Recent Advances in Veterinary Pharmacology and Toxicology

Recent Advances in Veterinary Pharmacology and Toxicology

Topic Editors

Chongshan Dai
Jichang Li

MDPI

Basel • Beijing • Wuhan • Barcelona • Belgrade • Novi Sad • Cluj • Manchester

Topic Editors
Chongshan Dai
Department of Pharmacology and Toxicology
China Agricultural University
Beijing
China

Jichang Li
College of Veterinary Medicine
Northeast Agricultural University
Harbin
China

Editorial Office
MDPI AG
Grosspeteranlage 5
4052 Basel, Switzerland

This is a reprint of the Topic, published open access by the journals *Antioxidants* (ISSN 2076-3921), *Biomolecules* (ISSN 2218-273X), *International Journal of Molecular Sciences* (ISSN 1422-0067), *Journal of Xenobiotics* (ISSN 2039-4713), *Metabolites* (ISSN 218-1989), *Toxics* (ISSN 2305-6304) and *Veterinary Sciences* (ISSN 2306-7381), freely accessible at: https://www.mdpi.com/topics/K91G9VC4Z1.

For citation purposes, cite each article independently as indicated on the article page online and as indicated below:

Lastname, A.A.; Lastname, B.B. Article Title. *Journal Name* **Year**, *Volume Number*, Page Range.

ISBN 978-3-7258-7643-3 (Hbk)
ISBN 978-3-7258-7644-0 (PDF)
https://doi.org/10.3390/books978-3-7258-7644-0

Contents

About the Editors

Chongshan Dai

Chongshan Dai, Ph.D., doctoral supervisor, postdoctoral fellow at Southwest Medical Center in the United States. He focuses on clinical highly drug-resistant pathogens and host immunity, conducting basic theoretical and applied research. He has hosted projects funded by the National Natural Science Foundation and National Key Research and Development, and has published over 100 papers as the first or corresponding author in journals such as *PNAS*, *Cell Reports*, and *Environment International* (3 highly cited ESI papers and 1 recommended F1000 paper), with over 4500 citations. He has applied for 24 national invention patents (11 have been granted) and 2 PCT patents. He has served as an Editorial Board Member for four SCI journals, including *Antioxidants*.

Jichang Li

Jichang Li, Ph.D., doctoral supervisor. Her main research directions include the development and application of new veterinary drugs, veterinary pharmacology and toxicology safety evaluation, and animal disease prevention and control. She has hosted over 30 scientific research projects, including the National Natural Science Foundation of China, the Natural Science Foundation of Heilongjiang Province, and the pre-industrialization research and development cultivation project for scientific and technological achievements in universities in Heilongjiang Province. She has published over 200 scientific papers, including more than 70 SCI indexed papers. She has authorized 7 national and international invention patents and published over 10 textbooks and works.

biomolecules

MDPI

Review

Antifungal Activity of Natural Thymol: Advances on Molecular Mechanisms and Therapeutic Potential

Chun Chen [1,2], Lu Liu [1,2], Shusheng Tang [1,2], Daowen Li [3,*] and Chongshan Dai [1,2,*]

[1] Technology Innovation Center for Food Safety Surveillance and Detection (Hainan), Sanya Institute of China Agricultural University, Sanya 572025, China
[2] State Key Laboratory of Veterinary Public Health and Safety, College of Veterinary Medicine, China Agricultural University, Beijing 100193, China
[3] Tianjin Key Laboratory of Agricultural Animal Breeding and Healthy Husbandry, College of Animal Science and Veterinary Medicine, Tianjin Agricultural University, Tianjin 300384, China
* Correspondence: daichongshan@cau.edu.cn (C.D.); lidaowen@tjau.edu.cn (D.L.)

Abstract

Currently, the increased incidence of invasive fungal infections globally is posing a significant challenge to public health. Due to drug resistance issues, the clinical efficacy of existing antifungal drugs is seriously insufficient, while new drug development progresses slowly. Consequently, there is an urgent need to discover and develop novel antifungal therapeutics. Natural products have the characteristics of wide sources and few adverse reactions and are one of the sources for developing antifungal drugs. Numerous studies have shown that many compounds isolated from plants and traditional Chinese medicine have antifungal activity and diverse antifungal mechanisms. Thymol, a monoterpene phenol compound from thyme (*Lamiaceae*), has multiple biological functions such as antibacterial, antioxidant, and anti-inflammatory. Recent research has found that thymol has strong antifungal activity, and its molecular mechanisms involve cell membrane rupture, interference with cell wall synthesis, disruption of mitochondrial function and energy metabolism, inhibition of biofilm, inhibition of virulence factor expression, inhibition of key enzymes, and induction of cell apoptosis. This review aimed to summarize the antifungal activity of thymol and the underlying molecular mechanisms, safety, and potential clinical applications. Emerging technologies in thymol delivery systems and future research directions are also discussed. The comprehensive analysis aims to provide a detailed understanding of fungal infections and the role of thymol in antifungal treatment, offering insights for further research and clinical practice.

Keywords: thymol; antifungal activity; molecular mechanisms; safety; clinical applications

1. Introduction

Fungal diseases are posing a significant threat to human and animal health [1–3]. In humans, invasive fungal infections have become increasingly common, especially in immunocompromised patients [4]. For instance, invasive pulmonary aspergillosis (IPA) is a severe manifestation of pulmonary aspergillosis. Among all clinical manifestations of pulmonary aspergillosis, IPA is the most acute presentation, caused by *Aspergillus* hyphae invading the pulmonary tissue [5]. In a study conducted at a Mexican hospital during the COVID-19 pre-vaccination era, the prevalence of possible COVID-19-associated pulmonary aspergillosis (CAPA) was 4.13% among 2080 individuals with severe COVID-19. All CAPA cases had a positive respiratory culture for *Aspergillus* species, with *Aspergillus fumigatus* being the most frequent isolate (64%) [6]. In organ transplant recipients, cryptococcosis is the third most commonly occurring invasive fungal disease [7]. Although kidney transplant recipients are at a relatively lower risk compared to other solid-organ transplant recipients, the mortality rates with cryptococcosis range from 10– to 25% and can be as high as 50% in those with central nervous system involvement [8]. In Japan, a nationwide observational study from 2015 to 2021 on disseminated cryptococcosis found that the age-adjusted incidence rate was higher in men, and the prevalence did not show a significant increase over the study period, with lower incidence observed in Northern Japan [9]. In animals, fungi can cause cutaneous mycoses, histoplasmosis, and coccidioidomycosis [10]. They can also contaminate animal feed via the toxic secondary metabolites, such as aflatoxins, T-2 toxin, and zearalenone, which can lead to mycotoxicosis [11–13]. Unfortunately, some fungal pathogens, such as *Candida*, *Cryptococcus*, and *Aspergillus*, exhibited potential resistance, resulting in a decreasing number of clinically available drugs [14–16]. Therefore, there is an urgent need to develop new antifungal drugs.

In recent decades, natural products have proven to be a diverse source of bioactive compounds with potential antifungal properties [17–19]. Multiple studies have shown that some natural products exhibited strong antifungal activity or reversed the resistance of some fungi to drugs, demonstrating potential clinical application value [20–22]. Thymol (the structure is shown in Figure 1) is a naturally occurring phenolic compound and can be found in multiple essential oils extracted from many plants, including oregano, thyme, sweet basil, black cumin, elsholtzia, and savory [23], which are widely distributed throughout China, Korea, and Europe [16]. Thymol has been reported to exhibit a broad spectrum of antifungal activities [24–27]. Importantly, thymol and its isomer carvacrol are the main active ingredients of these plant essential oils, and their compound proportion is high and accounts for up to about 20–60% [28,29]. In addition, some compounds, such as estragole, 1,8-cineole, terpineol-4, and γ-terpinene, found in these essential oils, also exhibited potent antifungal activity [30–34]. Previous studies have demonstrated that thymol has potent potential to inhibit biofilm formation of various fungal pathogens, such as *Fusarium* species and *Candida* species [35–37]. Gao et al. showed that the average median effective concentration (EC_{50}) value of thymol for 59 *Fusarium graminearum* isolates was about 26.3 μg/mL, and its antifungicidal activity is mainly attributed to the induction of lipid peroxidation and the disruption of ergosterol biosynthesis [38]. Thymol, in combination with other active ingredients in these essential oils at a special proportion, exhibited potent additive or synergistic effects. For example, Radocchia et al. reported that the combination of thymol and carvacrol exhibited potent synergistic effect against *C. albicans* (FICI is 0.24) and *C. tropicalis* (FICI is 0.37) [39]. In addition, it also has strong biological activities such as anti-inflammatory, antioxidant, and immune regulation and has a certain tissue repair effect [27]. Wang et al. found that thymol supplementation can markedly improve the inflammatory response caused by *Aspergillus fumigatus* infection by inhibiting lectin-type oxidized LDL receptor 1 (LOX-1)/interleukin-1β (IL-1β) signaling pathway-mediated inflammation activation and reducing the recruitment of neutrophils and macrophages in a mouse model [40]. It was also reported that thymol, in combination with other antifungal agents, such as posaconazole, fluconazole, and nystatin, can effectively reduce biofilm formation and disrupt mature biofilms, thereby enhancing the susceptibility of *Candida* species [41–44]. Collectively, these studies underscore the potential of thymol as a versatile antifungal agent with applications in agriculture and medicine.

https://doi.org/10.3390/biom16010149

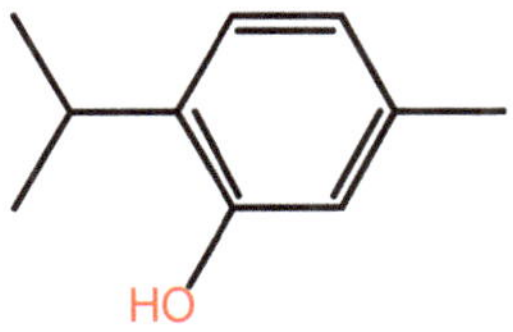

Figure 1. The chemical structure of thymol.

In the present study, we utilized keywords such as "thymol" or "thyme essential oil", and "antifungal activity" to gather research on the antifungal activity of thymol and the underlying molecular mechanisms from databases such as Web of Science and PubMed, up to November 2025. The collected information was then summarized and discussed. Then, we summarized the antifungal activity of thymol and the underlying molecular mechanisms. Additionally, the safety, potential clinical applications, emerging technologies in thymol delivery systems, and future research directions were also discussed. We hope this review paper can provide a detailed understanding of fungal infections and the role of thymol in antifungal treatment, offering insights for further research and clinical practice.

2. Epidemiology of Fungal Infections and Thymol's Role

2.1. *Prevalence and Impact of Fungal Infections*

Fungal infections have a significant global impact on public health, affecting various populations and causing a wide range of diseases [45]. In Greece, an estimate of the burden of serious fungal diseases showed that annually, out of a population of 10.8 million, about 1.79 cases per 100,000 individuals suffer from serious fungal diseases [46]. Approximately 1.93% of the population in Thailand (an estimated 1,254,562 people) is affected by serious fungal infections, including invasive aspergillosis and candidiasis [47]. In the Czech Republic, around 176,000 individuals (1.67% of the population) suffer from severe fungal infections each year, primarily manifesting as recurrent vaginitis and allergic respiratory diseases [48]. Additionally, nearly one million people in Ukraine are affected by fungal diseases annually [49]. Fungal infections range from mild superficial conditions (like athlete's foot) to severe systemic diseases (such as invasive aspergillosis or cryptococcosis). Among these, the following five globally prevalent fungal diseases cause significant morbidity in both humans and animals, i.e., candidiasis, aspergillosis, cryptococcosis, dermatophytosis, and Pneumocystis pneumonia (Table 1). Candidiasis is caused by fungi of the *Candida genus*, most commonly *Candida albicans*, though other clinically relevant species such as *Candida glabrata*, *Candida parapsilosis*, *Candida tropicalis*, and *Candida krusei* are increasingly significant in clinical settings [50,51]. Both humans and various animal species are susceptible, particularly immunocompromised individuals. The disease manifests as superficial infections, including cutaneous candidiasis, oral thrush, and vulvovaginal candidiasis, or as invasive infections that may lead to candidemia and multiorgan dissemination [52]. Aspergillosis refers to infections or allergic conditions caused by fungi of the *Aspergillus genus*, most commonly *Aspergillus fumigatus* [53,54]. This pathogen is highly transmissible, primarily via airborne spores, and manifests in several clinical forms including invasive aspergillosis, chronic necrotizing pulmonary aspergillosis, allergic bronchopulmonary aspergillosis, aspergilloma (fungus ball), as well as localized infections such as keratitis and otitis externa [55]. Cryptococcosis is a fungal disease caused by species of the *Cryptococcus genus*, primarily *Cryptococcus neoformans* and *Cryptococcus gattii*, and infection typically occurs through inhalation of airborne cryptococcal spores in humans and animals [56,57]. Cryptococcal meningitis represents the most significant clinical manifestation, particularly in individuals with HIV infection [58]. Dermatophytosis refers to a group of superficial fungal infections affecting keratinized tissues (skin, hair, and nails),

https://doi.org/10.3390/biom16010149

primarily caused by dermatophytes belonging to the genera *Trichophyton*, *Microsporum*, and *Epidermophyton* [59,60]. These fungi are widely prevalent across the globe and exhibit zoonotic potential, enabling transmission between humans and animals [61]. Infections may involve virtually any skin surface and hair [62,63]. Pneumocystis pneumonia (PCP) is a pulmonary infection caused by the fungus *Pneumocystis jirovecii* [64,65]. Transmitted via the respiratory route, the pathogen exhibits low-level circulation in human populations and primarily colonizes the human alveoli [65]. In developing countries, PCP remains a leading cause of mortality among individuals with HIV/AIDS [66,67]. In recent years, the emergence of resistance in *Aspergillus* and *Candida* spp. to azole drugs has been a growing issue [68]. This underscores the urgent need for novel antifungal agents.

Thymol has proven effective in diverse fungal infection scenarios, encompassing both in vitro and in vivo settings. Research highlights thymol's robust antibiofilm capabilities against *Cryptococcus neoformans* and *Cryptococcus laurentii*, where it curtails growth, biofilm development, and established biofilms in these fungi [69]. In animal-based investigations, thymol also exhibits significant potential. For example, in studies on Nile tilapia, initial in vitro work demonstrated thymol's antifungal properties, alongside other essential oils, targeting yeasts derived from infected fish. Subsequent in vivo experiments showed that incorporating thymol into diets markedly improved fish growth metrics, immune function, and survival rates (90%) when challenged with *Cryptococcus uniguttulatus* [70]. These outcomes indicate thymol's utility as a potent agent for combating fungal infections in human and animal contexts, emphasizing its prospects for refinement as an antifungal therapy.

Table 1. Top 5 globally prevalent fungal diseases.

Disease	Principal Pathogen(s)	Host Range	Epidemiological Features	References
Candidiasis	*Candida albicans*, etc.	Humans and various animals	High risk in immunocompromised individuals	[50–52,71,72]
Aspergillosis	*Aspergillus fumigatus*	Humans, birds	Airborne transmission	[53,54,73]
Cryptococcosis	*Cryptococcus neoformans*/*C. gattii*	Humans, cats, dogs	Associated with environmental exposure; high risk in patients with HIV	[55–58,74,75]
Dermatophytosis	*Trichophyton*, *Microsporum*, etc.	Humans, pets, livestock	More prevalent in moist, crowded environments	[59–63]
Pneumocystis pneumonia	*Pneumocystis jirovecii*	Humans	High risk in patients with HIV/AIDS	[64–66]

2.2. An Overview of Thymol's Antifungal Activity

Thymol exhibits broad-spectrum antifungal activity against a variety of fungi, with its effective concentration and mechanism of action varying by species. Against phytopathogenic fungi, thymol shows the highest activity against *Botrytis cinerea*, with a MIC as low as 12.30 μg/mL, primarily by disrupting cell membrane integrity [76]. The half-maximal effective concentrations EC_{50} against *Fusarium graminearum* and *Batrachochytrium salamandrivorans* are 26.3 μg/mL and 25 μg/mL [38,77], respectively, involving mechanisms such as the induction of lipid peroxidation and interference with ergosterol biosynthesis. For *Trichophyton rubrum*, the MIC of thymol is 32 μg/mL, while derivatives can significantly enhance the activity, reducing the MIC to 4 μg/mL [78]. It also demonstrates moderate to strong inhibitory effects against other common pathogens, such as *Alternaria alternata* (MIC = 250 μg/mL) [79] and *Rhizoctonia solani* (MIC = 250–500 μg/mL) [80], achieved by damaging the cell wall and cell membrane. Against human and animal

https://doi.org/10.3390/biom16010149

pathogenic yeasts, thymol is effective against various *Candida species*. Notably, it exhibits activity against non-albicans *Candida species*, including *Candida glabrata*, *Candida krusei*, and *Candida tropicalis*, with MICs ranging from 32 to 128 μg/mL, and effectively disrupts biofilm formation [81]. Importantly, it retains activity against fluconazole-resistant strains of *Candida dubliniensis* (MIC = 64 μg/mL) [81]. For *Candida albicans*, the MIC range is broader (32–256 μg/mL) [81], with mechanisms including the induction of reactive oxygen species (ROS) burst, inhibition of efflux pumps [82], and suppression of ergosterol biosynthesis [81]. Against *Malassezia pachydermatis*, thymol shows relatively lower activity (MIC = 800 μg/mL), where its hydrophobicity facilitates disruption of ion gradients across the cell membrane [83].

The core antifungal mechanisms of thymol are highly focused on disrupting cell membrane integrity (e.g., pore formation, inhibition of ergosterol biosynthesis) and inducing oxidative stress (ROS accumulation). Additionally, it interferes with cell wall metabolism (e.g., inhibition of chitin synthesis [84], activation of cell wall-degrading enzymes [85]) and ultimately leads to cell death by disrupting ion homeostasis and energy metabolism [86]. Formulation improvements, such as nanoparticle encapsulation, can further enhance its targeting and antifungal efficacy [76], demonstrating its potential against antifungal drug-resistant strains. A systematic summary of thymol on antifungal activity and potential effects is presented in Table 2.

Table 2. Antifungal activity and mechanisms of action of thymol.

Fungal Species	MIC/EC_{50} (μg/mL)	Effects	References
Alternaria alternata	250	It damages the cell wall and cell membrane, leading to cytoplasmic disorganization and organelle destruction; inhibits spore germination, hyphal growth, and early infection structure formation.	[79]
Aspergillus flavus	80	It induces ROS burst and NO production; the combined action of ROS and NO leads to spore lysis and cell death.	[87]
Batrachochytrium salamandrivorans	25	It interferes with ergosterol biosynthesis, increases cell membrane permeability, and degrades cell function.	[77]
Botrytis aclada	30 (Fumigation)	It likely acts by inhibiting the synthesis of chitin or β-1,3-glucan (fungal cell wall components).	[84]
Botrytis cinerea	12.30 (Thymol) 9.90 (Chitosan-Thymol NPs)	It disrupts cell membrane integrity; nanoparticles of thymol are pH-responsive and enable intelligent release in the acidic environment created by the pathogen, then enhance targeted inhibition.	[76]
Botrytis cinerea	65	It disrupts cell membrane integrity, leading to leakage of cellular contents, decrease in extracellular pH, and increase in conductivity, finally causing hyphal deformation, shrinkage, and rupture.	[88]
Candida albicans	32–256	It disrupts cell membrane integrity, inhibits ergosterol synthesis, induces ROS generation, causing oxidative damage, and significantly damages mature biofilms. It also inhibits efflux pump activity and exhibits synergy with fluconazole.	[81,82]
Candida dubliniensis	64	It is effective against fluconazole-resistant strains via disrupting cell membrane structure.	[81]

https://doi.org/10.3390/biom16010149

Table 2. *Cont.*

Fungal Species	MIC/EC$_{50}$ (μg/mL)	Effects	References
Candida glabrata *Candida krusei* *Candida tropicalis*	32–128	It effectively disturbs the growth of both planktonic cells and biofilms.	[81]
Colletotrichum chrysophillum *Colletotrichum nymphaeae*	100	It disrupts cell membrane integrity based on the membrane-interfering properties of monoterpenes.	[89]
Fusarium graminearum	EC$_{50}$ = 26.3	It can induce lipid peroxidation, disrupt ergosterol biosynthesis, cause cell membrane damage, and affect hyphal morphology and conidia production.	[38]
Fusarium oxysporum f. sp. *niveum*	EC$_{50}$ = 80	It inhibits cell wall synthesis and activates cell wall-degrading enzymes. It can also disrupt membrane integrity, induce the production of ROS, and disturb TCA cycle and fatty acid metabolism.	[85]
Fusarium tricinctum	90	It destroys cell ultrastructure, disrupts plasma membrane integrity, transiently increases energy metabolism, and induces lipid peroxidation.	[36]
Malassezia pachydermatis	800	Hydrophobicity allows distribution into the cell membrane; free hydroxyl group disrupts ion gradients, leading to membrane dysfunction.	[83]
Rhizoctonia solani	250–500	It can cause severe hyphal malformation, shrinkage, inhibits cell wall-degrading enzyme activity, and disrupt cell membrane integrity.	[80]
Trichophyton rubrum	32 (Thymol) 4 (Derivative)	It disrupts the structure and function of the fungal cell membrane via inhibiting the ergosterol biosynthesis pathway.	[78]
Zygosaccharomyces rouxii	62.5	It disrupts cell membrane integrity, causing membrane perforation, membrane potential hyperpolarization, and the decrease in intracellular ATP and pH, followed by disrupting cell homeostasis and inducing cell death.	[86]

3. The Underlying Molecular Mechanisms

3.1. *Disruption of the Fungal Cell Membrane and Mitochondrial Function*

Thymol's antifungal activity is partly attributed to its ability to disrupt the fungal cell membrane. Similarly to other membrane-active compounds, thymol can interact with the fungal cell membrane, leading to changes in its permeability and integrity [90]. Thymol primarily acts on the cell membrane surface. It can interact with the polar headgroups (especially the phosphate groups) of negatively charged phospholipids via hydrogen bonding or electrostatic interactions. This leads to monolayer expansion, reduced elasticity, and the induction of local aggregation. Concurrently, thymol mildly perturbs the packing of the alkyl chains, increasing their disorder and thereby enhancing membrane fluidity of fungal cells [91,92]. Collectively, these structural alterations compromise membrane stability and function, which may underlie its associated biological activities. For example, Kim et al. demonstrated that thymol induces the generation of ROS, which disrupts the cell membrane integrity of *Phytophthora quercina* and *Agrobacterium tumefaciens*, ultimately leading to cell death [93]. Treatment with the γ-cyclodextrin inclusion complex of thymol (γ-CDTL) for one hour resulted in disruption of cell membrane integrity in *Penicillium* spp. [94].

https://doi.org/10.3390/biom16010149

Fungal mitochondrial function can be disrupted by thymol, which may contribute to its antifungal activity. In many fungal pathogens, disruption of mitochondrial function can lead to cell death. Exposure of *Zygosaccharomyces rouxii* to subinhibitory concentrations of thymol induced intracellular accumulation of ROS, elevated Ca^{2+} levels, and decreased mitochondrial membrane potential (MMP), demonstrating its disruptive effect on the cell membrane [95]. Wang et al. revealed that thymol treatment induces increased membrane permeability in *Zygosaccharomyces rouxii*, leading to cytoplasmic leakage and mitochondrial membrane hyperpolarization, accompanied by the formation of pores in the cell membrane [86]. Treatment with thymol resulted in a significant increase in relative electrical conductivity and enhanced cell membrane permeability in *Fusarium graminearum*, indicating that thymol induces damage to the cell membrane [38]. In addition, some studies have shown that compounds affecting the ergosterol biosynthesis in the fungal cell membrane can have antifungal effects. Ergosterol is a key component of the fungal cell membrane, and disruption of its synthesis or interaction with it can lead to membrane dysfunction. Thymol interferes with the ergosterol biosynthesis or directly interacts with the membrane components, causing membrane disruption. This is supported by the fact that many natural products with antifungal activity target the fungal cell membrane, and thymol, being a natural monoterpene phenol, shares similar mechanisms of action [96]. Thymol inhibits ergosterol biosynthesis in *Fusarium graminearum* by downregulating the expression of genes FGSG_02771, FGSG_09031, and FGSG_11044 [38]. Thymol targets the mitochondria of fungal cells, inducing a reduction in MMP and an increase in ROS production. These effects lead to disruption of the cell membrane structure and inhibition of sterol lactone biosynthesis, thereby impairing cellular energy metabolism and ultimately resulting in cell death.

3.2. *Inhibition of Fungal Cell Wall Activity*

The fungal cell wall is a crucial target for antifungal agents [97], and thymol exerts its antifungal effects by interfering with cell wall synthesis or integrity. Thymol inhibits the growth of *Aspergillus niger* by increasing cell wall permeability [98]. Thymol may act in a similar way with poacic acid, a plant-derived antifungal agent, which can target β-1,3-glucan, a major component of the fungal cell wall [99]. It can also either directly interact with cell wall components or interfere with the enzymes involved in cell wall synthesis [94]. For example, it was reported that thymol can inhibit fungal growth in *Alternaria alternata* by simultaneously disrupting the cell wall and plasma membrane and damaging organelles within the cytoplasm [79]. Furthermore, Zhang et al. employed RNA-Seq analysis to investigate changes in gene expression in Fusarium species following thymol treatment. They observed significant downregulation in the majority of genes associated with glycosphingolipid biosynthesis and sphingolipid metabolism. In contrast, genes involved in chitin biosynthesis and cell wall remodeling were upregulated. These results demonstrate that thymol exerts its antifungal effects by suppressing genes critical to cell wall and membrane synthesis, inducing ROS accumulation, and ultimately disrupting cell wall integrity [85].

3.3. *Thymol Induces Apoptosis in Fungal Cells*

Apoptosis is a form of programmed cell death [100,101]. Previous studies have demonstrated that thymol can induce both intrinsic and extrinsic apoptotic pathways in fungi. Hu et al. demonstrated that exposure of *Aspergillus flavus* conidia to thymol (200 μg/mL) for 6 h induced chromatin condensation, DNA fragmentation, reduced MMP, and activation of caspase-9, indicating thymol-triggered caspase-dependent apoptosis. The antifungal effect was rescued by 4-aminopyridine, confirming that thymol targets potas-

https://doi.org/10.3390/biom16010149

sium channel subunit beta (KCNAB) to induce K^+ efflux, which subsequently promotes ROS accumulation and ultimately leads to apoptotic cell death in *Aspergillus* conidia. Treatment of *Zygosaccharomyces rouxii* with thymol resulted in increased cytochrome c release, caspase activation, and elevated DNA fragmentation. These effects were accompanied by upregulation of pro-apoptotic factors including yeast caspase-1 (YCA1), dynamin-1 (DNM1), nuclease-1 (NUC1), NADH dehydrogenase 1 subunit (NDI1), and microtubule and mitochondria interacting protein (MMI1), along with downregulation of anti-apoptotic regulators such as fission-1 (FIS1) and cell division control protein-48 (CDC48), collectively indicating thymol-induced apoptosis in *Zygosaccharomyces rouxii* [95]. Darvishi E et al. demonstrated that thymol, at a subinhibitory concentration of 50 μg/mL, accelerates telomere shortening by inhibiting expressed sequence tag-2 (EST2) transcription and thereby blocking telomerase activity, ultimately promoting fungal cellular senescence and apoptosis [102].

3.4. The Multifaceted Therapeutic Contributions of Thymol's Antioxidant and Anti-Inflammatory Properties in Antifungal Applications

In addition to its antifungal properties, thymol's antioxidant and anti-inflammatory activities significantly contribute to its therapeutic potential. The antioxidant activity of thymol is crucial in mitigating oxidative stress, a common pathway in fungal pathogenesis. Previous studies provide evidence of thymol's ability to reduce ROS production and lipid peroxidation, thereby protecting cells from oxidative damage [103]. This antioxidant effect is complemented by thymol's anti-inflammatory properties, which involve the modulation of key inflammatory pathways such as MAPK and NF-κB, leading to a reduction in pro-inflammatory cytokines [104,105]. These combined effects not only enhance thymol's antifungal efficacy but also contribute to its role in reducing inflammation and oxidative stress associated with fungal infections. The integration of thymol's antioxidant and anti-inflammatory activities into its antifungal therapy underscores the importance of a holistic approach to treating fungal infections. By targeting multiple pathways, thymol not only inhibits fungal growth but also alleviates the inflammatory and oxidative stress responses that often accompany such infections. This multifaceted approach is particularly relevant in the context of increasing antifungal resistance, as it offers a natural and potentially more effective alternative to synthetic antifungal agents [26].

Overall, the antifungal mechanism of thymol involves a multi-target synergistic cascade. It initiates by disrupting both the cell membrane (via ergosterol inhibition and pore formation) and cell wall (through interference with chitin and glycosphingolipid metabolism), compromising structural integrity. These primary damages trigger intracellular ion imbalance (e.g., K^+ efflux and Ca^{2+} elevation) and oxidative stress (ROS production). Subsequently, mitochondrial dysfunction occurs, characterized by membrane potential collapse and cytochrome C release, which ultimately activates the caspase cascade and modulates apoptosis-related genes, leading to programmed cell death. A schematic diagram to illustrate the multi-target synergistic effects and cascade pathways of thymol's antifungal action is shown in Figure 2.

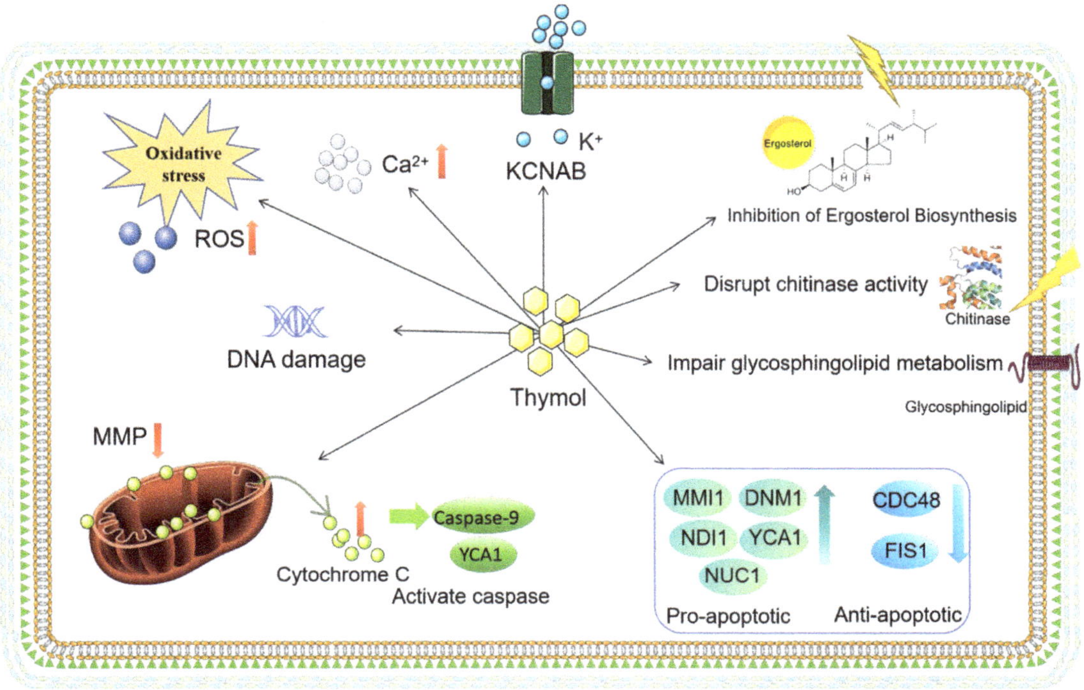

Figure 2. Thymol exerts its antifungal effect through multi-target synergistic actions. Firstly, it disrupts the peripheral cellular structures by increasing cell wall permeability, inhibiting chitinase activity, and interfering with glycosphingolipid metabolism. Concurrently, it enhances plasma membrane permeability, induces pore formation, and suppresses ergosterol biosynthesis, collectively leading to cytoplasmic leakage. Secondly, thymol targets mitochondria, causing a decrease or hyperpolarization of the MMP and resulting in mitochondrial dysfunction. Furthermore, thymol triggers substantial intracellular accumulation of ROS and elevated Ca^{2+} levels, leading to oxidative stress and disrupting calcium homeostasis. These multifaceted damages ultimately converge to induce potassium ion efflux, further promote ROS accumulation, and trigger cytochrome C release, thereby activating the caspase cascade (e.g., Caspase-9 and its yeast homolog YCA1). This process is accompanied by the upregulation of key pro-apoptotic factors (such as YCA1, DNM1, NUC1, NDI1, and MMI1) and downregulation of anti-apoptotic regulators (such as FIS1 and CDC48), culminating in DNA fragmentation and systematically driving fungal cells toward programmed cell death.

3.5. Comparative Analysis of Antifungal Efficacy Between Thymol and Other Monoterpenoids

Besides thymol, several other natural monoterpenoids also exhibit antifungal activity, such as carvacrol, linalool, D-limonene, and eucalyptol. However, each has certain limitations compared to thymol. Carvacrol, an isomer of thymol, shows comparable antifungal activity but is more toxic [23]. Linalool and D-limonene are only effective against specific fungi and lack broad-spectrum activity [106,107]. In contrast, eucalyptol possesses relatively weak antifungal activity and is commonly used in synergistic antifungal applications [108]. The comparative antifungal activities of thymol and other monoterpenoids are systematically summarized in Table 3.

https://doi.org/10.3390/biom16010149

Table 3. Comparative analysis of antifungal efficacy between thymol and other monoterpenoids.

Compound	Chemical Class	Major Natural Sources	Core Mechanism of Action	Characteristics and Status	Reference
Thymol	Phenolic monoterpene	Thyme	Multi-target synergy: disrupts cell membrane and wall, inhibits ergosterol biosynthesis, induces mitochondrial damage and oxidative stress	Broad-spectrum, highly potent representative; most extensively studied	[38,85,93]
Carvacrol	Phenolic monoterpene	Thyme	Similar to thymol	Synergistic with thymol; higher toxicity compared to thymol	[23,109,110]
Linalool	Alcoholic monoterpene	Lavender	Disrupts cell membrane structure, inhibits hyphal growth	Particularly effective against specific molds (e.g., *Aspergillus flavus*)	[106]
D-Limonene	Hydrocarbon monoterpene	Citrus	Membrane interference, induces mitochondrial dysfunction	Shows significant activity against certain yeasts	[107]
Cineole (Eucalyptol)	Oxide monoterpene	Eucalyptus	Primarily membrane interference	Generally moderate to weak activity; often used in combination therapies	[108]

4. Treatment Strategies and Potential Clinical Significance

In contemporary medical applications, thymol has gained recognition as a versatile antifungal agent with broad utility. Studies on Thymus vulgaris essential oil and thymol have confirmed its inhibitory action on *Candida albicans* growth [41]. Wang et al. further demonstrated that thymol not only exhibits direct antifungal effects against *Aspergillus fumigatus* but also provides defense against corneal infections by inhibiting the LOX-1/IL-1β signaling cascade, which minimizes neutrophil and macrophage infiltration [40].

Additionally, integrating thymol into antifungal protocols demands careful consideration of its properties and potential synergies. Advanced delivery systems, such as starch Pickering emulsions for oral candidiasis, improve thymol's bioavailability and activity against C. albicans when combined with enzymes like α-amylase, ensuring stability over weeks [111]. Chemical modifications, including sulfonamide group additions, have markedly increased thymol's antimicrobial potency against pathogens like *Phytophthora capsici* and *Aspergillus flavus*, outperforming commercial fungicides such as azoxystrobin and carbendazim [112]. Synthesized derivatives, such as naphthoquinone/thymol hybrids via Williamson ether synthesis and copper-catalyzed reactions, bind to fungal ergosterol. This disrupts potassium permeability, causing ionic imbalance and cell death, with efficacy comparable to thiabendazole against *Fusarium solani* [113]. Formulations combining thymol with other agents, like fluconazole-loaded biopolymer films for resistant vaginal candidiasis, enhance drug distribution in tissues and boost anti-*Candida* effects, reducing fluconazole doses by 50% against resistant strains [114]. Combinations with caprylic acid disrupt cell membrane integrity and inhibit efflux pumps, leading to intracellular accumulation and mortality [115]. Additionally, thymol paired with Folicur® effectively suppresses resistant *P. nodorum strains* [116], highlighting that integrated strategies can address resistant fungal infections effectively. Investigations into combination therapies reveal promising results; for instance, thymol synergizes with nystatin against Candida species in oral infections, reducing minimum inhibitory concentrations by 87.4% and achieving a fractional inhibitory concentration index of 0.25. This enhancement could

https://doi.org/10.3390/biom16010149

lead to superior clinical outcomes, but comprehensive trials are essential to validate thymol's standalone and combined efficacy in real-world scenarios [44]. This indicated thymol as a viable standalone or adjunct therapy for candidiasis in humans and animals.

5. Controversies and Challenges in Thymol's Antifungal Use

5.1. Debates on the Efficacy of Thymol in Resistant Fungal Strains

The efficacy of thymol against resistant fungal strains is a topic of debate. While some studies have shown promising results, others suggest limitations. In the context of Candida species, which are known to develop resistance to conventional antifungal drugs, thymol has been tested against resistant strains. A study evaluated the activity of thymol in combination with fluconazole against clinical isolates of *Candida albicans*, *Candida glabrata*, and *Candida kruseii* [43]. Thymol, in combination with fluconazole, exhibited synergistic effects against all tested species, with FICI values ranging from 0.366 to 0.607 for *Candida albicans* strains, 0.367 to 0.482 for *Candida glabrata* strains, and 0.375 to 0.563 for *Candida kruseii* strains. This indicates that thymol can be effective against some resistant Candida strains when used in combination. Conte et al. demonstrated that a biopolymer film incorporating thymol combined with half the conventional dose of fluconazole exhibited potent fungicidal efficacy against drug-resistant Candida glabrata [114].

However, there is a lack of comprehensive studies on a wide range of resistant fungal strains. Different fungal species and strains may have varying susceptibilities to thymol, and the mechanisms of resistance may also play a role. For example, in bacteria, resistance mechanisms such as enzymatic inhibition, penicillin-binding protein modification, porin mutations, efflux pumps, and molecular modifications of antibiotic targets can affect the efficacy of antimicrobial agents [117]. Fungi may have similar or unique resistance mechanisms that could potentially reduce the effectiveness of thymol. More research is needed to fully understand the efficacy of thymol against resistant fungal strains and to develop strategies to overcome potential resistance issues.

5.2. Safety and Toxicity Concerns of Thymol in Clinical Use

Safety assessment of thymol reveals an application- and dose-dependent profile, with a generally favorable safety margin observed across multiple routes of exposure. When used as a poultry feed additive, thymol is classified by the U.S. Food and Drug Administration as Generally Recognized as Safe. Pharmacokinetic studies indicate that thymol is rapidly and completely absorbed in the gastrointestinal tract, exhibits a short half-life, and is primarily excreted via urine, thereby minimizing the risk of long-term accumulation [118]. Studies have confirmed its absence of neurotoxicity while demonstrating neuroprotective effects in cortical neurons [27,119,120]. Thymol induces apoptosis in A549 cancer cells while exhibiting no significant cytotoxicity toward normal peripheral blood mononuclear cells [121]. In broiler chickens, thymol administered at doses of up to 400 mg/kg demonstrated comparable protective efficacy to the synthetic antioxidant butylated hydroxytoluene, with no adverse effects reported [122]. When co-administered with carvacrol at doses of up to 200 mg/kg, thymol not only elicited no toxic effects but also enhanced antioxidant enzyme activities and improved immune responses, thereby promoting overall health [123]. However, thymol exhibits a narrow level of safety in anesthesia applications for zebrafish, i.e., low concentrations (50–75 mg/L) safely and effectively induce anesthesia with no mortality, and higher concentrations ($\geq$100 mg/L) exhibit behavioral aversion and cardiorespiratory depression, with a 60% mortality rate observed at 200 mg/L, underscoring its significant toxic risk at elevated doses [124]. In chicken embryos, thymol demonstrated significantly lower toxicity compared to its isomer carvacrol, inducing only minor malformations at high doses without affecting embryonic growth [125]. Furthermore, thymol exhibits negligible

https://doi.org/10.3390/biom16010149

estrogenic activity and mutagenic potential, with only inconsistent weak effects observed at very low concentrations that do not reach established safety risk thresholds [125]. A six-month chronic inhalation toxicity study further established that even prolonged exposure of mice to thymol at doses up to 858-fold the maximum human dose revealed no toxicological alterations in pulmonary and respiratory tissues [126]. More clinical studies are required to fully assess the safety and toxicity of thymol, especially in long-term use and in different patient populations.

5.3. Regulatory Challenges in the Approval of Thymol-Based Treatments

The approval of thymol-based treatments faces several regulatory challenges: (i) It lacks comprehensive data on its safety, efficacy, and long-term effects. Regulatory agencies typically require extensive pre-marketing approval data, including detailed toxicology studies, clinical trials, and information on the manufacturing process to ensure the quality, safety, and effectiveness of the treatment. For thymol-based treatments, there is a need for more well-designed clinical trials to demonstrate their efficacy in treating specific fungal infections. (ii) The standardization of thymol-based products is crucial. Natural products can vary in their composition depending on factors such as the source plant, extraction method, and formulation. This variability can make it difficult to establish consistent quality control and regulatory standards. For example, in the case of essential oils containing thymol, the concentration of thymol and other components can vary significantly, which may affect the product's efficacy and safety. Regulatory agencies need to develop clear guidelines for the standardization of thymol-based treatments to ensure their reliable and reproducible performance. Moreover, the potential interactions of thymol with other medications also need to be investigated and regulated to avoid adverse drug–drug interactions in clinical use.

6. Future Directions and Innovations in Thymol's Antifungal Research

6.1. Emerging Technologies in Thymol Delivery Systems

Emerging technologies in thymol delivery systems offer promising opportunities to enhance thymol's antifungal efficacy. One such technology is the use of nano-encapsulation. For example, nanoemulsions can improve the solubility, stability, and bioavailability of thymol. A study on the nanoencapsulation of a formulation based on thymol, methyl cinnamate, and linalool (Ne-TML) showed that it had excellent antifungal and anti-aflatoxin B_1 potential [127]. The Ne-TML was able to completely inhibit the growth and Aflatoxin B1 (AFB_1) production of fungi at low concentrations. Nanoencapsulation can protect thymol from degradation, control its release, and improve its penetration into fungal cells, thereby enhancing its antifungal activity.

Another emerging technology is the development of responsive delivery systems. For instance, enzyme-responsive food packaging systems based on pectin-coated poly (lactic acid) nanofiber films have been fabricated to load and release thymol [128]. These films can respond to the presence of enzymes secreted by microorganisms from food contamination, triggering the release of thymol. This targeted release mechanism can improve the efficiency of thymol in inhibiting fungal growth in food products, reducing the need for high-dose or continuous application. Such technologies can be translated to clinical applications, where responsive delivery systems could potentially improve the treatment of fungal infections by delivering thymol precisely to the site of infection.

6.2. Prospective Studies on Thymol's Long-Term Antifungal Effects

Prospective studies on thymol's long-term antifungal effects are essential to fully understand its potential as a therapeutic agent. Long-term studies can help determine the

https://doi.org/10.3390/biom16010149

durability of thymol's antifungal activity, its impact on fungal resistance development, and its safety in extended use. For example, in the treatment of recurrent fungal infections, understanding how thymol performs over an extended period is crucial. A study on the long-term suppressive therapy of fungal infections in patients with artificial implants using fluconazole showed the importance of long-term data [129]. Similarly, for thymol, long-term studies could evaluate its effectiveness in preventing recurrence of fungal infections, especially in immunocompromised patients who are at a high risk of repeated infections.

In addition, studies have also investigated the impact of thymol on the host's microbiota. Fungal infections often occur in the context of a disrupted microbiota, and long-term use of antifungal agents can further affect the balance of microbiota. Prospective studies could explore how thymol affects the composition and function of the host microbiota over time and whether it has any beneficial or adverse effects on the overall health of the host. This information is vital for developing sustainable and effective antifungal treatment strategies using thymol.

6.3. Potential for Thymol in Combination Therapies for Fungal Infections

Combination therapies using thymol have significant potential in the treatment of fungal infections. Thymol has been shown to exhibit synergistic effects with various antifungal agents. This synergistic effect can potentially reduce the required dosage of individual drugs, minimizing the risk of toxicity and resistance development. In addition, combination therapies can target multiple aspects of fungal pathogenesis. Fungi often have multiple virulence factors and resistance mechanisms, and combination therapies can simultaneously target different pathways. For instance, thymol combined with agents that target different stages of biofilm formation or fungal cell metabolism could be more effective in treating biofilm-associated fungal infections. Future research could focus on identifying the most effective combinations, understanding the mechanisms of synergy, and optimizing the dosing regimens to maximize the therapeutic benefits of thymol-based combination therapies for fungal infections.

7. Conclusions

Thymol has strong antifungal activity, and its underlying molecular mechanisms involve cell membrane rupture, interference with cell wall synthesis, disruption of mitochondrial function and energy metabolism, inhibition of biofilm, inhibition of virulence factor expression, inhibition of key enzymes, and induction of cell apoptosis. Additionally, it also exhibited good safety. In the future, more clinical research for testing the therapeutic effect of thymol against fungal infection in humans and animals is needed.

Author Contributions: Writing—original draft preparation, C.C. and C.D.; methodology, L.L. and C.C.; writing—review and editing, S.T., C.D. and D.L.; funding acquisition, L.L. and C.D. All authors have read and agreed to the published version of the manuscript.

Funding: This research was supported by the Project of Sanya Yazhou Bay Science and Technology City, Grant No: SKJC-JYRC-2025-56. It was also funded by the National Natural Science Foundation of China (Award number 32102724).

Institutional Review Board Statement: Not applicable.

Informed Consent Statement: Not applicable.

Data Availability Statement: Not applicable.

Conflicts of Interest: The authors have read and approved this manuscript and declare no conflicts of interest.

https://doi.org/10.3390/biom16010149

Abbreviations

The following abbreviations are used in this manuscript:

AFB1	Aflatoxin B1
ATP	Adenosine Triphosphate
BHT	Butylated Hydroxytoluene
CAPA	COVID-19-associated Pulmonary Aspergillosis
CDC48	Cell Division Control protein-48
CNKI	China National Knowledge Infrastructure
DNA	Deoxyribonucleic Acid
DNM1	Dynamin-1
EC_{50}	Half Maximal Effective Concentration
EST2	Expressed Sequence Tag-2
FDA	U.S. Food and Drug Administration
FICI	Fractional Inhibitory Concentration Index
FIS1	Fission-1
γ-CDTL	γ-Cyclodextrin Inclusion Complex of Thymol
HIV/AIDS	Human Immunodeficiency Virus/Acquired Immunodeficiency Syndrome
IL-1β	Interleukin-1 Beta
IPA	Invasive Pulmonary Aspergillosis
KCNAB	Potassium Channel Subunit Beta
LOX-1	Lectin-type Oxidized LDL Receptor 1
MAPK	Mitogen-Activated Protein Kinase
MIC	Half Maximal Effective Concentration
MMP	Mitochondrial Membrane Potential
MMI1	Mitochondrial Mediator of Inheritance 1
NDI1	NADH Dehydrogenase (Internal) 1
NUC1	Nuclease 1
ROS	Reactive Oxygen Species
YCA1	Yeast Caspase-1

References

1. Xu, J.; Cai, L. Assessing global fungal threats to humans. *mLife* **2022**, *1*, 223–240. [CrossRef]
2. Fisher, M.C.; Gurr, S.J.; Cuomo, C.A.; Blehert, D.S.; Jin, H.; Stukenbrock, E.H.; Stajich, J.E.; Kahmann, R.; Boone, C.; Denning, D.W.; et al. Threats Posed by the Fungal Kingdom to Humans, Wildlife, and Agriculture. *mBio* **2020**, *11*, e00449-20. [CrossRef] [PubMed]
3. Anand, G.; Rajeshkumar, K.C. Challenges and Threats Posed by Plant Pathogenic Fungi on Agricultural Productivity and Economy. In *Fungal Diversity, Ecology and Control Management; Fungal Biology*; Springer: Singapore, 2022. [CrossRef]
4. Low, C.; Rotstein, C. Emerging fungal infections in immunocompromised patients. *F1000 Med. Rep.* **2011**, *3*, 14. [CrossRef] [PubMed]
5. Heylen, J.; Vanbiervliet, Y.; Maertens, J.; Rijnders, B.; Wauters, J. Acute Invasive Pulmonary Aspergillosis: Clinical Presentation and Treatment. *Semin. Respir. Crit. Care Med.* **2024**, *45*, 69–87. [CrossRef]
6. Hernández-Silva, G.; Corzo-León, D.E.; Becerril-Vargas, E.; Peralta-Prado, A.B.; Odalis, R.-G.; Morales-Villarreal, F.; Ríos-Ayala, M.A.; Alonso, T.-G.; Agustín, F.-L.D.; Ramón, A.-F.; et al. Clinical characteristics, bacterial coinfections and outcomes in COVID-19-associated pulmonary aspergillosis in a third-level Mexican hospital during the COVID-19 pre-vaccination era. *Mycoses* **2024**, *67*, e13693. [CrossRef] [PubMed]
7. Baddley, J.; Forrest, G. Cryptococcosis in Solid Organ Transplantation. *Am. J. Transplant.* **2013**, *13*, 242–249. [CrossRef] [PubMed]
8. Meena, P.; Bhargava, V.; Singh, K.; Sethi, J.; Prabhakar, A.; Panda, S. Cryptococcosis in kidney transplant recipients: Current understanding and practices. *World J. Nephrol.* **2023**, *12*, 120–131. [CrossRef] [PubMed]
9. Akazawa, H.; Hagiya, H.; Koyama, T.; Otsuka, F. Trends in the Incidence of Disseminated Cryptococcosis in Japan: A Nationwide Observational Study, 2015–2021. *Mycopathologia* **2024**, *189*, 8. [CrossRef]
10. Carpouron, J.E.; de Hoog, S.; Gentekaki, E.; Hyde, K.D. Emerging Animal-Associated Fungal Diseases. *J. Fungi* **2022**, *8*, 611. [CrossRef]

https://doi.org/10.3390/biom16010149

11. Zhu, F.; Zhu, L.; Xu, J.; Wang, Y.; Wang, Y. Effects of moldy corn on the performance, antioxidant capacity, immune function, metabolism and residues of mycotoxins in eggs, muscle, and edible viscera of laying hens. *Poult. Sci.* **2023**, *102*, 102502. [CrossRef]
12. Zuo, N.; Wang, R.T.; Bian, W.M.; Liu, X.; Han, B.Q.; Wang, J.J.; Shen, W.; Li, L. Vigor King mitigates spermatogenic disorders caused by environmental estrogen zearalenone exposure. *Ecotoxicol. Environ. Saf.* **2024**, *282*, 116757. [CrossRef]
13. Pei, X.; Ma, S.; Hong, L.; Zuo, Z.; Xu, G.; Chen, C.; Shen, Y.; Liu, D.; Li, C.; Li, D. Molecular insights of T-2 toxin exposure-induced neurotoxicity and the neuroprotective effect of dimethyl fumarate. *Food Chem. Toxicol.* **2025**, *196*, 115166. [CrossRef]
14. Zhang, Y.; Zeng, L.; Huang, X.; Wang, Y.; Chen, G.; Moses, M.; Zou, Y.; Xiong, S.; Xue, W.; Dong, Y.; et al. Targeting epigenetic regulators to overcome drug resistance in the emerging human fungal pathogen *Candida auris*. *Nat. Commun.* **2025**, *16*, 4668. [CrossRef]
15. Xie, J.L.; Polvi, E.J.; Shekhar-Guturja, T.; Cowen, L.E. Elucidating drug resistance in human fungal pathogens. *Future Microbiol.* **2014**, *9*, 523–542. [CrossRef]
16. Yu, S.-J.; Chang, Y.-L.; Chen, Y.-L. Deletion of ADA2 Increases Antifungal Drug Susceptibility and Virulence in *Candida glabrata*. *Antimicrob. Agents Chemother.* **2018**, *62*, e01924-17. [CrossRef] [PubMed]
17. Zuzarte, M.; Gonçalves, M.J.; Cruz, M.T.; Cavaleiro, C.; Canhoto, J.; Vaz, S.; Pinto, E.; Salgueiro, L. *Lavandula luisieri* essential oil as a source of antifungal drugs. *Food Chem.* **2012**, *135*, 1505–1510. [CrossRef]
18. Di Santo, R. Natural products as antifungal agents against clinically relevant pathogens. *Nat. Prod. Rep.* **2010**, *27*, 1084–1098. [CrossRef] [PubMed]
19. Zhang, X.; Liu, Y.; Shen, Z.; Wang, S.; Wu, C.; Liu, D.; Tang, S.; Dai, C. Osthole ameliorates myonecrosis caused by *Clostridium perfringens* type A infection in mice. *One Health Adv.* **2023**, *1*, 27. [CrossRef]
20. Dong, F.R.; Gao, L.; Wang, L.; Jiang, Y.Y.; Jin, Y.S. Natural Products as Antifungal Agents against Invasive Fungi. *Curr. Top. Med. Chem.* **2023**, *23*, 1859–1917. [CrossRef] [PubMed]
21. Luo, L.; Yang, J.; Wang, C.; Wu, J.; Li, Y.; Zhang, X.; Li, H.; Zhang, H.; Zhou, Y.; Lu, A.; et al. Natural products for infectious microbes and diseases: An overview of sources, compounds, and chemical diversities. *Sci. China Life Sci.* **2022**, *65*, 1123–1145. [CrossRef] [PubMed]
22. Tran, T.M.; Atanasova, V.; Tardif, C.; Richard-Forget, F. Stilbenoids as Promising Natural Product-Based Solutions in a Race against Mycotoxigenic Fungi: A Comprehensive Review. *J. Agric. Food Chem.* **2023**, *71*, 5075–5092. [CrossRef]
23. Salehi, B.; Mishra, A.P.; Shukla, I.; Sharifi-Rad, M.; Contreras, M.D.M.; Segura-Carretero, A.; Fathi, H.; Nasrabadi, N.N.; Kobarfard, F.; Sharifi-Rad, J. Thymol, thyme, and other plant sources: Health and potential uses. *Phytother. Res.* **2018**, *32*, 1688–1706. [CrossRef]
24. Dong, Y.; Wei, Z.; Yang, R.; Zhang, Y.; Sun, M.; Bai, H.; Mo, M.; Yao, C.; Li, H.; Shi, L. Chemical Compositions of Essential Oil Extracted from Eight Thyme Species and Potential Biological Functions. *Plants* **2023**, *12*, 4164. [CrossRef] [PubMed]
25. Kowalczyk, A.; Przychodna, M.; Sopata, S.; Bodalska, A.; Fecka, I. Thymol and Thyme Essential Oil—New Insights into Selected Therapeutic Applications. *Molecules* **2020**, *25*, 4125. [CrossRef]
26. Marchese, A.; Orhan, I.E.; Daglia, M.; Barbieri, R.; Di Lorenzo, A.; Nabavi, S.F.; Gortzi, O.; Izadi, M.; Nabavi, S.M. Antibacterial and antifungal activities of thymol: A brief review of the literature. *Food Chem.* **2016**, *210*, 402–414. [CrossRef]
27. Peng, X.; Zhang, X.; Sharma, G.; Dai, C. Thymol as a Potential Neuroprotective Agent: Mechanisms, Efficacy, and Future Prospects. *J. Agric. Food Chem.* **2024**, *72*, 6803–6814. [CrossRef] [PubMed]
28. Raal, A.; Gontova, T.; Ivask, A.; Orav, A.; Koshovyi, O. Yield, Composition, and Chemotypes of Essential Oils from *Origanum vulgare* L. Aerial Parts Cultivated in Different European Countries. *Agronomy* **2024**, *14*, 3046. [CrossRef]
29. de Sousa, D.P.; Damasceno, R.O.S.; Amorati, R.; Elshabrawy, H.A.; de Castro, R.D.; Bezerra, D.P.; Nunes, V.R.V.; Gomes, R.C.; Lima, T.C. Essential Oils: Chemistry and Pharmacological Activities. *Biomolecules* **2023**, *13*, 1144. [CrossRef]
30. Perestrelo, R.; Sousa, P.; Hontman, N.; Câmara, J.S. Composition of Different Species of Lamiaceae Plant Family: A Potential Biosource of Terpenoids and Antifungal Agents. In *Advances in Antifungal Drug Development: Natural Products with Antifungal Potential*; Manzoor, N., Ed.; Springer Nature: Singapore, 2024; pp. 41–63.
31. Stević, T.; Berić, T.; Šavikin, K.; Soković, M.; Gođevac, D.; Dimkić, I.; Stanković, S. Antifungal activity of selected essential oils against fungi isolated from medicinal plant. *Ind. Crops Prod.* **2014**, *55*, 116–122. [CrossRef]
32. Liang, L.; Zhang, W.; Hao, J.; Wang, Y.; Wei, S.; Zhang, S.; Hu, Y.-s.; Lv, Y. Estragole Inhibits Growth and Aflatoxin Biosynthesis of *Aspergillus flavus* by Affecting Reactive Oxygen Species Homeostasis. *Microbiol. Spectr.* **2023**, *11*, e0134823. [CrossRef] [PubMed]
33. Wang, K.; Li, Z.; Shen, S.; Wang, L.; Wang, H.; Li, D. Terpinen-4-ol triggers autophagy activation and metacaspase-dependent apoptosis against *Botrytis cinerea*. *Front. Microbiol.* **2025**, *16*, 1600831. [CrossRef] [PubMed]
34. Castro, J.A.M.; Monteiro, O.; Coutinho, D.; Rodrigues, A.A.C.; Silva, J.; Maia, J. Seasonal and Circadian Study of a Thymol/γ-Terpinene/p-Cymene Type Oil of *Ocimum gratissimum* L. and Its Antioxidant and Antifungal Effects. *J. Braz. Chem. Soc.* **2019**, *30*, 930–938. [CrossRef]
35. Jung, K.-W.; Chung, M.-S.; Bai, H.-W.; Chung, B.-Y.; Lee, S. Investigation of Antifungal Mechanisms of Thymol in the Human Fungal Pathogen, *Cryptococcus neoformans*. *Molecules* **2021**, *26*, 3476. [CrossRef] [PubMed]

https://doi.org/10.3390/biom16010149

36. Wang, W.; He, L.; Zhang, Z.; Li, W.; Chen, J.; Chen, T.; Long, Y. Activity of the botanical compound thymol against kiwifruit rot caused by *Fusarium tricinctum* and the underlying mechanisms. *Pest Manag. Sci.* **2023**, *79*, 2493–2502. [CrossRef]
37. Braga, P.C.; Culici, M.; Alfieri, M.; Dal Sasso, M. Thymol inhibits *Candida albicans* biofilm formation and mature biofilm. *Int. J. Antimicrob. Agents* **2008**, *31*, 472–477. [CrossRef]
38. Gao, T.; Zhou, H.; Zhou, W.; Hu, L.; Chen, J.; Shi, Z. The Fungicidal Activity of Thymol against *Fusarium graminearum* via Inducing Lipid Peroxidation and Disrupting Ergosterol Biosynthesis. *Molecules* **2016**, *21*, 770. [CrossRef]
39. Radocchia, G.; Giammarino, A.; Barberini, S.; Verdolini, L.; De Angelis, M.; Simonetti, G.; Pantanella, F.; Schippa, S.; Angiolella, L. Carvacrol and Thymol, a Synergistic Antimicrobial Activity Against Bacterial and *Candida* Species. *MicrobiologyOpen* **2025**, *14*, e70089. [CrossRef]
40. Wang, L.-M.; Yang, H.; Yan, H.-J.; Ge, R.-F.; Wang, Y.-X.; Xue, S.-S.; Li, L.; Lyu, L.-Y.; Che, C.-Y. Thymol Protects against *Aspergillus Fumigatus* Keratitis by Inhibiting the LOX-1/IL-1β Signaling Pathway. *Curr. Med. Sci* **2022**, *42*, 620–628. [CrossRef] [PubMed]
41. Jafri, H.; Ahmad, I. *Thymus vulgaris* essential oil and thymol inhibit biofilms and interact synergistically with antifungal drugs against drug resistant strains of *Candida albicans* and *Candida tropicalis*. *J. Mycol. Med.* **2020**, *30*, 100911. [CrossRef] [PubMed]
42. Blanc, A.R.; Sortino, M.A.; Butassi, E.; Svetaz, L.A. Synergistic effects of *Thymus vulgaris* essential oil in combination with antifungal agents and inhibition of virulence factors of *Candida albicans*. *Phytomed. Plus* **2023**, *3*, 100481. [CrossRef]
43. Sharifzadeh, A.; Khosravi, A.R.; Shokri, H.; Shirzadi, H. Potential effect of 2-isopropyl-5-methylphenol (thymol) alone and in combination with fluconazole against clinical isolates of *Candida albicans*, *C. glabrata* and *C. krusei*. *J. Mycol. Medicale* **2018**, *28*, 294–299. [CrossRef]
44. de Castro, R.D.; de Souza, T.M.P.A.; Bezerra, L.M.D.; Ferreira, G.L.S.; Costa, E.M.M.d.B.; Cavalcanti, A.L. Antifungal activity and mode of action of thymol and its synergism with nystatin against *Candida* species involved with infections in the oral cavity: An in vitro study. *BMC Complement. Altern. Med.* **2015**, *15*, 417. [CrossRef]
45. Denning, D.W. Global incidence and mortality of severe fungal disease. *Lancet Infect. Dis.* **2024**, *24*, e428–e438. [CrossRef]
46. Gamaletsou, M.N.; Drogari-Apiranthitou, M.; Denning, D.W.; Sipsas, N.V. An estimate of the burden of serious fungal diseases in Greece. *Eur. J. Clin. Microbiol. Infect. Dis.* **2016**, *35*, 1115–1120. [CrossRef] [PubMed]
47. Chayakulkeeree, M.; Denning, D.W. Serious fungal infections in Thailand. *Eur. J. Clin. Microbiol. Infect. Dis.* **2017**, *36*, 931–935. [CrossRef] [PubMed]
48. Chrdle, A.; Mallátová, N.a.; Vašáková, M.; Haber, J.; Denning, D.W. Burden of serious fungal infections in the Czech Republic. *Mycoses* **2015**, *58*, 6–14. [CrossRef] [PubMed]
49. Osmanov, A.; Denning, D.W. Burden of serious fungal infections in Ukraine. *Mycoses* **2015**, *58*, 94–100. [CrossRef] [PubMed]
50. Lass-Flörl, C.; Kanj, S.S.; Govender, N.P.; Thompson, G.R.; Ostrosky-Zeichner, L.; Govrins, M.A. Invasive candidiasis. *Nat. Rev. Dis. Primers* **2024**, *10*, 20. [CrossRef]
51. Mayer, F.L.; Wilson, D.; Hube, B. *Candida albicans* pathogenicity mechanisms. *Virulence* **2013**, *4*, 119–128. [CrossRef] [PubMed]
52. Pappas, P.G.; Kauffman, C.A.; Andes, D.R.; Clancy, C.J.; Marr, K.A.; Ostrosky-Zeichner, L.; Reboli, A.C.; Schuster, M.G.; Vazquez, J.A.; Walsh, T.J.; et al. Clinical Practice Guideline for the Management of Candidiasis: 2016 Update by the Infectious Diseases Society of America. *Clin. Infect. Dis.* **2015**, *62*, e1–e50. [CrossRef]
53. Cadena, J.; Thompson, G.R.; Patterson, T.F. Aspergillosis: Epidemiology, Diagnosis, and Treatment. *Infect. Dis. Clin. N. Am.* **2021**, *35*, 415–434. [CrossRef]
54. Jaggi, T.K.; Agarwal, R.; Tiew, P.Y.; Shah, A.; Lydon, E.C.; Hage, C.A.; Waterer, G.W.; Langelier, C.R.; Delhaes, L.; Chotirmall, S.H. Fungal lung disease. *Eur. Respir. J.* **2024**, *64*, 2400803. [CrossRef]
55. Barrs, V.R.; Talbot, J.J. Fungal Rhinosinusitis and Disseminated Invasive Aspergillosis in Cats. *Vet. Clin. N. Am. Small Anim. Pract.* **2019**, *50*, 331–357. [CrossRef]
56. Chaturvedi, V.; Chaturvedi, S. *Cryptococcus gattii*: A resurgent fungal pathogen. *Trends Microbiol.* **2011**, *19*, 564–571. [CrossRef] [PubMed]
57. May, R.C.; Stone, N.R.H.; Wiesner, D.L.; Bicanic, T.; Nielsen, K. *Cryptococcus*: From environmental saprophyte to global pathogen. *Nat. Rev. Microbiol.* **2015**, *14*, 106–117. [CrossRef] [PubMed]
58. Rajasingham, R.; Smith, R.M.; Park, B.J.; Jarvis, J.N.; Govender, N.P.; Chiller, T.M.; Denning, D.W.; Loyse, A.; Boulware, D.R. Global burden of disease of HIV-associated cryptococcal meningitis: An updated analysis. *Lancet Infect. Dis.* **2017**, *17*, 873–881. [CrossRef]
59. Osman, M.; Kasir, D.; Rafei, R.; Kassem, I.I.; Ismail, M.B.; El Omari, K.; Dabboussi, F.; Cazer, C.; Papon, N.; Bouchara, J.-P.; et al. Trends in the epidemiology of dermatophytosis in the Middle East and North Africa region. *Int. J. Dermatol.* **2021**, *61*, 935–968. [CrossRef] [PubMed]
60. Son, J.H.; Doh, J.Y.; Han, K.; Kim, Y.H.; Han, J.H.; Bang, C.H.; Park, Y.M.; Lee, J.H. Risk factors of dermatophytosis among Korean adults. *Sci. Rep.* **2022**, *12*, 13444. [CrossRef] [PubMed]

61. Aneke, C.I.; Otranto, D.; Cafarchia, C. Therapy and Antifungal Susceptibility Profile of *Microsporum canis*. *J. Fungi* **2018**, *4*, 107. [CrossRef]
62. Abdolrasouli, A.; Barton, R.C.; Borman, A.M. Spread of Antifungal-Resistant *Trichophyton indotineae*, United Kingdom, 2017–2024. *Emerg. Infect. Dis.* **2025**, *31*, 192–194. [CrossRef] [PubMed]
63. Baltazar, L.M.; Santos, D.A. Perspective on animal models of dermatophytosis caused by *Trichophyton rubrum*. *Virulence* **2015**, *6*, 372–375. [CrossRef]
64. Weyant, R.B.; Kabbani, D.; Doucette, K.; Lau, C.; Cervera, C. *Pneumocystis jirovecii*: A review with a focus on prevention and treatment. *Expert Opin. Pharmacother.* **2021**, *22*, 1579–1592. [CrossRef] [PubMed]
65. Nevez, G.; Hauser, P.M.; Le Gal, S. *Pneumocystis jirovecii*. *Trends Microbiol.* **2020**, *28*, 1034–1035. [CrossRef] [PubMed]
66. Catherinot, E.; Lanternier, F.; Bougnoux, M.-E.; Lecuit, M.; Couderc, L.-J.; Lortholary, O. *Pneumocystis jirovecii* Pneumonia. *Infect. Dis. Clin. N. Am.* **2010**, *24*, 107–138. [CrossRef]
67. McDonald, E.G.; Afshar, A.; Assiri, B.; Boyles, T.; Hsu, J.M.; Khuong, N.; Prosty, C.; So, M.; Sohani, Z.N.; Butler-Laporte, G.; et al. *Pneumocystis jirovecii* pneumonia in people living with HIV: A review. *Clin. Microbiol. Rev.* **2024**, *37*, e0010122. [CrossRef] [PubMed]
68. Busca, A.; Tortorano, A.M.; Pagano, L. Reviewing the importance and evolution of fungal infections and potential antifungal resistance in haematological patients. *J. Glob. Antimicrob. Resist.* **2015**, *3*, 237–241. [CrossRef]
69. Kumari, P.; Mishra, R.; Arora, N.; Chatrath, A.; Gangwar, R.; Roy, P.; Prasad, R. Antifungal and Anti-Biofilm Activity of Essential Oil Active Components against *Cryptococcus neoformans* and *Cryptococcus laurentii*. *Front. Microbiol.* **2017**, *8*, 2161. [CrossRef] [PubMed]
70. Mahboub, H.H.; Tartor, Y.H. Carvacrol essential oil stimulates growth performance, immune response, and tolerance of Nile tilapia to *Cryptococcus uniguttulatus* infection. *Dis. Aquat. Organ.* **2020**, *141*, 1–14. [CrossRef] [PubMed]
71. Poulain, D. *Candida albicans*, plasticity and pathogenesis. *Crit. Rev. Microbiol.* **2013**, *41*, 208–217. [CrossRef]
72. Han, S.-M. Successful treatment of *Candida albicans*-induced fungal cystitis in a dog using caspofungin: A case report. *Vet. Res. Forum* **2024**, *15*, 435–438. [CrossRef]
73. Arné, P.; Thierry, S.; Wang, D.; Deville, M.; Le Loc'h, G.; Desoutter, A.; Féménia, F.; Nieguitsila, A.; Huang, W.; Chermette, R.; et al. *Aspergillus fumigatus* in Poultry. *Int. J. Microbiol.* **2011**, *2011*, 746356. [CrossRef]
74. Headley, S.A.; Mota, F.C.D.; Lindsay, S.; de Oliveira, L.M.; Medeiros, A.A.; Pretto-Giordano, L.G.; Saut, J.P.E.; Krockenberger, M. *Cryptococcus neoformans* var. *grubii*-Induced Arthritis with Encephalitic Dissemination in a Dog and Review of Published Literature. *Mycopathologia* **2016**, *181*, 595–601. [CrossRef] [PubMed]
75. Li, Y.-H.; Liu, T.-B. Zinc Finger Proteins in the Human Fungal Pathogen *Cryptococcus neoformans*. *Int. J. Mol. Sci.* **2020**, *21*, 1361. [CrossRef]
76. Zhao, X.; Zhang, Y.; Chen, L.; Ma, Z.; Zhang, B. Chitosan-thymol nanoparticle with pH responsiveness as a potential intelligent botanical fungicide against *Botrytis cinerea*. *Pestic. Biochem. Physiol.* **2023**, *195*, 105571. [CrossRef]
77. Tompros, A.; Wilber, M.Q.; Fenton, A.; Carter, E.D.; Gray, M.J. Efficacy of Plant-Derived Fungicides at Inhibiting *Batrachochytrium salamandrivorans* Growth. *J. Fungi* **2022**, *8*, 1025. [CrossRef] [PubMed]
78. de Sousa Cutrim, T.A.; Eloy, M.A.; Barcelos, F.F.; Meireles, L.M.; de Freitas Ferreira, L.C.; Reis, T.A.; Gonçalves, S.S.; Lacerda, V.; Fronza, M.; Morais, P.A.B.; et al. New thymol-derived triazole exhibits promising activity against *Trichophyton rubrum*. *Braz. J. Microbiol.* **2024**, *55*, 1287–1295. [CrossRef] [PubMed]
79. Perina, F.J.; Amaral, D.C.; Fernandes, R.S.; Labory, C.R.; Teixeira, G.A.; Alves, E. *Thymus vulgaris* essential oil and thymol against *Alternaria alternata* (Fr.) Keissler: Effects on growth, viability, early infection and cellular mode of action. *Pest Manag. Sci.* **2014**, *71*, 1371–1378. [CrossRef] [PubMed]
80. Usha, I.; Pramesh, D.; Shrikanth, B.; Raghunandana, A.; Harish, M.N.; Prasannakumar, M.K.; Sharanabasav, H.; Manjunatha, C. Thymol—A plant-based monoterpenoid phenol compound of an essential oil for the management of sheath blight disease of rice. *Pest Manag. Sci.* **2024**, *80*, 6334–6343. [CrossRef]
81. Miranda-Cadena, K.; Marcos-Arias, C.; Mateo, E.; Aguirre-Urizar, J.M.; Quindós, G.; Eraso, E. In vitro activities of carvacrol, cinnamaldehyde and thymol against *Candida* biofilms. *Biomed. Pharmacother.* **2021**, *143*, 112218. [CrossRef]
82. Negro, M.F.; Bustos, P.S.; Bellezze, L.; Ortega, M.G.; Echeverría, J.; Silva, M.F.; Peralta, M.A. Menthol-thymol NADES as a fungicidal and chemosensitizing agent against multidrug-resistant *Candida albicans*: ROS induction, efflux pump inhibition, and synergy with fluconazole. *Front. Pharmacol.* **2025**, *16*, 1643472. [CrossRef] [PubMed]
83. Sim, J.X.F.; Khazandi, M.; Chan, W.Y.; Trott, D.J.; Deo, P. Antimicrobial activity of thyme oil, oregano oil, thymol and carvacrol against sensitive and resistant microbial isolates from dogs with otitis externa. *Vet. Dermatol.* **2019**, *30*, 524-e159. [CrossRef] [PubMed]
84. Ji, S.H.; Kim, T.K.; Keum, Y.S.; Chun, S.-C. The Major Postharvest Disease of Onion and Its Control with Thymol Fumigation During Low-Temperature Storage. *Mycobiology* **2018**, *46*, 242–253. [CrossRef] [PubMed]

https://doi.org/10.3390/biom16010149

85. Zhang, M.; Ge, J.; Yu, X. Transcriptome Analysis Reveals the Mechanism of Fungicidal of Thymol Against *Fusarium oxysporum* f. sp. *niveum*. *Curr. Microbiol.* **2018**, *75*, 410–419. [CrossRef]
86. Wang, H.; Peng, Z.; Sun, H. Antifungal activities and mechanisms of trans-cinnamaldehyde and thymol against food-spoilage yeast *Zygosaccharomyces rouxii*. *J. Food Sci.* **2022**, *87*, 1197–1210. [CrossRef]
87. Shen, Q.; Zhou, W.; Li, H.; Hu, L.; Mo, H. ROS Involves the Fungicidal Actions of Thymol against Spores of *Aspergillus flavus* via the Induction of Nitric Oxide. *PLoS ONE* **2016**, *11*, e0155647. [CrossRef] [PubMed]
88. Zhang, J.; Ma, S.; Du, S.; Chen, S.; Sun, H. Antifungal activity of thymol and carvacrol against postharvest pathogens *Botrytis cinerea*. *J. Food Sci. Technol.* **2019**, *56*, 2611–2620. [CrossRef] [PubMed]
89. Schorr, R.R.; Ballesteros Garcia, M.J.; Petermann, D.; Moreira, R.R.; Sales Maia, B.H.L.N.; Marques, F.A.; May-De Mio, L.L. Eugenol, Isoeugenol, Thymol, Carvacrol, and Ester Derivatives as an Ecofriendly Option to Control Glomerella Leaf Spot and Bitter Rot on Apple. *Plants* **2024**, *13*, 3196. [CrossRef]
90. Burt, S. Essential oils: Their antibacterial properties and potential applications in foods—A review. *Int. J. Food Microbiol.* **2004**, *94*, 223–253. [CrossRef]
91. Ferreira, J.V.N.; Lago, J.H.G.; Caseli, L. Thymol in cellular membrane models formed by negative charged lipids causes aggregation at the air-water interface. *Chem. Phys. Lett.* **2019**, *717*, 87–90. [CrossRef]
92. Karonen, M. Insights into Polyphenol–Lipid Interactions: Chemical Methods, Molecular Aspects and Their Effects on Membrane Structures. *Plants* **2022**, *11*, 1809. [CrossRef]
93. Kim, J.; Lee, J.-E.; Huh, M.-J.; Lee, S.-C.; Seo, S.-M.; Kwon, J.H.; Park, I.-K. Fumigant Antifungal Activity via Reactive Oxygen Species of *Thymus vulgaris* and *Satureja hortensis* Essential Oils and Constituents against *Raffaelea quercus-mongolicae* and *Rhizoctonia solani*. *Biomolecules* **2019**, *9*, 561. [CrossRef]
94. Zhang, Y.; Tan, Y.; OuYang, Q.; Duan, B.; Wang, Z.; Meng, K.; Tan, X.; Tao, N. γ-Cyclodextrin encapsulated thymol for citrus preservation and its possible mechanism against *Penicillium digitatum*. *Pestic. Biochem. Physiol.* **2023**, *194*, 105501. [CrossRef] [PubMed]
95. Wang, H.; Peng, Z.; Wang, C.; Zhu, Y.; Xia, F.; Sun, H. Thymol and trans-cinnamaldehyde induce apoptosis through a metacaspase-dependent mitochondrial pathway in food-spoilage yeast *Zygosaccharomyces rouxii*. *J. Food Sci.* **2022**, *87*, 4119–4136. [CrossRef]
96. Zida, A.; Bamba, S.; Yacouba, A.; Ouedraogo-Traore, R.; Guiguemdé, R.T. Anti-*Candida albicans* natural products, sources of new antifungal drugs: A review. *J. Mycol. Med.* **2017**, *27*, 1–16. [CrossRef]
97. Hasim, S.; Coleman, J.J. Targeting the fungal cell wall: Current therapies and implications for development of alternative antifungal agents. *Future Med. Chem.* **2019**, *11*, 869–883. [CrossRef] [PubMed]
98. Ranjbar, A.; Valero, D.; Shekarforoush, S.; Niakosari, M.; Eshghi, S. Antifungal activity of thymol against the main fungi causing pomegranate fruit rot by suppressing the activity of cell wall degrading enzymes. *LWT* **2022**, *161*, 113303. [CrossRef]
99. Piotrowski, J.S.; Okada, H.; Lu, F.; Li, S.C.; Hinchman, L.; Ranjan, A.; Smith, D.L.; Higbee, A.J.; Ulbrich, A.; Coon, J.J.; et al. Plant-derived antifungal agent poacic acid targets β-1,3-glucan. *Proc. Natl. Acad. Sci. USA* **2015**, *112*, E1490–E1497. [CrossRef]
100. Zhang, X.; Liu, Y.; Liu, M.; Ma, Q.; Hao, Z.; Tang, S.; Dai, S. Ellagic acid supplementation ameliorates cisplatin-induced liver injury in mice by inhibiting the NF-κB pathway and activating the Nrf2/HO-1 pathway. *One Health Adv.* **2024**, *2*, 20. [CrossRef]
101. Moyer, A.; Tanaka, K.; Cheng, E.H. Apoptosis in Cancer Biology and Therapy. *Annu. Rev. Pathol.* **2025**, *20*, 303–328. [CrossRef]
102. Darvishi, E.; Omidi, M.; Bushehri, A.A.; Golshani, A.; Smith, M.L. Thymol antifungal mode of action involves telomerase inhibition. *Med. Mycol.* **2013**, *51*, 826–834. [CrossRef] [PubMed]
103. Lanzarin, G.A.B.; Félix, L.M.; Monteiro, S.M.; Ferreira, J.M.; Oliveira, P.A.; Venâncio, C. Anti-Inflammatory, Anti-Oxidative and Anti-Apoptotic Effects of Thymol and 24-Epibrassinolide in Zebrafish Larvae. *Antioxidants* **2023**, *12*, 1297. [CrossRef]
104. Gago, C.; Serralheiro, A.; Miguel, M.D.G. Anti-Inflammatory Activity of Thymol and Thymol-Rich Essential Oils: Mechanisms, Applications, and Recent Findings. *Molecules* **2025**, *30*, 2450. [CrossRef] [PubMed]
105. Vervaeke, A.; Lamkanfi, M. MAP Kinase Signaling at the Crossroads of Inflammasome Activation. *Immunol. Rev.* **2025**, *329*, e13436. [CrossRef] [PubMed]
106. Li, Y.-N.; Zhang, S.-B.; Lv, Y.-Y.; Zhai, H.-C.; Cai, J.-P.; Hu, Y.-S. Mechanisms underlying the inhibitory effects of linalool on *Aspergillus flavus* spore germination. *Appl. Microbiol. Biot.* **2022**, *106*, 6625–6640. [CrossRef]
107. Leite-Andrade, M.C.; de Araújo Neto, L.N.; Buonafina-Paz, M.D.S.; de Assis Graciano Dos Santos, F.; da Silva Alves, A.I.; de Castro, M.C.A.B.; Mori, E.; de Lacerda, B.C.G.V.; Araújo, I.M.; Coutinho, H.D.M.; et al. Antifungal Effect and Inhibition of the Virulence Mechanism of D-Limonene against *Candida parapsilosis*. *Molecules* **2022**, *27*, 8884. [CrossRef] [PubMed]
108. Keymaram, M.; Falahati, M.; Farahyar, S.; Lotfali, E.; Abolghasemi, S.; Mahmoudi, S.; Sadeghi, F.; Khalandi, H.; Ghasemi, R.; Shamsaei, S.; et al. Anti-biofilm properties of eucalyptol in combination with antifungals against *Candida albicans* isolates in patients with hematological malignancy. *Arch. Microbiol.* **2022**, *204*, 295. [CrossRef] [PubMed]
109. Duan, W.-Y.; Zhu, X.-M.; Zhang, S.-B.; Lv, Y.-Y.; Zhai, H.-C.; Wei, S.; Ma, P.-A.; Hu, Y.-S. Antifungal effects of carvacrol, the main volatile compound in *Origanum vulgare* L. essential oil, against *Aspergillus flavus* in postharvest wheat. *Int. J. Food Microbiol.* **2023**, *410*, 110514. [CrossRef]

https://doi.org/10.3390/biom16010149

110. Zheng, H.; Li, J.; Guo, Z.; Smagghe, G.; Gui, S.; Wu, X.; Chen, X. Antifungal activities and underlying mechanisms of carvacrol against *Neopestalotiopsis saprophytica* causing leaf brown spot in *Rosa roxburghii*. *Pest Manag. Sci.* **2025**, *81*, 5848–5863. [CrossRef]
111. Cossu, A.; Wang, M.S.; Chaudhari, A.; Nitin, N. Antifungal activity against *Candida albicans* of starch Pickering emulsion with thymol or amphotericin B in suspension and calcium alginate films. *Int. J. Pharmaceut.* **2015**, *493*, 233–242. [CrossRef]
112. Yang, X.; Jiang, S.; Zhang, M.; Li, T.; Jin, Z.; Wu, X.; Chi, Y.R. Discovery of novel piperidine-containing thymol derivatives as potent antifungal agents for crop protection. *Pest Manag. Sci.* **2024**, *80*, 4906–4914. [CrossRef]
113. Alves Eloy, M.; Ribeiro, R.; Martins Meireles, L.; de Sousa Cutrim, T.A.; Santana Francisco, C.; Lirian Javarini, C.; Borges, W.d.S.; Costa, A.V.; Queiroz, V.T.d.; Scherer, R.; et al. Thymol as an Interesting Building Block for Promising Fungicides against *Fusarium solani*. *J. Agric. Food Chem.* **2021**, *69*, 6958–6967. [CrossRef]
114. Conte, J.; Saatkamp, R.H.; Sanches, M.P.; Argenta, D.F.; da Rosa Monte Machado, G.; Kretzer, I.F.; Parize, A.L.; Caon, T. Development of biopolymer films loaded with fluconazole and thymol for resistant vaginal candidiasis. *Int. J. Biol. Macromol.* **2024**, *275*, 133356. [CrossRef]
115. Bae, Y.S.; Rhee, M.S. Short-Term Antifungal Treatments of Caprylic Acid with Carvacrol or Thymol Induce Synergistic 6-Log Reduction of Pathogenic *Candida albicans* by Cell Membrane Disruption and Efflux Pump Inhibition. *Cell. Physiol. Biochem.* **2019**, *53*, 285–300. [CrossRef]
116. Shcherbakova, L.; Mikityuk, O.; Arslanova, L.; Stakheev, A.; Erokhin, D.; Zavriev, S.; Dzhavakhiya, V. Studying the Ability of Thymol to Improve Fungicidal Effects of Tebuconazole and Difenoconazole Against Some Plant Pathogenic Fungi in Seed or Foliar Treatments. *Front. Microbiol.* **2021**, *12*, 629429. [CrossRef] [PubMed]
117. Sadeer, N.B.; Mahomoodally, M.F. Antibiotic Potentiation of Natural Products: A Promising Target to Fight Pathogenic Bacteria. *Curr. Drug Targets* **2021**, *22*, 555–572. [CrossRef] [PubMed]
118. Michiels, J.; Missotten, J.; Dierick, N.; Fremaut, D.; Maene, P.; De Smet, S. In vitro degradation and in vivo passage kinetics of carvacrol, thymol, eugenol and trans-cinnamaldehyde along the gastrointestinal tract of piglets. *J. Sci. Food Agric.* **2008**, *88*, 2371–2381. [CrossRef]
119. Delgado-Marín, L.; Sánchez-Borzone, M.; García, D.A. Neuroprotective effects of gabaergic phenols correlated with their pharmacological and antioxidant properties. *Life Sci.* **2017**, *175*, 11–15. [CrossRef]
120. García, D.A.; Bujons, J.; Vale, C.; Suñol, C. Allosteric positive interaction of thymol with the GABAA receptor in primary cultures of mouse cortical neurons. *Neuropharmacology* **2005**, *50*, 25–35. [CrossRef]
121. Balan, D.J.; Rajavel, T.; Das, M.; Sathya, S.; Jeyakumar, M.; Devi, K.P. Thymol induces mitochondrial pathway-mediated apoptosis via ROS generation, macromolecular damage and SOD diminution in A549 cells. *Pharmacol. Rep.* **2021**, *73*, 240–254. [CrossRef] [PubMed]
122. Luna, A.; Lema-Alba, R.C.; Dambolena, J.S.; Zygadlo, J.A.; Labaque, M.C.; Marin, R.H. Thymol as natural antioxidant additive for poultry feed: Oxidative stability improvement. *Poult. Sci.* **2017**, *96*, 3214–3220. [CrossRef] [PubMed]
123. Hashemipour, H.; Kermanshahi, H.; Golian, A.; Veldkamp, T. Effect of thymol and carvacrol feed supplementation on performance, antioxidant enzyme activities, fatty acid composition, digestive enzyme activities, and immune response in broiler chickens. *Poult. Sci.* **2013**, *92*, 2059–2069. [CrossRef] [PubMed]
124. Félix, L.; Monteiro, S.M.; Venâncio, C. Evaluation of thymol application for anaesthesia of adult zebrafish. *Fish Physiol. Biochem.* **2025**, *51*, 155. [CrossRef]
125. Zhang, X.; Peng, Y.; Wu, C. Chicken embryonic toxicity and potential in vitro estrogenic and mutagenic activity of carvacrol and thymol in low dose/concentration. *Food Chem. Toxicol.* **2021**, *150*, 112038. [CrossRef]
126. Xie, K.; Tashkin, D.P.; Luo, M.Z.; Zhang, J.Y. Chronic toxicity of inhaled thymol in lungs and respiratory tracts in mouse model. *Pharmacol. Res. Perspect.* **2019**, *7*, e00516. [CrossRef] [PubMed]
127. Kumar, A.; Kujur, A.; Singh, P.P.; Prakash, B. Nanoencapsulated plant-based bioactive formulation against food-borne molds and aflatoxin B1 contamination: Preparation, characterization and stability evaluation in the food system. *Food Chem.* **2019**, *287*, 139–150. [CrossRef] [PubMed]
128. Min, T.; Zhou, L.; Sun, X.; Du, H.; Bian, X.; Zhu, Z.; Wen, Y. Enzyme-responsive food packaging system based on pectin-coated poly (lactic acid) nanofiber films for controlled release of thymol. *Food Res. Int.* **2022**, *157*, 111256. [CrossRef] [PubMed]
129. Penk, A.; Pittrow, L. Role of fluconazole in the long-term suppressive therapy of fungal infections in patients with artificial implants. *Mycoses* **1999**, *42*, 91–96. [CrossRef] [PubMed]

https://doi.org/10.3390/biom16010149

antioxidants

Article

Ellagic Acid Reduces Cadmium Exposure-Induced Apoptosis in HT22 Cells via Inhibiting Oxidative Stress and Mitochondrial Dysfunction and Activating Nrf2/HO-1 Pathway

Yue Liu [1,2,3], Chunhong Chen [1], Zhihui Hao [1,2,3], Jianzhong Shen [1,2,3], Shusheng Tang [1,2,3,*] and Chongshan Dai [1,2,3,*]

1 National Key Laboratory of Veterinary Public Health and Safety, College of Veterinary Medicine, China Agricultural University, Beijing 100193, China; s20203050760@cau.edu.cn (Y.L.); s20223050899@cau.edu.cn (C.C.); haozhihui@cau.edu.cn (Z.H.); sjz@cau.edu.cn (J.S.)
2 Technology Innovation Center for Food Safety Surveillance and Detection (Hainan), Sanya Institute of China Agricultural University, Sanya 572025, China
3 Key Biology Laboratory of Chinese Veterinary Medicine, Ministry of Agriculture and Rural Affairs, Beijing 100193, China
* Correspondence: tssfj@cau.edu.cn (S.T.); daichongshan@cau.edu.cn (C.D.)

Abstract: Exposure to cadmium sulfate ($CdSO_4$) can lead to neurotoxicity. Nevertheless, the precise molecular mechanisms underlying this phenomenon remain unclear, and effective treatment strategies are scarce. This study explored the protective effects of ellagic acid (EA), a natural polyphenolic compound, against $CdSO_4$ exposure-induced neurotoxicity in HT22 cells and the underlying molecular mechanisms. Our findings demonstrated that exposure of HT22 cells to $CdSO_4$ resulted in apoptosis, which was effectively reversed by EA in a dose-dependent manner. EA supplementation also decreased reactive oxygen species (ROS) and mitochondrial ROS production, reduced malondialdehyde (MDA) levels, and restored the activities of superoxide dismutase (SOD) and catalase (CAT). Additionally, EA supplementation at 5–20 μM significantly counteracted Cd-induced the loss of mitochondrial membrane potential and the decrease of ATP and reduced the ratio of Bax/Bcl-2 and cleaved-caspase-3 protein expression. Furthermore, EA supplementation resulted in the upregulation of Nrf2 and HO-1 protein and mRNAs while simultaneously downregulating the phosphorylation of JNK and p38 proteins. The pharmacological inhibition of c-Jun N-terminal kinase (JNK) partially attenuated the activation of the Nrf2/HO-1 pathway induced by $CdSO_4$ and exacerbated its cytotoxic effects. In conclusion, our findings suggest that ethyl acetate (EA) supplementation offers protective effects against $CdSO_4$-induced apoptosis in HT22 cells by inhibiting oxidative stress and activating the Nrf2 signaling pathway. Furthermore, the activation of the JNK pathway appears to play a protective role in $CdSO_4$-induced apoptosis in HT22 cells.

Keywords: cadmium; ellagic acid; JNK pathway; oxidative stress; apoptosis; Nrf2/HO-1 pathway

1. Introduction

Cadmium (Cd) is a prevalent non-essential heavy metal known to have highly toxic effects on both humans and animals [1]. Cd has been ranked 7th in the hazardous waste priority list according to the U.S. Environmental Protection Agency's Superfund, based on its level of toxic risk [1]. The World Health Organization and The International Agency for Research on Cancer (IARC) have both classified Cd as a human carcinogen [2]. Previous studies from animal models and in vitro cell models showed that Cd exposure could cause multiple toxic effects, including hepatoxicity, nephrotoxicity, reproductive toxicity, immunotoxicity, and gastrointestinal toxicity [3–7]. Cd has the potential to penetrate the blood-brain barrier (BBB), leading to its accumulation in the central nervous system and ultimately causing neurotoxicity. A recent epidemiological investigation showed that environmental exposure to Cd may be a risk factor for Alzheimer's disease in humans [8]. Epidemiologic studies have indicated that Cd exposure is positively corrected with Alzheimer's disease [9,10]. For example, a meta-analysis showed that Cd concentrations in whole blood, serum, or plasma samples of Alzheimer's disease patients were significantly higher than those in controls [10]. In a mouse model, Cd exposure could result in abnormal neurobehavioral changes (such as movement disorders, attention deficits, headaches, and cognitive dysfunction) and pathological injury in brain tissues [11–13]. Hence, investigating the potential health risks of environmental Cd exposure in the nervous system and understanding the underlying molecular mechanisms are of great importance.

The precise molecular mechanisms underlying neurotoxicity induced by Cd exposure are not yet fully elucidated. Deng et al. found that chronic exposure to Cd could result in cognitive function impairment in male mice by promoting mitochondrial fission and dysfunction within brain tissues [1]. It was reported that Cd exposure could affect the function of mitochondria by blocking the electron transport chain and reducing membrane potential in neuronal cells in vitro [8,14]. A prior study indicated that Cd exposure at 2.5–20 μM could induce excessive production of reactive oxygen species (ROS) and decrease the abilities of the antioxidant defense system, triggering oxidative stress damage and mitochondrial dysfunction in PC12 cells and rat primary cerebral cortical neurons [15]. Moreover, it was also documented that Cd exposure could also trigger the activation of p53, sirtuin-1 (SIRT1), mitogen-activated protein kinase (MAPK), PKR-like ER kinase (PERK)/activating transcription factor 4 (ATF4), and ferroptosis pathways in neuronal cells and human peripheral blood lymphocytes [13,15–19].

Ellagic acid (EA) is a polyphenolic compound (2,3,7,8-tetrahydroxychromeno (5,4,3-cde) chromene-5,10-dione, $C_{14}H_6O_8$; Figure 1), and it could be found in multiple gallnuts and fruits, such as pomegranates, black currants, raspberries, and mangoes [20]. The literature reports that EA exhibits various biological activities, such as anti-inflammatory, antiviral, antioxidant, antimicrobial, and immune-regulatory effects [21]. A mouse model has demonstrated that oral EA administration could effectively attenuate rotenone exposure-induced dopamine neuronal damage via targeting activation of the Nrf2 pathway [22]. Goudarzi et al. study showed that oral EA supplementation at the doses of 10 and 30 mg/kg per day for successive twenty-one days can effectively attenuate sodium arsenate-caused neurotoxicity by inhibiting oxidative stress and neuroinflammation in the brain tissue of rats [23]. The current study aims to examine the protective effects of EA supplementation against Cd exposure-induced toxicity in HT22 cells, a mouse hippocampal neuronal cell line. Additionally, we explore the underlying molecular mechanisms.

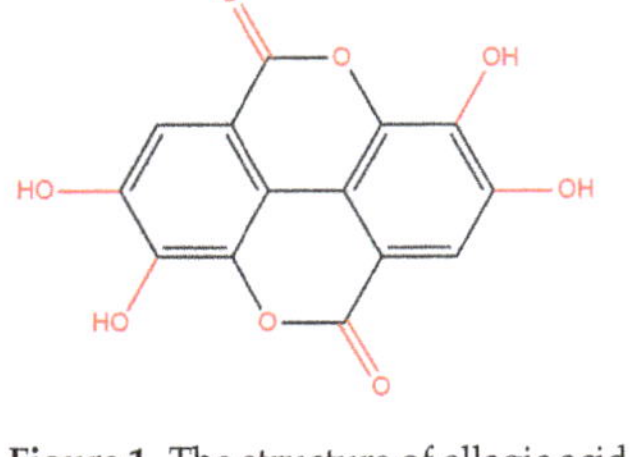

Figure 1. The structure of ellagic acid (2,3,7,8-tetrahydroxychromeno (5,4,3-cde) chromene-5,10-dione, $C_{14}H_6O_8$).

2. Materials and Methods

2.1. Chemical and Reagents

Cadmium sulfate ($CdSO_4$) (purity ≥ 98%) and ellagic acid (purity more than 96%) were both purchased from Aladdin (Shanghai, China). SB203580 (a special inhibitor of p38 MAPK), SP600125 (a special inhibitor of JNK), dimethyl sulfoxide (DMSO), 1% (*v*/*v*) penicillin-streptomycin, phenylmethanesulfonylfluoride (PMSF), pan-caspase inhibitor (Z-VAD-FMK), and N-acetylcysteine (NAC) were supplied by Beyotime Biotechnology

Company (Haimen, Jiangsu, China). ASN007 (a special inhibitor of ERK) was purchased from MCE (Shanghai, China). Mito-TEMPO was purchased from Santa Cruz Biotechnology (Dallas, TX, USA). All other chemical reagents used in the present study were at analysis grade.

2.2. Cell Viability and Treatment

The mouse HT22 and rat PC12 cell lines were both obtained from the Shanghai Cell Bank of the Chinese Academy of Sciences (Shanghai, China). Cells were cultured in Dulbecco's Modified Eagle's Medium (DMEM) supplemented with 10% (*v*/*v*) fetal bovine serum (FBS) and 1% (*v*/*v*) penicillin-streptomycin under controlled conditions in a cell culture incubator (Thermo Fisher Scientific, Shanghai, China) maintained at 5% CO_2 and 37 °C.

2.3. Measurement of Cell Viabilities

Cell viabilities were determined using a CCK-8 detection kit (Nanjing KeyGEN Biotech Co., Ltd., Nanjing, China), according to the instructions provided by the manufacturer. Briefly, A 96-well cell culture plate was used. A total of 1×10^4 cells were seeded per well. After 24 h cultivation, cells were treated with various concentrations of $CdSO_4$ (0.625, 1.25, 2.5, 5, 10, 20, and 40 μM) or EA (5, 10, 20, 40, and 80 μM) for another 24 h. After the treatment, cells were incubated with 10 μL of CCK-8 reagent for 1 h. Then, cell viabilities were determined at 450 nm using a microplate reader (Tecan Trading AG, Kanton Zürich, Switzerland).

We also tested the effects of EA, antioxidant NAC, and JNK inhibitor (i.e., SP600125) supplementation on Cd exposure-induced cytotoxicity in HT22 cells. In brief, cells were pre-treated with EA at the doses of 5, 10, and 20 μM, NAC at the dose of 2.5 mM, or the JNK inhibitor at a concentration of 10 μM, as well as SB203580 and ASN007 at concentrations of 5 and 10 μM for a duration of 2 h. Subsequently, the cells were co-treated with $CdSO_4$ at a concentration of 10 μM for an additional 24 h. Finally, cell viability was assessed.

2.4. LDH Measurement

Shown in Supplementary Material and Methods.

2.5. Measurement of Cell Apoptosis

Cell apoptosis analysis was performed using a commercial Annexin V-FITC/PI Apoptosis Detection Kit (Vazyme Biotech Co., Ltd., Nanjing, China) by flow cytometry analysis (Becton Dickinson, San Jose, CA, USA), according to the description in a previous study [24]. Meanwhile, the nuclear morphological alterations were assessed using the Hoechst 33,342 staining technique. The aggregation of chromosomes and the formation of apoptotic bodies were identified as indicators of cell apoptosis and were quantified accordingly.

2.6. Measurement of Intracellular ROS Levels and Oxidative Stress Biomarkers

Shown in Supplementary Material and Methods.

2.7. Measurement of Mitochondrial ROS

The levels of mitochondrial ROS were measured using the MitoSOX staining method (M36008, Invitrogen, Shanghai, China), according to our previous study [25]. In brief, cells were pre-treated with EA at the dose of 20 μM or Mito-TEMPO at 5 μM for 2 h, then co-treated with or without $CdSO_4$ at 10 μM for an additional 24 h. Then, cells were washed twice with PBS and incubated with MitoSOX Red (at 5 μM) plus 4′-6-diamidino-2-phenylindole (DAPI) for an additional 30 min. Finally, cells were washed using PBS twice, and a laser confocal microscope (Leica Microsystems, Hessen, Germany) was used to observe the changes in mitochondrial ROS.

2.8. Measurement of Intracellular ATP Levels

Show in Supplementary Material and Methods.

2.9. Western Blotting

The expression of proteins was analyzed using Western Blotting, according to the detailed protocol provided in a previous study [26]. Primary rabbit polyclonal antibodies targeting phosphorylated (p)-p38 (Thr180/Tyr182), p-JNK (Thr183/Tyr185), p-ERK1/2 (Thr202/Tyr204) (1:1000 dilution; CST company, Beverly, MA, USA), Nrf2, HO-1, Bax, Bcl-2, and caspase-3 (1:1000 dilution; Proteintech, Chicago, IL, USA), and mouse monoclonal antibody against β-actin (1:1000 dilution; Santa Cruz, CA, USA) were used. A detailed protocol is shown in the Supplementary Material and Methods.

2.10. qRT-PCR Analysis

The RNA samples were isolated using a commercial Fast Pure® Cell/Tissue Total RNA Isolation Kit V2 (Vazyme Biotech Co., Ltd.). The details are shown in Supplementary Material and Methods.

2.11. Statistical Analysis

All data were processed using GraphPad Prism 9.0 software. A nonlinear model-based approach from GraphPad Prism 9.0 software was employed for the calculation of IC_{50}. All data were shown as means ± standard deviation (SD) unless specially stated. The differences between any two groups were performed under Tukey's multiple comparison test of 1-way analysis of variance. A p-value < 0.05 was considered as considered statistically significant.

3. Results

3.1. EA Supplementation Attenuates $CdSO_4$-Induced Loss of Cell Viability and the Release of LDH in HT22 Cells

We first established the toxic model of $CdSO_4$ cytotoxicity in HT22 cells. The results showed that $CdSO_4$ treatment for 24 h dose-dependently decreased cell viability in HT22 cells. As shown in Figure 2A, $CdSO_4$ exposure at the concentrations of 2.5, 5, 10, 20, and 40 μM for 24 h reduced the cell viabilities to 86%, 77%, 65%, 33%, and 9% (all $p < 0.01$), respectively, and the half maximal inhibitory concentration (IC_{50}) was 12.4 μM. Additionally, we also assessed the toxic effects of EA. As shown in Figure 2B, EA treatment at 5–20 μM for 24 h did not change the cell viabilities of HT22 cells, but at 40 and 80 μM, the cell viabilities were decreased to 85% and 61% (both $p < 0.01$), respectively, compared to the untreated cells. Therefore, the concentrations of EA at 5–20 μM in the following investigation were used. Then, EA treatment at 5–20 μM could effectively decrease $CdSO_4$-induced cytotoxicity in HT22 cells, and it was in a dose-dependent manner. As shown in Figure 2C, EA treatment at concentrations of 5, 10, and 20 μM increased the cell viabilities to 77%, 84%, and 92% (all $p < 0.01$), respectively, compared to 10 μM $CdSO_4$ group, increased the cell viabilities to 43%, 49%, and 59% (all $p < 0.01$), respectively, compared to 20 μM $CdSO_4$ group; and increased cell viabilities to 20%, 26% ($p < 0.01$), and 29% ($p < 0.01$), respectively, compared to 40 μM $CdSO_4$ group. Correspondingly, $CdSO_4$ treatment at 10 μM for 24 h could result in marked morphology changes, such as cell shrinkage and lysis. EA co-treatment at 20 μM could improve these changes (Figure 2D). Similarly, in PC12 cells, $CdSO_4$ exposure at 5 μM significantly decreased the cell viabilities to 56% ($p < 0.01$) and morphological changes, which were ameliorated by EA supplementation at the concentrations of 10 and 20 μM (Supplementary Figure S1). The corresponding cell viability increased to 66% and 74% ($p < 0.05$ or 0.01), respectively.

Additionally, a sub-IC_{50} concentration of $CdSO_4$ at 10 μM was used to investigate the potential mechanisms of EA. The levels of LDH released by $CdSO_4$ exposure and the protective effects of EA were assessed. Compared to the control group, $CdSO_4$ exposure for 24 h could dose-dependently induce the release of LDH in the medium. The levels of

LDH in the 2.5, 5, and 10 μM $CdSO_4$ treatment groups were significantly increased to 2.2-, 3.7-, and 5.5-fold (all $p < 0.01$) (Figure 2E). Pre-treatment with EA at doses of 5, 10, and 20 μM significantly reduced LDH levels to 4.1-, 3.4-, and 1.7-fold (all $p < 0.01$), respectively, compared to the group treated with $CdSO_4$ only (Figure 2F). Treatment with EA alone at a concentration of 20 μM for 24 h did not result in changes to the LDH levels of HT22 cells, compared to the untreated control group.

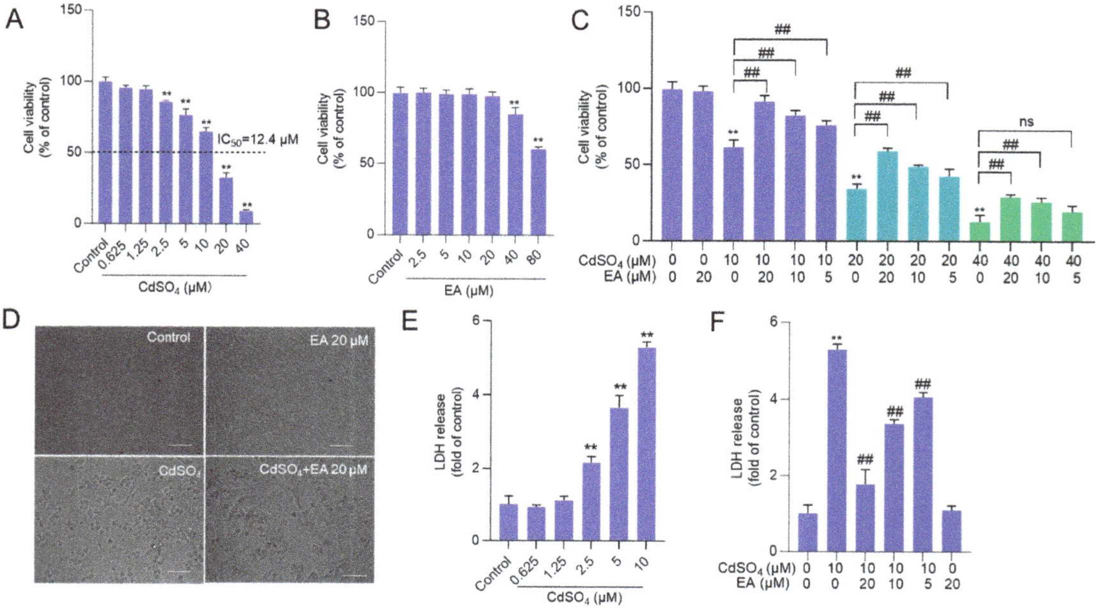

Figure 2. The cell viability and LDH level in HT22 cells exposed to ellagic acid (EA), $CdSO_4$, or combination. A and B cells were treated with $CdSO_4$ (at 0.625, 1.25, 2.5, 5, 10, 20, and 40 μM) (**A**) or EA at various concentrations (at 2.5, 5, 10, 20, and 40 μM) (**B**) for 24 h, the cell viability was examined using the CCK-8 method. (**C**) cells were pre-treated with EA at the doses of 5, 10, or 20 μM for 2 h; then, cells were co-treated with $CdSO_4$ at concentrations of 10, 20, or 40 μM for an additional 24 h. After treatment, the cell viability was examined using the CCK-8 method. (**D**) the morphology changes of HT22 cells exposed to $CdSO_4$ treatment at 10 μM with or without EA at 20 μM. Bar = 50 μm. (**E**,**F**), the levels of LDH in the medium. HT22 cells were treated with the same conditions as cells in (**A**,**B**), respectively. All results were shown in mean ± SD (n = 3). Compared to the untreated control group, ** $p < 0.01$; compared to the $CdSO_4$ only-treated group, ## $p < 0.01$. ns, no significant.

3.2. *EA Supplementation Ameliorates $CdSO_4$-Induced Cell Apoptosis in HT22 Cells*

Exposure to CdSO4, when compared to the control group, resulted in significant changes in nuclear morphology. Figure 3A shows the presence of chromosomal aggregation and nuclear fragmentation in the 10 μM $CdSO_4$ treatment group, effects that were effectively reversed by pre-treatment with EA. Similarly, EA pre-treatment at doses of 5, 10, and 20 μM markedly prevented Cd-induced cell apoptosis, leading to a decrease in apoptotic rates from 29% to 20%, 12%, and 8% (all $p < 0.01$) respectively, compared to that in the 10 μM $CdSO_4$ only group (Figure 3B). Treatment with EA alone at a concentration of 20 μM for 24 h did not result in alterations in apoptotic rates in HT22 cells compared to those in the untreated control group. Consistently, our results also found that caspase pan-inhibitor Z-VAD-FMK treatment at the dose of 20 μM could markedly inhibit $CdSO_4$-induced apoptosis (Supplementary Figure S2). These data suggested that apoptosis plays a critical role in $CdSO_4$-induced cell death and the protective effects of EA.

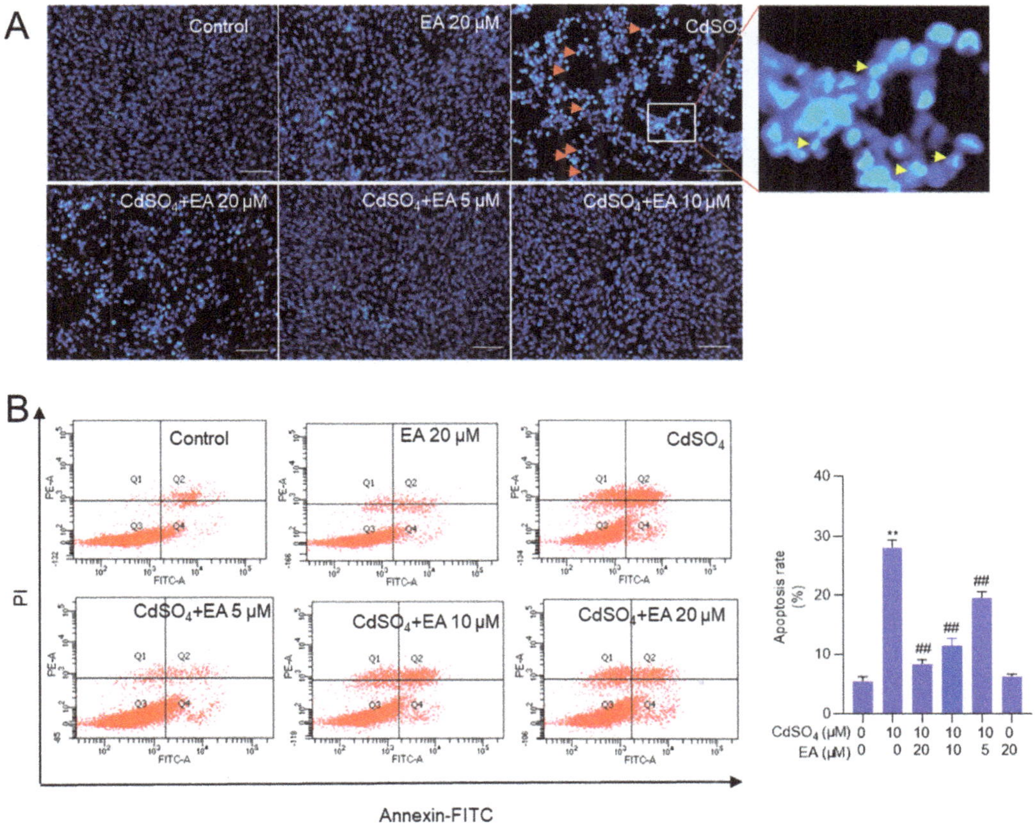

Figure 3. EA supplementation attenuates $CdSO_4$-induced cell apoptosis in HT22 cells. Cells were pre-treated with ellagic acid (EA) at final concentrations of 5, 10, or 20 μM for 2 h, then co-treated with $CdSO_4$ at 10 μM for 24 h. (**A**) nuclear morphology changes were assessed using the Hoechst 33,342 staining method. Bar = 50 μm. (**B**) apoptotic rates were quantified through Annexin V-FITC staining combined with flow cytometry analysis. Representative images of the flow cytometry analysis (left) and quantitative results (right) are displayed. All data are presented as mean ± SD (n = 3). ** $p < 0.01$ compared to the control group; ## $p < 0.01$ compared to the $CdSO_4$ only-treated group. The red arrow indicates chromosomal aggregation and the yellow arrow indicates nuclear fragmentation.

Similarly, we also found that Cd exposure at 5 μM increased the apoptosis rates to 43% ($p < 0.01$) compared to the control group. EA supplementation at 5, 10, and 20 μM decreased the apoptosis rates to 34%, 23%, and 17% (all $p < 0.01$), respectively (Supplementary Figure S3). EA supplementation at 20 μM for 24 h did not affect the apoptosis rates in PC12 cells compared to the untreated control group (Supplementary Figure S3).

3.3. EA Supplementation Attenuates $CdSO_4$-Induced ROS Production and Oxidative Stress Damage in HT22 Cells

Supplementary Figure S4A shows that the fluorescence intensities of DCFH-DA in HT22 cells treated with $CdSO_4$ alone were significantly heightened. This effect was partially alleviated by co-treatment with EA in a dose-dependent manner. A similar finding was also detected in PC12 cells (Supplementary Figure S4B). Then, the quantitative analysis using flow cytometry revealed a notable increase in fluorescence intensities to 46% in the 10 μM $CdSO_4$ treatment group in HT22 cells. Meanwhile, EA treatment at concentrations of 5, 10, and 20 μM led to significant reductions in fluorescence intensities to 30%, 26%, and 14% (all $p < 0.01$), respectively, compared to the group solely treated with $CdSO_4$ (Figure 4A). Furthermore, the MitoSOX staining was performed. The results showed that the mitochondrial ROS levels were significantly increased to 2.6-fold, which were effectively decreased to 1.3- and 1.1-fold (both $p < 0.01$) by EA (at 20 μM) and Mito-TEMPO (at 5 μM) (Figure 4B). Antioxidant N-acetylcysteine (NAC) supplementation (at 2.5 mM) could effectively mitigate $CdSO_4$-induced ROS production (Figure 4C) and partially alleviated $CdSO_4$-induced cytotoxicity, resulting in an increase in the cell viability from 61% to 90% ($p < 0.01$; Figure 4D).

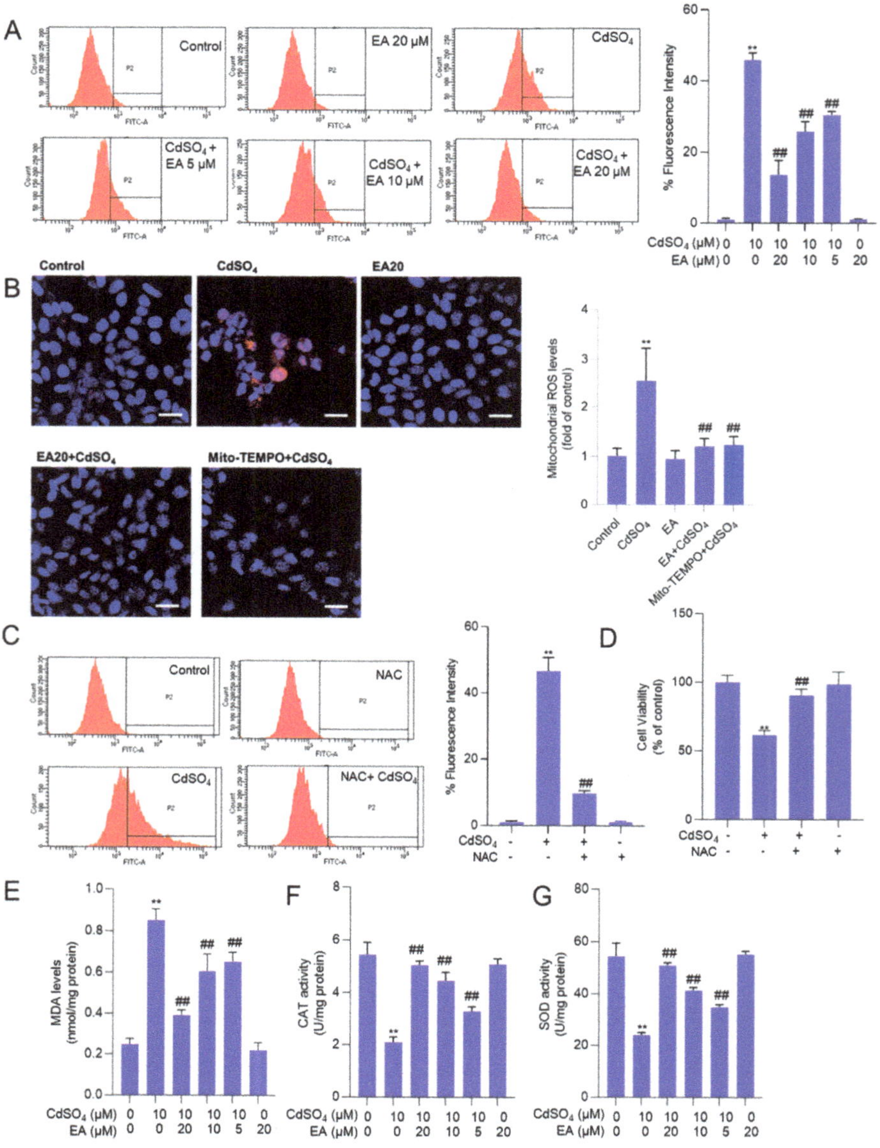

Figure 4. EA supplementation attenuates $CdSO_4$-induced ROS production and oxidative stress damage in HT22 cells. Cells were pre-treated with ellagic acid (EA) at 5, 10, or 20 μM for 2 h, followed

by co-treatment with $CdSO_4$ at a final concentration of 10 μM for 24 h. (**A**) the levels of reactive oxygen species (ROS) were assessed using DCFH-DA staining in conjunction with flow cytometry analysis, with representative images displayed (on the left) and quantitative analysis (on the right). (**B**) levels of mitochondrial ROS in HT22 cells. Cells were pre-treated with EA at 20 μM or Mito-TEMPO at 5 μM for 2 h, followed by co-treatment with $CdSO_4$ at 10 μM for an additional 24 h. Representative images (left) and quantitative results (right) are displayed. Bar = 25 μm. (**C**,**D**) antioxidant N-acetylcysteine (NAC) supplementation attenuates $CdSO_4$ exposure-induced the production of intracellular ROS and the decrease of cell viability (**D**). Cells were pre-treated with NAC at a final concentration of 2.5 mM for 2 h, followed by co-treatment with $CdSO_4$ at 10 μM for an additional 24 h. The ROS levels were quantified using a flow cytometry analysis (**C**). Cell viability was measured using the CCK-8 method (**D**). (**E**) malondialdehyde (MDA) levels. (**F**) the catalase (CAT) activities. (**G**) the superoxide dismutase (SOD) activities. All results are presented as mean ± SD (n = 3). ** $p < 0.01$ compared to that in the untreated control group; ## $p < 0.01$ compared to the $CdSO_4$ only-treated group.

Additionally, the oxidative stress biomarkers, including MDA levels and the activities of SOD and CAT in HT22 cells, were evaluated. The results demonstrated that supplementation with EA effectively ameliorated Cd-induced oxidative damage. In Figure 4B–D, compared to the group exposed to 10 μM $CdSO_4$ only, treatment with EA at concentrations of 5, 10 and 20 μM notably reduced MDA levels from 0.85 nmol/mg protein to 0.65, 0.60 and 0.39 nmol/mg protein (all $p < 0.01$), respectively. Moreover, EA treatment significantly increased CAT activities from 2.1 U/mg protein to 3.3, 4.5 and 5.1 U/mg protein (all $p < 0.05$ or 0.01), respectively, and elevated SOD activities from 24.2 U/mg protein increased to 35.2, 41.3 and 51.5 U/mg protein (all $p < 0.05$ or 0.01). Notably, EA treatment at the concentration of 20 μM did not alter the levels of MDA or the activities of CAT and SOD compared to the control group (Figure 4E–G).

3.4. EA Supplementation Attenuates $CdSO_4$-Caused Mitochondrial Dysfunction

Mitochondrial function was evaluated by measuring changes in mitochondrial membrane potential. In Figure 5A, EA treatment at concentrations of 5, 10, and 20 μM significantly increased the membrane potential from 56% to 70%, 81%, and 93% (all $p < 0.05$ or 0.01), respectively, compared to the group exposed to $CdSO_4$ alone. Consistently, NAC supplementation significantly attenuated the loss of mitochondrial membrane potential induced by $CdSO_4$ exposure (Supplementary Figure S5). Consistently, we also found that $CdSO_4$ exposure significantly decreased the intracellular ATP levels to 46% ($p < 0.01$), and EA supplementation at 10 and 20 μM effectively restored the intracellular ATP levels to 89% and 92% (both $p < 0.01$) (Figure 5B). Additionally, EA treatment at a dose of 20 μM did not alter the mitochondrial membrane potential and the intracellular ATP levels compared to the untreated control group (Figure 5).

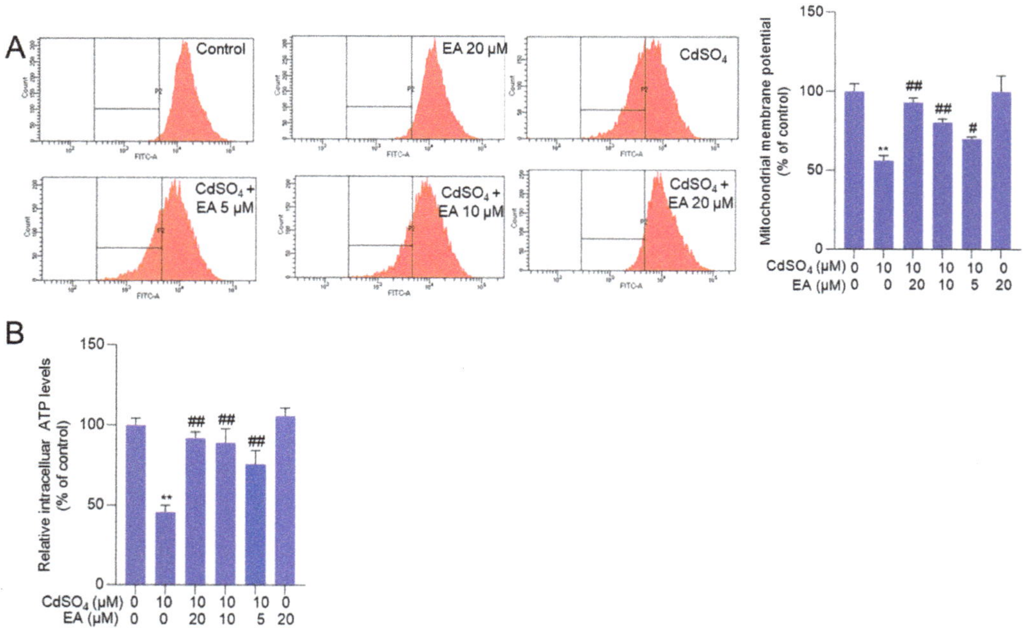

Figure 5. The changes in mitochondrial function. Cells were pre-treated with ellagic acid (EA) at 5, 10, or 20 μM for 2 h, followed by co-treatment with $CdSO_4$ at 10 μM for an additional 24 h. (**A**) the changes in mitochondrial membrane potential were evaluated using Rh123 staining and flow cytometry analysis. The representative images (left) and quantitative analysis (right) were shown. (**B**) the intracellular ATP levels. All results are presented as mean ± SD (n = 3). ** $p < 0.01$, compared to that in the untreated control group; # $p < 0.05$, ## $p < 0.01$ compared to the $CdSO_4$ only-treated group.

3.5. EA Supplementation Upregulates the Expression of Nrf2, HO-1, and Bcl-2 Proteins and Downregulates the Expression of Bax, p-JNK, p-p38, and p-ERK Proteins

Compared to the untreated control group, exposure to 10 μM $CdSO_4$ significantly upregulated the expression of Nrf2, HO-1, p-Erk1/2, p-JNK, p-p38, Bax/Bcl-2 ratio, cleaved caspase-3, and cleaved PARP-1 proteins by 1.9-, 10.1-, 2.4-, 4.3-, 3.7-, 1.9-, 4.0- and 2.6-fold, respectively (all $p < 0.01$) (Figure 6). These protein expressions were modulated by EA supplementation. EA supplementation at a dose of 20 μM plus $CdSO_4$ treatment significantly increased the expression of Nrf2 and HO-1 proteins by 3.1- and 16.3-fold, respectively, while notably decreasing the expression of p-Erk1/2, p-JNK, p-p38, Bax/Bcl-2 ratio, cleaved caspase-3, and cleaved PARP-1 proteins to 1.5-, 2.0-, 1.6-, 1.2-, 2.1-, and 1.0-fold (all $p < 0.01$; Figure 6), respectively. Furthermore, we also found that EA supplementation could significantly upregulate the expression of Nrf2, HO-1, and Bcl-2 mRNAs and significantly downregulate the expression of Bax mRNA. As shown in Supplementary Figure S6, EA supplementation at 20 μM significantly upregulated the expression of Nrf2 mRNA from 1.9-fold to 3.6-fold, upregulated the expression of HO-1 from 4.6-fold to 10.1-fold, upregulated the expression of Bcl-2 from 0.4-fold to 0.9-fold, and significantly downregulated the expression of Bax from 2.3-fold to 1.2-fold, compared to those in the $CdSO_4$-only group. EA alone treatment at 20 μM upregulated the expression of Nrf2 and HO-1 mRNAs but did not affect the expression of Bcl-2 and Bax mRNAs compared to those in the control group (Supplementary Figure S6).

We also confirmed the impact of oxidative stress on $CdSO_4$-induced activation of the JNK and mitochondrial apoptotic pathways. As shown in Supplementary Figure S7, NAC supplementation led to a substantial decrease in the expression of p-JNK, the Bax/Bcl-2 ratio, cleaved caspase-3, and cleaved PARP-1 proteins. In the NAC and $CdSO_4$ co-treatment group, the levels of p-JNK, the Bax/Bcl-2 ratio, cleaved caspase-3, and cleaved PARP-1 proteins nearly normalized, and they were decreased to 1.3-, 0.8-, 1.1-, and 1.1-fold (all $p < 0.01$) (Supplementary Figure S7), respectively. When compared to the control group, treatment with NAC alone did not affect the expression of the aforementioned proteins (Supplementary Figure S7).

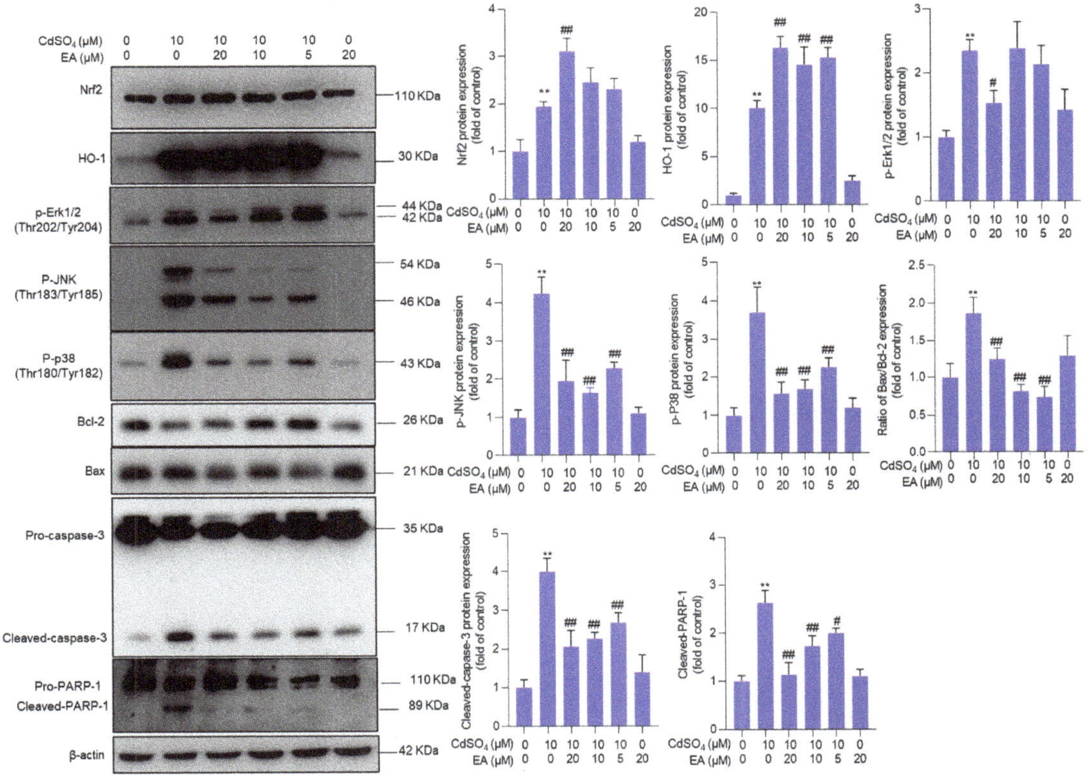

Figure 6. The expression of Nrf2, HO-1, p-Erk1/2, p-JNK, p-p38, Bax/Bcl-2 ratio, cleaved caspase-3, and cleaved PARP-1 proteins. Cells were pre-treated with ellagic acid (EA) at 5, 10, or 20 μM for 2 h, followed by co-treatment with $CdSO_4$ at 10 μM for an additional 24 h. Protein expressions were assessed using the Western Blotting method. All results were presented as Mean ± SD (n = 3). Statistical significance levels were delineated as follows: ** $p < 0.01$ compared to the untreated control group; # $p < 0.05$ and ## $p < 0.01$ compared to the $CdSO_4$ only-treated group.

3.6. The Pharmacological Inhibition of JNK Promotes $CdSO_4$-Induced Cytotoxicity

We further investigated the impact of the MAPK pathway on $CdSO_4$ exposure-induced cytotoxicity. We found that pharmacological inhibitions of p38 and ERK pathways by SB203580 and ASN007 did not affect $CdSO_4$-induced cytotoxicity in HT22 cells (Supplementary Figure S8). However, the pharmacological inhibition of JNK by SP600125 could significantly exacerbate $CdSO_4$ exposure-induced cytotoxicity (Figure 7A). Additionally, JNK inhibition led to a substantial decrease in the expression of Nrf2 and HO-1 proteins, which were decreased to 1.1- and 5.2-fold (both $p < 0.05$), respectively. On the contrary, compared to the group exposed to $CdSO_4$ only, JNK inhibition significantly increased the expression of the Bax/Bcl-2 ratio, cleaved caspase-3, and cleaved PARP-1 proteins to 2.4-fold ($p < 0.05$), 4.3- fold ($p < 0.01$), and 3.8-fold ($p < 0.05$) (Figure 7B).

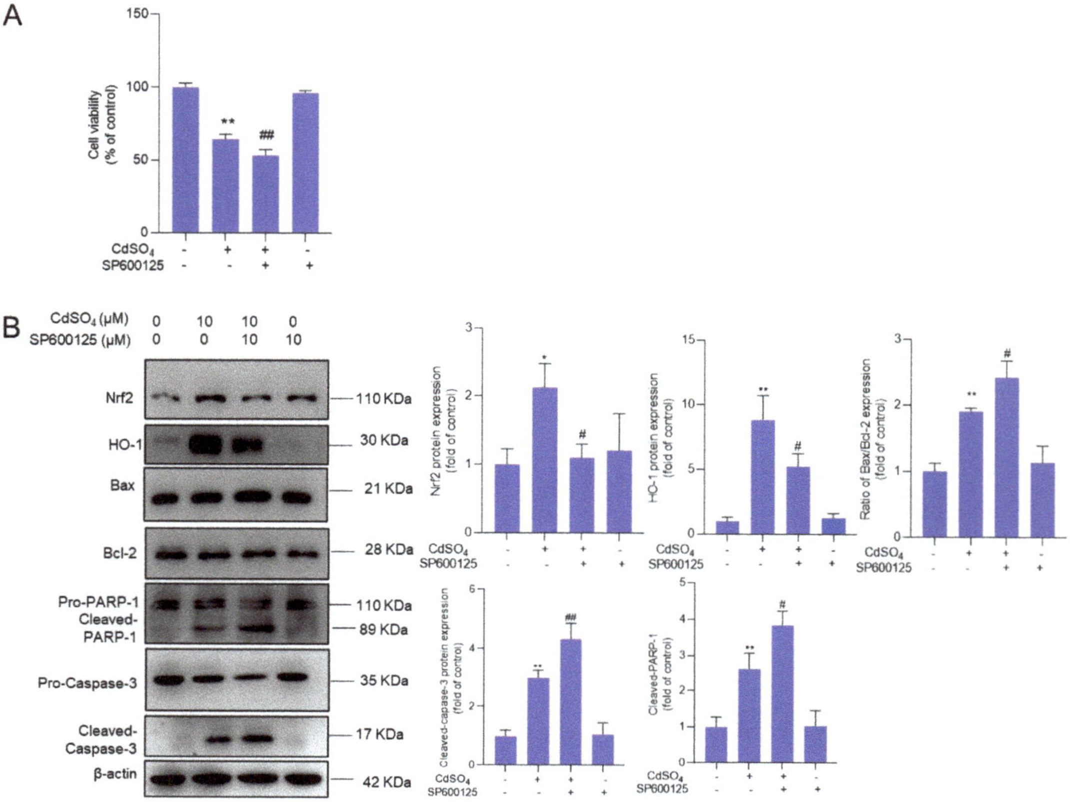

Figure 7. Pharmacological inhibition of JNK attenuates the activation of Nrf2/HO-1 and promotes $CdSO_4$ exposure-induced cytotoxicity. HT22 cells were treated with the JNK inhibitor SP600125 at 10 μM for 2 h, then co-treated with $CdSO_4$ at 10 μM for an additional 24 h. (**A**) the changes in cell viability were assessed. (**B**) the protein expressions were analyzed using the Western Blotting method. All results were presented as mean ± SD (n = 3). Significance levels were indicated as follows: * $p < 0.05$, ** $p < 0.01$ compared to the untreated control group; # $p < 0.05$, and ## $p < 0.01$ compared to the $CdSO_4$ only-treated group.

4. Discussion

Neurotoxicity is one of the multiple toxic effects caused by $CdSO_4$ exposure [18]. Previous studies demonstrated that $CdSO_4$ exposure at the range of 0.625–20 μM could induce marked decreases in the cell viability in PC12 cells and HT22 cells [13,27]. Consistent with a previous study [13], our current study reveals that exposure to $CdSO_4$ at concentrations ranging from 0.625 to 40 μM resulted in dose-dependent cell death in HT22 cells. Previous studies have reported that the different test methods have different sensitivity to Cd-induced cytotoxicity [28]. One non-negligible reason is that some heavy metals can affect the sensitivity of CCK-8 detection, and it might be increased with the increase of heavy metal concentration. In the present study, we used two methods, i.e., the CCK-8 and LDH methods, to assess the cytotoxicity and the protective effects of EA. Our results found that the LDH method is more sensitive than the CCK-8 method in Cd-treated HT22 cells (Figure 2). Meanwhile, we found that both two methods demonstrated the protective effects of EA on Cd-induced cytotoxicity (Figure 2). Similarly, EA supplementation at 5–20 μM could provide protection against Cd-induced cytotoxicity in PC12 cells (Supplementary Figure S1). Furthermore, we observed a dose-dependent inhibition by EA supplementation on $CdSO_4$ exposure-induced oxidative stress, cell apoptosis, and mitochondrial dysfunction. These effects may be linked to EA's ability to scavenge free radicals, enhance antioxidant defense function, inhibit the JNK pathway, and activate the Nrf2/HO-1 pathway (Figures 3–7 and Supplementary Figures S2–S8).

Apoptosis, a form of programmed cell death, can be induced by various drugs or environmental toxins, such as copper, cisplatin, and T-2 toxin [26,29–31]. In this study, we observed that exposure to $CdSO_4$ could initiate cell apoptosis (Figure 3), consistent with prior research [32,33]. Furthermore, our findings revealed that EA supplementation effectively ameliorated $CdSO_4$-induced cell apoptosis in HT22 and PC12 cells (Figure 3 and Supplementary Figures S2 and S3). Several previous studies demonstrated that EA supplementation can reduce drugs or toxins-induced apoptotic cell death in vitro and in vivo [34–36]. These findings suggest that EA's protective effects against $CdSO_4$-induced cytotoxicity involve the inhibition of apoptosis.

Apoptotic cell death could be triggered by multiple signals, including ROS, p53, and other apoptosis-related factors [26,37]. A prior study demonstrated that exposure to $CdSO_4$ can induce excessive ROS production, leading to direct damage to intracellular macromolecules (such as DNA, lipids, and proteins) and subcellular organelles (such as mitochondria, endoplasmic reticulum, and lysosomes), ultimately culminating in cell apoptosis [38]. Our current findings reveal that $CdSO_4$ exposure significantly elevates intracellular ROS in HT22 cells and PC12 cells (Figure 4 and Supplementary Figure S4). Meanwhile, $CdSO_4$ exposure significantly upregulated the MDA levels while simultaneously decreasing the activities of SOD and CAT in HT22 cells (Figure 4). We also found that $CdSO_4$ exposure could significantly upregulate the mitochondrial ROS levels (Figure 4). Similarly, Hyun et al. showed that Cd exposure at 10–40 μM for 24 h could significantly upregulate the mitochondrial ROS levels in human prostate stromal cells and mouse embryonic fibroblasts [39]. These data suggested that $CdSO_4$ exposure-induced excessive production is partly dependent on the production of mitochondrial ROS. SOD and CAT, as antioxidant enzymes, play crucial roles in combating oxidative stress by scavenging intracellular superoxide and hydroxyl radicals [26]. Increased MDA levels serve as an essential indicator of membrane lipid peroxidation [25]. These results indicate that $CdSO_4$ exposure elicits oxidative stress damage in HT22 cells. Notably, EA supplementation effectively attenuates $CdSO_4$-induced ROS and mitochondrial ROS productions and enhances SOD and CAT activities (Figure 4). Previous studies by Firdaus et al. demonstrated that EA pre-treatment at 10–20 μM can inhibit ROS production and apoptosis induced by arsenic trioxide in human neuroblastoma SH-SY5Y cells [40]. Similarly, Ding et al. reported that EA treatment at 15–30 μM reduces ROS production and MDA levels caused by high glucose exposure while improving SOD activities in HepG2 cells [41]. Moreover, Zhao et al. showed that EA supplementation at doses of 50–100 mg/kg/day for four weeks upregulates antioxidant enzymes, including CAT, SOD, and GSH-PX, to ameliorate ethanol exposure-induced liver injury in a mouse model [42]. Collectively, these findings suggest that EA supplementation effectively reduces $CdSO_4$-induced apoptosis by alleviating oxidative stress and mitochondrial ROS production. Furthermore, our study found that ATP production was significantly decreased after $CdSO_4$ exposure, indicating mitochondrial dysfunction (Figure 5). It is known that mitochondrial respiration mainly relies on the enzymatic activities of five mitochondrial complexes that couple electron transport with proton pumping, finally leading to ATP synthesis [43]. Several previous studies also reported that Cd is a potent uncoupling agent and can inhibit mitochondrial respiratory chain activities by interacting with CI at the Q site and NADH site and complexes I (CI), CII, and CIII at the Fe–S cluster [14,44–46]. Our results also showed that EA treatment for 24 h could effectively restore $CdSO_4$ exposure-induced loss of intracellular ATP (Figure 5). These data suggested that EA supplementation may provide protection for $CdSO_4$ exposure-induced mitochondrial dysfunction via the inhibition of oxidative stress. The inhibition of complexes I and II could induce ROS generation [47,48]. Khodaei et al.'s findings showed that EA supplementation could inhibit cuprizone exposure-induced loss of CII, CIII, and CIV proteins in mouse muscle tissue [49]. Therefore, the regulation of EA on mitochondrial complexes may participate in its protective effects against $CdSO_4$-induced mitochondrial dysfunction. This is a limitation of our current study. Further investigation is warranted to elucidate the underlying molecular mechanisms.

Mitochondria serve not only as producers but also as targets of ROS [50]. Bax, a pro-apoptotic protein, contrasts with Bcl-2, an anti-apoptotic protein [50]. The mitochondrial membrane potential assessment stands as a pivotal indicator of mitochondrial dysfunction [26]. Deng et al. observed a significant loss of mitochondrial membrane potential in N2a neuronal cells upon exposure to cadmium chloride ($CdCl_2$) [1]. The elevation in the Bax/Bcl-2 ratio can instigate mitochondrial dysfunction, triggering the release of CytC and subsequently activating caspases-9 and -3 in a cascading fashion, culminating in cell apoptosis [51]. Poly (ADP-ribose) polymerase-1 (PARP-1) acts as a target of activated caspases-3 in caspase-dependent apoptosis, with cleaved PARP-1 suppressing DNA repair and serving as a key marker of apoptosis [52]. Our study demonstrates that $CdSO_4$ exposure notably diminishes mitochondrial membrane potential, leading to an increase in the Bax/Bcl-2 ratio, ultimately resulting in the upregulation of cleaved caspase-3 and cleaved PARP-1 protein expression (Figures 5 and 6 and Supplementary Figure S6). Pan-caspase inhibitors could markedly revise $CdSO_4$-induced cell apoptosis (Supplementary Figure S2), indicating that $CdSO_4$-induced apoptotic cell death is caspase-dependent. EA supplementation partially reduces these $CdSO_4$-induced alterations (Figures 5 and 6 and Supplementary Figure S6). Furthermore, our data reveals that the antioxidant NAC supplementation significantly attenuates $CdSO_4$-induced cytotoxicity by suppressing ROS production and the mitochondrial apoptotic pathway (Figure 4 and Supplementary Figures S5 and S7). Therefore, these findings suggest that EA supplementation could effectively safeguard against $CdSO_4$-induced apoptosis by hindering mitochondrial dysfunction and the mitochondrial apoptotic pathway in HT22 cells, potentially attributable to its role in oxidative stress modulation.

Nrf2, a crucial transcription factor often referred to as a "housekeeper," responds to oxidative stress and inflammatory damage [53,54]. Studies have highlighted that Nrf2 activation exerts a protective effect against Cd-induced cytotoxicity and tissue damage [55,56]. Consistent with prior findings, our study demonstrated a significant upregulation of Nrf2 mRNA and protein expression and its downstream gene HO-1 following $CdSO_4$ exposure, with further enhancements observed upon co-treatment with EA (Figure 6 and Supplementary Figure S6). Recent research has underscored Nrf2 as a vital target of EA, showing that EA promotes nuclear Nrf2 expression by inhibiting Keap1 expression [57]. Notably, Nrf2 knockout partially attenuated the protective effects of EA against rotenone-induced neurotoxicity in a mouse model [22]. Our current data strongly suggest that Nrf2 activation contributes to the EA-mediated protection against $CdSO_4$-induced apoptosis.

Previous studies have shown that Cd exposure can promptly activate the MAPK pathway, leading to a significant upregulation of p-Erk1/2, p-JNK, and p-p38 proteins in various cell types, including PC12 and SH-SY5Y neuronal cells and human bronchial epithelial cells (BEAS-2B) [58–60]. In these cell models, researchers observed that the inhibition of p-Erk1/2 and p-JNK partially attenuated Cd exposure-induced cell apoptosis [58–60]. Conversely, Liu et al. reported that inhibiting JNK did not influence Cd exposure-induced cell death in BmE cells, a silkworm embryonic cell line [61]. Given its diverse functions, the MAPK pathway typically exhibits a dual role in regulating cell apoptosis [62]. For instance, Hu et al. demonstrated that JNK-mediated activation of the Nrf2 pathway contributed to the protective effects of coptisine against 2,2′-azodiisobutyramidine dihydrochloride-induced oxidative stress damage in zebrafish embryos [62]. In the present study, we observed that JNK inhibition, not p38 and ERK inhibition, could promote $CdSO_4$-induced cytotoxicity at 24 h (Supplementary Figure S8). Moreover, JNK inhibition by SP600125 markedly suppresses the expression of Nrf2 and HO-1 proteins, consequently exacerbating the mitochondrial apoptotic pathway and ultimately exacerbating $CdSO_4$ exposure-induced cell death (Figure 7). We also found that blocking oxidative stress with NAC substantially inhibited JNK activation (Supplementary Figure S7), suggesting that oxidative stress is a crucial upstream regulator of the JNK pathway in $CdSO_4$-induced cytotoxicity. For the MAPK pathway, the time profile of activation is usually responsible for the biological outcome. In the current study, we just tested the effects of the MAPK pathway on $CdSO_4$-induced cytotoxicity at 24 h. The dynamic changes of the MAPK pathway in response to $CdSO_4$ and how it mediates the protective effects of EA are still unknown. This is another limitation of the present study.

Importantly, EA has a higher safety for humans and animals and has been used in food production as a food additive [63,64]. Tasaki et al. found that the no-observed-effect levels (via the 90-day sub-chronic toxicity study) of EA in male or female rats are 3011 mg/kg body weight/day and 3254 mg/kg body weight/day, respectively [63]. Various studies indicate that oral EA supplementation at 10–100 mg/kg body weight per day could provide neuroprotection for drugs or environmental toxins (such as rotenone, 6-hydroxydopamine, and 1-methyl-4-phenyl 1,2,3,6 tetrahydropyridine)-induced neurotoxicity in rats or mice [22,36,65,66]. A recent human trial showed that EA intervention (90 mg twice a day for 12 weeks) could provide a favorable effect on depression in multiple sclerosis [67]. The information indicated that oral supplementation of EA is safer. Taken together, these findings suggest that EA may be a potential protective agent against Cd-induced neurotoxicity, and further commercial development could be considered.

5. Conclusions

In summary, our results demonstrate that EA supplementation effectively reduced $CdSO_4$-induced cytotoxicity in HT22 cells by inhibiting oxidative stress and apoptosis, potentially via the activation of the Nrf2/HO-1 pathway. Additionally, the activation of the JNK pathway in response to $CdSO_4$ exposure also confers protection via the Nrf2/HO-1 pathway. A proposed model illustrating the impact of EA supplementation on $CdSO_4$-induced apoptosis is presented in Figure 8. This study highlights EA as a promising candidate for combating Cd-induced cytotoxicity and neurotoxicity.

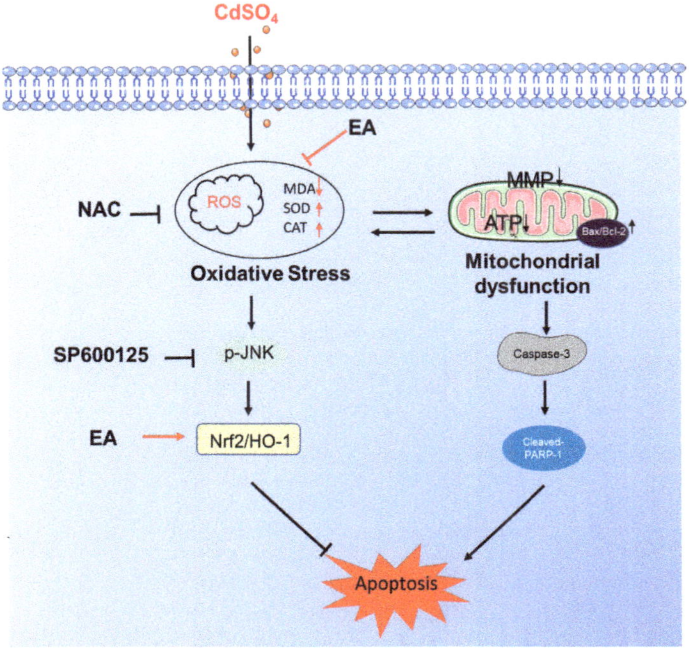

Figure 8. A potential proposed model of EA supplementation on $CdSO_4$ exposure-caused apoptosis. $CdSO_4$ exposure leads to the generation of reactive oxygen species (ROS) and lipid peroxidation while

reducing the activities of antioxidant enzymes SOD and CAT, resulting in oxidative stress damage. Elevated ROS levels may disrupt mitochondrial function, triggering the activation of the mitochondrial apoptotic pathway and eventually culminating in apoptosis. Furthermore, activated JNK exerts a protective effect by stimulating the Nrf2/HO-1 pathway.

Supplementary Materials: The following supporting information can be downloaded at: https://www.mdpi.com/article/10.3390/antiox13111296/s1, Supplementary Figure S1 The morphology changes (left) and the results of cell viabilities (right) in PC12 cells after CdSO4 treatment with or without ellagic acid (EA). Supplementary Figure S2 Pan-caspase inhibitor Z-VAD-FMK suppresses CdSO4-induced apoptosis in HT22 cells. Supplementary Figure S3 Ellagic acid (EA) supplementation attenuates CdSO4-induced cell apoptosis in PC12 cells. Supplementary Figure S4 Representative images of DCFH-DA staining in HT22 cells (A) and PC12 cells (B). Supplementary Figure S5 NAC supplementation attenuates CdSO4-induced alterations in mitochondrial membrane potential. Supplementary Figure S6 The mRNA expression of Nrf2 (A), HO-1 (B), Bcl-2 (C), and Bax (D) genes. Supplementary Figure S7 NAC supplementation attenuates CdSO4-induced activation of JNK and mitochondrial apoptotic pathways. Supplementary Figure S8 Effects of pharmacological inhibition of p38 (A) and Erk (B) pathways by SB203580 and ASN007 on CdSO4 exposure-induced cytotoxicity. References [24,26] are cited in the supplementary materials.

Author Contributions: Investigation, Y.L. and C.C.; data curation, Y.L., C.C. and Z.H.; writing—original draft preparation, Y.L., C.C., and C.D.; methodology, Y.L., C.D. and S.T.; formal analysis, Y.L., C.D. and S.T.; validation, Y.L., C.D. and S.T.; writing—review and editing, C.D. and J.S.; supervision, C.D. All authors have read and agreed to the published version of the manuscript.

Funding: This study was funded and supported by the Guangdong Major Project of Basic and Applied Basic Research (2020B0301030007). It was funded by the National Natural Science Foundation of China (Award number 32102724) and the Pinduoduo-China Agricultural University Research Fund (Grant No. PC2023A01002).

Institutional Review Board Statement: Not applicable.

Informed Consent Statement: Not applicable.

Data Availability Statement: The data are contained within the article and Supplementary Materials.

Conflicts of Interest: The authors declare no conflicts of interest.

References

1. Deng, P.; Zhang, H.; Wang, L.; Jie, S.; Zhao, Q.; Chen, F.; Yue, Y.; Wang, H.; Tian, L.; Xie, J.; et al. Long-term cadmium exposure impairs cognitive function by activating lnc-Gm10532/m6A/FIS1 axis-mediated mitochondrial fission and dysfunction. *Sci. Total Environ.* **2023**, *858*, 159950. [CrossRef]
2. Wang, M.; Chen, Z.; Song, W.; Hong, D.; Huang, L.; Li, Y. A review on Cadmium Exposure in the Population and Intervention Strategies Against Cadmium Toxicity. *Bull. Environ. Contam. Toxicol.* **2021**, *106*, 65–74. [CrossRef]
3. Tinkov, A.A.; Gritsenko, V.A.; Skalnaya, M.G.; Cherkasov, S.V.; Aaseth, J.; Skalny, A.V. Gut as a target for cadmium toxicity. *Environ. Pollut.* **2018**, *235*, 429–434. [CrossRef]
4. Xie, D.; Yan, J.; Zhang, H.; Zhang, H.; Nie, G.; Zhu, X.; Li, X. Cadmium exacerbates liver injury by remodeling ceramide metabolism: Multiomics and laboratory evidence. *Sci. Total Environ.* **2024**, *923*, 171405. [CrossRef]
5. Zhao, C.; Yu, D.; He, Z.; Bao, L.; Feng, L.; Chen, L.; Liu, Z.; Hu, X.; Zhang, N.; Wang, T.; et al. Endoplasmic reticulum stress-mediated autophagy activation is involved in cadmium-induced ferroptosis of renal tubular epithelial cells. *Free Radic. Biol. Med.* **2021**, *175*, 236–248. [CrossRef]
6. Qu, Z.; Liu, L.; Wu, X.; Guo, P.; Yu, Z.; Wang, P.; Song, Y.; Zheng, S.; Liu, N. Cadmium-induced reproductive toxicity combined with a correlation to the oogenesis process and competing endogenous RNA networks based on a Caenorhabditis elegans model. *Ecotoxicol. Environ. Saf.* **2023**, *268*, 115687. [CrossRef]
7. Wang, Z.; Sun, Y.; Yao, W.; Ba, Q.; Wang, H. Effects of Cadmium Exposure on the Immune System and Immunoregulation. *Front. Immunol.* **2021**, *12*, 695484. [CrossRef]
8. Min, J.Y.; Min, K.B. Blood cadmium levels and Alzheimer's disease mortality risk in older US adults. *Environ. Health* **2016**, *15*, 69. [CrossRef]
9. Bakulski, K.M.; Seo, Y.A.; Hickman, R.C.; Brandt, D.; Vadari, H.S.; Hu, H.; Park, S.K. Heavy Metals Exposure and Alzheimer's Disease and Related Dementias. *J. Alzheimer's Dis.* **2020**, *76*, 1215–1242. [CrossRef]

10. Xu, L.; Zhang, W.; Liu, X.; Zhang, C.; Wang, P.; Zhao, X. Circulatory Levels of Toxic Metals (Aluminum, Cadmium, Mercury, Lead) in Patients with Alzheimer's Disease: A Quantitative Meta-Analysis and Systematic Review. *J. Alzheimer's Dis.* **2018**, *62*, 361–372. [CrossRef]
11. Wang, H.; Abel, G.M.; Storm, D.R.; Xia, Z. Adolescent cadmium exposure impairs cognition and hippocampal neurogenesis in C57BL/6 mice. *Environ. Toxicol.* **2022**, *37*, 335–348. [CrossRef]
12. Wang, H.; Zhang, L.; Abel, G.M.; Storm, D.R.; Xia, Z. Cadmium Exposure Impairs Cognition and Olfactory Memory in Male C57BL/6 Mice. *Toxicol. Sci. Off. J. Soc. Toxicol.* **2018**, *161*, 87–102. [CrossRef]
13. Wang, D.; Wu, Y.; Zhou, X.; Liang, C.; Ma, Y.; Yuan, Q.; Wu, Z.; Hao, X.; Zhu, X.; Li, X.; et al. Cadmium exposure induced neuronal ferroptosis and cognitive deficits via the mtROS-ferritinophagy pathway. *Environ. Pollut.* **2024**, *349*, 123958. [CrossRef]
14. Branca, J.J.V.; Pacini, A.; Gulisano, M.; Taddei, N.; Fiorillo, C.; Becatti, M. Cadmium-Induced Cytotoxicity: Effects on Mitochondrial Electron Transport Chain. *Front. Cell Dev. Biol.* **2020**, *8*, 604377. [CrossRef]
15. Wen, S.; Xu, M.; Zhang, W.; Song, R.; Zou, H.; Gu, J.; Liu, X.; Bian, J.; Liu, Z.; Yuan, Y. Cadmium induces mitochondrial dysfunction via SIRT1 suppression-mediated oxidative stress in neuronal cells. *Environ. Toxicol.* **2023**, *38*, 743–753. [CrossRef]
16. Khan, A.; Ikram, M.; Muhammad, T.; Park, J.; Kim, M.O. Caffeine Modulates Cadmium-Induced Oxidative Stress, Neuroinflammation, and Cognitive Impairments by Regulating Nrf-2/HO-1 In Vivo and In Vitro. *J. Clin. Med.* **2019**, *8*, 680. [CrossRef]
17. Chandler, J.D.; Wongtrakool, C.; Banton, S.A.; Li, S.; Orr, M.L.; Barr, D.B.; Neujahr, D.C.; Sutliff, R.L.; Go, Y.M.; Jones, D.P. Low-dose oral cadmium increases airway reactivity and lung neuronal gene expression in mice. *Physiol. Rep.* **2016**, *4*, e12821. [CrossRef]
18. Al Olayan, E.M.; Aloufi, A.S.; AlAmri, O.D.; El-Habit, O.H.; Abdel Moneim, A.E. Protocatechuic acid mitigates cadmium-induced neurotoxicity in rats: Role of oxidative stress, inflammation and apoptosis. *Sci. Total Environ.* **2020**, *723*, 137969. [CrossRef]
19. Al-Assaf, A.H.; Alqahtani, A.M.; Alshatwi, A.A.; Syed, N.A.; Shafi, G.; Hasan, T.N. Mechanism of cadmium induced apoptosis in human peripheral blood lymphocytes: The role of p53, Fas and Caspase-3. *Environ. Toxicol. Pharmacol.* **2013**, *36*, 1033–1039. [CrossRef]
20. Kim, D.H.; Sim, Y.; Hwang, J.H.; Kwun, I.S.; Lim, J.H.; Kim, J.; Kim, J.I.; Baek, M.C.; Akbar, M.; Seo, W.; et al. Ellagic Acid Prevents Binge Alcohol-Induced Leaky Gut and Liver Injury through Inhibiting Gut Dysbiosis and Oxidative Stress. *Antioxidants* **2021**, *10*, 1386. [CrossRef]
21. Gupta, A.; Singh, A.K.; Kumar, R.; Jamieson, S.; Pandey, A.K.; Bishayee, A. Neuroprotective Potential of Ellagic Acid: A Critical Review. *Adv. Nutr.* **2021**, *12*, 1211–1238. [CrossRef]
22. Wei, Y.Z.; Zhu, G.F.; Zheng, C.Q.; Li, J.J.; Sheng, S.; Li, D.D.; Wang, G.Q.; Zhang, F. Ellagic acid protects dopamine neurons from rotenone-induced neurotoxicity via activation of Nrf2 signalling. *J. Cell Mol. Med.* **2020**, *24*, 9446–9456. [CrossRef]
23. Goudarzi, M.; Amiri, S.; Nesari, A.; Hosseinzadeh, A.; Mansouri, E.; Mehrzadi, S. The possible neuroprotective effect of ellagic acid on sodium arsenate-induced neurotoxicity in rats. *Life Sci.* **2018**, *198*, 38–45. [CrossRef]
24. Dai, C.; Ciccotosto, G.D.; Cappai, R.; Tang, S.; Li, D.; Xie, S.; Xiao, X.; Velkov, T. Curcumin Attenuates Colistin-Induced Neurotoxicity in N2a Cells via Anti-inflammatory Activity, Suppression of Oxidative Stress, and Apoptosis. *Mol. Neurobiol.* **2018**, *55*, 421–434. [CrossRef]
25. Dai, C.; Li, M.; Sun, T.; Zhang, Y.; Wang, Y.; Shen, Z.; Velkov, T.; Tang, S.; Shen, J. Colistin-induced pulmonary toxicity involves the activation of NOX4/TGF-β/mtROS pathway and the inhibition of Akt/mTOR pathway. *Food Chem. Toxicol.* **2022**, *163*, 112966. [CrossRef]
26. Li, M.; Tang, S.; Velkov, T.; Shen, J.; Dai, C. Copper exposure induces mitochondrial dysfunction and hepatotoxicity via the induction of oxidative stress and PERK/ATF4 -mediated endoplasmic reticulum stress. *Environ. Pollut.* **2024**, *352*, 124145. [CrossRef]
27. Ben, P.; Zhang, Z.; Xuan, C.; Sun, S.; Shen, L.; Gao, Y.; Cao, X.; Zhou, Y.; Lan, L.; Yin, Z.; et al. Protective Effect of L-Theanine on Cadmium-Induced Apoptosis in PC12 Cells by Inhibiting the Mitochondria-Mediated Pathway. *Neurochem. Res.* **2015**, *40*, 1661–1670. [CrossRef]
28. Fotakis, G.; Timbrell, J.A. In vitro cytotoxicity assays: Comparison of LDH, neutral red, MTT and protein assay in hepatoma cell lines following exposure to cadmium chloride. *Toxicol. Lett.* **2006**, *160*, 171–177. [CrossRef]
29. Chen, J.; Chu, Y.; Cao, J.; Yang, Z.; Guo, X.; Wang, Z. T-2 toxin induces apoptosis, and selenium partly blocks, T-2 toxin induced apoptosis in chondrocytes through modulation of the Bax/Bcl-2 ratio. *Food Chem. Toxicol.* **2006**, *44*, 567–573. [CrossRef]
30. Liu, J.; Wang, L.; Guo, X.; Pang, Q.; Wu, S.; Wu, C.; Xu, P.; Bai, Y. The role of mitochondria in T-2 toxin-induced human chondrocytes apoptosis. *PLoS ONE* **2014**, *9*, e108394. [CrossRef]
31. Mokhtari, Z.; Seyedhashemi, E.; Eftekhari, M.; Ghasemi, S.; Sabouri, A.; Abbaszadeh-Goudarzi, K.; Abuali, M.; Azimi, H.; Kesharwani, P.; Pourghadamyari, H.; et al. Enhancement of cisplatin-induced apoptosis by saffron in human lung cancer cells. *J. Trace Elem. Med. Biol.* **2023**, *79*, 127229. [CrossRef] [PubMed]
32. Wang, T.; Yan, L.; Wang, L.; Sun, J.; Qu, H.; Ma, Y.; Song, R.; Tong, X.; Zhu, J.; Yuan, Y.; et al. VPS41-mediated incomplete autophagy aggravates cadmium-induced apoptosis in mouse hepatocytes. *J. Hazard. Mater.* **2023**, *459*, 132243. [CrossRef] [PubMed]
33. Yi, L.; Shang, X.J.; Lv, L.; Wang, Y.; Zhang, J.; Quan, C.; Shi, Y.; Liu, Y.; Zhang, L. Cadmium-induced apoptosis of Leydig cells is mediated by excessive mitochondrial fission and inhibition of mitophagy. *Cell Death Dis.* **2022**, *13*, 928. [CrossRef]

34. Arab, H.H.; Gad, A.M.; Fikry, E.M.; Eid, A.H. Ellagic acid attenuates testicular disruption in rheumatoid arthritis via targeting inflammatory signals, oxidative perturbations and apoptosis. *Life Sci.* **2019**, *239*, 117012. [CrossRef]
35. Aslan, A.; Hussein, Y.T.; Gok, O.; Beyaz, S.; Erman, O.; Baspinar, S. Ellagic acid ameliorates lung damage in rats via modulating antioxidant activities, inhibitory effects on inflammatory mediators and apoptosis-inducing activities. *Environ. Sci. Pollut. Res. Int.* **2020**, *27*, 7526–7537. [CrossRef] [PubMed]
36. Ardah, M.T.; Eid, N.; Kitada, T.; Haque, M.E. Ellagic Acid Prevents α-Synuclein Aggregation and Protects SH-SY5Y Cells from Aggregated α-Synuclein-Induced Toxicity via Suppression of Apoptosis and Activation of Autophagy. *Int. J. Mol. Sci.* **2021**, *22*, 13398. [CrossRef]
37. Dai, C.; Li, J.; Tang, S.; Li, J.; Xiao, X. Colistin-induced nephrotoxicity in mice involves the mitochondrial, death receptor, and endoplasmic reticulum pathways. *Antimicrob. Agents Chemother.* **2014**, *58*, 4075–4085. [CrossRef]
38. Luo, H.; Gu, R.; Ouyang, H.; Wang, L.; Shi, S.; Ji, Y.; Bao, B.; Liao, G.; Xu, B. Cadmium exposure induces osteoporosis through cellular senescence, associated with activation of NF-κB pathway and mitochondrial dysfunction. *Environ. Pollut.* **2021**, *290*, 118043. [CrossRef]
39. Hyun, M.; Kim, H.; Kim, J.; Lee, J.; Lee, H.J.; Rathor, L.; Meier, J.; Larner, A.; Lee, S.M.; Moon, Y.; et al. Melatonin protects against cadmium-induced oxidative stress via mitochondrial STAT3 signaling in human prostate stromal cells. *Commun. Biol.* **2023**, *6*, 157. [CrossRef]
40. Firdaus, F.; Zafeer, M.F.; Waseem, M.; Anis, E.; Hossain, M.M.; Afzal, M. Ellagic acid mitigates arsenic-trioxide-induced mitochondrial dysfunction and cytotoxicity in SH-SY5Y cells. *J. Biochem. Mol. Toxicol.* **2018**, *32*, e22024. [CrossRef]
41. Ding, X.; Jian, T.; Wu, Y.; Zuo, Y.; Li, J.; Lv, H.; Ma, L.; Ren, B.; Zhao, L.; Li, W.; et al. Ellagic acid ameliorates oxidative stress and insulin resistance in high glucose-treated HepG2 cells via miR-223/keap1-Nrf2 pathway. *Biomed. Pharmacother.* **2019**, *110*, 85–94. [CrossRef] [PubMed]
42. Zhao, L.; Mehmood, A.; Soliman, M.M.; Iftikhar, A.; Iftikhar, M.; Aboelenin, S.M.; Wang, C. Protective Effects of Ellagic Acid Against Alcoholic Liver Disease in Mice. *Front. Nutr.* **2021**, *8*, 744520. [CrossRef]
43. Vercellino, I.; Sazanov, L.A. The assembly, regulation and function of the mitochondrial respiratory chain. *Nat. Rev. Mol. Cell Biol.* **2022**, *23*, 141–161. [CrossRef]
44. Wolin, M.S. Evidence for novel aspects of Nox4 oxidase regulation of mitochondrial function and peroxide generation in an endothelial cell model of senescence. *Biochem. J.* **2013**, *452*, e1–e2. [CrossRef] [PubMed]
45. Okoye, C.N.; MacDonald-Jay, N.; Kamunde, C. Effects of bioenergetics, temperature and cadmium on liver mitochondria reactive oxygen species production and consumption. *Aquat. Toxicol.* **2019**, *214*, 105264. [CrossRef]
46. Adiele, R.C.; Stevens, D.; Kamunde, C. Differential inhibition of electron transport chain enzyme complexes by cadmium and calcium in isolated rainbow trout (*Oncorhynchus mykiss*) hepatic mitochondria. *Toxicol. Sci. Off. J. Soc. Toxicol.* **2012**, *127*, 110–119. [CrossRef] [PubMed]
47. Okoye, C.N.; Koren, S.A.; Wojtovich, A.P. Mitochondrial complex I ROS production and redox signaling in hypoxia. *Redox Biol.* **2023**, *67*, 102926. [CrossRef]
48. Dröse, S. Differential effects of complex II on mitochondrial ROS production and their relation to cardioprotective pre- and postconditioning. *Biochim. Biophys. Acta* **2013**, *1827*, 578–587. [CrossRef]
49. Khodaei, F.; Rashedinia, M.; Heidari, R.; Rezaei, M.; Khoshnoud, M.J. Ellagic acid improves muscle dysfunction in cuprizone-induced demyelinated mice via mitochondrial Sirt3 regulation. *Life Sci.* **2019**, *237*, 116954. [CrossRef]
50. Dai, C.; Sharma, G.; Liu, G.; Shen, J.; Shao, B.; Hao, Z. Therapeutic detoxification of quercetin for aflatoxin B1-related toxicity: Roles of oxidative stress, inflammation, and metabolic enzymes. *Environ. Pollut.* **2024**, *345*, 123474. [CrossRef]
51. Green, D.R.; Reed, J.C. Mitochondria and apoptosis. *Science* **1998**, *281*, 1309–1312. [CrossRef] [PubMed]
52. Mashimo, M.; Onishi, M.; Uno, A.; Tanimichi, A.; Nobeyama, A.; Mori, M.; Yamada, S.; Negi, S.; Bu, X.; Kato, J.; et al. The 89-kDa PARP1 cleavage fragment serves as a cytoplasmic PAR carrier to induce AIF-mediated apoptosis. *J. Biol. Chem.* **2021**, *296*, 100046. [CrossRef] [PubMed]
53. Bello, M.; Morales-González, J.A. Molecular recognition between potential natural inhibitors of the Keap1-Nrf2 complex. *Int. J. Biol. Macromol.* **2017**, *105*, 981–992. [CrossRef]
54. Zhang, X.; Liu, Y.; Shen, Z.; Wang, S.; Wu, C.; Liu, D.; Tang, S.; Dai, C. Osthole ameliorates myonecrosis caused by Clostridium perfringens type A infection in mice. *One Health Adv.* **2023**, *1*, 27. [CrossRef]
55. Gao, M.; Li, C.; Xu, M.; Liu, Y.; Cong, M.; Liu, S. LncRNA MT1DP Aggravates Cadmium-Induced Oxidative Stress by Repressing the Function of Nrf2 and is Dependent on Interaction with miR-365. *Adv. Sci.* **2018**, *5*, 1800087. [CrossRef]
56. Ren, X.; Xu, Y.; Yu, Z.; Mu, C.; Liu, P.; Li, J. The role of Nrf2 in mitigating cadmium-induced oxidative stress of Marsupenaeus japonicus. *Environ. Pollut.* **2021**, *269*, 116112. [CrossRef] [PubMed]
57. Wang, Q.; Botchway, B.O.A.; Zhang, Y.; Liu, X. Ellagic acid activates the Keap1-Nrf2-ARE signaling pathway in improving Parkinson's disease: A review. *Biomed. Pharmacother.* **2022**, *156*, 113848. [CrossRef]
58. Cao, X.; Fu, M.; Bi, R.; Zheng, X.; Fu, B.; Tian, S.; Liu, C.; Li, Q.; Liu, J. Cadmium induced BEAS-2B cells apoptosis and mitochondria damage via MAPK signaling pathway. *Chemosphere* **2021**, *263*, 128346. [CrossRef]
59. Chen, L.; Liu, L.; Huang, S. Cadmium activates the mitogen-activated protein kinase (MAPK) pathway via induction of reactive oxygen species and inhibition of protein phosphatases 2A and 5. *Free Radic. Biol. Med.* **2008**, *45*, 1035–1044. [CrossRef]

60. Chen, L.; Liu, L.; Luo, Y.; Huang, S. MAPK and mTOR pathways are involved in cadmium-induced neuronal apoptosis. *J. Neurochem.* **2008**, *105*, 251–261. [CrossRef]
61. Liu, Y.; Chang, J.; Yang, C.; Zhang, T.; Chen, X.; Shi, R.; Liang, Y.; Xia, Q.; Ma, S. Genome-wide CRISPR-Cas9 screening in Bombyx mori reveals the toxicological mechanisms of environmental pollutants, fluoride and cadmium. *J. Hazard. Mater.* **2021**, *410*, 124666. [CrossRef] [PubMed]
62. Hu, Y.R.; Ma, H.; Zou, Z.Y.; He, K.; Xiao, Y.B.; Wang, Y.; Feng, M.; Ye, X.L.; Li, X.G. Activation of Akt and JNK/Nrf2/NQO1 pathway contributes to the protective effect of coptisine against AAPH-induced oxidative stress. *Biomed. Pharmacother.* **2017**, *85*, 313–322. [CrossRef] [PubMed]
63. Tasaki, M.; Umemura, T.; Maeda, M.; Ishii, Y.; Okamura, T.; Inoue, T.; Kuroiwa, Y.; Hirose, M.; Nishikawa, A. Safety assessment of ellagic acid, a food additive, in a subchronic toxicity study using F344 rats. *Food Chem. Toxicol.* **2008**, *46*, 1119–1124. [CrossRef]
64. Heilman, J.; Andreux, P.; Tran, N.; Rinsch, C.; Blanco-Bose, W. Safety assessment of Urolithin A, a metabolite produced by the human gut microbiota upon dietary intake of plant derived ellagitannins and ellagic acid. *Food Chem. Toxicol.* **2017**, *108*, 289–297. [CrossRef]
65. Ardah, M.T.; Bharathan, G.; Kitada, T.; Haque, M.E. Ellagic Acid Prevents Dopamine Neuron Degeneration from Oxidative Stress and Neuroinflammation in MPTP Model of Parkinson's Disease. *Biomolecules* **2020**, *10*, 1519. [CrossRef] [PubMed]
66. Baluchnejadmojarad, T.; Rabiee, N.; Zabihnejad, S.; Roghani, M. Ellagic acid exerts protective effect in intrastriatal 6-hydroxydopamine rat model of Parkinson's disease: Possible involvement of ERβ/Nrf2/HO-1 signaling. *Brain Res.* **2017**, *1662*, 23–30. [CrossRef]
67. Hajiluian, G.; Karegar, S.J.; Shidfar, F.; Aryaeian, N.; Salehi, M.; Lotfi, T.; Farhangnia, P.; Heshmati, J.; Delbandi, A.A. The effects of Ellagic acid supplementation on neurotrophic, inflammation, and oxidative stress factors, and indoleamine 2, 3-dioxygenase gene expression in multiple sclerosis patients with mild to moderate depressive symptoms: A randomized, triple-blind, placebo-controlled trial. *Phytomedicine* **2023**, *121*, 155094. [CrossRef]

antioxidants

MDPI

Review

Natural Products as Potentiators of β-Lactam Antibiotics: A Review of Mechanisms, Advances, and Future Directions

Wenjie Yang [1,2], Shuocheng Fan [1], Jie Luo [1], Yichu Zhou [1], Xingyang Dai [1], Jinhu Huang [1], Liping Wang [1,*] and Xiaoming Wang [1,*]

1 Sanya Institute of Nanjing Agricultural University, College of Veterinary Medicine, Nanjing Agricultural University, Nanjing 210095, China; 9201710317@stu.njau.edu.cn (W.Y.); shuochengfan3@stu.njau.edu.cn (S.F.); luojie-@stu.njau.edu.cn (J.L.); 2024807224@stu.njeu.edu.cn (Y.Z.); t2024002@njau.edu.cn (X.D.); jhuang@njau.edu.cn (J.H.)

2 Department of Comparative Biosciences, University of Illinois at Urbana-Champaign, Urbana, IL 61802, USA

* Correspondence: wlp71@njau.edu.cn (L.W.); wangxm@njau.edu.cn (X.W.)

Abstract

This review focuses on the research progress on natural products as β-lactam antibiotic adjuvants, aiming to address the escalating challenge of antibiotic resistance, particularly the inactivation of antibiotics caused by β-lactamases. The article provides an in-depth analysis of the mechanisms by which plant-derived (e.g., flavonoids, tannins, phenolics, terpenoids, and alkaloids) and microbial-derived (e.g., clavulanic acid, fungal metabolites, bacteriophages) natural products enhance antimicrobial efficacy. Key potentiation strategies discussed include efflux pump inhibition, membrane permeability alteration, biofilm disruption, PBP2a inhibition, and direct β-lactamase inhibition. Additionally, the review outlines in vitro methods (e.g., dilution and checkerboard assays) and in vivo models (e.g., mouse infection models) used to assess synergistic effects. It also addresses major challenges in identifying active compounds, elucidating mechanisms of action, and pharmacokinetic characterization. Looking forward, the article highlights the potential of multi-omics approaches, artificial intelligence, and nanotechnology to overcome existing bottlenecks, providing novel strategies for the development of effective and safe antibiotic adjuvants. These advances are expected to provide both theoretical insights and practical guidance for combating antibiotic-resistant bacterial infections.

Keywords: natural products; β-lactam antibiotics; antibiotic resistance; synergistic effects; potentiators

1. Introduction

Antibiotic resistance constitutes a significant global public health threat, with bacteria developing multiple mechanisms to evade the effects of antimicrobial agents [1]. β-lactam antibiotics, including penicillins, cephalosporins, carbapenems, and monobactams, are among the most extensively utilized and clinically effective classes of antibiotics [2]. In 2023, the global market for β-lactam and β-lactamase inhibitors was estimated at approximately USD 29 billion, representing nearly 57% of the total antibiotic market value, which stood at USD 50.91 billion [3]. These antibiotics exert their bactericidal activity by inhibiting bacterial cell wall synthesis. Specifically, they bind to penicillin-binding proteins (PBPs), enzymes responsible for cross-linking peptidoglycan, a crucial structural component of the bacterial cell wall [4]. This inhibition compromises cell wall integrity, resulting in osmotic imbalance and ultimately leading to bacterial cell lysis, making β-lactams exhibit broad-spectrum efficacy against diverse pathogens [5].

The therapeutic potential of β-lactam antimicrobial agents is increasingly compromised due to the global expansion of drug-resistant bacterial strains. A principal mechanism underlying this resistance is the bacterial production of β-lactamases—enzymes that hydrolyze the β-lactam ring, a four-membered lactam structure that constitutes the pharmacologically active core of these agents. Enzymatic cleavage of this moiety renders the antibiotic inactive [2]. This mechanism is particularly prevalent in Gram-negative bacteria, where β-lactamases are frequently encoded on mobile genetic elements, promoting horizontal gene transfer and rapid dissemination [6]. In addition to β-lactamases, bacteria have developed alternative resistance mechanisms, including mutations in PBPs (e.g., PBP2a in methicillin-resistant *Staphylococcus aureus*), which reduce β-lactam binding affinity, and the upregulation of efflux pumps that expel antibiotics from the intracellular environment. The emergence and global spread of multidrug-resistant (MDR) strains, resistant to three or more antibiotic classes—has further intensified this crisis, severely constraining available therapeutic options and posing a formidable challenge to public health systems worldwide.

The term 'natural products' refers to chemical compounds produced by living organisms. In the context of drug discovery, this term specifically denotes 'secondary metabolites'. Unlike primary metabolites (e.g., amino acids and sugars) essential for survival, secondary metabolites—such as flavonoids, alkaloids, and terpenoids—play crucial roles in an organism's defense, competition, and signaling. It is due to their vast chemical diversity and potent biological activities that natural products have long played a key role in drug discovery, serving as a prolific source of bioactive compounds, including antibiotics [7]. In recent years, increasing attention has been directed toward the use of natural products as potential adjuvants to β-lactam antibiotics, with the goal of restoring their efficacy against resistant bacterial strains [8]. These compounds can enhance β-lactam activity through diverse mechanisms, such as inhibiting β-lactamases, increasing bacterial membrane permeability, suppressing biofilm formation, and interfering with other resistance pathways [9].

Antibiotic resistance typically arises through genetic mutations or horizontal gene transfer, enabling bacteria to evade drug action via mechanisms such as target modification, enzymatic antibiotic degradation, reduced membrane permeability, or active efflux. These resistance traits are often rapidly selected under the strong selective pressure imposed by single-target, bactericidal antibiotics [10]. Natural products offer distinct advantages in combating resistance. Their vast chemical diversity provides a broad range of novel structures, which may reduce the likelihood of pre-existing bacterial resistance. However, it must also be considered that bacteria have long been exposed to these compounds in the natural environment, which may have facilitated the co-evolution of resistance mechanisms against certain natural products [11]. Nevertheless, many natural products exhibit multi-target or pleiotropic activity, simultaneously affecting multiple bacterial pathways. Such multi-site interference increases the evolutionary barrier to resistance by requiring concurrent adaptations at several loci, thereby reducing the probability of stable resistance emergence [12]. Clinically, the β-lactamase inhibitor clavulanic acid exemplifies the utility of natural products, having been effectively combined with β-lactam antibiotics such as amoxicillin [13]. Moreover, many natural products, particularly flavonoids and polyphenols, possess potent antioxidant properties that complement their antimicrobial activities [14]. These compounds can scavenge reactive oxygen species (ROS) generated during bacterial infections, thereby mitigating oxidative stress and preserving host tissue integrity. For instance, flavonoids like quercetin and epigallocatechin-3-gallate (EGCG) not only inhibit resistance mechanisms but also alleviate ROS-mediated cellular damage, thereby contributing to improved therapeutic outcomes in combination regimens [15].

https://doi.org/10.3390/antiox15020154

Similarly, the coumarin derivative osthole has been reported to ameliorate myonecrosis caused by *Clostridium perfringens* infection via the activation of the Nrf2/HO-1 antioxidant pathway, effectively mitigating oxidative stress and inflammatory responses in host tissues [16]. Furthermore, derived from renewable sources, natural products present a sustainable and potentially cost-effective strategy for drug development.

This review explores recent advances in natural products as potentiators of β-lactam antibiotics, with a focus on their mechanisms of action, representative findings, and prospective research directions. A comprehensive literature survey was conducted using the PubMed database and the Web of Science citation index, covering studies published between 2015 and 2025. Search terms included "β-lactamase", "synergy", "natural products", "synergistic effect", "potentiators", and "antibiotic synergism".

2. Mechanisms of Natural Product-Mediated Potentiation of β-Lactam Antibiotics

Natural products can enhance the efficacy of β-lactam antibiotics through various mechanisms, such as inhibiting drug efflux pumps, modulating cell membrane and wall permeability, disrupting biofilms, inhibiting PBP2a, suppressing β-lactamase activity, interfering with ATP synthesis, and other pathways (Figure 1).

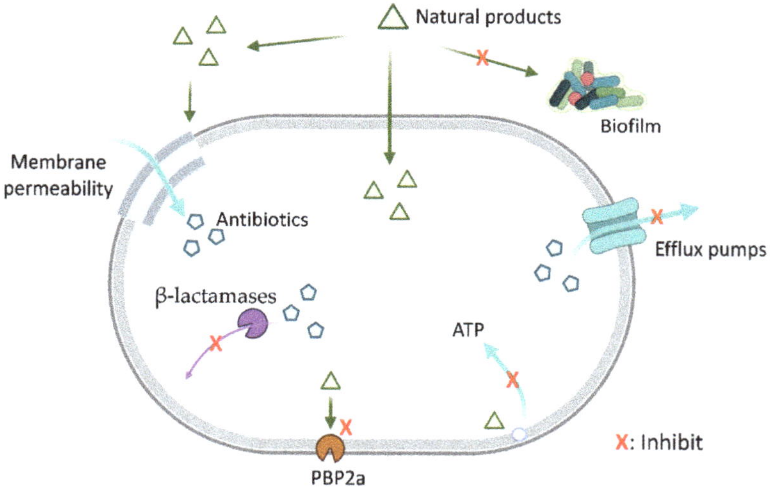

Figure 1. Natural Products Mitigate Bacterial Resistance Mechanisms. This schematic diagram illustrates the multifaceted strategies by which natural products (green triangles) may potentiate antibiotic activity by targeting bacterial resistance pathways. The diagram depicts mechanisms including inhibition of biofilm formation, alteration of bacterial membrane permeability, efflux pump inhibition, suppression of β-lactamase activity, inhibition of penicillin-binding protein 2a (PBP2a), and disruption of bacterial ATP generation (energy metabolism). Each red 'X' indicates a potential point of interference by natural products. Mechanisms are presented for conceptual completeness. Among them, β-lactamase inhibition and PBP2a inhibition are supported by substantial experimental evidence and, in some cases, clinical validation. In contrast, mechanisms such as modulation of ATP metabolism, membrane permeability, efflux activity, and biofilm-related processes are primarily supported by indirect or preclinical in vitro studies, and their roles in β-lactam potentiation remain less well established. Created in BioRender. Li, S. (2026) https://BioRender.com/c22k30u (accessed on 23 December 2025).

https://doi.org/10.3390/antiox15020154

2.1. *Efflux Pumps Inhibition*

Efflux pumps are transmembrane protein complexes that actively export a wide range of antibiotics out of the bacterial cell, thereby reducing intracellular drug concentrations and contributing to antimicrobial resistance [17]. This resistance mechanism is widely observed in both Gram-positive and Gram-negative bacteria, such as *Staphylococcus aureus*, *Escherichia coli*, and *Pseudomonas aeruginosa* [18].

Bacterial efflux pumps are primarily classified into several major superfamilies based on their structure and energy source [19]. Five of these are particularly prominent: the RND (Resistance Nodulation Division), MFS (Major Facilitator Superfamily), SMR (Small Multidrug Resistance), MATE (Multidrug and Toxic Compound Extrusion), and ABC (ATP-Binding Cassette) families [19]. ABC transporters utilize the energy from ATP hydrolysis, while the other four families are secondary transporters that rely on the proton motive force [20].

In Gram-negative bacteria, the RND family of efflux pumps is of paramount clinical importance due to its unique structure and high efficiency [19]. These pumps typically assemble into a tripartite complex that spans the inner membrane, the periplasm, and the outer membrane [20]. The AcrAB-TolC system in *E. coli* is a classic example, consisting of an inner membrane RND transporter (AcrB), a periplasmic adaptor protein (AcrA), and an outer membrane channel (TolC) [19]. This architecture allows the pump to expel captured drugs from the periplasm or inner membrane directly into the external medium, bypassing the periplasmic space and working synergistically with the low-permeability outer membrane barrier. In Gram-negative bacteria, this mechanism effectively protects targets on the inner membrane, such as PBPs, from antibiotics like many β-lactams that must first cross the outer membrane [18].

Natural products can act as Efflux Pump Inhibitors (EPIs) to subvert this defense system through several distinct mechanisms [17]. One primary mode of action is competitive inhibition, where the inhibitor acts as a substrate for the pump and competes with the antibiotic for the same binding site [17]. A more effective mechanism is often non-competitive or allosteric inhibition, where the inhibitor binds tightly to a distinct hydrophobic trap or "inhibitor-binding pit" within the pump protein [19]. This binding induces a conformational change that locks the pump in an inactive state, thereby blocking its functional rotation and transport cycle. Additionally, some compounds can indirectly inhibit proton-motive-force-dependent pumps by disrupting the proton gradient across the membrane, though this approach often lacks specificity and can be toxic to the cell [17].

Research has demonstrated that certain natural products are effective EPIs. For instance, EGCG enhances the susceptibility of *P. aeruginosa* to antibiotics like chloramphenicol and tetracyclines, likely by inhibiting the RND-family pump MexAB-OprM [15]. Similarly, berberine, an alkaloid found in traditional herbal medicines like *Corydalis* Tuber, has been identified as an efflux pump inhibitor in Gram-positive bacteria. Studies on *Staphylococcus aureus* demonstrated that berberine inhibits efflux activity, and molecular docking simulations suggest it binds to multiple efflux pump proteins, including MepA, NorA, NorB, and SdrM. This mechanism was shown to restore the efficacy of antibiotics such as ciprofloxacin and tobramycin against methicillin-resistant *S. aureus* (MRSA) [21]. These examples highlight the diverse strategies and potent potential of natural products to overcome efflux-mediated resistance, offering promising avenues for the development of combination therapies to re-sensitize resistant pathogens [17].

2.2. *Alteration of Membrane and Cell Wall Permeability*

The bacterial cell envelope, particularly the lipopolysaccharide (LPS)-rich outer membrane of Gram-negative bacteria, serves as a formidable permeability barrier that restricts

https://doi.org/10.3390/antiox15020154

the entry of many antibiotics [22]. Consequently, disrupting this barrier has emerged as a key therapeutic strategy to enhance antibiotic uptake and reverse resistance [22]. Natural products can function as permeability enhancers by directly interacting with membrane components, leading to destabilization, depolarization, and the formation of pores, which in turn facilitates the diffusion of antibiotics like β-lactams into the periplasm [13].

This principle is classically exemplified by polymyxins, such as colistin, which are natural peptides that specifically target the LPS of the Gram-negative outer membrane [23]. By electrostatically interacting with the anionic phosphate groups of lipid A, polymyxins competitively displace the divalent cations (Mg^{2+} and Ca^{2+}) that stabilize the LPS layer [23]. This disruption damages the integrity of the outer membrane, making it more permeable and creating a synergistic effect that allows other antibiotics, such as β-lactams, to reach their periplasmic targets [24].

Many plant-derived compounds operate on similar principles. For instance, cinnamaldehyde enhances the susceptibility of multidrug-resistant *Salmonella* to ceftriaxone by causing cell membrane depolarization and observable physical damage, which is evidenced by increased LPS release [25]. Similarly, compounds like gallic acid (GA) and methyl gallate (MG) have been shown to induce significant morphological changes in the cell membrane of MRSA, including surface wrinkles and the leakage of intracellular contents [14]. Although the PBP targets in Gram-positive bacteria are located outside the cell membrane, this NP-induced membrane damage compromises the cell's osmotic integrity, creating a synergistic lethal effect when combined with β-lactams that inhibit cell wall synthesis. While promising, membrane-targeting strategies also face challenges, notably ensuring selectivity for bacterial membranes over host cells to minimize toxicity, and the potential for bacteria to develop resistance by modifying their membrane structures, such as altering the net charge of their LPS [23].

2.3. *Suppression of Biofilm Formation*

Biofilms are structured bacterial communities embedded within a self-produced extracellular polymeric substance (EPS matrix composed of polysaccharides, proteins, and DNA). This matrix confers increased resistance by blocking antibiotic access to targets like PBPs and by harboring dormant bacteria that are less susceptible to cell wall inhibition [26,27]. Natural products can disrupt biofilms by inhibiting quorum sensing (QS), degrading the EPS matrix, or interfering with bacterial metabolism, thereby enhancing β-lactam efficacy.

Quorum sensing is a cell-to-cell communication system, mediated by signaling molecules, that coordinates gene expression and is crucial for biofilm maturation [28]. Natural products acting as QS inhibitors (QSIs) can block these signals, preventing mature biofilm formation and reducing resistance gene expression (e.g., *mecA*), thus increasing bacterial vulnerability. For example, baicalin synergizes with oxacillin against MRSA, lowering its minimum inhibitory concentration (MIC) four-fold, with a fractional inhibitory concentration index (FICI) ≤ 0.5, by inhibiting QS pathways, enhancing β-lactam penetration and bactericidal activity [29,30].

The EPS matrix acts as a physical shield, protecting embedded bacteria from antibiotics and host immune responses [27]. Degrading this layer with natural products, such as enzymes from bacteriophages, exposes bacterial targets. For instance, the lytic enzymes of bacteriophage F1Pa disrupt the EPS matrix of *Pseudomonas aeruginosa* biofilms, thereby facilitating piperacillin access to PBPs and resulting in a reduction in bacterial viability by over 2 $\log_{10}$ CFU/mL [31–33].

Bacteria within mature biofilms often enter a slow-growing or dormant metabolic state, rendering them tolerant to β-lactams that target actively dividing cells [26]. Natural

https://doi.org/10.3390/antiox15020154

products can disrupt this dormancy by interfering with essential metabolic processes and compromising membrane integrity. For instance, *Origanum vulgare* essential oil, rich in carvacrol, has been shown to damage the bacterial cell membrane, increase membrane permeability, and induce cytoplasmic leakage, as evidenced by potassium efflux, release of 260-nm–absorbing materials, and ultrastructural membrane alterations [34]. Such membrane disruption is widely associated with downstream energetic stress in bacteria, including impairment of cellular energy homeostasis and depletion of intracellular ATP, as reported for carvacrol and related essential oil components in independent metabolic studies [35]. This metabolic disruption reactivates dormant bacteria, enhancing their susceptibility to β-lactams. When combined with oxacillin, it reduces *S. aureus* biofilm biomass by approximately 50–60% (1–1.3 $\log_{10}$ CFU/mL) within 24 h, as shown by decreased viability [36]. Additionally, the oil induces reactive oxygen species (ROS) accumulation and membrane hyperpolarization, which may exacerbate metabolic stress, though the primary enhancement of β-lactam efficacy stems from ATP depletion and PBP targeting [35]. This synergistic action overcomes biofilm-mediated resistance, restoring bactericidal activity.

2.4. PBP2a Inhibition

Resistance to nearly all β-lactam antibiotics in MRSA is primarily mediated by the expression of Penicillin-Binding Protein 2a (PBP2a), an alternative transpeptidase encoded by the *mecA* gene [37]. Unlike the native PBPs of methicillin-susceptible *S. aureus* (MSSA), which are readily inactivated by β-lactams, PBP2a possesses a distorted active site that confers a low binding affinity for these antibiotics [38]. This structural distinction allows PBP2a to continue catalyzing the essential peptidoglycan cross-linking reactions required for cell wall biosynthesis, even when all other native PBPs are saturated by high concentrations of antibiotics, thereby enabling bacterial survival [37]. Therefore, inhibiting the function of PBP2a represents a critical strategy to overcome this resistance mechanism and restore the efficacy of β-lactam antibiotics against MRSA [39].

Natural products offer a diverse chemical arsenal to counteract PBP2a-mediated resistance through several distinct mechanisms, including the downregulation of gene expression, inhibition of protein translation, and direct binding to the enzyme [38]. For instance, the pomegranate-derived polyphenol punicalagin has been shown to suppress the transcription of the entire *mec* operon (*mecA*, *mecI*, and *mecR1*), leading to a significant reduction in PBP2a protein levels and restoring oxacillin susceptibility [40]. Similarly, glycyrrhizic acid nanoparticles (GA-NPs), formulated from the natural triterpenoid glycyrrhizic acid isolated from licorice, have been reported to potentiate the action of beta-lactams against MRSA by markedly suppressing PBP2a expression [41]. In this case, nanoparticle formulation primarily serves to improve the delivery and efficacy of the parent natural product, rather than representing a distinct class of antimicrobial agents.

Beyond transcriptional control, other natural compounds interfere at different stages of PBP2a production and function. Demethoxycurcumin (DMC), a natural curcuminoid from turmeric, hinders the translation of the PBP2a protein while also downregulating the transcription of its associated genes [37]. Furthermore, certain compounds exhibit multi-modal action; the non-membrane active antimicrobial peptide GN1, for example, resensitizes MRSA to β-lactams by both downregulating *mecA* gene expression and directly binding to an allosteric site on the PBP2a enzyme to inhibit its function [38]. In addition, a bioactive fraction from *Acalypha wilkesiana* has been shown to attenuate the production of PBP2a within the MRSA biofilm matrix [39].

Computational approaches have further expanded the list of potential PBP2a inhibitors. In silico molecular docking studies have predicted that flavonoids such as luteolin and kaempferol, as well as the microbial metabolite anisomycin, exhibit favorable binding

https://doi.org/10.3390/antiox15020154

affinity for PBP2a. Specifically, these compounds demonstrated calculated binding free energies of −8.47, −7.51, and −7.28 kcal/mol, respectively, which are comparable to or more negative than those of the reference antibiotic vancomycin (−6.95 kcal/mol) within the same docking framework [42]. These energetic profiles identify them as promising candidates for further experimental validation. Collectively, these findings demonstrate that targeting PBP2a with natural products—whether by suppressing gene expression, inhibiting protein synthesis, or through direct enzymatic binding—represents a highly promising and multifaceted strategy to overcome methicillin resistance and restore the clinical utility of β-lactam antibiotics against MRSA. It is essential to recognize, however, that most natural product–based strategies targeting the transcriptional or translational regulation of specific resistance genes, such as *mecA*, are currently restricted to the experimental or preclinical stages. To date, the clinical pharmacopoeia lacks approved beta-lactam adjuvants that primarily function through the selective modulation of resistance gene expression, emphasizing the forward-looking nature of these natural product investigations.

2.5. *β-Lactamase Inhibition*

The production of β-lactamase enzymes is the most significant mechanism of resistance to β-lactam antibiotics, particularly in Gram-negative bacteria [43,44]. These enzymes confer resistance by hydrolyzing the amide bond in the four-membered β-lactam ring, a core structural feature of all penicillins, cephalosporins, and carbapenems, thereby rendering the antibiotics pharmacologically inactive [6]. Based on amino acid sequence homology, β-lactamases are categorized into four molecular Ambler classes: A, B, C, and D [45]. This structural classification reflects fundamental mechanistic differences, as enzymes in classes A, C, and D utilize a serine residue in their active site for hydrolysis (serine β-lactamases), while class B enzymes are metallo-β-lactamases (MBLs) that require one or two zinc ions for their catalytic activity [44,45].

These enzymes present a formidable clinical challenge, especially those with broad hydrolytic spectrums that are encoded on mobile genetic elements, facilitating their rapid global dissemination [43]. Prominent examples include *Klebsiella pneumoniae* carbapenemase-2 (KPC-2), a class A serine carbapenemase, and New Delhi metallo-β-lactamase 1 (NDM-1), a class B metalloenzyme [43]. The widespread emergence of such enzymes, which can inactivate even last-resort carbapenem antibiotics, severely limits therapeutic options.

A clinically validated strategy to overcome this resistance is the co-administration of a β-lactamase inhibitor with a β-lactam antibiotic. Natural products serve as a rich source for such inhibitors, which can act by competitively or irreversibly binding to the enzyme's active site. The archetypal example is clavulanic acid, a natural product derived from *Streptomyces clavuligerus*. As a potent inhibitor of many class A enzymes, clavulanic acid has been successfully developed for clinical use and is routinely co-administered with β-lactam antibiotics, such as in amoxicillin-clavulanate formulations [46]. Furthermore, crude soy saponins have been found to exhibit an inhibitory effect against various β-lactamases, including the challenging New Delhi metallo-β-lactamase 1 (NDM-1), with soyasaponin V identified as having the highest inhibitory activity against NDM-1. The synergistic effect on the antimicrobial activity of β-lactam antibiotics by crude soy saponins is thought to result from the inhibition of β-lactamase activity [47]. This approach of targeting β-lactamases with natural compounds offers a promising avenue to restore the effectiveness of existing β-lactam antibiotics and combat the growing threat of antibiotic resistance.

https://doi.org/10.3390/antiox15020154

3. Advances in Natural Products as β-Lactam Antibiotic Potentiators

3.1. Advances in Research on Plant-Derived Natural Products

Plant-derived natural products possess rich chemical diversity, providing a broad candidate pool for the study of antibiotic potentiators [48]. These natural products enhance the antibacterial activity of β-lactam antibiotics through various mechanisms, including inhibiting efflux pumps, altering cell membrane and cell wall permeability, suppressing biofilm formation, inhibiting PBP2a, and suppressing β-lactamase activity (Figure 2). Key plant-derived natural products discussed in this section, their sources, representative structures, and primary mechanisms of potentiation are summarized in Table 1 (Plant-Derived section).

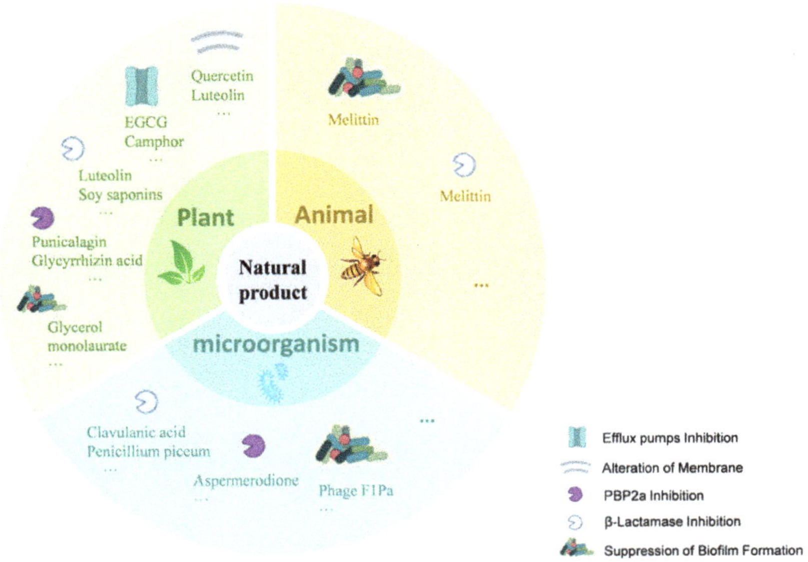

Figure 2. Mechanisms of Natural Products Enhancing β-Lactam Antibiotic Activity. This figure illustrates how natural products from plants, microorganisms, and animals enhance β-lactam antibiotic activity through various mechanisms. These include efflux pump inhibition (e.g., Quercetin, EGCG, Camphor); alteration of membrane permeability (e.g., Luteolin, Gallic acid, Eugenol); PBP2a inhibition (e.g., Punicalagin, Glycyrrhizin acid nanoparticles, Aspermerodione); β-lactamase inhibition (e.g., Soy saponins, Clavulanic acid); and suppression of biofilm formation (e.g., Glycerol monolaurate, Phage F1Pa, Melittin). The icons within the figure indicate the targets of these natural products, highlighting their multi-targeted strategies against bacterial resistance.

It should be noted that not all beneficial effects observed in combination studies constitute true pharmacological synergy or direct reversal of bacterial resistance. In particular, antioxidant and anti-inflammatory activities may improve therapeutic outcomes by mitigating host tissue damage or infection-associated stress but do not necessarily reflect direct modulation of bacterial resistance mechanisms.

https://doi.org/10.3390/antiox15020154

Table 1. Classification of Natural Product-Derived beta-Lactam Potentiators by Biological Origin. Mechanisms marked with an asterisk (*) indicate those categorized as "Validated," supported by direct biochemical evidence, genetic assays, and/or in vivo studies. Unmarked mechanisms are classified as "Postulated," primarily derived from in vitro phenotypic observations (e.g., checkerboard assays) and require further mechanistic confirmation. To ensure clarity within the table, evidence levels are integrated via these symbols. Chemical structures are standard 2D representations sourced from authoritative databases (e.g., PubChem). For heterogeneous mixtures or broad compound classes (e.g., soy saponins, cranberry proanthocyanidins, or essential oils), characteristic scaffolds or representative bioactive monomers are depicted for illustrative purposes and do not represent a single defined molecular entity.

Natural Product	Source (Origin)	Structure	Mechanism	Reference
Plant-Derived				
Quercetin	Ubiquitous in plants	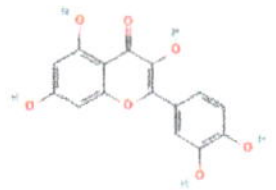	Enhance membrane permeability	[49]
Epigallocatechin-3-gallate (EGCG)	*Camellia sinensis* (green tea)	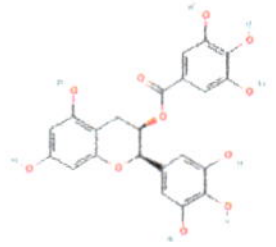	Inhibits/interferes with efflux pumps (e.g., MexAB-OprM) *	[15]
Theaflavin-3,3′-digallate	*Camellia sinensis* (black tea)	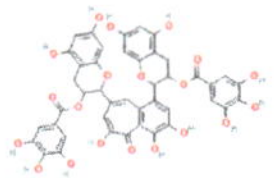	Binds PBP2a allosteric site; inhibits metallo-β-lactamases *	[50]
Luteolin	Ubiquitous in plants	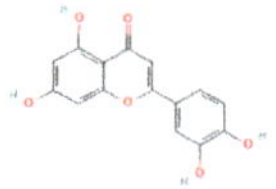	Synergy via membrane disruption or resistance gene suppression *	[51]
Galangin	*Alpinia officinarum* (Zingiberaceae)	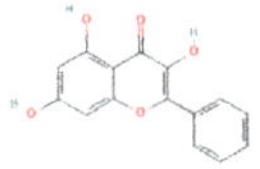	Enhances β-lactam activity	[52]
Baicalin	*Scutellaria baicalensis*	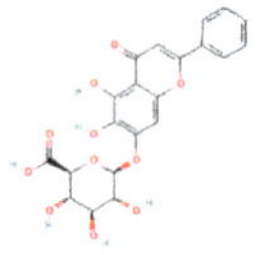	Disrupts bacterial membrane integrity (depolarization and leakage) and inhibits biofilm formation *	[29]
Liquiritigenin	*Glycyrrhiza glabra* (Fabaceae)	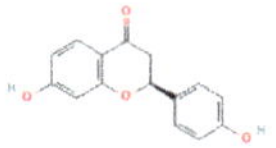	Reverses resistance by acting synergistically with beta-lactams *	[53]
Tannic acid	Leguminous plants	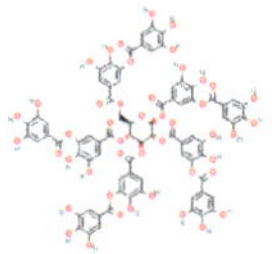	Sensitizes MRSA to β-lactams	[54]
Punicalagin	*Punica granatum* (Lythraceae)	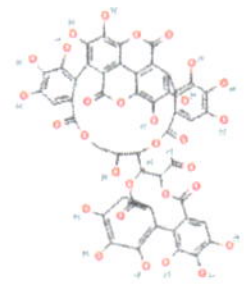	Sensitizes pathogens by inhibiting efflux pumps, and damaging cell membrane integrity *	[55]

https://doi.org/10.3390/antiox15020154

Table 1. *Cont.*

Natural Product	Source (Origin)	Structure	Mechanism	Reference
Cinnamic acid derivatives	*Cinnamomum* spp. (Lauraceae)	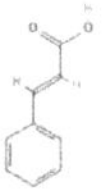	Acts synergistically to potentiate beta-lactam activity	[56]
Gallic acid	*Quercus* spp., *Camellia sinensis*	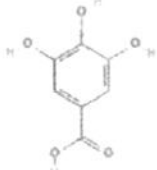	Modulates membrane permeability; putative β-lactamase inhibition	[14]
Eugenol	*Syzygium aromaticum* (Myrtaceae)	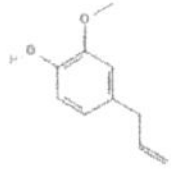	Increases membrane permeability	[57]
Cinnamaldehyde	*Cinnamomum* spp. (Lauraceae)	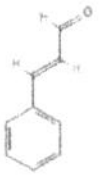	Disrupts membranes; Inhibits efflux pump; Disrupts energy metabolism *	[25]
Glycerol monolaurate	Coconut oil derivatives		Enhances betalactam efficacy *	[58]
Camphor	*Cinnamomum camphora* (Lauraceae)		Reduces carbapenem MIC via efflux inhibition	[59]
1,8-Cineole	*Eucalyptus* spp. (Myrtaceae)	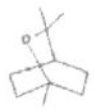	Potentiates beta-lactams by directly inhibiting ESBL enzymatic activity and increasing bacterial membrane permeability *	[60]
Carvacrol	Oregano, thyme (Lamiaceae)	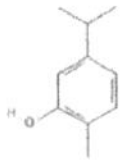	Membrane disruption; Enhances antibiotic penetration *	[61]
Clerodane diterpenoid	*Callicarpa americana* (Lamiaceae)	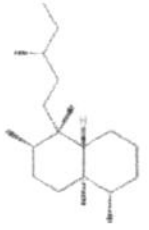	Restores β-lactam susceptibility in MRSA *	[62]
Berberine	*Berberis* spp. (Berberidaceae)	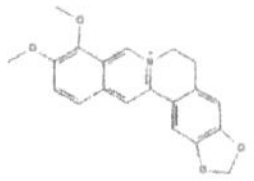	Inhibits efflux pumps; metabolic interference *	[63]
Ginsenoside Rg3	*Panax ginseng* (Araliaceae)	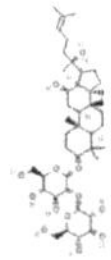	Membrane depolarization and increased permeability *	[64]
Soy saponins	*Glycine max* (Fabaceae)	-	Inhibits β-lactamases (including NDM-1) *	[47]

Table 1. *Cont.*

Natural Product	Source (Origin)	Structure	Mechanism	Reference
Glycyrrhizin acid	Chinese herb licorice	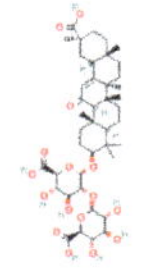	Suppress PBP2a *	[41]
Honokiol	*Magnolia officinalis* (Magnoliaceae)	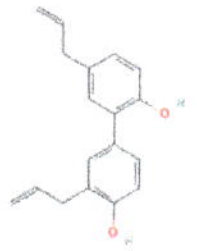	Inhibits bacterial cell division by directly binding to the FtsZ protei, suppressing its GTPase activity and polymerization, while also disrupting membrane integrity *	[65]
Cranberry proanthocyanidins	*Vaccinium macrocarpon* (Ericaceae)	-	Inhibits ESBL/MBL enzymes *	[66]
Lignin	Plant cell walls	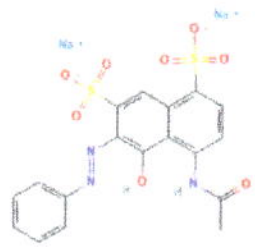	Disrupts cell envelope homeostasis by damaging the cell membrane *	[67]
Fucoxanthin	Brown algae (Phaeophyceae)	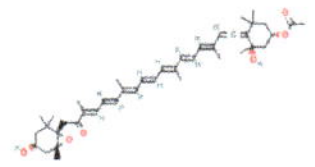	Enhances ceftriaxone efficacy	[68]
Hypericin	*Hypericum perforatum* (Hypericaceae)	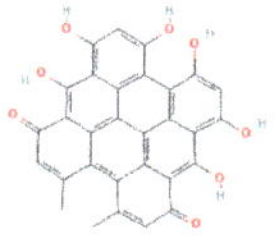	Suppresses sarA-mediated resistance regulation *	[68]
Origanum vulgare essential oil	Oregano plant	-	Disrupts cell membrane *	[34]
Microbial-Derived				
Clavulanic acid	*Streptomyces clavuligerus*	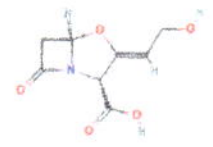	Irreversible inhibition of β-lactamases *	[46]
Penicillium piceum	Penicillium genus	-	Inhibits β-Lactamase *	[69]
Aspergillomarasmine A	*Aspergillus versicolor*	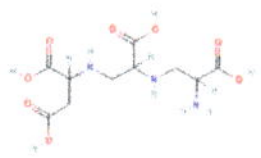	Potent metallo-β-lactamase (MBL) inhibitor *	[70]
Aspermerodione	*Aspergillus* sp. TJ23 (endophytic fungus)	-	Binds PBP2a allosteric site; inhibits MRSA resistance *	[71]
Tunicamycin (TUN) derivatives	*Streptomyces lysosuperificus*	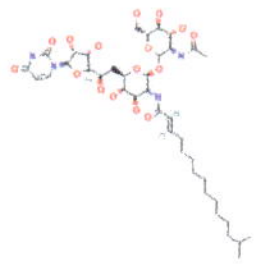	Acts as a non-toxic potentiator by inhibiting the bacterial wall teichoic acid (WTA) synthesis pathway *	[72]

https://doi.org/10.3390/antiox15020154

Table 1. *Cont.*

Natural Product	Source (Origin)	Structure	Mechanism	Reference
Plectasin	Fungal defensin peptide (saprophytic fungi)	-	Enhances antibiotic activity by binding to the cell wall precursor Lipid II to inhibit peptidoglycan biosynthesis *	[73]
Humimycin 17S	Human microbiome-derived lipopeptide	-	Inhibits the Lipid II flippase MurJ, thereby disrupting the transport of cell wall precursors and compromising cell wall integrity *	[74]
Probiotic metabolites	*Enterococcus faecium, Lactococcus lactis*	-	Reduces MRSA resistance via quorum sensing inhibition *	[75]
Phage F1Pa	Bacteriophages	-	Disrupts biofilms; enhances antibiotic penetration *	[33]
Animal-Derived				
Melittin	*Apis mellifera* (honeybee venom)		Disrupts the bacterial cell membrane (increasing permeability) and degrades the protective biofilm matrix *	[76]

3.1.1. Flavonoids

Flavonoids, widely distributed in plant families such as Fabaceae, Lamiaceae, and Asteraceae, have attracted significant attention due to their antioxidant, anti-inflammatory, and antimicrobial properties. For instance, flavonoids such as quercetin and epigallocatechin gallate significantly enhance the antibacterial efficacy of β-lactam antibiotics against MRSA when administered in combination. The co-administration of quercetin (MIC > 512 μg/mL alone) and EGCG demonstrates potent synergistic antibacterial activity (FICI: 0.06–0.31), producing a bactericidal effect with up to a 4–5 $\log_{10}$ CFU/mL reduction in bacterial counts within 24 h [49], potentially by inhibiting resistance-related enzymes or enhancing membrane permeability. Quercetin exhibits antimicrobial activity and antioxidant properties that may help mitigate infection-associated oxidative stress, typically achieving a 4- to 32-fold reduction in the MICs of clinical antibiotics, thereby supporting improved therapeutic outcomes when used as an adjunct to β-lactam antibiotics [77]. EGCG, a principal polyphenolic constituent of green tea, exhibits robust antibacterial and antioxidant activities and enhances the efficacy of β-lactam antibiotics against carbapenem-resistant *Acinetobacter baumannii* by reducing MIC values (e.g., meropenem) from 64 μg/mL to 8–4 μg/mL (8- to 16-fold reduction; FICI $\leq$ 0.5), specifically by inhibiting multidrug efflux pumps and increasing intracellular antibiotic accumulation [78]. Its potent antioxidant activity neutralizes ROS, mitigating infection-induced oxidative damage and contributing to enhanced antibacterial efficacy in combination with β-lactam antibiotics. Similarly, theaflavin-3,3′-digallate (TF3), a polyphenol from black tea, significantly enhances the antibacterial activity of carbapenems against NDM-1-producing Enterobacteriaceae by decreasing the MIC of meropenem from 64 μg/mL to 4 μg/mL (16-fold reduction; FICI $\leq$ 0.25), through directly inhibiting metallo-beta-lactamase activity. This compound demonstrates robust synergistic effects both in vitro and in a murine pneumonia model, effectively reducing bacterial load and improving survival rates [50]. Luteolin has also exhibited synergistic activity with ampicillin and oxacillin against MRSA, potentially by enhancing membrane permeability or downregulating the expression of resistance genes [51]. Other flavonoids, including galangin [52], baicalin [29], and liquiritigenin [53], have also been shown to potentiate β-lactam antibiotics through diverse mechanisms, providing a broad spectrum of plant-derived candidates for combating resistant bacteria.

3.1.2. Tannins

Tannins, a class of polyphenolic compounds widely distributed in plants, possess notable astringent and antibacterial properties and can effectively counteract bacterial resistance. Tannic acid, primarily derived from leguminous plants, has been reported to sensitize MRSA to β-lactam antibiotics, effectively reducing the MICs of agents such as oxacillin and ampicillin by 2- to 16-fold (with synergistic FICI values 0.174–0.477) when used at sub-inhibitory concentrations, likely through interference with cell wall synthesis or resistance gene expression [54]. Additionally, its antioxidant properties reduce ROS-mediated oxidative stress, contributing to improving therapeutic outcomes when combined with β-lactam antibiotics. Punicalagin, a key tannin compound found in pomegranates, reverses MRSA resistance to oxacillin by significantly decreasing the MIC from 256 μg/mL to 16 μg/mL (16-fold reduction; FICI = 0.31) through the inhibition of mecA and PBP2a expression [40]. It has also been demonstrated to potentiate the activity of β-lactams against MDR *Enterobacteriaceae*, where its combination with amoxicillin/clavulanic acid resulted in an increase of up to 54% in the growth inhibition area against *Salmonella typhi* isolates [55]. These findings highlight the potential of tannin-based compounds as adjunctive therapies to enhance β-lactam efficacy and provide novel strategies for overcoming bacterial resistance.

3.1.3. Phenolic Compounds

Phenolic compounds, recognized for their antibacterial and antioxidant properties, enhance the efficacy of β-lactam antibiotics by disrupting bacterial membranes or inhibiting resistance-associated enzyme activity. Cinnamic acid and its derivatives (such as ferulic acid and p-coumaric acid), in combination with ampicillin and cefazolin, exhibit synergistic effects against *Staphylococcus epidermidis* (FICI $\leq$ 0.5), possibly by compromising membrane integrity or increasing antibiotic permeability [56]. Gallic acid and its methyl ester enhance the antibacterial activity of ampicillin and penicillin G against MRSA, with in vitro studies and molecular docking analyses suggesting β-lactamase inhibition as a potential mechanism [14]. Additionally, its antioxidant effects diminish ROS accumulation during bacterial infections, augmenting the antibacterial efficacy of β-lactam antibiotics. Eugenol, a phenolic compound derived from cloves, has demonstrated synergistic bactericidal activity with cefotaxime against *S. aureus*, reducing the antibiotic's MICs by at least 4-fold (FICI = 0.375), likely by increasing membrane permeability and facilitating antibiotic uptake [57]. Cinnamaldehyde enhances ceftriaxone activity against MDR *Salmonella* by significantly decreasing the MIC from 128 μg/mL to 8 μg/mL (16-fold reduction; FICI = 0.375), primarily through downregulating extended-spectrum β-lactamase (ESBL) expression and disrupting bacterial membranes [25]. Moreover, glycerol monolaurate, when combined with ampicillin, exerts synergistic effects against MRSA with a 4- to 16-fold reduction in MIC values (FICI: 0.15–0.5), highlighting the broad applicability of phenolic compounds as β-lactam adjuvants [58].

3.1.4. Terpenoids

Terpenoids, primarily derived from Lamiaceae and Zingiberaceae plants, possess antibacterial and anti-inflammatory activities. These compounds have been shown to act synergistically with β-lactam antibiotics by increasing membrane permeability or inhibiting

https://doi.org/10.3390/antiox15020154

resistance mechanisms. Camphor, a monoterpenoid present in *Mentha* and *Thymus* species, reduces the MIC of carbapenem-resistant *Acinetobacter* spp. when combined with imipenem, achieving synergistic effects with FICI values as low as 0.37 and reducing the antibiotic's MIC by up to 4-fold, potentially by enhancing membrane permeability or inhibiting efflux pumps [59,62]. 1,8-Cineole, in combination with amoxicillin, significantly enhances efficacy against ESBL-producing *E. coli* and *K. pneumoniae*, reducing the antibiotic's MIC from 512 μg/mL to as low as 64 μg/mL (8-fold reduction; FICI $\leq$ 0.5), primarily by inhibiting enzymatic resistance mechanisms [60]. Carvacrol, another monoterpenoid, exhibits synergistic antibacterial activity against *E. coli* when combined with cefixim, demonstrating a synergistic FICI of 0.5, likely by disrupting the bacterial membrane and increasing antibiotic penetration [61]. Additionally, the clerodane diterpenoid 12(S), 16S-dihydroxycleroda-3,13-diene-15,16-olide, isolated from *Justicia* species, enhances the susceptibility of MRSA to β-lactam antibiotics, reducing the MIC of oxacillin from 16 μg/mL to 0.5 μg/mL (32-fold reduction; FICI = 0.125), demonstrating a remarkable synergistic effect [62].

3.1.5. Alkaloids

Alkaloids have demonstrated potentiation of beta-lactam antibiotics against MDR pathogens. Berberine, a protoberberine alkaloid, has surfaced as a promising antibiotic adjuvant. Experimental evidence indicates that berberine hydrochloride significantly restores the susceptibility of MDR *Acinetobacter baumannii* to carbapenems, reducing the MIC of meropenem from 64 μg/mL to 8 μg/mL (8-fold reduction; FICI = 0.375) [79]. It has also been reported to synergize with other beta-lactams against MRSA by inhibiting efflux pump systems and disrupting bacterial metabolic processes [63].

3.1.6. Other Classes of Phytochemicals

Beyond the major classes of natural products, a variety of plant-derived natural products, including saponins, lignans, and polyphenols, synergize with β-lactam antibiotics through various mechanisms, offering diverse options for antibacterial therapy. Saponin compounds, such as ginsenoside Rg3 and soybean saponins derived from soybeans, have demonstrated synergistic antibacterial activity by enhancing the sensitivity of MRSA to oxacillin and cefazolin, and by inhibiting β-lactamases such as NDM-1, respectively [47,64]. Glycyrrhizin nanoparticles reduce MRSA activity by inhibiting PBP2a [41]. Lignan compounds, including honokiol, exhibit synergistic antibacterial effects with β-lactam antibiotics against MRSA and vancomycin-resistant enterococci (VRE), potentially by disrupting cell division through interference with the FtsZ protein [65]. Polyphenol compounds, such as cranberry proanthocyanidins (cPAC), restore the activity of oxacillin against MRSA by inhibiting ESBLs and MBLs, thus overcoming PBP2a-mediated resistance [66]. Lignin, a primary component of plant cell walls, also restores MRSA susceptibility to β-lactam antibiotics when used in combination with oxacillin, by disrupting cell wall structure and resulting in ultrastructural disorganization [67]. Other phytochemicals, such as fucoxanthin and hypericin [68], exert their effects by enhancing ceftriaxone activity or inhibiting the expression of resistance genes, respectively. The structural and functional diversity of these compounds expands the repertoire of synergistic adjuvants and provides valuable leads for the development of novel antibacterial interventions.

3.2. Advances in Research on Microbial-Derived Natural Products

Microbial sources, including bacteria and filamentous fungi, have long been an essential font of antibiotics and their adjuvants. These sources produce small-molecule natural products that exhibit significant potential in combating resistant bacterial strains through mechanisms such as β-lactamase inhibition, enhanced targeting of cell wall synthesis, and inhibition of biofilm formation (Figure 2). In addition, other biological entities, such as

https://doi.org/10.3390/antiox15020154

bacteriophages, have emerged as powerful synergistic agents that operate through distinct mechanisms. Prominent examples are compiled in Table 1 (Microbial-Derived section).

3.2.1. Small-Molecule Potentiators from Bacteria and Fungi

This section focuses on small-molecule compounds produced by bacteria and filamentous fungi (e.g., *Penicillium, Aspergillus*), which act as sources of antibacterial agents, distinguishing them from pathogenic yeasts such as *Candida albicans*.

Actinomycetes are a crucial source of β-lactamase inhibitors. Clavulanic acid, produced by *Streptomyces clavuligerus*, remains the archetypal example [80]. It is a broad-spectrum, mechanism-based inhibitor that irreversibly inactivates many class A serine β-lactamases. Clinically, its combination with amoxicillin (e.g., Augmentin) is a cornerstone therapy for infections caused by β-lactamase-producing pathogens, and recent advancements in genetic and metabolic engineering continue to improve its biosynthetic yield [81].

Beyond clavulanic acid, other historically significant inhibitors have been discovered. The olivanic acids, a family of natural carbapenems also from *Streptomyces* species, demonstrated potent, broad-spectrum inhibition of both class A and C enzymes [82]. A mechanistically distinct molecule, lactivicin, isolated from *Lysobacter* and *Empedobacter* species, represented a paradigm shift; though it is not a β-lactam, its novel bicyclic structure mimics the transition state of β-lactam hydrolysis, enabling it to acylate and inhibit both PBPs and certain serine β-lactamases [83]. The discovery of these compounds broadened the chemical scaffolds known to inhibit β-lactamases and provided key insights for the development of modern synthetic inhibitors [84].

The fungal kingdom, particularly the genera *Penicillium* and *Aspergillus*, has also demonstrated strong potential. While the *Penicillium* genus is famed for producing penicillin, its potential as a source of β-lactamase inhibitors remains underexplored, though some metabolites from *Penicillium piceum* show inhibitory activity [69]. The *Aspergillus* genus, however, has yielded notable potentiators targeting different resistance pathways. Aspergillomarasmine A (AMA), a natural product from *Aspergillus versicolor*, acts as a potent MBL inhibitor. Recent efficacy studies demonstrate that the addition of AMA (8–16 μg/mL) effectively rescues the activity of meropenem against NDM-1-producing *Klebsiella pneumoniae*, restoring bacterial susceptibility to clinical breakpoints [70]. In contrast, aspermerodione, a metabolite from an endophytic *Aspergillus* sp., targets MRSA by binding to an allosteric site on PBP2a; it has been shown to reduce the MIC of oxacillin against MRSA ATCC 43300 from 128 μg/mL to 2–4 μg/mL(up to a 64-fold reduction) [71]. Fungi also produce antimicrobial peptides (AMPs), such as plectasin, a defensin that exhibits synergistic bactericidal activity against MRSA when combined with penicillins [73].

Other microbial ecosystems and compounds also serve as rich sources of adjuvants. Humimycin 17S, a lipopeptide from the human microbiome, shows potent synergy with ampicillin against MRSA and VRE [74]. Metabolites from probiotics like *Enterococcus faecium* can act as quorum-sensing inhibitors, reducing MRSA resistance to cefoxitin and attenuating its virulence [75]. Lastly, tunicamycin (TUN) and its less cytotoxic synthetic derivatives, such as TunR1, enhance beta-lactam activity by inhibiting phosphoglycosyl transferases. In models using *Bacillus subtilis*, TunR1 was found to potentiate the activity of various beta-lactams, including oxacillin and cephalosporins, by factors of up to 256-fold [72].

3.2.2. Bacteriophages as Synergistic Agents

Distinct from small-molecule natural products, bacteriophages—viruses that specifically infect bacteria—represent a powerful biological strategy to enhance antibiotic efficacy. Their synergy often stems from their ability to physically dismantle bacterial defenses,

https://doi.org/10.3390/antiox15020154

most notably biofilms. For instance, phage F1Pa exhibits synergistic activity with β-lactam antibiotics against multidrug-resistant *P. aeruginosa* by disrupting the biofilm matrix, which allows the antibiotic to reach its target cells [33]. This approach physically removes the protective barrier, restoring antibiotic susceptibility. Phage therapy combined with ceftaroline has also shown synergistic bactericidal activity against daptomycin-resistant VISA and MRSA strains while inhibiting biofilm formation [85]. Other studies have confirmed that bacteriophages and their lytic proteins enhance the activity of cephalosporins and piperacillin against pathogens like *Shigella dysenteriae* and *P. aeruginosa* PAO1 [86,87]. While the high specificity of phages offers a favorable safety profile by avoiding off-target effects, it also presents a challenge for broad-spectrum applications, and the potential for phage resistance remains an area of active investigation.

This growing body of research highlights the immense potential of products from diverse microbial sources as β-lactam antibiotic potentiators. Further exploration of these compounds and biological entities could provide novel strategies for overcoming antibiotic resistance.

3.3. Research Status and Limitations of Animal-Derived Natural Products

In contrast to plant- and microbial-derived natural products, research on animal-derived natural products remains relatively scarce. Current studies mainly focus on AMPs, such as melittin from bee venom, which have demonstrated certain antibacterial activities. However, their potential as β-lactam antibiotic adjuvants remains underexplored, and the available data are insufficient to support their widespread use. Moreover, animal-derived natural products face challenges such as difficulty in extraction, low yield, and complex composition. Mechanistic understanding and comprehensive safety evaluations are also lacking, further hindering their development in this field. Future research should focus on screening, isolating, and elucidating the mechanisms of animal-derived natural products to explore their potential as antimicrobial adjuvants.

AMPs are important components of animal-derived natural products and typically exert their effects by disrupting bacterial membranes, inhibiting bacterial metabolism, or interfering with biofilm formation. Among them, melittin—a cationic AMP derived from bee venom—has demonstrated synergistic antibacterial activity when combined with β-lactam antibiotics such as penicillin and oxacillin against MRSA and VRSA. In vitro experiments indicated that melittin can disrupt bacterial biofilms, thereby enhancing the bactericidal efficacy of β-lactam antibiotics [76].

In addition to AMPs, certain structural proteins and their derivatives from animals have also been found to possess antibacterial properties, though research on their role as β-lactamase inhibitors or adjuvants remains in its infancy. The antimicrobial protein Seroin, found in silkworm silk, has shown antibacterial activity, particularly against Gram-positive bacteria. Current research mainly focuses on its antimicrobial mechanisms and structural modifications, but no direct evidence has yet confirmed its role as a β-lactam antibiotic adjuvant [88].

It is important to acknowledge that many of the membrane-disrupting or membrane-active natural products discussed in this section possess the potential to enhance the activity of various antibiotic classes, rather than acting strictly as beta-lactam-specific adjuvants. Nevertheless, considering the paramount clinical significance of beta-lactams and the pivotal role that membrane permeability plays in their resistance mechanisms, this review maintains a dedicated focus on their combinational efficacy with this class of antibiotics.

https://doi.org/10.3390/antiox15020154

4. Research on the Combined Application of Natural Products and β-Lactam Antibiotics

4.1. Screening Methods for Natural Products as Antibiotic Potentiators or Inhibitors

The escalating challenge of antibiotic resistance necessitates the discovery of natural products that can act as potentiators to enhance the efficacy of β-lactam antibiotics against MDR bacteria [89]. High-throughput screening (HTS) and virtual screening (VS) represent two complementary discovery strategies widely employed to identify candidate β-lactam potentiators. At this early stage, the primary objective is to efficiently prioritize active candidates from large chemical or natural product libraries for downstream mechanism and pharmacological evaluation.

HTS-based discovery of β-lactam potentiators typically relies on synergy-oriented phenotypic assays or direct enzyme inhibition formats, enabling the rapid experimental evaluation of thousands to hundreds of thousands of compounds [90,91]. For example, a fluorescence-quenching assay based on the hydrolysis of the chromogenic substrate nitrocefin was used to screen nearly 190,000 compounds against the OXA-48 β-lactamase, leading to the identification of novel inhibitor scaffolds such as the pyrazolo[3,4-b]pyridine core [91]. Similarly, a near-infrared fluorescent probe (LXMB) was developed to selectively detect β-lactamase (Blac) activity from *Mycobacterium tuberculosis* and applied to the screening of herbal medicines, resulting in the identification of tannic acid as a Blac inhibitor that enhanced the activity of multiple β-lactam antibiotics against *M. tuberculosis* H37Ra (Figure 3) [92].

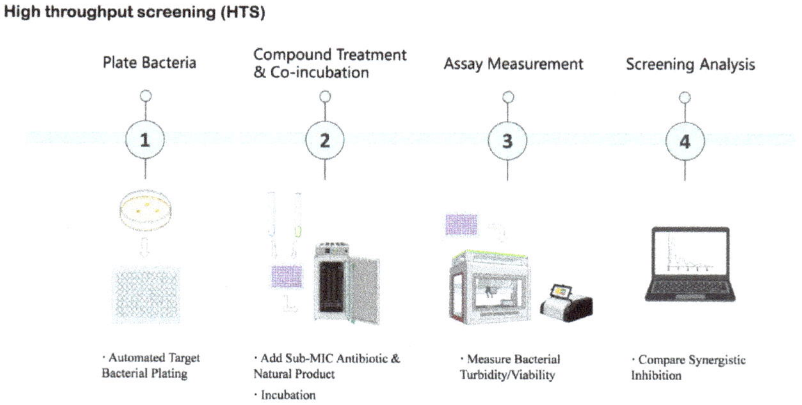

Figure 3. High-Throughput Screening (HTS) Workflow for Identifying Natural Product Antibiotic Potentiators. This schematic diagram illustrates the key steps involved in high-throughput screening to identify natural products that enhance antibiotic efficacy. The process begins with automated target bacterial plating, followed by the addition of sub-minimum inhibitory concentration (MIC) antibiotics and natural products for incubation. Subsequently, bacterial turbidity or viability is measured using automated readers. Finally, screening analysis is performed to compare synergistic inhibition effects, aiming to identify promising natural product hits. Created in BioRender. Li, S. (2026) https://BioRender.com/bpw3b6c (accessed on 17 January 2026).

When applied to natural products (NPs), however, HTS campaigns face distinct physicochemical challenges that differ from those encountered with conventional synthetic libraries. Many NPs, including polyphenols, terpenoids, and fatty acid derivatives, exhibit limited aqueous solubility and therefore require the use of organic solvents such as DMSO or surfactants (e.g., Tween 80) for screening. Importantly, accumulating evidence indicates that solvent composition itself may influence bacterial growth, membrane integrity, or

https://doi.org/10.3390/antiox15020154

biofilm formation, underscoring the necessity of rigorous vehicle controls to avoid false-positive potentiation effects during synergy screening [93,94]. These considerations are particularly relevant during primary HTS, where subtle solvent-dependent artifacts may be amplified across large datasets.

VS serves as a complementary in silico strategy that is especially valuable for prioritizing NP-derived candidates prior to experimental validation. Structure-based VS has been successfully applied to clinically relevant β-lactam resistance targets, including serine- and metallo-β-lactamases as well as altered penicillin-binding proteins (PBPs) [90,95]. For instance, large-scale VS campaigns have identified inhibitors of carbapenemases such as GES-5 by screening extensive libraries of drug-like compounds [90]. Parallel VS against multiple resistance enzymes, including KPC-2 and the metallo-β-lactamases NDM-1 and VIM-2, has further enabled the identification of broad-spectrum inhibitor scaffolds [89]. Mutated PBPs, most notably PBP2a in MRSA, have also emerged as important VS targets for rational discovery efforts [96] (Figure 4). Nevertheless, the application of VS to natural products presents additional challenges arising from their structural complexity, conformational flexibility, and frequent deviation from conventional "drug-like" chemical space. These features can complicate docking accuracy and scoring reliability, necessitating careful target selection, appropriate constraint strategies, and experimental validation to minimize false positives. Accordingly, the integration of HTS and VS—coupled with NP-specific controls and prioritization criteria—provides a pragmatic and efficient framework for the early-stage discovery of natural product–derived β-lactam potentiators.

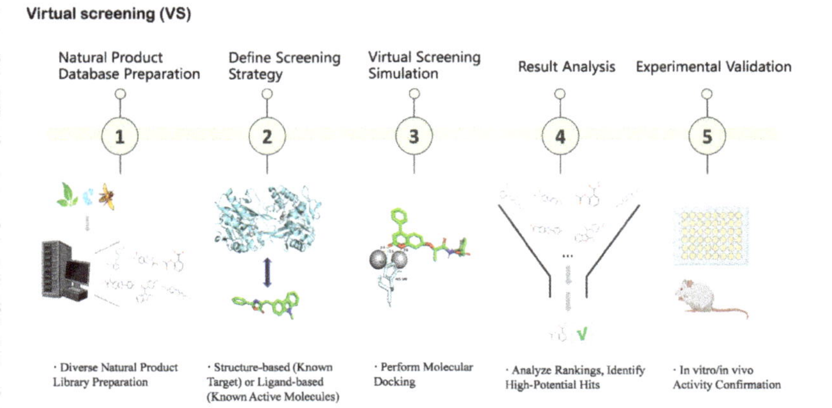

Figure 4. Virtual Screening (VS) Workflow for Natural Product Discovery. This schematic diagram outlines the key steps involved in leveraging virtual screening to identify potential natural product hits. The process commences with diverse natural product library preparation, followed by defining the screening strategy (either structure-based using a known target or ligand-based using known active molecules). Subsequently, molecular docking is performed to predict binding or activity. This leads to result analysis and ranking to identify high-potential "hit" molecules. Finally, these selected candidates undergo in vitro/in vivo activity confirmation. Created in BioRender. Li, S. (2026) https://BioRender.com/xlvn6h2 (accessed on 17 January 2026).

4.2. Methodological Adaptations for In Vitro Evaluation of Natural Product–Based Potentiators

Following the identification of candidate β-lactam potentiators, rigorous in vitro evaluation is essential to accurately assess antibacterial activity, quantify synergistic interactions, and assign bioactivity to specific compounds. Standard in vitro assays—such as MIC determination, checkerboard assays, and time-kill studies—remain foundational tools

https://doi.org/10.3390/antiox15020154

for evaluating antibiotic potentiation [97]. However, when applied to natural products, these methods require careful methodological adaptation to account for compound-specific properties that may confound assay readouts and data interpretation.

One major challenge in the in vitro evaluation of NP-based potentiators is interference with conventional optical or colorimetric readouts. Many plant-derived NPs, including polyphenols, flavonoids, and tannins, are intrinsically pigmented or redox-active and can artificially influence optical density (OD)–based growth measurements in microdilution assays. To mitigate such interference, modified protocols incorporating metabolic indicators such as resazurin have been widely adopted, enabling indirect assessment of bacterial viability through redox-dependent fluorescence or color change rather than turbidity alone [98,99]. These approaches have proven particularly valuable for evaluating crude extracts, essential oils, and antimicrobial peptides, although confirmatory CFU-based assays are often required to validate apparent synergistic effects.

Volatility represents an additional NP-specific consideration that is not adequately captured by standard liquid-phase in vitro assays. Volatile compounds, such as essential oils and low-molecular-weight terpenoids, may partially evaporate during incubation, leading to underestimation of antibacterial or potentiating activity in broth-based MIC or checkerboard formats. In such cases, vapor-phase or modified diffusion-based assays—such as inverted Petri dish or sealed-chamber systems—provide a more appropriate framework for evaluating antimicrobial efficacy and synergy under conditions that better reflect their physicochemical behavior [100,101]. These approaches are particularly relevant for assessing NP–β-lactam combinations intended for respiratory or topical applications.

A further methodological challenge arises from the chemical complexity of many NP preparations. Unlike single-molecule synthetic compounds, NPs are frequently encountered as partially characterized extracts or heterogeneous mixtures containing dozens to hundreds of constituents. In this context, activity-guided fractionation approaches, including thin-layer chromatography (TLC)–bioautography, offer an effective strategy for localizing antibacterial or potentiating activity directly on chromatographic media [102,103]. By coupling chemical separation with biological detection, TLC-bioautography facilitates the rapid identification of active fractions prior to full structural elucidation, thereby reducing the risk of misattributing activity to inactive or minor components.

Collectively, these methodological adaptations highlight the importance of tailoring in vitro evaluation strategies to the unique properties of natural products. Integrating standard synergy assays with NP-specific controls, alternative readouts, and activity-localization techniques enhances the reliability and interpretability of in vitro data, providing a more robust foundation for downstream mechanistic studies and in vivo validation of β-lactam potentiators.

4.3. *Models for In Vivo Validation of Natural Product–Based β-Lactam Potentiators*

Following promising in vitro results, in vivo validation represents a critical step in assessing the translational potential of natural product–based β-lactam potentiators. Animal infection models provide an integrated physiological context in which antibacterial efficacy, pharmacokinetics, toxicity, and host–pathogen interactions can be evaluated simultaneously. Commonly employed models include murine skin, pulmonary, and systemic infection models, which remain essential for establishing in vivo synergy and therapeutic relevance [104]. Additionally, the Caenorhabditis elegans infection model provides a high-throughput and ethically advantageous bridge between in vitro assays and mammalian studies for early-stage efficacy and β-lactam potentiation screening [105]; however, its limited pharmacokinetic resolution necessitates subsequent validation in higher vertebrate models to confirm translational relevance.

https://doi.org/10.3390/antiox15020154

When applied to NPs, however, vivo evaluation presents additional challenges that extend beyond those encountered with conventional small-molecule antibiotics. A major limitation arises from unfavorable pharmacokinetic (PK) properties, including poor solubility, low oral bioavailability, rapid metabolism, and limited tissue penetration. These factors can lead to discrepancies between in vitro potency and in vivo efficacy, complicating the interpretation of synergy observed in animal models. Consequently, careful PK characterization and dose optimization are essential to distinguish true pharmacological potentiation from exposure-limited efficacy [104].

Route of administration constitutes another NP-specific consideration in in vivo validation. While many β-lactam antibiotics are administered systemically, certain classes of NPs—such as essential oils or highly lipophilic compounds—may be more effectively delivered via alternative routes, including topical, inhalational, or localized administration. For example, inhalation-based infection models or aerosolized delivery systems may be more appropriate for evaluating volatile or respiratory-targeted NP–β-lactam combinations, whereas topical wound infection models may better capture the therapeutic potential of membrane-active or poorly absorbed compounds [100].

In addition, host-mediated effects must be carefully considered when interpreting in vivo outcomes for NP-based potentiators. Many natural products possess anti-inflammatory or antioxidant activities that can modulate disease severity, tissue damage, or immune responses independently of direct antibacterial action. While such effects may contribute to improved therapeutic outcomes, they do not necessarily reflect direct reversal of bacterial resistance mechanisms. Therefore, in vivo studies should integrate bacterial burden measurements (e.g., CFU reduction) with host-response markers to clearly delineate antibacterial potentiation from host-directed benefits [106].

Overall, in vivo validation of natural product–derived β-lactam potentiators requires tailored experimental design that accounts for NP-specific pharmacokinetic constraints, delivery strategies, and host effects. Integrating appropriate animal models with PK-informed dosing and mechanistic endpoints enhances the translational relevance of preclinical findings and supports more informed progression toward clinical development.

5. Challenges and Prospects in Natural Product Research

5.1. Challenges

The chemical complexity of natural products complicates the identification of active components that enhance antibacterial effects. Plant extracts contain numerous compounds with intricate interactions, requiring the integration of multiple chromatographic techniques (such as silica gel column and reversed-phase C18 column chromatography), high-performance liquid chromatography (HPLC), gas chromatography (GC), mass spectrometry (MS), and nuclear magnetic resonance (NMR) spectroscopy for repeated separation and purification. This process is time-consuming, yields low quantities of target compounds, and carries a high risk of component inactivation. Additionally, microbial metabolites are influenced by strain variability and culture conditions, leading to unstable production and structural diversity. The lag in new metabolite identification technologies further restricts the discovery and development of antibacterial potentiating compounds, consequently impeding progress in novel antimicrobial drug research [107].

The mechanisms of natural products involve multiple targets and pathways, rendering their elucidation particularly complex. Although certain studies have made progress, most remain confined to the cellular level, lacking detailed molecular-level characterization. Key knowledge gaps include the dynamic interactions between natural compounds and bacterial targets, the complexity of regulatory signaling networks, and an insufficient understanding of the deep molecular mechanisms underlying bacterial resistance reversal.

https://doi.org/10.3390/antiox15020154

Despite the generation of massive datasets through multi-omics approaches, the integration and interpretation of this information remain major obstacles, substantially limiting the comprehension of antibacterial potentiation mechanisms and impeding drug design, development, and innovation [108].

Pharmacokinetic studies of natural products face multiple difficulties. Their absorption is constrained by chemical structure, physicochemical properties, and gastrointestinal physiological conditions. For example, flavonoids and similar compounds possess high molecular weights and strong polarity, often undergoing glycosylation modifications that hinder absorption and reduce bioavailability. Studies on tissue distribution lack highly sensitive detection methods and reliable in vivo models. The metabolic pathways of natural products are highly complex, involving diverse enzymatic reactions, which complicates the identification of metabolite activity and potential toxicity. Additionally, research on pharmacokinetic–pharmacodynamic (PK-PD) correlations of natural products remains insufficient. Excretion profiling is further complicated by the structural complexity of natural products, while the lack of sensitive detection technologies constrains dosage form development and clinical application, ultimately hindering their clinical translation and broader therapeutic use [109,110].

The chemical complexity, mechanistic diversity, and pharmacokinetic challenges of natural products vary significantly across their structural and functional categories. Structurally, natural products can be classified into flavonoids (e.g., quercetin, EGCG), phenolics (e.g., gallic acid, cinnamaldehyde), terpenoids (e.g., carvacrol), alkaloids (e.g., berberine), microbial metabolites (e.g., clavulanic acid), and AMPs (e.g., melittin). Flavonoids, due to their high polarity and glycosylation, require complex chromatographic techniques (e.g., HPLC, C18 columns), which are time-consuming and risk activity loss [9]. Microbial metabolites face production instability due to strain variability and culture conditions, necessitating optimized fermentation processes [81]. AMPs are limited by low yields and extraction difficulties [76]. Mechanistically, β-lactamase inhibitors (e.g., soy saponins) have well-defined targets, whereas efflux pump inhibitors (e.g., berberine) involve complex dynamic interactions requiring advanced molecular imaging [17]. Pharmacokinetically, flavonoids and alkaloids suffer from low bioavailability largely due to their high polarity and associated poor solubility/absorption profiles [111,112], while terpenoids' high volatility and sensitivity to environmental factors (such as light, heat, and oxygen) compromise formulation stability and shelf-life [113]. These category-specific challenges underscore the need for tailored technological solutions.

5.2. Future Prospects

Based on the current body of evidence, different classes of natural products exhibit markedly distinct potentials for clinical translation as β-lactam antibiotic potentiators. Microbial-derived metabolites, particularly β-lactamase inhibitors and PBP-targeting compounds, currently represent the most clinically promising category, as exemplified by clavulanic acid and aspergillomarasmine A, which demonstrate well-defined mechanisms, robust in vivo efficacy [4,114], and, in some cases, established clinical use. These compounds typically act on specific and validated resistance targets, making their pharmacological optimization and regulatory development more feasible.

In contrast, many plant-derived natural products, including flavonoids, phenolics, and terpenoids, display broad-spectrum potentiating effects through mechanisms such as membrane disruption, efflux pump modulation, and biofilm suppression. While these compounds offer valuable chemical diversity and multifunctional biological activities, their clinical translation is often constrained by issues related to bioavailability, target specificity, and pharmacokinetic variability. Consequently, further structural optimization,

https://doi.org/10.3390/antiox15020154

formulation strategies, and in vivo validation are required before these agents can be advanced toward clinical application [115].

Collectively, these observations suggest that microbial-derived natural products with defined resistance targets are currently best positioned for near-term clinical development, whereas plant-derived compounds may serve as important long-term leads or adjunctive scaffolds pending further optimization.

Despite these differences in translational readiness, both microbial- and plant-derived natural products face common challenges, including incomplete mechanistic understanding, limited target validation, and insufficient pharmacokinetic characterization. Addressing these barriers requires systematic, mechanism-driven approaches that can more precisely link natural product activity to bacterial resistance pathways. To this end, the development of novel antibacterial potentiators can be facilitated by multi-omics technologies, including genomics, transcriptomics, proteomics, and metabolomics, to construct comprehensive molecular interaction networks of natural products within bacterial systems. This strategy enables the precise identification of critical targets and signaling pathways involved in antibacterial potentiation, thereby offering a clear framework for drug discovery and development [116]. Additionally, advances in artificial intelligence (AI) facilitate in-depth data mining and support virtual screening coupled with experimental validation. Such approaches significantly accelerate lead compound optimization, reduce development costs, and enhance research efficiency and success rates [117]. Furthermore, advancements in microfluidic chip technology have led to the development of miniaturized and integrated analytical platforms. These platforms enable rapid analysis of small samples, accurately simulate in vivo microenvironments, and allow for in-depth studies of natural product–bacteria interactions and pharmacokinetics, thereby introducing innovation into natural product research and advancing antimicrobial drug development [118]. Moreover, leveraging the antioxidant properties of natural products presents promising avenues for combating antibiotic resistance. By integrating multi-omics to identify the molecular targets of antioxidant compounds and applying nanotechnology to enhance their bioavailability, researchers can develop β-lactam potentiators that concurrently target bacterial infections and oxidative stress, providing novel therapeutic strategies for clinical applications.

Natural products exhibit substantial potential as β-lactam antibiotic potentiators in clinical settings. For instance, the combination of clavulanic acid and amoxicillin (Augmentin) has been clinically effective against infections caused by β-lactamase-producing bacteria, illustrating the practical value of this strategy. Other natural compounds, such as malleamycin A [70], have demonstrated promising preclinical efficacy by restoring bacterial susceptibility to β-lactam antibiotics. However, the clinical translation of these compounds faces several obstacles, including toxicity, side effects, dosage safety, and unfavorable pharmacokinetic properties. Many natural compounds exhibit low bioavailability, making it difficult to reach effective concentrations in vivo. Moreover, the absence of large-scale clinical trial data significantly restricts their widespread adoption. Future research should focus on conducting more preclinical and clinical studies, particularly human trials, to confirm the safety and therapeutic efficacy of these natural compounds.

Nanotechnology offers a promising strategy to overcome the pharmacokinetic limitations of natural compounds. The application of nanocarriers, such as nanoparticles and liposomes, enhances drug stability, facilitates targeted delivery, enables controlled release, and improves antibacterial potentiation while minimizing adverse effects. These advancements contribute to improved patient adherence and overall treatment outcomes [119,120].

In addition, the establishment of a clinical monitoring system for antibacterial potentiators derived from natural products is crucial. A comprehensive assessment of their efficacy, safety, and resistance characteristics, combined with real-time treatment adjust-

https://doi.org/10.3390/antiox15020154

ments informed by monitoring data, will support rational clinical application. Such a strategy will facilitate the efficient translation of natural product research from bench to bedside, thereby contributing to the fight against antibiotic resistance and the protection of global public health.

As discussed above, natural products exhibit substantial structural and functional diversity, which underlies their varied mechanisms of β-lactam potentiation. This diversity is reflected in the markedly different translational trajectories observed across major natural product classes. Microbial metabolites, such as clavulanic acid and aspergillomarasmine A, exhibit the highest potential due to their specific β-lactamase inhibition and proven clinical success (e.g., Augmentin), with multi-omics and genetic engineering enhancing yield and target identification [46,70]. Phenolics (e.g., cinnamaldehyde) and terpenoids (e.g., carvacrol), with simple structures and broad-spectrum activity, are ideal for rapid development as topical or local adjuvants, leveraging nanotechnology to improve stability [25,61]. Flavonoids (e.g., EGCG) provide notable anti-inflammatory benefits that may support improved therapeutic outcomes when used alongside β-lactam antibiotics, but their clinical translation is often limited by low bioavailability, necessitating AI-driven structural optimization [78]. Alkaloids (e.g., berberine) target complex resistance but face toxicity challenges [63]. AMPs, despite novel mechanisms, are limited by extraction costs [76]. Prioritizing microbial metabolites and accelerating phenolics and terpenoids via emerging technologies will drive innovative β-lactam potentiator development.

6. Conclusions

Significant progress has been made in the investigation of natural products as β-lactam antibiotic potentiators, offering a promising strategy to combat antibiotic resistance. Natural products derived from plants and microbes enhance the antibacterial activity of β-lactam antibiotics through diverse mechanisms, including efflux pump inhibition, increased membrane permeability, and direct β-lactamase inhibition. Representative compounds, such as clavulanic acid and malleamycin A, have demonstrated substantial value in both clinical applications and research. Both in vitro and in vivo experimental models provide reliable means for evaluating their synergistic antibacterial effects.

Nonetheless, several challenges persist, including difficulties in identifying active components, incomplete elucidation of mechanisms of action, and insufficient pharmacokinetic studies. While an increasing repertoire of phytochemicals has been identified as β-lactam potentiators, the field is still constrained by a paucity of translational studies bridging molecular discovery and therapeutic application. Bridging this gap will require a multidisciplinary approach integrating pharmacokinetic optimization, target-based screening, and structure-guided drug design, ultimately enabling the development of next-generation β-lactam adjuvants derived from natural scaffolds.

In summary, current evidence indicates that distinct classes of natural products differ substantially in their translational potential as β-lactam antibiotic potentiators. Microbial-derived metabolites, particularly β-lactamase inhibitors and penicillin-binding protein (PBP)-targeting compounds, emerge as the most compelling candidates for near-term clinical translation, supported by well-defined mechanisms of action, robust in vivo validation, and, in some cases, established clinical use. By contrast, many plant-derived natural products are better positioned as long-term leads, as further optimization is often required to overcome limitations related to bioavailability, pharmacokinetics, and target specificity.

Building on these translational priorities, future research integrating multi-omics approaches, artificial intelligence, and nanotechnology holds great promise for precisely identifying active compounds, elucidating their mechanisms of action, and optimizing their pharmacokinetic properties. These advancements are expected to accelerate the

https://doi.org/10.3390/antiox15020154

development of novel antibacterial potentiators and provide innovative strategies to combat resistant bacterial infections. Ultimately, these efforts will contribute to optimizing clinical antibacterial therapy, offering broad applications and substantial public health benefits.

An important but relatively underexplored question concerns the potential development and frequency of bacterial resistance to natural product–based antibiotic adjuvants. At present, systematic data on the emergence or prevalence of resistance to NP adjuvants remain limited, with most available evidence derived from short-term in vitro studies. Because many NP adjuvants act by targeting resistance mechanisms, stress response pathways, or membrane integrity rather than essential bacterial growth processes, they may impose different selective pressures compared with conventional antibiotics. Nevertheless, adaptive responses, compensatory mechanisms, or cross-resistance cannot be excluded, underscoring the need for long-term evolutionary and in vivo studies.

Author Contributions: L.W.: Supervision, Resources. X.W.: Writing—review & editing, Supervision, Funding acquisition, Conceptualization. W.Y., S.F. and J.L.: Validation, Investigation, Data curation. Y.Z., X.D. and J.H.: Investigation, Formal analysis. W.Y.: Writing—original draft, Methodology, Formal analysis, Data curation, Conceptualization. All authors have read and agreed to the published version of the manuscript.

Funding: This work was supported by Jiangsu Agriculture Science and Technology Innovation Fund (CX(24)3078), the Jiangsu Province Key Research and Development Program [BE2023332], the Fundamental Research Funds for the Central Universities (KJYQ2025042), the Natural Science Foundation of Jiangsu Province (BK20252056), and the grants from National Natural Science Foundation of China (32573433).

Institutional Review Board Statement: Not applicable.

Informed Consent Statement: Not applicable.

Data Availability Statement: No new data were created or analyzed in this study. Data sharing is not applicable to this article.

Acknowledgments: During the preparation of this manuscript, the authors used ChatGPT (OpenAI, https://www.openai.com/chatgpt) (accessed on 20 December 2025) for language editing and polishing. The authors have reviewed and edited the output and take full responsibility for the content of this publication.

Conflicts of Interest: The authors declare no conflicts of interest.

Abbreviations

The following abbreviations are used in this manuscript:

AMPs	Antimicrobial peptides
CFU	Colony-forming units
EGCG	Epigallocatechin-3-gallate
ESBLs	Extended-spectrum β-lactamases
FICI	Fractional inhibitory concentration index
MBLs	Metallo-β-lactamases
MDR	Multidrug resistance
MIC	Minimum inhibitory concentration
MRSA	Methicillin-resistant Staphylococcus aureus
PBPs	Penicillin-binding proteins
PBP2a	Penicillin-binding protein 2a
PK/PD	Pharmacokinetics/pharmacodynamics
QS	Quorum sensing
ROS	Reactive oxygen species
EPS	Extracellular polymeric substance

https://doi.org/10.3390/antiox15020154

References

1. Landecker, H. Antibiotic resistance and the biology of history. *Body Soc.* **2016**, *22*, 19–52. [CrossRef] [PubMed]
2. Khan, R.T.; Sharma, V.; Khan, S.S.; Rasool, S. Prevention and potential remedies for antibiotic resistance: Current research and future prospects. *Front. Microbiol.* **2024**, *15*, 1455759. [CrossRef]
3. Grand View Research. Antibiotics Market Size, Share & Trends Analysis Report by Action Mechanism, Drug Class, Type, and Segment Forecasts. Available online: https://www.grandviewresearch.com/industry-analysis/antibiotic-market (accessed on 20 January 2026).
4. Bush, K.; Bradford, P.A. β-Lactams and β-lactamase inhibitors: An overview. *Cold Spring Harb. Perspect. Med.* **2016**, *6*, a025247. [CrossRef]
5. Kohanski, M.A.; Dwyer, D.J.; Collins, J.J. How antibiotics kill bacteria: From targets to networks. *Nat. Rev. Microbiol.* **2010**, *8*, 423–435. [CrossRef]
6. Bush, K. Past and present perspectives on β-lactamases. *Antimicrob. Agents Chemother.* **2018**, *62*, e01076-18. [CrossRef] [PubMed]
7. Ojima, I. Modern natural products chemistry and drug discovery. *J. Med. Chem.* **2008**, *51*, 2587–2588. [CrossRef]
8. Liu, M.; El-Hossary, E.M.; Oelschlaeger, T.A.; Donia, M.S.; Quinn, R.J.; Abdelmohsen, U.R. Potential of marine natural products against drug-resistant bacterial infections. *Lancet Infect. Dis.* **2019**, *19*, e237–e245. [CrossRef] [PubMed]
9. Chen, S.; Zhi, Z.; Wong, W.-L.; Yuan, W.; Sun, N. Understanding the synergistic sensitization of natural products and antibiotics: An effective strategy to combat MRSA. *Eur. J. Med. Chem.* **2025**, *281*, 117012. [CrossRef]
10. Darby, E.M.; Trampari, E.; Siasat, P.; Gaya, M.S.; Alav, I.; Webber, M.A.; Blair, J.M.A. Molecular mechanisms of antibiotic resistance revisited. *Nat. Rev. Microbiol.* **2023**, *21*, 280–295. [CrossRef]
11. Liu, Y.; Tong, Z.; Shi, J.; Li, R.; Upton, M.; Wang, Z. Drug repurposing for next-generation combination therapies against multidrug-resistant bacteria. *Theranostics* **2021**, *11*, 4910–4928. [CrossRef]
12. Silver, L.L. Challenges of antibacterial discovery. *Clin. Microbiol. Rev.* **2011**, *24*, 71–109. [CrossRef]
13. Burillo, A.; Bouza, E. The evolution of knowledge for treating Gram-negative bacterial infections. *Curr. Opin. Infect. Dis.* **2024**, *37*, 573–581. [CrossRef] [PubMed]
14. Jiamboonsri, P.; Eurtivong, C.; Wanwong, S. Assessing the potential of gallic acid and methyl gallate to enhance the efficacy of β-lactam antibiotics against methicillin-resistant *Staphylococcus aureus* by targeting β-lactamase: In silico and in vitro studies. *Antibiotics* **2023**, *12*, 1622. [CrossRef]
15. Kanagaratnam, R.; Sheikh, R.; Alharbi, F.; Kwon, D.H. An efflux pump (MexAB-OprM) of *Pseudomonas aeruginosa* is associated with antibacterial activity of epigallocatechin-3-gallate (EGCG). *Phytomedicine* **2017**, *36*, 194–200. [CrossRef]
16. Zhang, X.; Liu, Y.; Shen, Z.; Wang, S.; Wu, C.; Liu, D.; Tang, S.; Dai, C. Osthole ameliorates myonecrosis caused by *Clostridium perfringens* type A infection in mice. *One Health Adv.* **2023**, *1*, 27. [CrossRef]
17. Sharma, A.; Gupta, V.K.; Pathania, R. Efflux pump inhibitors for bacterial pathogens: From bench to bedside. *Indian J. Med. Res.* **2019**, *149*, 129–145. [CrossRef] [PubMed]
18. Li, X.-Z.; Plésiat, P.; Nikaido, H. The challenge of efflux-mediated antibiotic resistance in Gram-negative bacteria. *Clin. Microbiol. Rev.* **2015**, *28*, 337–418. [CrossRef]
19. Nishino, K.; Yamasaki, S.; Nakashima, R.; Zwama, M.; Hayashi-Nishino, M. Function and inhibitory mechanisms of multidrug efflux pumps. *Front. Microbiol.* **2021**, *12*, 737288. [CrossRef]
20. Van Bambeke, F.; Balzi, E.; Tulkens, P.M. Antibiotic efflux pumps. *Biochem. Pharmacol.* **2000**, *60*, 457–470. [CrossRef]
21. Seo, Y.; Kim, M.; Kim, T.-J. Enhanced efficacy of ciprofloxacin and tobramycin against *Staphylococcus aureus* when combined with Corydalis tuber and berberine through efflux pump inhibition. *Antibiotics* **2024**, *13*, 469. [CrossRef]
22. Mehta, D.; Saini, V.; Aggarwal, B.; Khan, A.; Bajaj, A. Unlocking the bacterial membrane as a therapeutic target for next-generation antimicrobial amphiphiles. *Mol. Aspects Med.* **2021**, *81*, 100999. [CrossRef]
23. El-Sayed Ahmed, M.A.E.-G.; Zhong, L.-L.; Shen, C.; Yang, Y.; Doi, Y.; Tian, G.-B. Colistin and its role in the era of antibiotic resistance: An extended review (2000–2019). *Emerg. Microbes Infect.* **2020**, *9*, 868–885. [CrossRef] [PubMed]
24. Kaur, J.N.; Klem, J.F.; Liu, Y.; Boissonneault, K.R.; Holden, P.N.; Kreiswirth, B.; Chen, L.; Smith, N.M.; Tsuji, B.T. Maximally precise combinations to overcome metallo-β-lactamase-producing *Klebsiella pneumoniae*. *Antimicrob. Agents Chemother.* **2024**, *68*, e00770-24. [CrossRef]
25. Yin, L.; Gou, Y.; Dai, Y.; Wang, T.; Gu, K.; Tang, T.; Hussain, S.; Huang, X.; He, C.; Liang, X.; et al. Cinnamaldehyde restores ceftriaxone susceptibility against multidrug-resistant *Salmonella*. *Int. J. Mol. Sci.* **2023**, *24*, 9288. [CrossRef]
26. Høiby, N.; Bjarnsholt, T.; Givskov, M.; Molin, S.; Ciofu, O. Antibiotic resistance of bacterial biofilms. *Int. J. Antimicrob. Agents* **2010**, *35*, 322–332. [CrossRef]
27. Karygianni, L.; Ren, Z.; Koo, H.; Thurnheer, T. Biofilm matrixome: Extracellular components in structured microbial communities. *Trends Microbiol.* **2020**, *28*, 668–681. [CrossRef] [PubMed]
28. Rasmussen, T.B.; Givskov, M. Quorum-sensing inhibitors as anti-pathogenic drugs. *Int. J. Med. Microbiol.* **2006**, *296*, 149–161. [CrossRef]

https://doi.org/10.3390/antiox15020154

29. Meng, X.; Kang, M.; Yu, Z.; Li, C.; Chen, Y.; Jin, T.; Wang, K.; Guo, H. Synergistic antibacterial activity of baicalin in combination with oxacillin sodium against methicillin-resistant *Staphylococcus aureus*. *FEBS Open Bio* **2024**, *14*, e13952. [CrossRef]
30. Shang, D.; Han, X.; Du, W.; Kou, Z.; Jiang, F. Trp-containing antibacterial peptides impair quorum sensing and biofilm development in multidrug-resistant *Pseudomonas aeruginosa* and exhibit synergistic effects with antibiotics. *Front. Microbiol.* **2021**, *12*, 611009. [CrossRef] [PubMed]
31. Kong, L.; Hu, X.; Xia, D.; Wu, J.; Zhao, Y.; Guo, H.; Zhang, S.; Qin, C.; Wang, Y.; Li, L.; et al. Janus PEGylated CuS-engineered *Lactobacillus casei* combats biofilm infections via metabolic interference and innate immunomodulation. *Biomaterials* **2025**, *317*, 123060. [CrossRef]
32. Qin, M.; Zhang, X.; Ding, H.; Chen, Y.; He, W.; Wei, Y.; Chen, W.; Chan, Y.K.; Shi, Y.; Huang, D.; et al. Engineered probiotic bio-heterojunction with robust antibiofilm modality via "eating" extracellular polymeric substances for wound regeneration. *Adv. Mater.* **2024**, *36*, 2402530. [CrossRef]
33. Santamaría-Corral, G.; Aguilera-Correa, J.J.; Esteban, J.; García-Quintanilla, M. Bacteriophage therapy on an in vitro wound model and synergistic effects in combination with beta-lactam antibiotics. *Antibiotics* **2024**, *13*, 800. [CrossRef]
34. De Souza, E.L.; De Barros, J.C.; De Oliveira, C.E.V.; Da Conceição, M.L. Influence of *Origanum vulgare* L. essential oil on enterotoxin production, membrane permeability and surface characteristics of *Staphylococcus aureus*. *Int. J. Food Microbiol.* **2010**, *137*, 308–311. [CrossRef] [PubMed]
35. Luo, K.; Zhao, P.; He, Y.; Kang, S.; Shen, C.; Wang, S.; Guo, M.; Wang, L.; Shi, C. Antibacterial effect of oregano essential oil against *Vibrio vulnificus* and its mechanism. *Foods* **2022**, *11*, 403. [CrossRef]
36. Rosato, A.; Sblano, S.; Salvagno, L.; Carocci, A.; Clodoveo, M.L.; Corbo, F.; Fracchiolla, G. Anti-biofilm inhibitory synergistic effects of combinations of essential oils and antibiotics. *Antibiotics* **2020**, *9*, 637. [CrossRef]
37. Li, Q.-Q.; Kang, O.-H.; Kwon, D.-Y. Study on demethoxycurcumin as a promising approach to reverse methicillin resistance of *Staphylococcus aureus*. *Int. J. Mol. Sci.* **2021**, *22*, 3778. [CrossRef]
38. Shi, J.; Chen, C.; Kong, P.; Yu, F.; Lv, Q.; Wang, Z.; Liu, Y. Non-membrane-active peptide resensitizes MRSA to β-lactam antibiotics and inhibits *Staphylococcus aureus* virulence. *Adv. Sci.* **2025**, *12*, 2416260. [CrossRef] [PubMed]
39. Santiago, C.; Lim, K.-H.; Loh, H.-S.; Ting, K.N. Prevention of cell-surface attachment and reduction of penicillin-binding protein 2a (PBP2a) level in methicillin-resistant *Staphylococcus aureus* biofilms by *Acalypha wilkesiana*. *BMC Complement. Altern. Med.* **2015**, *15*, 79. [CrossRef] [PubMed]
40. Mun, S.-H.; Kang, O.-H.; Kong, R.; Zhou, T.; Kim, S.-A.; Shin, D.-W.; Kwon, D.-Y. Punicalagin suppresses methicillin resistance of *Staphylococcus aureus* to oxacillin. *J. Pharmacol. Sci.* **2018**, *137*, 317–323. [CrossRef]
41. Rijo, P.; Abuamara, T.M.M.; Ali Lashin, L.S.; Kamar, S.A.; Isca, V.M.S.; Mohammed, T.S.; Abdrabo, M.S.M.; Amin, M.A.; Abd El Maksoud, A.I.; Hassan, A. Glycyrrhizic acid nanoparticles subside the activity of methicillin-resistant *Staphylococcus aureus* by suppressing PBP2a. *Pharmaceuticals* **2024**, *17*, 589. [CrossRef]
42. Nandhini, P.; Gupta, P.K.; Mahapatra, A.K.; Das, A.P.; Agarwal, S.M.; Mickymaray, S.; Alothaim, A.S.; Rajan, M. In silico molecular screening of natural compounds as potential therapeutic inhibitors for methicillin-resistant *Staphylococcus aureus*. *Chem.-Biol. Interact.* **2023**, *374*, 110383. [CrossRef]
43. Bush, K.; Bradford, P.A. Epidemiology of β-lactamase-producing pathogens. *Clin. Microbiol. Rev.* **2020**, *33*, e00047-19. [CrossRef] [PubMed]
44. Tooke, C.L.; Hinchliffe, P.; Bragginton, E.C.; Colenso, C.K.; Hirvonen, V.H.A.; Takebayashi, Y.; Spencer, J. β-Lactamases and β-lactamase inhibitors in the 21st century. *J. Mol. Biol.* **2019**, *431*, 3472–3500. [CrossRef] [PubMed]
45. Bush, K.; Jacoby, G.A. Updated functional classification of β-lactamases. *Antimicrob. Agents Chemother.* **2010**, *54*, 969–976. [CrossRef] [PubMed]
46. Huttner, A.; Bielicki, J.; Clements, M.N.; Frimodt-Møller, N.; Muller, A.E.; Paccaud, J.-P.; Mouton, J.W. Oral amoxicillin and amoxicillin–clavulanic acid: Properties, indications and usage. *Clin. Microbiol. Infect.* **2020**, *26*, 871–879. [CrossRef]
47. Horie, H.; Chiba, A.; Wada, S. Inhibitory effect of soy saponins on the activity of β-lactamases, including New Delhi metallo-β-lactamase 1. *J. Food Sci. Technol.* **2018**, *55*, 1948–1952. [CrossRef]
48. Kolewe, M.E.; Gaurav, V.; Roberts, S.C. Pharmaceutically active natural product synthesis and supply via plant cell culture technology. *Mol. Pharm.* **2008**, *5*, 243–256. [CrossRef]
49. Betts, J.W.; Sharili, A.S.; Phee, L.M.; Wareham, D.W. In vitro activity of epigallocatechin gallate and quercetin alone and in combination versus clinical isolates of methicillin-resistant *Staphylococcus aureus*. *J. Nat. Prod.* **2015**, *78*, 2145–2148. [CrossRef]
50. Teng, Z.; Guo, Y.; Liu, X.; Zhang, J.; Niu, X.; Yu, Q.; Deng, X.; Wang, J. Theaflavin-3,3′-digallate increases the antibacterial activity of β-lactam antibiotics by inhibiting metallo-β-lactamase activity. *J. Cell. Mol. Med.* **2019**, *23*, 6955–6964. [CrossRef]
51. Liu, X.; Zuo, J.; Teng, J.; Yang, L.; Guo, J.; Liu, L.; Li, P. Antibiofilm potential of luteolin against multidrug-resistant *Staphylococcus aureus* isolated from dairy goats and farm environments. *Environ. Pollut.* **2023**, *335*, 122274. [CrossRef]

https://doi.org/10.3390/antiox15020154

52. Teethaisong, Y.; Pimchan, T.; Srisawat, R.; Hobbs, G.; Eumkeb, G. *Boesenbergia rotunda* (L.) Mansf. extract potentiates the antibacterial activity of some β-lactams against β-lactam-resistant staphylococci. *J. Glob. Antimicrob. Resist.* **2018**, *12*, 207–213. [CrossRef]
53. Gaur, R.; Gupta, V.K.; Singh, P.; Pal, A.; Darokar, M.P.; Bhakuni, R.S. Drug resistance reversal potential of isoliquiritigenin and liquiritigenin isolated from *Glycyrrhiza glabra* against methicillin-resistant *Staphylococcus aureus* (MRSA). *Phytother. Res.* **2016**, *30*, 1708–1715. [CrossRef] [PubMed]
54. Kırmusaoğlu, S. Sensitizing of β-lactam resistance by tannic acid in methicillin-resistant *Staphylococcus aureus*. *World J. Microbiol. Biotechnol.* **2019**, *35*, 57. [CrossRef] [PubMed]
55. Kiran, S.; Tariq, A.; Iqbal, S.; Naseem, Z.; Siddique, W.; Jabeen, S.; Bashir, R.; Hussain, A.; Rahman, M.; Habib, F.; et al. Punicalagin, a pomegranate polyphenol, sensitizes the activity of antibiotics against three MDR pathogens of the *Enterobacteriaceae*. *BMC Complement. Med. Ther.* **2024**, *24*, 93. [CrossRef]
56. Zawiła, T.; Swolana, D.; Zawiła, M.; Wojtyczka, R.D. Synergistic interactions between selected β-lactam antibiotics and cinnamic acid and its chosen derivatives. *Antibiotics* **2024**, *13*, 710. [CrossRef]
57. Lady, J.; Nurcahyanti, A.D.R.; Tjoa, E. Synergistic effect and time-kill evaluation of eugenol combined with cefotaxime against *Staphylococcus aureus*. *Curr. Microbiol.* **2023**, *80*, 244. [CrossRef]
58. Ghany, S.S.H.A.E.; Ibrahem, R.A.; El-Gendy, A.O.; El-Baky, R.M.A.; Mustafa, A.; Azmy, A.F. Novel synergistic interactions between monolaurin, a mono-acyl glycerol, and β-lactam antibiotics against *Staphylococcus aureus*: An in vitro study. *BMC Infect. Dis.* **2024**, *24*, 379. [CrossRef]
59. Mahmoudi, H.; Shokoohizadeh, L.; Zare Fahim, N.; Mohamadi Bardebari, A.; Moradkhani, S.; Alikhani, M.Y. Detection of *adeABC* efflux pump-encoding genes and antimicrobial effect of *Mentha longifolia* and menthol on MICs of imipenem and ciprofloxacin in clinical isolates of *Acinetobacter baumannii*. *BMC Complement. Med. Ther.* **2020**, *20*, 92. [CrossRef]
60. Akhmouch, A.A.; Hriouech, S.; Mzabi, A.; Tanghort, M.; Chefchaou, H.; Remmal, A.; Chami, N. Synergistic action of AMX associated with 1,8-cineole and its effect on the ESBL enzymatic resistance mechanism. *Antibiotics* **2022**, *11*, 1002. [CrossRef] [PubMed]
61. Asadi, S.; Nayeri-Fasaei, B.; Zahraei-Salehi, T.; Yahya-Rayat, R.; Shams, N.; Sharifi, A. Antibacterial and anti-biofilm properties of carvacrol alone and in combination with cefixime against *Escherichia coli*. *BMC Microbiol.* **2023**, *23*, 55. [CrossRef]
62. Dettweiler, M.; Melander, R.J.; Porras, G.; Risener, C.; Marquez, L.; Samarakoon, T.; Melander, C.; Quave, C.L. A clerodane diterpene from *Callicarpa americana* resensitizes methicillin-resistant *Staphylococcus aureus* to β-lactam antibiotics. *ACS Infect. Dis.* **2020**, *6*, 1667–1673. [CrossRef]
63. Zhou, H.; Wang, W.; Cai, L.; Yang, T. Potentiation and mechanism of berberine as an antibiotic adjuvant against multidrug-resistant bacteria. *Infect. Drug Resist.* **2023**, *16*, 7313–7326. [CrossRef] [PubMed]
64. Tsutamoto, S.; Iwasaki, Y.; Shinohara, A.; Imamiya, R.; Samukawa, K.; Kawada-Matsuo, M.; Komatsuzawa, H.; Yamada, Y.; Mandokoro, K.; Iwao, H.; et al. Triterpenoid saponin from *Panax ginseng* increases the sensitivity of methicillin-resistant *Staphylococcus aureus* to β-lactam and aminoglycoside antibiotics. *Microbiol. Spectr.* **2024**, *12*, e03227-23. [CrossRef]
65. Sun, N.; Zhi, Z.; Xiao, T.; Deng, X.; He, T.; Dong, W.; Feng, S.; Chen, S.; Wong, W.-L.; Yuan, W. The study of honokiol as a natural product-based antimicrobial agent and its potential interaction with FtsZ protein. *Front. Microbiol.* **2024**, *15*, 1361508. [CrossRef] [PubMed]
66. Gallique, M.; Wei, K.; Maisuria, V.B.; Okshevsky, M.; McKay, G.; Nguyen, D.; Tufenkji, N. Cranberry-derived proanthocyanidins potentiate β-lactam antibiotics against resistant bacteria. *Appl. Environ. Microbiol.* **2021**, *87*, e00127-21. [CrossRef] [PubMed]
67. Grossman, A.B.; Vermerris, W.; Rice, K.C. Dysregulation of cell envelope homeostasis in *Staphylococcus aureus* exposed to solvated lignin. *Appl. Environ. Microbiol.* **2022**, *88*, e00548-22. [CrossRef]
68. Wang, G.; Li, L.; Wang, X.; Li, X.; Zhang, Y.; Yu, J.; Jiang, J.; You, X.; Xiong, Y.Q. Hypericin enhances β-lactam antibiotic activity by inhibiting *sarA* expression in methicillin-resistant *Staphylococcus aureus*. *Acta Pharm. Sin. B* **2019**, *9*, 1174–1182. [CrossRef]
69. Dubey, S.; Patil, K.; Sharma, P.; Tandon, G.D. *Penicillium piceum*: A potential source for antimicrobial agents and β-lactamase inhibitors. *Biologia* **2015**, *70*, 19–26. [CrossRef]
70. Rotondo, C.M.; Wright, G.D. Efficacy of aspergillomarasmine A/meropenem combinations with and without avibactam against bacterial strains producing multiple β-lactamases. *Antimicrob. Agents Chemother.* **2024**, *68*, e00272-24. [CrossRef]
71. Qiao, Y.; Zhang, X.; He, Y.; Sun, W.; Feng, W.; Liu, J.; Hu, Z.; Xu, Q.; Zhu, H.; Zhang, J.; et al. Aspermerodione, a novel fungal metabolite with an unusual 2,6-dioxabicyclo[2.2.1]heptane skeleton, as an inhibitor of penicillin-binding protein 2a. *Sci. Rep.* **2018**, *8*, 5454. [CrossRef]
72. Price, N.P.J.; Jackson, M.A.; Singh, V.; Hartman, T.M.; Blackburn, J.A.; Dowd, P.F. Synergistic enhancement of β-lactam antibiotics by modified tunicamycin analogs TunR1 and TunR2. *J. Antibiot.* **2019**, *72*, 807–815. [CrossRef]

https://doi.org/10.3390/antiox15020154

73. Hu, Y.; Liu, A.; Vaudrey, J.; Vaiciunaite, B.; Moigboi, C.; McTavish, S.M.; Kearns, A.; Coates, A. Combinations of β-lactam or aminoglycoside antibiotics with plectasin are synergistic against methicillin-sensitive and methicillin-resistant *Staphylococcus aureus*. *PLoS ONE* **2015**, *10*, e0117664. [CrossRef]
74. Chu, J.; Vila-Farres, X.; Inoyama, D.; Gallardo-Macias, R.; Jaskowski, M.; Satish, S.; Freundlich, J.S.; Brady, S.F. Human microbiome-inspired antibiotics with improved β-lactam synergy against MDR *Staphylococcus aureus*. *ACS Infect. Dis.* **2018**, *4*, 33–38. [CrossRef]
75. Cella, M.A.; Coulson, T.; MacEachern, S.; Badr, S.; Ahmadi, A.; Tabatabaei, M.S.; Labbe, A.; Griffiths, M.W. Probiotic disruption of quorum sensing reduces virulence and increases cefoxitin sensitivity in methicillin-resistant *Staphylococcus aureus*. *Sci. Rep.* **2023**, *13*, 4373. [CrossRef] [PubMed]
76. Jalalifar, S.; Razavi, S.; Mirzaei, R.; Irajian, G.; Pooshang Bagheri, K. A hope for ineffective antibiotics to return to treatment: Investigating the anti-biofilm potential of melittin alone and in combination with penicillin and oxacillin against multidrug-resistant MRSA and VRSA. *Front. Microbiol.* **2024**, *14*, 1269392. [CrossRef] [PubMed]
77. Nguyen, T.L.A.; Bhattacharya, D. Antimicrobial activity of quercetin: An approach to its mechanistic principle. *Molecules* **2022**, *27*, 2494. [CrossRef] [PubMed]
78. Lee, S.; Razqan, G.S.A.; Kwon, D.H. Antibacterial activity of epigallocatechin-3-gallate (EGCG) and its synergism with β-lactam antibiotics sensitizing carbapenem-associated multidrug-resistant clinical isolates of *Acinetobacter baumannii*. *Phytomedicine* **2017**, *24*, 49–55. [CrossRef]
79. Li, X.; Song, Y.; Wang, L.; Kang, G.; Wang, P.; Yin, H.; Huang, H. A potential combination therapy of berberine hydrochloride with antibiotics against multidrug-resistant *Acinetobacter baumannii*. *Front. Cell. Infect. Microbiol.* **2021**, *11*, 660431. [CrossRef]
80. Tahlan, K.; Shaikh, A.A.; Liu, J.; Gupta, K.; AbuSara, N.; Srivastava, S.K.; Deng, A.; Rouah, A.; Swackhamer, M.J. Functional insights into *Streptomyces* isolates containing both clavulanic acid-like and carbapenem biosynthetic gene clusters. *mSphere* **2025**, *10*, e00188-25. [CrossRef]
81. Viana Marques, D.D.A.; Machado, S.E.F.; Ebinuma, V.C.S.; Duarte, C.D.A.L.; Converti, A.; Porto, A.L.F. Production of β-lactamase inhibitors by *Streptomyces* species. *Antibiotics* **2018**, *7*, 61. [CrossRef]
82. Basker, M.J.; Boon, R.J.; Box, S.J.; Prestige, E.A.; Smith, G.M.; Spear, S.R. Properties of MM 13902 and other carbapenem antibiotics in vitro and in vivo. *J. Antibiot.* **1983**, *36*, 416–422. [CrossRef] [PubMed]
83. Nozaki, Y.; Katayama, N.; Harada, S.; Ono, H.; Okazaki, H. Lactivicin, a naturally occurring non-β-lactam antibiotic having β-lactam-like action: Biological activities and mode of action. *J. Antibiot.* **1989**, *42*, 84–93. [CrossRef]
84. Wang, D.Y.; Abboud, M.I.; Markoulides, M.S.; Brem, J.; Schofield, C.J. The road to avibactam: The first clinically useful non-β-lactam working somewhat like a β-lactam. *Future Med. Chem.* **2016**, *8*, 1063–1084. [CrossRef]
85. Kunz Coyne, A.J.; Eshaya, M.; Bleick, C.; Vader, S.; Biswas, B.; Wilson, M.; Deschenes, M.V.; Alexander, J.; Lehman, S.M.; Rybak, M.J. Exploring synergistic and antagonistic interactions in phage–antibiotic combinations against ESKAPE pathogens. *Microbiol. Spectr.* **2024**, *12*, e00427-24. [CrossRef]
86. Lu, H.; Li, Z.; Elbaz, A.; Ni, S.-Q. Synergistic action of phages and lytic proteins with antibiotics: A combination strategy to target bacteria and biofilms. *BMC Microbiol.* **2023**, *23*, 149. [CrossRef]
87. Uchiyama, J.; Shigehisa, R.; Nasukawa, T.; Mizukami, K.; Takemura-Uchiyama, I.; Ujihara, T.; Murakami, H.; Imanishi, I.; Nishifuji, K.; Sakaguchi, M.; et al. Piperacillin and ceftazidime produce the strongest synergistic phage–antibiotic effect in *Pseudomonas aeruginosa*. *Arch. Virol.* **2018**, *163*, 1941–1948. [CrossRef] [PubMed]
88. Zhu, H.; Zhang, X.; Lu, M.; Chen, H.; Chen, S.; Han, J.; Zhang, Y.; Zhao, P.; Dong, Z. Antibacterial mechanism of silkworm seroins. *Polymers* **2020**, *12*, 2985. [CrossRef]
89. Spyrakis, F.; Santucci, M.; Maso, L.; Cross, S.; Gianquinto, E.; Sannio, F.; Verdirosa, F.; De Luca, F.; Docquier, J.-D.; Cendron, L.; et al. Virtual screening identifies broad-spectrum β-lactamase inhibitors with activity on clinically relevant serine- and metallo-carbapenemases. *Sci. Rep.* **2020**, *10*, 12763. [CrossRef]
90. Spyrakis, F.; Bellio, P.; Quotadamo, A.; Linciano, P.; Benedetti, P.; D'Arrigo, G.; Baroni, M.; Cendron, L.; Celenza, G.; Tondi, D. First virtual screening and experimental validation of inhibitors targeting GES-5 carbapenemase. *J. Comput.-Aided Mol. Des.* **2019**, *33*, 295–305. [CrossRef]
91. Garofalo, B.; Prati, F.; Buonfiglio, R.; Coletta, I.; D'Atanasio, N.; Molteni, A.; Carettoni, D.; Wanke, V.; Pochetti, G.; Montanari, R.; et al. Discovery of novel chemical series of OXA-48 β-lactamase inhibitors by high-throughput screening. *Pharmaceuticals* **2021**, *14*, 612. [CrossRef] [PubMed]
92. Yan, F.; He, S.; Han, X.; Wang, J.; Tian, X.; Wang, C.; James, T.D.; Cui, J.; Ma, X.; Feng, L. High-throughput fluorescent screening of β-lactamase inhibitors to improve antibiotic treatment strategies for tuberculosis. *Biosens. Bioelectron.* **2022**, *216*, 114606. [CrossRef] [PubMed]
93. Bubonja-Šonje, M.; Knežević, S.; Abram, M. Challenges to antimicrobial susceptibility testing of plant-derived polyphenolic compounds. *Arch. Ind. Hyg. Toxicol.* **2020**, *71*, 300–311. [CrossRef]

https://doi.org/10.3390/antiox15020154

94. Knap, K.; Kwiecień, K.; Ochońska, D.; Reczyńska-Kolman, K.; Pamuła, E.; Brzychczy-Włoch, M. Synergistic effect of antibiotics, α-linolenic acid and solvent type against *Staphylococcus aureus* biofilm formation. *Pharmacol. Rep.* **2024**, *76*, 1456–1469. [CrossRef]
95. Kang, J.S.; Zhang, A.L.; Faheem, M.; Zhang, C.J.; Ai, N.; Buynak, J.D.; Welsh, W.J.; Oelschlaeger, P. Virtual screening and experimental testing of B1 metallo-β-lactamase inhibitors. *J. Chem. Inf. Model.* **2018**, *58*, 1902–1914. [CrossRef]
96. Shi, C.; Bao, J.; Sun, Y.; Kang, X.; Lao, X.; Zheng, H. Discovery of baicalin as an NDM-1 inhibitor: Virtual screening, biological evaluation and molecular simulation. *Bioorg. Chem.* **2019**, *88*, 102953. [CrossRef] [PubMed]
97. Balouiri, M.; Sadiki, M.; Ibnsouda, S.K. Methods for in vitro evaluating antimicrobial activity: A review. *J. Pharm. Anal.* **2016**, *6*, 71–79. [CrossRef]
98. Sarker, S.D.; Nahar, L.; Kumarasamy, Y. Microtitre plate-based antibacterial assay incorporating resazurin as an indicator of cell growth, and its application in the in vitro antibacterial screening of phytochemicals. *Methods* **2007**, *42*, 321–324. [CrossRef] [PubMed]
99. Allen, J.L. Colorimetric assays for the rapid and high-throughput screening of antimicrobial peptide activity against diverse bacterial pathogens. In *Methods in Enzymology*; Elsevier: Amsterdam, The Netherlands, 2022; Volume 663, pp. 131–156.
100. Amat, S.; Baines, D.; Alexander, T.W. A vapour phase assay for evaluating the antimicrobial activities of essential oils against bovine respiratory bacterial pathogens. *Lett. Appl. Microbiol.* **2017**, *65*, 489–495. [CrossRef]
101. Pereira, M.A.A.; Sartoratto, A.; Miguel, C.M.T.S.; Murakami, C.; Silva, F.L.; Cordeiro, I.; Machado, K.N.; Young, M.C.M.; Kaneko, T.M.; Lopes, P.S.; et al. Essential oil vapor phase approach for antimicrobial evaluation using *Hesperozygis myrtoides* (A. St.-Hill. ex Benth.) Epling (Lamiaceae) as a model. *An. Acad. Bras. Cienc.* **2024**, *96*, e20231370. [CrossRef]
102. Suleiman, M.M.; McGaw, L.J.; Naidoo, V.; Eloff, J.N. Detection of antimicrobial compounds by bioautography of different extracts of leaves of selected South African tree species. *Afr. J. Tradit. Complement. Altern. Med.* **2010**, *7*, 64–78. [CrossRef]
103. Wang, M.; Zhang, Y.; Wang, R.; Wang, Z.; Yang, B.; Kuang, H. Thin-layer chromatography bioautography: An evolving technology integrating classical methods with continuous technological developments. *Molecules* **2021**, *26*, 4647. [CrossRef]
104. Andes, D.R.; Lepak, A.J. In vivo infection models in the preclinical pharmacokinetic/pharmacodynamic evaluation of antimicrobial agents. *Curr. Opin. Pharmacol.* **2017**, *36*, 94–99. [CrossRef] [PubMed]
105. Moy, T.I.; Ball, A.R.; Anklesaria, Z.; Casadei, G.; Lewis, K.; Ausubel, F.M. Identification of novel antimicrobials using a live-animal infection model. *Proc. Natl. Acad. Sci. USA* **2006**, *103*, 10414–10419. [CrossRef] [PubMed]
106. Brodin, P. Immune–microbe interactions early in life: A determinant of health and disease long term. *Science* **2022**, *376*, 945–950. [CrossRef] [PubMed]
107. Gonzalez-Pastor, R.; Carrera-Pacheco, S.E.; Zúñiga-Miranda, J.; Rodríguez-Pólit, C.; Mayorga-Ramos, A.; Guamán, L.P.; Barba-Ostria, C. Current landscape of methods to evaluate antimicrobial activity of natural extracts. *Molecules* **2023**, *28*, 1068. [CrossRef]
108. Hudson, M.A.; Lockless, S.W. Elucidating the mechanisms of action of antimicrobial agents. *mBio* **2022**, *13*, e02240-21. [CrossRef]
109. Schneider, Y.K. Bacterial natural product drug discovery for new antibiotics: Strategies for tackling antibiotic resistance by efficient bioprospecting. *Antibiotics* **2021**, *10*, 842. [CrossRef]
110. Vaou, N.; Stavropoulou, E.; Voidarou, C.; Tsigalou, C.; Bezirtzoglou, E. Towards advances in medicinal plant antimicrobial activity: Challenges and future perspectives. *Microorganisms* **2021**, *9*, 2041. [CrossRef]
111. Hu, L.; Luo, Y.; Yang, J.; Cheng, C. Botanical flavonoids: Efficacy, absorption, metabolism and advanced pharmaceutical technologies for improving bioavailability. *Molecules* **2025**, *30*, 1184. [CrossRef]
112. Haque, S.; Mathkor, D.M.; Bhat, S.A.; Musayev, A.; Khituova, L.; Ramniwas, S.; Phillips, E.; Swamy, N.; Kumar, S.; Yerer, M.B.; et al. Prospects of phytonutrient berberine as an anticancer agent: A comprehensive review. *J. Biochem. Mol. Toxicol.* **2025**, *39*, e70073. [CrossRef]
113. Kaspute, G.; Ivaskiene, T.; Ramanavicius, A.; Ramanavicius, S.; Prentice, U. Terpenes and essential oils in pharmaceutics: Therapeutic applications and advanced delivery systems for improved stability and bioavailability. *Pharmaceutics* **2025**, *17*, 793. [CrossRef]
114. King, A.M.; Reid-Yu, S.A.; Wang, W.; King, D.T.; De Pascale, G.; Strynadka, N.C.J.; Walsh, T.R.; Coombes, B.K.; Wright, G.D. Aspergillomarasmine A overcomes metallo-β-lactamase antibiotic resistance. *Nature* **2014**, *510*, 503–506. [CrossRef]
115. Malczak, I.; Gajda, A. Interactions of naturally occurring compounds with antimicrobials. *J. Pharm. Anal.* **2023**, *13*, 1452–1470. [CrossRef]
116. Leão, T.; Wang, M.; Moss, N.; Da Silva, R.; Sanders, J.; Nurk, S.; Gurevich, A.; Humphrey, G.; Reher, R.; Zhu, Q.; et al. Multi-omics characterization of the natural product potential of tropical filamentous marine cyanobacteria. *Mar. Drugs* **2021**, *19*, 20. [CrossRef]
117. Melo, M.C.R.; Maasch, J.R.M.A.; De La Fuente-Nunez, C. Accelerating antibiotic discovery through artificial intelligence. *Commun. Biol.* **2021**, *4*, 1050. [CrossRef] [PubMed]
118. Postek, W.; Pacocha, N.; Garstecki, P. Microfluidics for antibiotic susceptibility testing. *Lab Chip* **2022**, *22*, 3637–3662. [CrossRef]

119. Chakraborty, N.; Jha, D.; Roy, I.; Kumar, P.; Gaurav, S.S.; Marimuthu, K.; Ng, O.-T.; Lakshminarayanan, R.; Verma, N.K.; Gautam, H.K. Nanobiotics against antimicrobial resistance: Harnessing nanoscale materials and technologies. *J. Nanobiotechnol.* **2022**, *20*, 375. [CrossRef]
120. Zhu, X.; Tang, Q.; Zhou, X.; Momeni, M.R. Antibiotic resistance and nanotechnology: A narrative review. *Microb. Pathog.* **2024**, *193*, 106741. [CrossRef] [PubMed]

https://doi.org/10.3390/antiox15020154

toxics

MDPI

Article

Antimicrobial Peptide Sublancin Skin Sensitization and Irritation Assessment in Guinea Pigs and Rabbits

Yong Guo [1,2,3], Lin Zhang [4], Gantong Guo [2,3], Tao He [2,3], Yangke Liu [2,3] and Yujiao Lai [4,*]

1 School of Chemistry and Environmental Engineering, Wuhan Institute of Technology, Wuhan 420200, China; 11809010003@stu.wit.edu.cn

2 Sinagri YingTai Bio-peptide Co., Ltd., Linzhou 456550, China; znytggt@126.com (G.G.); znytht@126.com (T.H.); znytlyk@126.com (Y.L.)

3 Key Laboratory of Feed Antibiotics Replacement Technology, Ministry of Agriculture and Rural Affairs, Linzhou 456550, China

4 Hubei Key Laboratory of Animal Nutrition and Feed Science, Engineering Research Center of Feed Protein Resources on Agricultural By-Products, Ministry of Education, Wuhan Polytechnic University, Wuhan 430023, China; 17809057814@163.com

* Correspondence: 22113060@whpu.edu.cn; Tel.: +86-2783956175

Abstract

This study evaluated the skin sensitization of the antimicrobial peptide sublancin to support its safety assessment for topical application. Sensitization was assessed using the guinea pig maximization test (GPMT), in which animals received sublancin (2 mg/kg), vehicle (negative control), or 2,4-dinitrochlorobenzene (positive control) during induction and challenge phases. Skin reactions (erythema and edema) were recorded after challenge. Irritation was evaluated in rabbits following single and repeated applications of sublancin to intact and abraded skin, with observations made at multiple time points. In the GPMT, no erythema or edema was observed in the sublancin-treated group or negative control group at 24, 48, and 72 h post-challenge, corresponding to a sensitization rate of 0%. All animals in the positive control group exhibited moderate to severe erythema and edema (sensitization rate 100%). In both single- and repeated-dose rabbit irritation tests, sublancin induced no erythema or edema on intact or abraded skin at any observation point, resulting in a total irritation score of 0. Furthermore, no significant differences in the daily weight gain were observed between any experimental group and the negative group ($p > 0.05$). In conclusion, under the conditions of this study, sublancin showed no skin sensitization potential in guinea pigs and no irritant effects in rabbits, supporting its local tolerance for topical veterinary use.

Keywords: sublancin; antimicrobial peptide; skin sensitization; skin irritation; safety evaluation; Guinea Pig Maximization Test (GPMT)

1. Introduction

The intensification of modern animal husbandry has led to the widespread use of antibiotics, leading to serious challenges including bacterial resistance and drug residues that threaten animal health, food safety and public health [1–3]. In response, China has implemented policies promoting the "reduction and replacement" of antibiotics, explicitly encouraging the development of safe and efficient novel alternatives [4]. Antimicrobial peptides (AMPs) are considered one of the most promising candidates due to their broad-spectrum antimicrobial activity, unique mechanisms of action, and low propensity for inducing resistance [5,6].

Sublancin, an AMP produced by Bacillus subtilis 168 [7], exhibits potent inhibitory activity against various Gram-positive bacteria [8] and demonstrates additional biological functions such as immunomodulation and gut barrier protection [9–11]. Importantly, a recently completed systematic toxicology study established a foundational safety profile for systemic exposure: in subchronic and chronic toxicity studies in SD rats, doses as high as 50,000 mg/kg feed induced no drug-related systemic toxicity, confirming a No-Observed-Effect Level (NOEL) of 50,000 mg/kg [12]. This result has laid a solid foundation for the safety evaluation of systemic exposure.

However, systemic safety does not guarantee local tolerance for topical applications. For veterinary products intended for skin use (e.g., teat dips, wound sprays), assessing local effects is a critical and mandatory step in safety evaluation [13]. Skin sensitization tests evaluate a substance's potential to induce allergic contact dermatitis after repeated exposure [14], while irritation tests focus on reversible local inflammatory damage (such as erythema and edema) following a single or short-term application [13]. These assessments form the core safety basis for topical products and are explicit requirements for veterinary drug registration.

Therefore, a standardized evaluation of local tolerance is essential early in the development of new veterinary topical formulations. This study, following standardized toxicological guidelines, systematically evaluated the skin sensitization and irritation potential of sublancin using the Guinea Pig Maximization Test (GPMT) and rabbit skin irritation tests, respectively. The work addresses a key gap in its dermal safety assessment. Together with existing systemic toxicity data, it forms a complete safety evidence chain from local to systemic exposure. Furthermore, it provides a standardized practical framework for the preclinical safety evaluation of novel AMP-based preparations and supports their subsequent industrial development.

2. Materials and Methods

2.1. Test Substances and Characterization

The test substance, sublancin, is a bacteria-derived glycopeptide (lantibiotic antimicrobial peptide) produced and isolated by Bacillus subtilis 168. It is composed of 37 amino acid residues. Its molecular weight is approximately 3879.8 Da. The three-dimensional solution structure of this peptide was analyzed by nuclear magnetic resonance (NMR) spectroscopy and exhibited a spatial conformation characterized by intramolecular disulfide bonds and specific O-glycosylation (structural diagram is shown in Figure S1 of Supplementary Materials) [15]. The sublancin used in this study was a purified preparation, with a purity of 93.0% on a dry basis.

2.2. Chemicals and Reagents

Sublancin was supplied by Linzhou Sinagri Yingtai Bio-peptide Co., Ltd. (Linzhou, China). The positive control substance, 2,4-dinitrochlorobenzene (DNCB, purity > 99%), was purchased from Sigma-Aldrich (St. Louis, MO, USA).

2.3. Experimental Animals

Hartley strain albino guinea pigs (weight range: 260–320 g, half male and half female) were purchased from Beijing Haidian Xingwang Animal Farm. A total of 18 New Zealand White rabbits (weight range: 2.5–3.0 kg, half male and half female) were obtained from Beijing Haidian Xinglong Experimental Animal Farm. The specific housing conditions for each species were as follows. Guinea pigs were housed in plastic cages (80 cm × 50 cm × 20 cm) with five animals per cage in a standard conventional animal room. The ambient conditions were maintained at a temperature of 22 ± 4 °C, relative humidity of 55 ± 10%, with an

https://doi.org/10.3390/toxics14010069

artificial light/dark cycle, and noise levels below 60 dB. Bedding was changed twice a week. Rabbits were housed individually in stainless steel cages (50 cm × 30 cm × 30 cm) in a standard conventional animal room. The ambient conditions were maintained at a temperature of 18–22 °C, relative humidity of 40–60%, with a ventilation rate of ≥14 air changes per hour and adequate lighting.

All animals had ad libitum access to a standard diet (purchased from Beijing Ke'ao Xieli Feed Co., Ltd., Beijing, China) and purified water. All animal experiments were performed at the animal facility of the College of Veterinary Medicine, China Agricultural University. The experimental protocol was reviewed and approved by the Institutional Animal Care and Use Committee of Wuhan Polytechnic University (Approval No. WPU202211006), which authorized all procedures conducted at the collaborating site.

2.4. Experimental Procedures

2.4.1. Dose Preparation and Rationale

The test substance (sublancin) was formulated as an aqueous suspension for topical application. The dose selected for both the skin irritation and sensitization assays was 2 mg/kg.

Rationale for Dose Selection: This dose was determined to simulate a clinically relevant, worst-case single dermal exposure from the intended veterinary use (e.g., as a topical spray). The calculation was based on a two-step rationale: (1) estimation of the maximum probable dermal load of the active ingredient during a full application, derived from a proposed product concentration (0.1% *w*/*v*) and typical application volume; (2) translation of this total load (~5–6 mg) into a standardized dose for the rabbit model, normalized by the average animal weight (~2.7 kg). Testing at this exposure level allows for a conservative assessment of local tolerance and provides a direct basis for establishing a safety margin relative to the effective dose.

2.4.2. Skin Sensitization Test Protocol

Forty healthy guinea pigs were randomly divided into three groups (equal numbers of males and females per group): a test substance group (sublancin, $n = 20$), a negative control group ($n = 10$), and a positive control group (DNCB, $n = 10$). Twenty-four hours before the experiment, a 4 cm × 6 cm area on the dorsal neck region of each animal was closely clipped, taking care to keep the skin intact.

Induction Phase: Conducted on days 0, 7, and 14. For the test substance group, sublancin (2 mg/kg) was applied topically to the left clipped area. The positive control group received 0.2 mL of a 0.6% (*w*/*v*) DNCB solution, and the negative control group remained untreated. The applied materials were covered with gauze and non-irritating occlusive film, secured with adhesive tape for 6 h, after which the patches were removed. Skin reactions were observed and recorded at 1 h and 24 h after each exposure.

Challenge Phase: Conducted on day 28. A challenge dose was applied to the contralateral (right) clipped region. The test substance and the negative control groups received topical sublancin, while the positive control group received 0.2 mL of a 0.3% (*w*/*v*) DNCB solution, using the same application and occlusion procedure as in the induction phase. After 6 h, patches were removed. Erythema and edema were observed and recorded at 24, 48, and 72 h post-removal according to the criteria in Table 1. The sensitization rate was calculated based on Table 2.

https://doi.org/10.3390/toxics14010069

Table 1. Scoring criteria for skin allergic reactions.

Skin Reaction	Intensity	Score
Erythema	No erythema	0
	Mild erythema (barely visible)	1
	Moderate erythema (clearly visible)	2
	Severe erythema	3
	Purple-red erythema to slight eschar formation	4
Edema	No edema	0
	Mild edema (slight raising)	1
	Moderate edema (distinct raising)	2
	Edematous erythema	3

Table 2. Classification of skin sensitization intensity.

Skin Reaction (%)	Grade	Intensity
0–8	I	Weak
9–28	II	Mild
29–64	III	Moderate
65–80	IV	Strong
81–100	V	Extreme

Note: A sensitization rate of 0% is interpreted as no skin allergic reaction observed. The sensitization rate was calculated as follows: (Number of animals exhibiting erythema or edema/Total number of animals in the group) × 100%.

2.4.3. Skin Irritation Test Protocol

Twelve healthy rabbits were randomly assigned to four groups (n = 3 per group): single-application intact skin, single-application abraded skin, repeated-application intact skin, and repeated-application abraded skin. Twenty-four hours before dosing, fur on both sides of the dorsal spine (approximately 3 cm × 3 cm per side) was closely clipped. For animals in the abraded-skin groups, the stratum corneum was lightly breached with a sterile needle immediately before substance application to produce a non-occlusive abrasion with minimal bleeding.

Single-Application Irritation Test: A volume of 0.5 mL of sublancin suspension, equivalent to 2.0 mg/kg body weight, was applied to the designated test areas. An equal volume of vehicle was applied to the contralateral control areas. Sites were covered with two layers of gauze and a non-irritating plastic film, then secured. After 4 h, patches were removed and residual substance was gently cleansed with lukewarm water. Skin reactions at test and control sites were observed and scored at 1, 24, 48, and 72 h after patch removal.

Repeated-Application Irritation Test: The test substance was applied once daily to the same site for seven consecutive days. Reactions were observed and scored at 1 h after each daily removal. Following the final (7th) application, additional observations and scoring were performed at 1, 24, 48, and 72 h post-removal. Skin irritations were evaluated using the scoring criteria in Table 3, and the overall irritation intensity was classified according to the grading system in Table 4.

2.4.4. General Clinical Observations

The overall health status of all guinea pigs and rabbits was monitored daily by trained personnel. Detailed records were maintained for appearance, behavior, respiration, appetite, fur condition, and any abnormal secretions. Fecal consistency and color were noted. All animals were closely observed for toxic signs. In the event of severe illness or death, a post-mortem examination was conducted promptly. Skin conditions at applica-

https://doi.org/10.3390/toxics14010069

tion sites were meticulously observed and recorded at scheduled times before and after each administration.

Table 3. Scoring Criteria for Skin Irritation Reactions.

Irritation Response	Description of Intensity	Score
Erythema (A)	No erythema	0
	Mild erythema (barely visible)	1
	Moderate erythema (clearly visible)	2
	Severe erythema	3
	Erythema with eschar formation	4
Edema (B)	No edema	0
	Mild edema (barely visible)	1
	Moderate edema (distinct raising with clear outline)	2
	Severe edema (raised >1 mm)	3
	Severe edema (raised >1 mm and extensive)	4
Highest total score (A + B)		8

Table 4. Evaluation criteria for skin irritation intensity.

Mean Score (A + B)	Evaluation
0–0.5	Non-irritant
0.6–3.0	Mildly irritant
3.1–6.0	Moderately irritant
6.1–8.0	Severely irritant

Note: The Mean Score (A + B) is the average total irritation score (Erythema + Edema, from Table 3) per group across all animals and observation times. Classification thresholds are conventionally derived from the OECD Guideline 404 skin reaction scoring system.

2.4.5. Body Weight Measurement and Skin Reaction Assessment

Body Weight Measurement: For guinea pigs enrolled in the 28-day skin sensitization study, body weight was recorded as a general health indicator at the beginning of the study (Day 0) and on a weekly basis thereafter using a calibrated electronic balance. Weight gain was calculated and analyzed statistically.

Skin Reaction and Clinical Assessment: All scoring of dermal reactions (erythema and edema) according to the specified criteria (Tables 1 and 3), along with general clinical observations, was performed independently by two trained technicians. The assessors were blinded to the treatment groups and were experienced in conducting Guinea Pig Maximization Tests (GPMT) and acute dermal irritation studies in accordance with OECD guidelines. Any scoring discrepancies were resolved through consensus.

2.5. Statistical Analysis

Body weight data are presented as mean ± standard deviation (Mean ± SD). Differences between groups were analyzed using one-way analysis of variance (ANOVA). Statistical significance was set at $p < 0.05$. Data for skin irritation and sensitization endpoints were evaluated using descriptive statistics only and are presented as the incidence of response. All statistical analyses were performed using SPSS version 26.0 (IBM Corporation, Chicago, IL, USA). Graphical representations were generated using GraphPad Prism version 10.0 (GraphPad Software, San Diego, CA, USA).

3. Results

3.1. General Clinical Symptoms

Throughout the study, no remarkable changes in the general clinical sign, including behavior, mental state, appetite, feces, fur, respiration, and body weight, were observed

https://doi.org/10.3390/toxics14010069

in any treatment group of guinea pigs and rabbits compared to the negative control. No abnormal secretions from the nose, eyes, or oral cavity were noted. Critically, the skin at all application sites remained free of erythema, edema, or other signs of irritation for the duration of the administration and observation periods.

3.2. Body Weight

Body weight, a key indicator of general health, was monitored throughout the study (Figure 1 and Table S1). Statistical analysis showed no significant difference in body weight gain between the sublancin-treated group ($n = 20$) and the negative control group ($n = 10$; $p > 0.05$). Similarly, body weight changes in the positive control group ($n = 10$) did not differ significantly from those in the negative control group ($n = 10$; $p > 0.05$), indicating that neither the test nor the positive control substance markedly affected overall animal growth under the experimental conditions.

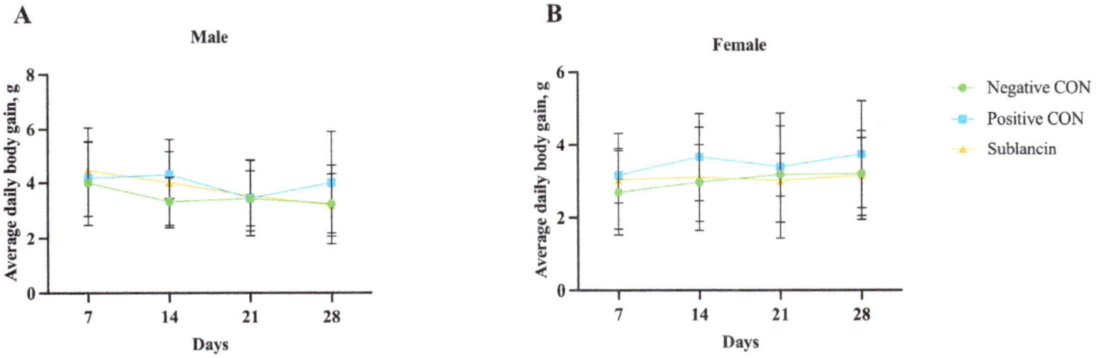

Figure 1. Body weight changes in guinea pigs over the experimental period. (**A**) Male. (**B**) Female.

3.3. Skin Sensitization Test

3.3.1. Induction Phase

Following each of the three sensitizing exposures, no skin reactions (erythema or edema) or other abnormalities were observed in the sublancin-treated (2 mg/kg) or negative control groups at the 1 h and 72 h observation points. Animals in these groups maintained normal dietary intake, excretion, and behavior throughout this phase. In sharp contrast, the positive control group (DNCB) exhibited typical sensitization signs after each exposure, including agitation, pain, scratching, and the development of moderate erythema with mild edema.

3.3.2. Challenge Phase

Skin reactions at 24, 48, and 72 h post-challenge are presented in Figure 2, with detailed scores and incidence rates provided in Table 5. No erythema or edema was observed at any time point in the sublancin-treated group ($n = 20$) or the negative control group ($n = 10$), resulting in a sensitization incidence of 0% for both. Conversely, all animals in the positive control group ($n = 10$) exhibited moderate to severe erythema and edema, corresponding to a 100% allergic reaction incidence. Animals in this group also displayed renewed signs of agitation, pain and itching during the challenge phase.

https://doi.org/10.3390/toxics14010069

Figure 2. Comparison of skin reactions between the test group and control groups in the guinea pig skin sensitization study.

Table 5. Skin allergy reaction scores and sensitization rates in guinea pigs.

Time Point (h)	Group	N	Erythema Score (No. of Animals)					Edema Score (No. of Animals)				Sensitization Rate (%)
			0	1	2	3	4	0	1	2	3	
24	Positive Control	10	0	0	4	6	0	0	0	3	7	100
	Negative Control	10	10	0	0	0	0	10	0	0	0	0
	Sublancin-Treated	20	20	0	0	0	0	20	0	0	0	0
48	Positive Control	10	0	0	7	3	0	0	0	4	6	100
	Negative Control	10	10	0	0	0	0	10	0	0	0	0
	Sublancin-Treated	20	20	0	0	0	0	20	0	0	0	0
72	Positive Control	10	0	0	7	2	1	0	0	4	6	100
	Negative Control	10	10	0	0	0	0	10	0	0	0	0
	Sublancin-Treated	20	20	0	0	0	0	20	0	0	0	0

3.4. *Skin Irritation Tests*

3.4.1. Single-Dose Application

In the single-dose test on intact rabbit skin, sublancin induced no irritant response. As detailed in Table 6 and visualized in Figure 3A, no erythema or edema was observed at the application sites at any observation time (1, 24, 48 and 72 h). The total irritation score remained 0 for all animals, indicating no difference from control sites or pre-application skin. This confirms that sublancin is non-irritating to intact rabbit skin.

Table 6. Skin irritation response scores in rabbits following single application on intact skin.

Time Point (h)	Rabbit ID	Erythema (A)		Edema (B)		Total Score (A + B)	
		Negative Control	Sublancin-Treated	Negative Control	Sublancin-Treated	Negative Control	Sublancin-Treated
1	1	0	0	0	0	0	0
	2	0	0	0	0	0	0
	3	0	0	0	0	0	0
24	1	0	0	0	0	0	0
	2	0	0	0	0	0	0
	3	0	0	0	0	0	0

https://doi.org/10.3390/toxics14010069

Table 6. *Cont.*

Time Point (h)	Rabbit ID	Erythema (A)		Edema (B)		Total Score (A + B)	
		Negative Control	Sublancin-Treated	Negative Control	Sublancin-Treated	Negative Control	Sublancin-Treated
48	1	0	0	0	0	0	0
	2	0	0	0	0	0	0
	3	0	0	0	0	0	0
72	1	0	0	0	0	0	0
	2	0	0	0	0	0	0
	3	0	0	0	0	0	0

Note: The scores for both the application and control sites of all rabbits were 0 at all time points. According to the criteria in Table 4, sublancin is determined to be non-irritating to the intact skin of rabbits. The same applies to the tables below.

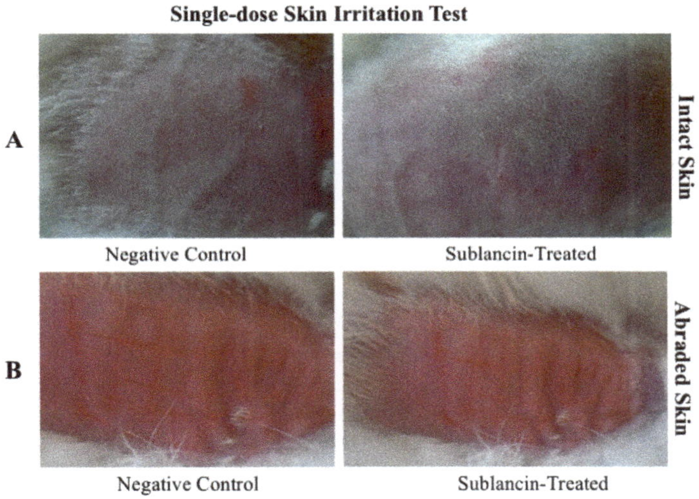

Figure 3. Representative photographs of rabbit skin from the single-dose irritation test. (**A**) intact and (**B**) abraded skin.

Table test on abraded skin yielded identical results. No signs of irritation were observed at any abraded application site during the entire observation period. The total irritation score remained 0 at all time points for all individuals (Table 7, Figure 3B), confirming that sublancin is also non-irritating to abraded rabbit skin.

Table 7. Skin irritation response scores in rabbits following single application on abraded skin.

Time Point (h)	Rabbit ID	Erythema (A)		Edema (B)		Total Score (A + B)	
		Negative Control	Sublancin-Treated	Negative Control	Sublancin-Treated	Negative Control	Sublancin-Treated
1	1	0	0	0	0	0	0
	2	0	0	0	0	0	0
	3	0	0	0	0	0	0
24	1	0	0	0	0	0	0
	2	0	0	0	0	0	0
	3	0	0	0	0	0	0

https://doi.org/10.3390/toxics14010069

Table 7. *Cont.*

Time Point (h)	Rabbit ID	Erythema (A)		Edema (B)		Total Score (A + B)	
		Negative Control	Sublancin-Treated	Negative Control	Sublancin-Treated	Negative Control	Sublancin-Treated
48	1	0	0	0	0	0	0
	2	0	0	0	0	0	0
	3	0	0	0	0	0	0
72	1	0	0	0	0	0	0
	2	0	0	0	0	0	0
	3	0	0	0	0	0	0

3.4.2. Repeated-Dose Application

Repeated application of sublancin over one week did not induce irritation on intact rabbit skin. No erythema or edema was observed following any application. Scores recorded after the final application confirmed a total irritation score of 0 for all animals at all time points (Table 8, Figure 4A).

Table 8. Skin irritation response scores in rabbits following repeated application on intact skin.

Time Point (h)	Rabbit ID	Erythema (A)		Edema (B)		Total Score (A + B)	
		Negative Control	Sublancin-Treated	Negative Control	Sublancin-Treated	Negative Control	Sublancin-Treated
1	1	0	0	0	0	0	0
	2	0	0	0	0	0	0
	3	0	0	0	0	0	0
24	1	0	0	0	0	0	0
	2	0	0	0	0	0	0
	3	0	0	0	0	0	0
48	1	0	0	0	0	0	0
	2	0	0	0	0	0	0
	3	0	0	0	0	0	0
72	1	0	0	0	0	0	0
	2	0	0	0	0	0	0
	3	0	0	0	0	0	0

Similarly, repeated application to abraded skin caused no irritation. As shown in Table 9 and Figure 4B, no erythema or edema was observed on test or control skin before or after any administration, or after the final dose. The total irritation score was consistently 0, confirming that repeated application of sublancin is non-irritating even to compromised skin.

Table 9. Skin irritation response scores in rabbits following repeated application on abraded skin.

Time Point (h)	Rabbit ID	Erythema (A)		Edema (B)		Total Score (A + B)	
		Negative Control	Sublancin-Treated	Negative Control	Sublancin-Treated	Negative Control	Sublancin-Treated
1	1	0	0	0	0	0	0
	2	0	0	0	0	0	0
	3	0	0	0	0	0	0

Table 9. *Cont.*

Time Point (h)	Rabbit ID	Erythema (A)		Edema (B)		Total Score (A + B)	
		Negative Control	Sublancin-Treated	Negative Control	Sublancin-Treated	Negative Control	Sublancin-Treated
24	1	0	0	0	0	0	0
	2	0	0	0	0	0	0
	3	0	0	0	0	0	0
48	1	0	0	0	0	0	0
	2	0	0	0	0	0	0
	3	0	0	0	0	0	0
72	1	0	0	0	0	0	0
	2	0	0	0	0	0	0
	3	0	0	0	0	0	0

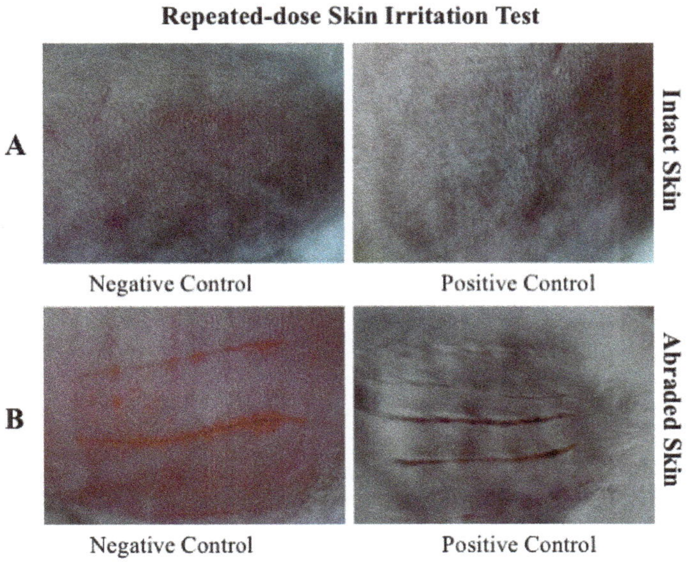

Figure 4. Representative photographs of rabbit skin from the repeated-dose irritation test. (**A**) Intact and (**B**) abraded skin.

4. Discussion

This study, in accordance with the Technical Guidelines for Veterinary Drug Research issued by the Ministry of Agriculture and Rural Affairs of China and the OECD Test Guidelines (TG 404 and 406) [16,17], systematically evaluated the topical safety profile of the antimicrobial peptide sublancin through standardized GPMT and acute dermal skin irritation test in rabbits. The core experimental outcomes consistently demonstrated an absence of skin irritation and sensitization potential under the tested conditions. The following discussion synthesizes these findings, moving beyond a mere restatement of results to explore their mechanistic underpinnings, regulatory significance, and translational implications for developing sublancin as a topical veterinary agent.

The reliability of the concluded safety profile is fundamentally underpinned by the appropriate responsiveness of the employed toxicological models. In the rabbit skin irritation test, sublancin elicited no erythema or edema on either intact or abraded skin across all observation periods, resulting in a Primary Irritation Index (PII) of 0. This outcome definitively classifies sublancin as a "non-irritant" substance according to OECD TG 404 criteria. Crucially, the concurrent positive control (e.g., DNCB) successfully induced expected irritant responses, validating the sensitivity and proper conduct of the assay system. Similarly, in the GPMT, the complete absence of allergic reactions in sublancin-treated guinea pigs (0% sensitization rate) stands in sharp contrast to the robust sensitization (100%) induced by the positive control, 2,4-dinitrochlorobenzene (DNCB). This clear dichotomy not only confirms the assay's capability to identify sensitizers but also provides high confidence in the negative result for sublancin. It is noteworthy that while DNCB provoked severe local skin reactions, it did not significantly affect systemic parameters like body weight in this model, underscoring the localized nature of contact hypersensitivity and reinforcing that the observed safety of sublancin is not due to a generalized insensitivity of the test animals [18,19].

The favorable safety outcome can be rationalized through an analysis of sublancin's physicochemical and structural properties, which likely minimize its interaction with skin biology pathways, leading to adverse effects. As a glycosylated peptide with a molecular weight of approximately 3.8 kDa [15,20], sublancin exceeds the typical size threshold (<500 Da) considered optimal for passive transdermal penetration [21]. Its complex, amphiphilic cyclic structure, stabilized by disulfide bonds, may further impose steric and diffusional barriers to permeation through the stratum corneum [22]. This inherently low potential for dermal absorption is a critical factor. For sensitization to occur, a compound must penetrate the skin, be haptenized or recognized as an antigen, and be presented by Langerhans cells to T-lymphocytes [23,24]. The limited bioavailability of sublancin in the skin significantly reduces its opportunity to engage this immunological cascade. Similarly, irritation often involves direct cytotoxic effects or pro-inflammatory interactions with epidermal cells; minimal penetration likely circumvents these processes. This perspective is conceptually supported by preliminary pharmacokinetic data (unpublished), which suggest low systemic bioavailability for sublancin via other routes and hint at a general pattern of low membrane permeability that may extend to the skin. These data are cited for mechanistic plausibility only and were not used for quantitative interpretation. Therefore, the observed safety is not coincidental but appears rooted in a predictable combination of its substantial molecular size and complex conformation.

From a regulatory standpoint, the results provide the necessary data to classify sublancin as a "non-irritant" and "non-sensitizer" under the OECD guidelines for skin irritation and sensitization testing and the corresponding criteria of the Globally Harmonized System of Classification and Labelling [25]. This classification is a pivotal component of the non-clinical safety dossier required for the registration of any new veterinary topical product.

It is essential to contextualize these dermal findings within the broader toxicological assessment of sublancin. The favorable local tolerance observed herein aligns coherently with and is reinforced by other toxicological studies. Notably, independent research has demonstrated an absence of systemic toxicity even at very high oral doses in rodent models [12]. Furthermore, evaluations for genotoxic potential (e.g., Ames test, micronucleus assay) have yielded negative results. This consistency across diverse endpoints, from local tolerance to systemic toxicity and genetic safety, creates a convergent and robust weight-of-evidence for the compound's overall low toxicological risk. The present study specifically addresses the relevant exposure route for intended applications, thereby filling a key data gap and complementing the existing safety database to sketch a more complete and reassuring non-clinical safety profile.

The demonstrated safety, particularly its maintenance on abraded skin, carries significant practical value for intended veterinary uses. In dairy farming, maintaining teat skin integrity is paramount for preventing mastitis. While traditional biocides like iodophors

https://doi.org/10.3390/toxics14010069

are effective, some can cause skin dryness or irritation with prolonged use, potentially compromising the physical barrier [26]. Sublancin's non-irritating property suggests its potential as an alternative that could provide effective antimicrobial activity while better preserving skin health, which is crucial for animal welfare and milk quality [27,28]. Similarly, in managing wounds in livestock (e.g., from hoof lesions or procedures), topical agents must avoid causing additional pain or contact dermatitis [26]. The safety demonstrated on abraded skin in this study supports the potential tolerability of sublancin in such wound care scenarios.

This application's potential is highly relevant within the global strategy of seeking antibiotic alternatives to combat resistance. Antimicrobial peptides (AMPs), as natural defense molecules, often possess mechanisms of action that make resistance development less likely compared to conventional antibiotics [29,30]. The progression of other AMPs, like the PL-5 spray for human wound infections [31], illustrates the translational pathway. Sublancin, with its documented in vitro antimicrobial efficacy against relevant pathogens [8–11,32,33], combined with the in vivo topical safety established here, presents a compelling candidate for development. However, it is critical to distinguish between demonstrated safety in a controlled test and proven efficacy in complex field conditions. The current data robustly support the former and justify further investment into the latter.

This work provides foundational safety data for the topical application of sublancin, yet certain limitations inherent to its design must be acknowledged to contextualize the findings and direct future veterinary product development. A primary consideration is the use of standard rodent and lagomorph models. While essential for regulatory safety screening, interspecies differences in skin thickness, immune response, and metabolism may affect the precise extrapolation of these results to all target livestock species, such as cattle or swine. Furthermore, the study evaluated sublancin as a purified entity. In a practical veterinary setting, its compatibility with necessary formulation excipients (e.g., emulsifiers, preservatives), its stability in final products, and its interaction with complex environmental factors on farms remain to be characterized.

Therefore, subsequent research should logically progress through a translational pipeline specific to veterinary medicine. The immediate step involves confirmatory safety and efficacy testing in relevant target species under conditions mimicking intended use. Concurrently, applied research must focus on developing stable, user-friendly formulations (e.g., long-acting teat dips, adhesive wound sprays) and rigorously assessing their local tolerance and physical stability [34]. Finally, to complete the safety dossier and define practical use parameters, systematic local pharmacokinetic and residue studies in target species are indispensable. Quantifying dermal absorption and depletion kinetics is critical for establishing scientifically justified withdrawal periods and maximum residue limits, which are fundamental for regulatory approval and consumer safety [35].

5. Conclusions

This study confirmed through guinea pig maximization tests and rabbit skin irritation tests that the antimicrobial peptide Sublancin does not have skin sensitization or irritation under the test conditions. Specifically, no skin lesions (erythema or edema) were observed in any of the 20 guinea pigs (sensitization rate 0%), and all 12 rabbits showed a total irritation score of 0 for both intact and abraded skin. According to relevant evaluation criteria, it can be determined that the skin sensitization and irritation test results of Sublancin are negative, which provides core experimental support for its local application safety. These findings support the progression of sublancin to safety and efficacy evaluations in target animal species, marking a key translational step toward its potential development as a topical veterinary agent.

https://doi.org/10.3390/toxics14010069

Supplementary Materials: The following supporting information can be downloaded at: https://www.mdpi.com/article/10.3390/toxics14010069/s1, Figure S1: Schematic representation of sublancin's structure; Table S1: The weight gain of guinea pigs in different groups over 4 weeks.

Author Contributions: Investigation, writing—original draft, data curation, supervision, project administration, Y.G.; writing, methodology, software, data curation, L.Z.; investigation, resources, G.G.; validation, investigation, resources T.H.; validation, investigation, resources Y.L. (Yangke Liu); writing—review and editing, supervision, Y.L. (Yujiao Lai). All authors have read and agreed to the published version of the manuscript.

Funding: This study was supported by Wuhan Polytechnic University Industry-Academia-Research Project (Project No.: whpu-2023-kj-2709).

Institutional Review Board Statement: The animal study protocol was approved by the Institutional Animal Care and Use Committee (IACUC) of Wuhan Polytechnic University (Approval No. WPU202211006, Approval Date: 1 November 2022), which authorized all procedures conducted at the collaborating site.

Informed Consent Statement: Not applicable.

Data Availability Statement: The data presented in this study are available on request from the corresponding author.

Acknowledgments: This study was supported by the research facilities and resources provided by Wuhan Textile University. The authors are especially grateful to Chen Rong for his attentive guidance and thorough revision of the manuscript, which greatly enhanced its linguistic quality and readability. They also extend their sincere thanks to China Agricultural University, Wuhan Institute of Technology, and Wuhan Polytechnic University for offering access to laboratory equipment and related resources. Additionally, during the writing process, the authors employed Kimi (https://www.kimi.com), developed by Moonshot AI (Beijing, China), to assist with grammatical refinement and improve the overall fluency of the text.

Conflicts of Interest: Yong Guo, Gantong Guo, Tao He, and Yangke Liu are employees of Sinagri YingTai Bio-peptide Co., Ltd. The other authors have no commercial or financial interests that could present a potential conflict of interest. The funders had no involvement in the study design, data collection, analysis, interpretation, manuscript writing, or the decision to publish the findings.

Abbreviations

The following abbreviations are used in this manuscript:

GPMT	Guinea Pig Maximization Test
AMPs	antimicrobial peptides
NOEL	No-Observed-Effect Level
NMR	nuclear magnetic resonance
DNCB	2,4-dinitrochlorobenzene
ANOVA	one-way analysis of variance
PII	primary irritation index

References

1. Liu, M.; Luan, H.; Qiu, W.; Zhang, Y.; Feng, W.; Xu, W.; Wang, F.; Xuan, H.; Song, P. Antibiotic alternatives in livestock feeding. *Sci. Total Environ.* **2025**, *989*, 179867. [CrossRef]
2. Qian, M.; Wang, J.; Ji, X.; Yang, H.; Tang, B.; Zhang, H.; Yang, G.; Bao, Z.; Jin, Y. Sub-chronic exposure to antibiotics doxycycline, oxytetracycline or florfenicol impacts gut barrier and induces gut microbiota dysbiosis in adult zebrafish (Daino rerio). *Ecotoxicol. Environ. Saf.* **2021**, *221*, 112464. [CrossRef]
3. Grabowski, L.; Choszcz, M.; Wisniewska, K.; Gaffke, L.; Namiotko, D.; Podlacha, M.; Wegrzyn, A.; Wegrzyn, G.; Pierzynowska, K. Induction of the mitochondrial pathway of apoptosis by enrofloxacin in the context of the safety issue of its use in poultry. *Apoptosis* **2024**, *29*, 1260–1270. [CrossRef] [PubMed]

https://doi.org/10.3390/toxics14010069

4. Gao, J.; Yang, Z.; Zhao, C.; Tang, X.; Jiang, Q.; Yin, Y. A comprehensive review on natural phenolic compounds as alternatives to in-feed antibiotics. *Sci. China Life Sci.* **2023**, *66*, 1518–1534. [CrossRef]
5. Oliveira Júnior, N.G.; Souza, C.M.; Buccini, D.F.; Cardoso, M.H.; Franco, O.L. Antimicrobial peptides: Structure, functions and translational applications. *Nat. Rev. Microbiol.* **2025**, *23*, 687–700. [CrossRef] [PubMed]
6. Liu, Y.; Gao, Z.; Chen, T.; Gao, Y.; Chen, H.; Ye, H.; Luo, Q.; Wu, D.; Zhang, X. Versatile Fabrication of Biocompatible Antimicrobial Materials Enabled by Cationic Peptide Bundles. *ACS Appl. Mater. Interfaces* **2024**, *16*, 43918–43932. [CrossRef] [PubMed]
7. Paik, S.H.; Chakicherla, A.; Hansen, J.N. Identification and characterization of the structural and transporter genes for, and the chemical and biological properties of, sublancin 168, a novel lantibiotic produced by Bacillus subtilis 168. *J. Biol. Chem.* **1998**, *273*, 23134–23142. [CrossRef]
8. Wang, S.; Zeng, X.F.; Wang, Q.W.; Zhu, J.L.; Peng, Q.; Hou, C.L.; Thacker, P.; Qiao, S.Y. The antimicrobial peptide sublancin ameliorates necrotic enteritis induced by Clostridium perfringens in broilers. *J. Anim. Sci.* **2015**, *93*, 4750–4760. [CrossRef]
9. Wang, S.; Wang, Q.; Zeng, X.; Ye, Q.; Huang, S.; Yu, H.; Yang, T.; Qiao, S. Use of the Antimicrobial Peptide Sublancin with Combined Antibacterial and Immunomodulatory Activities To Protect against Methicillin-Resistant Staphylococcus aureus Infection in Mice. *J. Agric. Food Chem.* **2017**, *65*, 8595–8605. [CrossRef]
10. Li, J.; Chen, J.; Yang, G.; Tao, L. Sublancin protects against methicillin-resistant Staphylococcus aureus infection by the combined modulation of innate immune response and microbiota. *Peptides* **2021**, *141*, 170533. [CrossRef]
11. Wang, Q.; Zeng, X.; Wang, S.; Hou, C.; Yang, F.; Ma, X.; Thacker, P.; Qiao, S. The Bacteriocin Sublancin Attenuates Intestinal Injury in Young Mice Infected With Staphylococcus aureus. *Anat. Rec.* **2014**, *297*, 1454–1461. [CrossRef]
12. Guo, Y.; Li, Z.; Xu, P.; Guo, G.; He, T.; Lai, Y. Subchronic and Chronic Toxicity Assessment of Sublancin in Sprague-Dawley Rats. *Toxics* **2025**, *13*, 413. [CrossRef]
13. Mustafa, N.; Ahmad, R.; Ahmad Khushaini, M.A.; Kamar Affendi, N.H.; Ab Ghani, S.M.; Tan, S.K.; Ismail, M.H.; Goo, C.L.; Kassim, M.Z.; Lim, T.W.; et al. Porous NiTi Dental Implant Fabricated by a Metal Injection Molding: An in Vivo Biocompatibility Evaluation in an Animal Model. *ACS Biomater. Sci. Eng.* **2024**, *10*, 405–419. [CrossRef]
14. Zhao, N.; Song, X.; Naito, H.; Li, H.; Huang, Y.; Liu, L.; Lu, F.; Cai, T.; Ito, Y.; Kamijima, M.; et al. Trichloroethylene and trichloroethanol induce skin sensitization with focal hepatic necrosis in guinea pigs. *J. Occup. Health* **2020**, *62*, e12142. [CrossRef]
15. Garcia De Gonzalo, C.V.; Zhu, L.; Oman, T.J.; van der Donk, W.A. NMR structure of the S-linked glycopeptide sublancin 168. *ACS Chem. Biol.* **2014**, *9*, 796–801. [CrossRef]
16. OECD. Test No. 404: Acute Dermal Irritation/Corrosion. In *OECD Guidelines for the Testing of Chemicals*; Section 4; OECD Publishing: Paris, France, 2015. [CrossRef]
17. OECD. Test No. 406: Skin Sensitisation. In *OECD Guidelines for the Testing of 1203 Chemicals*; Section 4: Health Effects; OECD Publishing: Paris, France, 1992.
18. Carlsson, J.; Dostberg, A.; Astrom, T.; Matyjasiak, J.; Kallin, A.; Juric, S.; Nilsson, U. Health risks from exposure to chemicals in clothing—Non-regulated halogenated aromatic compounds. *Chemosphere* **2024**, *363*, 142930. [CrossRef] [PubMed]
19. Qiao, W.; Duan, C.; Ma, J.; Hu, W.; Xie, Y.; Yang, L.; Wang, T.; Wu, S.; Li, X.; Wang, Y.; et al. Costimulatory Molecule CD226 Regulates Atopic Dermatitis in a Mouse Model. *J. Investig. Dermatol.* **2024**, *144*, 1743–1753.e4. [CrossRef] [PubMed]
20. Ji, S.; Li, W.; Baloch, A.R.; Wang, M.; Cao, B. Improved production of sublancin via introduction of three characteristic promoters into operon clusters responsible for this novel distinct glycopeptide biosynthesis. *Microb. Cell Fact.* **2015**, *14*, 17. [CrossRef]
21. Wang, Y.; Zeng, L.; Song, W.; Liu, J. Influencing factors and drug application of iontophoresis in transdermal drug delivery: An overview of recent progress. *Drug Deliv. Transl. Res.* **2022**, *12*, 15–26. [CrossRef] [PubMed]
22. Chen, Z.; Huang, Q.; Song, Y.; Feng, X.; Zeng, L.; Liu, Z.; Hu, X.; Tao, C.; Wang, L.; Qi, Y.; et al. Cubosomes-assisted transdermal delivery of doxorubicin and indocyanine green for chemo-photothermal combination therapy of melanoma. *Biomed. Pharmacother.* **2023**, *166*, 115316. [CrossRef]
23. Peterman, E.; Quitevis, E.J.A.; Goo, C.E.A.; Rasmussen, J.P. Rho-associated kinase regulates Langerhans cell morphology and responsiveness to tissue damage. *Cell Rep.* **2024**, *43*, 114208. [CrossRef]
24. Oltramare, C.; Burri, O.; Hopf, N.B.; Jaccoud, S.; Seitz, A.; Applegate, L.A.; Hirt-Burri, N.; Berthet, A. Development of image analysis tool to evaluate Langerhans cell migration after exposure to isothiazolinones. *Arch. Toxicol.* **2025**, *99*, 2463–2477. [CrossRef]
25. UNECE (United Nations Economic Commission for Europe). GHS (Rev.8) (2019): Globally Harmonized System of Classification and Labelling of Chemicals (GHS). Part 3.3. 2019. Available online: https://unece.org/ghs-rev8-2019 (accessed on 7 January 2026).
26. Lomax, S.; Dickson, H.; Sheil, M.; Windsor, P.A. Topical anaesthesia alleviates short-term pain of castration and tail docking in lambs. *Aust. Vet. J.* **2010**, *88*, 67–74. [CrossRef]
27. Tammineni, Y.; Kumar, P.; Reddy, S.; Muskan; Birader, K.; Kumar, K.; Arumugam, A.; Saha, S.; Nayee, N.; Suman, P. Portable and affordable device for on-site quantitative detection of somatic cells in milk. *NPJ Sci. Food* **2025**, *9*, 182. [CrossRef] [PubMed]

https://doi.org/10.3390/toxics14010069

28. Santamarina-Garcia, G.; Yap, M.; Crispie, F.; Amores, G.; Lordan, C.; Virto, M.; Cotter, P.D. Shotgun metagenomic sequencing reveals the influence of artisanal dairy environments on the microbiomes, quality, and safety of Idiazabal, a raw ewe milk PDO cheese. *Microbiome* **2024**, *12*, 262. [CrossRef] [PubMed]
29. Woodby, B.; Pambianchi, E.; Ferrara, F.; Therrien, J.P.; Pecorelli, A.; Messano, N.; Lila, M.A.; Valacchi, G. Cutaneous antimicrobial peptides: New "actors" in pollution related inflammatory conditions. *Redox Biol.* **2021**, *41*, 101952. [CrossRef]
30. de Alcantara Sica de Toledo, L.; Rosseto, H.C.; Buccini, D.F.; Franco, O.L. Synergy between Antimicrobial Peptides and Lipid Nanoparticles for Skin Infection Control. *ACS Appl. Mater. Interfaces* **2025**. [CrossRef]
31. Wei, Y.; Wu, J.; Chen, Y.; Fan, K.; Yu, X.; Li, X.; Zhao, Y.; Li, Y.; Lv, G.; Song, G.; et al. Efficacy and Safety of PL-5 (Peceleganan) Spray for Wound Infections: A Phase IIb Randomized Clinical Trial. *Ann. Surg.* **2023**, *277*, 43–49. [CrossRef] [PubMed]
32. Biswas, S.; Garcia De Gonzalo, C.V.; Repka, L.M.; van der Donk, W.A. Structure-Activity Relationships of the S-Linked Glycocin Sublancin. *ACS Chem. Biol.* **2017**, *12*, 2965–2969. [CrossRef]
33. Wang, S.; Ye, Q.; Wang, K.; Zeng, X.; Huang, S.; Yu, H.; Ge, Q.; Qi, D.; Qiao, S. Enhancement of Macrophage Function by the Antimicrobial Peptide Sublancin Protects Mice from Methicillin-Resistant Staphylococcus aureus. *J. Immunol. Res.* **2019**, *2019*, 3979352. [CrossRef]
34. Zhong, G.; Lei, P.; Guo, P.; Yang, Q.; Duan, Y.; Zhang, J.; Qiu, M.; Gou, K.; Zhang, C.; Qu, Y.; et al. A Photo-induced Cross-Linking Enhanced A and B Combined Multi-Functional Spray Hydrogel Instantly Protects and Promotes of Irregular Dynamic Wound Healing. *Small* **2024**, *20*, e2309568. [CrossRef] [PubMed]
35. Zarmpi, P.; Tsikritsis, D.; Watson, A.C.; Vorng, J.L.; Tyagi, V.; Belsey, N.A.; Rantou, E.; Ghosh, P.; Bunge, A.L.; Woodman, T.J.; et al. Raman spectroscopy as a tool to accelerate development of complex medicinal products. *J. Control. Release* **2025**, *387*, 114190. [CrossRef] [PubMed]

https://doi.org/10.3390/toxics14010069

Journal of
Xenobiotics

MDPI

Review

Hazards and Health Risks of the Antibacterial Agent Triclosan to Fish: A Review

Jiangang Wang [1,†], Nannan Ma [1,†], Gancong Mo [2], Xian Qin [3], Jin Zhang [1], Xiangping Yao [2], Jiahua Guo [4] and Zewei Sun [1,*]

1 Guangdong Provincial Key Laboratory of Marine Disaster Prediction and Prevention, Shantou University, Shantou 515063, China; 23jgwang@stu.edu.cn (J.W.); 21nnma@alumni.stu.edu.cn (N.M.); 24jzhang@stu.edu.cn (J.Z.)
2 College of Engineering, Guangzhou College of Technology and Business, Guangzhou 510850, China; mogancong@163.com (G.M.); xp_yao921@163.com (X.Y.)
3 State Key Laboratory of Marine Environmental Health, City University of Hong Kong, Kowloon, Hong Kong SAR, China; xianqin2@cityu.edu.hk
4 Shaanxi Key Laboratory of Earth Surface System and Environmental Carrying Capacity, College of Urban and Environmental Sciences, Northwest University, Xi'an 710127, China; jiahua_guo@nwu.edu.cn
* Correspondence: zwsun@stu.edu.cn; Tel.: +86-13923998896
† These authors contributed equally to this work.

Abstract: Triclosan (TCS) is a widely used antimicrobial agent found in personal care products and household cleaners. While valued since the 1960s for its ability to inhibit bacterial fatty acid synthesis, its environmental persistence, ecotoxicity, and bioaccumulative potential have raised significant global concern. The increased use of disinfectants during the COVID-19 pandemic has further exacerbated its prevalence as an aquatic pollutant. In the environment, TCS is distributed through water bodies and sediments, undergoing processes such as biodegradation and photochemical degradation. Its bioaccumulation poses a substantial threat to aquatic organisms, particularly fish. A growing body of research indicates that TCS acts as an endocrine disruptor and developmental toxicant, with documented adverse effects encompassing impaired embryonic and larval development, skeletal malformations, and induction of oxidative stress, mitochondrial dysfunction, DNA damage, and inflammatory responses. Furthermore, TCS exposure is linked to reproductive toxicity, including altered sex hormone levels and diminished reproductive capacity. This review consolidates current knowledge on the chemical properties, environmental fate, biodegradation pathways, and ecotoxicological impacts of TCS, with a specific emphasis on its multifaceted health risks to fish. The synthesis aims to provide a foundation for future research, inform environmental risk assessments, and support the development of evidence-based regulatory measures.

Keywords: pharmaceuticals and personal care products; triclosan; developmental toxicity; long-term effects; fish; ecological risk assessment

1. Introduction

Triclosan (TCS) is a chlorinated aromatic compound with broad-spectrum antibacterial activity. Its mechanism of action involves inhibiting bacterial fatty acid synthesis and disrupting cell membranes [1–4], which led to its initial adoption as a hospital bactericide. Historically, TCS was incorporated as a preservative and antimicrobial agent into a vast array of over 2000 consumer products, including soaps, toothpastes, cosmetics, textiles, kitchenware, and children's toys. This pervasive use resulted in TCS becoming a ubiquitous environmental contaminant, primarily entering waterways through wastewater treatment plant effluents and sewage sludge application on agricultural land [5].

Regulatory responses to health and environmental concerns have varied globally. The U.S. Food and Drug Administration and the European Union banned the use of TCS in over-the-counter antibacterial soaps in 2016. However, this ban is not worldwide, and TCS remains a key ingredient in many other personal care products, such as toothpaste and hand sanitizers, at concentrations typically between 0.1% and 1% (*w*/*w*) [2,3,6,7]. This variance in global regulation has led to known environmental hotspots, with elevated TCS concentrations detected in waters receiving effluent from manufacturing plants, urban centers, and hospitals [8,9]. This extensive use and environmental resilience have led to the widespread detection of TCS in rivers, lakes, and coastal waters globally. Notably, environmental concentrations surged during the COVID-19 pandemic due to heightened disinfectant consumption [10], underscoring the direct link between consumer use and environmental loading and highlighting its continued relevance as a pollutant of concern.

Structurally, TCS contains two benzene rings with a high chlorine content, conferring properties similar to persistent pollutants like polychlorinated biphenyls and dioxins [11]. This structural analogy suggests that TCS may also share the tendencies of these classic pollutants for bioaccumulation and biomagnification, raising concerns for long-term ecosystem exposure. Over 90% of used TCS enters wastewater systems, and conventional treatment processes remove it with limited efficiency; approximately 30% is ultimately discharged into water bodies on a global scale. It is routinely detected in wastewater influent and effluent, sludge, surface water, and sediments (Table 1). Characterized by low volatility (vapor pressure: 4×10^{-6} mm Hg at 20 °C), high lipophilicity (logKow: 4.8), and a long half-life up to several months, TCS persists in surface water at ng/L concentrations, where it remains stable and bioaccumulates [12–14]. In severely contaminated areas, concentrations can reach 40 μg/L, underscoring significant environmental concern [15].

Table 1. Reported concentrations of triclosan in water and sediments across different regions.

Location	Matrix	Concentration	Reference
Six lakes and Nag River, Nagpur, India	Water	74.3 μg/L	[16]
Gomati River, South India	Water	1.1–9.65 μg/L (Maximum: 9.56 μg/L)	[17,18]
Freshwater lake, Yangtze River Basin, Central China	Water	0.47 μg/L	[17]
Baiyangdian Lake, China	Water	26.1 ng/L	[19]
Laizhou Bay, China	Water	58.3 ng/L	[20]
Bouregreg River, Rabat, Morocco	Water	48–301 ng/L	[20]
Yangtze River, China	Water	1.0–20.6 ng/L	[20]
Yamuna River, India	Water	269.8 ± 1.0 ng/L	[20]
Lake Victoria, Uganda	Water	89–1400 ng/L	[20]
Tamiraparani River, India	Water	Up to 5.2 μg/L	[21]
Huangpu River, China	Water	1.48–89.76 ng/L	[22]
Shanghai, China	Water	533–774 ng/L	[23]

Table 1. *Cont.*

Location	Matrix	Concentration	Reference
Benedict River and lakes, India	Water	297–1761 ng/L	[24]
Tamiraparani River, India	Water	0.944 μg/L	[25]
Kaveri and Vellar Rivers, India	Water	3.8–5.16 μg/L	[25]
Xiaoqing River, China	Water	Up to 245 ng/L	[26]
Torsa River, India	Water	0.055–0.184 μg/L	[27]
Buffalo River, Eastern Cape, South Africa	Water	0–1264.2 ng/L	[28]
Estuarine system, Bangkok, Thailand	Water	Up to 185 ng/L	[29]
Freshwater, Switzerland	Water	18–98 ng/L	[2]
Yellow River, China	Water	Up to 64.7 ng/L	[30]
Atibaia River, São Paulo, Brazil	Water	Up to 0.34 mg/L	[31]
Paraíba do Sul River, São Paulo, Brazil	Water	Up to 0.78 mg/L	[31]
Pearl River Delta, South China	Water	1 μg/L	[32]
Campredo Lake, Spain	Water	0.285 μg/L	[32]
Hunting Creek, USA	Water	15.5 ± 3.71 ng/L	[33]
Ikpa River Basin, Nigeria	Water	55.1–297.7 ng/L	[34]
Jiulong River and estuary, China	Water	64 ng/L	[35]
Ton Canal, Japan	Water	134 ng/L	[35]
Ems estuary, Germany	Surface water	0.012–0.11 ng/L	[35]
Weser estuary, Germany	Surface water	0.018–0.620 ng/L	[35]
Elbe estuary, Germany	Surface water	1.20–6.87 ng/L	[35]
Vellar River, India	Surface water	Up to 516 μg/L	[36]
Tamiraparani River, Tamil Nadu, India	Surface water	Mean: 944 ng/L; Maximum: 5160 ng/L	[14]
United States	Surface water	250–850 ng/L	[18]
England	Surface water	58 μg/L	[20]
Guangzhou, China	Tap water	14.5 ng/L	[26]
Hunting Creek, USA	Sediment	72.5 ± 9.41 ng/g (d.w.)	[33]
Estuarine system, Bangkok, Thailand	Sediment	242 ng/g	[29]
Valiyar estuary, India	Sediment	132–3073 μg/kg	[25]
Baiyangdian Lake, China	Sediment	32.5 ng/g	[19]
Pearl River, China	Sediment	1329 μg/kg	[16]
Gomati River, South India	Sediment	5.11–50.36 μg/kg	[37]
Valiyar estuary, Tamil Nadu, India	Sediment	132–3073 ng/kg	[14]
Freshwater, Thailand	Sediment	Up to 0.726 mg/kg	[38]
Municipal WWTP, Savannah, USA	Effluent	86.161 μg/L	[23]

In aquatic systems, TCS primarily undergoes photodegradation and biodegradation. Photochemical reactions can transform TCS into hazardous derivatives, including 2,8-dichlorodibenzo-p-dioxin, 2,7-dichlorodibenzo-p-dioxin, and hydroxylated dichlorodibenzofurans, which may pose greater risks than the parent compound [39,40]. Notably, these dioxin and furan derivatives are of particular concern as they are generally more persistent, bioaccumulative, and significantly more toxic than the parent TCS, with well-documented potential for carcinogenic, endocrine-disrupting, and developmental effects [41]. Biodegradation represents another critical pathway. Fungal laccase/redox mediator systems can enzymatically transform TCS by cleaving its diphenyl ether bond, reducing its toxicity [42]. Certain bacterial species, such as *Pseudomonas putida* and *Alcaligenes xylosoxidans*, can utilize TCS as a carbon source for degradation with an efficiency up to 98% [43,44].

Due to its prevalence and persistence, TCS is ranked among the top ten pollutants in terrestrial and aquatic environments [45,46] and bioaccumulates in organisms across trophic levels, from algae to marine mammals [16]. The bioaccumulation of TCS in marine mammals indeed highlights a potential public health concern, as humans, particularly high-level seafood consumers, may be inadvertently exposed to this compound through the diet [47]. While the direct acute toxicity from dietary exposure is likely low, chronic intake has been linked to potential endocrine-disrupting effects and contributions to antimicrobial resistance in experimental models [48,49]. Crucially, maximum residue limits for TCS in food are not consistently established globally. Ecotoxicological studies (Tables 2 and 3) demonstrate that acute and sub-chronic TCS exposure induces a range of adverse effects in aquatic organisms. These include the induction of oxidative stress, disruption of lipid metabolism, DNA damage, and various histopathological changes. These effects can result in embryonic developmental abnormalities, endocrine disruption, neurotoxicity, behavioral alterations, and reproductive impairment (Figure 1) [50,51]. Notably, although many studies have revealed that triclosan exposure in fish causes thyroid dysfunction, metabolic disorders, reduced fertility, and liver damage, these adverse effects are mostly observed at high concentrations [50,52,53]. In contrast, multiple investigations, including Mihaich [54], demonstrate that at the lower (<90 μg/L), environmentally relevant concentrations of triclosan found in water and sediments (Table 1), no significant impacts on the HTP-axis, liver, metabolism, or reproductive outcomes are observed. Furthermore, TCS degradation products formed under oxidative conditions may also pose risks to aquatic life, though their toxicity to fish remains inadequately studied. Therefore, future research must prioritize the identification, monitoring, and toxicological assessment of these transformation products to fully understand the ecological risks posed by TCS throughout its lifecycle.

Table 2. Reported LC_{50} values for various fish species following exposure to triclosan.

Test Organism	Exposure Duration	LC_{50}	Reference
Aphaniops hormuzensis	96 h	0.924 mg/L	[55]
Carassius auratus	96 h	1111.9 μg/L, 1.839 mg/L	[56,57]
Catla catla	96 h	0.73 mg/L	[58]
Catla catla	96 h	0.36 mg/L	[3]
Cirrhinus mrigala	96 h	0.131 mg/L	[46]
Clarias gariepinus	96 h	16.04 mg/L	[52]
Ctenopharyngodon idella	96 h	0.116 mg/L	[46]
Cyprinus carpio	96 h	0.80 mg/L	[30]
Cyprinus carpio	96 h	0.80 mg/L	[56]
Cyprinus carpio	96 h	0.315 mg/L	[46]
Zebrafish embryos (*Danio rerio*)	120 h	217 μg/L	[59]
Zebrafish embryos (*Danio rerio*)	96 h	267.8 μg/L, 0.608 ± 0.064 mg/L, 420 μg/L	[20,60–62]
Zebrafish embryos (*Danio rerio*)	48 h	1.50 ± 0.48 mg/L	[63]

Table 2. *Cont.*

Test Organism	Exposure Duration	LC_{50}	Reference
Juvenile zebrafish (*Danio rerio*)	96 h	510 μg/L, 0.42 mg/L	[64–68]
Danio rerio	96 h	340 mg/L, 398.9 μg/L	[60,62,66,69]
Danio rerio	32 days	74.6 μg/L	[20]
Gambusia affinis	96 h	1.399 mg/L	[51]
Labeo rohita	96 h	126 μg/L	[25]
Labeo rohita	96 h	96 μg/L, 0.39 mg/L	[2,20,45]
Misgurnus anguillicaudatus	96 h	0.045 mg/L	[57]
Oreochromis mossambicus	96 h	715 μg/L, 740 μg/L	[20,70]
Oryzias latipes	96 h	1.7 mg/L, 210 mg/L	[71,72]
Oryzias latipes	21 days	330.6 μg/L	[73]
Oryzias latipes	96 h	169.78 μg/L, 399 μg/L	[73,74]
Oryzias latipes	96 h	117.9, 600, 602 μg/L	[73,74]
Oryzias latipes	48 h	0.352 mg/L	[35]
Oryzias latipes	96 h	1700 μg/L	[75]
Oryzias melastigma	-	300 μg/L	[20]
Oryzias sinensis	96 h	0.63 mg/L	[76]
Pangasianodon hypophthalmus	96 h	848.33 μg/L (25 °C) 1181.94 μg/L (30 °C) 1356.96 μg/L (35 °C)	[77]
Pangasianodon hypophthalmus	96 h	910 mg/L (pH 6.5) 1110 mg/L (pH 7.5) 1380 mg/L (pH 8.5)	[78]
Pangasianodon hypophthalmus	96 h	1458 μg/L	[79]
Pangasianodon hypophthalmus	96 h	1177 μg/L	[80]
Pimephales promelas	96 h	260 μg/L	[75,81]
Poecilia vivipara	96 h	0.6 mg/L	[82]
Pseudorasbora parva	96 h	0.071 mg/L	[57]
Tanichthys albonubes	96 h	0.889 mg/L	[57]
Xiphophorus helleri	96 h	1.47 mg/L	[83]
Lepomis macrochirus	96 h	370 μg/L	[81]

Table 3. Reported NOEC values for various fish species following exposure to triclosan.

Test Organism	Exposure Duration	NOEC	References
Zebrafish (*Danio rerio*) embryos	96 h	200 μg/L	[84]
Zebrafish (*Danio rerio*) embryos	144 h	160 μg/L	[19]
Zebrafish (*Danio rerio*) embryos	144 h	160 μg/L	[36]
Zebrafish (*Danio rerio*)	9 days	26 μg/L	[20]
Zebrafish (*Danio rerio*)	32 days	48.4 μg/L	[20]
Rainbow trout (*Oncorhynchus mykiss*)	35 days	71.3 μg/L	[20]
Medaka (*Oryzias latipes*)	182 days	11 μg/L	[54]

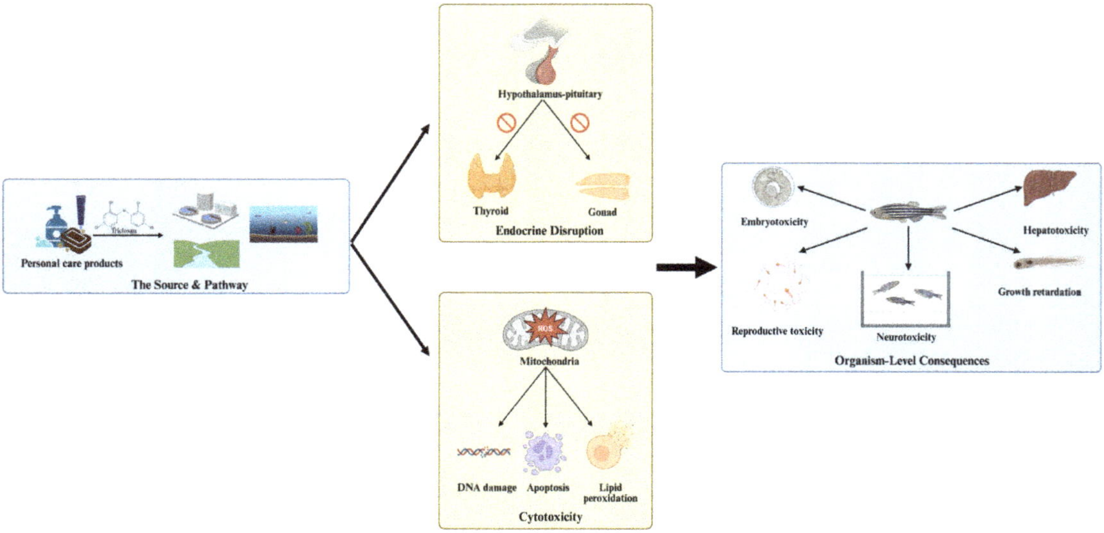

Figure 1. A schematic diagram showing the sources, environmental occurrence, and ecotoxicity of triclosan.

2. Methods

To comprehensively evaluate the latest advances in understanding the effects of triclosan on fish, literature searches were conducted by three coauthors. The search encompassed publications from January 1998 to December 2024. Primary databases, including PubMed, Web of Science (Figure 2), and Google Scholar, were queried using the key term "triclosan fish" or "triclosan toxicity fish". The resulting articles in English were selected for in-depth reading based on their relevance to the titles and abstracts. The search strategy was supplemented by a manual examination of reference lists from these relevant articles to identify additional studies. In total, 103 references were selected for final review.

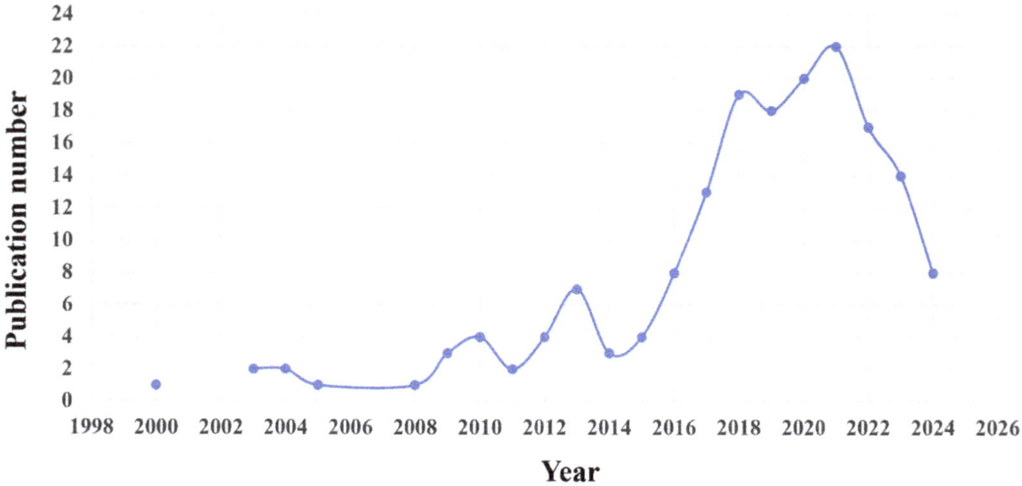

Figure 2. Publications related to triclosan-induced toxicity in fish in the literature. Literature searching was performed on Web of Science, and relevant papers published during 1998–2024 were sourced. The first related study recorded on Wed of Science was published in 2000.

3. Metabolism and Cytotoxicity of Triclosan

TCS can enter fish through dermal contact, diet, and aqueous exposure, undergoing a typical two-phase metabolic process: Phase I Metabolism (oxidation, reduction, hydrolysis) and Phase II Metabolism (conjugation reactions) [5,9,85]. Phase I Metabolism can be further divided into hydroxylation, ether bond cleavage, cyclization, and dechlorination. Hydroxylation involves the introduction of a hydroxyl (-OH) group onto the benzene ring of TCS, potentially occurring on the less chlorinated ring to form monohydroxy-triclosan (TP 302). Ether bond cleavage, one of the primary metabolic pathways, involves the breakage of the ether bond (C-O-C) at the C-6 position, generating metabolites such as 2,4-dichlorophenol (TP 160). TCS may also undergo cyclization to form dioxin-like derivatives, which are of significant concern due to their high ecotoxicological potential. These include compounds such as hydroxylated dichlorodibenzodioxin (DCDD, P4). Dechlorination at the C-3 position produces di-chloro-hydroxydiphenyl ether (TP 252). These processes significantly increase the polarity of TCS, preparing it for subsequent metabolism. Phase II Metabolism typically involves the addition of polar groups (e.g., sulfate, glucuronic acid, or glutathione) to Phase I metabolites, further enhancing their water solubility and promoting excretion with an efficiency larger than 80% [82]. Phase II reactions include sulfation and glucuronidation. TCS or its hydroxylated metabolites may undergo sulfation to form sulfated metabolites, such as 2,4-dichlorophenol-O-sulfate (TP 240), triclosan-O-sulfate (TP 366), and monohydroxy-triclosan-O-sulfate (TP 382). Glucuronidation involves the addition of a glucuronic acid group to hydroxylated metabolites, increasing their water solubility and facilitating excretion [7,86].

While these metabolic pathways are primarily detoxification mechanisms, they can also produce reactive intermediates whose presence directly connects to the observed cytotoxic effects. Mitochondria generate ATP via the electron transport chain, but TCS can interfere with the electron transport chain, increasing electron leakage and contributing to the generation of reactive oxygen species (ROS) such as superoxide. Simultaneously, it reduces the activity of key mitochondrial antioxidant enzymes (e.g., SOD, GPx), weakening the cell's ability to clean ROS. The resulting ROS accumulation triggers oxidative stress, further leading to mitochondrial dysfunction, decreased membrane potential, impaired ATP synthesis, and induction of apoptosis via activation of the MAPK/p53 and Bax/Bcl-2/caspase pathways. It also promotes DNA damage and inflammatory responses, constituting the primary cytotoxic mechanisms of TCS. ROS can activate multiple apoptotic signaling pathways, including intrinsic and extrinsic pathways. ROS promote the expression of Bax and Bak, increase mitochondrial membrane permeability, release cytochrome C, and activate the caspase cascade, leading to apoptosis. Experiments show that TCS exposure significantly increases the number of TUNEL-positive cells in the hearts of zebrafish larvae (250 μg/L from 30 to 120 days post fertilization under the conditions of a 14 h: 10 h light: dark cycle and temperature of 28 $\pm$ 0.5 °C) and in the olfactory epithelium and olfactory bulbs of goldfish, while decreasing Ca^{2+}-ATPase and Na^+/K^+-ATPase activity and increasing caspase-1 and caspase-3 activity [6,87]. Zebrafish embryos exposed to TCS (50 and 250 μg/L) from 4 h post-fertilization to 96 h also showed significant apoptosis [88]. In reproductive organs, high concentrations of TCS (85, 170 μg/L) led to TUNEL-positive signals in ovarian oocytes and testicular Sertoli cells and spermatocytes, along with downregulation of antioxidant genes (*sod*, *gpx1a*, and *cat*), upregulation of apoptosis-related genes (*bax* and *p53*), and a decrease in Bcl-2 [89,90]. These findings indicate that TCS can induce oxidative stress and apoptosis in key reproductive cells, which is highly likely to impair individual fertility and, at a population level, could lead to reduced reproductive success in wild fish populations. In vitro exposure of zebrafish liver cells to TCS (0.5–2 mg/L) for 24 h showed apoptotic cell proportions of 21%, 36%, and 80%, respectively [91]. Mechanistically,

TCS increases ROS to activate oxidative stress, further inducing MAPK (ERK, JNK, and p38) and p53 signaling pathways. The imbalance in Bax/Bcl-2 leads to cytochrome C release, activating effector caspases such as caspase-3, ultimately triggering apoptosis [11,91].

ROS can react with intracellular lipids, proteins, and DNA, causing oxidative damage. Lipid peroxidation (LPO) damages membrane fluidity and integrity. Malondialdehyde (MDA), a primary marker of LPO, reflects oxidative stress levels. Sustained LPO damages cell membranes by impairing their fluidity and integrity, which disrupts critical functions like ion transport, signal transduction, and organelle compartmentalization, ultimately leading to a loss of cellular homeostasis. Experiments show that TCS exposure significantly increases MDA and LPO levels in fish. In zebrafish intestinal tissue, 250 μg/L TCS elevated MDA [92]. In zebrafish larvae, MDA content increased positively with TCS concentration [93]. In zebrafish liver, 189 μg/L TCS increased LPO by approximately 40%, while the same concentration exposure in South American catfish gills resulted in a 13% increase in LPO [71]. In *Anabas testudineus* hepatocytes and goldfish liver homogenate, low concentrations of TCS (9 μg/L and 0.28–0.56 mg/L) also significantly elevated MDA levels [26]. Notably, the induction of oxidative stress in *Anabas testudineus* at a much lower concentration (9 μg/L) compared to the goldfish model (280–560 μg/L) suggests a potentially higher sensitivity in this species, which could be attributed to differences in metabolic capacity or antioxidant defense systems. These results indicate that TCS induces oxidative stress and cellular damage (Figure 3).

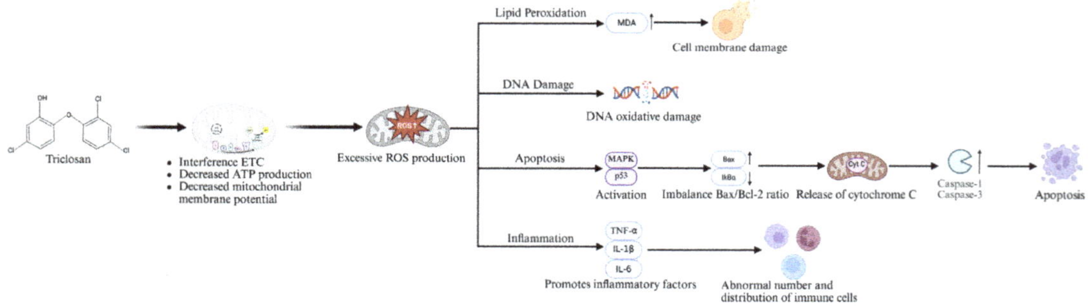

Figure 3. Molecular mechanism diagram of triclosan-induced cytotoxicity in fish. MDA: malondialdehyde, MAPK: mitogen-activated protein kinase, p53: tumor protein 53, TNF-α: tumor necrosis factor-alpha, IL-1β: interleukin-1 beta, IL-6: interleukin-6, ETC: electron transfer chain, ROS: reactive oxygen species. An upward and downward arrow indicates increase and decrease, respectively.

Mitochondria are both a primary site of ROS generation and a key target of ROS attack. TCS exposure can lead to mitochondrial dysfunction in fish (*Anabas testudineus*) and cell models (zebrafish liver cells ZF-L), manifested as decreased ATP production, enhanced oxidative stress, reduced antioxidant enzyme activity, decreased levels of reduced glutathione, and lowered mitochondrial membrane potential (MMP)—a particularly sensitive early indicator of toxicity that often precedes irreversible commitment to cell death [91]. In zebrafish, detection of MMP in larvae showed weakened red fluorescence in the 50 and 250 μg/L TCS treatment groups, indicating decreased MMP [53]. Excessive ROS formation also causes oxidative DNA damage, including base oxidation, single-strand and double-strand breaks, further activating apoptotic pathways. Micronucleus tests and comet assays show that TCS significantly increases micronucleus formation and DNA damage in fish erythrocytes and hepatocytes. For example, DNA damage significantly increased in yellow eels exposed to TCS under low pH conditions [78]. Micronucleus tests observed a significant increase in the number of micronuclei and binucleated cells in erythrocytes of

African sharptooth catfish (*Clarias gariepinus*) exposed to TCS [52]. In zebrafish, the extent of DNA damage was quantified by measuring the fluorescence intensity of comet tails. Following TCS treatment at 0.5596 mg/L, tail moment in hepatocytes profoundly increased compared to the solvent control group, indicating TCS-induced DNA damage. Tail length, tail moment, and tail DNA percentage in zebrafish hepatocyte comet assays increased with TCS concentration [56,69]. Comet assay results on zebrafish embryos showed that exposure to 1 μg/L of TCS led to a significant increase in the frequency of DNA damage at the end of 120 h of exposure, indicating slight primary genotoxic effects [94]. Furthermore, Indian major carp (*Labeo rohita*) exposed to TCS showed a significant increase in DNA damage, which remained higher than the control group even during the 10-day recovery period, suggesting the potential persistence of TCS genotoxicity [25,95]. These results indicate that TCS, by inducing mitochondrial dysfunction and ROS accumulation, not only damages cellular energy metabolism but also triggers oxidative stress and DNA damage, severely impacting cell survival and genetic stability.

Excessive ROS formation can promote the expression of inflammatory factors such as TNF-α, IL-1β, and IL-6. These factors interact with ROS, forming pathological positive feedback that exacerbates cellular damage and apoptosis. Studies have found that TCS exposure interferes with the development and function of immune cells in zebrafish, altering the balance between pro-inflammatory and anti-inflammatory factor expression [50,68]. Under 250 μg/L TCS treatment, the number and distribution of neutrophils, macrophages, and T cells in zebrafish quantified by bioimaging under a fluorescent microscope were abnormal, and the expression of immune-related genes and proteins significantly changed, including upregulation of pro-inflammatory factors and downregulation of anti-inflammatory factors [67]. Studies on striped catfish showed that TCS exposure led to decreased respiratory burst activity, myeloperoxidase activity, and phagocytic activity, indicating impaired immune defense function [80]. Additionally, mucus deposition on the body surface of the freshwater fish *Anabas testudineus* increased and rose with TCS concentration (0.009, 9, and 176.7 μg/L), reflecting the fish immune system's response to exogenous toxic substances by enhancing mucus secretion [14]. While mucus serves as a critical first-line immune barrier, its hyper-secretion in this context is more indicative of a chronic stress response. The sustained, dose-dependent increase signals a persistent toxic challenge that forces the fish to expend significant energy on defense, potentially leading to immunosuppression and reduced fitness over time. In summary, TCS exposure can perturb the fish immune system through ROS-mediated pro-inflammatory responses, manifesting as both decreased immune cell function and increased defensive mucus secretion, forming complex immunomodulatory effects.

ROS accumulation can affect cell survival and death by activating the MAPK signaling pathway (including JNK, p38, and ERK). Studies show that high-concentration TCS exposure (125–250 μg/L) causes chondrocyte swelling, deformation, reduced numbers, and nuclear displacement, along with a significant decrease in bone-derived alkaline phosphatase activity, suggesting inhibited osteoblast activity and weakened bone mineralization, thereby affecting larval bone formation. At the molecular level, TCS exposure downregulated skeletal development-related genes that play key roles in cartilage formation (e.g., *sox4a, sox11b, sox9a, col2a1a,*), bone mineralization (e.g., *spp1, bglap, bmp2a, bmp4*), and osteoblast differentiation (e.g., *hoxb3a, runx2b*) [13]. Furthermore, TCS exposure caused abnormal miRNA expression, specifically upregulation of miR-144, miR-34a, and miR-153a-5p, and downregulation of miR-34b, miR-200b-5p, and miR-21. These miRNA changes may participate in the regulation of oxidative stress responses and skeletal development by modulating their target genes through mRNA degradation or translation suppression, providing molecular mechanistic support for TCS-induced bone development abnormalities [26,93]. In summary, TCS interferes with fish skeletal development and mineralization processes through synergistic multi-pathway actions, including induction of oxidative stress, inhibition of osteogenesis-related genes, and alteration of miRNA expression.

4. Embryonic Development

4.1. Embryotoxicity

Fish embryos are highly sensitive to environmental pollutants, in addition to transparency, rapid development, ecological relevance, making it a common model for assessing the toxicity of chemical substances. Studies have shown that TCS exhibits endocrine-disrupting effects and significant toxicity to fish embryos [69]. Qian et al. (2024) found that 250 μg/L TCS significantly reduced the hatching rate of zebrafish embryos [68]. Ning Tang et al. further reported that zebrafish embryos treated with 30–900 ng/mL TCS showed a decline in hatching rate from 34.5% to 4.3%, with all larvae dying within 24 h at 900 μg/L [66]. Annamaria Iannetta observed a dose-dependent delay in zebrafish embryo development and hatching after 72 h post-fertilization (hpf) [84]. Under high-concentration exposure (1.60–4.16 mg/L), the hatching rate decreased to approximately 68–70%, while the proportion of developmental abnormalities reached 43–46% [52]. Similarly, Nile tilapia exposed to 2.092 mg/L TCS exhibited 100% mortality within 24 h, while no mortality was observed at lower concentrations (0.131–0.262 mg/L) [70]. This rapid and complete mortality at a relatively high concentration demonstrates a clear acute toxicological threshold and underscores the potential for TCS to cause severe, immediate population crashes in the event of a spill or point-source pollution event in aquatic environments. The mortality rate of zebrafish embryos increased significantly with time and dose, with significant increase after 96 h of exposure to 0.4 mg/L and higher concentrations of TCS [61].

In addition to changes in hatching and survival rates, TCS also caused noticeable embryonic morphological abnormalities, including melanin deficiency, yolk sac enlargement, spinal deformities, and cardiac hemorrhage. The malformation rate under acute exposure reached 27.4% [96,97]. Zebrafish embryos exposed to 0.9 mg/L TCS showed significant delays in otolith formation and ocular pigmentation [12]. The disruption of these processes is particularly significant; otoliths are vital for the vestibular system (governing balance and orientation), while timely ocular pigmentation is crucial for the development of functional vision. Furthermore, surviving larvae exposed to 0.5 mg/L TCS exhibited spinal deformities and pericardial edema [12]. The edema rate, an important indicator of TCS embryotoxicity, increased significantly as the exposure pH decreased from 8 to 6 (pH 6: 17.5%, pH 7: 14.7%, and pH 8: 14.1%) [98]. Furthermore, the severity of yolk sac edema was enhanced in TCS-exposed larvae (from 6 to 72 h post fertilization), and their average body length was significantly shorter compared to the control group [68].

4.2. Developmental Toxicity

Following the impact of TCS on embryonic development, fish growth and development also exhibit significant toxic effects. Studies have shown that TCS exposure increases mucus secretion in the skin and gills of fish such as Mozambique tilapia (*Oreochromis mossambicus*), striped catfish (*Pangasianodon hypophthalmus*), and guppy (*Poecilia reticulata*) [70,79,99]. Mucus secretion is a natural defense response of fish to environmental pollutants, pathogens, and physical damage. TCS exposure stimulates mucus secretion to form a physical barrier, reducing the penetration and absorption of harmful substances.

Furthermore, growth indicators of fish vary significantly under different experimental conditions. Qian et al. (2024) observed that zebrafish exposed to 125 and 250 μg/L TCS between 30 and 100 dpf showed significant reductions in body weight, length, and width [68]. In contrast, Danting Wang found that exposure to a mixture of TCS and its derivatives (TCS-DT) significantly increased the body mass index and width of zebrafish at 90 days [100]. In a 3-month chronic TCS exposure experiment on adult zebrafish, significant increases in body weight and length were observed [15]. These results indicate that the effects of TCS on fish development are complex and depend on dose, duration, and species.

The molecular structure of TCS resembles several nonsteroidal estrogens, such as diethylstilbestrol and bisphenol A, due to its two phenolic hydroxyl functional groups. This suggests that TCS has the potential to act as an endocrine disruptor [101]. Specifically, it can interfere with the hypothalamic–pituitary–gonadal (HPG) axis, leading to altered sex hormone levels, impaired gonadal development, reduced fertility, and the induction of vitellogenin in male fish—a clear marker of estrogenic exposure. These disruptions can ultimately result in skewed sex ratios, abnormal reproductive behavior, and diminished reproductive success at the population level. Indeed, some studies have shown that TCS exposure activates the HPG axis, increasing the levels of hormones such as GnRH, FSH, and LH, thereby affecting the production and regulation of sex hormones (Figure 4). This may ultimately impact fish reproductive health, including reproductive capacity and the sexual development of offspring [30,58]. TCS exposure upregulated GnRH mRNA expression in the hypothalamus, indicating that TCS may stimulate the hypothalamus to produce more GnRH [56,58]. Consistent with upregulation of GnRH mRNA expression, GnRH levels in the serum of Yellow River carp (*Cyprinus carpio*) and zebrafish also increased after TCS treatment [56]. However, in a 42-day exposure experiment on 7-week-old Yellow River carp using different concentrations of TCS (0, 0.04, 0.08, and 0.16 mg/L), a decrease in GnRH mRNA expression in the hypothalamus, particularly in the 0.08 and 0.16 mg/L TCS treatment groups was found, where cGnRH-II-2 mRNA levels and GnRH concentrations were reduced [30]. The observed contradiction, where low concentrations of TCS increase GnRH while higher concentrations decrease it, is a hallmark of endocrine disruption. A plausible mechanism involves a compensatory negative feedback loop. Initially, low-level disruption may stimulate a compensatory surge in GnRH release. However, at higher concentrations, the sustained disruption could trigger a powerful negative feedback response from downstream hormones (e.g., elevated cortisol or sex steroids), ultimately suppressing hypothalamic GnRH synthesis and release to prevent systemic overload.

Simultaneously, TCS exposure increased the mRNA expression of the β-subunits of follicle-stimulating hormone (FSH) and luteinizing hormone (LH) in the pituitary, indicating that TCS stimulates the anterior pituitary to produce more FSH and LH [56,58]. This heightened gonadotropin signaling directly promotes gonadal maturation by stimulating steroidogenesis (the production of sex steroids like estrogen and testosterone) and driving the progression of gametogenesis, including vitellogenesis in females and spermatogenesis in males. In *Catla catla* treated with 0.073 mg/L TCS, FSH and LH concentrations increased significantly by 35.11% and 40.69%, respectively, compared to the control group [58]. However, Chokki Veettil et al. (2024) observed a concentration-dependent significant decrease in FSH and LH. The observed contradiction, where lower TCS doses increase FSH/LH while higher doses decrease them, is a hallmark of a disrupted hypothalamic-pituitary-gonadal (HPG) axis feedback loop [14]. A plausible mechanism involves an initial compensatory stimulation of gonadotropin release at low doses. However, at higher concentrations, this may trigger a powerful negative feedback response from downstream hormones (e.g., elevated sex steroids or cortisol), or lead to the downregulation of pituitary GnRH receptors, ultimately suppressing FSH and LH synthesis to prevent systemic hormonal overload. Gonadal hormone levels and ratios were also affected. The E2/T ratio—the balance between the primary female hormone, estradiol (E2), and the primary male hormone, testosterone (T)—is a critical indicator of sexual phenotype. Increases or decreases in this ratio can lead to hormonal imbalance, potentially causing intersex conditions, impaired gonadal development, and reduced fertility [30,56]. Chokki Veettil et al. (2024) showed that serum estradiol levels significantly decreased in female fish and increased in male fish, while testosterone levels significantly decreased in both sexes [14]. These opposite changes suggest a feminizing effect on males and a de-feminizing or suppressive effect on females. In males, elevated E2 can lead to the inappropriate production of vitellogenin (an egg yolk protein) and disrupt courtship behaviors, while the overall suppression of testosterone in both sexes can impair libido, gametogenesis, and secondary sexual characteristics, collectively threatening reproductive success. The primary reason is that TCS affects the activity of aromatase and the mRNA expression of cyp19a1a and cyp19a1b in the gonads and hypothalamus. These genes encode for the aromatase enzyme, but with distinct tissue-specific functions: cyp19a1a is predominantly expressed in the ovaries and is essential for converting androgens to estrogens to support oocyte growth and vitellogenesis, while cyp19a1b is primarily neuronal, expressed in the hypothalamus, and crucial for regulating neuroendocrine pathways that control sexual behavior and the feedback of the reproductive axis. Changes in their activity may lead to imbalances in sex hormones [14,30,56,58].

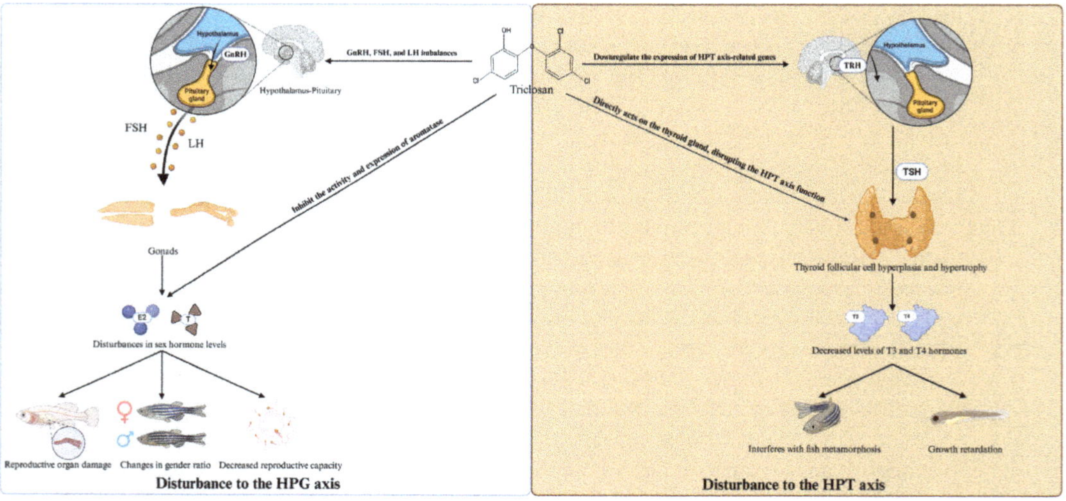

Figure 4. A schematic diagram illustrating triclosan-induced endocrine disruption in fish. GnRH: gonadotropin-releasing hormone, FSH: follicle-stimulating hormone, LH: luteinizing hormone, TSH: thyroid-stimulating hormone, HPG axis: hypothalamic–pituitary–gonadal axis, HPT axis: hypothalamic–pituitary–thyroid axis, T3: triiodothyronine, T4: thyroxine, E2: 17β-estradiol, T: testosterone.

Additionally, TCS exposure inhibits the activity of 3β-hydroxysteroid dehydrogenase (3β-HSD) and 17β-hydroxysteroid dehydrogenase (17β-HSD), enzymes critical for sex hormone biosynthesis. The inhibition of 3β-HSD disrupts the conversion of pregnenolone to progesterone, a key early step in the steroidogenic pathway, while the suppression of 17β-HSD directly impairs the final activation step of androgens (like androstenedione to testosterone). Since testosterone is both a crucial end-product and the primary precursor for estradiol synthesis via aromatase, this dual inhibition provides a direct mechanistic explanation for the observed decreases in both testosterone and estradiol [14]. TCS exposure also significantly increased Er protein levels and mRNA expression in the liver of female carp, suggesting that TCS may enhance Er levels by increasing the transcriptional activity of the Er gene [56]. This upregulation is likely a pathological response to the endocrine-disrupting activity of TCS. By increasing Er expression, TCS could sensitize the liver to even low levels of circulating estrogen (or estrogen-mimics), leading to an exaggerated and dysregulated vitellogenic response, which is energetically costly and can disrupt systemic homeostasis. However, Wang found that TCS exposure reduced mRNA expression of Er and Ar in the hepatopancreas. Specifically, the gene expression of Er and Ar was significantly downregulated in the 0.04, 0.08, and 0.16 mg/L TCS treatment groups, with the most pronounced reduction in the 0.04 mg/L treatment [30]. This discrepancy highlights that the endocrine-disrupting effects of TCS are not uniform and are likely highly dependent on specific experimental conditions. The opposing results could be attributed to differences in the target organ (liver vs. hepatopancreas), the exposure concentration, or the duration of exposure, all of which can influence the transcriptional and feedback responses of hormone receptors. Moreover, TCS exposure led to a significant increase in Vtg mRNA and protein levels in the liver of experimental carp, indicating that TCS may have estrogen-like effects and can induce Vtg expression [58]. The TCS-induced increase in Vtg levels was correlated with elevated serum E2 levels, suggesting that TCS may indirectly induce Vtg production by increasing endogenous E2 levels [30]. Quantitative real-time PCR analysis of zebrafish tissue samples exposed to TCS revealed changes in the expression levels of HPG axis-related genes [16]. Specifically, alterations were observed in hypothalamic releasing hormones (*gnrh2*, *gnrh3*), the pituitary gonadotropin subunit (*fshβ*), and key steroid and gonadotropin receptors (*erα*, *erβ*, *lhr*, *fshr*), demonstrating a multi-level disruption of the reproductive endocrine system. Transcriptome sequencing analysis indicated that TCS exposure affects sex hormone biosynthesis pathways, including steroid biosynthesis and steroid hormone biosynthesis pathways [16]. This broad dysregulation at the genomic level provides a mechanistic basis for the observed physiological effects of TCS, namely disrupted gonadal maturation and ultimately, reduced fertility.

Beyond its well-characterized effects on the reproductive (HPG) axis, TCS also poses a significant threat to thyroid homeostasis. Due to its structural similarity to thyroid hormones, TCS (3–300 μg/L) may disrupt the HPT axis function, affecting the synthesis and secretion of T3 and T4, thereby interfering with fish morphology, physiology, and metamorphic development [66,102–104]. TCS exposure led to an increase in follicular area in zebrafish thyroid tissue, accompanied by hyperplasia, nuclear hypertrophy, and angiogenesis. Morphological changes in thyroid follicular epithelial cells, including increased nuclear area and cell height, as well as thickening of the follicular cell layer, are classic histological indicators of impaired thyroid function. These alterations represent a compensatory hyperplastic response by the gland, which is attempting to overcome a blockade in hormone synthesis (e.g., inhibition of thyroperoxidase) and thus directly suggest that TCS acts as a thyroid-inhibiting agent [66,102,105]. It was found that TCS exposure (300 μg/L, embryos from 30 min to for up to 7 or 14 days post fertilization) reduced thyroid hormone levels in zebrafish, including total triiodothyronine, total thyroxine, free triiodothyronine, and free thyroxine [66]. Simultaneously, TCS exposure downregulated the expression of HPT axis-related genes, further disrupting thyroid hormone function. This includes genes essential for pituitary stimulation (*tshβ*), iodide uptake (*nis*), hormone synthesis (*tpo*, *tg*), and cellular signaling (*thrα*/*β*), representing a multi-level attack on the HPT axis that severely affects metamorphosis and development [103–105]. TCS exposure also inhibited the expression of the insulin-like growth factor-1 (IGF-1) gene, a key regulator of somatic growth, which may further impair fish growth and development [20].

Serum biochemical and metabolomic analyses revealed that TCS exposure significantly disrupts metabolic homeostasis and tissue function in fish embryos. Specifically, increases in serum glucose, aspartate aminotransferase and alanine aminotransferase, along with decreases in total protein and albumin, indicate metabolic disorders and tissue damage [38]. Meanwhile, decreased oxygen partial pressure and increased carbon dioxide partial pressure reflect impaired respiratory function and acid-base balance disturbances [17]. At the metabolite level, significant increases in urea, citric acid, D-glucose, D-galactose, stearic acid, L-proline, phenylalanine, and L-glutamate indicate disruptions in nitrogen metabolism, energy metabolism, and protein synthesis [106,107]. Elevated urea levels suggest abnormal nitrogen metabolism or impaired kidney function, while increased citric acid reflects disturbances in the TCA cycle, potentially affecting energy supply. The upregulation of key metabolic enzymes such as alanine aminotransferase and glutamate dehydrogenase in embryos indicates enhanced amino acid metabolism to compensate for energy stress induced by TCS. Increased monosaccharide levels suggest elevated energy demand or disrupted metabolic pathways, while increased amino acids may be associated with adaptive regulation of protein synthesis, neurotransmission, and antioxidant defense. Additionally, upregulation of glucose-6-phosphate dehydrogenase suggests enhanced pentose phosphate pathway activity to produce more NADPH in response to oxidative stress, while downregulation of phosphoglucomutase indicates inhibition of gluconeogenesis and glycolysis. These changes demonstrate that TCS exposure disrupts metabolic pathways at multiple levels, leading to comprehensive stress on energy supply, nitrogen metabolism, and antioxidant defense in embryos [106,107].

5. Reproductive Toxicity

In the reproductive system, a growing body of evidence demonstrates that TCS exposure inflicts pronounced damage on the reproductive organs of both sexes. In females, TCS induces ovarian pathologies such as ooplasmic atrophy, accumulation of proteinaceous fluid in the interstitial stroma, altered oocyte morphology, and structural abnormalities of the corona radiata, collectively impairing oocyte maturation and ovulation [89]. In males, TCS exposure not only elevates the gonadosomatic index in zebrafish [14] but also elicits histological lesions—including seminiferous-tubule fusion, hypertrophy of spermatogonia and Leydig cells, intratubular edema, and karyolysis of spermatogonia—that ultimately compromise spermatogenesis and sperm quality [14,89].

Beyond direct gonadal injury, TCS markedly perturbs sexual differentiation and reproductive function. TCS up-regulates male-promoting genes (e.g., *sox9a*, *dmrt1a*, and *amh*) while down-regulating female-promoting genes (e.g., *wnt4a*, *cyp19a1b*, *cyp19a1a*, and *vtg2*), shifting sex ratios even at environmentally relevant concentrations (2 μg/L) [21]. Male mosquitofish exposed to 350 nM TCS exhibit a pronounced upregulation of hepatic Vtg mRNA, indicating estrogen-mimetic activity that drives feminization and skews population sex ratios [101]. This apparent contradiction, where TCS can induce both masculinizing and feminizing effects, is a recognized feature of endocrine-disrupting chemicals. The divergent outcomes likely result from differences in species sensitivity, exposure concentration, and the specific biological pathways affected. This duality highlights that TCS is a multifaceted endocrine disruptor, with its net effect being highly context-dependent. Concomitant histological deterioration of ovaries and testes reduces gametogenesis and reproductive output. A decline in the gonadosomatic index following TCS treatment reflects gonadal underdevelopment and impaired fecundity [14]. Sperm counts in TCS-treated males are significantly reduced [14,101], whereas TCS-exposed females show markedly lower daily fecundity (eggs per female per day), underscoring compromised oviposition [73]. Importantly, TCS effects are not confined to exposed individuals; epigenetic or cytoplasmic factors transmitted via oocytes or sperm can precipitate transgenerational reproductive dysfunction in offspring [23].

6. Neurotoxicity

In the nervous system, TCS exposure has been shown to markedly disrupt neurodevelopment and function in larval zebrafish. Morphological analyses reveal that TCS treatment reduces neuron number, induces vascular malformations, impairs hair-cell development, and perturbs neuromast formation along the lateral line, accompanied by an increased V-shaped body angle and discontinuous neuromuscular junctions, ultimately leading to aberrant motor and sensory behaviors [61,93,108]. Behavioral assays further demonstrate that TCS exposure significantly alters spontaneous locomotion and the response patterns to light–dark and acoustic stimuli [109].

Mechanistic studies at the molecular level indicate that TCS dysregulates the expression of key neurodevelopmental genes—including *olig2*, *gfap*, *mbp*, *bdnf*, *ngn1*, *syn2a*, and *cd36*—that govern neuronal maturation, myelination, and synaptic function [109]. For instance, down-regulation of myelin basic protein (*mbp*) can slow nerve impulse conduction, whereas aberrant synapsin IIa (*syn2a*) expression compromises synapse formation and neurotransmitter release [60]. Moreover, TCS interferes with cholinergic and dopaminergic signaling by inhibiting acetylcholinesterase (AChE) and dopamine activity, promoting neuronal apoptosis and diencephalic injury and thereby disrupting neurotransmission and behaviors [69,79,92]. Evidence suggests that TCS can directly bind to the catalytic site of AChE, exerting competitive inhibition [61]. At the post-transcriptional level, TCS modulates microRNA expression to further impair neurodevelopment. Exposure markedly up-regulates miR-137, thereby disturbing neural stem-cell proliferation, differentiation, and synaptic maturation [65], while altered miR-219 expression impedes oligodendrocyte precursor cell differentiation into mature oligodendrocytes, resulting in defective myelination [110].

7. Hepatotoxicity

In the liver, TCS exposure likewise elicits pronounced toxic effects. Chronic exposure significantly elevates liver weight and the hepatosomatic index in adult zebrafish, implying hepatic injury or steatosis [6,58]. Histological examination reveals cytoplasmic vacuolation, hepatocellular dissociation, and inflammatory-cell infiltration in TCS-treated livers, indicating disruption of hepatic architecture and function [15]. Cytoplasmic vacuolation is thought to reflect compromised membrane integrity and impaired energy metabolism [51].

At the metabolic level, TCS exposure leads to massive hepatic lipid accumulation and is accompanied by elevated serum triglycerides, total cholesterol, and low-density lipoprotein cholesterol, together with decreased high-density lipoprotein cholesterol, denoting disorders of lipid metabolism [15]. Mechanistic studies show that TCS binds avidly to PPARγ and markedly up-regulates PPARγ and its downstream targets (e.g., *scd* and *fads2*) while down-regulating *lpl*, thereby disrupting lipid metabolism via PPARγ pathway activation [6,15]. Additionally, TCS exposure reduces miR-101a expression, whereas miR-101a overexpression mitigates hepatic injury and metabolic imbalance, suggesting a protective role for miR-101a [6].

Inflammatory responses also contribute to TCS-induced hepatotoxicity. Exposure increases the pro-inflammatory cytokine TNF-α and promotes formation of Mallory–Denk bodies and intracellular hyaline bodies in hepatocytes; these protein aggregates are classic histopathological markers of hepatic injury and inflammation [51]. Moreover, TCS causes excessive hepatic accumulation of vitellogenin, leading to hepatocyte hypertrophy and further compromising liver function [15]. At the epigenetic level, TCS exposure markedly decreases global m6A-RNA methylation in zebrafish liver. This reduction is associated with enhanced demethylase FTO activity, suppressed expression of the methyltransferase METTL3, and dysregulated expression of the m6A reader protein YTHDF1. Collectively, this disruption of the m6A machinery leads to widespread instability and dysregulation of messenger RNAs involved in critical liver functions, such as metabolism and detoxification. This indicates that TCS exacerbates liver injury by perturbing m6A-mediated post-transcriptional regulation [15].

8. Summary and Perspectives

TCS, a widely used antimicrobial agent, persists in aquatic environments and bioaccumulates in organisms, raising significant concerns regarding its ecological and human health risks. Emerging research indicates that beyond acute toxicity, TCS induces a range of complex biological effects even under chronic, low-dose exposure conditions. It disrupts mitochondrial function, resulting in excessive ROS formation, oxidative stress, and apoptosis. Concurrently, it interferes with multiple signaling pathways, leading to energy-metabolic imbalances and DNA damage. At the organismal level, TCS exposure has severe multigenerational impacts on fish. It markedly impairs embryonic development, reducing hatchability and larval survival while inducing malformations and growth restrictions. Furthermore, TCS alters sex ratios and diminishes reproductive capacity in adulthood. Critically, these effects are dose- and time-dependent; even environmentally relevant concentrations can elicit significant developmental and reproductive toxicity, underscoring that the risks of TCS extend well beyond those identified by traditional acute toxicity evaluations.

The adverse impacts of TCS are cross-hierarchical and multi-systemic. In the liver, it disrupts hepatic structure and function, precipitating lipid metabolic disorders and inflammatory responses. Within the immune system, TCS disturbs the balance of pro- and anti-inflammatory cytokines, thereby weakening host defenses. In the nervous system, it induces neuronal apoptosis, impairs myelination, and alters behavioral patterns. Finally, for reproductive health, TCS causes histopathological damage in gonadal tissues, modulates genes involved in sex determination, and ultimately reduces fecundity and gamete quality. Notably, growing evidence suggests that TCS acts not only through established pathways of endocrine disruption and oxidative stress, but also via epigenetic mechanisms—including DNA methylation, miRNA regulation, and m6A RNA modification—which can reprogram gene expression. Such epigenetic alterations may lead to functional impairments in exposed individuals and can be transmitted to subsequent generations via gametes, resulting in transgenerational phenotypic effects that significantly amplify ecological risks. Critically, this potential for heritable, epigenetic damage challenges current risk assessment frameworks, which largely focus on direct, acute toxicity, and underscores the urgent need for regulatory policies to account for these long-term, multigenerational hazards.

From an ecological standpoint, the persistence and bioaccumulation of TCS in water, sediment, and biota facilitate its trophic transfer through food webs, threatening the sustainability of fish populations. This biomagnification poses a direct human health risk, as TCS enters the human food chain primarily through the consumption of contaminated fish and seafood, potentially exposing higher trophic levels to concerning concentrations. For human health, TCS exposure has been associated with thyroid dysfunction, metabolic disorders, and reduced fertility. Its multi-route exposure via personal care products, drinking water, and food constitutes a considerable public health concern. These findings represent a considerable public health concern, necessitating stricter regulatory measures such as limiting the use of TCS in consumer products and enhancing environmental monitoring to protect ecosystem and human health.

Despite recent advances, critical knowledge gaps remain. The current understanding of TCS toxicity, while substantial, is constrained by several critical gaps that limit a complete ecological risk assessment. Foremost among these is the scarcity of long-term, multigenerational studies that reflect realistic, low-dose exposure scenarios, which is crucial for uncovering chronic health impacts and true population-level threats. Furthermore, the mechanistic basis of TCS action remains partially elucidated, particularly concerning its epigenetic influences and the potential for these changes to cause stable, transgenerational effects. Finally, the environmental reality of TCS co-occurring with a mixture of other pollutants is poorly represented in toxicological studies, leaving a significant blind spot regarding potential synergistic interactions that could amplify its toxicity. Addressing these interconnected gaps—through integrated multi-omics approaches and multi-stressor experimental designs—is essential for developing accurate risk models and effective regulatory policies.

Moving forward, research on TCS should pursue several key directions: (1) Elucidate core mechanisms: mechanistic studies must further dissect mitochondrial dysfunction, pathway interactions, and the integrated impacts on the immune and nervous systems to identify critical molecular targets. (2) Decipher epigenetic regulation: epigenetic analyses provide a promising framework for understanding chronic and transgenerational risks and warrant integrated transcriptomic, methylomic, and non-coding RNA investigations. (3) Enhance ecological relevance: ecological and environmental studies should prioritize long-term, low-dose, and multi-pollutant exposure regimes, combining laboratory data with field monitoring to improve risk assessment accuracy. In summary, TCS exemplifies how a potential endocrine disruptor can elicit effects spanning from acute toxicity to chronic, systemic, and transgenerational impairments, highlighting the profound hazards pollutants pose across multiple levels of biological organization. Effectively addressing this complex threat demands a deeply interdisciplinary approach, integrating the expertise of molecular toxicologists, ecologists, geneticists, and environmental modelers. This collaborative framework is essential to leverage advances in detection technologies, omics platforms, and systems biology, ultimately elucidating the complete mechanistic and risk profile of TCS and providing the robust scientific foundation needed to protect environmental and human health.

Author Contributions: J.W. and N.M.: Writing—original draft, Methodology, Investigation, Data curation. G.M., X.Q., J.Z., X.Y. and J.G.: Data curation, Visualization, Writing—review and editing. Z.S.: Writing—original draft, Supervision, Data curation, Conceptualization. All authors have read and agreed to the published version of the manuscript.

Funding: This work was supported by Research on Industrial Innovation Technology for Guangdong Modern Marine Ranching (2024-MRI-001).

Institutional Review Board Statement: Not applicable.

Informed Consent Statement: Not applicable.

Data Availability Statement: No new data were created or analyzed in this study.

Conflicts of Interest: The authors declare no conflicts of interest.

References

1. Narrowe, A.B.; Albuthi-Lantz, M.; Smith, E.P.; Bower, K.J.; Roane, T.M.; Vajda, A.M.; Miller, C.S. Perturbation and restoration of the fathead minnow gut microbiome after low-level triclosan exposure. *Microbiome* **2015**, *3*, 1–18. [CrossRef]
2. Hemalatha, D.; Nataraj, B.; Rangasamy, B.; Shobana, C.; Ramesh, M. DNA damage and physiological responses in an Indian major carp *Labeo rohita* exposed to an antimicrobial agent triclosan. *Fish Physiol. Biochem.* **2019**, *45*, 1463–1484. [CrossRef]
3. Hemalatha, D.; Rangasamy, B.; Nataraj, B.; Ramesh, M. Assessment of triclosan impact on enzymatic biomarkers in an Indian major carp, *Catla catla*. *J. Basic Appl. Zool.* **2019**, *80*, 1–8. [CrossRef]
4. Chen, X.; Mou, L.; Qu, J.; Wu, L.; Liu, C. Adverse effects of triclosan exposure on health and potential molecular mechanisms. *Sci. Total Environ.* **2023**, *879*, 163068. [CrossRef]
5. Luo, N.; Chen, J.; Chen, X.; Wang, M.; Niu, X.; Chen, G.; An, T. Toxicity evolution of triclosan during environmental transformation and human metabolism: Misgivings in the post-pandemic era. *Environ. Int.* **2024**, *190*, 108927. [CrossRef]

6. Yan, J.; Li, J.; Wang, Y.; Song, J.; Ni, A.; Fang, L.; Xi, M.; Qian, Q.; Wang, Z.; Wang, H. Deciphering the molecular mediators of triclosan-induced lipid accumulation: Intervention via short-chain fatty acids and miR-101a. *Environ. Pollut.* **2024**, *343*, 123153. [CrossRef]
7. Yao, L.; Liu, Y.H.; Zhou, X.; Yang, J.H.; Zhao, J.L.; Chen, Z.Y. Uptake, tissue distribution, and biotransformation pattern of triclosan in tilapia exposed to environmentally-relevant concentrations. *Sci. Total Environ.* **2024**, *922*, 171270. [CrossRef]
8. Quan, B.; Li, X.; Zhang, H.; Zhang, C.; Ming, Y.; Huang, Y.; Xi, Y.; Weihua, X.; Yunguo, L.; Tang, Y. Technology and principle of removing triclosan from aqueous media: A review. *Chem. Eng. J.* **2019**, *378*, 122185. [CrossRef]
9. Dar, O.I.; Aslam, R.; Pan, D.; Sharma, S.; Andotra, M.; Kaur, A.; Faggio, C. Source, bioaccumulation, degradability and toxicity of triclosan in aquatic environments: A review. *Environ. Technol. Innov.* **2022**, *25*, 102122. [CrossRef]
10. Sun, X.; Guo, Y.; Luo, D.; Xu, Z.; Liu, Z. Tradeoffs between hygiene behaviors and triclosan loads from rivers to coastal seas in the post COVID-19 era. *Mar. Pollut. Bull.* **2024**, *203*, 116507. [CrossRef]
11. Wan, X.; Wang, H.; Qian, Q.; Yan, J. *MiR*-133b as a crucial regulator of TCS-induced cardiotoxicity via activating β-adrenergic receptor signaling pathway in zebrafish embryos. *Environ. Pollut.* **2023**, *334*, 122199. [CrossRef] [PubMed]
12. Oliveira, R.; Domingues, I.; Koppe Grisolia, C.; Soares, A.M. Effects of triclosan on zebrafish early-life stages and adults. *Environ. Sci. Pollut. Res.* **2009**, *16*, 679–688. [CrossRef]
13. Wang, W.; Wang, D.; Li, X.; Ai, W.; Wang, X.; Wang, H. Toxicity mechanisms regulating bone differentiation and development defects following abnormal expressions of miR-30c targeted by triclosan in zebrafish. *Sci. Total Environ.* **2022**, *850*, 158040. [CrossRef]
14. Chokki Veettil, P.; Nikarthil Sidhick, J.; Kavungal Abdulkhader, S.; Ms, S.P.; Kumari Chidambaran, C. Triclosan, an antimicrobial drug, induced reproductive impairment in the freshwater fish, *Anabas testudineus* (Bloch, 1792). *Toxicol. Ind. Health* **2024**, *40*, 254–271. [CrossRef]
15. Li, J.; Fang, L.; Xi, M.; Ni, A.; Qian, Q.; Wang, Z.; Yan, J. Toxic effects of triclosan on hepatic and intestinal lipid accumulation in zebrafish via regulation of m6A-RNA methylation. *Aquat. Toxicol.* **2024**, *269*, 106884. [CrossRef]
16. Wang, F. Reproductive endocrine disruption effect and mechanism in male zebrafish after life cycle exposure to environmental relevant triclosan. *Aquat. Toxicol.* **2024**, *270*, 106899. [CrossRef]
17. Dar, O.I.; Vinothkanna, A.; Aslam, B.; Furkh, A.; Sharma, S.; Kaur, A.; Gao, Y.-A.; Jia, A.Q. Dynamic alterations in physiological and biochemical indicators of *Cirrhinus mrigala* hatchlings: A sublethal exposure of triclosan. *Sci. Total Environ.* **2024**, *924*, 171701. [CrossRef]
18. Wang, F.; Liu, F. Mechanism-based understanding of the potential cellular targets of triclosan in zebrafish larvae. *Environ. Toxicol. Pharmacol.* **2023**, *102*, 104255. [CrossRef]
19. Lin, D.; Hamilton, C.; Hobbs, J.; Miller, E.; Sutton, R. Triclosan and methyl triclosan in prey fish in a wastewater-influenced estuary. *Environ. Toxicol. Chem.* **2023**, *42*, 620–627. [CrossRef]
20. Horie, Y. Environmentally relevant concentrations of triclosan induce lethality and disrupt thyroid hormone activity in zebrafish (*Danio rerio*). *Environ. Toxicol. Pharmacol.* **2023**, *100*, 104151. [CrossRef]
21. Wang, F.; Liu, H.; Liu, F. Analysis of the effect of triclosan on gonadal differentiation of zebrafish based on metabolome. *Chemosphere* **2023**, *331*, 138856. [CrossRef]
22. Jin, Y.; Yuan, T.; Li, J.; Shen, Z.; Tian, Y. Occurrence, health risk assessment and water quality criteria derivation of six personal care products (PCPs) in Huangpu River, China. *Environ. Monit. Assess.* **2022**, *194*, 577. [CrossRef]
23. Qiao, Y.; He, J.; Han, P.; Qu, J.; Wang, X.; Wang, J. Long-term exposure to environmental relevant triclosan induces reproductive toxicity on adult zebrafish and its potential mechanism. *Sci. Total Environ.* **2022**, *826*, 154026. [CrossRef]
24. Gopal, C.M.; Bhat, K.; Ramaswamy, B.R.; Kumar, V.; Singhal, R.K.; Basu, H.; Udayashankar, H.N.; Vasantharaju, S.G.; Praveenkumarreddy, Y.; Shailesh; et al. Seasonal occurrence and risk assessment of pharmaceutical and personal care products in Bengaluru rivers and lakes, India. *J. Environ. Chem. Eng.* **2021**, *9*, 105610. [CrossRef]
25. Sharma, S.; Dar, O.I.; Singh, K.; Thakur, S.; Kesavan, A.K.; Kaur, A. Genomic markers for the biological responses of Triclosan stressed hatchlings of *Labeo rohita*. *Environ. Sci. Pollut. Res.* **2021**, *28*, 67370–67384. [CrossRef]
26. Bao, S.; He, C.; Ku, P.; Xie, M.; Lin, J.; Lu, S.; Nie, X. Effects of triclosan on the RedoximiRs/Sirtuin/Nrf2/ARE signaling pathway in mosquitofish (*Gambusia affinis*). *Aquat. Toxicol.* **2021**, *230*, 105679. [CrossRef]
27. Das Sarkar, S.; Nag, S.K.; Kumari, K.; Saha, K.; Bandyopadhyay, S.; Aftabuddin, M.; Das, B.K. Occurrence and safety evaluation of antimicrobial compounds triclosan and triclocarban in water and fishes of the multitrophic niche of River Torsa, India. *Arch. Environ. Contam. Toxicol.* **2020**, *79*, 488–499. [CrossRef] [PubMed]
28. Olaniyan, L.W.; Okoh, A.I. Determination and ecological risk assessment of two endocrine disruptors from River Buffalo, South Africa. *Environ. Monit. Assess.* **2020**, *192*, 1–15. [CrossRef] [PubMed]
29. Juksu, K.; Zhao, J.L.; Liu, Y.S.; Yao, L.; Sarin, C.; Sreesai, S.; Klomjek, P.; Jiang, Y.-X.; Ying, G.G. Occurrence, fate and risk assessment of biocides in wastewater treatment plants and aquatic environments in Thailand. *Sci. Total Environ.* **2019**, *690*, 1110–1119. [CrossRef]

30. Wang, F.; Liu, F.; Chen, W.; Xu, R.; Wang, W. Effects of triclosan (TCS) on hormonal balance and genes of hypothalamus-pituitary-gonad axis of juvenile male Yellow River carp (*Cyprinus carpio*). *Chemosphere* **2018**, *193*, 695–701. [CrossRef] [PubMed]
31. Silva, D.C.; Araújo, C.V.; López-Doval, J.C.; Neto, M.B.; Silva, F.T.; Paiva, T.C.; Pompêo, M.L. Potential effects of triclosan on spatial displacement and local population decline of the fish *Poecilia reticulata* using a non-forced system. *Chemosphere* **2017**, *184*, 329–336. [CrossRef]
32. Ho, J.C.; Hsiao, C.D.; Kawakami, K.; William, K.F. Triclosan (TCS) exposure impairs lipid metabolism in zebrafish embryos. *Aquat. Toxicol.* **2016**, *173*, 29–35. [CrossRef]
33. Arya, G. Endocrine Disrupting Chemicals in Water, Sediments and Fish Observed in Urban Tributaries of the Freshwater Tidal Potomac River: Occurrence, Bioaccumulation and Tissue Distribution. Ph.D. Thesis, George Mason University, Fairfax, VA, USA, 2015.
34. Inam, E.; Offiong, N.A.; Kang, S.; Yang, P.; Essien, J. Assessment of the occurrence and risks of emerging organic pollutants (EOPs) in Ikpa River Basin freshwater ecosystem, Niger Delta-Nigeria. *Bull. Environ. Contam. Toxicol.* **2015**, *95*, 624–631. [CrossRef]
35. Hedrick-Hopper, T.L.; Koster, L.P.; Diamond, S.L. Accumulation of triclosan from diet and its neuroendocrine effects in Atlantic croaker (*Micropogonias undulatus*) under two temperature regimes. *Mar. Environ. Res.* **2015**, *112*, 52–60. [CrossRef]
36. Macedo, S.; Torres, T.; Santos, M.M. Methyl-triclosan and triclosan impact embryonic development of *Danio rerio* and *Paracentrotus lividus*. *Ecotoxicology* **2017**, *26*, 482–489. [CrossRef] [PubMed]
37. Sharma, S.; Dar, O.I.; Thakur, S.; Kesavan, A.K.; Kaur, A. Environmentally relevant concentrations of triclosan cause transcriptomic and biomolecular alterations in the hatchlings of *Labeo rohita*. *Environ. Toxicol. Pharmacol.* **2022**, *96*, 104004. [CrossRef]
38. Kantal, D.; Kumar, S.; Shukla, S.P.; Karmakar, S.; Jha, A.K.; Singh, A.B.; Kumar, K. Chronic toxicity of sediment-bound triclosan on freshwater walking catfish *Clarias magur*: Organ level accumulation and selected enzyme biomarker responses. *Environ. Pollut.* **2024**, *351*, 124108. [CrossRef]
39. Boreen, A.L.; Arnold, W.A.; McNeill, K. Photodegradation of pharmaceuticals in the aquatic environment: A review. *Aquat. Sci.* **2003**, *65*, 320–341. [CrossRef]
40. Latch, D.E.; Packer, J.L.; Stender, B.L.; VanOverbeke, J.; Arnold, W.A.; McNeill, K. Aqueous photochemistry of triclosan: Formation of 2,4-dichlorophenol, 2,8-dichlorodibenzo-*p*-dioxin, and oligomerization products. *Environ. Toxicol. Chem.* **2005**, *24*, 517–525. [CrossRef]
41. Greim, L.W.W.H. The toxicity of brominated and mixed-halogenated dibenzo-p-dioxins and dibenzofurans: An overview. *J. Toxicol. Environ. Health Part A* **1997**, *50*, 195–216. [CrossRef]
42. Murugesan, K.; Chang, Y.Y.; Kim, Y.M.; Jeon, J.R.; Kim, E.J.; Chang, Y.S. Enhanced transformation of triclosan by laccase in the presence of redox mediators. *Water Res.* **2010**, *44*, 298–308. [CrossRef]
43. Hundt, K.A.I.; Martin, D.; Hammer, E.; Jonas, U.; Kindermann, M.K.; Schauer, F. Transformation of triclosan by *Trametes versicolor* and *Pycnoporus cinnabarinus*. *Appl. Environ. Microbiol.* **2000**, *66*, 4157–4160. [CrossRef]
44. Balakrishnan, P.; Mohan, S. Treatment of triclosan through enhanced microbial biodegradation. *J. Hazard. Mater.* **2021**, *420*, 126430. [CrossRef]
45. Dar, O.I.; Sharma, S.; Singh, K.; Sharma, A.; Bhardwaj, R.; Kaur, A. Biochemical markers for prolongation of the acute stress of triclosan in the early life stages of four food fishes. *Chemosphere* **2020**, *247*, 125914. [CrossRef]
46. Dar, O.I.; Sharma, S.; Singh, K.; Sharma, A.; Bhardwaj, R.; Kaur, A. Biomarkers for the toxicity of sublethal concentrations of triclosan to the early life stages of carps. *Sci. Rep.* **2020**, *10*, 17322. [CrossRef] [PubMed]
47. Klaiber, M.; Moreno-Gordaliza, E.; Gómez-Gómez, M.M.; Marazuela, M.D. Human intake assessment of triclosan associated with the daily use of polypropylene-made antimicrobial food packaging. *Food Chem.* **2024**, *451*, 139475. [CrossRef]
48. Morgan, M.K.; Clifton, M.S. Dietary exposures and intake doses to bisphenol A and triclosan in 188 duplicate-single solid food items consumed by US adults. *Int. J. Environ. Res. Public Health* **2021**, *18*, 4387. [CrossRef]
49. Zhang, D.; Lu, S. A holistic review on triclosan and triclocarban exposure: Epidemiological outcomes, antibiotic resistance, and health risk assessment. *Sci. Total Environ.* **2023**, *872*, 162114. [CrossRef]
50. Zhao, C.; Xie, R.; Qian, Q.; Yan, J.; Wang, H.; Wang, X. Triclosan induced zebrafish immunotoxicity by targeting miR-19a and its gene socs3b to activate IL-6/STAT3 signaling pathway. *Sci. Total Environ.* **2022**, *815*, 152916. [CrossRef] [PubMed]
51. Song, X.; Wang, X.; Li, X.; Yan, X.; Liang, Y.; Huang, Y.; Huang, L.; Zeng, H. Histopathology and transcriptome reveals the tissue-specific hepatotoxicity and gills injury in mosquitofish (*Gambusia affinis*) induced by sublethal concentration of triclosan. *Ecotoxicol. Environ. Saf.* **2021**, *220*, 112325. [CrossRef]
52. Jimoh, R.O.; Sogbanmu, T.O. Sublethal and environmentally relevant concentrations of triclosan and triclocarban induce histological, genotoxic, and embryotoxic effects in *Clarias gariepinus* (Burchell, 1822). *Environ. Sci. Pollut. Res.* **2021**, *28*, 31071–31083. [CrossRef]
53. Fu, L.; Sun, Y.; Zhou, J.; Li, H.; Liang, S.X. Parabens, triclosan and bisphenol A in surface waters and sediments of Baiyang Lake, China: Occurrence, distribution, and potential risk assessment. *Toxics* **2023**, *12*, 31. [CrossRef]

54. Mihaich, E.; Capdevielle, M.; Urbach-Ross, D.; Gallagher, S.; Wolf, J. Medaka (*Oryzias latipes*) multigeneration test with triclosan. *Environ. Toxicol. Chem.* **2019**, *38*, 1770–1783. [CrossRef]
55. Saemi-Komsari, M.; Pashaei, R.; Abbasi, S.; Esmaeili, H.R.; Dzingelevičienė, R.; Hadavand, B.S.; Kalako, M.P.; Szultka-Mlynska, M.; Gadzała-Kopciuch, R.; Buszewski, B.; et al. Accumulation of polystyrene nanoplastics and triclosan by a model tooth-carp fish, *Aphaniops hormuzensis* (Teleostei: Aphaniidae). *Environ. Pollut.* **2023**, *333*, 121997. [CrossRef]
56. Wang, F.; Guo, X.; Chen, W.; Sun, Y.; Fan, C. Effects of triclosan on hormones and reproductive axis in female Yellow River carp (*Cyprinus carpio*): Potential mechanisms underlying estrogen effect. *Toxicol. Appl. Pharmacol.* **2017**, *336*, 49–54. [CrossRef]
57. Wang, X.N.; Liu, Z.T.; Yan, Z.G.; Zhang, C.; Wang, W.L.; Zhou, J.L.; Pei, S.W. Development of aquatic life criteria for triclosan and comparison of the sensitivity between native and non-native species. *J. Hazard. Mater.* **2013**, *260*, 1017–1022. [CrossRef]
58. Adhikari, A.; Das, B.K.; Ganguly, S.; Nag, S.K.; Sadhukhan, D.; Raut, S.S. Emerging contaminant triclosan incites endocrine disruption, reproductive impairments and oxidative stress in the commercially important carp, Catla (*Labeo catla*): An insight through molecular, histopathological and bioinformatic approach. *Comp. Biochem. Physiol. Part C Toxicol. Pharmacol.* **2023**, *268*, 109605. [CrossRef]
59. Kraft, M.; Gölz, L.; Rinderknecht, M.; Koegst, J.; Braunbeck, T.; Baumann, L. Developmental exposure to triclosan and benzophenone-2 causes morphological alterations in zebrafish (*Danio rerio*) thyroid follicles and eyes. *Environ. Sci. Pollut. Res.* **2023**, *30*, 33711–33724. [CrossRef]
60. Pullaguri, N.; Nema, S.; Bhargava, Y.; Bhargava, A. Triclosan alters adult zebrafish behavior and targets acetylcholinesterase activity and expression. *Environ. Toxicol. Pharmacol.* **2020**, *75*, 103311. [CrossRef]
61. Pullaguri, N.; Grover, P.; Abhishek, S.; Rajakumara, E.; Bhargava, Y.; Bhargava, A. Triclosan affects motor function in zebrafish larva by inhibiting ache and syn2a genes. *Chemosphere* **2021**, *266*, 128930. [CrossRef]
62. Dong, G.; Zhang, R.; Huang, H.; Lu, C.; Xia, Y.; Wang, X.; Du, G. Exploration of the developmental toxicity of TCS and PFOS to zebrafish embryos by whole-genome gene expression analyses. *Environ. Sci. Pollut. Res.* **2021**, *28*, 56032–56042. [CrossRef]
63. Zhou, S.; Chen, Q.; Di Paolo, C.; Shao, Y.; Hollert, H.; Seiler, T.B. Behavioral profile alterations in zebrafish larvae exposed to environmentally relevant concentrations of eight priority pharmaceuticals. *Sci. Total Environ.* **2019**, *664*, 89–98. [CrossRef]
64. Liu, J.; Sun, L.; Zhang, H.; Shi, M.; Dahlgren, R.A.; Wang, X.; Wang, H. Response mechanisms to joint exposure of triclosan and its chlorinated derivatives on zebrafish (*Danio rerio*) behavior. *Chemosphere* **2018**, *193*, 820–832. [CrossRef]
65. Liu, J.; Xiang, C.; Huang, W.; Mei, J.; Sun, L.; Ling, Y.; Wang, C.; Wang, X.; Dahlgren, R.A.; Wang, H. Neurotoxicological effects induced by up-regulation of miR-137 following triclosan exposure to zebrafish (*Danio rerio*). *Aquat. Toxicol.* **2019**, *206*, 176–185. [CrossRef]
66. Tang, N.; Fan, P.; Chen, L.; Yu, X.; Wang, W.; Wang, W.; Ouyang, F. The effect of early life exposure to triclosan on thyroid follicles and hormone levels in zebrafish. *Front. Endocrinol.* **2022**, *13*, 850231. [CrossRef]
67. Li, X.; Wang, W.; Wang, X.; Wang, H. Differential immunotoxicity effects of triclosan and triclocarban on larval zebrafish based on RNA-Seq and bioinformatics analysis. *Aquat. Toxicol.* **2023**, *262*, 106665. [CrossRef]
68. Qian, Q.; Pu, Q.; Li, X.; Liu, X.; Ni, A.; Han, X.; Wang, Z.; Wang, X.; Yan, J.; Wang, H. Acute/chronic triclosan exposure induces downregulation of m6A-RNA methylation modification via mettl3 suppression and elicits developmental and immune toxicity to zebrafish. *Chemosphere* **2024**, *352*, 141395. [CrossRef]
69. Gyimah, E.; Dong, X.; Qiu, W.; Zhang, Z.; Xu, H. Sublethal concentrations of triclosan elicited oxidative stress, DNA damage, and histological alterations in the liver and brain of adult zebrafish. *Environ. Sci. Pollut. Res.* **2020**, *27*, 17329–17338. [CrossRef]
70. Deepika, S.; Padmavathy, P.; Srinivasan, A.; Sugumar, G.; Jawahar, P. Effect of triclosan (TCS) on the protein content and associated histological changes on tilapia, *Oreochromis mossambicus* (Peters, 1852). *Environ. Sci. Pollut. Res.* **2021**, *28*, 59899–59907. [CrossRef]
71. Sager, E.; Scarcia, P.; Marino, D.; Mac Loughlin, T.; Rossi, A.; de La Torre, F. Oxidative stress responses after exposure to triclosan sublethal concentrations: An integrated biomarker approach with a native (*Corydoras paleatus*) and a model fish species (*Danio rerio*). *J. Toxicol. Environ. Health* **2022**, *85*, 291–306. [CrossRef]
72. Gomes, M.F.; de Paula, V.D.C.S.; Martins, L.R.R.; Garcia, J.R.E.; Yamamoto, F.Y.; de Freitas, A.M. Sublethal effects of triclosan and triclocarban at environmental concentrations in silver catfish (*Rhamdia quelen*) embryos. *Chemosphere* **2021**, *263*, 127985. [CrossRef]
73. Horie, Y.; Yamagishi, T.; Takahashi, H.; Iguchi, T.; Tatarazako, N. Effects of triclosan on Japanese medaka (*Oryzias latipes*) during embryo development, early life stage and reproduction. *J. Appl. Toxicol.* **2018**, *38*, 544–551. [CrossRef]
74. Ishibashi, H.; Matsumura, N.; Hirano, M.; Matsuoka, M.; Shiratsuchi, H.; Ishibashi, Y.; Takao, Y.; Arizono, K. Effects of triclosan on the early life stages and reproduction of medaka *Oryzias latipes* and induction of hepatic vitellogenin. *Aquat. Toxicol.* **2004**, *67*, 167–179. [CrossRef]
75. Falisse, E.; Voisin, A.S.; Silvestre, F. Impacts of triclosan exposure on zebrafish early-life stage: Toxicity and acclimation mechanisms. *Aquat. Toxicol.* **2017**, *189*, 97–107. [CrossRef]
76. Cui, L.; Fan, M.; Belanger, S.; Li, J.; Wang, X.; Fan, B.; Li, W.; Gao, X.; Chen, J.; Liu, Z. *Oryzias sinensis*, a new model organism in the application of eco-toxicity and water quality criteria (WQC). *Chemosphere* **2020**, *261*, 127813. [CrossRef]

77. Paul, T.; Shukla, S.P.; Kumar, K.; Poojary, N.; Kumar, S. Effect of temperature on triclosan toxicity in *Pangasianodon hypophthalmus* (Sauvage, 1878): Hematology, biochemistry and genotoxicity evaluation. *Sci. Total Environ.* **2019**, *668*, 104–114. [CrossRef]
78. Paul, T.; Kumar, S.; Shukla, S.P.; Pal, P.; Kumar, K.; Poojary, N.; Biswal, A.; Mishra, A. A multi-biomarker approach using integrated biomarker response to assess the effect of pH on triclosan toxicity in *Pangasianodon hypophthalmus* (Sauvage, 1878). *Environ. Pollut.* **2020**, *260*, 114001. [CrossRef]
79. Sahu, V.K.; Karmakar, S.; Kumar, S.; Shukla, S.P.; Kumar, K. Triclosan toxicity alters behavioral and hematological parameters and vital antioxidant and neurological enzymes in *Pangasianodon hypophthalmus* (Sauvage, 1878). *Aquat. Toxicol.* **2018**, *202*, 145–152. [CrossRef]
80. Bera, K.K.; Kumar, S.; Paul, T.; Prasad, K.P.; Shukla, S.P.; Kumar, K. Triclosan induces immunosuppression and reduces survivability of striped catfish *Pangasianodon hypophthalmus* during the challenge to a fish pathogenic bacterium *Edwardsiella tarda*. *Environ. Res.* **2020**, *186*, 109575. [CrossRef]
81. Nassef, M.; Matsumoto, S.; Seki, M.; Khalil, F.; Kang, I.J.; Shimasaki, Y.; Oshima, Y.; Honjo, T. Acute effects of triclosan, diclofenac and carbamazepine on feeding performance of Japanese medaka fish (*Oryzias latipes*). *Chemosphere* **2010**, *80*, 1095–1100. [CrossRef]
82. Escarrone, A.L.V.; Caldas, S.S.; Primel, E.G.; Martins, S.E.; Nery, L.E.M. Uptake, tissue distribution and depuration of triclosan in the guppy Poecilia vivipara acclimated to freshwater. *Sci. Total Environ.* **2016**, *560*, 218–224. [CrossRef]
83. Liang, X.; Nie, X.; Ying, G.; An, T.; Li, K. Assessment of toxic effects of triclosan on the swordtail fish (*Xiphophorus helleri*) by a multi-biomarker approach. *Chemosphere* **2013**, *90*, 1281–1288. [CrossRef]
84. Iannetta, A.; Caioni, G.; Di Vito, V.; Benedetti, E.; Perugini, M.; Merola, C. Developmental toxicity induced by triclosan exposure in zebrafish embryos. *Birth Defects Res.* **2022**, *114*, 175–183. [CrossRef]
85. Kumar, S.; Paul, T.; Shukla, S.P.; Kumar, K.; Karmakar, S.; Bera, K.K. Biomarkers-based assessment of triclosan toxicity in aquatic environment: A mechanistic review. *Environ. Pollut.* **2021**, *286*, 117569. [CrossRef]
86. Araújo, M.J.; Quintaneiro, C.; Soares, A.M.V.M.; Monteiro, M.S. Effects of triclosan on early development of *Solea senegalensis*: From biochemical to individual level. *Chemosphere* **2019**, *235*, 885–899. [CrossRef] [PubMed]
87. Huang, L.; Zhang, W.; Tong, D.; Lu, L.; Zhou, W.; Tian, D.; Liu, G.; Shi, W. Triclosan and triclocarban weaken the olfactory capacity of goldfish by constraining odorant recognition, disrupting olfactory signal transduction, and disturbing olfactory information processing. *Water Res.* **2023**, *233*, 119736. [CrossRef] [PubMed]
88. Liu, F.; Zhang, Y.; Wang, F. Environmental relevant concentrations of triclosan affected developmental toxicity, oxidative stress, and apoptosis in zebrafish embryos. *Environ. Toxicol.* **2022**, *37*, 848–857. [CrossRef] [PubMed]
89. Akbulut, C. Histopathological and apoptotic examination of zebrafish (*Danio rerio*) gonads exposed to triclosan. *Arch. Biol. Sci.* **2021**, *73*, 465–472. [CrossRef]
90. Liu, M.; Ai, W.; Sun, L.; Fang, F.; Wang, X.; Chen, S.; Wang, H. Triclosan-induced liver injury in zebrafish (*Danio rerio*) via regulating MAPK/p53 signaling pathway. *Comp. Biochem. Physiol. Part C Toxicol. Pharmacol.* **2019**, *222*, 108–117. [CrossRef]
91. Guidony, N.S.; Scaini, J.L.R.; Oliveira, M.W.B.; Machado, K.S.; Bastos, C.; Escarrone, A.L.; Souza, M.M. ABC proteins activity and cytotoxicity in zebrafish hepatocytes exposed to triclosan. *Environ. Pollut.* **2021**, *271*, 116368. [CrossRef]
92. Wang, Y.; Song, J.; Wang, X.; Qian, Q.; Wang, H. Study on the toxic-mechanism of triclosan chronic exposure to zebrafish (*Danio rerio*) based on gut-brain axis. *Sci. Total Environ.* **2022**, *844*, 156936. [CrossRef]
93. Diao, W.; Qian, Q.; Sheng, G.; He, A.; Yan, J.; Dahlgren, R.A.; Wang, X.; Wang, H. *Triclosan targets miR*-144 abnormal expression to induce neurodevelopmental toxicity mediated by activating PKC/MAPK signaling pathway. *J. Hazard. Mater.* **2022**, *431*, 128560. [CrossRef] [PubMed]
94. Parenti, C.C.; Ghilardi, A.; Della Torre, C.; Mandelli, M.; Magni, S.; Del Giacco, L.; Binelli, A. Environmental concentrations of triclosan activate cellular defence mechanism and generate cytotoxicity on zebrafish (*Danio rerio*) embryos. *Sci. Total Environ.* **2019**, *650*, 1752–1758. [CrossRef]
95. Sharma, S.; Dar, O.I.; Singh, K.; Kaur, A.; Faggio, C. Triclosan elicited biochemical and transcriptomic alterations in *Labeo rohita* larvae. *Environ. Toxicol. Pharmacol.* **2021**, *88*, 103748. [CrossRef] [PubMed]
96. Sun, L.; Ling, Y.; Jiang, J.; Wang, D.; Wang, J.; Li, J.; Wang, X.; Wang, H. Differential mechanisms regarding triclosan vs. bisphenol A and fluorene-9-bisphenol induced zebrafish lipid-metabolism disorders by RNA-Seq. *Chemosphere* **2020**, *251*, 126318. [CrossRef]
97. Song, X.; Wang, X.; Bhandari, R.K. Developmental abnormalities and epigenetic alterations in medaka (*Oryzias latipes*) embryos induced by triclosan exposure. *Chemosphere* **2020**, *261*, 127613. [CrossRef]
98. Fu, J.; Bae, S. The pH-dependent toxicity of triclosan on developing zebrafish (*Danio rerio*) embryos using metabolomics. *Aquat. Toxicol.* **2020**, *226*, 105560. [CrossRef]
99. Çelebi, H.; Gök, O. Effect of triclosan exposure on mortality and behavioral changes of *Poecilia reticulata* and *Danio rerio*. *Hum. Ecol. Risk Assess.* **2018**, *24*, 1327–1341. [CrossRef]
100. Wang, D.; Zhang, Y.; Li, J.; Dahlgren, R.A.; Wang, X.; Huang, H.; Wang, H. Risk assessment of cardiotoxicity to zebrafish (*Danio rerio*) by environmental exposure to triclosan and its derivatives. *Environ. Pollut.* **2020**, *265*, 114995. [CrossRef]

101. Raut, S.A.; Angus, R.A. Triclosan has endocrine-disrupting effects in male western mosquitofish, *Gambusia affinis*. *Environ. Toxicol. Chem.* **2010**, *29*, 1287–1291. [CrossRef]
102. Pinto, P.I.; Guerreiro, E.M.; Power, D.M. Triclosan interferes with the thyroid axis in the zebrafish (*Danio rerio*). *Toxicol. Res.* **2013**, *2*, 60–69. [CrossRef]
103. Schnitzler, J.G.; Frederich, B.; Dussenne, M.; Klaren, P.H.; Silvestre, F.; Das, K. Triclosan exposure results in alterations of thyroid hormone status and retarded early development and metamorphosis in *Cyprinodon variegatus*. *Aquat. Toxicol.* **2016**, *181*, 1–10. [CrossRef]
104. Araújo, M.J.; Quintaneiro, C.; Rocha, R.J.; Pousão-Ferreira, P.; Candeias-Mendes, A.; Soares, A.M.; Monteiro, M.S. Single and combined effects of ultraviolet radiation and triclosan during the metamorphosis of *Solea senegalensis*. *Chemosphere* **2022**, *307*, 135583. [CrossRef]
105. Guerreiro, E.M. Thyroid Axis Disruption by Goitrogens: A Molecular and Functional Approach. Master's Thesis, Universidade do Algarve, Faro, Portugal, 2018.
106. Fu, J.; Gong, Z.; Kelly, B.C. Metabolomic profiling of zebrafish (*Danio rerio*) embryos exposed to the antibacterial agent triclosan. *Environ. Toxicol. Chem.* **2019**, *38*, 240–249. [CrossRef]
107. Fu, J.; Tan, Y.X.R.; Gong, Z.; Bae, S. The toxic effect of triclosan and methyl-triclosan on biological pathways revealed by metabolomics and gene expression in zebrafish embryos. *Ecotoxicol. Environ. Saf.* **2020**, *189*, 110039. [CrossRef]
108. Diao, W.; Yan, J.; Wang, X.; Qian, Q.; Wang, H. Mechanisms regarding cardiac toxicity triggered by up-regulation of miR-144 in larval zebrafish upon exposure to triclosan. *J. Hazard. Mater.* **2023**, *443*, 130297. [CrossRef]
109. Wang, W.; Li, X.; Qian, Q.; Yan, J.; Huang, H.; Wang, X.; Wang, H. Mechanistic exploration on neurodevelopmental toxicity induced by upregulation of alkbh5 targeted by triclosan exposure to larval zebrafish. *J. Hazard. Mater.* **2023**, *457*, 131831. [CrossRef] [PubMed]
110. Ling, Y.; Sun, L.; Wang, D.; Jiang, J.; Sun, W.; Ai, W.; Wang, X.; Wang, H. Triclosan induces zebrafish neurotoxicity by abnormal expression of miR-219 targeting oligodendrocyte differentiation of central nervous system. *Arch. Toxicol.* **2020**, *94*, 857–871. [CrossRef] [PubMed]

toxics

MDPI

Review

Use of Formaldehyde in Poultry Production for the Treatment of Litter, Hatching Eggs, Hatcheries, and Feed: An Updated Mini Review

Pedro Henrique Gomes de Sá Santos [1], Gabriel da Silva Oliveira [1,2,*], Concepta McManus [3] and Vinícius Machado dos Santos [1,2,*]

1 Faculty of Agronomy and Veterinary Medicine, University of Brasília, Brasília 70910-900, Brazil
2 Federal Institute of Brasília—Campus Planaltina, Brasília 73380-900, Brazil
3 Center for Nuclear Energy in Agriculture (CENA), University of São Paulo, São Paulo 13416-000, Brazil
* Correspondence: gabriels.unb@gmail.com (G.d.S.O.); vinicius.santos@ifb.edu.br (V.M.d.S.)

Abstract: Bacteria such as *Salmonella* spp. are primarily transmitted through contaminated eggs and infected poultry; however, other routes, including the movement of personnel, vehicles, and lapses in biosecurity protocols, also play a significant role in their dissemination within poultry systems. Control of a wide range of microorganisms, including bacteria, is often carried out using chemical agents, such as formaldehyde, applied in its solid, liquid, or gaseous forms. Reports on the use of formaldehyde in poultry production date back more than a century. However, it continues to attract research interest due to growing concerns about bacterial resistance, embryotoxicity, occupational exposure, the generation of toxic byproducts, and the search for safer alternatives in poultry production systems. It remains widely used worldwide, but comprehensive and updated evaluations of its efficacy, toxicity, and risks to both poultry and workers are still limited. This review aims to synthesize the current knowledge on the use of formaldehyde in poultry production. Overall, the synthesis shows that formaldehyde remains an effective but high-risk sanitizer whose continued use in poultry systems requires rigorous control and monitoring protocols, and that the development and adoption of efficient and safer alternatives is recommended.

Keywords: bacterial control; *Escherichia coli*; formalin; hatchery eggs; natural alternatives; paraformaldehyde; poultry health; toxic products; *Salmonella* spp.; sanitization

1. Introduction

Salmonella spp., *Campylobacter* spp., *Escherichia coli*, *Staphylococcus aureus*, and *Clostridium perfringens* are among the pathogenic bacteria commonly found in poultry systems, which can be spread through various routes of contamination, primarily via horizontal transmission [1,2] (Figure 1). These bacteria have impacted the poultry industry in the past, continue to do so today, and will likely become an even greater concern in the future. Colibacillosis and clostridiosis, caused by *Escherichia coli* and *Clostridium perfringens*, respectively, are major bacterial diseases that can lead to severe outbreaks, resulting in high mortality rates in poultry and substantial economic losses. This highlights the importance of prioritizing prevention and control in both broiler and laying hen production systems [3,4].

 https://doi.org/10.3390/toxics13111003

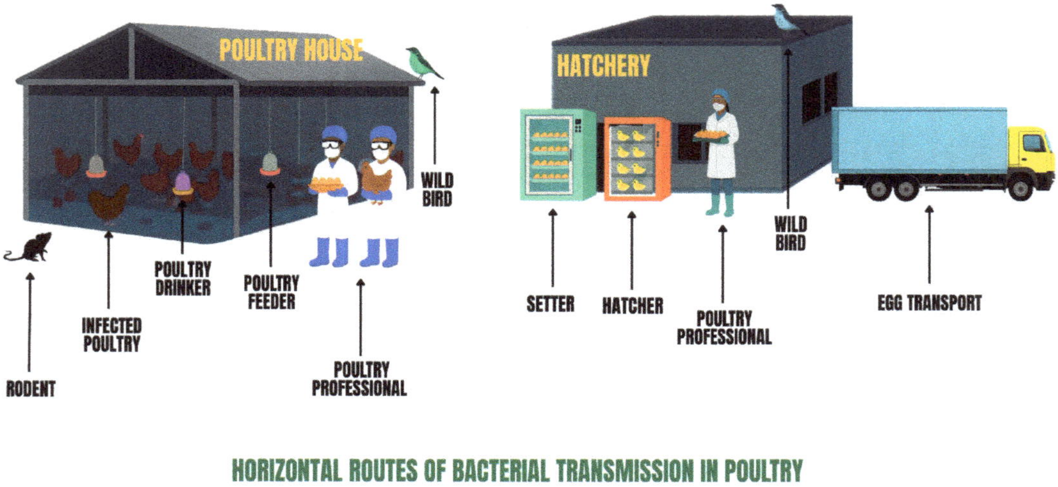

Figure 1. Horizontal routes of bacterial transmission in poultry systems. This figure illustrates some pathways by which pathogenic bacteria can spread within and between poultry facilities. In the poultry house, bacteria can be transmitted through direct contact with infected poultry, contaminated feeders and drinkers, and via fomites carried by poultry workers. Rodents and wild birds serve as mechanical vectors, introducing and disseminating bacteria within the poultry environment. In the hatchery, eggs and chicks can become contaminated during incubation in the setter and hatcher machines. Handling of eggs or chicks by poultry farmers, as well as the movement of vehicles transporting poultry products in and out of the facility, can further spread bacteria. Wild birds around the hatchery also pose a risk of introducing external bacteria.

Bacterial control requires the effective application of sanitizers to achieve performance and quality goals in poultry production facilities. Formaldehyde is a sanitizing agent with strong antibacterial efficacy, ease of application, and low cost, making it readily available and straightforward to prepare and apply. However, strict safety measures are required due to its toxicity. These attributes have led to its global use in poultry production, and despite its long-standing use, it is still considered more advantageous than other sanitizers in the poultry industry. It is recognized as one of the most effective bactericidal agents for controlling bacteria in poultry operations in various countries [5–8]. It exhibits antibacterial activity against various bacteria, is cost-effective compared to other products, and is readily available in the poultry sanitizer market. Its use for bacterial control in poultry production has been associated with toxicity issues and may cause unintended impacts on the production environment [9]. This review aims to highlight the current knowledge of the effects of formaldehyde in poultry production environments, thereby contributing to an understanding of the interactions among formaldehyde, poultry, eggs, surfaces, and bacteria in hatcheries and farms.

For this purpose, publications investigating the use of formaldehyde in poultry production were identified through Google Scholar. Between 29 March 2025 and 7 November 2025, a bibliographic search was conducted in Google Scholar for research articles, reviews, books, and book chapters published in English, Portuguese, or Spanish, with no restrictions on publication date. The search strategy used different combinations of keywords related to formaldehyde, paraformaldehyde, formalin, toxicity, poultry, bacteria, contamination, infection, antibacterial, bacterial control, antibacterial resistance, laying hens, broilers, eggs, hatching eggs, embryo, chicks, sanitization, disinfection, sterilization, sanitizers, natural sanitizers, synthetic sanitizers, alternative sanitizers, disinfectants, synthetic disinfectants, natural disinfectants, chemical products, hatcheries, poultry workers and humans, employing the Boolean operators AND and OR. To complement the review, the sources of gray literature, such as conference papers, monographs, dissertations, and theses available online, were also consulted. For each search combination, the first ten results were screened based on their titles and abstracts to identify publications suitable for inclusion in the review. For those who passed this stage, the content of each selected study was critically analyzed. All works that met the criteria for the review, including language, thematic relevance, and alignment with the established topics and subtopics, were included. Studies that did not meet the objectives defined for each section were excluded, for example, in the topic "antibacterial effects", this excluded those focusing exclusively on other microorganisms (such as fungi), as well as publications in languages other than English, Portuguese, or Spanish, duplicate articles, or those without full-text availability. Non-peer-reviewed materials were carefully evaluated by all authors, considering the credibility of the source, the clarity of the information, and its consistency with the evidence presented in peer-reviewed publications. Google Scholar was selected for its ability to integrate studies from different databases, a feature particularly relevant given the scarcity of research in this specific area. To maintain consistency in this review, the term "formaldehyde" has been used to encompass all its forms.

2. Formaldehyde in Poultry Field

Formaldehyde (CH_2O) is the simplest aliphatic aldehyde, an electrophilic compound containing an aldehyde group (–CHO). At room temperature, it occurs as a colorless gas with a pungent odor. It can also exist as a white polymeric solid (paraformaldehyde) or as a colorless aqueous solution known as formalin (37–50% formaldehyde, often stabilized with 0–15% methanol) [10]. In aqueous media, formaldehyde is highly reactive and exists in dynamic equilibrium with methylene glycol ($HO–CH_2–OH$) and its oligomeric forms (polyoxymethylene glycols). Its chemical reactivity arises from the electrophilic nature of the carbonyl carbon, which enables addition and condensation reactions with nucleophiles such as amines and phenols, as well as disproportionation under basic conditions (Cannizzaro reaction) to yield methanol and formic acid [10]. Formaldehyde can be applied by spraying (liquid dispersion), immersion (submersion in solution), or fumigation (vapor release in a closed space) [11]. In poultry production, formaldehyde can be applied in liquid form by spraying or in solid form. Most reports describe its use by fumigation, particularly for sanitizing hatching eggs, which remains the primary protocol in many countries, such as Brazil, Egypt, and Saudi Arabia. The success of fumigation depends on several factors, including environmental conditions, concentration, and exposure time. In this process, formaldehyde concentrations typically range from 5 to 14 g/m^3, occurring at temperatures of 24 to 30 °C and relative humidity between 65 and 75%, resulting in exposure times of 15 to 40 min, depending on the study and experimental context [8,12–17]. These parameters act in combination, as temperature and humidity influence the gas's activity, release, and diffusion, and together with concentration and exposure time determine its antibacterial effectiveness [9]. Additionally, effective containment of the gas within enclosed and well-sealed chambers is crucial to prevent leaks, maintain fumigation efficiency, and minimize occupational exposure risks [9,11]. The operational conditions can be adjusted according to the specific conditions of the farm or hatchery, ensuring maximum antibacterial efficiency while minimizing impacts on poultry, human, and environmental health.

Formaldehyde, regardless of its physical state, has a documented antibacterial potential [18–20]. It appears that the antibacterial potential of formaldehyde is related to alkylation reactions with cellular macromolecules [21]. Formaldehyde reacts with amino ($–NH_2$), carboxyl (–C=O), hydroxyl (–OH), and sulfhydryl (–SH) groups of proteins, thereby disrupting disulfide bonds and altering their tertiary structure, as well as with purine bases in DNA and RNA [21,22]. Such covalent modifications ultimately result in bacterial inactivation. This mechanism of action, which has been reviewed and discussed for over 25 years, remains widely accepted and continues to be actively debated [22,23].

The main reason for the use of formaldehyde in industry is its antibacterial activity, characterized by its ability to reach inaccessible spaces, especially in protocols involving the release of the gas, similar to other antibacterial gases such as chlorine dioxide and ozone [15,24]. This allows for simultaneous bacterial control of materials, surfaces, and raw materials within a closed space. Antibacterial protocols involving formaldehyde have been used as one strategy in biosecurity programs aimed at interrupting reservoirs and horizontal transmission routes of pathogenic bacteria, such as those belonging to the Enterobacteriaceae family. These programs seek to reduce the dissemination of pathogens among poultry, between poultry and environmental surfaces, and between poultry products and humans, ultimately preventing disease outbreaks, such as salmonellosis, in the poultry industry. The potential spread of these bacteria beyond the production environment poses a risk to public health. For example, *Salmonella enterica* serovar Typhi can have a fatality rate of up to 80% in humans [25]. However, the likelihood of this scenario arising from poultry practices is minimal when sanitary standards, including those related to the use of formaldehyde, are correctly implemented and monitored.

2.1. Antibacterial Effects

In poultry practice, formaldehyde has been used beyond bacterial control in broiler and layer houses [26], with applications including the treatment of poultry litter, sanitization of hatching eggs and hatcheries, and even the incorporation into poultry feed (Figure 2).

Figure 2. Some forms of formaldehyde application for bacterial control in poultry production, (**A**) house sanitization, (**B**) litter treatment, (**C**) hatching eggs fumigation and (**D**) incorporation of formaldehyde into feed.

2.1.1. Poultry Litter

According to Toledo [27], the levels of *Pseudomonas* spp. and *Staphylococcus* spp. present in poultry litter were effectively controlled by the application of 15 g of 90% formaldehyde. After the first fumigation, *Pseudomonas* spp. counts dropped from

5.16 $\log_{10}$ CFU/g to undetectable levels within 24 h and remained undetectable throughout the 15-day evaluation period. Similarly, *Staphylococcus* spp. showed a marked reduction from 4.62 to 4.19 $\log_{10}$ CFU/g after 24 h, reaching only 2.03 $\log_{10}$ CFU/g after 15 days. Similarly, Argueta [28] reported that 8 g of formaldehyde (90–92%) can be used to eliminate coliform bacteria in poultry litter. Bampi et al. [29] experimentally evaluated the sanitization efficiency of 37% formaldehyde diluted to 4% in poultry litter contaminated with *Salmonella* spp. and *Escherichia coli*. After two hours of contact, formaldehyde significantly reduced *Salmonella* spp. by 1.4 $\log_{10}$ and *Escherichia coli* by 0.5 $\log_{10}$. Saleh et al. [30] mixed 5 g of 92% formaldehyde into the nest litter material and observed a significant reduction in the bacterial load. Total bacterial counts decreased from 8.28 to 7.70 $\log_{10}$ CFU/g after 24 h and to 7.59 $\log_{10}$ CFU/g after 72 h, increasing again to 8.11 $\log_{10}$ CFU/g after 120 h. Coliform counts declined from 8.18 to 7.55 $\log_{10}$ CFU/g after 24 h and to 7.23 $\log_{10}$ CFU/g after 72 h, with a slight increase to 7.30 $\log_{10}$ CFU/g after 120 h. According to the authors, this practice would likely decrease bacterial contamination of hatching eggs.

2.1.2. Hatching Eggs

Luna [31] reported that fumigation with 2% formaldehyde achieved approximately 91% efficiency in reducing the total aerobic bacterial load on the eggshell surface. According to Rojas [32], fumigation with formaldehyde (11 mL/m^3) removed approximately 90% of the organic matter on eggshells as measured by luminometry. Concurrently, microbiological analysis of the eggshells confirmed the elimination of *Escherichia coli*, *Klebsiella* spp., and *Pseudomonas* spp., demonstrating that formaldehyde can effectively remove organic matter that could otherwise protect microorganisms, thereby enhancing its sanitizing action on the eggshell surface against bacteria that naturally occur in poultry farms. Similarly, Vale et al. [33] demonstrated that applying a 1.5% formaldehyde solution (36.5–38% concentration) to eggs significantly reduced the total aerobic mesophilic bacterial count from 4.96 ± 0.52 to 2.03 ± 0.47 $\log_{10}$ CFU/mL, indicating a satisfactory reduction from a sanitary perspective. More recently, dos Santos et al. [34] estimated that fumigation with formaldehyde at a concentration of 6.07 g/m^3 is effective in achieving minimal counts of mesophilic bacterial colony-forming units on the eggshell surface.

2.1.3. Poultry Hatcheries

Harlia et al. [35] demonstrated that fumigation with a 40% formaldehyde solution eliminated bacterial contamination in the hatching airspace of poultry setters. This was also observed by Selby et al. [36], who conducted a controlled hatchery experiment in which 19-day-old embryonated eggs were exposed to a bacterial challenge designed to simulate natural contamination in commercial hatcheries. The inoculum, prepared from isolates of *Escherichia coli*, *Staphylococcus aureus*, and *Enterococcus* spp., was applied as a suspension containing approximately 10^7–10^8 CFU per 100 µL per egg over a 28 mm^2 area located on the blunt end of the shell. Following this bacterial challenge, the treated hatchers were subjected to formaldehyde fumigation through drip application of 6 mL of formaldehyde every three hours, continuing until 12 h before hatching. The efficacy of formaldehyde was evaluated by quantifying Gram-negative bacteria, *Staphylococcus aureus*, and *Enterococcus* spp. recovered from the incubation environment (air and fluff samples) and from the gastrointestinal tract of newly hatched chicks. Air samples were collected using the open-agar plate method, in which selective agar plates were placed open on the top tray of the hatcher through a sampling port and exposed for 1 or 5 min, depending on the medium. In Experiment 1, for example, at approximately 80% hatch, airborne recovery of Gram-negative bacteria ranged from 1.62 to 2.16 $\log_{10}$ CFU/plate in the challenged or naturally contaminated groups, whereas formaldehyde treatment reduced this value to

1.11 $\log_{10}$ CFU/plate. Similar patterns were observed for *Enterococcus* spp. (1.73–1.86 versus 1.08 $\log_{10}$ CFU/plate) and *Staphylococcus aureus* (1.26–1.57 $\log_{10}$ CFU/plate versus undetectable), and the same trend occurred in Experiments 2 and 3. Formaldehyde fumigation significantly reduced the bacterial load across the other sampling sites evaluated, without affecting hatchability, seven-day body weight gain, feed conversion, and mortality. On the other hand, Espinosa [37] reported that fumigation of a commercial poultry hatcher, using a conventional system with 38% formaldehyde concentration (6.9 mL/m^3), was insufficient to prevent an increase in the number of microorganisms in the air during the hatching period. Microbiological analysis of the down revealed that it was heavily contaminated.

2.1.4. Feed Poultry

Salmonella enterica serovar Enteritidis can be eliminated from poultry feed within a short period of time after exposure to formaldehyde gas [38]. Avila et al. [39] reported that the feed for breeder hens containing 720 mg of formaldehyde per kilogram was free of *Salmonella* spp. Still, the initial levels of this pathogen were non-detectable in the untreated samples. The treated feed also had lower counts of total aerobic bacteria, Enterobacteriaceae, and *Clostridium perfringens*. The eggshells from these hens were also free of *Salmonella* spp. and showed reduced total aerobic bacterial counts.

The applicability of formaldehyde, due to its antibacterial activity, at various points along the production chain demonstrates that it can be adapted to diverse situations, making it a strategic differentiator for poultry production. This flexibility not only contributes to maintaining avian health but also supports the optimization of productive processes.

2.2. *Bacterial Resistance to Formaldehyde*

Aarestrup and Hasman [40] tested formaldehyde in vitro at concentrations ranging from 0.0008 to 0.4% against *Salmonella*, *Escherichia coli*, *Staphylococcus aureus*, *Staphylococcus hyicus*, *Enterococcus faecalis*, and *Enterococcus faecium*, which were isolated from broilers, cattle, and pigs. Most isolates exhibited minimal inhibitory concentrations ranging from 0.003 to 0.006%, with no significant differences observed among the bacterial species tested. Alijani et al. [41] tested formaldehyde at different concentrations, from a 37% stock solution, against Avian Pathogenic *Escherichia coli* isolates from broiler chickens. All strains exhibited minimal inhibitory concentration and minimal bactericidal concentrations at $\leq$0.009%. However, studies have shown that some bacterial strains can tolerate much higher levels of formaldehyde. Hoseinzadeh et al. [42] isolated *Escherichia coli* from cloacal and pericardial swabs collected from apparently healthy and colibacillosis-affected broiler chickens. The isolates were subjected to in vitro susceptibility testing using serial twofold dilutions of formaldehyde. According to the authors, isolates capable of growth at a concentration of 0.929% (9.29 mg/mL) formaldehyde were defined as resistant. Among the isolates, 15% were classified as resistant and 85% as sensitive to formaldehyde. Molecular screening revealed that 61% of isolates carried the class 1 integron, 94% harbored the formaldehyde dehydrogenase gene, and 8% contained the *orfF* gene.

Choroszy-Król et al. [20] prepared formaldehyde solutions from a 37% stock solution and tested them at concentrations of 0.02, 0.2, and 2.0% for 1 and 15 min against *Salmonella enterica* serovar Senftenberg and *Escherichia coli* isolated from poultry farms. The concentration of 2.0% completely eradicated the biofilms of all *Salmonella enterica* serovar Senftenberg strains after 1 min of exposure. At lower concentrations, 0.2 and 0.02%, only partial inhibition of bacterial growth was observed. The *Escherichia coli* strains showed higher tolerance, as some isolates were inactivated entirely by 2.0% after 1 min, while others with greater biofilm-forming ability required 15 min of exposure for complete eradication. Although formaldehyde concentrations of 0.2 and 0.02% produced some reduction in viable cell counts for certain *Salmonella enterica* serovar Senftenberg and *Escherichia coli* strains, these effects were strain-dependent and insufficient to ensure effective biofilm eradication. Brazilian researchers observed that all 63 avian pathogenic *Escherichia coli* isolates from poultry exhibited resistance to formaldehyde [43], underscoring the presence of formaldehyde-resistant bacteria in production environments.

Various studies have advanced our understanding of the bacterial molecular responses that confer resistance to formaldehyde. For example, Klein et al. [44] identified a novel formaldehyde resistance mechanism associated with the *yycR* gene, which encodes a zinc- and NAD^+-dependent formaldehyde dehydrogenase. This enzyme, termed YycR, functions independently of thiols and can oxidize formaldehyde to formic acid, thereby contributing to cellular detoxification. The deletion of the *yycR* gene markedly reduced tolerance to formaldehyde, demonstrating that, in addition to previously recognized pathways mediated by 3-hexulose-6-phosphate synthase (*hps*), 6-phospho-3-hexuloisomerase (*phi*), and bacillithiol-dependent aldehyde dehydrogenase (*adhA*), an additional detoxification route based on YycR activity is also present.

Complementing these enzymatic detoxification mechanisms, Bazurto et al. [45] demonstrated that bacterial resistance to formaldehyde can also involve regulatory responses mediated by intracellular sensors. The study identified the protein EfgA (enhanced formaldehyde growth) as the first described bacterial formaldehyde sensor that does not act through enzymatic degradation of the compound. Instead, EfgA directly senses elevated intracellular formaldehyde levels and triggers immediate arrest of cell growth and protein translation, thereby preventing macromolecular damage caused by the aldehyde's high reactivity with amino and thiol groups. In the same study, the authors also reported that, in addition to *efgA*, other genes identified through experimental evolution, such as *efgB* and hrcA, as well as genes involved in protein quality control, suggest that formaldehyde resistance is strongly associated with protection against protein damage rather than solely with chemical detoxification.

These molecular insights help explain the observed patterns of resistance across diverse bacterial species. For example, *Pseudomonas putida* strains can survive formaldehyde exposure through multiple protective mechanisms, including detoxification via glutathione-dependent formaldehyde and formate dehydrogenases that oxidize formaldehyde to CO_2, efflux of the toxic molecule via the MexEF-OprN pump, and protection of cellular structures (DNA and proteins) [46]. Similarly, *Escherichia coli* may exhibit resistance to formaldehyde due to enzymes capable of degrading it into less toxic compounds, such as glutathione- and NAD^+-dependent formaldehyde dehydrogenases encoded by a plasmid-mediated gene [47,48].

The prevalence of formaldehyde-resistant bacteria has significant consequences for poultry production. As reviewed by Oliveira et al. [49], *Escherichia coli*, for example, is a Gram-negative bacterium commonly isolated from eggs, dead-in-shell embryos, newly hatched chicks, and poultry. It is associated with septicemia, omphalitis, and congenital embryonic deformities, frequently resulting in high mortality rates. A survey conducted in just four Brazilian states revealed that contamination and diseases associated with pathogenic Escherichia coli result in million-dollar economic losses [50]. In this context, the bacterium's increasing resistance to formaldehyde is particularly concerning, as *Escherichia coli* can be present at various points along the poultry production chain, where formaldehyde is widely used as a sanitizing agent. Therefore, a crucial point that cannot be overlooked is that the antibacterial efficacy of formaldehyde may be limited under certain application conditions. For example, Badr and Yoseif [26] reported that a 37% formaldehyde solution at a 10% concentration was unable to eliminate pathogenic bacteria in poultry houses completely. Moreover, the continuous use of formaldehyde in poultry production may favor the selection of resistant bacterial strains, such as *Pseudomonas aeruginosa* [51]. Therefore, careful planning of the protocol and continuous monitoring of process efficiency are essential.

2.3. Effects on Hatchability

The use of formaldehyde in poultry production is recognized for its low cost and ease of application, making it accessible to producers of various scales. Its application occurs primarily at points in the production chain preceding the hatching process, aiming to ensure high hatchability rates and optimize post-hatch productivity. For example, hatchability rates above 80% and no adverse effects after hatching on quality, mortality, weight gain, feed intake, or feed conversion ratio were observed in poultry originating from eggs sanitized by fumigation with a 40% formaldehyde solution [52]. Although there are reports evaluating post-hatch productive parameters, such as overall poultry performance, the primary production indicator associated with formaldehyde use remains hatchability. Reports comparing the hatchability of eggs fumigated with formaldehyde to that obtained with other sanitizing agents, such as ozone, hydrogen peroxide, and essential oils, are available (Table 1).

Table 1. Comparison of hatchability rates between formaldehyde and other sanitizers.

Formaldehyde	Ammonia	Ozone	Hydrogen Peroxide	Essential Oils	Propolis	Garlic Oil	p Value	Reference
			Hatchability (%)					
45.00		40.00					ns	Souza et al. [17]
86.42		86.64	86.37				ns	Melo et al. [15]
81.87				84.69			ns	Oliveira et al. [8]
78.23		64.32	82.50				$p < 0.05$	Wlazlo et al. [53]
97.22			96.01				$p < 0.05$	Badran et al. [54]
91.10		91.79					ns	Hrnčár et al. [55]
75.34				87.83			$p < 0.05$	Oliveira et al. [56]
95.32					95.76		ns	Shahein and Sedeek [57]
93.46						94.07	ns	Rizk et al. [58]
93.1	93.5						ns	Cony et al. [13]

Badran et al. [54] demonstrated that treating hatching eggs with formaldehyde for 30 min, prepared from a mixture of 60 mL of formaldehyde, 30 mL of water, and 48 g of potassium permanganate, resulted in higher hatchability compared to 5% hydrogen peroxide. No significant differences in chick weight at hatch were observed between the two treatments. However, based on the analysis of blood components (total protein, albumin, globulin, glucose, triiodothyronine, uric acid, creatinine, aspartate aminotransferase, and alanine aminotransferase) and the relative organ weights, the authors suggested that chicks hatched from eggs fumigated with formaldehyde exhibited signs of hepatotoxicity, evidenced by hepatic hypertrophy and possible impairment of renal function.

Shahein and Sedeek [57] demonstrated that the hatchability of eggs fumigated with formaldehyde, obtained from a mixture of 119.8 mL formaldehyde and 59.9 g potassium permanganate per 2.83 m^3 for 20 min, did not differ from that of eggs sprayed with a 14% propolis solution. Similarly, embryo weights at 18 days of incubation and at hatch were not significantly different between the two treatments. However, embryo and chick weights tended to be higher in eggs sanitized with propolis.

Oliveira et al. [56] reported that hatching eggs sanitized with essential oils of *Citrus aurantifolia* (9.38 mg/mL), *Ocimum basilicum* (4.69 mg/mL), and *Allium sativum* (1.17 mg/mL) showed superior hatchability performance compared to fumigation with formaldehyde, applied at a concentration of 5 g/m^3 for 15 min. While embryos from eggs sanitized with essential oils showed no reports of tracheal or lung damage, embryos from eggs fumigated with formaldehyde exhibited tracheal lesions such as goblet cell hyperplasia, lymphocytic inflammation, and lung congestion. The weight of newly hatched chicks was comparable between the groups treated with essential oils and the formaldehyde group; however, formaldehyde significantly reduced chick weight compared with the non-sanitized control group.

Melo et al. [15] observed that the hatchability of fertile eggs sanitized with formaldehyde at a concentration of 5.03 g/m^3 for 30 min was similar to that of eggs treated with ozone (5–15 ppm for 30 min), ultraviolet light at an average intensity of 8.09 mW/cm^2 for 120 s, hydrogen peroxide at 3%, and peracetic acid at 0.3%. None of these sanitization methods harmed chick weight at hatching or the percentage of saleable chicks. From a production standpoint, formaldehyde can still be used to preserve hatchability potential, depending on the application conditions. This factor likely explains its continued use in poultry management despite growing pressure to replace it due to its high toxicity.

It is important to emphasize, however, that the apparent competitiveness of formaldehyde in maintaining hatchability rates should be interpreted with caution. Despite studies reporting hatchability performance similar to that observed with other sanitizers, such as hydrogen peroxide, ozone, or essential oils, these results are not entirely consistent (Table 1). In some cases, formaldehyde resulted in slight reductions in hatchability, while in others, it maintained or even exceeded the rates achieved with alternatives treatments (Table 1). This variability may be associated with differences in experimental protocols across studies, including concentration, exposure time, and even the quality and purity of the commercial product. Even though formaldehyde continues to show promise for sanitization and hatchability maintenance under specific controlled conditions, its efficacy must be evaluated contextually, accounting for methodological variability and, above all, the balance between antimicrobial efficacy and potential embryonic toxicity.

Formaldehyde is not exempt from causing reductions in hatchability under certain circumstances, as poultry are sensitive to toxic chemicals such as it. Its toxicity represents a central point of discussion within the scientific community [9,11,56,59]. The risks may extend beyond production losses, including the potential to compromise the quality of hatched chicks, the poultry's healthy growth, and, above all, the safety of the professionals who handle and apply this chemical. Therefore, the hatchability results observed with formaldehyde are not universally guaranteed and may vary depending on management practices and experimental conditions. This emphasizes the need also to consider both its potential toxicity to embryos and the risks to poultry workers when making decisions about its use in poultry operations.

2.4. *Toxicity in Poultry*

Formaldehyde can cause a range of effects, from mild to severe outcomes, including poultry mortality, which occurs during both the embryonic phase and the post-hatch period. For instance, studies have shown that fumigation of hatching eggs with formaldehyde can lead to embryonic mortality rates exceeding 20% [60,61]. Additionally, other outcomes that may or may not progress to embryonic mortality have been reported depending on the study focus, encompassing fumigation of eggs and hatchery environments as well as poultry feeding, thus demonstrating varied impacts depending on the route, concentration and context of formaldehyde exposure [62–64].

These varied outcomes further demonstrate that poultry organs and tissues are particularly vulnerable to formaldehyde, exhibiting significant alterations across all developmental stages. The main observed effects were categorized by organ or tissue, with emphasis on the route of exposure (Table 2). In poultry, formaldehyde exposure can manifest as clinical signs such as apathy, depression, lethargy, anorexia, respiratory distress, and diarrhea, accompanied by reductions in body weight, weight gain, and feed intake, as well as changes in hematological and biochemical parameters [63–65]. Collectively, these manifestations reflect complex systemic impairments involving the liver, kidneys, heart, lungs, and gastrointestinal tract (Table 2), directly affecting the poultry's clinical condition and growth performance.

Table 2. Effects of formaldehyde exposure on organs of poultry, regardless of their developmental stage.

Structure	Observed Effects	Exposure Route	Reference
Cornea	Corneoscleral junction congestion Iris heterophil infiltration Basal cell degeneration	Hatchery fumigation	Espinosa [37]
Brain and head	Microcephaly Microphthalmia Exencephaly Brain degeneration Collapsed cavities Absence of segmentation Forebrain and hindbrain non-closure Severe ventricular reduction	Egg fumigation	Bekhet and Khalifa [66]
Spinal cord	Severe layer degeneration Central canal hemorrhage Central canal closure Zig-zag deformity Gray/white matter degeneration	Egg fumigation	Bekhet and Khalifa [66]
Trachea and lungs	Goblet cell hyperplasia Lymphocytic inflammation Deciliation Ciliary membrane rupture Ciliary agglutination Epithelial desquamation Heterophil infiltration Necrosis Congestion	Egg fumigation or hatchery fumigation	Di Matteo et al. [67], Oliveira et al. [56], de Freitas [68]
Proventriculus and Gizzard	Proventricular ulcers Hardened mucosa in both	Feed exposure	Khan et al. [64]
Liver	Congestion Sinusoidal dilation Minor hemorrhages Lymphoid infiltration Central vein congestion Tissue degeneration Reduced size	Poultry exposure or feed exposure	Khan et al. [64], Albaghdady et al. [69]
Kidneys	Atrophied glomeruli Ruptured glomeruli Leukocyte infiltration Tissue degeneration Hemorrhagic congestion Reduced size	Poultry exposure or feed exposure	Khan et al. [64], Albaghdady et al. [69]
Skeletal muscles	Breast hemorrhages Thigh hemorrhages	Feed exposure	Khan et al. [64]
Heart	Hemorrhage Congestion Inflammatory infiltration Reduced size	Feed exposure or poultry exposure	Khan et al. [64], Al-Saeed et al. [70]
Digestive tract	Lesions	Feed exposure	Khan et al. [64]
Reproductive system	Oviduct gland degeneration (nuclear vacuolation) Reduced ovary and oviduct size	Feed exposure	Khan et al. [64]

Several adverse outcomes in embryos and chicks originating from eggs fumigated with formaldehyde, as reported in studies published between 2015 and 2021, were summarized in a review by Oliveira et al. [11]. The concerning findings include embryonic anomalies, such as congenital malformations, reduced chick survival, and impaired chick quality. Chicks hatched from eggs sanitized with formaldehyde may exhibit compromised quality, evidenced by visible lesions, redness on the legs and beak, a swollen, edematous, and reddened abdomen, as well as poor navel healing (omphalitis) (Figure 3). In addition, severe eggshell wear, characterized by extensive structural damage and significant compromise, has also been reported [56]. The toxic effects of formaldehyde fumigation are related to its ability to penetrate the eggshell's structure. The compound can react with structural components, such as proteins and minerals, and even reach the internal contents of the egg [9,71]. Consequently, toxic effects on embryos may occur even at relatively low concentrations, becoming progressively more severe with increasing concentration and exposure time [71]. In addition, its adverse effects may also be associated with the limited long-term antibacterial persistence, which favors subsequent bacterial recontamination of the eggshell surface and the embryonic development environment [71], allowing infectious processes to occur.

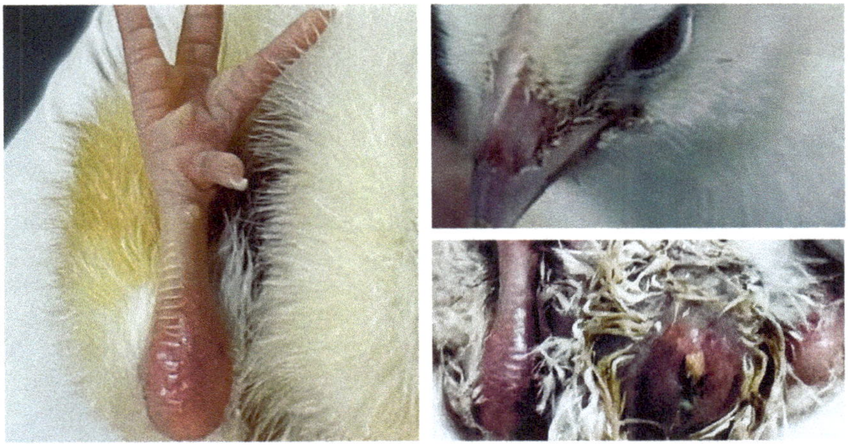

Figure 3. Morphological alterations observed in newly hatched chicks originating from eggs fumigated with formaldehyde in an experiment evaluating chicks from untreated eggs, eggs treated with formaldehyde, and essential oils. (**Left**) lesion on the hock, evidenced by swelling and erythema. (**Upper right**) lesion on the beak, characterized by areas of deformity. (**Lower right**) unhealed navel, presenting signs of omphalitis, with inflammation and exudate.

Valverde-Santiago and Pontel [72] reviewed the mechanisms involved in formaldehyde toxicity. They demonstrated that its harmful effects arise from its high reactivity with nucleophilic groups in DNA, RNA, and proteins, leading to the formation of adducts and crosslinks (DNA–protein, RNA–protein, and DNA–DNA). These lesions generate replicative and transcriptional stress, as well as single- and double-strand breaks, leading to genomic instability. Such damage activates multiple repair pathways, including the Fanconi anemia pathway, homologous recombination, nucleotide excision repair, and specific proteolytic mechanisms for the degradation of DNA–protein crosslinks mediated by the protease SPRTN and the proteasome. The review further highlights that formaldehyde reacts with glutathione, causing a redox imbalance and the accumulation of reactive oxygen species, and can inhibit enzymes involved in one-carbon metabolism, resulting in epigenetic dysregulation. At the cellular level, exposure to formaldehyde triggers ribotoxic stress and the integrated stress response through the formation of RNA–protein crosslinks, ultimately impairing translation and reducing cell viability.

The application of formaldehyde in poultry production, particularly through fumigation processes, can lead to the formation of toxic byproducts, including carbon monoxide and formic acid. These compounds have been associated with embryotoxic and oxidative effects in poultry. Matias et al. [73] observed that exposure of chicken embryos to carbon monoxide concentrations ranging from 3 to 18 ppm during the first 10 days of development resulted in a significant reduction in heart weight and a dose-dependent thickening of the ventricular, septal, and atrial walls, indicating that even very low levels of carbon monoxide may compromise embryonic cardiac development. Amini et al. [74] observed that exposure of chicken embryos to formic acid markedly reduced hatchability rates and significantly increased the levels of malondialdehyde, total lipid hydroperoxides, and total nitrite–nitrate, while decreasing total antioxidant capacity and superoxide dismutase activity in the liver, heart, and brain tissues. The authors concluded that formic acid, even at low concentrations ($\geq$2 mM), can induce oxidative damage and embryolethality in avian embryos. Therefore, the potential conversion of formaldehyde into these reactive molecules during fumigation processes raises concerns regarding toxicology and biosafety. It has been reported that formic acid has limited use in poultry feed due to its corrosive nature toward the gastrointestinal tract [75]. Similarly, adverse effects associated with the inhalation of formic acid by animals have ranged from eyelid closure to mortality [76].

2.5. Risks for Poultry Workers

According to da Silva [77], the occupational exposure limit for formaldehyde allowed in Brazil is 2.3 mg/m^3, while in other countries these values are considerably more restrictive, such as in Japan (0.24 mg/m^3), Denmark (0.4 mg/m^3), and Finland (1.2 mg/m^3). Only the United States presents a slightly more permissive limit of 2.5 mg/m^3. In contrast, the sanitization of hatching eggs frequently exceeds these limits by multiple times. In Brazil, for example, concentrations ranging from 5 to 13 g/m^3 have been reported, approximately 2000 to 5000 times higher than the national occupational limit [8,13,14,56]. Although the procedure is conducted in closed fumigation chambers or rooms, the seal may not be absolute. Furthermore, exposure can occur intermittently and silently, especially during chamber opening, handling the product before fumigation, and manipulating newly fumigated eggs, creating toxic microenvironments. Under these conditions, local concentration may significantly exceed occupational safety limits. In the case of formaldehyde application to eggs using the spraying method, the procedure is usually performed by professionals.

The sanitization of hatching eggs with formaldehyde, typically performed by the worker during their daily 8 h shift, may continue over several working years. Additionally, these workers are often responsible for handling and applying this chemical to vehicles entering the farm and to the poultry houses [78]. This repetitive and prolonged exposure over many years can lead to chronic poisoning, causing severe neurological and respiratory damage, including cerebral atrophy, epileptic seizures, signs of dementia, as well as chronic rhinosinusitis with nasal septum deviation and perforation [79]. The intense routine and constant handling in environments where formaldehyde is manipulated and applied put the health and survival of workers at risk. Castellani et al. [80] reviewed the adverse effects of formaldehyde on human health, reporting bodily irritations, respiratory problems, allergic reactions, cancer risk, and even death.

2.6. Mitigating Formaldehyde Hazards in Poultry Systems

Mitigating formaldehyde hazards in poultry systems requires a combination of immediate protective measures, strict occupational protocols, and a progressive transition to safer alternatives. This involves the use of fully enclosed, automated fumigation systems, which control temperature and humidity, ensure adequate ventilation, isolate treated areas for a specified period, and properly dispose of product residues. It also requires adopting standardized protocols that define appropriate concentrations, exposure times, and application methods, based on research, manufacturer instructions, and regulatory guidelines. Proper worker training is essential to ensure safe handling [81], with a particular emphasis on the consistent use of personal protective equipment, including masks, gloves, goggles, aprons, and protective clothing. Occupational exposure limits vary widely between countries, from highly restrictive thresholds in Japan to more permissive values in Brazil and the United States [77]. Given that formaldehyde concentrations used in egg fumigation often exceed these limits by several orders of magnitude, aligning international regulations and enforcing stricter compliance are urgent priorities.

Promoting a transition to poultry farming that is less reliant on synthetic products, such as formaldehyde, and more focused on green alternatives, is the most effective way to prevent the harmful effects of this compound on poultry production and the professionals responsible for its management. As long as the intensive use of toxic synthetic products persists, ensuring the safety of poultry and humans will remain a significant challenge, both in large and small poultry systems. In the face of anticipated severely deleterious events, this challenge may, at some point in the future, lead to production failure. Positive gains should not be limited to productivity and economy alone, but should also encompass environmental safety, poultry health, and human health, ensuring that all these aspects are secured. For this reason, numerous studies have focused on developing and testing environmentally friendly, sustainable solutions for poultry systems. This list includes those tested and recommended for litter treatment, the sanitization of hatching eggs, and feed additives (Table 3).

Table 3. Some green products to replace formaldehyde in poultry management.

Study Area	Green Product	Efficacy	Reference
Litter treatment	*Thymus vulgaris* essential oil	Reduced bacterial counts	Galgano et al. [82]
Litter treatment	Neem leaves	Improved poultry performance	Bishnoi et al. [83]
Litter treatment	Cinnamon essential oil	Bacterial reduction and the incidence of footpad lesions	Marchioro et al. [84]
Litter treatment	Plant parts of *Satureja hortensis*, *Origanum vulgare*, *Melissa officinalis*, *Salvia officinalis*, and *Thymus vulgaris*	Reduced bacterial counts	Gontar et al. [85]
Sanitization of hatching eggs	Tahiti lemon juice	Inhibited bacterial growth	de Jesus et al. [86]
Sanitization of hatching eggs	Cherry leaf extract	Improved hatchability rate	Ayuningtyas et al. [87]
Sanitization of hatching eggs	Tea tree and lavender essential oils	Reduced the bacterial load and improved hatchability	Iraqi et al. [88]
Sanitization of hatching eggs	Cemele pepper extract	Reduced the bacterial load and improved hatchability	Ergün et al. [89]
Feed additive	Coffee pulp extract	Improved growth performance and intestinal structure	Huanhong et al. [90]
Feed additive	*Curcuma xanthorrhiza* extract	Improved growth performance and intestinal microbial balance	Sinurat et al. [91]
Feed additive	*Minthostachys verticillata* essential oil	Improved growth performance and influenced the composition of the gut microflora, without inducing genotoxic or cytotoxic effects.	Escobar et al. [92]
Feed additive	Rosemary essential oil	Improved growth performance, nutrient digestibility, and carcass traits	Adil et al. [93]

Despite the availability of various green products, including those listed in Table 3, for routine poultry management practices, large-scale implementation of these compounds still faces obstacles that limit their official and widespread adoption in commercial production. Protocols involving essential oils and plant extracts often require adjustments to minimize limitations associated with high volatility, low solubility, instability, and variability in chemical composition. These characteristics directly affect production costs, which represent another limiting factor. The cost of using these natural compounds is usually higher than that of several synthetic sanitizers, such as formaldehyde, which reduces their economic feasibility, especially for small- and medium-sized producers. For this reason, recent research has focused on developing protocols using green products that employ minimal effective concentrations to reduce the amount required to achieve satisfactory antibacterial effects and, consequently, lower application costs. This approach is particularly relevant because some conventional chemical products require higher concentrations to achieve the same level of sanitization. Another significant challenge lies in defining standardized application protocols that ensure consistent antibacterial efficacy without compromising egg hatchability and/or poultry performance. In summary, large-scale applications still depend on improvements in formulation stability, cost reductions to levels comparable to those of synthetic products, and the establishment of standardized usage protocols.

3. Study Field Limitation

One of the major current challenges in the field of poultry production is the scarcity of detailed investigations into formaldehyde, despite its well-established role across different production sectors, both for its recognized antibacterial activity and its toxic potential. For instance, integrated research linking histopathology, genetics, and formaldehyde in poultry, as well as case studies involving professionals who handle this compound, is practically nonexistent. A few reviews have been published in recent decades, including one in 2009 that focused on egg sanitization with formaldehyde [9] and another in 2019 that addressed its incorporation into poultry diets [94]. However, it remains evident that the topic still needs to be explored in greater depth in the poultry field despite consistent efforts over the years. Among the numerous existing knowledge gaps are those related to bacterial resistance to formaldehyde in hatchery and farm environments, dose–response relationships associated with histological and genetic damage and malformations in poultry, residues and byproducts in poultry products, and the cumulative risks for workers exposed to fumigation vapors in poultry systems. Over the years, there has been a growing interest in alternatives to formaldehyde, as evidenced by the significant increase in publications on new sanitizers and substitute strategies. However, this focus on other options has essentially treated formaldehyde as a control treatment, without fostering innovative evaluations, resulting in reduced attention to its continuous and detailed study. Although limited in scope, the present mini review aims to update the current state of knowledge on the application of formaldehyde in poultry production, encouraging new investigations that more thoroughly explore the topics discussed herein.

4. Conclusions

The current literature indicates that formaldehyde has multiple applications in the poultry industry, ranging from its use as a feed additive to the sanitization of hatching eggs. This is due to its strong antibacterial efficacy, low cost, and practicality. When managed within well-studied and properly tested protocols, it remains an effective tool for controlling bacterial contamination in hatcheries and poultry farms. However, its well-documented toxicity to embryos, poultry, and workers raises significant safety concerns. The findings summarized here emphasize that the advantages of formaldehyde are closely linked to its proper handling and application, considering both concentration and exposure time. At the same time, its risks arise mainly from improper use and excessive exposure. Among the alternatives for antibacterial treatment in poultry production, oxidative sanitizers, such as hydrogen peroxide, and plant-based sanitizers, including essential oils and botanical extracts, are potential alternatives. Protocols involving these compounds need to be designed to minimize possible limitations and to suit each stage of the production chain, ensuring their routine use and meeting the demands of industrial-scale production.

Author Contributions: Conceptualization, P.H.G.d.S.S., G.d.S.O. and V.M.d.S.; writing—original draft preparation, P.H.G.d.S.S.; writing—review and editing, P.H.G.d.S.S., G.d.S.O., C.M. and V.M.d.S.; visualization, G.d.S.O., C.M. and V.M.d.S.; supervision, V.M.d.S. All authors have read and agreed to the published version of the manuscript.

Funding: This study was funded by Conselho Nacional de Desenvolvimento Científico e Tecnológico (CNPq) and the Coordenação de Aperfeiçoamento de Pessoal de Nível Superior (CAPES; financing code 001). The APC was funded by Fundação de Apoio à Pesquisa do Distrito Federal (FAPDF).

Institutional Review Board Statement: Not applicable.

Informed Consent Statement: Not applicable.

Data Availability Statement: No new data were created or analyzed in this study. Data sharing does not apply to this paper.

Acknowledgments: The authors thank CNPq, CAPES, and FAPDF for their support.

Conflicts of Interest: The authors declare that they have no conflicts of interest.

References

1. Castañeda-Gulla, K.; Sattlegger, E.; Mutukumira, A.N. Persistent contamination of *Salmonella*, *Campylobacter*, *Escherichia coli*, and *Staphylococcus aureus* at a broiler farm in New Zealand. *Can. J. Microbiol.* **2020**, *66*, 171–185. [CrossRef] [PubMed]
2. Xu, W.; Wang, H.; Liu, L.; Miao, Z.; Huo, Y.; Zhong, Z. Prevalence and characterization of *Clostridium perfringens* isolated from different chicken farms in China. *Anaerobe* **2021**, *72*, 102467. [CrossRef]
3. Jones, P.J.; Niemi, J.; Christensen, J.P.; Tranter, R.B.; Bennett, R.M. A review of the financial impact of production diseases in poultry production systems. *Anim. Prod. Sci.* **2018**, *59*, 1585–1597. [CrossRef]
4. Wibisono, F.; Sumiarto, B.; Kusumastuti, T. Economic Losses Estimation of Pathogenic *Escherichia coli* Infection in Indonesian Poultry Farming. *Bul. Peternak.* **2018**, *42*, 341–346. [CrossRef]
5. Batkowska, J.; Al-Shammari, K.I.A.; Gryzinska, M.M.; Brodacki, A.; Wlazlo, L.; Nowakowicz-Debek, B. Effect of Using Colloidal Silver in the Disinfection of Hatching Eggs on Some Microbial, Hatchability and Performance Traits in Japanese Quail (*Coturnix cot. japonica*). *Eur. Poult. Sci.* **2017**, *81*, 1–11. [CrossRef]
6. Abo-Samaha, M.I.; Basha, H.A. Effect of Spraying Japanese Quail Eggs with Garlic Oil on Hatching Performance and Hatch Weight. *Adv. Anim. Vet. Sci.* **2021**, *9*, 156–161. [CrossRef]
7. Hrnčár, C.; Hanusová, E.; Hanus, A.; Arpášová, H.; Kokoszyński, D.; Bujko, J. The Effect of Various Disinfectants on Hatching Results in Chickens. *Sci. Pap. Anim. Sci. Biotechnol.* **2021**, *54*, 193–196.
8. Oliveira, G.S.; Nascimento, S.T.; dos Santos, V.M.; Silva, M.G. Clove Essential Oil in the Sanitation of Fertile Eggs. *Poult. Sci.* **2020**, *99*, 5509–5516. [CrossRef]
9. Cadirci, S. Disinfection of Hatching Eggs by Formaldehyde Fumigation-a Review. *Eur. Poult. Sci.* **2009**, *73*, 116–123. [CrossRef]
10. Kowatsch, S. Formaldehyde. In *Phenolic Resins: A Century of Progress*; Springer: Berlin/Heidelberg, Germany, 2010; pp. 25–40.
11. Oliveira, G.d.S.; McManus, C.; Salgado, C.B.; dos Santos, V.M. Effects of Sanitizers on Microbiological Control of Hatching Eggshells and Poultry Health during Embryogenesis and Early Stages after Hatching in the Last Decade. *Animals* **2022**, *12*, 2826. [CrossRef]
12. Williams, J.E.; Siegel, H.S. Formaldehyde levels on and in chicken eggs following preincubation fumigation. *Poult. Sci.* **1969**, *48*, 552–558. [CrossRef]
13. Cony, H.C.; Vieira, S.L.; Berres, J.; Gomes, H.A.; Coneglian, J.L.B.; Freitas, D.M.D. Técnicas de pulverização e imersão com distintos desinfetantes sobre ovos incubáveis. *Ciênc. Rural* **2008**, *38*, 1407–1412. [CrossRef]

14. Clímaco, W.L.D.S.; Melo, É.D.F.; Vaz, D.P.; Saldanha, M.M.; Pinto, M.F.V.D.S.; Fernandes, L.C.C.; Baião, N.C.; Oliveira, L.G.D.; Sant'Anna, F.M.D.; Souza, M.R.D.; et al. Eggshell Microbiology and Quality of Hatching Eggs Subjected to Different Sanitizing Procedures. *Pesq. Agrop. Bras.* **2018**, *53*, 1177–1183. [CrossRef]
15. Melo, E.F.; Clímaco, W.L.S.; Triginelli, M.V.; Vaz, D.P.; de Souza, M.R.; Baião, N.C.; Pompeu, M.A.; Lara, L.J.C. An Evaluation of Alternative Methods for Sanitizing Hatching Eggs. *Poult. Sci.* **2019**, *98*, 2466–2473. [CrossRef]
16. Branco, J.R.O.; Dallago, B.S.L.; Bernal, F.E.M. Efficiency of Ultraviolet Light for Disinfection of Fertile Broiler Eggs. *Arq. Bras. Med. Vet. Zootec.* **2021**, *73*, 1137–1146. [CrossRef]
17. Souza, L.C.; Silva, J.L.D.; Rufino, J.P.F.; Nazaré, S.C.D.; Mendonça, M.A.D.F. Efficiency of ozone compared to commercial sanitizers for hatching eggs from older breeders. *Acta Sci. Anim. Sci.* **2024**, *46*, e65262. [CrossRef]
18. Graziano, K.U.; Figueiredo, L. Paraformaldehyde Tablets microbiological activity reproducing the use conditions of the Brazilian health Institutions. *Rev. Esc. Enferm. USP* **2003**, *37*, 90–96. [CrossRef]
19. Nikolic, P.; Mudgil, P.; Whitehall, J. The in vitro antibacterial effect of permethrin and formaldehyde on *Staphylococcus aureus*. *MicrobiologyOpen* **2020**, *9*, e1054. [CrossRef] [PubMed]
20. Choroszy-Król, I.; Futoma-Kołoch, B.; Kuźnik, K.; Wojnicz, D.; Tichaczek-Goska, D.; Frej-Mądrzak, M.; Jama-Kmiecik, A.; Sarowska, J. Exposing *Salmonella* Senftenberg and *Escherichia coli* Strains Isolated from Poultry Farms to Formaldehyde and Lingonberry Extract at Low Concentrations. *Int. J. Mol. Sci.* **2023**, *24*, 14579. [CrossRef]
21. McKeen, L. Introduction to Food Irradiation and Medical Sterilization. In *The Effect of Sterilization on Plastics and Elastomers*, 3rd ed.; McKeen, L., Ed.; Plastics Design Library, William Andrew Publishing: Boston, MA, USA, 2012; pp. 1–40.
22. McDonnell, G.; Russell, A.D. Antiseptics and disinfectants: Activity, action, and resistance. *Clin. Microbiol. Rev.* **1999**, *12*, 147–179. [CrossRef] [PubMed]
23. Amponin, D.E.; Przybek-Skrzypecka, J.; Zyablitskaya, M.; Takaoka, A.; Suh, L.H.; Nagasaki, T.; Trokel, S.L.; Paik, D.C. Ex Vivo anti-microbial efficacy of various formaldehyde re-leasers against antibiotic resistant and antibiotic sensitive microorganisms involved in infectious keratitis. *BMC Ophthalmol.* **2020**, *20*, 28. [CrossRef]
24. Chung, H.; Kim, H.; Myeong, D.; Kim, S.; Choe, N.H. Effect of Chlorine Dioxide Gas Application to Egg Surface: Microbial Reduction Effect, Quality of Eggs, and Hatchability. *Korean J. Food Sci. Anim. Resour.* **2018**, *38*, 487–497.
25. Mantovam, V.B.; Dos Santos, D.F.; Giola Junior, L.C.; Landgraf, M.; Pinto, U.M.; Todorov, S.D. *Listeria monocytogenes*, *Salmonella* spp., and *Staphylococcus aureus*: Threats to the food industry and public health. *Foodborne Pathog. Dis.* **2025**. [CrossRef]
26. Badr, J.M.; Yoseif, A.I. Evaluation of Two Disinfection Programs in Disinfecting Poultry Houses Under Field Conditions. *Assiut Vet. Med. J.* **2014**, *60*, 119–127.
27. Toledo, A.M.R. Efecto de Cuatro Productos Desinfectantes Aplicados en el Material de Nido de Aves Reproductoras Pesadas. Ph.D. Thesis, Universidad de San Carlos de Guatemala, Guatemala City, Guatemala, 2005.
28. Argueta, F.M.G. Comparación de Dos Productos (Formaldehído y Paraformaldehído) Usados en la Desinfección de Cama de Nidos en Granja de Aves Reproductoras y El Efecto de Cada Uno Sobre El Porcentaje de Incubabilidad. Bachelor's Thesis, Universidad de San Carlos de Guatemala, Guatemala City, Guatemala, 2005.
29. Bampi, J.; Bampi, M.C.; Antonucci, A.M. Effectiveness of Formol and Ammonia on Disinfection of Chicken Bed. *Interfaces Científicas Saúde Ambiente* **2020**, *8*, 195–205. [CrossRef]
30. Saleh, M.; Laban, S.E.; Mostafa, G.Z.; Moubarak, S.T. An evaluation of some disinfectants effectiveness in reducing nest bedding contamination in broiler breeder farms. *J. Anim. Health Prod.* **2022**, *10*, 390–395. [CrossRef]
31. Luna, C.D.C. Comparación de Tres Protocolos de Desinfección en Huevo Fertil, SU Relación Con la Disminución en la Carga Bacteriana y Viabilidad Del Pollo de Engorde. Bachelor's Thesis, Universidad de La Salle, Bogotá, Colombia, 2014.
32. Rojas, C.E.J. Evaluación Microbiológica de Huevo Fértil con dos Procesos de Desinfección y el Uso de Cuatro Productos Comerciales en la Granja Guacata en el Municipio de Fusagasugá, Cundinamarca. Bachelor's Thesis, Universidad de Cundinamarca, Fusagasugá, Colombia, 2017.
33. Vale, I.R.R.; Oliveira, G.d.S.; de Jesus, L.M.; de Castro, M.B.; McManus, C.; dos Santos, V.M. Sustainable Bacterial Control of Hatching Eggshells Using Essential Oils. *Antibiotics* **2024**, *13*, 1025. [CrossRef] [PubMed]
34. Dos Santos, I.P.; Rieger, G.; Garcia, R.G.; Valentim, J.K.; Caldara, F.R.; de Oliveira Troguilho, A.; Sgavioli, S. Optimal paraformaldehyde levels for disinfection of eggs used in vaccine production. *Poult. Sci.* **2025**, *104*, 104614. [CrossRef] [PubMed]
35. Harlia, E.; Andrianto; Marlina, E.T.; Suryanto, D. Utilization of Bee Nest Waste as a Natural Disinfectant on Hatching Eggs Poultry. In Proceedings of the International Seminar on Tropical Animal Production, Yogyakarta, Indonesia, 20–22 October 2015; pp. 770–773.
36. Selby, C.M.; Beer, L.C.; Forga, A.J.; Coles, M.E.; Graham, L.E.; Teague, K.D.; Tellez-Isaias, G.; Hargis, B.M.; Vuong, C.N.; Graham, B.D. Evaluation of the impact of formaldehyde fumigation during the hatching phase on contamination in the hatch cabinet and early performance in broiler chickens. *Poult. Sci.* **2023**, *102*, 102584. [CrossRef] [PubMed]
37. Espinosa, H.V.G. Efecto de la Fumigación de Nacedoras con Formaldehído en el Tracto Respiratorio del Pollo Recién Nacido (Estudio Patológico). Bachelor's Thesis, Universidad Nacional Autónoma de México, Ciudad de México, Mexico, 1992.

38. Qadyanloo, B.; Rahimi, S.; Torshizi, M.A.K. Effect of organic acids and formaldehyde on morphology of broiler intestine and *Salmonella* reduction in feed. *J. Vet. Res.* **2009**, *64*, 2015–2020.
39. Avila, L.P.; Sweeney, K.M.; Schaeffer, C.; Holcombe, N.; Selby, C.; Montiel, E.; Wilson, J.L. Broiler breeder feed treatment with a formaldehyde-based sanitizer and its consequences on reproduction, feed and egg contamination, and offspring livability. *J. Appl. Poult. Res.* **2023**, *32*, 100330. [CrossRef]
40. Aarestrup, F.M.; Hasman, H. Susceptibility of different bacterial species isolated from food animals to copper sulphate, zinc chloride and antimicrobial substances used for disinfection. *Vet. Microbiol.* **2004**, *100*, 83–89. [CrossRef] [PubMed]
41. Alijani, N.; Khakpour, M.; Sabzi, S.; Fardsanei, F.; Sayfzadeh, S.; Shamlu, A.A.; Nikkhahi, F. Virulence Genes Profile, Biofilm Formation, and Resistance to Disinfectant Agents in Avian Pathogenic *Escherichia coli* (APEC) Isolated from Broiler Farms. *Iran. J. Med. Microbiol.* **2025**, *19*, 171–181. [CrossRef]
42. Hoseinzadeh, F.; Ghaemi, M.; Sharifiyazdi, H.; Hachesoo, B.A. Class 1 integron causes vulnerability to formaldehyde in *Escherichia coli*. *Arch. Microbiol.* **2021**, *203*, 4509–4515. [CrossRef]
43. Pilati, G.V.T.; Cadamuro, R.D.; Filho, V.B.; Dahmer, M.; Elois, M.A.; Savi, B.P.; Salles, G.B.C.; Muniz, E.C.; Fongaro, G. Bacteriophage-Associated Antimicrobial Resistance Genes in Avian Pathogenic *Escherichia coli* Isolated from Brazilian Poultry. *Viruses* **2023**, *15*, 1485. [CrossRef]
44. Klein, V.J.; Troøyen, S.H.; Fernandes Brito, L.; Courtade, G.; Brautaset, T.; Irla, M. Identification and characterization of a novel formaldehyde dehydrogenase in *Bacillus subtilis*. *Appl. Environ. Microbiol.* **2024**, *90*, e02181-23. [CrossRef] [PubMed]
45. Bazurto, J.V.; Nayak, D.D.; Ticak, T.; Davlieva, M.; Lee, J.A.; Hellenbrand, C.N.; Lambert, L.B.; Benski, O.J.; Quates, C.J.; Johnson, J.L.; et al. EfgA Is a Conserved Formaldehyde Sensor That Leads to Bacterial Growth Arrest in Response to Elevated Formalde-hyde. *PLoS Biol.* **2021**, *19*, e3001208. [CrossRef]
46. Roca, A.; Rodríguez-Herva, J.J.; Duque, E.; Ramos, J.L. Physiological responses of *Pseudomonas putida* to formaldehyde during detoxification. *Microb. Biotechnol.* **2008**, *1*, 158–169. [CrossRef]
47. Kummerle, N.; Feucht, H.H.; Kaulfers, P.M. Plasmid-mediated formaldehyde resistance in *Escherichia coli*: Characterization of resistance gene. *Antimicrob. Agents Chemother.* **1996**, *40*, 2276–2279. [CrossRef]
48. Roedel, A.; Vincze, S.; Projahn, M.; Roesler, U.; Robé, C.; Hammerl, J.A.; Noll, M.; Al Dahouk, S.; Dieckmann, R. Genetic but No Phenotypic Associations between Biocide Tolerance and Antibiotic Resistance in *Escherichia coli* from German Broiler Fattening Farms. *Microorganisms* **2021**, *9*, 651. [CrossRef] [PubMed]
49. Oliveira, G.d.S.; McManus, C.; dos Santos, V.M. Control of *Escherichia coli* in Poultry Using the In Ovo Injection Technique. *Antibiotics* **2024**, *13*, 205. [CrossRef]
50. Ebling, P.D.; Basurco, V. Análise das perdas econômicas oriundas da condenação de carcaças nos principais estados brasileiros produtores de frangos de corte. *Cien. Agrovet. Aliment.* **2016**, *1*, 1–11.
51. Sondossi, M.; Rossmoore, H.W.; Wireman, J.W. Induction and selection of formaldehyde-based resistance in *Pseudomonas aeruginosa*. *J. Ind. Microbiol.* **1986**, *1*, 97–103. [CrossRef]
52. Jabbar, A.; Yousaf, A.; Hameed, A.; Riaz, A.; Ditta, Y.A. Influence of fumigation strength on hatchery parameters and later life of chicks. *J. Hol. Vet. Sci. Anim. Care* **2020**, *1*, 101–106.
53. Wlazlo, L.; Drabik, K.; Al-Shammari, K.I.A.; Batkowska, J.; Nowakowicz-Debek, B.; Gryzińska, M. Use of Reactive Oxygen Species (Ozone, Hydrogen Peroxide) for Disinfection of Hatching Eggs. *Poult. Sci.* **2020**, *99*, 2478–2484. [CrossRef]
54. Badran, A.M.M.; Osman, A.M.R.; Yassein, D.M.M. Comparative Study of the Effect of Some Disinfectants on E-bryonic Mortality, Hatchability, and Some Blood Components. *Egypt. Poult. Sci. J.* **2018**, *38*, 1069–1081. [CrossRef]
55. Hrnčár, C.; Prachárová, S.; Bujko, J. The Effect of Disinfection of Hatching Eggs on Hatchability of Oravka Chickens. *Sci. Pap. Anim. Sci. Biotechnol.* **2012**, *45*, 411–414.
56. Oliveira, G.D.S.; McManus, C.; Santos, P.H.G.D.S.; de Sousa, D.E.R.; Jivago, J.L.D.P.R.; de Castro, M.B.; dos Santos, V.M. Hatching egg sanitizers based on essential oils: Microbiological parameters, hatchability, and poultry health. *Antibiotics* **2024**, *13*, 1066. [CrossRef]
57. Shahein, E.H.A.; Sedeek, E.K. Role of Spraying Hatching Eggs with Natural Disinfectants on Hatching Characteristics and Eggshell Bacterial Counts. *Egypt. Poult. Sci. J.* **2014**, *34*, 213–230. [CrossRef]
58. Rizk, Y.S. Effects of spraying eggs with garlic oil on hatching traits, post-hatch chick growth and physiological response of hatched sinai chicks. *Egypt. Poult. Sci. J.* **2022**, *42*, 187–198. [CrossRef]
59. Oliveira, G.d.S.; Vale, I.R.R.; de Jesus, L.M.; McManus, C.; Sousa, H.A.d.F.; Pires, P.G.d.S.; Jivago, J.L.d.P.R.; dos Santos, V.M. Bacterial Counts and Physical Properties of Hatching Eggshells Sprayed with a Formaldehyde Solution. *Antibiotics* **2025**, *14*, 972. [CrossRef] [PubMed]
60. Tezcan, Z.; Çelen, M. The Effects of Different Formaldehyde Fumigation for Broiler Hatching Eggs on Hatching Results. *J. Eng. Technol.* **2020**, *4*, 14–18.
61. Pagala, M.A.; Has, H.; Sandiah, N. The Effect of Fumigation Duration Using Potassium Permanganate (KMnO4) and Formalin on the Fertility and Mortality of Quail Eggs (*Coturnix coturnix japonica*). *Indones. J. Anim. Agric. Sci. IJAAS* **2024**, *6*, 196–200.

62. Zulkifli, I.; Fauziah, O.; Omar, A.R.; Shaipullizan, S.; Selina, A.H.S. Respiratory epithelium, production performance and behaviour of formaldehyde-exposed broiler chicks. *Vet. Res. Commun.* **1999**, *23*, 91–99. [CrossRef]
63. Taşdemir, A.N.; Onbaşılar, E.E.; Yalçın, S.; Boyalı, B.; Aygören, H.; Tülek, E.; Sarıçam, S.; Akan, M. Effects of Oregano Juice on Eggshell Microbial Load, Layer Embryo Development, Hatching Results, and Growth at the First 2 Weeks after Hatch. *Trop. Anim. Health Prod.* **2021**, *53*, 404. [CrossRef] [PubMed]
64. Khan, A.; Bachaya, H.A.; Khan, M.Z.; Mahmood, F. Pathological effects of formalin (37% formaldehyde) feeding in female Japanese quails (*Coturnix coturnix japonica*). *Hum. Exp. Toxicol.* **2005**, *24*, 415–422. [CrossRef]
65. Dalave, P.R.; Khan, M.A.; Gaikwad, N.Z.; Bhonsle, A.V.; Manwar, S.J.; Chavhan, S.G. Effect of formaldehyde spraying in broiler pens on growth and haemato-biochemical parameters. *Indian Vet. J.* **2018**, *95*, 11–14.
66. Bekhet, G.; Khalifa, A.Y.Z. Essential Oil Sanitizers to Sanitize Hatching Eggs. *J. Appl. Anim. Res.* **2022**, *50*, 695–701. [CrossRef]
67. Di Matteo, A.M.; Soñez, M.C.; Plano, C.M.; von Lawzewitsch, I. Morphologic Observations on Respiratory Tracts of Chickens after after Hatchery Infectious Bronchitis Vaccination and Formaldehyde Fumigation. *Avian Dis.* **2000**, *44*, 507–518. [CrossRef]
68. De Freitas, A.G. Efeito da Fumigação de Nascedouros com Formaldeído Sobre o Trato Respiratório e Desempenho de Frangos de Corte. Bachelor's Thesis, Federal University of Uberlândia, Minas Gerais, Brazil, 2007.
69. Albaghdady, E.F.; Alhachem, E.I.; Kadhim, A.B. Effects of Thirty-Day Exposure of Quails (Coturnix Coturnix) to Formalin Vapor on Biochemical, Immunological, and Histological Features of Liver and Kidney. *Asia Pac. J. Med. Toxicol.* **2023**, *12*, 137–144.
70. Al-Saeed, M.H.; Hamza, B.S.; Ahmed, H.A. Evaluation the exposure of formalin as a disinfectant for poultry house on hematological, biochemical parameters and histopathological examination in broiler. *Al-Qadisiya J. Vet. Med. Sci.* **2013**, *12*, 144–150. [CrossRef]
71. Santos, P.H.G.d.S.; Oliveira, G.d.S.; Cerqueira, L.d.A.; Jivago, J.L.d.P.R.; Paixão, S.S.R.M.; de Castro, M.B.; McManus, C.; Santos, V.M.d. Formaldehyde Fumigation: Antibacterial Profile and Toxic Effects on Hatching Eggs. *Toxics* **2025**, *13*, 851. [CrossRef] [PubMed]
72. Valverde-Santiago, M.; Pontel, L.B. Emerging mechanisms underlying formaldehyde toxicity and response. *Mol. Cell* **2025**, *85*, 2068–2079. [CrossRef] [PubMed]
73. Matias, F.R.; Groves, I.; Durrans, J.; Herigstad, M. Carbon monoxide affects early cardiac development in an avian model. *Birth Defects Res.* **2024**, *116*, e2330. [CrossRef] [PubMed]
74. Amini, M.; Zadeh-Hashem, E.; Allymehr, M. Assessment of the effect of kinetin against formic acid toxicity in chicken embryo model. *J. Anim. Physiol. Anim. Nutr.* **2023**, *107*, 238–247. [CrossRef]
75. Ragaa, N.M.; Korany, R.M.S. Studying the Effect of Formic Acid and Potassium Diformate on Performance, Immunity and Gut Health of Broiler Chickens. *Anim. Nutr.* **2016**, *2*, 296. [CrossRef]
76. EFSA Panel on Additives and Products or Substances used in Animal Feed (FEEDAP). Scientific Opinion on the safety and efficacy of formic acid when used as a technological additive for all animal species. *EFSA J.* **2014**, *12*, 3827. [CrossRef]
77. Da Silva, R. Efeitos tóxicos do formaldeído e a regulamentação aplicada em atividades ocupacionais. *RevistaFT* **2024**, *28*, 1–15.
78. Velasco, N.J.H.; Ríos, Y.A.O. Principales Usos y Sus Posibles Consecuencias en los Trabajadores Avícolas Expuestos al Formaldehído 2020/2021. Ph.D. Thesis, Corporación Universitaria Minuto de Dios, Programa de Ciencias Empresariales, Bogotá, Colombia, 2021.
79. Idrobo-Avila, E.H.; Vasquez-López, J.A.; Vargas-Cañas, R. La exposición ocupacional al formol y la nueva tabla de enfermedades laborales. *Rev. Salud Pública* **2017**, *19*, 382–385. [CrossRef]
80. Castellani, F.; Vitali, M.; Antonucci, A.; Cofone, L.; D'Ancona, G.; Pindinello, I.; Buomprisco, G.; Petyx, M.; Ursini, C.L.; Protano, C. Effective mitigation strategies for reducing workers' exposure to formaldehyde: A systematic review. *Air Qual. Atmos. Health* **2024**, *17*, 877–890. [CrossRef]
81. Oliveira, G.D.S.; McManus, C.; dos Santos, V.M. Garlic as Active Principle of Sanitiser for Hatching Eggs. *Worlds Poult. Sci. J.* **2022**, *78*, 1037–1052. [CrossRef]
82. Galgano, M.; Pellegrini, F.; Fracchiolla, G.; Mrenoshki, D.; Zarea, A.A.K.; Bianco, A.; Del Sambro, L.; Capozzi, L.; Schiavone, A.; Saleh, M.S.; et al. Pilot Study on the Action of *Thymus vulgaris* Essential Oil in Treating the Most Common Bacterial Contaminants and *Salmonella enterica* Subsp. *enterica* Serovar Derby in Poultry Litter. *Antibiotics* **2023**, *12*, 436.
83. Bishnoi, P.; Sharma, S.; Singh, N.; Karnani, M.; Mishra, G.; Dhakad, G.S.; Sharma, R.A. Effect of neem leaves (*Azadirachta indica*) as a supplementary litter material on performance of broiler birds. *Indian J. Poult. Sci.* **2021**, *56*, 155–159. [CrossRef]
84. Marchioro, T.; Stefani, L.M.; Zampar, A.; Strapazzon, J.V.; Araújo, D.N.; Boiago, M.M. The Use of *a* Compound Based on Phyllosilicates and Cinnamon Essential Oil for Chicken Broiler's Litter. *Braz. J. Poult. Sci.* **2022**, *24*, eRBCA-2022. [CrossRef]
85. Gontar, Ł.; Sitarek-Andrzejczyk, M.; Kochański, M.; Buła, M.; Drutowska, A.; Zych, D.; Markiewicz, J. Dynamics and diversity of microbal contamination in poultry bedding materials containing parts of medical plants. *Materials* **2022**, *15*, 1290. [CrossRef]
86. De Jesus, L.M.; Oliveira, G.D.S.; Vale, I.R.R.; McManus, C.; Sousa, H.A.D.F.; dos Santos, V.M. Tahiti Lemon Juice: A Natural Alternative to Reduce Bacteria from Eggshells. *Microbiol. Res.* **2024**, *15*, 1406–1411. [CrossRef]

87. Ayuningtyas, G.; Martini, R.; Yulianti, W. The role of dipping duck hatching eggs with cherry leaf extract as natural sanitizers on hatching performance and eggshell bacterial counts. *E3S Web Conf.* **2022**, *348*, 00023. [CrossRef]
88. Iraqi, E.E.; El-Sahn, A.A.; El-Barbary, A.M.; Ahmed, M.M.; Elkomy, A.E. Antimicrobial activity of tea tree and lavender essential oils and their effects on hatching performance and eggshell bacterial count of Japanese quail eggs. *BMC Vet. Res.* **2025**, *21*, 176. [CrossRef] [PubMed]
89. Ergun, F.; Taskin, A.; Ergun, D. Effects of spraying cemele pepper (*Capsicum annuum* L.) extracts onto hatching eggs of atak-s layer hens on hatching results, serum glutathione and malondialdehyde levels. *J. Anim. Plant Sci.* **2024**, *34*, 1210–1219. [CrossRef]
90. Huanhong, K.; Lumsangkul, C.; Arjin, C.; Sirilun, S.; Tangpao, T.; Wang, Y.-L.; Mektrirat, R.; Lin, C.-S.; Sommano, S.R.; Sringarm, K. Dietary supplementation of coffee pulp extract enhances growth performance and intestinal morphology in broiler chicken. *Poult. Sci.* **2025**, *104*, 104873. [CrossRef]
91. Sinurat, A.P.; Haryati, T.; Purba, M.; Pasaribu, T.; Irawan, Y.; Setyawan, R.H.; Saputra, F.; Ilyas, M.; Miraya, N.; Sumiati, S. Evaluation of Curcuma xanthorrhiza Roxb. extract as a functional alternative to antibiotic growth promoters in broiler chicken nutrition. *Vet. World* **2025**, *18*, 1944–1954. [CrossRef] [PubMed]
92. Escobar, F.M.; Magnoli, A.; Barrios, R.B.; Montenegro, M.A.; Vergara, L.I.D.; Luna, M.J.; Parada, J.; Cavaglieri, L.R. Effect of microencapsulated *Minthostachys verticillata* essential oil as an antibiotic substitute on broiler performance. *Iraqi J. Vet. Sci.* **2025**, *39*, 439–449. [CrossRef]
93. Adil, S.; Banday, M.T.; Hussain, S.A.; Wani, M.A.; Al-Olayan, E.; Patra, A.K.; Rasool, S.; Gani, A.; Sheikh, I.U.; Khan, A.A.; et al. Impact of Nanoencapsulated Rosemary Essential Oil as a Novel Feed Additive on Growth Performance, Nutrient Utilization, Carcass Traits, Meat Quality and Gene Expression of Broiler Chicken. *Foods* **2024**, *13*, 1515. [CrossRef]
94. Ricke, S.C.; Richardson, K.; Dittoe, D.K. Formaldehydes in feed and their potential interaction with the poultry gastrointestinal tract microbial community—A review. *Front. Vet. Sci.* **2019**, *6*, 188. [CrossRef] [PubMed]

antioxidants

MDPI

Review

Oxalis corniculata L. As a Source of Natural Antioxidants: Phytochemistry, Bioactivities, and Application Potential

Tao Zhong [1,2], Junying He [1,2], Hao Zhao [1,2], Chang Tan [1,2], Wenjing Zhou [1,2], Congming Wu [1,2] and Jijun Kang [1,2,*]

1 Technology Innovation Center for Food Safety Surveillance and Detection (Hainan), Sanya Institute of China Agricultural University, Sanya 572025, China; sy20243051241@cau.edu.cn (T.Z.); sy20233051122@cau.edu.cn (J.H.); b20243050624@cau.edu.cn (H.Z.); tanchang@cau.edu.cn (C.T.); s20233050999@cau.edu.cn (W.Z.); wucm@cau.edu.cn (C.W.)

2 State Key Laboratory of Veterinary Public Health and Safety, College of Veterinary Medicine, China Agricultural University, Beijing 100193, China

* Correspondence: kangjijun@cau.edu.cn

Abstract: *Oxalis corniculata* L. (*O. corniculata*) has attracted increasing attention as a natural source of antioxidants with diverse pharmacological potential. Phytochemical studies have identified a diverse spectrum of metabolites, dominated by flavonoids, polysaccharides, and organic acids. These compounds exhibit antioxidant properties as well as related biological activities, including anti-inflammatory, antimicrobial, neuroprotective, hypoglycemic, and anticancer effects. Its long-standing use in traditional remedies, along with its incorporation into approved Chinese patent medicines, underscores its safety and translational value. This review synthesizes recent advances in the chemical composition, bioactivities, and molecular mechanisms of *O. corniculata*, emphasizing its antioxidant-driven pharmacological prospects. The review highlights *O. corniculata* as a sustainable and accessible botanical resource with significant potential for the development of pharmaceuticals, dietary supplements, and health-promoting applications.

Keywords: *Oxalis corniculata* L.; phytochemistry; antioxidant activity; biological functions

1. Introduction

Plant-derived natural products have long served as an indispensable source of bioactive compounds, owing to their structural diversity and broad pharmacological potential [1]. In the context of increasing demand for sustainable resources in food and health industries, herbs with wide adaptability, ease of cultivation, and abundant bioactive constituents are attracting renewed scientific and industrial interest [2,3].

O. corniculata, commonly known as creeping woodsorrel, belongs to the family Oxalidaceae and is one of the most widely distributed vascular plants worldwide [4,5]. Traditionally, it has been used in various medical systems to treat digestive, dermatological, and urinary disorders, and in modern practice it is also included in several approved Chinese patent medicines for related indications [6]. Pharmacological studies have demonstrated that extracts of *O. corniculata* exhibit diverse biological activities, including antioxidant, anti-inflammatory, antimicrobial, hypoglycemic, and anticancer effects [7]. Flavonoids, polysaccharides, and organic acids are recognized as the major classes of bioactive compounds in *O. corniculata* [8,9]. Building on these findings, subsequent studies suggest that these compounds play key roles in mediating its pharmacological effects through antioxidant enhancement and regulation of inflammation, metabolism, and cellular protection [10]. In addition, emerging formulation and delivery strategies, such as nanomaterial-based systems and exosome-mediated approaches [11], have been explored to improve stability, bioavailability, and functional efficacy. Preliminary toxicological evaluations in animal models also indicate a favorable safety margin, supporting the translational potential of *O. corniculata* [12].

Although extensive research has been conducted, a comprehensive and up-to-date synthesis of the phytochemistry, antioxidant mechanisms, and health-related applications of *O. corniculata* remains lacking. By systematically summarizing recent advances, this review aims to clarify its antioxidant basis, link bioactive constituents to biological activities, and provide new perspectives on its potential roles in dietary supplements and therapeutic applications. *O. corniculata* also offers a sustainable source for the development of natural antioxidant-based products.

2. Botanical Features and Geographic Distribution

O. corniculata is a small herbaceous plant, annual to perennial in habit. It is characterized by profusely branched [13], slender, and creeping stems with reddish nodes that readily produce adventitious roots, enabling rapid vegetative propagation in environments (Figure 1) [14]. Plants typically reach 10–40 cm in height and are sparsely covered with fine hairs. The leaves are trifoliolate, bearing obcordate leaflets (4–16 mm long, 4–22 mm wide) notched at the apex and ciliate along the margins. Flowers are axillary, solitary, or borne in small umbels, each comprising five yellow petals (6–8 mm) and five lanceolate sepals (3–5 mm). The floral structure includes ten stamens of unequal lengths and a five-loculed ovary topped with capitate stigmas. The fruit is a cylindrical capsule (1–2.5 cm) that dehisces explosively at maturity, releasing ovoid seeds (1–1.5 mm) with transverse ridged-reticulate ornamentation [15]. Flowering and fruiting occur from March to September. (https://www.iplant.cn/, accessed on 1 September 2025)

Figure 1. Morphology of *O. corniculata*. (**A**): Field photograph of *O. corniculata* displaying its characteristic trifoliate leaves and yellow flowers; (**B**): Diagrammatic illustration of *O. corniculata*, displaying the whole plant (III) with detailed depictions of the flower (I), leaf (II), fruit (IV), and seed (V).

O. corniculata exhibits a nearly global distribution, thriving across tropical, subtropical, and temperate regions (Figure 2). Native to Asia, particularly East, South, and Southeast Asia, it has since naturalized across Europe, the Mediterranean, Oceania, and the Americas [16]. The species grows most vigorously at elevations of 500–1500 m, with higher density and frequency observed during the rainy season [17]. It commonly colonizes roadsides, grasslands, field margins, and forest edges, especially in moist or human-disturbed habitats. Its ecological success is largely attributed to adaptive growth habits and versatile reproductive strategies, which support survival and dispersal across diverse environments [18].

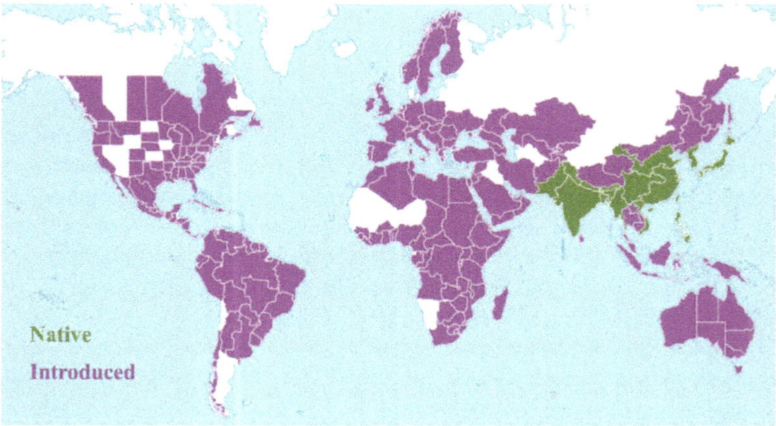

Figure 2. Global distribution of *O. corniculata*. Green areas indicate regions where the species is native, while purple areas represent regions where it has been introduced (https://powo.science.kew.org/taxon/urn:lsid:ipni.org:names:177893-2 (accessed on 1 July 2025)).

3. Phytochemistry

Phytochemical studies have identified 227 chemical constituents (Table S1) in *O. corniculata*, primarily isolated from its whole herb and leaves. Flavonoids are the predominant class of compounds, with approximately 84 identified in the plant [19], contributing to a total flavonoid content of up to 13.5 mg/g [10]. Organic acids, particularly phenolic acids, represent another major class of active constituents, with 51 organic acids identified, including 23 phenolic acids [20,21]. The total phenolic content in dry plant samples is quantified at 98.6 μg/g, with notable concentrations of hydroxybenzoic acid, p-coumaric acid, caffeic acid, and others [22]. The terpenoid fraction, primarily monoterpenes, consists of approximately 20 different compounds, with the highest concentration found in the leaves. Terpenoids such as geraniol and linalool have demonstrated antioxidant, antimicrobial, and anti-inflammatory activities [23].

Notably, compounds 47 and 48 (new flavones), compound 160 (a novel alkaloid, Aspergillus triazolate A), and compound 161 (a lignan named corniculin A) are among the novel constituents first isolated from *O. corniculata*. Additional constituents include coumarins, simple phenolics, fatty acid esters, anthraquinones, aldehydes, glycosides, sterols, carbohydrates, amino acids, carotenoids, pigments, and other specialized metabolites. Typical compounds are shown in Figure 3.

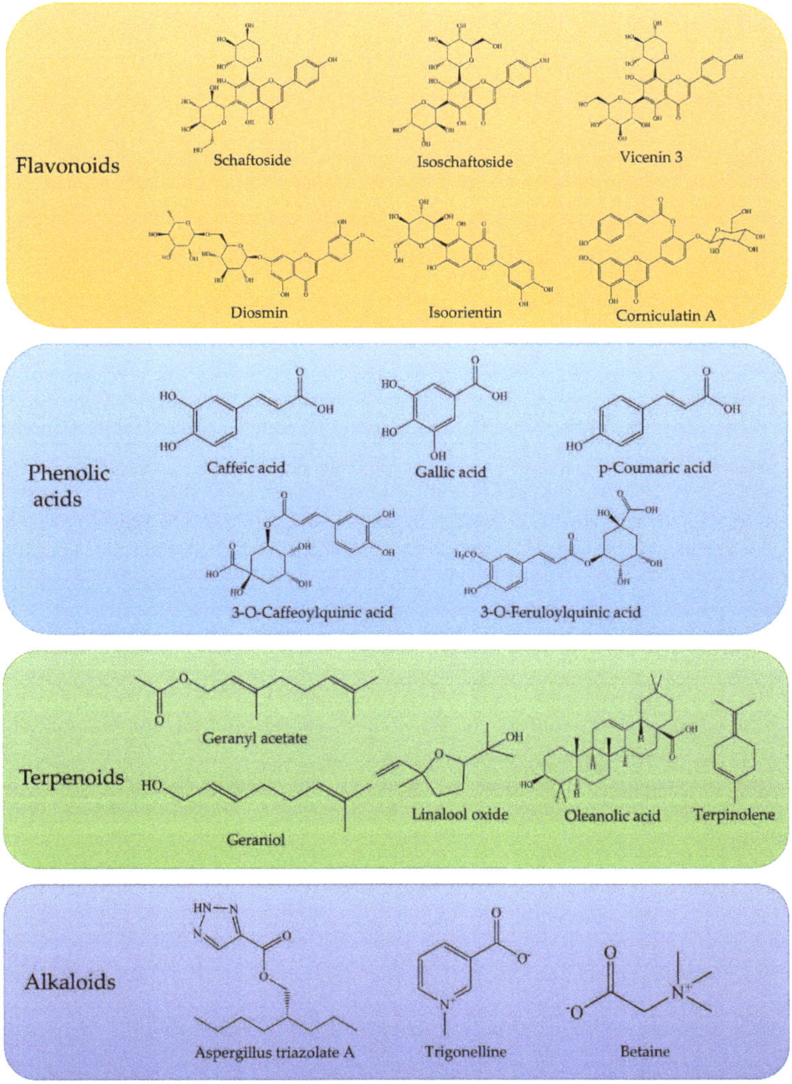

Figure 3. Representative flavonoids, phenolic acids, terpenoids, and alkaloids identified in *O. corniculata*.

3.1. Flavonoids

Flavonoids are representative compounds in *O. corniculata*, featuring a basic 2-phenylchromen-4-one skeleton. They are categorized into several subclasses, including flavones, flavonols, dihydroflavones, isoflavones, chalcone, dihydrochalcones, and flavanols (Figures S1 and S2). The antioxidant activity of these compounds depends on both the number and position of hydroxyl groups in their structure. Hydroxylation at the 3, 5, and 7 positions and the presence of a C2=C3 double bond in the C-ring enhance radical-scavenging activity and anti-inflammatory effects [24]. For instance, flavonols possess a hydroxyl or other oxygen-containing substituent at the 3-position of the flavonoid backbone, a structural feature that has been associated with enhanced antioxidant activity [25]. In addition, glycosylation generally improves water solubility and bioavailability, which may enhance physiological efficacy [26]. These insights underscore the importance of structure-activity relationships in determining the functional properties of flavonoids.

In *O. corniculata*, flavonoid content may reach optimal levels when harvested in June or October, periods likely corresponding to peaks in secondary metabolite accumulation [27]. Flavonoids occur mainly as C-glycosyl and O-glycosyl flavones. C-glycosyl flavones (schaftoside, isoschaftoside, and isovitexin) demonstrate strong radical-scavenging activity and superior structural stability [28]. In contrast, O-glycosyl flavones (orientin, isoorientin and diosmin) exhibit enhanced solubility and absorption [29]. These flavonoids collectively represent the major contributors to the antioxidant and anti-inflammatory activities of *O. corniculata*.

3.2. Organic Acids

Organic acids constitute an important class of phytochemicals in *O. corniculata*, contributing both to its characteristic sour taste and to its diverse biological activities [20]. Among them, phenolic acids (85–107) are dominated by caffeic acid and its derivatives (Figure S3A). Caffeic acid content peaks in dry-area samples, reaching up to 1.28 μg/g [22], with lower concentrations observed in samples from marshy and moist areas. Caffeic acid derivatives, characterized by a catechol moiety, exhibit strong free radical scavenging capacity and protect cells against oxidative stress. Their antioxidant potential is largely determined by structural factors such as the number and position of hydroxyl groups on the aromatic ring, while modifications like methoxylation can further modulate activity [30].

In addition to phenolic acids, *O. corniculata* also contains low-molecular-weight organic acids such as malic and citric acids [20]. Fatty acid-type organic acids have been identified as well, with gas chromatography-mass spectrometry (GC/MS) analyses revealing oleic acid and 6-octadecenoic acid as predominant components (31.08%), followed by palmitic acid (2.55%) (Figure S3B) [31]. These fatty acids are associated with metabolic regulation and potential cardiovascular protective effects [32].

3.3. Terpenoids

Terpenoids, biosynthesized from isoprene units, constitute a structurally diverse class of natural products that contribute to plant defense, volatile signaling, and physiological regulation [33]. To date, 20 terpenoids (136–155) have been identified in *O. corniculata*, comprising monoterpenes (136–150), a cycloether monoterpene (151), and triterpenes (152–155).

Supercritical CO_2 extraction followed by GC/MS analysis of leaf samples from Guangdong Province, China, revealed that monoterpenes dominate the essential oil fraction [23]. The most abundant constituents were geranyl acetate (13.3%), terpinolene (9.2%), linalool oxide (7.4%), and geraniol (6.4%). These compounds impart characteristic aromatic properties and exhibit antioxidant, antimicrobial, and anti-inflammatory activities [34]. In addition to volatile monoterpenes, several triterpenoids have been detected in the non-volatile fractions, including oleanolic acid, eburicoic acid, squalene, and phytol [31,35]. The chemical structures of the identified terpenoids are shown in Figure S4.

3.4. Alkaloids

Alkaloids are nitrogen-containing natural compounds with diverse chemical structures [36]. Five alkaloids (156–160) have been reported from *O. corniculata*. Among them, trigonelline (156), a classical pyridine alkaloid with an N^+-methylated pyridine ring, displays high water solubility due to its quaternary ammonium structure [37]. Betaine (158), a naturally occurring zwitterion with a trimethylated amino group, serves as an important osmolyte and methyl donor, and has been associated with antioxidant, hepatoprotective, and metabolic regulatory effects [38].

Particularly noteworthy is compound 160, Aspergillus triazolate A (ATA), a novel triazole alkaloid first isolated from the dried whole plant of *O. corniculata*. Structurally, ATA consists of a hydrophilic triazole ring and a hydrophobic C_8 alkyl chain, forming an amphiphilic scaffold. This structural balance is proposed to facilitate stable interactions with target enzymes, thereby enhancing inhibitory activity. Bioassays have demonstrated that ATA lowers blood glucose levels by inhibiting α-glucosidase (α-Glu), highlighting its potential as a natural lead compound for α-Glu inhibitor development [39]. The chemical structures of these alkaloids are shown in Figure S5.

3.5. *Polysaccharides*

Polysaccharides are natural macromolecules consisting of more than ten monosaccharide units linked through glycosidic bonds, often comprising hundreds to thousands of residues [40]. In *O. corniculata*, crude polysaccharides (OCP) are typically extracted using hot water and ethanol precipitation, followed by purification steps such as deproteinization and ethanol fractionation, with a total yield of 9.45% [7]. The major fraction, OCP-3, has a molecular weight of 31.5 kDa and is mainly composed of arabinose (47.83%) and galacturonic acid (17.81%), indicating its acidic nature. OCP-3 demonstrates strong free radical scavenging capacity, suppresses lipid peroxidation, and confers protection against oxidative damage in both cell-based and in vivo models [7].

3.6. *Nutrients*

On a dry-weight basis, *O. corniculata* contains considerable amounts of essential minerals, including calcium (5.63%), potassium (3.15%), and magnesium (2.63%) [41]. Trace elements are also present [41], with iron (89.16 ppm), manganese (4.21 ppm), zinc (1.59 ppm), and copper (0.12 ppm) detected. The vitamin C content reaches 0.139 mg/g. Fresh methanolic extracts provide approximately 710 μg vitamin C equivalent antioxidant capacity per gram of fresh leaves [42], reflecting strong antioxidant capacity attributable to vitamin C. This effect is likely enhanced by synergistic interactions with flavonoids, organic acids, and other bioactive constituents [43]. In addition, dried *O. corniculata* powder contains appreciable levels of crude protein (12.25%), crude fiber (10.64%), and ash (9.98%). The total antioxidant capacity (TAC) has been measured at 31.60 mmol ascorbic acid equivalent/kg dry weight and 31.22 mmol vitamin E equivalent/kg dry weight, indicating strong overall antioxidant potential [44].

4. Traditional Uses

O. corniculata has a long-standing history in traditional medicine, first documented in the Xin Xiu Ben Cao in 659 AD [45]. Over centuries, it has been incorporated into diverse ethnomedical systems, including Traditional Chinese Medicine (TCM), Ayurveda, and Unani. In oral applications, decoctions, teas, or fresh juices prepared from the whole plant or leaves are commonly used to relieve gastrointestinal discomfort, fever, urinary disorders, and liver ailments [6,46]. Topical preparations, such as poultices or expressed juices, are applied to treat skin conditions including warts, eczema, wounds, and abscesses, reflecting recognition of its antimicrobial and wound-healing effects [47,48].

In TCM, *O. corniculata* is traditionally categorized as a heat-clearing and detoxifying herb, used to eliminate damp-heat, relieve strangury, and reduce swelling and pain. It has been incorporated into several approved Chinese patent medicines, such as Mi Lin Qing Capsules (for damp-heat urinary tract infections/strangury), Fu Yan Xiao Capsules (for gynecological inflammation), and Gu Kang Capsules (for traumatic injury and swelling). These commercial formulations illustrate its compatibility within compound prescriptions and its translational potential from traditional use to standardized products. These applications reflect its traditional recognition for antimicrobial, antioxidant, and anti-inflammatory properties (Table 1).

Table 1. Traditional uses of *O. corniculata*.

Country	Components and Preparation Methods	Applications	Prescription Source
China	Harvested in April and May, shade-dried, and pounded for juice.	Treats sores; pounded and applied to kill parasites.	Xin Xiu Ben Cao (Tang Dynasty, 659 AD) [45]
China	Pounded for juice and prepared with vinegar	Indicated for strangury, hematuria, dysuria, and pediatric gingival ulceration with bleeding and fetor.	Zheng Lei Ben Cao (Song Dynasty, 1082 AD) [45]
China	Pounded for juice and prepared with wine	Treats strangury with painful reddish urination and bladder heat; for postpartum uterine prolapse.	Shi Yi De Xiao Fang (Yuan Dynasty, 1337 AD) [45]
China	Decocted with Bassia scoparia seeds and Plantago asiatica seeds	Treats strangury and leukorrhea.	Ben Cao Gang Mu (Ming Dynasty, 1578 AD) [49]
China	Leaves harvested in summer and shade-dried. Juice prepared with wine, honey, or sugar.	For sores, thirst, strangury, urinary retention, and difficult labor; for red and white vaginal discharge in women.	Zheng Zhi Zhun Sheng (Ming Dynasty, 1602 AD) [45]
China	Decocted with rock sugar and taken orally	For chronic diarrhea and various types of dysentery.	Dian Nan Ben Cao (Ming Dynasty, 1436 AD) [49]
China	Pounded for juice and prepared with wine, vinegar, honey, or sugar	For strangury, carbuncles, heat toxins, burns, and venomous bites; promotes urination and defecation. for phlegm disorders and anal swelling.	De Pei Ben Cao (Qing Dynasty, 1761 AD) [45]
India	With turmeric and Emblica officinalis in decoction; leaf juice with honey as mouth rinse; also consumed and used as fodder.	For diarrhea, oral ulcers, skin infections, and lowering blood glucose.	Ayurvedic Tradition [6,50,51]
Tanzania	Whole plant decoction; crushed fresh leaves applied topically	Treatment of skin infections, wounds, conjunctivitis, sore throat, fever, gastrointestinal discomfort	Ethnobotanical practice in Tanzania [52]
/	Used fresh, juiced, made into paste or decoction; administered orally or topically	Treat liver diseases, jaundice, skin disorders, and urinary tract infections	Traditional medical systems: Ayurveda, Unani, and Siddha [53]
Pakistan	Fresh leaves boiled in water for 2 h and placed for cooling.	For vitamin C deficiency and mouth smell	Traditional Persian Medicine [54]
China	Combined with Musa basjoo rhizome, Psoralea corylifolia, Dipsacus asper, etc	For fractures, coxarthritis, and osteoporosis associated with liver-kidney deficiency and meridian obstruction.	Gu Kang Capsule (www.nmpa.gov.cn)
China	Combined with Alternanthera philoxeroides, Agrimonia pilosa, Plantago asiatica, etc	For urinary tract infections, including cystitis, strangury, dysuria, and painful urination.	Mi Lin Qing Capsule (www.nmpa.gov.cn/)

Table 1. *Cont.*

Country	Components and Preparation Methods	Applications	Prescription Source
China	Combined with Boenninghausenia albiflora, Polygonum capitatum and Plantago asiatica	For treating stranguria caused by damp-heat accumulation, characterized by difficult urination, dribbling, and painful urination.	Mi Ling Granule/Capsule (www.nmpa.gov.cn/)
China	Combined with Patrinia scabiosaefolia, Rheum palmatum, Paeonia suffruticosa bark, ect	Adjuvant treatment for chronic gynecological inflammation; also regulates menstruation and relieves pain.	Fu Yan Xiao Capsule (www.nmpa.gov.cn/)
China	Combined with Root of Musa basjoo and Curcuma longa	For traumatic injuries, contusions, and soft tissue damage with swelling and pain.	Zhong Tong Shu Spray (www.nmpa.gov.cn/)

5. Biological Activities of *O. corniculata*

O. corniculata exhibits a wide range of biological activities, increasingly supported by experimental evidence. These effects are largely attributed to the synergistic actions among its diverse phytochemicals, particularly flavonoids, polysaccharides, and phenolic acids. The plant demonstrates notable potential in antioxidant, anti-inflammatory, antimicrobial, hepatoprotective, neuroprotective, hypoglycemic, and anticancer applications. An overview of these biological activities is presented in Figure 4, providing a foundation for the detailed discussions that follow.

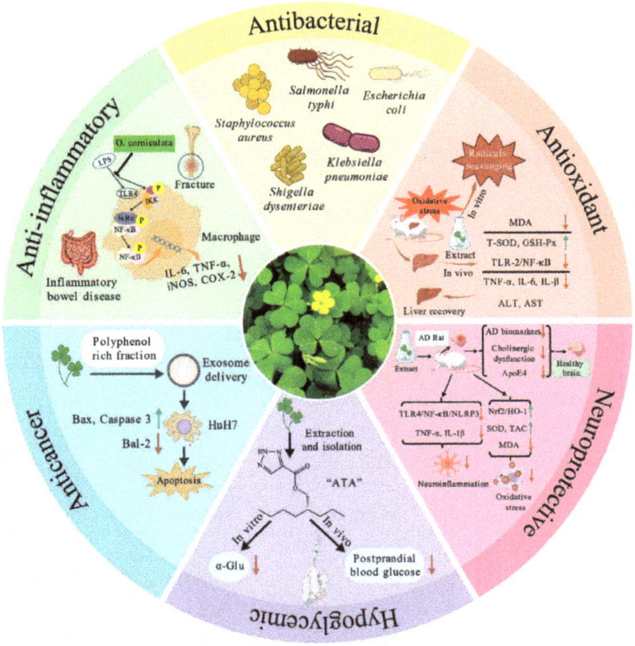

Figure 4. Biological Activities of *O. corniculata*. Upward-pointing arrows denote upregulation of gene/protein expression or protective markers, while downward-pointing arrows indicate downregulation of expression levels or damage markers. Several elements in the image were obtained from https://biogdp.com (accessed on 1 September 2025) [55].

5.1. Antioxidant Activity

Oxidative stress caused by excessive reactive oxygen species (ROS) is a major contributor to lipid peroxidation, protein oxidation, and DNA damage, processes implicated in chronic diseases such as hepatic dysfunction, neurodegeneration, and cancer [56]. It suppresses enzymatic antioxidants, including superoxide dismutase (SOD), catalase (CAT), and glutathione peroxidase (GSH-Px), as well as non-enzymatic antioxidants such as glutathione (GSH). This imbalance leads to the accumulation of oxidative products such as malondialdehyde (MDA) and protein carbonyls (PC), which further aggravate cellular injury and inflammation [7].

O. corniculata contains abundant bioactive constituents, notably flavonoids, phenolic acids, and polysaccharides, that contribute to strong free radical scavenging and anti-oxidative stress effects (Table 2) [4]. The polysaccharide fraction OCP-3 inhibits lipid peroxidation and protects DNA from oxidative damage. In both HEK-293 cells and *C. elegans*, OCP-3 reduces MDA and PC levels, while enhancing antioxidant enzymes including SOD, CAT, and GSH-Px [7]. Methanol extracts of *O. corniculata* activate the Nrf2/HO-1 signaling pathway, enhancing antioxidant enzyme activities [57]. In carbon tetrachloride-induced hepatic and renal injury models, the extracts restore the activities of multiple antioxidant enzymes, including SOD, CAT, GSH-Px, glutathione-S-transferase (GST), glutathione reductase (GSR), peroxidase (POD), and quinone reductase (QR). The extracts also lower MDA levels, suppress lipid peroxidation, improve serum biochemical indices such as alanine aminotransferase (ALT) and aspartate aminotransferase (AST), and mitigate histopathological damage [58,59].

Dietary supplementation with *O. corniculata* powder has been shown to enhance antioxidant capacity under heat stress in broilers by increasing SOD, CAT, and GSH-Px activities while reducing MDA levels [60]. It also improved oxidative stability and nutritional quality of broiler meat by lowering lipid peroxidation and increasing polyunsaturated fatty acid (PUFA) content [4,5]. These findings indicate that *O. corniculata* supplementation effectively mitigates oxidative stress and maintains redox balance in animal systems, supporting its use as a natural antioxidant source in animal nutrition [44].

Table 2. Summary of antioxidant activity studies of *O. corniculata*.

No.	Testing Subjects	Application Part or Compounds	Doses/Duration	Effects	Ref.
1	Free radicals	Polyphenol-rich methanol extract	0.5–50 μg/mL	DPPH·scavenging	[61]
2	pUC18 plasmid DNA	Polyphenol-rich methanol extract	100–500 ng/mL	Protected DNA from·OH-induced strand breaks	[61]
3	BSA	Polyphenol-rich methanol extract	100–500 μg/mL	Inhibited PC formation	[61]
4	Liver cells	Polyphenol-rich methanol extract	10–50 μg/mL, 45 min incubation	Protected liver cells from oxidative damage by OH	[61]
5	Free radicals; Fe^{3+}	Hot water extract	50–400 μg/mL	Free radicals scavenging and iron-reducing capacity	[62]
6	HEK-293 cells; *C. elegans*	Acidic polysaccharide OCP-3	50–800 μg/mL; 4–8 mg/mL	SOD ↑, CAT ↑, GSH-Px ↑, MDA ↓, PC ↓, radicals scavenging, DNA protection	[7]

Table 2. *Cont.*

No.	Testing Subjects	Application Part or Compounds	Doses/Duration	Effects	Ref.
7	Free radicals	Biogenic silver nanoparticles synthesized using aqueous extract of *O. corniculata*	50–400 μg/mL	DPPH·/ABTS·scavenging	[31]
8	Free radicals, Fe^{2+}, Fe^{3+}, lipid	Ethanol extract	DPPH: 26.2 ± 2 μg/mL; Ferrous Ion: 74.3 ± 0.4 μg/mL; NO: 75.4 ± 7.3 μg/mL; ABTS: 59.9 ± 5.2 μg/mL; NBT: 118.2 ± 2.3 μg/mL; FRAP: 152.1 ± 9.5 μg/mL; LPO: 1.8 ± 0.4 μg/mL	The antioxidant effect was as demonstrated by IC_{50} values in seven assays.	[63]
9	Free radicals	TFO	100–500 μg/mL	DPPH/OH/O_2-scavenging	[10]
10	Pork lard	*O. corniculata* polyphenols	0.02–0.04%, 50 °C, 21 days	Significantly inhibited lipid peroxidation	[64]
11	Fractured rats	*O. corniculata* (whole plant, 47.5% of mixture) aqueous extract	150–600 mg/kg, percutaneous and p.o., 2 weeks	SOD ↑, CAT ↑, GSH ↑, enhances antioxidant enzyme activity and alleviates oxidative stress in bone tissue	[65]
12	$AlCl_3$-induced AD rats	Methanol extract	150 mg/kg, p.o., 5 weeks	TAC ↑, SOD ↑, MDA ↓; upregulates Nrf2/HO-1 pathway, enhances antioxidant defense, reduces oxidative damage in brain	[57]
13	SD rats (CCl_4-induced acute liver injury)	Aqueous extract	4–16 g/kg, p.o., 10 days	ALT ↓, AST ↓, MDA ↓, T-SOD ↑, GSH-Px ↑; downregulates TLR2/NF-κB pathway, inhibits oxidative stress	[66]
14	CCl_4-induced nephrotoxic rats	Methanol extract	100, 200 mg/kg, p.o., 7 days	CAT ↑, POD ↑, SOD ↑, GSH-Px ↑, GST ↑, GSR ↑, QR ↑, GSH ↑, MDA ↓, protein oxidation ↓, renal injury markers ↓, improved renal histology	[58]
15	CCl_4-induced hepatotoxic rats	Methanol extract	100, 200 mg/kg, p.o., 7 days	CAT ↑, POD ↑, SOD ↑, GSH-Px ↑, GST ↑, GSR ↑, QR ↑, GSH ↑, TBARS ↓, AST ↓, ALT ↓, ALP ↓, improved liver histology	[59]

Table 2. *Cont.*

No.	Testing Subjects	Application Part or Compounds	Doses/Duration	Effects	Ref.
16	Paracetamol-induced hepatotoxic rats	Ethanol extract	100–500 mg/kg, p.o., 4 days	AST ↓, ALT ↓, ALP ↓, MDA ↓, improved liver histology; enhanced antioxidant defense and reduced oxidative stress	[67]
17	Streptozotocin (STZ)-induced diabetic rats	Ethanol extract	100 and 300 mg/kg, p.o., 28 days	Reduced fasting glucose and MDA levels, enhanced SOD and GSH-Px activities, and improved pancreatic β-cell morphology.	[8]
18	Heat-stressed broilers	*O. corniculata* powder	10 g/kg diet, 28–42 days	MDA ↓ in muscle, TAC ↑, improves lipid oxidative stability and mitigates oxidative damage	[60]
19	Heat-stressed broilers	*O. corniculata* powder + Chromium picolinate (CrPic)	10 g *O. corniculata* powder + 0.2 mg CrPic/kg diet, 1–42 days.	Improvement of microbiota balance, enhancing gut health in heat-stressed broilers, with more significant effects in the early growth stage	[44]
20	Heat-stressed broilers	*O. corniculata* powder + CrPic	10 g *O. corniculata* powder + 0.2 mg CrPic/kg diet, 28–42 days	Improves meat quality (crude protein ↑, fat in breast meat ↓) and enhances antioxidant defense against heat stress.	[3]

5.2. Anti-Inflammatory Activity

Inflammation is an essential host defense mechanism, but chronic or excessive responses contribute to the pathogenesis of cancer, cardiovascular disorders, and diabetes [68,69]. Recent studies indicate that *O. corniculata* exhibits anti-inflammatory activity through multiple pathways involving its diverse phytochemicals, particularly flavonoids and phenolic acids [70]. These compounds act primarily by inhibiting nuclear factor kappa B (NF-κB)-mediated transcriptional activation, downregulating pro-inflammatory cytokines and enzymes such as cyclooxygenase-2 (COX-2), inducible nitric oxide synthase (iNOS), and phospholipase A_2 (PLA_2), and modulating oxidative stress-associated immune responses (Table 3).

In vitro, hot water extracts enriched in 3-O-caffeoylquinic acid and ellagic acid suppressed LPS-induced nitric oxide (NO) production in RAW264.7 cells [62], accompanied by downregulation of *iNOS* and *COX-2* mRNA expression, and suppressed the protein levels of Interleukin-6 (IL-6) and tumor necrosis factor (TNF-α) [62]. Ethanol extracts inhibited NF-κB activation in PC-3 cells by downregulating phosphorylated p65 and IκBα, while upregulating IκBα expression [71]. In vivo, ethanol extracts significantly alleviated colonic inflammation in a rat model of acetic acid-induced inflammatory bowel disease (IBD) [72]. In a LPS-induced acute lung injury (ALI) rat model, the extracts also protected against lung damage by reducing inflammatory cell infiltration and tissue injury [73]. Serum metabolomics further identified flavonoids such as vicenin-3, isovitexin, and isoschaftoside as major absorbed constituents, which were proposed to act via suppression of IL-17/NF-κB signaling and concomitant antioxidant effects [73].

Among isolated compounds, diosmin demonstrated specific anti-inflammatory effects by inhibiting PLA_2 activity [29]. Evidence from experimental studies suggests that this compound can attenuate venom-induced inflammation and protect against associated tissue injury, further supporting the role of *O. corniculata* flavonoids in modulating inflammatory pathways [29].

Table 3. Summary of studies on anti-inflammatory, antipyretic and analgesic activity of *O. corniculata*.

No.	Testing Subjects	Application Part or Compounds	Doses/Duration	Effects	Ref.
1	RAW264.7 cells (LPS 2 μg/mL)	Hot water extract	100–400 μg/mL, 24 h	NO ↓, iNOS ↓, COX-2 ↓, IL-6 ↓, TNF-α ↓	[62]
2	RAW264.7 cells (LPS 0.25 μg/mL)	Ethanol extract	10–50 μg/mL, 24 h	NO ↓, IL-17 ↓; NF-κB pathway ↓	[73]
3	Human RBCs	Diosmin (isolated from *O. corniculata*)	15–120 μM, 2 h	PLA_2 inhibition ↑, anti-hemolysis, inflammation mediators ↓	[29]
4	Human PC-3 prostate cancer cells	Ethanol extract	50–200 μg/mL, 24 h	Cell proliferation ↓, migration, invasion; apoptosis ↑; NF-κB pathway activity ↓ (p-p65 ↓, p-IκBα ↓, IκBα ↑); anti-inflammatory mechanism via NF-κB inhibition	[71]
5	Swiss albino mice	Diosmin (for snake venom toxicity)	Diosmin (1:200 *w*/*w*) pre-incubated for 1 h (i.p. injection), observation for 5 h	Myotoxicity ↓, pulmonary hemorrhage ↓, CPK ↓, LDH ↓, tissue damage ↓	[29]
6	Mice (acute peritonitis model)	Ethanol extract	6% and 12% extract, p.o., 5 days	Abdominal writhing response ↓ (analgesic); abdominal capillary permeability ↓ (anti-inflammatory)	[74]
7	Swiss mice/Wistar rats	β-sitosterol (extract from leaves)	5–20 mg/kg, i.p., single dose	Analgesia (hot plate latency ↑, writhing ↓), anti-inflammation (paw edema ↓); opioid receptor involvement, PGE inhibition	[70]
8	Rats (acetic acid-induced IBD)	Ethanol extract	200–400 mg/kg, p.o., 7 days	Colon weight ↓, visible lesion score ↓, histopathological score ↓	[72]
9	SD rats (LPS-induced ALI)	Ethanol extract	0.8–3.2 g/kg, p.o., bid × 7 d	Lung edema ↓, inflammatory cell infiltration ↓; serum TNF-α ↓, IL-6 ↓, IL-1β ↓, IL-18 ↓	[73]
10	Rats (pylorus ligation model, indomethacin-induced ulcer model)	Methanol extract	125–500 mg/kg, p.o., single dose	Gastric secretion ↓, acidity ↓, ulcer index ↓; protection against NSAID-induced gastric injury; anti-inflammatory and gastroprotective effects	[75]

Table 3. *Cont.*

No.	Testing Subjects	Application Part or Compounds	Doses/Duration	Effects	Ref.
11	Rats (CNP model)	Compound prescription containing *O. corniculata*	1 g/kg, p.o., 28 days	Inflammatory infiltration ↓, MCP-1 ↓, ROS ↓, GSH ↑, 4-HNE ↓, ALDH2 ↓, FGF2 ↓; improved tissue pathology;	[48]
12	Rats (CNP model)	Compound prescription containing *O. corniculata*	9.37 g/kg, p.o., 49 days	Prostate index ↓, TNF-α ↓, IL-1β ↓, IL-6 ↓; improved histopathology; cGAS ↓, STING ↓, TRAF6 ↓, HSP70 ↑	[76]

5.3. Antibacterial Activity

O. corniculata extracts exhibit broad-spectrum antibacterial activity. Methanol extracts containing compounds such as rutin, p-hydroxybenzoic acid, and ferulic acid have been reported to contribute to both antimicrobial and antioxidant effects [61]. Recent findings suggest that the antibacterial effects of *O. corniculata* are associated with the inhibition of biofilm formation and interference with bacterial adhesion and colonization (Table 4) [77].

In vitro, ethanol and aqueous extracts inhibited the growth of both Gram-positive and Gram-negative bacteria, including *S. aureus* and *E. coli*, and were shown to suppress biofilm formation [77,78]. In vivo, methanol extracts reduced intestinal colonization by pathogenic *Shigella* strains in a suckling mouse diarrhea model, with stronger inhibition observed against *S. dysenteriae* type 1 compared to *S. flexneri* 2a [79]. High-performance liquid chromatography (HPLC) analysis suggested that polyphenols are the principal active constituents. The study further showed that efficacy depended on timing of administration, suggesting that the extract may interfere with bacterial adhesion or early colonization processes.

Recent advances in nanotechnology have expanded the antibacterial applications of *O. corniculata* (Table 5). Extracts have been utilized for the eco-friendly synthesis of silver nanoparticles (AgNPs), which display stronger and broader antibacterial activity compared with crude extracts [80]. Hybrid nanomaterials, including silver nanoparticles on graphene oxide (AgNPs@GO) [81] and AgNP-graphene nanocomposites, provide enhanced efficacy by synergistically disrupting bacterial membranes, inhibiting biofilm formation, and interfering with bacterial metabolism [82]. Furthermore, ZnO nanoflowers (ZnO NFs) synthesized using *O. corniculata* extracts damage bacterial membranes and intensify oxidative stress [83]. Flavonoids such as quercetin contribute both antibacterial effects and electron-donating capacity, thereby enhancing the overall activity of these nanostructures [83].

Table 4. Summary of studies on the antimicrobial activity of *O. corniculata*.

No.	Testing Subjects	Application Part or Compounds	Methods	Effects	Ref.
1	*S. aureus*, *E. coli*	Aqueous extract	Disk diffusion	At 20% concentration, inhibition zones: *S. aureus* (21 mm), *E. coli* (19.33 mm)	[78]

Table 4. *Cont.*

No.	Testing Subjects	Application Part or Compounds	Methods	Effects	Ref.
2	*E. coli*, *S. Typhi*, MDR *S. Typhi*, *K. pneumoniae*, MDR *C. koseri*	Methanol extract	Agar well diffusion, MIC	At 50 mg/mL concentration, inhibition zones: *E. coli* (17 mm), *S. Typhi* (13 mm), MDR *S. Typhi* (16 mm), *K. pneumoniae* (11 mm), *C. koseri* (12 mm). MIC for *E. coli*, *K. pneumoniae*, *C. koseri*: 25 mg/mL; *S. Typhi*: 100 mg/mL; MDR *S. Typhi*: 50 mg/mL.	[84]
3	*S. aureus*	Ethanol extract	Microdilution, biofilm inhibition and eradication assays	At 1% (w/v), antibacterial activity against *S. aureus* was 76.23%; biofilm inhibition rate 71.32% (mid-phase), 69.33% (mature-phase); biofilm eradication rate 64.1%.	[77]
4	*E. coli*, *S. dysenteriae*, *S. typhi*, *B. subtilis*	5-hydroxy-6,7,8,4′-tetramethoxyflavone and 5,7,4′-trihydroxy-6,8-dimethoxyflavone	Agar diffusion assay	The inhibition zones of these two flavonoids isolated from *O. corniculata* ranged from 10 to 16.5 mm.	[85]
5	Suckling mice infected with *S. dysenteriae* 1 (NT4907) and *S. flexneri* 2a (2457T)	Methanol extract	20 mg/kg, single oral administration (simultaneous or 3 h post-infection)	Reduced intestinal colonization of Shigella strains.	[79]
6	Pomfret fish (food preservation model)	Methanol leaf extract (rutin, p-hydroxybenzoic acid, ferulic acid)	Food storage assay (bacterial growth and oxidative stability)	Significantly inhibited *S. aureus* growth during 10 °C storage (48 h) and reduced lipid oxidation, suggesting potential as a natural preservative.	[61]

Table 5. Antimicrobial activity of nanomaterials synthesized using *O. corniculata*.

No.	Nanomaterial Type	Biosynthesis Agent	Test Microorganisms	Methods	Effects	Ref.
1	Silver nanoparticles (AgNPs)	*O. corniculata* extract + $AgNO_3$	*S. aureus*, *E. coli*, *B. subtilis*, *K. pneumoniae*, *S. typhi*, *S. pyogenes*, *P. aeruginosa*, *B. cereus*, *S. typhimurium*, *E. faecalis*, *A. baumannii*, *P. mirabilis*	Disk diffusion, broth microdilution, biofilm inhibition assay	Inhibition zones 10–20 mm (25–50 μg/mL); MIC 0.11–11.5 μg/mL; disrupts membrane integrity, inhibits biofilm, induces ROS, interferes with bacterial metabolic pathways and interacts with intracellular targets	[31,80,86–88]

Table 5. *Cont.*

No.	Nanomaterial Type	Biosynthesis Agent	Test Microorganisms	Methods	Effects	Ref.
2	AgNPs@GO nanocomposite	*O. corniculata* extract + $AgNO_3$ + GO	*B. subtilis, E. coli*	Agar well diffusion	Inhibition zones: *B. subtilis* 27 mm, *E. coli* 21 mm. Enhanced antibacterial activity due to synergistic interaction.	[81]
3	AgNPs-graphene nanocomposite	*O. corniculata* extract + $AgNO_3$ + graphene	*E. coli, S. aureus, B. cereus, S. typhimurium*	MIC, MBC, inhibition zone	MIC and MBC both as low as 10 μg/mL for AgNPs-graphene composites; disrupts membranes and inhibits metabolism.	[82]
4	ZnO nanoflowers (ZnO NFs)	*O. corniculata* extract + $Zn(NO_3)_2$	*S. aureus, E. faecium, P. aeruginosa*	Broth dilution	Dose-dependent antibacterial activity (40–120 μg/mL); Gram-positive bacteria more sensitive; induces membrane damage, ROS, and synergizes with flavonoids.	[83]

5.4. Anticancer Activity

Research on the anticancer potential of *O. corniculata* has gained momentum in recent years, with investigations employing cell-based, animal, and molecular docking approaches to elucidate its activity across diverse cancer models (Table 6) [63]. Ethanol extracts have been shown to induce apoptosis in MCF-7 breast cancer cells by promoting oxidative stress and modulating apoptosis-related genes, including tumor protein p53 (p53), Fas cell surface death receptor (CD95), and B-cell lymphoma 2 (Bcl-2) [89]. Importantly, these extracts displayed low cytotoxicity toward normal cells, suggesting selective activity. Ethanol and ethyl acetate extracts also exhibited significant cytotoxicity against HepG2 cells, with IC_{50} values of 34.49 μg/mL and 30.25 μg/mL, respectively [63]. Molecular docking identified apigenin, a flavonoid constituent, as a potential inhibitor of epidermal growth factor receptor tyrosine kinase (EGFR-TK), with a binding energy of −7.90 kcal/mol, implicating interference with EGFR signaling [90]. In addition, corniculin, a newly identified lignan from *O. corniculata*, exhibited moderate cytotoxicity against SMMC-7721, MCF-7, HCT-15, and A549 cells, with potency comparable to cisplatin. In vivo, ethanol extracts inhibited tumor progression and prolonged survival in Ehrlich ascites carcinoma (EAC) mice, again without significant toxicity to normal tissues.

Advances in nanotechnology and drug delivery systems have further expanded its anticancer potential. Green-synthesized silver nanoparticles (O-AgNPs) using O. *corniculata* extract exhibited cytotoxicity in MCF-7 and AGS (human gastric carcinoma) cells by inducing apoptosis, while showing low toxicity toward normal cells [88]. Moreover, exosome-mediated delivery of *O. corniculata* polyphenols improved cellular uptake and cytotoxicity in HuH7 hepatocellular carcinoma cells [11]. This strategy promoted apoptosis by upregulating pro-apoptotic genes, including Bcl-2-associated X protein (*Bax*) and *caspase-3*, while downregulating the anti-apoptotic gene *Bcl-2*, thereby enhancing therapeutic efficacy and selectivity.

Antioxidant and anti-inflammatory properties of *O. corniculata* contribute to reducing the risk of cancer development. Selective induction of apoptosis in cancer cells is another important mechanism of its anticancer activity. Evidence suggests that the bioactive constituents of *O. corniculata* possess potential anticancer activity, and further elucidation of mitochondrial signaling, autophagy, and immune modulation may provide valuable insights into its underlying mechanisms.

Table 6. Summary of studies on anticancer activity of *O. corniculata*.

No.	Testing Subjects	Application Part or Compounds	Doses/Duration	Effects	Ref.
1	MCF-7 cell line	Ethanol extract	31.25–2000 μg/mL, 24 and 72 h	*p53* ↑, *CD95* ↑, *Bcl-2* ↓; selective cytotoxicity against cancer cells	[89]
2	HepG2 cell line	Ethanol extract and ethyl acetate fraction	35–45 μg/mL, 48 h	Inhibited proliferation of HepG2 cells	[63]
3	SMMC-7721, MCF-7, HCT-15, A549 cancer cell lines	Corniculin	concentration range 0–100 μM, 48 h	Moderate cytotoxicity against multiple cell lines: IC_{50} for SMMC-7721, MCF-7, HCT-15, and A549 cells were 29.0, 35.6, 31.3, and 25.7 μM, respectively	[90]
4	EAC mice	Ethanol extract	100 mg/kg and 400 mg/kg, p.o., 9 days	Inhibited tumor growth; prolonged survival; improved hematological and biochemical parameters; CAT ↑, GSH ↑, MDA ↓; no significant toxicity to normal cells	[91]
5	MCF-7 and AGS cell lines	O-AgNPs	0.1–50 μg/mL, 24 h	Significant cytotoxicity and growth inhibition; induced apoptosis; low toxicity to normal cells	[88]
6	HuH7 cell line	Polyphenols from *O. corniculata* loaded in exosomes	10–100 μg/mL, 24 h	Exosome delivery enhanced cytotoxicity and cellular uptake; promoted apoptosis via upregulation of *Bax* and *caspase-3*, downregulation of *Bcl-2*	[11]

5.5. Neuroprotective Activity

Neurodegenerative diseases, including Alzheimer's disease, Parkinson's disease, and depression, are commonly associated with oxidative stress and neuroinflammation. These processes drive protein aggregation, neuronal injury, and neurotransmitter imbalance, ultimately resulting in cognitive decline and motor dysfunction [92].

In an $AlCl_3$-induced rat model of AD, methanol extracts of *O. corniculata* activated the Nrf2/HO-1 antioxidant pathway, elevated total antioxidant capacity (TAC) and SOD activity, and decreased MDA levels [57]. The extracts restored neurotransmitter balance by increasing dopamine (DA), norepinephrine (NE), and 5-hydroxytryptamine (5-HT), while inhibiting acetylcholinesterase (AChE). They also suppressed neuroinflammation by downregulating TLR4/NF-κB/NLRP3 signaling and reducing pro-inflammatory cytokines. Furthermore, the extracts attenuated endoplasmic reticulum stress by inhibiting PERK/CHOP-mediated apoptosis, upregulated anti-apoptotic *Bcl-2* and autophagy-related *Beclin-1*, and activated Wnt3a/β-catenin signaling. These changes decreased *GSK-3β*, ApoE4, and AD-associated proteins (APP, Aβ, p-Tau), while enhancing low-density lipoprotein receptor-related protein 1 (LRP1)-mediated Aβ clearance. Phytochemical profiling indicated that the extract is rich in flavonoids, organic acids, terpenoids. Behavioral tests demonstrated that the extract significantly improved spatial learning and memory in AD rats, partly due to increased brain-derived neurotrophic factor (BDNF) levels [57].

O. corniculata extracts exhibit notable neuroprotective potential by mitigating oxidative stress, neuroinflammation, and neuronal apoptosis, while modulating autophagy and key intracellular signaling cascades. Evidence from multiple neurodegenerative models (Table 7) indicates that these effects are primarily mediated through activation of the Nrf2/HO-1 antioxidant pathway [93,94], inhibition of TLR4/NF-κB/NLRP3-driven inflammatory signaling, and regulation of apoptosis- and autophagy-associated proteins. Collectively, these integrated mechanisms contribute to the alleviation of oxidative stress, preservation of neuronal structure, and improvement of cognitive and motor functions.

Table 7. Summary of neuroprotective studies of *O. corniculata.*

No.	Disease Model	Animal Subjects	Application Part or Compounds	Doses/Duration	Effects	Ref.
1	Parkinson's disease (MPTP-induced)	C57 black male mice	Ethanol extract	250, 500 mg/kg, p.o., co-administered with MPTP	Restored SOD, CAT, reduced LPO; attenuated oxidative stress; improved locomotor activity, muscle coordination, and cognitive performance	[93,94]
2	Parkinson's disease (Rotenone-induced)	Swiss albino mice	Ethanol extract	500 mg/kg, p.o., for 21 days	Improved motor performance; SOD ↑, CAT ↑, GSH ↑, DA ↑; MDA ↓, NO ↓, Glu ↓; reduced neuronal loss and neuroinflammation	[95]
3	Dementia	Albino mice	Methanol extract	100, 200 mg/kg, p.o., for 21 days	Significantly improved learning and memory abilities	[96]
4	Epilepsy (MES/PTZ-induced)	Wistar rats	Methanol extract	200, 400 mg/kg, i.p., single dose	Restored antioxidant enzymes (SOD, GPx, GR, CAT), reduced LPO; increased DA, NA, 5-HT, GABA; delayed seizures and reduced severity	[97]
5	Alzheimer's disease ($AlCl_3$-induced)	Male SD rats	Methanol extract	150 mg/kg, p.o., weeks	Improved spatial learning and memory; TAC ↑, SOD ↑, MDA ↓; restored DA, NE, 5-HT levels, AChE ↓; upregulation of Nrf2/HO-1, *Bcl-2*, *Beclin-1*, Wnt3a, *β-catenin*, LRP1; downregulation of TLR4/NF-κB/NLRP3, PERK/CHOP, *GSK-3β*, ApoE4; reduced APP, BACE1, Aβ, p-Tau; alleviated neuronal damage and brain pathology.	[57]

6. Safety Studies on *O. corniculata*

O. corniculata has a long history of medicinal use and is generally considered to have a good safety record. In modern clinical practice, several formulations derived from *O. corniculata* have received approval from the China National Medical Products Administration (NMPA). These include the prescription formulations Gu Kang Capsules, Mi Lin Granules, and Fu Yan Xiao Capsules, all of which are widely used in clinical practice, and no major adverse effects have been reported to date.

Evidence from experimental models further supports the potentially safety profile of *O. corniculata*. A brine shrimp lethality assay indicated low cytotoxicity, with an LC_{50} of 156 μg/mL [52]. In Swiss albino mice [75], oral administration of *O. corniculata* extract at doses up to 2000 mg/kg caused no observable behavioral abnormalities or toxic effects. Li et al. [12] reported that daily oral administration of 0.432 g/kg of crude herb in rats for seven consecutive days caused no observable toxic effects. In a rabbit fracture model, daily dosing of 0.224 g/kg for 30 days produced no adverse reactions [98].

Despite these findings, most studies to date have focused on acute or short-term exposure. Comprehensive data on long-term toxicity, genotoxicity, and reproductive safety remain limited. Moreover, because *O. corniculata* contains calcium oxalate [20], excessive or prolonged intake may pose a theoretical risk of oxalate accumulation, particularly in individuals susceptible to kidney stones [99]. Modern extraction and processing techniques may help to mitigate this risk, but further systematic and quantitative evaluation is necessary to confirm its safety under long-term or high-dose exposure conditions. Preliminary serum metabolomics studies have detected absorbed bioactive constituents after oral administration of *O. corniculata* extract, including flavonoids (e.g., vicenin-3, isovitexin, isoschaftoside) and phenolic acids (e.g., 3-O-, 4-O-, 5-O-p-coumaroylquinic acid) [73]. However, the interactions between different bioactive constituents and their potential effects when combined with other medications have not been systematically evaluated. Further investigation of toxicokinetics, tissue distribution, and maximum tolerated dose (MTD) would be meaningful for confirming safe dosage ranges and supporting its clinical applications.

7. Conclusions and Future Prospective

O. corniculata exhibits antioxidant, antimicrobial, and anti-inflammatory activities, and its wide distribution, ease of cultivation, and chemical diversity underscore its potential as a sustainable bioresource. Flavonoids are the principal absorbed constituents in vivo and likely mediate antioxidant and anti-inflammatory effects through multiple signaling pathways [73]. Recent studies have also identified structurally unique compounds, such as the alkaloid ATA, which represents a promising lead for α-glucosidase inhibition [39], and the lignan corniculin A, which demonstrates moderate anticancer activity comparable to cisplatin [100]. Advances in nanotechnology have further expanded its therapeutic potential. Green-synthesized nanomaterials derived from *O. corniculata* extracts exhibit enhanced antioxidants, antibacterial and anticancer activities. Exosome-mediated delivery systems for polyphenols also improve bioavailability, addressing limitations of conventional herbal extracts [11]. Collectively, these findings highlight the broad application potential of *O. corniculata*, as outlined below:

(1) Dietary supplements and nutraceuticals: Flavonoids and polysaccharides enhance antioxidant defenses and help mitigate disorders associated with oxidative stress, supporting their potential development as dietary supplements and nutraceuticals [7].

(2) Antimicrobial applications: Extracts exhibit antibacterial activity against *S. aureus* and *E. coli*, with topical formulations such as antibacterial creams showing efficacy [78]. Green-synthesized nanomaterials further enhance antibacterial activity, offering opportunities for wound-care and infection-control products.

(3) Hypoglycemic potential: Ethanol extracts exhibit antihyperglycemic effects in STZ-induced diabetic rats, mainly through the enhancement of antioxidant defenses and protection of pancreatic β-cells [8]. The alkaloid ATA has been identified as one of the contributors to this activity [37,39]. These findings suggest that the plant's bioactive constituents hold promise for the development of natural hypoglycemic agents.
(4) Food preservation: Polyphenols delay lipid peroxidation in animal fats and extend shelf life. Oil-soluble extracts show stronger antioxidant effects than water-soluble forms, with optimal activity at 0.02–0.04% concentrations, comparable to butylated hydroxyanisole (BHA) and butylated hydroxytoluene (BHT) [77]. Methanol extracts of *O. corniculata* effectively inhibit bacterial growth and oxidative spoilage in fish meat, indicating its potential as a natural preservative [61].
(5) Poultry production: Dietary supplementation with *O. corniculata* improves antioxidant stability of broiler meat and increases the levels of bioactive compounds such as lutein, zeaxanthin, and PUFAs [4]. It also optimizes the fatty acid profile, helps maintain gut microbiota balance, alleviates heat stress, and supports growth performance [44,60].

Future research should prioritize standardized cultivation and harvesting practices, as geographic origin, environmental conditions, and plant varieties strongly influence the accumulation and composition of bioactive compounds. Standardized extraction and quality-control protocols, particularly those based on antioxidant-rich components such as flavonoids and polyphenols, will be essential to ensure reproducibility. Mechanistic studies are also needed to clarify how these constituents exert their biological effects and to guide practical applications. Systematic evaluation of pharmacokinetics and long-term safety is also required to support clinical translation.

Considering its multi-target pharmacological properties, *O. corniculata* may have potential for synergistic or modulatory effects when combined with antibacterial, anti-cancer, or neuroprotective drugs. However, such interactions have not yet been systematically investigated and warrant further study. Moreover, mechanistic investigations focusing on the antioxidant, anti-inflammatory, and gut microbiota-modulating effects of *O. corniculata* in poultry production will be valuable to clarify its roles and optimize its practical application as a feed additive. With its low cost, broad availability, and multi-target bioactivities, *O. corniculata* holds considerable promise as a sustainable botanical resource for improving animal health, food preservation, and the development of novel antioxidant-based therapeutics.

Supplementary Materials: The following supporting information can be downloaded at: https://www.mdpi.com/article/10.3390/antiox14111352/s1. Figure S1: Structures of flavones (1–48) isolated from *O. corniculata*. Figure S2: Structures of other flavonoid subclasses isolated from *O. corniculata*, including flavonols (49–67), dihydroflavones (68–70), isoflavones (71–79), chalcone (80), dihydrochalcones (81–82), and flavanols (83–84). Figure S3: Structures of organic acids identified from *O. corniculata*. (A): Phenolic acids (85–107); (B): Other organic acids (108–135). Figure S4: Structures of terpenoids (136–155) identified from *O. corniculata*. Figure S5: Structures of alkaloids identified from *O. corniculata*. Table S1: Chemical compounds isolated from *O. corniculata*. References [19–23,27,29,31,35,37,39,42,57,61–63,68,70,83,85,90,100–109] are cited in the supplementary materials.

Author Contributions: T.Z.: conceptualization, literature search, software, figure preparation, validation and original draft preparation; J.K.: resources, supervision, project administration, funding acquisition, review and editing; C.W.: conceptualization, resources, supervision, project administration; H.Z.: validation and funding acquisition; J.H.: proofreading the manuscript and editing; C.T.: data curation and editing; W.Z.: software and data curation. All authors have read and agreed to the published version of the manuscript.

Funding: This research was funded by the 2115 Talent Development Program of China Agricultural University (NO. 00109015); Joint Project of the Scientific Research Innovation Fund for PhD Students (NO. HSPHDSRF-2024-05-013).

Institutional Review Board Statement: Not applicable.

Informed Consent Statement: Not applicable.

Data Availability Statement: The original contributions presented in the study are included in the article; further inquiries can be directed to the corresponding author.

Conflicts of Interest: The authors declare no conflicts of interest.

References

1. Zhang, X.; Liu, Y.; Shen, Z.; Wang, S.; Wu, C.; Liu, D.; Tang, S.; Dai, C. Osthole ameliorates myonecrosis caused by Clostridium perfringens type A infection in mice. *One Health Adv.* **2023**, *1*, 27. [CrossRef]
2. Gliozheni, E.; Salem, Y.; Cho, E.; Wahlstrom, S.; Olbrich, D.; Shams, B.; Alexander, M.; Ichii, H. Food-Derived Phytochemicals: Multicultural Approaches to Oxidative Stress and Immune Response. *Int. J. Mol. Sci.* **2025**, *26*, 7316. [CrossRef]
3. Saracila, M.; Panaite, T.D.; Mironeasa, S.; Untea, A.E. Dietary Supplementation of Some Antioxidants as Attenuators of Heat Stress on Chicken Meat Characteristics. *Agriculture* **2021**, *11*, 638. [CrossRef]
4. Saracila, M.; Untea, A.E.; Panaite, T.D.; Varzaru, I.; Oancea, A.; Turcu, R.P.; Vlaicu, P.A. Creeping wood sorrel and chromium picolinate effect on the nutritional composition and lipid oxidative stability of broiler meat. *Antioxidants* **2022**, *11*, 780. [CrossRef] [PubMed]
5. Groom, Q.J.; Straeten JVder Hoste, I. The origin of *Oxalis corniculata* L. *PeerJ* **2019**, *7*, e6384. [CrossRef] [PubMed]
6. Bharti, R.; Priyanka, P.; Bhargava, P.; Khatri, N. Ethnopharmacology and therapeutic potentials of *Oxalis corniculata*: An in-depth study. *Beni-Suef Univ. J. Basic Appl. Sci.* **2024**, *13*, 81. [CrossRef]
7. Gao, T.; Hu, W.; Zhang, Z.; Tang, Z.; Chen, Y.; Zhang, Z.; Yuan, S.; Chen, T.; Huang, Y.; Feng, S.; et al. An acidic polysaccharide from *Oxalis corniculata* L. and the preliminary study on its antioxidant activity. *J. Food Biochem.* **2022**, *46*, e14235. [CrossRef]
8. Ansari Khan, M.; Khan, N.; Ahmed, D.; Khan, M.; Lall, A. Exploring antihyperglycemic and histopathological analysis of *Oxalis corniculata* L. hydroethanolic extract in STZ induced diabetic rats with in silico α-amylase inhibitor assessment. *Ann. Phytomedicine* **2024**, *13*, 612–626. [CrossRef]
9. Sarfraz, I.; Rasul, A.; Hussain, G.; Shah, M.A.; Nageen, B.; Jabeen, F.; Selamoğlu, Z.; Uçak, İ.; Asrar, M.; Adem, S. A review on phyto-pharmacology of *Oxalis corniculata*. *Comb. Chem. High Throughput Screen.* **2022**, *25*, 1181–1186. [CrossRef]
10. Jiang, D.; Yu, D.; Zeng, M.; Liu, W.; Li, D.; Liu, K. Optimization of ultrasonic-assisted extraction of total flavonoids from *Oxalis corniculata* by a hybrid response surface methodology-artificial neural network-genetic algorithm (RSM-ANN-GA) approach, coupled with an assessment of antioxidant activities. *RSC Adv.* **2024**, *14*, 39069–39080. [CrossRef]
11. Begum, S.; Mazumder, P.B.; Talukdar, A.D. In vitro evaluation of anticancer activity of *Oxalis corniculata* L., by harnessing exosome for effective drug delivery against hepatocellular carcinoma. *Pharmacol. Res.-Nat. Prod.* **2025**, *8*, 100298. [CrossRef]
12. Li, J.; Yang, J.; Liu, T.; Li, Y.; Shen, Y.; Yang, C.; He, B.; Wang, A. Treatment of Soft Tissue Injury in Rats with Gukang Capsule, *Oxalis Corniculata* and *Musa Basjoo* Sieb. *J. Guizhou Med. Univ.* **2020**, *45*, 1015–1019.
13. Shamso, E.; Draz, A.; Hosni, H.; Hussein, S. Taxonomic revision of genus *Oxalis* L. (Oxalidaceae) in the flora of Egypt. *Taeckholmia* **2022**, *41*, 56–69. [CrossRef]
14. Yang, Y.; He, L.; Liu, Y.; Liu, Y.; Chen, J. Identification and Analysis of KNOX Gene Family Members Based on RNA-Seq in *Oxalis corniculata*. *Mol. Plant Breed.* **2022**, *23*, 4320.
15. Loi, V.D.; Nguyet, L.T.; Duong, L.T.H.; Mai, N.T.; Huong, N.T.T.; Hai, P.T.M. Morphological and Microscopical Characteristics of *Oxalis corniculata* L. *VNU J. Sci. Med. Pharm. Sci.* **2021**, *37*, 1.
16. Al Sheikh, B.; Gedeon, J. Newly Documented Invasive Alien Plant Species in the West Bank, Palestine. *Feddes Repert.* **2025**, *136*, 136–145. [CrossRef]
17. Rokaya, M.B.; Münzbergová, Z.; Shrestha, M.R.; Timsina, B. Distribution pattern, ecological status and ethnomedicinal uses of medicinal plants along an altitudinal gradient among Jaunsar tribes of Western himalaya. *J. Mt. Sci.* **2025**, *31*, 1–25.
18. Liu, Z.; Wang, L.; Ding, S.; Xiao, H. Enhancer assisted-phytoremediation of mercury-contaminated soils by *Oxalis corniculata* L., and rhizosphere microorganism distribution of *Oxalis corniculata* L. *Ecotoxicol. Environ. Saf.* **2018**, *160*, 171–177. [CrossRef]
19. Imran, M.; Irfan, A.; Ibrahim, M.; Assiri, M.A.; Khalid, N.; Ullah, S.; Al-Sehemi, A.G. Carbonic anhydrase and cholinesterase inhibitory activities of isolated flavonoids from *Oxalis corniculata* L. and their first-principles investigations. *Ind. Crops Prod.* **2020**, *148*, 112285. [CrossRef]
20. Badwaik, H.; Singh, M.; Thakur, D.; Giri, T.; Tripathi, D. The botany, chemistry, pharmacological and therapeutic application of *Oxalis corniculata* Linn-a review. *Int. J. Phytomed.* **2011**, *3*, 1.

21. Zhang, B.; Kuang, W.; Liu, J.; Jiang, L.; Li, Y.; Ma, X. Chemical Constituents Analysis of *Oxalis Corniculata* L. Based on UPLC-Q-Exactive-Plus-Orbitrap-MS. *Chin. J. Mod. Appl. Pharm.* **2025**, *42*, 611–622.
22. Zeb, A.; Imran, M. Carotenoids, pigments, phenolic composition and antioxidant activity of *Oxalis corniculata* leaves. *Food Biosci.* **2019**, *32*, 100472. [CrossRef]
23. Zhang, S.; Dong, J.; Cheng, H. Essential oil composition of the leaves of *Oxalis corniculata* from China. *Chem. Nat. Compd.* **2018**, *54*, 380–381. [CrossRef]
24. Al-Khayri, J.M.; Sahana, G.R.; Nagella, P.; Joseph, B.V.; Alessa, F.M. Flavonoids as potential anti-inflammatory molecules: A review. *Molecules* **2022**, *27*, 2901. [CrossRef]
25. Dai, W.; Pan, M.; Peng, L.; Zhang, D.; Ma, Y.; Wang, M.; Wang, N. Integrated Transcriptome and Metabolome Analysis Reveals Insights into Flavone and Flavonol Biosynthesis in Salicylic Acid-Induced Citrus Huanglongbing Tolerance. *J. Agric. Food Chem.* **2024**, *73*, 919–937. [CrossRef]
26. Tang, S.; Wang, B.; Liu, X.; Xi, W.; Yue, Y.; Tan, X.; Bai, J.; Huang, L. Structural insights and biological activities of flavonoids: Implications for novel applications. *Food Front.* **2025**, *6*, 218–247. [CrossRef]
27. Yang, J.; Jiang, J.; Zong, D.; Peng, Y.; Chen, J.; Zhao, J. Chemical Composition from *Oxalis corniculata* L., Content of Total Flavonoids in *Oxalis corniculata* L. in Different Harvest Times. *Guangzhou Chem. Ind.* **2021**, *49*, 64–65.
28. Zhang, B.; Sun, X.; Tang, W.; Jiang, L.; Liu, J.; Wang, Y.; Li, Y.; Li, Y.; Ma, X. An effective high-performance liquid chromatography method for fingerprint and multi-component quantification of *Oxalis corniculata*. *Curr. Pharm. Anal.* **2025**, *21*, 434–438. [CrossRef]
29. Kiran, K.S.; Kameshwar, V.H.; Mudnakudu Nagaraju, K.K.; Nagalambika, P.; Varadaraju, K.R.; Karthik, N.A. Diosmin: A Daboia russelii venom PLA_2s inhibitor-purified, and characterized from *Oxalis corniculata* L medicinal plant. *J. Ethnopharmacol.* **2024**, *318*, 116977. [CrossRef]
30. Khan, F.; Bamunuarachchi, N.I.; Tabassum, N.; Kim, Y.M. Caffeic acid and its derivatives: Antimicrobial drugs toward microbial pathogens. *J. Agric. Food Chem.* **2021**, *69*, 2979–3004. [CrossRef]
31. Karimzadeh, K.; Bakhshi, N.; Ramzanpoor, M. Biogenic silver nanoparticles using *Oxalis corniculata* characterization and their clinical implications. *J. Drug Deliv. Sci. Technol.* **2019**, *54*, 101263. [CrossRef]
32. Vari, F.; Bisconti, E.; Serra, I.; Stanca, E.; Friuli, M.; Vergara, D.; Giudetti, A.M. Exploring the Role of Oleic Acid in Muscle Cell Differentiation: Mechanisms and Implications for Myogenesis and Metabolic Regulation in C2C12 Myoblasts. *Biomedicines* **2025**, *13*, 1568. [CrossRef]
33. Peng, B.; Wei, S. Synthetic Engineering of Microbes for Production of Terpenoid Food Ingredients. *J. Agric. Food Chem.* **2025**, *73*, 10052–10068. [CrossRef]
34. Jaramillo, S.P.; Calva, J.; Jiménez, A.; Armijos, C. Isolation of Geranyl Acetate and Chemical Analysis of the Essential Oil from Melaleuca armillaris (Sol. ex Gaertn.) Sm. *Appl. Sci.* **2024**, *14*, 1864. [CrossRef]
35. Durgawale, P.P.; Hendre, A.S.; Phatak, R.S. GC/MS characterization, antioxidant and free radical scavenging capacities of methanolic extract of *Oxalis corniculata* LINN: An ayurvedic herb. *Rasayan J. Chem.* **2015**, *8*, 271–278.
36. Ramakrishna, G.V.; Latif, Z.; Romiti, F. Enantioselective Total Syntheses of Vallesamidine and Schizozygane Alkaloids. *J. Am. Chem. Soc.* **2025**, *147*, 4613–4623. [CrossRef]
37. Feng, Q.; Yang, W.; Peng, Z.; Wang, G. Utilizing bio-affinity ultrafiltration combined with UHPLC Q-Exactive Plus Orbitrap HRMS to detect potential α-glucosidase inhibitors in *Oxalis corniculate* L. *Int. J. Biol. Macromol.* **2023**, *252*, 126490. [CrossRef]
38. Yang, Z.-J.; Huang, S.-Y.; Cheng, J.; Zeng, J.-W.; Wusiman, M.; Li, H.-B.; Zhu, H.-L. Betaine: A comprehensive review on dietary sources, health benefits, mechanisms of action, and application. *Trends Food Sci. Technol.* **2025**, *159*, 104993. [CrossRef]
39. Feng, Q.; Yang, W.; Ma, X.; Peng, Z.; Wang, G. Investigation on the anti-α-glucosidase mechanism of aspergillus triazolate A from *Oxalis corniculate* L. *Int. J. Biol. Macromol.* **2024**, *279*, 135457. [CrossRef]
40. Hu, W. Ultrasound-Enzyme Assisted Extraction, Physicochemical Properties and Antioxidant Activities of Polysaccharides from *Oxalis corniculata* L. Master's Thesis, Sichuan Agricultural University, Ya'an, China, 2019.
41. Aruna, K.; Devi, P.; Rajeswari, R.; Prabu, K.; Ramkumar, M.; Chidambaram, R.; Raja Sankar, S. Quantitative phytochemical analysis of *Oxalis corniculata* L. (Oxalidaceae). *World J. Pharm. Sci* **2014**, *3*, 711–716.
42. Bordoloi, M.; Bordoloi, P.K.; Dutta, P.P.; Singh, V.; Nath, S.; Narzary, B.; Bhuyan, P.D.; Rao, P.G.; Barua, I.C. Studies on some edible herbs: Antioxidant activity, phenolic content, mineral content and antifungal properties. *J. Funct. Foods* **2016**, *23*, 220–229. [CrossRef]
43. Chowdhury, M.M.K.; Islam, M.A.; Yasmin, F.; Ryhan, M.A. Biochemical potentials and stability of *Oxalis corniculata* L. leaf extracts. *Dhaka Univ. J. Biol. Sci.* **2025**, *34*, 75–83. [CrossRef]
44. Saracila, M.; Panaite, T.D.; Tabuc, C.; Soica, C.; Untea, A.; Varzaru, I.; Wojdyło, A.; Criste, R.D. Maintaining intestinal microflora balance in heat-stressed broilers using dietary creeping wood sorrel (*Oxalis corniculata*) powder and chromium (chromium picolinate). *Span. J. Agric. Res.* **2020**, *18*, 19. [CrossRef]
45. Jiang, X.; Zhao, R.; Zhou, X.; Fang, X.; Miao, M. Herbal research and analysis of ancient and modern clinical application of fresh *Oxalis corniculata*. *China J. Tradit. Chin. Med. Pharm.* **2024**, *39*, 3761–3767.

46. Lou, Q.; Zhan, M.W.; Lai, Y.Q.; Zhan, X.X.; Xiao, Y.P.; Shang, X.J. Study on the treatment of chronic nonbacterial prostatitis caused by dampness-heat stasis with Oxalis Formula combined with transacupuncture. *Natl. J. Androl.* **2025**, *31*, 165–171.
47. Zargar, S.A.; Ganie, A.H.; Reshi, Z.A.; Shah, M.A.; Sharma, N.; Khuroo, A.A. *Oxalis corniculata* L.(Oxalidaceae), an addition of an alien plant species to the flora of Ladakh, India. *Vegetos* **2024**, *37*, 373–378. [CrossRef]
48. Lai, Y.; Yu, Y.; Liu, P.; Zhan, M.; Ma, M.; Wang, L.; Lou, Q.; Shang, X. Research on the mechanism of Miao ethnicity medicine formula of *Oxalis corniculata* against chronic non-bacterial prostatitis. *Natl. J. Androl.* **2023**, *29*, 783–789.
49. He, L.; Sun, W.; Suo, X.; Shi, S.; Zhang, C.; Chen, Y. Herbal textual research of *Oxalis corniculata* L. *Chin. J. Ethnomed. Ethnopharm.* **2025**, *34*, 66–73.
50. Kubade, M.; Shetti, P. Medicinal importance of traditional Indian herbal plant *Oxalis corniculata*: A review. *Int. J. Pharm. Sci. Res.* **2024**, *15*, 1015–1024.
51. Noor, N.; Satapathy, K.B. Leafy vegetable diversity and their ethnomedicinal uses against gastrointestinal disorders in the Balasore district of Odisha, India. *J. Appl. Biol. Biotechnol.* **2023**, *11*, 259–267.
52. Tibuhwa, D.D. *Oxalis corniculata* L. in Tanzania: Traditional use, cytotoxicity and antimicrobial activities. *J. Appl. Biosci.* **2016**, *105*, 10055–10063. [CrossRef]
53. Silalahi, M. Utilization of *Oxalis corniculata* Linn as a traditional medicine and its bioactivity. *Magna Sci. Adv. Res. Rev.* **2022**, *5*, 27–33. [CrossRef]
54. Rahman, I.U.; Afzal, A.; Iqbal, Z.; Hart, R.; Abd_Allah, E.F.; Hashem, A.; Alsayed, M.F.; Ijaz, F.; Ali, N.; Shah, M.; et al. Herbal teas and drinks: Folk medicine of the Manoor valley, Lesser Himalaya, Pakistan. *Plants* **2019**, *8*, 581. [CrossRef]
55. Jiang, S.; Li, H.; Zhang, L.; Mu, W.; Zhang, Y.; Chen, T.; Wu, J.; Tang, H.; Zheng, S.; Liu, Y.; et al. Generic Diagramming Platform (GDP): A comprehensive database of high-quality biomedical graphics. *Nucleic Acids Res.* **2025**, *53*, D1670–D1676. [CrossRef]
56. Glorieux, C.; Liu, S.; Trachootham, D.; Huang, P. Targeting ROS in cancer: Rationale and strategies. *Nat. Rev. Drug Discov.* **2024**, *23*, 583–606. [CrossRef]
57. Abu-Elfotuh, K.; Hamdan, A.M.E.; Mohamed, S.A.; Bakr, R.O.; Ahmed, A.H.; Atwa, A.M.; Hamdan, A.M.; Alanzai, A.G.; Alnahhas, R.K.; Gowifel, A.M.H.; et al. The potential anti-Alzheimer's activity of *Oxalis corniculata* Linn. Methanolic extract in experimental rats: Role of APOE4/LRP1, TLR4/NF-κβ/NLRP3, Wnt 3/β-catenin/GSK-3β, autophagy and apoptotic cues. *J. Ethnopharmacol.* **2024**, *324*, 117731. [CrossRef]
58. Khan, M.R.; Zehra, H. Amelioration of CCl_4-induced nephrotoxicity by *Oxalis corniculata* in rat. *Exp. Toxicol. Pathol.* **2013**, *65*, 327–334. [CrossRef]
59. Khan, M.R.; Marium, A.; Shabbir, M.; Saeed, N.; Bokhari, J. Antioxidant and hepatoprotective effects of *Oxalis corniculata* against carbon tetrachloride (CCl_4) induced injuries in rat. *Afr. J. Pharm. Pharmacol.* **2012**, *6*, 2255–2267. [CrossRef]
60. Saracila, M.; Tatiana, P.; Untea, A.; Varzaru, I. Effect of dietary supplementation of some antioxidant combinations on nutrient digestibility in heat-stressed broilers. *Arch. Zootech.* **2022**, *25*, 116–129. [CrossRef]
61. Mukherjee, S.; Pal, S.; Chakrabarti, R.; Koley, H.; Dhar, P. Biochemical assessment of extract from *Oxalis corniculata* L.: Its role in food preservation, antimicrobial and antioxidative paradigms using in situ and in vitro models. *Indian J. Exp. Biol.* **2018**, *56*, 230–243.
62. Kim, J.H.; Hong, M.; Han, J.H.; Lee, H.J.; Choi, D.H.; Hoang, K.; Van Dung, L.; Kwon, T.; Ahn, Y. Antioxidant and Anti-inflammatory Effects of *Oxalis Corniculata* Hot Water Extract. *Korean J. Med. Crop Sci.* **2022**, *30*, 419–429. [CrossRef]
63. Gudasi, S.; Gharge, S.; Koli, R.; Patil, K. Antioxidant properties and cytotoxic effects of *Oxalis corniculata* on human Hepatocarcinoma (Hep-G2) cell line: An in vitro and in silico evaluation. *Future J. Pharm. Sci.* **2023**, *9*, 25. [CrossRef]
64. Sun, X. Antioxidant Effect of *Oxalis corniculata* Polyphenols on Animal Fats. *Heilongjiang Anim. Husb. Vet. Med.* **2015**, 129–130. [CrossRef]
65. Ngueguim, F.; Jouonzo, J.; Donfack, J.; Gounoue, R.; Djientcheu Tientcheu, J.P.; Clarice, D.; Fifen, R.; Dzeufiet, P.; Théophile, D. The Mixture Aqueous Extracts from *Oxalis corniculata* L. and Acmella caulirhiza Delile Accelerates Bone Healing in Fractured Rats. *J. Complement. Altern. Med.* **2022**, *17*, 18–28. [CrossRef]
66. Chen, C.; Chen, Y.; Cao, H.; Wang, G.; Guo, D.; Zhang, K. Protective effect and mechanism of Oxalidis Corniculatase Herba on acute hepatic injury induced by carbon tetrachloride in rats. *Chin. J. Exp. Tradit. Med. Formulae* **2018**, *24*, 141–145.
67. Sreejith, G.; Jayasree, M.; Latha, P.G.; Suja, S.R.; Shyamal, S.; Shine, V.J.; Anuja, G.I.; Sini, S.; Shikha, P.; Krishnakumar, N.M.; et al. Hepatoprotective activity of *Oxalis corniculata* L. ethanolic extract against paracetamol induced hepatotoxicity in Wistar rats and its in vitro antioxidant effects. *Indian J. Exp. Biol.* **2014**, *52*, 147–152.
68. Zhang, J.; Shen, W.; He, H. Exploring the action mechanism of *Oxalis corniculata* L. decoction in treating osteoarthritis utilizing liquid chromatography–mass spectrometry technology combined with network pharmacology. *Medicine* **2024**, *103*, e39515. [CrossRef]
69. Ding, Q.; Huang, X.; Yang, X.; Zhang, M.; Qin, P.; Wu, H.; Wang, X. *Study of the Mechanism of Oxalis corniculata (L.) in the Treatment of Hepatitis Based on Network Pharmacology*; IOP Conference Series: Earth and Environmental Science; IOP Publishing: Bristol, UK, 2021.

70. Dighe, S.B.; Kuchekar, B.S.; Wankhede, S.B. Analgesic and anti-inflammatory activity of β-sitosterol isolated from leaves of *Oxalis corniculata*. *Int. J. Pharmacol. Res.* **2016**, *6*, 109–113.
71. Zhang, G.W.; Zhan, M.W.; Zhan, X.X.; Yu, Y.; Liu, P.F.; Wang, L.; Wu, H.; He, C.Q.; Lou, Q.; Shang, X.J. Biological mechanisms of *Oxalis corniculata* regulating human prostate cancer PC-3 cells: An investigation based on the NF-κB pathway. *Natl. J. Androl.* **2023**, *29*, 202–209.
72. Dutta, A.; Handique, C.; Lahkar, M. Evaluation of anti-inflammatory activity of *Oxalis corniculata* in experimentally induced inflammatory bowel disease in rats. *Int. J. Basic Clin. Pharmacol.* **2015**, *4*, 744–749. [CrossRef]
73. Ma, X.; Zhang, B.; Jiang, L.; Kuang, W.; Wang, M.; Wang, Y.; Sun, J.; Li, Y. The Activity, Composition, and Molecular Mechanism of *Oxalis corniculata* in Alleviating Acute Lung Injury. *Nat. Prod. Commun.* **2025**, *20*, 1934578X251361195. [CrossRef]
74. Guo, M.; Wang, Y.; Shi, G.; Shen, L. Anti-inflammatory and analgesic effects of *Oxalis corniculata* L. on acute peritonitis in mice. *J. Dali Univ.* **2014**, *13*, 6–8.
75. Sakat, S.S.; Tupe, P.; Juvekar, A. Gastroprotective effect of *Oxalis corniculata* (whole plant) on experimentally induced gastric ulceration in Wistar rats. *Indian J. Pharm. Sci.* **2012**, *74*, 48.
76. Lou, Q.; Zhan, M.-W.; Lai, Y.-Q.; Zhan, X.-X.; Shang, X.-J. Mechanism of regulation of CNP rat model by *Oxalis* decoction via cGAS-STING signaling pathway. *Natl. J. Androl.* **2023**, *29*, 973–979.
77. Hamzah, H.; Siregar, K.A.A.K.; Nurwijayanto, A.; Wahyuningrum, R.; Sari, S. Effectiveness of *Oxalis corniculata* L. ethanol extract against mono-species of biofilm staphylococcus aureus. *Borneo J. Pharm.* **2021**, *4*, 184–191. [CrossRef]
78. Handali, S.; Hosseini, H.; Ameri, A. Formulation and evaluation of an antibacterial cream from *Oxalis corniculata* aqueous extract. *Jundishapur J. Microbiol.* **2011**, *4*, 255–260.
79. Mukherjee, S.; Koley, H.; Barman, S.; Mitra, S.; Datta, S.; Ghosh, S.; Paul, D.; Dhar, P. *Oxalis corniculata* (Oxalidaceae) leaf extract exerts in vitro antimicrobial and in vivo anticolonizing activities against Shigella dysenteriae 1 (NT4907) and Shigella flexneri 2a (2457T) in induced diarrhea in suckling mice. *J. Med. Food* **2013**, *16*, 801–809. [CrossRef]
80. Singh, B.; Kumar, D.; Soni, M.; Sinha, J.; Saroha, P.; Sharma, K.; Singh, K.; Gupta, S.K.; Singh, H. Larvicidal and anti-bacterial efficacy of silver nanoparticles derived from *Oxalis corniculata*. *Biosci. Nanotechnol.* **2025**, *1*, 2. [CrossRef]
81. Perumal, S.; Atchudan, R.; Ramalingam, S.; Edison, T.N.J.I.; Lee, H.M.; Cheong, I.W.; Devarajan, N.; Lee, Y.R. Comparative investigation on antibacterial studies of *Oxalis corniculata* and silver nanoparticle stabilized graphene surface. *J. Mater. Sci.* **2022**, *57*, 11630–11648. [CrossRef]
82. Jakhar, V.; Sharma, D.K. A sustainable approach for graphene–oxide surface decoration using *Oxalis corniculata* leaf extract–derived silver nanoparticles: Their antibacterial activities and electrochemical sensing. *Dalton Trans.* **2020**, *49*, 8625–8635. [CrossRef]
83. Badgujar, H.F.; Bora, S.; Kumar, U. Eco-benevolent synthesis of ZnO nanoflowers using *Oxalis corniculata* leaf extract for potential antimicrobial application in agriculture and cosmeceutical. *Biocatal. Agric. Biotechnol.* **2021**, *38*, 102216. [CrossRef]
84. Manandhar, S.; Luitel, S.; Dahal, R.K. In vitro antimicrobial activity of some medicinal plants against human pathogenic bacteria. *J. Trop. Med.* **2019**, *2019*, 1895340. [CrossRef]
85. Rehman, A.; Rehman, A.; Ahmad, I. Antibacterial, antifungal, and insecticidal potentials of *Oxalis corniculata* and its isolated compounds. *Int. J. Anal. Chem.* **2015**, *2015*, 842468. [CrossRef]
86. Sreya, K.R.; Uma, R.; Jayavardhanan, K.K. Antibacterial potential of silver nanoparticles synthesized using *Oxalis corniculata* leaf extract. *Pharma Innov. J.* **2022**, *11*, 1362–1366.
87. Das Mahapatra, A.; Patra, C.; Pal, K.; Mondal, J.; Sinha, C.; Chattopadhyay, D. Green synthesis of AgNPs from aqueous extract of *Oxalis corniculata* and its antibiofilm and antimicrobial activity. *J. Indian Chem. Soc.* **2022**, *99*, 100529. [CrossRef]
88. Ebrahimzadeh, M.A.; Alizadeh, S.R.; Hashemi, Z. Discovery of high antibacterial and antitumor effects against multi-drug resistant clinically isolated bacteria and MCF-7 and AGS cell lines by biosynthesized silver nanoparticles using *Oxalis corniculata* extract. *Eur. J. Chem.* **2023**, *14*, 202–210. [CrossRef]
89. Gholipour, A.R.; Jafari, L.; Ramezanpour, M.; Evazalipour, M.; Chavoshi, M.; Yousefbeyk, F.; Kargar Moghaddam, S.J.; Yekta Kooshali, M.H.; Ramezanpour, N.; Daei, P.; et al. Apoptosis Effects of *Oxalis corniculata* L. Extract on Human MCF-7 Breast Cancer Cell Line. *Galen Med. J.* **2022**, *11*, e2484.
90. Zhang, B.; Jiang, L.; Ma, X.; Wang, A.M.; Liu, T.; Zhou, M.; Liao, S.G.; Wang, Y.L.; Huang, Y.; Li, Y.J. A New arylnaphthalide Lignan from *Oxalis corniculata*. *Nat. Prod. Commun.* **2019**, *14*, 1934578X19875885. [CrossRef]
91. Kathiriya, A.; Das, K.; Kumar, E.P.; Mathai, K.B. Evaluation of antitumor and antioxidant activity of *Oxalis corniculata* Linn. against Ehrlich ascites carcinoma on mice. *Iran. J. Cancer Prev.* **2010**, *3*, 157–165.
92. Wilson, D.M.; Cookson, M.R.; Van Den Bosch, L.; Zetterberg, H.; Holtzman, D.M.; Dewachter, I. Hallmarks of neurodegenerative diseases. *Cell* **2023**, *186*, 693–714. [CrossRef]
93. Aruna, K.; Rajeswari, P.D.R.; Sankar, S.R. The effects of *Oxalis corniculata* L. extract against MPTP induced oxidative stress in mouse model of Parkinson's disease. *J. Pharm. Sci. Res.* **2016**, *8*, 1136.

94. Aruna, K.; Rajeswari, P.D.R.; Sankar, S.R. The effect of *Oxalis corniculata* extract against the behavioral changes induced by 1-methyl-4-phenyl-1, 2, 3, 6-tetrahydropyridine (MPTP) in mice. *J. Appl. Pharm. Sci.* **2017**, *7*, 148–153.
95. Kumar, P.; Sachdeva, H.; Kalra, S.; Singh, B.; Gupta, R.; Singh, G. *Oxalis corniculata* and Ficus religiosa mitigates rotenone-induced Parkinson's disease in Swiss Albino mice: Mechanistic insights and therapeutic potential. *J. Appl. Pharm. Sci.* **2024**, *14*, 253–262.
96. Das, M.; Gohain, K. Evaluation of memory enhancing activity of methanolic Extract of *Oxalis corniculata* Linn on dementia in experimental animals. *Int. J. Sci. Eng. Res.* **2018**, *9*, 922–928.
97. Al-Snafi, A.E. Medicinal plants with anticonvulsant activities with emphasis on their mechanisms of action. *Int. J. Biol. Pharm. Sci. Arch.* **2021**, *1*, 177–189. [CrossRef]
98. Tang, Y.; Yang, J.; Li, J.; Wang, Y.; Li, Y.; Yang, C.; Sun, J.; Liu, C.; Liu, T. Effects of *Oxalis corniculata* and *musa basjoo sieb* of gukang capsule on fracture healing in rabbits and its mechanism. *J. Guizhou Med. Univ.* **2022**, *47*, 635–639.
99. Wang, X.; Wang, Q. Current dietary and medical prevention of renal calcium oxalate stones. *Int. J. Gen. Med.* **2024**, *17*, 1635–1649. [CrossRef]
100. Ibrahim, M.; Hussain, I.; Imran, M.; Hussain, N.; Hussain, A.; Mahboob, T. Corniculatin A, a new flavonoidal glucoside from *Oxalis corniculata. Rev. Bras. Farmacogn.* **2013**, *23*, 630–634. [CrossRef]
101. Zhang, B.; Peng, X.; He, Y. Chemical constituents from *Oxalis corniculata. J. Chin. Med. Mat.* **2018**, *41*, 1883–1886.
102. Prasad Pandey, B.; Prakash Pradhan, S.; Adhikari, K. LC-ESI-QTOF-MS for the profiling of the metabolites and in vitro enzymes inhibition activity of Bryophyllum pinnatum and *Oxalis corniculata* collected from Ramechhap District of Nepal. *Chem. Biodivers.* **2020**, *17*, e2000155. [CrossRef]
103. Mondal, S.; Talukdar, P.; Mondal, T.K. Study of molecular docking to detect antihypertensive phytochemicals of *Oxalis corniculata* Linn. against angiotensin converting enzyme. *World Sci. News* **2018**, *110*, 42–55.
104. Ahamad, T.; Khan, M.A.; Khan, M.F.; Ahmad, R.; Rahman, M.A.; Siddiqui, S. *Oxalis corniculata*-Derived Bioactive Compounds Target Hormone Receptors in Breast Cancer: HPLC-ESI-MS/MS Analysis, Cytotoxicity, and Computational Studies. *ChemistrySelect* **2025**, *10*, e202404547. [CrossRef]
105. Jain, B. An evidence-based ethnomedicinal study on *Oxalis corniculata*: Review of decade study. *Int. J. Green Pharm.* **2023**, *17*, 3371.
106. Duc, L.V.; Thi, M.N.; Le Hong, D.; Le Huong, G. Antioxidant Activity, Inhibition of No Production and Cytotoxicity of Chemical Compounds Isolated from *Oxalis corniculata* L. *Pharm. Chem. J.* **2023**, *57*, 388–394.
107. Lei, Y.; Zhang, B.; Ma, X.; Zheng, L.; Gong, Z.P.; Wang, Y.L.; Li, Y.J. Chemical constituents from *Oxalis corniculata* L. *Chin. Pharm. J.* **2021**, *56*, 1378–1383.
108. Loi, V.D.; Nga, D.T.Q.; Huong, D.T.M.; Huy, N.Q. Compounds Isolated from the Ethyl Acetate Fraction of the Aerial Parts of *Oxalis corniculata* L. *VNU J. Sci. Med. Pharm. Sci.* **2018**, *34*, 4112.
109. Absar, K.M.B.; Rifat, H.B.S.M.; Das, S.; Eisha, J.A.; Das, T.R.; Dash, P.R. Phytochemical and Pharmacological Properties of *Oxalis corniculata*: A Review. *Trop. J. Phytochem. Pharm. Sci.* **2024**, *3*, 364–374. [CrossRef]

International Journal of
Molecular Sciences

MDPI

Review

Progress of AI-Driven Drug–Target Interaction Prediction and Lead Optimization

Qiqi Wang [1,2], Boyan Sun [1,2], Yunpeng Yi [3], Tony Velkov [4], Jianzhong Shen [1,2], Chongshan Dai [1,2,*] and Haiyang Jiang [1,2,*]

[1] State Key Laboratory of Veterinary Public Health and Safety, College of Veterinary Medicine, China Agricultural University, Beijing 100193, China
[2] Technology Innovation Center for Food Safety Surveillance and Detection (Hainan), Sanya Institute of China Agricultural University, Sanya 572025, China
[3] Shandong Provincial Animal and Poultry Green Health Products Creation Engineering Laboratory, Institute of Poultry Science, Shandong Academy of Agricultural Science, 202 Gongyebeilu Jinan, Jinan 250023, China
[4] Department of Pharmacology, Biodiscovery Institute, Monash University, Clayton, VIC 3800, Australia
* Correspondence: daichongshan@cau.edu.cn (C.D.); haiyang@cau.edu.cn (H.J.)

Abstract: In modern pharmaceutical research and development (R&D), drug discovery remains a challenging process. Artificial intelligence (AI) has been extensively incorporated into various phases of drug discovery and development. AI enable effectively extract molecular structural features, perform in-depth analysis of drug–target interactions, and systematically model the relationships among drugs, targets, and diseases. These approaches improve prediction accuracy, accelerate discovery timelines, reduce costs from trial and error methods, and enhance success probabilities. This review summarizes recent advances in AI applications for drug design, including target identification, synthetic accessibility prediction, lead optimization, and ADMET property evaluation. Furthermore, it introduces various deep learning tools to guide researchers in selecting and implementing the most appropriate AI-driven strategies throughout the drug discovery process. We hope it can establish a conceptual framework intended to advance AI-driven methodologies in pharmaceutical research by comprehensively organizing novel perspectives and critical insights.

Keywords: artificial Intelligence; drug discovery; machine learning; virtual screening

1. Introduction

The traditional drug discovery paradigm faces formidable challenges characterized by lengthy development cycles, prohibitive costs, and high preclinical trial failure rate [1]. The process from lead compound identification to regulatory approval typically spans over 12 years with cumulative expenditures exceeding $2.5 billion [2]. Clinical trial success probabilities decline precipitously from Phase I (52%) to Phase II (28.9%), culminating in an overall success rate of merely 8.1% [3–7]. Global efforts are intensifying to address the persistent inefficiency challenges in drug development. The strategy involves diversifying therapeutic targets to overcome the limitations of traditional approaches. This aims to reduce the preclinical attrition rate of candidate drugs, improve R&D efficiency, and optimize the cost-effectiveness ratio to alleviate the burden of high investment.

Computational molecular modeling has catalyzed a paradigm shift in pharmaceutical research. It enables the precise simulation of receptor–ligand interactions and the optimization of lead compounds [8,9]. For example, Talukder et al. integrated molecular docking, QSAR, and molecular dynamics to identify phytochemical inhibitors targeting

EGFR in non-small cell lung cancer [10]. Similarly, Kaur et al. designed blood–brain barrier (BBB) permeable β-secretase enzyme (BACE-1) inhibitors for Alzheimer's disease using 2D-QSAR and ADMET profiling. The novel molecules exhibited good potency, BBB permeability, excellent binding affinity, and stable conformations with BACE-1 [11]. Souza et al. investigated the ability of Artificial Intelligence (AI) to balance binding affinity with drug-likeness in the design of SARS-CoV-2 protease inhibitors using machine learning (ML) [12]. Moreover, Maliyakkal et al. employed QSAR-driven virtual screening to identify *Trypanosoma cruzi* inhibitors with high predicted efficacy [13]. This paradigm shift has catalyzed the rise of AI-driven drug discovery (AIDD). ML integrates multiple omics data and structural biology insights to provide information for experimental design. Modern drug development workflows increasingly rely on these predictive systems for critical tasks including target prioritization [14–16], high-throughput compound screening [17–20], synthetic route planning [21–23], and polymorph screening [24–26] (Figure 1). A representative research achievement from Insilico Medicine is rentosertib, an AI-discovered drug that has completed Phase II trials for pulmonary fibrosis, showcasing the power of its AI platform [27]. Parallel advancements in antimicrobial discovery reveal that computational drug design platforms can systematically decode structural determinants of antibiotic efficacy, particularly against drug resistant pathogens [28]. The field's transformative potential was further validated by the 2024 Nobel Prize in Chemistry, awarded for AI-powered innovations in protein engineering [29].

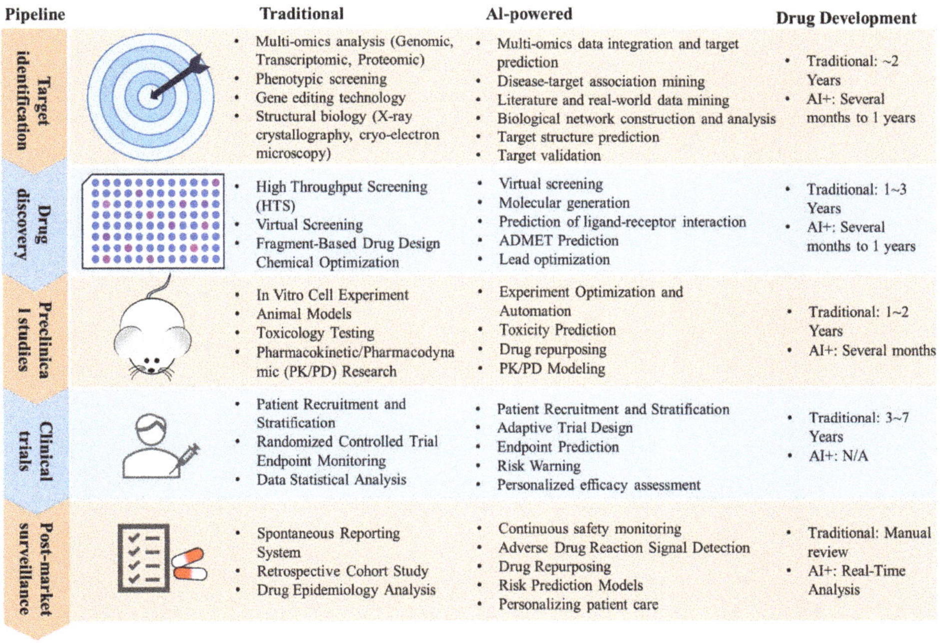

Figure 1. Comparative analysis of AI applications across multistage drug development pipelines.

AIDD has experienced rapid advances, and Table 1 presents representative AI-designed small molecules currently progressing through clinical trials. A key technological strength lies in the capacity to decode intricate structure-activity relationships, facilitating de novo generation of bioactive compounds with optimized pharmacokinetic properties. The efficacy of these algorithms is intrinsically linked to the quality and volume of training data, particularly in deciphering latent patterns within complex biological datasets [30–33].

As deep learning architectures continue to evolve, the integration of AIDD into pharmaceutical pipelines is poised to significantly enhance development efficiency and substantially reduce costs.

Table 1. Selected examples of AI-discovered/designed small molecules in clinical stages.

Small Molecule	Company	Target	Stage	Indication
REC-1245 [34]	Recursion	RBM39	Phase 1	Biomarker-enriched solid Tumors and lymphoma
REC-3565 [34]	Recursion	MALT1	Phase 1	B-Cell Malignancies
REC-4539 [34]	Recursion	LSD1	Phase 1/2	Small-Cell Lung Cancer
REC-4881 [34]	Recursion	MEK Inhibitor	Phase 2	Familial adenomatous polyposis
REC-3964 [34]	Recursion	Selective C. diff Toxin Inhibitor	Phase 2	Clostridioides difficile Infection
REC-7735 [34]	Recursion	PI3Kα H1047R	Preclinical	HER2−HR+ Breast cancer
REV102 [34]	Recursion	ENPP1	Candidate profiling	Hypophosphatasia
ISM-6631 [35]	Insilico Medicine	Pan-TEAD	Phase 1	Mesothelioma, and Solid Tumors
ISM-3412 [35]	Insilico Medicine	MAT2A	Phase 1	MTAP−/− Cancers
INS018-055 [35]	Insilico Medicine	TNIK	Phase 2a	IPF
ISM-3091 [35]	Insilico Medicine	USP1	Phase 1	BRCA mutant cancer
ISM-8207 [35]	Insilico Medicine	QPCTL	Phase 1	Solid Tumors
ISM-5043 [35]	Insilico Medicine	KAT6	Phase 1	Breast cancer
ISM-5939 [35]	Insilico Medicine	ENPP1	IND Clearance	Solid Tumors
ISM-5411 [35]	Insilico Medicine	PHD	Phase 1	IBD/Anemia of CKD
ISM-3312 [35]	Insilico Medicine	3CLpro	Phase 1	COVID-19
RLY-4008 [36]	Relay Therapeutics	FGFR2	Phase 1/2	FGFR2-altered cholangiocarcinoma
RLY-8161 [36]	Relay Therapeutics	NRAS	Preclinical	Solid Tumors
RLY-2608 [36]	Relay Therapeutics	PI3Kα	Phase 1/2	Advanced Breast Cancer
EXS4318 [37]	Exscientia	PKC-theta	Phase 1	Inflammatory and immunologic diseases
GTAEXS617 [37]	Exscientia	CDK7	Phase 1/2	Solid Tumors
SIGX-1094 [38]	Signet Therapeutics	FAK	Phase 1/2	Solid Tumors
BG-89894 [39]	BeiGene	Mat2A	Phase 1	Advanced Solid Tumors
H002 [40]	RedCloud Bio	EGFR	Phase 1	Non-Small Cell Lung Cancer
AC0682 [41]	Accutar Biotech	ER Degrader	Phase 1	Breast Cancer
AC0682 [41]	Accutar Biotech	ER Degrader	Phase 1	Breast Cancer
AC0176 [41]	Accutar Biotech	AR Degrader	Phase 1	Prostate Cancer
AC0676 [41]	Accutar Biotech	BTK Degrader	Phase 1	Hematology Oncology Indications

Table 1. *Cont.*

Small Molecule	Company	Target	Stage	Indication
MDR-001 [42]	MindRank	GLP-1	Phase 1/2	Obesity/Type 2 Diabetes Mellitus
DF-006 [43]	Drug Farm	ALPK1	Phase 1	Hepatitis B/Hepatocellular cancer
LAM-001 [44]	OrphAI Therapeutics	mTOR	Phase 2	PH/BOS
HLX-1502 [45]	Healx	N/A [b]	Phase 2	Neurofibromatosis Type 1
BGE-105 [46]	BioAge	APJ agonist	Phase 2	Obesity/Type 2 diabetes
BXCL501 [47]	BioXcel Therapeutics	alpha-2 adrenergic	Phase 2/3	Neurological Disorders
EVX-01 [48]	Evaxion Biotech	N/A [b]	Phase 2	Metastatic melanoma
EVX-02 [48]	Evaxion Biotech	N/A [b]	Phase 1	Adjuvant melanoma

[b], no marked information.

This review provides a systematic overview of recent advances in AI for medicinal chemistry design. While previous surveys have broadly outlined AI applications across the entire drug discovery pipeline, this work focuses specifically on two pivotal stages: drug–target interaction (DTI) prediction and lead optimization. We summarize a range of deep learning (DL) tools to assist researchers in selecting and implementing suitable AI-driven strategies within the drug discovery process. In addition to the core focus on DTI and lead optimization, discussions were included on target identification, synthetic feasibility prediction, virtual screening, and ADMET evaluation. It demonstrates that integrating these phases with AI reduces false positives, improves compound prioritization, and accelerates therapeutic design. The discussion also highlights the translational impact of current methods and identifies promising directions for future development.

2. AI Technology

AI develops systems capable of human-like reasoning and decision-making. Contemporary AI systems integrate ML and DL to address pharmaceutical challenges ranging from target validation to formulation optimization.

ML employs algorithmic frameworks to analyze high-dimensional datasets, identify latent patterns, and construct predictive models through iterative optimization processes [49]. ML has evolved into four principal paradigms: supervised learning, unsupervised learning, semi-supervised learning, and reinforcement learning [50]. Supervised learning employs labeled datasets for classification via algorithms like support vector machines (SVMs) and for regression via algorithms like support vector regression (SVR) and random forests (RFs) [51,52]. Unsupervised learning identifies latent data structures through clustering and dimensionality reduction techniques (such as principal component analysis and K-means clustering) to reveal underlying pharmacological patterns and streamline chemical descriptor analysis [53,54]. In contrast, t-distributed stochastic neighbor embedding (t-SNE) primarily serves as a nonlinear visualization tool, effectively mapping high-dimensional molecular features into low-dimensional spaces to facilitate the interpretation of chemical similarity and class separation [55]. Semi-supervised learning boosts drug–target interaction prediction by leveraging a small set of labeled data alongside a large pool of unlabeled data. This is achieved through model collaboration and by generating simulated data, which enhances prediction reliability [56,57]. Reinforcement learning optimizes molecular design via Markov decision processes, where agents iteratively refine policies to generate inhibitors and balance pharmacokinetic properties through reward-driven strategies [58,59]. A comparison of ML models are shown in Table 2.

Table 2. Comparison of machine learning models.

Comparison Dimension	Supervised Learning	Unsupervised Learning	Semi-Supervised Learning	Reinforcement Learning
Data Requirements	Fully labeled datasets (input–output pairs)	Unlabeled data	Small amount of labeled data coupled with a large amount of unlabeled data	Dynamic environment interaction (state-action-reward signals)
Core Functions	Classification, Regression	Clustering, Dimensionality reduction	Label expansion & data augmentation	Strategy optimization & sequential decision-making
Typical Algorithms	Support Vector Machine (SVM); Logistic Regression; Random Forest [53]	Principal Component Analysis (PCA); t-SNE; K-means [53,54]	Graph-based Semi-supervised Learning [56,57]	Q-learning; Policy Gradient [53,60]
Advantages	High prediction accuracy; Clear objective functions	Reveals intrinsic data structures; No labeling cost	Mitigates labeled data scarcity; Enhances model generalization	Adapts to dynamic environments; Optimizes long-term goals
Disadvantages	Relies on high-quality labels; High overfitting risk	Poor interpretability of results; Requires manual validation of clusters	High model complexity; Performance depends on initial label quality	Slow training convergence; High computational cost
Drug Discovery Applications	Molecular toxicity prediction; Activity classification [51,52]	Chemical space visualization; Compound library deduplication [55]	Drug–target interaction prediction [61]	Inhibitor reaction path optimization; Multi attribute molecular design [58,59]

Building upon these models, a broad spectrum of shallow learning models (such as SVM, decision Trees, RF, and artificial neural networks) has historically served as the computational backbone of data-driven drug discovery. These models provide strong interpretability, efficient training on moderate-sized datasets, and robust performance for QSAR modeling, activity prediction, and ADMET estimation. Their comparative strengths and limitations are summarized in Table 3, which outlines their respective mechanisms, advantages, and representative applications across medicinal chemistry tasks.

Table 3. Comparative overview of commonly used shallow machine learning algorithms.

Model	Core Mechanism	Advantages	Disadvantages	Drug Discovery Applications
Support Vector Machine (SVM)	Finds optimal hyperplane maximizing the margin between classes using kernel functions.	High classification accuracy for small and high-dimensional datasets; robust to overfitting with suitable kernel.	Sensitive to kernel and hyperparameter selection; limited scalability for large datasets.	Drug–target interaction prediction; compound activity classification; ADMET and toxicity modeling.

Table 3. *Cont.*

Model	Core Mechanism	Advantages	Disadvantages	Drug Discovery Applications
Decision Tree	Builds hierarchical decision rules by recursively splitting features that maximize information gain.	Simple, interpretable, and requires minimal preprocessing; handles nonlinear relationships.	Prone to overfitting; small data perturbations can change structure.	Compound classification; QSAR modeling; structure–activity relationship visualization.
Random Forest (RF)	Ensemble of multiple decision trees via bootstrap aggregation (bagging) and majority voting.	Reduces variance and overfitting; robust to noise; provides feature importance ranking.	Decreased interpretability; slower inference for large forests.	Bioactivity prediction; virtual screening; toxicity classification.
k-Nearest Neighbors (KNN)	Classifies samples by majority vote of k nearest data points in feature space.	Simple implementation; adaptable to diverse datasets.	Computationally expensive for large datasets; sensitive to irrelevant features and scaling.	Molecular similarity searching; ligand-based virtual screening; QSAR/QSPR modeling.
Artificial Neural Network (ANN)	Composed of interconnected layers of neurons performing weighted summations and nonlinear activation.	Learns nonlinear mappings; flexible architecture; high predictive performance with proper tuning.	Prone to overfitting; requires large labeled data; limited interpretability compared to tree models.	QSAR/QSPR modeling; bioactivity prediction; pharmacokinetic property estimation.

Despite the success of shallow learning models, the representational capacity of shallow models is limited when dealing with complex, nonlinear biological data or multimodal molecular information. This limitation has driven the evolution toward DL architectures. DL, as a subset of ML, processes information through multilayer neural networks. These networks automatically learn a hierarchy of features directly from raw data via a sequence of functional layers (e.g., convolutional layers, self-attention layers). The process begins with initial layers capturing low-level, local patterns (such as atomic properties in a molecule or edges in an image). These elementary features are then progressively combined and transformed through subsequent layers to form higher-level, more abstract representations. This multi-stage, hierarchical processing is powered by nonlinear transformations introduced by activation functions at each step, enabling DL models to approximate highly complex, nonlinear relationships that are inherent in multidimensional biological and chemical data [62]. Different models have different core mechanisms. Convolutional neural networks (CNNs) are designed to analyze grid-like data, owing to their unique convolutional architecture. This capability makes them particularly valuable for biomedical image processing tasks, such as cell morphology analysis [63]. Recurrent neural networks (RNNs), with their internal recurrent connections, effectively model sequential data. For instance, they can process SMILES strings for automated molecular generation [64]. Although RNNs once represented the state-of-the-art in this domain, they have gradually been supplanted by Transformer-based architectures. Transformers excel because they more effectively capture long-range dependencies and contextual relationships in complex datasets. Nonetheless, RNNs still demonstrate advantages in modeling local sequential patterns [65]. Graph neural networks (GNNs) analyze graph structures by propagating neighborhood information, excelling at molecular property prediction (Figure 2A) [66,67]. In contrast, Transformer models use self-attention to capture long-range dependencies in sequences, showing great promise for innovative tasks like designing antiviral analogs (Figure 2B) [66–70].In terms of new molecule generation, Generative adversarial networks (GANs) synthesize structurally diverse candidate molecules through an adversarial training framework involving a generator and a discriminator [71–74]. Variational autoencoders (VAEs) explore molecular structures within a latent space using an encoder–decoder architecture (Figure 2C) [75–79]. Large language models (LLMs) have shown remarkable capabilities in handling massive multimodal data, and their potential in generative drug design is being gradually uncovered. These models are expected to assist drug discovery by integrating structured knowledge extracted from scientific literature (Figure 2D) [80–83]. It should be noted that their direct application to generative drug design remains nascent. In the short term, their more established potential lies in supporting rational drug discovery workflows through literature mining, hypothesis generation, and knowledge integration [84]. These diverse DL architectures collectively aim to address critical bottlenecks in drug research, powerfully supporting core tasks such as toxicity prediction, molecular property optimization, and de novo drug design (for a detailed comparative summary of various models, see Table 4).

In practice, many studies employ the same public datasets (such as MoleculeNet, TDC) to ensure fair model assessment and to compare results with prior work [85,86]. However, differences in research objectives, model architectures, and evaluation strategies often lead to diverse reporting practices. Increasingly, hybrid or ensemble models combine complementary strengths. For instance, GNN to capture 2D/3D molecular structures, recurrent architectures to handle sequential representations such as SMILES, and transformer-based models for long-range contextual dependencies [87]. These integrative approaches underscore the value of multi-perspective feature extraction over relying on a single model. Consequently, future benchmarking should evaluate not only raw performance but also the complementary strengths of different models.

Table 4. Comparison of deep learning models.

Model	Core Mechanism	Advantages	Disadvantages	Drug Discovery Applications
CNNs	Convolution-pooling stack [88]	Convolutional layers extract local features; pooling layers reduce dimensionality [89]	Relies on local receptive fields; struggles with global dependencies	Cellular morphology analysis, drug–drug interaction prediction [90]
RNNs	Recurrent connections	Process sequential data via cyclic connections [73,91–93]	Gradient vanishing/explosion; weak long-range dependency handling	SMILES-based molecular generation [94]
GNNs	Graph convolution	Aggregate neighborhood information through graph message-passing [95–101]	High computational complexity; hyperparameter sensitivity	Solubility prediction, binding affinity calculation [51,102,103]
Transformers	Self-attention mechanism [104]	Capture global dependencies via self-attention mechanisms [105,106]	High memory consumption; prone to overfitting on small datasets	Derivative design, multimodal data integration [69,107]
GANs	Generator-discriminator adversarial	Adversarial training between generator and discriminator	Training instability, mode collapse risk	Novel chemical entity synthesis [71]
GANs	Generator-discriminator adversarial	Adversarial training between generator and discriminator	Training instability, mode collapse risk	Novel chemical entity synthesis [71]

Table 4. *Cont.*

Model	Core Mechanism	Advantages	Disadvantages	Drug Discovery Applications
VAEs	Encoder–decoder latent space	Compress molecular features via encoder–decoder architecture	Blurry generated samples, limited diversity	Targeted drug design [78]
LLMs	Multi-layer Transformer [108,109]	Multimodal Transformer-based joint reasoning	High training cost, requires domain-specific fine-tuning	Literature knowledge-driven molecular optimization [110]

Convolutional neural networks, CNNs. Recurrent neural networks, RNNs. Graph neural networks, GNNs. Generative adversarial networks, GANs. Variational autoencoders, VAEs. Large language models, LLMs.

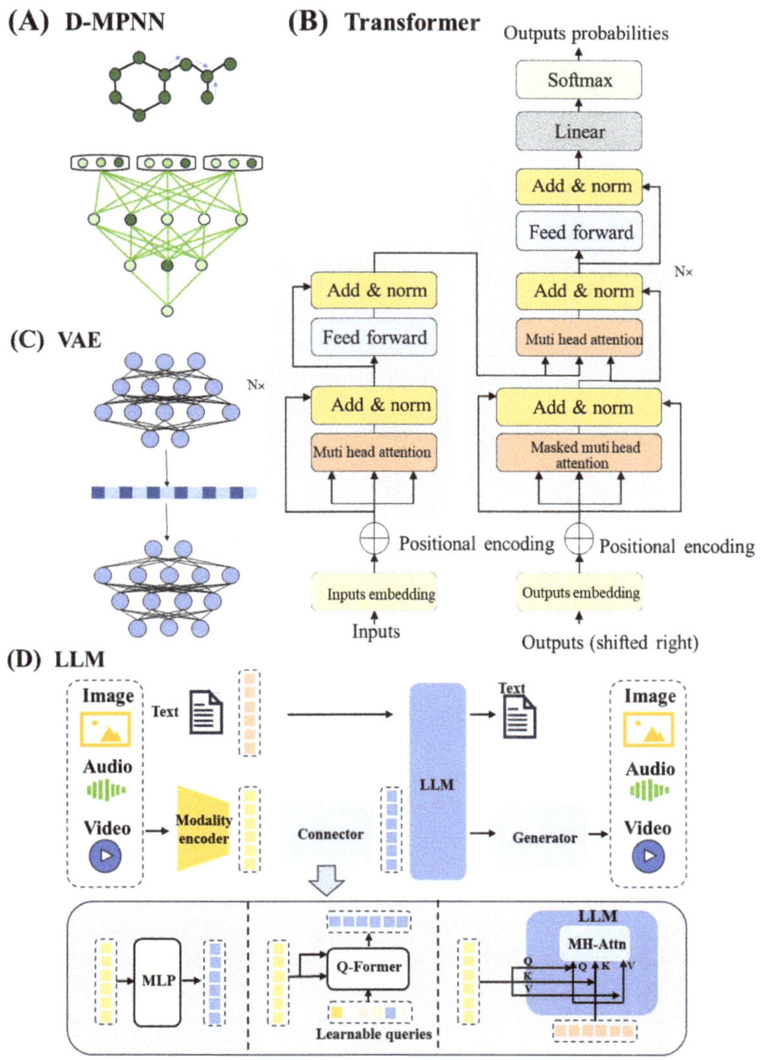

Figure 2. Deep learning model architectures. (**A**), Directed message-passing neural network architecture [102]. (**B**), Transformer model architecture. Key hyperparameters include: N denotes the number

of encoder/decoder layers, Muti-head attention represents the attention heads, and the dimension of input embeddings represents the length of the continuous, dense vector to which each discrete input token [104]. (**C**), VAE model architecture. (**D**), LLM model architecture. LLM processes diverse inputs through modality-specific encoders that convert heterogeneous data into latent representations. Cross-modal alignment is achieved via a query-based transformer (Q-Former) employing multi-head attention (MH-Attn) mechanisms. The central LLM performs joint reasoning enhanced by MLP-based feature fusion, and multimodal generators synthesize coordinated outputs such as text or images, completing the comprehension–generation cycle.

3. Applications in New Drug Development

The rapid expansion of bioactive compounds and biomedical multi-omics data, accelerated by technologies such as high-throughput virtual screening, has necessitated advanced computational approaches in drug discovery. AI plays a key role in multiple drug development stages, including molecular characterization, structural analysis of target proteins, and virtual screening of lead compounds. ML algorithms, with their predictive modeling capabilities, continuously optimize the entire drug design process. The following sections will highlight innovative applications of AI in these areas, illustrating how technological breakthroughs are driving further advances in new drug development.

3.1. Applications in Tertiary Structure of Target Proteins

The tertiary structure of a protein is formed by the folding of its amino acid sequence through complex atomic interactions, directly determining its biological function [111]. Since protein dysfunction is a key pathological mechanism in many diseases, accurately determining its three-dimensional structure is critical for structure-based drug discovery [112,113]. However, the immense computational cost of modeling complex molecular folding and dynamics renders traditional simulation-based methods a bottleneck in early drug development.

AlphaFold has revolutionized the field of structural biology by enabling data-driven prediction of protein conformations directly from sequence information [114,115]. AlphaFold2 achieves remarkable accuracy in predicting monomeric protein folds but is inherently limited to static conformations. It does not capture dynamic processes, such as allosteric transitions, conformational flexibility, or ligand-induced rearrangements. Consequently, its application in protein–ligand and protein–protein interaction studies remain constrained. On the other hand, more recent frameworks like AlphaFold-Multimer and AlphaFold3 can predict complexes of proteins, nucleic acids, and small molecules. However, they still cannot fully model dynamic interactions, which remain dependent on molecular dynamics simulations [116,117]. Despite these limitations, such models have shown practical value in structure-based drug discovery. For instance, through computational prioritization and preclinical validation, researchers successfully identified a potent lead compound that acts as a trace amine-associated receptor 1 (TAAR1) agonist [118]. Moreover, integration of structure prediction with generative AI systems further empowers de novo drug design. For example, the PCMol model enables simultaneous multi-target candidate molecule optimization while maintaining synthetic feasibility [119]. In summary, the synergy between structure prediction and AI-driven chemistry platforms can be directly applied to hit identification in drug discovery, offering viable development pathways for understudied targets [120].

Beyond AlphaFold, several other DL frameworks contribute to tertiary structure prediction and protein design. The RoseTTAFold model can design entirely novel proteins not observed in nature, expanding the druggable structural space [121,122]. Similarly, the trRosetta (transform-restrained Rosetta) server is a deep learning-driven platform for rapid and accurate de novo protein structure prediction. Unlike traditional homology modeling approaches, trRosetta excels in de novo prediction when no experimentally resolved templates are available [123]. Meanwhile, ESMFold, a Transformer-based protein language model, achieves rapid sequence to structure prediction through training at various scales. It is up to ten times faster than traditional methods and maintains accuracy even for understudied proteins with limited structural data [113]. Although its precision is slightly below mainstream benchmarks, this speed advantage makes such models critical for streamlining early-stage drug design. Overall, these technological advancements are accelerating structure-guided therapeutic development by enhancing computational efficiency and enabling structural innovation.

3.2. Applications in Target Identification

In precision medicine, the growing complexity of multifactorial diseases has created an urgent need for new molecular targets. However, the number of clinically validated candidate drug targets is still limited, patient applicability is narrow, and clinical-trial failure rates remain high [124]. Until 2022, fewer than 500 drug targets had been successfully established worldwide, a figure that falls far short of the needs for complex disease treatment [125,126]. Consequently, discovering and evaluating new therapeutic targets has become a central focus of drug development. AI has emerged as a key enabler in this area by systematically integrating multi-scale biomedical data to perform probabilistic target identification with enhanced mechanistic interpretability [14].

Modern target discovery relies on three core strategies: experimental validation, multi-omics integration, and computational modeling. Experimental approaches use functional genomics tools like CRISPR-based screens to directly validate connections between targets and diseases [127–131]. Multi-omics platforms integrate genomic, proteomic, and metabolomic data to systematically identify how genes influence disease characteristics [132–137]. Computational methods use ML and structural biology to prioritize drug targets. Techniques like inverse docking (which screens a single ligand against multiple protein targets) and pharmacophore analysis help identify promising candidates, reducing the need for experimental resources [138–140].

As biomedical datasets have grown, AI techniques leveraging large scale data analysis, pattern recognition and network construction have found broad application in target discovery. For example, AI analysis of transcriptome profiles from young and aged skeletal muscle identified key drivers of muscle aging [141]. At the same time, Transformer-based architectures have been particularly impactful for integrating multi-omics datasets. In single-cell transcriptomics, pre-trained transformer models learn gene relationships through attention mechanisms to capture gene–gene interactions. This enables researchers to accelerate the discovery of key network regulators and candidate therapeutic targets (Figure 3A) [142]. In parallel, building on pretrained protein language models have transformed target prioritization by integrating data across human, murine, and cellular contexts. Frameworks such as protein essentiality predictors use a multi-module structure that encodes protein sequences, weights biological context information, and classifies essentiality probabilities. These models generate protein essentiality scores that assist in identifying key therapeutic targets, cancer prognostic markers, and microprotein functions (Figure 3B) [143].

Alongside these strategies, by applying RefMap to human induced pluripotent stem cells (iPSC) derived motor neurons, the researchers mapped noncoding regulatory regions to target genes, identifying 690 ALS-associated candidates and achieving a fivefold increase in recovered heritability [144]. Finally, AI-powered drug-repurposing platforms (such as deepDTnet, which embeds chemical, genomic, phenotypic, and cellular network data) have successfully predicted topotecan as a related orphan receptor-gamma t inhibitor and demonstrated its efficacy in a mouse model of multiple sclerosis [145].

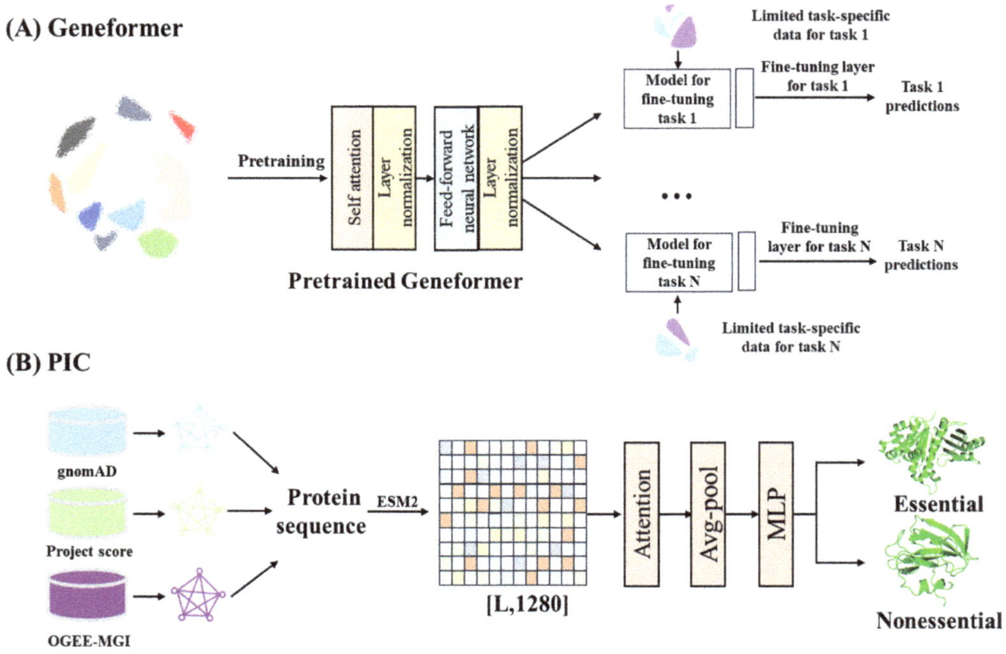

Figure 3. Applications of AI in target discovery. (**A**), Transformer-based gene network model for single-cell transcriptomics [142]. (**B**), Protein essentiality prediction framework integrating multi-source biological data [143]. ESM2, Evolutionary scale modeling 2. MLP, Multilayer perceptron. Avg-pool, Average pooling. Attention, Attention mechanism.

As algorithmic sophistication converges with expanding biomedical datasets, AI-driven target discovery is transitioning from exploratory research to clinical implementation. Through their bridging of mechanistic insights and therapeutic design, these technologies promise to streamline drug development cycles and advance precision medicine.

3.3. *Applications in Drug–Target Interaction Prediction*

Accurate prediction of drug–target interactions is a cornerstone of modern drug discovery, providing key insights to understand binding mechanisms and directly guiding rational therapeutic design. Molecular docking is a core computational approach in this field, aiming to predict the preferred orientation and conformation of small-molecule ligands when bound to protein targets, while estimating their binding affinity through scoring functions. By combining search algorithms and energy-based scoring, docking enables virtual screening of large chemical libraries to identify potential binders efficiently [146]. Thanks to the close integration of ML and structural biology, this field has made significant progress in recent years. One of the prominent directions is the innovation of virtual screening methods. For example, molecular docking approaches optimized through active learning employ intelligent sampling strategies to effectively balance chemical space exploration with the identification of potentially highly active molecules. Such methods have demonstrated the ability to recover over 80% of experimentally confirmed active compounds while reducing computational costs by a factor of 14. Moreover, over 90% of active scaffolds were retained within the top 5% of model-predicted compounds, preserving the diversity of confirmed actives [147].

Innovative techniques for drug–target interaction (DTI) prediction are overcoming traditional limitations. To address the challenge of limited training data and poor generalization, ML based on pre-trained protein language models have significantly enhanced extrapolation capabilities by analyzing the structural determinants of molecular recognition. These systems have been shown to accurately distinguish true binders from decoy compounds. Furthermore, they offer interpretable visualizations of interactions, enabling the use of embeddings to represent the functional characteristics of human cell surface proteins [148].

The deep integration of DL and molecular docking has catalyzed a transformation in virtual screening platforms. These platforms have been extended to multi-target pharmacological studies and to the creation of customized, synergistic kinase inhibition regimens for aggressive cancers through transcriptome-driven models (Figure 4A) [149]. Traditional docking involves generating candidate poses and scoring them based on binding energy. DL now learns from docking outcomes to accelerate screening across vast chemical libraries. For instance, Deep Docking utilizes QSAR deep models to estimate the docking outcome for yet unprocessed entries. In this approach, a small subset of compounds is first docked. A deep neural network is then trained on this subset to predict the docking scores for the remaining library, which dramatically reduces computation time while maintaining high accuracy (Figure 4B) [150,151]. Similarly, receptor activation pattern analysis could guide the development of dual-target agonists with enhanced pharmacological properties [152].

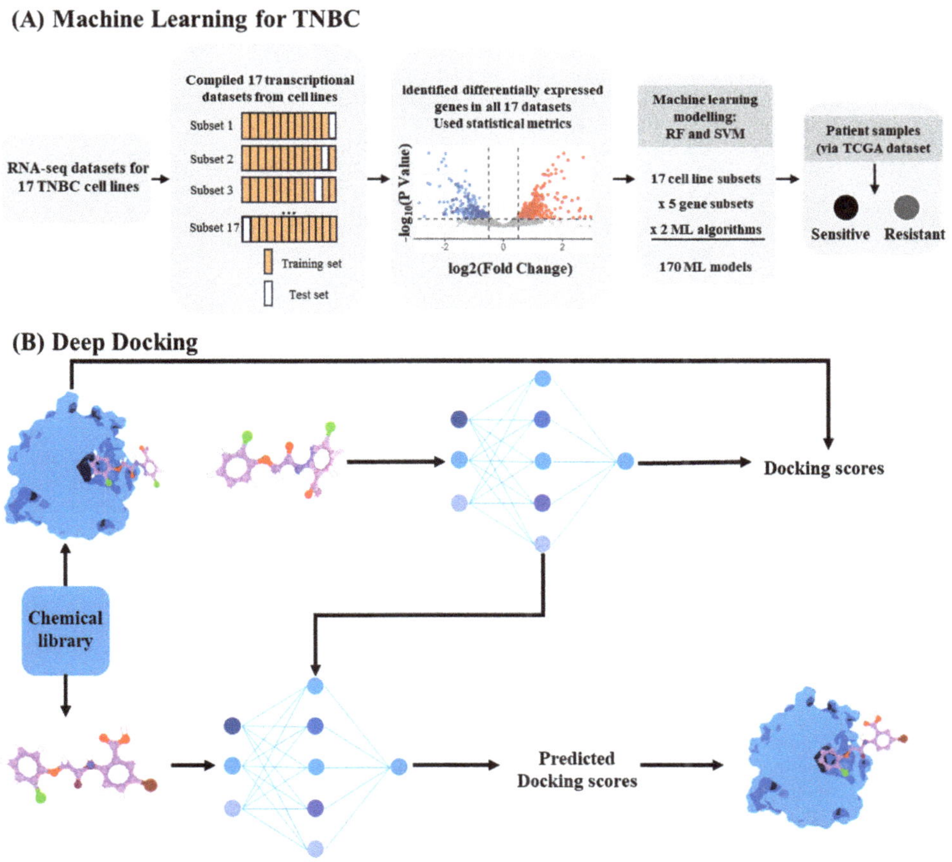

Figure 4. AI applications in drug–target interaction prediction. (**A**), Transcriptome-guided framework for target prioritization [149]. (**B**), Deep Docking approach combining molecular docking and neural prediction [150].

However, systemic challenges still need to be overcome. Rare diseases and emerging pathogens often lack sufficient labeled data, limiting model generalizability. Current predictors often suffer from reduced reliability due to dataset bias. To mitigate these issues, transfer learning and zero-shot learning strategies have been increasingly adopted. For example, TxGNN, a graph neural network model developed for zero-shot drug repurposing, leverages large-scale medical knowledge graphs to predict indications and contraindications even for diseases without existing therapies [95]. Future advances will likely require systematic curation of balanced interaction datasets combined with bias-resistant and transfer-learning-enabled modeling architectures to improve predictive reliability. Recent DL models in the target prediction tasks were collected in Table 5.

Table 5. DL Models for target prediction frameworks.

No.	Model Name	Framework Description	Web
1	**Geneformer**	A deep learning model based on a six-layer Transformer architecture. It is self-supervised pretrained on 30 million single-cell transcriptomes to learn gene interaction dynamics. Its attention mechanisms autonomously capture hierarchical gene relationships, enabling transfer to data-scarce scenarios such as disease gene prediction, chromatin interaction analysis, and therapeutic target discovery [142].	https://huggingface.co/ctheodoris/Geneformer (accessed on 20 August 2025)
2	**PIC**	A framework built on the pretrained Evolutionary Scale Modeling 2 (ESM2) protein language model. It extracts sequence features via an embedding module, captures residue importance through an attention module, and outputs probability-based predictions of protein essentiality across human tissues, cell lines, and mouse. The resulting Protein Essentiality Score quantifies protein importance, supports breast-cancer prognostic marker identification, and evaluates microprotein function [143].	https://github.com/KangBoming/PIC (accessed on 20 August 2025)
3	**Deep Docking**	A binary classifier deep neural network trained iteratively on 1024-bit circular Morgan fingerprint features. It dynamically predicts and filters high-docking-score compounds from ultra-large libraries, reducing a multi-billion-member library to a manageable size while retaining over 90% of active candidates and drastically lowering compute requirements [150].	https://github.com/jamesgleave/DD_protocol (accessed on 20 August 2025)
4	**EMCIP**	An ensemble model combining multiple machine-learning algorithms with a multi-instance 3D graph neural network. Using stacked generalization and soft-voting, it integrates predictions from all base models for efficient forecasting of Candida drug resistance inhibitors [153].	https://github.com/trinhthechuong/Cdr1_inhibitors (accessed on 20 August 2025) https://huggingface.co/spaces/thechuongtrinh/EMCIP_Cdr1_inhibitor_prediction (accessed on 20 August 2025)
5	**CSNN**	A graph-neural network leveraging Chemical Space Networks (CSNs) and label information as node features. This method achieves zero-training predictions by querying a compound's chemical neighbors and exploiting network homophily, significantly enhancing drug–target interaction accuracy for human G protein-coupled receptors [154].	https://github.com/cansyl/TransferLearning4DTI/ (accessed on 20 August 2025)
6	**GraphBAN**	A graph-based knowledge-distillation model combining a graph autoencoder and a bilinear attention network. It fuses multi-source features of compounds and proteins, along with a cross-domain adaptation module, to predict interactions between unseen nodes, improving both accuracy and generalization in compound–protein interaction prediction for drug discovery [155].	https://github.com/HamidHadipour/GraphBAN/blob/main/README.md (accessed on 20 August 2025)

Table 5. *Cont.*

No.	Model Name	Framework Description	Web
7	DTIAM	A unified framework that uses self-supervised pretraining for drug molecules (multi-task learning) and Transformer-based protein representations. It incorporates automated machine-learning to predict drug–target interactions, binding affinities, and activation/inhibition mechanisms [156].	https://github.com/CSUBioGroup/DTIAM (accessed on 20 August 2025)
8	KinGen	A recurrent neural network (RNN)-based molecule generator combined with reinforcement learning and transfer learning. By optimizing reward functions and transferring knowledge, it could generate high-activity compounds targeting specific kinases [157].	https://github.com/Shawn-Lau-lxm/KinGen (accessed on 20 August 2025)
9	MINDG	A hybrid architecture that extracts drug and target sequence features via deep networks, captures higher-order structural information through graph-attention convolutional layers, and integrates multi-view predictions with an adaptive decision module. This approach markedly improves drug–target interaction prediction accuracy [158].	https://github.com/AGI-FBHC/MINDG (accessed on 20 August 2025)
10	PIGNet2	A DL model combining physics-informed graph neural networks with novel data-augmentation strategies. It accurately predicts protein–ligand binding affinities and efficiently screens candidate drugs, demonstrating exceptional cross-task performance in hit identification and lead optimization [159].	https://github.com/ACE-KAIST/PIGNet2 (accessed on 20 August 2025)
11	CarsiDock	A DL-based protein–ligand docking model pretrained on 9 million predicted complexes. Coupled with geometric optimization strategies, it greatly enhances pose-prediction accuracy, reliably reproduces key interactions observed in crystal structures, and preserves ligand topology for high-throughput virtual screening [160].	https://github.com/carbonsilicon-ai/CarsiDock (accessed on 20 August 2025)

PIC, Protein Importance Calculator. EMCIP, Ensemble Model for Cdr1 Inhibitor Prediction. CSNN, Chemical Space Neural Network. MINDG, Multi-view Integrated Learning Network.

3.4. Applications in Virtual Screening of Lead Compounds

Virtual screening has emerged as a transformative approach in lead compound identification, overcoming the cost and scalability limitations of traditional high-throughput screening (HTS). This capability is achieved through the AI-driven prioritization of bioactive molecules from extensive chemical libraries [161]. The integration of structural informatics with DL has proven particularly impactful in antimicrobial discovery. For instance, directed message-passing neural network (D-MPNN) captures complex atomic relationships by transmitting messages along chemical bonds, generating molecular embeddings that encode both local and global structural information. This approach achieves high predictive accuracy for antimicrobial and anticancer lead discovery (Figure 5A) [51]. This computational framework has been further extended to develop narrow-spectrum therapeutics selectively targeting high-risk pathogens like *Acinetobacter baumannii*, demonstrating precision in addressing antimicrobial resistance [162]. Subsequent innovations combining structural feature analysis with bioactivity prediction enabled the discovery of *Helicobacter pylori* growth inhibitors [163]. Despite these advances, the limited interpretability of DL remains a critical challenge in virtual screening. Recent explainable AI frameworks address this gap by incorporating chemical substructure recognition and structural analysis into graph-based neural networks, thereby maintaining interpretable design principles while exploring novel molecular scaffolds [28].

The therapeutic applications of such models span multiple disease domains, with notable progress in anticancer and metabolic drug development. ML approaches combining molecular feature encoding with dimensionality reduction techniques have identified validated antidiabetic natural [164]. In oncology, proteomics-driven prediction frameworks that integrate protein interaction networks demonstrate robust performance through noise-resistant algorithms. However, while these approaches show promise, their broader adoption requires addressing dataset limitations to mitigate overfitting risks inherent to proteomic-based deep learning models [165].

Finally, bioactivity prediction has been revolutionized through meta-learning architectures that mitigate data scarcity and experimental variability. A representative framework trained on cross-platform bioactivity data achieves robust generalization capabilities, enabling reliable predictions for novel drug development even with limited training samples (Figure 5B) [166]. Recent DL models in the virtual screening and bioactivity prediction tasks were collected in Table 6.

Table 6. DL models for virtual screening and bioactivity prediction.

No.	Model Name	Framework Description	Web
1	Chemprop	A deep learning package implementing Directed Message Passing Neural Networks (D-MPNNs) for molecular property prediction. It efficiently predicts physicochemical properties (such as logP, reaction barriers) and bioactivity, enabling rapid ADME/efficacy assessment in lead-compound screening to guide candidate optimization and reduce experimental cost [102].	https://github.com/chemprop/chemprop (accessed on 20 August 2025)
2	iPADD	Screening molecular fingerprint features through feature selection strategies to predict the activity of anti-diabetic compounds using an XGBoost model [164].	https://github.com/llllxw/iPADD/blob/main/README.md (accessed on 20 August 2025)
3	DRUMLR	An ensemble machine learning framework leverages proteomic and phosphoproteomic data to rank over 400 anticancer drugs by efficacy, enabling rapid prioritization of high-potential leads [165].	https://github.com/CutillasLab/DRUMLR (accessed on 20 August 2025)
4	ActFound	A meta-learning and pairwise-learning bioactivity model that rapidly adapts to small assay datasets to predict relative activity differences with high precision, requiring minimal fine-tuning [166].	https://github.com/BFeng14/ActFound (accessed on 20 August 2025)
5	TOML-BERT	This dual-level pretrained model combines self-supervised learning on molecular structures with domain-knowledge transfer using pseudo-labels. It integrates atomic and molecular-level tasks to achieve state-of-the-art ADMET prediction accuracy across ten drug datasets, especially when labeled data is scarc [167].	https://github.com/yanjing-duan/TOML-BERT (accessed on 20 August 2025)
6	FP-GNN	A hybrid GNN model that fuses molecular-graph structural information with fingerprint-based substructure features, markedly improving prediction accuracy for molecular properties [168].	https://github.com/idrugLab/FP-GNN (accessed on 20 August 2025)
7	ChemBERTa2	A model pretrained on chemical molecular structures (SMILES) that efficiently predicts a wide range of physicochemical and bioactivity properties [169].	https://github.com/miservilla/ChemBERTa (accessed on 20 August 2025)
8	Uni-Mol	This 3D molecular deep-learning framework is pretrained on extensive datasets of small molecules and protein binding pockets. It can directly predict molecular physicochemical properties, generate accurate 3D conformations, and simulate drug-target binding modes [170].	https://github.com/deepmodeling/Uni-Mol/tree/main/unimol (accessed on 20 August 2025)
9	SPMM	A multimodal molecular model that jointly learns from both structural representations and associated properties, enabling bidirectional prediction and generation [171].	https://github.com/jinhojsk515/SPMM/ (accessed on 20 August 2025)
10	MoLFormer	An efficient Transformer model pretrained on large-scale chemical SMILES datasets, capable of accurately predicting molecular properties to aid both drug discovery and materials design [172].	https://github.com/IBM/molformer (accessed on 20 August 2025)

TOML-BERT, Task-oriented multilevel learning based on BERT. FP-GNN, fingerprints and graph neural network. SPMM, Structure–Property Multi-Modal foundation model.

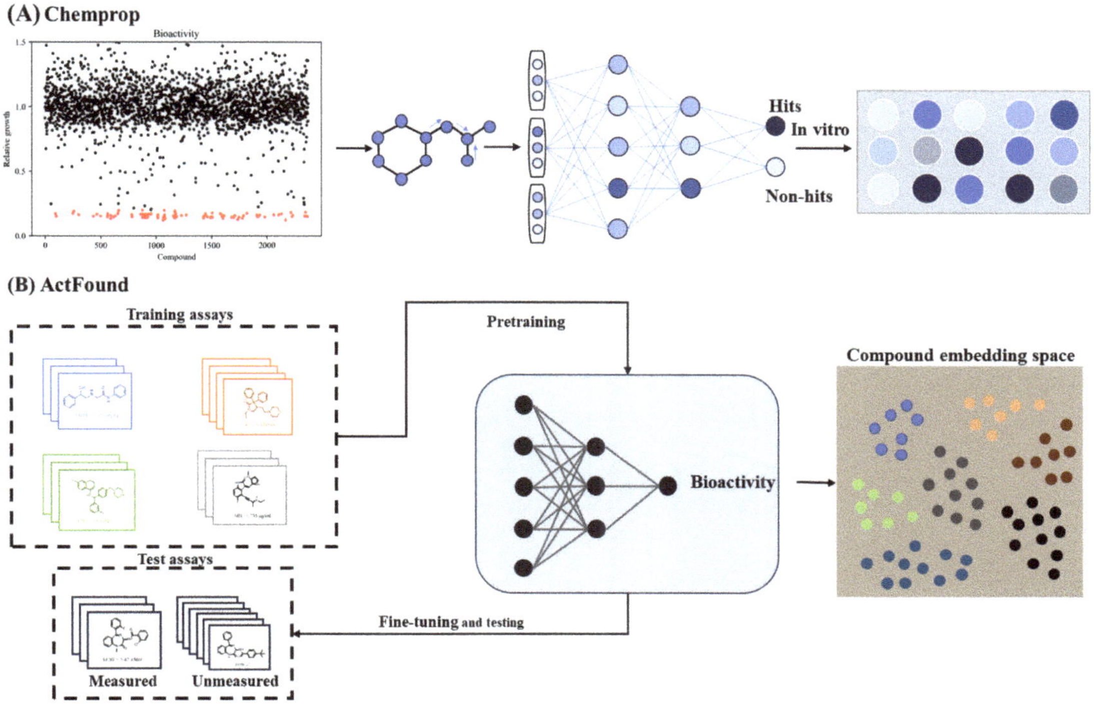

Figure 5. AI applications in lead compound virtual screening. (**A**), Graph-based neural network architecture for antimicrobial discovery [51]. (**B**), Meta-learning framework addressing data scarcity across assays [166].

3.5. Application in the Generation of Lead Compounds

The identification and generation of lead compounds remain pivotal challenges in early drug discovery. Current estimates suggest the chemical space contains 10^{23}–10^{60} potential drug-like molecules, yet only ~10^{8} have been synthesized to date [173,174]. Lead compound generation and optimization constitute one of the most critical bottlenecks in early-stage drug discovery, since even subtle structural modifications can profoundly influence potency, selectivity, pharmacokinetics, and safety profiles [175]. Traditional approaches such as molecular docking, and free-energy perturbation (FEP) have provided valuable guidance but often rely on predefined chemical rules or approximated scoring functions, limiting their capacity to explore the vast chemical space and to capture complex structure–activity relationships [176].

Deep generative models trained on structural patterns and bioactivity relationships enable de novo design of novel molecular entities with optimized drug-like properties, effectively expanding the scope of virtual screening libraries. Equivariant diffusion models enable the generation of novel ligands conditioned on protein pockets [177]. Meanwhile, toxicity-controlled ligand generation frameworks integrate safety considerations directly into the design process [178]. Additionally, DiffGui facilitates the simultaneous generation of atoms and bonds, producing molecules that exhibit high structural rationality and desirable molecular properties [179]. Compared with classical QSAR or docking-based strategies, these models not only enhance hit rates and novelty but also improve synthetic feasibility and safety awareness, thereby redefining the state of the art in lead optimization.

VAEs encode molecules into a latent space and reconstruct new structures while optimizing predicted bioactivity. This approach has successfully developed novel compounds with high potency against parasitic enzymes, exhibiting IC_{50} values of 0.023 μM and 0.025 μM against *P. falciparum* 3D7 (Figure 6A) [180]. The SyntheMol platform, through combining a validated library of chemical transformations with Monte Carlo tree search, generated structurally novel antibiotics with strong activity against multidrug-resistant *Acinetobacter baumannii* [181].For tuberculosis treatment, a target-aware generative system successfully identified protease inhibitors with measurable efficacy against *Mycobacterium tuberculosis*, discovering 14 compounds with significant inhibitory activity against the ClpP protease of *Mycobacterium tuberculosis*. The most potent compound had an IC_{50} of 1.9 μM [182].

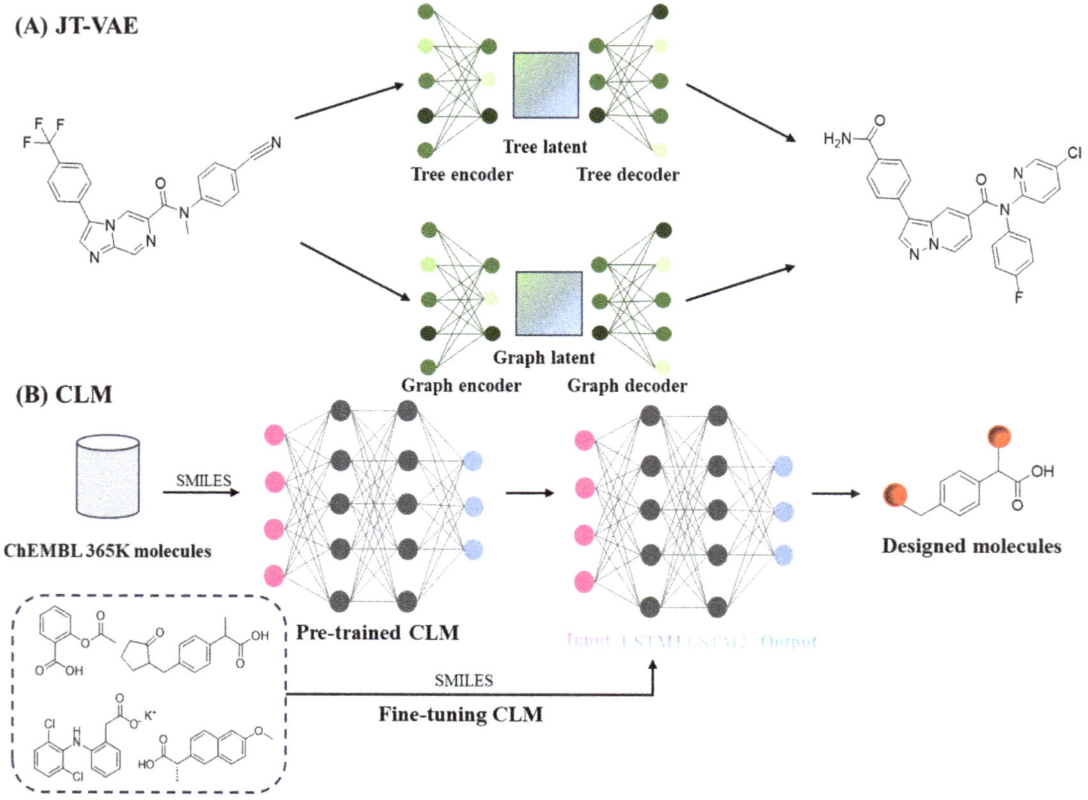

Figure 6. Application of AI in lead compound generation. (**A**), Variational autoencoder-based generative framework [180]. (**B**), Chemical language model (CLM) for SMILES-based molecular design [94]. LSTM, long short-term memory.

Recent advances in chemical language modeling have further enhanced generative capabilities [183–185]. Chemical language models (CLMs), inspired by natural language processing, treat SMILES strings as sequences of tokens. Pretrained on large chemical datasets, these models learn the "grammar" of chemistry and can generate valid, synthetically accessible molecules. When fine-tuned on small datasets, CLMs can design dual-target ligands, such as multitarget compounds for metabolic disorders (Figure 6B) [94]. In a complementary approach, transcriptomic signature-driven generative models are designed to reconstruct drug-induced biological profiles. For example, they have been applied to repurpose therapies for pancreatic cancer by efficiently screening clinical compound libraries [186]. These synergistic innovations establish generative AI as a versatile platform for chemical space exploration. Their integration into high-throughput experimental workflows is accelerating therapeutic development across infectious diseases, metabolic disorders, and oncology. A summary of recent deep learning models for lead compound generation is provided in Table 7.

Table 7. DL models for the generation of lead compounds.

No.	Model Name	Framework Description	Web
1	DRLinker	A deep reinforcement learning framework for fragment-based drug design that optimizes linker generation, controls physicochemical properties, enhances bioactivity, and promotes structural innovation during lead optimization [187].	https://github.com/biomed-AI/Drlinker (accessed on 20 August 2025)
2	Link-INVENT	Link-INVENT uses an RNN to generate novel linkers as SMILES strings between two molecular subunits. Unlike database-search methods, it creates linkers token-by-token, enabling exploration of new chemical space. Its key advantage is a customizable Scoring Function that uses reinforcement learning to optimize linkers for multiple specific properties, as demonstrated in fragment linking, scaffold hopping, and proteolysis targeting chimeras design case studies [188].	https://github.com/MolecularAI/Reinvent (accessed on 20 August 2025)
3	DeepHop	Integrates 3D molecular conformations and target protein information to efficiently generate compounds with novel 2D scaffolds, improved bioactivity, and preserved 3D similarity. It is specifically used for scaffold-hopping optimization in drug design [189].	https://github.com/prokia/deepHops (accessed on 20 August 2025)
4	ScaffoldGVAE	A deep generative model that smartly modifies the molecular core scaffold while retaining key side chains, enabling the design of novel active compounds. It has been successfully applied to develop Leucine-rich repeat ki-nase 2 inhibitors for Parkinson's disease, offering a tool for innovative chemical-space exploration [75].	https://github.com/ecust-hc/ScaffoldGVAE (accessed on 20 August 2025)
5	MolGPT	A Transformer-based molecular generator that assembles chemically valid drug molecules and provides controllable optimization of structure and molecular properties [190].	https://github.com/devalab/molgpt (accessed on 20 August 2025)
6	Scaffold Decorator	A SMILES-based generative model that decorates a given scaffold to pro-duce novel compounds with potential activity. It has experimentally generated predicted Dopamine Receptor D2 ligands while ensuring synthetic feasibility, offering an effective scaffold-decoration tool [191].	https://github.com/undeadpixel/reinvent-scaffold-decorator (accessed on 20 August 2025)
7	SAMOA	A deep learning framework for lead optimization that maintains a specified core scaffold while improving bioactivity and drug-likeness under structural constraints [192].	https://github.com/maxime-langevin/scaffold-constrained-generation (accessed on 20 August 2025)
8	LibINVENT	A reaction-guided molecular generation platform that constructs libraries of compounds sharing a common scaffold, streamlining lead optimization and improving synthetic accessibility [193].	https://github.com/MolecularAI/Lib-INVENT (accessed on 20 August 2025)
9	DiffDec	A structure-aware molecule optimization tool based on diffusion models. Given a protein binding pocket's 3D structure, it intelligently generates side-chain R-groups to enhance binding affinity, offering an efficient solution for lead optimization [194].	https://github.com/biomed-AI/DiffDec (accessed on 20 August 2025)
10	DeepFrag	An AI method using diffusion models that, guided by a protein's 3D structure, generates chemically compatible R-groups for lead molecules, significantly boosting binding affinity and drug-likeness [195].	https://durrantlab.pitt.edu/deepfragmodel/ (accessed on 20 August 2025)

Table 7. *Cont.*

No.	Model Name	Framework Description	Web
11	STRIFE	A structure-guided fragment expansion tool that designs optimized fragments based on target protein structures, improving efficiency for antiviral and anti-inflammatory drug discovery [196].	https://github.com/oxpig/STRIFE (accessed on 20 August 2025)
12	DEVELOP	A 3D-pharmacophore-guided generative framework (DEVELOP) that combines graph neural networks with 3D pharmacophoric features for linker and R-group design, improving target specificity and molecular quality [197].	https://github.com/oxpig/DEVELOP (accessed on 20 August 2025)
13	DrugEX-v3	A graph-transformer and reinforcement-learning model that generates high-affinity compounds from user-defined fragments [198].	https://github.com/CDDLeiden/DrugEx (accessed on 20 August 2025)
16	REINVENT4	A hybrid RNN/Transformer generative framework enhanced with reinforcement learning for de novo molecule generation, R-group replacement, linker design, and property optimization in an end-to-end workflow [199].	https://github.com/MolecularAI/REINVENT4 (accessed on 20 August 2025)
17	TamGen	A GPT-based chemical language model that generates new drug-like molecules targeting disease proteins and optimizes existing scaffolds for improved activity and synthetic accessibility [182].	https://github.com/SigmaGenX/TamGen/tree/main/data (accessed on 20 August 2025)
18	ClickGen	An AI model combining modular reaction chemistry with reinforcement learning to rapidly design easily synthesizable and bioactive drug molecules [200].	https://github.com/mywang1994/cligen_gen (accessed on 20 August 2025)
19	JAEGER	A deep generative model integrating structure generation with activity prediction, successfully designing low-toxicity antimalarials with nanomolar experimental efficacy [180].	https://github.com/Novartis/JAEGER (accessed on 20 August 2025)
20	SyntheMol	An AI–synthetic-chemistry hybrid system that rapidly proposes easily synthesizable novel antibiotic scaffolds. Experimentally, it identified six new compounds effective against drug-resistant Acinetobacter baumannii and other pathogens [181].	https://github.com/swansonk14/SyntheMol (accessed on 20 August 2025)
21	CLM	An RNN-based generative model trained on limited molecular data to produce structures containing specified motifs, applied to discover bacterial, plant, and fungal metabolites as potential drug leads [183].	https://github.com/skinnider/low-data-generative-models (accessed on 20 August 2025)

SAMOA, Scaffold constrained molecular generation. STRIFE, Structure informed fragment elaboration. DEVELOP, DEep Vision-Enhanced Lead OPtimisation. TamGen, Target-aware molecular generation. JAEGER, JT-VAE Generative Modeling. CLM, Chemical language model.

3.6. Applications in Synthesis Prediction in Drug Discovery

The integration of computational synthesis prediction into drug discovery pipelines has emerged as a transformative strategy for addressing the inherent complexity and cost challenges in therapeutic development [201]. Researchers embed AI within the design-make-test-analyze (DMTA) cycle to accelerate synthetic route planning, while also ensuring practical feasibility through reaction condition optimization and outcome validation [23]. For instance, Thakkar et al. employ a Monte Carlo tree search guided by learned policies to break down target molecules into purchasable precursors, typically completing full routes in under one minute. Building on that approach, a lightweight classifier predicts retrosynthetic accessibility thousands of times faster than a full search by estimating whether a viable route exists [202].

Beyond route finding, predicting synthesis feasibility via learned metrics has matured significantly. DeepSA, trained on 3.6 million SMILES from public and proprietary sources, achieves an AUROC of 0.896 for synthesis accessibility. This model enables chemists to prioritize cost-effective and readily synthesizable molecules, reducing experimental attrition and accelerating lead optimization [22]. AI-driven advances in reaction prediction prioritize spatial molecular interactions. In late-stage functionalization, where selective modification of complex molecular scaffolds is required, geometric deep learning plays a crucial role. With researchers integrating 3D atomic coordinates into GNNs, these models quantify both steric and electronic interactions to predict regioselectivity and reaction outcomes. This enables chemists to design site-selective transformations with improved efficiency and precision (Figure 7A) [203].

Complementary innovations address forward reaction prediction. Transformer-based models (e.g., T5Chem) treat reactions as text-to-text problems, achieving state-of-the-art accuracy in classification, retrosynthesis, and yield estimation on USPTO_500_MT data (SHAP analysis providing functional group-level interpretability) [204]. Graph-based reaction prediction systems, such as GraphRXN, represent another major innovation. Instead of relying on molecular descriptors, they directly encode reactant, reagent, and product graphs, node and edge features to learn reaction patterns. This end-to-end learning enables real-time product yield and regioselectivity prediction and allows seamless integration with robotic synthesis platforms, supporting fully automated chemical workflows (Figure 7B) [205].

With the maturation of multimodal data integration, AI-powered synthesis prediction stands poised to redefine pharmaceutical chemistry, transforming it into a fully data-driven discipline capable of intelligent molecular design and automated synthetic planning. Recent DL models in synthesis prediction tasks are collected in Table 8.

Table 8. DL Models for the synthesis prediction tasks.

No.	Model Name	Framework Description	Web
1	GSETransformer	An integrated model that combines graph neural networks and sequence processing to predict biosynthetic pathways of natural products. It accelerates route design in drug development and provides a visual interface that supports research workflow efficiency [206].	https://github.com/momozhangcn/GSETRetro (accessed on 20 August 2025)
2	Molecular Transformer	An attention-based model that unifies reaction prediction and retrosynthetic analysis for drug molecules. It maintains strong accuracy on novel compounds and serves as an effective tool for planning pharmaceutical syntheses [207].	https://github.com/pschwllr/MolecularTransformer (accessed on 20 August 2025)
3	DeepSA	A deep-learning chemical language model that predicts synthetic accessibility from molecular structure. It prioritizes easily synthesizable compounds and reduces both development time and cost [22].	https://github.com/Shihang-Wang-58/DeepSA (accessed on 20 August 2025)
4	Retro-MTGR	A multitask learning framework that uses molecular structure features to predict key bond disconnections and leaving groups in single-step retrosynthesis. It provides an efficient tool for planning synthetic routes [208].	https://github.com/zpczaizheli/Retro-MTGR (accessed on 20 August 2025)
5	LocalRetro	A retrosynthesis prediction model that integrates local molecular structure analysis with global attention mechanisms. It accurately designs synthetic routes for a broad range of drug-like molecules [209].	https://github.com/kaist-amsg/LocalRetro (accessed on 20 August 2025)
6	RetroTRAE	An atom-environment-aware model that predicts reactants directly for single-step retrosynthesis. It learns chemical fragment patterns and achieves 61.6% accuracy on benchmark datasets, surpassing SMILES-based methods [210].	https://github.com/knu-lcbc/RetroTRAE (accessed on 20 August 2025)
7	ReroSub	An end-to-end retrosynthesis model that automatically identifies conserved molecular substructures to simplify reaction prediction. It improves accuracy by over 5% compared with template-based approaches and removes the need for predefined reaction templates [211].	https://github.com/fangleigit/RetroSub (accessed on 20 August 2025)

Table 8. *Cont.*

No.	Model Name	Framework Description	Web
8	G2GT	A hybrid framework combining graph networks and Transformer architectures in an encoder–decoder design. It uses self-training to predict required reactants for a target molecule and guides retrosynthetic route construction with high precision [212].	https://github.com/ZaiyunLin/G2GT_2 (accessed on 20 August 2025)
9	MolecularGET	A fusion model that combines graph neural networks with Transformer encoders to integrate structural and sequential chemical information. It enhances retrosynthesis prediction accuracy and accelerates route design [213].	https://github.com/papercodekl/MolecularGET (accessed on 20 August 2025)
10	Graph2SMILES	A template-free neural model that predicts reactants or products directly from molecular graphs. It improves the accuracy of both retrosynthesis and forward reaction prediction without complex preprocessing [214].	https://github.com/coleygroup/Graph2SMILES (accessed on 20 August 2025)
11	Graph2Edits	An edit-based architecture that applies stepwise graph modifications to infer feasible reactants from products. It achieves 55.1% accuracy in single-step prediction and performs well in complex multi-center transformations [208].	https://github.com/Jamson-Zhong/Graph2Edits (accessed on 20 August 2025)
12	RetroPrime	A two-stage Transformer model that first decomposes a target molecule and then generates plausible reactant sets. It improves retrosynthesis accuracy while increasing diversity and chemical feasibility of predicted routes [215].	https://github.com/wangxr0526/RetroPrime (accessed on 20 August 2025)
13	RAscore	A machine learning classifier that estimates synthetic accessibility scores for molecules. It enables large-scale prescreening to enrich chemical libraries with easily synthesizable compounds [202].	https://github.com/reymond-group/Rascore (accessed on 20 August 2025)
14	T5Chem	A sequence-to-sequence model based on SMILES representation that unifies multiple chemistry tasks, including reaction prediction, retrosynthesis, classification, and yield estimation. It also supports model interpretability studies [204].	https://github.com/HelloJocelynLu/t5chem (accessed on 20 August 2025)
15	GraphRXN	A graph neural network framework that analyzes 2D molecular structures of reactants and products to predict reaction outcomes. It demonstrates strong performance when trained with high-throughput experimental data [205].	https://github.com/jidushanbojue/GraphRXN (accessed on 20 August 2025)
16	Geometric deep learning	A 3D-geometry-based framework that integrates structural information with high-throughput experimentation to predict late-stage functionalization outcomes such as yield and regioselectivity. It accelerates physicochemical property optimization for complex drug candidates [203].	https://github.com/ETHmodlab/lsfml (accessed on 20 August 2025)

GSETransformer, graph-sequence enhanced transformer. Retro-MTGR, a Multi-Task Graph Representation learning framework for Retrosynthesis prediction. LocalRetro, a local retrosynthesis framework. G2T2, a graph-to-graph transformation model. Molecular GET, molecular graph enhanced transformer. RAscore, retrosynthetic accessibility score. T5Chem, "Text-to-Text Transfer Transformer" (T5) framework.

3.7. Absorption, Distribution, Metabolism, Excretion, and Toxicity (ADMET) Prediction in Drug Discovery

The accurate prediction of ADMET properties has become indispensable in modern drug discovery. AI has transformed this domain by enabling systematic analysis of molecular structure-pharmacokinetic relationships, overcoming the cost limitation of traditional experimental approaches. For instance, Bayesian pharmacokinetic (bPK) frameworks combine physicochemical and biological descriptors into composite metrics that quantitatively estimate molecular developability. These models outperform single-endpoint predictors by enabling multi-parameter optimization of absorption, bioavailability, and clearance profiles, effectively prioritizing compounds likely to succeed in downstream stages (Figure 8A) [216]. To address the interpretability–accuracy trade-off, fragment-based explainable AI (FBXAI) models have emerged. These methods decompose molecules into functional fragments and apply attention-based deep networks to learn how each substructure contributes to ADMET scores. Using domain-aware fragmentation and visualization tools such as Grad-CAM, the framework finds key functional domains responsible for properties like toxicity or metabolic stability (Figure 8B) [217].

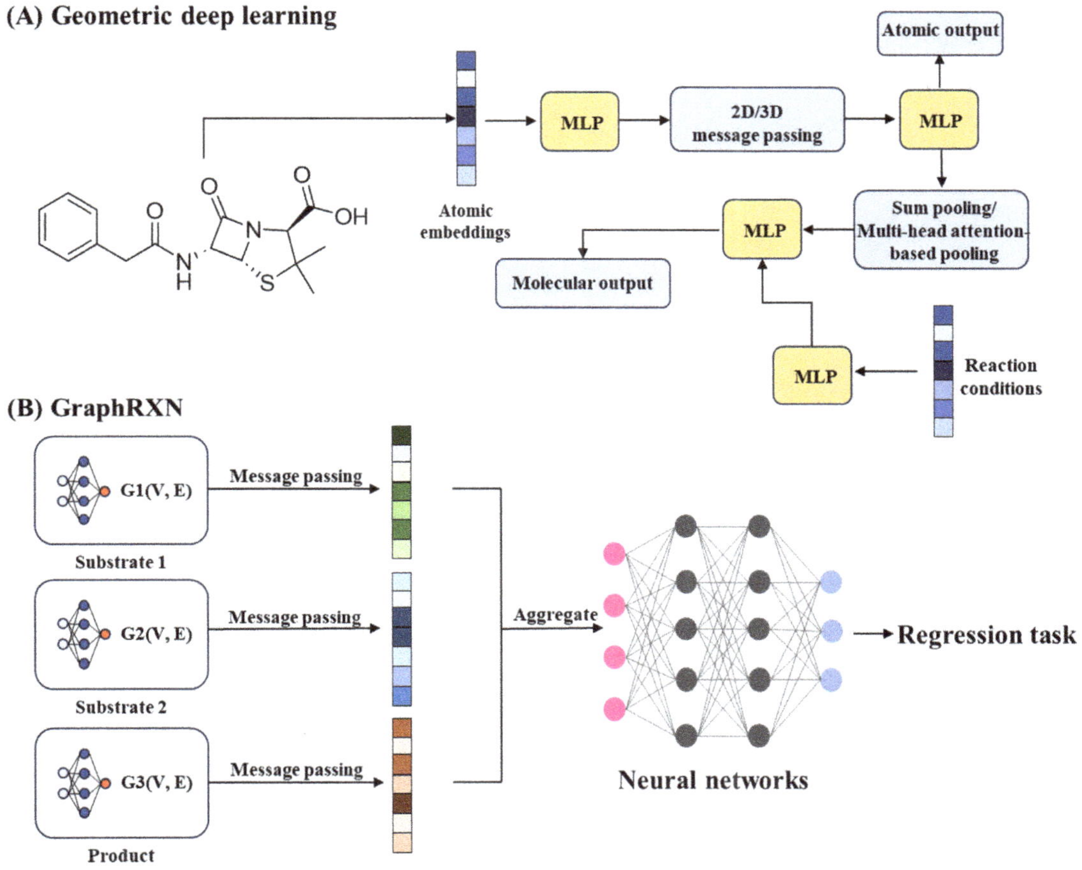

Figure 7. Applications of AI in chemical synthesis. (**A**), Geometric deep learning for regioselectivity prediction and late-stage functionalization [203]. (**B**), Graph-based and transformer models for reaction and yield prediction [205]. MLP, Multilayer perceptron.

Recent advances in toxicity prediction emphasize the integration of mechanistic interpretability with computational precision. Unified modeling approaches achieve superior discriminative power across diverse toxicity scores through DL-enhanced molecular descriptor analysis [218,219]. For example, specialized cardiac safety frameworks rigorously benchmark structural feature representations to elucidate drivers of multi-ion channel cardiotoxicity, significantly enhancing prediction reliability through structurally dissimilar validation sets [52]. Complementary platforms employ interpretable pattern recognition to decode structural determinants across diverse ADMET endpoints [220]. Furthermore, the field continues to evolve through large-scale open-source data initiatives that systematically address historical limitations in dataset diversity and balance, establishing comprehensive benchmarks for next-generation ADMET prediction [221]. Recent DL models in ADMET prediction tasks are collected in Table 9.

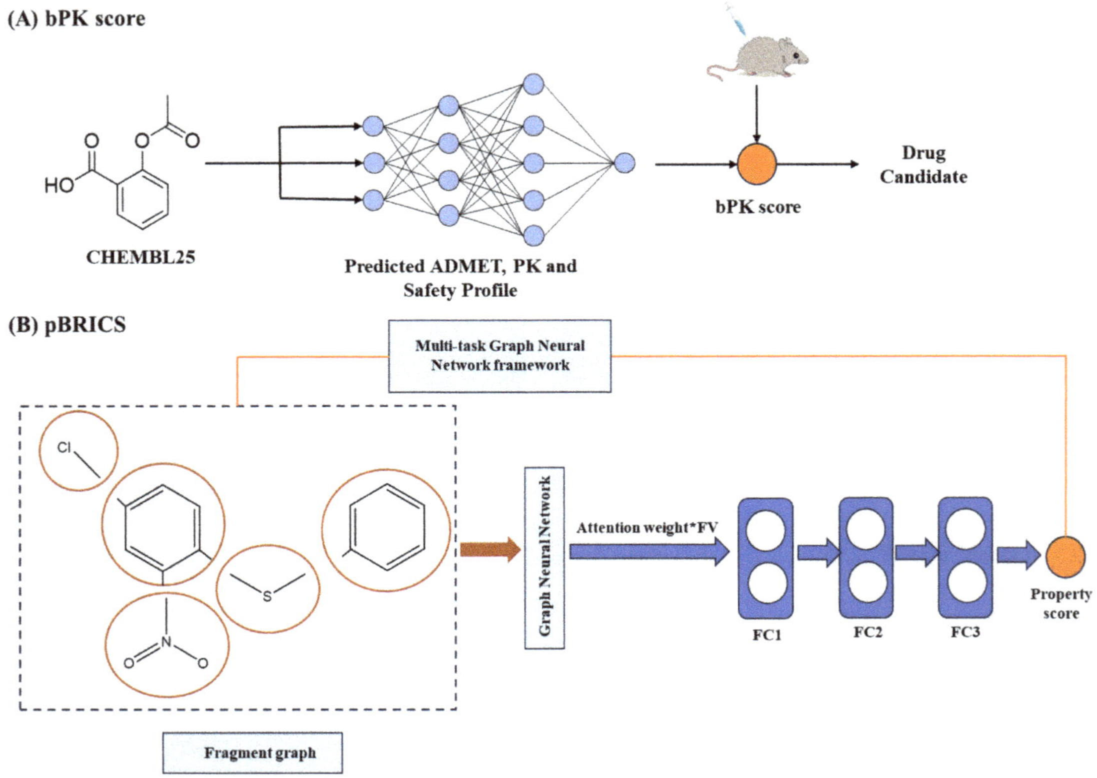

Figure 8. Applications of AI in ADMET. (**A**), Bayesian pharmacokinetic (bPK) framework integrating multi-descriptor data for compound prioritization [216]. (**B**), Fragment-based explainable model linking molecular substructures to pharmacokinetic properties [217]. FV, feature vector. FC, fully connected layers.

Table 9. DL Models for the ADMET prediction tasks.

No.	Model Name	Framework Description	Web
1	ADMET-PrInt	An online platform that integrates machine learning models with interpretability modules to predict major ADMET properties. It provides visual analytics of structural features to guide molecular optimization and accelerate in silico lead evaluation [218].	https://github.com/JamEwe/ADMET-PrInt (accessed on 20 August 2025)
2	PharmaBench	A benchmark dataset built from more than 14,000 bioassay entries covering over 52,000 compounds. It offers a standardized training and evaluation framework for AI models in absorption, distribution, metabolism, excretion, and toxicity prediction [221].	https://github.com/mindrank-ai/PharmaBench (accessed on 20 August 2025)
3	ADMET-AI	A hybrid platform combining graph neural networks and molecular descriptors for rapid ADMET property prediction. It supports both web and local deployment for high-throughput virtual screening focused on metabolism and toxicity endpoints [222].	https://admet.ai.greenstonebio.com/ (accessed on 20 August 2025)
4	SwissADME	A free web service that evaluates small-molecule ADME properties, drug-likeness, and synthetic accessibility. It assists researchers in early-stage drug candidate optimization [223].	http://www.swissadme.ch/ (accessed on 20 August 2025)

Table 9. *Cont.*

No.	Model Name	Framework Description	Web
5	ADMETlab 3.0	A deep learning platform that predicts key ADMET parameters using integrated molecular features and neural network models. It streamlines early screening and reduces experimental workload [224].	https://admetlab3.scbdd.com/server/screening (accessed on 20 August 2025)
6	pkCSM	A graph-based predictive model that estimates metabolic stability and potential toxicity directly from molecular structure. It uses graph signatures to rapidly identify high-risk compounds during early ADMET assessment [225].	https://biosig.lab.uq.edu.au/pkcsm/ (accessed on 20 August 2025)
7	vNN-ADMET	A similarity-based prediction algorithm that applies nearest-neighbor statistics in chemical space. It provides ADMET predictions with confidence intervals to support risk-aware decision-making in candidate selection [226].	https://vnnadmet.bhsai.org/vnnadmet/login.xhtml (accessed on 20 August 2025)
8	ChemMORT	An automated optimization platform that combines reversible molecular representations with reinforcement learning. It refines ADME and safety properties while preserving bioactivity [227].	https://cadd.nscc-tj.cn/deploy/chemmort/ (accessed on 20 August 2025)
9	Conformal ADMET Prediction	A graph-neural-network model enhanced with conformal prediction calibration that generates both point estimates and prediction intervals for ADMET endpoints. It offers statistical confidence for decision-making under uncertainty [101].	https://github.com/peiyaoli/Conformal-ADMET-Prediction (accessed on 20 August 2025)

vNN-ADMET, variable nearest neighbor-ADMET. ChemMORT, Chemical Molecular Optimization, Representation and Translation.

4. Future Challenges

Over the past decade, AI has driven transformative advances in drug discovery, particularly in compound property prediction and virtual screening, demonstrating its potential to shorten drug discovery cycles and reduce costs. However, AI is not a panacea for pharmaceutical R&D, as its integration faces multifaceted technical limitations, safety concerns, and ethical dilemmas that demand urgent resolution.

AI model performance is highly dependent on high-quality and sufficient datasets. Despite the vast scale of existing chemical libraries, rigorously curated datasets with robust biological, pharmacological, and clinical annotations remain scarce. This challenge is particularly acute in drug discovery, where the procurement of high-quality training data is hindered by extreme costs, stringent privacy regulations, and restrictive data-sharing agreements, especially for rare diseases and novel target studies [183]. Furthermore, biological variability poses a major obstacle to reproducibility and generalization. This variability, stemming from differences in experimental conditions, patient demographics, and genetic backgrounds, introduces significant noise and can lead to substantial fluctuations in molecular response data. Without careful normalization and multi-omics integration, AI models may learn these variations instead of the underlying biological signals, resulting in predictions that are biased, non-reproducible, and poorly generalizable to new populations or experimental settings.

Another critical issue is the lack of negative and failed experimental data in public databases. Because most published datasets overrepresent successful experiments, AI models are often unable to learn from failure modes, which are crucial for predicting adverse effects, toxicity, and off-target interactions [228]. Therefore, future efforts should emphasize the systematic curation of both positive and negative results, along with the establishment of shared, standardized data infrastructures to enhance cross-study comparability and robustness [229].

The black-box nature of DL models represents another major barrier to their adoption in regulated biomedical environments [230]. While DL architectures such as convolutional and Transformer-based networks have significantly advanced molecular representation learning, their lack of interpretability remains a fundamental limitation [228]. In particular, large models are prone to high-risk modeling illusions due to the inherent bias of the attention mechanism that leads to the generation process. In addition, the black-box nature of large models makes it difficult to meet the interpretable requirements of vertical applications. Developing inherently explainable AI frameworks capable of elucidating structure-activity relationships and toxicity pathways is essential to bridge this gap. Recent advances in attention mechanisms and causal inference models offer promising pathways toward interpretable predictions that align with domain expertise [28].

Computational efficiency represents another major obstacle. Although modern graphics processing units (GPUs) have improved training performance, state-of-the-art deep learning architectures, particularly large Transformer models, demand massive computational and energy resources, creating both environmental and practical challenges [231]. To mitigate these costs, future research should prioritize green computing strategies. This includes developing lightweight yet high-performing architectures to reduce carbon footprints without sacrificing predictive power. Furthermore, decentralized frameworks like federated learning can enhance sustainability by minimizing redundant data transfers and enabling secure, collaborative model development across institutions without centralizing sensitive data.

In summary, while AI-driven drug discovery has achieved significant progress, its continued success depends on addressing data scarcity, biological variability, model interpretability, and sustainability challenges. With the ongoing evolution of algorithmic efficiency, interdisciplinary data integration, and open-source model development, AI holds great promise to fundamentally reshape drug discovery—making it faster, greener, and more reliable for global health advancement.

Author Contributions: Q.W. and B.S.: Data curation, Writing—original draft; Y.Y., T.V. and J.S.: Writing—review & editing; C.D. and H.J.: Conceptualization, Funding acquisition, Writing—review & editing, Supervision. All authors have read and agreed to the published version of the manuscript.

Funding: This study was funded supported by the National Key Research and Development Program of China (Grant No. 2023YFD1802300) and the National High-Level Talents Special Support Program.

Institutional Review Board Statement: Not applicable.

Informed Consent Statement: Not applicable.

Data Availability Statement: Not applicable. No new data were created or analyzed in this study. Data sharing is not applicable to this article.

Conflicts of Interest: The authors declare no conflicts of interest.

References

1. Jung, Y.L.; Yoo, H.S.; Hwang, J. Artificial intelligence-based decision support model for new drug development planning. *Expert Syst. Appl.* **2022**, *198*, 116825. [CrossRef] [PubMed]
2. Wouters, O.J.; McKee, M.; Luyten, J. Estimated Research and Development Investment Needed to Bring a New Medicine to Market, 2009–2018. *JAMA* **2020**, *323*, 844–853. [CrossRef] [PubMed]
3. Harrison, R.K. Phase II and phase III failures: 2013–2015. *Nat. Rev. Drug Discov.* **2016**, *15*, 817–818. [CrossRef]
4. Dowden, H.; Munro, J. Trends in clinical success rates and therapeutic focus. *Nat. Rev. Drug Discov.* **2019**, *18*, 495–496. [CrossRef]
5. Smietana, K.; Siatkowski, M.; Møller, M. Trends in clinical success rates. *Nat. Rev. Drug Discov.* **2016**, *15*, 379–380. [CrossRef]
6. Sun, D.; Gao, W.; Hu, H.; Zhou, S. Why 90% of clinical drug development fails and how to improve it? *Acta Pharm. Sin. B* **2022**, *12*, 3049–3062. [CrossRef] [PubMed]
7. Biotechnology Innovation Organization; Informa Pharma Intelligence; Quantitative Life Sciences Advisors. Clinical Development Success Rates and Contributing Factors 2011–2020. 2021. Available online: https://www.bio.org/clinical-development-success-rates-and-contributing-factors-2011-2020 (accessed on 20 February 2025).

8. Peng, C.; Zhao, S.; Tang, L.; Wang, K.; Wang, Y.; Ding, L. A simplified and reliable LC-tandem mass spectrometry method for determination of ulipristal acetate in human plasma and its application to a pharmacokinetic study in healthy Chinese volunteers. *Biomed. Chromatogr.* **2020**, *34*, e4908. [CrossRef]
9. Lyu, J.; Wang, S.; Balius, T.E.; Singh, I.; Levit, A.; Moroz, Y.S.; O'Meara, M.J.; Che, T.; Algaa, E.; Tolmachova, K.; et al. Ultra-large library docking for discovering new chemotypes. *Nature* **2019**, *566*, 224–229. [CrossRef]
10. Talukder, M.E.K.; Atif, M.F.; Siddiquee, N.H.; Rahman, S.; Rafi, N.I.; Israt, S.; Shahir, N.F.; Islam, M.T.; Samad, A.; Wani, T.A.; et al. Molecular docking, QSAR, and simulation analyses of EGFR-targeting phytochemicals in non-small cell lung cancer. *J. Mol. Struct.* **2025**, *1321*, 139924. [CrossRef]
11. Kaur, N.; Gupta, S.; Pal, J.; Bansal, Y.; Bansal, G. Design of BBB permeable BACE-1 inhibitor as potential drug candidate for Alzheimer disease: 2D-QSAR, molecular docking, ADMET, molecular dynamics, MMGBSA. *Comput. Biol. Chem.* **2025**, *116*, 108371. [CrossRef]
12. Souza, A.S.; Amorim, V.M.F.; Soares, E.P.; de Souza, R.F.; Guzzo, C.R. Antagonistic Trends Between Binding Affinity and Drug-Likeness in SARS-CoV-2 Mpro Inhibitors Revealed by Machine Learning. *Viruses* **2025**, *17*, 935. [CrossRef]
13. Maliyakkal, N.; Kumar, S.; Bhowmik, R.; Vishwakarma, H.C.; Yadav, P.; Mathew, B. Two-dimensional QSAR-driven virtual screening for potential therapeutics against *Trypanosoma cruzi*. *Front. Chem.* **2025**, *13*, 1600945. [CrossRef] [PubMed]
14. Pun, F.W.; Ozerov, I.V.; Zhavoronkov, A. AI-powered therapeutic target discovery. *Trends Pharmacol. Sci.* **2023**, *44*, 561–572. [CrossRef]
15. Cassan, O.; Lèbre, S.; Martin, A. Inferring and analyzing gene regulatory networks from multi-factorial expression data: A complete and interactive suite. *BMC Genom.* **2021**, *22*, 387. [CrossRef] [PubMed]
16. Nogales, C.; Mamdouh, Z.M.; List, M.; Kiel, C.; Casas, A.I.; Schmidt, H.H.H.W. Network pharmacology: Curing causal mechanisms instead of treating symptoms. *Trends Pharmacol. Sci.* **2022**, *43*, 136–150. [CrossRef]
17. Zhuo, C.; Gao, J.; Li, A.; Liu, X.; Zhao, Y. A Machine Learning Method for RNA–Small Molecule Binding Preference Prediction. *J. Chem. Inf. Model.* **2024**, *64*, 7386–7397. [CrossRef]
18. Zhou, J.-B.; Tang, D.; He, L.; Lin, S.; Lei, J.H.; Sun, H.; Xu, X.; Deng, C.-X. Machine learning model for anti-cancer drug combinations: Analysis, prediction, and validation. *Pharmacol. Res.* **2023**, *194*, 106830. [CrossRef]
19. He, D.; Liu, Q.; Mi, Y.; Meng, Q.; Xu, L.; Hou, C.; Wang, J.; Li, N.; Liu, Y.; Chai, H.; et al. De Novo Generation and Identification of Novel Compounds with Drug Efficacy Based on Machine Learning. *Adv. Sci.* **2024**, *11*, 2307245. [CrossRef] [PubMed]
20. Huang, L.; Xu, T.; Yu, Y.; Zhao, P.; Chen, X.; Han, J.; Xie, Z.; Li, H.; Zhong, W.; Wong, K.-C.; et al. A dual diffusion model enables 3D molecule generation and lead optimization based on target pockets. *Nat. Commun.* **2024**, *15*, 2657. [CrossRef]
21. Zhou, G.; Rusnac, D.-V.; Park, H.; Canzani, D.; Nguyen, H.M.; Stewart, L.; Bush, M.F.; Nguyen, P.T.; Wulff, H.; Yarov-Yarovoy, V.; et al. An artificial intelligence accelerated virtual screening platform for drug discovery. *Nat. Commun.* **2024**, *15*, 7761. [CrossRef]
22. Wang, S.; Wang, L.; Li, F.; Bai, F. DeepSA: A deep-learning driven predictor of compound synthesis accessibility. *J. Cheminform.* **2023**, *15*, 103. [CrossRef] [PubMed]
23. Struble, T.J.; Alvarez, J.C.; Brown, S.P.; Chytil, M.; Cisar, J.; DesJarlais, R.L.; Engkvist, O.; Frank, S.A.; Greve, D.R.; Griffin, D.J.; et al. Current and Future Roles of Artificial Intelligence in Medicinal Chemistry Synthesis. *J. Med. Chem.* **2020**, *63*, 8667–8682. [CrossRef] [PubMed]
24. Ye, Z.; Wang, N.; Zhou, J.; Ouyang, D. Organic crystal structure prediction via coupled generative adversarial networks and graph convolutional networks. *Innovation* **2024**, *5*, 100562. [CrossRef]
25. Yang, Z.; Zhao, Y.-M.; Wang, X.; Liu, X.; Zhang, X.; Li, Y.; Lv, Q.; Chen, C.Y.-C.; Shen, L. Scalable crystal structure relaxation using an iteration-free deep generative model with uncertainty quantification. *Nat. Commun.* **2024**, *15*, 8148. [CrossRef]
26. Ryan, K.; Lengyel, J.; Shatruk, M. Crystal Structure Prediction via Deep Learning. *J. Am. Chem. Soc.* **2018**, *140*, 10158–10168. [CrossRef] [PubMed]
27. Ren, F.; Aliper, A.; Chen, J.; Zhao, H.; Rao, S.; Kuppe, C.; Ozerov, I.V.; Zhang, M.; Witte, K.; Kruse, C.; et al. A small-molecule TNIK inhibitor targets fibrosis in preclinical and clinical models. *Nat. Biotechnol.* **2025**, *43*, 63–75. [CrossRef] [PubMed]
28. Wong, F.; Zheng, E.J.; Valeri, J.A.; Donghia, N.M.; Anahtar, M.N.; Omori, S.; Li, A.; Cubillos-Ruiz, A.; Krishnan, A.; Jin, W.; et al. Discovery of a structural class of antibiotics with explainable deep learning. *Nature* **2024**, *626*, 177. [CrossRef]
29. Nobel Prize in Chemistry 2024. Available online: https://www.nature.com/collections/edjcfdihdi (accessed on 16 December 2024).
30. Van de Sande, B.; Lee, J.S.; Mutasa-Gottgens, E.; Naughton, B.; Bacon, W.; Manning, J.; Wang, Y.; Pollard, J.; Mendez, M.; Hill, J.; et al. Applications of single-cell RNA sequencing in drug discovery and development. *Nat. Rev. Drug Discov.* **2023**, *22*, 496–520. [CrossRef]
31. Yang, F.; Wang, W.; Wang, F.; Fang, Y.; Tang, D.; Huang, J.; Lu, H.; Yao, J. scBERT as a large-scale pretrained deep language model for cell type annotation of single-cell RNA-seq data. *Nat. Mach. Intell.* **2022**, *4*, 852–866. [CrossRef]
32. Chen, J.; Wang, X.; Ma, A.; Wang, Q.-E.; Liu, B.; Li, L.; Xu, D.; Ma, Q. Deep transfer learning of cancer drug responses by integrating bulk and single-cell RNA-seq data. *Nat. Commun.* **2022**, *13*, 6494. [CrossRef]

33. Hou, R.; Xie, C.; Gui, Y.; Li, G.; Li, X. Machine-Learning-Based Data Analysis Method for Cell-Based Selection of DNA-Encoded Libraries. *ACS Omega* **2023**, *8*, 19057–19071. [CrossRef] [PubMed]
34. Recursion. Recursion's Drug Pipeline. Available online: https://www.recursion.com/pipeline (accessed on 20 February 2025).
35. Insilico Medicine. Insilico Medicine's Drug Pipeline. Available online: https://insilico.com/ (accessed on 20 February 2025).
36. Relay Therapeutics. Relay Therapeutics's Drug Pipeline. Available online: https://relaytx.com/pipeline/ (accessed on 20 February 2025).
37. Exscientia. Exscientia's Drug Pipeline. Available online: https://www.exscientia.com/pipeline/ (accessed on 20 February 2025).
38. Signet Therapeutics. Signet Therapeutics' Drug Pipeline. Available online: https://www.signettx.com/about/ (accessed on 20 February 2025).
39. BeiGene. BeiGene's Drug Pipeline. Available online: https://www.beonemedicines.com.cn/science/pipeline/ (accessed on 20 February 2025).
40. RedCloud Bio. RedCloud Bio's Drug Pipeline. Available online: http://www.redcloudbio.com/en/h-col-105.html (accessed on 20 February 2025).
41. Accutar Biotech. Accutar Biotech's Drug Pipeline. Available online: https://www.accutarbio.com/workflow/ (accessed on 20 February 2025).
42. MindRank. MindRank's Drug Pipeline. Available online: https://www.mindrank.ai/zh-CN/pipeline (accessed on 20 February 2025).
43. Drug Farm. Drug Farm's Drug Pipeline. Available online: https://www.drug-farm.com/pipeline (accessed on 20 February 2025).
44. OrphAI Therapeutics. OrphAI Therapeutics's Drug Pipline. Available online: https://www.orphai-therapeutics.com/pipeline (accessed on 20 February 2025).
45. Healx. Healx's Drug Pipeline. Available online: https://healx.ai/pipeline/ (accessed on 20 February 2025).
46. BioAge. BioAge's Drug Pipeline. Available online: https://bioagelabs.com/apj (accessed on 20 February 2025).
47. BioXcel Therapeutics. BioXcel Therapeutics's Drug Pipeline. Available online: https://www.bioxceltherapeutics.com/our-pipeline/ (accessed on 20 February 2025).
48. Evaxion Biotech. Evaxion Biotech's Drug Pipeline. Available online: https://evaxion.ai/pipeline (accessed on 20 February 2025).
49. Deo, R.C. Machine Learning in Medicine. *Circulation* **2015**, *132*, 1920–1930. [CrossRef] [PubMed]
50. Jiang, T.; Gradus, J.L.; Rosellini, A.J. Supervised Machine Learning: A Brief Primer. *Behav. Ther.* **2020**, *51*, 675–687. [CrossRef]
51. Stokes, J.M.; Yang, K.; Swanson, K.; Jin, W.; Cubillos-Ruiz, A.; Donghia, N.M.; MacNair, C.R.; French, S.; Carfrae, L.A.; Bloom-Ackerman, Z.; et al. A deep learning approach to antibiotic discovery. *Cell* **2020**, *180*, 688–702. [CrossRef]
52. Arab, I.; Egghe, K.; Laukens, K.; Chen, K.; Barakat, K.; Bittremieux, W. Benchmarking of Small Molecule Feature Representations for hERG, Nav1.5, and Cav1.2 Cardiotoxicity Prediction. *J. Chem. Inf. Model.* **2024**, *64*, 2515–2527. [CrossRef]
53. Singh, S.; Kaur, N.; Gehlot, A. Application of artificial intelligence in drug design: A review. *Comput. Biol. Med.* **2024**, *179*, 108810. [CrossRef]
54. Glielmo, A.; Husic, B.E.; Rodriguez, A.; Clementi, C.; Noé, F.; Laio, A. Unsupervised Learning Methods for Molecular Simulation Data. *Chem. Rev.* **2021**, *121*, 9722–9758. [CrossRef]
55. Cihan Sorkun, M.; Mullaj, D.; Koelman, J.M.V.A.; Er, S. ChemPlot, a Python Library for Chemical Space Visualization. *Chem. –Methods* **2022**, *2*, e202200005. [CrossRef]
56. van Engelen, J.E.; Hoos, H.H. A survey on semi-supervised learning. *Mach. Learn.* **2020**, *109*, 373–440. [CrossRef]
57. Niu, Q.; Li, H.; Tong, L.; Liu, S.; Zong, W.; Zhang, S.; Tian, S.; Wang, J.; Liu, J.; Li, B.; et al. TCMFP: A novel herbal formula prediction method based on network target's score integrated with semi-supervised learning genetic algorithms. *Brief. Bioinform.* **2023**, *24*, bbad102. [CrossRef] [PubMed]
58. Jiang, X.; Lu, L.; Li, J.; Jiang, J.; Zhang, J.; Zhou, S.; Wen, H.; Cai, H.; Luo, X.; Li, Z.; et al. Synthetically Feasible De Novo Molecular Design of Leads Based on a Reinforcement Learning Model: AI-Assisted Discovery of an Anti-IBD Lead Targeting CXCR4. *J. Med. Chem.* **2024**, *67*, 10057–10075. [CrossRef] [PubMed]
59. Wang, Y.; Hu, Z.; Chang, J.; Yu, B. Thinking on the Use of Artificial Intelligence in Drug Discovery. *J. Med. Chem.* **2025**, *68*, 4996–4999. [CrossRef] [PubMed]
60. Botvinick, M.; Ritter, S.; Wang, J.X.; Kurth-Nelson, Z.; Blundell, C.; Hassabis, D. Reinforcement Learning, Fast and Slow. *Trends Cogn. Sci.* **2019**, *23*, 408–422. [CrossRef] [PubMed]
61. Xie, Z.; Tu, S.; Xu, L. Multilevel Attention Network with Semi-supervised Domain Adaptation for Drug-Target Prediction. *Proc. AAAI Conf. Artif. Intell.* **2024**, *38*, 329–337. [CrossRef]
62. LeCun, Y.; Bengio, Y.; Hinton, G. Deep learning. *Nature* **2015**, *521*, 436–444. [CrossRef]
63. Kusumoto, D.; Seki, T.; Sawada, H.; Kunitomi, A.; Katsuki, T.; Kimura, M.; Ito, S.; Komuro, J.; Hashimoto, H.; Fukuda, K.; et al. Anti-senescent drug screening by deep learning-based morphology senescence scoring. *Nat. Commun.* **2021**, *12*, 257. [CrossRef]
64. Grebner, C.; Matter, H.; Plowright, A.T.; Hessler, G. Automated De Novo Design in Medicinal Chemistry: Which Types of Chemistry Does a Generative Neural Network Learn? *J. Med. Chem.* **2020**, *63*, 8809–8823. [CrossRef]

65. Chen, Y.; Wang, Z.; Zeng, X.; Li, Y.; Li, P.; Ye, X.; Sakurai, T. Molecular language models: RNNs or transformer? *Brief. Funct. Genom.* **2023**, *22*, 392–400. [CrossRef]
66. Shor, B.; Schneidman-Duhovny, D. DockFormer: Efficient Multi-Modal Receptor-Ligand Interaction Prediction using Pair Transformer. *bioRxiv* **2024**. [CrossRef]
67. Su, X.; Hu, P.; You, Z.-H.; Yu, P.S.; Hu, L. Dual-Channel Learning Framework for Drug-Drug Interaction Prediction via Relation-Aware Heterogeneous Graph Transformer. *Proc. AAAI Conf. Artif. Intell.* **2024**, *38*, 249–256. [CrossRef]
68. Teng, S.; Yin, C.; Wang, Y.; Chen, X.; Yan, Z.; Cui, L.; Wei, L. MolFPG: Multi-level fingerprint-based Graph Transformer for accurate and robust drug toxicity prediction. *Comput. Biol. Med.* **2023**, *164*, 106904. [CrossRef] [PubMed]
69. Wei, G.-W.; Chen, D.; Liu, J. TopoFormer: Multiscale Topology-enabled Structure-to-Sequence Transformer for Protein-Ligand Interaction Predictions. *Nat. Mach. Intell.* **2024**, *6*, 799–810.
70. Jiang, L.; Jiang, C.; Yu, X.; Fu, R.; Jin, S.; Liu, X. DeepTTA: A transformer-based model for predicting cancer drug response. *Brief. Bioinform.* **2022**, *23*, bbac100. [CrossRef]
71. Zhang, O.; Lin, H.; Zhang, H.; Zhao, H.; Huang, Y.; Hsieh, C.-Y.; Pan, P.; Hou, T. Deep Lead Optimization: Leveraging Generative AI for Structural Modification. *J. Am. Chem. Soc.* **2024**, *146*, 31357–31370. [CrossRef]
72. Wang, F.; Feng, X.; Kong, R.; Chang, S.; Wang, F.; Feng, X.; Kong, R.; Chang, S. Generating new protein sequences by using dense network and attention mechanism. *Math. Biosci. Eng.* **2023**, *20*, 4178–4197. [CrossRef]
73. Zeng, X.; Wang, F.; Luo, Y.; Kang, S.-g.; Tang, J.; Lightstone, F.C.; Fang, E.F.; Cornell, W.; Nussinov, R.; Cheng, F. Deep generative molecular design reshapes drug discovery. *Cell Rep. Med.* **2022**, *3*, 100794. [CrossRef]
74. Guimaraes, G.L.; Sanchez-Lengeling, B.; Outeiral, C.; Farias, P.L.C.; Aspuru-Guzik, A. Objective-Reinforced Generative Adversarial Networks (ORGAN) for Sequence Generation Models. *arXiv* **2017**, arXiv:1705.10843.
75. Hu, C.; Li, S.; Yang, C.; Chen, J.; Xiong, Y.; Fan, G.; Liu, H.; Hong, L. ScaffoldGVAE: Scaffold generation and hopping of drug molecules via a variational autoencoder based on multi-view graph neural networks. *J. Cheminform.* **2023**, *15*, 91. [CrossRef]
76. Wang, M.; Hsieh, C.-Y.; Wang, J.; Wang, D.; Weng, G.; Shen, C.; Yao, X.; Bing, Z.; Li, H.; Cao, D.; et al. RELATION: A Deep Generative Model for Structure-Based De Novo Drug Design. *J. Med. Chem.* **2022**, *65*, 9478–9492. [CrossRef] [PubMed]
77. Tong, X.; Liu, X.; Tan, X.; Li, X.; Jiang, J.; Xiong, Z.; Xu, T.; Jiang, H.; Qiao, N.; Zheng, M. Generative Models for De Novo Drug Design. *J. Med. Chem.* **2021**, *64*, 14011–14027. [CrossRef] [PubMed]
78. Ragoza, M.; Masuda, T.; Koes, D.R. Generating 3D molecules conditional on receptor binding sites with deep generative models. *Chem. Sci.* **2022**, *13*, 2701–2713. [CrossRef]
79. Skalic, M.; Jiménez, J.; Sabbadin, D.; De Fabritiis, G. Shape-Based Generative Modeling for de Novo Drug Design. *J. Chem. Inf. Model.* **2019**, *59*, 1205–1214. [CrossRef] [PubMed]
80. DeepSeek-AI; Guo, D.; Yang, D.; Zhang, H.; Song, J.-M.; Zhang, R.; Xu, R.; Zhu, Q.; Ma, S.; Wang, P.; et al. DeepSeek-R1: Incentivizing Reasoning Capability in LLMs via Reinforcement Learning. *arXiv* **2025**, arXiv:abs/2501.12948.
81. Anthropic, S. Model Card Addendum: Claude 3.5 Haiku and Upgraded Claude 3.5 Sonnet. 2024. Available online: https://assets.anthropic.com/m/1cd9d098ac3e6467/original/Claude-3-Model-Card-October-Addendum.pdf (accessed on 20 February 2025).
82. Dubey, A.; Jauhri, A.; Pandey, A.; Kadian, A.; Al-Dahle, A.; Letman, A.; Mathur, A.; Schelten, A.; Yang, A.; Fan, A.; et al. The Llama 3 Herd of Models. *arXiv* **2024**, arXiv:2407.21783. [CrossRef]
83. El-Kishky, A.; Wei, A.; Saraiva, A.; Minaev, B.; Selsam, D.; Dohan, D.; Song, F.; Lightman, H.; Clavera, I.; Pachocki, J.W.; et al. Competitive Programming with Large Reasoning Models. *arXiv* **2025**, arXiv:2502.06807.
84. Lin, A.; Ye, J.; Qi, C.; Zhu, L.; Mou, W.; Gan, W.; Zeng, D.; Tang, B.; Xiao, M.; Chu, G.; et al. Bridging artificial intelligence and biological sciences: A comprehensive review of large language models in bioinformatics. *Brief. Bioinform.* **2025**, *26*, bbaf357. [CrossRef]
85. Wu, Z.; Ramsundar, B.; Feinberg, E.N.; Gomes, J.; Geniesse, C.; Pappu, A.S.; Leswing, K.; Pande, V. MoleculeNet: A benchmark for molecular machine learning. *Chem. Sci.* **2018**, *9*, 513–530. [CrossRef] [PubMed]
86. Huang, K.; Fu, T.; Gao, W.; Zhao, Y.; Roohani, Y.; Leskovec, J.; Coley, C.W.; Xiao, C.; Sun, J.; Zitnik, M. Artificial intelligence foundation for therapeutic science. *Nat. Chem. Biol.* **2022**, *18*, 1033–1036. [CrossRef]
87. Xu, L.-C.; Tang, M.-J.; An, J.; Cao, F.; Qi, Y. A unified pre-trained deep learning framework for cross-task reaction performance prediction and synthesis planning. *Nat. Mach. Intell.* **2025**, *7*, 1561–1571. [CrossRef]
88. He, K.; Zhang, X.; Ren, S.; Sun, J. Deep Residual Learning for Image Recognition. In Proceedings of the IEEE Conference on Computer Vision and Pattern Recognition (CVPR), Las Vegas, NV, USA, 27–30 June 2016.
89. Dou, B.; Zhu, Z.; Merkurjev, E.; Ke, L.; Chen, L.; Jiang, J.; Zhu, Y.; Liu, J.; Zhang, B.; Wei, G.-W. Machine Learning Methods for Small Data Challenges in Molecular Science. *Chem. Rev.* **2023**, *123*, 8736–8780. [CrossRef]
90. Asfand-e-yar, M.; Hashir, Q.; Shah, A.A.; Malik, H.A.M.; Alourani, A.; Khalil, W. Multimodal CNN-DDI: Using multimodal CNN for drug to drug interaction associated events. *Sci. Rep.* **2024**, *14*, 4076. [CrossRef]
91. Chen, S.; Li, T.; Yang, L.; Zhai, F.; Jiang, X.; Xiang, R.; Ling, G. Artificial intelligence-driven prediction of multiple drug interactions. *Brief. Bioinform.* **2022**, *23*, bbac427. [CrossRef] [PubMed]

92. Abbasi, M.; Carvalho, F.G.; Ribeiro, B.; Arrais, J.P. Predicting drug activity against cancer through genomic profiles and SMILES. *Artif. Intell. Med.* **2024**, *150*, 102820. [CrossRef]
93. Hochreiter, S.; Schmidhuber, J. Long Short-Term Memory. *Neural Comput.* **1997**, *9*, 1735–1780. [CrossRef] [PubMed]
94. Isigkeit, L.; Hörmann, T.; Schallmayer, E.; Scholz, K.; Lillich, F.F.; Ehrler, J.H.M.; Hufnagel, B.; Büchner, J.; Marschner, J.A.; Pabel, J.; et al. Automated design of multi-target ligands by generative deep learning. *Nat. Commun.* **2024**, *15*, 7946. [CrossRef]
95. Huang, K.; Chandak, P.; Wang, Q.; Havaldar, S.; Vaid, A.; Leskovec, J.; Nadkarni, G.N.; Glicksberg, B.S.; Gehlenborg, N.; Zitnik, M. A foundation model for clinician-centered drug repurposing. *Nat. Med.* **2024**, *30*, 3601–3613. [CrossRef]
96. Lim, J.; Ryu, S.; Park, K.; Choe, Y.J.; Ham, J.; Kim, W.Y. Predicting Drug–Target Interaction Using a Novel Graph Neural Network with 3D Structure-Embedded Graph Representation. *J. Chem. Inf. Model.* **2019**, *59*, 3981–3988. [CrossRef]
97. Sun, Y.; Li, Y.Y.; Leung, C.K.; Hu, P. iNGNN-DTI: Prediction of drug–target interaction with interpretable nested graph neural network and pretrained molecule models. *Bioinformatics* **2024**, *40*, btae135. [CrossRef] [PubMed]
98. Zhang, Z.; Chen, L.; Zhong, F.; Wang, D.; Jiang, J.; Zhang, S.; Jiang, H.; Zheng, M.; Li, X. Graph neural network approaches for drug-target interactions. *Curr. Opin. Struct. Biol.* **2022**, *73*, 102327. [CrossRef]
99. Ng, S.S.S.; Lu, Y. Evaluating the Use of Graph Neural Networks and Transfer Learning for Oral Bioavailability Prediction. *J. Chem. Inf. Model.* **2023**, *63*, 5035–5044. [CrossRef]
100. Yang, Z.; Zhong, W.; Zhao, L.; Chen, C.Y.-C. MGraphDTA: Deep multiscale graph neural network for explainable drug–target binding affinity prediction. *Chem. Sci.* **2022**, *13*, 816–833. [CrossRef] [PubMed]
101. Li, P.; Hua, L.; Ma, Z.; Hu, W.; Liu, Y.; Zhu, J. Conformalized Graph Learning for Molecular ADMET Property Prediction and Reliable Uncertainty Quantification. *J. Chem. Inf. Model.* **2024**, *64*, 8705–8717. [CrossRef]
102. Heid, E.; Greenman, K.P.; Chung, Y.; Li, S.-C.; Graff, D.E.; Vermeire, F.H.; Wu, H.; Green, W.H.; McGill, C.J. Chemprop: A Machine Learning Package for Chemical Property Prediction. *J. Chem. Inf. Model.* **2024**, *64*, 9–17. [CrossRef]
103. Bai, P.; Miljković, F.; John, B.; Lu, H. Interpretable bilinear attention network with domain adaptation improves drug–target prediction. *Nat. Mach. Intell.* **2023**, *5*, 126–136. [CrossRef]
104. Vaswani, A. Attention is all you need. *Adv. Neural Inf. Process. Syst.* **2017**, *30*, 5998–6008.
105. Radford, A.; Narasimhan, K. Improving Language Understanding by Generative Pre-Training. 2018. Available online: https://www.semanticscholar.org/paper/Improving-Language-Understanding-by-Generative-Radford-Narasimhan/cd18800a0fe0b668a1cc19f2ec95b5003d0a5035 (accessed on 15 February 2025).
106. Devlin, J.; Chang, M.-W.; Lee, K.; Toutanova, K. BERT: Pre-training of Deep Bidirectional Transformers for Language Understanding. In Proceedings of the 2019 Conference of the North American Chapter of the Association for Computational Linguistics: Human Language Technologies, Volume 1 (Long and Short Papers), Minneapolis, MN, USA, 2–7 June 2019; pp. 4171–4186.
107. Mao, J.; Wang, J.; Zeb, A.; Cho, K.-H.; Jin, H.; Kim, J.; Lee, O.; Wang, Y.; No, K.T. Transformer-Based Molecular Generative Model for Antiviral Drug Design. *J. Chem. Inf. Model.* **2024**, *64*, 2733–2745. [CrossRef] [PubMed]
108. OpenAi; Achiam, J.; Adler, S.; Agarwal, S.; Ahmad, L.; Akkaya, I.; Aleman, F.L.; Almeida, D.; Altenschmidt, J.; Altman, S.; et al. GPT-4 Technical Report. *arXiv* **2024**, arXiv:2303.08774.
109. Brown, T.B.; Mann, B.; Ryder, N.; Subbiah, M.; Kaplan, J.; Dhariwal, P.; Neelakantan, A.; Shyam, P.; Sastry, G.; Askell, A.; et al. Language models are few-shot learners. In Proceedings of the 34th International Conference on Neural Information Processing Systems, Vancouver, BC, Canada, 6–12 December 2020; p. 159.
110. Liu, S.; Wang, J.; Yang, Y.; Wang, C.; Liu, L.; Guo, H.; Xiao, C. Chatgpt-powered conversational drug editing using retrieval and domain feedback. *arXiv* **2023**, arXiv:2305.18090.
111. Greenblatt, J.F.; Alberts, B.M.; Krogan, N.J. Discovery and significance of protein-protein interactions in health and disease. *Cell* **2024**, *187*, 6501–6517. [CrossRef]
112. Lim, Y.; Tamayo-Orrego, L.; Schmid, E.; Tarnauskaite, Z.; Kochenova, O.V.; Gruar, R.; Muramatsu, S.; Lynch, L.; Schlie, A.V.; Carroll, P.L.; et al. In silico protein interaction screening uncovers DONSON's role in replication initiation. *Science* **2023**, *381*, eadi3448. [CrossRef]
113. Lin, Z.; Akin, H.; Rao, R.; Hie, B.; Zhu, Z.; Lu, W.; Smetanin, N.; Verkuil, R.; Kabeli, O.; Shmueli, Y.; et al. Evolutionary-scale prediction of atomic-level protein structure with a language model. *Science* **2023**, *379*, 1123–1130. [CrossRef] [PubMed]
114. Senior, A.W.; Evans, R.; Jumper, J.; Kirkpatrick, J.; Sifre, L.; Green, T.; Qin, C.; Žídek, A.; Nelson, A.W.R.; Bridgland, A.; et al. Improved protein structure prediction using potentials from deep learning. *Nature* **2020**, *577*, 706–710. [CrossRef]
115. Jumper, J.; Evans, R.; Pritzel, A.; Green, T.; Figurnov, M.; Ronneberger, O.; Tunyasuvunakool, K.; Bates, R.; Žídek, A.; Potapenko, A.; et al. Highly accurate protein structure prediction with AlphaFold. *Nature* **2021**, *596*, 583–589. [CrossRef]
116. Evans, R.; O'Neill, M.; Pritzel, A.; Antropova, N.; Senior, A.; Green, T.; Žídek, A.; Bates, R.; Blackwell, S.; Yim, J.; et al. Protein complex prediction with AlphaFold-Multimer. *bioRxiv* **2021**. [CrossRef]
117. Abramson, J.; Adler, J.; Dunger, J.; Evans, R.; Green, T.; Pritzel, A.; Ronneberger, O.; Willmore, L.; Ballard, A.J.; Bambrick, J.; et al. Accurate structure prediction of biomolecular interactions with AlphaFold 3. *Nature* **2024**, *630*, 493–500. [CrossRef]

118. Díaz-Holguín, A.; Saarinen, M.; Vo, D.D.; Sturchio, A.; Branzell, N.; Cabeza de Vaca, I.; Hu, H.; Mitjavila-Domènech, N.; Lindqvist, A.; Baranczewski, P.; et al. AlphaFold accelerated discovery of psychotropic agonists targeting the trace amine–associated receptor 1. *Sci. Adv.* **2024**, *10*, eadn1524. [CrossRef]
119. Bernatavicius, A.; Šícho, M.; Janssen, A.P.A.; Hassen, A.K.; Preuss, M.; van Westen, G.J.P. AlphaFold Meets De Novo Drug Design: Leveraging Structural Protein Information in Multitarget Molecular Generative Models. *J. Chem. Inf. Model.* **2024**, *64*, 8113–8122. [CrossRef]
120. Ren, F.; Ding, X.; Zheng, M.; Korzinkin, M.; Cai, X.; Zhu, W.; Mantsyzov, A.; Aliper, A.; Aladinskiy, V.; Cao, Z.; et al. AlphaFold accelerates artificial intelligence powered drug discovery: Efficient discovery of a novel CDK20 small molecule inhibitor. *Chem. Sci.* **2023**, *14*, 1443–1452. [CrossRef]
121. Michaud, J.M.; Madani, A.; Fraser, J.S. A language model beats alphafold2 on orphans. *Nat. Biotechnol.* **2022**, *40*, 1576–1577. [CrossRef]
122. Baek, M.; DiMaio, F.; Anishchenko, I.; Dauparas, J.; Ovchinnikov, S.; Lee, G.R.; Wang, J.; Cong, Q.; Kinch, L.N.; Schaeffer, R.D.; et al. Accurate prediction of protein structures and interactions using a three-track neural network. *Science* **2021**, *373*, 871–876. [CrossRef] [PubMed]
123. Du, Z.; Su, H.; Wang, W.; Ye, L.; Wei, H.; Peng, Z.; Anishchenko, I.; Baker, D.; Yang, J. The trRosetta server for fast and accurate protein structure prediction. *Nat. Protoc.* **2021**, *16*, 5634–5651. [CrossRef] [PubMed]
124. Zhou, Y.; Zhang, Y.; Lian, X.; Li, F.; Wang, C.; Zhu, F.; Qiu, Y.; Chen, Y. Therapeutic target database update 2022: Facilitating drug discovery with enriched comparative data of targeted agents. *Nucleic Acids Res.* **2022**, *50*, D1398–D1407. [CrossRef]
125. Finan, C.; Gaulton, A.; Kruger, F.A.; Lumbers, R.T.; Shah, T.; Engmann, J.; Galver, L.; Kelley, R.; Karlsson, A.; Santos, R.; et al. The druggable genome and support for target identification and validation in drug development. *Sci. Transl. Med.* **2017**, *9*, eaag1166. [CrossRef] [PubMed]
126. Kana, O.; Brylinski, M. Elucidating the druggability of the human proteome with eFindSite. *J. Comput. Aided Mol. Des.* **2019**, *33*, 509–519. [CrossRef]
127. Li, X.; Wang, S.; Xie, Y.; Jiang, H.; Guo, J.; Wang, Y.; Peng, Z.; Hu, M.; Wang, M.; Wang, J.; et al. Deacetylation induced nuclear condensation of HP1gamma promotes multiple myeloma drug resistance. *Nat. Commun.* **2023**, *14*, 1290. [CrossRef]
128. Wang, Y.; Gao, S.; Chen, L.; Liu, S.; Ma, J.; Cao, Z.; Li, Q. DUT enhances drug resistance to proteasome inhibitors via promoting mitochondrial function in multiple myeloma. *Carcinogenesis* **2022**, *43*, 1030–1038. [CrossRef]
129. Qi, T.F.; Tang, F.; Yin, J.; Miao, W.; Wang, Y. Parallel-reaction monitoring revealed altered expression of a number of epitranscriptomic reader, writer, and eraser proteins accompanied with colorectal cancer metastasis. *Proteomics* **2023**, *23*, e2200059. [CrossRef] [PubMed]
130. Nidhi, S.; Anand, U.; Oleksak, P.; Tripathi, P.; Lal, J.A.; Thomas, G.; Kuca, K.; Tripathi, V. Novel CRISPR-Cas Systems: An Updated Review of the Current Achievements, Applications, and Future Research Perspectives. *Int. J. Mol. Sci.* **2021**, *22*, 3327. [CrossRef]
131. Ramkumar, P.; Abarientos, A.B.; Tian, R.; Seyler, M.; Leong, J.T.; Chen, M.; Choudhry, P.; Hechler, T.; Shah, N.; Wong, S.W.; et al. CRISPR-based screens uncover determinants of immunotherapy response in multiple myeloma. *Blood Adv.* **2020**, *4*, 2899–2911. [CrossRef]
132. Raivola, J.; Dini, A.; Karvonen, H.; Piki, E.; Salokas, K.; Niininen, W.; Kaleva, L.; Zhang, K.; Arjama, M.; Gudoityte, G.; et al. Multiomics characterization implicates PTK7 in ovarian cancer EMT and cell plasticity and offers strategies for therapeutic intervention. *Cell Death Dis.* **2022**, *13*, 714. [CrossRef]
133. Gulfidan, G.; Soylu, M.; Demirel, D.; Erdonmez, H.B.C.; Beklen, H.; Ozbek Sarica, P.; Arga, K.Y.; Turanli, B. Systems biomarkers for papillary thyroid cancer prognosis and treatment through multi-omics networks. *Arch Biochem. Biophys.* **2022**, *715*, 109085. [CrossRef] [PubMed]
134. Assum, I.; Krause, J.; Scheinhardt, M.O.; Muller, C.; Hammer, E.; Borschel, C.S.; Volker, U.; Conradi, L.; Geelhoed, B.; Zeller, T.; et al. Tissue-specific multi-omics analysis of atrial fibrillation. *Nat. Commun.* **2022**, *13*, 441. [CrossRef]
135. Abell, N.S.; DeGorter, M.K.; Gloudemans, M.J.; Greenwald, E.; Smith, K.S.; He, Z.; Montgomery, S.B. Multiple causal variants underlie genetic associations in humans. *Science* **2022**, *375*, 1247–1254. [CrossRef]
136. Namba, S.; Konuma, T.; Wu, K.H.; Zhou, W.; Global Biobank Meta-analysis, I.; Okada, Y. A practical guideline of genomics-driven drug discovery in the era of global biobank meta-analysis. *Cell Genom.* **2022**, *2*, 100190. [CrossRef]
137. Deelen, J.; Evans, D.S.; Arking, D.E.; Tesi, N.; Nygaard, M.; Liu, X.; Wojczynski, M.K.; Biggs, M.L.; van der Spek, A.; Atzmon, G.; et al. Publisher Correction: A meta-analysis of genome-wide association studies identifies multiple longevity genes. *Nat. Commun.* **2021**, *12*, 2463. [CrossRef]
138. Vamathevan, J.; Clark, D.; Czodrowski, P.; Dunham, I.; Ferran, E.; Lee, G.; Li, B.; Madabhushi, A.; Shah, P.; Spitzer, M.; et al. Applications of machine learning in drug discovery and development. *Nat. Rev. Drug. Discov.* **2019**, *18*, 463–477. [CrossRef]
139. Lo, Y.C.; Senese, S.; Damoiseaux, R.; Torres, J.Z. 3D Chemical Similarity Networks for Structure-Based Target Prediction and Scaffold Hopping. *ACS Chem. Biol.* **2016**, *11*, 2244–2253. [CrossRef]

140. Wolber, G.; Seidel, T.; Bendix, F.; Langer, T. Molecule-pharmacophore superpositioning and pattern matching in computational drug design. *Drug Discov. Today* **2008**, *13*, 23–29. [CrossRef]
141. Mamoshina, P.; Volosnikova, M.; Ozerov, I.V.; Putin, E.; Skibina, E.; Cortese, F.; Zhavoronkov, A. Machine Learning on Human Muscle Transcriptomic Data for Biomarker Discovery and Tissue-Specific Drug Target Identification. *Front. Genet.* **2018**, *9*, 242. [CrossRef] [PubMed]
142. Theodoris, C.V.; Xiao, L.; Chopra, A.; Chaffin, M.D.; Al Sayed, Z.R.; Hill, M.C.; Mantineo, H.; Brydon, E.M.; Zeng, Z.; Liu, X.S.; et al. Transfer learning enables predictions in network biology. *Nature* **2023**, *618*, 616–624. [CrossRef] [PubMed]
143. Kang, B.; Fan, R.; Cui, C.; Cui, Q. Comprehensive prediction and analysis of human protein essentiality based on a pretrained large language model. *Nat. Comput. Sci.* **2025**, *5*, 196–206. [CrossRef] [PubMed]
144. Zhang, S.; Cooper-Knock, J.; Weimer, A.K.; Shi, M.; Moll, T.; Marshall, J.N.G.; Harvey, C.; Nezhad, H.G.; Franklin, J.; Souza, C.D.S.; et al. Genome-wide identification of the genetic basis of amyotrophic lateral sclerosis. *Neuron* **2022**, *110*, 992–1008.e11. [CrossRef]
145. Zeng, X.; Zhu, S.; Lu, W.; Liu, Z.; Huang, J.; Zhou, Y.; Fang, J.; Huang, Y.; Guo, H.; Li, L.; et al. Target identification among known drugs by deep learning from heterogeneous networks. *Chem. Sci.* **2020**, *11*, 1775–1797. [CrossRef]
146. Zhou, L.; Lian, G.; Zhou, T.; Cai, Z.; Yang, S.; Li, W.; Cheng, L.; Ye, Y.; He, M.; Lu, J.; et al. Palmitoylation of GPX4 via the targetable ZDHHC8 determines ferroptosis sensitivity and antitumor immunity. *Nat. Cancer* **2025**, *6*, 768–785. [CrossRef]
147. Yang, Y.; Yao, K.; Repasky, M.P.; Leswing, K.; Abel, R.; Shoichet, B.K.; Jerome, S.V. Efficient Exploration of Chemical Space with Docking and Deep Learning. *J. Chem. Theory Comput.* **2021**, *17*, 7106–7119. [CrossRef]
148. Singh, R.; Sledzieski, S.; Bryson, B.; Cowen, L.; Berger, B. Contrastive learning in protein language space predicts interactions between drugs and protein targets. *Proc. Natl. Acad. Sci. USA* **2023**, *120*, e2220778120. [CrossRef]
149. Schade, A.E.; Perurena, N.; Yang, Y.; Rodriguez, C.L.; Krishnan, A.; Gardner, A.; Loi, P.; Xu, Y.; Nguyen, V.T.M.; Mastellone, G.M.; et al. AKT and EZH2 inhibitors kill TNBCs by hijacking mechanisms of involution. *Nature* **2024**, *635*, 755–763. [CrossRef]
150. Gentile, F.; Yaacoub, J.C.; Gleave, J.; Fernandez, M.; Ton, A.T.; Ban, F.; Stern, A.; Cherkasov, A. Artificial intelligence-enabled virtual screening of ultra-large chemical libraries with deep docking. *Nat. Protoc.* **2022**, *17*, 672–697. [CrossRef] [PubMed]
151. Acharya, A.; Agarwal, R.; Baker, M.B.; Baudry, J.; Bhowmik, D.; Boehm, S.; Byler, K.G.; Chen, S.Y.; Coates, L.; Cooper, C.J.; et al. Supercomputer-Based Ensemble Docking Drug Discovery Pipeline with Application to Covid-19. *J. Chem. Inf. Model.* **2020**, *60*, 5832–5852. [CrossRef] [PubMed]
152. Puszkarska, A.M.; Taddese, B.; Revell, J.; Davies, G.; Field, J.; Hornigold, D.C.; Buchanan, A.; Vaughan, T.J.; Colwell, L.J. Machine learning designs new GCGR/GLP-1R dual agonists with enhanced biological potency. *Nat. Chem.* **2024**, *16*, 1436–1444. [CrossRef]
153. Trinh, T.C.; Falson, P.; Tran-Nguyen, V.K.; Boumendjel, A. Ligand-Based Drug Discovery Leveraging State-of-the-Art Machine Learning Methodologies Exemplified by Cdr1 Inhibitor Prediction. *J. Chem. Inf. Model.* **2025**, *65*, 4027–4042. [CrossRef]
154. Hansson, F.G.; Madsen, N.G.; Hansen, L.G.; Jakociunas, T.; Lengger, B.; Keasling, J.D.; Jensen, M.K.; Acevedo-Rocha, C.G.; Jensen, E.D. Labels as a feature: Network homophily for systematically annotating human GPCR drug-target interactions. *Nat. Commun.* **2025**, *16*, 4121. [CrossRef] [PubMed]
155. Hadipour, H.; Li, Y.Y.; Sun, Y.; Deng, C.; Lac, L.; Davis, R.; Cardona, S.T.; Hu, P. GraphBAN: An inductive graph-based approach for enhanced prediction of compound-protein interactions. *Nat. Commun.* **2025**, *16*, 2541. [CrossRef] [PubMed]
156. Lu, Z.; Song, G.; Zhu, H.; Lei, C.; Sun, X.; Wang, K.; Qin, L.; Chen, Y.; Tang, J.; Li, M. DTIAM: A unified framework for predicting drug-target interactions, binding affinities and drug mechanisms. *Nat. Commun.* **2025**, *16*, 2548. [CrossRef]
157. Liu, X.; Li, Q.; Yan, X.; Wang, L.; Qiu, J.; Yao, X.; Liu, H. A Specialized and Enhanced Deep Generation Model for Active Molecular Design Targeting Kinases Guided by Affinity Prediction Models and Reinforcement Learning. *J. Chem. Inf. Model.* **2025**, *65*, 3294–3308. [CrossRef]
158. Yang, H.; Chen, Y.; Zuo, Y.; Deng, Z.; Pan, X.; Shen, H.B.; Choi, K.S.; Yu, D.J. MINDG: A drug-target interaction prediction method based on an integrated learning algorithm. *Bioinformatics* **2024**, *40*, btae147. [CrossRef]
159. Moon, S.; Hwang, S.-Y.; Lim, J.; Kim, W.Y. PIGNet2: A versatile deep learning-based protein–ligand interaction prediction model for binding affinity scoring and virtual screening. *Digit. Discov.* **2024**, *3*, 287–299. [CrossRef]
160. Cai, H.; Shen, C.; Jian, T.; Zhang, X.; Chen, T.; Han, X.; Yang, Z.; Dang, W.; Hsieh, C.Y.; Kang, Y.; et al. CarsiDock: A deep learning paradigm for accurate protein-ligand docking and screening based on large-scale pre-training. *Chem. Sci.* **2024**, *15*, 1449–1471. [CrossRef] [PubMed]
161. Jin, W.; Stokes, J.M.; Eastman, R.T.; Itkin, Z.; Zakharov, A.V.; Collins, J.J.; Jaakkola, T.S.; Barzilay, R. Deep learning identifies synergistic drug combinations for treating COVID-19. *Proc. Natl. Acad. Sci. USA* **2021**, *118*, e2105070118. [CrossRef]
162. Liu, G.; Catacutan, D.B.; Rathod, K.; Swanson, K.; Jin, W.; Mohammed, J.C.; Chiappino-Pepe, A.; Syed, S.A.; Fragis, M.; Rachwalski, K.; et al. Deep learning-guided discovery of an antibiotic targeting *Acinetobacter baumannii*. *Nat. Chem. Biol.* **2023**, *19*, 1342–1350. [CrossRef]
163. Guo, X.; Zhao, X.; Lu, X.; Zhao, L.; Zeng, Q.; Chen, F.; Zhang, Z.; Xu, M.; Feng, S.; Fan, T.; et al. A deep learning-driven discovery of berberine derivatives as novel antibacterial against multidrug-resistant *Helicobacter pylori*. *Signal Transduct. Target. Ther.* **2024**, *9*, 183. [CrossRef] [PubMed]

164. Liu, X.-W.; Shi, T.-Y.; Gao, D.; Ma, C.-Y.; Lin, H.; Yan, D.; Deng, K.-J. iPADD: A Computational Tool for Predicting Potential Antidiabetic Drugs Using Machine Learning Algorithms. *J. Chem. Inf. Model.* **2023**, *63*, 4960–4969. [CrossRef]
165. Gerdes, H.; Casado, P.; Dokal, A.; Hijazi, M.; Akhtar, N.; Osuntola, R.; Rajeeve, V.; Fitzgibbon, J.; Travers, J.; Britton, D.; et al. Drug ranking using machine learning systematically predicts the efficacy of anti-cancer drugs. *Nat. Commun.* **2021**, *12*, 1850. [CrossRef] [PubMed]
166. Feng, B.; Liu, Z.; Huang, N.; Xiao, Z.; Zhang, H.; Mirzoyan, S.; Xu, H.; Hao, J.; Xu, Y.; Zhang, M.; et al. A bioactivity foundation model using pairwise meta-learning. *Nat. Mach. Intell.* **2024**, *6*, 962–974. [CrossRef]
167. Duan, Y.; Yang, X.; Zeng, X.; Wang, W.; Deng, Y.; Cao, D. Enhancing Molecular Property Prediction through Task-Oriented Transfer Learning: Integrating Universal Structural Insights and Domain-Specific Knowledge. *J. Med. Chem.* **2024**, *67*, 9575–9586. [CrossRef]
168. Cai, H.; Zhang, H.; Zhao, D.; Wu, J.; Wang, L. FP-GNN: A versatile deep learning architecture for enhanced molecular property prediction. *Brief. Bioinform.* **2022**, *23*, bbac408. [CrossRef]
169. Ahmad, W.; Simon, E.; Chithrananda, S.; Grand, G.; Ramsundar, B. Chemberta-2: Towards chemical foundation models. *arXiv* **2022**, arXiv:2209.01712. [CrossRef]
170. Zhou, G.; Gao, Z.; Ding, Q.; Zheng, H.; Xu, H.; Wei, Z.; Zhang, L.; Ke, G. Uni-mol: A universal 3d molecular representation learning framework. *ChemRxiv* **2023**. [CrossRef]
171. Chang, J.; Ye, J.C. Bidirectional generation of structure and properties through a single molecular foundation model. *Nat. Commun.* **2024**, *15*, 2323. [CrossRef]
172. Ross, J.; Belgodere, B.; Chenthamarakshan, V.; Padhi, I.; Mroueh, Y.; Das, P. Large-scale chemical language representations capture molecular structure and properties. *Nat. Mach. Intell.* **2022**, *4*, 1256–1264. [CrossRef]
173. Virshup, A.M.; Contreras-García, J.; Wipf, P.; Yang, W.; Beratan, D.N. Stochastic Voyages into Uncharted Chemical Space Produce a Representative Library of All Possible Drug-Like Compounds. *J. Am. Chem. Soc.* **2013**, *135*, 7296–7303. [CrossRef]
174. Polishchuk, P.G.; Madzhidov, T.I.; Varnek, A. Estimation of the size of drug-like chemical space based on GDB-17 data. *J. Comput.-Aided Mol. Des.* **2013**, *27*, 675–679. [CrossRef] [PubMed]
175. Tropsha, A.; Isayev, O.; Varnek, A.; Schneider, G.; Cherkasov, A. Integrating QSAR modelling and deep learning in drug discovery: The emergence of deep QSAR. *Nat. Rev. Drug Discov.* **2024**, *23*, 141–155. [CrossRef] [PubMed]
176. Carvalho Martins, L.; Cino, E.A.; Ferreira, R.S. PyAutoFEP: An Automated Free Energy Perturbation Workflow for GROMACS Integrating Enhanced Sampling Methods. *J. Chem. Theory Comput.* **2021**, *17*, 4262–4273. [CrossRef]
177. Schneuing, A.; Harris, C.; Du, Y.; Didi, K.; Jamasb, A.; Igashov, I.; Du, W.; Gomes, C.; Blundell, T.L.; Lio, P.; et al. Structure-based drug design with equivariant diffusion models. *Nat. Comput. Sci.* **2024**, *4*, 899–909. [CrossRef]
178. Li, P.; Zhang, K.; Liu, T.; Lu, R.; Chen, Y.; Yao, X.; Gao, L.; Zeng, X. A deep learning approach for rational ligand generation with toxicity control via reactive building blocks. *Nat. Comput. Sci.* **2024**, *4*, 851–864. [CrossRef]
179. Hu, Q.; Sun, C.; He, H.; Xu, J.; Liu, D.; Zhang, W.; Shi, S.; Zhang, K.; Li, H. Target-aware 3D molecular generation based on guided equivariant diffusion. *Nat. Commun.* **2025**, *16*, 7928. [CrossRef]
180. Godinez, W.J.; Ma, E.J.; Chao, A.T.; Pei, L.; Skewes-Cox, P.; Canham, S.M.; Jenkins, J.L.; Young, J.M.; Martin, E.J.; Guiguemde, W.A. Design of potent antimalarials with generative chemistry. *Nat. Mach. Intell.* **2022**, *4*, 180–186. [CrossRef]
181. Swanson, K.; Liu, G.; Catacutan, D.B.; Arnold, A.; Zou, J.; Stokes, J.M. Generative AI for designing and validating easily synthesizable and structurally novel antibiotics. *Nat. Mach. Intell.* **2024**, *6*, 338–353. [CrossRef]
182. Wu, K.; Xia, Y.; Deng, P.; Liu, R.; Zhang, Y.; Guo, H.; Cui, Y.; Pei, Q.; Wu, L.; Xie, S.; et al. TamGen: Drug design with target-aware molecule generation through a chemical language model. *Nat. Commun.* **2024**, *15*, 9360. [CrossRef]
183. Skinnider, M.A.; Stacey, R.G.; Wishart, D.S.; Foster, L.J. Chemical language models enable navigation in sparsely populated chemical space. *Nat. Mach. Intell.* **2021**, *3*, 759–770. [CrossRef]
184. Flam-Shepherd, D.; Zhu, K.; Aspuru-Guzik, A. Language models can learn complex molecular distributions. *Nat. Commun.* **2022**, *13*, 3293. [CrossRef]
185. Grisoni, F. Chemical language models for de novo drug design: Challenges and opportunities. *Curr. Opin. Struct. Biol.* **2023**, *79*, 102527. [CrossRef]
186. Tong, X.; Qu, N.; Kong, X.; Ni, S.; Zhou, J.; Wang, K.; Zhang, L.; Wen, Y.; Shi, J.; Zhang, S.; et al. Deep representation learning of chemical-induced transcriptional profile for phenotype-based drug discovery. *Nat. Commun.* **2024**, *15*, 5378. [CrossRef]
187. Tan, Y.; Dai, L.; Huang, W.; Guo, Y.; Zheng, S.; Lei, J.; Chen, H.; Yang, Y. DRlinker: Deep Reinforcement Learning for Optimization in Fragment Linking Design. *J. Chem. Inf. Model.* **2022**, *62*, 5907–5917. [CrossRef]
188. Guo, J.; Knuth, F.; Margreitter, C.; Janet, J.P.; Papadopoulos, K.; Engkvist, O.; Patronov, A. Link-INVENT: Generative linker design with reinforcement learning. *Digit. Discov.* **2023**, *2*, 392–408. [CrossRef]
189. Zheng, S.; Lei, Z.; Ai, H.; Chen, H.; Deng, D.; Yang, Y. Deep scaffold hopping with multimodal transformer neural networks. *J. Cheminform.* **2021**, *13*, 87. [CrossRef] [PubMed]

190. Bagal, V.; Aggarwal, R.; Vinod, P.K.; Priyakumar, U.D. MolGPT: Molecular Generation Using a Transformer-Decoder Model. *J. Chem. Inf. Model.* **2022**, *62*, 2064–2076. [CrossRef] [PubMed]
191. Arus-Pous, J.; Patronov, A.; Bjerrum, E.J.; Tyrchan, C.; Reymond, J.L.; Chen, H.; Engkvist, O. SMILES-based deep generative scaffold decorator for de-novo drug design. *J. Cheminform.* **2020**, *12*, 38. [CrossRef]
192. Langevin, M.; Minoux, H.; Levesque, M.; Bianciotto, M. Scaffold-Constrained Molecular Generation. *J. Chem. Inf. Model.* **2020**, *60*, 5637–5646. [CrossRef]
193. Fialkova, V.; Zhao, J.; Papadopoulos, K.; Engkvist, O.; Bjerrum, E.J.; Kogej, T.; Patronov, A. LibINVENT: Reaction-based Generative Scaffold Decoration for in Silico Library Design. *J. Chem. Inf. Model.* **2022**, *62*, 2046–2063. [CrossRef] [PubMed]
194. Xie, J.; Chen, S.; Lei, J.; Yang, Y. DiffDec: Structure-Aware Scaffold Decoration with an End-to-End Diffusion Model. *J. Chem. Inf. Model.* **2024**, *64*, 2554–2564. [CrossRef]
195. Green, H.; Koes, D.R.; Durrant, J.D. DeepFrag: A deep convolutional neural network for fragment-based lead optimization. *Chem. Sci.* **2021**, *12*, 8036–8047. [CrossRef]
196. Hadfield, T.E.; Imrie, F.; Merritt, A.; Birchall, K.; Deane, C.M. Incorporating Target-Specific Pharmacophoric Information into Deep Generative Models for Fragment Elaboration. *J. Chem. Inf. Model.* **2022**, *62*, 2280–2292. [CrossRef] [PubMed]
197. Imrie, F.; Hadfield, T.E.; Bradley, A.R.; Deane, C.M. Deep generative design with 3D pharmacophoric constraints. *Chem. Sci.* **2021**, *12*, 14577–14589. [CrossRef]
198. Liu, X.; Ye, K.; van Vlijmen, H.W.T.; AP, I.J.; van Westen, G.J.P. DrugEx v3: Scaffold-constrained drug design with graph transformer-based reinforcement learning. *J. Cheminform.* **2023**, *15*, 24. [CrossRef]
199. Loeffler, H.H.; He, J.; Tibo, A.; Janet, J.P.; Voronov, A.; Mervin, L.H.; Engkvist, O. Reinvent 4: Modern AI-driven generative molecule design. *J. Cheminform.* **2024**, *16*, 20. [CrossRef]
200. Wang, M.; Li, S.; Wang, J.; Zhang, O.; Du, H.; Jiang, D.; Wu, Z.; Deng, Y.; Kang, Y.; Pan, P.; et al. ClickGen: Directed exploration of synthesizable chemical space via modular reactions and reinforcement learning. *Nat. Commun.* **2024**, *15*, 10127. [CrossRef]
201. Segler, M.H.S.; Preuss, M.; Waller, M.P. Planning chemical syntheses with deep neural networks and symbolic AI. *Nature* **2018**, *555*, 604–610. [CrossRef] [PubMed]
202. Thakkar, A.; Chadimová, V.; Bjerrum, E.J.; Engkvist, O.; Reymond, J.-L. Retrosynthetic accessibility score (RAscore)—rapid machine learned synthesizability classification from AI driven retrosynthetic planning. *Chem. Sci.* **2021**, *12*, 3339–3349. [CrossRef]
203. Nippa, D.F.; Atz, K.; Hohler, R.; Muller, A.T.; Marx, A.; Bartelmus, C.; Wuitschik, G.; Marzuoli, I.; Jost, V.; Wolfard, J.; et al. Enabling late-stage drug diversification by high-throughput experimentation with geometric deep learning. *Nat. Chem.* **2024**, *16*, 239–248. [CrossRef]
204. Lu, J.; Zhang, Y. Unified Deep Learning Model for Multitask Reaction Predictions with Explanation. *J. Chem. Inf. Model.* **2022**, *62*, 1376–1387. [CrossRef]
205. Li, B.; Su, S.; Zhu, C.; Lin, J.; Hu, X.; Su, L.; Yu, Z.; Liao, K.; Chen, H. A deep learning framework for accurate reaction prediction and its application on high-throughput experimentation data. *J. Cheminform* **2023**, *15*, 72. [CrossRef] [PubMed]
206. Cong, S.; Zhang, M.; Song, Y.; Chang, S.; Tian, J.; Zeng, H.; Ji, H. Graph-sequence enhanced transformer for template-free prediction of natural product biosynthesis. *Patterns* **2025**, *6*, 101259. [CrossRef] [PubMed]
207. Lee, A.A.; Yang, Q.; Sresht, V.; Bolgar, P.; Hou, X.; Klug-McLeod, J.L.; Butler, C.R. Molecular Transformer unifies reaction prediction and retrosynthesis across pharma chemical space. *Chem. Commun.* **2019**, *55*, 12152–12155. [CrossRef]
208. Zhao, P.C.; Wei, X.X.; Wang, Q.; Wang, Q.H.; Li, J.N.; Shang, J.; Lu, C.; Shi, J.Y. Single-step retrosynthesis prediction via multitask graph representation learning. *Nat. Commun.* **2025**, *16*, 814. [CrossRef]
209. Chen, S.; Jung, Y. Deep Retrosynthetic Reaction Prediction using Local Reactivity and Global Attention. *JACS Au* **2021**, *1*, 1612–1620. [CrossRef]
210. Ucak, U.V.; Ashyrmamatov, I.; Ko, J.; Lee, J. Retrosynthetic reaction pathway prediction through neural machine translation of atomic environments. *Nat. Commun.* **2022**, *13*, 1186. [CrossRef]
211. Fang, L.; Li, J.; Zhao, M.; Tan, L.; Lou, J.G. Single-step retrosynthesis prediction by leveraging commonly preserved substructures. *Nat. Commun.* **2023**, *14*, 2446. [CrossRef] [PubMed]
212. Lin, Z.; Yin, S.; Shi, L.; Zhou, W.; Zhang, Y.J. G2GT: Retrosynthesis Prediction with Graph-to-Graph Attention Neural Network and Self-Training. *J. Chem. Inf. Model.* **2023**, *63*, 1894–1905. [CrossRef]
213. Mao, K.; Xiao, X.; Xu, T.; Rong, Y.; Huang, J.; Zhao, P. Molecular graph enhanced transformer for retrosynthesis prediction. *Neurocomputing* **2021**, *457*, 193–202. [CrossRef]
214. Tu, Z.; Coley, C.W. Permutation Invariant Graph-to-Sequence Model for Template-Free Retrosynthesis and Reaction Prediction. *J. Chem. Inf. Model.* **2022**, *62*, 3503–3513. [CrossRef]
215. Wang, X.; Li, Y.; Qiu, J.; Chen, G.; Liu, H.; Liao, B.; Hsieh, C.-Y.; Yao, X. Retroprime: A diverse, plausible and transformer-based method for single-step retrosynthesis predictions. *Chem. Eng. J.* **2021**, *420*, 129845. [CrossRef]
216. Beckers, M.; Sturm, N.; Sirockin, F.; Fechner, N.; Stiefl, N. Prediction of Small-Molecule Developability Using Large-Scale In Silico ADMET Models. *J. Med. Chem.* **2023**, *66*, 14047–14060. [CrossRef]

217. Vangala, S.R.; Krishnan, S.R.; Bung, N.; Srinivasan, R.; Roy, A. pBRICS: A Novel Fragmentation Method for Explainable Property Prediction of Drug-like Small Molecules. *J. Chem. Inf. Model.* **2023**, *63*, 5066–5076. [CrossRef]
218. Jamrozik, E.; Śmieja, M.; Podlewska, S. ADMET-PrInt: Evaluation of ADMET Properties: Prediction and Interpretation. *J. Chem. Inf. Model.* **2024**, *64*, 1425–1432. [CrossRef] [PubMed]
219. Mamada, H.; Takahashi, M.; Ogino, M.; Nomura, Y.; Uesawa, Y. Predictive Models Based on Molecular Images and Molecular Descriptors for Drug Screening. *ACS Omega* **2023**, *8*, 37186–37195. [CrossRef] [PubMed]
220. Komissarov, L.; Manevski, N.; Groebke Zbinden, K.; Schindler, T.; Zitnik, M.; Sach-Peltason, L. Actionable Predictions of Human Pharmacokinetics at the Drug Design Stage. *Mol. Pharm.* **2024**, *21*, 4356–4371. [CrossRef]
221. Niu, Z.; Xiao, X.; Wu, W.; Cai, Q.; Jiang, Y.; Jin, W.; Wang, M.; Yang, G.; Kong, L.; Jin, X.; et al. PharmaBench: Enhancing ADMET benchmarks with large language models. *Sci. Data* **2024**, *11*, 985. [CrossRef] [PubMed]
222. Swanson, K.; Walther, P.; Leitz, J.; Mukherjee, S.; Wu, J.C.; Shivnaraine, R.V.; Zou, J. ADMET-AI: A machine learning ADMET platform for evaluation of large-scale chemical libraries. *Bioinformatics* **2024**, *40*, btae416. [CrossRef]
223. Daina, A.; Michielin, O.; Zoete, V. SwissADME: A free web tool to evaluate pharmacokinetics, drug-likeness and medicinal chemistry friendliness of small molecules. *Sci. Rep.* **2017**, *7*, 42717. [CrossRef]
224. Fu, L.; Shi, S.; Yi, J.; Wang, N.; He, Y.; Wu, Z.; Peng, J.; Deng, Y.; Wang, W.; Wu, C.; et al. ADMETlab 3.0: An updated comprehensive online ADMET prediction platform enhanced with broader coverage, improved performance, API functionality and decision support. *Nucleic Acids Res.* **2024**, *52*, W422–W431. [CrossRef]
225. Pires, D.E.; Blundell, T.L.; Ascher, D.B. pkCSM: Predicting Small-Molecule Pharmacokinetic and Toxicity Properties Using Graph-Based Signatures. *J. Med. Chem.* **2015**, *58*, 4066–4072. [CrossRef]
226. Schyman, P.; Liu, R.; Desai, V.; Wallqvist, A. vNN Web Server for ADMET Predictions. *Front. Pharmacol.* **2017**, *8*, 889. [CrossRef]
227. Yi, J.C.; Yang, Z.Y.; Zhao, W.T.; Yang, Z.J.; Zhang, X.C.; Wu, C.K.; Lu, A.P.; Cao, D.S. ChemMORT: An automatic ADMET optimization platform using deep learning and multi-objective particle swarm optimization. *Brief. Bioinform.* **2024**, *25*, bbae008. [CrossRef] [PubMed]
228. Zhang, K.; Yang, X.; Wang, Y.; Yu, Y.; Huang, N.; Li, G.; Li, X.; Wu, J.C.; Yang, S. Artificial intelligence in drug development. *Nat. Med.* **2025**, *31*, 45–59. [CrossRef] [PubMed]
229. Ji, Z.; Lee, N.; Frieske, R.; Yu, T.; Su, D.; Xu, Y.; Ishii, E.; Bang, Y.J.; Madotto, A.; Fung, P. Survey of Hallucination in Natural Language Generation. *ACM Comput. Surv.* **2023**, *55*, 248. [CrossRef]
230. Wellawatte, G.P.; Gandhi, H.A.; Seshadri, A.; White, A.D. A Perspective on Explanations of Molecular Prediction Models. *J. Chem. Theory Comput.* **2023**, *19*, 2149–2160. [CrossRef]
231. Frye, L.; Bhat, S.; Akinsanya, K.; Abel, R. From computer-aided drug discovery to computer-driven drug discovery. *Drug Discov. Today Technol.* **2021**, *39*, 111–117. [CrossRef]

metabolites

MDPI

Article

Yigong San Extract Modulates Metabolism, Antioxidant Status, and Immune Function to Improve Health in Diarrheic Calves

Sijuan Huang [1,2,3], Chao Han [1,2,3], Jianyu Lv [1,2,3], Xiaosong Zhang [1,2,3], Xuan Ni [1,2,3], Xin Wang [4], Jianfei Wang [4], Yunfei Ma [1,2,3,*] and Zhihui Hao [1,2,3,*]

1 State Key Laboratory of Veterinary Public Health and Safety, College of Veterinary Medicine, China Agricultural University, Beijing 100193, China; b20223050398@cau.edu.cn (S.H.); s20233051040@cau.edu.cn (C.H.); s20223050978@cau.edu.cn (J.L.); zhangxsgs@cau.edu.cn (X.Z.); b20243050542@cau.edu.cn (X.N.)
2 Innovation Centre of Chinese Veterinary Medicine, College of Veterinary Medicine, China Agricultural University, Beijing 100193, China
3 Key Biology Laboratory of Chinese Veterinary Medicine, Ministry of Agriculture and Rural Affairs, Beijing 100193, China
4 Dairy Cattle Center, Beijing Shounong Animal Husbandry Development Co., Ltd. (Capital Agribusiness Group), Beijing 100029, China; b20223050447@cau.edu.cn (X.W.); s20243051163@cau.edu.cn (J.W.)
* Correspondence: yunfeima@cau.edu.cn (Y.M.); haozhihui@cau.edu.cn (Z.H.)

Abstract: Background/Objectives: Calf diarrhea is a serious health problem in dairy farming, which seriously affects their production performance in adulthood. Diarrhea in calves is usually treated with antibiotics, which may lead to issues of antibiotic residue and resistance. Therefore, finding antibiotic alternatives is of critical importance. Yigong San (YGS) is a traditional Chinese medicine formula traditionally indicated for inflammatory gastrointestinal disorders. This study aimed to investigate whether YGS can be used as an alternative to antibiotics for the effective treatment of calf diarrhea and the underlying mechanisms. Methods: Ten healthy Holstein calves served as the control (Ctrl) group, while twenty diarrheic calves were randomly assigned to either a natural diarrhea (ND) group or a YGS treatment (YGS) group, which received YGS for seven days. Serum samples were collected post-treatment and analyzed for biochemical parameters, oxidative stress indicators, cytokine levels, and metabolomic profiles. Results: The results showed that YGS effectively alleviated diarrhea and improved abnormal biochemical indicators. YGS significantly increased serum levels of CAT, GSH-PX, and SOD, while reducing MDA levels. In addition, YGS also suppressed the expression of multiple proinflammatory cytokines, including IL-1α, TNF-α, IL-1β, IL-17A, IL-18, IL-21, IFN-γ, as well as chemokines CXCL9 and CXCL10. Metabolomic analysis revealed that YGS treatment significantly improved metabolic disorders and enriched the Arginine and Proline Metabolism pathways. The metabolites 1-methylhydantoin and ornithine were significantly and negatively correlated with pro-inflammatory cytokine levels. Conclusions: YGS effectively treats calf diarrhea by enhancing antioxidant capacity, reducing inflammatory factor levels, regulating immune function, and modulating serum metabolites. It provided valuable insights for the development of safe and effective antibiotic alternatives for preventing and treating calf diarrhea.

Keywords: traditional Chinese medicine; calf diarrhea; therapeutic effect; metabolomics

1. Introduction

Calf diarrhea represents a critical global challenge to livestock health and productivity, imposing substantial economic burdens due to its high morbidity and mortality rates. Data

 https://doi.org/10.3390/metabo15090618

from the United States highlights the severity of this issue: the 2018 National Animal Health Monitoring System (NAHMS) report indicates that diarrhea accounts for 39% of calf deaths within the first three weeks of life [1]. Beyond acute mortality, surviving calves face long-term repercussions, including impaired growth (manifested as reduced pre-weaning average daily gain), delayed reproductive maturity (extending the interval to first calving by 3–6 months), and a 12–15% reduction in milk yield during their first lactation, accompanied by inferior milk components [2]. These persistent effects arise from compromised intestinal barrier function, which disrupts nutrient partitioning, induces epigenetic metabolic reprogramming, and triggers chronic immune dysregulation. Cumulatively, each diarrheic calf incurs an average lifetime economic loss of $1800, encompassing treatment expenses, diminished productivity, and premature culling [3]. Addressing this disease is therefore pivotal for enhancing the sustainability of dairy operations and improving animal welfare.

Factors contributing to calf diarrhea include infectious factors, such as intestinal pathogen infections, and non-infectious factors such as poor feeding management and low immunity [4]. Although antibiotics remain a conventional clinical strategy for the prevention and treatment of calf diarrhea, growing evidence highlights the significant risks they pose: exacerbating the spread of antimicrobial resistance in zoonotic pathogens and contributing to drug residues within the food chain [3]. Consequently, identifying effective antibiotic alternatives capable of improving gut health and controlling diarrhea is imperative for promoting sustainable and healthy livestock production.

Traditional Chinese Medicine (TCM) has emerged as a promising candidate for such an alternative, exhibiting significant anti-inflammatory, antibacterial, and immunomodulatory properties that contribute to effective disease management [5]. TCM extracts have a positive impact on livestock production when used as drugs or feed additives, effectively treating diarrhea or reducing its incidence [6,7]. Oral administration of plant extracts or natural compounds from them can also be effective in treating calf diarrhea [8,9]. Calves with diarrhea may experience inflammation and oxidative stress [10], and metabolic disorders have also been reported to be associated with calf diarrhea. [11].Metabolites from the host and gut are crucial in regulating host energy metabolism and immune responses [11]. According to research reports, calf diarrhea alters the gut metabolite profile, causing growth impairment [12].

Therefore, therapeutic agents that can ameliorate metabolic disturbances while enhancing antioxidant defenses and immune function represent a viable strategy for mitigating diarrhea and improving calf survival. Yigong San (YGS), a renowned traditional Chinese herbal formula composed of *Panax ginseng* C.A.Mey., *Citrus reticulata* Blanco, *Glycyrrhiza uralensis* Fisch. ex DC., *Poria cocos* (Schw.) Wolf and *Atractylodes macrocephala* Koidz., exhibits significant antioxidant and anti-inflammatory properties [13,14]. Key components of YGS have also been reported to confer beneficial effects in livestock, including improved weight gain, enhanced antioxidant capacity, boosted immunity, and reduced disease incidence [15,16]. However, the mechanism of action of YGS remains inadequately elucidated, particularly concerning its impact on the metabolic profile of calves. Hence, this study aimed to evaluate the therapeutic efficacy of YGS in diarrheic calves. We specifically assessed its effects on oxidative stress indicators, inflammatory cytokines, chemokines, and serum metabolite profiles. Our objective was to elucidate the underlying mechanisms by which YGS alleviates calf diarrhea. The findings are anticipated to contribute to the development of effective and safe antibiotic alternatives for the prevention and management of this economically significant disease.

2. Materials and Methods

2.1. Preparation and Chemical Composition Analysis of YGS

All herbs were commercially sourced from Beijing Meikang Hall Pharmaceutical Technology Co., Ltd. (Beijing, China), accompanied by inspection certificates of conformity (Supplemental Information S3). The herbal formulation YGS, whose complete composition is detailed in Table 1, was used for subsequent experiments. The dried herbs underwent two 2.5-h boiling extractions, each using a 12-fold volume of water (v/w). Concentrate the mixed filtrate under reduced pressure to 1 g/mL and store at 4 °C for subsequent use. Qualitative analysis of the active ingredients in YGS was conducted based on previous research [17], with the detailed mass spectrometry conditions provided in the Supplementary Information S1.

Table 1. The composition of YGS.

Chinese Name	Latin Name	Batch Number	Ratio
Dang Shen	*Codonopsis pilosula* (Franch.) Nannf.	220925002	1
Bai Zhu	*Atractylodes macrocephala* Koidz.	220911004	1
Fu Ling	*Poria cocos* (Schw.) Wolf	221221002	1
Chen Pi	*Citrus reticulata* Blanco	230615001	1
Gan Cao	*Glycyrrhiza uralensis* Fisch. ex DC.	230627001	1

2.2. Animals

Twenty diarrheic Holstein calves and ten healthy calves (5–7 days old, 45–55 kg) were selected from Shounong Farm (Beijing, China). Diarrheic calves met the inclusion criteria: (1) fecal score > 2 and (2) age 1–30 days. The twenty diarrheic calves finally selected were all from the 5–7 days of age segment, matching the age of the healthy calves. All diarrheal calves were screened using antigen test cards for common infectious pathogens, including Bovine Coronavirus, Rotavirus, Cryptosporidium, Giardia, *E. coli* K99, and Bovine Viral Diarrhea Virus. Only calves that tested negative for all these infectious agents were included in the study. Calves were raised individually in calf pens. Overall, feeding management followed standard operating procedures for commercial farms. All experimental protocols were reviewed and approved by the Animal Ethics Committee of China Agricultural University (approval number: AW31605202-2-01).

2.3. Drug Administration

Twenty diarrheic calves were randomly allocated into two groups: a natural diarrhea (ND) group and a YGS-treated (YGS) group, with ten calves each. Details regarding random grouping procedures were provided in Supplementary Information S1. The clinical routine dosage of YGS for humans is 75 g/d [14], and the administration dose for calves, converted based on body surface area, is 0.915 g/kg. Detailed calculation steps were provided in Supplementary Information S1. Ten healthy calves serving as the control (Ctrl) group and the ND group received daily oral gavage with equivalent volumes of water. The calf did not receive any other medication during the YGS treatment. Disease remission was defined as a fecal score ≤ 1 persisting for two consecutive days. Detailed supportive care protocols and resuscitation criteria are provided in Supplementary Information S1. In short, calves received fluid and electrolyte support, nutritional support, and environmental and nursing care. After the experiment, collect blood samples from each group. During sampling, calves are kept in a standing position to minimize stress caused by the procedure. Blood was collected into plain tubes (without anticoagulant). Following centrifugation at 3200 rpm for 6 min, the resulting serum was aliquoted and stored at −20 °C for subsequent analysis.

2.4. Therapeutic Efficacy Observation

During the experimental period (Day 0–Day 7), as shown in Table S1, the calves' mental status, dehydration status, and fecal consistency were observed and recorded in accordance with a previous study [18]. Therapeutic efficacy was subsequently calculated.

2.5. Analysis of Serum Biochemical Indices

Serum levels of aspartate aminotransferase (AST), globulin (GLB), alkaline phosphatase (ALP), lactate dehydrogenase (LDH), total protein (TP), alanine aminotransferase (ALT), creatine kinase (CK), total bile acids (TBA), total cholesterol (CHOL), uric acid (UA), amylase (AMY), phosphorus (P), iron (Fe), chloride (Cl), potassium (K), calcium (Ca), and magnesium (Mg) were measured using a Roche Cobas 6000 c501 automated clinical chemistry analyzer.

2.6. Oxidative Stress Indicator Detection

According to the manufacturer's instructions, we analyzed serum levels of catalase (CAT), malondialdehyde (MDA), glutathione peroxidase (GSH-PX), and superoxide dismutase (SOD) using commercial kits (Nanjing Jiancheng Bioengineering Institute, Nanjing, China).

2.7. Cytokine Detection

The serum levels of IFN-α, IFN-γ, IL-13, IL-1α, IL-1F5, IL-21, CXCL10, CXCL9, CCL4, and TNF-α in calf serum were quantified using enzyme-linked immunosorbent assays (ELISA) (Catalog #QAB-CYT-1-2, RayBiotech, Norcross, GA, USA). Levels of ANG-1, Decorin, IL-1β, IL-17A, LIF, CD40 Ligand, IFN-β, IL-10, IL-18, and CCL5 were quantified using ELISA (Catalog #QAB-CYT-3-2, RayBiotech, Norcross, GA, USA). All steps were performed strictly according to the manufacturer's protocols.

2.8. Preparation of Serum Samples for Metabolomic Analysis

To evaluate system stability, a quality control (QC) sample was prepared by combining equal volumes of serum from each of the three groups. For metabolite extraction, 200 μL of serum was combined with 800 μL of ice-cold methanol-acetonitrile (1:1, *v*/*v*) and vortex-mixed. The mixture was then sonicated for 1 h in an ice-water bath, incubated at −20 °C for 1.5 h, and finally centrifuged at 13,000× *g* for 25 min at 4 °C.

2.9. Metabolomics Analysis of Serum Samples

Serum metabolomic profiling was conducted using ultra-performance liquid chromatography coupled with quadrupole-Orbitrap mass spectrometry (UPLC-Q-Orbitrap MS). The system integrated a Nexera X2 LC-30AD UHPLC (Shimadzu, Kyoto, Japan) with a Q-Exactive Plus mass spectrometer (Thermo Scientific, San Jose, CA, USA). Chromatographic separation was performed on a HSS T3 column (2.1 × 100 mm, 1.8 μm; Waters, Milford, MA, USA) with a gradient elution program. Mass spectrometric detection was carried out in both positive and negative ionization modes with a heated electrospray ionization (HESI) source. Full MS and data-dependent MS^2 scans were acquired at resolutions of 70,000 and 17,500, respectively.

2.10. Data Processing and Statistical Analysis

Raw data were processed using MS-DIAL for peak detection, alignment, and metabolite identification based on accurate mass and MS/MS spectra matching against HMDB, MassBank, and an in-house database. Features detected in more than 50% of samples in at least one group were retained. Multivariate analyses were performed in R. Metabolites with VIP > 1.0 and $p < 0.05$ were considered significant. Pathway enrichment analysis was performed using KEGG. Detailed experimental procedures and data processing workflows are provided in the Supplementary Information S1.

Data distribution normality was evaluated using GraphPad Prism 10 and SPSS (version 26.0). Data are presented as mean ± standard deviation (SD). One-way analysis of variance (ANOVA) was employed if the data were normally distributed with homogeneous variance, and a non-parametric test was used if not. Correlations were analysed using Spearman's method via chiplot.online. Significance was set at $p < 0.05$.

3. Results

3.1. Identification of the Main Chemical Components in YGS by UPLC-MS/MS

The main active components in YGS were identified using the UPLC-MS/MS analytical method (Figure 1). A total of 147 primary compounds were identified (Table S2), among which the most abundant categories include flavonoids and their glycosides, phenolic acids and their derivatives, amino acids and their derivatives, terpenoids, organic acids, and alkaloids.

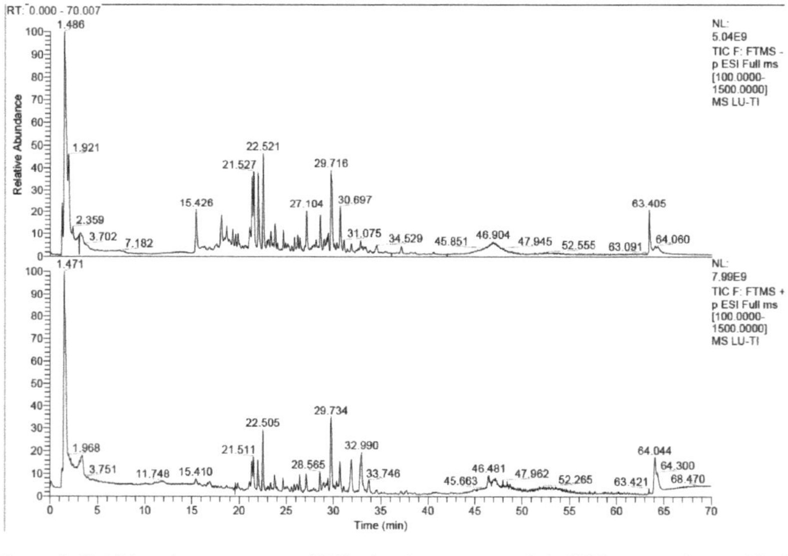

Figure 1. Total ion chromatograms (TIC) of active compounds in YGS extract detected by UPLC-MS/MS. The upper panel shows the TIC in negative ion mode, and the lower panel shows the TIC in positive ion mode.

3.2. The Therapeutic Effects of YGS on Diarrheal Calves

As shown in Table 2, diarrheic calves exhibited significantly higher fecal status, mental status and dehydration scores compared to the Ctrl group ($p < 0.05$). Following YGS treatment, fecal status, mental status, and dehydration scores showed a decreasing trend on Day 7, ultimately showing no significant difference compared to the Ctrl group ($p > 0.05$). The clinical efficacy rate of YGS treatment, defined as the ratio of calves cured within 7 days to the total number of diarrheic calves treated, was 100%. We present the daily trajectories of fecal status, mental status, and dehydration scores for all 30 calves in the Table S3. We characterized the cumulative remission rate over the observation period. Results showed the YGS group achieved a 100% cumulative remission rate by D7 (Figure S1). The median time to remission was 6.5 days, with a 95% confidence interval (95% CI) of 6.13–6.88 days, confirming consistent and timely therapeutic effects. These results indicate that YGS was associated with improved clinical scores compared with ND under the conditions tested.

Table 2. Effects of YGS treatment on clinical symptom scores of diarrheic calves (Day 0 to Day 7).

Items	Treatment	Group			p-Value
		Ctrl	ND	YGS	
Fecal scores					
	D0	0.10 ± 0.32 [b]	3.70 ± 0.48 [a]	3.50 ± 0.53 [a]	<0.001
	D1	0.00 ± 0.00 [b]	3.60 ± 0.52 [a]	3.50 ± 0.53 [a]	<0.001
	D2	0.10 ± 0.32 [b]	3.60 ± 0.52 [a]	3.20 ± 0.42 [a]	<0.001
	D3	0.00 ± 0.00 [b]	3.00 ± 0.00 [a]	2.70 ± 0.48 [a]	<0.001
	D4	0.20 ± 0.42 [b]	3.00 ± 0.82 [a]	2.20 ± 0.63 [a]	<0.001
	D5	0.00 ± 0.00 [b]	2.60 ± 0.52 [a]	1.50 ± 0.53 [a]	<0.001
	D6	0.10 ± 0.32 [b]	2.50 ± 0.53 [a]	0.20 ± 0.42 [b]	<0.001
	D7	0.10 ± 0.32 [b]	2.60 ± 0.52 [a]	0.30 ± 0.48 [b]	<0.001
Dehydration scores					
	D0	0.00 ± 0.00 [b]	3.60 ± 0.52 [a]	3.60 ± 0.52 [a]	<0.001
	D1	0.00 ± 0.00 [b]	3.40 ± 0.52 [a]	3.00 ± 0.47 [a]	<0.001
	D2	0.00 ± 0.00 [b]	3.40 ± 0.52 [a]	2.70 ± 0.48 [a]	<0.001
	D3	0.00 ± 0.00 [c]	3.50 ± 0.53 [a]	2.20 ± 0.63 [b]	<0.001
	D4	0.00 ± 0.00 [c]	2.80 ± 0.42 [a]	1.60 ± 0.52 [b]	<0.001
	D5	0.00 ± 0.00 [b]	2.40 ± 0.52 [a]	0.90 ± 0.57 [b]	<0.001
	D6	0.00 ± 0.00 [b]	2.60 ± 0.52 [a]	0.20 ± 0.42 [b]	<0.001
	D7	0.00 ± 0.00 [b]	2.70 ± 0.67 [a]	0.00 ± 0.00 [b]	<0.001
Mental scores					
	D0	0.00 ± 0.00 [b]	3.50 ± 0.53 [a]	3.60 ± 0.52 [a]	<0.001
	D1	0.30 ± 0.48 [b]	3.50 ± 0.53 [a]	2.90 ± 0.32 [a]	<0.001
	D2	0.10 ± 0.32 [b]	3.70 ± 0.48 [a]	2.60 ± 0.52 [a]	<0.001
	D3	0.30 ± 0.48 [b]	3.00 ± 0.00 [a]	2.00 ± 0.47 [a]	<0.001
	D4	0.10 ± 0.32 [b]	3.00 ± 0.00 [a]	1.60 ± 0.52 [a]	<0.001
	D5	0.00 ± 0.00 [b]	2.40 ± 0.52 [a]	0.70 ± 0.48 [b]	<0.001
	D6	0.00 ± 0.00 [b]	2.40 ± 0.52 [a]	0.20 ± 0.42 [b]	<0.001
	D7	0.10 ± 0.32 [b]	2.50 ± 0.53 [a]	0.00 ± 0.00 [b]	<0.001

[a–c] Values in a row with no common letters differ significantly ($p < 0.05$) with mean ± SD. Ctrl: healthy control group; ND: natural diarrhea group; YGS: YGS treatment group.

3.3. The Effects of YGS on the Biochemical Blood Indexes

We further investigated the effect of YGS administration on blood biochemical indices of diarrhea calves (Table 3). Compared to the Ctrl group, diarrheic calves exhibited significantly lower ($p < 0.05$) serum levels of ALT, ALP, and CHOL, Ca and Mg, while levels of TP, GLB, CK, AMY and UREA were significantly higher ($p < 0.05$). YGS treatment reversed previously altered indices to baseline levels ($p < 0.05$). Furthermore, no significant alterations were found in the concentrations of AST, LDH, TBA, UA, P, Cl, K, or Fe, despite the occurrence of sporadic variations ($p > 0.05$).

Table 3. Effects of YGS treatment on serum biochemical indicators of diarrheic calves.

Items	Group			p-Value
	Ctrl	ND	YGS	
ALT (U/L)	11.16 ± 3.17 [a]	6.30 ± 0.55 [b]	8.49 ± 1.58 [a]	<0.001
AST (U/L)	46.34 ± 6.65 [a]	50.52 ± 6.46 [a]	47.18 ± 5.62 [a]	0.341

Table 3. *Cont.*

Items	Group			p-Value
	Ctrl	ND	YGS	
TP (g/dL)	5.40 ± 0.56 [b]	6.65 ± 0.61 [a]	5.21 ± 0.65 [b]	<0.001
GLB (g/dL)	2.54 ± 0.50 [b]	3.97 ± 0.48 [a]	2.34 ± 0.57 [b]	<0.001
ALP (U/L)	446.43 ± 187.82 [a]	230.13 ± 65.38 [b]	512.96 ± 198.97 [a]	0.001
LDH (U/L)	779.71 ± 76.95 [a]	744.27 ± 90.07 [a]	768.66 ± 64.00 [a]	0.381
CK (U/L)	75.70 ± 19.62 [c]	188.13 ± 47.30 [a]	127.27 ± 36.79 [b]	<0.001
TBA (μmol/L)	19.43 ± 6.39 [a]	17.76 ± 9.05 [a]	18.78 ± 3.66 [a]	0.869
UREA (mg/dL)	10.44 ± 2.12 [b]	19.16 ± 5.29 [a]	13.64 ± 1.95 [b]	<0.001
UA (mg/dL)	1.06 ± 0.21 [a]	1.03 ± 0.19 [a]	0.89 ± 0.10 [a]	0.129
CHOL (mg/dL)	83.09 ± 4.31 [b]	42.26 ± 12.02 [c]	105.29 ± 19.47 [a]	<0.001
AMY (U/L)	21.20 ± 6.74 [b]	38.06 ± 3.69 [a]	24.50 ± 5.01 [b]	<0.001
P (mmol/L)	10.86 ± 1.30 [a]	10.78 ± 1.09 [a]	11.33 ± 1.15 [a]	0.573
CL (mmol/L)	94.93 ± 2.01 [a]	97.76 ± 3.81 [a]	96.90 ± 1.47 [a]	0.085
K (mmol/L)	6.22 ± 0.80 [a]	6.20 ± 0.34 [a]	6.39 ± 0.52 [a]	0.744
Fe (μg/dL)	140.98 ± 73.35 [a]	157.00 ± 110.06 [a]	172.13 ± 100.02 [a]	0.862
Ca (mg/dL)	11.30 ± 0.77 [a]	10.01 ± 0.27 [b]	10.82 ± 0.31 [a]	0.003
Mg (mmol/L)	1.05 ± 0.10 [a]	0.84 ± 0.05 [b]	0.97 ± 0.03 [b]	0.009

[a–c] Values in a row with no common letters differ significantly ($p < 0.05$) with mean ± SD. alanine aminotransferase (ALT), aspartate aminotransferase (AST), total protein (TP), globulin (GLB), alkaline phosphatase (ALP), lactate dehydrogenase (LDH), creatine kinase (CK), total bile acids (TBA), UREA, uric acid (UA), total cholesterol (CHOL), amylase (AMY), calcium (Ca), phosphorus (P), iron (Fe), magnesium (Mg), potassium (K) and chloride (Cl). Ctrl: healthy control group; ND: natural diarrhea group; YGS: YGS treatment group.

3.4. The Effects of YGS on the Oxidative Stress Indicators

Serum levels of oxidative stress indicators in calves are closely associated with intestinal health [19]. Subsequently, we evaluated the effect of YGS treatment on serum oxidative stress indicators in diarrheic calves. The results revealed that serum levels of CAT, SOD and GSH-PX were significantly lower in diarrheic calves compared to healthy controls ($p < 0.05$), YGS treatment significantly restored the serum levels of these antioxidant enzymes ($p < 0.05$) (Figure 2A–C). Moreover, the administration of YGS significantly reduced the increase in serum MDA levels induced by calf diarrhea. ($p < 0.05$) (Figure 2D). This indicates that YGS administration significantly improves oxidative stress levels in calves with diarrhea.

3.5. The Effects of YGS on the Cytokine Levels

To further investigate the effects of YGS on inflammatory responses in diarrheic calves, we comprehensively analysed serum inflammatory mediators. Compared to the Ctrl group, diarrheic calves exhibited significantly elevated levels of the pro-inflammatory cytokines IL-1α, TNF-α, IL-1β, IL-17A, IL-18, IL-21, LIF, and IFN-γ ($p < 0.05$), indicating a severe systemic inflammatory response. YGS treatment significantly reduced the levels of these cytokines ($p < 0.05$) (Figure 2E–H and 3A–D). Levels of CD40 ligand and IFN-α were increased in diarrheic calves and decreased after YGS treatment, although these changes were non-significant (Figure 3E,F). Concurrently, the chemokines CXCL9, CXCL10, and CCL5 were significantly upregulated ($p < 0.05$) in diarrheic calves. Following YGS intervention, CXCL9 and CXCL10 levels decreased significantly ($p < 0.05$) compared to the ND group (Figure 4A–C), while CCL5 levels decreased but the change was non-significant. Notably, levels of the anti-inflammatory cytokines IL-10, IL-13, and IL-1F5 were significantly reduced ($p < 0.05$) in diarrheic calves. YGS administration increased their levels, but the results were non-significant (Figure 4D–F).

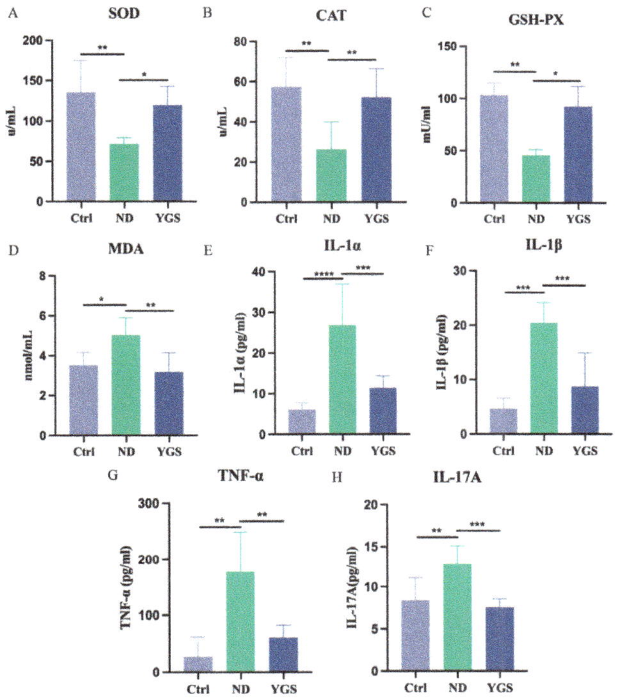

Figure 2. Effects of YGS treatment on serum oxidative stress indicators and pro-inflammatory cytokines in diarrheic calves. (**A**) Superoxide dismutase (SOD), (**B**) Catalase (CAT), (**C**) Glutathione Peroxidase (GSH-PX), (**D**) Malondialdehyde (MDA), (**E**) Interleukin-1α (IL-1α), (**F**) Interleukin-1β (IL-1β), (**G**) Tumor necrosis factor-α (TNF-α), (**H**) Interleukin-17A (IL-17A). Ctrl: healthy control group; ND: natural diarrhea group; YGS: YGS treatment group. The data is presented as mean ± SD (n = 7). * $p < 0.05$, ** $p < 0.01$, *** $p < 0.001$, **** $p < 0.0001$.

3.6. The Effects of YGS on Metabolomics

To further investigate the changes in serum metabolites of diarrheic calves following YGS administration, we conducted serum metabolomic profiling. QC were employed to assess sample reproducibility and analytical system stability throughout the experimental process. Following normalization, principal component analysis (PCA) was performed on peaks extracted from all experimental and QC. QC clustered tightly, indicating excellent experimental reproducibility (Figure 5A). QC were further subjected to Pearson correlation analysis. Correlation coefficients all exceeding 0.9 confirmed good correlation (Figure 5B). Following rigorous quality assessment, the data were analyzed further.

PCA, partial least squares-discriminant analysis (PLS-DA), and Orthogonal PLS-DA (OPLS-DA) were employed to assess inter-group metabolic differences. Pronounced clustering of diarrheic calves versus controls in score plots (Figure 5C–J) evidenced significant endogenous metabolite reorganization during calf diarrhea.

Notably, significant metabolic changes were observed following YGS group compared to the ND group. The model parameters obtained through 200 permutation tests confirmed the high reliability and accuracy of the two comparative analyses. The comparison model between the ND group and the Ctrl group performed exceptionally well [$Q^2 = 0.876$; $R^2X = 0.265$; $R^2Y = 0.993$; CV-ANOVA $p = 0.005$], while the comparison model between the YGS group and the ND group demonstrated strong reliability and predictive capability [$Q^2 = 0.95$; $R^2X = 0.409$; $R^2Y = 0.999$; CV-ANOVA $p = 0.005$].

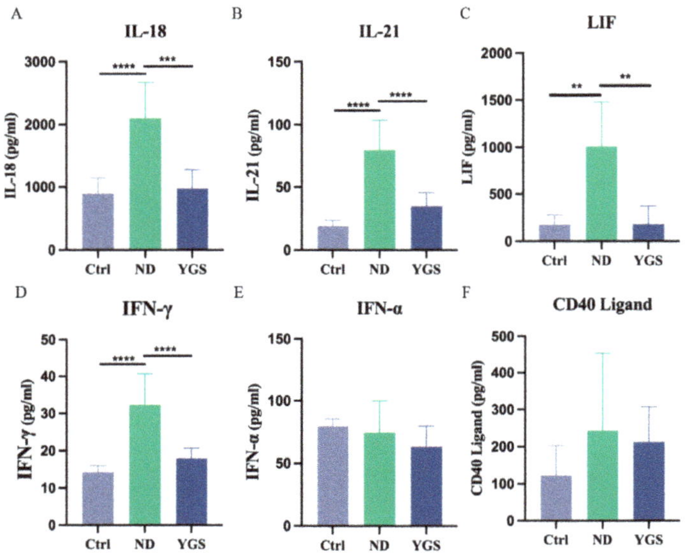

Figure 3. Effects of YGS treatment on serum pro-inflammatory cytokines in diarrheic calves. (**A**) Interleukin-18 (IL-18), (**B**) Interleukin-21 (IL-21), (**C**) Leukemia Inhibitory Factor (LIF), (**D**) Interferon-γ (IFN-γ), (**E**) Interferon-α (IFN-α), (**F**) CD40 Ligand. Ctrl: healthy control group; ND: natural diarrhea group; YGS: YGS treatment group. The data is presented as mean ± SD (n = 7). ** $p < 0.01$, *** $p < 0.001$, **** $p < 0.0001$.

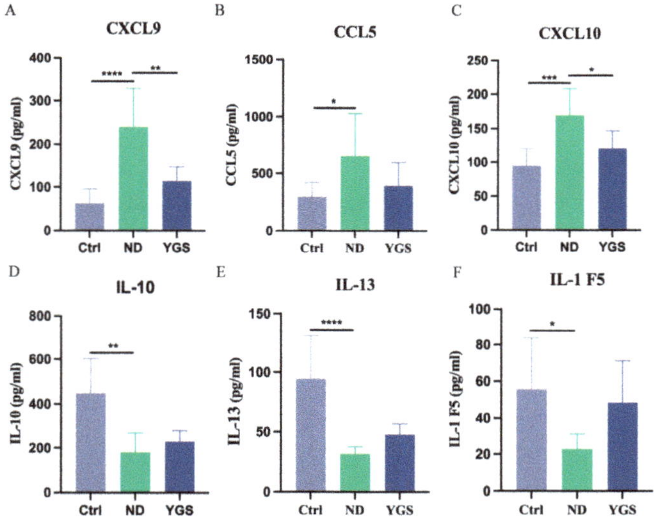

Figure 4. Effects of YGS treatment on serum chemokines and anti-inflammatory cytokines in diarrheic calves. (**A**) C-X-C Motif Chemokine Ligand 9 (CXCL9), (**B**) C-C Motif Chemokine Ligand 5 (CCL5), (**C**) C-X-C Motif Chemokine Ligand 10 (CXCL10), (**D**) Interleukin-10 (IL-10), (**E**) Interleukin-13 (IL-13), (**F**) Interleukin-1 Family Member 5 (IL-1F5). ND: natural diarrhea group; YGS: YGS treatment group. The data is presented as mean ± SD (n = 7). * $p < 0.05$, ** $p < 0.01$, *** $p < 0.001$, **** $p < 0.0001$.

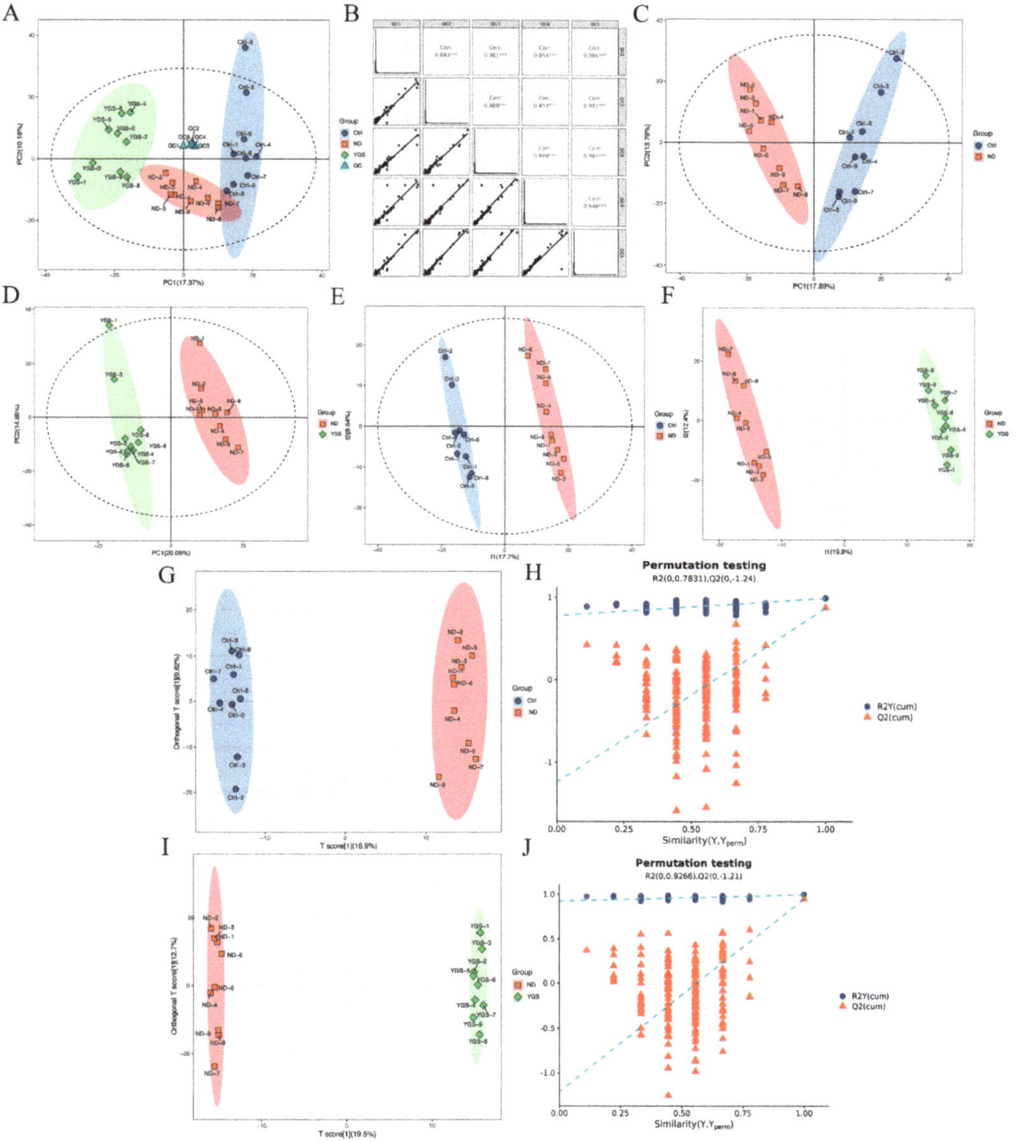

Figure 5. Evaluation of serum metabolomic profiling differences in diarrheic calves before and after YGS treatment. (**A**) Principal Component Analysis (PCA) of experimental samples (Ctrl, ND, YGS groups) and quality control (QC) samples. (**B**) Pearson correlation analysis of QC samples. (**C**) PCA score plot of the ND group vs. Ctrl group; (**D**) PCA score plot of the YGS group vs. ND group. (**E**) Partial Least Squares-Discriminant Analysis (PLS-DA) score plot of the ND group vs. the Ctrl group; (**F**) PLS-DA score plot of the YGS group vs. the ND group. (**G**,**H**) Orthogonal PLS-DA (OPLS-DA) score plot of the ND group vs. the Ctrl group; (**I**,**J**) OPLS-DA score plot of the YGS group vs. the ND group. *** $p < 0.001$.

Compared to the Ctrl group, 164 metabolites were significantly up-regulated and 196 were down-regulated in the ND group. Compared to the ND group, the YGS treatment group exhibited 192 upregulated metabolites and 277 downregulated metabolites (Figure 6A,B). Detailed information on differential metabolites was available in

Supplementary Information S2. The top 50 significantly differential metabolites were identified as potential biomarkers distinguishing the ND and YGS groups and are displayed in the heatmap in Figure 6C. Most of the significantly altered metabolites could be categorized into lipids, steroids and derivatives, organic acids and derivatives, alkaloids and derivatives, and phenolic compounds, among others. To further investigate the metabolic pathways altered in diarrheic calves following YGS administration, enrichment analysis was performed on the differentially expressed metabolites. Compared to the ND group, YGS administration induced significant alterations in multiple metabolic pathways (Figure 6D), including Biosynthesis of amino acids, Arginine biosynthesis, Arginine and proline metabolism, Pyrimidine metabolism, and beta−Alanine metabolism. Furthermore, correlation analysis revealed that upregulated metabolites in the arginine biosynthesis and arginine-proline metabolism pathways were negatively correlated with pro-inflammatory cytokines (including IL-1α, TNF-α, IL-1β, IL-17A, IL-18, IL-21 and IFN-γ) and MDA, and positively correlated with antioxidant indicators (CAT, SOD and GSH-PX). Notably, the strongest correlations were observed for the metabolites 1-methylhydantoin, ornithine, and citrulline (Figure 7).

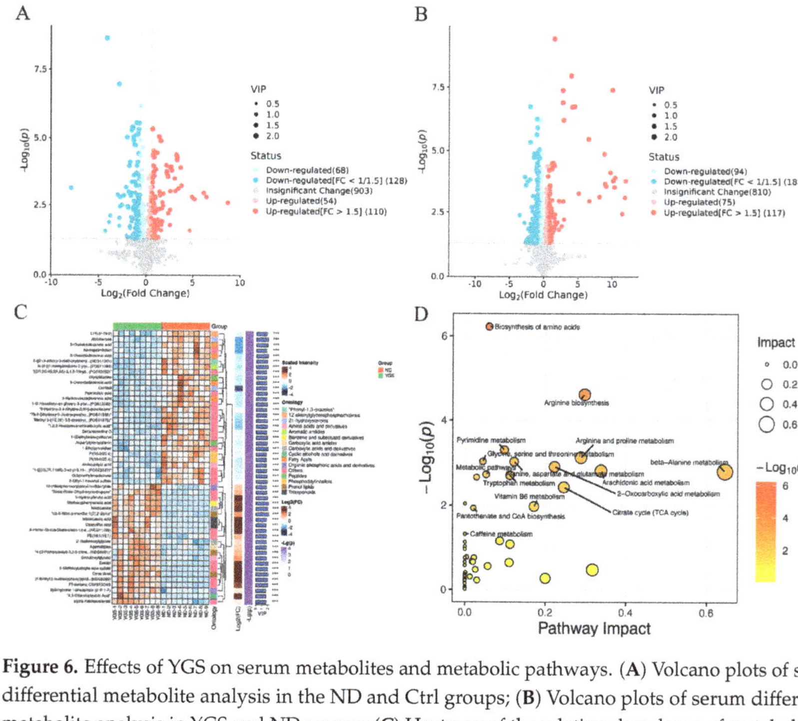

Figure 6. Effects of YGS on serum metabolites and metabolic pathways. (**A**) Volcano plots of serum differential metabolite analysis in the ND and Ctrl groups; (**B**) Volcano plots of serum differential metabolite analysis in YGS and ND groups; (**C**) Heatmap of the relative abundance of metabolites in the top 50 with significantly differences between the YGS and ND groups (VIP > 1.0, and $p < 0.05$ with 95 % confidence intervals); (**D**) The metabolic pathway impact prediction between the YGS and ND groups based on the KEGG. *** $p < 0.001$.

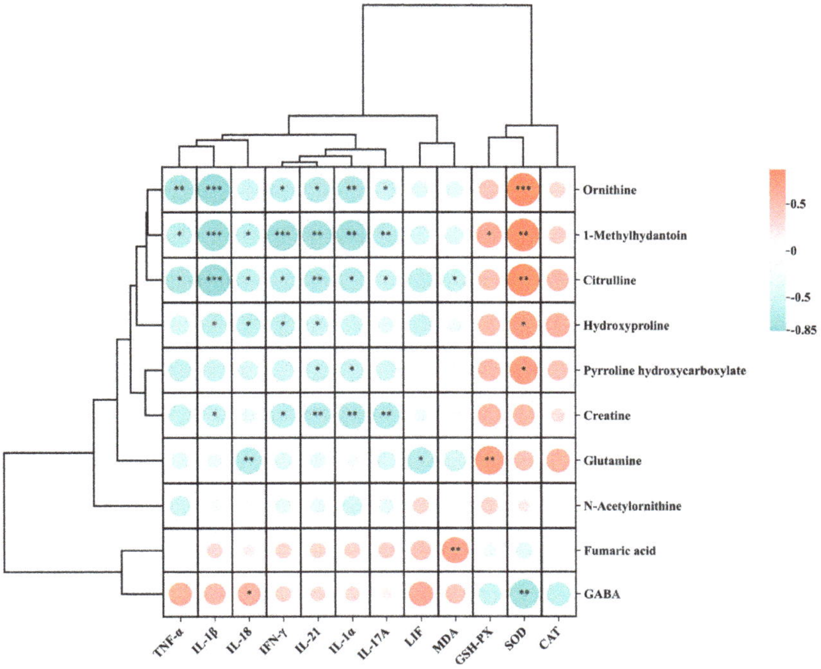

Figure 7. Correlation analysis between differential metabolites in Arginine and proline metabolism and Arginine biosynthesis pathways with serum oxidative stress indicators and cytokines. * $p < 0.05$, ** $p < 0.01$, *** $p < 0.001$.

4. Discussion

Antimicrobial agents have long been the primary treatment for calf diarrhea. However, the long-term use of antibiotics contributes to antimicrobial resistance and residue concerns, driving the search for alternative therapies. This study investigated the therapeutic effects of YGS on diarrheic calves. Our findings demonstrate that YGS administration significantly ameliorated diarrhea, improved abnormal biochemical parameters, enhanced antioxidant and anti-inflammatory capacity, and alleviated metabolic disturbances in diarrheic calves.

YGS treatment effectively reduced fecal scores, dehydration scores, and mental status scores in diarrheic calves, indicating successful therapeutic intervention. Serum biochemical parameters serve as crucial indicators of animal health status. Furthermore, diarrheic calves exhibited elevated serum GLB levels. Given the close association between GLB and immune function [20], the normalization of GLB levels following YGS treatment suggests its efficacy in modulating the immune response in diarrheic calves.

Antioxidant capacity serves as a critical indicator of animal health status. It has been reported that diarrheic calves exhibit oxidative stress [21]. Consistent with this, the present study observed that the levels of SOD, CAT and GSH-PX were significantly decreased, while the level of MDA was significantly increased in diarrheic calves. Antioxidant capacity can be directly assessed by the activities of SOD, CAT and GSH-PX. SOD acts as the primary scavenger of oxygen free radicals, while CAT catalyzes the decomposition of hydrogen peroxide into water and oxygen, thereby mitigating cellular oxidative damage [22]. The observed decline likely reflects depletion of these antioxidant enzymes in the bloodstream, as the organism attempts to counteract excessive free radical production and restore redox homeostasis during oxidative stress [23]. MDA, a biomarker of oxidative stress, is produced through free radical-induced lipid peroxidation and serves as an indicator of biomembrane peroxidative damage [24]. Administration of YGS significantly elevated SOD, CAT, and GSH-PX levels, consequently enhancing antioxidant activity. This synergistic enhancement of the primary antioxidant enzyme cascade ensures the efficient and complete neutralization of reactive oxygen species (ROS). Consequently, the diminished substrate for lipid peroxidation reactions directly leads to the decreased production of MDA, a reliable end-product biomarker of such damage. This finding aligns with previous research demonstrating that the principal components of YGS possess notable antioxidant properties [25–29].

Beyond biochemical abnormalities and oxidative stress, diarrhea triggers inflammatory responses in calves. Cytokines critically orchestrate inflammatory responses and gut barrier homeostasis [30]. Previous studies have reported that serum levels of pro-inflammatory cytokines were significantly higher in calves with E. coli-induced diarrhea compared with those in healthy controls [31]. Our results also demonstrate that YGS significantly suppressed the expression of pro-inflammatory cytokines (IL-1α, IL-1β, TNF-α, IL-17A, IL-18, IL-21, and IFN-γ), effectively reducing inflammation and alleviating diarrhea. These cytokines are key mediators in immune and inflammatory responses. Consistent with our findings, YGS has been shown to inhibit macrophage inflammatory factors and modulate NF-κB and caspase-1 activation, contributing to its anti-inflammatory effects [13].

Chemokines, a family of >50 structurally conserved 8–10 kDa chemoattractant proteins, orchestrate leukocyte trafficking to direct inflammatory and immune responses through targeted recruitment [32]. The CC-chemokine CCL5 (RANTES) drives leukocyte migration, and its pathological elevation is implicated in chronic inflammation and associated disorders [33]. IFN-γ-inducible chemokines (CXCL9 and CXCL10) are crucial for immune cell maturation and migration, modulating inflammation by inducing, sustaining, and amplifying inflammatory/immune responses [34]. Our study confirmed significantly elevated levels of CXCL9, CXCL10, and CCL5 in diarrheic calves, further validating severe inflammation. YGS treatment effectively reduced CXCL9 and CXCL10 levels, suggesting its mechanism may involve modulation of the IFN signaling pathway. Concurrently, we detected downregulation of IL-10, IL-13, and IL-1F5 in diarrheic calves, indicating an imbalance between TH1 and TH2 cells and suppressed secretion of anti-inflammatory factors by TH2 cells, potentially compromising humoral immunity [35]. While YGS showed a tendency to restore these anti-inflammatory cytokine levels, the effect was not statistically significant, suggesting its primary anti-diarrheal action may be through suppressing pro-inflammatory mediators rather than potently elevating anti-inflammatory ones.

Liquid chromatography-mass spectrometry (LC-MS) enables the identification of specific metabolites and prediction of disease phenotype-associated metabolic pathways, facilitating the elucidation of mechanistic links through integrated analysis of the metabolome and host phenotype [36]. In this study, serum metabolomic profiling using UPLC-Q-Exactive Orbitrap-MS was performed to investigate the mechanism by which YGS alleviates calf diarrhea. The results revealed that YGS significantly reduced serum levels of Aldosterone and Aminoadipic acid, metabolites known to promote inflammation and oxidative stress. Aldosterone enhances the expression of inflammatory mediators and chemokine receptors, inducing inflammation and oxidative stress [37]. Aminoadipic acid increases serum inflammatory cytokines and hepatic malondialdehyde levels while altering SOD activity, promoting inflammation via the ROS/TXNIP/NLRP3 signaling pathway [38]. Importantly, metabolic pathway enrichment analysis of differential metabolites indicated that YGS alleviates diarrhea potentially by modulating pathways related to Biosynthesis of amino acids, Arginine biosynthesis, Arginine and proline metabolism, and beta−Alanine metabolism. Amino acid metabolism underpins not only protein synthesis but also controls immune cell function, regulates T cell fate, supports metabolic reprogramming (e.g., glycolysis, mitochondrial metabolism) [39]. Dysfunctional amino acid metabolism is a key driver of inflammatory bowel disease (IBD) progression. Amino acid repletion attenuates IBD-related inflammation and oxidative stress, whereas deficiency exacerbates colitic pathology [39]. Previous studies confirm that disturbances in Arginine and Proline Metabolism can trigger inflammatory responses [40]. Correlation analysis revealed statistically significant negative correlations between the levels of two metabolites in this pathway—1-Methylhydantoin and Ornithine—and the expression of inflammatory cytokines. Both compounds display significant anti-inflammatory properties with concurrent augmentation of antioxidant defenses [41,42]. Therefore, we propose that the anti-diarrheal effect of YGS may involve modulation of the Arginine and Proline Metabolism pathway, elevating levels of 1-Methylhydantoin and Ornithine, which merits further investigation. However, our study still suffers from the problem of small sample size, and the next step will be to further expand the number of test animals to further determine the efficacy of YGS.

5. Conclusions

YGS demonstrates significant therapeutic efficacy in the treatment of diarrhea calves. Administration of YGS effectively alleviated clinical symptoms, normalized serum biochemical parameters, enhanced antioxidant capacity by elevating levels of SOD, CAT, and GSH-PX while reducing MDA, and suppressed the expression of key pro-inflammatory cytokines and chemokines. Metabolomic analysis revealed that YGS treatment notably modulated metabolic disorders, particularly enriching the Arginine and Proline Metabolism pathway. Key metabolites such as 1-methylhydantoin and ornithine were significantly correlated with reduced inflammation and improved antioxidant status. These results suggest that YGS ameliorates calf diarrhea through mechanisms involving antioxidant, anti-inflammatory, and metabolic regulatory actions. This study supports the potential of YGS as a safe and effective antibiotic alternative for managing calf diarrhea, contributing to sustainable dairy farming practices. Further research with larger sample sizes is warranted to validate these findings and explore underlying molecular mechanisms.

Supplementary Materials: The following supporting information can be downloaded at https://www.mdpi.com/article/10.3390/metabo15090618/s1, Table S1. The scoring criteria of clinical examination [1]; Table S2. Biologically active components in YGS extract identified by UPLC-MS/MS. Table S3. Daily clinical symptom trajectories of individual calves in the Ctrl group, ND group and YGS group. Figure S1: Kaplan–Meier survival curves for clinical remission of diarrhea in calves from the YGS group and ND group. Statistical comparison between the two groups was performed using the log-rank test ($\chi^2 = 18.81$, $p < 0.0001$), indicating an extremely significant difference in remission dynamics between the YGS group and ND group; Supplementary Information S1: Supplementary experimental methods and results; Supplementary Information S2: Differences in metabolites between the ND group and the Ctrl group; Supplementary Information S3: Inspection certificates for component herbs of YGS.

Author Contributions: Z.H. and Y.M. designed and directed the project. S.H., X.Z. and C.H. planned experiments. S.H., C.H., X.Z., X.N., X.W. and J.W. carried out the experiments. J.L. and S.H. drafted the manuscript. All authors helped with the research and analysis. J.L., C.H., Z.H. and Y.M. critically revised the manuscript and provided supervision. All authors contributed to the writing of manuscript and approved the final version. All authors have read and agreed to the published version of the manuscript.

Funding: This study was supported by the Major Special Project of the Autonomous Region-Xinjiang Livestock and Poultry Disease Prevention and Control System Quality Improvement Project (2023A02007), The 2115 Talent Development Program of China Agricultural University (No. 00109015), the State Key Research and Development Plan, China (No. 2022YFD1801105); the

China Agriculture Research System of MOF and MARA(CARS-37) and the Shandong Province Agricultural Major Technology Collaborative Promotion Plan (No. SDNYXTTG-2024-06).

Institutional Review Board Statement: All experimental protocols were reviewed and approved by the Animal Ethics Committee of China Agricultural University (approval number: AW31605202-2-01, approved 13 June 2025).

Informed Consent Statement: Not appliable.

Data Availability Statement: The original contributions presented in this study are included in the article/supplementary material. Further inquiries can be directed to the corresponding authors.

Acknowledgments: We acknowledge Shanghai Bioprofile for technical assistance in metabolomics.

Conflicts of Interest: Author Xin Wang and Jianfei Wang are employed by the company Beijing Shounong Animal Husbandry Development Co. The remaining authors declare that the research was conducted in the absence of any commercial or financial relationships that could be construed as a potential conflict of interest.

Abbreviations

The following abbreviations are used in this manuscript:

ALP	alkaline phosphatase
ALT	alanine aminotransferase
AMY	amylase
AST	aspartate aminotransferase
Ca	calcium
CAT	catalase
CHOL	total cholesterol
CK	creatine kinase
Cl	chloride
ELISA	enzyme-linked immunosorbent assays
Fe	iron
GLB	globulin
GSH-PX	glutathione peroxidase
IBD	inflammatory bowel disease
K	potassium
KEGG	Kyoto Encyclopedia of Genes and Genomes
LDH	lactate dehydrogenase
MDA	malondialdehyde
Mg	magnesium
NAHMS	National Animal Health Monitoring System
OPLS-DA	orthogonal partial least squares-discriminant analysis
P	phosphorus
PCA	principal component analysis
PLS-DA	partial least squares-discriminant analysis
QC	quality control
SD	standard deviation
SOD	superoxide dismutase
TBA	total bile acids
TCM	Traditional Chinese medicines
TP	total protein
UA	uric acid
VIP	Variable Importance in Projection
YGS	Yigong San

References

1. Kim, H.S.; Whon, T.W.; Sung, H.; Jeong, Y.S.; Jung, E.S.; Shin, N.R.; Hyun, D.W.; Kim, P.S.; Lee, J.Y.; Lee, C.H. Longitudinal evaluation of fecal microbiota transplantation for ameliorating calf diarrhea and improving growth performance. *Nat. Commun.* **2021**, *12*, 161. [CrossRef] [PubMed]
2. Abuelo, A.; Cullens, F.; Brester, J.L. Effect of preweaning disease on the reproductive performance and first-lactation milk production of heifers in a large dairy herd. *J. Dairy Sci.* **2021**, *104*, 7008–7017. [CrossRef] [PubMed]
3. Bernal-Córdoba, C.; Branco-Lopes, R.; Latorre-Segura, L.; de Barros-Abreu, M.; Fausak, E.D.; Silva-Del-Río, N. Use of antimicrobials in the treatment of calf diarrhea: A systematic review. *Anim. Health Res. Rev.* **2022**, *23*, 101–112. [CrossRef]
4. Yong-Il, C.; Kyoung-Jin, Y. An overview of calf diarrhea-infectious etiology, diagnosis, and intervention. *J. Vet. Sci.* **2014**, *15*, 1.
5. Jia, J.; Chen, J.; Wang, G.; Li, M.; Zheng, Q.; Li, D. Progress of research into the pharmacological effect and clinical application of the traditional Chinese medicine Rehmanniae Radix. *Biomed. Pharmacother.* **2023**, *168*, 115809. [CrossRef]
6. Wang, M.; Huang, H.; Hu, Y.; Liu, Y.; Zeng, X.; Zhuang, Y.; Yang, H.; Wang, L.; Chen, S.; Yin, L.; et al. Effects of dietary supplementation with herbal extract mixture on growth performance, organ weight and intestinal morphology in weaning piglets. *J. Anim. Physiol. Anim. Nutr.* **2020**, *104*, 1462–1470. [CrossRef]
7. Zhao, H.; Huang, Y.; Yang, W.; Huang, C.; Ou, Z.; He, J.; Yang, M.; Wu, J.; Yao, H.; Yang, Y.; et al. Viola yedoensis Makino alleviates lipopolysaccharide-induced intestinal oxidative stress and inflammatory response by regulating the gut microbiota and NF-κB-NLRP3/ Nrf2-MAPK signaling pathway in broiler. *Ecotoxicol. Environ. Saf.* **2024**, *282*, 116692. [CrossRef]
8. Bonelli, F.; Turini, L.; Sarri, G.; Serra, A.; Mele, M. Oral administration of chestnut tannins to reduce the duration of neonatal calf diarrhea. *BMC Vet. Res.* **2018**, *14*, 227. [CrossRef]
9. Wang, S.; Cui, D.; Lv, Y.; Yan, Z.; Zhang, J. Cangpu Oral Liquid as a Possible Alternative to Antibiotics for the Control of Undifferentiated Calf Diarrhea. *Front. Vet. Sci.* **2022**, *9*, 879857. [CrossRef] [PubMed]
10. Han, H.; Liu, K.; Wang, Y.; Song, M.; Lian, H.; Gao, T.; Zhang, L.; Fu, T. Supplementation of the probiotic Lactobacillus rhamnosus GG to pre-weaning calves decreases diarrhea incidence by modulating gut bacteria and associated metabolites. *Anim. Nutr.* **2025**, *21*, 234–244. [CrossRef]
11. Van Treuren, W.; Dodd, D. Microbial Contribution to the Human Metabolome: Implications for Health and Disease. *Annu. Rev. Pathol.* **2020**, *15*, 345–369. [CrossRef]
12. Cao, P.P.; Hu, C.L.; Li, M.J.; An, Y.H.; Feng, X.; Ma, X.H.; Wang, D.Z.; Song, Z.H.; Ji, G.S.; Yang, D.; et al. 16S rRNA and metabolomics reveal the key microbes and key metabolites that regulate diarrhea in Holstein male calves. *Front. Microbiol.* **2024**, *15*, 1521719. [CrossRef]
13. Kim, S.J.; Shin, H.J.; Lee, B.J.; Kim, D.S.; Lee, J.H.; Jeong, M.Y.; Kim, H.L.; Park, J.; Lim, H.; Kim, S.H.; et al. The antiinflammatory mechanism of Igongsan in mouse peritoneal macrophages via suppression of NF-κB/Caspase-1 activation. *Phytother. Res.* **2014**, *28*, 736. [CrossRef]
14. Zheng, Q.; Guan, Y.; Xia, L.; Wang, Z.; Jiang, Y.; Zhang, X.; Wang, J.; Wang, G.; Pu, Y.; Xia, J.; et al. Effect of Yi Gong San Decoction on Iron Homeostasis in a Mouse Model of Acute Inflammation. *Evid.-Based Complement. Altern. Med.* **2016**, *2016*, 2696480. [CrossRef] [PubMed]
15. Liu, J.; Wang, H.; Luo, J.; Chen, T.; Xi, Q.; Sun, J.; Wei, L.; Zhang, Y. Synergism of fermented feed and ginseng polysaccharide on growth performance, intestinal development, and immunity of Xuefeng black-bone chickens. *BMC Vet. Res.* **2024**, *20*, 13. [CrossRef]
16. Zhu, J.; Lian, J.; Deng, H.; Luo, J.; Chen, T.; Sun, J.; Zhang, Y.; Yang, Y.; Liu, P.; Xi, Q. Effects of Spinach Extract and Licorice Extract on Growth Performance, Antioxidant Capacity, and Gut Microbiota in Weaned Piglets. *Animals* **2024**, *14*, 321. [CrossRef]
17. Lv, J.; Fu, Y.; Ga, Y.; Han, C.; Fan, Y.; Wei, Y.; Hao, S.; Hao, Z. Lianweng Granules Alleviate Intestinal Barrier Damage via the IL-6/STAT3/PI3K/AKT Signaling Pathway with Dampness-Heat Syndrome Diarrhea. *Antioxidants* **2024**, *13*, 661. [CrossRef]
18. Shen, L.; Shen, Y.; You, L.; Zhang, Y.; Su, Z.; Peng, G.; Deng, J.L.; Zhong, Z.; Yu, S.; Zong, X.; et al. Blood metabolomics reveals the therapeutic effect of Pueraria polysaccharide on calf diarrhea. *BMC Vet. Res.* **2023**, *19*, 98. [CrossRef] [PubMed]
19. Fu, Z.L.; Yang, Y.; Ma, L.; Malmuthuge, N.; Guan, L.L.; Bu, D.P. Dynamics of oxidative stress and immune responses in neonatal calves during diarrhea. *J. Dairy Sci.* **2024**, *107*, 1286–1298. [CrossRef] [PubMed]
20. Zhang, L.; Zhao, M.; Zhang, M.; Zhang, H.; Huang, Q.; Jian, F.; Yin, F.; Zhang, L.; Du, Z.; Yang, R.; et al. Effects of Sodium Butyrate and Ampelopsis grossedentata Flavonoidson Production Performance, Diarrhea and Serum Indicators in Australian-Canadian F1 Holstein Weaned Calves. *Chin. Anim. Husb. Vet. Med.* **2025**, *52*, 3651–3660.
21. Zhang, Y.T.; Yang, Y.; Bu, D.P.; Ma, L. The effects of gamma-aminobutyric acid on growth performance, diarrhoea, ruminal fermentation, and antioxidant capacity in pre-weaned calves. *Animal* **2025**, *19*, 101493. [CrossRef] [PubMed]
22. Sammari, H.; Jedidi, S.; Selmi, H.; Rtibi, K.; Jabri, M.A.; Jridi, M.; Zouari, N.; Toumi, L.; Sebai, H. Protective effects of *Crataegus azarolus* L. berries aqueous extract against castor oil–induced diarrhea, oxidative stress, and inflammation in rat. *Neurogastroenterol. Motil.* **2020**, *33*, e14065. [CrossRef] [PubMed]

23. Wu, T.; Wang, X.; Xiong, H.; Deng, Z.; Peng, X.; Xiao, L.; Jiang, L.; Sun, Y. Bioactives and their metabolites from *Tetrastigma hemsleyanum* leaves ameliorate DSS-induced colitis via protecting the intestinal barrier, mitigating oxidative stress and regulating the gut microbiota. *Food Funct.* **2021**, *12*, 11760–11776. [CrossRef]
24. Ma, X.; Liu, Y.; Ding, B.; Liu, Y.; Zhang, Y.; Wang, Y.; Yang, L.; Yang, Y.; Liu, X. Blueberry anthocyanins ameliorate arsenic-induced cognitive impairment in rats: Mitigating mitochondrial damage and dysregulation. *Phytomedicine* **2025**, *145*, 157062. [CrossRef]
25. Liu, M.; Zhang, G.; Zhou, K.; Wen, J.; Zheng, F.; Sun, L.; Ren, X. Structural characterization, antioxidant activity, and the effects of *Codonopsis pilosula* polysaccharides on the solubility and stability of flavonoids. *J. Pharm. Biomed. Anal.* **2023**, *229*, 115368. [CrossRef]
26. Zhu, B.; Zhang, Q.L.; Hua, J.W.; Cheng, W.L.; Qin, L.P. The traditional uses, phytochemistry, and pharmacology of *Atractylodes macrocephala* Koidz.: A review. *J. Ethnopharmacol.* **2018**, *226*, 143–167. [CrossRef]
27. Zhang, Y.; Huang, J.; Sun, M.; Duan, Y.; Wang, L.; Yu, N.; Peng, D.; Chen, W.; Wang, Y. Preparation, characterization, antioxidant and antianemia activities of Poria cocos polysaccharide iron (III) complex. *Heliyon* **2023**, *9*, e12819. [CrossRef]
28. Mastrodi Salgado, J.; Baroni Ferreira, T.R.; de Oliveira Biazotto, F.; Dos Santos Dias, C.T. Increased antioxidant content in juice enriched with dried extract of pomegranate (*Punica granatum*) peel. *Plant Foods Hum. Nutr.* **2012**, *67*, 39–43. [CrossRef]
29. Zhou, Q.; Zhang, S.; Geng, X.; Jiang, H.; Dai, Y.; Wang, P.; Hua, M.; Gao, Q.; Lang, S.; Hou, L.; et al. Antioxidant Effects of Roasted Licorice in a Zebrafish Model and Its Mechanisms. *Molecules* **2022**, *27*, 7743. [CrossRef]
30. Yang, B.; Huang, Z.; He, Z.; Yue, Y.; Zhou, Y.; Ross, R.P.; Stanton, C.; Zhang, H.; Zhao, J.; Chen, W. Protective effect of Bifidobacterium bifidum FSDJN7O5 and Bifidobacterium breve FHNFQ23M3 on diarrhea caused by enterotoxigenic *Escherichia coli*. *Food Funct.* **2021**, *12*, 7271–7282. [CrossRef] [PubMed]
31. Caliskan, M.; Dabak, M.; Tumer, K.C. The relationship between serum cytokine profile and vitamin D in calves with neonatal diarrhea. *Cytokine* **2023**, *165*, 156173. [CrossRef]
32. Jiang, B.C.; Liu, T.; Gao, Y.J. Chemokines in chronic pain: Cellular and molecular mechanisms and therapeutic potential. *Pharmacol. Ther.* **2020**, *212*, 107581. [CrossRef]
33. Yu, L.; Dong, J.; Wang, Y.; Zhang, P.; Liu, Y.; Zhang, L.; Liang, P.; Wang, L.; Song, C. Porcine epidemic diarrhea virus nsp4 induces pro-inflammatory cytokine and chemokine expression inhibiting viral replication in vitro. *Arch. Virol.* **2019**, *164*, 1147–1157. [CrossRef]
34. Xu, W.; Fang, S.; Wang, Y.; Chi, X.; Ma, X.; Zhang, T.; Hu, S. Receptor and signaling pathway involved in bovine lymphocyte activation by Atractylodis macrocephalae polysaccharides. *Carbohydr. Polym.* **2020**, *234*, 115906. [CrossRef]
35. He, L.; Wang, C.; Simujide, H.; Aricha, H.; Zhang, J.; Liu, B.; Zhang, C.; Cui, Y.; Aorigele, C. Effect of Early Pathogenic *Escherichia coli* Infection on the Intestinal Barrier and Immune Function in Newborn Calves. *Front. Cell. Infect. Microbiol.* **2022**, *12*, 818276. [CrossRef]
36. Ong, E.S. Urine Metabolites and Bioactive Compounds from Functional Food: Applications of Liquid Chromatography Mass Spectrometry. *Crit. Rev. Anal. Chem.* **2024**, *54*, 3196–3211. [CrossRef]
37. Ferreira, N.S.; Tostes, R.C.; Paradis, P.; Schiffrin, E.L. Aldosterone, Inflammation, Immune System, and Hypertension. *Am. J. Hypertens.* **2021**, *34*, 15–27. [CrossRef]
38. Wang, T.; Wu, H.; Shi, X.; Dai, M.; Liu, Y. Aminoadipic acid aggravates atherosclerotic vascular inflammation through ROS/TXNIP/NLRP3 pathway, a harmful microbial metabolite reduced by paeonol. *Int. J. Biochem. Cell Biol.* **2024**, *177*, 106678. [CrossRef]
39. Zheng, X.; Zhu, Y.; Zhao, Z.; Chu, Y.; Yang, W. The role of amino acid metabolism in inflammatory bowel disease and other inflammatory diseases. *Front. Immunol.* **2023**, *14*, 1284133. [CrossRef]
40. Song, G.; Gan, Q.; Qi, W.; Wang, Y.; Xu, M.; Li, Y. Fructose Stimulated Colonic Arginine and Proline Metabolism Dysbiosis, Altered Microbiota and Aggravated Intestinal Barrier Dysfunction in DSS-Induced Colitis Rats. *Nutrients* **2023**, *15*, 782. [CrossRef]
41. Ding, Y.; Koda, Y.; Shashni, B.; Takeda, N.; Zhang, X.; Tanaka, N.; Nishikawa, Y.; Nagasaki, Y. An orally deliverable ornithine-based self-assembling polymer nanomedicine ameliorates hyperammonemia in acetaminophen-induced acute liver injury. *Acta Biomater.* **2023**, *168*, 515–528. [CrossRef] [PubMed]
42. Gao, L.; Yuan, H.; Xu, E.; Liu, J. Toxicology of paraquat and pharmacology of the protective effect of 5-hydroxy-1-methylhydantoin on lung injury caused by paraquat based on metabolomics. *Sci. Rep.* **2020**, *10*, 1790. [CrossRef] [PubMed]

veterinary sciences

MDPI

Article

Influence of Untreated and Microbially Degraded Mangrove Sediment Microplastics on *Zebrafish* (*Danio rerio*) Intestinal Histology and Immune and Antioxidant Biomarkers

Xin-Yu Zheng †, Wan Wei †, Asim Muhammad †, Min Zhang, Yan-Jun Chen, Jia-Hong Xie, Dan-Ju Kang and Jin-Jun Chen *

Department of Veterinary Medicine, School of Coastal Agriculture, Guangdong Ocean University, Zhanjiang 524088, China; 18604068009@stu.gdou.edu.cn (X.-Y.Z.); 19806873590@163.com (W.W.); asimmuhammad9862@gmail.com (A.M.); 13634322028@163.com (M.Z.); chenyanjun2@stu.gdou.edu.cn (Y.-J.C.); 18758019771@stu.gdou.edu.cn (J.-H.X.); kangdanju@126.com (D.-J.K.)

* Correspondence: chenjj@gdou.edu.cn; Tel.: +86-759-8128600

† These authors contributed equally to this work.

Simple Summary: Plastic pollution poses an increasing threat to marine life and human health, particularly in accumulation zones such as mangrove wetlands. This study examined the effects of microplastics (MPs) collected from mangrove sediments in Zhanjiang, China, on zebrafish intestinal health. We compared the impacts of untreated MPs with those degraded by microorganisms. Zebrafish were exposed to both types of MPs for 21 days. Results indicated that fish exposed to untreated MPs exhibited more severe intestinal damage and inflammation compared to those exposed to degraded particles, with damage intensifying at higher concentrations and longer exposure durations. These findings suggest that microbial degradation can mitigate the harmful effects of MPs on aquatic organisms, supporting the development of safer bioremediation strategies to address plastic pollution and protect marine ecosystems.

Abstract: MPs are pervasive pollutants in marine ecosystems, posing risks to aquatic organisms due to their small size and bioaccumulation potential. This study investigated the intestinal toxicity of MP particles extracted from mangrove sediments in *zebrafish*, comparing the effects before and after microbial degradation. *Zebrafish* were exposed to either undegraded MPs or microbially degraded MP extracts at concentrations of 0 (control), 2, 10, and 50 mg/L for 21 days in 10 L tanks (stocking density: 10 fish/L), with three replicate tanks per concentration. MPs were dispersed ultrasonically before addition to the water. Intestinal samples were collected on 7, 14, and 21 days for the analysis of immune response (tumor necrosis factor-alpha, TNF-α; interleukin-1 beta, IL-1β; interleukin-6, IL-6; interleukin-8, IL-8) and antioxidant activity (superoxide dismutase, SOD; catalase, CAT). Histopathological analysis revealed intestinal wall thinning, villus damage, and epithelial cell detachment in *zebrafish* exposed to both undegraded and degraded MP extracts; however, undegraded MPs induced more severe intestinal damage. Results indicated dynamic changes in cytokine expression: TNF-α decreased initially before increasing, while IL-1β and IL-8 first rose then declined. IL-6 peaked on day 7, dropped by day 14, and increased again on day 21. CAT expression decreased, whereas SOD increased only in the pre-degradation group. Microbial degradation reduced intestinal damage severity, with effects intensifying at higher MP exposure levels. These findings demonstrate that MPs can impair *zebrafish* digestive systems, but microbial degradation mitigates their toxicity. This study underscores the importance of biodegradation as a potential environmental remediation strategy and provides experimental evidence on MPs' impact on aquatic organisms.

Keywords: microplastics; degradation; mangrove sediments; oxidative stress; gene expression; immune response; intestinal toxicology; *zebrafish*

1. Introduction

Plastics have become one of the most widespread and persistent pollutants in marine environments, from the poles to the equator [1]. Plastic particles smaller than 5 mm in diameter, found in oceans or on beaches, are classified as microplastics (MPs) [2]. MPs are characterized by their small size, slow degradation rate, and capacity for long-range transport via natural forces such as rivers and ocean currents [3]. Numerous studies have confirmed the global distribution of MPs in oceans, rivers, and sediments [4–6]. Sometimes referred to as the "PM2.5 of the ocean," MPs are readily ingested by marine organisms due to their small size [7].

While larger MPs are generally not absorbed through the intestine wall, their prolonged retention in the digestive tract can cause localized inflammation and disrupt the gut microbiota [8,9]. However, MPs with small particle sizes mainly accumulate in the intestine and are further transferred to the blood. MP exposure can impair growth, development, and reproduction, and can also induce oxidative stress and immune responses [10]. The primary route of exposure is the ingestion of contaminated food and water [11,12]. Therefore, the effects of MP exposure on animal intestines deserve attention. *Zebrafish* are an ideal model for studying intestinal inflammation and host–microbiome interactions due to their small size, high fecundity, fully sequenced genome, and physiological similarities to mammals in intestinal and immune function [13]. Several studies have confirmed that MP exposure over a period of time can induce different levels of inflammation and oxidative stress in the gut of *zebrafish* [8,14].

The extensive root systems of mangrove plants not only mitigate wave energy but also trap pollutants [15]. Consequently, mangrove sediments are considered major sinks for contaminants [16]. Plastics constitute up to 70% of pollutants found in mangrove areas [17]. The abundance of MPs in mangroves is higher than in non-mangrove areas, and MPs are trapped in every layer of sediment [18]. Furthermore, various studies have shown that microorganisms isolated from mangrove sediments can degrade MPs [19–21]. Mostly, research on the risks of MP exposure focuses on standard single plastics in the laboratory, which differ from the MPs found in the natural environment. Presently, there are few reports on the effects of extracted MP particles from sediments on the gut of *zebrafish* before and after degradation.

In this study, MPs were extracted from mangrove sediments in Mazhang District, Zhanjiang, and divided into two portions: one subjected to microbial degradation for 120 days, and the other kept undegraded. *Zebrafish* were exposed to various concentrations of degraded MP products or undegraded MPs. Intestinal tissues were collected on days 7, 14, and 21 to compare the effects of degraded and undegraded MPs on intestinal histopathology. We measured the relative expression levels of SOD, CAT, TNF-α, IL-1β, IL-6, and IL-8 mRNA in different MP concentration groups. Our findings provide a scientific basis for bioremediation strategies aimed at mitigating the intestinal toxicity of environmentally sourced MPs.

2. Materials and Methods

2.1. Extraction and Pretreatment of Microplastics from Mangrove Sediments

Sediment samples were collected from mangrove forests in Mazhang District, Zhanjiang City, Guangdong Province, China. MPs were extracted from the sediments using a wet oxidation method, which included drying, flotation, digestion, and filtration [22]. Polyethylene (PE) powder, identified as the predominant type of MPs in the samples, was spread on a sterile Petri dish and sterilized under ultraviolet light for 6 h with the dish rotated every 30 min. The sterilized PE plastic was then inoculated onto Luria–Bertani (LB) solid medium—a nutrient-rich agar used for bacterial cultivation—and incubated at 30 °C for 3 days. The absence of microbial growth around the powder confirmed sterility.

2.2. Zebrafish Acclimation and Rearing Conditions

A total of 2500 healthy adult *zebrafish*, with an average body length of 3.0 ± 0.5 cm and wet weight of 0.34 ± 0.05 g, were purchased from the Zhanjiang Flower and Bird Market. They were identified as *zebrafish* by the Species Identification Group of the Fisheries College of Guangdong Ocean University. The *zebrafish* were subjected to 30 days of adaptive feeding in a 250 L fish tank with an environmental density of 10 tails/L. A circulating water pump was installed to ensure a sufficient oxygen supply, and the water temperature was maintained at 25–28 °C. The light duration was set to 12 h, and the light avoidance duration was set to 12 h. The water was filtered through reverse osmosis, maintaining a pH range of 7.4–7.6 and a conductivity of 500–550 uS/cm. The *zebrafish* were fed twice a day with a commercial pellet feed (Biofish-F100, Shanghai Feixi Biotechnology Co., Ltd., Shanghai, China). According to the manufacturer, the proximate composition of the diet was as follows: crude protein $\geq$ 48%, crude fat $\geq$ 10%, crude fiber $\leq$ 5%, crude ash $\leq$ 15%, moisture $\leq$ 12%, phosphorus $\geq$ 1.0%, lysine $\geq$ 2.2%. The feeding ration was 3% body weight per feeding with a 12 h interval between feedings. The same diet and feeding regime were maintained throughout the subsequent MP exposure experiment. *Zebrafish* mortality was monitored daily, with immediate removal of deceased individuals to prevent decomposition. Mortality was monitored but not statistically analyzed, as it was not a primary focus of this study.

2.3. Microbial Degradation of MPs

Extracted MP particles degradation group: Based on the previous research of our laboratory, K1-1 *Acinetobacter venetianus*—a bacterial strain originally isolated from mangrove sediments—was screened out for degradation experiments due to its significant efficacy in MP degradation. In a 40 mL wide-mouth conical bottle, 4 mL of bacterial solution and 36 mL of MM30 liquid medium were cultured with extracted MP particles. In triplicate, 20, 100, and 500 mg of MPs were added to each bottle and cultured for 120 d in a 180 r/min incubator at 30 °C and were maintained at 40 mL by adding MM30 every 10 days.

The bacterial strain *Acinetobacter venetianus* K1-1, previously isolated from mangrove sediments for its MP-degrading capability, was used for degradation experiments. In 40 mL wide-mouth conical bottles, 4 mL of bacterial suspension and 36 mL of MM30 liquid medium were mixed with extracted MP particles. Triplicate bottles containing 0 (control), 20, 100, or 500 mg of MPs were incubated at 30 °C with shaking at 180 rpm for 120 days. The volume was maintained at 40 mL by adding fresh MM30 medium every 10 days. For the non-degraded MP group, the same amounts of MPs were added to sterile MM30 medium without bacteria and incubated under identical conditions.

2.4. Zebrafish Exposure Experiment

Zebrafish were randomly distributed into seven experimental groups, each maintained in triplicate tanks (n = 3 tanks per group). The groups consisted of the following: a single blank control group (0 mg/L MPs), shared for both degradation conditions; three groups exposed to untreated, pre-degradation MPs at concentrations of 2, 10, and 50 mg/L; and three groups exposed to microbially degraded MPs at the same concentrations of 2, 10, and 50 mg/L.

Each tank consisted of 100 *zebrafish* in 10 L of fully aerated tap water. To ensure homogeneous exposure, the MPs were dispersed using an ultrasonic bath (10 s pulse, 5 s pause, 40 cycles) before being added to the water. The entire volume of water in each tank was replaced every 48 h, and the respective MPs were re-added at that time to maintain the target exposure concentrations throughout the 21-day experimental period. All holding conditions—including water temperature (25–28 °C), pH (7.4–7.6), conductivity (500–550 μS/cm), photoperiod (12 h:12 h light–dark), stocking density (10 fish/L), and feeding regimen (twice daily with commercial pellets at 3% body weight)—were identical to those described for the acclimation period (Section 2.2) to ensure that any observed effects were attributable solely to MP exposure and not to environmental variability.

2.5. Sample Collection

Sampling was conducted at 7, 14, and 21 days of exposure. At each time point, 10 *zebrafish* were randomly sampled from each of the three replicate tanks per treatment group, resulting in a total of 30 intestinal samples per group per time point. Intestinal tissues were dissected on a sterile bench, rinsed with phosphate-buffered saline (PBS), and processed for subsequent analysis.

For the 7- and 14-day time points, all 30 samples per group were immediately frozen and stored at −80 °C for subsequent gene expression analysis. For the 21-day time point, the sampling was subdivided: 2 fish from each tank (totaling 6 per treatment group) were fixed in 4% paraformaldehyde for histopathological sectioning, while the remaining 8 fish from each tank (totaling 24 per treatment group) were stored at −80 °C for biochemical and molecular analyses. All samples were processed within 6 months of collection. All animal procedures were approved by the Animal Ethics Committee of Guangdong Ocean University and were conducted in accordance with the relevant guidelines to minimize suffering.

2.6. Histological Analysis

Intestinal samples fixed in 4% paraformaldehyde were dehydrated, embedded in paraffin, sectioned at 4 μm thickness, and stained with hematoxylin and eosin (HE). A total of 42 fish (6 per group across 7 groups) were analyzed. For each fish, one representative slide was prepared, and two non-overlapping fields per section were examined under a light microscope (Nikon ECLIPSE E200; Nikon Corporation, Tokyo, Japan) at 200× magnification. Normal intestinal morphology was defined by intact villi, continuous epithelium, visible goblet cells, and absence of necrosis or inflammation. Abnormalities included villus shortening, epithelial detachment, wall thinning, inflammatory infiltration, and cellular necrosis.

2.7. Relative mRNA Expression in Intestinal Tissues of Zebrafish

Total RNA was extracted individually from each intestinal sample (30 individuals per group for the 7- and 14-day time points; 24 individuals per group for the 21-day time point) without pooling, using a total RNA extraction kit (TRIZOL Reagent) purchased from Invitrogen, and the concentration of the extracted RNA samples was detected using a nucleic acid protein analyzer. Samples with OD260/280 values of 1.8~2.1 were selected, and the extracted 2 μL RNA was reverse-transcribed into cDNA using a reverse transcription kit. The qPCR reaction system: Green Taq Mix 10 μL of upper and downstream primers 0.4 μL of cDNA 2 μL of ddH2O supplementation 20 μL. The qPCR amplification procedure was pre-denatured at 94 °C for 30 s. The samples were denatured at 94 °C for 5 s, annealed at 55 °C for 15 s, and extended at 72 °C for 10 s for a total of 40 cycles. To determine the changes in the relative expression of target genes, the results were evaluated using the $2^{-\Delta\Delta CT}$ method, with β-actin as the reference gene. Each sample was analyzed in triplicate. The primers used are listed in Table 1.

Table 1. Primer sequences for real-time PCR.

Gene	Forward (5′–3′)	Reverse (3′–5′)
IL-1β	GATTCGCAGATGGTGGAGATGGAC	TCGTCTTTGGATGGAAGCACAGC
IL-6	TCAACTTCTCCAGCGTGATG	TCTTTCCCTCTTTTCCTCCTG
IL-8	AACATGGAGGTCATTGCCACTGTG	GAGGTAGAATTTGGAGGGAGGGTA
TNF-α	GTCGGGTGTATGGAGGGTGTTTG	CTGGTCTTATGGAGCGTGAAGC
SOD	GTCCGCACTTCAACCCTCA	TCCTCATTGCCACCCTTCC
CAT	AGGGCAACTGGGATCTTACA	TTTATGGGACCAGACCTTGG
β-actin	GTGATGGACTCTGGTGATGGTGTG	CACGCTCGGTCAGGATCTTCATC

2.8. Statistical Analysis

Statistical analyses were conducted using SPSS Statistics 27.0. For each treatment group and time point, data from individual *zebrafish* within the same tank were averaged to obtain a single value per tank, as the tank was considered the independent experimental unit. Differences among treatment groups, including the control, undegraded MP groups (2, 10, and 50 mg/L), and degraded MP groups at the same concentrations—were analyzed using one-way ANOVA at each exposure time point (7, 14, and 21 days). A significance threshold of $p < 0.05$ was applied. Results are expressed as mean ± standard error of the tank means ($n = 3$).

3. Results

3.1. Effects of Microbial Degradation Products of Extracted MP Particles on Intestinal Tract of Zebrafish

After 21 days of exposure, the histological analysis of intestinal tissues was performed (Figure 1). In the control group (0 mg/L), applicable to both degradation conditions, intestinal morphology remained intact, with no noticeable pathological changes throughout the experiment. In the low-dose groups (2 mg/L), both pre- and post-degradation treatments caused only minor villus damage. In contrast, significant alterations were observed in the medium-dose group (10 mg/L). Both treatments exhibited epithelial shedding at villus tips; however, the undegraded MP group displayed markedly thinner intestinal walls, severe villus destruction, and greater cell loss and necrosis. In the high-dose group (50 mg/L), lesions were most pronounced: villus height in the mid-intestine decreased, mucosal epithelium became hypertrophic, and the villous epithelial boundary in the large intestine appeared blurred or absent. Columnar epithelial cells were necrotic and detached, and lymphocytes and goblet cells were scattered. Similar lesions occurred in the degraded MP group, but the intestinal wall was generally more preserved compared to the undegraded group.

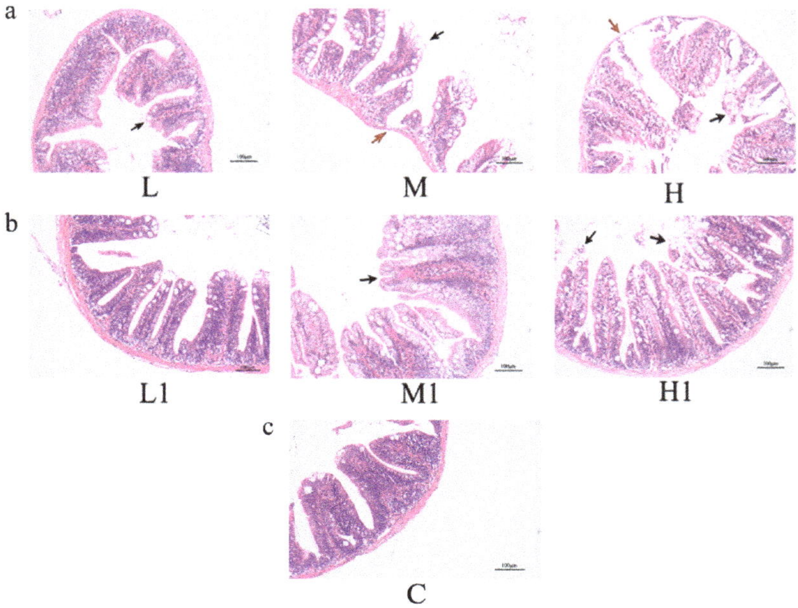

Figure 1. Histological changes in the gut of *zebrafish* after 21 days of exposure. (**a**) MP group before degradation; (**b**) MP group after degradation; (**c**) blank control (0 mg/L); L.L1. Low dose group; M.M1. Medium dose group; H.H1. High dose group; HE, 200×; red arrows indicate thinning of the intestinal wall, while black arrows indicate cell detachment.

3.2. Relative mRNA Expression of Oxidative Stress-Related Genes

The relative mRNA expression of SOD and CAT in *zebrafish* intestines was analyzed after 7, 14, and 21 days of exposure using qRT-PCR (Figure 2a). In the undegraded MP group, SOD expression increased progressively with exposure duration, while microbial degradation attenuated this effect. Significant differences were detected at medium and high concentrations between the two treatments ($p < 0.01$, $p < 0.001$). After 21 days, SOD expression remained significantly elevated in the undegraded MP group but continued to decline in the degraded MP group, even at low doses ($p < 0.001$). Conversely, CAT expression decreased in all treated groups compared with the control at all time points (Figure 2b). Although CAT levels were consistently lower in the undegraded MP group than in the degraded MP group, the difference between treatments was not statistically significant.

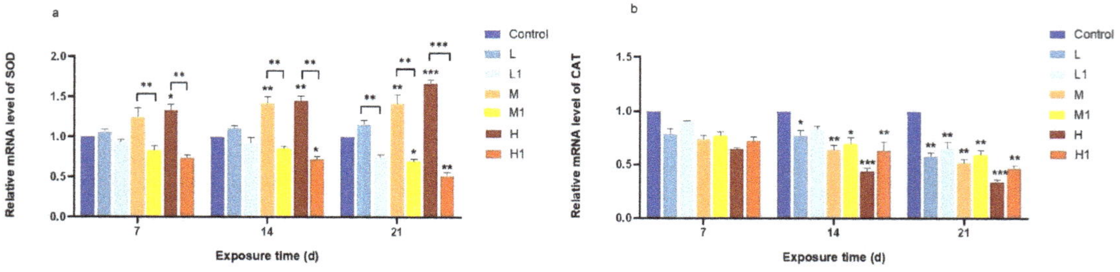

Figure 2. Relative mRNA expression of oxidative stress markers in the intestines of *zebrafish* exposed to extracted MP particles before and after microbial degradation. (**a**) SOD; (**b**) CAT. MPs after degradation are marked with 1 in the lower corner. Data are presented as mean ± SEM; * $p < 0.05$, ** $p < 0.01$, *** $p < 0.001$.

3.3. Relative mRNA Expression of Inflammatory Response-Related Genes

IL-1β expression increased significantly in all treatment groups after 7 days, with the undegraded MP group showing the highest levels ($p < 0.001$). However, expression dropped below control levels by days 14 and 21, with a more pronounced decline in the degraded MP group (Figure 3a). IL-6 exhibited a biphasic pattern, increasing at days 7 and 21 but decreasing on day 14, with high-dose effects becoming more evident over time in both treatments ($p < 0.01$) (Figure 3b). IL-8 followed a similar trend to IL-1β, peaking at day 7 before decreasing at later time points; significant differences between undegraded and degraded MPs were only detected at medium and high doses on day 7 ($p < 0.001$) (Figure 3c). TNF-α expression decreased initially but rose again by day 21, with significant differences observed only at the medium dose after 21 days ($p < 0.05$) (Figure 3d).

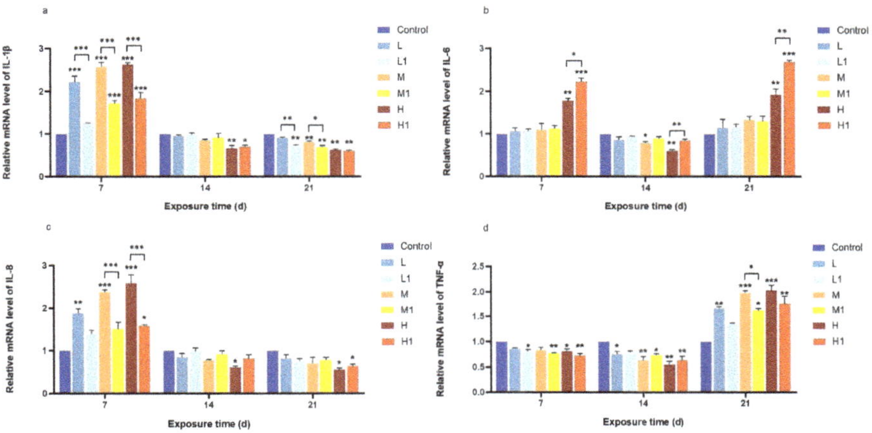

Figure 3. Changes in the relative expression of cytokine mRNA in *zebrafish* before and after degradation of extracted MP particles. (**a**) IL-1β; (**b**) IL-6; (**c**) IL-8; (**d**) TNF-α; MPs after degradation are marked with 1 in the lower corner; * $p < 0.05$, ** $p < 0.01$, *** $p < 0.001$.

4. Discussion

The *zebrafish* intestine plays a crucial role in digestion, nutrient absorption, waste elimination, and immune defense [23]. Contaminated food is a major source of MP exposure, and the accumulation of MPs in intestinal tissues may lead to local inflammation and affect gut microbes, resulting in various adverse reactions [10,24]. Exposure to polystyrene (PS), a common type of plastic, with a particle size of 5 μm at concentrations of 50 μg/L and 500 μg/L resulted in intestinal damage after 21 d of exposure. This damage includes villus cracking and intestinal cell division [14]. This study showed that exposure of *zebrafish* to MPs can cause intestinal oxidative stress and intestinal inflammation. With an increase in concentration and extension of exposure time, the intestinal inflammation and antioxidant damage in *zebrafish* were more pronounced. However, after microbial degradation, the MP-exposed group had significantly lower intestinal toxicity to *zebrafish* than the MP-exposed group without degradation. Histological analysis revealed intestinal wall thinning, villus damage, and epithelial shedding in both undegraded and degraded MP groups, but lesions were more severe with undegraded MPs. The possible reason for this result is attributed to the fact that MPs are extremely difficult to degrade, and large plastic particles cannot be divided into countless smaller plastic particles in a short period of time [25] and are eventually completely discharged with excreta. However, compared with MPs without degradation, the amount of plastic particles after degradation was relatively reduced. These results demonstrated that the histological sections of different treatment groups showed different degrees of intestinal damage, and the degree of intestinal lesions was not only related to the dose of exposure but also to the degradation of MPs, indicating that microbial degradation of MPs requires further research.

Under external stressors, organisms produce excess reactive oxygen species (ROS), leading to oxidative stress when ROS levels exceed antioxidant capacity [26]. SOD and CAT are the most common metabolic enzymes in the antioxidant stress system of organisms [27]. The main function of CAT is to remove excess hydrogen peroxide from the body and break it down into oxygen and water to protect the cells from hydrogen peroxide damage. SOD is an important metal enzyme, and its main function is to remove excess oxygen free radicals produced in the body, especially superoxide anions (O^{2-}). While some studies reported increased SOD and CAT mRNA levels under MP exposure [14], our study found a decrease in CAT expression, with pre-degradation MPs inducing lower CAT levels than post-degradation MPs. However, the relative mRNA expression of the SOD increased only in the group before MP degradation but decreased in the group after degradation, and the reaction became more obvious with the increase in MP exposure dose in both treatment groups. A possible reason for this result is that the MPs used in this study were extracted from the natural environment, and it is difficult to ensure their purity and particle size. For example, the size of MPs determines the extent to which they are transported in the circulatory system, leading to differences in tissue accumulation [28]. Moreover, the extracted MP particles after bacterial degradation, whose components remain to be verified, may be released after bacterial degradation to inhibit the activity of SOD and CAT. As the energy consumed to combat oxidative stress is exhausted, MP exposure leads to a decrease in antioxidant enzymes [29]. Exposure to PS-MPs inhibited CAT activity in *zebrafish* larvae, suggesting that oxidative damage leads to an imbalance in the antioxidant defense system [30]. CAT activity also decreased with increasing MP concentration, suggesting oxidative damage and stress. The composition of the products after plastic biodegradation is complex and may be toxic to the ecosystem. However, there is still a lack of a summary of the environmental behavior of the degradation products. Therefore, it is very necessary to study the composition analysis of MP degradation products for the subsequent work of improving biodegradation.

Zebrafish, as aquatic vertebrates, possess a well-developed immune system comprising various tissues and cytokines that contribute to maintaining internal homeostasis. Studies have indicated that cytokine expression levels serve as effective biomarkers of inflammation in *zebrafish* [8]. To further investigate the effects of MP exposure on the intestinal tract of *zebrafish*, this study measured the relative mRNA expression levels of four immune-related cytokines: IL-1β, TNF-α, IL-6, and IL-8. IL-6 is a pivotal cytokine implicated in the development of chronic inflammation [31]. In this study, it was found that the mRNA expression level of IL-6 in both groups before and after MP degradation displayed an initial increase, followed by a decrease and then an increase. However, only the high-dose group showed statistically significant differences as compared to others. These findings imply that with an increase in exposure time, MPs can enter the interior of cells, interact with intracellular proteins, and affect IL-6 production by regulating cell-signaling pathways [32]. TNF-α, secreted primarily by monocytes, acts as an important cellular activator that regulates tissue metabolism and stimulates the synthesis and release of other cytokines [33]. Its principal functions include inhibiting and killing tumor cells, combating infections, and promoting inflammatory responses. In this study, the mRNA expression of TNF-α in *zebrafish* initially decreased and then increased relative to the control group. Significant differences from the control were observed at low, medium, and high MP concentrations after 7 and 14 days of exposure; however, there was no significant difference observed before and after the degradation of MPs. Notably, a significant difference was observed only after 21 d of exposure. Exposure to

0.5 μm polystyrene microplastics significantly increased the mRNA level of TNF-α in *zebrafish* viscera compared to the control group [8], a result consistent with the findings in the 21-day exposed group of this study, which may be related to the abnormal activation of the immune system. This suggests that MP pollution may negatively affect the immune system of *zebrafish*, impairing their immune function and leading to chronic inflammation and disease. IL-1β is an important pro-inflammatory cytokine produced by activated macrophages in response to exogenous stimuli [34], which can regulate the immune response, and cell metabolism and induce the gradual release of other pro-inflammatory factors. MP exposure significantly increased the expression and release of IL-1β in *zebrafish* [35]. This may be because MPs activated the immune system after entering the *zebrafish*, leading to the increased secretion of interleukin IL-1β. In contrast, our results demonstrated an increase in IL-1β levels only after 7 days of MP exposure, with a decline observed after 14 and 21 days. IL-8 is a chemokine secreted by mononuclear macrophages and can induce neutrophils to release active products that participate in inflammatory responses [36]. This study found that the relative expression of IL-8 mRNA in response to MP exposure initially increased and then decreased compared to the control group, with no significant differences among exposure groups. These results suggest that MPs may reduce the resistance of white blood cells to pathogenic bacteria, leading to diminished IL-8 expression. This suggests that MP contamination may affect the immune system of *zebrafish*, making them more vulnerable to infections and diseases.

In this study, both degraded and non-degraded MP groups caused immune and oxidative stress damage in *zebrafish*. However, the relative mRNA expression levels of TNF-α, IL-1β, and IL-8 decreased after degradation compared to the non-degraded group. This may be attributed to the reduced particle size of MPs following bacterial degradation, facilitating their excretion in feces, or to the diminished toxicity of certain harmful substances, such as associated additives, thereby reducing the overall harm to *zebrafish*. Previous studies have indicated that MPs of smaller particle sizes can be eliminated via fecal excretion after ingestion by fish [37], resulting in reduced toxicity—a finding consistent with the results of this experiment. The effects of MPs on IL-1β, IL-6, IL-8, TNF-α, SOD, and CAT have not been fully studied. Moreover, the influence of MPs varying in type, shape, and concentration on these biomarkers may differ significantly, and their impacts in real-world environments remain to be elucidated. Therefore, further research is necessary to examine the toxic effects of MPs under varying conditions such as particle size, polymer type, and exposure concentration.

5. Conclusions

This study demonstrated that the intestinal toxicity of MPs extracted from mangrove sediments decreased significantly after microbial degradation compared with undegraded MPs. The severity of toxicity was influenced by both concentration and exposure duration, with higher doses and longer exposure times causing more pronounced intestinal inflammation and oxidative stress. Although microbial degradation mitigated these adverse effects, toxicity was not completely eliminated. These findings offer important insights into the potential of microbial degradation as a strategy to reduce MP-induced toxicity in aquatic organisms and provide a foundation for future bioremediation research.

Author Contributions: X.-Y.Z. and W.W.: Methodology, Investigation, Data analysis and curation, Writing—original draft. A.M.: Methodology and Investigation. M.Z.: Microbil Investigation, Writing—review and editing. Y.-J.C.: C-Methodology. J.-H.X.: Conceptualization. D.-J.K.: Methodology and Investigation. J.-J.C.: Conceptualization, Methodology, Data curation, Supervision, Funding Acquisition, Writing—review and editing. All authors have read and agreed to the published version of the manuscript.

Funding: Marine Economic Development Project of Guangdong Province Special Fund for Promoting High-quality Economic Development (Contract grant number: GDOE2019A52), and Guangdong Province Postgraduate Education Innovation Program Grant Project-CP Group (Zhanjiang) Modern Agricultural Investment Co., Ltd. Joint training of graduate students demonstration base 2020 (Contract grant number: 230420081).

Institutional Review Board Statement: All experimental protocols were approved by the Animal Ethics Committee of Guangdong Ocean University (IACUC No. GDOU-LAE-2020-007), and the experiment was performed according to the ethical guidelines of the European Community guidelines.

Informed Consent Statement: Not applicable.

Data Availability Statement: The data presented in this study are available on request from the corresponding author. The data are not publicly available due to privacy protection.

Acknowledgments: The authors are grateful to the staff of the School of Coastal Agriculture, Guangdong Ocean University, Zhanjiang, China, for their technical assistance and support in the course of the study.

Conflicts of Interest: The authors declare no conflicts of interest.

Abbreviations

MPs	Microplastics
SOD	Superoxide dismutase
CAT	Catalase
HE	Hematoxylin and eosin
PE	polyethylene
LB	Luria–Bertani
TNF-α	tumor necrosis factor-alpha
IL-1β	interleukin-1 beta
IL-6	interleukin-6
IL-8	interleukin-8
PBS	phosphate-buffered saline
PS	polystyrene
ROS	reactive oxygen species

References

1. Jambeck, J.R.; Geyer, R.; Wilcox, C.; Siegler, T.R.; Perryman, M.; Andrady, A.; Narayan, R.; Law, K.L. Plastic waste inputs from land into the ocean. *Science* **2015**, *347*, 768–771. [CrossRef] [PubMed]
2. Bao, M.; Xiang, X.; Huang, J.; Kong, L.; Wu, J.; Cheng, S. Microplastics in the Atmosphere and Water Bodies of Coastal Agglomerations: A Mini-Review. *Int. J. Environ. Res. Public Health* **2023**, *20*, 2466. [CrossRef] [PubMed]
3. Auta, H.S.; Emenike, C.U.; Fauziah, S.H. Distribution and importance of microplastics in the marine environment: A review of the sources, fate, effects, and potential solutions. *Environ. Int.* **2017**, *102*, 165–176. [CrossRef] [PubMed]
4. Ter Halle, A.; Jeanneau, L.; Martignac, M.; Jarde, E.; Pedrono, B.; Brach, L.; Gigault, J. Nanoplastic in the North Atlantic Subtropical Gyre. *Environ. Sci. Technol.* **2017**, *51*, 13689–13697. [CrossRef]
5. Phuong, N.N.; Fauvelle, V.; Grenz, C.; Ourgaud, M.; Schmidt, N.; Strady, E.; Sempéré, R. Highlights from a review of microplastics in marine sediments. *Sci. Total Environ.* **2021**, *777*, 146225. [CrossRef]
6. Uddin, S.; Fowler, S.W.; Uddin, M.F.; Behbehani, M.; Naji, A. A review of microplastic distribution in sediment profiles. *Mar. Pollut. Bull.* **2021**, *163*, 111973. [CrossRef]
7. Sun, X.; Li, Q.; Shi, Y.; Zhao, Y.; Zheng, S.; Liang, J.; Liu, T.; Tian, Z. Characteristics and retention of microplastics in the digestive tracts of fish from the Yellow Sea. *Environ. Pollut.* **2019**, *249*, 878–885. [CrossRef]
8. Jin, Y.; Xia, J.; Pan, Z.; Yang, J.; Wang, W.; Fu, Z. Polystyrene microplastics induce microbiota dysbiosis and inflammation in the gut of adult zebrafish. *Environ. Pollut.* **2018**, *235*, 322–329. [CrossRef]
9. Lu, L.; Wan, Z.; Luo, T.; Fu, Z.; Jin, Y. Polystyrene microplastics induce gut microbiota dysbiosis and hepatic lipid metabolism disorder in mice. *Sci. Total Environ.* **2018**, *631–632*, 449–458. [CrossRef]

10. Dong, X.; Liu, X.; Hou, Q.; Wang, Z. From natural environment to animal tissues: A review of microplastics (nanoplastics) translocation and hazards studies. *Sci. Total Environ.* **2023**, *855*, 158686. [CrossRef] [PubMed]
11. Zhang, J.; Wang, L.; Kannan, K. Polyethylene Terephthalate and Polycarbonate Microplastics in Pet Food and Feces from the United States. *Environ. Sci. Technol.* **2019**, *53*, 12035–12042. [CrossRef]
12. Wagner, M.; Scherer, C.; Alvarez-Munoz, D.; Brennholt, N.; Bourrain, X.; Buchinger, S.; Fries, E.; Grosbois, C.; Klasmeier, J.; Marti, T.; et al. Microplastics in freshwater ecosystems: What we know and what we need to know. *Environ. Sci. Eur.* **2014**, *26*, 12. [CrossRef]
13. Brugman, S. The zebrafish as a model to study intestinal inflammation. *Dev. Comp. Immunol.* **2016**, *64*, 82–92. [CrossRef]
14. Qiao, R.; Sheng, C.; Lu, Y.; Zhang, Y.; Ren, H.; Lemos, B. Microplastics induce intestinal inflammation, oxidative stress, and disorders of metabolome and microbiome in zebrafish. *Sci. Total Environ.* **2019**, *662*, 246–253. [CrossRef] [PubMed]
15. Liu, X.; Liu, H.; Chen, L.; Wang, X. Ecological interception effect of mangroves on microplastics. *J. Hazard. Mater.* **2022**, *423*, 127231. [CrossRef] [PubMed]
16. Martin, C.; Baalkhuyur, F.; Valluzzi, L.; Saderne, V.; Cusack, M.; Almahasheer, H.; Krishnakumar, P.K.; Rabaoui, L.; Qurban, M.A.; Arias-Ortiz, A.; et al. Exponential increase of plastic burial in mangrove sediments as a major plastic sink. *Sci. Adv.* **2020**, *6*, eaaz5593. [CrossRef] [PubMed]
17. Deng, H.; He, J.; Feng, D.; Zhao, Y.; Sun, W.; Yu, H.; Ge, C. Microplastics pollution in mangrove ecosystems: A critical review of current knowledge and future directions. *Sci. Total Environ.* **2021**, *753*, 142041. [CrossRef]
18. Duan, J.; Han, J.; Cheung, S.G.; Chong, R.K.Y.; Lo, C.M.; Lee, F.W.; Xu, S.J.; Yang, Y.; Tam, N.F.; Zhou, H.C. How mangrove plants affect microplastic distribution in sediments of coastal wetlands: Case study in Shenzhen Bay, South China. *Sci. Total Environ.* **2021**, *767*, 144695. [CrossRef]
19. Auta, H.S.; Emenike, C.U.; Fauziah, S.H. Screening of Bacillus strains isolated from mangrove ecosystems in Peninsular Malaysia for microplastic degradation. *Environ. Pollut.* **2017**, *231*, 1552–1559. [CrossRef]
20. Auta, H.S.; Emenike, C.U.; Jayanthi, B.; Fauziah, S.H. Growth kinetics and biodeterioration of polypropylene microplastics by *Bacillus* sp. and *Rhodococcus* sp. isolated from mangrove sediment. *Mar. Pollut. Bull.* **2018**, *127*, 15–21. [CrossRef]
21. Yuan, J.; Ma, J.; Sun, Y.; Zhou, T.; Zhao, Y.; Yu, F. Microbial degradation and other environmental aspects of microplastics/plastics. *Sci. Total Environ.* **2020**, *715*, 136968. [CrossRef]
22. Kesavan, S.; Martin Xavier, K.A.; Nair, M.M.; Gurjar, U.R.; Sukla, S.P.; Jaiswar, A.K.; Bhusan, S.; Abdul Azeez, S.; Deshmukhe, G. Assessment of secondary microplastics trapped in mangrove ecosystem of a highly populated tropical megacity, India. *J. Hazard. Mater. Adv.* **2025**, *17*, 100587. [CrossRef]
23. Zhao, Y.; Qiao, R.; Zhang, S.; Wang, G. Metabolomic profiling reveals the intestinal toxicity of different length of microplastic fibers on zebrafish (*Danio rerio*). *J. Hazard. Mater.* **2021**, *403*, 123663. [CrossRef]
24. Zhang, Q.; Xu, E.G.; Li, J.; Chen, Q.; Ma, L.; Zeng, E.Y.; Shi, H. A Review of Microplastics in Table Salt, Drinking Water, and Air: Direct Human Exposure. *Environ. Sci. Technol.* **2020**, *54*, 3740–3751. [CrossRef]
25. Arpia, A.A.; Chen, W.H.; Ubando, A.T.; Naqvi, S.R.; Culaba, A.B. Microplastic degradation as a sustainable concurrent approach for producing biofuel and obliterating hazardous environmental effects: A state-of-the-art review. *J. Hazard. Mater.* **2021**, *418*, 126381. [CrossRef]
26. Umamaheswari, S.; Priyadarshinee, S.; Kadirvelu, K.; Ramesh, M. Polystyrene microplastics induce apoptosis via ROS-mediated p53 signaling pathway in zebrafish. *Chem. Biol. Interact.* **2021**, *345*, 109550. [CrossRef]
27. Ighodaro, O.; Akinloye, O. First line defence antioxidants-superoxide dismutase (SOD), catalase (CAT) and glutathione peroxidase (GPX): Their fundamental role in the entire antioxidant defence grid. *Alex. J. Med.* **2018**, *54*, 287–293. [CrossRef]
28. Lu, Y.; Zhang, Y.; Deng, Y.; Jiang, W.; Zhao, Y.; Geng, J.; Ding, L.; Ren, H. Uptake and Accumulation of Polystyrene Microplastics in Zebrafish (*Danio rerio*) and Toxic Effects in Liver. *Environ. Sci. Technol.* **2016**, *50*, 4054–4060. [CrossRef]
29. Hamed, M.; Soliman, H.A.M.; Osman, A.G.M.; Sayed, A.E.H. Antioxidants and molecular damage in Nile Tilapia (*Oreochromis niloticus*) after exposure to microplastics. *Environ. Sci. Pollut. Res. Int.* **2020**, *27*, 14581–14588. [CrossRef] [PubMed]
30. Wan, Z.; Wang, C.; Zhou, J.; Shen, M.; Wang, X.; Fu, Z.; Jin, Y. Effects of polystyrene microplastics on the composition of the microbiome and metabolism in larval zebrafish. *Chemosphere* **2019**, *217*, 646–658. [CrossRef] [PubMed]
31. Singh, M.; Prasad, N.; Agarwal, V.; Rai, M. # 5108 sustained IL-6 secretion can lead to an amplified inflammatory response, mediated by the activation of stat3 and nfkb via the IL-6 amplifier loop. *Nephrol. Dial. Transplant.* **2023**, *38*, gfad063c_5108.
32. Huang, J.N.; Wen, B.; Zhu, J.G.; Zhang, Y.S.; Gao, J.Z.; Chen, Z.Z. Exposure to microplastics impairs digestive performance, stimulates immune response and induces microbiota dysbiosis in the gut of juvenile guppy (*Poecilia reticulata*). *Sci. Total Environ.* **2020**, *733*, 138929. [CrossRef] [PubMed]
33. Chen, J.W.; Chen, Y.H.; Lin, S.J. Long-term exposure to oxidized low-density lipoprotein enhances tumor necrosis factor-alpha-stimulated endothelial adhesiveness of monocytes by activating superoxide generation and redox-sensitive pathways. *Free Radic. Biol. Med.* **2006**, *40*, 817–826. [CrossRef] [PubMed]

34. Kim, J.Y.; Choi, C.Y.; Lee, K.J.; Shin, D.W.; Jung, K.S.; Chung, Y.C.; Lee, S.S.; Shin, J.G.; Jeong, H.G. Induction of inducible nitric oxide synthase and proinflammatory cytokines expression by o,p'-DDT in macrophages. *Toxicol. Lett.* **2004**, *147*, 261–269. [CrossRef]
35. Yang, H.; Lai, H.; Huang, J.; Sun, L.; Mennigen, J.A.; Wang, Q.; Liu, Y.; Jin, Y.; Tu, W. Polystyrene microplastics decrease F-53B bioaccumulation but induce inflammatory stress in larval zebrafish. *Chemosphere* **2020**, *255*, 127040. [CrossRef] [PubMed]
36. Zitouni, N.; Bousserrhine, N.; Missawi, O.; Boughattas, I.; Chevre, N.; Santos, R.; Belbekhouche, S.; Alphonse, V.; Tisserand, F.; Balmassiere, L.; et al. Uptake, tissue distribution and toxicological effects of environmental microplastics in early juvenile fish Dicentrarchus labrax. *J. Hazard. Mater.* **2021**, *403*, 124055. [CrossRef]
37. Ding, J.; Zhang, S.; Razanajatovo, R.M.; Zou, H.; Zhu, W. Accumulation, tissue distribution, and biochemical effects of polystyrene microplastics in the freshwater fish red tilapia (*Oreochromis niloticus*). *Environ. Pollut.* **2018**, *238*, 1–9. [CrossRef]

veterinary sciences

MDPI

Article

Metabolic Profiling, Tissue Distribution, and Tolerance Assessment of Bopu Powder in Laying Hens Following Long-Term Dietary Administration

Hongting Wang [1,2,†], Xinhao Wang [1,2,†], Jiaxin Xu [1,2], Zihui Yang [1,2], Zhen Dong [1,2], Jianguo Zeng [1,2,*] and Hua Liu [1,2,*]

1 HunanProvince Key Laboratory of Traditional Chinese Veterinary Medicine, College of Veterinary Medicine, Hunan Agricultural University, Changsha 410128, China; 322860@stu.hunau.edu.cn (H.W.); wxh995651051@stu.hunau.edu.cn (X.W.); 15588521035@163.com (J.X.); yangzihui_2021@hunau.edu.cn (Z.Y.); dongzhen2022@stu.hunau.edu.cn (Z.D.)

2 Chinese Medicinal Materials Breeding Innovation Centre of Yuelushan Laboratory, Changsha 410128, China

* Correspondence: zengjianguo@hunau.edu.cn (J.Z.); 13875829927@163.com (H.L.)

† These authors contributed equally to this work.

Simple Summary: This study assessed the safety and metabolism of Bopu Powder—a herbal veterinary product—in laying hens following 56 days of dietary supplementation. Its major alkaloids, protopine and allocryptopine, were extensively metabolized. Residual levels were low and limited to the liver and kidneys, with occasional detection in eggs only at the highest dose (500 mg/kg). No residues were found in muscle, fat, or other edible tissues. Serum biochemistry and histopathology revealed no adverse effects, supporting the safety of Bopu Powder for long-term use in laying hens.

Abstract: This study aimed to elucidate the metabolic profile and safety of Bopu Powder in laying hens, focusing on its principal alkaloids protopine (PRO) and allocryptopine (ALL). Nine hens received intravenous PRO (1 mg/kg·bw), ALL (1 mg/kg·bw), or saline. Blood (5–120 min post-dose) and feces (48 h collection) were analyzed via LC-Q-TOF-MS, revealing 9 ALL and 12 PRO metabolites, including two novel structures. In a separate 56-day trial, 324 hens were assigned to control, 50 mg/kg Bopu Powder (BP), or 500 mg/kg Bopu Powder (BPX) groups. Post-trial analyses showed PRO and ALL residues primarily in their kidneys (BP: PRO 11.21 ng/g, ALL 6.59 ng/g; BPX: PRO 23.62 ng/g, ALL 7.92 ng/g) and livers (BP: PRO 15.52 ng/g; BPX: PRO 269.49 ng/g, ALL 56.14 ng/g). Residues in eggs occurred exclusively in the BPX group (PRO 26.86 ng/g, ALL 12.29 ng/g). No residues were detected in other tissues (jejunum, ileum, ovaries, oviducts, uterus, muscle, fat, gizzard). Serum biochemical indicators and histopathological examinations were used to evaluate the long-term effects of Bopu Powder. The results confirmed that supplementation at doses up to 500 mg/kg did not induce any significant physiological or histopathological alterations.

Keywords: Bopu Powder; laying hens; metabolism; residues; safety evaluation

1. Introduction

The poppy family plant *Macleaya cordata* (Willd.) R. Br. is rich in alkaloid compounds, with sanguinarine (SAN), chelerythrine (CHE), allocryptopine (ALL), and protopine (PRO) being the most abundant [1]. Modern pharmacological studies have demonstrated that Macleaya cordata extracts exhibit diverse biological activities, including antibacterial [2], anti-inflammatory [3], and antitumor effects [4,5]. Bopu Powder, a new veterinary drug formulation primarily composed of PRO and ALL as its main active ingredients, has been approved for treating *Escherichia coli*-induced diarrhea in chickens [6]. Furthermore, evidence suggests that dietary supplementation with Bopu Powder can enhance growth performance in broilers [7] and improve egg quality in laying hens [8]. However, evidence indicates that PRO/ALL exhibit dose-dependent toxicity, with acute exposure causing pulmonary hemorrhage and hepatorenal injury in mice [9,10], although the mechanisms underlying their chronic toxicity remain unclear. Given the potential cardiotoxicity of PRO, phytopharmaceutical preparations containing PRO/ALL may pose potential health risks to both humans and animals [11].

Guo et al. [12] investigated the tissue distribution of PRO in *Dactylicapnos scandens* using LC-ESI-MS/MS. Following a single oral administration of D. scandens extract (126.6 mg/kg) to rats, PRO residues were highest in the small intestine (9.31 ± 2.45 μg/g), followed by the stomach (2.91 ± 0.19 μg/g) and kidneys (2.33 ± 0.71 μg/g). PRO concentrations fell below the limit of quantification (LOQ, 10 ng/mL) within 8 h post-administration, indicating rapid systemic elimination. Huang et al. [13] employed HPLC-QTOF-MS to study the residues of PRO and ALL in rats. After PRO administration, the highest residual levels were observed in the spleen at 24 h (44.04 ng/g) and the liver at 48 h (20.64 ng/g). Following ALL administration, residues were detected in urine, feces, plasma, and various tissues, albeit at consistently low levels (< 3 ng/g). The highest residue level after PRO administration was found in the liver at 48 h (61.8 ng/g), with levels in other tissues below 6.5 ng/g. Liu et al. [14] examined the residual distribution of PRO and ALL in laying hens' tissues following 7 days of dietary supplementation with Bopu Powder. At a dose of 400 mg/kg feed, PRO residues in eggs peaked on day 7 (2.8 μg/kg) but became undetectable 5 days post-withdrawal; ALL remained undetectable throughout. At the higher dose of 2000 mg/kg feed, residual levels of PRO and ALL reached 16.72 μg/kg and 8.61 μg/kg, respectively, 1 day after withdrawal. PRO was undetectable 7 days post-withdrawal, while ALL residues persisted. Tissue analysis revealed significantly elevated concentrations of both compounds 2 h after withdrawal, followed by a rapid decline within 24 h. No residues were detected in any tissues by 7 days post-withdrawal. The study indicated a low residue risk associated with the 400 mg/kg dose. However, the 2000 mg/kg dose may elevate the residue risk, necessitating a withdrawal period of at least 7 days to ensure food safety.

Currently, research on residues of Bopu Powder in laying hens has focused solely on short-term exposure (7 days) for therapeutic purposes. Short-term exposure studies, while valid for establishing therapeutic withdrawal periods, are critically limited for two reasons: Firstly, they fail to account for the potential of low-level residue accumulation in metabolic tissues (e.g., liver, kidneys) during continuous intake. Secondly, they cannot assess the steady-state residue deposition in eggs or the potential for delayed physiological effects under long-term, zootechnical use conditions. Currently, there remains a lack of systematic evaluation regarding the safety of long-term dietary administration of Bopu Powder throughout the entire laying cycle of hens. To address this data gap, we hypothesize that even prolonged exposure at doses as high as 500 mg/kg will not lead to toxicologically significant residue accumulation of its primary active constituents, PRO and ALL, in edible tissues and eggs, nor will it cause significant adverse physiological effects in laying hens. To test this hypothesis, this study systematically characterizes, for the first time, the metabolic profiles of PRO and ALL in laying hens; quantitatively assesses their residual distribution across multiple tissues; and comprehensively evaluates their safety and tolerability following 56 days of continuous administration through integrated serum biochemical parameters and histopathological examinations. Ultimately, this research aims to elucidate the transformation and residue patterns of PRO and ALL from Bopu Powder in laying hens, thereby providing critical scientific evidence to support the safety of its long-term use.

2. Materials and Methods

2.1. Chemical Reagent

Bopu Powder (Production Batch No.: 190701; specification: total PRO and ALL content not less than 1.5%, with PRO content not less than 1.0%; supplied by Hunan MICOLTA Biological Resources Co., Ltd., Changsha, China.). PRO (Batch No.: 140701, purity $\geq$ 98.0%; Hunan MICOLTA Biological Resources Co., Ltd., Changsha, China.). ALL (Batch No.: 170601, purity $\geq$ 98.0%; Hunan MICOLTA Biological Resources Co., Ltd., Changsha, China.). Sodium barbiturate (purity $\geq$ 97.0%, Merck KGaA, Darmstadt, Germany). Dimethyl sulfoxide (DMSO; Tedia, Fairfield, OH, USA). Acetonitrile and methanol (HPLC grade; Merck KgaA, Germany). Formic acid (LC-MS grade; Tedia, Fairfield, OH, USA). Tetrahydropalmatine standard (TET, Batch No.: 110726-201819, purity 99.8%; National Institutes for Food and Drug Control, NIFDC, Beijing, China). Basal diet (Hunan Jinpai Haotian Biotechnology Co., Ltd., Changsha, China). Physiological saline (Anhui Shuanghe Pharmaceutical Co., Ltd., Wuhu, China). Ultrapure water was prepared in-house using a Milli-Q system (MilliporeSigma, Burlington, MA, USA).

2.2. Solution Preparation

Saline Solution (0.75%): A 0.75% saline solution was prepared by diluting 250 mL of 0.9% saline solution with 50 mL of sterile ultrapure water to a final volume of 300 mL. Dosing Solution: PRO and ALL monomers (2.00 mg each) were accurately weighed and dissolved in 200 μL of dimethyl sulfoxide (DMSO) via sonication. Subsequently, 2 mL of the 0.75% saline solution was added to dilute the mixture. Control Solution: The control solution was prepared by combining 2 mL of the 0.75% saline solution with 200 μL of DMSO, followed by thorough mixing.

2.3. Birds and Grouping

A total of 333 sixty-five-week-old Roman Gray laying hens, exhibiting a laying rate of 85.75 $\pm$ 8.35%, were obtained from the Hunan Institute of Animal Science and Veterinary Medicine. Throughout the experimental period, the hens were housed in cages under standard laboratory conditions, with ad libitum access to feed and water. Following a 7-day acclimatization period, standardized feeding regimens and routine health management procedures, including biosecurity protocols, were implemented. The nutritional composition of the diet was formulated in accordance with the Chinese National Standard GB/T 5916-2020 [15] ("Formula Feed for Laying Hens and Broilers") and contained no added medications; its detailed nutritional profile is presented in Table 1. Nine hens were allocated to the metabolic study. These birds were randomly divided into three groups (n = 3 per group): the blank control group, the PRO group, and the ALL group. Test compounds were administered via intravenous injection at a dose of 1 mg/kg body weight (bw).The remaining 324 hens were utilized for tolerance and residue studies, conducted in compliance with the "Guidelines for Target Animal Tolerance Testing of Feed and Feed Additives in Livestock and Poultry" issued by the Ministry of Agriculture and Rural Affairs [16], P.R. China. Three experimental groups were established: CN (control, no additive), BP (50 mg/kg additive), and BPX (500 mg/kg additive). The study employed a randomized design with 6 replicates per group and 18 hens per replicate. The feeding trial duration was 56 days. This study received approval from the Animal Ethics Committee of Hunan Agricultural University (Approval No.: HUNAULSK 20230807–03).

Table 1. Composition and nutrient levels of basal diets (air-dry basis, %).

Items	Diets
Ingredients, %	
Corn	58
Soybean meal	27
Soybean oil	2.2
Wheat bran	1.5
Limestone	8.28
$CaHPO_4{\cdot}2H_2O$	1.5
NaCl	0.2
Na_2SO_4	0.2
Met	0.12
50%	0.1
Premix 1)	1
Total	100.00
Nutrient Contents 2)	
Metabolic energy, MJ/kg	11.37
Crude protein, %	16.86
Crude fiber, %	3.03
Ca, %	3.63
Total phosphorus, %	0.61
Available phosphorus, %	0.35
Lys, %	0.87
Met, %	0.38
Met + Cys, %	0.66

1) The premix provided the following (per kilogram of complete diet) micronutrients: VA 6 000 IU, VD_3 2 500 IU, VE 25 mg, VK_3 2.25 mg, VB_1 1.8 mg, VB_2 7 mg, VB_6 4 mg, VB_{12} 0.2 mg, *D*-pantothenic acid 12 mg, nicotinic acid 35 mg, biotin 0.14 mg, folic acid 0.8 mg, Cu (as copper sulfate) 11 mg, Zn (as zinc sulfate) 70 mg, Fe (as ferrous sulfate) 60 mg, Mn (as manganese sulfate) 115 mg, Se (as sodium selenite) 0.30 mg, I (as potassium iodide) 0.4 mg. 2) Nutrient levels are calculated values.

2.4. Sample Collection

2.4.1. Identification of Metabolites for Allocryptopine and Protopine in Laying Hens

Plasma: Blood samples were collected from the subcutaneous ulnar vein at 5, 10, 20, 30, 60, and 120 min post-administration. Samples were centrifuged at 4 °C and 3200× *g* for 10 min. The supernatant (plasma) was collected and stored at −80 °C for subsequent metabolite analysis. Additionally, at the end of week 8, from each replicate group, one hen was randomly selected. A 10 mL blood sample was collected from the subcutaneous ulnar vein; 5 mL was transferred into an EDTA-K2 anticoagulant tube, gently mixed, and centrifuged at 3000× *g* for 10 min. The separated plasma was transferred into EP tubes (Eppendorf tubes) and stored at −20 °C for residue analysis. The remaining 5 mL was transferred into a plain serum tube (without additives). Following a 30 min clotting period at room temperature (with the tube inclined), the sample was centrifuged at 3000× *g* for 10 min. The supernatant (serum) was collected into EP tubes and stored at −20 °C for serum biochemical parameter analysis.

Feces: Feces were collected over the intervals 0–6 h, 6–12 h, 12–24 h, and 24–48 h following intravenous administration. All fecal samples were immediately stored at −80 °C.

2.4.2. Quantification of Protopine and Allocryptopine in Laying Hens' Tissues and Eggs

Eggs: During weeks 4 and 8 of the trial, two eggs per replicate group, approximating the mean egg weight, were collected and stored at 4 °C until analysis.

Tissues: At week 8 of the trial, one laying hen per replicate was euthanized via intravenous sodium barbiturate injection following blood collection, in strict adherence to the *AVMA Guidelines for the Euthanasia of Animals: 2020 Edition* [17]. Necropsy was performed to isolate cardiac, hepatic, splenic, renal, ovarian, oviductal, and intestinal tissues. Segments of jejunum, ileum, liver, ovary, oviduct, uterus, pectoral muscle, thigh muscle, gizzard, abdominal fat, and subcutaneous fat were individually wrapped in aluminum foil, flash-frozen in liquid nitrogen, and stored at −80 °C pending analysis. Concurrently, samples of hepatic, renal, oviductal, and uterine tissues were preserved in 10% neutral buffered formalin for histopathological processing.

2.4.3. Serum Biochemistry

At the end of the 8th week, one laying hen was randomly selected from each replicate, and 5 mL of blood was collected from the wing vein using additive-free blood collection tubes. The tubes were inclined at an angle and allowed to clot for 30 min, followed by centrifugation at 3000 r/min for 10 min. The supernatant serum was carefully collected, transferred into EP tubes, and stored at −20 °C for subsequent serum biochemical analysis.

2.5. Sample Pretreatment

Plasma: Plasma samples stored at −80 °C were thawed at room temperature. A pooled plasma sample was prepared by combining equal volumes of plasma from each designated collection timepoint. A 200 μL aliquot of this pooled plasma was precisely transferred into a 1.5 mL microcentrifuge tube. Subsequently, 800 μL of acetonitrile was added to precipitate proteins, and the mixture was vortex-mixed vigorously for 2 min. Following protein precipitation, the tube was sealed and sonicated for 20 min. The sample was then centrifuged at 12,000× *g* for 10 min. The resulting supernatant was carefully collected and evaporated to dryness under a gentle stream of nitrogen at room temperature. The dried residue was reconstituted in 200 μL of acetonitrile, vortex-mixed for 2 min, and centrifuged again at 12,000× *g* for 5 min. Finally, the supernatant was filtered through a 0.22 μm organic membrane filter prior to LC-MS/MS analysis.

Feces: Fecal samples stored at −80 °C were thawed at room temperature and homogenized. A 2.00 g aliquot of the homogenized feces was accurately weighed into a 50 mL centrifuge tube. Ten milliliters of acetonitrile, serving as the extraction solvent, was added to the sample, which was then vortex-mixed for 5 min. The tube was sealed and subjected to sonication for 20 min. The mixture was subsequently centrifuged at 10,000× *g* and 4 °C for 8 min. The supernatant was collected and dried under a nitrogen stream at room temperature. The residue was reconstituted in 0.5 mL of acetonitrile, vortex-mixed for 1 min, and centrifuged at 12,000× *g* for 10 min. The obtained supernatant was filtered through a 0.22 μm organic membrane filter before instrumental analysis.

Eggs: Two grams of homogenized whole egg liquid was transferred into a 50 mL centrifuge tube containing homogenization beads. Twenty microliters of the 1 mg/L internal standard (IS) working solution was precisely added, followed by vortex mixing for 30 s. Subsequently, 10 mL of acetonitrile was added and the mixture was vortex-mixed for 5 min to facilitate extraction. The sample was then centrifuged at 8000× *g* for 5 min. Five milliliters of the resulting supernatant was transferred to a ten-milliliter centrifuge tube, combined with three milliliters of n-hexane, and vortex-mixed for 1 min. After phase separation, the upper n-hexane layer was carefully removed using a rubber-bulb pipette. Then, 1 mL of the lower aqueous acetonitrile phase was collected, evaporated to dryness under a nitrogen stream, and reconstituted in 0.5 mL of acetonitrile with 30 s of vortex mixing. The final extract was filtered through a 0.22 μm organic solvent-compatible membrane prior to LC-MS/MS analysis. Only two quality control (QC) measures were implemented: parallel duplicate samples, and blank/carryover checks.

Tissues: Tissue samples stored at −80 °C were thawed to room temperature. Approximately 2 g of tissue was weighed into a 50 mL screw-cap centrifuge tube. Twenty microliters of freshly prepared 20 mg·L^{-1} IS solution and 1 mL of physiological saline were added. The mixture was homogenized using a high-speed homogenizer for 30 s to form a tissue homogenate. Twenty milliliters of ice-cold extraction solvent consisting of 1% formic acid in acetonitrile–water (95:5, v/v) was added to precipitate proteins, followed by 1 min of vortex mixing. The tube opening was sealed with plastic wrap, and the sample was sonicated for 20 min. Centrifugation was performed at 4 °C and 3200× g for 10 min. Five milliliters of the supernatant was transferred to a ten-milliliter centrifuge tube, mixed with two milliliters of n-hexane, and vortexed for 30 s. After centrifugation at 4 °C and 3200× g for 5 min, the upper lipid layer was aspirated and discarded using a pipette. One milliliter of the supernatant was transferred to a clean ten-milliliter tube and evaporated to dryness under nitrogen at 35 °C. The residue was reconstituted in 1 mL of 1% formic acid in methanol, vortex-mixed for 30 s, and centrifuged at 4 °C and 3200× g for 10 min. The clarified solution was filtered through a 0.22 μm membrane before instrumental analysis. Only two QC measures were implemented: parallel duplicate samples, and blank/carryover checks.

2.6. LC-MS Conditions

2.6.1. LC-QTOF-MS Conditions

Chromatographic separation was performed on a Waters BEH C18 column (2.1 mm × 100 mm, 1.7 μm particle size) using a mobile phase consisting of 0.1% formic acid in water (A) and acetonitrile (B). The column temperature was maintained at 35 °C, and the sample tray was set to 4 °C. The following gradient elution program was employed: 8% B (0–10 min), linearly increased to 20% B (10–15 min), ramped to 85% B (15–25 min), stepped to 95% B (25–25.1 min), held at 95% B (25.1–29.5 min), returned to 8% B (29.5–30 min), and re-equilibrated at 8% B (30–35 min). The injection volume was 5 μL.

Mass spectrometric detection was conducted in positive electrospray ionization (ESI+) mode with the following optimized parameters: capillary voltage, 4000 V; nebulizer gas pressure, 35 psi; drying gas temperature, 350 °C; drying gas flow rate, 12 L·min^{-1}; sheath gas temperature, 350 °C; sheath gas flow rate, 11 L·min^{-1}. High-purity nitrogen served as both the drying gas (≥99.9%) and the collision gas (≥99.999%).

2.6.2. LC-QQQ-MS Conditions

Quantification of ALL and PRO in laying hens' tissues and eggs was performed according to the method established by Liu et al. [14,18]. Chromatographic separation was achieved using an Agilent ZORBAX SB-C18 column (2.1 mm × 50 mm i.d., 1.8 μm particle size) maintained at 35 °C. The mobile phase consisted of 0.1% aqueous formic acid (A) and 0.1% formic acid in acetonitrile (B), delivered at a flow rate of 0.3 mL/min. The gradient elution program was executed as follows: 15% B (0–4 min), increased linearly to 17% B (4–5 min), ramped to 20% B (5–6 min), elevated to 35% B (6–12 min), stepped to 95% B (12–12.1 min), maintained at 95% B (12.1–14 min), returned to 15% B (14–14.1 min), and re-equilibrated at 15% B (14.1–17 min). A 1 μL aliquot was injected for analysis.

Mass spectrometric detection employed an electrospray ionization (ESI) source operating in positive ion mode under the following optimized parameters: capillary voltage, 4000 V; electron multiplier voltage (EMV), 200 V; nebulizer gas pressure, 30 psi; desolvation gas flow rate, 13 L/min; desolvation temperature, 350 °C. Both nebulizing and drying gases were supplied by a nitrogen generator. Multiple reaction monitoring (MRM) mode was utilized for quantitative analysis, with specific precursor–product ion transitions detailed in Table 2.

Table 2. Mass spectrometry parameters and retention times for TET, PRO, and ALL.

Analyte	Retention Time (min)	Monitored Ion Pair (m/z)	Fragmentor (V)	Collision Energy (eV)
TET	5.5	356.0→192 * 356.0→165.0	142	26
PRO	3.8	354.1→189.0 * 354.1→149	128	34
ALL	5.0	370.1→188.1 * 370.1→290	114	22

Note: * Quantify ions.

2.7. Comprehensive Metabolic Panel

The colorimetric methods, employed on a Mindray BS-420 automated biochemical analyzer (Mindray Bio-Medical Electronics Co., Ltd., Shenzhen, China), were used to measure the serum levels of nine biochemical and physiological parameters in laying hens. These parameters included aspartate transaminase (AST), alanine aminotransferase (ALT), alkaline phosphatase (ALP), globulin (GLB), albumin (ALB), uric acid (UA), glucose (GLU), total cholesterol (TC), and triacylglycerol (TG), adhering to the protocols outlined in the reagent kits provided by BioSino Bio-Technology & Science Inc. (Beijing, China).

2.8. Histopathological Examination

Liver, kidney, oviduct, and uterus tissue samples that had been preserved in a fixative solution were embedded in paraffin and then sectioned. After dewaxing, the sections were stained using hematoxylin and eosin (H&E) and subsequently analyzed under a light microscope. In our study, the histopathological examination of liver, kidney, oviduct, and uterus tissues was conducted by two experienced pathologists from the College of Veterinary Medicine, Hunan Agricultural University. The assessment was performed in a blinded manner, meaning that the pathologists were not informed of the group assignments (CN, BP, BPX) of the samples that they were evaluating, so as to prevent any potential bias during the observation and scoring process. Regarding the semi-quantitative scoring criteria, the examination focused on key pathological changes such as inflammatory cell infiltration, steatosis (fat vacuoles), and cellular integrity/architecture. These changes were scored based on their severity and distribution: Inflammatory Cell Infiltration: Scored as 0 (absent), 1 (mild, focal), 2 (moderate, multifocal), or 3 (severe, diffuse). Steatosis (hepatocytes): Scored based on the percentage of affected cells: 0 (<5%), 1 (5–25%), 2 (26–50%), or 3 (>50%). Tubular Epithelial Detachment (Kidney): Scored as 0 (absent), 1 (mild), 2 (moderate), or 3 (severe). General tissue architecture and integrity were assessed qualitatively as either 'normal' or 'altered'.

2.9. Data Processing

Residue data from laying hens were recorded and organized using Microsoft Excel. The replicate was consistently treated as the experimental unit throughout the study. All statistical analyses were performed using the mean values of each replicate (n = 6). Data are expressed as replicate means ± standard deviation (SD). A one-way analysis of variance (ANOVA) followed by Duncan's multiple comparison test was applied, with statistical significance defined at $p < 0.05$.

For the identification of metabolites in laying hens, potential metabolite chromatographic peaks were initially discerned by comparative analysis of MS1 spectra between blank and test samples. Subsequently, a comprehensive investigation of candidate metabolites was conducted through the integrated application of precisely targeted, widely targeted, and untargeted mass spectrometric approaches. Key parameters of all reported ALL and PRO metabolites—including mass-to-charge ratios (m/z), structural formulae, molecular formulae, and characteristic fragmentation patterns—were systematically compiled through extensive literature curation. Plausible metabolite structures were hypothesized based on established biotransformation pathways, with theoretical m/z values and molecular formulae calculated using ChemBioDraw Ultra 14.0. Ions demonstrating chromatographic peak intensities $\geq 10^4$ in blank-subtracted samples were prioritized for subsequent MS/MS analysis. To ensure high-confidence identifications, these candidate ions were subjected to fragmentation only when exhibiting a mass error of <5 ppm for the precursor ion. Structural proposals were inferred through comprehensive interpretation of product ion spectra and characteristic fragmentation pathways, leading to putative annotations at MSI Level 2 confidence for all reported metabolites.

3. Results

3.1. Metabolite Identification

3.1.1. Metabolic Profile of ALL in Laying Hens

Fragmentation Pathways of ALL

Under the present chromatographic conditions, ALL exhibited a retention time of 11.3 min and was detected as the protonated ion $[M + H]^+$ at m/z 370.1634 (molecular formula $C_{21}H_{24}NO_5^+$). As a protopine-type alkaloid, ALL undergoes characteristic fragmentation dominated by retro-Diels–Alder (RDA) cleavage and α-bond fission. The molecular ion fragmented to yield product ions at m/z 206.0794 and 165.0897. Subsequent losses of OH and CH_4 from these ions generated fragments at m/z 189.0772 and 149.0582, respectively. The ion at m/z 206.0794 further eliminated H_2O to form m/z 188.0692. Concurrently, the precursor ion underwent rearrangement followed by sequential neutral losses: dehydration produced m/z 352.1527, demethylation yielded m/z 339.1201, and further dehydration generated m/z 321.1249. The proposed fragmentation pathways are illustrated in Figure 1 (MS1/MS2 spectra) and Figure 2 (fragmentation scheme). Due to the unavailability of authentic standards for the identified metabolites, structural elucidation was based on high-resolution mass spectrometry data and characteristic fragmentation patterns.

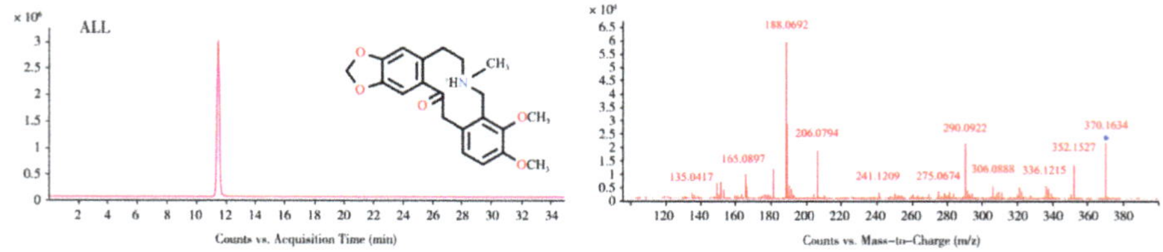

Figure 1. MS and MS/MS spectra of ALL.

Tentative Identification of ALL Metabolites in Laying Hens

Subsequent comparative analysis of dosed versus blank control samples using UPLC-Q-TOF-MS tentatively identified nine ALL metabolites in laying hen plasma and fecal samples, designated as ALL-M1 to ALL-M9. Representative MS1 and MS2 spectra and proposed metabolic pathways are presented in Figures 3 and 4, respectively, with Table 3 summarizing critical identification parameters, including retention times, molecular formulae, observed m/z values, and mass accuracy errors.

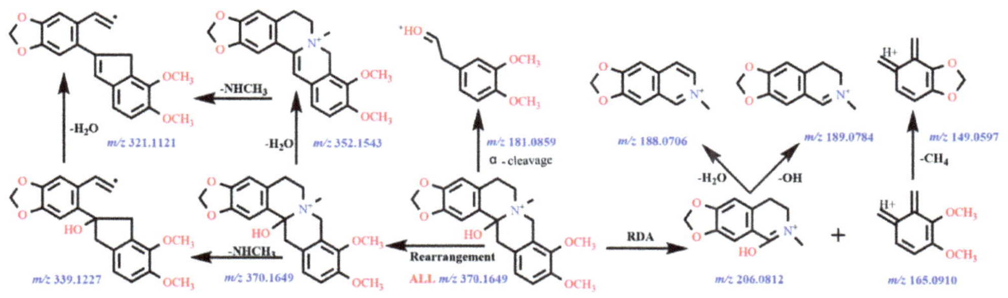

Figure 2. The proposed fragmentation pathways of ALL.

Table 3. Metabolite identification results of ALL in vivo.

Metabolite	Retention Time	Proposed Formula	Measured	Calculated	Error (ppm)	Fragment Ions	Fragmentation Pathway
ALL	11.3	$C_{21}H_{24}NO_5{}^+$	370.1634	370.1649	−4.05	352.1527, 206.079, 165.0897, 188.0692	Absent
ALL-M1	6.7	$C_{20}H_{22}NO_5{}^+$	356.1427	356.1492	−18.25	338.1212, 275.055, 188.0662, 189.0785	Demethylation
ALL-M2	4.1	$C_{26}H_{30}NO_{11}{}^+$	532.1784	532.1813	−5.45	356.1403, 338.1277, 188.0666, 189.0783	Glucuronidation
ALL-M3	9.2	$C_{21}H_{26}NO_5{}^+$	372.1765	372.1805	−10.75	354.1721, 208.0862, 208.0862, 149.0607	Ring Cleavage
ALL-M4	6.4	$C_{20}H_{24}NO_5{}^+$	358.1627	358.1649	−6.14	340.1428, 194.0762, 165.0854, 176.0685	Ring Cleavage and Demethylation
ALL-M5	5.1	$C_{26}H_{32}NO_{11}{}^+$	534.1958	534.1970	−2.25	358.1717, 340.1412, 176.0701	Glucuronidation
ALL-M6	9.7	$C_{21}H_{24}NO_6{}^+$	386.1543	386.1593	−12.95	368.1453, 297.1032, 222.0790, 204.0682	Oxidation
ALL-M7	11.1	$C_{22}H2_8NO_5{}^+$	386.1949	386.1962	−3.37	368.1857, 306.1114, 222.1243, 204.0965	Methylation
ALL-M8	9.5	$C_{21}H_{26}NO_5{}^+$	372.1727	372.1805	−20.96	354.1720, 208.0797, 189.0758, 149.0607	Ring Cleavage
ALL-M9	7.9	$C_{20}H_{22}NO_5{}^+$	356.1445	356.1492	−13.20	338.1220, 275.0637, 188.0701, 189.0780	Demethylation

(1) Plasma

In addition to the prototype compound ALL, seven metabolites designated as ALL-M1 through ALL-M7 were detected in pooled plasma samples collected from 5 to 120 min post-administration.

The metabolite ALL-M1 exhibited a retention time of 6.7 min. Detected in positive ion mode as the $[M + H]^+$ ion at m/z 356.1427, it displayed a mass decrease of 14 Da compared to the prototype compound ALL. This mass shift indicates the loss of a methylene group, suggesting that ALL-M1 is likely a demethylated metabolite of ALL. Fragmentation of the precursor ion produced a fragment at m/z 338.1212, corresponding to dehydration. Further sequential losses of methylamine and methanol from this fragment yielded the ion observed at m/z 275.0552. A distinct fragment ion at m/z 206.0741 originated from RDA cleavage of the precursor ion. Dehydration of this RDA-derived fragment generated the ion at m/z 188.0662, while loss of a hydroxyl group produced the ion at m/z 189.0785. The specific RDA fragmentation pattern and the resulting fragment ions localized the hydroxyl and methoxy substituents to the C9 and C10 positions. Consequently, the structural evidence identifies ALL-M1 as a demethylated derivative of ALL at either the C9 or C10 position.

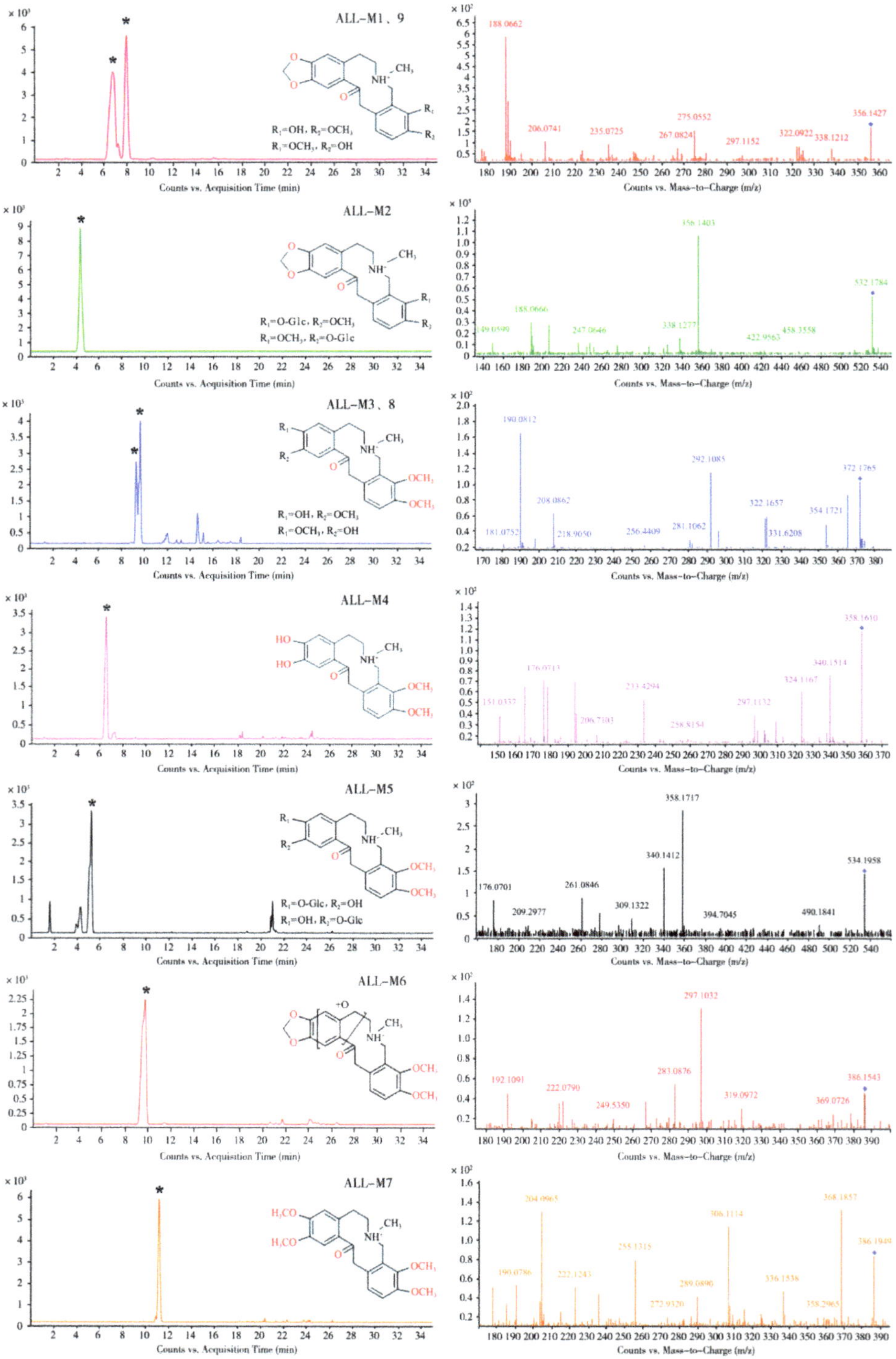

* Indicates the chromatographic peak of the metabolite of interest.

Figure 3. MS and MS/MS spectra for metabolites of ALL.

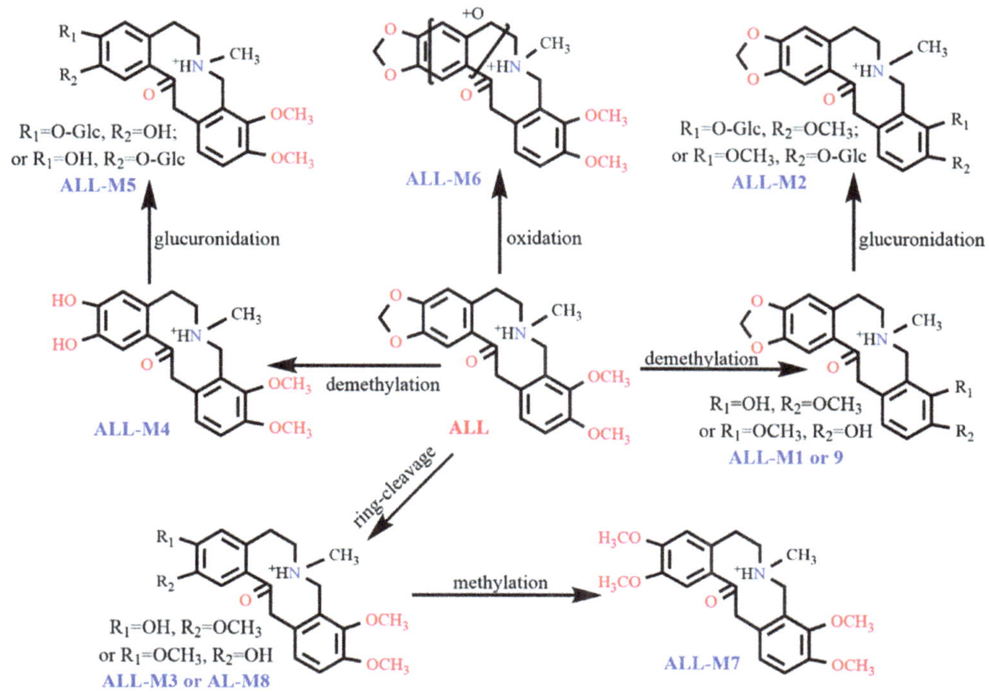

Figure 4. The proposed biotransformation patterns of ALL in vivo.

The metabolite ALL-M2 exhibited a retention time of 4.1 min and an $[M + H]^+$ ion at m/z 532.1784, representing a mass increase of 176 Da compared to ALL-M1. Its major fragment ions at m/z 356.1403, 338.1277, 188.0666, and 189.0783 were identical to those of ALL-M1. Therefore, ALL-M2 was inferred to be the glucuronide conjugate of ALL-M1.

ALL-M3 displayed a retention time of 9.2 min and an $[M + H]^+$ ion at m/z 372.1765, corresponding to a mass increase of 2 Da relative to the prototype compound ALL. Dehydration of the precursor ion yielded the fragment ion at m/z 354.1721. The fragment ion at m/z 208.0862 originated from an RDA cleavage of the parent nucleus; subsequent dehydration of this RDA fragment generated the ion at m/z 190.0812. The fragment ion at m/z 149.0607 resulted from the loss of methane from the RDA-derived fragment ion m/z 165.0897. The fragment ions at m/z 208.0862 and 149.0607 collectively indicate methylenedioxy ring cleavage at positions C2 and C3 on the ALL scaffold.

ALL-M4 had a retention time of 6.4 min and an $[M + H]^+$ ion at m/z 358.1627, representing a mass decrease of 12 Da relative to ALL and a loss of 14 Da compared to ALL-M3. Dehydration of the precursor ion produced the fragment ion at m/z 340.1428. The appearance of fragment ions at m/z 165.0854 and m/z 194.0762 signifies RDA cleavage of the ALL-M4 structure, with hydroxyl groups located at C2 and C3 and methoxy groups at C9 and C10. The fragment ion at m/z 176.0685 arose from dehydration of the m/z 194.0762 ion. Consequently, ALL-M4 was identified as a demethylated derivative of ALL-M3.

ALL-M5 showed a retention time of 5.1 min and an $[M + H]^+$ ion at m/z 534.1958, indicating a mass increase of 176 Da relative to ALL-M4. Its major fragment ions corresponded to those of ALL-M4. Thus, ALL-M5 was inferred to be the glucuronide conjugate of ALL-M4, with the glucuronyl moiety likely substituting a hydroxyl group at either the C2 or C3 position.

ALL-M6 eluted at 9.7 min with an $[M + H]^+$ ion at m/z 386.1543, corresponding to a mass increase of 16 Da compared to ALL. Dehydration of the precursor ion yielded the fragment ion at m/z 368.1453. This ion subsequently lost a molecule of C_3H_5NO to generate the fragment ion at m/z 297.1032. Concurrently, the ion at m/z 368.1453 underwent RDA cleavage analogous to ALL, producing the fragment ion at m/z 222.0790, which underwent further dehydration to yield the ion at m/z 204.0682. Therefore, ALL-M6 is proposed to be an epoxidized derivative of ALL on ring A or B [19].

ALL-M7 exhibited a retention time of 11.1 min and an $[M + H]^+$ ion at m/z 386.1949, representing a mass increase of 14 Da relative to ALL-M3. Combined with its fragmentation pattern, this suggests ALL-M7 is the methylated derivative of ALL-M3. Fragment ions at m/z 368.1857, 306.1114, 222.1243, and 204.0965 were generated through dehydration of the precursor ion, loss of methylamine and methanol, RDA cleavage of the precursor ion, and subsequent dehydration of the RDA fragment, respectively.

(2) Fecal Samples

Analysis of fecal samples collected 0–6 h post-administration revealed the presence of seven metabolites. Four of these metabolites—specifically, ALL-M1 through ALL-M4—were also detected in plasma. The remaining three metabolites were designated as ALL-M8 and ALL-M9; their respective primary and secondary mass spectra are presented in Figure 3.

Within the fecal samples, two compounds were detected with identical m/z values of 372.1727 and 372.1765, exhibiting retention times of 9.2 min and 9.5 min, respectively. One compound corresponded to ALL-M3, while the other represented an isomeric form, designated as ALL-M8. Both isomers exhibited identical fragmentation pathways, with all fragment ions matching precisely. This observation collectively indicates that ALL-M3 and ALL-M8 are positional isomers differing in the locations of one hydroxyl and one methoxy group on the C2 and C3 positions.

Similarly, metabolites at m/z 356.1427 and 356.1430 were observed with retention times of 6.7 min and 7.9 min, respectively. One metabolite was identified as ALL-M1, and the other as its isomeric counterpart, designated as ALL-M9. The fragmentation patterns of ALL-M1 and ALL-M9 were indistinguishable, with complete agreement in their fragment ion profiles. Consequently, this data suggests that ALL-M9 and ALL-M1 are positional isomers resulting from the alternate positioning of one hydroxyl and one methoxy group on the C9 and C10 positions.

3.1.2. Metabolism of PRO in Laying Hens

Fragmentation Pathways of PRO

Under the chromatographic conditions employed, PRO exhibited a retention time of 10.3 min and was detected in its protonated form at m/z 354.1343, corresponding to the molecular formula $C_{20}H_{20}NO_5^+$. As a representative protopine alkaloid, PRO underwent characteristic RDA and alpha-cleavage reactions, generating product ions at m/z 149.0594, 206.0797, and 165.0539. The ion at m/z 206.0797 subsequently lost a hydroxyl group and a water molecule, yielding fragment ions at m/z 189.0773 and 188.0696, respectively. Further fragmentation involved the loss of methylamine from PRO, producing the fragment ion at m/z 323.0933. Concurrently, the presence of a carbonyl group facilitated an intramolecular rearrangement yielding a hydroxyl intermediate, culminating in the loss of water to form the fragment ion at m/z 336.1242, corresponding to a mass loss of 18.0106 Da from the precursor ion. Figures 5 and 6 depict the primary and secondary mass spectra of PRO, respectively, alongside the proposed fragmentation scheme.

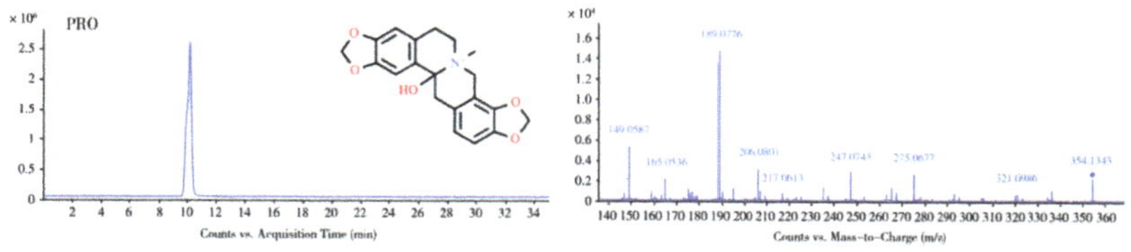

Figure 5. MS and MS/MS spectra of PRO.

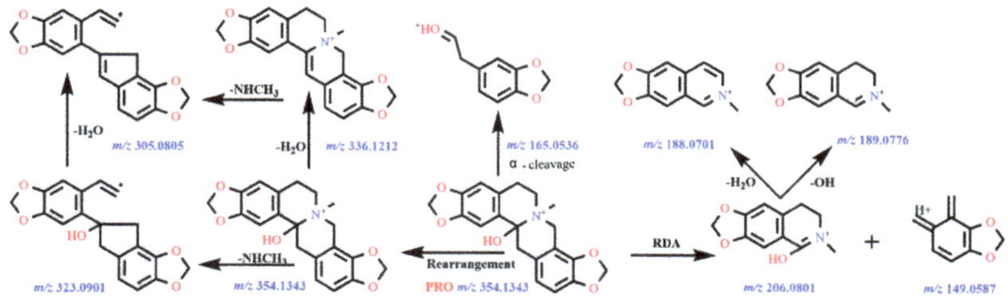

Figure 6. The proposed fragmentation pathways of PRO.

Identification of PRO Metabolites in Laying Hens

Employing UPLC-Q-TOF-MS technology, comparative analysis of dosed samples against blank control samples enabled the inference of twelve PRO metabolites in plasma and fecal matrices of laying hens. These metabolites were designated PRO-M1 through PRO-M12. Primary and secondary mass spectra for these metabolites are presented in Figures 7 and 8, respectively, while Table 4 comprehensively summarizes detailed metabolite characteristics, including retention times, molecular formulae, m/z values, and mass errors.

Table 4. Metabolite identification results of PRO in vivo.

Metabolite	Retention Time	Proposed Formula	Measured	Calculated	Error (ppm)	Fragment Ions	Fragmentation Pathway
PRO	10.3	$C_{20}H_{20}NO_5^+$	354.1343	354.1336	1.98	336.124, 149.0594, 206.0797 165.053, 189.0773, 188.0696	Absent
PRO-M1	5.6	$C_{19}H_{20}NO_5^+$	342.1342	342.1336	1.75	324.126, 194.0732, 149.1772, 176.0626, 177.0768	Ring cleavage, demethylation, or rearrangement; ring cleavage and demethylation
PRO-M2	11.1	$C_{22}H_{28}NO_5^+$	386.1941	386.1962	−5.44	368.197, 222.1093, 165.0916, 204.0964, 205.1066	Methylation
PRO-M3	6.1	$C_{19}H_{20}NO_5^+$	342.1300	342.1336	−10.52	324.121, 206.0793, 188.0703, 189.0786	Ring cleavage and demethylation
PRO-M4	6.4	$C_{19}H_{20}NO_5^+$	342.1365	342.1336	8.48	324.123, 194.0730, 149.1692, 176.061, 177.0765	Ring cleavage, demethylation, or rearrangement; ring cleavage and demethylation
PRO-M5	3.3	$C_{25}H_{28}NO_{11}^+$	518.1633	518.1657	−4.63	500.155, 342.1335, 324.1236, 176.0701	Glucuronidation
PRO-M6	4.3	$C_{25}H_{28}NO_{11}^+$	518.1626	518.1657	−5.98	342.130, 324.1193, 194.0785, 176.0668	Glucuronidation
PRO-M7	4.6	$C_{25}H_{28}NO_{11}^+$	518.1656	518.1657	−0.19	342.134, 324.1229, 206.0816, 188.0714	Glucuronidation
PRO-M8	4.2	$C_{20}H_{22}NO_6^+$	372.1389	372.1442	−14.24	206.120, 189.0965, 151.0659, 167.0703	Ring cleavage and hydroxylation
PRO-M9	4.3	$C_{26}H_{30}NO_{11}^+$	532.1796	532.1813	−3.19	356.148, 338.1295, 206.0812, 188.0654, 189.0785	Glucuronidation
PRO-M10	5.4	$C_{26}H_{30}NO_{11}^+$	532.1816	532.1813	0.56	356.148, 338.1295, 208.0968, 190.0862, 191.0941	Glucuronidation
PRO-M11	5.6	$C_{26}H_{32}NO_{11}^+$	534.1925	534.1970	−8.42	358.1566	Glucuronidation
PRO-M12	7.2	$C_{20}H_{22}NO_5^+$	356.1556	356.1492	17.97	238.1479, 220.1079, 202.0968, 188.0983	Methylation, etc.

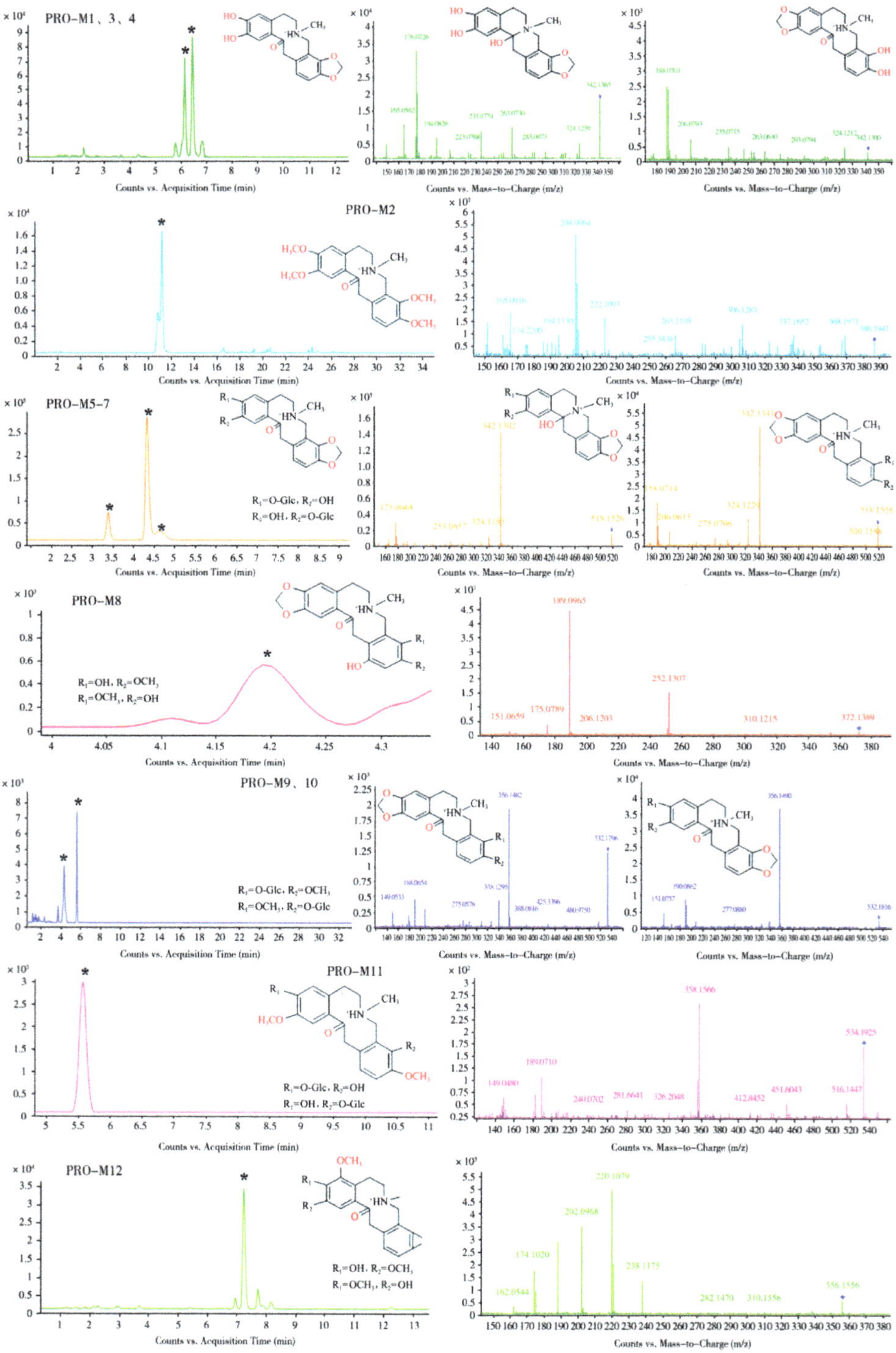

* Indicates the chromatographic peak of the metabolite of interest.

Figure 7. MS and MS/MS spectra for metabolites of PRO.

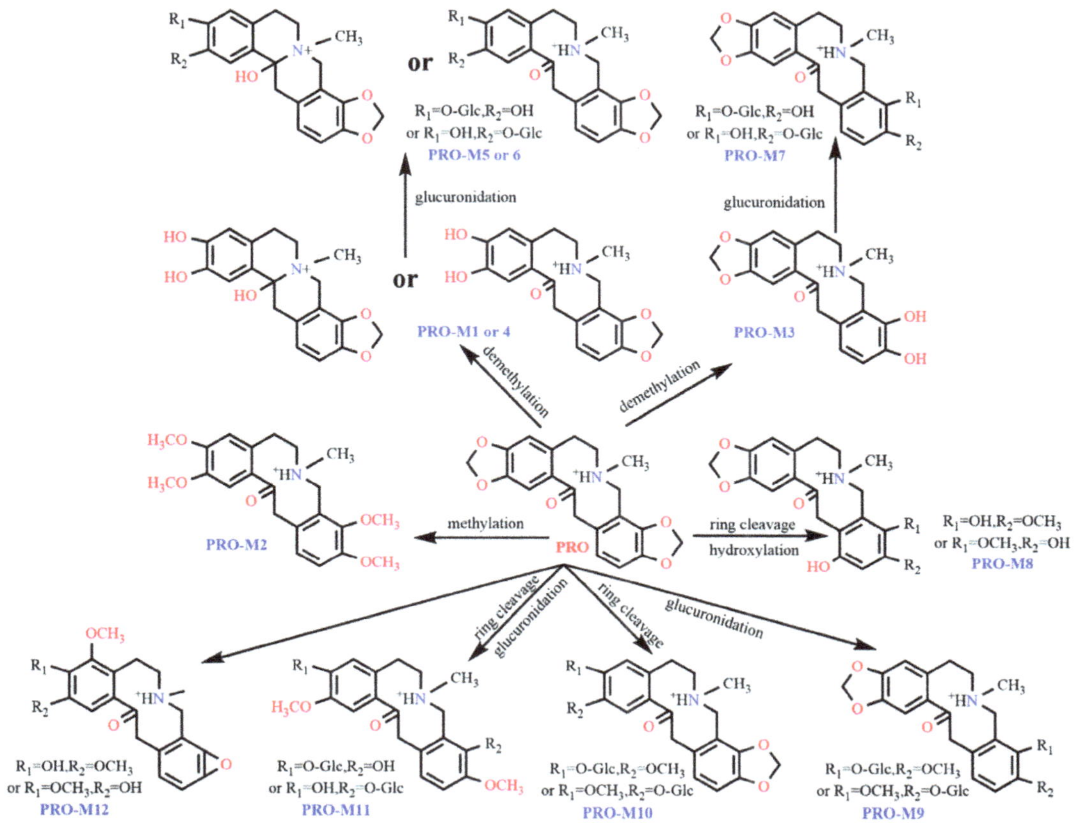

Figure 8. The proposed biotransformation patterns of PRO in vivo.

(1) Plasma

Analysis of pooled plasma samples collected 5 to 120 min post-administration detected the PRO prototype alongside two metabolites, designated PRO-M1 and PRO-M2.

The metabolite PRO-M1 exhibited a retention time of 5.6 min and an *m/z* value of 342.1313, which is 12 Da lower than that of the protonated molecular ion of the parent compound PRO. The parent nucleus structure initially underwent dehydration (−18 Da) at a specified collision energy to form the product ion at *m/z* 324.1263. Concurrently, PRO-M1 underwent RDA cleavage, yielding product ions at *m/z* 194.0732 and *m/z* 149.1772. The ion at *m/z* 194.0732 subsequently lost H_2O and OH, forming fragment ions at *m/z* 176.0626 and 177.0768, respectively. Analysis of the two RDA fragments suggests that the methylenedioxy bridge at C2 and C3 of PRO-M1 cleaved, resulting in the loss of a CH_2 molecule and the formation of two neighboring dihydroxy groups. If, following ring cleavage and demethylation, the parent nucleus structure underwent reductive cleavage of an oxygen double bond, the resulting fragments and cleavage patterns were identical. Therefore, PRO-M1 may undergo either of these two reactions, leading to its identification as either the ring-opened demethylated product at C2 and C3 of PRO or the ring-opened, demethylated, and reduced product at C2 and C3 of PRO.

PRO-M2 showed a retention time of 11.1 min and an *m/z* value of 386.1941. Based on the fragment ions at *m/z* 222.1093 and *m/z* 165.0916, it was inferred that this metabolite underwent ring cleavage and methylation at both the C2,3 and C9,10 positions. During mass spectrometric fragmentation, additional ions were formed, including $[M–H_2O + H]^+$

at m/z 368.1971 and $[M–NH_2CH_3 + OCH_3]^+$ at m/z 306.1283. The ions at m/z 204.0964 and 205.1066 originated from the product ion at m/z 222.1093 through dehydration and dehydroxylation, respectively. Based on the MS/MS fragment information, PRO-M2 was identified as the tetramethylated product of PRO at the C2,3,9,10 positions.

(2) Feces

Eleven metabolites were detected in fecal samples collected 0–6 h post-dosing. One metabolite, PRO-M1, had been previously identified in plasma. The remaining ten metabolites were designated PRO-M3 through PRO-M12.

PRO-M1, PRO-M3, and PRO-M4 exhibited m/z values of 342.1342, 342.1347, and 342.1365, with retention times of 5.6 min, 6.1 min, and 6.4 min, respectively, suggesting that they may be isomers. The fragment ion of PRO-M3 at m/z 324.1212, 18 Da lower than its precursor ion, indicates a potential dehydration reaction. Characteristic RDA fragment ions at m/z 206.0793, 188.0703, and 189.0786 suggest ring-opening demethylation of the methylenedioxy bridge at the C9,10 positions. PRO-M3 is therefore proposed as the ring-opened demethylated product of PRO at C9,10. PRO-M4 shared identical major fragment ions and mass spectrometric fragmentation patterns with PRO-M1, indicating that it is either the ring-opened demethylated product or the ring-opened, demethylated, reduced product of PRO at C2,3.

PRO-M5, PRO-M6, and PRO-M7 displayed m/z values of 518.1633, 518.1626, and 518.1606, with retention times of 3.3 min, 4.3 min, and 4.6 min, respectively. The 176 Da mass increase relative to m/z 342.1336 suggests the addition of a glucosyl group. As PRO-M7 produced fragment ions identical to those of PRO-M3, it was identified as the glucuronide conjugate of PRO-M3. PRO-M5 and PRO-M6 yielded secondary fragments fully consistent with PRO-M1 and PRO-M4, identifying them as glucuronide conjugates of either PRO-M1 or PRO-M4.

PRO-M8 (m/z 372.1389, retention time 4.2 min) exhibited an 18 Da increase compared to PRO, tentatively suggesting ring-opening (+2 Da) and hydroxylation (+16 Da). Fragment ions at m/z 206.1203 and its dehydroxylated product at m/z 189.0965 indicate that methylenedioxy bridge cleavage did not occur at C9,10 and hydroxylation did not take place on this aromatic ring. The product ion at m/z 151.0659 likely originated from the RDA fragment m/z 167.0703 via cyclization (+2 Da) and dehydration (-18 Da), confirming methylenedioxy bridge cleavage at C2,3. The hydroxyl group was positioned at either C11 or C12. Based on this evidence, PRO-M8 is proposed as a ring-opened hydroxylated metabolite of PRO.

PRO-M9 and PRO-M10 (m/z 532.1796 and 532.1816, retention times 4.3 min and 5.4 min, respectively) were isomers but exhibited distinct fragment ions, indicating different reaction sites. Shared fragment ions at m/z 356.1482 and m/z 338.1295 revealed glucosylation (+176 Da) of m/z 356.1482 and subsequent dehydration in both. Additional fragments of PRO-M9 included m/z 206.0812, 188.0654, and 189.0785. The RDA fragmentation patterns suggested that the methylenedioxy bridge at C9,10 remained intact. The fragment at m/z 149.0533 was likely derived from dehydrogenative cyclization of the RDA fragment m/z 151.0754, indicating substitution at C2,3 by a methoxy group and a hydroxyl group. Therefore, PRO-M9 is proposed as the glucuronide conjugate at either C2 or C3 following ring-opening of PRO. Fragments of PRO-M10 (m/z 208.0968, 190.0862, 191.0941) suggested modification at the C9,10 methylenedioxy bridge, identifying it as the glucuronide conjugate at either C9 or C10 following ring-opening of PRO.

PRO-M11 (m/z 534.1925, retention time 5.6 min) yielded a major fragment at m/z 358.1566. This suggests ring-opening at both C2,3 and C9,10 of PRO, forming two hydroxyl and two methoxy groups, with glucosylation replacing a hydrogen on one hydroxyl group at an undetermined position.

PRO-M12 (m/z 356.1556, retention time 7.2 min) exhibited secondary fragments similar to those of PRO-M8 and PRO-M9, with an 18 Da mass increase. It underwent RDA cleavage, yielding a product ion at m/z 238.1479, followed by sequential loss of water to m/z 220.1079. Fragment m/z 202.0968 further lost CH2 to form m/z 188.0983. This fragmentation pattern suggests that, after ring-opening demethylation of the C2,3 methylenedioxy bridge, PRO-M12 underwent dehydration, forming an oxo bridge. The C9,10 methylenedioxy bridge underwent ring-opening, forming a hydroxyl and a methoxy group, concurrent with hydroxylation at C4 and subsequent methylation to form a methoxy group.

3.2. Residues of Bopu Powder in Laying Hens' Tissues

As shown in Table 5, after feeding laying hens a diet supplemented with Bopu Powder for 56 days in the late laying period, no residues of PRO or ALL were detected in the BP group within egg, breast muscle, thigh muscle, skin with fat, abdominal fat, gizzard, plasma, jejunum, ileum, ovary, oviduct, or uterus tissues, with levels below the limit of detection (LOD) of 1 ng/g wet weight. Similarly, in the BPX group, no detectable residues of PRO or ALL were found in breast muscle, thigh muscle, skin with fat, abdominal fat, gizzard, plasma, jejunum, ileum, ovary, oviduct, or uterus tissues. However, residues of PRO at 26.86 ng/g and ALL at 12.29 ng/g were detected in eggs from the BPX group. Low levels of residues were detected in liver and kidney tissues. Specifically, PRO residues were measured at 15.52 ng/g in liver tissue from the BP group, while the BPX group showed PRO residues at 269.49 ng/g and ALL residues at 56.14 ng/g in liver tissue. In kidney tissue, residues of PRO and ALL were 11.21 ng/g and 6.59 ng/g in the BP group, respectively, and 23.62 ng/g and 7.92 ng/g in the BPX group, respectively.

Table 5. The residues of PRO and ALL in the tissues of laying hens (ng/g, n = 6, mean ± sd).

Tissues	Analytes	CN	BP	BPX
Eggs	PRO	ND	ND	26.86 ± 15.12
	ALL	ND	ND	12.29 ± 4.69
Breast muscle	PRO	ND	ND	ND
	ALL	ND	ND	ND
Thigh muscle	PRO	ND	ND	ND
	ALL	ND	ND	ND
Skin with fat	PRO	ND	ND	ND
	ALL	ND	ND	ND
Abdominal fat	PRO	ND	ND	ND
	ALL	ND	ND	ND
Gizzard	PRO	ND	ND	ND
	ALL	ND	ND	ND
Plasma	PRO	ND	ND	ND
	ALL	ND	ND	ND
Liver	PRO	ND	ND	ND
	ALL	ND	ND	56.14 ± 9.39
Kidney	PRO	ND	11.21 ± 8.87	23.62 ± 6.94
	ALL	ND	6.59 ± 4.61	7.92 ± 4.12
Jejunum	PRO	ND	ND	ND
	ALL	ND	ND	ND
Ileum	PRO	ND	ND	ND
	ALL	ND	ND	ND

Table 5. *Cont.*

Tissues	Analytes	CN	BP	BPX
Ovary	PRO	ND	ND	ND
	ALL	ND	ND	ND
Oviduct	PRO	ND	ND	ND
	ALL	ND	ND	ND
Uterus	PRO	ND	ND	ND
	ALL	ND	ND	ND

ND: not detected.

3.3. Serum Biochemistry and Histopathological Examination

As shown in Table 6, serum ALT and AST levels in laying hens were significantly decreased in both the BP and BPX groups compared to the CN group. Dietary supplementation with 50 mg/kg Bopu Powder significantly reduced serum TG and TC levels.

Table 6. Effects of Bopu Powder on plasma biochemical parameter indices of laying hens.

Item	CN	BP	BPX	SEM	p-Value
ALT, U/L	62.48 ^{b}	38.30 ^{a}	32.65 ^{a}	7.738	0.032
AST, U/L	194.10 ^{b}	130.62 ^{a}	126.62 ^{a}	5.58	0.035
LDH, U/L	347.38	236.50	235.80	17.202	0.161
T-Bil, μmol/L	46.40	47.90	49.26	0.506	0.077
ALB, g/L	5.97	9.22	8.63	0.669	0.192
GLU, mmol/L	9.04	11.69	9.71	0.624	0.320
UREA, mmol/L	1.06	2.06	0.97	0.221	0.205
TG, mmol/L	16.67 ^{b}	5.71 ^{a}	10.01 ^{ab}	1.722	0.039
TC, mmol/L	6.36 ^{b}	3.82 ^{a}	5.44 ^{ab}	0.305	0.008
HDL-C, mmol/L	0.156 ^{b}	0.216 ^{a}	0.164 ^{b}	0.007	<0.001
LDL-C, mmol/L	0.664	0.501	0.434	0.098	0.713

^{a,b} Different superscript letters indicate statistically significant differences at $p < 0.05$.

Histological analysis of liver sections stained with HE (Figure 9) showed mild inflammatory cell infiltration and a few fat vacuoles present in the CN group. Conversely, the BP and BPX groups demonstrated normal liver architecture, characterized by densely packed hepatocytes and an absence of noticeable lesions. The evaluation of HE-stained kidney sections (Figure 9) revealed an intact renal structure, healthy cells, and distinct nucleoli across all groups (CN, BP, BPX), without any significant lesions. Nonetheless, the CN group exhibited slight detachment of tubular epithelial cells alongside mild inflammatory cell infiltration, while these alterations were less evident in the BP and BPX groups. The analysis of oviduct (Figure 9) and uterine (Figure 9) sections stained with HE showed typical tissue structure, cellular integrity, and closely arranged epithelial layers, with no visible pathological alterations such as inflammatory infiltration in any of the groups. No noteworthy differences were identified among the three experimental groups regarding these reproductive tissues.

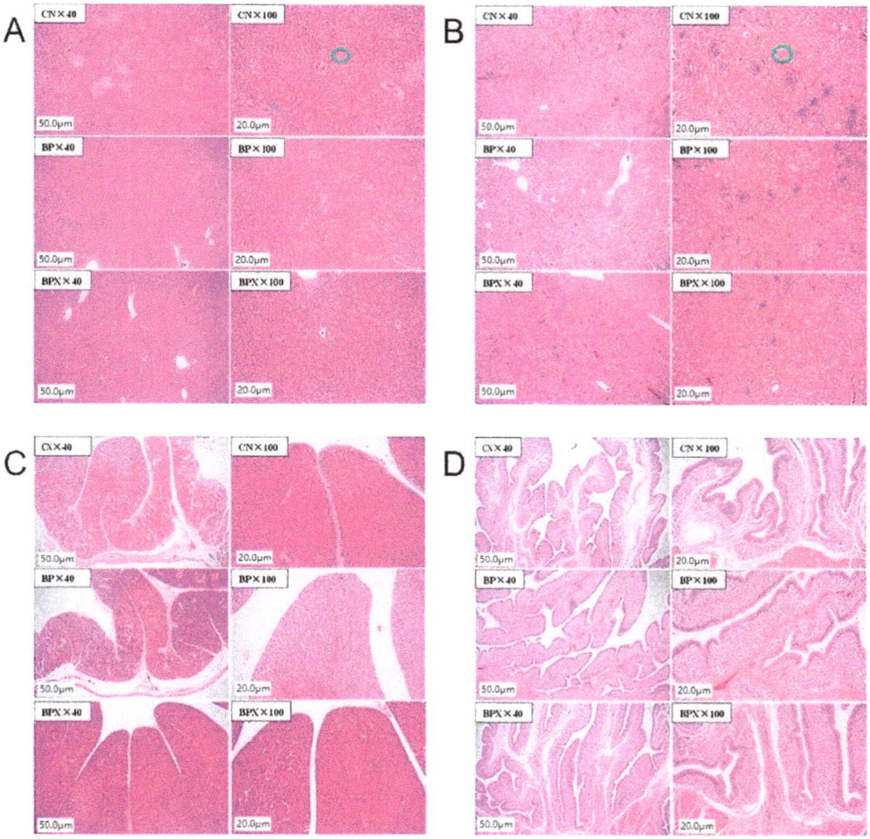

The circular area represents the pathological changes in key structures.

Figure 9. Layer tissue HE sectioning: (**A**) liver; (**B**) kidney; (**C**) oviduct; (**D**) uterus.

4. Discussion

4.1. Numbers and Types of Metabolites

Comparative analysis of metabolite profiles revealed interspecies differences between laying hens and rats. Following oral administration of ALL to rats, Huang et al. identified 25 metabolites, including isomers. While Phase I metabolic reactions were nearly identical to those observed in our study, Phase II metabolism in rats yielded glucuronidated, glucosylated, and sulfated conjugates. In contrast, the latter two conjugates were undetectable in laying hens [19]. In vitro metabolic data corroborate this pattern: Huang et al. detected eight and five metabolites for ALL and PRO, respectively, in rat liver S9 fractions, compared to only four and three metabolites in porcine liver S9 fractions, respectively, with substantially lower metabolite abundance in the latter system [13]. Wu et al. further characterized ALL metabolism in broiler hepatic microsomes using stable isotope tracing, identifying five key metabolites, with demethylated derivatives predominating [20].

Two factors may explain these discrepancies: First, the relatively low dosage administered in our study resulted in limited abundance of potential metabolites. Many candidates exhibited insufficient signal intensity in primary mass spectrometry for confirmatory secondary mass spectrometry analysis, reducing the number of identifiable metabolites. Minor mass-to-charge ratio deviations occurred in the few detectable metabolites due to low ion abundance. Second, consistent evidence from in vivo and in vitro studies indicates substantial interspecies variation in both metabolite diversity and quantity.

4.2. Metabolic Pathways

Structural elucidation of secondary metabolite fragments revealed that both PRO and ALL metabolites share an identical parent nucleus skeleton. These compounds exclusively underwent RDA fragmentation characteristic of protopine alkaloids, followed by intramolecular rearrangement with water elimination via carbonyl groups. Integrating published data on ALL and PRO metabolism with our experimental results [19], the primary metabolic pathways include methylation, demethylation, ring cleavage, oxidation, hydroxylation, and glucuronidation. Notably, these biotransformations predominantly occur at the methylenedioxy or methoxy groups positioned at C2/C3 and C9/C10 of the parent nucleus, and on their associated benzene rings.

4.3. Residues of Bopu Powder in Laying Hens' Tissues

In this study, laying hens were fed diets supplemented with Bopu total alkaloids powder at 50 mg/kg or 500 mg/kg for 56 days. The 50 mg/kg dose was selected based on its established efficacy in improving egg quality and antioxidant capacity without adverse effects [8]. The 500 mg/kg dose was included not only to investigate residue deposition patterns under exaggerated exposure conditions but also to comprehensively evaluate tolerance. The results demonstrated a favorable safety profile for Bopu Powder. No residues of PRO or ALL were detected in edible tissues (eggs, breast muscle, or thigh muscle) at the 50 mg/kg dose. Residues were confined to metabolizing organs (liver and kidneys), consistent with distribution patterns observed in previous studies [14]. At 500 mg/kg, low but detectable residue levels were found in eggs (PRO: 26.86 ng/g; ALL: 12.29 ng/g), indicating potential accumulation at excessive doses, and highlighting the need for a withdrawal period when dosages exceed recommended levels.

These findings confirm a 10-fold safety margin for the 50 mg/kg dose and provide practical guidance for the safe use of Bopu Powder under field conditions. This study supports its application as a long-term feed additive at recommended concentrations and emphasizes adherence to proper dosage regimens to minimize potential risks.

4.4. Tolerance Evaluation of Bopu Powder in Laying Hens

During their late production phase, laying hens frequently develop hepatic lipid metabolism disorders and fatty liver syndrome due to intensive egg-laying activity and physiological aging [21]. The serum biochemical indices observed in this study, including reduced ALT and AST levels and improved TG and TC profiles, suggest a potential positive effect of dietary supplementation with 50 mg/kg Bopu Powder on hepatic lipid metabolism and liver function. These improvements may be attributed to the known antioxidant and anti-inflammatory properties of protopine alkaloids, which could help mitigate oxidative stress and modulate lipid metabolism pathways in the liver. Notably, all experimental groups received the same basal diet with identical premix supplementation, ensuring that micronutrients such as selenium and copper did not confound the assessment of hepatic safety.

Histopathological examination of liver and kidney sections revealed no treatment-induced tissue damage at either 50 or 500 mg/kg supplementation levels. Moreover, the supplementation reduced hepatic inflammatory infiltration and steatosis while decreasing renal tubular epithelial cell detachment, indicating potential protective effects on hepatic and renal tissues. These observations align with documented hepatorenal-protective properties of active constituents including PRO and ALL [22,23]. Previous studies established that PRO administration at 11–22 mg/kg body weight effectively counteracts carbon tetrachloride, acetaminophen, and simvastatin-induced hepatotoxicity [24]. In complementary research, Liu et al. demonstrated in broilers that 40 mg/kg dietary supplementation improves hepatic health through inhibition of the TLR4/MyD88/NF-κB/NLRP3 signaling pathway, consequently reducing caspase-1-mediated hepatic pyroptosis [7].

Laying hens' production involves continuous ovulation and egg-laying activity, where reproductive system health and functionality are critical for maintaining high production performance. In this study, dietary supplementation with 50 or 500 mg/kg Bopu Powder induced no significant pathological alterations in ovarian, oviductal, or uterine tissues based on histological examination, indicating the absence of toxic effects on reproductive organs. Supporting these findings, Yang et al. demonstrated that administering 400 mg/kg (clinical therapeutic dose) or 2000 mg/kg (5-fold clinical dose) to breeder hens for 10 consecutive days negatively impacted neither reproductive performance nor egg fertility and hatchability rates [25]. Furthermore, long-term supplementation at 500 mg/kg (10-fold recommended dose) caused no significant adverse effects on serum biochemical parameters or tissue integrity in laying hens, confirming excellent tolerance at this dosage level.

5. Conclusions

This study employed LC-QTOF-MS to identify, for the first time, nine ALL and twelve PRO metabolites in laying hens' plasma and feces. Among these, two novel metabolites—PRO-M1 (4) and PRO-M12—were characterized, which have not been previously reported in poultry species based on an exhaustive review of the existing literature. Dietary supplementation with 50 mg/kg or 500 mg/kg Bopu Powder demonstrated favorable safety profiles, exhibiting no significant adverse effects on serum biochemical parameters or major organs in late-phase laying hens. No residues were detected in eggs or muscle tissues at the 50 mg/kg dose, demonstrating its safety for long-term use. However, low but detectable residue levels were observed in eggs at the 500 mg/kg dose, suggesting a potential risk of accumulation under exaggerated exposure conditions. These findings highlight the need for further studies to evaluate residue depletion kinetics and establish an appropriate withdrawal period when high-dose supplementation is used.

Author Contributions: H.W.: Investigation, Formal Analysis, Writing—Original Draft, Visualization, Validation. X.W.: Investigation, Formal Analysis, Visualization, Validation. J.X.: Investigation, Methodology, Data Curation. Z.Y.: Methodology, Validation, Formal Analysis. Z.D.: Resources, Conceptualization. J.Z.: Funding Acquisition, Supervision, Writing—Review and Editing. H.L.: Conceptualization, Project Administration, Writing—Review and Editing. All authors have read and agreed to the published version of the manuscript.

Funding: This work was supported by the National Key Research and Development Program (2022YFD1801105).

Institutional Review Board Statement: This study received approval from the Animal Ethics Committee of Hunan Agricultural University (Approval No.: HUNAULSK 20230807–03).

Informed Consent Statement: Not applicable.

Data Availability Statement: The original contributions presented in this study are included in the article. Further inquiries can be directed to the corresponding author.

Conflicts of Interest: The authors declare no conflicts of interest.

References

1. Chen, W.; Zhang, R.; Lei, S.; Zuo, H.; Chen, S.; Chang, Y. Advances in Chemical Constituents and Bioactivities of Macleaya Cordata. *Chin. J. Exp. Tradit. Med. Formulae* **2019**, *26*, 243–250. [CrossRef]
2. Wang, X.; Min, C.; Han, P. Antibacterial Effect of the Extracts from Different Parts of Macleya cordata Against Coliforms. *Nat. Prod. Res. Dev.* **2016**, *28*, 247–250+288. [CrossRef]
3. Li, J.; Zhang, M.; Liu, Y.; Ji, X.; Liang, F. An overview of natural plant-derived anti-inflammatory agents. *J. Exp. Tradit. Vet. Med.* **2016**, *35*, 24–26. [CrossRef]
4. Zhang, Y.; Zhang, T.; Ma, W.; Liu, P.; Wu, Y.; Liu, Y. Advances in the Anticancer Activities of Sanguinarine from Macleaya cordata (Bo Luo Hui) in Shennongjia. *J. Hubei Univ. Med.* **2018**, *37*, 476–483. [CrossRef]
5. Yang, S.; Liu, Y.; Yang, Q.; Xiang, J.; Tang, Y.; Xv, G. Antitumor effect of Macleaya cordata and its molecular mechanism on inducement of human telomeric DNA to form G-quadruplex. *Chin. Tradit. Herb. Drugs* **2011**, *42*, 738–742.
6. Dong, Z.; Xu, Y.; Liu, Z.; Zeng, J. Evaluation of the Efficacy of the Protopine Total Alkaloids of *Macleaya Cordata* (Willd), R. Br. in Controlling *E. Coli* Infection in Broiler Chickens. *bioRxiv* **2024**. 2024.07.03.601902. [CrossRef]
7. Liu, Y.; Wang, Q.; Liu, H.; Niu, J.-I.; Jiao, N.; Huang, L.; Jiang, S.; Guan, Q.; Yang, W.; Li, Y. Effects of dietary Bopu Powder supplementation on intestinal development and microbiota in broiler chickens. *Front. Microbiol.* **2022**, *13*, 1019130. [CrossRef]
8. Liu, H.; Lin, Q.; Liu, X.; Huang, P.; Yang, Z.; Cao, M.; Liu, M.; Li, X.; Zeng, J.; He, J. Effects of Dietary Bopu Powder Supplementation on Serum Antioxidant Capacity, Egg Quality, and Intestinal Microbiota of Laying Hens. *Front. Physiol.* **2022**, *13*, 902784. [CrossRef]
9. Dong, Z.; Tang, S.-S.; Ma, X.-L.; Tan, B.; Tang, Z.-S.; Li, C.-H.; Yang, Z.-H.; Zeng, J.-G. Acute, Chronic, and Genotoxic Studies on the Protopine Total Alkaloids of the *Macleaya Cordata* (Willd.) R. Br. in Rodents. *Front. Pharmacol.* **2022**, *13*, 987800. [CrossRef]
10. Hu, W.; Yang, F.; Liu, W.; Guo, L.; Ai, L.; Zhang, X.; Sheng, Z.; Gao, C. Potential Toxicity Evaluation of Protopine in *Macleaya Cordata* (Willd.) R. Br.—A Bioactivity Guided Approach. *Front. Vet. Sci.* **2021**, *8*, 752767. [CrossRef]
11. Xu, Y.; Sun, J.; Li, W.; Zhang, S.; Yang, L.; Teng, Y.; Lv, K.; Liu, Y.; Su, Y.; Zhang, J.; et al. Analgesic Effect of the Main Components of Corydalis Yanhusuo (Yanhusuo in Chinese) Is Caused by Inhibition of Voltage Gated Sodium Channels. *J. Ethnopharmacol.* **2021**, *280*, 114457. [CrossRef]
12. Guo, C.; Jiang, Y.; Li, L.; Hong, L.; Wang, Y.; Shen, Q.; Lou, Y.; Hu, H.; Zhou, H.; Yu, L.; et al. Application of a Liquid Chromatography–Tandem Mass Spectrometry Method to the Pharmacokinetics, Tissue Distribution and Excretion Studies of Dactylicapnos Scandens in Rats. *J. Pharm. Biomed. Anal.* **2013**, *74*, 92–100. [CrossRef]
13. Huang, Y.-J.; Cheng, P.; Zhang, Z.-Y.; Tian, S.-J.; Sun, Z.-L.; Zeng, J.-G.; Liu, Z.-Y. Biotransformation and Tissue Distribution of Protopine and Allocryptopine and Effects of Plume Poppy Total Alkaloid on Liver Drug-Metabolizing Enzymes. *Sci. Rep.* **2018**, *8*, 537. [CrossRef] [PubMed]
14. Liu, M.; Xu, J.; Jing, Y.; Yao, K.; Tang, Z.; Zeng, J.; Yang, Z. Residue of Allocryptopine and Protopine in Tissues and Eggs of Laying Hens Using Ultra-High Performance Liquid Chromatography–Tandem Mass Spectrometry. *Poult. Sci.* **2025**, *104*, 105124. [CrossRef]
15. *GB/T 5916-2020*; Formula Feed for Chicks, Laying Hens and Broilers. Standardization Administration of China (SAC): Beijing, China, 2020. Available online: http://c.gb688.cn/bzgk/gb/showGb?type=online&hcno=7071118571C7E3C4412977B5392522E3 (accessed on 2 August 2025).
16. *Guidelines for Target Animal Tolerance Testing of Feed and Feed Additives in Livestock and Poultry*; Ministry of Agriculture and Rural Affairs Announcement No. 424; Ministry of Agriculture and Rural Affairs: Beijing, China, 2021. Available online: https://www.moa.gov.cn/govpublic/XMYS/201107/t20110720_2064934.htm (accessed on 2 August 2025).
17. *AVMA Guidelines for the Euthanasia of Animals: 2020 Edition*; American Veterinary Medical Association: Schaumburg, IL, USA, 2020. Available online: https://www.avma.org/sites/default/files/2020-02/Guidelines-on-Euthanasia-2020.pdf (accessed on 2 August 2025).
18. Liu, M.; Hu, N.; Zou, X.; Yang, Z.; Zeng, J. Simultaneous Determination of 4 Isoquinoline Alkaloids in Broiler Tissues by Ultra-Performance Liquid Chromatography-Tandem Mass Spectrometry. *China Anim. Husb. Vet. Med.* **2022**, *49*, 4843–4853. [CrossRef]
19. Huang, A.; Chi, Y.; Ou, L.; Chen, W.; Zhan, R.; Xv, H. Analysis of Metabolites of α-Allocryptopine in Rats Based on UPLC-Q-TOF-MS. *Tradit. Chin. Drug Res. Clin. Pharmacol.* **2020**, *31*, 1454–1461. [CrossRef]
20. Wu, Z.; Wang, X.; Wang, L.; Sun, N.; Yang, Z.; Zeng, J. Research on the Metabolites and Key Metabolic Enzymes of Allocryptopine in Chicken Liver Microsomes via Stable Isotope Tracing Technology. *J. Pharm. Biomed. Anal.* **2025**, *255*, 116667. [CrossRef] [PubMed]
21. Song, J.; Huang, M.; Shi, X.; Li, X.; Chen, X.; He, Z.; Li, J.; Xu, G.; Zheng, J. T329S Mutation in the *FMO3* Gene Alleviates Lipid Metabolic Diseases in Chickens in the Late Laying Period. *Animals* **2021**, *12*, 48. [CrossRef] [PubMed]

22. Zhang, B.; Zeng, M.; Li, M.; Kan, Y.; Li, B.; Xu, R.; Wu, Y.; Wang, S.; Zheng, X.; Feng, W. Protopine Protects Mice against LPS-Induced Acute Kidney Injury by Inhibiting Apoptosis and Inflammation via the TLR4 Signaling Pathway. *Molecules* **2019**, *25*, 15. [CrossRef] [PubMed]
23. Huailing Wei, G.L. Protective Effects of Corydaline, Acetylcorydaline, and Protopine on Experimental Liver Injury in Mice. *Acta Pharm. Sin.* **1997**, *32*, 12–17. [CrossRef]
24. Janbaz, K.H.; Saeed, S.A.; Gilani, A.H. An Assessment of the Potential of Protopine to Inhibit Microsomal Drug Metabolising Enzymes and Prevent Chemical-Induced Hepatotoxicity in Rodents. *Pharmacol. Res.* **1998**, *38*, 215–219. [CrossRef] [PubMed]
25. Yang, N.; Zhao, Z.; Tang, S.; Liu, H.; Wang, Q.; Xie, X.; Zhang, Y.; Song, L.; Xv, X. The reproductive toxicity test of Bopu Total Alkali Powder on breeding chickens. *J. Yangzhou Univ. (Agric. Life Sci. Ed.)* **2019**, *40*, 65–69. [CrossRef]

toxics

MDPI

Review

Mycotoxin-Caused Intestinal Toxicity: Underlying Molecular Mechanisms and Further Directions

Tian Li [1,2,†], Weidong Qiao [1,2,†], Jiehong Zhou [1,2], Zhihui Hao [1,2], Gea Oliveri Conti [3], Tony Velkov [4], Shusheng Tang [1,2], Jianzhong Shen [1,2] and Chongshan Dai [1,2,*]

[1] State Key Laboratory of Veterinary Public Health and Safety, Key Laboratory for Detection of Veterinary Drug Residues and Illegal Additives of Ministry of Agriculture and Rural Affairs, College of Veterinary Medicine, China Agricultural University, Beijing 100193, China
[2] Technology Innovation Center for Food Safety Surveillance and Detection (Hainan), Sanya Institute of China Agricultural University, Sanya 572025, China
[3] Department of Medical, Surgical Sciences and Advanced Technologies "G.F. Ingrassia", University of Catania, Catania 95123, Italy
[4] Department of Pharmacology, Biodiscovery Institute, Monash University, Clayton, VIC 3800, Australia
* Correspondence: daichongshan@cau.edu.cn
† These authors contributed equally to this work.

Abstract: Mycotoxins represent a group of highly toxic secondary metabolites produced by diverse fungal pathogens. Mycotoxin contaminations frequently occur in foods and feed and pose significant risks to human and animal health due to their carcinogenic, mutagenic, and immunosuppressive properties. Notably, deoxynivalenol, zearalenone, fumonisins (mainly including fumonisins B1, B2, and FB3), aflatoxin B1 (AFB1), and T-2/HT-2 toxins are the major mycotoxin contaminants in foods and feed. Undoubtedly, exposure to these mycotoxins can disrupt gut health, particularly damaging the intestinal epithelium in humans and animals. In this review, we summarized the detrimental effects caused by these mycotoxins on the intestinal health of humans and animals. The fundamental molecular mechanisms, which cover the induction of inflammatory reaction and immune dysfunction, the breakdown of the intestinal barrier, the triggering of oxidative stress, and the intestinal microbiota imbalance, were explored. These signaling pathways, such as MAPK, Akt/mTOR, TNF, TGF-β, Wnt/β-catenin, PKA, NF-kB, NLRP3, AHR, TLR2, TLR4, IRE1/XBP1, Nrf2, and MLCK pathways, are implicated. The abnormal expression of micro-RNA also plays a critical role. Finally, we anticipate that this review can offer new perspectives and theoretical foundations for controlling intestinal health issues caused by mycotoxin contamination and promote the development of prevention and control products.

Keywords: mycotoxin; intestinal health; toxicity mechanisms; micro-RNA

1. Introduction

Filamentous fungi are extensively spread throughout nature and they have a high species diversity and exhibit strong environmental adaptability [1]. Secondary metabolites produced by filamentous fungi, commonly known as mycotoxins, have been extensively documented. To date, over 400 mycotoxins have been identified. Among these, aflatoxins (AFs), ochratoxins (OTA), type-A trichothecenes (such as T-2, HT-2, neosolaniol [NEO], and trichodermin, 4,15-diacetoxyscirpenol [DAS]), type-B trichothecenes (including nivalenol [NIV], and trichothecin [TCN], and deoxynivalenol [DON]), and zearalenone (ZEN), and fumonisins (FUMs), citrinin, ergot alkaloids, and patulin stand out as the most prominent compounds associated with various health disorders in humans and animals [2,3]. Currently, mycotoxin-contaminated animal feed is frequently detected across the world. A global survey of mycotoxin contamination in 4601 samples in 2023 showed that DON is the most frequently found in all samples and finished feed, with the corresponding detection rates being as high as 70% and 75%, respectively, among the emerging regulated or guideline mycotoxins; the top five in all samples from various countries are DON, ZEN, FUMs (including fumonisin B1 [FB1], FB2, and FB3), aflatoxin B1 (AFB1), and T-2/HT-2 toxin [4]. In detail, DON, known as vomitoxin, is a harmful secondary metabolite produced by *Fusarium* fungi, especially *F. graminearum*. These fungi frequently infest multiple field crops, thereby threatening agricultural yields and food safety [5]. ZEN, one of the most detected contaminating mycotoxins usually produced by *Fusarium graminearum*, is classified as an endocrine-disrupting substance due to its capacity to directly bind to and stimulate estrogen receptors (ERs), thus showing an estrogen-like effect [6,7]. FBs are heat-resistant metabolites found in cereal grains (including maize) and crop products, including pelleted feed. The FB1 corresponds to 70% of fumonisins and is produced by *Fusarium verticillioides* [8]. AFs are mainly generated by *Aspergillus flavu* and *Aspergillus parasiticus*, which can contaminate a wide range of food commodities, including nuts (coconut, walnut, pistachio, Brazil nut, and almond), spices (ginger, coriander, turmeric, chilies, and black pepper), oilseeds (cotton, sunflower, soybean, and peanut), cereals (wheat, rice, pearl millet, sorghum, and maize), yam and various milk products [9,10]. The most common metabolic pathway of T-2 is fast deacetylation of the C-4 site, which is then converted into HT-2 toxin and they are all common contaminants in oats, wheat, and corn [11]. These mycotoxins can result in multiple harmful effects, such as genotoxicity, neurotoxicity, nephrotoxicity, hepatotoxicity, gastrointestinal toxicity, reproductive toxicity, and embryotoxicity [2,12–15]. These detrimental impacts can significantly impair production efficiency and, in severe cases, lead to livestock mortality [16–19]. Notably, some mycotoxins, such as DON, ZEN, and AFB1, can be detected in various animal-derived products, such as eggs, meat, milk, and processed foods, thereby posing a significant risk to human health through the food chain [20]. A recent comprehensive meta-analysis revealed that feeding DON to broilers can significantly impair their small intestine health and productive performance [16]. Comprehending the precise molecular mechanisms of mycotoxin-induced harmful effects is crucial for advancing potential treatments or detoxification strategies.

Farm animals such as chickens, pigs, ducks, cows, and sheep are mainly exposed to mycotoxins through the consumption of contaminated feed [14]. The gastrointestinal tract usually serves as the first physiological defense against food-borne mycotoxin contaminants [21,22]. Mucosa in the intestinal tissue constitutes a significant functional part of intestinal integrity; an increasing amount of evidence suggests that other elements, such as the microbiota and mucus, are also involved [23]. Multiple studies have shown that exposure to DON, ZEN, FUMs (including FB1, FB2, and FB3), AFB1, and T-2/HT-2 can cause abnormal changes in the function of the intestinal barrier. Additionally, they can also harm the enteric nervous system, disturb the balance of the host's gut microbiota, and cause the death of intestinal epithelial cells [24–29]. The molecular mechanisms of these mycotoxin-induced multiple harmful effects on animal intestinal function are complex and context-dependent [20,23]. They may involve the induction of inflammatory response, cell apoptosis, as well as cell autophagy, the activation of oxidative stress and endoplasmic reticulum stress, the suppression of immune function, and imbalance of gut microbiota in the intestinal tissues [14,29–31]. Currently, there is a deficiency in systematic summary and analysis concerning intestinal damage caused by these prominent mycotoxins in farm animals. Thus, in this present review, we mainly summarized the detrimental effects caused by

DON, ZEN, FUMs, AFB1, and T-2/HT-2 exposure on the intestinal function in farm animals. Additionally, the potential molecular mechanisms are discussed. Our aim is to put forward new viewpoints regarding intestinal toxicity in farm animals due to mycotoxin exposure, thereby offering a theoretical basis for the development of detoxifying substances.

2. An Overview of Mycotoxin-Induced Intestinal Toxicity Effects

Intestinal epithelial cells, together with tight junction proteins and various components, create a physical and biochemical barricade. The intestinal mucosa serves as the first line of defense against these external contaminants, playing a crucial role in maintaining overall health by acting as a barrier to harmful substances. The impact of mycotoxins on the intestinal mucosa is multifaceted, affecting various components of the intestinal barrier, including the mechanical, chemical, immunological, and microbial barriers [32]. The mechanical barrier of the intestinal mucosa is primarily composed of epithelial cells that are tightly joined by proteins such as claudins, which form tight junctions. Mycotoxins can disrupt this barrier by altering the expression and function of these proteins, leading to increased intestinal permeability and compromised barrier integrity. This disruption can facilitate the translocation of pathogens and toxins, potentially leading to systemic infections and inflammation [32]. In addition to the mechanical barrier, mycotoxins can also affect the chemical barrier of the intestinal mucosa, which includes the mucus layer composed of mucins. Mucins are glycoproteins that protect the epithelial surface and maintain mucosal homeostasis. The immunological barrier of the intestinal mucosa is another critical component that can be affected by mycotoxins [21]. These toxins can modulate immune responses by altering the balance of inflammatory cytokines and immune cells, such as lymphocytes, within the intestinal mucosa [21,29,33,34].

It has been shown that DON exposure via oral administration at 2.4 mg/kg body weight for one week significantly decreased the crypt depth and villus height in the intestinal tissues of mice [35]. Pierron et al. showed that DON, not DOM-1 (the diepoxy-metabolite of DON), exposure by gavage at 0.5 nmol per body weight per day for 21 days significantly decreased the villus height in the jejunum of pigs [36]. Recent research has demonstrated that dietary DON exposure at 0.83 mg/kg feed (below the EU-proposed threshold of 0.9 mg/kg feed) can induce villous atrophy and fusion in the intestinal tissues of pigs [37]. Pasternak et al. reported that dietary DON at 3.8 mg/kg body weight for 24 days significantly upregulated the expression of claudin-7, a critical regulator of tight junctions and intestinal homeostasis in weaned piglets [38]. An in vitro proteomic analysis further revealed that 5 μM DON exposure for 24 h substantially modified protein profiles in human Caco-2 cells (a colorectal adenocarcinoma line), particularly affecting cell junction/adhesion molecules. These alterations were partially mediated through suppressed extracellular regulated protein kinase (ERK) and protein kinase A (PKA) pathways [39]. Taken together, this evidence indicates that tight junction proteins may be the potential target of DON exposure. Mechanistically, it has been reported that DON exposure can cause cell apoptosis and inflammation in intestinal tissues through multiple signaling pathways, such as the mitogen-activated protein kinase (MAPK), phosphatidylinositol 3-kinase/Akt (PI3K/Akt), and Janus kinase/signal transducer and activator of transcription (JAK/STAT) pathways [40]. Additionally, DON exposure decreased the expression of β-catenin, and its downstream targets, such as cyclin D1 and Lgr5 proteins in the jejunum tissues of mice, followed by causing intestinal homeostasis dysfunction [41]. This has indicated that DON exposure can damage intestinal mucosal function and the stability of intestinal stem cells.

It is well known that estrogen has various modulatory roles in regulating cellular proliferation and differentiation, modulating intestinal peristalsis, and maintaining normal intestinal physiological functions through the interaction with nuclear estrogen receptor

(ER) α/β [6]. ZEN exposure can induce significant toxic effects in the intestines of animals and humans by directly binding to and stimulating ER expression. In a mouse model, exposure to ZEN at 40 mg/kg body weight for 12 days can cause a marked morphological change, which was evident by a reduction in the villus height, an increase in the crypt depth, and a decrease in the ratio of villus length/crypt depth [42]. Very recently, Huangfu et al. found that ZEN exposure at 5 mg/kg body weight per day via the gavage administration for 21 days can significantly disrupt the intestinal barrier, increase intestinal inflammation, disrupt gut microbial diversity, and induce ferroptosis of intestinal epithelial cells in rats [43]. Additionally, ZEN exposure also decreased the expression of claudin-1, ZO-1, and occludin proteins, followed by disturbing the tight-junction structure in the jejunum, and ultimately resulting in the increased permeability of intestinal tissues [42,43]. Lahjouji et al. reported that the exposure to ZEN at 100 μM for 4 h can increase the expression of ERα protein and upregulate the Wnt/β-catenin signaling pathway in the explant culture of jejunum tissue of piglets [6].

It has been demonstrated that exposure to fumonisins can cause various detrimental effects on the intestinal histomorphometry and the expression of intestinal tight junction proteins in farm animals (such as chickens and pigs) [44–47]. For example, Tomaszewska et al. reported that when hens were orally administered an extract containing FB1 and FB2 at doses of 1, 4, and 10.9 mg/kg body weight for 21 days, it could markedly damage the epithelial integrity of the duodenum and jejunum. Additionally, it significantly reduced the villus height in hens, the ratio of the thickness to the width of duodenal villi, and the ratio of the thickness to the width as well as the depth of the duodenal crypts [44]. Consistently, a recent study reported that oral administration of FB1 (at 5 mg/kg body weight) or hydrolyzed FB1 (at 2.8 mg/kg body weight) for 21 days can both result in compromised integrity, atrophy of intestinal villus, elevated levels of inflammatory factors, and decreased total short-chain fatty acids (SCFAs) in the intestinal tissues in mice [48]. Yu et al. found that FB1 exposure at 10–40 μM for 24 or 48 h can induce dose-dependently cytotoxicity and elevate the ROS accumulation, increase the production of inflammatory factors, disrupt the tight connectivity and permeability, and result in cell apoptosis in IPEC-J2 cells (a porcine intestinal epithelial cell line) [49]. The Nrf2 signal pathway, an endogenous antioxidant stress signaling pathway, was significantly inhibited [50]. Additionally, research has demonstrated that FB1 primarily affects proliferating intestinal cells, leading to lipid peroxidation, inhibition of cell proliferation, and immunomodulatory effects, as evidenced by changes in IL-8 secretion and membrane microviscosity [51].

Around the world, it has been clearly demonstrated that AF contamination poses a significant threat to food safety as well as public health security [3,52]. Among these AFs, AFB1 is the most toxic one [15]. It has been reported that AFB1 exposure can disrupt intestinal microbiota, induce immune responses and oxidative damage, and impair intestinal barrier function and morphology in farm animals [53]. For example, a 21-day exposure to AFB1 at a dose of 0.1 mg/kg body weight can induce ileum tissue damage in ducks. This is manifested by a significantly increased crypt depth and a decreased villus height [54]. Li et al. reported that AFB1 exposure at 0.75 mg/kg body weight once daily for 4 weeks can markedly damage the intestinal barrier and induce an inflammatory response via the induction the dysbiosis of the gut microbiota and inhibition of the AHR signaling pathway in the intestinal tissues of mice [55]. Ding et al. reported that AFB1 exposure can induce pyroptosis and mitochondrial dynamics imbalance via activating NLRP3 and AMPK pathways in IPEC-J2 cells [56].

It was reported that HT-2 toxin exposure at 6.25 nM or T-2 toxin at 3.125 nM for 24 h can significantly decrease the expression of claudin-1, occludin, and ZO-1 mRNAs and proteins in IPEC-J2 cells, indicating damage to the intestinal barrier function [57]. Additionally, the exposure of HT-2 is accompanied by T-2 toxin, and it has been confirmed that the combined exposure of HT-2 and T-2 can induce a significant synergistic toxic effect [57]. Liu et al. reported that T-2 toxin exposure at 1.0, 3.0, and 6.0 mg/kg body weight via the diet for 2 weeks significantly decreased the activities of glutathione peroxidase (GPX), thioredoxin reductase and total antioxidant capacity but increased the concentrations of protein carbonyl and malondialdehyde (MDA) in the duodenum in a dose-dependent manner in the intestinal tissues of chicks. This study also indicated that T-2 toxin-induced intestinal damage is involved in the regulation of nucleotide and glycerophospholipid metabolism, redox homeostasis, inflammation, and apoptosis [58].

The combined exposure to multiple mycotoxins, such as DON, FB1, and ZE, further complicates the toxicological landscape. Studies using metabolomics and lipidomics approaches have shown that these mycotoxins can induce metabolic disorders and damage antioxidant capacity in intestinal cells, with synergistic and antagonistic interactions observed in co-exposure scenarios [59]. It was also reported that the exposure of FB1 at 100 μM or in combination with DON (at 10 μM) can both damage the intestinal morphology (such as the reduction in villus height and the expression of E-cadherin, a junction protein) and the number of goblet cells in jejunal explants from piglets [47,60]. Such findings underscore the importance of considering the cumulative effects of mycotoxin mixtures in risk assessments, as they may exacerbate intestinal toxicity beyond the effects of individual toxins.

In conclusion, these mycotoxin exposures can disrupt the intestinal structure of animals, mainly reflected in the destruction of the intestinal barrier and integrity, as well as its immune function. The intricate molecular mechanisms underlying mycotoxin-induced intestinal toxicity involve the induction of inflammatory reactions and immune dysfunction, the breakdown of the intestinal barrier, the triggering of oxidative stress, and the intestinal microbiota imbalance. These signaling pathways, such as MAPK, Akt/mTOR, TNF, TGF-β, Wnt/β-catenin, PKA, NF-kB, NLRP3, AHR, TLR2, TLR4, IRE1/XBP1, Nrf2, and MLCK pathways, are also implicated. The abnormal expression of micro-RNA also plays a critical role. Continued research in this area is essential for advancing our understanding of mycotoxin-related health risks and for the development of effective interventions to protect human and animal health. In the following section, the detailed molecular mechanisms of mycotoxin exposure-induced intestinal toxicity will be thoroughly discussed.

3. Mycotoxin Exposure Triggers Oxidative Stress and Cell Apoptosis

Oxidative stress is commonly regarded as a mechanism in toxic-pathology associated with a variety of adverse environmental agents, like organic compounds, veterinary drugs, and heavy metals [61]. The appearance of oxidative stress is directly related to a rise in the amount of reactive nitrogen species (RNS)/reactive oxygen species (ROS) in cells or tissues [61]. One of the principal sites where ROS/RNS is formed is the gastrointestinal tract [7]. Generally, ROS and RNS encompass a series of oxygen or nitrogen containing free radicals, mainly generated by cell respiration, like singlet oxygen ($^1O_2$), superoxide anion ($O_2^{-}\bullet$), peroxynitrite (ONOO–), nitrous acid (HNO_2), hydroxyl radical (OH•), nitrotyrosine, nitric oxide (NO), hydrogen peroxide (H_2O_2), and nitrosyl anion (NO−) [62,63]. To keep free radical levels low inside cells, these ROS/RNS can be swiftly cleared or neutralized by the cell's self-defense antioxidant enzymes (such as GPX, glutathione S transferase [GST], superoxide dismutase [SOD], and catalase [CAT]) or antioxidants (such as reduced glutathione [GSH], selenium, and vitamin C) [3,15]. In normal physiological circumstances, the intracellular states of oxidation and antioxidation remain in an equilibrium. When this balance gets disrupted, there will be a significant rise in intracellular ROS/RNS, which eventually causes oxidative stress damage [2,64].

Similar to other toxic compounds, mycotoxins can also induce oxidative stress damage through conventional pathophysiological mechanisms. Multiple studies have reported that farm animals exposed to mycotoxins might lead to an excess production of ROS/RNS, thereby initiating oxidative stress, finally leading to mitochondrial malfunction and apoptosis, ultimately resulting in intestinal damage [65–67]. De Souza et al. reported that DON exposure at 19.3 mg per body weight for seven days markedly promoted the production of ROS and reduced the intracellular GSH levels, subsequently inducing lipid peroxidation in the chickens' jejunum and ileum tissues [68]. Xu et al. discovered that DON treatment at 0.5 μg/mL for 6–48 h significantly increased the levels of MDA and significantly diminished the activities of CAT and SOD enzymes, then promoted the apoptotic cell death in IPEC-J2 cells (an immortalized intestinal porcine enterocyte cell line) [69]. Consistently, multiple other mycotoxins like HT-2 toxin, AFB1, and ZEN are reported to be able to induce the production of ROS/RNS and interfere with the antioxidant system [70,71]. Sun et al. found that ZEN treatment at 20 μg/mL for 24 h markedly reduced the activities of CAT, GPX, and SOD while markedly increasing intracellular MDA and ROS levels in IPEC-J2 cells [71]. At the same time, it is also reported that ZEN exposure can markedly decrease thioredoxin reductase (TrxR) activities and reduce the levels of intracellular GSH, followed by causing oxidative stress in the intestinal tissues [71]. These results indicated that the overproduction of ROS/RNS or the dysfunction of the antioxidant system-induced oxidative stress plays a critical role in mycotoxin exposure-induced intestinal tissue damage.

When it comes to oxidative stress, the nuclear factor E2-related factor 2 (Nrf2) acts as a housekeeping gene. Additionally, Nrf2 is a crucial transcription factor that directly or indirectly controls the expression of over two thousand genes involved in various biological processes, such as xenobiotic metabolism, phase II detoxification enzyme systems, anti-oxidative stress, anti-inflammatory response, redox-reduction regulation, and drug detoxification [72,73]. A large number of studies have shown that several mycotoxins like T-2, DON, AFB1, and ZEN are able to target and inhibit Nrf2 transcription, thus promoting the downregulation of SOD, GPXs, CAT, and GST at the genetic levels, finally worsening the oxidative stress damage [66,71,74–76]. Researchers found that DON exposure at 3 mg/kg body weight for 10 days significantly upregulated the expression of Keap1 and significantly downregulated the expression of Nrf2, then significantly downregulated the expression of heme oxygenase-1 (HO-1) and GPX4 in the jejunum tissues, finally inducing oxidative stress damage [66]. Consistently, Li et al. found that DON exposure at 4 mg/kg via the diet can markedly downregulate the expressions of Nrf2 and its downstream genes glutamate-cysteine ligase modifier subunit (GCLM), glutamate cysteine ligase catalytic subunit (GCLC), HO-1, and NAD (P)H Quinone Dehydrogenase 1 (NQO-1) in the jejunum tissues of piglets [77]. Notably, it was reported that T-2 toxin and AFB1 exposure at the low dose levels can both markedly inhibit the transcription activation of Nrf2 and its downstream antioxidant components in intestinal epithelial cells [74,78]. Nrf2 knockout or knockdown markedly exacerbated AFB1 or T-2 toxin exposure-induced intracellular oxidative stress damage [79,80]. Additionally, it was reported that AFB1 can also selectively impede the Nrf2's transcriptional activation, then reduce the expression of the downstream detoxification and antioxidant genes, including HO-1, CAT, SOD, GPX, and GST [81–83]. Moreover, it was found that T-2 toxin-induced ubiquitination and degradation of Nrf2 is dependent on the activation of activating transcription factor 3 (ATF3) [84]. Unlike T-2 toxin, Xu et al. demonstrated that AFB1-induced the decrease in Nrf2 is dependent on the activation of caveolin-1 protein [85]. This evidence indicate that Nrf2-mediated antioxidant defense plays a vital role in mycotoxin-caused intestinal toxicity. But the underlying mechanisms of different mycotoxins on Nrf2 may be different.

Excessive ROS/RNS production in the animal body can damage intracellular biological macromolecules such as DNA, RNA, proteins, and lipids, ultimately causing programmed cell death [86,87]. Apoptosis, a type of programmed cell death, can be induced in various mycotoxins (such as DON, ZEN, and AFB1)-treated intestinal epithelial cells. It involves the downregulation of B-cell lymphoma-2 (Bcl-2) gene and the upregulation of caspases-3, 8, and 9, Bcl-2-associated x gene (Bax), and tumor suppressor p53 genes and proteins [69,88]. It is widely recognized that a raised Bax/Bcl-2 ratio has the potential to induce mitochondrial dysfunction and initiate the mitochondrial apoptotic pathway through the creation of mitochondrial outer membrane permeabilization (MOMP) [69,88]. The enhanced MOMP can trigger the release of cytochrome C (CytC), then lead to the activation of caspases-9 and -3, and finally induce cell apoptosis [89,90]. Caspases-9 and -3 are usually regarded as the biomarkers of the mitochondrial apoptosis pathway and apoptotic cell death, respectively [89,90]. Xu et al. found that DON treatment at 0.5 μg/mL for six hours markedly increased the Bcl-2 mRNA expression and decreased Bax and caspase-3 mRNA expressions, then induced cell apoptosis in IPEC-J2 cells [69]. In another study, it was found that AFB1 exposure at 30 μg/mL for 24 h markedly increased the mRNA expression of Bax and caspase-3 and decreased the Bcl-2 mRNA expression in IPEC-J2 cells. Similar findings were also observed in AFB1-treated intestinal tissues in mice [91].

Antioxidant supplements can effectively suppress ROS production and oxidative stress resulting from mycotoxin exposure. Subsequently, they can effectively correct mitochondrial dysfunction and the mitochondrial apoptosis pathway caused by mycotoxin exposure. This verifies that oxidative stress is of crucial importance in apoptosis in the intestinal tissues induced by mycotoxin exposure [66,75,92,93]. Additionally, studies have demonstrated that exposure to mycotoxins can trigger cell autophagy to lower ROS levels and relieve the damage caused by oxidative stress. In the end, it safeguards against intestinal damage induced by mycotoxin exposure [94]. It was reported that mycotoxin exposure-induced cell autophagy involves the protein kinase B (Akt) pathway, the p38/MAPK pathway, the IκB kinase (IKK) pathway, and the mammalian target of rapamycin (mTOR) pathway [94–96]. For instance, Liu et al. discovered that DON treatment is capable of triggering autophagy through reducing the expression of phosphor-Akt (p-Akt) and p-mTOR while elevating the expression of microtubule-associated protein 1 light chain 3B (LC3-II), Beclin-1, and autophagy-related protein 5 (ATG5) mRNAs and proteins in IPEC-J2 cells [96]. Consistently, ATG5 deletion markedly boosts the ROS production, then significantly increases the expression of cleaved-caspase-3 and CytC proteins, finally promoting cell DON exposure-induced cell apoptosis in IPEC-J2 cells. Moreover, it was shown that DON-induced autophagy depends on the activation of IKK and AMPK pathways [95]. In addition, it was also found that the inhibition of ROS production could also significantly block Akt/mTOR-mediated autophagy activation, indicating that DON-induced autophagy is also partly due to the induction of ROS [96].

In short, oxidative stress is extremely important in intestinal damage caused by exposure to mycotoxins. Mycotoxins can lead to mitochondrial malfunction and cell apoptosis by inducing oxidative stress, and this process can be controlled by the Nrf2, Akt/mTOR, IKK, and AMPK pathways (Figure 1). Relieving oxidative stress can efficiently reduce the damage to the intestine caused by mycotoxins, and it has been considered an effective approach to lessen the deleterious effects of mycotoxin exposure on the intestinal tissues of animals.

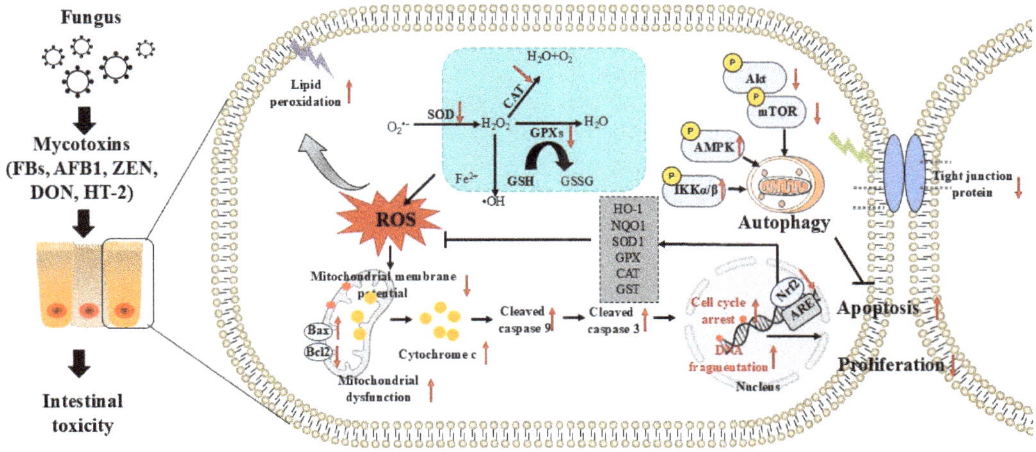

Figure 1. A proposal molecular mechanism model of mycotoxin exposure-induced cell apoptosis and oxidative stress in the intestines of animals. Mycotoxin exposure can promote ROS production via inhibiting SOD, CAT, GPXs activities and reducing GSH levels, then induce lipid peroxidation and oxidative stress. Excessive ROS production can damage mitochondria, cause DNA fragmentation, and cell cycle arrest, finally blocking cell proliferation and inducing cell apoptosis in intestinal epithelial cells. Additionally, mycotoxin exposure can also activate autophagy via the activation of AMPK and IKK pathways and the inhibition of the Akt/mTOR pathway. The inhibition of Nrf2 at the transcriptional level caused by mycotoxin exposure exacerbated oxidative stress damage. Bcl-2, B-cell lymphoma-2; Bax, Bcl-2-associated x gene; AMPK, AMP-activated protein kinase; Akt, protein kinase B; AFB1, aflatoxin B1; CytC, cytochrome C; CAT, catalase; HO-1, heme oxygenase-1; H_2O_2, hydrogen peroxide; GST, glutathione S transferase; GSH, glutathione; DON, deoxynivalenol; GPX, glutathione peroxidase; mTOR, mammalian target of rapamycin; IKK, IκB Kinase; SOD: superoxide dismutase; ROS, reactive oxygen species; Nrf2, nuclear factor E2-related factor 2; ZEN, zearalenone.

4. Mycotoxin Exposure Triggers the Intestinal Immune Dysfunction and Inflammatory Responses

It is widely recognized that the intestine ranks among the largest immune organs. Moreover, it acts as the chief barrier to ensure animals' ability to resist natural toxins, which are ingested via the diet or other methods [97]. The immune-regulatory function of the intestine is intricate, and this process involves numerous immune cells and immune-active factors, including T cells producing IL-17 and IL-22, innate lymphoid cells, intraepithelial T cells with innate and cytolytic effector functions, and the production of antimicrobial peptides, among others [98]. Macrophages take in pathogens, display antigens, control the interaction between B lymphocytes and T lymphocytes, and are of central importance in immune regulation [99]. Numerous studies have shown that exposure to mycotoxins can damage intestinal immune function in vitro and animal experiments [100–103]. It is reported that contact with DON, AFB1, ZEN, FBs, and HT-2 toxin can remarkably decrease cellular immune responses, such as the overall amount of white blood cells, neutrophil phagocytosis activity, macrophage function, and antibody concentration, etc. [104–108]. This is the main reason for the reduction in production capacity and disease resistance among farm animals.

The epoxy group structure of DON can react with the nucleotides present in ribosomal RNA (rRNA). This reaction then disrupts the spatial structure of the ribosome and inhibits the activity of the peptidyl transferase on the 60 S subunit [109]. Ribotoxic stressors can activate the pro-apoptotic as well as pro-inflammatory signaling pathways that are downstream of p38 and JNK signaling pathways [110]. Hu et al. found that DON exposure at 12 mg/kg feed via the diet for 5 weeks significantly increased the expression of p-p38 and p-ERK1/2 proteins and the mRNA expression of several inflammatory factors, including IL-1β, IL-6, IL-8, and TNF-α, and significantly decreased the expression of IL-10 mRNA in the jejunum tissues of mice [111]. Zhang et al. found that DON treatment at 0.5–2 μg/mL can dose-dependently increase expression of p-p38, p-JNK, p-ERK1/2, TNF-α, IL-6, and IL-1α in IPEC-J2 cells. Furthermore, inhibition of p38 can markedly diminish DON exposure-induced upregulation of pro-inflammatory factors, including IL-6, TNF-α, IL-1α, CCL4, CXCL8, CXCL2, IL-12α, CCL20, and IL-15. And the inhibition of ERK 1/2 could notably alleviate the production of DON-induced IL-15 and IL-6, indicating that the activation of p38 and ERK signaling pathways is critical in DON-mediated inflammatory response [112]. Moreover, research has also shown that exposure to the different levels of DON can trigger two different immune effects in vivo as well as in vitro [99]. For instance, DON exposure at the low dose led to intestinal mucosal immune-stimulation, which entailed an increment in the expression of IL-6 and TNF-α mRNA and protein levels, along with a marked increase in the number of T cells and goblet cells. Nevertheless, exposure at the high dose triggered intestinal mucosal immunosuppression, involving an elevation in the mRNA and protein expression of IL-10 and TGF-β genes, and a reduction in the number of T cells and goblet cells in weaned piglets [113]. Moreover, researchers have found that DON exposure at a low concentration can activate the TLR4/NFκB signaling pathway, while DON exposure at high concentration can result in immunosuppression via mitochondrial dysfunction by inhibiting mitophagy [113]. In addition, DON exposure at low doses can also promote inducible nitric oxide synthase (iNOS) ubiquitinoylation and its degradation, resulting in a decrease in NO production in Caco-2 cells, which might potentially account for the enhanced susceptibility of animals to bacterial or viral infection in the intestinal tissues [114]. This is an important factor in explaining DON exposure-induced immune suppression. In another study, it was observed that the humoral and cellular immunity were significantly diminished in pigs fed a DON naturally contaminated diet (i.e., 3.5 mg DON per kg feed) for twenty-eight days in comparison with pigs fed an uncontaminated diet. Additionally, they found that DON exposure significantly upregulated the expression of IL-4 and CXCL10 in the jejunum tissue and the expression of interferon gamma (IFNG) and CXCL10 mRNA in the ileum tissue [115]. Van De et al. discovered that DON treatment at 0.5–5 μg/mL can trigger the activation of NF-κB and the secretion of IL-8 in a dose-dependent manner in Caco-2 cells. Moreover, under pro-inflammatory stimulation, this effect became more intense. This shows that exposure to DON has the potential to lead to or exacerbate intestinal inflammation [116]. Consistently, a recent study showed that DON exposure at the lower dose can trigger the activation of the inflammatory response in IPEC-J2 cells through the activation of the TNF-α/NF-κB/myosin light chain kinase (MLCK) pathway and it may be attributed to the inhibition of the aryl hydrocarbon receptor (AHR) pathway [117].

Jiang et al. found that when broilers are exposed to AFB1 at 0.6 mg/kg feed via the diet, the proportions of CD3+/CD8+, CD3+/CD4+, and CD3+ in the small intestinal T lymphocytes can be decreased. Meanwhile, the mRNA expressions of IL-4, IL-6, IL-10, IL-17, and TNF-α in the duodenal, jejunal, and ileal tissues will also be affected [118]. Furthermore, it was also discovered by them that exposure to AFB1 resulted in a marked decrease in the levels of immunoglobulin A (IgA), immunoglobulin M (IgM), and polymeric immunoglobulin receptor (pIgR) in the serum sample of broilers. Moreover, the mRNA expression of these immunoglobulins, the quantity of mature T cells, and the mRNA expression of IL-6 and IL-2 in the intestinal tissues were also decreased [118]. The evidence presented indicates that the decline in the content of IgA and the expression of inflammatory cytokines could be strongly related to the reduction in the proportion of T-cell subsets brought about by AFB1 exposure. Additionally, AFB1 has been regarded as a non-canonical AHR ligand [55,119]. Zhang et al. discovered that when mice were exposed to AFB1 at 5, 25, and 50 μg/kg body weight daily for thirty days, it could dose-dependently activate the inflammatory response and cause colitis in mice, and the process was partly dependent on the activation of the macrophage AHR/TLR4/STAT3 pathway [119]. In addition, AFB1 treatment can also result in a marked increase in the levels of CD68+ myeloid cells and CD80+ M1 macrophages, and a marked decrease in the levels of CD11b+ cells [119]. Moreover, studies also found that AFB1 exposure can remarkably upregulate the expression of immune-functional proteins, such as indoleamine 2,3-dioxygenase-1 (IDO-1), iNOS, NLRP3, NF-κB, ICAM-1, and COX-2 in the intestinal tissues [119]. Guo et al. discovered that AFB1 exposure at 40 μg/kg body weight via the diet significantly increased the expression of TNF-α, IL-8, IL-6, iNOS, NF-κB, NOD1, and TLR2 mRNAs in the intestinal tissues of broilers [120]. Zhang et al. discovered that a 5-day exposure to AFB1 at a dose of 0.3 mg/kg body weight is capable of triggering an inflammatory reaction in the jejunum and enhancing the expression of TNF-α, IL-6, and IL-1β mRNAs and proteins in the rabbit intestinal tissues [78]. In an another study, it was found that oral AFB1 exposure at 3 mg/kg body weight or AFM1 (a metabolite of AFB1) exposure at 0.3 mg/kg body weight daily for 28 days can both cause a significant increase in the crypt depth, and a decrease in the ratio of villus length/crypt depth in the jejunum tissues of mice [121]. Furthermore, a proteomic analysis revealed that a combination of AFM1 or AFB1 exposure can exacerbate the intestinal barrier dysfunction, and this is positively correlated with several signaling pathways, including the TNF signaling pathway, the microRNAs in cancer, and the IL-17 signaling pathway [121]. These pieces of evidence suggest that immune dysfunction and inflammation in AFB1-treated intestinal tissues are related to multiple signaling pathways, including TLR4, TLR2, NLRP3, AHR, and NF-κB signaling pathways.

Additionally, it has also been reported that ZEN and HT-2 toxin can inhibit intestinal immune function, leading to intestinal damage in farm animals. For example, Wang et al. found that treating with ZEN at 20 mg/kg body weight daily through oral administration for seven days markedly enhanced the IL-1β, IL-10, TNF-α, and IFNG mRNA expressions in the jejunum tissues of mice [122]. Girish et al. showed that dietary Fusarium mycotoxins (including deoxynivalenol at 3.9 μg/g feed, ZEN at 0.67–0.75 μg/g feed, 15-acetyl-DON at 0.34 μg/g feed, and HT-2 toxin at 0.078–0.085 μg/g feed) for 21 days significantly increased the percentage of B-lymphocytes in ileum, and reduced the percentages of CD8(+)-lymphocytes in cecal tonsil in turkey poults [123].

It is also reported that exposure to certain mycotoxins can disrupt intestinal homeostasis and heighten intestinal permeability, which is also referred to as "leaky gut". Subsequently, this facilitates the translocation of pathogens and lipopolysaccharides (LPS) across the epithelial cell barrier, ultimately leading to gut syndrome [22,124,125]. For example, Awad et al. reported that DON exposure at 5 or 10 mg/kg feed via the diet for five weeks significantly increases the intestinal paracellular permeability in broiler chickens, then promotes the translocation of *Escherichia coli* (*E. coli*) [126]. Verbrugghe et al. found that T-2 toxin exposure can promote the transepithelial passage of *Salmonella* Typhimurium through the intestinal epithelium [127]. Ye et al. reported that AFB1 exposure at low doses can increase serum LPS levels via disrupting the balance of gut microbiota and intestinal barrier structure [128].

Briefly, these mycotoxins can damage the intestinal immune system and induce an inflammatory response, and different mycotoxins have different effects on the expression of inflammation-associated cytokines. They can inhibit the expression of relevant cytokines and genes, reduce the immune response in an inflammatory situation, and increase the susceptibility to diseases. Additionally, mycotoxin can also promote the translocation of intestinal pathogens and LPS through the epithelial cell barrier, leading to induce gut syndrome. Mycotoxin-regulated immune modulation and inflammatory response in the intestinal tissues may be implicated in multiple signaling pathways, including TGF-β, NF-kB, NLRP3, AHR, TLR2, TLR4, MAPK, IRE1/XBP1, and MLCK pathways. These findings provide important intervention targets for the development of intervention strategies to alleviate intestinal damage due to mycotoxin exposure.

5. Mycotoxin Exposure Causes Disruption of Intestinal Microbiota

The intestine is a vital digestive organ, and its stable, functional mucosal barrier serves as the biological interface separating the body's internal environment from food antigens and microorganisms. The gut microbiota, the largest microecological system, plays a crucial role in maintaining a stable internal and external environment. Disruption of the intestinal microbiota is primarily characterized by an imbalance in the growth of symbiotic and pathogenic bacteria and is closely associated with disease progression [129]. Generally, the composition and function of intestinal microbiota can be disturbed by various factors, including genetic factors, antibiotics, diet, and environmental chemicals [130]. Intestinal microbiota can modulate the endocrine functions and energy absorption; therefore, it plays a key role in host metabolism [131–135]. It has been reported that mycotoxins can be biologically inactivated or degraded by the intestinal microbiota [136,137]. Nevertheless, excessive exposure to mycotoxins can disrupt the homeostasis of the intestinal microbiota and impede the absorption of intestinal nutrients.

DON exposure at the low dose (i.e., at 10 μg/kg body weight daily) via the diet for 280 days can markedly induce the changes in the microbial composition in the mouse intestine; a remarkable rise was observed in the relative abundances of *Cyanobacteria*, *Tenericutes*, *Verrucomicrobia*, TM7, *Proteobacteria*, *Deferribacteres*, and a substantial decline was observed in the relative abundances of *Actinobacteria* and *Bacteroidetes* [138]. Furthermore, exposure to DON can significantly raise the quantity of *Mucispirillum* as well. These bacteria are capable of degrading mucin and may play a role in the already-known impacts of DON on the intestinal barrier [138]. It has also been demonstrated that *Lactobacillus* and *Bacteroides* can also eliminate DON and ZEN in the intestinal tissue [139,140]. Therefore, leveraging the antibacterial properties of beneficial bacteria can effectively relieve the harm of mycotoxins to intestinal functions.

Piotrowska et al. found that ZEN treatment separately or together with DON can both cause a marked effect on mesophilic aerobic bacteria, and significantly reduces the relative enrichment of *E. coli* and *Clostridium perfringens* (*C. perfringens*) in the intestinal tissues of porcine [141]. The enrichment of bacteria containing lipopolysaccharide (LPS) may be an important cause of systemic inflammation and immune response. In addition, a recent study showed that T-2 toxin treatment can promote *E. coli* to develop stable resistance to multiple clinical antibiotics, such as cephalosporins, carbapenems, tigecycline, and colistin [142]. This indicates that mycotoxin exposure not only increases the susceptibility of the intestine to pathogenic bacteria but may also enhance bacterial resistance, thereby reducing the effectiveness of clinical antibiotic treatment.

It is widely acknowledged that intestinal microbiota dysbiosis can affect gut health through the metabolites such as SCFAs (i.e., acetate, propionate, and butyrate), LPS, bile acid, and small peptides [143]. Notably, alterations in the levels of SCFAs and LPS play a vital role in regulating intestinal NLRP3 signaling pathways [144]. In the above-mentioned, the intestinal toxicity caused by NLRP3 activation upon exposure to T-2 and ZEN has been described. Therefore, we believe that exposure to these mycotoxins may lead to a reduction in SCFAs and an increase in LPS by inhibiting the growth of probiotics and promoting the growth of pathogenic bacteria, which triggers excessive NLRP3 activation and results in intestinal inflammation. It has also been reported that DON exposure could cause the disruption of intestinal microbiota, and then lead to excessive secretion of peptide YY (PYY) and 5-HT, which are two key neurotransmitters in controlling emetic [145]. It is known that this abnormal PYY and 5-HT secretions can directly or indirectly inhibit or activate multiple signaling pathways, such as NF-κB, MAPK, TLR pathways, and cell apoptosis and autophagy in intestinal epithelial cells [146,147]. This indicated that the specific intestinal toxicity of DON may be closely related to its effect on intestinal nerve cells. Additionally, it reported that DON exposure can inhibit intestinal bile acid reabsorption and lead to bile acid malabsorption in the intestine [148]. Bile acids can stimulate mitochondria to produce excessive ROS by disrupting the respiratory complexes and electron chain transfer in mitochondria [149]. This evidence indicated that DON can induce intestinal oxidative damage via the induction of the accumulation of bile acids in the intestinal tissues.

In brief, mycotoxin exposure can disrupt the intestinal micro-ecological balance, subsequently result in the enrichment of harmful metabolites or reduction in beneficial metabolites, leading to disturbance of the intestinal barrier and permeability and ultimately result in inflammatory responses, oxidative stress, immune dysfunction, and other adverse health effects (Figure 2). Moreover, the mechanisms by which different mycotoxins disrupt the homeostasis of intestinal microbiota suggest special relationships between mycotoxins and intestinal microbiota. This may be due to different mycotoxins having specific antibacterial properties and the interactions between different mycotoxin entities and the intestinal microbiota in certain specific environments. Currently, some studies have confirmed that certain intestinal microbiota, such as commercially available lactic acid bacteria or probiotics isolated from animal intestines, can effectively ameliorate intestinal microbiota dysbiosis induced by mycotoxin exposure and other health-damaging effects [150–153]. In addition, future research exploring the effects of mycotoxins on a specific bacterial genus to understand its specific mechanism of action is required.

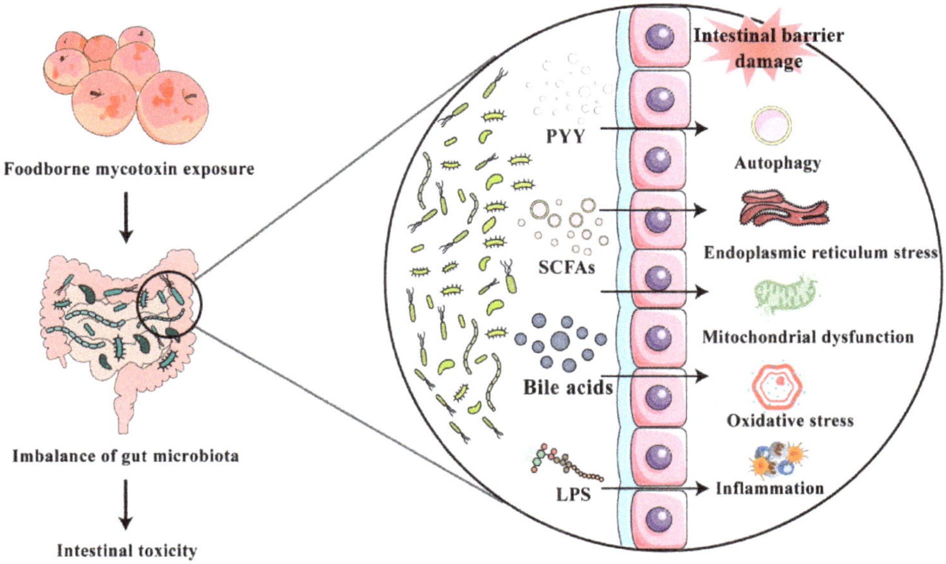

Figure 2. Mycotoxin exposure can disrupt the intestinal micro-ecological balance, subsequently resulting in a reduction in short-chain fatty acid (SCFA) production, an increase in LPS, an excessive secretion of peptide YY (PYY), and an accumulation of bile acids, initiating a cascade to

induce endoplasmic reticulum stress, mitochondrial dysfunction, oxidative stress, and inflammatory response, finally resulting in intestinal barrier injury.

6. Role of MiRNAs in Mycotoxin-Induced Intestine Toxicity

MicroRNAs (miRNAs) belong to a type of endogenous non-coding small molecular substances, and their lengths are approximately 21–28 nt. They can regulate gene expression post-transcriptionally, influencing various biological processes, including those related to toxicology. [154]. miRNAs are regulators whose up- or downregulation acts on the development of pathological dysregulation. MiRNAs also participate in regulating multiple signaling pathways to govern various mycotoxin exposure-induced toxicity and tissue damage [155].

The role of miRNAs in mycotoxin-induced intestinal toxicity is an emerging area of research that underscores the complex interplay between genetic regulation and toxicological responses. A great number of studies have shown that miRNAs can act as nodes within signaling networks and play a crucial role in maintaining homeostasis and controlling various types of cell death, including pyroptosis, ferroptosis, and apoptosis, as well as chronic diseases like fibrosis, metastasis, neurodegenerative diseases, cardiovascular diseases, and cancer [156,157]. Variations in microRNA expression have been detected following exposure to different mycotoxins, primarily resulting from altered regulatory mechanisms post-toxin contact and interspecies disparities in miRNA functions [158]. It was reported that ZEN exposure can upregulate the expression of miR-452-3p, miR-424-5p, and miR-1 in piglets [159]. Chuturgoon et al. reported that FB1 exposure can significantly upregulate the expression of CYP1B1 via the target inhibition of miR-27b [160]. Rieswijk et al. demonstrated that AFB1 exposure can upregulate the expression of miR-301b-3p to regulate cell cycle arrest and DNA damage [161]. Additionally, it also reported that AFB1 exposure can upregulate the expression of miR-24, miR-33a, miR-34a, miR-34a-5p, and downregulate the expression of miR-138-1 and miR-122, following to regulate cell proliferation, DNA damage, or cancer generation [158,162–164]. A systematic review on the impact of several different mycotoxins (e.g., DON, ZEN, and AFB1) on miRNAs and the underlying toxicology mechanisms has been performed by Chen et al. [155] and Rong et al. [158].

Collectively, these studies provide a comprehensive view of the multifaceted roles of miRNAs in mycotoxin-induced intestinal toxicity. This highlights the potential of miRNAs as diagnostic biomarkers and therapeutic targets, offering new avenues for research and intervention in the management of mycotoxin-related health risks. By integrating insights from various studies, it becomes evident that miRNAs play a crucial role in mediating the toxicological and protective responses to mycotoxin exposure, thereby shaping the future of food safety and toxicology research.

7. Summary and Prospect

Mycotoxin exposure can damage the integrity of the intestinal morphological structure, cause intestinal dysfunction, and finally lead to a decrease in the production performance or products of farm animals. Various mycotoxins like DON, ZEN, FBs, AFB1, and HT-2 toxin may have different levels of impact (both molecular and tissue level), and their mechanisms of action are complex. Studies indicate that the possible mechanisms behind mycotoxin exposure-induced abnormal intestinal function involve the breakdown of the intestinal barrier, the triggering of an inflammatory response and immune malfunction, the activation of oxidative stress, and the imbalance of the intestinal microbiota. Multiple signaling pathways, including TGF-β, PKA, MAPK, TNF, NF-κB, NLRP3, AHR, TLR2, TLR4, IRE1/XBP1, Nrf2, Akt/mTOR, Wnt/β-catenin, and MLCK pathways, are also implicated. These findings provide important theoretical foundations for intervention strategies against these mycotoxin-induced intestinal toxicities. The targeting of the reduction in mycotoxin exposure from agricultural practice to food/and feed additives is also required.

The general body of literature highlights several aspects for future research: (i) It is crucial to investigate the potential toxic effects and the exact molecular correlations between exposure to multiple mycotoxins and other food contaminant-caused intestinal toxicity in animals and humans. (ii) Importantly, the abnormal expression of miRNA can disturb various signal transduction and seems to play a critical role in mycotoxin-caused intestinal dysregulation. Currently, the precise molecular mechanisms are incompletely understood, and more investigations are required. (iii) As demonstrated by large-scale investigations, both humans and animals are simultaneously exposed to several mycotoxins, such as AFB1, DON, ZEN, HT-2/T-2, or FUNs. The exposure to several mycotoxin combinations may lead to synergistic, cumulative, or antagonistic effects. Certain mycotoxins, such as DON and AFB1 combined, might produce significant synergistic toxic effects. Consequently, molecular mechanism studies focused on single mycotoxins might not be suitable for the joint exposure of multiple mycotoxins. (iii) Previous studies have mainly centered on the influence of these mycotoxins on gut microbiota but might have overlooked the effect of mycotoxins on the formation and dissemination of bacterial resistance, which could directly affect clinical drug treatment and even pose a threat to public health safety. More attention is needed in future research. (iv) There are still many unknown areas in the research around this 'mycotoxin–gut microbiota–metabolite-signaling pathways' metabolic axis, and in-depth research is urgently needed.

Author Contributions: Investigation: C.D., J.Z., T.L., Z.H., W.Q.; methodology, S.T., J.Z., C.D., T.V., Z.H.; funding acquisition, J.S., C.D.; Writing—original draft preparation, T.L. C.D., T.V., Z.H.; writing—review and editing, G.O.C., J.S., C.D. and T.V. All authors have read and agreed to the published version of the manuscript.

Funding: This study was supported by Beijing Natural Science Foundation (L252139 and 6252017) and the National Natural Science Foundation of China (Award number 32102724).

Institutional Review Board Statement: Not applicable.

Informed Consent Statement: Not applicable.

Data Availability Statement: All data are included in this manuscript.

Conflicts of Interest: The authors declare no conflicts of interest.

Abbreviations

The following abbreviations are used in this manuscript:

Akt, protein kinase B; AFs, aflatoxins; AHR, aryl hydrocarbon receptor; ATG5, autophagy-related protein 5; AMPK: AMP-activated protein kinase; Bax, Bcl-2-associated x gene; Bcl-2, B-cell lymphoma-2; CytC, cytochrome C; CAT, catalase; Caco-2, human epithelial colorectal adenocarcinoma; DAS, 4,15-diacetoxyscirpenol; DON, deoxynivalenol; ER, estrogen receptors; eNOS, endothelial nitric oxide synthase; ERK1/2, extracellular regulated protein kinase 1/2; FUMs, fumonisins; GST, glutathione S-transferase; GSH, glutathione; GR, glutathione reductase; GPX, glutathione peroxidase; H_2O_2, hydrogen peroxide; HNO2, nitrous acid; HO-1, heme oxygenase-1; IgA, immunoglobulin A; IgM, immunoglobulin M; IKK, IκB Kinase; iNOS, inducible nitric oxide synthase; LPS, lipopolysaccharide; LC3-II, microtubule-associated protein 1 light chain 3B; mTOR, mammalian target of rapamycin; MDA, malondialdehyde; MOMP, mitochondrial outer membrane permeabilization; MLCK, myosin light chain kinase; MAPK, mitogen-activated protein kinase; NEO, neosolaniol; NIV, nivalenol; NO, nitric oxide; NO−, nitrosyl anion; Nrf2, nuclear factor E2-related factor

2; ONOO–, peroxynitrite; $O_2^{-}\bullet$, superoxide anion; OTA, ochratoxins; OH•, hydroxyl radical; $^1O_2$, singlet oxygen; pIgR, polymeric immunoglobulin receptor; PYY, peptide YY; PKA, protein kinase A; rRNA, ribosomal RNA; RNS, reactive nitrogen species; ROS, reactive oxygen species; TCN, trichothecin; TrxR, thioredoxin reductase; TNF, tumor necrosis factor; VIP-LI, vasoactive intestinal polypeptide-like immunoreactive; ZEN, zearalenone; ZO-1, zonula occludens-1; XBP1: X-box-binding protein 1; SCFAs, short-chain fatty acids; SOD, superoxide dismutase.

References

1. Khan, R.; Anwar, F.; Ghazali, F.M. A comprehensive review of mycotoxins: Toxicology, detection, and effective mitigation approaches. *Heliyon* **2024**, *10*, e28361. [CrossRef]
2. Li, M.; Tang, S.; Peng, X.; Sharma, G.; Yin, S.; Hao, Z.; Li, J.; Shen, J.; Dai, C. Lycopene as a therapeutic agent against aflatoxin B1-related toxicity: Mechanistic insights and future directions. *Antioxidants* **2024**, *13*, 452. [CrossRef]
3. Dai, C.; Tian, E.; Hao, Z.; Tang, S.; Wang, Z.; Sharma, G.; Jiang, H.; Shen, J. Aflatoxin B1 toxicity and protective effects of curcumin: Molecular mechanisms and clinical implications. *Antioxidants* **2022**, *11*, 2031. [CrossRef]
4. dsm-firmenich World Mycotoxin Survey. Available online: https://www.dsm.com/content/dam/dsm/anh/en/documents/REP_MTXsurvey_Q4_2023_EN_0124_AUE_doublePage.pdf (accessed on 23 July 2025).
5. Yue, J.; Guo, D.; Gao, X.; Wang, J.; Nepovimova, E.; Wu, W.; Kuca, K. Deoxynivalenol (vomitoxin)-induced anorexia is induced by the release of intestinal hormones in mice. *Toxins* **2021**, *13*, 512. [CrossRef]
6. Lahjouji, T.; Bertaccini, A.; Neves, M.; Puel, S.; Oswald, I.P.; Soler, L. Acute exposure to zearalenone disturbs intestinal homeostasis by modulating the Wnt/β-Catenin signaling pathway. *Toxins* **2020**, *12*, 113. [CrossRef]
7. Yan, W.K.; Liu, Y.N.; Song, S.S.; Kang, J.W.; Zhang, Y.; Lu, L.; Wei, S.W.; Xu, Q.X.; Zhang, W.Q.; Liu, X.Z.; et al. Zearalenone affects the growth of endometriosis via estrogen signaling and inflammatory pathways. *Ecotoxicol. Environ. Saf.* **2022**, *241*, 113826. [CrossRef]
8. Dilkin, P.; Direito, G.; Simas, M.M.; Mallmann, C.A.; Corrêa, B. Toxicokinetics and toxicological effects of single oral dose of fumonisin B1 containing Fusarium verticillioides culture material in weaned piglets. *Chem. Biol. Interact.* **2010**, *185*, 157–162. [CrossRef]
9. Fouad, A.M.; Ruan, D.; El-Senousey, H.K.; Chen, W.; Jiang, S.; Zheng, C. Harmful effects and control strategies of aflatoxin b_1 produced by Aspergillus flavus and Aspergillus parasiticus strains on poultry: Review. *Toxins* **2019**, *11*, 176. [CrossRef]
10. Kumar, A.; Pathak, H.; Bhadauria, S.; Sudan, J. Aflatoxin contamination in food crops: Causes, detection, and management: A review. *Food Prod. Process. Nutr.* **2021**, *3*, 17. [CrossRef]
11. Shanakhat, H.; Sorrentino, A.; Raiola, A.; Romano, A.; Masi, P.; Cavella, S. Current methods for mycotoxins analysis and innovative strategies for their reduction in cereals: An overview. *J. Sci. Food Agric.* **2018**, *98*, 4003–4013. [CrossRef]
12. Li, M.; Yu, R.; Bai, X.; Wang, H.; Zhang, H. Fusarium: A treasure trove of bioactive secondary metabolites. *Nat. Prod. Rep.* **2020**, *37*, 1568–1588. [CrossRef]
13. Rai, A.; Das, M.; Tripathi, A. Occurrence and toxicity of a fusarium mycotoxin, zearalenone. *Crit. Rev. Food Sci. Nutr.* **2020**, *60*, 2710–2729. [CrossRef]
14. Ruan, H.; Huang, Y.; Yue, B.; Zhang, Y.; Lv, J.; Miao, K.; Zhang, D.; Luo, J.; Yang, M. Insights into the intestinal toxicity of foodborne mycotoxins through gut microbiota: A comprehensive review. *Compr. Rev. Food Sci. Food Saf.* **2023**, *22*, 4758–4785. [CrossRef]
15. Dai, C.; Sharma, G.; Liu, G.; Shen, J.; Shao, B.; Hao, Z. Therapeutic detoxification of quercetin for aflatoxin B1-related toxicity: Roles of oxidative stress, inflammation, and metabolic enzymes. *Environ. Pollut. (Barking Essex 1987)* **2024**, *345*, 123474. [CrossRef]
16. Adugna, C.; Wang, K.; Du, J.; Li, C. Deoxynivalenol mycotoxin dietary exposure on broiler performance and small intestine health: A comprehensive meta-analysis. *Poult. Sci.* **2024**, *103*, 104412. [CrossRef]
17. Duarte, S.C.; Lino, C.M.; Pena, A. Ochratoxin A in feed of food-producing animals: An undesirable mycotoxin with health and performance effects. *Vet. Microbiol.* **2011**, *154*, 1–13. [CrossRef]
18. Costamagna, D.; Gaggiotti, M.; Smulovitz, A.; Abdala, A.; Signorini, M. Mycotoxin sequestering agent: Impact on health and performance of dairy cows and efficacy in reducing AFM(1) residues in milk. *Environ. Toxicol. Pharmacol.* **2024**, *105*, 104349. [CrossRef]
19. Kolawole, O.; Graham, A.; Donaldson, C.; Owens, B.; Abia, W.A.; Meneely, J.; Alcorn, M.J.; Connolly, L.; Elliott, C.T. Low doses of mycotoxin mixtures below eu regulatory limits can negatively affect the performance of broiler chickens: A longitudinal study. *Toxins* **2020**, *12*, 433. [CrossRef]
20. Yang, C.; Song, G.; Lim, W. Effects of mycotoxin-contaminated feed on farm animals. *J. Hazard. Mater.* **2020**, *389*, 122087. [CrossRef]

21. Kozieł, M.J.; Ziaja, M.; Piastowska-Ciesielska, A.W. Intestinal barrier, claudins and mycotoxins. *Toxins* **2021**, *13*, 758. [CrossRef]
22. Gao, Y.; Meng, L.; Liu, H.; Wang, J.; Zheng, N. The Compromised intestinal barrier induced by mycotoxins. *Toxins* **2020**, *12*, 619. [CrossRef]
23. Song, C.; Chai, Z.; Chen, S.; Zhang, H.; Zhang, X.; Zhou, Y. Intestinal mucus components and secretion mechanisms: What we do and do not know. *Exp. Mol. Med.* **2023**, *55*, 681–691. [CrossRef]
24. Gao, J.; Song, G.; Shen, H.; Wu, Y.; Zhao, C.; Zhang, Z.; Jiang, Q.; Li, X.; Ma, X.; Tan, B.; et al. Allicin Improves intestinal epithelial barrier function and prevents LPS-induced barrier damages of intestinal epithelial cell monolayers. *Front. Immunol.* **2022**, *13*, 847861. [CrossRef]
25. Kang, R.; Li, S.; Perveen, A.; Shen, J.; Li, C. Effects of maternal T-2 toxin exposure on microorganisms and intestinal barrier function in young mice. *Ecotoxicol. Environ. Saf.* **2022**, *247*, 114252. [CrossRef]
26. Bastos-Amador, P.; Duarte, E.L.; Torres, J.; Caldeira, A.T.; Silva, I.; Salvador, C.; Assunção, R.; Alvito, P.; Ferreira, M. Maternal dietary exposure to mycotoxin aflatoxin B(1) promotes intestinal immune alterations and microbiota modifications increasing infection susceptibility in mouse offspring. *Food Chem. Toxicol.* **2023**, *173*, 113596. [CrossRef]
27. Xue, D.; Cheng, Y.; Pang, T.; Kuai, Y.; An, Y.; Wu, K.; Li, Y.; Lai, M.; Wang, B.; Wang, S. Sodium butyrate alleviates deoxynivalenol-induced porcine intestinal barrier disruption by promoting mitochondrial homeostasis via PCK2 signaling. *J. Hazard. Mater.* **2023**, *459*, 132013. [CrossRef]
28. Du, K.; Wang, C.; Liu, P.; Li, Y.; Ma, X. Effects of dietary mycotoxins on gut microbiome. *Protein Pept. Lett.* **2017**, *24*, 397–405. [CrossRef]
29. Zhou, B.; Xiao, K.; Guo, J.; Xu, Q.; Xu, Q.; Lv, Q.; Zhu, H.; Zhao, J.; Liu, Y. Necroptosis contributes to the intestinal toxicity of deoxynivalenol and is mediated by methyltransferase SETDB1. *J. Hazard. Mater.* **2024**, *474*, 134601. [CrossRef]
30. Lin, J.; Zuo, C.; Liang, T.; Huang, Y.; Kang, P.; Xiao, K.; Liu, Y. Lycopene alleviates multiple-mycotoxin-induced toxicity by inhibiting mitochondrial damage and ferroptosis in the mouse jejunum. *Food Funct.* **2022**, *13*, 11532–11542. [CrossRef]
31. Singh, N.; Dev, I.; Pal, S.; Yadav, S.K.; Idris, M.M.; Ansari, K.M. Transcriptomic and proteomic insights into patulin mycotoxin-induced cancer-like phenotypes in normal intestinal epithelial cells. *Mol. Cell. Biochem.* **2022**, *477*, 1405–1416. [CrossRef]
32. Miner-Williams, W.M.; Moughan, P.J. Intestinal barrier dysfunction: Implications for chronic inflammatory conditions of the bowel. *Nutr. Res. Rev.* **2016**, *29*, 40–59. [CrossRef]
33. Wu, Y.; Xiao, W.; Xiao, B.; Wang, Y.; Li, Y.; Wu, A.; Zhang, Q.; Liu, X.; Liu, S.; Yuan, Z.; et al. Melatonin alleviates t-2 toxin-induced intestinal injury by enhancing gut barrier function and modulating microbiota in weaned piglets. *J. Agric. Food Chem.* **2025**, *73*, 6903–6916. [CrossRef]
34. Liu, Y.; Yang, Y.; Dong, R.; Zhang, Z.; Jia, F.; Yu, H.; Wang, Y.; Zhang, Z. Protective effect of selenomethionine on intestinal injury induced by T- 2 toxin. *Res. Vet. Sci.* **2020**, *132*, 439–447. [CrossRef]
35. Cao, Z.; Gao, J.; Huang, W.; Yan, J.; Shan, A.; Gao, X. Curcumin mitigates deoxynivalenol-induced intestinal epithelial barrier disruption by regulating Nrf2/p53 and NF-κB/MLCK signaling in mice. *Food Chem. Toxicol.* **2022**, *167*, 113281. [CrossRef]
36. Pierron, A.; Bracarense, A.; Cossalter, A.M.; Laffitte, J.; Schwartz-Zimmermann, H.E.; Schatzmayr, G.; Pinton, P.; Moll, W.D.; Oswald, I.P. Deepoxy-deoxynivalenol retains some immune-modulatory properties of the parent molecule deoxynivalenol in piglets. *Arch. Toxicol.* **2018**, *92*, 3381–3389. [CrossRef]
37. Pierron, A.; Balbo, L.C.; Soler, L.; Pinton, P.; Puel, S.; Laffitte, J.; Albin, M.; Bracarense, A.; Rodriguez, M.A.; Oswald, I.P. Deoxynivalenol Induces local inflammation and lesions in tissues at doses recommended by the EU. *Int. J. Mol. Sci.* **2024**, *25*, 9790. [CrossRef]
38. Pasternak, J.A.; Aiyer, V.I.A.; Hamonic, G.; Beaulieu, A.D.; Columbus, D.A.; Wilson, H.L. Molecular and physiological effects on the small intestine of weaner pigs following feeding with deoxynivalenol-contaminated feed. *Toxins* **2018**, *10*, 40. [CrossRef]
39. Wang, X.; Li, L.; Zhang, G. A proteomic study on the protective effect of kaempferol pretreatment against deoxynivalenol-induced intestinal barrier dysfunction in a Caco-2 cell model. *Food Funct.* **2020**, *11*, 7266–7279. [CrossRef]
40. Zhang, Z.Q.; Wang, S.B.; Wang, R.G.; Zhang, W.; Wang, P.L.; Su, X.O. Phosphoproteome analysis reveals the molecular mechanisms underlying deoxynivalenol-induced intestinal toxicity in IPEC-J2 Cells. *Toxins* **2016**, *8*, 270. [CrossRef]
41. Zhou, J.Y.; Zan, G.X.; Zhu, Q.J.; Gao, C.Q.; Yan, H.C.; Wang, X.Q. Recombinant porcine r-spondin 1 facilitates intestinal stem cell expansion along the crypt-villus axis through potentiating Wnt/β-Catenin signaling in homeostasis and deoxynivalenol injury. *J. Agric. Food Chem.* **2022**, *70*, 10644–10653. [CrossRef]
42. Xia, S.; Yan, C.; Gu, J.; Yuan, Y.; Zou, H.; Liu, Z.; Bian, J. Resveratrol alleviates zearalenone-induced intestinal dysfunction in mice through the NF-κB/Nrf2/HO-1 signalling pathway. *Foods* **2024**, *13*, 1217. [CrossRef]
43. Huangfu, B.; Li, J.; Xu, T.; Ren, X.; Zhang, R.; Chen, Y.; Wang, J.; Huang, K.; He, X. Zearalenone induces intestinal damage and flora disturbance in rats by triggering ferroptosis via the system Xc(-)-GSH-GPX4 signaling pathway. *Ecotoxicol. Environ. Saf.* **2025**, *302*, 118600. [CrossRef]

44. Tomaszewska, E.; Rudyk, H.; Dobrowolski, P.; Donaldson, J.; Świetlicka, I.; Puzio, I.; Kamiński, D.; Wiącek, D.; Kushnir, V.; Brezvyn, O.; et al. Changes in the intestinal histomorphometry, the expression of intestinal tight junction proteins, and the bone structure and liver of pre-laying hens following oral administration of Fumonisins for 21 Days. *Toxins* **2021**, *13*, 375. [CrossRef]
45. Antonissen, G.; Van Immerseel, F.; Pasmans, F.; Ducatelle, R.; Janssens, G.P.; De Baere, S.; Mountzouris, K.C.; Su, S.; Wong, E.A.; De Meulenaer, B.; et al. Mycotoxins deoxynivalenol and fumonisins alter the extrinsic component of intestinal barrier in broiler chickens. *J. Agric. Food Chem.* **2015**, *63*, 10846–10855. [CrossRef]
46. Shanmugasundaram, R.; Kappari, L.; Pilewar, M.; Jones, M.K.; Olukosi, O.A.; Pokoo-Aikins, A.; Applegate, T.J.; Glenn, A.E. Exposure to subclinical doses of fumonisins, deoxynivalenol, and zearalenone affects immune response, amino acid digestibility, and intestinal morphology in broiler chickens. *Toxins* **2025**, *17*, 16. [CrossRef]
47. da Silva, E.O.; Gerez, J.R.; Hohmann, M.S.N.; Verri, W.A., Jr.; Bracarense, A. Phytic acid decreases oxidative stress and intestinal lesions induced by Fumonisin B_1 and deoxynivalenol in intestinal explants of pigs. *Toxins* **2019**, *11*, 18. [CrossRef]
48. Ye, Y.; Yang, D.; Huang, H.; Li, Y.; Ji, J.; Wang, J.S.; Sun, X. Effect of fumonisin b1 and hydrolyzed fb1 exposure on intestinal and hepatic toxicity in BALB/c mice. *J. Agric. Food Chem.* **2025**, *73*, 10603–10614. [CrossRef]
49. Yu, S.; Zou, L.; Zhao, J.; Zhu, Y. Resveratrol alleviates fumonisin-induced intestinal cytotoxicity by modulating apoptosis, tight junction, and inflammation in IPEC-J2 porcine intestinal epithelial cells. *Environ. Toxicol.* **2024**, *39*, 905–914. [CrossRef]
50. Peng, X.; Zhang, X.; Sharma, G.; Dai, C. Thymol as a potential neuroprotective agent: Mechanisms, efficacy, and future prospects. *J. Agric. Food Chem.* **2024**, *72*, 6803–6814. [CrossRef]
51. Minervini, F.; Garbetta, A.; D'Antuono, I.; Cardinali, A.; Martino, N.A.; Debellis, L.; Visconti, A. Toxic mechanisms induced by fumonisin b1 mycotoxin on human intestinal cell line. *Arch. Environ. Contam. Toxicol.* **2014**, *67*, 115–123. [CrossRef]
52. Eskola, M.; Kos, G.; Elliott, C.T.; Hajšlová, J.; Mayar, S.; Krska, R. Worldwide contamination of food-crops with mycotoxins: Validity of the widely cited 'FAO estimate' of 25. *Crit. Rev. Food Sci. Nutr.* **2020**, *60*, 2773–2789. [CrossRef]
53. Choi, H.; Garavito-Duarte, Y.; Gormley, A.R.; Kim, S.W. Aflatoxin B1: Challenges and strategies for the intestinal microbiota and intestinal health of monogastric animals. *Toxins* **2025**, *17*, 43. [CrossRef]
54. Pan, H.; Hu, T.; He, Y.; Zhong, G.; Wu, S.; Jiang, X.; Rao, G.; You, Y.; Ruan, Z.; Tang, Z.; et al. Curcumin attenuates aflatoxin B1-induced ileum injury in ducks by inhibiting NLRP3 inflammasome and regulating TLR4/NF-κB signaling pathway. *Mycotoxin Res.* **2024**, *40*, 255–268. [CrossRef]
55. Li, J.; Shi, M.; Wang, Y.; Liu, J.; Liu, S.; Kang, W.; Liu, X.; Chen, X.; Huang, K.; Liu, Y. Probiotic-derived extracellular vesicles alleviate AFB1-induced intestinal injury by modulating the gut microbiota and AHR activation. *J. Nanobiotechnol.* **2024**, *22*, 697. [CrossRef]
56. Ding, J.; Cheng, X.; Zeng, C.; Zhao, Q.; Xing, C.; Zhang, C.; Cao, H.; Guo, X.; Hu, G.; Zhuang, Y. Aflatoxin B1 promotes pyroptosis in ipec-j2 cells by disrupting mitochondrial dynamics through the AMPK/NLRP3 pathway. *J. Agric. Food Chem.* **2024**, *72*, 28093–28108. [CrossRef]
57. He, W.; Wang, J.; Han, M.; Wang, L.; Li, L.; Zhang, J.; Chen, S.; Guo, J.; Zhai, X.; Yang, J. Potential toxicity and mechanisms of T-2 and HT-2 individually or in combination on the intestinal barrier function of porcine small intestinal epithelial cells. *Toxins* **2023**, *15*, 682. [CrossRef]
58. Liu, M.; Zhao, L.; Wei, J.T.; Huang, Y.X.; Khalil, M.M.; Wu, W.D.; Kuča, K.; Sun, L.H. T-2 toxin-induced intestinal damage with dysregulation of metabolism, redox homeostasis, inflammation, and apoptosis in chicks. *Arch. Toxicol.* **2023**, *97*, 805–817. [CrossRef]
59. Wang, X.; Xu, Y.; Yu, H.; Lu, Y.; Qian, Y.; Wang, M. Metabolomics and lipidomics reveal the metabolic disorders induced by single and combined exposure of Fusarium mycotoxins in IEC-6 cells. *Foods* **2025**, *14*, 230. [CrossRef]
60. Basso, K.; Gomes, F.; Bracarense, A.P. Deoxynivanelol and fumonisin, alone or in combination, induce changes on intestinal junction complexes and in E-cadherin expression. *Toxins* **2013**, *5*, 2341–2352. [CrossRef]
61. Peters, A.; Nawrot, T.S.; Baccarelli, A.A. Hallmarks of environmental insults. *Cell* **2021**, *184*, 1455–1468. [CrossRef]
62. Jomova, K.; Raptova, R.; Alomar, S.Y.; Alwasel, S.H.; Nepovimova, E.; Kuca, K.; Valko, M. Reactive oxygen species, toxicity, oxidative stress, and antioxidants: Chronic diseases and aging. *Arch. Toxicol.* **2023**, *97*, 2499–2574. [CrossRef]
63. Poprac, P.; Jomova, K.; Simunkova, M.; Kollar, V.; Rhodes, C.J.; Valko, M. Targeting free radicals in oxidative stress-related human diseases. *Trends Pharmacol. Sci.* **2017**, *38*, 592–607. [CrossRef]
64. Li, M.; Tang, S.; Velkov, T.; Shen, J.; Dai, C. Copper exposure induces mitochondrial dysfunction and hepatotoxicity via the induction of oxidative stress and PERK/ATF4 -mediated endoplasmic reticulum stress. *Environ. Pollut. (Barking Essex 1987)* **2024**, *352*, 124145. [CrossRef]
65. Ben Salah-Abbès, J.; Mannai, M.; Belgacem, H.; Zinedine, A.; Abbès, S. Efficacy of lactic acid bacteria supplementation against Fusarium graminearum growth in vitro and inhibition of Zearalenone causing inflammation and oxidative stress in vivo. *Toxicon Off. J. Int. Soc. Toxinol.* **2021**, *202*, 115–122. [CrossRef]
66. Zhu, C.; Liang, S.; Zan, G.; Wang, X.; Gao, C.; Yan, H.; Wang, X.; Zhou, J. Selenomethionine alleviates don-induced oxidative stress via modulating Keap1/Nrf2 signaling in the small intestinal epithelium. *J. Agric. Food Chem.* **2023**, *71*, 895–904. [CrossRef]

67. Liang, S.J.; Wang, X.Q. Deoxynivalenol induces intestinal injury: Insights from oxidative stress and intestinal stem cells. *Environ. Sci. Pollut. Res. Int.* **2023**, *30*, 48676–48685. [CrossRef]
68. de Souza, M.; Baptista, A.A.S.; Valdiviezo, M.J.J.; Justino, L.; Menck-Costa, M.F.; Ferraz, C.R.; da Gloria, E.M.; Verri, W.A., Jr.; Bracarense, A. *Lactobacillus* spp. reduces morphological changes and oxidative stress induced by deoxynivalenol on the intestine and liver of broilers. *Toxicon Off. J. Int. Soc. Toxinol.* **2020**, *185*, 203–212. [CrossRef]
69. Xu, X.; Yan, G.; Chang, J.; Wang, P.; Yin, Q.; Liu, C.; Liu, S.; Zhu, Q.; Lu, F. Astilbin ameliorates deoxynivalenol-induced oxidative stress and apoptosis in intestinal porcine epithelial cells (IPEC-J2). *J. Appl. Toxicol. JAT* **2020**, *40*, 1362–1372. [CrossRef]
70. Chen, F.; Wang, Y.; Chen, Y.; Fan, J.; Zhang, C.; He, X.; Yang, X. JNK molecule is a toxic target for IPEC-J2 cell barrier damage induced by T-2 toxin. *Ecotoxicol. Environ. Saf.* **2023**, *263*, 115247. [CrossRef]
71. Sun, H.; Zhang, M.; Li, J.; Shan, A. DL-Selenomethionine alleviates oxidative stress induced by zearalenone via Nrf2/Keap1 signaling pathway in IPEC-J2 Cells. *Toxins* **2021**, *13*, 557. [CrossRef]
72. He, F.; Ru, X.; Wen, T. NRF2, a transcription factor for stress response and beyond. *Int. J. Mol. Sci.* **2020**, *21*, 4777. [CrossRef]
73. McCord, J.M.; Gao, B.; Hybertson, B.M. The complex genetic and epigenetic regulation of the Nrf2 pathways: A Review. *Antioxidants* **2023**, *12*, 366. [CrossRef]
74. Yu, T.; Deng, X.; Yang, X.; Yin, Y.; Liu, Y.; Xu, S. New insights into evodiamine attenuates IPEC-J2 cells pyroptosis induced by T-2 toxin—Activating Keap1-Nrf2/NF-κB signaling pathway through binding with Keap1. *J. Environ. Manag.* **2024**, *370*, 122605. [CrossRef]
75. Jiang, X.; Liu, H.; You, Y.; Zhong, G.; Ruan, Z.; Liao, J.; Zhang, H.; Pan, J.; Tang, Z.; Hu, L. Multi-omics reveals the protective effects of curcumin against AFB1-induced oxidative stress and inflammatory damage in duckling intestines. *Comp. Biochem. physiology. Toxicol. Pharmacol. CBP* **2024**, *276*, 109815. [CrossRef]
76. Zhang, Z.; Wang, J.; Wang, J.; Xie, H.; Zhang, Z.; Shi, L.; Zhu, X.; Lv, Q.; Chen, X.; Liu, Y. Selenomethionine attenuates ochratoxin A-induced small intestinal injury in rabbits by activating the Nrf2 pathway and inhibiting NF-κB activation. *Ecotoxicol. Environ. Saf.* **2023**, *256*, 114837. [CrossRef]
77. Li, R.; Tan, B.; Jiang, Q.; Chen, F.; Liu, K.; Liao, P. Eucommia ulmoides flavonoids alleviate intestinal oxidative stress damage in weaned piglets by regulating the Nrf2/Keap1 signaling pathway. *Ecotoxicol. Environ. Saf.* **2024**, *288*, 117373. [CrossRef]
78. Zhang, Z.; Zhang, Q.; Li, M.; Xu, J.; Wang, J.; Li, M.; Wei, L.; Lv, Q.; Chen, X.; Wang, Y.; et al. SeMet attenuates AFB1-induced intestinal injury in rabbits by activating the Nrf2 pathway. *Ecotoxicol. Environ. Saf.* **2022**, *239*, 113640. [CrossRef]
79. Taguchi, K.; Takaku, M.; Egner, P.A.; Morita, M.; Kaneko, T.; Mashimo, T.; Kensler, T.W.; Yamamoto, M. Generation of a new model rat: Nrf2 knockout rats are sensitive to aflatoxin B1 Toxicity. *Toxicol. Sci. Off. J. Soc. Toxicol.* **2016**, *152*, 40–52. [CrossRef]
80. Pang, Y.; Zhang, L.; Liu, Q.; Peng, H.; He, J.; Jin, H.; Su, X.; Zhao, J.; Guo, J. NRF2/PGC-1α-mediated mitochondrial biogenesis contributes to T-2 toxin-induced toxicity in human neuroblastoma SH-SY5Y cells. *Toxicol. Appl. Pharmacol.* **2022**, *451*, 116167. [CrossRef]
81. Zhao, L.; Deng, J.; Xu, Z.J.; Zhang, W.P.; Khalil, M.M.; Karrow, N.A.; Sun, L.H. Mitigation of Aflatoxin B(1) hepatoxicity by dietary hedyotis diffusa is associated with activation of NRF2/ARE signaling in chicks. *Antioxidants* **2021**, *10*, 878. [CrossRef]
82. Rajput, S.A.; Shaukat, A.; Wu, K.; Rajput, I.R.; Baloch, D.M.; Akhtar, R.W.; Raza, M.A.; Najda, A.; Rafał, P.; Albrakati, A.; et al. Luteolin alleviates AflatoxinB(1)-induced apoptosis and oxidative stress in the liver of mice through activation of Nrf2 signaling pathway. *Antioxidants* **2021**, *10*, 1268. [CrossRef]
83. Wang, Y.; Liu, F.; Zhou, X.; Liu, M.; Zang, H.; Liu, X.; Shan, A.; Feng, X. Alleviation of oral exposure to aflatoxin B1-induced renal dysfunction, oxidative stress, and cell apoptosis in mice kidney by curcumin. *Antioxidants* **2022**, *11*, 1082. [CrossRef]
84. Chen, X.; Mu, P.; Zhu, L.; Mao, X.; Chen, S.; Zhong, H.; Deng, Y. T-2 toxin induces oxidative stress at low doses via Atf3ΔZip2a/2b-mediated ubiquitination and degradation of Nrf2. *Int. J. Mol. Sci.* **2021**, *22*, 7936. [CrossRef]
85. Xu, Q.; Shi, W.; Lv, P.; Meng, W.; Mao, G.; Gong, C.; Chen, Y.; Wei, Y.; He, X.; Zhao, J.; et al. Critical role of caveolin-1 in aflatoxin B1-induced hepatotoxicity via the regulation of oxidation and autophagy. *Cell Death Dis.* **2020**, *11*, 6. [CrossRef]
86. Jakubczyk, K.; Dec, K.; Kałduńska, J.; Kawczuga, D.; Kochman, J.; Janda, K. Reactive oxygen species—sources, functions, oxidative damage. *Pol. Merkur. Lek. Organ Pol. Tow. Lek.* **2020**, *48*, 124–127.
87. Dos Santos, L.; Bertoli, S.R.; Ávila, R.A.; Marques, V.B. Iron overload, oxidative stress and vascular dysfunction: Evidences from clinical studies and animal models. *Biochim. Biophys. Acta Gen. Subj.* **2022**, *1866*, 130172. [CrossRef]
88. Wan, D.; Wu, Q.; Qu, W.; Liu, G.; Wang, X. Pyrrolidine dithiocarbamate (pdtc) inhibits DON-induced mitochondrial dysfunction and apoptosis via the NF-κB/iNOS pathway. *Oxidative Med. Cell. Longev.* **2018**, *2018*, 1324173. [CrossRef]
89. Brunelle, J.K.; Letai, A. Control of mitochondrial apoptosis by the Bcl-2 family. *J. Cell Sci.* **2009**, *122*, 437–441. [CrossRef]
90. Youle, R.J.; Strasser, A. The BCL-2 protein family: Opposing activities that mediate cell death. *Nat. Rev. Mol. Cell Biol.* **2008**, *9*, 47–59. [CrossRef]

91. Zhang, M.; Li, Q.; Wang, J.; Sun, J.; Xiang, Y.; Jin, X. Aflatoxin B1 disrupts the intestinal barrier integrity by reducing junction protein and promoting apoptosis in pigs and mice. *Ecotoxicol. Environ. Saf.* **2022**, *247*, 114250. [CrossRef]
92. Xu, Y.; Xie, Y.; Wu, Z.; Wang, H.; Chen, Z.; Wang, J.; Bao, W. Protective effects of melatonin on deoxynivalenol-induced oxidative stress and autophagy in IPEC-J2 cells. *Food Chem. Toxicol.* **2023**, *177*, 113803. [CrossRef]
93. Chen, J.; Huang, Z.; Cao, X.; Chen, X.; Zou, T.; You, J. Plant-derived polyphenols as Nrf2 activators to counteract oxidative stress and intestinal toxicity induced by deoxynivalenol in swine: An Emerging Research Direction. *Antioxidants* **2022**, *11*, 2379. [CrossRef]
94. Shen, T.; Miao, Y.; Ding, C.; Fan, W.; Liu, S.; Lv, Y.; Gao, X.; De Boevre, M.; Yan, L.; Okoth, S.; et al. Activation of the p38/MAPK pathway regulates autophagy in response to the CYPOR-dependent oxidative stress induced by zearalenone in porcine intestinal epithelial cells. *Food Chem. Toxicol.* **2019**, *131*, 110527. [CrossRef]
95. Tang, Y.; Li, J.; Li, F.; Hu, C.A.; Liao, P.; Tan, K.; Tan, B.; Xiong, X.; Liu, G.; Li, T.; et al. Autophagy protects intestinal epithelial cells against deoxynivalenol toxicity by alleviating oxidative stress via IKK signaling pathway. *Free Radic. Biol. Med.* **2015**, *89*, 944–951. [CrossRef]
96. Liu, S.; Mao, X.; Ge, L.; Hou, L.; Le, G.; Gan, F.; Wen, L.; Huang, K. Phenethyl isothiocyanate as an anti-nutritional factor attenuates deoxynivalenol-induced IPEC-J2 cell injury through inhibiting ROS-mediated autophagy. *Anim. Nutr. (Zhongguo Xu Mu Shou Yi Xue Hui)* **2022**, *8*, 300–309. [CrossRef]
97. Takiishi, T.; Fenero, C.I.M.; Câmara, N.O.S. Intestinal barrier and gut microbiota: Shaping our immune responses throughout life. *Tissue Barriers* **2017**, *5*, e1373208. [CrossRef]
98. Mowat, A.M.; Agace, W.W. Regional specialization within the intestinal immune system. *Nat. Rev. Immunol.* **2014**, *14*, 667–685. [CrossRef]
99. Fessler, M.B. The intracellular cholesterol landscape: Dynamic integrator of the immune response. *Trends Immunol.* **2016**, *37*, 819–830. [CrossRef]
100. Watson, S.; Gong, Y.Y.; Routledge, M. Interventions targeting child undernutrition in developing countries may be undermined by dietary exposure to aflatoxin. *Crit. Rev. Food Sci. Nutr.* **2017**, *57*, 1963–1975. [CrossRef]
101. Gu, M.J.; Song, S.K.; Lee, I.K.; Ko, S.; Han, S.E.; Bae, S.; Ji, S.Y.; Park, B.C.; Song, K.D.; Lee, H.K.; et al. Barrier protection via Toll-like receptor 2 signaling in porcine intestinal epithelial cells damaged by deoxynivalnol. *Vet. Res.* **2016**, *47*, 25. [CrossRef]
102. Liu, M.; Gao, R.; Meng, Q.; Zhang, Y.; Bi, C.; Shan, A. Toxic effects of maternal zearalenone exposure on intestinal oxidative stress, barrier function, immunological and morphological changes in rats. *PLoS ONE* **2014**, *9*, e106412. [CrossRef]
103. Gu, M.J.; Han, S.E.; Hwang, K.; Mayer, E.; Reisinger, N.; Schatzmayr, D.; Park, B.C.; Han, S.H.; Yun, C.H. Hydrolyzed fumonisin B(1) induces less inflammatory responses than fumonisin B(1) in the co-culture model of porcine intestinal epithelial and immune cells. *Toxicol. Lett.* **2019**, *305*, 110–116. [CrossRef]
104. Hao, S.; Hu, J.; Song, S.; Huang, D.; Xu, H.; Qian, G.; Gan, F.; Huang, K. Selenium alleviates aflatoxin B_1-induced immune toxicity through improving glutathione peroxidase 1 and selenoprotein S expression in primary porcine splenocytes. *J. Agric. Food Chem.* **2016**, *64*, 1385–1393. [CrossRef]
105. Sharma, R.P. Immunotoxicity of mycotoxins. *J. Dairy Sci.* **1993**, *76*, 892–897. [CrossRef]
106. Pierron, A.; Alassane-Kpembi, I.; Oswald, I.P. Impact of mycotoxin on immune response and consequences for pig health. *Anim. Nutr. (Zhongguo Xu Mu Shou Yi Xue Hui)* **2016**, *2*, 63–68. [CrossRef]
107. Burel, C.; Tanguy, M.; Guerre, P.; Boilletot, E.; Cariolet, R.; Queguiner, M.; Postollec, G.; Pinton, P.; Salvat, G.; Oswald, I.P.; et al. Effect of low dose of fumonisins on pig health: Immune status, intestinal microbiota and sensitivity to Salmonella. *Toxins* **2013**, *5*, 841–864. [CrossRef]
108. Bulgaru, C.V.; Marin, D.E.; Pistol, G.C.; Taranu, I. Zearalenone and the immune response. *Toxins* **2021**, *13*, 248. [CrossRef]
109. Wang, S.; Wu, K.; Xue, D.; Zhang, C.; Rajput, S.A.; Qi, D. Mechanism of deoxynivalenol mediated gastrointestinal toxicity: Insights from mitochondrial dysfunction. *Food Chem. Toxicol.* **2021**, *153*, 112214. [CrossRef]
110. Sauter, K.A.; Magun, E.A.; Iordanov, M.S.; Magun, B.E. ZAK is required for doxorubicin, a novel ribotoxic stressor, to induce SAPK activation and apoptosis in HaCaT cells. *Cancer Biol. Ther.* **2010**, *10*, 258–266. [CrossRef]
111. Hu, P.; Zong, Q.; Zhao, Y.; Gu, H.; Liu, Y.; Gu, F.; Liu, H.Y.; Ahmed, A.A.; Bao, W.; Cai, D. Lactoferrin attenuates intestinal barrier dysfunction and inflammation by modulating the MAPK pathway and gut microbes in mice. *J. Nutr.* **2022**, *152*, 2451–2460. [CrossRef]
112. Zhang, H.; Deng, X.; Zhou, C.; Wu, W.; Zhang, H. Deoxynivalenol induces inflammation in IPEC-J2 cells by activating P38 Mapk And Erk1/2. *Toxins* **2020**, *12*, 180. [CrossRef]
113. Liu, D.; Wang, Q.; He, W.; Chen, X.; Wei, Z.; Huang, K. Two-way immune effects of deoxynivalenol in weaned piglets and porcine alveolar macrophages: Due mainly to its exposure dosage. *Chemosphere* **2020**, *249*, 126464. [CrossRef]
114. Graziani, F.; Pujol, A.; Nicoletti, C.; Pinton, P.; Armand, L.; Di Pasquale, E.; Oswald, I.P.; Perrier, J.; Maresca, M. The food-associated ribotoxin deoxynivalenol modulates inducible NO synthase in human intestinal cell model. *Toxicol. Sci. Off. J. Soc. Toxicol.* **2015**, *145*, 372–382. [CrossRef]

115. Chen, J.; Zhang, X.; He, Z.; Xiong, D.; Long, M. Damage on intestinal barrier function and microbial detoxification of deoxynivalenol: A review. *J. Integr. Agric.* **2024**, *23*, 2507–2524. [CrossRef]
116. Van De Walle, J.; Romier, B.; Larondelle, Y.; Schneider, Y.-J. Influence of deoxynivalenol on NF-κB activation and IL-8 secretion in human intestinal Caco-2 cells. *Toxicol. Lett.* **2008**, *177*, 205–214. [CrossRef]
117. Hu, Z.Y.; Yang, S.J.; Chang, Y.H.; Wang, X.Q.; Liu, R.Q.; Jiang, F.W.; Chen, M.S.; Wang, J.X.; Liu, S.; Zhu, H.M.; et al. AHR activation relieves deoxynivalenol-induced disruption of porcine intestinal epithelial barrier functions. *J. Hazard. Mater.* **2024**, *480*, 136095. [CrossRef]
118. Jiang, M.; Peng, X.; Fang, J.; Cui, H.; Yu, Z.; Chen, Z. Effects of aflatoxin b1 on T-cell subsets and mRNA expression of cytokines in the intestine of broilers. *Int. J. Mol. Sci.* **2015**, *16*, 6945–6959. [CrossRef]
119. Zhang, L.; Cheng, D.; Zhang, J.; Tang, H.; Li, F.; Peng, Y.; Duan, X.; Meng, E.; Zhang, C.; Zeng, T.; et al. Role of macrophage AHR/TLR4/STAT3 signaling axis in the colitis induced by non-canonical AHR ligand aflatoxin B1. *J. Hazard. Mater.* **2023**, *452*, 131262. [CrossRef]
120. Guo, H.; Wang, P.; Liu, C.; Zhou, T.; Chang, J.; Yin, Q.; Wang, L.; Jin, S.; Zhu, Q.; Lu, F. Effects of compound mycotoxin detoxifier on alleviating aflatoxin B(1)-induced inflammatory responses in intestine, liver and kidney of broilers. *Toxins* **2022**, *14*, 665. [CrossRef]
121. Gao, Y.N.; Wang, Z.W.; Su, C.Y.; Wang, J.Q.; Zheng, N. Omics analysis revealed the intestinal toxicity induced by aflatoxin B1 and aflatoxin M1. *Ecotoxicol. Environ. Saf.* **2024**, *278*, 116336. [CrossRef]
122. Wang, X.; Yu, H.; Shan, A.; Jin, Y.; Fang, H.; Zhao, Y.; Shen, J.; Zhou, C.; Zhou, Y.; Fu, Y.; et al. Toxic effects of Zearalenone on intestinal microflora and intestinal mucosal immunity in mice. *Food Agric. Immunol.* **2018**, *29*, 1002–1011. [CrossRef]
123. Girish, C.K.; Smith, T.K.; Boermans, H.J.; Anil Kumar, P.; Girgis, G.N. Effects of dietary Fusarium mycotoxins on intestinal lymphocyte subset populations, cell proliferation and histological changes in avian lymphoid organs. *Food Chem. Toxicol.* **2010**, *48*, 3000–3007. [CrossRef]
124. Stewart, A.S.; Pratt-Phillips, S.; Gonzalez, L.M. Alterations in intestinal permeability: The role of the "leaky gut" in health and disease. *J. Equine Vet. Sci.* **2017**, *52*, 10–22. [CrossRef]
125. Park, S.H.; Kim, D.; Kim, J.; Moon, Y. Effects of Mycotoxins on mucosal microbial infection and related pathogenesis. *Toxins* **2015**, *7*, 4484–4502. [CrossRef]
126. Awad, W.A.; Ruhnau, D.; Hess, C.; Doupovec, B.; Schatzmayr, D.; Hess, M. Feeding of deoxynivalenol increases the intestinal paracellular permeability of broiler chickens. *Arch. Toxicol.* **2019**, *93*, 2057–2064. [CrossRef]
127. Verbrugghe, E.; Vandenbroucke, V.; Dhaenens, M.; Shearer, N.; Goossens, J.; De Saeger, S.; Eeckhout, M.; D'Herde, K.; Thompson, A.; Deforce, D.; et al. T-2 toxin induced Salmonella Typhimurium intoxication results in decreased Salmonella numbers in the cecum contents of pigs, despite marked effects on Salmonella-host cell interactions. *Vet. Res.* **2012**, *43*, 22. [CrossRef]
128. Ye, L.; Chen, H.; Wang, J.; Tsim, K.W.K.; Wang, Y.; Shen, X.; Lei, H.; Liu, Y. Aflatoxin B(1)-induced liver pyroptosis is mediated by disturbing the gut microbial metabolites: The roles of pipecolic acid and norepinephrine. *J. Hazard. Mater.* **2024**, *474*, 134822. [CrossRef]
129. Neish, A.S. The gut microflora and intestinal epithelial cells: A continuing dialogue. *Microbes Infect.* **2002**, *4*, 309–317. [CrossRef]
130. Zhang, L.; Zhang, Z.; Xu, L.; Zhang, X. Maintaining the Balance of Intestinal Flora through the Diet: Effective Prevention of Illness. *Foods* **2021**, *10*, 2312. [CrossRef]
131. Chen, J.; Li, Y.; Tian, Y.; Huang, C.; Li, D.; Zhong, Q.; Ma, X. Interaction between Microbes and Host Intestinal Health: Modulation by Dietary Nutrients and Gut-Brain-Endocrine-Immune Axis. *Curr. Protein Pept. Sci.* **2015**, *16*, 592–603. [CrossRef]
132. Cai, J.; Sun, L.; Gonzalez, F.J. Gut microbiota-derived bile acids in intestinal immunity, inflammation, and tumorigenesis. *Cell Host Microbe* **2022**, *30*, 289–300. [CrossRef]
133. Singh, R.K.; Chang, H.W.; Yan, D.; Lee, K.M.; Ucmak, D.; Wong, K.; Abrouk, M.; Farahnik, B.; Nakamura, M.; Zhu, T.H.; et al. Influence of diet on the gut microbiome and implications for human health. *J. Transl. Med.* **2017**, *15*, 73. [CrossRef]
134. Paone, P.; Cani, P.D. Mucus barrier, mucins and gut microbiota: The expected slimy partners? *Gut* **2020**, *69*, 2232–2243. [CrossRef]
135. Dang, A.T.; Marsland, B.J. Microbes, metabolites, and the gut-lung axis. *Mucosal Immunol.* **2019**, *12*, 843–850. [CrossRef]
136. Xia, D.; Mo, Q.; Yang, L.; Wang, W. Crosstalk between mycotoxins and intestinal microbiota and the alleviation approach via microorganisms. *Toxins* **2022**, *14*, 859. [CrossRef]
137. Jin, J.; Beekmann, K.; Ringø, E.; Rietjens, I.M.C.M.; Xing, F. Interaction between food-borne mycotoxins and gut microbiota: A review. *Food Control* **2021**, *126*, 107998. [CrossRef]
138. Vignal, C.; Djouina, M.; Pichavant, M.; Caboche, S.; Waxin, C.; Beury, D.; Hot, D.; Gower-Rousseau, C.; Body-Malapel, M. Chronic ingestion of deoxynivalenol at human dietary levels impairs intestinal homeostasis and gut microbiota in mice. *Arch. Toxicol.* **2018**, *92*, 2327–2338. [CrossRef]
139. Charlet, R.; Bortolus, C.; Sendid, B.; Jawhara, S. Bacteroides thetaiotaomicron and Lactobacillus johnsonii modulate intestinal inflammation and eliminate fungi via enzymatic hydrolysis of the fungal cell wall. *Sci. Rep.* **2020**, *10*, 11510. [CrossRef]

140. Ragoubi, C.; Quintieri, L.; Greco, D.; Mehrez, A.; Maatouk, I.; D'Ascanio, V.; Landoulsi, A.; Avantaggiato, G. Mycotoxin Removal by *Lactobacillus* spp. Their Application in Animal Liquid Feed. *Toxins* **2021**, *13*, 185. [CrossRef]
141. Piotrowska, M.; Sliżewska, K.; Nowak, A.; Zielonka, L.; Zakowska, Z.; Gajęcka, M.; Gajęcki, M. The effect of experimental fusarium mycotoxicosis on microbiota diversity in porcine ascending colon contents. *Toxins* **2014**, *6*, 2064–2081. [CrossRef]
142. Deng, F.; Zhao, L.; Wei, P.; Mai, E.; Chen, M.; Yang, H.; Mu, P.; Wu, J.; Wen, J.; Deng, Y. Role and mechanism of the outer membrane porin LamB in T-2 mycotoxin-mediated extensive drug resistance in Escherichia coli. *J. Hazard. Mater.* **2024**, *480*, 136437. [CrossRef]
143. Chen, C.; Liu, C.; Zhang, K.; Xue, W. The role of gut microbiota and its metabolites short-chain fatty acids in food allergy. *Food Sci. Hum. Wellness* **2023**, *12*, 702–710. [CrossRef]
144. Liu, P.; Wang, Y.; Yang, G.; Zhang, Q.; Meng, L.; Xin, Y.; Jiang, X. The role of short-chain fatty acids in intestinal barrier function, inflammation, oxidative stress, and colonic carcinogenesis. *Pharmacol. Res.* **2021**, *165*, 105420. [CrossRef]
145. Wei, B.; Xiao, H.; Xu, B.; Kuca, K.; Qin, Z.; Guo, X.; Wu, W.; Wu, Q. Emesis to trichothecene deoxynivalenol and its congeners correspond to secretion of peptide YY and 5-HT. *Food Chem. Toxicol.* **2023**, *178*, 113874. [CrossRef]
146. Larraufie, P.; Doré, J.; Lapaque, N.; Blottière, H.M. TLR ligands and butyrate increase Pyy expression through two distinct but inter-regulated pathways. *Cell Microbiol.* **2017**, *19*, e12648. [CrossRef]
147. Zhou, L.; Ni, C.; Liao, R.; Tang, X.; Yi, T.; Ran, M.; Huang, M.; Liao, R.; Zhou, X.; Qin, D.; et al. Activating SRC/MAPK signaling via 5-HT1A receptor contributes to the effect of vilazodone on improving thrombocytopenia. *eLife* **2024**, *13*, RP94765. [CrossRef]
148. Wang, J.; Bakker, W.; Zheng, W.; de Haan, L.; Rietjens, I.; Bouwmeester, H. Exposure to the mycotoxin deoxynivalenol reduces the transport of conjugated bile acids by intestinal Caco-2 cells. *Arch. Toxicol.* **2022**, *96*, 1473–1482. [CrossRef]
149. Fang, Y.; Han, S.I.; Mitchell, C.; Gupta, S.; Studer, E.; Grant, S.; Hylemon, P.B.; Dent, P. Bile acids induce mitochondrial ROS, which promote activation of receptor tyrosine kinases and signaling pathways in rat hepatocytes. *Hepatology* **2004**, *40*, 961–971. [CrossRef]
150. Zhang, F.L.; Ma, H.H.; Dong, P.Y.; Yan, Y.C.; Chen, Y.; Yang, G.M.; Shen, W.; Zhang, X.F. Bacillus licheniformis ameliorates Aflatoxin B1-induced testicular damage by improving the gut-metabolism-testis axis. *J. Hazard. Mater.* **2024**, *468*, 133836. [CrossRef]
151. Guo, H.; Liu, T.; Li, J.; Li, E.; Wen, X.; Chen, F.; Li, S.; Li, Y.; Yin, Q.; Zhu, Q. Compound probiotics regulate the NRF2 antioxidant pathway to inhibit aflatoxin B(1)-induced autophagy in mouse Sertoli TM4 cells. *Ecotoxicol. Environ. Saf.* **2024**, *281*, 116619. [CrossRef]
152. Fan, J.; Zhang, Y.; Zuo, M.; Ding, S.; Li, J.; Feng, S.; Xiao, Y.; Tao, S. Novel mechanism by which extracellular vesicles derived from Lactobacillus murinus alleviates deoxynivalenol-induced intestinal barrier disruption. *Environ. Int.* **2024**, *185*, 108525. [CrossRef]
153. Jiang, S.; Du, L.; Zhao, Q.; Su, S.; Huang, S.; Zhang, J. Tropical postbiotics alleviate the disorders in the gut microbiota and kidney damage induced by ochratoxin A exposure. *Food Funct.* **2024**, *15*, 3980–3992. [CrossRef]
154. Saliminejad, K.; Khorram Khorshid, H.R.; Soleymani Fard, S.; Ghaffari, S.H. An overview of microRNAs: Biology, functions, therapeutics, and analysis methods. *J. Cell Physiol.* **2019**, *234*, 5451–5465. [CrossRef]
155. Chen, J.; Yang, S.; Li, P.; Wu, A.; Nepovimova, E.; Long, M.; Wu, W.; Kuca, K. MicroRNA regulates the toxicological mechanism of four mycotoxins in vivo and in vitro. *J. Anim. Sci. Biotechnol.* **2022**, *13*, 37. [CrossRef]
156. Inui, M.; Martello, G.; Piccolo, S. MicroRNA control of signal transduction. *Nat. Rev. Mol. Cell Biol.* **2010**, *11*, 252–263. [CrossRef]
157. Zuo, Y.B.; Zhang, Y.F.; Zhang, R.; Tian, J.W.; Lv, X.B.; Li, R.; Li, S.P.; Cheng, M.D.; Shan, J.; Zhao, Z.; et al. Ferroptosis in cancer progression: Role of noncoding RNAs. *Int. J. Biol. Sci.* **2022**, *18*, 1829–1843. [CrossRef]
158. Rong, X.; Sun-Waterhouse, D.; Wang, D.; Jiang, Y.; Li, F.; Chen, Y.; Zhao, S.; Li, D. The significance of regulatory micrornas: Their roles in toxicodynamics of mycotoxins and in the protection offered by dietary therapeutics against mycotoxin-induced toxicity. *Compr. Rev. Food Sci. Food Saf.* **2019**, *18*, 48–66. [CrossRef]
159. Grenier, B.; Hackl, M.; Skalicky, S.; Thamhesl, M.; Moll, W.D.; Berrios, R.; Schatzmayr, G.; Nagl, V. MicroRNAs in porcine uterus and serum are affected by zearalenone and represent a new target for mycotoxin biomarker discovery. *Sci. Rep.* **2019**, *9*, 9408. [CrossRef]
160. Chuturgoon, A.A.; Phulukdaree, A.; Moodley, D. Fumonisin B_1 modulates expression of human cytochrome P450 1b1 in human hepatoma (Hepg2) cells by repressing Mir-27b. *Toxicol. Lett.* **2014**, *227*, 50–55. [CrossRef]
161. Rieswijk, L.; Brauers, K.J.; Coonen, M.L.; van Breda, S.G.; Jennen, D.G.; Kleinjans, J.C. Evaluating microRNA profiles reveals discriminative responses following genotoxic or non-genotoxic carcinogen exposure in primary mouse hepatocytes. *Mutagenesis* **2015**, *30*, 771–784. [CrossRef]
162. Wang, Y.; Zhang, Z.; Wang, H.; Zhang, Y.; Ji, M.; Xu, H.; Wang, C.; Sun, Z.; Gao, W.; Wang, S.L. miR-138-1* regulates aflatoxin B1-induced malignant transformation of BEAS-2B cells by targeting PDK1. *Arch. Toxicol.* **2016**, *90*, 1239–1249. [CrossRef]

163. Liu, Y.X.; Long, X.D.; Xi, Z.F.; Ma, Y.; Huang, X.Y.; Yao, J.G.; Wang, C.; Xing, T.Y.; Xia, Q. MicroRNA-24 modulates aflatoxin B1-related hepatocellular carcinoma prognosis and tumorigenesis. *BioMed Res. Int.* **2014**, *2014*, 482926. [CrossRef]
164. Marrone, A.K.; Tryndyak, V.; Beland, F.A.; Pogribny, I.P. Microrna responses to the genotoxic carcinogens aflatoxin B1 and benzo[a]pyrene in human HepaRG Cells. *Toxicol. Sci. Off. J. Soc. Toxicol.* **2016**, *149*, 496–502. [CrossRef]

veterinary sciences

MDPI

Article

Comparison of Intravenous and Oral Meloxicam Pharmacokinetics in Female and Male Saanen Goats

Zeynep Ozdemir Kutahya [1], Busra Aslan Akyol [2], Selen Mamuk [3], Petek Piner Benli [1] and Cengiz Gokbulut [2,4,*]

[1] Department of Veterinary Pharmacology and Toxicology, Faculty of Ceyhan Veterinary Medicine, Cukurova University, Adana 01330, Türkiye; zkutahya@cu.edu.tr (Z.O.K.); ppinerbenli@cu.edu.tr (P.P.B.)
[2] Department of Veterinary Pharmacology and Toxicology, Institute of Health Sciences, Balikesir University, Balikesir 10145, Türkiye; busra.aslan@baun.edu.tr
[3] Department of Veterinary Pharmacology and Toxicology, Institute of Health Sciences, Cukurova University, Adana 01330, Türkiye; smamuk@student.cu.edu.tr
[4] Department of Medical Pharmacology, Faculty of Medicine, Balikesir University, Balikesir 10145, Türkiye
* Correspondence: cgokbulut@balikesir.edu.tr; Tel.: +90-535-297-36-62

Simple Summary: Meloxicam is a medication commonly used in animals to reduce pain and inflammation, helping them recover more comfortably after illness or surgery. However, the way this medication is processed in the body can differ between females and males, which may affect how long it stays in the system and how well it works. In this study, we examined how female and male goats handle meloxicam after it was given by mouth and administered directly into the bloodstream. We found that when the medicine was given by mouth, it stayed in male goats' bodies longer and at higher levels than in females. These differences are important because they can affect how often the medication should be given and determine how long we should wait before using goat products for human consumption to ensure safety. Understanding these differences can help veterinarians make better decisions about treating pain in goats and may guide medication practices in other animals raised for food, ensuring both animal welfare and food safety.

Abstract: This study aimed to investigate the effect of gender on the pharmacokinetics of meloxicam in goats following intravenous (IV, 0.5 mg/kg) and oral (PO, 1.0 mg/kg) administration. A crossover design was used with 12 clinically healthy Saanen goats (six females and six males). Plasma samples were collected up to 96 h post-administration and analyzed with an HPLC for meloxicam concentrations. Pharmacokinetic parameters were calculated and statistically compared between genders and administration routes. The results show that male goats exhibited significantly longer terminal half-life ($T_{1/2\lambda z}$), a greater mean residence time ($MRT_{0-\infty}$), and higher systemic exposure ($AUC_{0-\infty}$) than females, particularly after oral administration. Oral bioavailability was calculated as 77.43% in females and 104.73% in males. These differences may be linked to gender-based variations in hepatic metabolism, enterohepatic recirculation, and the hormone-mediated modulation of cytochrome P450 activity. The findings are consistent with previous research demonstrating that gender can influence drug disposition through hormonal and enzymatic mechanisms. This study underscores the importance of considering gender as a biological variable in pharmacokinetic assessments of veterinary drugs, especially those used in food-producing animals, to optimize dosing strategies and ensure both therapeutic efficacy and food safety.

Keywords: meloxicam; goat; gender; pharmacokinetics; intravenous; oral

1. Introduction

Non-steroidal anti-inflammatory drugs (NSAIDs), which are widely used in veterinary medicine, play a significant role, particularly in managing pain, inflammation, and fever. They are often preferred in conditions such as postoperative pain, musculoskeletal diseases (e.g., osteoarthritis and laminitis), visceral pain (e.g., colic), inflammatory diseases (e.g., pneumonia and mastitis), and fever reduction [1]. NSAIDs mainly suppress prostaglandin synthesis by inhibiting cyclooxygenase (COX) enzymes, thereby reducing pain, fever, and inflammation [2]. While COX-1 is involved in normal physiological functions, COX-2 is induced during inflammation. Consequently, COX-2 selective NSAIDs (e.g., meloxicam and firocoxib) have fewer gastrointestinal side effects [1,2]. NSAIDs may cause adverse effects, particularly in the gastrointestinal and renal systems [2]. Although NSAIDs are highly effective and essential drugs widely used in veterinary medicine, they should be administered with caution and awareness. Meloxicam belongs to the enolic acid group of NSAIDs and exhibits analgesic, anti-inflammatory, and antipyretic effects [3]. It preferentially inhibits COX-2 over COX-1, but it is not entirely COX-2 selective; thus, at higher doses, its selectivity for the COX-2 isoenzyme diminishes [4]. Meloxicam is extensively metabolized in the liver and undergoes significant enterohepatic recirculation [5]. Its metabolites have not been found to possess any pharmacological activity [6]. Meloxicam is an approved NSAID for cattle, horses, cats, and dogs, while it is used off-label in sheep and goats. Pharmacological studies increasingly highlight gender-based differences in drug absorption, distribution, metabolism, and excretion [7–12]. Incorporating gender as a crucial variable in dosage adjustment is important. Moyer et al. [13] discussed how gender chromosomes and endogenous steroid hormones modulate drug transporters, metabolic enzymes, and receptors, thereby influencing pharmacokinetics and adverse drug responses. Moreover, Bosch et al. [14] stated that female sex hormones, particularly estrogen, significantly impact drug metabolism mediated by CYP and UGT enzyme activity. These factors are especially relevant for orally administered drugs due to the influence of first-pass hepatic metabolism affected by estrogenic activity. Meloxicam is known to have high oral bioavailability (72–96%) across various species including goats, sheep, cattle, and horses, indicating minimal first-pass hepatic metabolism [15]. Data on the gender-dependent pharmacokinetics of drugs used in ruminants remain limited. It is scientifically and clinically valuable to explore how the pharmacokinetics of meloxicam, a commonly used NSAID in veterinary medicine, is affected by gender in food-producing species like goats. While the existing literature addresses gender differences in humans and laboratory animals, no prior studies have focused on meloxicam in goats. Therefore, our research aimed to assess whether and how the pharmacokinetic parameters of meloxicam differ between female and male Saanen goats following intravenous (0.5 mg/kg) and oral (1.0 mg/kg) administration.

2. Materials and Methods

2.1. Chemicals

Analytical standards of meloxicam (99.2%, CAS: 71125-38-7) and piroxicam (99.7%, CAS: 36322-90-4) were obtained from LGC Standards (Wesel, Germany). HPLC-grade acetonitrile (99.9%), methanol (99.9%), magnesium sulphate heptahydrate (99.9%), and phosphoric acid (89%) were supplied by Sigma-Aldrich (Steinheim, Germany).

2.2. Animals

This study involved a total of 12 Saanen goats, comprising six females and six males, all approximately one year of age. The average body weight was 28.19 ± 3.35 kg for the female goats and 36.87 ± 4.58 kg for the males. Before the experiment, each animal underwent a comprehensive clinical evaluation along with hematological testing to ensure they were in good health. None of the goats had been treated with any medication for two months prior to the experiment. The animals were divided into two groups according to gender. Throughout the study period, they were provided with commercial concentrate feed, and both dry forage and water were available *ad libitum*. The experiments with the males were performed in late autumn as the endogenous testosterone levels are at a maximum at this time. The experiments with the females were undertaken during the same period. The female goats used in this study were approximately one year old and had not yet been bred; thus, their estrous cycles were not active or regular, and monitoring cycle stages was not considered necessary. The study protocol was approved by the Ethics Committee of Cukurova University, Health Sciences Experimental Application and Research Centre.

2.3. Study Design

The study was conducted using a crossover design consisting of two experimental phases separated by a 10-day washout period [16–18] between administration routes in both female and male Saanen goats. Meloxicam was administered in two commercially available formulations: oral tablets (Melox Fort, 15 mg/tablet, Nobel Pharmaceuticals, Istanbul, Türkiye) at a dose of 1.0 mg/kg and an injectable solution (Metacam, 20 mg/mL, Boehringer Ingelheim, Istanbul, Türkiye) at a single intravenous dose of 0.5 mg/kg. The doses used in this study (0.5 mg/kg IV and 1.0 mg/kg PO) were selected based on previous pharmacokinetic studies in ruminants, aligning with commonly reported ranges while reflecting clinically relevant dosing practices [16,17,19]. In this crossover design, each animal received both the IV and PO formulations in two separate phases, with a washout period in between. All treatments were administered in the morning before feeding with a meloxicam-free commercial concentrate. Animals were closely observed for any adverse effects (behavior, emesis, salivation, diarrhea, weakness, and anorexia) following both IV and PO administration of meloxicam. Blood samples (3 mL) were collected in heparinized tubes before the treatment (0) and subsequently at 5, 15, 30, and 45 min and 1, 1.5, 2, 4, 6, 8, 10, 16, 24, 36, 48, 72, and 96 h post-administration. Blood samples were subjected to centrifugation at 4000× *g* for 7 min within 60 min of collection, and the plasma samples were stored at −80 °C until further analysis.

2.4. Analytical Procedure

2.4.1. Instrumentation

Chromatographic analyses were performed by using an Agilent 1260 HPLC system (Agilent, Waldbronn, Germany) consisting of a binary high-pressure gradient system used for the analysis of meloxicam. An Agilent binary pump (G1312B) was used to deliver the mobile phase to the analytical column. Sample injection was performed via an Agilent autosampler (G1367E) coupled with an injection valve (Rheodyne®, Rohnert Park, CA, USA) equipped with a 100 μL variable loop. Detection was achieved using a diode array detector (G4212B) in compliance with the data acquisition ChemStation Software by Agilent (Germany). Degassing of the mobile phase was achieved using an Agilent vacuum degasser unit (G4225A). Operations and functions of the whole HPLC system were controlled using ChemStation® Software (C.01.08, Agilent, Waldbronn, Germany).

All evaporations following sample extraction were performed at 50 °C using a vacuum evaporator (Maxi-Dry Plus, Heto Lab. Equipment, Allerod, Denmark). A vortex mixer (622, Isolab, Wertheim, Germany) and a centrifuge (Hettich Rotina® 380R, Kirchlengern, Germany) were used for the extraction procedure.

2.4.2. Chromatographic Conditions

The separation of meloxicam was achieved through the use of an analytical column (Eclipse XDB-C18, 5 μm, 250 mm × 4.6 mm, Agilent) with a Nucleosil C18 guard column

(Phenomenex, Macclesfield, UK), which was maintained at 50 °C during analysis in a column oven (G1316A). The mobile phase was composed of ultra-pure water (H_2O), methanol, and acetonitrile in a ratio of 40:30:30 (*v*/*v*/*v*) and adjusted to a pH of 3.5 using phosphoric acid. It was delivered isocratically at a flow rate of 1 mL/min. A photodiode array detector (G4212B) was used at a wavelength of 355 nm. The injection volume was consistently 50 μL, and each chromatographic analysis took 9 min after the injection.

2.4.3. Preparation of Standard Solution

Stock analytical standard solutions (100 μg/mL) of meloxicam and the internal standard piroxicam (5 μg/mL) were prepared in a mixture of acetonitrile/ultra-pure water (50:50) and stored in glass bottles at 4 °C. The stock solution of meloxicam was diluted with a mixture of acetonitrile/ultra-pure water (50:50) to give solutions for plasma samples (0.1, 0.5, 1, 5, 10, and 50 μg/mL). These standards were used to spike drug-free samples at various concentrations to generate standard curves and determine extraction recovery rates.

2.4.4. Sample Preparations and Extraction

The plasma levels of meloxicam were analyzed using HPLC with a diode array detector following liquid–liquid phase extraction procedures according to Karademir et al. [16] and Bae et al. [20] with small modifications described below. Accordingly, blank plasma samples (0.25 mL) were spiked with 25 μL of meloxicam standard solution to reach the following final concentrations: 0, 0.01, 0.05, 0.1, 0.5, 1, 5, and 10 μg/mL for the IV route and 0, 0.01, 0.05, 0.1, 0.5, and 1 μg/mL for the PO route. The Spike concentrations used in the analysis of both intravenous and oral administration samples were determined by considering the peak concentrations obtained in previous studies in the literature. The plasma samples were combined with 50 μL of internal standard (piroxicam 5 μg/mL). Subsequently, magnesium sulphate heptahydrate (0.1 g) was added, and the mixture was vortexed for 1 min. Then, 1 mL of acetonitrile was added to deproteinize the plasma samples, and the mixture was vortexed for an additional 1 min. The mixture was then centrifuged at 14,500× *g* for 5 min. The upper phase was transferred to a 10 mL tube and evaporated using a vacuum concentrator set at 50 °C. Once the residue was completely dried, it was dissolved in 200 μL of the mobile phase and vortexed for 15 s. Finally, 50 μL of this solution was injected into the chromatographic system for analysis.

2.4.5. Validation of Analytical Method

Analytical validation was conducted to determine meloxicam levels in plasma samples in accordance with the International Conference on Harmonization (ICH) guidelines for validating analytical procedures [21]. The analyte was identified based on the retention times of a pure reference standard. Recoveries of the substance under study were assessed by comparing the peak areas from spiked plasma samples with those obtained from direct injections of standards prepared in a mixture of acetonitrile and ultra-pure water. The recoveries were calculated with five multiple concentrations and six replications. To evaluate the inter- and intra-assay precision of the extraction and chromatography procedures, replicate aliquots of drug-free goat plasma samples containing known amounts of meloxicam were processed on different days. A calibration graph for meloxicam was created, demonstrating a linear range of 0.01 to 10 μg/mL for IV administration and 0.01 to 1 μg/mL for PO administration. The slope of the line relating peak areas to drug concentration was determined using least squares linear regression, and both the correlation coefficient (*r*) and the coefficient of variation (CV) were calculated. Linearity was established to confirm the relationship between the meloxicam concentration and detector response. The detection limit (LOD) for meloxicam was determined through an HPLC analysis of blank plasma samples fortified with the standard. This involved measuring the baseline noise at the retention time of the peak. The LOD was defined as the mean baseline noise at the peak retention time plus three standard deviations (SDs). Additionally, the limit of quantification (LOQ) was defined as the mean baseline noise plus six SDs.

2.5. Pharmacokinetics and Statistical Analysis of Data

The plasma concentrations vs. time curves obtained following each administration route in individual animals were fitted with the WinNonlin software program (Version 5.2, Pharsight Corporation, Mountain View, CA, USA). The pharmacokinetic parameters for each animal were analyzed using non-compartmental model analysis. The maximum plasma concentration (C_{max}) and time to reach maximum concentration (T_{max}) were obtained from the plotted concentration–time curve of the drug in each animal. The trapezoidal rule was used to calculate the area under the plasma concentration–time curve (AUC). The fraction of dose absorbed (i.e., *F*) was calculated with the use of mean AUCs calculated for each route of administration using the following equation after dose normalization:

$$F = (AUC_{PO}/AUC_{IV}) \times (0.5/1) \times 100$$

Pharmacokinetic parameters are presented as mean ± SD. Harmonic means were calculated for $T_{1/2\lambda z}$, MRT_{last}, and $MRT_{0-\infty}$. For the comparison of administration routes, non-normally distributed data were analyzed using the non-parametric Wilcoxon signed-rank test, while normally distributed data were assessed using the paired *t*-test. When comparing gender, non-normally distributed data were compared with the non-parametric Mann–Whitney U test, and normally distributed data were analyzed using the independent *t*-test. All statistical analyses were performed using SPSS Statistics version 23.0 (IBM Corp).

3. Results

The validation parameters of meloxicam for the analysis of plasma samples of goats are given in Table 1.

Table 1. Validation parameters of the analytical method used to determine meloxicam in plasma samples.

Parameters	Meloxicam
LOD (μg/mL)	0.0031
LOQ (μg/mL)	0.01
Range of linearity (μg/mL)	0.01–10.00
Linearity (r^2)	0.9987–1.0000
Recovery (%)	85.85 (2.91)
Coefficient of variation (%)	4.51 (0.73)

LOD: limit of detection, LOQ: limit of quantification, *r*: correlation coefficient. The values in the brackets represent the standard deviations for the recovery assays ($n = 6$).

Animals were closely observed for adverse effects, including behavior, emesis, salivation, diarrhea, weakness, and anorexia, throughout the 20-day study period following both IV and PO administration of meloxicam. Gender-related differences were observed in several pharmacokinetic parameters following IV and PO administration of meloxicam at doses of 0.5 mg/kg and 1.0 mg/kg, respectively, in Saanen goats (Table 2). Although the initial plasma concentration (C_0), area under the curve ($AUC_{0-\infty}$), clearance (Cl), mean residence time (MRT), and area under the first moment curve (AUMC) values were similar between female and male goats, the terminal elimination half-life ($T_{1/2\lambda z}$) was notably longer in males (10.09 ± 0.97 h vs. 11.72 ± 0.74 h) after IV administration. Following

PO administration of meloxicam at a dose of 1.0 mg/kg, notable pharmacokinetic differences were observed between female and male Saanen goats. Although the maximum plasma concentration (C_{max}) and time to reach C_{max} (T_{max}) were similar across genders, significant differences emerged in elimination and exposure parameters. The $T_{1/2\lambda z}$ was significantly longer in male goats (13.10 ± 2.01 h) compared to females (9.87 ± 0.85 h). In parallel, $MRT_{0-\infty}$ was also significantly prolonged in males (22.18 ± 3.47 h) relative to females (17.12 ± 1.73 h). Significant differences were observed between the pharmacokinetic parameters obtained after intravenous (IV, 0.5 mg/kg) and oral (PO, 1.0 mg/kg) administration of meloxicam to female and male goats. No difference in mean $T_{1/2\lambda z}$ values was observed between IV and PO meloxicam administration in both female and male goats. The AUC, AUMC, and MRT parameters were determined to be significantly higher after PO administration. It was also determined that meloxicam has a high bioavailability (*F*) when administered orally in female (77.43%) and especially male goats (104.73%).

Table 2. Pharmacokinetic parameters of meloxicam after intravenous (IV, 0.5 mg/kg) and oral (PO, 1.0 mg/kg) administration to female and male goats (n = 6).

Parameters	IV Administration		PO Administration	
	Female	Male	Female	Male
$T_{1/2\lambda z}$ (h) [a]	10.09 ± 0.97	11.72 ± 0.74 *	9.87 ± 0.85	13.10 ± 2.01 *
C_0 (μg/mL)	3.76 ± 0.82	3.47 ± 0.39	-	-
T_{max} (h)	-	-	7.33 ± 1.03	8.33 ± 1.51
C_{max} (μg/mL)	-	-	1.87 ± 0.38	1.90 ± 0.50
AUC_{last} (μg.h/mL)	25.35 ± 6.65	26.07 ± 4.14	39.24 ± 7.36	53.60 ± 21.12
$AUC_{0-\infty}$ (μg.h/mL)	25.57 ± 6.76	26.43 ± 4.30	39.59 ± 7.45	55.36 ± 22.38
Vz (mL/kg)	302.08 ± 66.15	325.47 ± 36.97	-	-
Cl (mL/h/kg)	20.69 ± 5.20	19.30 ± 2.83	-	-
$AUMC_{last}$ (μg.h^2/mL)	319.66 ± 113.72	351.23 ± 82.09	654.61 ± 141.85	1147.71 ± 556.79
$AUMC_{0-\infty}$ (μg.h^2/mL)	338.73 ± 124.71	383.57 ± 97.28	684.96 ± 153.68	1311.62 ± 685.47
MRT_{last} (h) [a]	12.22 ± 1.50	13.30 ± 0.97	16.55 ± 1.52	20.35 ± 2.55 *
$MRT_{0-\infty}$ (h) [a]	12.77 ± 1.74	14.27 ± 1.26	17.12 ± 1.73	22.18 ± 3.47 *
V_{ss} (L/kg)	263.27 ± 52.26	274.50 ± 24.70	-	-
F (%)	-	-	77.43	104.73

$T_{1/2\lambda z}$: terminal half-life; C_0: calculated concentration at time zero of IV administration; T_{max}: time to reach peak plasma concentration; C_{max}: peak plasma concentration; AUC: area under (zero moment) curve from time 0 to last detectable concentration; Vz: volume of distribution; Cl: clearance of drug; AUMC: area under moment curve from time 0 to t last detectable concentration; MRT: mean residence time; V_{ss}: apparent volume of distribution at steady state; *F*: bioavailability. [a] Harmonic mean. * Kinetic parameters in male goats are significantly different ($p < 0.05$) from those in female goats.

Meloxicam was detected in plasma for approximately 70 h in female and male goats after IV (Figure 1) and PO administration (Figure 2).

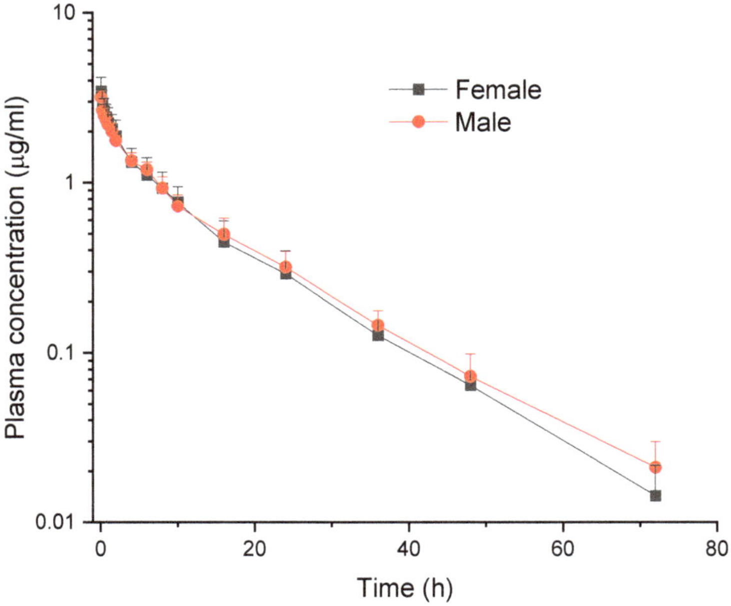

Figure 1. Comparative mean (±SD) plasma concentration vs. time curves of meloxicam in female and male goats following intravenous administration at a dose of 0.5 mg/kg (n = 6).

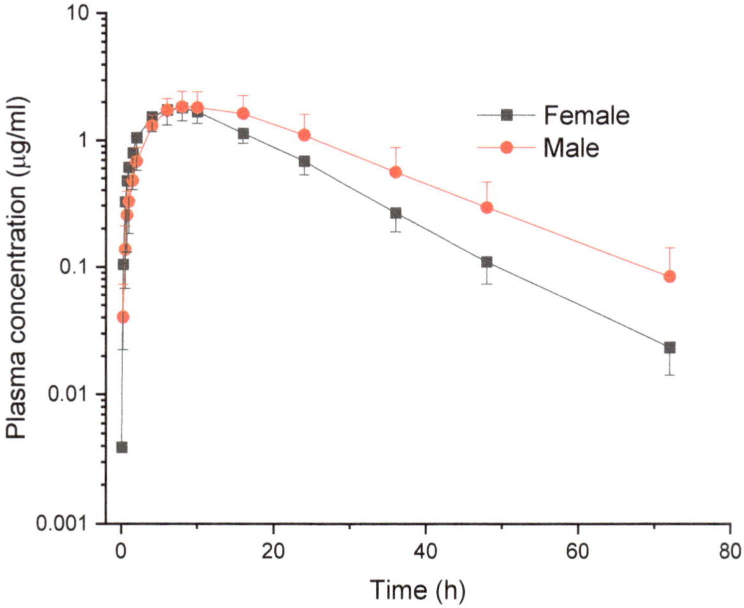

Figure 2. Comparative mean (±SD) plasma concentration vs. time curves of meloxicam in female and male goats following per os administration at a dose of 1.0 mg/kg (n = 6).

4. Discussion

It is imperative to comprehend how gender influences drug disposition to optimize therapeutic strategies, ensure animal welfare, and refine dosage recommendations, particularly in food-producing species such as goats. In both genders, pharmacokinetic variability can have both clinical and regulatory implications. The present study was conducted to evaluate the pharmacokinetics of meloxicam, a commonly used NSAID in veterinary medicine, in male and female Saanen goats following both IV and PO administration.

Following IV administration of meloxicam at a dose of 0.5 mg/kg, a statistically significant difference was observed in the $T_{1/2\lambda z}$ values between female and male goats. The $T_{1/2\lambda z}$ was notably longer in males (11.72 ± 0.74 h) than in females (10.09 ± 0.97 h), suggesting that meloxicam is cleared more slowly in male goats. This prolonged elimination could reflect gender-based differences in hepatic metabolism, protein binding, or renal excretion rates. Although the estrous cycle was not active in the female goats used in this study due to their age and breeding status, gender-based differences in pharmacokinetics can still occur. These differences may result from baseline hormonal variations, including higher testosterone levels in males during late autumn, and from physiological factors such as plasma protein binding, body composition, and renal clearance, all of which can influence drug disposition independently of the estrous cycle. Despite the absence of statistically significant differences in other pharmacokinetic parameters, such as C_0, $AUC_{0-\infty}$, Cl, and volume of distribution (V_{ss}), between the genders, the consistently higher $MRT_{0-\infty}$ and AUMC values observed in males provide further evidence that supports the hypothesis of slower drug disposition in males. These findings indicate that, despite similar initial plasma concentrations and exposure levels, meloxicam remains in the system for longer in male goats. It is imperative to consider the implications of this prolonged systemic persistence on both therapeutic efficacy and withdrawal time considerations. A recent study investigated the pharmacokinetic differences in meloxicam between male and female sheep following IV administration. The results show that the total clearance (Cl_T) and Vdss were significantly higher in male sheep, while the $T_{1/2\lambda z}$ was significantly shorter compared to female sheep [11]. In our study, meloxicam was administered at a dose of 0.5 mg/kg IV to Saanen goats, whereas Corum et al. [11] employed a 1.0 mg/kg IV dose in Romanov sheep. Despite this difference in dosing, both studies reveal that gender significantly influences meloxicam disposition, particularly in terms of $T_{1/2\lambda z}$. In male goats, the $T_{1/2\lambda z}$ (11.72 ± 0.74 h) was longer than that observed in male sheep (9.47 ± 0.25 h), while female goats had a slightly shorter $T_{1/2\lambda z}$ (10.09 ± 0.97 h) than female sheep (11.96 ± 0.33 h). The fact that the $T_{1/2\lambda z}$ is longer in male goats than in male sheep is a surprising result because studies have shown that goats metabolize and eliminate compounds faster than sheep [22–25]. Although male sheep exhibited faster Cl (8.72 ± 1.34 mL/h/kg) and larger Vd (100.95 ± 14.73 mL/kg) than females, such pronounced differences were not observed in the goats. In our study, both Cl and Vd values were similar across genders, suggesting that species-related physiological or metabolic factors may play a more prominent role in drug disposition than gender alone in goats. Another notable distinction concerns the extent of systemic exposure. In sheep, female animals showed significantly higher $AUC_{0-\infty}$ values (224.68 ± 28.22 μg·h/mL) compared to males (117.25 ± 20.38 μg·h/mL), indicating a marked gender-based difference in meloxicam exposure. In contrast, in goats, AUC values were comparable between genders (females: 25.57 ± 6.76 vs. males: 26.43 ± 4.30 μg·h/mL), further underscoring the interspecies variability in meloxicam pharmacokinetics. These differences could be attributed not only to species-specific hepatic enzyme activity (such as CYP2C9-mediated metabolism) but also to variations in plasma protein binding, tissue distribution, and renal excretion patterns between goats and sheep. Nevertheless, direct comparisons should be interpreted with caution due to differences in administered doses (0.5 mg/kg in goats vs. 1.0 mg/kg in sheep), species-related physiological variations (Saanen goats vs. Romanov sheep), and potential differences in study design, all of which may influence pharmacokinetic parameters. Both studies highlight the importance of gender in determining the pharmacokinetics of meloxicam; however, the results demonstrate that these effects are species dependent.

In the present study, gender-related pharmacokinetic differences were also observed in Saanen goats following a single PO administration of meloxicam at a dose of 1.0 mg/kg. Male goats exhibited a significantly longer $T_{1/2\lambda z}$ (13.10 ± 2.01 h) compared to females (9.87 ± 0.85 h), indicating slower elimination. This was further supported by the longer $MRT_{0-\infty}$ observed in males (22.18 ± 3.47 h vs. 17.12 ± 1.73 h in females), suggesting an extended systemic persistence of meloxicam in male animals. In addition, although C_{max} was similar between genders, the total drug exposure represented by $AUC_{0-\infty}$ was higher in males (55.36 ± 22.38 μg·h/mL) than in females (39.59 ± 7.45 μg·h/mL), although not statistically significant, but with a high degree of inter-individual variability. These findings suggest that meloxicam is more systemically available in males than females following PO administration in goats. Meloxicam is subject to complete metabolic conversion to four pharmacologically inactive metabolites, primarily through the cytochrome P450 2C pathway [26]. Thus, the observed differences in elimination and exposure in male and female animals may be attributable to gender-based variations in hepatic enzyme activity, enterohepatic recirculation, gastrointestinal transit times, or hormonal influences affecting drug metabolism and excretion.

In this study, the IV and PO routes were selected to evaluate the pharmacokinetics and calculate the absolute *F* of meloxicam in goats, as the IV route serves as the reference for *F* calculations. Although the IM route is commonly used in veterinary practice, it was not included in this study to maintain methodological clarity for *F* assessment and to focus on the establishment of pharmacokinetic parameters relevant to dosage regimen design in food-producing animals. A thorough evaluation of meloxicam pharmacokinetics after IV and PO administration in both female and male Saanen goats shows significant differences in drug disposition based on the route of administration and gender. While the $T_{1/2\lambda z}$ values remained generally consistent across genders and administration routes—suggesting similar elimination kinetics once meloxicam enters the systemic circulation—statistically significant differences emerged in drug exposure and MRT, especially following PO administration. In both genders, PO administration of meloxicam at 1.0 mg/kg resulted in significantly higher systemic exposure compared to 0.5 mg/kg IV administration. This was evident in the $AUC_{0-\infty}$ values, which increased from 25.57 ± 6.76 to 39.59 ± 7.45 μg·h/mL in females and from 26.43 ± 4.30 to 55.36 ± 22.38 μg·h/mL in males. Correspondingly, $AUMC_{0-\infty}$ and $MRT_{0-\infty}$ values were significantly elevated after PO dosing in both genders, indicating a prolonged systemic presence of meloxicam via the oral route. Notably, the increase in $MRT_{0-\infty}$ from IV to PO was more pronounced in males (14.27 h vs. 22.18 h) than in females (12.77 h vs. 17.12 h), suggesting that male goats may exhibit a slower drug turnover and more sustained drug retention after PO administration. Interestingly, the calculated absolute *F* of oral meloxicam was 77.43% in females and 104.73% in males. While the high *F* value in males may partly reflect enterohepatic recirculation, it may also be influenced by the sex hormone-mediated modulation of hepatic enzyme activity, differences in gastrointestinal physiology affecting absorption efficiency, and potential variations in plasma protein binding between genders in addition to inter-individual variability. Although dose normalization was applied in *F* calculation, potential nonlinear pharmacokinetics at the administered dose cannot be completely excluded as a contributing factor. The consistent T_{max} values (7.33 ± 1.03 h in females vs. 8.33 ± 1.51 h in males) suggest a delayed yet efficient absorption profile for oral meloxicam across both genders, with slightly slower absorption kinetics in males. From a clinical standpoint, these findings underscore that although both genders achieve therapeutic plasma levels of meloxicam after PO administration, male goats may maintain these levels for a longer duration due to prolonged MRT and higher AUC. While $T_{1/2\lambda z}$ remained unaffected by the administration route, the overall pharmacokinetic behavior was modulated by both the administration route and gender, reinforcing the need to account for these variables in designing dosage regimens, determining dosing intervals, and establishing withdrawal periods for food-producing animals. The therapeutic plasma concentration of meloxicam in goats has not been definitively established in this study, although data from related ruminant species suggest that maintaining plasma levels above a certain threshold is essential for effective anti-inflammatory and analgesic effects [27–29]. The prolonged MRT and higher systemic exposure observed in male goats suggest that dosing intervals might be extended compared to females to maintain therapeutic levels while minimizing drug accumulation and potential toxicity.

The gender-related pharmacokinetic differences observed in meloxicam disposition in goats may be partially attributed to hormonal modulation of drug metabolism. Previous studies in goats have shown that females can eliminate compounds like antipyrine more rapidly, while males may exhibit slower elimination, suggesting a gender influence on hepatic enzyme activity [30]. Additionally, testosterone administration in dwarf goats has been shown to suppress the formation of specific oxidative antipyrine metabolites, indicating hormonal modulation of hepatic metabolism [31]. Similarly, studies with ivermectin reported longer $T_{1/2}$ and MRT in male goats despite similar C_{max} and AUC values, indicating slower drug elimination in males [12]. Hormonal influences, particularly testosterone in males, may suppress hepatic oxidative metabolism and contribute to prolonged drug exposure, while estrogen in females may enhance CYP3A activity and drug clearance [31,32]. Differences in gastrointestinal physiology such as motility and pH, as well as body composition and renal function differences between genders, may further explain the observed differences in *F* and MRT [13,14,33,34]. These physiological and hormonal factors can impact the pharmacokinetics of orally administered drugs, aligning with our findings of higher bioavailability, prolonged MRT, and elevated AUC in male goats, especially after PO meloxicam administration. Our results, together with previous studies highlighting the clinical impact of gender-related pharmacokinetic differences, support the importance of considering gender when designing dosing regimens and withdrawal times for meloxicam in goats to ensure optimal efficacy and food safety [35–37].

While this study provides valuable insights into the gender-related pharmacokinetics of meloxicam in Saanen goats, certain limitations should be noted. The investigation was limited to single-dose administration under controlled conditions in a single breed, which may affect extrapolation to other breeds or field settings. Meloxicam metabolites and hepatic enzyme activities were not assessed, which could clarify the mechanisms underlying the observed gender differences. Additionally, potential influences of body weight-related physiological differences were not separately analyzed. Future studies incorporating repeated dosing, pharmacodynamic evaluation, and residue depletion data will be essential to translate these pharmacokinetic findings into clinical and regulatory practice.

Taken together, these findings underscore the need to incorporate gender as a biological variable in pharmacokinetic studies, particularly for veterinary drugs administered orally in food-producing animals. In particular, the longer MRT and higher systemic exposure observed in male goats indicate that male animals may require extended withdrawal periods or adjusted dosing intervals to avoid violative residues while maintaining therapeutic efficacy. The integration of the findings in the literature with our data suggests that sex hormones can affect the activity of hepatic enzymes involved in meloxicam metabolism, which may lead to clinically relevant differences in drug exposure and clearance. These findings could have significant implications for adjusting dosage regimens and withdrawal periods to ensure both therapeutic efficacy and food safety.

Author Contributions: Conceptualization, Z.O.K., C.G. and P.P.B.; Methodology, C.G. and B.A.A.; Data Curation, Z.O.K., S.M. and P.P.B.; Writing—Original Draft Preparation, Z.O.K., C.G. and P.P.B.; Writing—Review and Editing, Z.O.K., C.G. and P.P.B.; Project Administration, Z.O.K. All authors have read and agreed to the published version of the manuscript.

Funding: This study was supported by the Coordination Unit of Scientific Research Projects, the University of Cukurova (Project No TKB-2024-16859).

Institutional Review Board Statement: The animal study protocol was approved by the Ethics Committee of Cukurova University (protocol code: 5-1, approval date: 28 May 2024) for studies involving animals.

Informed Consent Statement: Informed consent was obtained from all subjects involved in the study.

Data Availability Statement: The data presented in this study are available within the article's figures and tables.

Conflicts of Interest: The authors declare that they have no conflicts of interest.

References

1. Mathews, K.A. Non-steroidal anti-inflammatory analgesics: A review of current practice. *J. Vet. Emerg. Crit. Care* **2002**, *12*, 89–97. [CrossRef]
2. Curry, S.L.; Cogar, S.M.; Cook, J.L. Nonsteroidal antiinflammatory drugs: A review. *J. Am. Anim. Hosp. Assoc.* **2005**, *41*, 298–309. [CrossRef] [PubMed]
3. Luger, P.; Daneck, K.; Engel, W.; Trummlitz, G.; Wagner, K. Structure and physicochemical properties of meloxicam, a new NSAID. *Eur. J. Pharm. Sci.* **1996**, *4*, 175–187. [CrossRef]
4. Del Tacca, M.; Colucci, R.; Fornai, M.; Blandizzi, C. Efficacy and tolerability of meloxicam, a COX-2 preferential nonsteroidal anti-inflammatory drug: A review. *Clin. Drug Investig.* **2002**, *22*, 799–818. [CrossRef]
5. Gates, B.J.; Nguyen, T.T.; Setter, S.M.; Davies, N.M. Meloxicam: A reappraisal of pharmacokinetics, efficacy and safety. *Expert. Opin. Pharmacother.* **2005**, *6*, 2117–2140. [CrossRef]
6. Furst, D.E. Meloxicam: Selective COX-2 inhibition in clinical practice. *Semin. Arthritis Rheum.* **1997**, *26*, 21–27. [CrossRef]
7. Gorski, J.C.; Vannaprasaht, S.; Hamman, M.A.; Ambrosius, W.T.; Bruce, M.A.; Haehner-Daniels, B.; Hall, S.D. The effect of age, sex, and rifampin administration on intestinal and hepatic cytochrome P450 3A activity. *Clin. Pharmacol. Ther.* **2003**, *74*, 275–287. [CrossRef]
8. Overholser, B.R.; Kays, M.B.; Forrest, A.; Sowinski, K.M. Sex-related differences in the pharmacokinetics of oral ciprofloxacin. *J. Clin. Pharmacol.* **2004**, *44*, 1012–1022. [CrossRef]
9. Carcas, A.; Guerra, P.; Frias, J.; Soto, A.; Fernandez-Aijón, A.; Montuenga, C.; Govantes, C. Gender differences in the disposition of metronidazole. *Int. J. Clin. Pharmacol. Ther.* **2001**, *39*, 213–218. [CrossRef]
10. Konieczna, L.; Chmielewska, A.; Lamparczyk, H. Influence of sex on the pharmacokinetics of ciprofloxacin and ofloxacin. *Chemotherapy* **2006**, *52*, 111–121. [CrossRef]
11. Corum, O.; Uney, K.; Durna Corum, D.; Coskun, D.; Akin, F.; Oguz, H.; Elmas, M. Effect of Gender on the pharmacokinetics of meloxicam in sheep. *J. Vet. Pharmacol. Ther.* **2025**. [CrossRef]
12. Gokbulut, C.; Bilgili, A.; Hanedan, B.; Aksit, D.; Aksoy, A.M.; Turgut, C. Sex-related plasma disposition of ivermectin following pour-on administration in goats. *Vet. Parasitol.* **2009**, *162*, 342–345. [CrossRef]
13. Moyer, A.M.; Matey, E.T.; Miller, V.M. Individualized medicine: Sex, hormones, genetics, and adverse drug reactions. *Pharmacol. Res. Perspect.* **2019**, *7*, e00541. [CrossRef] [PubMed]
14. Bosch, E.; Sommer, I.; Touw, D. The influence of female sex and estrogens on drug pharmacokinetics: What is the evidence? *Expert. Opin. Drug Metab. Toxicol.* **2025**, *21*, 637–647. [CrossRef] [PubMed]
15. de la Puente, R.; Diez, R.; Diez, M.J.; Fernandez, N.; Sahagun, A.M.; Rodriguez, J.M.; Garcia, J.J.; Lopez, C. Pharmacokinetics of meloxicam in different animal species: A comprehensive review. *Vet. Sci.* **2024**, *11*, 519. [CrossRef] [PubMed]
16. Karademir, U.; Erdogan, H.; Boyacioglu, M.; Kum, C.; Sekkin, S.; Bilgen, M. Pharmacokinetics of meloxicam in adult goats: A comparative study of subcutaneous, oral and intravenous administration. *N. Z. Vet. J.* **2016**, *64*, 165–168. [CrossRef]
17. Stock, M.L.; Coetzee, J.F.; KuKanich, B.; Smith, B.I. Pharmacokinetics of intravenously and orally administered meloxicam in sheep. *Am. J. Vet. Res.* **2013**, *74*, 779–783. [CrossRef]
18. Woodland, A.N.; Van der Saag, D.; Kimble, B.; White, P.J.; Govendir, M.; Lomax, S. Plasma pharmacokinetic profile and efficacy of meloxicam administered subcutaneously and intramuscularly to sheep. *PLoS ONE* **2019**, *14*, e0215842. [CrossRef]

19. Kreuder, A.J.; Coetzee, J.F.; Wulf, L.W.; Schleining, J.A.; KuKanich, B.; Layman, L.L.; Plummer, P.J. Bioavailability and pharmacokinetics of oral meloxicam in llamas. *BMC Vet. Res.* **2012**, *8*, 85. [CrossRef]
20. Bae, J.-W.; Kim, M.-J.; Jang, C.-G.; Lee, S.-Y. Determination of meloxicam in human plasma using a HPLC method with UV detection and its application to a pharmacokinetic study. *J. Chromatogr. B Analyt. Technol. Biomed. Life Sci.* **2007**, *859*, 69–73. [CrossRef]
21. International Conference on Harmonization (ICH). Technical Requirements for the Registration of Pharmaceutical for Human Use, Validation of Analytical Procedures; Text and Methodology Q2 (R1). 2005. Available online: https://www.ema.europa.eu/en/documents/scientific-guideline/ich-guideline-q2r1-validation-analytical-procedures-text-methodology-step-5-first-version_en.pdf (accessed on 23 June 2025).
22. Bogan, J.; Benoit, E.; Delatour, P. Pharmacokinetics of oxfendazole in goats: A comparison with sheep. *J. Vet. Pharmacol. Ther.* **1987**, *10*, 305–309. [CrossRef] [PubMed]
23. Aksit, D.; Yalinkilinc, H.S.; Sekkin, S.; Boyacioğlu, M.; Cirak, V.Y.; Ayaz, E.; Gokbulut, C. Comparative pharmacokinetics and bioavailability of albendazole sulfoxide in sheep and goats, and dose-dependent plasma disposition in goats. *BMC Vet. Res.* **2015**, *11*, 124. [CrossRef] [PubMed]
24. Gokbulut, C.; Yalinkilinc, H.S.; Aksit, D.; Veneziano, V. Comparative pharmacokinetics of levamisole-oxyclozanide combination in sheep and goats following per os administration. *Can. J. Vet. Res.* **2014**, *78*, 316–320. [PubMed]
25. Fadel, C.; Łebkowska-Wieruszewska, B.; Serih, F.; Lisowski, A.; Poapolathep, A.; Giorgi, M. Comparative pharmacokinetic evaluation of metronidazole in sheep and goats. *Res. Vet. Sci.* **2025**, *184*, 105507. [CrossRef]
26. Fleischmann, R.; Iqbal, I.; Slobodin, G. Meloxicam. *Expert. Opin. Pharmacother.* **2002**, *3*, 1501–1512. [CrossRef]
27. Toutain, P.; Autefage, A.; Legrand, C.; Alvinerie, M. Plasma concentrations and therapeutic efficacy of phenylbutazone and flunixin meglumine in the horse: Pharmacokinetic/pharmacodynamic modelling. *J. Vet. Pharmacol. Ther.* **1994**, *17*, 459–469. [CrossRef]
28. Lees, P.; Landoni, M.F.; Giraudel, J.; Toutain, P.-L. Pharmacodynamics and pharmacokinetics of nonsteroidal anti-inflammatory drugs in species of veterinary interest. *J. Vet. Pharmacol. Ther.* **2004**, *27*, 479–490. [CrossRef]
29. Landoni, M.; Cunningham, F.; Lees, P. Pharmacokinetics and pharmacodynamics of ketoprofen in calves applying PK/PD modelling. *J. Vet. Pharmacol. Ther.* **1995**, *18*, 315–324. [CrossRef]
30. Witkamp, R.; Lohuis, J.; Nijmeijer, S.; Kolker, H.; Noordhoek, J.; Van Miert, A. Species-and sex-related differences in the plasma clearance and metabolite formation of antipyrine. A comparative study in four animal species: Cattle, goat, rat and rabbit. *Xenobiotica* **1991**, *21*, 1483–1492. [CrossRef]
31. Witkamp, R.; Nijmeijer, S.; Kolker, H.; Noordhoek, J.; van Miert, A.S. Effect of gonadal hormones on the plasma clearance and metabolite formation of antipyrine in the dwarf goat. *J. Vet. Pharmacol. Ther.* **1993**, *16*, 164–173. [CrossRef]
32. Gleiter, C.; Gundert-Remy, U. Gender differences in pharmacokinetics. *Eur. J. Drug Metab. Pharmacokinet.* **1996**, *21*, 123–128. [CrossRef]
33. Scandlyn, M.J.; Stuart, E.C.; Rosengren, R.J. Sex-specific differences in CYP450 isoforms in humans. *Expert. Opin. Drug Metab. Toxicol.* **2008**, *4*, 413–424. [CrossRef]
34. Schwartz, J.B. The influence of sex on pharmacokinetics. *Clin. Pharmacokinet.* **2003**, *42*, 107–121. [CrossRef]
35. Meibohm, B.; Beierle, I.; Derendorf, H. How important are gender differences in pharmacokinetics? *Clin. Pharmacokinet.* **2002**, *41*, 329–342. [CrossRef]
36. Soldin, O.P.; Mattison, D.R. Sex differences in pharmacokinetics and pharmacodynamics. *Clin. Pharmacokinet.* **2009**, *48*, 143–157. [CrossRef]
37. Spoletini, I.; Vitale, C.; Malorni, W.; Rosano, G.M. Sex differences in drug effects: Interaction with sex hormones in adult life. *Handb. Exp. Pharmacol.* **2012**, *214*, 91–105.

antioxidants

MDPI

Review

Puerarin as a Phytochemical Modulator of Gastrointestinal Homeostasis in Livestock: Molecular Mechanisms and Translational Applications

Jiehong Zhou [1,2,†], Jianyu Lv [2,†], Xin Chen [1,2], Tian Li [1,2], Jianzhong Shen [1,2], Zhanhui Wang [1,2], Chongshan Dai [1,2,*] and Zhihui Hao [1,2,*]

1 Technology Innovation Center for Food Safety Surveillance and Detection (Hainan), Sanya Institute of China Agricultural University, Sanya 572025, China
2 National Key Laboratory of Veterinary Public Health and Safety, College of Veterinary Medicine, China Agricultural University, Beijing 100193, China
* Correspondence: daichongshan@cau.edu.cn (C.D.); haozhihui@cau.edu.cn (Z.H.)
† These authors contributed equally to this work.

Abstract: The gut serves as the main site for nutrient digestion and absorption. Simultaneously, it functions as the body's largest immune organ, playing a dual role in sustaining physiological equilibrium and offering immunological defense against intestinal ailments. Maintaining the structural and functional integrity of the intestine is paramount for ensuring animal health and productivity. Puerarin, a naturally derived isoflavonoid from the *Pueraria* species, exhibits multifaceted bioactivities, such as antioxidant, anti-inflammatory, antimicrobial, and immunomodulatory properties. Emerging evidence highlights puerarin's capacity to enhance gut health in farm animals through four pivotal mechanisms: (1) optimization of intestinal morphology via crypt-villus architecture remodeling, (2) augmentation of systemic and mucosal antioxidant defenses through Nrf2/ARE pathway activation, and (3) reinforcement of intestinal barrier function by regulating tight junction proteins (e.g., ZO-1, occludin), mucin secretion, intestinal mucosal immune barrier, the composition of microbiota, and the derived beneficial metabolites; (4) regulating the function of the intestinal nervous system via reshaping the distribution of intestinal neurons and neurotransmitter secretion function. This review synthesizes current knowledge on puerarin's protective effects on intestinal physiology in farm animals, systematically elucidates its underlying molecular targets (including TLR4/NF-κB, MAPK, and PI3K/Akt signaling pathways), and critically evaluates its translational potential in mitigating enteric disorders such as post-weaning diarrhea and inflammatory bowel disease in agricultural practices.

Keywords: puerarin; intestinal health; intestinal barrier function; livestock and poultry production

1. Introduction

The intestine fulfills a dual physiological mandate: it functions as the principal organ for nutrient assimilation while simultaneously serving as the primary immunological interface against luminal and systemic pathogen incursion [1]. Consequently, intestinal homeostasis is intricately linked to systemic health and organismal vitality. The intestinal mucosal barrier system is composed of four coordinated components, i.e., the mechanical barrier (such as epithelial tight junctions), the biochemical barrier (such as mucins and enzymes), the immunological barrier (such as gut-associated lymphoid tissues-associated defenses), and the microbial barrier (such as commensal microbiota). They act collectively to prevent pathogenic infiltration and inhibit the translocation of luminal-derived endotoxins, microbial fragments, and immunogenic antigens [2,3]. Preservation of this barrier's multidimensional integrity, encompassing both architectural continuity (via epithelial regeneration) and functional competence, constitutes a critical determinant in sustaining intestinal physiological equilibrium and preventing enteropathy-associated systemic dysregulation [4,5].

In intensive livestock production systems, animals are chronically exposed to multifactorial stressors, including pathogen infections, thermal stress, and premature weaning, which collectively damage the intestinal epithelial barrier and compromise growth performance [6]. Historically, antibiotics were widely used prophylactically against enteric disorders, but this could result in drug residue and antimicrobial resistance issues, causing public health risks [7]. Consequently, developing sustainable alternatives to antibiotics, such as probiotic formulations, phytogenic bioactive compounds, and immunomodulatory nutraceuticals, has become a strategic priority in animal husbandry [8].

Emerging as promising alternatives to antibiotics, plant-derived bioactive compounds have garnered significant research attention owing to their diverse botanical origins, non-residual safety profiles, and multi-target mechanisms of action [9]. Among these, *Pueraria lobata (Willd.)* Ohwi is a perennial leguminous vine indigenous to Asian ecosystems (*Pueraria thomsonii* and *Pueraria lobata*). It has a rich repository of phytochemicals, including isoflavonoids (such as puerarin, daidzein, daidzin, and puerarin-7-xyloside), amino acids, polysaccharides, and trace elements with demonstrated bioactivity [10]. As the principal bioactive constituent, puerarin (an 8-C-glycosylated isoflavone) exhibits multimodal pharmacological properties, such as reactive oxygen species (ROS) scavenging, antioxidative stress, anti-inflammatory, anti-infection, immune regulation, anti-aging, improving intestinal function and maintaining the balance of gut microbiota effects [11–21]. These activities highlight puerarin's therapeutic versatility across multiple disease spectrums, with demonstrated efficacy in ameliorating cardiovascular dysfunction, diabetes, hepatic pathologies, infective diseases, cancer, and neurodegenerative diseases [22–27]. Recent studies indicate that puerarin supplementation can effectively improve intestinal function and increase feed conversion efficiency, meat quality metrics (e.g., intramuscular fat deposition), and enteroprotective efficacy, finally improving livestock productivity [28–30]. For instance, Li et al. showed that dietary puerarin supplementation can markedly reduce the intestinal injury and oxidative stress in the small intestines of diquat-challenged piglets [31]. Wu et al. showed that puerarin supplementation at 0.5 mg/kg body weight (dissolved in a liquid milk replacer) for nine days can effectively reduce oxidative stress and inflammatory response, promote beneficial intestinal microbiota, and improve the function of the mucosal barrier in the intestine tissues of neonatal piglets, ultimately reducing morbidity caused by porcine epidemic diarrhea virus (PEDV) infection [21]. In this review, we summarized the current evidence on puerarin supplementation on its beneficial impact on animal intestinal health. Additionally, the underlying regulatory mechanisms in improving animal intestinal morphology, enhancing intestinal antioxidant capacity, and preserving the intestinal mucosal barrier were also discussed. Collectively, this review provides a comprehensive perspective for the intestinal health regulatory effects of puerarin in livestock, and it will promote its conversion and application in livestock production systems.

2. The Physicochemical Properties of Puerarin

Puerarin (also known as daidzein-8-C-glucoside) (Figure 1), is a predominant isoflavonoid in the root of *Pueraria lobata (Willd.)* Ohwi presents unique molecular characteristics that govern its pharmacokinetic behavior [32]. Pharmacopeial standards mandate $\geq$13% puerarin content in *Pueraria lobata*'s root extracts. With the molecular formula

$C_{21}H_{20}O_9$ (with a molecular weight of 416 Da), this crystalline compound exhibits pH-dependent solubility: freely soluble in methanol, sparingly soluble in ethanol, and poorly soluble in water (it is about 2.46 g/L at 25 °C) due to its amphiphilic nature arising from the hydrophobic isoflavone core and hydrophilic C-8 β-D-glucopyranosyl moiety [33–35]. The C-glycosidic linkage confers metabolic stability against β-glucosidase hydrolysis, a critical advantage over O-glycosylated analogs [27,36].

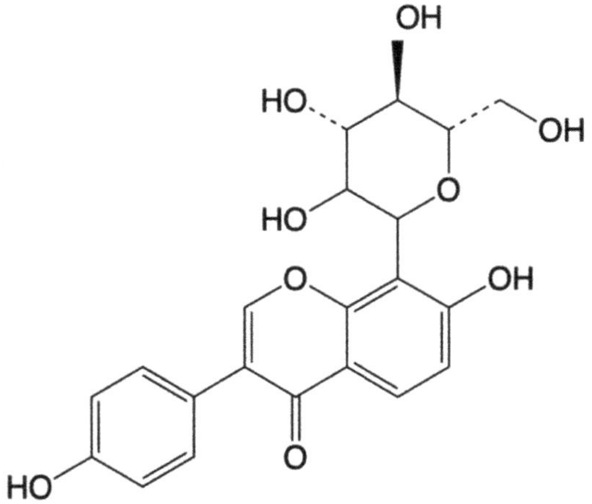

Figure 1. Chemical structure of puerarin.

3. Absorption and Metabolism of Puerarin

Puerarin's pharmacokinetics are characterized by limited absorption, broad tissue distribution, multi-phase metabolism, and enterohepatic-dependent elimination [37]. While negligible gastric absorption occurs, intestinal uptake in the duodenum and jejunum is rapid yet hindered by P-glycoprotein (P-gp) efflux and its low water solubility and permeability, resulting in the poor oral bioavailability (~7%) of puerarin [38]. However, the lower oral bioavailability (less than 1%) was also observed following oral higher dosing. This may be corrected by its low dissolution state in gastrointestinal fluid [38]. The poor intestinal absorption also resulted in a short half-life ($t_{1/2}$) and low peak plasma concentration (C_{max}) of puerarin [39,40]. Therefore, selecting appropriate strategies to improve the physicochemical properties of puerarin is of great significance for further enhancing its oral bioavailability and promoting the development of its oral formulations. Several formulation innovations and drug delivery systems, such as nanoemulsions, phospholipid complexes, and solid self-microemulsifying, can effectively improve the oral bioavailability of puerarin in animal models [41–43]. For instance, Wu et al. reported that microemulsion-based puerarin exhibited a 15.8-fold increase in relative bioavailability compared to a suspension in mice [44]. Tang et al. prepared puerarin-loaded oil-in-water- and water-in-oil-microemulsions, which significantly increased the solubility to 11.3 and 23.1 mg/mL, respectively; and the oral relative bioavailability both increased 2-fold, compared to the puerarin suspension administration in rats [45]. Tu et al. prepared puerarin nanocrystals and improved the oral bioavailability of puerarin. In a rat model, the area under the curve (AUC) of the puerarin nanocrystals was 7.6-fold that of puerarin suspension, with an absolute bioavailability of 21.4% [46]. A systematic induction and review of the **drug delivery systems of** puerarin have been carried out by Zhang [37].

Following absorption, puerarin can distribute extensively to the heart, liver, kidney, lung, stomach, bone marrow, and mammary gland [38]. Although the concentration is limited, puerarin can still penetrate the blood–brain barrier (BBB) at pharmacologically relevant levels, and the glucose transporter type 1 (GLUT1) transporter promotes pharmacology-related BBB permeability [47–49]. Wu et al. found that co-administration with borneol or α-asarone (both at 25 mg/kg body weight) amplifies brain exposure of puerarin in rats, the AUC_{0-12} increased from 48.1 ng/mL·h to 64.8 ng/mL·h or 86.0 ng/mL·h, respectively, thereby enhancing its targeting efficacy for cerebral ischemia-induced brain damage [47,50]. In the study of Anukunwithaya et al., C_{max} of puerarin was reached within 5 minutes when rats were orally administered with 5 or 10 mg/kg body weight, and systemic clearance occurred at 8–12 h [38]. Several studies also showed that oral administration of a puerarin suspension resulted in C_{max} at 0.45–5.00 hours [35,51–53].

Extensive studies on puerarin metabolism in vitro and in vivo have revealed pathways including oxidation, glucuronidation, sulfation, creatinine conjugation, and gut microbiota-mediated metabolic transformation [38]. The major hydrolysis metabolite of puerarin was identified as daidzein, which was formed by cytochrome P450 proteins (CYP 450s) in the liver microsomes, and subsequently reduced to dihydrodaidzein and equol. Daidzein could also form daidzein-4′,7-di-O-sulfate and daidzein-4′-O-sulfate through the sulfation progress [54]. Glucuronides were the main metabolites of puerarin and were excreted in the urine and feces [38]. The glucuronide conjugates of puerarin, such as puerarin-7-O-glucuronide and puerarin-4′-O-glucuronide, were the main metabolites of the conjugation reaction [38]. It has been reported that seven UDP-glucuronosyl-transferase (UGT) isoforms (e.g., UGT1A1, 1A9, 1A10, 1A3, 1A6, 1A7, and 1A8) can catalyze the conversion of puerarin to form two water-soluble metabolites, i.e., puerarin-7-O-glucuronide and puerarin-4-O-glucuronide [55–57]. It has been demonstrated that puerarin-7-O-glucuronide can also contribute to potent activities in treating cardiovascular diseases, which is similar to its parent compound, puerarin [58]. Additionally, based on the data in rat liver and intestine, the metabolic profiles of puerarin were similar in liver and intestine, with no metabolic regioselectivity [37].

After absorption in the intestine, puerarin is rapidly excreted through urine and feces, mainly in the form of gluconates [37]. It is mainly excreted in urine, and approximately 50% and 15% of the intravenous dose is excreted via its gluconate metabolites through the urethra and the feces, respectively [38]. This process can be influenced by multiple factors, such as water solubility, bioavailability, gender, and pathological condition [37]. In addition, the pharmacokinetic characteristics of puerarin showed marked differences in its absorption and excretion patterns in vivo between *Pueraria lobata*′ root extract and pure puerarin after oral administration [59]. These findings suggest that puerarin's metabolism is highly susceptible to external factors, necessitating further investigation into its complex metabolic pathways.

Correctly, puerarin undergoes enterohepatic recycling following oral administration. It is absorbed in the small intestine via glucose transporters and metabolized in the liver during first-pass effects by UGTs and sulfotransferases (SULTs) into sulfated and glucuronidated derivatives. A fraction of these metabolites is excreted into bile and is then hydrolyzed back to the original compound by gut microbiota, thereby initiating enterohepatic recycling. For example, gut microbiota-derived bioactive metabolite daidzein can be reabsorbed into liver via the portal vein, then provide a hepatoprotective effect (shown in Figure 2) [27,37]. The enterohepatic recycling mechanism underlies puerarin's sustained efficacy and interpatient variability, highlighting its role as a dual-edged determinant for precision dosing strategies.

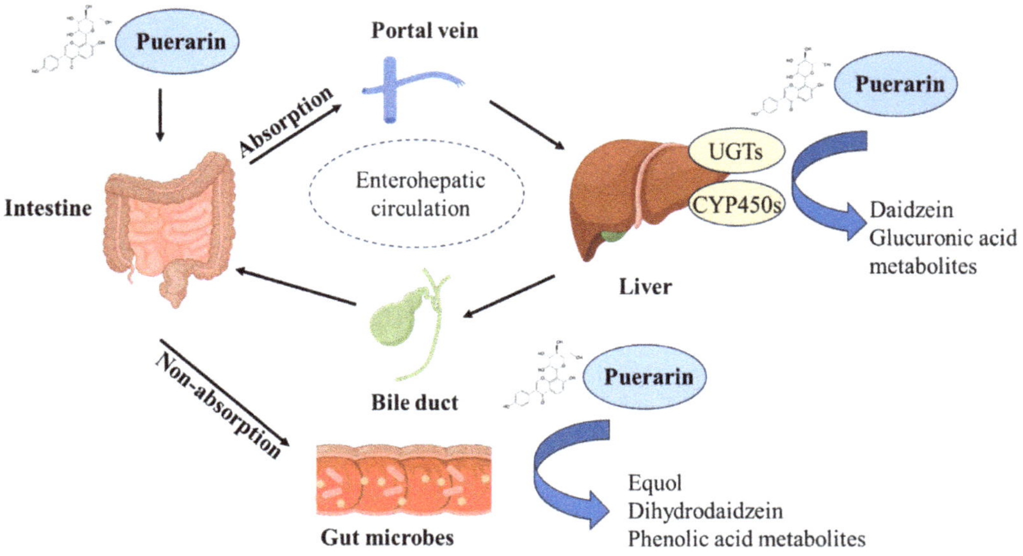

Figure 2. The absorption of puerarin and its metabolic pathways in the intestine and liver tissues. UGTs, UDP-glucuronosyltransferases; CYP450s, cytochrome P450s.

4. The Regulatory Effect of Puerarin on Intestinal Health of Farm Animals

4.1. Effects of Puerarin on Intestinal Morphology

The villus height (VH), crypt depth (CD), and VH/CD ratio serve as quantitative anatomical indices for evaluating intestinal mucosal integrity and digestive-absorptive capacity. In the small intestine, an increase in villus height (VH) expands the luminal surface area, enhancing nutrient absorption. Meanwhile, a decrease in crypt depth (CD) indicates improved epithelial regeneration efficiency [60]. Intestinal stem cells (ISCs) residing in the crypt base undergo continuous proliferation and upward migration along the crypt-villus axis, differentiating into functional enterocytes to maintain homeostatic turnover—a critical balance between apoptosis and regeneration that underpins mucosal barrier function [61]. Mechanistic studies reveal puerarin's therapeutic potential in preserving intestinal morphology under pathological stress. For instance, Dong found that the oral administration of puerarin at 5 mg/kg body weight for six days can enhance the values of VH and the ratio of VH/CD, and reduced oxidative stress and inflammation in piglets infected with Enterotoxigenic *Escherichia coli* (ETEC), thereby mitigating intestinal morphological damage and improving the production performance (such as growth rate and feed conversion rate) [62]. In another study, dietary supplementation with puerarin at 750 mg/kg feed in yellow-feathered broilers (*Gallus gallus* domesticus) can effectively reverse oxidized soybean oil-induced ileal atrophy, restoring duodenal VH to 428 μm (vs. 312 μm in controls) and VH/CD to 2.9 (vs. 1.7 in controls) [63]. The equilibrium of intestinal morphology is intricately linked to the balance between apoptosis and proliferation of intestinal cells. Proliferating cell nuclear antigen (PCNA) serves as a marker for cell proliferation [64,65]. Li et al. indicated that incorporating 0.1% puerarin into the diet of weaned piglets can increase the proportion of PCNA-positive cells in the intestinal mucosa, thereby promoting intestinal epithelial cell proliferation and alleviating the intestinal and ileal morphological damage induced by diquat [31].

These findings collectively demonstrate that puerarin stabilizes crypt-villus dynamics by reducing oxidative stress-induced apoptosis (via ROS scavenging) and promoting epithelial renewal (via Wnt/β-catenin activation), thereby preserving intestinal architectural homeostasis under diverse pathological insults.

4.2. Effect of Puerarin on Intestinal Antioxidant Capacity

Modern high-density livestock operations expose animals to cumulative oxidative stressors—heat stress, environmental toxins, infectious diseases, and weanling stress—that overwhelm endogenous antioxidant defenses. This redox imbalance drives pathological ROS accumulation, culminating in macromolecular oxidative lesions across tissues [66,67].

Currently, exogenous supplementation of antioxidants represents a critical strategy for enhancing systemic antioxidative capacity and mitigating intestinal oxidative stress in animal health management. A recent study showed that oral administration of puerarin at 5 mg/kg body weight for six days can significantly augment enzymatic defenses in ETEC-challenged piglets, as evidenced by significantly increased total superoxide dismutase (T-SOD) and catalase (CAT) activities in the ileum tissues of ETEC-challenged piglets [21]. In another study, it was found that puerarin supplementation improved antioxidant activity by significantly increasing the activity of total antioxidant capacity, glutathione peroxidase (GSH-Px), and CAT, while decreasing malondialdehyde (MDA) levels in the serum, jejunum, and ileum in aged laying hens, finally improving intestinal morphology with increased villus height and the ratio of villus height to crypt depth [30].

Nuclear factor erythroid 2-related factor 2 (Nrf2), encoded by the NFE2L2 gene, belongs to the Cap'n'Collar (CNC) basic leucine zipper transcription factor family. These evolutionarily conserved regulators are ubiquitously expressed in eukaryotic organisms and function as master regulators of cytoprotective gene networks that mediate adaptation to environmental stressors [68]. Upon activation, Nrf2 translocates to the nucleus, where it binds to antioxidant response elements (AREs) and triggers the expression of detoxifying enzymes such as heme oxygenase-1 (HO-1), NAD(P)H quinone oxidoreductase 1 (NQO1), glutamate cysteine ligase (GCLC), SOD, CAT, and glutathione S-transferase (GST) [69]. Puerarin has been identified as a potent Nrf2 activator, functioning through the Keap1-Nrf2-ARE signaling axis to upregulate downstream cytoprotective genes including HO-1, SOD, CAT, and NQO1, and GCLC [11,70–72]. Consistently, Nrf2 knockdown or knockout significantly reduces puerarin-induced oxidative stress damage to various environmental toxins or drugs in tissues or cells [11,70,73]. Li et al. demonstrated that dietary inclusion of 0.1% puerarin markedly upregulated the expression of Nrf2 protein in both jejunal and ileal mucosa tissue of piglets, then further induced the transcriptional expression of downstream gene-mediated protein expression (e.g., HO-1, GSH-Px, and CAT) [31]. In a separate intervention, oral administration of puerarin (0.5 mg/kg body weight) induced a marked increase in the levels of Nrf2 mRNA in the intestinal tissues, then significantly increased the levels of T-SOD, GSH-Px, and CAT activities, and finally, effectively reduced the damage in PEDV-infected piglets [21].

Collectively, these findings demonstrate that puerarin supplementation alleviates stress-induced intestinal damage in animals by inhibiting lipid peroxidation, enhancing antioxidant enzyme activity, or directly activating the Nrf2-mediated antioxidant defense pathway (Figure 3).

4.3. Effect of Puerarin on Intestinal Barrier Function

The intestinal barrier represents a sophisticated multidimensional defense system comprising four interactive components, including the mechanical barrier (also called the physical barrier), the chemical barrier, the immunological barrier, and the microbiological barrier [74–76]. Puerarin supplementation can integrate and regulate these barriers synergistically to prevent systemic dissemination of luminal pathogens and toxins through mechanisms including mucin secretion, tight junction regulation, IgA-mediated immune surveillance, and commensal microbiota modulation (shown in Figure 4). The detailed actional mechanisms will be discussed in the following sections.

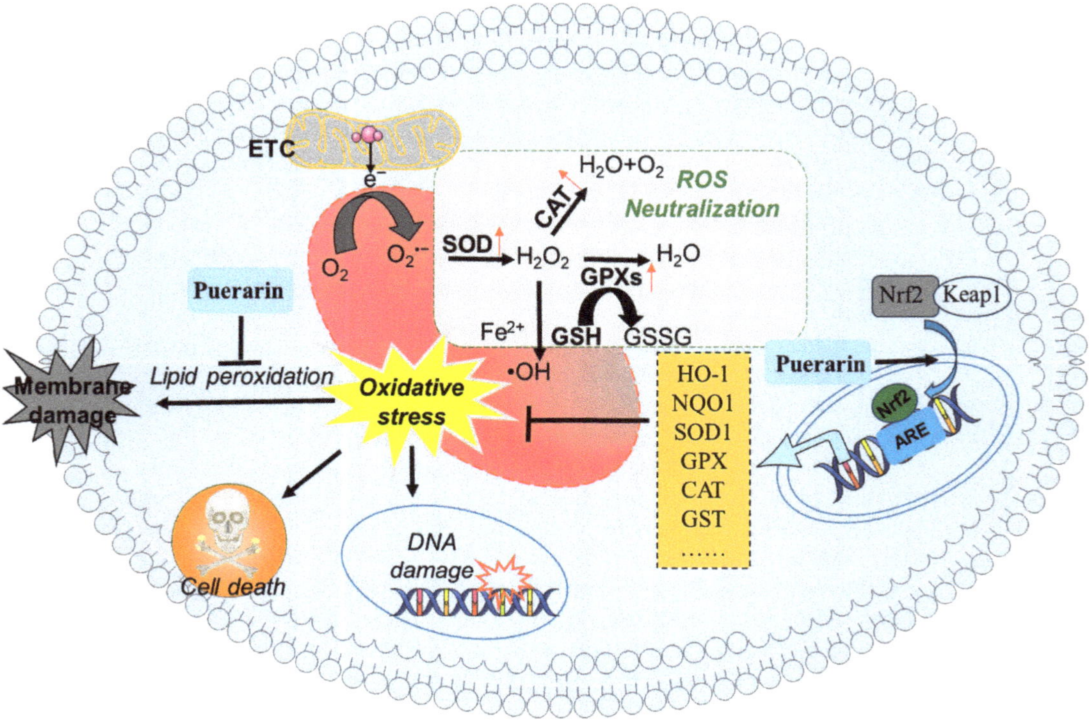

Figure 3. A working model of puerarin protecting against oxidative stress in the intestinal tissues. GSSG, oxidized glutathione; SOD, superoxide dismutase; SOD, superoxide dismutase; ROS, reactive oxygen species; H_2O_2, hydrogen peroxide; •OH, hydroxyl radical; $O_2^{\bullet-}$, superoxide anion; $O_2^{\bullet-}$, superoxide anion; Nrf2, nuclear factor erythroid 2-related factor 2; NQO1, NAD (P)H quinone oxidoreductase 1; NQO1, NAD (P)H quinone oxidoreductase 1; HO-1, heme oxygenase 1; GSH, glutathione; CAT, catalase; ETC, electron transport chain; ARE, antioxidant response element.

4.3.1. Intestinal Microbiological Barrier

The intestinal microbial barrier, consisting of the intestinal microbiota, forms a relatively stable ecosystem through mutual balance. The gut microbiota is instrumental in breaking down indigestible dietary fiber into short-chain fatty acids (SCFAs) within the gastrointestinal tract; these SCFAs provide energy, inhibit pathogenic microorganisms, and enhance the intestinal mucosal barrier [77]. Disruptions in the gut microbiota can impair mechanical barriers and intestinal morphology, precipitating intestinal inflammation [78]. The gut microbiota analysis revealed a significant reduction in Bacteroidetes coupled with *Akkermansia muciniphila* (AKK) enrichment in the puerarin-treated intestinal tissue, suggesting enhanced mucin utilization capacity. Notably, puerarin demonstrated selective elevation of fecal indole-3-propionic acid (IPA) concentrations while maintaining indoleacrylic acid and butyrate at physiological baselines [79]. Moreover, IPA can activate the aryl hydrocarbon receptor (AhR) signaling pathway, which can regulate barrier integrity and immune function by influencing the expression of tight junction proteins and modulating inflammatory responses [80]. It indicated that the marked upregulation of IPA in the intestinal tissues induced by puerarin may provide protection via the activation of AhR [79,81]. Intravenous administration of puerarin increased the number of lactic acid bacteria in the cecum and colon of ETEC-infected piglets, alleviating bacterial dysbiosis. Furthermore, puerarin enhances the relative abundance of Lactobacillus in the gut of mice with irritable bowel syndrome and decreases the relative abundance of *Desulfovibrionaceae* [82]. In addition, research also showed that oral administration of puerarin increases the relative abundance of AKK bacteria in mice under normal physiological conditions, colitis rats, and obese mice [83,84]. AKK bacteria can stimulate intestinal stem cell proliferation and IL-10 secretion, repairing *Salmonella*-induced intestinal damage [85]. Studies revealed that puerarin exerts a direct inhibitory effect on ETEC, significantly diminishing the expression of ETEC subunit proteins and virulence factors, and curbing the secretion of exotoxins [86]. Consistently, it also found that the intervention of silymarin and polyherbal extract (it includes silymarin, salvianolic acids B, and puerarin) significantly increased the levels of *Akkermansia* and *Blautia*, and reduced the levels of *Clostridium* in the intestinal tissues, then improved high-fat diet-induced liver injury [87].

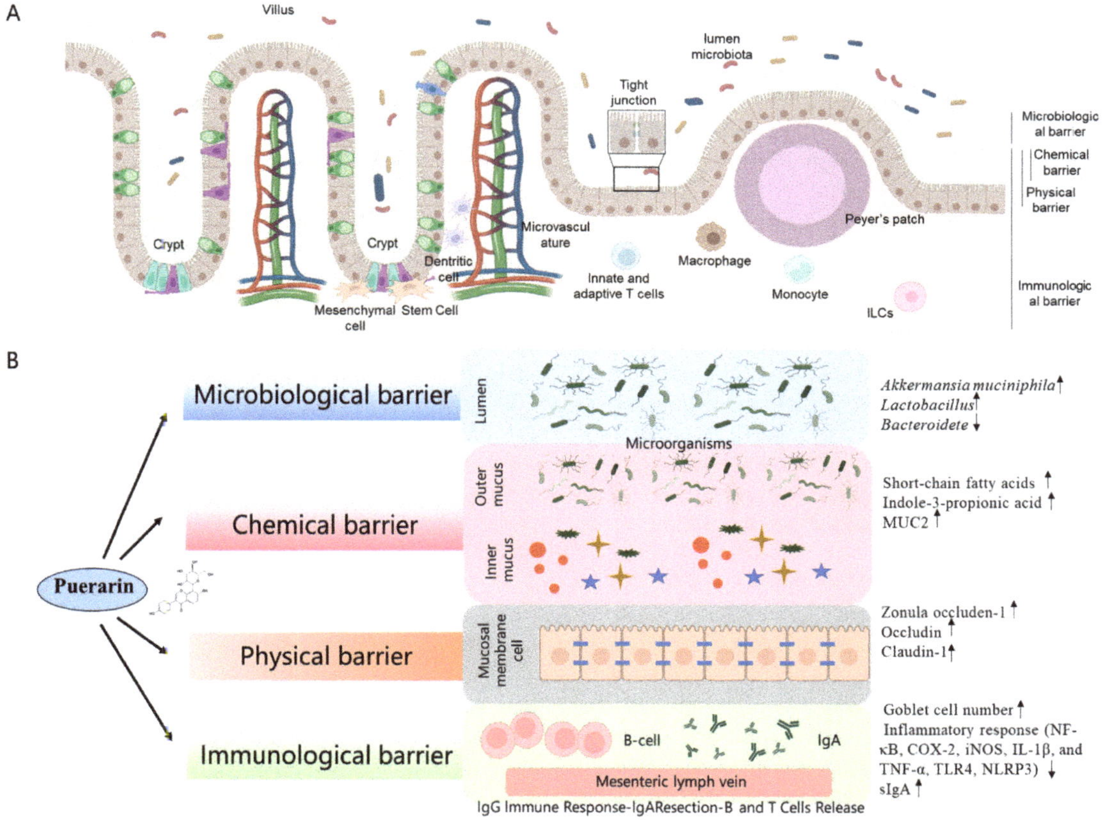

Figure 4. (**A**), intestinal barrier structure. (**B**), a working model of puerarin in maintaining the intestinal barrier function in the intestinal tissues. Puerarin supplementation can integrate and regulate the microbiological barrier (e.g., increase the levels of *Akkermansia muciniphila* and *Lactobacillus*, and decrease the levels of *Bacteroidete*), chemical barrier (e.g., increase the levels of short-chain fatty acids, indole-3-propionic acid, and secreted MUC2), physical barrier (improve the tight junctions in intestinal epithelial cells via upregulating the expression of zonula occluden-1, occludin, and claudin-1 proteins), and immunological barrier (e.g., increase the number of goblet cells and the levels of sIgA, and reduce the inflammatory response), and synergistically maintain the intestinal barrier function.

In short, puerarin can promote microecological reprogramming via enriching AKK while suppressing *Bacteroidete*, synergistically elevating IPA to AhR signaling, and finally contributing to its protective effect on the dysfunction of the intestinal barrier. Moreover, this dual mechanism, i.e., enhancing the integrity of the mucus layer through goblet cell hyperplasia and simultaneously optimizing the ecology of mucinophilic microbiota, demonstrates puerarin's therapeutic potential as a nutraceutical agent for precisely restoring the intestinal barrier by modulating host–microbiota crosstalk.

4.3.2. Intestinal Chemical Barrier

The intestinal chemical barrier, a multifunctional defense system comprising enzymatic secretions, antimicrobial peptides (e.g., defensins), and microbiota-derived metabolites (e.g., SCFAs and IPA), relies critically on the mucin 2 (MUC2)-mediated mucus layer (>300 μm) to prevent pathogen adhesion (such as *Salmonella*) while maintaining luminal pH homeostasis (6.0–7.4) [88–91]. Notably, SCFAs, mainly including acetate, propionate, and butyrate, are metabolites produced during the bacterial fermentation of dietary fiber in the intestinal tract. Butyrate is the principal substrate and energy source for colonocytes and provides at least 60–70% of colonic mucosal energy requirements [92]. Concurrently, SCFAs can reinforce barrier integrity by energizing colonocytes (i.e., ~70% ATP from butyrate β-oxidation) and upregulating Claudin-5/MUC2 expression via GPR43/HDAC modulation [93]. It has been demonstrated that physiological concentrations of SCFAs immediately suppress colonic epithelial permeability and reduce the invasion of exogenous toxins and pathogens [93]. Several studies have illustrated that puerarin can enhance the chemical barrier through three synergistic mechanisms: (1) stimulating goblet cell MUC2 secretion via Ca^{2+}/PKC-dependent exocytosis, (2) elevating protective metabolites, including fecal IPA and butyrate through bifidobacterium enrichment, and (3) suppressing inflammation via IPA-mediated NF-κB inhibition, downregulating the expression of tumor necrosis factor-α (TNF-α), interleukin-8 (IL-8), and IL-6 [61,94,95]. Consistently, Liu et al. demonstrated that oral administration of 10 mg/kg body weight puerarin can increase the activity of sucrase, protease, and amylase in the intestinal tract of mice, then promote the digestion and metabolism of carbohydrates [96]. Collectively, these actions establish puerarin as a multimodal enhancer of chemical barriers, integrating mucus reinforcement, regulation of microbial metabolites, and optimization of digestive processes to strengthen intestinal mucosal defenses.

4.3.3. Intestinal Mechanical Barrier

Tight junctions (TJs), the principal intercellular junctional complexes of the intestinal mechanical barrier, are dynamically regulated by transmembrane proteins (occludin, claudin family) and scaffolding proteins (Zonula occludens, ZO). These junctions function to seal intercellular spaces and selectively facilitate the passage of nutrients, electrolytes, and water from the intestinal lumen through the intestinal mucosa, while simultaneously preventing the infiltration of harmful substances [97]. A compromise in the integrity of this mechanical barrier results in increased intestinal permeability, allowing pathogens from the intestinal lumen to translocate into the submucosal tissue. This translocation activates the immune system, prompting the release of inflammatory cytokines and leading to intestinal inflammation [98,99]. In rat models of colitis induced by dextran sulfate sodium (DSS) and 2,4,6-trinitrobenzenesulfonic acid (TNBS), puerarin has been demonstrated to enhance TJ mRNA and protein expression levels, thereby restoring the integrity of the mechanical barrier [15]. Dietary supplementation with 1% puerarin has been shown to significantly elevate the mRNA expression levels of occludin and E-cadherin in the jejunum and ileum of piglets under oxidative stress [100]. Similarly, the supplementation of 750 mg/kg puerarin in the diet markedly enhances the mRNA expression levels of occludin, claudin-1, and ZO-1 within the intestinal tract of broiler chickens, thereby mitigating the mechanical barrier damage caused by oxidized soybean oil [38,99]. Additionally, puerarin has been shown to ameliorate intestinal mechanical barrier damage and reduce intestinal permeability in obese mice induced by a high-fat diet and in ovariectomized rats [83,101]. These findings suggest that puerarin plays a crucial role in restoring the integrity of the intestinal mechanical barrier in animals subjected to stress by upregulating the expression of TJ-relative proteins, thus safeguarding intestinal health (Figure 4).

4.3.4. Intestinal Mucosal Immune Barrier

The intestinal mucosal immune barrier constitutes a sophisticated interface integrating structural, immunological, and neural components to maintain gastrointestinal homeostasis. Central to this system, the gut-associated lymphoid tissue (GALT) orchestrates immune surveillance through specialized structures including Peyer's patches and mesenteric lymph nodes, coordinated by regulatory T cells that enforce immune tolerance and reduce the inflammatory response [102–104]. The epithelial barrier, comprising enterocytes and goblet cells, forms a selective physical blockade while actively participating in immune signaling through pattern recognition receptors [105]. Disruptions in the mucosal immune barrier can lead to increased intestinal permeability, allowing translocation of bacteria and their products, which can trigger systemic inflammation and contribute to the pathogenesis of various diseases, including inflammatory bowel disease, bacterial diarrhea, and virus infection-related gut dysfunction [106,107]. Li et al. found that puerarin supplementation via oral administration at 200 mg/kg body weight for 14 days can markedly restore mucus barrier integrity by upregulating mucin synthesis (e.g., upregulated the expression of MUC2 and MUC4 expression) and increasing goblet cell density in intestinal barrier dysfunction in rats exposed with 2,4,6 trinitrobenzene sulfonic acid (TNBS)-indued intestinal barrier dysfunction in rats [79].

The dysfunction of the intestinal mucosal immune barrier is a critical aspect of various inflammatory diseases, where the integrity of epithelial and endothelial barriers is compromised, leading to increased permeability and subsequent inflammation [108]. This dysfunction is often mediated by complex inflammatory pathways involving a variety of cellular and molecular mechanisms. One of the key players in barrier dysfunction is the nuclear factor kappa B (NF-κB) signaling pathway. Activation of NF-κB leads to the transcription of genes (such as cyclooxygenase-2 [COX-2], nitric oxide synthase [iNOS], IL-1β, and TNF-α) involved in inflammation and immune responses, contributing to the disruption of tight junction proteins and increased barrier permeability [109]. Studies have demonstrated that puerarin supplementation can effectively inhibit the transcriptional expression of NF-κB, then markedly downregulate the pro-inflammatory mediators such as COX-2, iNOS, IL-1β, and TNF-α, thereby alleviating DSS-induced colitis in mice [15]. Similarly, Wu et al. reported that puerarin supplementation can effectively inhibit the NF-κB signaling pathway in the ileum, decrease the levels of IL-1β, IL-6, and IL-8, and alleviate the damage of PEDV to the intestine [21,110]. In another study, it was reported that puerarin supplementation also inhibited the expression of NF-κB inflammatory signaling pathway-related genes in IPEC-J2 cells, reduced the number of ETEC adhesions, and mitigated ETEC-induced cell damage [111]. Consistently, puerarin supplementation can also markedly inhibit PEDV replication in both in vitro and in vivo settings and significantly reduce NF-κB activation-mediated inflammation response, effectively mitigating the decline in growth performance among piglets [21,110]. Zhou et al. found that dietary puerarin supplementation at 750 mg/kg feed via the diet can markedly inhibit the Toll-like Receptors 4 (TLR4)/NF-κB signaling pathway in the intestinal tract of broiler chickens, upregulate the expression of negative regulatory factors and anti-inflammatory cytokine genes, suppress the expression of pro-inflammatory cytokine genes, elevate the levels of secretory immunoglobulin A (sIgA), and ameliorate intestinal inflammation triggered by oxidized soybean oil [28]. These findings indicate that puerarin can effectively inhibit the inflammatory response to maintain the intestinal mucosal immune barrier via the inhibition of the NF-κB pathway.

Inflammasomes, particularly the NLRP3 inflammasome, play a crucial role in mediating inflammatory responses that lead to barrier dysfunction. Activation of the NLRP3 inflammasome results in the production of pro-inflammatory cytokines such as IL-1β, which can further exacerbate barrier disruption and inflammation [109,112]. Peng et al. discovered that puerarin activates the AMP-activated protein kinase (AMPK)/Sirtuin 1 (SIRT1) signaling pathway, inhibits the formation of NOD-like receptor protein 3 (NLRP3) inflammasomes, reduces IL-1β and IL-18 levels, and enhances the resistance of gastric epithelial cells to lipopolysaccharide (LPS)-induced apoptosis [113]. In addition, puerarin supplementation also exerts a mitigating effect on intestinal inflammation associated with obesity, food allergies, and irritable bowel syndrome [82,83,114]. Puerarin can also significantly increase the expression of TRPM5, a necessary protein for mucin secretion [81], but the precise molecular mechanisms are still unclear, and further investigations are required.

Of note, several studies showed that puerarin combined with tryptophan or its metabolites (e.g., IPA) can markedly improve the mucus layer barrier of intestinal tissues in dextran sodium sulfate-treated rats, which was mainly achieved by increasing the number of goblet cells and promoting the secretion of MUC2 [81]. This synergistic effect may also be related to the different regulatory effects of puerarin and tryptophan on the expression of TRPM5 protein in the intestinal tissues [81].

Collectively, puerarin orchestrates intestinal mucosal repair through the synergistic effects on mucosal reinforcement and anti-inflammatory circuitry. It can upregulate mucin synthesis and goblet cell density and dually inhibit the NF-κB and NLRP3 inflammasome pathways through AMPK/SIRT1 activation. These mechanistically stratified effects validate puerarin's potential as a phytoceutical intervention for barrier-related pathologies ranging from infectious enteritis to metabolic-associated gut disorders.

4.4. Enteric Nervous System

The enteric nervous system (ENS) has a close interaction with the intestinal epithelial barrier and the surrounding immune cells and plays a crucial part in maintaining the homeostasis of the gastrointestinal tract [115]. It has been reported that the ENS is a key factor in the pathogenesis of various diseases, such as cancer, diabetes, and neurodegenerative diseases, and regulates gut barrier function and intestinal homeostasis [116].

The ENS is divided into two nerve plexuses. In the myenteric plexus, the cell bodies are located between the circular and longitudinal muscle layers of the intestine, and they coordinate the contraction and relaxation required for gastrointestinal peristalsis. Meanwhile, the cell bodies of the submucosal plexus are closer to the lumen and penetrate deep into the mucosa [116]. They coordinate secretion, absorption, and local blood flow in the intestinal tissues [116]. ENS can sense the stimuli from the lumen and respond with appropriate effector signals to promote digestion, movement, and host defense. It can also develop and influence the mucosal immune system and the gut microbiota, jointly regulating gut homeostasis [117]. For instance, Lactobacillus strains can directly regulate colonic motility by activating calcium-dependent potassium channels in the intrinsic sensory neurons of the ENS [118].

A recent study reported that puerarin exerts inhibitory effects on the motor dorsal nucleus–vagus nerve axis by binding to gamma-aminobutyric acid receptors containing alpha 1 subunit, thereby controlling the length of microvilli in the jejunum to regulate the absorption of fat by the intestine [119]. In another study, it was found that puerarin treatment can significantly decrease the nitric oxide, acetylcholine, and serotonin levels, reduce the contraction response of the intestine, and thereby improving diarrhea-predominant irritable bowel syndrome in rats [120]. Huang reported that puerarin supplementation can upregulate the proportions and quantities of inhibitory neurons (e.g., neuronal nitric oxide synthase [nNOS]- and vasoactive intestinal peptide [VIP]- immunoreactive myenteric neurons) and downregulate proportions and quantities of the excitatory neurons (e.g., neocortical choline acetyltransferase [ChAT]- and substance P [SP] immunoreactive myenteric neurons) in myenteric plexus, maintaining the homeostasis of colonic interneurons and smooth muscle contraction function, and finally, improving intestinal function in rat models of diarrhea-predominant irritable bowel syndrome [121].

Collectively, puerarin can regulate the intestinal nervous system by affecting the number and proportion of intestinal neurons and the secretion of neurotransmitters, thereby regulating intestinal contraction, nutrient absorption, and the response to harmful exogenous substances. However, the precise mechanism still needs further research.

4.5. Others

Intestinal stem cells are the driving force behind intestinal epithelial renewal, and when damaged, they quickly replenish lost stocks to support epithelial repair. This plasticity mechanism is crucial for intestinal health [122]. Tao et al. found that puerarin supplementation can increase Paneth cell numbers and stimulate the expression of Wnt ligand Wnt3a and Lgr5 and PCNA in the intestinal tissues of sleep disorder-treated aged mice [123]. Of note, Lgr5 serves as a marker for intestinal stem cells and increased levels of PCNA signify cell proliferation. This information indicated that puerarin supplementation can enhance the ability of intestinal damage repair via modulating the number and functions of intestinal stem cells [123]. Additionally, it also reported that puerarin supplementation can increase the expression of phosphorylation-extracellular signal-regulated kinase (p-Erk) and p-Akt protein, and improved the wound healing ability [124]. It is known that the activities of Erk and Akt play a critical role in intestinal modification, cell proliferation, immune regulation, and stress resistance [125]. These modulations may also contribute to its effects on intestinal repair. The precise molecular mechanisms still need more investigation.

5. Pharmacological Modulation of Production Performance in Livestock by Puerarin

5.1. Poultry Health

The application of puerarin has gained attention in recent years for its potential benefits in poultry production. One of the primary benefits of puerarin in poultry production is its ability to improve growth performance. Puerarin can enhance feed intake and conversion efficiency, leading to improved weight gain and overall growth in poultry. For instance, Lu Yu et al. discovered that adding 10 mg/kg of total flavonoids from *Pueraria lobata* (with a puerarin content of 79%) in the feed can significantly enhance the average daily weight gain of broiler chickens and decrease the feed conversion ratio via the enhancement of mucin secretion and the upregulation of short-chain fatty acids derived from gut microbiota [63]. The use of natural additives to bolster the immune system and reduce the prevalence of pathogens is a growing area of interest in poultry science. For instance, the inclusion of probiotics, prebiotics, and other natural compounds in poultry diets has been shown to improve gut health and immune responses, thereby reducing the incidence of diseases [126]. Puerarin, with its antioxidant and anti-inflammatory properties, could play a similar role in enhancing the immune resilience of poultry, potentially reducing the need for antibiotics and other chemical interventions.

In addition to growth optimization, puerarin may also contribute to disease resilience in poultry. For example, it was shown that puerarin treatment of 100 mg/kg in the diet can markedly improve disease resistance by increasing antioxidant enzyme activity, reducing lipid peroxidation and inflammatory response, and improving caspase-8-mediated cell apoptosis in broiler chickens infected with *Salmonella* [63,127]. Zhou et al found that puerarin supplementation at 750 mg/kg body weight via the diet significantly increased the serum immunoglobulin content, reduced the HMGB1, MAPK, and TLR4 signaling pathway-mediated inflammatory response, and mitigated intestinal barrier damage caused by oxidized soybean oil in the intestine tissue of broiler chickens [28].

Furthermore, puerarin's potential role in modulating the gut microbiota is another avenue through which it could enhance poultry health. The gut microbiome is crucial for nutrient absorption and immune function, and its modulation through dietary interventions can lead to improved health outcomes in poultry. Research has highlighted the importance of maintaining a balanced gut microbiota for optimal poultry health and performance, with strategies such as the use of bacteriophage cocktails showing promise in this regard [128]. Gut microbiome analysis reveals a dual effect: suppression of *Enterobacteriaceae* and enrichment of butyrate-producing *Clostridium* clusters IV/XIVa, synergistically enhancing barrier function. This finding indicated that puerarin could contribute to this balance, supporting a healthy gut environment that fosters better nutrient utilization and disease resistance.

In short, puerarin supplementation can effectively improve intestinal health by reducing inflammatory response and remodeling gut microbiota. It presents a promising opportunity to enhance growth optimization and disease resilience in poultry production by improving intestinal health. By improving growth performance, bolstering immune function, and supporting gut health, puerarin could contribute to more sustainable and efficient poultry farming practices. Further research is needed to fully elucidate its mechanisms of action and to optimize its use dosage in commercial poultry operations, but the potential benefits make it a valuable area of exploration for the future of poultry production.

5.2. Porcine Health

Currently, research on puerarin in pig production is somewhat limited, with a primary focus on its impact on stressed piglets. Zeng et al. reported that puerarin supplementation at 5 mg/kg body weight can markedly enhance the antioxidant capacity and alleviate intestinal inflammation in *Escherichia coli* K88-infected piglets [129]. Furthermore, oral administration of 0.5 mg/kg puerarin has been shown to mitigate intestinal inflammation and oxidative stress caused by PEDV challenge [21,110]. Li et al. discovered that incorporating 0.1% puerarin into the feed can ameliorate the decline in growth performance of piglets infected with the enemy grass fast attack toxin [31]. Liu et al. found that *Pueraria lobata* root crude extract supplementation at 0.5 kg/ton feed in the diet can improve the gut microbial diversity of finishing pigs, enhancing the intestinal digestion and absorption of nutrients [130].

Collectively, these studies suggested that puerarin not only enhanced the intestinal and general health of stressed piglets and their stress resistance but also exerted a positive effect on the growth of piglets and fattening pigs under normal physiological conditions.

5.3. Cattle Health

Puerarin's rumen-modulating properties stem from its bifunctional capacity as both a fibrolytic enhancer and thermoresistant. Nie et al. discovered that the supplementation of 0.04% puerarin in beef cattle feed can significantly enhance rumen fermentation patterns and increase the degradation rate of feed nutrients and rumen bacterial protein synthesis [131]. This information indicates that puerarin supplementation via the diet can promote the growth of beneficial bacteria that enhance fiber digestion via modulation of the microbial population in the rumen in cattle. Under heat stress conditions, the inclusion of 0.04% puerarin in beef cattle feed significantly enhanced the apparent digestibility of dry matter, crude protein, and neutral detergent fiber by 3.23%, 3.19%, and 2.19%, respectively [132]. This improvement in digestibility was accompanied by a significant increase in fecal microbiota diversity and the relative abundance of beneficial bacteria, which in turn improved the growth performance, immune function, and antioxidant capacity of beef cattle [132]. Several studies have shown that the addition of 400-900 mg/kg puerarin to feed can improve the digestion rate of nutrients in beef cattle by regulating rumen fermentation mode and microbial community structure, enhancing the antioxidant capacity of muscle tissues [133,134].

Collectively, these findings indicate that puerarin supplementation can effectively improve the intestinal microbiota and antioxidant capacity of cattle, thereby enhance their digestive ability and improving their productivity. Since cows are ruminants, their digestive systems may differ from those of other animals such as chickens or pigs. Therefore, their precise mechanisms still need further investigation.

Although these data suggest that supplementing with puerarin can partly improve the intestinal health of poultry, porcine, and cattle, thereby enhancing their production performance, it still has some limitations. For example, the sample size, experimental period, and scale in the above-mentioned studies are very limited and insufficient. In addition, current experimental studies on the dosage of puerarin use are diverse, with some studies using relatively high doses, which may increase costs. Therefore, a more in-depth, systematic, and comprehensive evaluation of the clinical efficacy of puerarin supplementation in intestinal infections is still highly needed.

6. Conclusions and Future Perspectives

As a pleiotropic phytochemical, puerarin was demonstrated to have multidimensional gastrointestinal protective efficacy based on its multimodal bioactivities, such as anti-inflammatory, antioxidant, regulating gut microbiota, and immune regulation. These mechanisms synergistically enhance intestinal villus architecture, reinforce mucosal barrier integrity, and stabilize gut–liver axis homeostasis, positioning it as a viable antibiotic-alternative feed additive for sustainable livestock production.

Moreover, the environmental and economic benefits of using natural additives like puerarin in livestock production cannot be overlooked. As the industry moves towards more sustainable practices, the reduction in antibiotic use and the implementation of natural growth promoters are becoming increasingly important. The use of puerarin aligns with these goals, offering a natural and potentially cost-effective alternative to traditional growth promoters and disease control measures. Importantly, a healthy gut can reduce harmful gas emissions from feces. It has been demonstrated that the essential microbes may effectively reduce the enteric methane emission and help to mitigate the global methane crisis [135].

To advance translational applications, three strategic research priorities emerge: (1) in terms of life-stage-specific pharmacology, we need to delineate dose–response relationships across poultry growth phases (broiler starter/grower/finisher) and lactating versus gestating swine models to optimize feeding regimens; (2) in terms of mechanistic investigation, the precise molecular targets puerarin in barrier regulation, particularly its interplay with TLR4 signaling and occludin phosphorylation dynamics, need to be further addressed.

Some new technologies such as artificial intelligence, CRISPR-interference screening or multi-omics approaches may effectively accelerate our mechanism exploration; (3) in terms of bioavailability engineering, new formulations, such as β-cyclodextrin-encapsulated nano-formulations or others need to be investigated to overcome inherent solubility limitations; (4) in terms of safety assessments, the long-term endocrine effects and microbiota adaptation patterns caused by puerarin under industrial farming conditions should be further considered; (5) more systematic and comprehensive clinical studies (such as sufficient sample size, longer-term, and large-scale clinical trials) are still needed; (6) some investigations have shown that puerarin with its metabolites or other natural products may have a certain synergistic effect, thus maximizing its benefits for the intestinal health of livestock, which should be paid attention to in the future. By bridging the gap between growth promotion and disease prevention, with demonstrated benefits across poultry, swine, and ruminant systems, it promotes the translational applications to realize its full economic and ecological potential.

Author Contributions: Investigation, J.Z. and J.L.; writing-original draft preparation, J.Z. and J.L., X.C., T.L. and C.D.; methodology, C.D. and Z.H.; writing—review and editing, C.D., Z.W., Z.H. and J.S.; funding acquisition, C.D., Z.H. and J.S. All authors have read and agreed to the published version of the manuscript.

Funding: This study is supported by the National Key Research and Development Plan of China (No. 2022YFD1801105), the Beijing Natural Science Foundation (6252017), Beijing Natural Science Foundation (No. L252139), and the National Natural Science Foundation of China (32102724).

Data Availability Statement: All data in this article were obtained based on the author's request to the corresponding authors.

Conflicts of Interest: The authors declare no conflicts of interest.

References

1. Bany Bakar, R.; Reimann, F.; Gribble, F.M. The intestine as an endocrine organ and the role of gut hormones in metabolic regulation. *Nat. Rev. Gastroenterol. Hepatol.* **2023**, *20*, 784–796. [CrossRef] [PubMed]
2. Mörbe, U.M.; Jørgensen, P.B.; Fenton, T.M.; von Burg, N.; Riis, L.B.; Spencer, J.; Agace, W.W. Human gut-associated lymphoid tissues (GALT); diversity, structure, and function. *Mucosal Immunol.* **2021**, *14*, 793–802. [CrossRef] [PubMed]
3. Song, C.; Chai, Z.; Chen, S.; Zhang, H.; Zhang, X.; Zhou, Y. Intestinal mucus components and secretion mechanisms: What we do and do not know. *Exp. Mol. Med.* **2023**, *55*, 681–691. [CrossRef]
4. Zhang, L.; Wang, J.; Piao, X. Potential Effects of 25-Hydroxycholecalciferol on the Growth Performance, Blood Antioxidant Capacity, Intestinal Barrier Function and Microbiota in Broilers under Lipopolysaccharide Challenge. *Antioxidants* **2022**, *11*, 2094. [CrossRef] [PubMed]
5. Chelakkot, C.; Ghim, J.; Ryu, S.H. Mechanisms regulating intestinal barrier integrity and its pathological implications. *Exp. Mol. Med.* **2018**, *50*, 1–9. [CrossRef]
6. Tang, X.; Xiong, K.; Fang, R.; Li, M. Weaning stress and intestinal health of piglets: A review. *Front. Immunol.* **2022**, *13*, 1042778. [CrossRef]
7. Abreu, R.; Semedo-Lemsaddek, T.; Cunha, E.; Tavares, L.; Oliveira, M. Antimicrobial Drug Resistance in Poultry Production: Current Status and Innovative Strategies for Bacterial Control. *Microorganisms* **2023**, *11*, 953. [CrossRef]
8. O'Connell, L.M.; Coffey, A.; O'Mahony, J.M. Alternatives to antibiotics in veterinary medicine: Considerations for the management of Johne's disease. *Anim. Health Res. Rev.* **2023**, *24*, 12–27. [CrossRef]
9. Zhi, W.; Liu, Y.; Wang, X.; Zhang, H. Recent advances of traditional Chinese medicine for the prevention and treatment of atherosclerosis. *J. Ethnopharmacol.* **2023**, *301*, 115749. [CrossRef]
10. Chen, T.; Lin, Y.; Cao, Z.; Xue, Y.; Wang, W.; Wang, X. Network pharmacology analysis and experimental study strategy reveals the potential mechanism of puerarin against rotavirus. *Ann. Transl. Med.* **2022**, *10*, 14. [CrossRef]
11. Yang, M.; Xia, L.; Song, J.; Hu, H.; Zang, N.; Yang, J.; Zou, Y.; Wang, L.; Zheng, X.; He, Q.; et al. Puerarin ameliorates metabolic dysfunction-associated fatty liver disease by inhibiting ferroptosis and inflammation. *Lipids Health Dis.* **2023**, *22*, 202. [CrossRef] [PubMed]

12. Zhang, Y.; Yang, X.; Ge, X.; Zhang, F. Puerarin attenuates neurological deficits via Bcl-2/Bax/cleaved caspase-3 and Sirt3/SOD2 apoptotic pathways in subarachnoid hemorrhage mice. *Biomed. Pharmacother.* **2019**, *109*, 726–733. [CrossRef]
13. Li, H.T.; Zhong, K.; Xia, Y.F.; Song, J.; Chen, X.Q.; Zhao, W.; Zeng, X.H.; Chen, T.X. Puerarin improves busulfan-induced disruption of spermatogenesis by inhibiting MAPK pathways. *Biomed. Pharmacother.* **2023**, *165*, 115231. [CrossRef]
14. Fang, X.; Lan, X.; Zhu, M.; He, M.; Sun, M.; Cao, Y.; Zhu, D.; Guo, D.; Luo, H. Puerarin Induces Macrophage M2 Polarization to Exert Antinonalcoholic Steatohepatitis Pharmacological Activity via the Activation of Autophagy. *J. Agric. Food Chem.* **2024**, *72*, 7187–7202. [CrossRef] [PubMed]
15. Jeon, Y.D.; Lee, J.H.; Lee, Y.M.; Kim, D.K. Puerarin inhibits inflammation and oxidative stress in dextran sulfate sodium-induced colitis mice model. *Biomed. Pharmacother.* **2020**, *124*, 109847. [CrossRef]
16. Sun, S.; Gong, D.; Liu, R.; Wang, R.; Chen, D.; Yuan, T.; Wang, S.; Xing, C.; Lv, Y.; Du, G.; et al. Puerarin Inhibits NLRP3-Caspase-1-GSDMD-Mediated Pyroptosis via P2X7 Receptor in Cardiomyocytes and Macrophages. *Int. J. Mol. Sci.* **2023**, *24*, 13169. [CrossRef]
17. Lin, S.P.; Zhu, L.; Shi, H.; Ye, S.; Li, Q.; Yin, X.; Xie, Q.; Xu, Q.; Wei, J.X.; Mei, F.; et al. Puerarin prevents sepsis-associated encephalopathy by regulating the AKT1 pathway in microglia. *Phytomedicine* **2023**, *121*, 155119. [CrossRef] [PubMed]
18. Hu, Z.; Chen, D.; Yan, P.; Zheng, F.; Zhu, H.; Yuan, Z.; Yang, X.; Zuo, Y.; Chen, C.; Lu, H.; et al. Puerarin suppresses macrophage M1 polarization to alleviate renal inflammatory injury through antagonizing TLR4/MyD88-mediated NF-κB p65 and JNK/FoxO1 activation. *Phytomedicine* **2024**, *132*, 155813. [CrossRef]
19. Zhou, Y.X.; Zhang, H.; Peng, C. Puerarin: A review of pharmacological effects. *Phytother. Res.* **2014**, *28*, 961–975. [CrossRef]
20. Kulczyński, B.; Gramza-Michałowska, A.; Suliburska, J.; Sidor, A. Puerarin-an isoflavone with beneficial effects on bone health. *Front. Biosci. (Landmark Ed.)* **2021**, *26*, 1653–1667. [CrossRef]
21. Wu, M.; Yi, D.; Zhang, Q.; Wu, T.; Yu, K.; Peng, M.; Wang, L.; Zhao, D.; Hou, Y.; Wu, G. Puerarin enhances intestinal function in piglets infected with porcine epidemic diarrhea virus. *Sci. Rep.* **2021**, *11*, 6552. [CrossRef] [PubMed]
22. Shaojun, Z.; Yanyan, X.; Jian, C.; Xia, Z.; Qiang, F.; Saiping, J. Effects of puerarin on lipopolysaccharide-induced myocardial dysfunction in isolated rat hearts. *Pak. J. Pharm. Sci.* **2017**, *30*, 1195–1202. [PubMed]
23. Wang, Y.S.; Li, B.Y.; Xing, Y.F.; Huang, J.C.; Chen, Z.S.; Yue, L.; Zou, Y.G.; Guo, B. Puerarin Ameliorated PCOS through Preventing Mitochondrial Dysfunction Dependent on the Maintenance of Intracellular Calcium Homeostasis. *J. Agric. Food Chem.* **2024**, *72*, 2963–2976. [CrossRef] [PubMed]
24. Tan, C.; Wang, A.; Liu, C.; Li, Y.; Shi, Y.; Zhou, M.S. Puerarin Improves Vascular Insulin Resistance and Cardiovascular Remodeling in Salt-Sensitive Hypertension. *Am. J. Chin. Med.* **2017**, *45*, 1169–1184. [CrossRef]
25. Wei, H.; Sun, M.; Wang, R.; Zeng, H.; Zhao, B.; Jin, S. Puerarin mitigated LPS-ATP or HG-primed endothelial cells damage and diabetes-associated cardiovascular disease via ROS-NLRP3 signalling. *J. Cell. Mol. Med.* **2024**, *28*, e18239. [CrossRef]
26. Li, X.; Lin, Y.; Zhou, H.; Li, Y.; Wang, A.; Wang, H.; Zhou, M.S. Puerarin protects against endothelial dysfunction and end-organ damage in Ang II-induced hypertension. *Clin. Exp. Hypertens.* **2017**, *39*, 58–64. [CrossRef]
27. Wang, D.; Bu, T.; Li, Y.; He, Y.; Yang, F.; Zou, L. Pharmacological Activity, Pharmacokinetics, and Clinical Research Progress of Puerarin. *Antioxidants* **2022**, *11*, 2121. [CrossRef]
28. Zhou, H.; Yu, Y.; Shi, X.; Zou, T.; Yang, L.; OuYang, J.; Bian, Y.; Liu, Y.; Li, G. Dietary puerarin supplementation improves immune function in the small intestines of oxidized oil-challenged broilers. *Anim. Sci. J.* **2023**, *94*, e13895. [CrossRef]
29. Wang, R.; Li, T.; Pan, Z.; Chen, H.; Xu, S.; Lu, X.; Shi, K.; Bian, B.; Wu, G. Effect of Dietary Puerarin Supplementation on Growth Performance, Immune Response, Antioxidant Capacity, and Intestinal Morphology in Domestic Pigeons (*Columba livia*). *J. Poult. Sci.* **2024**, *61*, 2024003. [CrossRef]
30. Du, X.; Zhao, D.; Pian, H.; Li, Y.; Wu, X.; Liu, F.; Yu, D. Effects of puerarin as a feed additive on the laying performance, egg quality, endocrine hormones, antioxidant capacity, and intestinal morphology of aged laying hens. *Poult. Sci.* **2024**, *103*, 103420. [CrossRef]
31. Li, M.; Yuan, D.; Liu, Y.; Jin, H.; Tan, B. Dietary Puerarin Supplementation Alleviates Oxidative Stress in the Small Intestines of Diquat-Challenged Piglets. *Animals* **2020**, *10*, 631. [CrossRef] [PubMed]
32. Yeung, D.K.; Leung, S.W.; Xu, Y.C.; Vanhoutte, P.M.; Man, R.Y. Puerarin, an isoflavonoid derived from Radix puerariae, potentiates endothelium-independent relaxation via the cyclic AMP pathway in porcine coronary artery. *Eur. J. Pharmacol.* **2006**, *552*, 105–111. [CrossRef] [PubMed]
33. Muhammad, Z.; Ramzan, R.; Zhang, R.; Zhao, D.; Khalid, N.; Deng, M.; Dong, L.; Aziz, M.; Batool, R.; Zhang, M. Enhanced Bioaccessibility of Microencapsulated Puerarin Delivered by Pickering Emulsions Stabilized with OSA-Modified Hydrolyzed Pueraria montana Starch: In Vitro Release, Storage Stability, and Physicochemical Properties. *Foods* **2022**, *11*, 3591. [CrossRef] [PubMed]
34. Liga, S.; Paul, C. Puerarin-A Promising Flavonoid: Biosynthesis, Extraction Methods, Analytical Techniques, and Biological Effects. *Int. J. Mol. Sci.* **2024**, *25*, 5222. [CrossRef]

35. Quan, D.Q.; Xu, G.X.; Wu, X.G. Studies on preparation and absolute bioavailability of a self-emulsifying system containing puerarin. *Chem. Pharm. Bull.* **2007**, *55*, 800–803. [CrossRef]
36. Kato, E.; Kawabata, J. Glucose uptake enhancing activity of puerarin and the role of C-glucoside suggested from activity of related compounds. *Bioorg. Med. Chem. Lett.* **2010**, *20*, 4333–4336. [CrossRef]
37. Zhang, L. Pharmacokinetics and drug delivery systems for puerarin, a bioactive flavone from traditional Chinese medicine. *Drug Deliv.* **2019**, *26*, 860–869. [CrossRef]
38. Anukunwithaya, T.; Poo, P.; Hunsakunachai, N.; Rodsiri, R.; Malaivijitnond, S.; Khemawoot, P. Absolute oral bioavailability and disposition kinetics of puerarin in female rats. *BMC Pharmacol. Toxicol.* **2018**, *19*, 25. [CrossRef]
39. He, Y.X.; Liu, M.N.; Wu, H.; Lan, Q.; Liu, H.; Mazhar, M.; Xue, J.Y.; Zhou, X.; Chen, H.; Li, Z. Puerarin: A hepatoprotective drug from bench to bedside. *Chin. Med.* **2024**, *19*, 139. [CrossRef]
40. Song, X.; Bai, X.; Liu, S.; Dong, L.; Deng, H.; Wang, C. A Novel Microspheres Formulation of Puerarin: Pharmacokinetics Study and In Vivo Pharmacodynamics Evaluations. *Evid. Based Complement. Altern. Med.* **2016**, *2016*, 4016963. [CrossRef]
41. Ye, Q.; Li, J.; Li, T.; Ruan, J.; Wang, H.; Wang, F.; Zhang, X. Development and evaluation of puerarin-loaded controlled release nanostructured lipid carries by central composite design. *Drug Dev. Ind. Pharm.* **2021**, *47*, 113–125. [CrossRef] [PubMed]
42. Xu, H.; Hu, M.; Liu, M.; An, S.; Guan, K.; Wang, M.; Li, L.; Zhang, J.; Li, J.; Huang, L. Nano-puerarin regulates tumor microenvironment and facilitates chemo- and immunotherapy in murine triple negative breast cancer model. *Biomaterials* **2020**, *235*, 119769. [CrossRef]
43. Cheng, G.; Hu, R.; Ye, L.; Wang, B.; Gui, Y.; Gao, S.; Li, X.; Tang, J. Preparation and In Vitro/In Vivo Evaluation of Puerarin Solid Self-Microemulsifying Drug Delivery System by Spherical Crystallization Technique. *AAPS PharmSciTech* **2016**, *17*, 1336–1346. [CrossRef]
44. Wu, H.; Lu, C.; Zhou, A.; Min, Z.; Zhang, Y. Enhanced oral bioavailability of puerarin using microemulsion vehicle. *Drug Dev. Ind. Pharm.* **2009**, *35*, 138–144. [CrossRef]
45. Tang, T.T.; Hu, X.B.; Liao, D.H.; Liu, X.Y.; Xiang, D.X. Mechanisms of microemulsion enhancing the oral bioavailability of puerarin: Comparison between oil-in-water and water-in-oil microemulsions using the single-pass intestinal perfusion method and a chylomicron flow blocking approach. *Int. J. Nanomed.* **2013**, *8*, 4415–4426. [CrossRef] [PubMed]
46. Tu, L.; Yi, Y.; Wu, W.; Hu, F.; Hu, K.; Feng, J. Effects of particle size on the pharmacokinetics of puerarin nanocrystals and microcrystals after oral administration to rat. *Int. J. Pharm.* **2013**, *458*, 135–140. [CrossRef] [PubMed]
47. Wu, J.Y.; Li, Y.J.; Yang, L.; Hu, Y.Y.; Hu, X.B.; Tang, T.T.; Wang, J.M.; Liu, X.Y.; Xiang, D.X. Borneol and A-asarone as adjuvant agents for improving blood-brain barrier permeability of puerarin and tetramethylpyrazine by activating adenosine receptors. *Drug Deliv.* **2018**, *25*, 1858–1864. [CrossRef]
48. Zhao, L.X.; Liu, A.C.; Yu, S.W.; Wang, Z.X.; Lin, X.Q.; Zhai, G.X.; Zhang, Q.Z. The permeability of puerarin loaded poly(butylcyanoacrylate) nanoparticles coated with polysorbate 80 on the blood-brain barrier and its protective effect against cerebral ischemia/reperfusion injury. *Biol. Pharm. Bull.* **2013**, *36*, 1263–1270. [CrossRef]
49. Kong, H.; Zhang, G.; Cheng, J.; Shi, R.; Zhang, M.; Cao, P.; Zhao, Y.; Qu, H.; Wang, Q. Distribution kinetics of puerarin in rat hippocampus after acute local cerebral ischemia. *J. Pharm. Biomed. Anal.* **2019**, *164*, 196–201. [CrossRef]
50. Gao, L.; Ji, X.; Song, J.; Liu, P.; Yan, F.; Gong, W.; Dang, S.; Luo, Y. Puerarin protects against ischemic brain injury in a rat model of transient focal ischemia. *Neurol. Res.* **2009**, *31*, 402–406. [CrossRef]
51. Li, Y.; Pan, W.S.; Chen, S.L.; Xu, H.X.; Yang, D.J.; Chan, A.S. Pharmacokinetic, tissue distribution, and excretion of puerarin and puerarin-phospholipid complex in rats. *Drug Dev. Ind. Pharm.* **2006**, *32*, 413–422. [CrossRef] [PubMed]
52. Ren, F.; Jing, Q.; Shen, Y.; Ma, H.; Cui, J. Quantitative determination of puerarin in dog plasma by HPLC and study on the relative bioavailability of sustained release tablets. *J. Pharm. Biomed. Anal.* **2006**, *41*, 549–553. [CrossRef] [PubMed]
53. Liu, X.; Zhi, H.; Du, F.; Ye, Z.; Wang, N.; Qin, W.; Li, J. A HPLC-UV method for the determination of puerarin in rat plasma after intravenous administration of PEGylated puerarin conjugate. *J. Chromatogr. B Anal. Technol. Biomed. Life Sci.* **2010**, *878*, 3297–3302. [CrossRef]
54. Prasain, J.K.; Jones, K.; Brissie, N.; Moore, R.; Wyss, J.M.; Barnes, S. Identification of puerarin and its metabolites in rats by liquid chromatography-tandem mass spectrometry. *J. Agric. Food Chem.* **2004**, *52*, 3708–3712. [CrossRef]
55. Luo, C.F.; Cai, B.; Hou, N.; Yuan, M.; Liu, S.M.; Ji, H.; Xiong, L.G.; Xiong, W.; Luo, J.D.; Chen, M.S. UDP-glucuronosyltransferase 1A1 is the principal enzyme responsible for puerarin metabolism in human liver microsomes. *Arch. Toxicol.* **2012**, *86*, 1681–1690. [CrossRef] [PubMed]
56. Luo, C.F.; Yuan, M.; Chen, M.S.; Liu, S.M.; Ji, H. Metabolites of puerarin identified by liquid chromatography tandem mass spectrometry: Similar metabolic profiles in liver and intestine of rats. *J. Chromatogr. B Anal. Technol. Biomed. Life Sci.* **2010**, *878*, 363–370. [CrossRef]
57. Luo, C.F.; Hou, N.; Tian, J.; Yuan, M.; Liu, S.M.; Xiong, L.G.; Luo, J.D.; Chen, M.S. Metabolic profile of puerarin in rats after intragastric administration of puerarin solid lipid nanoparticles. *Int. J. Nanomed.* **2013**, *8*, 933–940. [CrossRef]

58. Hou, N.; Cai, B.; Ou, C.W.; Zhang, Z.H.; Liu, X.W.; Yuan, M.; Zhao, G.J.; Liu, S.M.; Xiong, L.G.; Luo, J.D.; et al. Puerarin-7-O-glucuronide, a water-soluble puerarin metabolite, prevents angiotensin II-induced cardiomyocyte hypertrophy by reducing oxidative stress. *Naunyn-Schmiedeberg's Arch. Pharmacol.* **2017**, *390*, 535–545. [CrossRef]
59. Zhang, G.; Ji, J.; Sun, M.; Ji, Y.; Ji, H. Comparative Pharmacokinetic Profiles of Puerarin in Rat Plasma by UHPLC-MS/MS after Oral Administration of Pueraria lobata Extract and Pure Puerarin. *J. Anal. Methods Chem.* **2020**, *2020*, 4258156. [CrossRef]
60. Adelman, D.C.; Murray, J.; Wu, T.T.; Mäki, M.; Green, P.H.; Kelly, C.P. Measuring Change In Small Intestinal Histology In Patients With Celiac Disease. *Am. J. Gastroenterol.* **2018**, *113*, 339–347. [CrossRef]
61. Wu, Y.; Li, Y.; Ruan, Z.; Li, J.; Zhang, L.; Lu, H.; Xu, Z. Puerarin Rebuilding the Mucus Layer and Regulating Mucin-Utilizing Bacteria to Relieve Ulcerative Colitis. *J. Agric. Food Chem.* **2020**, *68*, 11402–11411. [CrossRef] [PubMed]
62. Dong, Y. The Protective Effects of Bacillus Coagulans, Puerarin, and Zinc Oxide on Intestinal Function in Piglets Infected with *Escherichia coli*. Master's Thesis, Wuhan Polytechnic University, Wuhan, China, 2020. (In Chinese)
63. Lu, Y.; Ge, S.; Zhang, H.; Lu, W.; Bao, X.; Pan, S.; Pang, Q. Metabolomic and microbiome analysis of the protective effects of Puerarin against Salmonella Enteritidis Infection in chicks. *BMC Vet. Res.* **2023**, *19*, 242. [CrossRef] [PubMed]
64. Li, Y.; Zhang, H.; Su, W.; Ying, Z.; Chen, Y.; Zhang, L.; Lu, Z.; Wang, T. Effects of dietary Bacillus amyloliquefaciens supplementation on growth performance, intestinal morphology, inflammatory response, and microbiota of intra-uterine growth retarded weanling piglets. *J. Anim. Sci. Biotechnol.* **2018**, *9*, 22. [CrossRef]
65. Phoophitphong, D.; Wangnaitham, S.; Srisuwatanasagul, S.; Tummaruk, P. The use of proliferating cell nuclear antigen (PCNA) immuno-staining technique to determine number and type of follicles in the gilt ovary. *Livest. Sci.* **2012**, *150*, 425–431. [CrossRef]
66. Hu, R.; Wu, S.; Li, B.; Tan, J.; Yan, J.; Wang, Y.; Tang, Z.; Liu, M.; Fu, C.; Zhang, H.; et al. Dietary ferulic acid and vanillic acid on inflammation, gut barrier function and growth performance in lipopolysaccharide-challenged piglets. *Anim. Nutr.* **2022**, *8*, 144–152. [CrossRef]
67. Wu, Y.; Wang, W.; Kim, I.H.; Yang, Y. Dietary hydrolyzed wheat gluten supplementation ameliorated intestinal barrier dysfunctions of broilers challenged with *Escherichia coli* O78. *Poult. Sci.* **2022**, *101*, 101615. [CrossRef]
68. Dai, C.; Sharma, G.; Liu, G.; Shen, J.; Shao, B.; Hao, Z. Therapeutic detoxification of quercetin for aflatoxin B1-related toxicity: Roles of oxidative stress, inflammation, and metabolic enzymes. *Environ. Pollut.* **2024**, *345*, 123474. [CrossRef]
69. Peng, X.; Zhang, X.; Sharma, G.; Dai, C. Thymol as a Potential Neuroprotective Agent: Mechanisms, Efficacy, and Future Prospects. *J. Agric. Food Chem.* **2024**, *72*, 6803–6814. [CrossRef]
70. Li, X.; Zhang, J.; Zhang, X.; Dong, M. Puerarin suppresses MPP(+)/MPTP-induced oxidative stress through an Nrf2-dependent mechanism. *Food Chem. Toxicol.* **2020**, *144*, 111644. [CrossRef]
71. Chen, X.; Huang, C.; Sun, H.; Hong, H.; Jin, J.; Bei, C.; Lu, Z.; Zhang, X. Puerarin suppresses inflammation and ECM degradation through Nrf2/HO-1 axis in chondrocytes and alleviates pain symptom in osteoarthritic mice. *Food Funct.* **2021**, *12*, 2075–2089. [CrossRef]
72. Li, C.; Liu, Y. Puerarin reduces cell damage from cerebral ischemia-reperfusion by inhibiting ferroptosis. *Biochem. Biophys. Res. Commun.* **2024**, *693*, 149324. [CrossRef]
73. Zhang, W.; Xu, M.; Wen, S.; Wang, L.; Zhang, K.; Zhang, C.; Zou, H.; Gu, J.; Liu, X.; Bian, J.; et al. Puerarin alleviates cadmium-induced rat neurocyte injury by alleviating Nrf2-mediated oxidative stress and inhibiting mitochondrial unfolded protein response. *Ecotoxicol. Environ. Saf.* **2022**, *247*, 114239. [CrossRef]
74. Camilleri, M.; Madsen, K.; Spiller, R.; Greenwood-Van Meerveld, B.; Verne, G.N. Intestinal barrier function in health and gastrointestinal disease. *Neurogastroenterol. Motil.* **2012**, *24*, 503–512. [CrossRef]
75. Sánchez de Medina, F.; Romero-Calvo, I.; Mascaraque, C.; Martínez-Augustin, O. Intestinal inflammation and mucosal barrier function. *Inflamm. Bowel Dis.* **2014**, *20*, 2394–2404. [CrossRef] [PubMed]
76. Turner, J.R. Intestinal mucosal barrier function in health and disease. *Nat. Rev. Immunol.* **2009**, *9*, 799–809. [CrossRef] [PubMed]
77. Xia, B.; Wu, W.; Zhang, L.; Wen, X.; Xie, J.; Zhang, H. Gut microbiota mediates the effects of inulin on enhancing sulfomucin production and mucosal barrier function in a pig model. *Food Funct.* **2021**, *12*, 10967–10982. [CrossRef]
78. Ma, X.; Xu, T.; Qian, M.; Zhang, Y.; Yang, Z.; Han, X. Faecal microbiota transplantation alleviates early-life antibiotic-induced gut microbiota dysbiosis and mucosa injuries in a neonatal piglet model. *Microbiol. Res.* **2021**, *255*, 126942. [CrossRef] [PubMed]
79. Li, J.; Zhang, L.; Li, Y.; Wu, Y.; Wu, T.; Feng, H.; Xu, Z.; Liu, Y.; Ruan, Z.; Zhou, S. Puerarin improves intestinal barrier function through enhancing goblet cells and mucus barrier. *J. Funct. Foods* **2020**, *75*, 104246. [CrossRef]
80. Liu, D.; Zhang, S.; Li, S.; Zhang, Q.; Cai, Y.; Li, P.; Li, H.; Shen, B.; Liao, Q.; Hong, Y.; et al. Indoleacrylic acid produced by Parabacteroides distasonis alleviates type 2 diabetes via activation of AhR to repair intestinal barrier. *BMC Biol.* **2023**, *21*, 90. [CrossRef]
81. Wu, Y.; Li, J.; Ding, W.; Ruan, Z.; Zhang, L. Enhanced Intestinal Barriers by Puerarin in Combination with Tryptophan. *J. Agric. Food Chem.* **2021**, *69*, 15575–15584. [CrossRef]
82. Wang, Y.; Quan, F.; Cao, Q.; Lin, Y.; Yue, C.; Bi, R.; Cui, X.; Yang, H.; Yang, Y.; Birnbaumer, L.; et al. Quercetin alleviates acute kidney injury by inhibiting ferroptosis. *J. Adv. Res.* **2021**, *28*, 231–243. [CrossRef] [PubMed]

83. Wang, L.; Wu, Y.; Zhuang, L.; Chen, X.; Min, H.; Song, S.; Liang, Q.; Li, A.D.; Gao, Q. Puerarin prevents high-fat diet-induced obesity by enriching Akkermansia muciniphila in the gut microbiota of mice. *PLoS ONE* **2019**, *14*, e0218490. [CrossRef]
84. Gao, X.; Yue, C.; Tian, R.; Yu, L.; Tian, F.; Zhao, J.; Chen, W.; Zhai, Q. The regulatory effects of specific polyphenols on Akkermansia are dependent on uridine. *Food Chem.* **2023**, *410*, 135367. [CrossRef] [PubMed]
85. Zhu, L.; Lu, X.; Liu, L.; Voglmeir, J.; Zhong, X.; Yu, Q. Akkermansia muciniphila protects intestinal mucosa from damage caused by *S. pullorum* by initiating proliferation of intestinal epithelium. *Vet. Res.* **2020**, *51*, 34. [CrossRef]
86. Koupaei, M.; Saderi, H.; Amin Marashi, S.M.; Fathizadeh, H.; Owlia, P. Evaluation of the effect of Saccharomyces cerevisiae on the expression of enterotoxin genes in *Escherichia coli* O157: H7 (EHEC) and *Escherichia coli* H10407 (ETEC). *Microb. Pathog.* **2022**, *164*, 105450. [CrossRef]
87. Wang, X.; Jin, Y.; Di, C.; Zeng, Y.; Zhou, Y.; Chen, Y.; Pan, Z.; Li, Z.; Ling, W. Supplementation of Silymarin Alone or in Combination with Salvianolic Acids B and Puerarin Regulates Gut Microbiota and Its Metabolism to Improve High-Fat Diet-Induced NAFLD in Mice. *Nutrients* **2024**, *16*, 1169. [CrossRef]
88. Paone, P.; Cani, P.D. Mucus barrier, mucins and gut microbiota: The expected slimy partners? *Gut* **2020**, *69*, 2232–2243. [CrossRef] [PubMed]
89. Johansson, M.E.; Hansson, G.C. Immunological aspects of intestinal mucus and mucins. *Nat. Rev. Immunol.* **2016**, *16*, 639–649. [CrossRef]
90. Johansson, M.E.; Phillipson, M.; Petersson, J.; Velcich, A.; Holm, L.; Hansson, G.C. The inner of the two Muc2 mucin-dependent mucus layers in colon is devoid of bacteria. *Proc. Natl. Acad. Sci. USA* **2008**, *105*, 15064–15069. [CrossRef]
91. Heazlewood, C.K.; Cook, M.C.; Eri, R.; Price, G.R.; Tauro, S.B.; Taupin, D.; Thornton, D.J.; Png, C.W.; Crockford, T.L.; Cornall, R.J.; et al. Aberrant mucin assembly in mice causes endoplasmic reticulum stress and spontaneous inflammation resembling ulcerative colitis. *PLoS Med.* **2008**, *5*, e54. [CrossRef] [PubMed]
92. Suzuki, T.; Yoshida, S.; Hara, H. Physiological concentrations of short-chain fatty acids immediately suppress colonic epithelial permeability. *Br. J. Nutr.* **2008**, *100*, 297–305. [CrossRef] [PubMed]
93. Liu, P.; Wang, Y.; Yang, G.; Zhang, Q.; Meng, L.; Xin, Y.; Jiang, X. The role of short-chain fatty acids in intestinal barrier function, inflammation, oxidative stress, and colonic carcinogenesis. *Pharmacol. Res.* **2021**, *165*, 105420. [CrossRef] [PubMed]
94. Zhao, Z.H.; Xin, F.Z.; Xue, Y.; Hu, Z.; Han, Y.; Ma, F.; Zhou, D.; Liu, X.L.; Cui, A.; Liu, Z.; et al. Indole-3-propionic acid inhibits gut dysbiosis and endotoxin leakage to attenuate steatohepatitis in rats. *Exp. Mol. Med.* **2019**, *51*, 1–14. [CrossRef]
95. Feng, Y.; Wang, Y.; Wang, P.; Huang, Y.; Wang, F. Short-Chain Fatty Acids Manifest Stimulative and Protective Effects on Intestinal Barrier Function Through the Inhibition of NLRP3 Inflammasome and Autophagy. *Cell. Physiol. Biochem.* **2018**, *49*, 190–205. [CrossRef]
96. Liu, H.; Chang, G.; Wang, W.; Ji, Z.; Cui, J.; Peng, Y. Pharmacokinetics, Prostate Distribution and Metabolic Characteristics of Four Representative Flavones after Oral Administration of the Aerial Part of Glycyrrhiza uralensis in Rats. *Molecules* **2022**, *27*, 3245. [CrossRef]
97. Luissint, A.C.; Parkos, C.A.; Nusrat, A. Inflammation and the Intestinal Barrier: Leukocyte-Epithelial Cell Interactions, Cell Junction Remodeling, and Mucosal Repair. *Gastroenterology* **2016**, *151*, 616–632. [CrossRef]
98. Li, H.; Han, R.; Yong, F.; Fan, Y.; Zhao, B.; Hu, X.; Zhang, T.; Che, D. The protective effect of Eleutheroside E against the mechanical barrier dysfunction triggered by lipopolysaccharide in IPEC-J2 cells. *Res. Vet. Sci.* **2023**, *154*, 1–7. [CrossRef] [PubMed]
99. Awad, W.A.; Hess, C.; Hess, M. Enteric Pathogens and Their Toxin-Induced Disruption of the Intestinal Barrier through Alteration of Tight Junctions in Chickens. *Toxins* **2017**, *9*, 60. [CrossRef]
100. Runzhi, C.; Peng, W.; Zheng, C.; Lan, T.; Ling, Y.; Chuqi, H.; Na, Y.; Jie, Z. The combination of Puerariae Lobatae Radix and Chuanxiong Rhizoma enhanced the absorption and pharmacokinetics of puerarin by modulating the intestinal barrier and influenced gut microbiota. *J. Funct. Foods* **2018**, *47*, 72–82.
101. Li, B.; Liu, M.; Wang, Y.; Gong, S.; Yao, W.; Li, W.; Gao, H.; Wei, M. Puerarin improves the bone micro-environment to inhibit OVX-induced osteoporosis via modulating SCFAs released by the gut microbiota and repairing intestinal mucosal integrity. *Biomed. Pharmacother.* **2020**, *132*, 110923. [CrossRef]
102. Obata, Y.; Takahashi, D.; Ebisawa, M.; Kakiguchi, K.; Yonemura, S.; Jinnohara, T.; Kanaya, T.; Fujimura, Y.; Ohmae, M.; Hase, K.; et al. Epithelial cell-intrinsic Notch signaling plays an essential role in the maintenance of gut immune homeostasis. *J. Immunol.* **2012**, *188*, 2427–2436. [CrossRef] [PubMed]
103. Fishman, J.E.; Levy, G.; Alli, V.; Zheng, X.; Mole, D.J.; Deitch, E.A. The intestinal mucus layer is a critical component of the gut barrier that is damaged during acute pancreatitis. *Shock* **2014**, *42*, 264–270. [CrossRef]
104. Fenton, T.M.; Jørgensen, P.B.; Niss, K.; Rubin, S.J.S.; Mörbe, U.M.; Riis, L.B.; Da Silva, C.; Plumb, A.; Vandamme, J.; Jakobsen, H.L.; et al. Immune Profiling of Human Gut-Associated Lymphoid Tissue Identifies a Role for Isolated Lymphoid Follicles in Priming of Region-Specific Immunity. *Immunity* **2020**, *52*, 557–570.e556. [CrossRef]
105. Jacobse, J.; Li, J.; Rings, E.; Samsom, J.N.; Goettel, J.A. Intestinal Regulatory T Cells as Specialized Tissue-Restricted Immune Cells in Intestinal Immune Homeostasis and Disease. *Front. Immunol.* **2021**, *12*, 716499. [CrossRef]

106. Ilchmann-Diounou, H.; Menard, S. Psychological Stress, Intestinal Barrier Dysfunctions, and Autoimmune Disorders: An Overview. *Front. Immunol.* **2020**, *11*, 1823. [CrossRef] [PubMed]
107. Bhattacharyya, A.; Chattopadhyay, R.; Mitra, S.; Crowe, S.E. Oxidative stress: An essential factor in the pathogenesis of gastrointestinal mucosal diseases. *Physiol. Rev.* **2014**, *94*, 329–354. [CrossRef]
108. Niec, R.E.; Rudensky, A.Y.; Fuchs, E. Inflammatory adaptation in barrier tissues. *Cell* **2021**, *184*, 3361–3375. [CrossRef]
109. Liu, C.; Chi, K.; Yang, M.; Guo, N. Staphylococcal Enterotoxin A Induces Intestinal Barrier Dysfunction and Activates NLRP3 Inflammasome via NF-κB/MAPK Signaling Pathways in Mice. *Toxins* **2022**, *14*, 29. [CrossRef] [PubMed]
110. Wu, M.; Zhang, Q.; Yi, D.; Wu, T.; Chen, H.; Guo, S.; Li, S.; Ji, C.; Wang, L.; Zhao, D.; et al. Quantitative Proteomic Analysis Reveals Antiviral and Anti-inflammatory Effects of Puerarin in Piglets Infected With Porcine Epidemic Diarrhea Virus. *Front. Immunol.* **2020**, *11*, 169. [CrossRef]
111. Liu, X.; Liu, F.; Ma, Y.; Li, H.; Ju, X.; Xu, J. Effect of Puerarin, Baicalin and Berberine Hydrochloride on the Regulation of IPEC-J2 Cells Infected with Enterotoxigenic *Escherichia coli*. *Evid. Based Complement. Altern. Med.* **2019**, *2019*, 7438593. [CrossRef]
112. Jensen, S.B.; Sheikh, M.A.; Akkouh, I.A.; Szabo, A.; O'Connell, K.S.; Lekva, T.; Engh, J.A.; Agartz, I.; Elvsåshagen, T.; Ormerod, M.; et al. Elevated Systemic Levels of Markers Reflecting Intestinal Barrier Dysfunction and Inflammasome Activation Are Correlated in Severe Mental Illness. *Schizophr. Bull.* **2023**, *49*, 635–645. [CrossRef] [PubMed]
113. Peng, Z.T.; Liu, H. Puerarin attenuates LPS-induced inflammatory injury in gastric epithelial cells by repressing NLRP3 inflammasome-mediated apoptosis. *Toxicol. Vitr.* **2022**, *81*, 105350. [CrossRef] [PubMed]
114. Yamamoto, T.; Matsunami, E.; Komori, K.; Hayashi, S.; Kadowaki, M. The isoflavone puerarin induces Foxp3(+) regulatory T cells by augmenting retinoic acid production, thereby inducing mucosal immune tolerance in a murine food allergy model. *Biochem. Biophys. Res. Commun.* **2019**, *516*, 626–631. [CrossRef]
115. Abdullah, N.; Defaye, M.; Altier, C. Neural control of gut homeostasis. *Am. J. Physiol. Gastrointest. Liver Physiol.* **2020**, *319*, G718–G732. [CrossRef]
116. Niesler, B.; Kuerten, S.; Demir, I.E.; Schäfer, K.H. Disorders of the enteric nervous system—A holistic view. *Nat. Reviews Gastroenterol. Hepatol.* **2021**, *18*, 393–410. [CrossRef]
117. Pellegrini, C.; Antonioli, L.; Calderone, V.; Colucci, R.; Fornai, M.; Blandizzi, C. Microbiota-gut-brain axis in health and disease: Is NLRP3 inflammasome at the crossroads of microbiota-gut-brain communications? *Prog. Neurobiol.* **2020**, *191*, 101806. [CrossRef] [PubMed]
118. Kunze, W.A.; Mao, Y.K.; Wang, B.; Huizinga, J.D.; Ma, X.; Forsythe, P.; Bienenstock, J. Lactobacillus reuteri enhances excitability of colonic AH neurons by inhibiting calcium-dependent potassium channel opening. *J. Cell. Mol. Med.* **2009**, *13*, 2261–2270. [CrossRef]
119. Lyu, Q.; Xue, W.; Liu, R.; Ma, Q.; Kasaragod, V.B.; Sun, S.; Li, Q.; Chen, Y.; Yuan, M.; Yang, Y.; et al. A brain-to-gut signal controls intestinal fat absorption. *Nature* **2024**, *634*, 936–943. [CrossRef]
120. Lin, Y.; Yi, S.; Yang, M.; Li, X.; Li, J.; Chen, T.; Tan, B.; Zhang, J. Study on the effect of puerarin on the regulation mechanism of neurotransmitters in intestinal smooth muscle of diarrhea predominant irritable bowel syndrome. *Int. Med. Health Her.* **2022**, *28*, 2919–2923. (In Chinese) [CrossRef]
121. Huang, C. Study on the Regulation of Intestinal Nerve Homeostasis in IBS-D Rats by Puerarin. Master's Thesis, Guangzhou University of Chinese Medicine, Guangzhou, China, 2024. (In Chinese).
122. Lin, Y.; Lu, Y.; Wang, Y.; Lv, C.; Chen, J.; Luo, Y.; Quan, H.; Yu, W.; Chen, L.; Huang, Z.; et al. The Regeneration of Intestinal Stem Cells Is Driven by miR-29-Induced Metabolic Reprogramming. *Engineering* **2024**, *42*, 39–58. [CrossRef]
123. Tao, Q.; Zhang, J.; Liang, Q.; Song, S.; Wang, S.; Yao, X.; Gao, Q.; Wang, L. Puerarin alleviates sleep disorders in aged mice related to repairing intestinal mucosal barrier. *Nat. Prod. Bioprospect.* **2023**, *13*, 29. [CrossRef]
124. Nguyen, L.T.H.; Ahn, S.-H.; Choi, M.-J.; Yang, I.-J.; Shin, H.-M. Puerarin Improves Dexamethasone-Impaired Wound Healing In Vitro and In Vivo by Enhancing Keratinocyte Proliferation and Migration. *Appl. Sci.* **2021**, *11*, 9343. [CrossRef]
125. Mehmood, T.; Pichyangkura, R.; Muanprasat, C. Chitosan Oligosaccharide Promotes Junction Barrier through Modulation of PI3K/AKT and ERK Signaling Intricate Interplay in T84 Cells. *Polymers* **2023**, *15*, 1681. [CrossRef]
126. Abd El-Hack, M.E.; El-Saadony, M.T.; Salem, H.M.; El-Tahan, A.M.; Soliman, M.M.; Youssef, G.B.A.; Taha, A.E.; Soliman, S.M.; Ahmed, A.E.; El-Kott, A.F.; et al. Alternatives to antibiotics for organic poultry production: Types, modes of action and impacts on bird's health and production. *Poult. Sci.* **2022**, *101*, 101696. [CrossRef]
127. Lu, Y.; Guo, X.; Xu, F.; Wang, F.; Wu, H.; Bai, Y.; Li, W.; Zhang, G.; Yuan, J.; Pang, Q. Protective effects of puerarin on liver tissue in Salmonella-infected chicks: A proteomic analysis. *Poult. Sci.* **2024**, *103*, 103281. [CrossRef]
128. Upadhaya, S.D.; Ahn, J.M.; Cho, J.H.; Kim, J.Y.; Kang, D.K.; Kim, S.W.; Kim, H.B.; Kim, I.H. Bacteriophage cocktail supplementation improves growth performance, gut microbiome and production traits in broiler chickens. *J. Anim. Sci. Biotechnol.* **2021**, *12*, 49. [CrossRef] [PubMed]
129. Zeng, Y.; Li, R.; Dong, Y.; Yi, D.; Wu, T.; Wang, L.; Zhao, D.; Zhang, Y.; Hou, Y. Dietary Supplementation with Puerarin Improves Intestinal Function in Piglets Challenged with *Escherichia coli* K88. *Animals* **2023**, *13*, 1908. [CrossRef] [PubMed]

130. Li, L.; Sun, X.; Zhao, D.; Dai, H. Pharmacological Applications and Action Mechanisms of Phytochemicals as Alternatives to Antibiotics in Pig Production. *Front. Immunol.* **2021**, *12*, 798553. [CrossRef]
131. Fróes, R.; Bezerra, L.; Missasse, J.; Castro, D.; Barbosa, A.; Arce-Cordero, J.; Silva, T.; Portela, R.; Cunha, T.; Oliveira, R. Effects of yeast and exogenous fibrolytic enzyme additives on lamb performance and feed efficiency. *Trop. Anim. Health Prod.* **2024**, *56*, 235. [CrossRef]
132. Nie, C.; Shang, X.; Yang, Z.; Chen, H.; Qu, M.; You, J.; Song, X. Effects of Puerarian at Different Levels on Rumen Fermentation Indices and Nutrient Degradability of Beef Cattle in Vitro. *Acta Agric. Univ. Jiangxiensis* **2021**, *43*, 1381–1387. [CrossRef]
133. Lyu, C.; Yuan, B.; Meng, Y.; Cong, S.; Che, H.; Ji, X.; Wang, H.; Chen, C.; Li, X.; Jiang, H.; et al. Puerarin Alleviates H(2)O(2)-Induced Oxidative Stress and Blood-Milk Barrier Impairment in Dairy Cows. *Int. J. Mol. Sci.* **2023**, *24*, 7742. [CrossRef] [PubMed]
134. Dai, C.; Chen, X.; Li, J.; Comish, P.; Kang, R.; Tang, D. Transcription factors in ferroptotic cell death. *Cancer Gene Ther.* **2020**, *27*, 645–656. [CrossRef] [PubMed]
135. Thacharodi, A.; Hassan, S.; Ahmed, Z.H.T.; Singh, P.; Maqbool, M.; Meenatchi, R.; Pugazhendhi, A.; Sharma, A. The ruminant gut microbiome vs enteric methane emission: The essential microbes may help to mitigate the global methane crisis. *Environ. Res.* **2024**, *261*, 119661. [CrossRef] [PubMed]

biomolecules

MDPI

Review

Role of Nutritional Elements in Skin Homeostasis: A Review

Nansong Jiang [1,2,3], Tao Quan [1], Ran Li [1], Yaoxing Chen [1] and Ting Gao [1,*]

1 College of Veterinary Medicine, China Agricultural University, Beijing 100083, China; jiangnansong@faas.cn (N.J.); s20243051072@cau.edu.cn (T.Q.); 1214120476@ybu.edu.cn (R.L.); yxchen@cau.edu.cn (Y.C.)
2 Research Center for Poultry Diseases of Institute of Animal Husbandry and Veterinary Medicine, Fujian Academy of Agricultural Sciences, Fuzhou 350003, China
3 Fujian Key Laboratory for Prevention and Control of Avian Diseases, Fuzhou 350003, China
* Correspondence: gt1994@cau.edu.cn

Abstract: Skin aging is the most prominent phenotype of human aging and is the result from the interaction of genetic and environmental factors. Improving skin aging is essential for maintaining the normal physiological function of the skin and the mental health of the human body. Existing studies have proposed many theories of aging, such as the free radical theory, mitochondrial theory of aging, etc., and accordingly have also proposed appropriate interventions. It is worth noting that many nutrients contained in the skin itself play an important role in maintaining skin health. During the aging process, a decrease in the levels of these nutrients is closely related to a weakening of the related functions. Therefore, supplementing these nutrients offers potential therapeutic options for improving skin aging. In this review, we summarize the important roles of nutrients in the skin and their potential in anti-aging, with the expectation of providing a therapeutic basis and clinical reference for delaying skin aging.

Keywords: skin; aging; nutrient; antioxidant

1. Introduction

Human skin is one of the largest organs of the body and plays many critical roles in maintaining body homeostasis, protecting the organism from external damage, maintaining body temperature and water balance, and transmitting sensory information, among other functions [1]. In order to be able to effectively fulfill these complex and important functions, the skin has developed the ability to provide a high degree of plasticity in response to changing environmental conditions while maintaining its structural integrity. It plays an important role in water and electrolyte homeostasis (epidermal barrier and sweat glands), thermoregulation (in vitro receptors), and immune response (skin-associated lymphoid tissue (SALT)). It is also an important sensory organ (with free nerve endings, Pacini vesicles, Meissner vesicles, Ruffini vesicles, Krause telomeres, and Merkel cells). It is also involved in metabolism and homeostasis in the body and is responsible for the removal, selective absorption, and storage of substances [2]. In addition, the skin is the body's bearer of aesthetic functions, and its appearance, health, and well-being are inextricably linked to the body's mental health [3].

The skin functions largely rely on the composition of nutrients in the host skin ecological environment. Recently, however, growing attention has been given to the effect of nutrition on skin homeostasis, with selected food components emerging as an alternative way to improve photoaging phenotypes [4]. Food and nutrition are associated with skin homeostasis in a two-way relationship. While nutrient deficiency has a vital effect on skin status, skin homeostasis also influences nutritional intake. Moreover, unhealthy diets result in a range of skin disorders, influence all the skin's status from youth to aging, and induce injury to the skin, while a balanced diet is not only essential for the restoration of impaired skin but also for the skin's phenotype and homeostasis [5,6]. Thus, the goal of the review is to research the different structures of the skin and their functions, the role played by nutrients contained in the skin itself versus nutrients from outside (fruits or vegetables) in the maintenance of skin health, and several measures that are effective in improving skin aging today. The views and data were collated from a range of literature databases and resources such as ScienceDirect, PubMed, Wiley, Springer, Taylor & Francis, INFLIBNET, Scopus, Google, and Google Scholar. The keywords and corresponding Relevant Medical Subject Headings (MeSH) terms used for searching databases include aging, impaired skin, nutrition, nutritious foods, balanced diet, unhealthy diet, fruits and vegetables, antioxidants, probiotics, and phytoestrogens. The selection process for relevant articles involved various titles obtained from the electronic database, then an evaluation of the abstracts of all analyzed and selected titles based on predetermined inclusion criteria.

2. Structures of the Skin

2.1. Epidermis

The epidermis is a highly cellular structure that is free of blood vessels and provides the body with a major barrier against environmental damage [7], protecting it from pathogens, ultraviolet radiation (UVR), and environmental pollution, while regulating the release of water from the body [8]. The epidermis is a continuously renewed structure that is mainly regenerated by epidermal stem cells. Cells located in the epidermis are divided into two main types: keratin-forming cells and non-keratin-forming cells, in which stem cells from the basement membrane differentiate to form keratin-forming cells, which are the main epidermal cell type forming the superficial layer of the epidermis. These cells are enriched in waveform proteins, junctional proteins, α-endoproteins, and nestin proteins, in addition to the family of keratins, which are alternatively and abundantly expressed in the different stages of keratin-forming cell differentiation. Non-keratin-forming cells consist of three main types of cells: melanocytes, Langerhans cells, and Merkel cells. Furthermore, the epidermis itself can be divided into five layers. From the outside to the inside, they are the stratum corneum, stratum pellucidum, stratum granulosum, stratum spinosum, and stratum basale.

2.2. Dermis

The dermis is the deeper area located below the epidermis and is the main component of the skin, providing the scaffolding and structure for the skin. The dermis is a dense and irregular layer of connective tissue, 2–3 mm thick, accounting for the majority of the skin's thickness. It consists primarily of the fibroblast-produced extracellular matrix (ECM) that supports the skin's vascular system, sensory nerve cells, and organelles such as sweat glands and hair follicles. Although these glands and organelles contain multiple cell populations, the dermal ECM itself consists primarily of sparsely distributed fibroblasts. These cells synthesize a complex structural matrix composed of collagen, elastin fibers, negatively charged hydrophilic proteoglycans, and hyaluronic acid. In addition to these major components, there are many less abundant ECM components, including soluble proteases, protease inhibitors, and cytokines. The dermis contains two main parts: the papillary layer and the reticular layer. The papillary dermis is the layer closest to the epidermis, directly beneath it, and has dermal papillae, while the reticular layer is located deep within the dermis and contains a large number of collagen fibers, elastic fibers, and glycosaminoglycans. Among them, hyaluronic acid has the ability to bind water and thus maintain skin moisture [9]. Fibroblasts are the most abundant cells in the dermis, and an important feature of these cells is their ability to synthesize and remodel the ECM, with remodeling supported by the synthesis of cleaved metalloproteinases and their inhibitors.

2.3. Epidermal-Dermal Junction

The epidermal-dermal junction is a basement membrane synthesized by basal keratinocytes and dermal fibroblasts and consists of a highly specialized extracellular matrix (composed of proteins such as laminin and type IV collagen) that mechanically anchors the two tissues while also playing a regulatory role in development and wound healing [10,11]. The DEJ is able to act as a mechanical support for dermal-epidermal adhesion and as a barrier to macromolecules, inflammatory cells, and tumor cells. In addition, it serves as an obligatory channel for nutrient delivery from the dermis to the epidermis, which can be accomplished to enable the penetration of nutrients and metabolic wastes, thus maintaining the overall homeostasis of the skin. Typically, in young, healthy human skin, the DEJ consists of embedded protrusions (papillae), similar to the villi of the small intestine, that maximize the surface area between the two layers, which improves the efficiency of nutrient delivery from the dermis to the epidermis [10]. On a mechanical level, this structure enhances adhesion between the dermis and epidermis and protects the dermis from physical trauma by diffusing external forces along the contact zone [12]. Furthermore, the absence of these papillae is a key marker of aging [13]. The structure of the epidermal-dermal junction is complex and consists of four different layers, namely the cytomembrane layer, the thylakoid layer, the dense layer, and the sub-dense layer region [10].

2.4. Subcutaneous Tissue

The subcutaneous tissue is a structure consisting of loose connective and adipose tissue located deep within the dermis [14], which connects the skin to the muscles. The subcutaneous tissue gives the skin some mobility, and the thickness of the subcutaneous tissue varies considerably depending on the individual, age, sex, nutrition, and disease; in general, it is thickest in the abdomen and buttocks and thinnest on the eyelids, dorsum of the hands, and dorsum of the feet. The subcutaneous tissue is rich in fat, which serves to cushion underlying tissues and organs from physical damage, provide insulation, and act as an energy reserve [15].

3. Nutrients Contained in the Different Structures of the Skin and Their Functions

3.1. Lipids

Skin lipids are mainly produced by sebocytes, keratin-forming cells, and the skin microbiome [16] (Table 1). Cortical cell-derived lipids are synthesized by sebaceous glands and secreted to the skin surface, with squalene, triglycerides (TG), fatty acids (FA), wax esters, cholesterol (CHOL), and cholesteryl esters as the main components. Lipid precursors are synthesized by keratin-forming cells and secreted in the form of lamellipodia, which ultimately release lipid precursors and lipid synthase into the extracellular space of the stratum corneum (SC). Lipid precursors are then catalyzed by lipid synthase to produce lipids, which mainly consist of ceramides, fatty acids, and cholesterol; additionally, the skin microbiome is an important source of skin lipids, that should not be ignored [17–19]. The skin is colonized by a wide range of microorganisms, most of which produce skin lipids, such as short-chain fatty acids (SCFA), during metabolic processes [20]. Lipids are important metabolites with many important cellular functions [21], including energy metabolism, signal transduction, enzyme activation, transmembrane transport, cell proliferation, development, differentiation, and apoptosis. In the classification system provided by

Lipid MAPS, lipids can be categorized into eight groups, including fatty acids (FA), glycerol lipids (GL), glycerophospholipids (GP), sphingolipids (SP), sterol lipids (ST), isoprenoid lipids (PR), glycolipid lipids (GL), and polyketides (PK). Importantly, lipid composition provides a direct readout of the cellular metabolic state, and changes in lipid metabolism are important factors in the pathogenesis of respective diseases [22–28], such as metabolic disorders, several cancers, neurodegenerative diseases, and skin diseases.

In addition, skin condition is significantly influenced by skin lipids, and one of the earliest and widely studied aspects of skin lipid function is the maintenance of barrier function. The human skin is one of the largest organs that protects the body from irritants that affect human health, and skin lipids are among the most important compounds that maintain this barrier function; however, the influence of skin lipids on skin condition is not limited to this. They include (1) physicochemical functions, as they act as "mantle" that establishes the skin's physicochemical barrier; (2) biochemical functions, as they signal in a complex network of signals originating at the epidermal level; and (3) microecological functions, as sebocytes and keratinocytes derive lipids that alter the composition of the microbial dermatoflora. Microbial metabolism produces lipids as signal initiators and signal transducers.

3.1.1. Ceramides

It is always known that in addition to providing structural support to cell membranes, lipids produce cellular messengers (e.g., arachidonic acid, diacylglycerol, and ceramides) [29]. Ceramides belong to the sphingolipid class of compounds, and sphingolipids are structurally based on sphingomyelin or related bases. Ceramides are poorly soluble in water, but they are usually considered amphiphilic due to the presence of hydroxyl and amide bonds, which provide hydrophilicity to them.

Ceramides are mainly found in the epidermis, where they account for approximately 50% of the total epidermal lipid weight. Together with free fatty acids and cholesterol, ceramides constitute a hydrophilic extracellular lipid matrix, which is important for permeability barrier function [30]. In addition, ceramides act as active second messengers, regulate the proliferation and differentiation of keratinocytes, enhance the production of pro-inflammatory cytokines, and modulate the immune response. Epidermal ceramides are predominantly found in the stratum corneum and exhibit great molecular specificity. Several types of ceramides have been identified in the stratum corneum [31]. Normal-phase liquid chromatography-electrospray mass spectrometry analysis revealed more than 340 ceramides in the human stratum corneum [32]. The heterogeneity of the ceramide molecular structure allows for significant differences in the functions of different classes of ceramides, suggesting that they have different roles in skin homeostasis. Ceramides, fatty acids, and cholesterol constitute the extracellular lipid matrix of the epidermal barrier in approximately equimolar ratios and are assembled in dense orthogonal lateral stacks within the SC interstitium. Of these, ceramides are central to the structure and function of the permeability barrier. In certain skin diseases, such as atopic dermatitis and psoriasis, where skin barrier function is impaired, decreased ceramide levels are commonly observed. Reduced barrier function may be due to increased activity of enzymes such as ceramidase, sphingomyelin deacetylase, and glucosylceramide deacetylase, which enhance the degradation of ceramides and thus reduces the amount of ceramide in the skin. Patients with atopic dermatitis have lower concentrations of ceramide in the stratum corneum, which may be due to the increased activity of sphingosine deacetylase, an enzyme that degrades sphingolipids and produces sphingomyelin phosphorylcholine rather than ceramide [33]. As a signaling lipid that regulates epidermal function, ceramide activates several downstream signals, including ceramide-activated serine/threonine phosphatases

(CAPS) such as protein phosphatase 1A [34], protein kinase C (PKC), histone D [35], and the kinase inhibitor RAS (KSR) [36]. Activation of these proteins induces a variety of responses, including cell proliferation, differentiation, and apoptosis in the epidermis [37,38]. Among these effects, ceramide-induced apoptosis, including apoptosis and autophagy, has been extensively studied and has attracted much attention.

3.1.2. Squalenes

Squalene is secreted only by sebaceous glands and can be used as a marker to differentiate sebaceous gland lipids from stratum corneum lipids [39]. Squalene is an unsaturated triterpene consisting of six isoprenoids linked together, and due to its specific chemical structure, it possesses strong moisturizing and antioxidant properties. Analysis of sebaceous gland lipids by thin-layer chromatography (TLC) revealed that patients with atopic dermatitis (AD) had a twofold reduction in the amount of sebaceous gland lipids (especially squalene and wax esters) compared to healthy control subjects [40,41]. Additionally, the potential role of squalene and its peroxides in preventing sunburn and UV radiation has been reported [42]. Furthermore, waxes are more antioxidant than other lipids and may improve the water resistance of surfaces [43].

3.1.3. Triglycerides

Triglycerides (TGs) are the most abundant components of sebaceous gland lipids. Li et al. showed that there is a correlation between the altered composition of TGs and the subtypes of AD [44]. A group of triglyceride levels (TG46:1, TG48:1, TG48:2, TG50:1, TG50:2, and TG50:3) was significantly lower in participants with AD and was not adjusted for age and sex. In contrast, TG46:2 and TG56:2 were significantly lower in AD participants after adjustment for age and gender [44].

3.1.4. Free Fatty Acids

The ordered lipid accumulation in the stratum corneum is induced by free fatty acids (FFAs) with predominantly saturated chains with carbon chain lengths ranging from 12 to 30, with the majority of FFAs having chain lengths of 22, 24, or 26 carbon atoms [45,46]. Together with cholesterol sulfates, free fatty acids are the only lipids with charged/ionizable head groups in the stratum corneum [47], and thus may be necessary for the formation of the SC lipid bilayer. The presence of free fatty acids causes the SC epidermis to exhibit an acidic pH of approximately 4–5.5 [15].

3.1.5. Cholesterol

Cholesterol is the most abundant lipid in the whole body and is part of the stratum corneum, plasma membrane, and intercellular lipid lamellae. Although basal cells can reabsorb cholesterol from the circulation, most of the cholesterol in the epidermis is synthesized in situ from acetate [48]. It is likely that the function of cholesterol in the epidermal barrier is to provide a degree of mobility and flexibility to the rigid and fragile membrane system.

Table 1. Functions performed by different lipids of the skin and their machine.

Items	Functionality	Lipids	Machine
Physicochemical functions	Barrier function	Squalene, wax esters, free fatty acids, cholesterol esters, cholesterol, phospholipids, triglycerides	Carbon chains form an epidermal permeability barrier [49,50]
	Protection from sunburn and UV radiation	Squalene, triglycerides	
	Waterproof	Wax ester	
	Antioxidant	Wax ester	
Biochemical function	Stimulates lipid synthesis	Unsaturated fatty acids, e.g., AA, LA	PPARγ activation [51,52]
	Promotes terminal sebocyte differentiation	Unsaturated fatty acids, e.g., AA, LA	PPARγ activation [51]
	Stimulates cell proliferation and migration	Ceramide-1-phosphate, Sphingosine-1-phosphate	Inhibition of TNFα activation and Toll-like receptor 4-induced NF-kB activation [53]
	Inhibits proliferation of keratinocytes	Unsaturated fatty acid	PPARγ activation [51]
	Apoptosis	Ceramide	Activation of TNF receptors
	Inhibition of apoptosis	Ceramide-1-phosphate; Sphingosine-1-phosphate	Inhibition of TNFα activation and Toll-like receptor 4-induced NF-kB activation [53,54]
	Inflammations	AA	Production of leukotriene B4 and prostaglandin E2
	Anti-inflammatory	Unsaturated fatty acid	PPARγ activation [51] and inhibition of Langerhans cell function; decreased expression of adhesion molecules
	Anti-aging	Unsaturated fatty acid	Inhibition of TNFα activation and Toll-like receptor 4-induced NF-kB activation [53,54]
Microecological function	Against some sensitive skin microbiomes such as Staphylococcus aureus	Short-chain fatty acid	Inhibition of histone deacetylase (HDAC) [55]
	Glucococcus antiglucosus	Sapienic acid	Membrane depolarization [56]
	Gram-positive bacteria	Palmitoleic acid	Membrane depolarization [56]
	Promotes differentiation of preadipocytes to adipocytes	Butyrate	Inhibition of histone deacetylase (HDAC) [55]

3.2. Proteins

The proteins in the skin are mainly composed of fibrous proteins. Fibrous proteins are primarily found in the dermis. The properties of fibers made up of different proteins vary and can be classified as follows:

3.2.1. Collagen Fibers

Collagen fibers are the main component of dermal fibers, accounting for about 95%. Collagen fibers are tough and tensile, giving the skin tension and toughness against external mechanical damage while storing large amounts of water. Collagen fibers located in the superficial plexiform papillary layer are thin and have different directions, while collagen fibers in the deep reticular layer become thicker and intertwine parallel to the skin to form a network. Of these, collagen type I is the most common collagen, which accounts for approximately 50–80% of the dry weight of the entire dermal matrix [57]. It reduces skin aging caused by UV light and is present throughout the dermis in large bundles, larger in diameter than the embryonic-derived collagen type III. Type III collagen accounts for 10–15%. In addition to this, other important collagen types in the dermis include collagen types IV, VI, and VII. Type IV collagen is connected to laminin-6 and is found mainly in the basement membrane, whereas type V collagen is ubiquitous in connective tissues (about 10% of the dry weight of the dermis), where it establishes a separate network structure in the reticulation of type I and type III collagen [58]. Type VII collagen forms anchoring fibers at the dermal-epidermal junction, and additionally, its amount is significantly reduced in aging skin exposed to prolonged ultraviolet light [59]. X Type VII, called the motile version, is present at hemibridged granules; it connects to cenadherin-5, and it is also the main ligand for the integrins α-6-β-4. Type V collagen accounts for approximately 5% of the dermal matrix [60].

3.2.2. Elastic Fibers

Elastin is a gelatinous protein and is the main component of elastic fibers. In addition, the elastin molecules constituting elastic fibers have the structural characteristic of curling; when subjected to external forces, the curled elastin molecules stretch and lengthen, and after removing the external forces, the elongated elastin molecules return to the curled state, as if they were springs. Therefore, elastic fibers are rich in elasticity but poor in toughness, and elastin accounts for only 2% of the total dry weight of the skin [61]. Elastic fibers are more than collagen fibers intertwined and twisted together. The direction of the papillary layer of elastic fibers is perpendicular to the epidermis so that the skin can be bounced back to its original position after being touched and pressed; the direction of the reticular layer of elastic fibers is the same as that of the collagen fibers and is parallel to the skin surface so that the collagen fibers can be restored to their original state after being pulled so that the skin has lateral elasticity and compliance, and it has a protective effect against external mechanical damage.

Elastin fibers exist in a different state compared to collagen, which exists only in a mature state throughout the dermis. Elastin is first synthesized as a water-soluble precursor substance and then cross-linked. The maturity of elastin increases through extensive cross-linking; that is, the molecular solubility decreases and then increases, with only mature elastic fibers present deep in the dermis, whereas immature fibers are still present in the superficial dermis, i.e., the papillary layer. The mature fibers consist of an irregular core of elastin as well as surrounding microprimary fibers that are rich in elastin (90%), and the most mature elastic fibers are found at the very bottom of the dermal reticular layer. They are all closely related to the elasticity of the skin.

Elastin is rich in glycine and proline, as is collagen, but they differ in that elastin contains only small amounts of hydroxyproline and is free of hydroxylysine [62]. Its amino acid composition is highly nonpolar. Prolonged exposure to ultraviolet light leads to the degradation of elastic fibers, and some irregularities can be clearly observed under the light microscope, a condition known as elastic tissue degeneration.

3.2.3. Netted Fibers

They are more naïve fibrillar collagen fibers and are found in small numbers in the dermis. Reticular fibers are usually observed as a meshwork of fine protofibrils stained black by silver impregnation, which are usually located subcutaneously and cover the surface of muscle cells, adipocytes, etc. Under the electron microscope, the reticulated fibers are observed as individual collagen fibers or as small bundles of protofibrils, which are very thin in diameter, about 30 nm, and uniformly thick or thin. In silver-impregnated specimens, individual protofibrils in the reticular fibers were densely encapsulated by coarse metal particles, possibly be due to the high content of glycoproteins surrounding the protofibrils [63].

3.2.4. Polyfilament Proteins

The physical barrier of the skin consists mainly of the SC, while the nucleated epidermis provides additional support. In particular, cell-cell junctions and associated cytoskeletal proteins provide other important components. Among these, keratins are arranged into highly ordered and condensed arrays through interactions with the matrix protein polyfilament proteins during the final stages of normal differentiation of keratin-forming cells. The role of polyfilament proteins in skin barrier homeostasis is only partially known. Polyfilament proteins play an important role in retaining moisture in the stratum corneum and are located in the granular keratin-forming cells and hypokeratinocytes of the stratum corneum [64]. Polyfilament proteins, together with keratin intermediates, aggregate during the terminal differentiation of the epidermis of malignant tumors and are an integral part of the envelope of the keratinized cells.

Polyfilament proteins aggregate keratin filaments into tight bundles, which promotes cell collapse into the flattened shape that characterizes keratinocytes in the keratinized layer [65]. Together, keratin and polyfilament proteins account for 80–90% of the protein mass of mammalian epidermis [66,67]. The importance of polyfilament proteins to epidermal barrier homeostasis has been shown in mouse models and human diseases, where this protein is aberrantly expressed in patients with common ichthyosis and atopic dermatitis [65,68,69]. Overexpression of filament polyprotein on mouse basal epidermis results in delayed barrier repair [70].

3.2.5. Keratins

Keratins are the major structural proteins synthesized in keratin-forming cells; they assemble into a reticular pattern of intermediate filaments that extend from the perinuclear ring throughout the cytoplasm and terminate at the bridging and hemibridging granules. In keratin disorders, the network of filaments around the nucleus collapses, and thus interactions between neighboring cells are altered. In humans, genetic defects in suprabasal keratins result in hyperkeratosis and mild barrier defects manifested by a fragile and easily damaged epidermis. Keratin fundamentally affects epithelial cell structure (e.g., cell polarity and cell shape) and mitotic activity. The primary function of keratin and keratin filaments is to provide scaffolding for epithelial cells and tissues to maintain mechanical stress, preserve their structural integrity, and ensure mechanical elasticity [71].

3.3. *Amino Acids*

Each hundred grams of dried human skin tissue contains 71.9% collagen and 0.9% elastin. Skin collagen consists of glycine, proline, hydroxyproline, lysine, and hydroxylysine; elastin consists of glycine, alanine, valine, proline, and hydroxyproline. The properties and functions of each amino acid are described below in turn.

3.3.1. Glycines (Gly)

Glycine is a major component of skin collagen, accounting for 1/3 of the amino acids in collagen and elastin and plays an important role in nutrition and metabolism. Glycine can be synthesized in the body and is a non-essential amino acid. However, increasing evidence suggests that the amount of glycine synthesized in vivo is insufficient to meet the metabolic needs of animals (including the synthesis of collagen, glutathione, and hemoglobin) [72]. Glycine has a facilitating effect on epidermal cell wandering and facilitates wound healing. Elastin fibers are essential extracellular matrix components of the skin that help maintain skin elasticity. During skin aging, elastin synthesis decreases, and activity is reduced, leading to skin laxity and loss of elasticity. Glycine induces the activation of elastin promoter activity and inhibits the degradation of elastin fibers by elastase in vitro [73].

3.3.2. Prolines (Pro)

Proline, along with glycine, is the main amino acid involved in collagen synthesis, accounting for approximately 15% of the amino acid composition of collagen. It can be converted to hydroxylysine and hydroxyproline in the body to aid in the formation of collagen [74], Proline is beneficial for skin health and wound healing, stimulating cell migration and contributing to new tissue development, with blood proline levels near wounds being at least 50% higher than plasma levels during the early stages of wound healing [75]. Collagen supplements containing proline can improve skin barrier function, induce hyaluronic acid synthesis, and promote fibroblast growth and migration. In addition to its anti-aging and restorative effects, other functions of proline include support of the immune system, improvement of antioxidant status, improvement of intestinal health and nutrient absorption in the body, support of the metabolism, protection of the cardiovascular system, and participation in the skin's metabolism, which is important for the overall health of the skin [76].

3.3.3. Hydroxyprolines (Hpro)

Hydroxyproline is the main component of collagen, accounting for 14% of the total amino and imino content of collagen. Unlike glycine and proline, hydroxyproline is produced from proline-containing collagen. Specifically, hydroxyproline residues are formed by post-translational hydroxylation of proline in newly synthesized collagen [77]. Hydroxyproline is a structurally and physiologically important sub-amino acid in animals, which can be converted to glycine, enhancing the production of glutathione, DNA, hemoglobin, and proteins [78]. It has been shown that hydroxyproline can upregulate collagen biosynthesis and degradation, activate the collagen turnover cycle, and stimulate collagen metabolism [79].

3.3.4. Lysines (Lys)

Lysine is one of the essential amino acids that can promote human development, enhance immune function, improve central nervous system function, act as an antiviral, promote fat oxidation, relieve anxiety, and more. At the same time, it can also promote the absorption of certain nutrients and can synergize with some nutrients to better facilitate the physiological functions of various nutrients. The recommended daily intake of lysine is 41 mg/kg body weight for adults and 89 mg/kg body weight for children. Lysine is involved in the synthesis of many proteins, such as skeletal muscle, enzymes, serum proteins, and peptide hormones. It is particularly important for the normal function of collagen and elastin. Lysine is not only found in the dermis, but some studies have found lysine in the epidermis as well [80]. Lysine plays an important role in the synthesis of fibronectin, collagen, and elastin [81].

3.3.5. Cystines

Cystine is found in keratin in nails and hair and can be synthesized artificially, Cystine is a stable dimer consisting of two cysteine molecules linked by a disulfide bond and is obtained by food and food supplements. Purified cystine is 100% usable as a food supplement and is rapidly absorbed by the body and provides cells with enough cysteine for glutathione production [82]. Cystine assists in skin formation and is important for detoxification, protecting cells from copper toxicity by reducing the body's ability to absorb copper. When metabolized, it releases sulfuric acid, which chemically interacts with other substances to increase the detoxification function of the entire metabolic system. In addition, it aids in the supply of insulin, which is necessary for the body to utilize sugar and starch. It also promotes cellular redox, enables vigorous liver function, promotes leukocyte proliferation, and prevents the development of pathogenic bacteria. It is used for hepatitis, alopecia areata, and leukopenia.

3.3.6. Mercaptoethyl Amine

Cysteine is typically the limiting amino acid synthesized from glutathione, and increasing the supply of cysteine or its precursors enhances glutathione synthesis. Moreover, the main source of cysteine at the cellular level is cystine. While glutathione is a tripeptide composed of cysteine, glycine, and glutamic acid. It is a powerful intracellular antioxidant, and its antimelanogenic activity has been demonstrated [83–85].

3.3.7. Arginines (Arg)

Arginine can be formed as an intermediate in the urea cycle in the mammalian liver, but there is no net synthesis of arginine in the liver [86]. There are no reports on the cutaneous urea cycle, and in humans, arginine is synthesized from glutamine, glutamate, and proline via the long renal axis. Arginine is a conditionally essential amino acid for children and adults [87]. It has the ability to repair damaged skin [88].

3.4. *Vitamins*

3.4.1. Vitamin C

Normal skin contains high levels of vitamin C, comparable to other body tissues and much higher than plasma concentrations, indicating active circulatory accumulation, and most of the vitamin C in the skin is located in intracellular compartments at concentrations in the millimolar range [89,90]. It is transported from blood vessels in the dermis into the cells. Several reports have shown low levels of vitamin C in aging or photodamaged skin, and it has also been reported that overexposure to oxidative stress through pollutants or UV radiation is associated with the depletion of vitamin C levels in the epidermis [91,92]. In fact, there is more vitamin C in the epidermis than in the dermis, with a 2- to 5-fold difference between the two layers [90]. Levels of vitamin C in the skin are similar to those of other water-soluble antioxidants such as glutathione [90,92]. There are indications that vitamin C is present in the epidermal stratum corneum in a concentration gradient [93]. Vitamin C concentrations are lowest on the outer surface of the epidermis and increase dramatically in the deeper layers of the stratum corneum, reflecting exocytotic cellular depletion due to prolonged exposure to the environment [93].

Vitamin C is a cofactor for proline and lysine hydroxylases, stabilizes the tertiary structure of collagen molecules, and also promotes collagen gene expression [94,95], which protects against skin oxidation, contributes to the fight against skin aging, and plays a role in signaling pathways that play a role in cell growth and differentiation. Additionally. In vitro studies have shown that vitamin C can play a role in the differentiation of keratinocytes [96]. It promotes the synthesis of barrier lipids during differentiation and increases the formation of the keratinized envelope. There is evidence that vitamin C increases the proliferation and migration of dermal fibroblasts [97,98], functions that are essential for wound healing. Vitamin C deficiency can lead to significant skin problems; for example, early signs of scurvy include fragile skin and poor wound healing [99,100]. In contrast, signs of aging in human skin can be improved by providing vitamin C, which is supported by many studies [94,101].

3.4.2. Vitamins D

Vitamin D is a fat-soluble vitamin that is not only a nutrient required for bone health but also an essential nutrient needed to regulate many physiological functions [102]. Vitamin D can be obtained from the diet or its cholesterol precursor in the skin through the effects of UV exposure. This biologically active form requires two subsequent hydroxylation reactions to obtain active hormonal activity. Vitamin D can also be synthesized in the skin, starting with the fact that the rate of vitamin D synthesis in the granular layer is regulated by the dual processes of pigmentation and keratinization of the overlying stratum corneum, which allow only a defined amount of solar UV irradiation to penetrate the outer layers of the skin and reach the areas where vitamin D is synthesized.

3.4.3. Vitamins B

Thiamine (vitamin B1), riboflavin (vitamin B2), niacin (vitamin B3), pyridoxine (vitamin B6), and biotin (vitamin B7) are water-soluble vitamins that are essential for normal cellular metabolism and red blood cell synthesis [103–105]. These vitamins are stored in small quantities but play a role in the skin that cannot be replaced by other vitamins. The properties of the different B vitamins are described separately below.

Vitamin B3

Niacin (vitamin B3) is a water-soluble B vitamin that has a variety of physiological functions. It acts as a hydrogen electron donor and acceptor and is essential for adenosine triphosphate production, macronutrient metabolism, DNA repair, and cell signaling. Niacin deficiency can lead to dermatitis, diarrhea, and dementia, and the initial stages of deficiency are usually characterized by nonspecific symptoms such as lethargy, weakness, loss of appetite, mild digestive disturbances, and mental or emotional distress [106].

Vitamin B6

Vitamin B6, also known as pyridoxine, is a multifunctional coenzyme involved in many biochemical reactions, including amino acid metabolism, carbohydrate metabolism, and lipid metabolism. It is also effective in cognitive development for neurotransmitter synthesis, immune function for interleukin-2 production, hemoglobin formation, and gene expression [107].

Vitamin B7

Vitamin B7, also known as biotin, is an essential cofactor for mammalian carboxylase enzymes, which are involved in important metabolic pathways in humans. Biotin deficiency is rare in humans, but its deficiency leads to neuromuscular dysfunction, alopecia, and dermatitis in humans [108]. Additionally, despite the fact that there is no conclusive evidence to support that biotin can be used to treat skin disorders, many physicians frequently recommend biotin for the treatment of skin disorders, with 66% of dermatologists reporting that they recommend dietary supplements to their patients to improve their skin, hair, and nails [109].

Vitamin B12

Vitamin B12, also known as cobalamin, is a water-soluble vitamin that is important in hematology and the nervous system, and it has a complex relationship with the skin. Altered levels of cobalamin can lead to disease. The biosynthesis and metabolism of cobalamin are complex, and disease may be associated with alterations in this metabolic pathway. Skin manifestations of cobalamin deficiency include hyperpigmentation, changes in hair and nails, and changes in the oral cavity, including glossitis. In addition, several skin disorders, including vitiligo, stomatitis, atopic dermatitis, and acne, have been associated with cobalamin excess or deficiency [110,111].

3.4.4. Vitamin E

Vitamin E is a group of lipophilic compounds comprising 76 meso2tocopherols (α-, β-, γ-, and δ-tocopherols) and 77 tocotrienols (α-, β-, γ-, and δ-tocotrienols) [112,113]. Vitamin E is synthesized exclusively by plants [113], and all forms are supplied to the body through food. α-Tocopherol is the most important form, as it shows an affinity for a specific protein, α-TTP, which binds and transports only this form of the vitamin, while the rest of the dietary forms are metabolized in the liver and excreted in the bile [114,115].

Vitamin E plays an important role in maintaining skin health by contributing to the skin's antioxidant defenses and protecting the epidermis and dermis from oxidative stress induced by environmental factors. Vitamin E is the major fat-soluble antioxidant in humans [116]. Due to the antioxidant properties of vitamin E and its ability to scavenge free radicals and become part of the lipid structure, it prevents lipid peroxidation and slows down skin aging [115]. α tocopherol has been shown to reduce the amount of 8-hydroxydeoxyguanosine indirectly produced by reactive oxygen species, so it reduces the amount of ROS-induced DNA damage, thus helping to slow down the development of skin cancer. Several studies have also shown that vitamin E has strong photoprotective, firming, moisturizing, and anti-aging properties and improves the elasticity, structure, and suppleness of the epidermis and dermis. It is believed that vitamin E bound to intercellular and lipid structures protects the skin from solar UVB radiation, thereby preventing redness and swelling [116,117].

3.5. Minerals

Minerals, along with vitamins, are essential trace elements that cannot be synthesized by the body and therefore must be obtained through food. A healthy diet should ensure adequate intake of macronutrients (calcium, phosphorus, potassium, sodium, and magnesium) and trace minerals (iodine, sulfur, zinc, iron, cobalt, copper, manganese, and selenium). Minerals are responsible for the functions of the skeletal, circulatory, nervous, and endocrine systems. They also have many health benefits as cofactors and coenzymes in various enzyme systems and help regulate and harmonize biochemical and physiological functions. Micronutrient deficiencies can adversely affect human development and health, including the function and appearance of the skin [118]. In the context of skin aging, we also need to focus on the roles of minerals such as selenium, zinc, copper, and silicon (Figure 1).

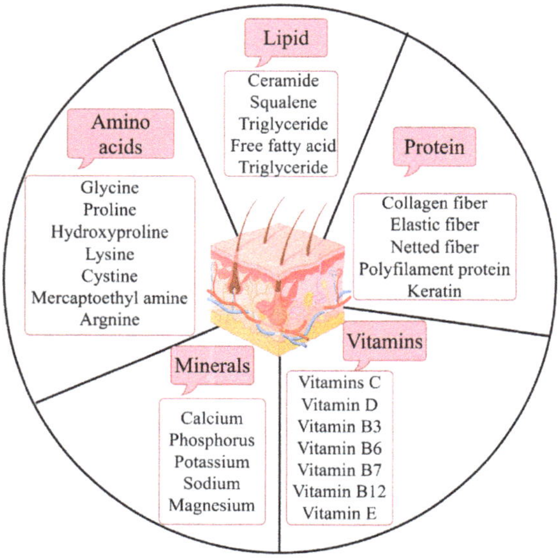

Figure 1. Nutrients contained in the different structures of the skin. The nutrients contained in the different structures and cells of the skin include lipids, proteins, amino acids, vitamins, and minerals.

4. Nutrients Contained in Fruits and Vegetables and Their Functions

As part of the daily diet, fruits and vegetables play a vital role in human health. They are rich in polyphenols, polysaccharides, dietary fiber, vitamins, minerals, natural pigments, and other active ingredients with antioxidant properties that protect the skin from free radical-induced cellular damage, maintain skin radiance, and prevent acne, pigmentation, and fine lines [119]. Consumption of fruits and vegetables also helps to reduce inflammation, brighten skin tone, and even lighten discoloration. Thus, bioactive compounds from a variety of plants are increasingly being used not only for body care, but also as ingredients that provide nutrients and influence the biological functions of the skin [120]. For example, polysaccharides are one of the most important active constituents in fruits and vegetables, with a large number of functional groups, high bioavailability, no toxic side effects, structural stability not easy to be denatured due to temperature and pH changes, as well as extraordinary bioactivities such as anti-aging, anti-inflammatory, antioxidant, antimicrobial, anticancer, digestive aid, immune modulation, anti-fatigue, lipid-lowering, and glucose-lowering [119]. The seeds of various fruits are also a source of oils, extracts, powders (flours), and a wide range of beneficial bioactive compounds [121]. Increased intake of green leafy vegetables and yellow vegetables was significantly associated with reduced wrinkles [122]. The Mediterranean diet, which is mainly characterized by a high intake of fruits, vegetables, legumes, nuts, and whole grains, is one of the most widely evaluated and described dietary patterns in the literature with proven health benefits [123]. The unique balanced composition of unsaturated fatty acids and vitamins in sea buckthorn oil has made it a popular ingredient in cosmetic and biomedical products, as it imparts a healthy and beautiful appearance to the skin and helps in caring for dry, rapidly aging skin, wounds, skin pigmentation, and infections [124].

4.1. Healthy Fats

Omega-3 and omega-6 polyunsaturated fatty acids are usually considered essential because humans cannot synthesize them on their own. Therefore, these fatty acids are obtained through various biochemical reactions in the diet [125]. Omega-3 polyunsaturated fatty acids significantly contribute to the production and circulating levels of specific anti-inflammatory and pro-resolving mediators i.e., autoreactive substances, including the D-series of antispasmodulins, protectins, and maresins derived from docosahexaenoic acid (DHA, C22:6 n-3), as well as the D-series from eicosapentaenoic acid (EPA, C20:5 n-3), an E-series antispasmodic [126]. Common plant sources of polyunsaturated fatty acids are vegetable oils such as corn oil, canola oil, flaxseed oil, safflower oil, soybean oil, and walnuts. DHA and EPA are omega-3 fatty acids found in fish and shellfish, whereas common plant sources of monounsaturated fatty acids are canola oil, safflower oil high in oleic acid, olive oil, sunflower oil, and nuts [127].

Omega-3 and omega-6 polyunsaturated fatty acids (PUFAs) are essential for maintaining of cellular function and are important components of cellular phospholipid membranes. Omega-3 polyunsaturated fatty acids and their metabolites also confer a variety of biological properties in the body, such as influencing neuronal function and reducing systemic inflammation. The commonly recommended optimal dietary ratio of omega-6 to omega-3 is 4 to 1, while higher ratios have been associated with autoimmune and cardiovascular diseases [128]. Anti-inflammatory PUFA-based supplementation and balanced nutrition are not only critical for maintaining and improving overall skin health, but also for the management of various skin disorders [129].

4.2. Carotenoids

Carotenoids are mainly tetraterpenoid compounds synthesized by plants, algae, fungi, and bacteria, with lipophilic pigments that are important for the diet, with greater variation and concentration of plant carotenoids. Fruits and vegetables rich in carotenoids have mainly antioxidant properties that help fight the incidence of certain diseases. The most common carotenoids that act in a dose- and time-dependent manner include α-carotene, β-carotene, β-cryptoxanthin, saffronin, curcumin, lycopene, lutein, and zeaxanthin [130]. α-carotene, β-carotene, and β-cryptoxanthin are known as vitamin A precursor carotenoids because they can be converted to retinol, whereas lutein, lycopene, and zeaxanthin are non-provitamin A precursor carotenoids with no vitamin A activity. However, β-carotene is the main and readily available carotenoid in the human diet and is found mainly in dark green, red, purple, yellow and orange fruits and vegetables, such as apricots, carrots, mangoes, papaya, spinach, pumpkin and sweet potatoes [131,132]. Immunomodulatory vitamin A plays an important role in the regulation, proliferation, and differentiation of a wide range of cells, including skin cells such as fibroblasts and keratin-forming cells, thereby stimulating the synthesis of collagen and elastin by fibroblasts, smoothing out deep wrinkles, and reversing photoaging [133]. Some of the roles of carotenoids include antioxidant, immune enhancement, inhibition of precancerous lesions, inhibition of mutation and transformation, quenching of non-photochemical fluorescence, reduction of the risk of certain cancers and cardiovascular events, and decreasing the risk of cataracts and macular degeneration by acting as pigments in the macula of primates [131].

5. Intervention of Different Nutrients in Skin Aging

5.1. Mechanisms of Skin Aging

Since birth, human beings are inevitably aging, and skin aging is a multifactorial process that can be categorized into endogenous aging (DNA damage, telomere shortening, mitochondrial impairment, etc.) and exogenous aging (ultraviolet rays, air pollution, smoking, etc.) according to the triggers. The former triggers aging mainly in non-sun-exposed areas of the skin, including the inner sides of the arms and buttocks, while the latter is mainly in skin that is frequently exposed to the sun, such as the face. Intrinsically aged skin appears thin on the outside and exhibits fine wrinkles, reduced elasticity, and marked dryness, often accompanied by itching. In contrast to these observations is extrinsically photoaged skin with epidermal thickening, dullness, and roughness. Capillary dilatation and pigment discoloration may also be observed in advanced and severe degrees of photoaging, which are the main skin aging-related changes found in Asian populations [134–136]. Skin aging is not only detrimental to physical health but also causes social impairment and can predispose to associated diseases, such as a significant increase in the incidence of dryness, itching, and skin irritation in older adults [137], alongside an increased risk of infections, chronic wounds (e.g., venous, pressure, or diabetic foot ulcers), as well as various types of dermatitis and malignant neoplasms (including melanoma) [138,139].

The aging process affects all skin layers and manifests itself in structural and functional changes [138]. The aged epidermis exhibits reduced barrier function and recovery after injury [140]. In addition, with age, the proliferative capacity of basal layer cells decreases, epidermal thickness decreases, and the contact surface area between the dermis and epidermis decreases, resulting in a smaller exchange surface for epidermal nutrient supply and a further diminution in the proliferative capacity of basal layer cells [141,142]. In addition to changes in the epidermis, the epidermal-dermal junction flattens, and the dermis thins. The flattening of the epidermal-dermal junction results in fewer cells, less nutrients, and reduced oxygen, leading to wrinkle formation. The dermal extracellular matrix (ECM) also exhibits structural and functional changes in intrinsically and extrinsically aged skin.

These include changes in the accumulation of labile collagen and type III collagen, changes in the ratio of type I/III collagen [143], impaired synthesis of these ECM molecules [144], and changes in elastic fiber organization [145]. One study found that as the COLI/COLIII ratio increases, protofiber width and stiffness increase, while the affinity for cell-matrix binding decreases. Mixed matrices had the best activation of fibroblasts, with high cell polarization, production of rigid fiber bundles, high α-SMA expression, and higher levels of collagen synthesis. It is evident that the proportion of COLI and COLIII in the matrix is important, and together they determine the nature of the matrix and direct different cell biological responses [146]. Decrease in the number of fibroblasts also leads to alterations and degradation of the ECM, which is manifested by thinning of the skin, increased wrinkles, and loss of elasticity [147]. Aging also leads to anatomical changes affecting the skin's microcirculation loop [112], which is characterized by a decrease in the size and number of blood vessels, associated with defective angiogenesis. The loss of dermal nutrient vascular density and exchange surface is accompanied by disorganization of the vascular network [148], which may also reveal changes in certain physiological features of aging skin, including pallor, decreased skin temperature, decreased UV-induced erythema, decreased nutrient intake, and decreased skin vascular reactivity [149]. These histologic changes are ultimately manifested as an overall thinning of the skin and diminished adipomuscular support, leading to skin laxity and increased wrinkles.

There are many mechanisms of skin aging, and researchers have proposed different models to explain the molecular basis of skin aging, including the theory of cellular senescence, the decline in cellular DNA repair and telomere loss, extranuclear mitochondrial DNA point mutations, oxidative stress, increased frequency of chromosomal anomalies, single-gene mutations, chronic inflammation, and others [13]. Some scientists have proposed that most of the effects are due to extrinsic factors and only 3% of aging is caused by intrinsic factors [102]. This paper focuses on the doctrine related to oxidative stress and telomere shortening.

It is believed that reactive oxygen species (ROS) play a key role in the dermal extracellular matrix alterations that are intrinsic to aging and photoaging. ROS can be produced by different sources, including the mitochondrial electron transport chain, peroxisomes, and endoplasmic reticulum (ER) localization proteins, as well as enzymes such as cyclooxygenases, lipoxygenases, xanthine oxidases, and nicotinamide adenine dinucleotide phosphate (NADPH) oxidases [101]. Under common conditions in the absence of ligands, cell surface receptor tyrosine kinase (RTK) activity is inhibited by the receptor protein tyrosine phosphatase RPTP, which dephosphorylates RTK. However, under UV radiation, cellular chromophores absorb energy and are excited to produce oxidation products and ROS. ROS inhibit RPTP activity by binding to cysteine at the catalytic site of RPTP [150], increasing the level of phosphorylated RTK and triggering downstream signaling pathways, including mitogen-activated protein kinase (MAPK) and subsequent nuclear factor-κB (NK-κB) and activation of transcription factor activator protein-1 (AP-1). Activated NF-κB and AP-1 inhibit collagen production and increase MMP gene transcription, leading to a decrease in collagen content in photoaged skin [151]. Notably, NF-κB has recently been found to be activated by the mammalian rapamycin complex 2/Akt/IκB kinase α pathway in aging and photoaging [152].

Telomere shortening is one of the earliest and most characteristic mechanisms that induce cellular senescence. Telomeres are located at the ends of linear chromosomes and consist of thousands of bases that contain tandemly repeated DNA sequences. Telomeres maintain the stability of chromosome structure by protecting chromosomes from degradation by nucleases and preventing end linkage between chromosomes [153]. When these telomeres remain intact, the life cycle of the cell is prolonged. However, with each cell division, the telomeres at the ends of the chromosomes are lost by about 30–200 base pairs (bp), and after multiple cell divisions the telomeres become very short, resulting in the loss of important DNA fragments during subsequent divisions, leading to loss of cellular function. The cell enters a state of value-added senescence, resulting in telomere loops or T-loops (TLoop) breaks and exposure of the 3′ end single-stranded overhangs. This exposure activates the tumor suppressor protein p53 and other DNA damage response proteins through interaction with the Werner protein, thereby inducing value-added cellular senescence or apoptosis, ultimately contributing to the onset of skin aging. It has also been noted that telomere loops can also naturally undergo disruption due to UV irradiation or other DNA damage [154].

Telomeres in skin cells may be particularly susceptible to accelerated shortening due to proliferation and the DNA-damaging agent ROS. Accelerated telomere loss after oxidative stress may be mediated by intracellular DNA repair processes, DNA damage accelerates telomere shortening in the skin, and telomeres also mediate melanin formation in senescent skin [155]. Keratin-forming cells are more susceptible to apoptosis and senescence than fibroblasts as a result of DNA damage induced by UV and ionizing radiation [4]. When keratinocytes are removed as apoptotic cells due to telomere shortening and DNA damage, fibroblasts may become senescent, affecting the epidermal growth process in the form of paracrine secretion and extracellular matrix deposition in the dermis.

Current anti-aging measures consist mainly of limiting the effects of senescence, particularly at the dermal and epidermal levels, and these approaches are diverse and involve a variety of strategies, including antioxidants, stem cell therapy, vitamin A-like acids, hormone replacement therapy, telomere modification, and dietary restrictions.

5.2. Antioxidants

Antioxidants act as reducing agents that can mitigate skin aging by neutralizing already formed ROS. ROS activates the MAPK pathway and subsequently increases MMP production, which degrades collagen. This can be prevented by antioxidants such as vitamin C and vitamin E or antioxidant enzymes such as superoxide dismutase, catalase, glutathione peroxidase, and coenzyme Q10 [156,157]. Some plants can also be used as natural sources of antioxidants, such as green tea and aloe vera [158]. It has been shown that a catechin (EGCG) in green tea prevents skin aging through the epidermal growth factor receptor (EGFR) pathway in a mouse model of aging, resulting in better skin structure than controls [159,160]. In addition, N-acetylcysteine, a form of the antioxidant glutathione, is capable of treating vascular and non-vascular neurological disorders as well as counteracting the age-related decline in tissue regeneration [161], which suggests its promising anti-aging applications in the skin.

However, it is worth noting that some researchers have suggested that antioxidant supplements have no preventive effect on chronic diseases and that excessive supplementation with β-carotene and vitamins A and E may be harmful and produce unwanted side effects [162,163]. Especially in well-nourished populations, the nearest source of antioxidants comes from our diet rather than from antioxidant supplements in pills or tablets [164], and a previous study found that EGCG induced significant death and DNA damage in normal cells of human lungs and skin through a reducing mechanism [165]. The aim of antioxidant therapy is to restore oxygen homeostasis rather than to completely eliminate all oxidants, as they have physiological functions [166]. Therefore, for the necessary clinical application of antioxidants, physicians should assess the patient's status before prescribing them; complete inactivation of all ROS is not desirable, and antioxidant therapy appears to be beneficial in aging, including skin aging, only when ROS levels are reduced to those of healthy cells.

5.3. Stem Cell Therapy

Stem cell transplantation is a promising therapy for treating skin aging, and adipose tissue grafting improves the quality of skin at the subject site in addition to increasing skin volume [167]. Further experiments have shown that adipose-derived stem cells (ADSC) contribute to skin regeneration during aging [168,169]. In recent clinical trials, autologous fat grafting rejuvenated aging skin and increased periocular and perioral skin volume in subjects with an average age of 50 years [170,171]. The data show that ADSCs produce a range of growth factors, including vascular endothelial growth factor (VEGF), basic fibroblast growth factor (bFGF), transforming growth factor (TGF)-β1, TGF-β2, hepatocyte growth factor (HGF), keratinocyte growth factor (KGF), platelet-derived growth factor AA (PDGF-AA), and placental growth factor (PGF) [172]. ADSCs may affect the surrounding skin cells through these secretions, which reveals the mechanism by which fat grafting rejuvenates the overlying skin.

5.4. Retinoids

Vitamin A analogs, including retinol, retinaldehyde, retinoic acid, and synthetic analogs (tazarotene, etc.). Vitamin A analogs are chemically similar to vitamin A, including retinoids, etc. Retinoids are the first retinoids approved for clinical use, and the topical application of retinoids inhibits AP-1, which inhibits MMP expression and prevents collagen degradation [173]. Increases in epidermal thickness and anchoring protofibers observed, and intrinsically aged skin may also benefit from topical application of vitamin A analogs [174]. Additionally, 0.02% retinoic acid cream is now FDA-approved for the treatment of photoaging. Griffiths and colleagues reported a 48-week controlled trial comparing the clinical efficacy and tolerability of 0.025% and 0.1% retinoic acid creams. Both 0.025% and 0.1% retinoic acid significantly improved photoaging compared to control but there was no significant difference between the two concentrations of retinoic acid. However, 0.025% retinoic acid had fewer side effects. Moreover, Kligman conducted a 9 month experiment in which a 0.025% retinoic acid cream was applied once a day to the skin of the inner thighs of 6 women (average age 74). The results of the experiment demonstrated a significant improvement in skin condition, so that retinoic acid can combat natural aging in addition to photoaging. For significant skin improvement, retinoic acid was used for at least 6 months. No significant changes in dermal papillae were observed in the retinoic acid-treated group (whites) at 6 months, and new collagen formation was observed after 12 months. However, the experimental sample is small, and further validation is awaited.

5.5. Hormone Replacement Therapy

Hormone replacement therapy (HRT) is a medical treatment method that is often used for hormone replacement when specific hormones are missing from a patient's body. This method involves: replacing the missing hormone with an agent containing the missing hormone that is injected into the patient's body through an IV. In addition to being used to treat symptoms caused by menopause, HRT is also used to slows down the aging process of the skin; it improves skin thickness, collagen content and elasticity, and enhances hydration. However, some studies have shown that HRT increases the risk of breast cancer [175].

Skin aging is accelerated during menopause; however, many women and healthcare professionals are insufficiently informed on the impact of menopause on skin [176]. Declining estrogen detrimentally affects the skin's ECM, which provides strength, elasticity, and resilience [177]. The relationship between dermal structural changes and altered skin function has been demonstrated by measuring the biomechanical properties of young, aged, and photoaged skin in parallel with histological analyses [178]. Estrogen levels also influence skin hydration, vascularity, and pigmentation [179], and women report a greater incidence of skin dryness and sensitivity in relation to the menstrual cycle and menopause [180,181]. However, the specific biological mechanisms underlying these observations are poorly understood, as are the respective contributions of epidermal and dermal changes to skin biomechanics. This indicates that epidermal lipid synthesis is altered postmenopause and is susceptible to regulation by HRT. In summary, increased viscoelastic skin distension post-menopause is reversible by HRT and may result from altered epidermal homeostasis and dermal morphology. Smaller R6 values are associated with improved skin condition, which is evident where decreased R6 is measured following the application of anti-aging skin care routines. Skewed CD44 expression and ceramide synthesis may contribute to declining epidermal barrier function and could represent key targets, beyond ECM rejuvenation, for improving skin function and appearance in midlife.

5.6. Telomere Modification

Cutaneous aging represents a multifaceted biological phenomenon that is influenced by an interplay of genetic predisposition, environmental factors, and lifestyle choices [182]. The length and integrity of telomeres exert a significant impact on the skin aging process, constituting one of the numerous influential factors [183], which has emerged as a prominent subject in the realm of anti-aging research and skin health. Although telomerase activation appears to be ideal for reversing skin aging, it is true that high levels of telomerase reverse transcriptase (TERT) expression in skin fibroblasts and keratinocytes lead to a significant enhancement of proliferative capacity [184] but at the same time an increased risk of epidermal carcinogenesis [185]. Therefore, further studies are needed to assess the safety of telomere lengthening.

Telomeres are specialized structures located at the ends of eukaryotic chromosomes, composed of repetitive DNA sequences that serve to safeguard against chromosomal injury and fusion [186]. Yet as cells undergo successive divisions, telomeres gradually shorten until they reach a critical length, thereby causing cell cycle arrest referred to as replicative senescence [187]. Telomere shortening is widely regarded as a priority indicator of cellular aging and closely related to the onset and progression of many age-related disorders. Recently, an increasing body of evidence has suggested a direct connection between telomere dynamics and skin aging, indicating that telomere length could be regarded as a valuable predictor for the process of skin aging and evaluation of anti-aging interventions [188,189]. The latest data advancements have suggested that telomere shortening is intricately associated not only with the conventional cellular aging process but also with biological pathways such as epigenetics, cell metabolism, and inflammatory response. For instance, it has been discovered that telomere shortening can impact gene expression patterns, thereby regulating cellular metabolism and inflammatory pathways [190], these alterations exert a vital effect on the process of skin aging. In addition, telomerase, an enzyme renowned for its ability to elongate telomeres, has garnered significant attention due to its regulation of activity and gene expression in retarding skin aging [191].

5.7. Dietary Interventions

Since it is still not technically feasible to reverse glycated proteins to their original state, the main strategy at present remains on the prevention of protein glycosylation. The problem, however, is that the diet provides not only sugars such as glucose and fructose, but also pre-formed angiogenic growth factors (AGFs), the latter of which are high in grilled, fried, and baked foods, but very low in foods prepared by water-based cooking such as poaching and steaming [192]. Thus, low-glycemic foods prepared with water reduce the intake of pre-formed exogenous AGFs and the endogenous production of physiologically glycated proteins. In the future, the search for drugs with deglycosylation ability will be a breakthrough discovery.

Some scientists believe that certain culinary herbs and spices, such as cinnamon, cloves, and oregano, inhibit fructose-induced glycosylation [193], and a number of compounds, including ginger, inositol, α-lipoic acid, carnitine, taurine, carnosine, flavonoids (e.g., green tea catechins), ammonium benzyl phosphate, α-tocopherol, niacinamide, pyridoxal, sodium selenite, riboflavin, zinc, and manganese, are also involved in inhibiting AGE formation [194,195]. More studies are needed to further validate these findings and reveal their inhibitory mechanisms.

In contrast, it is widely recognized that the nutritional status of macronutrients and micronutrients is important for skin health and appearance. Skin aging can be more effectively mitigated by supplementing nutrients that the skin synthesizes itself. Evidence for this is provided by the many vitamin-deficiency disorders that lead to serious skin diseases [196]. For example, dermatologic signs of B vitamin deficiency include blotchy red rashes, seborrheic dermatitis, and fungal skin and nail infections. Vitamin C deficiency scurvy is characterized by fragile skin, bleeding gums, and impaired wound healing [197,198].

5.8. Nutritional Interventions in Skin Aging

Nutritional status is critical for maintaining normal skin function during collagen synthesis and keratinocyte differentiation [118]. In addition, many of the body's antioxidant components, such as vitamins C and E and selenium, are obtained from the diet, and these components are important for protection against UV-induced damage [199–201].

The epidermis, on the other hand, is a challenging environment for nutrient delivery because of its lack of vasculature to deliver nutrients to cells. Epidermal nutrient delivery depends on diffusion from the vascularized dermis [202]. Whereas diffusion of nutrients into the outermost layers of the epidermis is limited, in addition, there is also little movement of extracellular fluid between cells due to the complex lipid/protein cross-linking structure that forms the skin barrier, all of which makes it difficult for nutrients to reach the cells in the outermost layers of the epidermis, resulting in these cells to receive little nutritional support.

The skin can target nutrient delivery by topical application. However, because the stratum corneum acts as a barrier that prevents the passage of many substances, it is unlikely that nutrients delivered by topical application will readily penetrate deeper into the dermis, although some uncharged, lipid-soluble molecules can pass through the surface layer. Therefore, the function of the dermis is best supported by nutrients delivered through the bloodstream [203].

Many researchers have also emphasized the relationship between a properly balanced diet and the condition of the body, including the appearance and function of the skin [204]. The intake of essential nutrients in the daily diet is extremely important for the biological processes in young and aging skin [205]. Skin is a tissue with high proliferative potential, which explains why adequate intake of proteins, carbohydrates, and fats that contribute to cell production is so important [205–207]. The overall condition of the skin—including surface texture, color, and physiological properties—is a result of factors such as hydration, i.e., the presence of sufficient water in the stratum corneum, sebum content, and surface acidity. Natural moisturizing factors (NMFs), composed mainly of amino acids, play an important role in hydration and acidity [208]. Specific fatty acids are also important for maintaining the function of the skin barrier and the integrity of the stratum corneum [209]. A growing body of research suggests that a well-balanced diet significantly influences the skin aging process. Functional anti-aging components in foods include substances involved in the synthesis and metabolism of skin components (e.g., protein peptides and essential fatty acids), as well as substances that inhibit the degradation of skin components and maintain their structural integrity (e.g., substances that regulate the expression of enzymes, such as matrix metalloproteinases (MMPs) and activated protein 1 (AP-1)) [205]. Due to their antimutagenic, antioxidant, and free radical-scavenging properties, which protect the skin from UV-induced deleterious effects, a number of dietary plants may be useful supplements for skin care [210,211]. These include carotenoids and polyphenols such as quercetin, curcumin, apigenin, proanthocyanidins, and resveratrol. Other key elements of an anti-aging diet are vitamins and minerals with antioxidant properties [212].

Substances that protect the skin from oxidative and UV damage, dehydration, and loss of elasticity include vitamins A, C, and E; selenium, zinc, copper, silica, and polyphenols; carotenoids; and essential polyunsaturated omega-3 and omega-6 fatty acids. A growing body of research suggests that a well-balanced diet significantly affects the skin aging process. These substances protect and restore the epidermal barrier, ensure proper levels of skin hydration, and protect against damage caused by external factors and inflammation. All of these nutrients have an intervening effect on skin aging (Figure 2).

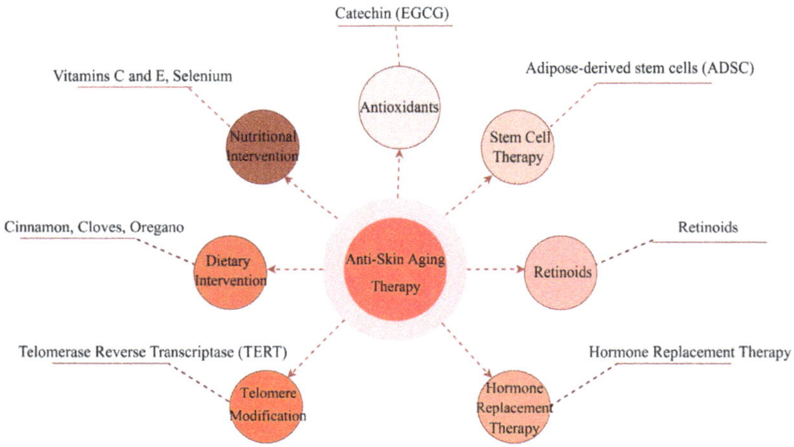

Figure 2. Several interventions against skin aging. These include antioxidant stem cell therapy, vitamin-likes treatments, hormone replacement therapy, telomere modification, dietary interventions, and nutrient interventions.

6. Pharmaceutical or Cosmetic Topical in Skin Aging

Various offerings of plant origin have been observed as anti-aging offerings regulating one or more mediators associated with skin elasticity and wrinkle production. Over the past few years, potential anti-aging isolated compounds from plants were reported to increase skin elasticity by various pharmacological actions. Some xanthones from *Garcinia mangostana*, mainly α-mangostin, decreased UVB-caused skin wrinkles and suppressed epidermal thickening in hairless mice, in addition to other pharmacological influences [213]. Phenol compounds from *Dendrobium loddigesii*, including batatasin III, up-regulated collagen synthesis [214], whereas various limonoids from *Carapa guainensis*, mainly genudin, acetoxygenudin, andirolide H, hydroxygenudin methyl angolensate, and carapanoside C, improved collagen synthesis without cytotoxicity [215]. Triterpenoids from *Eriobotrya japonica*, mainly ursolic, pomolic, colosolic, and methyl colosolic acids, exhibited anti-aging activity by stimulating collagen and hyaluronic acid synthesis [216]. A plant hormone derived from the jasmonate-caused expression of proteoglycan in the human reconstituted epidermis [217], and a by-product polysaccharide obtained from red ginseng (*Panax ginseng*) inhibited UV-induced MMP-1 through activator protein-1 (AP-1) [218]. Finally, ferulic acid improved skin elasticity with significant bleaching, anti-redness, smoothing, and moisturizing activity in a clinical study [219].

On the other hand, plant-based extracts or differential herbal formulations have also been exhaustively studied in several in vitro models, with different pharmacological actions on several mediators of skin elasticity, according to the following classification:

1. Elastase suppress activity was determined after the use of the plant extracts prepared with *Eugenia dysenterica* [219], *Gastrodia elata* [220], *Litchi sinensis* [221], *Magnolia officinalis* [222], *Malaxis acuminata* [223], *Manilkara zapota* [224], *Nephelium lappaceum* [221], *Phyllanthus emblica* [224], *Sclerocarya birrea* [225], *Sylibum marianum* [224], *Spatholobus suberectus* [226], *Tamarindus indica* [221], and a polyherbal formulation containing *Nyctanthes arbor-tristis* leaf, unripe and ripe *Aegle marmelos* fruit pulp, and the terminal meristem of *Musa paradisiaca* extracts [227];
2. An up-regulation in elastin synthesis or gene expression was reported after treatments with plant extracts of *Piper cambodianum* [228] and *Bidens pilosa* [229];
3. An up-regulation in pro-collagen expression or synthesis was reported after treatments with plant extracts prepared from *Alchemilla mollis* [230], *Azadirachta indica* [231], *Camellia sinensis* [232], *Citrus junus* [233], *Trapa japonica* [234], and a mixture of plant extracts of *Kochia scoparia* and *Rosa multiflora* [235];
4. An up-regulation in collagen synthesis or expression was reported after treatments with plant extracts of *Andrographis paniculata* [236], *Cassia fistula* [237], *Camelia sinensis* [238], *Passiflora tarminiana* [239], *Physalis peruviana* [240], *Piper cambodianum* [228], *Solanum tuberosum* [241], and *Bidens pilosa* [229];
5. The regulation of hyaluronic acid synthesis was reported after treatments with plant extracts of *Cassia fistula* [237], *Penthorium chinense* [241], and *Salvia officinalis* [242];
6. The reduction of MMP-1 level was observed after treatments with plant extracts of *Alchemilla mollis* [230], *Allium sativum* [243], *Azadirachta indica* [231], *Camellia sinensis* [232], *Gastrodia elata* [221], *Kochia scoparia* [235], *Magnolia officinalis* [223], *Passiflora tarminiana* [239], *Penthorium chinense* [242], *Rosa multiflora* [235], and *Syzygium aromaticum* [244];
7. The reduction of MMP-2 level was observed after treatments with plant extracts of *Cassia fistula* [237], *Magnolia officinalis* [223], and *Pourthiaea villosa* [245];
8. A decrease in other metalloproteinases was reported after treatments with plant extracts of *Penthorium chinense* [241], *Piper cambodianum* [228], *Pousthiaea villosa* [245] and *Syzygium aromaticum* [244];
9. The suppression of the unspecific collagenase activity appeared after treatments with plant extracts of *Cassia fistula* [237], *Curcuma heyneana* [246], *Eugenia dysenterica* [220], *Hibiscus sabdariffa* [247], *Litchi chinens* [232], *Magnolia officinalis* [223], *Malaxis acuminate* [224], *Manilkara zapota* [225], *Nephelium lappaceum* [222], *Passiflora tarminiana* [240], *Phyllanthus emblica* [226], *Piper cambodianum* [228], *Sclerocarya birrea* [226], *Sylibum marianum* [225], and *Tamarindus indica* [222].

In addition, two clinical experiments were performed to research herbal formulas on skin elasticity. The research explored the anti-aging potential in mammalian skin of topical therapies containing a mixture of *Camellia sinensis* leaf, *Polygonum cuspidatum* root, *Ginkgo biloba* leaf, *Cynara scolymus* leaf, and *Selaginella tamariscina* and reported helpful influences on facial skin as well as an improvement in ECM protein production, although the reports are not statistically obvious [248]. An herbal preparation containing *Vaccinum vitis-idaea* and *Phyllanthus emblica* was orally administered to volunteers in a double-blind, placebo-controlled clinical experiment, generating obvious helpful influences on skin elasticity [249]. The effects were associated with improvements in skin thickness and stratum corneum water content and a decrease in facial wrinkles in a dose-dependent manner [249]. Furthermore, a phytocosmetic formulation containing *Thymus vulgaris*, after its topical application in volunteers, produced a decrease in perioral and crow's feet wrinkles, accompanied by a decrease in nasolabial and smile lines, and an improvement in face oval remodeling [250].

7. Conclusions and Future Prospects

Skin aging is the direct manifestation of human aging, which is the result of the interaction of genetic and environmental factors. The improvement of skin aging is imperative and important to maintain normal physiological functions of the skin and people's psychological health. In this paper, we review the various nutrients contained in the skin and their important role in maintaining skin health. Furthermore, we also analyze the mechanisms of skin aging as well as current interventions to slow down skin aging, which seems to be a promising approach by supplementing the nutrients contained in the skin itself. However, basic and clinical research to verify whether these additives can be applied in skin anti-aging is still very limited and deserves further exploration.

Author Contributions: Conceptualization, methodology, formal analysis, writing—original draft preparation, visualization, N.J.; resources, supervision, project administration, funding acquisition, T.Q. and R.L.; conceptualization, formal analysis, validation, writing—review and editing, project administration, funding acquisition, supervision, Y.C. and T.G. All authors have read and agreed to the published version of the manuscript.

Funding: This work was supported by the Chinese National Natural Science Foundation (32402845) and Beijing Natural Science Foundation (5244041); Fujian Provincial Department of Science and Technology Project (2022R10260010); Fujian Academy of Agricultural Sciences "5511" Collaborative Innovation Project (XTCXGC2021012); Natural Science Foundation of Fujian Province (2024J01332); the Earmarked Fund for China Agriculture Research System (CARS-42); and Fujian Academy of Agricultural Sciences Outstanding Talents Program (YCZX202509).

Institutional Review Board Statement: Not applicable.

Informed Consent Statement: Not applicable.

Data Availability Statement: Data will be made available on request.

Conflicts of Interest: The authors declare that they have no known competing financial interests or personal relationships that could have appeared to influence the work reported in this paper.

Abbreviations

The following abbreviations are used in this manuscript:

SALT	Skin-associated lymphoid tissue
UVR	Ultraviolet radiation
ECM	Extracellular matrix
TG	Triglycerides
FA	Fatty acids
CHOL	Cholesterol
SC	Stratum corneum
SCFA	Short-chain fatty acids
GL	Glycerol lipids
GP	Glycerophospholipids
SP	Sphingolipids
ST	Sterol lipids

PR	Isoprenoid lipids
SL	Glycolipid lipids
PKC	Protein kinase C
KSR	Kinase inhibitor RAS
TLC	Thin-layer chromatography
AD	Atopic dermatitis
TGs	Triglycerides
FFAs	Free fatty acids
HDAC	Histone deacetylase
Gly	Glycine
Pro	Proline
Hpro	Hydroxyproline
Lys	Lysine
Arg	Arginine
ROS	Reactive oxygen species
ER	Endoplasmic reticulum
NADPH	Nicotinamide adenine dinucleotide phosphate
RTK	Receptor tyrosine kinase
MAPK	Mitogen-activated protein kinase
NK-kB	Nuclear factor-kB
AP-1	Activator protein-1
EGFR	Epidermal growth factor receptor
ADSC	Adipose-derived stem cells
VEGF	Vascular endothelial growth factor
TGF	Transforming growth factor
HGF	Hepatocyte growth factor
KGF	Keratinocyte growth factor
PDGF-AA	Platelet-derived growth factor AA
PGF	Placental growth factor
HRT	Hormone replacement therapy
TERT	Telomerase reverse transcriptase
NMFs	Natural moisturizing factors
MMPs	Matrix metalloproteinases
AP-1	Activated protein 1

References

1. Lephart, E.D. Skin aging and oxidative stress: Equol's anti-aging effects via biochemical and molecular mechanisms. *Ageing Res. Rev.* **2016**, *31*, 36–54. [CrossRef] [PubMed]
2. Ndiaye, M.A.; Nihal, M.; Wood, G.S.; Ahmad, N. Skin, reactive oxygen species, and circadian clocks. *Antioxid. Redox Signal.* **2014**, *20*, 2982–2996. [CrossRef] [PubMed]
3. Montagna, W.; Carlisle, K. Structural changes in aging human skin. *J. Investig. Dermatol.* **1979**, *73*, 47–53. [CrossRef]
4. Geng, R.; Kang, S.G.; Huang, K.; Tong, T. Boosting the photoaged skin: The potential role of dietary components. *Nutrients* **2021**, *13*, 1691. [CrossRef]
5. Greenberg, S.A. Diet and skin: A primer. *Cutis.* **2020**, *106*, E31–E32. [CrossRef]
6. Zohoori, F.V. Chapter 1: Nutrition and Diet. *Monogr. Oral Sci.* **2020**, *28*, 1–13. [PubMed]
7. D'Errico, M.; Lemma, T.; Calcagnile, A.; Proietti De Santis, L.; Dogliotti, E. Cell type and DNA damage specific response of human skin cells to environmental agents. *Mutat. Res.* **2007**, *614*, 37–47. [CrossRef]
8. Fluhr, J.W.; Feingold, K.R.; Elias, P.M. Transepidermal water loss reflects permeability barrier status: Validation in human and rodent in vivo and ex vivo models. *Exp. Dermatol.* **2006**, *15*, 483–492. [CrossRef]
9. Ni, C.; Zhang, Z.; Wang, Y.; Zhang, Z.; Guo, X.; Lv, H. Hyaluronic acid and HA-modified cationic liposomes for promoting skin penetration and retention. *J Control Release.* **2023**, *357*, 432–443. [CrossRef]
10. Waller, J.M.; Maibach, H.I. Age and skin structure and function, a quantitative approach (II): Protein, glycosaminoglycan, water, and lipid content and structure. *Ski. Res. Technol.* **2006**, *12*, 145–154. [CrossRef]
11. Xiong, X.; Wu, T.; He, S. Physical forces make rete ridges in oral mucosa. *Med. Hypotheses* **2013**, *81*, 883–886. [CrossRef]

12. Naylor, E.C.; Watson, R.E.; Sherratt, M.J. Molecular aspects of skin ageing. *Maturitas* **2011**, *69*, 249–256. [CrossRef]
13. Sharma, A.; Kuhad, A.; Bhandari, R. Novel nanotechnological approaches for treatment of skin-aging. *J. Tissue Viability* **2022**, *31*, 374–386. [CrossRef] [PubMed]
14. Baroni, A.; Buommino, E.; De Gregorio, V.; Ruocco, E.; Ruocco, V.; Wolf, R. Structure and function of the epidermis related to barrier properties. *Clin. Dermatol.* **2012**, *30*, 257–262. [CrossRef] [PubMed]
15. Cui, L.; Jia, Y.; Cheng, Z.W.; Gao, Y.; Zhang, G.L.; Li, J.Y.; He, C.F. Advancements in the maintenance of skin barrier/skin lipid composition and the involvement of metabolic enzymes. *J. Cosmet. Dermatol.* **2016**, *15*, 549–558. [CrossRef] [PubMed]
16. Grice, E.A.; Kong, H.H.; Conlan, S.; Deming, C.B.; Davis, J.; Young, A.C. NISC Comparative Sequencing Program; Bouffard GG, Blakesley RW, Murray PR, Green ED, Turner ML, Segre JA. Topographical and temporal diversity of the human skin microbiome. *Science* **2009**, *324*, 1190–1192. [CrossRef]
17. Lai, Y.; Di Nardo, A.; Nakatsuji, T.; Leichtle, A.; Yang, Y.; Cogen, A.L.; Wu, Z.R.; Hooper, L.V.; Schmidt, R.R.; von Aulock, S.; et al. Commensal bacteria regulate Toll-like receptor 3-dependent inflammation after skin injury. *Nat. Med.* **2009**, *15*, 1377–1382. [CrossRef]
18. Naik, S.; Bouladoux, N.; Wilhelm, C.; Molloy, M.J.; Salcedo, R.; Kastenmuller, W.; Deming, C.; Quinones, M.; Koo, L.; Conlan, S.; et al. Compartmentalized control of skin immunity by resident commensals. *Science* **2012**, *37*, 1115–1119. [CrossRef]
19. Achermann, Y.; Goldstein, E.J.; Coenye, T.; Shirtliff, M.E. Propionibacterium acnes: From commensal to opportunistic biofilm-associated implant pathogen. *Clin. Microbiol. Rev.* **2014**, *27*, 419–440. [CrossRef]
20. Han, X.; Gross, R.W. The foundations and development of lipidomics. *J. Lipid Res.* **2022**, *63*, 100164. [CrossRef]
21. Graessler, J.; Schwudke, D.; Schwarz, P.E.; Herzog, R.; Shevchenko, A.; Bornstein, S.R. Top-down lipidomics reveals ether lipid deficiency in blood plasma of hypertensive patients. *PLoS ONE* **2009**, *4*, e6261. [CrossRef] [PubMed]
22. Meikle, P.J.; Christopher, M.J. Lipidomics is providing new insight into the metabolic syndrome and its sequelae. *Curr. Opin. Lipidol.* **2011**, *22*, 210–215. [CrossRef] [PubMed]
23. Pietiläinen, K.H.; Róg, T.; Seppänen-Laakso, T.; Virtue, S.; Gopalacharyulu, P.; Tang, J.; Rodriguez-Cuenca, S.; Maciejewski, A.; Naukkarinen, J.; Ruskeepää, A.L.; et al. Association of lipidome remodeling in the adipocyte membrane with acquired obesity in humans. *PLoS Biol.* **2011**, *9*, e1000623. [CrossRef] [PubMed]
24. Hu, Q.; Wang, M.; Cho, M.S.; Wang, C.; Nick, A.M.; Thiagarajan, P.; Aung, F.M.; Han, X.; Sood, A.K.; Afshar-Kharghan, V. Lipid profile of platelets and platelet-derived microparticles in ovarian cancer. *BBA Clin.* **2016**, *6*, 76–81. [CrossRef]
25. Ahn, B.; Soundarapandian, M.M.; Sessions, H.; Peddibhotla, S.; Roth, G.P.; Li, J.L.; Sugarman, E.; Koo, A.; Malany, S.; Wang, M.; et al. MondoA coordinately regulates skeletal myocyte lipid homeostasis and insulin signaling. *J. Clin. Investig.* **2016**, *126*, 3567–3579. [CrossRef]
26. Wang, M.; Han, X. Advanced shotgun lipidomics for characterization of altered lipid patterns in neurodegenerative diseases and brain injury. *Methods Mol. Biol.* **2016**, *1303*, 405–422.
27. Bechmann, L.P.; Hannivoort, R.A.; Gerken, G.; Hotamisligil, G.S.; Trauner, M.; Canbay, A. The interaction of hepatic lipid and glucose metabolism in liver diseases. *J. Hepatol.* **2012**, *56*, 952–964. [CrossRef]
28. Spiegel, S.; Foster, D.; Kolesnick, R. Signal transduction through lipid second messengers. *Curr. Opin. Cell Biol.* **1996**, *8*, 159–167. [CrossRef]
29. Narangifard, A.; den Hollander, L.; Wennberg, C.L.; Lundborg, M.; Lindahl, E.; Iwai, I.; Han, H.; Masich, S.; Daneholt, B.; Norlén, L. Human skin barrier formation takes place via a cubic to lamellar lipid phase transition as analyzed by cryo-electron microscopy and EM-simulation. *Exp. Cell Res.* **2018**, *366*, 139–151. [CrossRef]
30. Meckfessel, M.H.; Brandt, S. The structure, function, and importance of ceramides in skin and their use as therapeutic agents in skin-care products. *J. Am. Acad. Dermatol.* **2014**, *71*, 177–184. [CrossRef]
31. YMasukawa, Y.; Narita, H.; Shimizu, E.; Kondo, N.; Sugai, Y.; Oba, T.; Homma, R.; Ishikawa, J.; Takagi, Y.; Kitahara, T.; et al. Characterization of overall ceramide species in human stratum corneum. *J. Lipid Res.* **2008**, *49*, 1466–1476. [CrossRef] [PubMed]
32. Imokawa, G.A. possible mechanism underlying the ceramide deficiency in atopic dermatitis: Expression of a deacylase enzyme that cleaves the N-acyl linkage of sphingomyelin and glucosylceramide. *J. Dermatol. Sci.* **2009**, *55*, 1–9. [CrossRef] [PubMed]
33. Chalfant, C.E.; Szulc, Z.; Roddy, P.; Bielawska, A.; Hannun, Y.A. The structural requirements for ceramide activation of serine-threonine protein phosphatases. *J. Lipid Res.* **2004**, *45*, 496–506. [CrossRef]
34. Lozano, J.; Berra, E.; Municio, M.M.; Diaz-Meco, M.T.; Dominguez, I.; Sanz, L.; Moscat, J. Protein kinase C zeta isoform is critical for kappa B-dependent promoter activation by sphingomyelinase. *J. Biol. Chem.* **1994**, *269*, 19200–19202. [CrossRef] [PubMed]
35. Heinrich, M.; Wickel, M.; Schneider-Brachert, W.; Sandberg, C.; Gahr, J.; Schwandner, R.; Weber, T.; Saftig, P.; Peters, C.; Brunner, J.; et al. Cathepsin D targeted by acid sphingomyelinase-derived ceramide. *EMBO J.* **1999**, *18*, 5252–5263. [CrossRef]
36. Zhang, Y.; Yao, B.; Delikat, S.; Bayoumy, S.; Lin, X.H.; Basu, S.; McGinley, M.; Chan-Hui, P.Y.; Lichenstein, H.; Kolesnick, R. Kinase suppressor of Ras is ceramide-activated protein kinase. *Cell* **1997**, *89*, 63–72. [CrossRef]
37. Lee, K.H.; Zeng, H. Aptamer-based ELISA assay for highly specific and sensitive detection of Zika NS1 protein. *Anal. Chem.* **2017**, *89*, 12743–12748. [CrossRef]

38. Smith, K.R.; Thiboutot, D.M. Thematic review series: Skin lipids. Sebaceous gland lipids: Friend or foe? *J. Lipid Res.* **2008**, *49*, 271–281. [CrossRef]
39. De Luca, C.; Valacchi, G. Surface lipids as multifunctional mediators of skin responses to environmental stimuli. *Mediat. Inflamm.* **2010**, *2010*, 321494. [CrossRef]
40. Shi, V.Y.; Leo, M.; Hassoun, L.; Chahal, D.S.; Maibach, H.I.; Sivamani, R.K. Role of sebaceous glands in inflammatory dermatoses. *J. Am. Acad. Dermatol.* **2015**, *73*, 856–863. [CrossRef]
41. Ohsawa, K.; Watanabe, T.; Matsukawa, R.; Yoshimura, Y.; Imaeda, K. The possible role of squalene and its peroxide of the sebum in the occurrence of sunburn and protection from the damage caused by U.V. irradiation. *J. Toxicol. Sci.* **1984**, *9*, 151–159. [CrossRef]
42. Wertz, P.W. Epidermal surface lipids. *Dermatoendocrinol* **2009**, *1*, 72–76.
43. Li, S.; Villarreal, M.; Stewart, S.; Choi, J.; Ganguli-Indra, G.; Babineau, D.C.; Philpot, C.; David, G.; Yoshida, T.; Boguniewicz, M.; et al. Altered composition of epidermal lipids correlates with Staphylococcus aureus colonization status in atopic dermatitis. *Br. J. Dermatol.* **2017**, *177*, e125–e127. [CrossRef] [PubMed]
44. Norlén, L.; Nicander, I.; Lundsjö, A.; Cronholm, T.; Forslind, B. A new HPLC-based method for the quantitative analysis of inner stratum corneum lipids with special reference to the free fatty acid fraction. *Arch. Dermatol. Res.* **1998**, *290*, 508–516. [CrossRef] [PubMed]
45. van Smeden, J.; Boiten, W.A.; Hankemeier, T.; Rissmann, R.; Bouwstra, J.A.; Vreeken, R.J. Combined LC/MS-platform for analysis of all major stratum corneum lipids, and the profiling of skin substitutes. *Biochim. Biophys. Acta* **2014**, *1841*, 70–79. [CrossRef]
46. Wertz, P.W. Epidermal lipids. *Semin. Dermatol.* **1992**, *11*, 106–113.
47. Karim, N.; Phinney, B.S.; Salemi, M.; Wu, P.W.; Naeem, M.; Rice, R.H. Human stratum corneum proteomics reveals cross-linking of a broad spectrum of proteins in cornified envelopes. *Exp. Dermatol.* **2019**, *28*, 618–622. [CrossRef]
48. Chandrasekaran, S.K.; Shaw, J.E. Factors influencing the percutaneous absorption of drugs. *Curr. Probl. Dermatol.* **1978**, *7*, 142–155.
49. Chandrasekaran, S.K.; Bayne, W.; Shaw, J.E. Pharmacokinetics of drug permeation through human skin. *J. Pharm. Sci.* **1978**, *67*, 1370–1374. [CrossRef]
50. Fluhr, J.W.; Man, M.Q.; Hachem, J.P.; Crumrine, D.; Mauro, T.M.; Elias, P.M.; Feingold, K.R. Topical peroxisome proliferator activated receptor activators accelerate postnatal stratum corneum acidification. *J. Investig. Dermatol.* **2009**, *129*, 365–374. [CrossRef]
51. Tyagi, S.; Gupta, P.; Saini, A.S.; Kaushal, C.; Sharma, S. The peroxisome proliferator-activated receptor: A family of nuclear receptors role in various diseases. *J. Adv. Pharm. Technol. Res.* **2011**, *2*, 236–240. [CrossRef] [PubMed]
52. Kim, T.J.; Kang, Y.J.; Lim, Y.; Lee, H.W.; Bae, K.; Lee, Y.S.; Yoo, J.M.; Yoo, H.S.; Yun, Y.P. Ceramide 1-phosphate induces neointimal formation via cell proliferation and cell cycle progression upstream of ERK1/2 in vascular smooth muscle cells. *Exp. Cell Res.* **2011**, *317*, 2041–2051. [CrossRef]
53. Gangoiti, P.; Granado, M.H.; Arana, L.; Ouro, A.; Gomez-Muñoz, A. Activation of protein kinase C-alpha is essential for stimulation of cell proliferation by ceramide 1-phosphate. *FEBS Lett.* **2010**, *584*, 517–524. [CrossRef]
54. Moissl-Eichinger, C.; Probst, A.J.; Birarda, G.; Auerbach, A.; Koskinen, K.; Wolf, P.; Holman, H.N. Human age and skin physiology shape diversity and abundance of Archaea on skin. *Sci. Rep.* **2017**, *7*, 4039. [CrossRef]
55. Fearon, D.T.; Locksley, R.M. The instructive role of innate immunity in the acquired immune response. *Science* **1996**, *27*, 50–53. [CrossRef] [PubMed]
56. Oikarinen, A. The aging of skin: Chronoaging versus photoaging. *Photodermatol. Photoimmunol. Photomed.* **1990**, *7*, 3–4. [PubMed]
57. Berardesca, E. Disorders of skin barriers: Clinical implications. *J. Eur. Acad. Dermatol. Venereol.* **2002**, *16*, 559–561. [CrossRef]
58. Craven, N.M.; Watson, R.E.; Jones, C.J.; Shuttleworth, C.A.; Kielty, C.M.; Griffiths, C.E. Clinical features of photodamaged human skin are associated with a reduction in collagen VII. *Br. J. Dermatol.* **1997**, *137*, 344–350. [CrossRef] [PubMed]
59. Glaser, D.A. *Cosmetic Dermatology: Principles and Practice: Leslie Baumann*; McGraw-Hill Companies Inc.: New York, NY, USA, 2003; Volume 49, p. 560. 226p, ISBN 0-07-136281-9.
60. Weinstein, G.D.; Boucek, R.J. Collagen and elastin of human dermis. *J. Investig. Dermatol.* **1960**, *35*, 227–229. [CrossRef]
61. Mostafavi Yazdi, S.J.; Baqersad, J. Mechanical modeling and characterization of human skin: A review. *J. Biomech.* **2022**, *130*, 110864. [CrossRef]
62. Ushiki, T. Collagen fibers, reticular fibers and elastic fibers. A comprehensive understanding from a morphological viewpoint. *Arch. Histol. Cytol.* **2002**, *65*, 109–126. [CrossRef] [PubMed]
63. Hsu, C.Y.; Henry, J.; Raymond, A.A.; Méchin, M.C.; Pendaries, V.; Nassar, D.; Hansmann, B.; Balica, S.; Burlet-Schiltz, O.; Schmitt, A.M.; et al. Deimination of human filaggrin-2 promotes its proteolysis by calpain 1. *J. Biol. Chem.* **2011**, *286*, 23222–23233. [CrossRef]
64. Palmer, C.N.; Irvine, A.D.; Terron-Kwiatkowski, A.; Zhao, Y.; Liao, H.; Lee, S.P.; Goudie, D.R.; Sandilands, A.; Campbell, L.E.; Smith, F.J.; et al. Common loss-of-function variants of the epidermal barrier protein filaggrin are a major predisposing factor for atopic dermatitis. *Nat. Genet.* **2006**, *38*, 441–446. [CrossRef] [PubMed]

65. Roop, D. Defects in the barrier. *Science* **1995**, *267*, 474–475. [CrossRef]
66. Nemes, Z.; Steinert, P.M. Bricks and mortar of the epidermal barrier. *Exp. Mol. Med.* **1999**, *31*, 5–19. [CrossRef]
67. Smith, F.J.; Irvine, A.D.; Terron-Kwiatkowski, A.; Sandilands, A.; Campbell, L.E.; Zhao, Y.; Liao, H.; Evans, A.T.; Goudie, D.R.; Lewis-Jones, S.; et al. Loss-of-function mutations in the gene encoding filaggrin cause ichthyosis vulgaris. *Nat. Genet.* **2006**, *38*, 337–342. [CrossRef]
68. Irvine, A.D.; McLean, W.H. Breaking the (un)sound barrier: Filaggrin is a major gene for atopic dermatitis. *J. Investig. Dermatol.* **2006**, *126*, 1200–1202. [CrossRef] [PubMed]
69. Presland, R.B.; Coulombe, P.A.; Eckert, R.L.; Mao-Qiang, M.; Feingold, K.R.; Elias, P.M. Barrier function in transgenic mice overexpressing K16, involucrin, and filaggrin in the suprabasal epidermis. *J. Investig. Dermatol.* **2004**, *123*, 603–606. [CrossRef] [PubMed]
70. Bragulla, H.H.; Homberger, D.G. Structure and functions of keratin proteins in simple, stratified, keratinized and cornified epithelia. *J. Anat.* **2009**, *214*, 516–559.
71. Hou, Y.; Yao, K.; Yin, Y.; Wu, G. Endogenous synthesis of amino acids limits growth, lactation, and reproduction in animals. *Adv. Nutr.* **2016**, *7*, 331–342. [CrossRef]
72. Zhao, R.; Bruning, E.; Rossetti, D.; Starcher, B.; Seiberg, M.; Iotsova-Stone, V. Extracts from Glycine max (soybean) induce elastin synthesis and inhibit elastase activity. *Exp. Dermatol.* **2009**, *18*, 883–886. [CrossRef]
73. Karna, E.; Szoka, L.; Huynh, T.Y.L.; Palka, J.A. Proline-dependent regulation of collagen metabolism. *Cell Mol. Life Sci.* **2020**, *77*, 1911–1918. [CrossRef] [PubMed]
74. Rauscher, S.; Baud, S.; Miao, M.; Keeley, F.W.; Pomès, R. Proline and glycine control protein self-organization into elastomeric or amyloid fibrils. *Structure* **2006**, *14*, 1667–1676. [CrossRef] [PubMed]
75. Saibi, W.; Feki, K.; Yacoubi, I.; Brini, F. Bridging between proline structure, functions, metabolism, and involvement in organism physiology. *Appl. Biochem. Biotechnol.* **2015**, *176*, 2107–2119. [CrossRef] [PubMed]
76. Gorres, K.L.; Raines, R.T. Prolyl 4-hydroxylase. *Crit. Rev. Biochem. Mol. Biol.* **2010**, *45*, 106–124. [CrossRef]
77. Wu, Z.; Hou, Y.; Dai, Z.; Hu, C.A.; Wu, G. Metabolism, nutrition, and redox signaling of hydroxyproline. *Antioxid. Redox Signal.* **2019**, *30*, 674–682. [CrossRef]
78. Aoki, M.; Suto, K.; Komatsu, M.; Kamimura, A.; Morishita, K.; Yamasaki, M.; Takao, T. Increasing effect of an oral intake of L-hydroxyproline on the soluble collagen content of skin and collagen fragments in rat serum. *Biosci. Biotechnol. Biochem.* **2012**, *76*, 1242–1244. [CrossRef]
79. Awuchi, C.G.; Igwe, V.S.; Amagwula, I.O.; Echeta, C.K. Health Benefits of Micronutrients (Vitamins and Minerals) and their Associated Deficiency Diseases: A Systematic Review. *Int. J. Food Sci.* **2020**, *3*, 1–32.
80. Dyer, D.G.; Dunn, J.A.; Thorpe, S.R.; Bailie, K.E.; Lyons, T.J.; McCance, D.R.; Baynes, J.W. Accumulation of Maillard reaction products in skin collagen in diabetes and aging. *J. Clin. Investig.* **1993**, *91*, 2463–2469. [CrossRef]
81. Wu, G.; Fang, Y.Z.; Yang, S.; Lupton, J.R.; Turner, N.D. Glutathione metabolism and its implications for health. *J. Nutr.* **2004**, *134*, 489–492. [CrossRef]
82. Imokawa, G. Analysis of initial melanogenesis including tyrosinase transfer and melanosome differentiation through interrupted melanization by glutathione. *J. Investig. Dermatol.* **1989**, *93*, 100–107. [PubMed]
83. del Marmol, V.; Solano, F.; Sels, A.; Huez, G.; Libert, A.; Lejeune, F.; Ghanem, G. Glutathione depletion increases tyrosinase activity in human melanoma cells. *J. Investig. Dermatol.* **1993**, *101*, 871–874. [CrossRef]
84. del Marmol, V.; Ito, S.; Bouchard, B.; Libert, A.; Wakamatsu, K.; Ghanem, G.; Solano, F. Cysteine deprivation promotes eumelanogenesis in human melanoma cells. *J. Investig. Dermatol.* **1996**, *107*, 698–702. [CrossRef] [PubMed]
85. Wu, G.; Morris Jr, S.M. Arginine metabolism: Nitric oxide and beyond. *Biochem. J.* **1998**, *336*, 1–17. [CrossRef]
86. Wu, G.; Bazer, F.W.; Davis, T.A.; Kim, S.W.; Li, P.; Marc Rhoads, J.; Carey Satterfield, M.; Smith, S.B.; Spencer, T.E.; Yin, Y. Arginine metabolism and nutrition in growth, health and disease. *Amino Acids* **2009**, *37*, 153–168. [CrossRef] [PubMed]
87. Stechmiller, J.K.; Childress, B.; Cowan, L. Arginine supplementation and wound healing. *Nutr. Clin. Pract.* **2005**, *20*, 52–61. [CrossRef]
88. Shindo, Y.; Witt, E.; Han, D.; Packer, L. Dose-response effects of acute ultraviolet irradiation on antioxidants and molecular markers of oxidation in murine epidermis and dermis. *J. Investig. Dermatol.* **1994**, *102*, 470–475. [CrossRef]
89. Rhie, G.; Shin, M.H.; Seo, J.Y.; Choi, W.W.; Cho, K.H.; Kim, K.H.; Park, K.C.; Eun, H.C.; Chung, J.H. Aging- and photoaging-dependent changes of enzymic and nonenzymic antioxidants in the epidermis and dermis of human skin in vivo. *J. Investig. Dermatol.* **2001**, *117*, 1212–1217.
90. Shindo, Y.; Witt, E.; Han, D.; Epstein, W.; Packer, L. Enzymic and non-enzymic antioxidants in epidermis and dermis of human skin. *J. Investig. Dermatol.* **1994**, *102*, 122–124. [CrossRef]
91. Shindo, Y.; Witt, E.; Packer, L. Antioxidant defense mechanisms in murine epidermis and dermis and their responses to ultraviolet light. *J. Investig. Dermatol.* **1993**, *100*, 260–265. [CrossRef]

92. Weber, S.U.; Thiele, J.J.; Cross, C.E.; Packer, L. Vitamin C, uric acid, and glutathione gradients in murine stratum corneum and their susceptibility to ozone exposure. *J. Investig. Dermatol.* **1999**, *113*, 1128–1132. [CrossRef] [PubMed]
93. Hinek, A.; Kim, H.J.; Wang, Y.; Wang, A.; Mitts, T.F. Sodium L-ascorbate enhances elastic fibers deposition by fibroblasts from normal and pathologic human skin. *J. Dermatol. Sci.* **2014**, *75*, 173–182. [CrossRef]
94. Ivanov, V.; Ivanova, S.; Kalinovsky, T.; Niedzwiecki, A.; Rath, M. Inhibition of collagen synthesis by select calcium and sodium channel blockers can be mitigated by ascorbic acid and ascorbyl palmitate. *Am. J. Cardiovasc. Dis.* **2016**, *6*, 26–35.
95. Pasonen-Seppänen, S.; Suhonen, T.M.; Kirjavainen, M.; Suihko, E.; Urtti, A.; Miettinen, M.; Hyttinen, M.; Tammi, M.; Tammi, R. Vitamin C enhances differentiation of a continuous keratinocyte cell line (REK) into epidermis with normal stratum corneum ultrastructure and functional permeability barrier. *Histochem. Cell Biol.* **2001**, *116*, 287–297. [CrossRef]
96. Takahashi, Y.; Takahashi, S.; Shiga, Y.; Yoshimi, T.; Miura, T. Hypoxic induction of prolyl 4-hydroxylase alpha (I) in cultured cells. *J. Biol. Chem.* **2000**, *275*, 14139–14146. [CrossRef]
97. Kishimoto, Y.; Saito, N.; Kurita, K.; Shimokado, K.; Maruyama, N.; Ishigami, A. Ascorbic acid enhances the expression of type 1 and type 4 collagen and SVCT2 in cultured human skin fibroblasts. *Biochem. Biophys. Res. Commun.* **2013**, *430*, 579–584. [CrossRef]
98. Ellinger, S.; Stehle, P. Efficacy of vitamin supplementation in situations with wound healing disorders: Results from clinical intervention studie. *Curr. Opin. Clin. Nutr. Metab. Care* **2009**, *12*, 588–595. [CrossRef]
99. Talarico, V.; Aloe, M.; Barreca, M.; Galati, M.C.; Raiola, G. Do you remember scurvy? *Clin. Ter.* **2014**, *165*, 253–256.
100. Rinnerthaler, M.; Bischof, J.; Streubel, M.K.; Trost, A.; Richter, K. Oxidative stress in aging human skin. *Biomolecules* **2015**, *5*, 545–589. [CrossRef]
101. Kechichian, E.; Ezzedine, K. Vitamin D and the skin: An update for dermatologists. *Am. J. Clin. Dermatol.* **2018**, *19*, 223–235. [CrossRef]
102. Powers, H.J. Riboflavin (vitamin B-2) and health. *Am. J. Clin. Nutr.* **2003**, *77*, 1352–1360. [CrossRef]
103. Sriram, K.; Manzanares, W.; Joseph, K. Thiamine in nutrition therapy. *Nutr. Clin. Pract.* **2012**, *27*, 41–50. [CrossRef]
104. Stover, P.J.; Field, M.S. Vitamin B-6. *Adv. Nutr.* **2015**, *6*, 132–133. [CrossRef]
105. Hegyi, J.; Schwartz, R.A.; Hegyi, V. Pellagra: Dermatitis, dementia, and diarrhea. *Int. J. Dermatol.* **2004**, *43*, 1–5. [CrossRef]
106. Spinneker, A.; Sola, R.; Lemmen, V.; Castillo, M.J.; Pietrzik, K.; González-Gross, M. Vitamin B6 status, deficiency and its consequences—An overview. *Nutr. Hosp.* **2007**, *22*, 7–24.
107. Daft, F.S.; Ashburn, L.L.; Sebrell, W.H. Biotin deficiency and other changes in rats given sulfanilylguanidine or succinyl sulfathiazole in purified diets. *Science* **1942**, *96*, 321–322. [CrossRef]
108. Dickinson, A.; Shao, A.; Boyon, N.; Franco, J.C. Use of dietary supplements by cardiologists, dermatologists and orthopedists: Report of a survey. *Nutr. J.* **2011**, *10*, 20. [CrossRef]
109. Aaron, S.; Kumar, S.; Vijayan, J.; Jacob, J.; Alexander, M.; Gnanamuthu, C. Clinical and laboratory features and response to treatment in patients presenting with vitamin B12 deficiency-related neurological syndromes. *Neurol. India* **2005**, *53*, 55–58; discussion 59.
110. Graells, J.; Ojeda, R.M.; Muniesa, C.; Gonzalez, J.; Saavedra, J. Glossitis with linear lesions: An early sign of vitamin B12 deficiency. *J. Am. Acad. Dermatol.* **2009**, *60*, 498–500. [CrossRef]
111. Poljšak, B.; Dahmane, R.G.; Godić, A. Intrinsic skin aging: The role of oxidative stress. *Acta Dermatovenerol. Alp. Pannonica Adriat.* **2012**, *21*, 33–36. [PubMed]
112. Khadangi, F.; Azzi, A. Vitamin E—The Next 100 Years. *IUBMB Life* **2019**, *71*, 411–415. [CrossRef]
113. Mustacich, D.J.; Bruno, R.S.; Traber, M.G. Vitamin E. *Vitam Horm.* **2007**, *76*, 1–21.
114. Lee, G.Y.; Han, S.N. The role of vitamin E in immunity. *Nutrients* **2018**, *10*, 1614. [CrossRef]
115. Manela-Azulay, M.; Bagatin, E. *Cosmeceuticals vitamins. Clin. Dermatol.* **2009**, *27*, 469–474. [CrossRef]
116. Keen, M.A.; Hassan, I. Vitamin E in dermatology. *Indian. Dermatol. Online J.* **2016**, *7*, 311–315. [CrossRef]
117. Park, K. Role of micronutrients in skin health and function. *Biomol. Ther.* **2015**, *23*, 207–217. [CrossRef]
118. Nouveau-Richard, S.; Yang, Z.; Mac-Mary, S.; Li, L.; Bastien, P.; Tardy, I.; Bouillon, C.; Humbert, P.; de Lacharrière, O. Skin ageing: A comparison between Chinese and European populations. A pilot study. *J. Dermatol. Sci.* **2005**, *40*, 187–193. [CrossRef]
119. Gao, Y.; Tan, J.; Sang, Y.; Tang, J.; Cai, X.; Xue, H. Preparation, structure, and biological activities of the polysaccharides from fruits and vegetables: A review. *Food Biosci.* **2023**, *54*, 102909. [CrossRef]
120. Hoang, H.T.; Moon, J.Y.; Lee, Y.C. Natural antioxidants from plant extracts in skincare cosmetics: Recent applications, challenges and perspectives. *Cosmetics* **2021**, *8*, 106. [CrossRef]
121. Sławińska, N.; Olas, B. Selected seeds as sources of bioactive compounds with/diverse biological activities. *Nutrients* **2022**, *15*, 187. [CrossRef]
122. Dreher, M.L.; Davenport, A.J. Hass avocado composition and potential health effects. *Crit. Rev. Food Sci. Nutr.* **2013**, *53*, 738–750. [CrossRef]
123. Zielinska, A.; Nowak, I. Abundance of active ingredients in sea-buckthorn oil. *Lipids Health Dis.* **2017**, *16*, 95. [CrossRef]
124. Knaggs, H.; Lephart, E.D. Enhancing skin anti-aging through healthy lifestyle factors. *Cosmetics* **2023**, *10*, 142. [CrossRef]

125. Tureck, C.; Barboza, B.P.; Bricarello, L.P.; Retondario, A.; Alves, M.d.A.; Souza, A.d.M.; Fernandes, R.; Vasconcelos, F.d.A.G.d. Scientific evidence of the association between oral intake of OMEGA-3 and OMEGA-6 fatty acids and the metabolicsyndrome in adolescents: A systematic review. *Nutr. Metab. Cardiovasc. Dis.* **2022**, *32*, 2689–2704. [CrossRef]
126. Sorokin, A.V.; Arnardottir, H.; Svirydava, M.; Ng, Q.; Baumer, Y.; Berg, A.; Pantoja, C.J.; Florida, E.M.; Teague, H.L.; Yang, Z.H.; et al. Comparison of the dietary omega-3 fatty acids impact on murine psoriasis-like skin inflammation and associated lipid dysfunction. *J. Nutr. Biochem.* **2023**, *117*, 109348. [CrossRef]
127. Dhaka, V.; Gulia, N.; Ahlawat, K.S.; Khatkar, B.S. Trans fats-sources, health risks and alternative approach—A review. *J. Food Sci. Technol.* **2011**, *48*, 534–541. [CrossRef]
128. Britten-Jones, A.C.; Craig, J.P.; Downie, L.E. Omega-3 polyunsaturated fatty acids and corneal nerve health: Current evidence and future directions. *Ocul. Surf.* **2023**, *27*, 1–12. [CrossRef]
129. Opálka, L.; Meyer, J.M.; Ondrejčeková, V.; Svatošová, L.; Radner, F.P.W.; Vávrová, K. ω-O-Acylceramides but not ω-hydroxy ceramides are required for healthy lamellar phase architecture of skin barrier lipids. *J Lipid Res.* **2022**, *63*, 100226. [CrossRef]
130. Priyadarshini Pradhan, S.; Padhi, S.; Dash, M.; Heena Mittu, B. Behera A: Chapter 7—Carotenoids. In *Nutraceuticals and Health Care*; Kour, J., Nayik, G.A., Eds.; Academic Press: Nayik, GA, USA, 2022; pp. 135–157.
131. Elvira-Torales, L.I.; García-Alonso, J.; Periago-Caston, M.J. Nutritional importance/of carotenoids and their effect on liver health: A Review. *Antioxidants* **2019**, *8*, 299. [CrossRef] [PubMed]
132. Maoka, T. Carotenoids as natural functional pigments. *J. Nat. Med.* **2020**, *74*, 1–16. [CrossRef] [PubMed]
133. Zerres, S.; Stahl, W. Carotenoids in human skin. *Biochim. Biophys. Acta Mol. Cell Biol. Lipids* **2020**, *1865*, 158588. [CrossRef] [PubMed]
134. Rawlings, A.V. Ethnic skin types: Are there differences in skin structure and function? *Int. J. Cosmet. Sci.* **2006**, *28*, 79–93. [CrossRef]
135. Zouboulis, C.C.; Makrantonaki, E.; Nikolakis, G. When the skin is in the center of interest: An aging issue. *Clin. Dermatol.* **2019**, *9*, 296–305. [CrossRef] [PubMed]
136. Kligman, A.M. Perspectives and problems in cutaneous gerontology. *J. Investig. Dermatol.* **1979**, *73*, 39–46. [CrossRef] [PubMed]
137. Kurban, R.S.; Kurban, A.K. Common skin disorders of aging: Diagnosis and treatment. *Geriatrics* **1993**, *48*, 30–31.
138. Blume-Peytavi, U.; Kottner, J.; Sterry, W.; Hodin, M.W.; Griffiths, T.W.; Watson, R.E.; Hay, R.J.; Griffiths, C.E. Age-associated skin conditions and diseases: Current perspectives and future options. *Gerontologist* **2016**, *2*, S230–S242. [CrossRef]
139. Choi, E.H.; Man, M.Q.; Xu, P.; Xin, S.; Liu, Z.; Crumrine, D.A.; Jiang, Y.J.; Fluhr, J.W.; Feingold, K.R.; Elias, P.M.; et al. Stratum corneum acidification is impaired in moderately aged human and murine skin. *J. Investig. Dermatol.* **2007**, *127*, 2847–2856. [CrossRef]
140. López-Otín, C.; Blasco, M.A.; Partridge, L.; Serrano, M.; Kroemer, G. The hallmarks of aging. *Cell* **2013**, *153*, 1194–1217. [CrossRef]
141. Makrantonaki, E.; Zouboulis, C.C.; William, J. Cunliffe scientific awards. characteristics and pathomechanisms of endogenously aged skin. *Dermatology* **2007**, *214*, 352–360. [CrossRef]
142. Lovell, C.R.; Smolenski, K.A.; Duance, V.C.; Light, N.D.; Young, S.; Dyson, M. Type I and III collagen content and fibre distribution in normal human skin during ageing. *Br. J. Dermatol.* **1987**, *117*, 419–428. [CrossRef]
143. Autio, P.; Risteli, J.; Haukipuro, K.; Risteli, L.; Oikarinen, A. Collagen synthesis in human skin in vivo: Modulation by aging, ultraviolet B irradiation and localization. *Photodermatol. Photoimmunol. Photomed.* **1994**, *10*, 212–216. [PubMed]
144. Braverman, I.M.; Fonferko, E. Studies in cutaneous aging: I. The elastic fiber network. *J. Investig. Dermatol.* **1982**, *78*, 434–443. [CrossRef]
145. Wang, A.S.; Dreesen, O. Biomarkers of cellular senescence and skin aging. *Front. Genet.* **2018**, *9*, 247. [CrossRef]
146. Bentov, I.; Reed, M.J. The effect of aging on the cutaneous microvasculature. *Microvasc. Res.* **2015**, *100*, 25–31. [CrossRef]
147. Li, W.; Chi, N.; Rathnayake, R.A.C.; Wang, R. Distinctive roles of fibrillar collagen I and collagen III in mediating fibroblast-matrix interaction: A nanoscopic study. *Biochem. Biophys. Res. Commun.* **2021**, *560*, 66–71. [CrossRef]
148. Johnson, J.M.; Minson, C.T.; Kellogg, D.L., Jr. Cutaneous vasodilator and vasoconstrictor mechanisms in temperature regulation. *Compr. Physiol.* **2014**, *4*, 33–89. [CrossRef]
149. Rittié, L.; Fisher, G.J. Natural and sun-induced aging of human skin. *Cold Spring Harb. Perspect. Med.* **2015**, *5*, a015370. [CrossRef]
150. Kammeyer, A.; Luiten, R.M. Oxidation events and skin aging. *Ageing Res. Rev.* **2015**, *21*, 16–29. [CrossRef]
151. Choi, Y.J.; Moon, K.M.; Chung, K.W.; Jeong, J.W.; Park, D.; Kim, D.H.; Yu, B.P.; Chung, H.Y. The underlying mechanism of proinflammatory NF-κB activation by the mTORC2/Akt/IKKα pathway during skin aging. *Oncotarget* **2016**, *7*, 52685–52694. [CrossRef]
152. Puizina-Ivić, N. Skin aging. *Acta Dermatovenerol. Alp. Pannonica Adriat.* **2008**, *17*, 47–54. [PubMed]
153. Li, G.Z.; Eller, M.S.; Firoozabadi, R.; Gilchrest, B.A. Evidence that exposure of the telomere 3′ overhang sequence induces senescence. *Proc. Natl. Acad. Sci. USA* **2003**, *100*, 527–531. [CrossRef]
154. Gilchrest, B.A.; Eller, M.S.; Yaar, M. Telomere-mediated effects on melanogenesis and skin aging. *J. Investig. Dermatol. Symp. Proc.* **2009**, *14*, 25–31. [CrossRef]

155. Ganceviciene, R.; Liakou, A.I.; Theodoridis, A.; Makrantonaki, E.; Zouboulis, C.C. Skin anti-aging strategies. *Dermato-Endocrinology* **2012**, *4*, 308–319. [CrossRef] [PubMed]
156. Veret, D.; Brondello, J.M. Senotherapy: Advances and new clinical perspectives. *Med. Sci.* **2020**, *36*, 1135–1142.
157. Tanuja, Y.; Mishra, S.; Das, S.; Aggarwal, S.; Rani, V. Anticedants and natural prevention of environmental toxicants induced accelerated aging of skin. *Environ. Toxicol. Pharmacol.* **2015**, *39*, 384–391. [CrossRef] [PubMed]
158. Geissler, S.; Textor, M.; Schmidt-Bleek, K.; Klein, O.; Thiele, M.; Ellinghaus, A.; Jacobi, D.; Ode, A.; Perka, C.; Dienelt, A.; et al. In serum veritas-in serum sanitas? Cell non-autonomous aging compromises differentiation and survival of mesenchymal stromal cells via the oxidative stress pathway. *Cell Death Dis.* **2013**, *4*, e970. [CrossRef]
159. Chen, J.; Li, Y.; Zhu, Q.; Li, T.; Lu, H.; Wei, N.; Huang, Y.; Shi, R.; Ma, X.; Wang, X.; et al. Anti-skin-aging effect of epigallocatechin gallate by regulating epidermal growth factor receptor pathway on aging mouse model induced by d-Galactose. *Mech. Ageing Dev.* **2017**, *164*, 1–7. [CrossRef]
160. Bavarsad Shahripour, R.; Harrigan, M.R.; Alexandrov, A.V. N-acetylcysteine (NAC) in neurological disorders: Mechanisms of action and therapeutic opportunities. *Brain Behav.* **2014**, *4*, 108–122. [CrossRef]
161. Bjelakovic, G.; Nikolova, D.; Gluud, L.L.; Simonetti, R.G.; Gluud, C. Antioxidant supplements for prevention of mortality in healthy participants and patients with various diseases. *Cochrane Database Syst. Rev.* **2012**, *2012*, CD007176.
162. Marosz, A.; Chlubek, D. The risk of abuse of vitamin supplements. *Pomeranian J. Life Sci.* **2014**, *60*, 60–64. [CrossRef]
163. Bjelakovic, G.; Nikolova, D.; Gluud, C. Antioxidant supplements and mortality. *Curr. Opin. Clin. Nutr. Metab. Care* **2014**, *17*, 40–44. [CrossRef] [PubMed]
164. Lu, L.Y.; Ou, N.; Lu, Q.B. Antioxidant induces DNA damage, cell death and mutagenicity in human lung and skin normal cells. *Sci. Rep.* **2013**, *3*, 3169. [CrossRef]
165. Rhee, S.G. Cell signaling. H2O2, a necessary evil for cell signaling. *Science* **2006**, *312*, 1882–1883. [CrossRef] [PubMed]
166. Mojallal, A.; Lequeux, C.; Shipkov, C.; Breton, P.; Foyatier, J.L.; Braye, F.; Damour, O. Improvement of skin quality after fat grafting: Clinical observation and an animal study. *Plast. Reconstr. Surg.* **2009**, *124*, 765–774. [CrossRef]
167. Kim, W.S.; Park, B.S.; Kim, H.K.; Park, J.S.; Kim, K.J.; Choi, J.S.; Chung, S.J.; Kim, D.D.; Sung, J.H. Evidence supporting antioxidant action of adipose-derived stem cells: Protection of human dermal fibroblasts from oxidative stress. *J. Dermatol. Sci.* **2008**, *49*, 133–142. [CrossRef] [PubMed]
168. Zhang, S.; Dong, Z.; Peng, Z.; Lu, F. Anti-aging effect of adipose-derived stem cells in a mouse model of skin aging induced by D-galactose. *PLoS ONE* **2014**, *9*, e97573. [CrossRef]
169. Bernardini, F.P.; Gennai, A.; Izzo, L.; Zambelli, A.; Repaci, E.; Baldelli, I.; Fraternali-Orcioni, G.; Hartstein, M.E.; Santi, P.L.; Quarto, R. Superficial enhanced fluid fat injection (SEFFI) to correct volume defects and skin aging of the face and periocular region. *Aesthet. Surg. J.* **2015**, *35*, 504–515. [CrossRef]
170. Gennai, A.; Zambelli, A.; Repaci, E.; Quarto, R.; Baldelli, I.; Fraternali, G.; Bernardini, F.P. Skin rejuvenation and volume enhancement with the micro superficial enhanced fluid fat injection (M-SEFFI) for skin aging of the periocular and perioral regions. *Aesthet. Surg. J.* **2017**, *37*, 14–23. [CrossRef]
171. Park, B.S.; Jang, K.A.; Sung, J.H.; Park, J.S.; Kwon, Y.H.; Kim, K.J.; Kim, W.S. Adipose-derived stem cells and their secretory factors as a promising therapy for skin aging. *Dermatol. Surg.* **2008**, *34*, 1323–1326.
172. Fisher, G.J.; Datta, S.C.; Talwar, H.S.; Wang, Z.Q.; Varani, J.; Kang, S.; Voorhees, J.J. Molecular basis of sun-induced premature skin ageing and retinoid antagonism. *Nature* **1996**, *379*, 335–339. [CrossRef]
173. Kafi, R.; Kwak, H.S.; Schumacher, W.E.; Cho, S.; Hanft, V.N.; Hamilton, T.A.; King, A.L.; Neal, J.D.; Varani, J.; Fisher, G.J.; et al. Improvement of naturally aged skin with vitamin A (retinol). *Arch. Dermatol.* **2007**, *143*, 606–612. [PubMed]
174. Verdier-Sévrain, S.; Bonté, F. Skin hydration: A review on its molecular mechanisms. *J. Cosmet. Dermatol.* **2007**, *6*, 75–82. [CrossRef] [PubMed]
175. Ramirez, R.D.; Morales, C.P.; Herbert, B.S.; Rohde, J.M.; Passons, C.; Shay, J.W.; Wright, W.E. Putative telomere-independent mechanisms of replicative aging reflect inadequate growth conditions. *Genes Dev.* **2001**, *15*, 398–403.
176. González-Suárez, E.; Geserick, C.; Flores, J.M.; Blasco, M.A. Antagonistic effects of telomerase on cancer and aging in K5-mTert transgenic mice. *Oncogene* **2005**, *24*, 2256–2270. [CrossRef]
177. LePillouer-Prost, A.; Kerob, D.; Nielsen, M.; Taieb, C.; Maitrot Mantelet, L. Skin and menopause: Women's point of view. *J. Eur. Acad. Dermatol. Venereol.* **2020**, *34*, e267–e269. [CrossRef]
178. Quan, T. Molecular insights of human skin epidermal and dermal aging. *J Dermatol Sci.* **2023**, *112*, 48–53. [CrossRef]
179. Asavasupreechar, T.; Saito, R.; Miki, Y.; Edwards, D.P.; Boonyaratanakornkit, V.; Sasano, H. Systemic distribution of progesterone receptor subtypes in human tissues. *J. Steroid Biochem. Mol. Biol.* **2020**, *199*, 105599.
180. Schmidt, J.B.; Lindmaier, A.; Spona, J. Hormone receptors in pubic skin of premenopausal and postmenopausal females. *Gynecol. Obstet. Investig.* **1990**, *30*, 97–100.
181. Pelletier, G.; Ren, L. Localization of sex steroid receptors in human skin. *Histol. Histopathol.* **2004**, *19*, 629–636.
182. Huang, A.H.; Chien, A.L. Photoaging: A review of current literature. *Curr. Dermatol. Rep.* **2020**, *9*, 22–29.

183. Tran, D.; Townley, J.P.; Barnes, T.M.; Greive, K.A. An antiaging skin care system containing alpha hydroxy acids and vitamins improves the biomechanical parameters of facial skin. *Clin. Cosmet. Investig. Dermatol.* **2014**, *8*, 9–17. [PubMed]
184. Ogluszka, M.; Lipinski, P.; Starzynski, R.R. Effect of Omega-3 fatty acids on telomeres-are they the elixir of youth? *Nutrients* **2022**, *14*, 3723. [CrossRef] [PubMed]
185. Danby, F.W. Nutrition and aging skin: Sugar and glycation. *Clin. Dermatol.* **2010**, *28*, 409–411.
186. Dearlove, R.P.; Greenspan, P.; Hartle, D.K.; Swanson, R.B.; Hargrove, J.L. Inhibition of protein glycation by extracts of culinary herbs and spices. *J. Med. Food* **2008**, *11*, 275–281. [PubMed]
187. Buckingham, E.M.; Klingelhutz, A.J. The role of telomeres in the ageing of human skin. *Exp. Dermatol.* **2011**, *20*, 297–302.
188. Jacczak, B.; Rubis, B.; Toton, E. Potential of naturally derived compounds in telomerase and telomere modulation in skin senescence and aging. *Int. J. Mol. Sci.* **2021**, *22*, 6381.
189. Cai, Z.; Zhang, J.; Li, H. Selenium, aging and aging-related diseases. *Aging Clin. Exp. Res.* **2019**, *31*, 1035–1047. [CrossRef]
190. Lopez-Otin, C.; Blasco, M.A.; Partridge, L.; Serrano, M.; Kroemer, G. Hallmarks of aging: An expanding universe. *Cell* **2023**, *186*, 243–278.
191. Vinci, M.C.; Costantino, S.; Damiano, G.; Rurali, E.; Rinaldi, R.; Vigorelli, V.; Sforza, A.; Carulli, E.; Pirola, S.; Mastroiacovo, G.; et al. Persistent epigenetic signals propel a senescence-associated secretory phenotype and trained innate immunity in CD34(+) hematopoietic stem cells from diabetic patients. *Cardiovasc. Diabetol.* **2024**, *23*, 107. [CrossRef]
192. Gao, Z.; Santos, R.B.; Rupert, J.; Van Drunen, R.; Yu, Y.; EckelMahan, K.; Kolonin, M.G. Endothelial-specific telomerase inactivation causes telomere-independent cell senescence and multi-organ dysfunction characteristic of aging. *Aging Cell* **2024**, *23*, e14138. [CrossRef]
193. Thirunavukkarasu, V.; Nandhini, A.T.; Anuradha, C.V. Fructose diet-induced skin collagen abnormalities are prevented by lipoic acid. *Exp. Diabesity Res.* **2004**, *5*, 237–244. [CrossRef] [PubMed]
194. Tarwadi, K.V.; Agte, V.V. Effect of micronutrients on methylglyoxal-mediated in vitro glycation of albumin. *Biol. Trace Elem. Res.* **2011**, *143*, 717–725. [CrossRef]
195. Boelsma, E.; van de Vijver, L.P.; Goldbohm, R.A.; Klöpping-Ketelaars, I.A.; Hendriks, H.F.; Roza, L. Human skin condition and its associations with nutrient concentrations in serum and diet. *Am. J. Clin. Nutr.* **2003**, *77*, 348–355. [CrossRef]
196. Alqanatish, J.T.; Alqahtani, F.; Alsewairi, W.M.; Al-kenaizan, S. Childhood scurvy: An unusual cause of refusal to walk in a child. *Pediatr. Rheumatol. Online J.* **2015**, *13*, 23. [CrossRef] [PubMed]
197. Trapani, S.; Rubino, C.; Indolfi, G.; Lionetti, P. A narrative review on pediatric scurvy: The last twenty years. *Nutrients* **2022**, *14*, 684. [CrossRef]
198. Stewart, M.S.; Cameron, G.S.; Pence, B.C. Antioxidant nutrients protect against UVB-induced oxidative damage to DNA of mouse keratinocytes in culture. *J. Investig. Dermatol.* **1996**, *106*, 1086–1089. [CrossRef] [PubMed]
199. Placzek, M.; Gaube, S.; Kerkmann, U.; Gilbertz, K.P.; Herzinger, T.; Haen, E.; Przybilla, B. Ultraviolet B-induced DNA damage in human epidermis is modified by the antioxidants ascorbic acid and D-alpha-tocopherol. *J. Investig. Dermatol.* **2005**, *124*, 304–307. [CrossRef]
200. Zussman, J.; Ahdout, J.; Kim, J. Vitamins and photoaging: Do scientific data support their use? *J. Am. Acad. Dermatol.* **2010**, *63*, 507–525. [CrossRef]
201. Langton, A.K.; Sherratt, M.J.; Griffiths, C.E.; Watson, R.E. A new wrinkle on old skin: The role of elastic fibres in skin ageing. *Int. J. Cosmet. Sci.* **2010**, *32*, 330–339. [CrossRef]
202. Baumann, L. Skin ageing and its treatment. *J. Pathol.* **2007**, *211*, 241–251. [CrossRef]
203. Kaimal, S.; Thappa, D.M. Diet in dermatology: Revisited. *Indian J. Dermatol. Venereol. Leprol.* **2010**, *76*, 103–115. [PubMed]
204. Cao, C.; Xiao, Z.; Wu, Y.; Ge, C. Diet and skin aging-from the perspective of food nutrition. *Nutrients* **2020**, *12*, 870. [CrossRef] [PubMed]
205. Strasser, B.; Volaklis, K.; Fuchs, D.; Burtscher, M. Role of dietary protein and muscular fitness on longevity and aging. *Aging Dis.* **2018**, *9*, 119–132. [CrossRef] [PubMed]
206. Balić, A.; Vlašić, D.; Žužul, K.; Marinović, B.; Bukvić Mokos, Z. Omega-3 versus omega-6 polyunsaturated fatty acids in the prevention and treatment of inflammatory skin diseases. *Int. J. Mol. Sci.* **2020**, *21*, 741. [CrossRef]
207. Rawlings, A.V.; Scott, I.R.; Harding, C.R.; Bowser, P.A. Stratum corneum moisturization at the molecular level. *J. Investig. Dermatol.* **1994**, *103*, 731–741. [CrossRef]
208. Schürer, N.Y.; Plewig, G. Elias PM Stratum corneum lipid function. *Dermatologica* **1991**, *183*, 77–94. [CrossRef]
209. Reuter, J.; Merfort, I.; Schempp, C.M. Botanicals in dermatology: An evidence-based review. *Am. J. Clin. Dermatol.* **2010**, *11*, 247–267. [CrossRef]
210. Tungmunnithum, D.; Thongboonyou, A.; Pholboon, A.; Yangsabai, A. Flavonoids and other phenolic compounds from medicinal plants for pharmaceutical and medical aspects: An overview. *Medicines* **2018**, *5*, 93. [CrossRef]
211. Anunciato, T.P.; da Rocha Filho, P.A. Carotenoids and polyphenols in nutricosmetics, nutraceuticals, and cosmeceuticals. *J. Cosmet. Dermatol.* **2012**, *11*, 51–54. [CrossRef]

212. Im, A.R.; Kim, Y.M.; Chin, Y.C.; Chae, S. Protective effects of compounds from *Garcinia mangostana* L. (mangosteen) against UVB damage in HaCaT cells and hairless mice. *Int. J. Mol. Med.* **2017**, *40*, 1941–1949. [CrossRef]
213. Ma, R.J.; Yang, L.; Bai, X.; Yuan, M.Y.; Wang, Y.K.; Xie, Y.; Hu, J.M.; Zhou, J. Phenolic constituents with antioxidative, tyrosinase inhibitory and anti-aging activities from *Dendrobium loddigesii* Rolfe. *Nat. Prod. Bioprospecting* **2019**, *9*, 329–336. [CrossRef] [PubMed]
214. Morikawa, T.; Nagatomo, A.; Kitazawa, K.; Muraoka, O.; Kikuchi, T.; Yamada, T.; Tanaka, R.; Ninomiya, K. Collagen synthesispromoting effects of andiroba oil and its limonoid constituents in normal human dermal fibroblasts. *J. Oleo Sci.* **2018**, *67*, 1271–1277. [CrossRef] [PubMed]
215. Tan, H.; Sonam, T.; Shimizu, K. The potential of triterpenoids from loquat leaves (*Eryobotrya japonica*) for prevention and treatment of skin disorder. *Int. J. Mol. Sci.* **2017**, *18*, 1030. [CrossRef]
216. Henriet, E.; Jäger, S.; Tran, C.; Bastien, P.; Michelet, J.F.; Minondo, A.M.; Formanek, F.; Dalko-Csiba, M.; Lortat-Jacob, H.; Breton, L.; et al. A jasmonic acid derivative improves skin healing and induces changes in proteoglycan expression and glycosaminoglycan structure. *Biochim. Biophys. Acta (BBA)—Gen. Subj.* **2017**, *1861*, 2250–2260. [CrossRef]
217. Kim, J.E.; Jang, S.G.; Lee, C.H.; Lee, J.Y.; Park, H.; Kim, J.H.; Lee, S.; Kim, S.H.; Park, E.Y.; Lee, K.W.; et al. Beneficial effects on skin health using polysaccharides from red ginseng by-product. *J. Food Biochem.* **2019**, *43*, e12961. [CrossRef]
218. Zdunska-Peciak, K.; Rotsztejn, H. The effectiveness of ferulic acid and microneedling in reducing signs of photoaging: A split-face comparative study. *Dermatol. Ther.* **2020**, *33*, e14000.
219. Moreira, L.C.; Ávila, R.I.; Veloso, D.F.M.C.; Pedrosa, T.N.; Lima, E.S.; Couto, R.O.; Lima, E.M.; Batista, A.C.; Paula, J.R.; Valadares, M.C. In vitro safety and efficacy of a complex botanical mixture of *Eugenia dysenterica* Dc. (Myrtaceae): Prospects for developing a new dermocosmetic product. *Toxicol. In Vitr.* **2017**, *45*, 394–408. [CrossRef]
220. Song, E.; Chung, H.; Shin, E.; Jeong, J.K.; Han, B.K.; Choi, H.J.; Hwang, J. *Gastrodea elata* Blume extract modulates antioxidant activity and ultraviolet A-irradiated skin aging in human dermal fibroblast cells. *J. Med. Food* **2016**, *19*, 1057–1064. [CrossRef]
221. Lourith, N.; Kanlayavattanakul, M.; Chaikul, P.; Chansriniyom, C.; Bunwatcharaphansakun, P. In vitro and cellular activities of the selected fruit residues for skin aging treatment. *An. Acad. Bras. Ciências* **2017**, *89*, 577–589. [CrossRef]
222. Wu, L.; Chen, C.; Cheng, C.; Dai, H.; Ai, Y.; Lin, C.; Chung, Y. Evaluation of tyronisase inhibitory, antioxidant, antimicrobial, and antiaging activities of *Magnolia officinalis* extracts after *Aspergillus niger* fermentation. *BioMed Res. Int.* **2018**, *2018*, 5201786. [CrossRef]
223. Bose, B.; Choudhury, H.; Tandon, P.; Kumaria, S. Studies on secondary metabolite profiling, anti-inflammatory potential, in vitro photoprotective and skin-aging related enzyme inhibitory activities of *Malaxis acuminate*, a threatened orchid of nutraceutical importance. *J. Photochem. Photobiol. B* **2017**, *173*, 686–695. [CrossRef] [PubMed]
224. Pientaweeratch, S.; Panapisal, V.; Tansirikongkol, A. Antioxidant, anti-collagenase and anti-elastase activities of *Phyllanthus emblica*, *Manilkara zapota* and sylimarin: An in vitro comparative study for anti-aging applications. *Pharm. Biol.* **2016**, *54*, 1865–1872. [CrossRef]
225. Shoko, T.; Naharaj, V.J.; Naidoo, D.; Tselanyane, M.; Nthambeleni, R.; Khorombi, E.; Apostolides, Z. Anti-aging potential of extracts from *Sclerocarya birrea* (A. Rich.) Hochst and its chemical profiling by UPLC-Q-TOF-MS. *BMC Complement. Altern. Med.* **2018**, *18*, 54. [CrossRef]
226. Kwon, K.R.; Alam, M.B.; Park, J.H.; Kim, T.H.; Lee, S.H. Attenuation of UVB-induced photo-aging by polyphenolic-rich *Spatholobus suberectus* stem extract via modulation of MAPK/AP-1/MMPs signaling in human keratinocytes. *Nutrients* **2019**, *11*, 1341. [CrossRef]
227. Sundaran, I.K.; Sarangi, D.D.; Sundararajan, V.; George, S.; Mohideen, S.S. Poly herbal formulation with anti-elastase and antioxidant properties for skin anti-aging. *BMC Complement. Altern. Med.* **2018**, *18*, 33.
228. Lee, H.; Hong, Y.; Kwon, S.H.; Park, J.; Park, J. Anti-aging effects of *Piper cambodianum* P. Fourn. extract on normal human dermal fibroblast cells and a wound-healing model in mice. *Clin. Interv. Aging* **2016**, *11*, 1017–1026. [PubMed]
229. Dieament, G.; Pereda, M.D.C.V.; Nogueira, C.; Eberlin, S.; Facchini, G.; Checon, J.T.; Cesar, C.K.; Mussi, L.; Polezel, M.A.; Martins-Oliveira, D., Jr.; et al. Antiageing mechanisms of a standardized supercritical CO2 preparation of Black Jack (*Bidens Pilosa* L.) in human fibroblasts and skin fragments. *Altern. Med.* **2015**, *2015*, 280529.
230. Hwang, E.; Ngo, H.T.T.; Seo, S.A.; Park, B.; Zhang, M.; Yi, T.H. Protective effect of dietary *Alchemilla mollis* on UVB-irradiated premature skin aging through regulation of transcription factor NFATc1 and Nrf2/ARE pathways. *Phytomedicine* **2018**, *39*, 125–136. [CrossRef]
231. Ngo, H.T.T.; Hwang, E.; Seo, S.A.; Park, B.; Sun, Z.W.; Zhang, M.; Shin, Y.K.; Yi, T.H. Topical application of neem leaves prevents wrinkles formation in UVB-exposed hairless mice. *J. Photochem. Photobiol. B* **2017**, *169*, 161–170. [CrossRef]
232. Zhao, P.; Alam, M.B.; Lee, S.H. Protection of UVB-induced photoaging by Fuzhuan-brick tea aqueous extract via MAPKs/Nrf2-mediated down-regulation of MMP-1. *Nutrients* **2019**, *11*, 60. [CrossRef]
233. Adhikari, D.; Panthi, V.K.; Pangeni, R.; Kim, H.J.; Park, J.W. Preparation, characterization, and biological activities of topical anti-aging ingredients in a *Citrus junos* callus extract. *Molecules* **2017**, *22*, 2198. [CrossRef] [PubMed]

234. Nam, G.H.; Kawk, H.W.; Kim, S.Y.; Kim, Y.M. Solvent fraction of fermented *Trapa japonica* fruit extract stimulates collagen synthesis through TGF-β1/GSK-3β/β-catenin pathway in human dermal fibroblasts. *J. Cosmet. Dermatol.* **2020**, *19*, 226–233. [CrossRef]
235. Jeon, H.; Kim, D.H.; Nho, Y.H.; Park, J.E.; Kim, S.N.; Choi, E.H. A mixture of *Kochia scoparia* and *Rosa multiflora* with PPAR α/γ dual agonistic effects prevents photoaging in hairless mice. *Int. J. Mol. Sci.* **2016**, *17*, 1919. [CrossRef]
236. You, J.; Roh, K.B.; Li, Z.; Liu, G.; Tang, J.; Shin, S.; Park, D.; Jung, E. The antiaging properties of *Andrographs paniculata* by activation epidermal cell stemness. *Molecules* **2015**, *20*, 17557–17569. [CrossRef] [PubMed]
237. Limtrakul, P.; Yodkeeree, S.; Thippraphan, P.; Punfa, W.; Srisomboom, J. Anti-aging and tyrosinase inhibition effects of *Cassia fistula* flower butanolic extract. *BMC Complement. Altern. Med.* **2016**, *16*, 497. [CrossRef] [PubMed]
238. Pakravan, N.; Mahmoudi, E.; Hashemi, S.A.; Kamali, J.; Hajiaghayi, R.; Rahimzadeh, M. Cosmeceutical effect of ethyl acetate fraction of kombucha tea by intradermal administration in the skin of aged mice. *J. Cosmet. Dermatol.* **2017**, *17*, 1216–1224. [CrossRef]
239. Bravo, K.; Duque, L.; Ferreres, F.; Moreno, D.A.; Osorio, E. *Passiflora tarminiana* fruits reduce UVB-induced photoaging in human skin fibroblasts. *J. Photochem. Photobiol. B* **2017**, *168*, 78–88. [CrossRef]
240. Cicchetti, E.; Duroure, L.; Le Borgne, E.; Laville, R. Upregulation of skin-aging biomarkers in aged NHDF cells by a sucrose ester extract from agroindustrial waste of *Physalis peruviana* calyces. *J. Nat. Prod.* **2018**, *81*, 1946–1955. [CrossRef]
241. Jeong, D.; Lee, J.; Park, S.H.; Kim, Y.A.; Park, B.J.; Oh, J.; Sung, G.H.; Aravithan, A.; Kim, J.H.; Kang, H.; et al. Antiphotoaging and antimelanogenic effects of *Penthorum chinense* pursh ethanol extract due to antioxidant- and autophagy-inducing properties. *Oxidative Med. Cell. Longev.* **2019**, *2019*, 9679731. [CrossRef]
242. Khare, R.; Upmanyu, N.; Jha, M. Exploring the potential of methanolic extract of *Salvia officinalis* against UV exposed skin aging: In vivo and in vitro model. *Curr. Aging Sci.* **2021**, *12*, 46–55. [CrossRef]
243. Kim, H.K. Protective effect of garlic on cellular senescence in UVB-exposed HaCaT human keratinocytes. *Nutrient* **2016**, *8*, 464. [CrossRef]
244. Hwang, E.; Lin, P.; Ngo, T.T.; Yi, T.H. Clove attenuates UVB-induced photodamage and repairs skin barrier function in hairless mice. *Food Nutr.* **2018**, *9*, 4936–4947. [CrossRef] [PubMed]
245. Choi, S.I.; Lee, J.S.; Lee, S.; Cho, B.Y.; Choi, S.H.; Han, X.; Sin, W.S.; Kim, Y.C.; Lee, B.Y.; Kang, I.J.; et al. Protective effects and mechanisms of *Pourthiaea villosa* (Thumb.) Decne. extract on hydrogen peroxide-induced skin aging in human dermal fibroblasts. *J. Med. Food* **2019**, *22*, 841–850. [CrossRef] [PubMed]
246. Kusumawati, I.; Kurniawan, K.O.; Rullyansyah, S.; Prijo, T.A.; Widyowati, R.; Ekowati, J.; Hestianah, E.P.; Maat, S.; Matsunami, K. Anti-aging of *Curcuma heyneana* Valeton & Zipj: A scientific approach use in Javanese tradition. *J. Ethnopharmacol.* **2018**, *225*, 64–70. [PubMed]
247. Li, J.; Lu, Y.R.; Lin, I.F.; Kang, W.; Chen, H.B.; Lu, H.F.; Wang, H.M.D. Reversing UVB-induced photoaging with *Hibiscus sabdariffacalyx* aqueous extract. *J. Sci. Food Agric.* **2020**, *100*, 672–681. [CrossRef]
248. Choi, M.; Oh, J.H.; Shin, M.K.; Lee, S.R. Beneficial effects of blood group antigen synthesis-increasing natural plant extracts and monosaccharides on extracellular matrix protein production in vivo. *J. Dermatol. Sci.* **2015**, *80*, 152–155. [CrossRef]
249. Uchiyama, T.; Tsunenaga, M.; Miyanaga, M.; Ueda, U.; Ogo, M. Oral intake of lingonberry and amla fruit extract improves skin conditions in healthy female subjects: A randomized, double-blind placebo-controlled clinical trial. *Biotechnol. Appl. Biochem.* **2018**, *66*, 870–879. [CrossRef]
250. Caversan, J.; Mussi, L.; Sufi, B.; Padovani, G.; Nazaro, L.; Camargo-Junior, F.B.; Magalhães, W.V.; Di Stasi, L.C. A new phytocosmetic preparation from *Thymus vulgaris* stimulates adipogenesis and controls skin aging process: In vitro studies and topical effects in a double-blind placebo-controlled clinical trial. *J. Cosmet. Dermatol.* **2021**, *20*, 2190–2202. [CrossRef]

International Journal of
Molecular Sciences

MDPI

Review

Regulatory Mechanism of Intestinal Stem Cells Based on Hippo Pathway and Signaling Crosstalk in Chicken

Tao Quan, Ran Li, Yaoxing Chen and Ting Gao *

College of Veterinary Medicine, China Agricultural University, Beijing 100083, China;
s20243051072@cau.edu.cn (T.Q.); 18246269672@163.com (R.L.); yxchen@cau.edu.cn (Y.C.)
* Correspondence: gt1994@cau.edu.cn

Abstract: Recently, there has been a gradual increase in the demand for chicken and eggs. The gut, as the vital place of nutrient digestion and absorption, is highly associated with the development of livestock and poultry and the quality of meat, eggs, and milk. Intestinal stem cells, as an important source of intestinal cell proliferation and renewal, exert a vital effect on repairing injured intestinal epithelial cells and keeping homeostasis. Intestinal stem cell-regulated intestinal epithelial balance is closely controlled and modulated by interlinked developmental loops that maintain cell proliferation and differentiation processes in balance. Some conservative signaling pathways, including the Wnt, Notch, hedgehog, and bone morphogenetic protein (BMP) loops, have been proved to modulate intestinal health in poultry. Meanwhile, studies have revealed the importance of the Hippo pathway in gastrointestinal tract physiology by regulating intestinal stem cells. Moreover, crosstalk between Hippo and other signaling pathways provides tight, yet versatile, regulation of tissue homeostasis. In this review, we summarize studies on the role of the Hippo pathway in the intestine in these physiological processes and the underlying mechanisms responsible via interacting with these signaling pathways and discuss future research directions and potential therapeutic strategies targeting Hippo signaling in intestinal disease. A comprehensive understanding of how these signaling pathways regulate stem cell proliferation, differentiation, and self-renewal will help to understand the regulation of intestinal homeostasis. In addition, it has the capacity for creative ways to govern intestinal damage, enteritis, and associated disorders induced by different factors.

Keywords: intestinal stem cells; intestinal mucosa; Hippo pathway; intestinal epithelial cell

1. Introduction

In recent years, there has been a gradual increase in the demand for chicken and eggs; the production of chicken meat will up-regulate from 117 million tons to 132 million tons in 2026 [1]. Several studies suggested that intestinal health is highly associated with metabolite state [2], immunocompetence homeostasis [3], host development [4], and general health [5], suggesting that intestinal homeostasis is a vital factor of homeostasis. Poultry digestive tract discomfort will not only affect the digestion and absorption of nutrients, but also undermine the health of the body, directly causing economic losses to the livestock and poultry industry. In contrast, in mammals, an imbalance of intestinal homeostasis not only induces a variety of intestinal-related disorders, including Crohn's disease, ulcerative colitis, etc., but also has the potential to induce lesions in other organs and tissues, such as Alzheimer's disease, type I diabetes, muscular dystrophy, and cancers, etc. [6] (Figure 1). Thus, maintaining intestinal homeostasis is essential for livestock and host homeostasis.

Anatomy divides the gut into the small intestines at the front and the large intestine at the posterior end, and the large intestine is dominated by the cecum in poultry. The small intestines are the main position of nutrient digestion and absorption in chickens. Nutrient transporters and digestive enzymes contained in the small intestinal pouch can resist pathogen invasion by secreting hormones, glycoproteins, and defensins [7,8]. The small intestinal epithelium folds internally to form the intestinal crypts and protrudes externally to produce intestinal villi, which are straight exposed to the intestinal lumen [9]. The intestinal epithelial lining is derived from the embryonic endoderm. Stem cells located in the intestinal crypt develop specialized types of epithelial cells that constitute the mature intestinal epithelium, and the crypt ecological niche provides essential signals and regulators for this process. These mature intestinal epithelial cells are distributed bottom-up along the intestinal villi, exert a vital effect on absorptive and secretory [10,11]. In chickens, proliferating cells are also distributed on the intestinal villi, which may be due to the kinetics of cell production from humans [12]. The chicken are widely used to study vertebrate embryonic development. While few studies have reported the molecular basis of epithelial cell development and differentiation in the chicken small intestine. Despite the limited knowledge of the characteristics of the chicken intestinal epithelium, recent researches have suggested that the chicken small intestinal epithelium has a number of similarities, although it is markedly different and uniquely characterized from that of humans and mice. Therefore, it is crucial to understand specific characteristics of chicken small intestinal epithelial cells.

Intestinal epithelial health is highly modulated by interrelated developmental loops and maintain cell proliferation and differentiation processes in balance. Many conserved signaling loops, including Wnt, Notch, hedgehog and bone morphogenetic protein (BMP) loops, have been indicated to modulate gut health and regeneration in mammals [13]. Recognizing how the Hippo pathway interacts with these signaling pathways could facilitate the understanding of the phenotypes that emerge in mouse models of Hippo signaling malfunctions and would help in the development of new therapeutic measures.

The Hippo pathway plays an important role in the regulation of cell proliferation, differentiation and death, and directly determines cell fate, such as tissue homeostasis and organ size [14,15], and disruption of this pathway can trigger the development of a variety of malignant diseases [16,17]. Several studies have demonstrated that the Hippo pathway plays a key role in gastrointestinal health [18]. Intestinal regeneration is co-regulated by multiple signaling pathways, and there is also a correlation between these different signaling pathways. The role of the Hippo pathway in intestinal regeneration has been reported. Meanwhile, interactions between the Hippo pathway and signaling pathways, including Wnt, Notch, and others, also play a key role in regulating intestinal homeostasis. Here, we outline how the Hippo pathway is involved in intestinal physiological functions and the potential mechanisms by which it plays a role, and accordingly summarize possible future research directions and propose therapeutic options in the treatment of intestinal diseases.

The paper reviews the development and growth of the chicken intestine and compares the cellular features and signaling loops, including the Notch and Wnt signaling pathways that control small intestinal epithelial cell differentiation, between chicken and mammals. Evidence suggests that the chicken intestinal epithelium has unique cellular structures and proliferative zones. The establishment of an in vitro cell culture model of chicken will offer a new artifact for exploring the molecular modulation of chicken intestinal development and progression and may also result in novel ideas for enhancing small intestinal health and nutrition in the poultry industry.

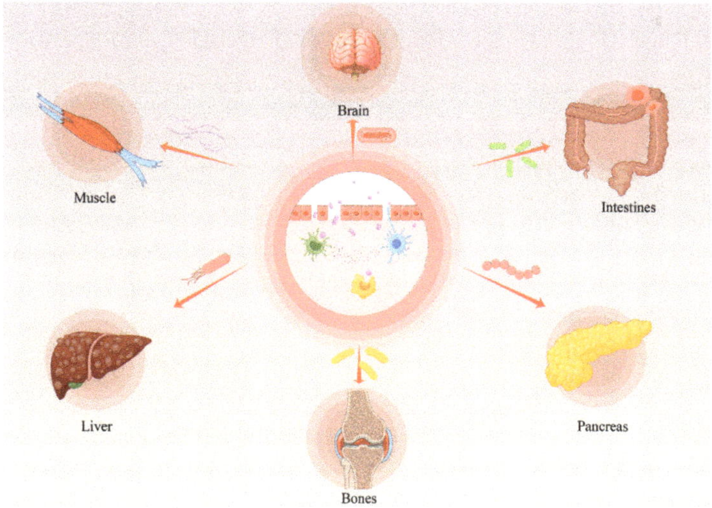

Figure 1. Intestinal health is closely related to the homeostasis of other organs in the body. Intestinal health is not only about the gut itself, but also about the transmission of gut microbiota and its metabolites through the bloodstream, which in turn affects the health status of other organs in the body. When the intestinal mucosal barrier is damaged and the intestinal microbiota is disturbed, it may be closely related to multi-organ health including brain, muscle, liver, bones and pancreas etc.

2. The Development of Chicken Small Intestine

The small intestines of the chicken consist of intestinal tubes, which are derived from the internal folding of the foregut and caudal intestinal entrances of the embryonic endoderm [19]. Intestinal tube formation is accomplished and modulated by diverse signals caused by Sonic hedgehog or Wnt [20,21], when the yolk stalk connecting the yolk sac is only lightly open around embryonic day e5 [22]. Between e5 and hatching day, the length of the small intestine increases [23]. Peristalsis within the chick small intestine was also observed at e5, suggesting the emergence of mechanical intestinal function during early embryogenesis [24]. By contrast, the expression of nutrient transporters, mineral transporters, and digestive enzymes is detected only at the final stage of embryonic development, while physiological structures such as the intestinal villi and crypts are formed on the day of hatching [25]. Mature and functional epithelial cells of the chicken small intestine, including goblet cells and intestinal epithelial cells, have been histologically examined by e18 [26]. As chickens are precocious species, the intestines are thoroughly mature at the end of embryogenesis and they are able to feed independently promptly after hatching. Thus, the production of the crypt–villus framework and functional epithelial cells is a precondition before hatching.

3. Cellular Proliferation in the Chicken Intestinal Epithelium

The intestinal crypt, first identified by Jonathan Lieberkühn, is a position where ISCs have the proliferative capacity to proliferate and differentiate, thereby maintaining homeostasis of the intestinal epithelium. For mammals, the proliferation and differentiation of intestinal stem cells are controlled by complex cellular cascades of signals that are regulated by Paneth cells in the crypt or by intercellular communication between stromal cells around the crypt [27]. Differentiation of intestinal stem cells initially forms transport amplification (TA). Proliferating cells rapidly propagate epithelial cell populations and migrate into the intestinal crypts along the villi generating ultimately differentiated cells (enterocytes, cisternae, enteroendocrine cells) or deeper into the crypts differentiating into Paneth cells [28]. Thus, the proliferative zone of mammalian intestinal epithelial cells is located within the crypt (Figure 2).

However, in addition to the crypts, chicken epithelial cells also proliferated extended to the villi, a conclusion reached by measuring the rate of uptake of thymidine at 3 hr. Immunohistochemical tests observed that 5-bromo-2-deoxyuridine (BrdU) and proliferating cell nuclear antigen (PCNA) were also localized to the villi in chicken small intestine. However, most of the proliferative activity of intestinal stem cells still occurred in the crypts [29]. Bar-Shira and Friedman also observed a small number of chromosome-concentrated proliferating columnar cells in chicken duodenal villi by hematoxylin and eosin staining. They also observed that lysozyme-secreting cells, probably chicken Paneth cells, were distributed in both crypts and villi. Considering that Paneth cells play an important role in the proliferation and differentiation of intestinal stem cells, it has been suggested that the proliferating cells found in chicken intestinal villi may be either intestinal stem cells or TA proliferating cells [30]. Yet another study identified cells expressing the leucine-rich repeat-containing G protein-coupled receptor 5 (*Lgr5*) (leucine-rich G protein-coupled receptor 5) and olfactory toxin 4 (*Olfm4*) genes in the crypts of chicken small intestine [31]. Therefore, chicken ISCs may be confined to the crypts, and the proliferating cells observed extend the villi may be delayed-differentiated TA cells. In addition, partially differentiated TA cells may also be present in the region above Olfm4-positive stem cells in the crypts and below PepT1 (oligopeptide transporter 1)-positive enterocytes extend the villi [32].

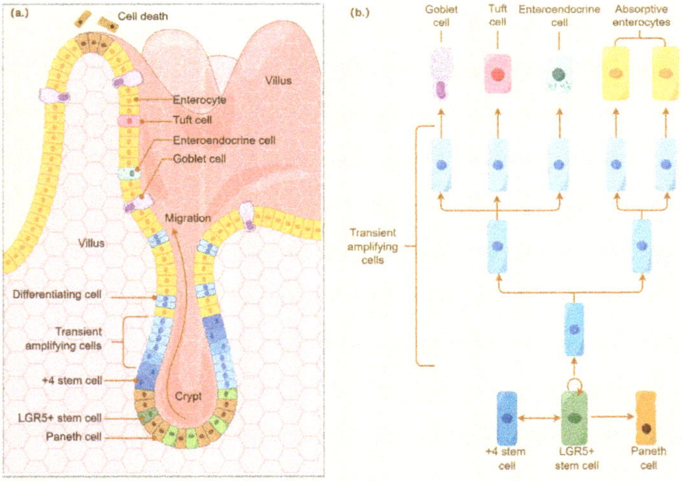

Figure 2. Proliferation and differentiation of intestinal stem cells. ISC is capable of differentiating into progenitor cells, and these newly formed cells proliferate and differentiate along the crypt–villus axis of the small intestine and colon. Lgr5 is an important marker of active ISC identified to generate differentiated epithelium cell types over a long period of time. In addition, another population of quiescent "reserve" ISC is located at the so-called '+4' position. In the small intestine, ISC can differentiate into five major cell types, including enterocytes, goblet cells, enteroendocrine cells, tuft cells, and Paneth cells, but only differentiate into three major cell types in the colon. (**a**) Different cell types in the intestinal epithelium; (**b**) Differentiation of stem cells into different types of intestinal epithelial cell processes.

4. The Importance of Intestinal Stem Cells

Under normal physiological balanced conditions, our intestinal mucosa, immune system, intestinal microorganisms, nutrients, and metabolites interact with each other to form a dynamic equilibrium. Disturbances in the intestinal microenvironment or endogenous pathogens can disrupt the homeostatic balance and lead to intestinal epithelial damage and lesions. Therefore, ISC must actively play the functions of regeneration and repair to replenish the damaged and lost cells. In this process, ISC need to precisely regulate the homeostasis between proliferation and differentiation. When the self-renewal capacity of ISC is unregulated and over-enhanced, it may cause the growth of ISCs and their progeny to be unregulated; on the other hand, when the differentiation process of ISC is inhibited, or impaired, the proliferative capacity of the cells may also become stronger. Both of these conditions can lead to imbalances in gut homeostasis in poultry and thus induce gut-related diseases.

On the contrary, when the self-renewal ability of ISC is weakened or impaired, the stem cells age prematurely and cannot be replenished with new stem cells, leading to a decrease of the ISCs number, which results in a down-regulation in the functional cells number in tissues and organs due to the lack of replenished cells; in addition, if the differentiation of the ISCs is impaired, it also results in a down-regulation of functional cells number in tissues and organs. When the number of functional cells cannot meet the wear and tear of tissues and organs, it will lead to incomplete renewal and repair of tissues and organs, which will lead to tissue and organ failure. A recent study has revealed the mechanism by which genomic instability in ISC triggers stem cell necrosis, leading to spontaneous intestinal inflammation. The research suggested that targeting ISC necrosis may be an approach to treat severe intestinal inflammation. ISC, as important players in intestinal renewal and maintenance, are highly associated with the pathogenesis of several disorders. More and more studies are now linking intestinal diseases to stem cell defects, providing additional dimensions and possibilities for us to be able to better understand and treat intestinal diseases.

5. Homeostatic Maintenance of Intestinal Stem Cells

The proliferation and differentiation of ISC from progenitor cells into many specialized typologies of epithelial cells allows the intestinal epithelium to be continuously renewed to maintain intestinal homeostasis [33,34]. In this process, the ISC niche offers the microecology for proliferation and differentiation. In the ISC niche, the presence of many different types of cells provides different signals for the normal functioning of ISC to ensure that these ISC are able to differentiate into normal intestinal epithelium and to prevent the progress of tumor cells. The current development of gut tissue has greatly improved comprehension of the formation of the ecological niche of ISC [35]. Existing studies have identified two main components of the ISC ecosystem: the extracellular matrix (ECM) and the cellular microecology, which encompasses all resident cells embedded in the outer matrix, such as immune cells, fibroblasts, smooth muscle cells, etc. These cells can nourish ISC by secreting different matrix components and growth factors to promote their renewal and differentiation [36,37]. From the available findings, cells in the gut microecology, including Paneth cells and mesenchymal telangiectasia, can offer growth factors including Wnt and Notch ligands or epidermal growth factor (EGF) to promote epithelial cell proliferation and self-renewal [38]. However, when the renewal process of the intestinal epithelium is impaired, such as ecological dysregulation or immune dysfunction, including necrotizing enterocolitis, can occur (Figure 3).

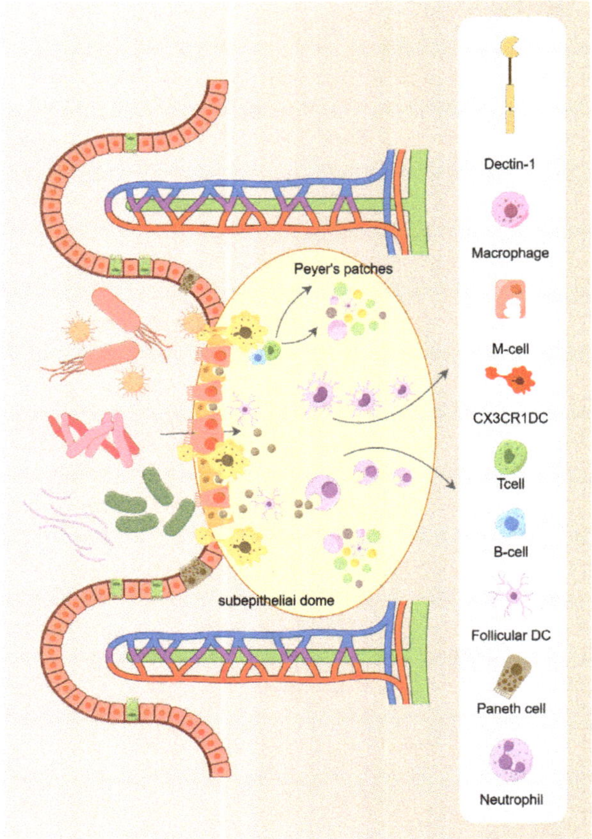

Figure 3. Maintenance of intestinal homeostasis. The intestinal tract plays a vital role in both digestion and immunity, making its equilibrium crucial for overall health. This equilibrium relies on the dynamic interplay between various intestinal cells, including intestinal epithelial cells, immune cells, crypt stem cells, and intestinal microbiota. Intestinal epithelial cells play a pivotal role in protecting and regulating the gut. They form vital barriers, modulate immune responses, and engage in pathogen defense and cytokine secretion. Moreover, they supervise the regulation of ISC. Immune cells actively influence the immune response through the phagocytosis of pathogens and the release of cytokines.

5.1. Intestinal Mesenchymal Cells

Intestinal mesenchymal stromal cells, including fibroblasts, muscles, fibroblasts, endothelial cells, and smooth muscle cells, not only provide structured supports for ISC, but also secrete Wnt ligands and bone morphogenetic proteins (BMP) antagonists to regulate ISC activity [39]. New research indicated that a down-regulation of the Foxl1+ mesenchymal stem cells (MSCs) number resulted in a decrease in the secretion of Wnt signaling, which induced dysfunction of ISC and ultimately led to intestinal inactivation in mice. Another related study showed that Wnt2B can also be derived from Gli1+ MSCs, which are also critical for the functional maintenance of ISC. Various types of MSCs originate from ISC ecological niches (e.g., Wnt2B, R-spondin 1, Gremlin 1, and CD34+), which is an important feature of ISC [40]. The above studies establish the relevance of MSCs providing Wnt ligands, BMP suppressant, and r-responsive proteins.

5.2. Paneth Cells

Results from studies of intestinal ISC derived from ISC indicate that the ISC niche can function independently in the absence of mesenchymal cells [41]. The most important component of this small intestinal crypt is the Paneth cells, which are located close to the ISC and provide niche signals that are essential for the survival and functioning of the ISC [11]. The antimicrobial peptides secreted by Paneth cells were found to form an essential part of the defense function of the ISC niche [42]. Moreover, Paneth cells express a variety of signaling molecules, including Wnt and Notch ligands, which are vital for the keep of the ISC niche [43]. While these studies only confirm the importance of these signals, the significance of Paneth cells in the ISC niche is critical. It was found that Atoh1 (also called Math1 and Hath1 in mouse and human separately) in the Notch signaling loop also exert a very important effect on Notch-based ISC destinies determination [44]. The data of the research suggested that in Atoh1$^{-/-}$ mice, the absence of Paneth cells did not have an effect on the proliferation of ISC [45,46]. Therefore, it has been suggested that EGF and Wnt could serve as a replacement for Paneth cells and mediate Notch signaling to exert a vital effect on the keep of the ISC niche. However, knockdown of Sox9 or Gfli1 genes depletes Paneth cells, which in turn leads to the loss of ISC. Another study found that insertion of the diphtheria toxin receptor gene into the Reg4 locus in mice caused desiccation of the crypt basal cup cells [47–49]. In the large intestines, Reg4+ crypt basal goblet cells, in turn, can be regarded as a landmark of Paneth cell function, and depletion of this cell results in the decrease of ISC in the colonic crypts, which in turn harasses intestinal health and affects the growth of organoids [50]. However, it has been suggested that knockout mice are not an excellent model for studying whether Paneth cells exert a decisive effect on the functioning of the ISC niche, and the actual situation remains to be hoped that future trials will be carried out.

5.3. Enteroendocrine Cells (EECs)

Enteroendocrine cells also play an important role in the maintenance of intestinal homeostasis, one of the mechanisms of which is through the regulation of intestinal stem cell function. Intestinal stem cells act as rapidly circulating cells, which work together to maintain intestinal health by proliferating and differentiating to form different intestinal epithelial cells. Since they are located at the root of the intestinal crypts and only have access to a small supply of nutrients, intestinal stem cells are highly sensitive to their surrounding microenvironment and different diets, including high-fat diets and fasting [51,52]. After a few days on a high-fat diet, intestinal stem cells change their metabolic pathway from glycolysis to fatty acid-like oxidation, at which point intestinal stem cell proliferative activity increases [53]. Similarly, when mice were fasted, intestinal stem cell activity increased and their metabolism was predominantly fatty acid-like, suggesting the ability of intestinal stem cells to rapidly capture nutrient alterations in the microenvironment and respond accordingly [54]. Together, these results suggest that enteroendocrine cells can act as nutrient-sensing cells to transmit signals from the intestinal lumen to the microenvironment of intestinal stem cells in a timely manner to regulate intestinal stem cell activity.

Moreover, it was found that the lipid metabolism pathway of intestinal stem cells was upregulated in animals with insufficient enteroendocrine cells, which is consistent with the results obtained in the fasting model [55,56]. Meanwhile, enhanced mitochondrial activity and increased oxygen consumption in the intestinal crypts were also seen, demonstrating the role of enteroendocrine cells in the regulation of intestinal stem cells [55]. Together, these results suggest that enteroendocrine cells can influence the function of intestinal stem cells by regulating the metabolic activity of the intestinal crypts [56]. In addition, derivatives of enteroendocrine cells can also influence intestinal stem cell activity, and this is achieved through the combined action of multiple derivatives, whereas the effect of a single derivative is limited [57]. For example, on a standard chow diet that was as deficient in neurotensin as possible, the mice were still better tolerated and had no significant differences in total intestinal length, villus height, and crypt depth. However, under fasting conditions, cell proliferative activity and Wnt signaling in the distal small intestinal crypts were significantly reduced if neurotensin was lacking [58]. These results suggest that neurotensin can affect cell proliferative activity in the intestinal crypts by modulating Wnt signaling during nutritional deficiencies, further emphasizing the role of enteroendocrine cells in the regulation of intestinal stem cells.

In addition, enterochromaffin cells in the enteroendocrine cells can promote the proliferative activity of intestinal epithelial cells through their secretion of 5-HT [59]. Human organoid models from in vitro have also shown that 5-HT produced by enterochromaffin cells significantly promotes the proliferation of intestinal epithelial cells [60]. The results provide a disorder mechanism via which patients Cronkhite–Canada syndrome show up-regulated gut 5-HT level and gut polyposis [60]. Studies have shown that GLP-2 is one of the most effective derivatives secreted by enteroendocrine cells to regulate the proliferation of intestinal epithelial cells and that it has been used to ameliorate intestinal-related disorders such as short bowel syndrome [61]. Although GLP-2 is not required for intestinal development and homeostasis maintenance, exogenous supplementation with GLP-2 promotes crypt cell proliferation and increases villus and microvillus length [62]. This process is regulated by several complex mechanisms, including the Wnt signaling pathway [63–65]. Specifically, GLP-2 is used to promote the proliferative activity of intestinal stem cells by accelerating their entry into the S phase of the cell cycle. In addition, GLP-2 may act to nourish the gut [66,67]. Further studies showed that in Drosophila, EECs can regulate the proliferation of intestinal stem cells by signaling to cells within the subepithelial ecological niche [68,69].

6. Mechanisms of Intestinal Stem Cells Promoting Intestinal Mucosal Repair

Overall, the process of ISC to daughter cell transition is in a dynamic equilibrium. At the molecular level, multiple signaling pathways work in concert to facilitate intercellular communication and thus maintain intestinal homeostasis. Dysfunction of these signaling pathways, on the other hand, can lead to various intestinal diseases. Therefore, we next summarize several signaling pathways that exert a vital effect on the proliferation and differentiation of ISC.

6.1. Integrity of the Epithelium as a Barrier

The intestinal epithelial cell layer is made up of a combination of different types of cells, such as enterocytes, Paneth cells, absorptive cells, goblet cells, enteroendocrine cells, and microfold cells, which work together to maintain the homeostasis of the intestinal epithelial barrier [70,71]. Goblet cells exert defense and repair by secreting mucus, trehalose peptides, and resistin-like molecule-β. Enteroendocrine secretion not only senses nutrients and gut flora stimulation, but also delivers nutritional signals to the nervous system [72]. Paneth cells, in turn, provide intestinal stem cells with their essential growth factors to maintain the stemness of intestinal stem cells [11]. Paneth cells also secrete antimicrobial peptides including α-defensin and lysozyme C to clear bacteria [73]. Absorptive enterocytes, on the other hand, can play an immunological role by expressing polymeric immunoglobulin receptors (pIgR) that mediate IgA secretion by B cells [74] (Table 1).

Table 1. Different intestinal epithelial cells and their functions. Intestinal epithelial cells include goblet cells, EECs, microfold cells, Paneth cells, and absorptive enterocytes. They function to protect the gastrointestinal tract through different pathways of action.

Cell Type	Mechanism of Action	Functionality
Goblet cells	Mucus trefoil peptides, and resistin-like molecule-β [70]	crucial for epithelial defense and repair
EECs	Triggers secretion of peptide or amine hormones	sense nutrients and microbial stimuli, transmitting nutritional information to the nervous system [72]
Microfold cells	IgA	presenting bacterial antigens to dendritic cells [46,74]
Paneth cells	secreting antimicrobial peptides, such as α-defensins, lysozyme C, regenerating islet-derived protein IIIβ (RegIIIβ), regenerating islet-derived protein IIIγ, and angiopoietin-4	support stemness in the intestinal stem cell niche by providing essential growth factors to cryptbasal columnar stem cells [75]
Absorptive enterocytes	express polymerized immunoglobulin receptors (pIgR) and transmembrane mucins, forming a glycocalyx	pIgR mediates the transcellular transport of IgA released by B cells from the basolateral to the apical surface and to the glycosyl surface facing the lamina propria [76]

6.1.1. Mucus Layer

The mucus layer consists of a highly dynamic extracellular matrix located outside the intestinal epithelial cells and is considered the first mechanical barrier in the gastrointestinal tract. As a thicker gel-like substance, the mucus layer covers the surface of the gastrointestinal tract and separates the contents of the intestinal lumen from the intestinal epithelium and the cells beneath it [75,76]. This barrier effectively prevents harmful substances in the intestinal lumen from coming into direct contact with the intestinal epithelium and the cells beneath it, thus exerting a protective effect on the intestinal mucosa. The main component of mucus is MUC2, and they are interconnected to form a reticular structure [77]. The study by Bergstrom et al. revealed that colonic mucus contains two different O-glycosylated forms of Muc2 from the proximal and distal colon, and that they wrap fecal matter and microbiota tightly. Notably, some microbiota induce cup cells in the proximal colon to encapsulate themselves by producing MUC2. Not only that, MUC2 production in the proximal colon alters the structure and function of the microbiota [78]. The mucus layer not only separates the harmful flora, but also provides a place for beneficial bacteria. In addition, the network formed by the mucus layer traps and immobilizes pathogens, including bacteria and viruses, to prevent them from coming into direct contact with the epithelial cells [77]. The gut microbiota can also influence the mucus layer: it secretes enzymes that degrade mucins to obtain nutrients to support its growth [79]. However, the presence of sialic acid on mucin protects it from degradation by degrading enzymes secreted by the gut microbiota [80]. Thus, similar to the promising data achieved by targeting ECM proteins in tumors, therapeutics targeting mucins or their glycosylation may serve as an effective approach to prevent or ameliorate inflammatory bowel disease (IBD) [81,82].

6.1.2. Antimicrobial Peptides

In the gastrointestinal tract, AMPs are mainly derived from epithelial cells, such as Pan cells and cup cells. Some AMPs from the gut exert their debacterizing effect by acting on the cell wall of bacteria [83,84]. The correlation between intestinal epithelial cells and AMPs is important for elucidating the functions and mechanisms of the body's digestive system [85]. First, the intestinal epithelium acts as a guard, capable of recognizing pathogens and triggering a timely immune response. Immediately afterward, these cells initiate the secretion of AMPs, which in turn remove bacteria and viruses [83].

Of greatest interest in this process is the selective action of AMPs. This selectivity is critical for maintaining the dynamic balance of the gut flora. The homeostasis of the gut flora benefits from the interplay of harmful and beneficial bacteria, and if all bacteria were removed, this would be equally disruptive to the gut and even host health [84]. In addition, lysozyme secreted by Paneth cells, as a glycosidase, catalyzes the hydrolysis of the 1,4-β-glycosidic bond between the N-acetylglucosamine and N-acetylcytidylic acid components of the peptidoglycan, which in turn destroys the bacterial cell wall [86]. In addition, another antimicrobial peptide synthesized by Paneth cells can be utilized to remove bacteria by an enzymatic mechanism using its secreted phospholipase A2. These antimicrobial peptides include defensins, C-type lectins of the REG3 family and neutral endotoxin [87]. Most of these AMPs disrupt bacterial membranes, causing membrane potential collapse, loss of metabolites and ions, and osmotic lysis [88].

6.1.3. Tight Junction

Tight junctions are key components that make up the intestinal epithelial barrier, and they seal the gaps between these specialized intestinal epithelial cells. The components that make up the tight junction structure include transmembrane proteins, articulin, and cytoskeleton [61,89]. Major participants include closure proteins, tight junction proteins, adhesion banding proteins, and closure banding proteins [90]. These proteins work in concert with each other to form a physical barrier that selectively regulates the entry and exit of molecules and ions. Thus, tight junctions primarily form a physical barrier between the intestinal lumen and the tissues of the body. This physical barrier is what prevents harmful substances, including pathogens, toxins, and undigested food particles, from entering the bloodstream [91]. In addition, this physical barrier is also involved in the nutrient absorption process, ensuring that adapted nutrients can be utilized by the body [91]. However, damage to the tight junctions can disrupt intestinal permeability, a phenomenon known as "leaky gut". In this case, toxic and harmful substances take advantage of the opportunity to enter the body through the bloodstream, causing intestinal disorders and even systemic health problems [92].

6.2. BMP Signaling Pathway

A gradient of BMP signaling was observed along the crypt–villus axis, and BMP signaling could antagonize Wnt signaling and thus inhibit the renewal and proliferation of small ISC. This process plays a key role in inhibiting the over-activation of small ISC. In the small intestine of mice, overexpression of Noggin resulted in the development of a large number of neonatal crypts, whereas downregulation of Bmpr1a led to crypt hyperplasia [93]. It was shown that BMP signaling could antagonize the proliferation and renewal process of small ISC by directly preventing the Wnt signaling pathway [94]. Therefore, the knockdown of BMP in $LGR5^+$ small ISC leads to rapid expansion of ISC, whereas inactivation of BMP signaling induces unrestricted expansion of $LGR5^+$ small ISC, which in turn leads to small intestinal epithelial polyp development, a phenomenon observed in small intestines with radiation-induced inactivation of BMP signaling [95]. It was found that radiation induces BMP signaling inactivation, which in turn causes apoptosis in some $LGR5^+$ small ISC, while healthy $LGR5^+$ small ISC undergo damage repair [96]. The study emphasizes that if BMP signaling is inactivated at a young age [97], it can lead to other cells becoming desensitized to BMP signaling, ultimately resulting in malignant hyperplastic intestinal polyps [98]. However, it cannot be determined whether BMP signaling in epithelial or mesenchymal cells exerts a vital effect on inhibiting the growth of small intestinal polyps [99]. Although previous studies have used the overexpression of Noggin and knockout of Bmpr1a mouse models, they have not been able to resolve this issue [100].

6.3. Wnt/β-Catenin Signaling Pathway

Wnt signaling is progressively weakened along the small intestinal crypt–villus axis. Thus, the crypt region is an area of high Wnt signaling expression, which is also an area of high ISC density. In contrast, Wnt signaling intensity is relatively low in the villus region, which also corresponds to the remains of differentiated cells in this region. In the small intestine, Wnt signaling exerts a vital effect on the self-renewal and proliferation of small ISC. Study indicated that proliferating cells and ISCs were lost in large numbers in the small intestine of adult mice overexpressing Dkk1 [100] virus or specifically knocking out the Ctnnb1 (encoding β-catenin protein) gene, which in turn lead to the stopping of self-renewal of the entire intestinal epithelium [101]. In addition, Wnt overexpression leads to excessive proliferation of small intestinal crypt cells and expansion of stem cell regions, which increases the risk of colorectal cancer. The marker genes for intestinal stem and progenitor cells, LGR5 and Olfm4, are both modulated by Wnt [102]. Specifically, in crypts, Wnt molecules secreted by epithelial or stromal cells first bind to LRP5/6 receptors, which in turn activate β-catenin and upregulate its expression, followed by binding of β-catenin to the nuclear transcription factor TCF4, which ultimately maintains stem cell stemness and promotes their proliferation and differentiation [103]. However, suppression of the Wnt/β-catenin signaling loop hinders the proliferation of ISC. It was shown that knockdown of the TCF4 gene in mouse embryonic intestinal epithelial cells lead to stagnant proliferation of small intestinal villus cells in newborn mice. Similarly, knockdown of TCF4 with Ctnnb1 in the intestinal epithelial cells of adult mice caused blocked cell proliferation in the crypt region of mice [104]. Therefore, Wnt signaling exerts a vital effect on stemness maintenance, proliferation and differentiation of ISC [105].

6.4. Notch Signaling Pathway

In chickens, the mechanism by which the Notch signaling loop mediates intestinal stem cell differentiation has not been investigated and can only be obtained by comparison with mammals. The Notch signaling pathway exists in almost all postnatal animals and is a closely conserved signaling mechanism that functions by cell–cell communication [106]. Within the small intestinal epithelium, the several ligands and receptors for Notch are expressed only in the crypts, and thus their activity exists only in that region [107]. Notch signaling exerts a vital effect on regulating the proliferation, differentiation, and self-renewal of small ISC, especially during the differentiation stage of small ISC [108]. Activation of the Notch signaling loop induces the differentiation of stem cells to absorptive cells and inhibits their differentiation to secretory cells [109]. Moreover, Notch signaling also plays an important role in modulating the renewal process of small ISC. In the microenvironment of ISC, Paneth cells can provide ligands for Notch signaling to be expressed abundantly in this microenvironment [110]. Lineage tracing trials have clarified that small ISC are a subset of cells characterized by high Notch signaling, and that this subset of cells has a unique ability to differentiate and give rise to all of the different cell types found in intestinal epithelial cells [111].

6.5. Hedgehog Signaling Pathway

Hedgehog (Hh) signaling is primarily vital during host embryonic development, which in turn continues to influence proliferation and differentiation after cytogenesis [112]. Three family members of Hh exist in mammals: the Sonic Hh, Indian Hh (Ihh), and Desert Hh. Ihh is the predominantly present Hh protein in the gut and mediates paracrine signaling initiated by differentiated epithelial cells to regulate mesenchymal cells. Notably, Ihh also inhibits the proliferation of columnar cells in the crypt by activating BMP signaling [113]. Not only that, Ihh has the ability to negatively regulate immune function within the lamina propria. Knockdown of the Ihh gene activates an immune process similar to the wound healing process, which involves remodeling of epithelial cells and recruitment of fibroblasts and macrophages, a process that does not damage the epithelial cells themselves [114]. Thus, reduced expression of Ihh in epithelial cell injury or dysfunction may be one of the important mechanisms for activating the Leymic cells wound healing response [115].

6.6. EGF Signaling Pathway

EGF receptors, such as EGFR, ErbB-2, ErbB-3, and ErbB-4, are collectively referred to as the receptor complex kinase family, and these receptors play important roles in cellular signaling transduction. They first bind to the growth factor EGF and subsequently activate range of downstream signaling loops, including PI3K/Akt, PLCγ/PKC, and Ras/Raf/Mek/Erk signaling loops, etc., which ultimately control cell proliferation and differentiation, especially of small intestinal ISCs and TA cells [116,117]. The downstream Ras/Raf/Mek/Erk signaling loop of EGF is a region of crypt a particularly active region [118]. To combat the over-activation of small ISC, small intestinal epithelial cells have also derived a series of negative feedback regulatory mechanisms to antagonize the EGF signaling pathway. Among them, Lrig1, as one of the many negative feedback regulators of EGF signaling, is mostly expressed in proliferating cells in the crypts of the small intestine. Specific knockdown of the Lrig1 gene in the small intestine induces expansion of the proliferative zone of the crypts, which in turn induces small bowel tumorigenesis [118,119].

7. Hippo Signaling Pathway

Hippo signaling is a highly conserved cascade signaling mechanism commonly found in drosophila and mammals, which participates in several biological processes, including cell proliferation, differentiation, survival, and ultimately determines cell fate and maintains tissue homeostasis, by regulating its key target genes [120,121]. In contrast, an aberrant Hippo pathway induces the onset and development of cancer or other diseases, in which it participates in the process of tumor formation via modulating cell proliferation and apoptosis [122]. In addition, blocking the nuclear translocation of YES-associated protein (YAP) and PDZ-binding motif (TAZ) inhibits tumor development by reducing cell proliferation and increasing apoptosis [123].

Not only that, the Hippo signaling loop also exerts a vital effect on keeping intestinal health, as evidenced by the fact that inhibition of the Hippo signaling loop results in organism overgrowth and induces tumorigenesis. When MST1 and MST2 were specifically knocked down in intestinal epithelial cells, the villus structure of mice was homeostatic and showed expanded undifferentiated cells [124]. Not only that, abnormal epithelium and adenomas develop in the colon of these mice. Similarly, the small and large intestine in sav1 gene-deficient mice also exhibit expanded crypt structures with increased polyp formation. However, a number of studies have found that the gut does not show adverse phenotypes when depleted of YAP1 and/or TAZ protein expression in the gut in the healthy state, suggesting that the importance of YAP1 and TAZ is not emphasized in the healthy state [125–127]. In contrast, MST1, MST2, and SAV1 in the Hippo pathway are essential for healthy intestinal tissue [126]. Overall, an impaired Hippo loop results in over-activation of YAP1 and TAZ, resulting in lost ISCs proliferation and increased tumorigenesis. The current study was only conducted in mice with gut-given knockout of MST1, MST2, SAV1, YAP1, and/or TAZ, and more studies are needed to corroborate this.

It has been shown that YAP1 and TAZ can maintain and promote ISCs activity in several organs not limited to the gut, including the liver, skin, and nerves [128]. In the intestine, the endogenous YAP1 protein is expressed only in the ISCs ecological niche at the base of the crypt, implying that he may be associated with the regulation of ISC activity [129]. It has been found that highly active YAP1 promotes the proliferation of ISC [126]. Overexpression of YAP1 in the intestine results in expansion of undifferentiated cells and self-renewal of differentiated cells present at the tip of the villi, a phenomenon that is similar to the phenotype of intestines lacking MST1 and MST2 genes. In addition, intestinal gene transfer techniques revealed that YAP1 with TAZ promotes the binding of TEAD transcription factors to stem and progenitor cells and further causes progenitor cells to differentiate into cup cells [130]. In the healthy state, the tumor suppressor gene protein kinase C zeta type (PKCζ) in ISC can directly inhibit YAP1 activity by phosphorylation [131]. Universally, active expression of YAP1 and TAZ promotes ISC activity. However, it was found that when YAP1 expression was specifically induced in the mouse intestine, it disrupted the normal intestinal tissue structure, resulting in phenotypes such as mislocalization of Paneth cells or reduction in the number of ISC, and ultimately inhibited the proliferation of ISC in the crypt [127]. This result indicates that YAP1 functions as a growth inhibitor in the balanced intestine. The reason for the inconsistency of this result with previous results may lie in the difference between the experimental setup and the mouse model [127]. Specifically, researches suggested that mice with inducible systemic overexpression of YAP1 were used, whereas Barry specifically induced YAP1 expression in intestinal epithelial cells. Intestinal epithelial cells continuously work with their neighboring stromal and immune cells, and in the process can modulate YAP1 activity through chemical or physical pathways. Thus, the different phenotypes detected in different models may be due to autonomous and non-autonomous influences. In the nucleus, YAP1 activates the Wnt signaling loop, whereas in the cytoplasm, YAP1 inhibits the Wnt signaling pathway, both of which can have opposite effects on the maintenance and proliferation of progenitor cells. In addition, YAP1 can also initiate a negative feedback regulatory mechanism via causing the expression of LATS2, which in turn acts as a growth inhibitor [132–134].

Intestinal epithelial balance is closely controlled and modulated by interrelated developmental loops, allowing the cell proliferation and differentiation processes to remain relatively balanced. Signaling loops such as Wnt, Notch, Hh, and BMP have been indicated to exert a vital effect on maintaining mammalian intestinal homeostasis and regeneration by interacting with the Hippo pathway [135]. Understanding how the Hippo pathway interacts with these signaling loops to regulate intestinal homeostasis will help develop appropriate therapeutic strategies (Figure 4).

7.1. Hippo and Wnt Signalling

Wnt signaling exerts a vital effect on the keep of gut homeostasis and stem cell activity [136]. The classical Wnt signaling pathway uses β-catenin as its essential inductor, and β-catenin together with the TCF and LEF families as transcription factors, which in turn trigger the transcription of marker genes [137]. Nuclear localization of β-catenin has been observed in mouse ISC located at the base of the crypt, which represents active Wnt signaling [138]. Thus, in mice, suppression of classical Wnt signaling results in loss of transporter-expanded cells and damage to the crypt structure, and further leads to an imbalance of homeostasis and associated dysfunction in the intestine [124]. The interplay between the Hippo pathway and the Wnt signaling pathway has been intensively investigated in the context of keep of ISC activity and damage restoration. Our results indicate that YAP1 can maintain ISC activity and promote stem cell prolifertion in the mammalian intestine via activating the Wnt signaling loop [124].

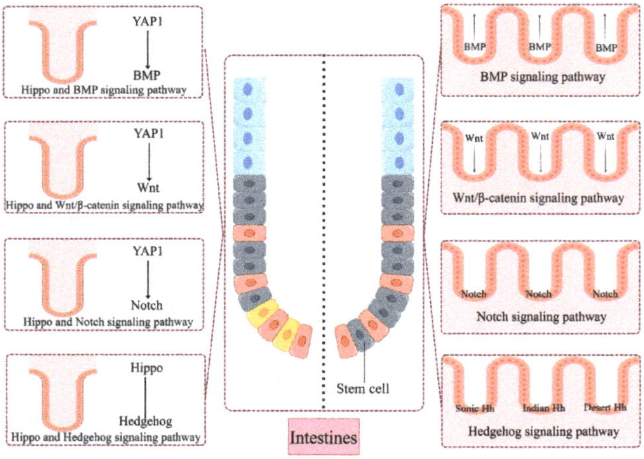

Figure 4. Potential mechanisms for regulating intestinal stem cells. Intestinal stem cell-regulated intestinal epithelial homeostasis is tightly controlled and regulated by interrelated developmental pathways that keep cell proliferation and differentiation processes in balance. Several conserved signaling pathways, such as the Wnt, Notch, hedgehog, and bone morphogenetic protein pathways, have been shown to regulate intestinal homeostasis and regeneration in mammals. The Hippo signaling system, as an exceptionally well-preserved cascade signaling mechanism, can play key roles in cell proliferation, cell survival, cell differentiation, cell fate determination, organ size, and tissue homeostasis by interacting with these signaling pathways.

Specifically, overexpression of YAP1 induced the accumulation of β-catenin and the expression of its target genes in the nucleus. Moreover, the expression of Wnt target genes and ISCs markers Lgr5 and Ascl2 was significantly increased in the intestines of mice specifically knocked down for MST1 versus MST2 genes [125]. In vitro findings showed that down-regulation of YAP1 expression in the SW480 human colon cancer cell line further resulted in decreased transcription of β-catenin [125]. Overall, YAP1 expression upregulated Wnt signaling activity. In addition, it has also been shown that YAP1 and TAZ in the cytoplasm can lead to restriction of classical Wnt signaling, and many regulatory mechanisms have been suggested to reveal this modulation based on the current findings. First, YAP1 and TAZ in the cytoplasm can bind to disheveled homologue DVL-1 (be famed as DVL), a positive modulator of Wnt, and further inhibit DVL expression [139]. In addition, YAP1 with TAZ can directly inhibit nuclear translocation of β-catenin; secondly, YAP1 with TAZ can join in a complex that destroys β-catenin and recruits F-box/WD repeat protein 1A (also known as β-transducing repeat protein, β-TrCP), which ubiquitinates β-catenin and results in its degradation. In the mouse intestine, overexpression of YAP1 means that there is a concomitant increase in cytoplasmic levels of YAP1, which in turn further reduces the expression of β-catenin and its target genes [127]. However, many researches have indicated that YAP1 and TAZ can also be inversely triggered by Wnt signaling and that this is still essential for a number of Wnt responses in the intestine. In particular, the non-classical Wnt signaling pathway, which has also been found to play a crucial role in intestinal tissue regeneration, is a process that is independent of β-catenin [129,140,141]. In addition, Wnt signaling of the nonclassical pathway can activate YAP1 and TAZ signaling, which in turn can inhibit the typical Wnt signaling pathway in vitro [142]. Therefore, it remains unclear whether the nonclassical Wnt signaling loop is mediated by YAP1 and Wnt signaling during intestinal regeneration [143]. The crosstalk between Hippo and Wnt signaling loops is complicated and interesting and will need to be explored in depth in future studies.

7.2. Hippo and Notch Signalling

Apart from the Wnt signaling pathway, the Notch pathway is also involved in maintaining ISC activity and proliferation and differentiation. There are four Notch receptors in mammals, including Notch1/2/3/4, each of which contains range of extracellular ligand-binding structural domain, a transmembrane structural domain, and an intracellular structural domain. Upon activation of the ligand, the Notch intracellular structural domain (NICD) is dissociated from the receptor and translocated to the nucleus [144]. Subsequently, the nuclear NICD binds to the recombining binding protein suppressor of hairless (RBPJ, also regarded as CSL (CBF1/RBPjκ/Su(H)/Lag-1)) and induces the expression of target genes including HES transcription factors and others. In mice, inhibition of Notch expression further leads to a down-regulation in the expression of Lgr5 and olfactomedin 4 (Olfm4) and the ISCs number located at the bottoms of the crypt [145]. In addition to maintaining stem cell activity, Notch is also associated with the fate decision of stem cell differentiation. High Notch activity significantly accelerates the differentiation of transit-expanded cells, whereas suppression of Notch signaling leads to loss of ISCs activity [44]. Notably, the Notch signaling pathway is also modulated by the Hippo pathway [124]. The results showed that specific knockdown of MST1 and MST2 genes in the intestine resulted in increased nuclear localization of NICD and upregulation of Hes1 gene expression in the intestine [146]. In contrast, suppression of Notch signaling attenuated the extension of undifferentiated intestinal progenitor cells caused by YAP1 overexpression [125]. Collectively, these findings indicate that YAP1 positively regulates Notch signaling in the intestine and that YAP1 induces the expansion of ISC by mediating Notch. Although how Hippo regulates Notch is unknown, some members of the Notch family, including Notch1/2/9, and Hes1, were up-regulated in YAP1-overexpressing hepatocytes, and Notch2 was regarded as a direct marker gene of the YAP1-tead complex [147]. The phenomenon may offer a theoretical reference for a deeper understanding of the interaction mechanism between Hippo and Notch in the intestine.

Notch signaling regulates intestinal health by controlling ISCs and inducing absorptive cells differentiation. Active Notch stimulates differentiation of TA cells along the uptake spectrum [148], whereas low non-active Notch results in dysfunction of stem cell properties and inducing differentiation of the secretory spectrum [149]. In addition, the Hippo pathway also regulates Notch signaling. If Notch signaling is positively regulated by YAP1, YAP1 can mediate Notch to promote the amplification of ISCs [150,151]. Hippo and Notch signaling loops cause the secretion of target genes, including *Wnt5a* and *Wnt9b* [124]. The Wnts associated with many Frizzled receptors to stimulate Wnt signaling in ISCs, and ultimately, the proliferation and differentiation activities of ISCs were significantly enhanced. Interaction of Notch signaling, Notch1, and Dll1 between Paneth cells and ISCs regulates small intestinal health by controlling ISC and inducing absorptive cells [152]. It is also clear that Notch may not directly regulate the differentiation of small intestinal epithelial cells through the Hippo signaling pathway. Yet the molecular mechanisms by which the Hippo pathway modulates intestinal Notch signaling remain unclear. Our overview offers understanding into the interplay between Hippo and Notch signaling loops in regulating ISCs activity. Some degree of intestinal epithelial regeneration is effective in improving IBD. YAP1 modulation has been regarded as an effective mechanism for intestinal epithelial regeneration in IBD in DSS-caused colitis mouse model [153]. The finding of nAChRs in non-neuronal cells and their effect on non-nervous disorder, such as IBD is indispensable for researching treatments [154]. It was concluded that a three-part mechanism of nAChR, Hippo, and Notch signaling would allow for revision of the rate of ISC and its daughter cells to keep ISCs and daughter cell density in adult mice over time.

7.3. Hippo and BMP Signalling

BMPs belong to the transforming growth factor β (TGF-β) superfamily, and they bind first to mitogen-activated protein kinase receptors and then through activation of Smad-dependent or non-dependent pathways [155]. BMP signaling plays a key role in embryonic development and adult stem cell homeostasis by regulating cell proliferation, differentiation and apoptosis, and its dysfunction induces diseases [156]. BMPs are secreted proteins that bind to BMP receptors (BMPRs) as homodimers or heterodimers. BMP-2/4 are the major BMPs present in the gut [157]. The BMP signaling pathway plays a crucial role in gut development and homeostasis maintenance from Drosophila to vertebrates. In Drosophila, Bmp-2/4 homologs Dpp and Gbb are essential for midgut development, differentiation, and regeneration [158]. BMP and its receptors are expressed in both epithelial and mesenchymal cells of the mammalian intestine, suggesting that there is an exchange of information between the epithelium and the mesenchyme, but the signaling of BMP in them is complex [159]. In addition, BMP activity is modulated by the antagonists, including Noggin and Gremlin 1, which are mainly expressed in mesenchymal cells around the crypt, and their aberrant expression has been associated with colon tumorigenesis. Through transmembrane receptors, BMP could trigger responses via Smad or non-Smad pathways.

Signal communication between Hippo and BMP or Hh has also been studied and confirmed, although this study was not conducted in intestinal cells. Studies have demonstrated that BMP exerts a vital effect on regulating cell differentiation, proliferation and growth [160]. In the intestine, inhibition of the BMP signaling loop using Noggin, an antagonist that overexpresses BMP, resulted in the overproliferation of ISC, an up-regulation in the crypts number, and ultimately the induction of intestinal polyposis. The data indicate that BMP negatively modulates the proliferation of ISC in intestinal crypts [93,94]. Activation of the BMP signaling loop occurs first with the binding of BMP ligands to type I and type II serine/threonine kinase receptors, followed by phosphorylation of the C-terminus of the activated receptors on SMADs (SMAD1, SMAD1, SMAD5, and SMAD8), which then binds to SMAD4 and ultimately translocates to the nucleus. The transcriptional complexes of SMAD1, SMAD5, SMAD7, and SMAD4 are completely activated by phosphorylation of the linker region of cyclin-dependent kinase (CDK)8 and/or CDK9. The ISCs in mouse embryo, YAP1 is recruited by the phosphorylated SMAD1-SMAD4 complex to the accelerator regions of BMP-responsive genes, thereby inducing the expression of BMP target genes [161].

7.4. Hippo and Hedgehog Signalling

The Hh signaling pathway, which is closely related to intestinal homeostasis, has been described previously, and Indian hedgehog and Sonic hedgehog, two important Hh signaling proteins, have been shown to be expressed in the intestinal epithelium of mice [162]. It was found that depletion of Indian hedgehog in the mouse small intestines resulted in an overexpansion of the ecological niche of ISC and a reduction in ISCs differentiation [163]. Although signaling crosstalk between the Hippo pathway and the Hh pathway in the intestine has not been identified, it has been demonstrated in other organs. In medulloblastoma, where the tumor originates from cerebellar granule neuron precursors (CGNPs), sonic hedgehog signaling promotes their expression [164,165]. Not only that, activation of sonic hedgehog signaling in CGNPs also upregulates the protein level of YAP1, which directly binds to insulin receptor substrate 1 (IRS1), which in turn leads to nuclear localization of YAP1 [166]. Moreover, YAP1 deficiency counteracts sonic hedgehog-induced proliferation of CGNP. sonic hedgehog signaling downstream of the factor GLI2, which has been regarded as a target gene of the YAP1–tead complex, suggests a mechanism by which YAP1 drives adult neural tube cell tumorigenesis. In the intestine, however, the predominant Hh protein is Indian hedgehog. therefore, crosstalk between Hh and Hippo signaling pathways in the intestine may be controlled by Indian hedgehog, but this needs to be further explored. Overall, the Hippo signaling loop has been suggested to crosstalk with the BMP and Hh loops, and one could hypothesize that this crosstalk also exists in the intestine, which in turn maintains intestinal homeostasis.

In general, hedgehog signaling modulates many organs' differentiation, proliferation, and morphogenesis, including patched (PTCH) and smoothened (SMO) trans membrane proteins. Apart from this, the loops include Indian hedgehog (IHH), desert hedgehog (DHH), Sonic hedgehog (Shh), suppressor of fused protein (SUFU), and glioma-associated oncogene (GLI) transcription factors. Dysfunction of this signaling pathway has been reported to be associated with a variety of cancers. Binding of ligand Hh to the PTCH receptor further internalizes and degrades the receptor protein, which in turn releases SMO and leads to dissociation of the SUFU-GLI complex. Ultimately, GLI transcriptional activators translocate to the nucleus and induce the expression of hedgehog target genes including GLI1 and PTCH. In addition, in the absence of ligand, PTCH hinders SMO expression and causes pathway blockage [167].

Studies have shown that abnormalities in the Hh signaling pathway are associated with the development of several intestinal cancers [168]. Dysfunctional Hh signaling pathways can be classified into three types, each of which can induce different types of cancers separately: autonomous and ligand-independent types of Hh signaling, ligand-dependent oncogenic Hh signaling in autocrine/juxtacrine manner, and ligand-dependent Hh signaling in paracrine or reverse paracrine manner [169]. Mutations often result in non-dependent ligand activation. Either SMO mutations (activation) or PTCH mutations (inactivation) activate the Hh signaling pathway, which in turn promotes tumor growth [170]. During autocrine/paracrine signaling, similar tumor cells produce and take up a ligand. Overactivation of Hh ligands promotes tumor growth during the pathogenesis of several cancers [171]. Studies have shown that paracrine activation of the Hh pathway can exacerbate cancer by binding to PTCH receptors on tumor cells, leading to increased proliferation and differentiation [172].

The Hippo pathway, a highly conserved pathway, can regulate biological processes including cell proliferation and differentiation by activating target genes, and dysregulation of this pathway induces the development of a variety of intestinal diseases. The Hippo pathway is composed of mammalian Ste20-like kinases 1/2, large tumor suppressor 1/2, yes association protein, and its paralog transcriptional coactivator with PDZ-binding motif (TAZ) (also called be WWTR1) [173]. Activation of the Hippo pathway induces phosphorylation of MST1/2, which further activates LATS1/2 and inhibits YAP/TAZ activity [174]. Subsequently, YAP/TAZ is confined to the cytoplasm and degraded by the proteasome, which further triggers signaling pathway dysregulation that ultimately promotes cancer progression [175].

8. Concluding Remarks and Future Perspectives

Multifunctional intestinal epithelial cells have organized heterogeneity. The single layer of epithelial cells is responsible for the digestion and absorption of nutrients and the secretion of digestive enzymes and other enzymes and regarded as a protective barrier stopping harmful substances. Chickens are precocious animals; thus, the intestines are thoroughly matured in the postembryonic stage. Several basic studies over the past decades have examined chicken embryonic gut evolution. Yet, the detailed cytoarchitecture and regulatory mechanisms of chicken intestinal epithelial cell differentiation are currently unknown. In this paper, we present detailed knowledge of chicken intestinal epithelial dynamics and physiology by describing the intestinal developmental process, differentiation, modulation, and cellular dynamics. In addition, we also have similarities and differences between birds and mammals. Finally, we concisely describe recent in vitro models for the detection of chicken small intestinal epithelium, in particular to offer new ways for future studies of the molecular signaling pathways that regulate cell proliferation and differentiation of chicken ISCs.

This review offers a basis for further study of the influences of pathogen incidence and nutritive supplementation on the differentiation or development of the small intestinal epithelium in chickens. Based on the prior literature, chickens were observed to differ from mammals and rodents in (1) the proliferation of intestinal epithelial cells from the crypt villi, and (2) the exists of Paneth cells that are controversial and may appear in both crypts and villi. These results may indicate that there are distinct underlying mechanisms for the differentiation and proliferation of chicken intestinal epithelial cells. Therefore, it is important to conduct further studies to understand the unique knowledge of the chicken intestinal epithelium so as to correctly analyze the nutritional or microbial connections that influence the health of the chicken intestinal epithelium.

Author Contributions: Conceptualization, methodology, formal analysis, writing—original draft preparation, visualization, T.Q.; resources, supervision, project administration, funding acquisition, R.L.; conceptualization, formal analysis, validation, writing—review and editing, project administration, funding acquisition, supervision, Y.C. and T.G. All authors have read and agreed to the published version of the manuscript.

Funding: This work was supported by the Chinese National Natural Science Foundation (32402845, 32172801, 32372954) and Beijing Natural Science Foundation (5244041, 6222019).

Institutional Review Board Statement: Not applicable.

Data Availability Statement: Data will be made available on request.

Conflicts of Interest: The authors declare that they have no known competing financial interests or personal relationships that could have appeared to influence the work reported in this paper.

Abbreviations

ISC	Intestinal stem cells
TA	Transport amplification
BrdU	5-bromo-2-deoxyuridine
PCNA	Proliferating cell nuclear antigen
IBD	Inflammatory bowel disease
ECM	Extracellular matrix
EGF	Epidermal growth factor
BMP	Bone morphogenetic proteins
MSCs	Mesenchymal stem cells
Hh	Hedgehog
Ihh	Indian Hh
YAP	YES-associated protein
TAZ	PDZ-binding motif
PKCζ	Protein kinase C zeta type
NICD	Notch intracellular structural domain
Olfm4	Olfactomedin 4
CDK	Cyclin-dependent kinase
CGNPs	Cerebellar granule neuron precursors
IRS1	Insulin receptor substrate 1

References

1. Van Boeckel, T.P.; Brower, C.; Gilbert, M.; Grenfell, B.T.; Levin, S.A.; Robinson, T.P.; Teillant, A.; Laxminarayan, R. Global trends in antimicrobial use in food animals. *Proc. Natl. Acad. Sci. USA* **2015**, *112*, 5649–5654. [CrossRef] [PubMed]
2. Cui, Y.; Wang, Q.; Liu, S.; Sun, R.; Zhou, Y.; Li, Y. Age-related variations in intestinal microflora of free-range and caged hens. *Front. Microbiol.* **2017**, *8*, 1310. [CrossRef]
3. Yeoman, C.J.; Chia, N.; Jeraldo, P.; Sipos, M.; Goldenfeld, N.D.; White, B.A. The microbiome of the chicken gastrointestinal tract. *Anim. Health Res. Rev.* **2012**, *13*, 89–99. [CrossRef]
4. Pan, D.; Yu, Z. Intestinal microbiome of poultry and its interaction with host and diet. *Gut Microbes* **2014**, *5*, 108–119. [CrossRef] [PubMed]
5. Diaz Carrasco, J.M.; Casanova, N.A.; Fernández Miyakawa, M.E. Microbiota, gut health and chicken productivity: What is the connection? *Microorganisms* **2019**, *7*, 374. [CrossRef]
6. Garrett, W.S.; Gordon, J.I.; Glimcher, L.H. Homeostasis and inflammation in the intestine. *Cell* **2010**, *140*, 859–870. [CrossRef]
7. Peterson, L.W.; Artis, D. Intestinal epithelial cells: Regulators of barrier function and immune homeostasis. *Nat. Rev. Immunol.* **2014**, *14*, 141–153. [CrossRef]
8. Gilbert, E.R.; Li, H.; Emmerson, D.A.; Webb, K.E.; Wong, E.A. Developmental regulation of nutrient transporter and enzyme mRNA abundance in the small intestine of broilers. *Poult. Sci.* **2007**, *86*, 1739–1753. [CrossRef] [PubMed]
9. Crosnier, C.; Stamataki, D.; Lewis, J. Organizing cell renewal in the intestine: Stem cells, signals and combinatorial control. *Nat. Rev. Genet* **2006**, *7*, 349–359. [CrossRef]
10. Quan, T.; Li, R.; Gao, T. The intestinal macrophage-intestinal stem cell axis in inflammatory bowel diseases: From pathogenesis to therapy. *Int. J. Mol. Sci.* **2025**, *26*, 2855. [CrossRef]
11. Jasper, H. Intestinal Stem Cell Aging: Origins and Interventions. *Annu. Rev. Physiol.* **2020**, *82*, 203–226. [CrossRef] [PubMed]
12. Uni, Z.; Platin, R.; Sklan, D. Cell proliferation in chicken intestinal epithelium occurs both in the crypt and along the villus. *J. Comp. Physiol. B Biochem. Syst. Environ. Physiol.* **1998**, *168*, 241–247. [CrossRef] [PubMed]
13. Vanuytsel, T.; Senger, S.; Fasano, A.; Shea-Donohue, T. Major signaling pathways in intestinal stem cells. *Biochim. Biophys. Acta* **2013**, *1830*, 2410–2426. [CrossRef] [PubMed]
14. Grapin-Botton, A.; Melton, D.A. Endoderm development: From patterning to organogenesis. *Trends Genet.* **2000**, *16*, 124–130. [CrossRef]
15. Barry, E.R.; Camargo, F.D. The Hippo superhighway: Signaling crossroads converging on the Hippo/Yap pathway in stem cells and development. *Curr Opin Cell Biol.* **2013**, *25*, 247–253. [CrossRef]
16. Piccolo, S.; Dupont, S.; Cordenonsi, M. The biology of YAP/TAZ: Hippo signaling and beyond. *Physiol Rev.* **2014**, *94*, 1287–1312. [CrossRef]
17. Moroishi, T.; Park, H.W.; Qin, B.; Chen, Q.; Meng, Z.; Plouffe, S.W.; Taniguchi, K.; Yu, F.X.; Karin, M.; Pan, D.; et al. A YAP/TAZ-induced feedback mechanism regulates Hippo pathway homeostasis. *Genes. Dev.* **2015**, *29*, 1271–1284. [CrossRef]
18. Harvey, K.F.; Zhang, X.; Thomas, D.M. The Hippo pathway and human cancer. *Nat. Rev. Cancer* **2013**, *13*, 246–257. [CrossRef]
19. Roberts, D.J.; Johnson, R.L.; Burke, A.C.; Nelson, C.E.; Morgan, B.A.; Tabin, C. Sonic hedgehog is an endodermal signal inducing Bmp-4 and Hox genes during induction and regionalization of the chick hindgut. *Development* **1995**, *121*, 3163–3174. [CrossRef]
20. Yu, F.X.; Meng, Z.; Plouffe, S.W.; Guan, K.L. Hippo pathway regulation of gastrointestinal tissues. *Annu. Rev. Physiol.* **2015**, *77*, 201–227. [CrossRef]
21. Theodosiou, N.A.; Tabin, C.J. Wnt signaling during development of the gastrointestinal tract. *Dev. Biol.* **2003**, *259*, 258–271. [CrossRef] [PubMed]
22. Southwell, B.R. Staging of intestinal development in the chick embryo. *Anat. Rec. Part. A Discov. Mol. Cell. Evol. Biol.* **2006**, *288*, 909–920. [CrossRef] [PubMed]
23. McBride, H.J.; Fatke, B.; Fraser, S.E. Wnt signaling components in the chicken intestinal tract. *Dev. Biol.* **2003**, *256*, 18–33. [CrossRef] [PubMed]
24. Chevalier, N.R.; Fleury, V.; Dufour, S.; Proux-Gillardeaux, V.; Asnacios, A. Emergence and development of gut motility in the chicken embryo. *PLoS ONE* **2017**, *12*, e0172511. [CrossRef]
25. Uni, Z.; Smirnov, A.; Sklan, D. Pre- and posthatch development of goblet cells in the broiler small intestine: Effect of delayed access to feed. *Poult. Sci.* **2003**, *82*, 320–327. [CrossRef]
26. Uni, Z.; Tako, E.; Gal-Garber, O.; Sklan, D. Morphological, molecular, and functional changes in the chicken small intestine of the late-term embryo. *Poult. Sci.* **2003**, *82*, 1747–1754. [CrossRef]
27. Kabiri, Z.; Greicius, G.; Madan, B.; Biechele, S.; Zhong, Z.; Zaribafzadeh, H.; Aliyev, J.; Wu, Y.; Bunte, R.; Williams, B.O.; et al. Stroma provides an intestinal stem cell niche in the absence of epithelial Wnts. *Development* **2014**, *141*, 2206–2215. [CrossRef]
28. Barker, N. Adult intestinal stem cells: Critical drivers of epithelial homeostasis and regeneration. *Nat. Rev. Mol. Cell Biol.* **2014**, *15*, 19–33. [CrossRef]

29. Uni, Z.; Geyra, A.; Ben-Hur, H.; Sklan, D. Small intestinal development in the young chick: Crypt formation and enterocyte proliferation and migration. *Br. Poult. Sci.* **2000**, *41*, 544–551. [CrossRef]
30. Shira, E.B.; Friedman, A. Innate immune functions of avian intestinal epithelial cells: Response to bacterial stimuli and localization of responding cells in the developing avian digestive tract. *PLoS ONE* **2018**, *13*, e0200393.
31. Zhang, H.; Wong, E.A. Identification of cells expressing OLFM4 and LGR5 mRNA by in situ hybridization in the yolk sac and small intestine of embryonic and early post-hatch chicks. *Poult. Sci.* **2018**, *97*, 628–633. [CrossRef] [PubMed]
32. Zhang, H.; Wong, E.A. Spatial transcriptional profile of PepT1 mRNA in the yolk sac and small intestine in broiler chickens. *Poult. Sci.* **2017**, *96*, 2871–2876. [CrossRef]
33. Barker, N.; van Es, J.H.; Kuipers, J.; Kujala, P.; van den Born, M.; Cozijnsen, M.; Haegebarth, A.; Korving, J.; Begthel, H.; Peters, P.J.; et al. Identification of stem cells in small intestine and colon by marker gene Lgr5. *Nature* **2007**, *449*, 1003–1007. [CrossRef]
34. van der Flier, L.G.; Clevers, H. Stem cells, self-renewal, and differentiation in the intestinal epithelium. *Annu. Rev. Physiol.* **2009**, *71*, 241–260. [CrossRef] [PubMed]
35. Date, S.; Sato, T. Mini-gut organoids: Reconstitution of the stem cell niche. *Annu. Rev. Cell Dev. Biol.* **2015**, *31*, 269–289. [CrossRef] [PubMed]
36. Santos, A.J.M.; Lo, Y.H.; Mah, A.T.; Kuo, C.J. The intestinal stem cell niche: Homeostasis and adaptations. *Trends Cell Biol.* **2018**, *28*, 1062–1078. [CrossRef]
37. McCarthy, N.; Kraiczy, J.; Shivdasani, R.A. Cellular and molecular architecture of the intestinal stem cell niche. *Nat. Cell Biol.* **2020**, *22*, 1033–1041. [CrossRef]
38. Beumer, J.; Clevers, H. Cell fate specification and differentiation in the adult mammalian intestine. *Nat. Rev. Mol. Cell Biol.* **2021**, *22*, 39–53. [CrossRef]
39. Aoki, R.; Shoshkes-Carmel, M.; Gao, N.; Shin, S.; May, C.L.; Golson, M.L.; Zahm, A.M.; Ray, M.; Wiser, C.L.; Wright, C.V.; et al. Foxl1-expressing mesenchymal cells constitute the intestinal stem cell niche. *Cell Mol. Gastroenterol. Hepatol.* **2016**, *2*, 175–188. [CrossRef]
40. Valenta, T.; Degirmenci, B.; Moor, A.E.; Herr, P.; Zimmerli, D.; Moor, M.B.; Hausmann, G.; Cantù, C.; Aguet, M.; Basler, K. Wnt ligands secreted by subepithelial mesenchymal cells are essential for the survival of intestinal stem cells and gut homeostasis. *Cell Rep.* **2016**, *15*, 911–918. [CrossRef]
41. Sato, T.; Vries, R.G.; Snippert, H.J.; van de Wetering, M.; Barker, N.; Stange, D.E.; van Es, J.H.; Abo, A.; Kujala, P.; Peters, P.J.; et al. Single Lgr5 stem cells build crypt-villus structures in vitro without a mesenchymal niche. *Nature* **2009**, *459*, 262–265. [CrossRef] [PubMed]
42. Porter, E.M.; Bevins, C.L.; Ghosh, D.; Ganz, T. The multifaceted Paneth cell. *Cell Mol. Life Sci.* **2002**, *59*, 156–170. [CrossRef] [PubMed]
43. Clevers, H. The intestinal crypt, a prototype stem cell compartment. *Cell* **2013**, *154*, 274–284. [CrossRef]
44. Ludikhuize, M.C.; Meerlo, M.; Gallego, M.P.; Xanthakis, D.; Burgaya Julià, M.; Nguyen, N.T.B.; Brombacher, E.C.; Liv, N.; Maurice, M.M.; Paik, J.H.; et al. Mitochondria Define Intestinal Stem Cell Differentiation Downstream of a FOXO/Notch Axis. *Cell Metab.* **2020**, *32*, 889–900.e7. [CrossRef]
45. Durand, A.; Donahue, B.; Peignon, G.; Letourneur, F.; Cagnard, N.; Slomianny, C.; Perret, C.; Shroyer, N.F.; Romagnolo, B. Functional intestinal stem cells after Paneth cell ablation induced by the loss of transcription factor Math1 (Atoh1). *Proc. Natl. Acad. Sci. USA* **2012**, *109*, 8965–8970. [CrossRef] [PubMed]
46. Kim, T.H.; Escudero, S.; Shivdasani, R.A. Intact function of Lgr5 receptor expressing intestinal stem cells in the absence of Paneth cells. *Proc. Natl. Acad. Sci. USA* **2012**, *109*, 3932–3937. [CrossRef]
47. Shroyer, N.F.; Wallis, D.; Venken, K.J.; Bellen, H.J.; Zoghbi, H.Y. Gfi1 functions downstream of Math1 to control intestinal secretory cell subtype allocation and differentiation. *Genes. Dev.* **2005**, *19*, 2412–2417. [CrossRef]
48. Bastide, P.; Darido, C.; Pannequin, J.; Kist, R.; Robine, S.; Marty-Double, C.; Bibeau, F.; Scherer, G.; Joubert, D.; Hollande, F.; et al. Sox9 regulates cell proliferation and is required for Paneth cell differentiation in the intestinal epithelium. *J. Cell Biol.* **2007**, *178*, 635–648. [CrossRef]
49. Mori-Akiyama, Y.; van den Born, M.; van Es, J.H.; Hamilton, S.R.; Adams, H.P.; Zhang, J.; Clevers, H.; de Crombrugghe, B. SOX9 is required for the differentiation of paneth cells in the intestinal epithelium. *Gastroenterology* **2007**, *133*, 539–546. [CrossRef]
50. Sasaki, N.; Sachs, N.; Wiebrands, K.; Ellenbroek, S.I.; Fumagalli, A.; Lyubimova, A.; Begthel, H.; van den Born, M.; van Es, J.H.; Karthaus, W.R.; et al. Reg4+ deep crypt secretory cells function as epithelial niche for Lgr5+ stem cells in colon. *Proc. Natl. Acad. Sci. USA* **2016**, *113*, E5399–E5407. [CrossRef]
51. Kraiczy, J.; McCarthy, N.; Malagola, E.; Tie, G.; Madha, S.; Boffelli, D.; Wagner, D.E.; Wang, T.C.; Shivdasani, R.A. Graded BMP signaling within intestinal crypt architecture directs self-organization of the Wnt-secreting stem cell niche. *Cell Stem Cell* **2023**, *30*, 433–449.e8. [CrossRef]

52. He, X.C.; Zhang, J.; Tong, W.G.; Tawfik, O.; Ross, J.; Scoville, D.H.; Tian, Q.; Zeng, X.; He, X.; Wiedemann, L.M.; et al. BMP signaling inhibits intestinal stem cell self-renewal through suppression of Wnt-beta-catenin signaling. *Nat. Genet.* **2004**, *36*, 1117–1121. [CrossRef] [PubMed]
53. Beyaz, S.; Mana, M.D.; Roper, J.; Kedrin, D.; Saadatpour, A.; Hong, S.J.; Yilmaz, O.H. High-fat diet enhances stemness and tumorigenicity of intestinal progenitors. *Nature* **2016**, *531*, 53. [CrossRef]
54. Yilmaz, O.H.; Katajisto, P.; Lamming, D.W.; Gültekin, Y.; Bauer-Rowe, K.E.; Sengupta, S.; Sabatini, D.M. mTORC1 in the Paneth cell niche couples intestinal stem-cell function to calorie intake. *Nature* **2012**, *486*, 490. [CrossRef]
55. Enriquez, J.R.; McCauley, H.A.; Zhang, K.X.; Sanchez, J.G.; Kalin, G.T.; Lang, R.A.; Wells, J.M. A dietary change to a high-fat diet initiates a rapid adaptation of the intestine. *Cell Rep.* **2022**, *41*, 111641. [CrossRef] [PubMed]
56. Mihaylova, M.M.; Cheng, C.W.; Cao, A.Q.; Tripathi, S.; Mana, M.D.; Bauer-Rowe, K.E.; Lewis, C.A. Fasting activates fatty acid oxidation to enhance intestinal stem cell function during homeostasis and aging. *Cell Stem Cell* **2018**, *22*, 769–778.e4. [CrossRef]
57. Blot, F.; Marchix, J.; Ejarque, M.; Jimenez, S.; Meunier, A.; Keime, C.; Gradwohl, G. Gut microbiota remodeling and intestinal adaptation to lipid malabsorption after enteroendocrine cell loss in adult mice. *Cell. Mol. Gastroenterol. Hepatol.* **2023**, *15*, 1443–1461. [CrossRef] [PubMed]
58. McCauley, H.A.; Riedman, A.M.; Enriquez, J.R.; Zhang, X.; Watanabe-Chailland, M.; Sanchez, J.G.; Wells, J.M. Enteroendocrine cells protect the stem cell niche by regulating crypt metabolism in response to nutrients. *Cell. Mol. Gastroenterol. Hepatol.* **2023**, *15*, 1293–1310. [CrossRef]
59. McCauley, H.A. Enteroendocrine regulation of nutrient absorption. *J. Nutr.* **2019**, *150*, 10–21. [CrossRef]
60. Rock, S.A.; Jiang, K.; Wu, Y.; Liu, Y.; Li, J.; Weiss, H.L.; Evers, B.M. Neurotensin regulates proliferation and stem cell function in the small intestine in a nutrientdependent manner. *Cell. Mol. Gastroenterol. Hepatol.* **2022**, *13*, 501–516. [CrossRef]
61. Gross, E.R.; Gershon, M.D.; Margolis, K.G.; Gertsberg, Z.V.; Li, Z.; Cowles, R.A. Neuronal serotonin regulates growth of the intestinal mucosa in mice. *Gastroenterology* **2012**, *143*, 408–417.e2. [CrossRef]
62. Poplaski, V.; Bomidi, C.; Kambal, A.; Nguyen-Phuc, H.; Di Rienzi, S.C.; Danhof, H.A.; Blutt, S.E. Human intestinal organoids from Cronkhite-Canada syndrome patients reveal link between serotonin and proliferation. *J. Clin. Investig.* **2023**, *133*, e166884. [CrossRef] [PubMed]
63. Lee, B.; Moon, K.M.; Kim, C.Y. Tight junction in the intestinal epithelium: Its association with diseases and regulation by phytochemicals. *J. Immunol. Res.* **2018**, *2018*, 2645465. [CrossRef]
64. Drucker, D.J.; Erlich, P.; Asa, S.L.; Brubaker, P.L. Induction of intestinal epithelial proliferation by glucagon-like peptide 2. *Proc. Natl. Acad. Sci. USA* **1996**, *93*, 7911–7916. [CrossRef]
65. Fesler, Z.; Mitova, E.; Brubaker, P.L. GLP-2, EGF, and the intestinal epithelial IGF-1 receptor interactions in the regulation of crypt cell proliferation. *Endocrinology* **2020**, *161*, bqaa040. [CrossRef] [PubMed]
66. Markovic, M.A.; Brubaker, P.L. The roles of glucagon-like peptide-2 and the intestinal epithelial insulin-like growth factor-1 receptor in regulating microvillus length. *Sci. Rep.* **2019**, *9*, 13010. [CrossRef] [PubMed]
67. Yusta, B.; Holland, D.; Koehler, J.A.; Maziarz, M.; Estall, J.L.; Higgins, R.; Drucker, D.J. ErbB signaling is required for the proliferative actions of GLP-2 in the murine gut. *Gastroenterology* **2009**, *137*, 986–996. [CrossRef]
68. Yusta, B.; Huang, L.; Munroe, D.; Wolff, G.; Fantaske, R.; Sharma, S.; Drucker, D.J. Enteroendocrine localization of GLP-2 receptor expression in humans and rodents. *Gastroenterology* **2000**, *119*, 744–755. [CrossRef]
69. Yusta, B.; Matthews, D.; Koehler, J.A.; Pujadas, G.; Kaur, K.D.; Drucker, D.J. Localization of glucagon-like peptide-2 receptor expression in the mouse. *Endocrinology* **2019**, *160*, 1950–1963. [CrossRef]
70. Amcheslavsky, A.; Song, W.; Li, Q.; Nie, Y.; Bragatto, I.; Ferrandon, D.; Ip, Y.T. Enteroendocrine cells support intestinal stem-cell-mediated homeostasis in Drosophila. *Cell Rep.* **2014**, *9*, 32–39. [CrossRef]
71. Scopelliti, A.; Cordero, J.B.; Diao, F.; Strathdee, K.; White, B.H.; Sansom, O.J.; Vidal, M. Local control of intestinal stem cell homeostasis by enteroendocrine cells in the adult Drosophila midgut. *Curr. Biol.* **2014**, *24*, 1199–1211. [CrossRef]
72. Groschwitz, K.R.; Hogan, S.P. Intestinal barrier function: Molecular regulation and disease pathogenesis. *J. Allergy Clin. Immunol.* **2009**, *124*, 3–20; quiz 21–22. [CrossRef]
73. Kagnoff, M.F. The intestinal epithelium is an integral component of a communications network. *J. Clin. Investig.* **2014**, *124*, 2841–2843. [CrossRef] [PubMed]
74. Birchenough, G.M.; Johansson, M.E.; Gustafsson, J.K.; Bergström, J.H.; Hansson, G.C. New developments in goblet cell mucus secretion and function. *Mucosal Immunol.* **2015**, *8*, 712–719. [CrossRef] [PubMed]
75. Bevins, C.L.; Salzman, N.H. Paneth cells, antimicrobial peptides and maintenance of intestinal homeostasis. *Nat. Rev. Microbiol.* **2011**, *9*, 356–368. [CrossRef] [PubMed]
76. Kaetzel, C.S.; Robinson, J.K.; Chintalacharuvu, K.R.; Vaerman, J.P.; Lamm, M.E. The polymeric immunoglobulin receptor (secretory component) mediates transport of immune complexes across epithelial cells: A local defense function for IgA. *Proc. Natl. Acad. Sci. USA* **1991**, *88*, 8796–8800. [CrossRef]

77. Johansson, M.E.; Sjovall, H.; Hansson, G.C. The gastrointestinal mucus system in health and disease. *Nat. Rev. Gastroenterol. Hepatol.* **2013**, *10*, 352–361. [CrossRef]
78. Herath, M.; Hosie, S.; Bornstein, J.C.; Franks, A.E.; Hill-Yardin, E.L. The role of the gastrointestinal mucus system in intestinal homeostasis: Implications for neurological disorders. *Front. Cell. Infect. Microbiol.* **2020**, *10*, 248. [CrossRef]
79. Nystrom, E.E.L.; Martinez-Abad, B.; Arike, L.; Birchenough, G.M.H.; Nonnecke, E.B.; Castillo, P.A.; Svensson, F.; Bevins, C.L.; Hansson, G.C.; Johansson, M.E.V. An intercrypt subpopulation of goblet cells is essential for colonic mucus barrier function. *Science* **2021**, *372*, eabb1590. [CrossRef]
80. Bergstrom, K.; Shan, X.; Casero, D.; Batushansky, A.; Lagishetty, V.; Jacobs, J.P.; Hoover, C.; Kondo, Y.; Shao, B.; Gao, L.; et al. Proximal colon-derived O-glycosylated mucus encapsulates and modulates the microbiota. *Science* **2020**, *370*, 467–472. [CrossRef]
81. La Rosa, S.L.; Ostrowski, M.P.; Vera-Ponce de Leon, A.; McKee, L.S.; Larsbrink, J.; Eijsink, V.G.; Lowe, E.C.; Martens, E.C.; Pope, P.B. Glycan processing in gut microbiomes. *Curr. Opin. Microbiol.* **2022**, *67*, 102143. [CrossRef]
82. Yao, Y.; Kim, G.; Shafer, S.; Chen, Z.; Kubo, S.; Ji, Y.; Luo, J.; Yang, W.; Perner, S.P.; Kanellopoulou, C.; et al. Mucus sialylation determines intestinal host-commensal homeostasis. *Cell* **2022**, *185*, 1172–1188.e28. [CrossRef] [PubMed]
83. Shang, W.; Gottlieb, S.; Yao, Y. Targeting the ECM component HA: A new route for mCRC treatment. *Sci. China Life Sci.* **2023**, *66*, 1944–1946. [CrossRef] [PubMed]
84. Martinez-Ordonez, A.; Duran, A.; Ruiz-Martinez, M.; Cid-Diaz, T.; Zhang, X.; Han, Q.; Kinoshita, H.; Muta, Y.; Linares, J.F.; Kasashima, H.; et al. Hyaluronan driven by epithelial aPKC deficiency remodels the microenvironment and creates a vulnerability in mesenchymal colorectal cancer. *Cancer Cell* **2023**, *41*, 252–271.e9. [CrossRef] [PubMed]
85. Mukherjee, S.; Hooper, L.V. Antimicrobial defense of the intestine. *Immunity* **2015**, *42*, 28–39. [CrossRef]
86. Caballero, S.; Pamer, E.G. Microbiota-mediated inflammation and antimicrobial defense in the intestine. *Annu. Rev. Immunol.* **2015**, *33*, 227–256. [CrossRef]
87. Huan, Y.; Kong, Q.; Mou, H.; Yi, H. Antimicrobial peptides: Classification, design, application and research progress in multiple fields. *Front. Microbiol.* **2020**, *11*, 582779. [CrossRef]
88. Erdem Büyükkiraz, M.; Kesmen, Z. Antimicrobial peptides (AMPs): A promising class of antimicrobial compounds. *J. Appl. Microbiol.* **2022**, *132*, 1573–1596. [CrossRef]
89. Koprivnjak, T.; Peschel, A. Bacterial resistance mechanisms against host defense peptides. *Cell. Mol. Life Sci.* **2011**, *68*, 2243–2254. [CrossRef]
90. Mukherjee, S.; Zheng, H.; Derebe, M.G.; Callenberg, K.M.; Partch, C.L.; Rollins, D.; Propheter, D.C.; Rizo, J.; Grabe, M.; Jiang, Q.X.; et al. Antibacterial membrane attack by a pore-forming intestinal C-type lectin. *Nature* **2014**, *505*, 103–107. [CrossRef]
91. Chelakkot, C.; Ghim, J.; Ryu, S.H. Mechanisms regulating intestinal barrier integrity and its pathological implications. *Exp. Mol. Med.* **2018**, *50*, 1–9. [CrossRef]
92. Paradis, T.; Begue, H.; Basmaciyan, L.; Dalle, F.; Bon, F. Tight junctions as a key for pathogens invasion in intestinal epithelial cells. *Int. J. Mol. Sci.* **2021**, *22*, 2506. [CrossRef]
93. Portincasa, P.; Bonfrate, L.; Khalil, M.; Angelis, M.; Calabrese, F.M.; D'Amato, M.; Wang, D.Q.; Di Ciaula, A. Intestinal barrier and permeability in health, obesity and NAFLD. *Biomedicines* **2021**, *10*, 83. [CrossRef] [PubMed]
94. Hollander, D.; Kaunitz, J.D. The "leaky gut": Tight junctions but loose associations? *Dig. Dis. Sci.* **2020**, *65*, 1277–1287. [CrossRef]
95. Katoh, M. Canonical and non-canonical WNT signaling in cancer stem cells and their niches: Cellular heterogeneity, omics reprogramming, targeted therapy and tumor plasticity (Review). *Int. J. Oncol.* **2017**, *51*, 1357–1369. [CrossRef] [PubMed]
96. Wang, S.; Chen, Y.G. BMP signaling in homeostasis, transformation and inflammatory response of intestinal epithelium. *Sci. China Life Sci.* **2018**, *61*, 800–807. [CrossRef] [PubMed]
97. Ye, W.; Takabayashi, H.; Yang, Y.; Mao, M.; Hibdon, E.S.; Samuelson, L.C.; Eaton, K.A.; Todisco, A. Regulation of gastric Lgr5+ve cell homeostasis by bone morphogenetic protein (BMP) signaling and inflammatory stimuli. *Cell Mol. Gastroenterol. Hepatol.* **2018**, *5*, 523–538. [CrossRef]
98. Li, Y.; Liu, Y.; Liu, B.; Wang, J.; Wei, S.; Qi, Z.; Wang, S.; Fu, W.; Chen, Y.G. A growth factor-free culture system underscores the coordination between Wnt and BMP signaling in Lgr5+ intestinal stem cell maintenance. *Cell Discov.* **2018**, *4*, 49. [CrossRef]
99. Bayne, R.A.; Donnachie, D.J.; Kinnell, H.L.; Childs, A.J.; Anderson, R.A. BMP signalling in human fetal ovary somatic cells is modulated in a gene-specific fashion by GREM1 and GREM2. *Mol. Hum. Reprod.* **2016**, *22*, 622–633. [CrossRef]
100. Sunkara, R.R.; Mehta, D.; Sarate, R.M.; Waghmare, S.K. BMP-AKT-GSK3β signaling restores hair follicle stem cells decrease associated with loss of sfrp1. *Stem Cells* **2022**, *40*, 802–817. [CrossRef]
101. Fevr, T.; Robine, S.; Louvard, D.; Huelsken, J. Wnt/beta-catenin is essential for intestinal homeostasis and maintenance of intestinal stem cells. *Mol. Cell Biol.* **2007**, *27*, 7551–7559. [CrossRef] [PubMed]
102. Kim, S.; Shin, Y.C.; Kim, T.Y.; Kim, Y.; Lee, Y.S.; Lee, S.H.; Kim, M.N.; O, E.; Kim, K.S.; Kweon, M.N. Mucin degrader Akkermansia muciniphila accelerates intestinal stem cell-mediated epithelial development. *Gut Microbes* **2021**, *13*, 1–20. [CrossRef]
103. Zhang, H.; Lin, M.; Dong, C.; Tang, Y.; An, L.; Ju, J.; Wen, F.; Chen, F.; Wang, M.; Wang, W.; et al. An MST4-pβ-CateninThr40 signaling axis controls intestinal stem cell and tumorigenesis. *Adv. Sci.* **2021**, *8*, e2004850. [CrossRef]

104. Chen, S.; Zheng, Y.; Ran, X.; Du, H.; Feng, H.; Yang, L.; Wen, Y.; Lin, C.; Wang, S.; Huang, M.; et al. Integrin αEβ7+ T cells direct intestinal stem cell fate decisions via adhesion signaling. *Cell Res.* **2021**, *31*, 1291–1307. [CrossRef]
105. Young, M.A.; Daly, C.S.; Taylor, E.; James, R.; Clarke, A.R.; Reed, K.R. Subtle deregulation of the Wnt-signaling pathway through loss of Apc2 reduces the fitness of intestinal stem cells. *Stem Cells* **2018**, *36*, 114–122. [CrossRef]
106. Artavanis-Tsakonas, S.; Rand, M.D.; Lake, R.J. Notch signaling: Cell fate control and signal integration in development. *Science* **1999**, *284*, 770–776. [CrossRef] [PubMed]
107. Schröder, N.; Gossler, A. Expression of Notch pathway components in fetal and adult mouse small intestine. *Gene Expr. Patterns* **2002**, *2*, 247–250. [CrossRef] [PubMed]
108. Ogaki, S.; Shiraki, N.; Kume, K.; Kume, S. Wnt and Notch signals guide embryonic stem cell differentiation into the intestinal lineages. *Stem Cells* **2013**, *31*, 1086–1096. [CrossRef]
109. Teratani-Ota, Y.; Yamamizu, K.; Piao, Y.; Sharova, L.; Amano, M.; Yu, H.; Schlessinger, D.; Ko, M.S.; Sharov, A.A. Induction of specific neuron types by overexpression of single transcription factors. *In Vitro Cell Dev. Biol. Anim.* **2016**, *52*, 961–973. [CrossRef]
110. Cheng, C.W.; Biton, M.; Haber, A.L.; Gunduz, N.; Eng, G.; Gaynor, L.T.; Tripathi, S.; Calibasi-Kocal, G.; Rickelt, S.; Butty, V.L.; et al. Ketone body signaling mediates intestinal stem cell homeostasis and adaptation to diet. *Cell* **2019**, *178*, 1115–1131.e15. [CrossRef]
111. Kurokawa, K.; Hayakawa, Y.; Koike, K. Plasticity of intestinal epithelium: Stem cell niches and regulatory signals. *Int. J. Mol. Sci.* **2020**, *22*, 357. [CrossRef] [PubMed]
112. Skoda, A.M.; Simovic, D.; Karin, V.; Kardum, V.; Vranic, S.; Serman, L. The role of the Hedgehog signaling pathway in cancer: A comprehensive review. *Bosn. J. Basic. Med. Sci.* **2018**, *18*, 8–20. [CrossRef] [PubMed]
113. Zeng, X.; Ju, D. Hedgehog signaling pathway and autophagy in cancer. *Int. J. Mol. Sci.* **2018**, *19*, 2279. [CrossRef]
114. Ingham, P.W. Hedgehog signaling. *Curr. Top. Dev. Biol.* **2022**, *149*, 1–58.
115. Quaglio, D.; Infante, P.; Di Marcotullio, L.; Botta, B.; Mori, M. Hedgehog signaling pathway inhibitors: An updated patent review (2015-present). *Expert. Opin. Ther. Pat.* **2020**, *30*, 235–250. [CrossRef]
116. Normanno, N.; De Luca, A.; Bianco, C.; Strizzi, L.; Mancino, M.; Maiello, M.R.; Carotenuto, A.; De Feo, G.; Caponigro, F.; Salomon, D.S. Epidermal growth factor receptor (EGFR) signaling in cancer. *Gene* **2006**, *366*, 2–16. [CrossRef] [PubMed]
117. De Luca, A.; Carotenuto, A.; Rachiglio, A.; Gallo, M.; Maiello, M.R.; Aldinucci, D.; Pinto, A.; Normanno, N. The role of the EGFR signaling in tumor microenvironment. *J. Cell Physiol.* **2008**, *214*, 559–567. [CrossRef]
118. Wong, V.W.; Stange, D.E.; Page, M.E.; Buczacki, S.; Wabik, A.; Itami, S.; van de Wetering, M.; Poulsom, R.; Wright, N.A.; Trotter, M.W.; et al. Lrig1 controls intestinal stem-cell homeostasis by negative regulation of ErbB signalling. *Nat. Cell Biol.* **2012**, *14*, 401–408. [CrossRef]
119. Powell, A.E.; Wang, Y.; Li, Y.; Poulin, E.J.; Means, A.L.; Washington, M.K.; Higginbotham, J.N.; Juchheim, A.; Prasad, N.; Levy, S.E.; et al. The pan-ErbB negative regulator Lrig1 is an intestinal stem cell marker that functions as a tumor suppressor. *Cell* **2012**, *149*, 146–158. [CrossRef]
120. An, Y.; Tan, S.; Yang, J.; Gao, T.; Dong, Y. The potential role of Hippo pathway regulates cellular metabolism via signaling crosstalk in disease-induced macrophage polarization. *Front. Immunol.* **2024**, *14*, 1344697. [CrossRef]
121. Wu, Z.; Guan, K.L. Hippo signaling in embryogenesis and development. *Trends Biochem. Sci.* **2021**, *46*, 51–63. [CrossRef]
122. Meng, Z.; Moroishi, T.; Guan, K.L. Mechanisms of hippo pathway regulation. *Genes. Dev.* **2016**, *30*, 1–17. [CrossRef] [PubMed]
123. Misra, J.R.; Irvine, K.D. The hippo signaling network and its biological functions. *Annu. Rev. Genet.* **2018**, *52*, 65–87. [CrossRef] [PubMed]
124. Zhou, D.; Zhang, Y.; Wu, H.; Barry, E.; Yin, Y.; Lawrence, E.; Dawson, D.; Willis, J.E.; Markowitz, S.D.; Camargo, F.D.; et al. Mst1 and Mst2 protein kinases restrain intestinal stem cell proliferation and colonic tumorigenesis by inhibition of Yes-associated protein (Yap) overabundance. *Proc. Natl. Acad. Sci. USA* **2011**, *108*, E1312–E1320. [CrossRef]
125. Gregorieff, A.; Liu, Y.; Inanlou, M.R.; Khomchuk, Y.; Wrana, J.L. Yap-dependent reprogramming of Lgr5(+) stem cells drives intestinal regeneration and cancer. *Nature* **2015**, *526*, 715–718. [CrossRef] [PubMed]
126. Cai, J.; Zhang, N.; Zheng, Y.; de Wilde, R.F.; Maitra, A.; Pan, D. The Hippo signaling pathway restricts the oncogenic potential of an intestinal regeneration program. *Genes. Dev.* **2010**, *24*, 2383–2388. [CrossRef]
127. Barry, E.R.; Morikawa, T.; Butler, B.L.; Shrestha, K.; de la Rosa, R.; Yan, K.S.; Fuchs, C.S.; Magness, S.T.; Smits, R.; Ogino, S.; et al. Restriction of intestinal stem cell expansion and the regenerative response by YAP. *Nature* **2013**, *493*, 106–110. [CrossRef]
128. Mo, J.S.; Park, H.W.; Guan, K.L. The Hippo signaling pathway in stem cell biology and cancer. *EMBO Rep.* **2014**, *15*, 642–656. [CrossRef]
129. Azzolin, L.; Panciera, T.; Soligo, S.; Enzo, E.; Bicciato, S.; Dupont, S.; Bresolin, S.; Frasson, C.; Basso, G.; Guzzardo, V.; et al. YAP/TAZ incorporation in the β-catenin destruction complex orchestrates the Wnt response. *Cell* **2014**, *158*, 157–170. [CrossRef]
130. Imajo, M.; Ebisuya, M.; Nishida, E. Dual role of YAP and TAZ in renewal of the intestinal epithelium. *Nat. Cell Biol.* **2015**, *17*, 7–19. [CrossRef]

131. Llado, V.; Nakanishi, Y.; Duran, A.; Reina-Campos, M.; Shelton, P.M.; Linares, J.F.; Yajima, T.; Campos, A.; Aza-Blanc, P.; Leitges, M.; et al. Repression of intestinal stem cell function and tumorigenesis through direct phosphorylation of β-catenin and Yap by PKCζ. *Cell Rep.* **2015**, *10*, 740–754. [CrossRef] [PubMed]
132. Qin, M.; Geng, E.; Wang, J.; Yu, M.; Dong, T.; Li, S.; Zhang, X.; Lin, J.; Shi, M.; Li, J.; et al. LATS2 condensates organize signalosomes for Hippo pathway signal transduction. *Nat. Chem. Biol.* **2024**, *20*, 710–720. [CrossRef] [PubMed]
133. Chen, Q.; Zhang, N.; Xie, R.; Wang, W.; Cai, J.; Choi, K.S.; David, K.K.; Huang, B.; Yabuta, N.; Nojima, H.; et al. Homeostatic control of Hippo signaling activity revealed by an endogenous activating mutation in YAP. *Genes. Dev.* **2015**, *29*, 1285–1297. [CrossRef]
134. Dai, X.; Liu, H.; Shen, S.; Guo, X.; Yan, H.; Ji, X.; Li, L.; Huang, J.; Feng, X.H.; Zhao, B. YAP activates the Hippo pathway in a negative feedback loop. *Cell Res.* **2017**, *27*, 1073. [CrossRef]
135. Gehart, H.; Clevers, H. Tales from the crypt: New insights into intestinal stem cells. *Nat. Rev. Gastroenterol. Hepatol.* **2019**, *16*, 19–34. [CrossRef]
136. Pinto, D.; Gregorieff, A.; Begthel, H.; Clevers, H. Canonical Wnt signals are essential for homeostasis of the intestinal epithelium. *Genes. Dev.* **2003**, *17*, 1709–1713. [CrossRef] [PubMed]
137. Batlle, E.; Henderson, J.T.; Beghtel, H.; van den Born, M.M.; Sancho, E.; Huls, G.; Meeldijk, J.; Robertson, J.; van de Wetering, M.; Pawson, T.; et al. Beta-catenin and TCF mediate cell positioning in the intestinal epithelium by controlling the expression of EphB/ephrinB. *Cell* **2002**, *111*, 251–263. [CrossRef]
138. Korinek, V.; Barker, N.; Moerer, P.; van Donselaar, E.; Huls, G.; Peters, P.J.; Clevers, H. Depletion of epithelial stem-cell compartments in the small intestine of mice lacking Tcf-4. *Nat. Genet.* **1998**, *19*, 379–383. [CrossRef]
139. Varelas, X.; Miller, B.W.; Sopko, R.; Song, S.; Gregorieff, A.; Fellouse, F.A.; Sakuma, R.; Pawson, T.; Hunziker, W.; McNeill, H.; et al. The Hippo pathway regulates Wnt/beta-catenin signaling. *Dev. Cell* **2010**, *18*, 579–591. [CrossRef]
140. Azzolin, L.; Zanconato, F.; Bresolin, S.; Forcato, M.; Basso, G.; Bicciato, S.; Cordenonsi, M.; Piccolo, S. Role of TAZ as mediator of Wnt signaling. *Cell* **2012**, *151*, 1443–1456. [CrossRef]
141. Konsavage, W.M.; Kyler, S.L., Jr.; Rennoll, S.A.; Jin, G.; Yochum, G.S. Wnt/β-catenin signaling regulates Yes-associated protein (YAP) gene expression in colorectal carcinoma cells. *J. Biol. Chem.* **2012**, *287*, 11730–11739. [CrossRef] [PubMed]
142. Miyoshi, H.; Ajima, R.; Luo, C.T.; Yamaguchi, T.P.; Stappenbeck, T.S. Wnt5a potentiates TGF-β signaling to promote colonic crypt regeneration after tissue injury. *Science* **2012**, *338*, 108–113. [CrossRef]
143. Park, H.W.; Kim, Y.C.; Yu, B.; Moroishi, T.; Mo, J.S.; Plouffe, S.W.; Meng, Z.; Lin, K.C.; Yu, F.X.; Alexander, C.M.; et al. Alternative Wnt signaling activates YAP/TAZ. *Cell* **2015**, *162*, 780–794. [CrossRef] [PubMed]
144. Kopan, R.; Ilagan, M.X. The canonical Notch signaling pathway: Unfolding the activation mechanism. *Cell* **2009**, *137*, 216–233. [CrossRef]
145. Van Dussen, K.L.; Carulli, A.J.; Keeley, T.M.; Patel, S.R.; Puthoff, B.J.; Magness, S.T.; Tran, I.T.; Maillard, I.; Siebel, C.; Kolterud, Å.; et al. Notch signaling modulates proliferation and differentiation of intestinal crypt base columnar stem cells. *Development* **2012**, *139*, 488–497. [CrossRef] [PubMed]
146. Liang, S.J.; Li, X.G.; Wang, X.Q. Notch signaling in mammalian intestinal stem cells: Determining cell fate and maintaining homeostasis. *Curr. Stem Cell Res. Ther.* **2019**, *14*, 583–590. [CrossRef]
147. Yimlamai, D.; Christodoulou, C.; Galli, G.G.; Yanger, K.; Pepe-Mooney, B.; Gurung, B.; Shrestha, K.; Cahan, P.; Stanger, B.Z.; Camargo, F.D. Hippo pathway activity influences liver cell fate. *Cell* **2014**, *157*, 1324–1338. [CrossRef]
148. Wang, R.N.; Green, J.; Wang, Z.; Deng, Y.; Qiao, M.; Peabody, M.; Zhang, Q.; Ye, J.; Yan, Z.; Denduluri, S.; et al. Bone Morphogenetic Protein (BMP) signaling in development and human diseases. *Genes. Dis.* **2014**, *1*, 87–105. [CrossRef]
149. Alarcón, C.; Zaromytidou, A.I.; Xi, Q.; Gao, S.; Yu, J.; Fujisawa, S.; Barlas, A.; Miller, A.N.; Manova-Todorova, K.; Macias, M.J.; et al. Nuclear CDKs drive Smad transcriptional activation and turnover in BMP and TGF-beta pathways. *Cell* **2009**, *139*, 757–769. [CrossRef]
150. Sancho, R.; Cremona, C.A.; Behrens, A. Stem cell and progenitor fate in the mammalian intestine: Notch and lateral inhibition in homeostasis and disease. *EMBO Rep.* **2015**, *16*, 571–581. [CrossRef]
151. van Es, J.H.; van Gijn, M.E.; Riccio, O.; van den Born, M.; Vooijs, M.; Begthel, H.; Cozijnsen, M.; Robine, S.; Winton, D.J.; Radtke, F.; et al. Notch/gamma-secretase inhibition turns proliferative cells in intestinal crypts and adenomas into goblet cells. *Nature* **2005**, *435*, 959–963. [CrossRef] [PubMed]
152. Camargo, F.D.; Gokhale, S.; Johnnidis, J.B.; Fu, D.; Bell, G.W.; Jaenisch, R.; Brummelkamp, T.R. YAP1 increases organ size and expands undifferentiated progenitor cells. *Curr. Biol.* **2007**, *17*, 2054–2060. [CrossRef] [PubMed]
153. Takahashi, T.; Shiraishi, A.; Murata, J. The Coordinated Activities of nAChR and Wnt Signaling Regulate Intestinal Stem Cell Function in Mice. *Int. J. Mol. Sci.* **2018**, *19*, 738. [CrossRef]
154. Sato, T.; van Es, J.H.; Snippert, H.J.; Stange, D.E.; Vries, R.G.; van den Born, M.; Barker, N.; Shroyer, N.F.; van de Wetering, M.; Clevers, H. Paneth cells constitute the niche for Lgr5 stem cells in intestinal crypts. *Nature* **2011**, *469*, 415–418. [CrossRef]

155. Takahashi, T.; Shiraishi, A. Stem Cell Signaling Pathways in the Small Intestine. *Int. J. Mol. Sci.* **2020**, *21*, 2032. [CrossRef] [PubMed]
156. Takahashi, T. Roles of nAChR and Wnt signaling in intestinal stem cell function and inflammation. *Int. Immunopharmacol.* **2020**, *81*, 106260. [CrossRef]
157. Feng, X.H.; Derynck, R. Specificity and versatility in TGF-β signaling through Smads. *Annu. Rev. Cell. Dev. Biol.* **2005**, *21*, 659–693. [CrossRef]
158. Lowery, J.W.; Rosen, V. Bone morphogenetic protein-based therapeutic approaches. *Cold Spring Harb. Perspect. Biol.* **2017**, *10*, a022327. [CrossRef]
159. Haramis, A.P.G.; Begthel, H.; van den Born, M.; van Es, J.; Jonkheer, S.; Offerhaus, G.J.A.; Clevers, H. De novo crypt formation and juvenile polyposis on BMP inhibition in mouse intestine. *Science* **2004**, *303*, 1684–1686. [CrossRef]
160. Tian, A.; Jiang, J. Dual role of BMP signaling in the regulation Of Drosophila intestinal stem cell self-renewal. *Fly* **2017**, *11*, 297–302. [CrossRef]
161. Li, X.; Madison, B.B.; Zacharias, W.; Kolterud, A.; States, D.; Gumucio, D.L. Deconvoluting the intestine: Molecular evidence for a major role of the mesenchyme in the modulation of signaling cross talk. *Physiol. Genom.* **2007**, *29*, 290–301. [CrossRef] [PubMed]
162. Batts, L.E.; Polk, D.B.; Dubois, R.N.; Kulessa, H. Bmp signaling is required for intestinal growth and morphogenesis. *Dev. Dyn.* **2006**, *235*, 1563–1570. [CrossRef] [PubMed]
163. Kosinski, C.; Stange, D.E.; Xu, C.; Chan, A.S.; Ho, C.; Yuen, S.T.; Mifflin, R.C.; Powell, D.W.; Clevers, H.; Leung, S.Y.; et al. Indian hedgehog regulates intestinal stem cell fate through epithelial-mesenchymal interactions during development. *Gastroenterology* **2010**, *139*, 893–903. [CrossRef]
164. Reifenberger, J.; Wolter, M.; Weber, R.G.; Megahed, M.; Ruzicka, T.; Lichter, P.; Reifenberger, G. Missense mutations in SMOH in sporadic basal cell carcinomas of the skin and primitive neuroectodermal tumors of the central nervous system. *Cancer Res.* **1998**, *58*, 1798–1803.
165. Wetmore, C. Sonic hedgehog in normal and neoplastic proliferation: Insight gained from human tumors and animal models. *Curr. Opin. Genet. Dev.* **2003**, *13*, 34–42. [CrossRef] [PubMed]
166. Fernandez-L., A.; Northcott, P.A.; Dalton, J.; Fraga, C.; Ellison, D.; Angers, S.; Taylor, M.D.; Kenney, A.M. YAP1 is amplified and up-regulated in hedgehog-associated medulloblastomas and mediates Sonic hedgehog-driven neural precursor proliferation. *Genes Dev.* **2009**, *23*, 2729–2741. [CrossRef]
167. Kumar, V.; Vashishta, M.; Kong, L.; Wu, X.; Lu, J.J.; Guha, C.; Dwarakanath, B.S. The Role of Notch, Hedgehog, and Wnt Signaling Pathways in the Resistance of Tumors to Anticancer Therapies. *Front. Cell Dev. Biol.* **2021**, *9*, 650772. [CrossRef]
168. Taipale, J.; Beachy, P.A. The Hedgehog and Wnt signalling pathways in cancer. *Nature* **2001**, *411*, 349–354. [CrossRef]
169. Rubin, L.L.; de Sauvage, F.J. Targeting the Hedgehog pathway in cancer. *Nat. Rev. Drug Discov.* **2006**, *5*, 1026–1033. [CrossRef]
170. Reifenberger, J.; Wolter, M.; Knobbe, C.B.; Köhler, B.; Schönicke, A.; Scharwächter, C.; Kumar, K.; Blaschke, B.; Ruzicka, T.; Reifenberger, G. Somatic mutations in the PTCH, SMOH, SUFUH and TP53 genes in sporadic basal cell carcinomas. *Br. J. Dermatol.* **2005**, *152*, 43–51. [CrossRef]
171. Szkandera, J.; Kiesslich, T.; Haybaeck, J.; Gerger, A.; Pichler, M. Hedgehog signaling pathway in ovarian cancer. *Int. J. Mol. Sci.* **2013**, *14*, 1179–1196. [CrossRef] [PubMed]
172. Ingham, P.W.; McMahon, A.P. Hedgehog signaling in animal development: Paradigms and principles. *Genes Dev.* **2001**, *15*, 3059–3087. [CrossRef] [PubMed]
173. Zhu, C.; Li, L.; Zhao, B. The regulation and function of YAP transcription co-activator. *Acta Biochim. Biophys. Sin.* **2015**, *47*, 16–28. [CrossRef] [PubMed]
174. Ma, S.; Meng, Z.; Chen, R.; Guan, K.L. The Hippo Pathway: Biology and Pathophysiology. *Annu. Rev. Biochem.* **2019**, *88*, 577–604. [CrossRef]
175. Calses, P.C.; Crawford, J.J.; Lill, J.R.; Dey, A. Hippo pathway in cancer: Aberrant regulation and therapeutic opportunities. *Trends Cancer* **2019**, *5*, 297–307. [CrossRef]

toxics

MDPI

Article

Subchronic and Chronic Toxicity Assessment of Sublancin in Sprague–Dawley Rats

Yong Guo [1,2,3], Zhihao Li [4], Penglong Xu [5], Gantong Guo [2,3], Tao He [2,3] and Yujiao Lai [4,*]

1 School of Chemistry and Environmental Engineering, Wuhan Institute of Technology, Wuhan 420200, China; 11809010003@stu.wit.edu.cn
2 Sinagri YingTai Bio-peptide Co., Ltd., Linzhou 456550, China; znytht@126.com (T.H.)
3 Key Laboratory of Feed Antibiotics Replacement Technology, Ministry of Agriculture and Rural Affairs, Linzhou 456550, China
4 Hubei Key Laboratory of Animal Nutrition and Feed Science, Engineering Research Center of Feed Protein Resources on Agricultural By-Products, Ministry of Education, Wuhan Polytechnic University, Wuhan 430023, China; zhhhli623@163.com
5 Beijing Key Laboratory of Detection Technology for Animal-Derived Food Safety, College of Veterinary Medicine, China Agricultural University, Beijing 100193, China
* Correspondence: 22113060@whpu.edu.cn; Tel.: +86-2783956175

Abstract: Sublancin, an S-linked antimicrobial (glycol) peptide produced by *Bacillus subtilis*, has emerged as a novel and promising veterinary drug due to its unique antibacterial mechanism, low risk of resistance, and properties that modulate the immune system, reduce inflammation, and promote gut health. This study comprehensively assessed the subchronic (90-day) and chronic (180-day) toxicity of Sprague–Dawley (SD) rats, following the guidelines issued by the Ministry of Agriculture of China. Rats were orally administered sublancin at doses of 2000, 10,000, or 50,000 mg/kg feed, representing 1666–5000 times the efficacious dose (1.0–1.2 mg/kg) reported in mice via the same administration route. Throughout this study, a wide range of physiological and behavioral parameters were monitored to access the toxicity of sublancin, including appetite, water intake, body weight gain, and organ weights. Hematological and biochemical analyses, as well as histopathological examinations of the major organs, were conducted at the end of each study period. The results indicated no adverse effects on any measured parameters at any dose level, with no significant differences observed between the sublancin-treated groups and the control group ($p > 0.05$). Notably, even the highest dose of 50,000 mg/kg did not induce growth inhibition or physiological dysfunction. A histopathological examination also revealed no tissue abnormalities in the major organs. The no-observed-effect level (NOEL) was determined to be 50,000 mg/kg for both study periods. These results demonstrate the long-term safety of sublancin in Sprague–Dawley rats, with no adverse effects during 180 days of oral administration at doses 1666–5000-fold the documented antimicrobially effective and immune-enhancing doses.

Keywords: sublancin; SD rat; subchronic toxicity; chronic toxicity; novel veterinary drugs; safety evaluation

1. Introduction

The rapid growth of modern animal husbandry has led to larger-scale and more intensive livestock and poultry farming, resulting in extensive antibiotic use in animal production. However, misusing and irregularly using antibiotics have increased bacterial resistance and drug residue issues, affecting animal health and meat quality, while also posing potential threats to human health [1]. To tackle these challenges, the Chinese government has banned antibiotics in feed and reduced their use in veterinary drugs. The goal is to reduce antibiotic use, minimize the risk of resistance, and ensure the safety of animal products. Thus, developing novel, safe, and efficient veterinary drugs, especially those with antimicrobial properties that leave no residues, has become a critical research priority in animal husbandry and food safety.

Sublancin, a 37-amino acid antimicrobial (glycol) peptide produced by *Bacillus subtilis 168*, featured two α-helices (Helix A and Helix B) connected by a loop and stabilized by two nested disulfide bonds between four cysteine residues, forming its characteristic helix–loop–helix hairpin structure [2,3]. This S-linked (glycol) peptide contains a crucial glucose modification at Cys22, with both glycosylation and the disulfide-bonded structure being essential for its antimicrobial activity [3]. Functionally, sublancin exhibits broad-spectrum efficacy against Gram-positive bacteria in vitro at a minimum inhibitory concentration (MIC) of 15 μM [4] (e.g., *Staphylococcus aureus*). In vivo efficacy has been demonstrated in mice via intraperitoneal injection (2–4 mg/kg) [4,5] and oral administration (1.0–1.2 mg/kg feed) [6]. Sublancin also inhibits *Clostridium perfringens* (a Gram-positive bacteria) at a MIC of 8 μM, and controls necrotic enteritis in broilers in vivo at dose of 5.76 mg/L via water, which is much lower than that of lincomycin at 75 mg activity/L [7]. Beyond direct antimicrobial effects, it modulates the gut microbiota [8], enhances intestinal barrier function [5], and suppresses excessive inflammation [4]. Notably, it also potentiates innate immunity by upregulating host defense peptides [4,6,8,9]. These multifaceted properties, combined with undetectable drug residues (according to our pharmacokinetic studies in rats) and low resistance risk (consistent with other AMPs like nisin [10]), position sublancin as an ideal antibiotic alternative in livestock production, aligning with global antimicrobial stewardship goals [3,11–13].

Chronic toxicity tests evaluate long-term drug/toxin effects on animal health. They help determine the maximum no-observed-adverse-effect level (NOAEL) of a drug and are crucial for establishing a comprehensive safety profile, especially in assessing long-term impacts [14]. Chronic toxicity tests typically last for extended periods to simulate long-term drug exposure [14–16]. Tests are categorized by duration: subchronic (90-day) and chronic (180-day) toxicity tests. Tests are usually performed on specific animal models, such as C57 mice [14], Wistar rats [15], and Sprague–Dawley rats [16–18]. Various countries and regions have specific regulations and guidelines for chronic toxicity and carcinogenicity tests of veterinary drugs. Notable examples include guidelines from the US Environmental Protection Agency (USEPA) [19], the European Food Safety Authority (EFSA), and the United States Food and Drug Administration (FDA) [18]. To evaluate the safety of sublancin as a novel veterinary drug in practical applications, this study conducted 90-day subchronic and 180-day chronic toxicity tests in accordance with the Guidelines for Chronic Toxicity and Carcinogenicity Testing of Veterinary Drugs issued by the Ministry of Agriculture and Rural Affairs of the People's Republic of China. In this study, varying doses of sublancin, equivalent to 1666–5000 times the dose shown to be efficacious in antibacterial function and immune enhancement, were administered to SD rats of different ages and sexes to systematically assess its effects on growth performance, hematological parameters, serum biochemistry, organ coefficients, and histopathology.

2. Materials and Methods

2.1. Chemicals and Reagents

Sublancin (purity > 93%), supplied by Sinagri Yingtai bio-peptide Co., Ltd. (Linzhou, Henan, China), has the following amino acid sequence: ESKICTWLGTGQCNCYK CKWEQSLQCRN.

2.2. Animals and Treatments

This study was approved by the Animal Care and Use Committee of China Agricultural University, which approved all the animal experiments in this study (approval number AW42800202-3-5). All the experimental protocols followed the guidelines for animal experiments provided by the Animal Care and Use Committee of China Agricultural University.

A total of 240 four-week-old male and female SPF-grade SD rats (70–90 g) were purchased from Beijing Weitonglihua Experimental Animal Technology Co., Ltd. (Beijing, China). For the subchronic (90-day) and chronic (180-day) toxicity tests, 80 and 160 rats were allocated, respectively. Rats were housed under standard conditions with a controlled temperature of 22–24 °C, humidity of 50–60%, and a light/dark cycle. All the rats had free access to a standard diet and drinking water throughout the experiment. On the basis of the acute toxicity test results (LD50 > 5000 mg/kg body weight) and regulatory guidelines (maximum 5% of feed weight) [20], we selected four dose levels: 0 mg/kg (control group), 2000 mg/kg (low-dose group), 10,000 mg/kg (medium-dose group), and 50,000 mg/kg (high-dose group). These doses represent 1666–5000 times the efficacious antibacterial dose (1.0–1.2 mg/kg) previously established in mice [6]. Each group consisted of 20 SD rats (10 males and 10 females), with the weight variation within each sex not exceeding 20% of the average body weight. For the subchronic toxicity test and chronic toxicity tests, the animals were organized into the same groups, with the number of animals doubled. Sublancin was added to the maintenance feed for rats at four dose levels: 0 mg/kg (control), 2000 mg/kg, 10,000 mg/kg, and 50,000 mg/kg. The feed was thoroughly mixed in a feed mixer to ensure uniform distribution and then pelleted. The maintenance feed was purchased from Beijing Keao Xieli Feed Co., Ltd. (Beijing, China).

2.3. Experiment 1: Observation and Measurement of Indicators of the Subchronic Toxicity Test for 90 Days

2.3.1. General Observations

The overall health status of the experimental animals was monitored daily. Detailed records were kept of the rats' behavior, including eye, mouth, and nose secretions; fecal consistency; signs of diarrhea or constipation; fecal color; and fur condition (e.g., dullness, piloerection). Each group of rats receiving a specific dose were closely monitored for toxicity signs, including the timing and appearance of symptoms. If an animal died or showed severe illness, immediate post-mortem examinations were performed. Detailed records were kept of any abnormal conditions and the time of death. Tissues from the heart, liver, spleen, lungs, kidneys, and duodenum were collected for histopathological analysis. Body weight and food intake were measured every five days, while water consumption was recorded every two days.

2.3.2. Hematology

Blood samples were collected for hematological analysis from 5 males and 5 females in each experimental group on Day 45 (midpoint) and Day 90 (endpoint) of the study. The following hematological parameters were analyzed: hemoglobin (HGB), red blood cell count (RBC), white blood cell count (WBC), platelet count (PLT), hematocrit (HCT), eosinophils (EOS), basophils (BAS), monocytes (MO), and lymphocytes (LY).

2.3.3. Serum Biochemical Parameters

Simultaneously with the hematological analysis, we assessed serum biochemical parameters, including albumin (ALB), alanine aminotransferase (ALT), aspartate aminotransferase (AST), blood urea nitrogen (BUN), total cholesterol (TC), creatinine (Cr), glucose (Glu), and total protein (TP).

2.3.4. Pathological Examination

On Day 45 (midpoint) and Day 90 (endpoint) of the study, five male and five female rats from each experimental group were euthanized. A gross autopsy was then conducted. Any suspicious lesions observed were quickly fixed and preserved for histopathological analysis. During the gross examination, body and major organ weights were recorded, with organ-to-body weight ratios calculated for the heart, liver, kidneys, spleen, lungs, gastrointestinal tract, testes, and ovaries.

Following the gross autopsy, the heart, liver, spleen, lungs, kidneys, and duodenum were collected and fixed in a 10% neutral formalin solution. Routine paraffin-embedded sections were prepared and stained with hematoxylin and eosin (H&E) for a histopathological examination of these organs in both the high-dose and control groups. If lesions were observed, the corresponding organs from the lower-dose groups were also examined via histopathological analysis.

2.4. Experiment 2: Observation and Measurement of Indicators of the Chronic Toxicity Test for 180 Days

The experimental groups were organized similarly to Experiment 1, but with double the number of animals in each group. General observations were consistent with Experiment 1. However, histopathological assessments were expanded to the brain, testes, epididymis, ovaries, and uterus. Water intake was recorded every 5 days, consistent with the frequency of body weight and food intake measurements. Blood samples were collected on Days 60, 120, and 180 for hematological and biochemical analyses, following the same parameters as Experiment 1. On Days 60, 120, and 180 of the chronic toxicity studies, rats were sacrificed for necropsy, including a uterine assessment. Other indices and procedural steps were consistent with those in Experiment 1. Additionally, histopathological examinations of rat tissues included brain tissue, with other indices and procedural steps consistent with those in Experiment 1.

2.5. Statistical Analysis

Differences in food intake, water intake, daily weight gain, hematological and serum biochemical parameters, and organ coefficients between the treatment groups and negative controls were analyzed by *t*-tests. The data are presented as the mean ± SD. Incidence rates of organ lesions, tumors, and mortality were analyzed with chi-square (χ^2) tests. We defined a *p*-value of less than 0.05 as statistically significant.

3. Results

3.1. Clinical Symptoms

In the subchronic (90-day) and chronic (180-day) toxicity studies, rats administered sublancin at three doses (2000 mg/kg, 10,000 mg/kg, and 50,000 mg/kg) showed normal behavior compared with the negative controls. All animals maintained normal physical conditions, including smooth fur without ocular, oral, or nasal discharge. Fecal characteristics (morphology and color) remained consistent across all groups, with no deviations from the typical light gray appearance observed in healthy rats.

3.2. Food Intake

Statistical analysis of food intake (Figure 1; Tables S1 and S2-1–S2-3) revealed no significant differences ($p > 0.05$) between the control (CON) and sublancin-treated groups (2000, 10,000, or 50,000 mg/kg feed) in either the 90-day subchronic or 180-day chronic toxicity studies. This consistency was observed in both male and female rats, indicating that prolonged oral administration of sublancin at these doses did not alter feeding behavior.

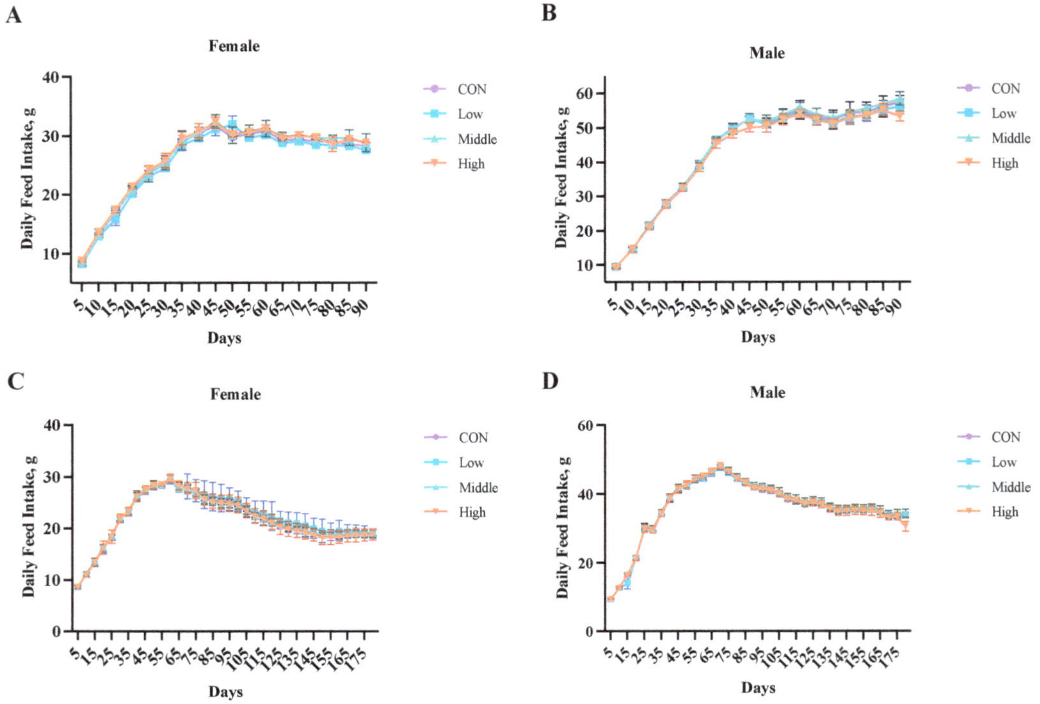

Figure 1. Effects of sublancin on the daily food intake of SD rats fed for the subchronic ((**A**,**B**); 90-day; n = 10/sex/group) and chronic ((**C**,**D**); 180-day; n = 20/sex/group) toxicity trials (g/day/rat). Each point represents the mean ± SD. CON group, basal diet; Low group, basal diet + 2000 mg/kg; Middle group, basal diet + 10,000 mg/kg; High group, basal diet + 50,000 mg/kg.

3.3. Water Intake

The water consumption analysis presented no significant differences ($p > 0.05$) between the control and sublancin-treated groups (2000–50,000 mg/kg feed) in either the 90-day subchronic or 180-day chronic toxicity studies (Figure 2; Tables S3 and S4-1–S4-3). These findings demonstrate that prolonged ad libitum administration of sublancin at these doses (2000 mg/kg, 10,000 mg/kg, and 50,000 mg/kg of feed) did not affect water intake in rats of either sex.

3.4. Average Daily Weight Gain

The body weight gain analysis (Figure 3; Tables S5 and S6-1–S6-3) showed no significant differences ($p > 0.05$) between the control and sublancin-treated groups (2000–50,000 mg/kg feed) during the 90-day subchronic or 180-day chronic toxicity studies. These findings indicate that continuous exposure to sublancin at these doses ranging from 2000 to 50,000 mg/kg of feed did not affect growth performance in either male or female rats.

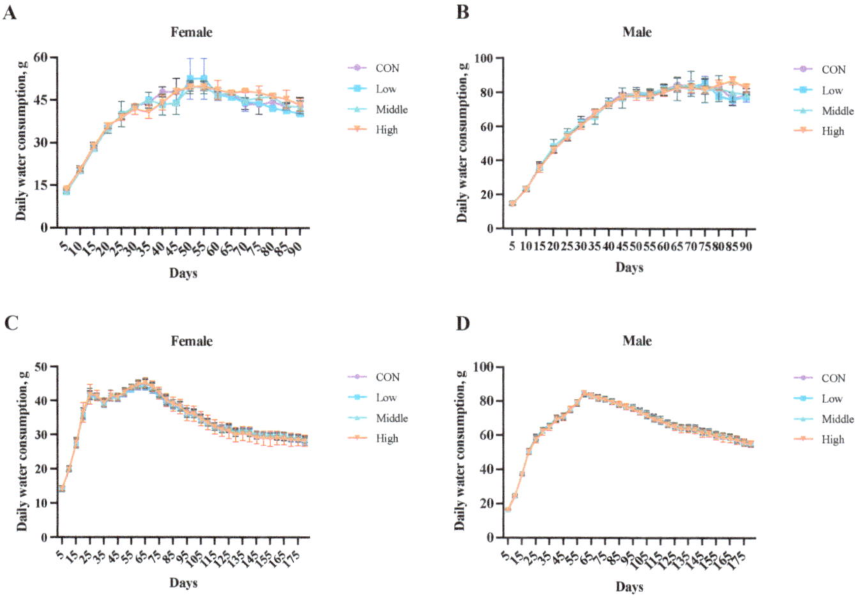

Figure 2. Effects of sublancin on the daily water intake of SD rats fed for the subchronic toxicity trial ((**A**,**B**); 90-day; n = 10/sex/group) and chronic ((**C**,**D**); 180-day; n = 20/sex/group) toxicity trials (g/day/rat). Each point represents the mean ± SD. CON group, basal diet; Low group, basal diet + 2000 mg/kg; Middle group, basal diet + 10,000 mg/kg; High group, basal diet + 50,000 mg/kg.

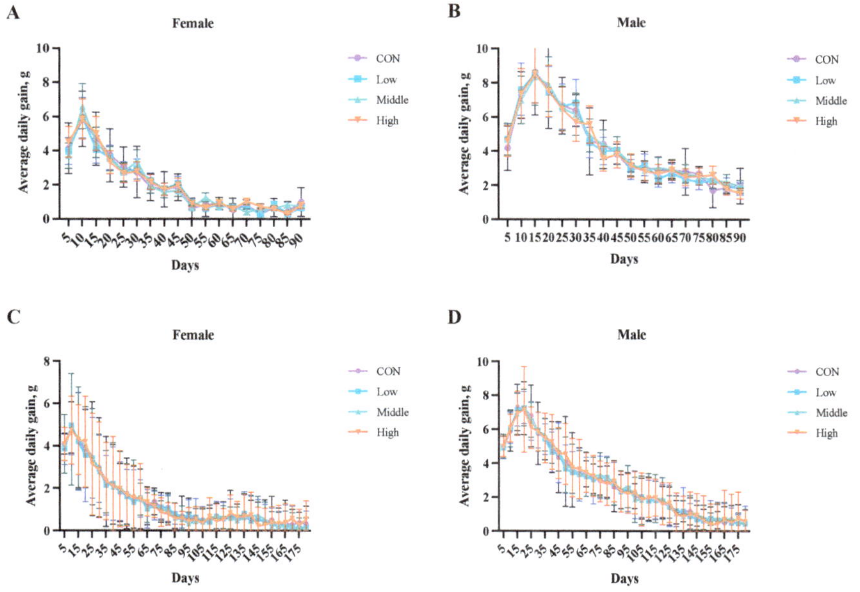

Figure 3. Effects of sublancin on the average daily gain of SD rats fed for the subchronic ((**A**,**B**); 90-day; n = 10/sex/group) and chronic ((**C**,**D**); 180-day; n = 20/sex/group) toxicity trials (g/day/rat). Each point represents the mean ± SD. CON group, basal diet; Low group, basal diet + 2000 mg/kg; Middle group, basal diet + 10,000 mg/kg; High group, basal diet + 50,000 mg/kg.

3.5. Hematological Parameters

Hematological analysis was performed on blood samples collected from five male and five female rats per group at (1) Days 45 and 90 (subchronic study endpoint), and (2) Days 60, 120, and 180 (chronic study endpoint). Analysis of the key parameters (HGB, RBC, WBC, PLT, HCT, EOS, BAS, MO, and LY) revealed no significant differences ($p > 0.05$) between the control and sublancin-treated groups (2000–50,000 mg/kg feed) at any time point (Figures 4 and 5; Tables S7 and S8). These findings demonstrate that prolonged sublancin exposure at these doses did not induce hematological alterations in either sex.

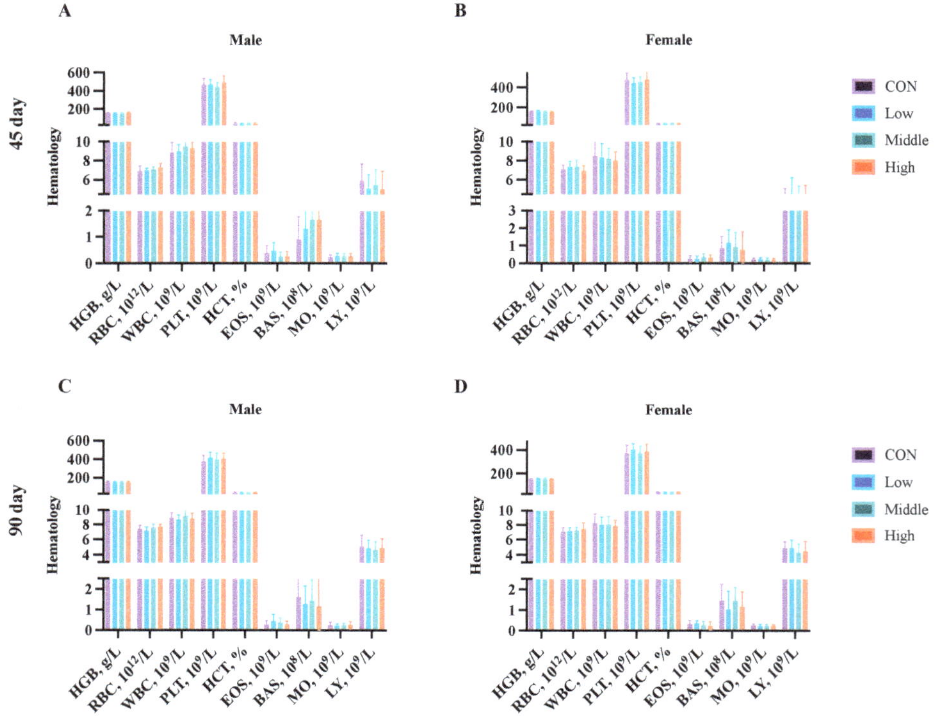

Figure 4. Effects of sublancin on the hematology of SD rats fed for 45 (**A**,**B**) days and 90 (**C**,**D**) days during the subchronic toxicity trial (n = 5/sex/group). Each point represents the mean ± SD. CON group, basal diet; Low group, basal diet + 2000 mg/kg; Middle group, basal diet + 10,000 mg/kg; High group, basal diet + 50,000 mg/kg. HGB, hemoglobin; RBC, red blood cell count; WBC, white blood cell count; PLT, platelets; HCT, hematocrit; EOS, eosinophils; BAS, basophils; MO, monocytes; LY, lymphocytes.

3.6. Clinical Biochemistry

Blood biochemical analysis was conducted at multiple time points: Day 45 (mid-study) and Day 90 (endpoint) in the subchronic study, and Days 60, 120, and 180 (end-point) in the chronic study. Evaluation of the key parameters (ALB, ALT, AST, BUN, TC, Cr, Glu, and TP) revealed no significant differences ($p > 0.05$) between the control and sublancin-treated groups (2000–50,000 mg/kg feed) at any time point (Figures 6 and 7; Tables S9 and S10). These results demonstrate that prolonged dietary exposure to sublancin at these doses did not induce clinically relevant alterations in blood biochemistry in either male or female rats.

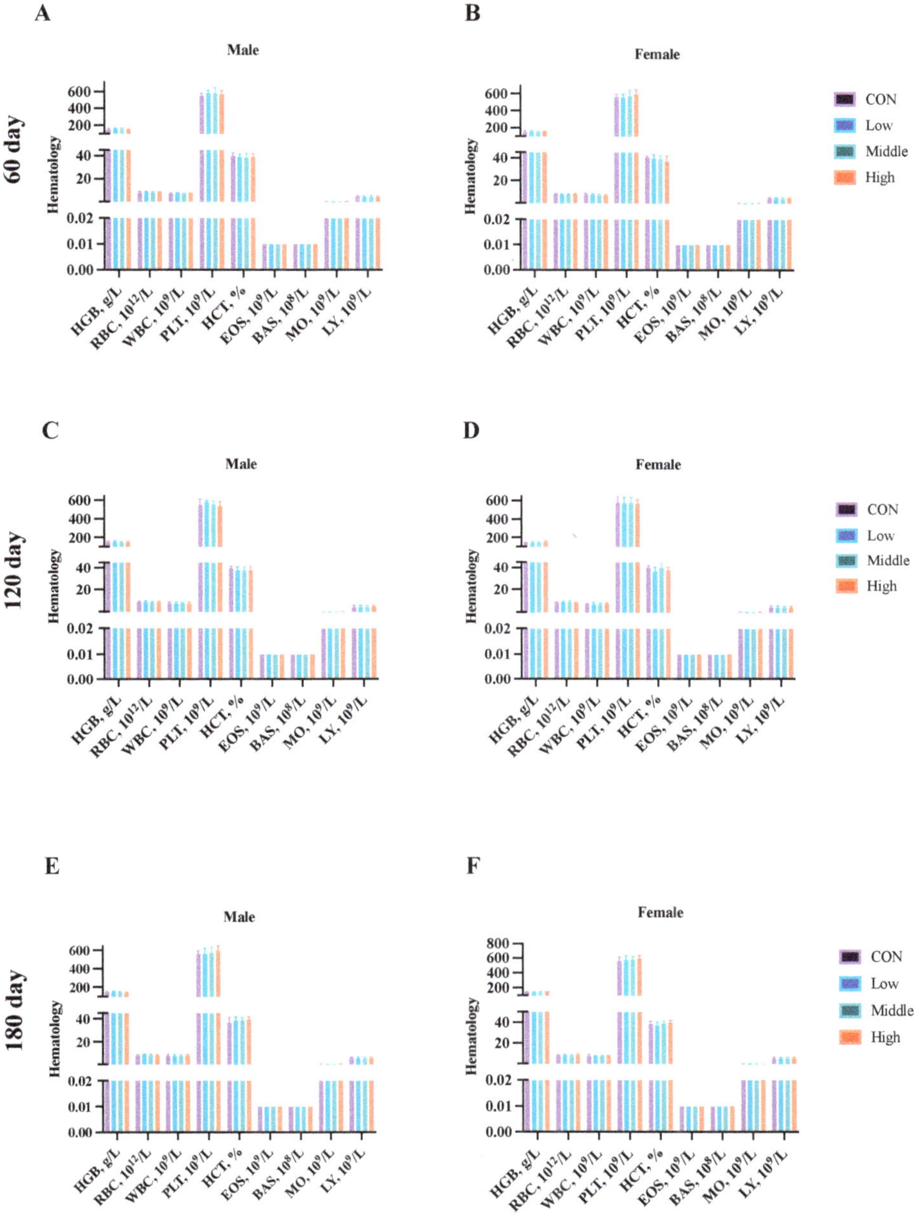

Figure 5. Effects of sublancin on the hematology of SD rats fed for 60 (**A**,**B**), 120 (**C**,**D**) and 180 (**E**,**F**) days during the chronic toxicity trial (n = 5/sex/group). Each point represents the mean ± SD. CON group, basal diet; Low group, basal diet + 2000 mg/kg; Middle group, basal diet + 10,000 mg/kg; High group, basal diet + 50,000 mg/kg. HGB, hemoglobin; RBC, red blood cell count; WBC, white blood cell count; PLT, platelet count; HCT, hematocrit; EOS, eosinophils; BAS, basophils; MO, monocytes; LY, lymphocytes.

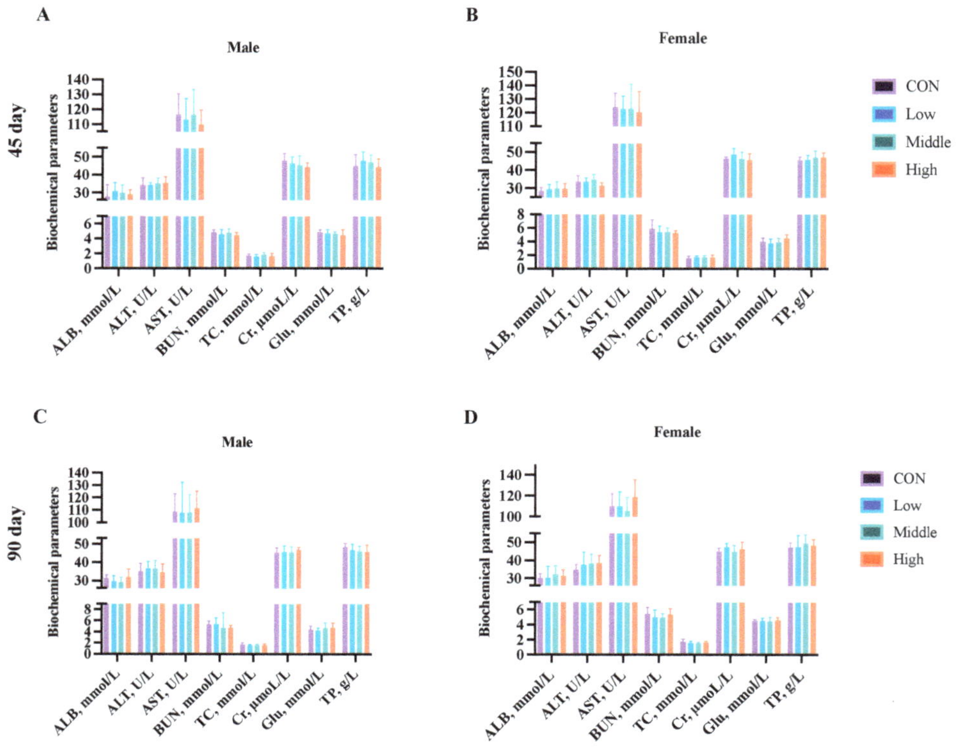

Figure 6. Effects of sublancin on the biochemical parameters of SD rats fed for 45 (**A**,**B**) and 90 (**C**,**D**) days during the subchronic toxicity trial (n = 5/sex/group). Each point represents the mean ± SD. The data were compared for statistical significance with the same-sex control group. CON group, basal diet; low group, basal diet + 2000 mg/kg; middle group, basal diet + 10,000 mg/kg; high group, basal diet + 50,000 mg/kg. ALB, albumin; ALT, alanine aminotransferase; AST, aspartate aminotransferase; BUN, blood urea nitrogen; TC, total cholesterol; Cr, creatinine; Glu, glucose; TP, total protein.

3.7. Organ Index

Organ weight analysis was performed at the subchronic study's midpoint (Day 45) and endpoint (Day 90), and during the chronic study (Days 60, 120, and 180) using five rats per sex per group. No significant differences ($p > 0.05$) were observed in relative organ weights (heart, liver, kidneys, spleen, lungs, gastrointestinal tract, brain, and reproductive organs) between the control and sublancin-treated groups (2000–50,000 mg/kg feed) at any time point (Figures 8 and 9; Tables S11 and S12). These findings demonstrate that chronic ad libitum exposure for up to 180 days to sublancin at doses from 2000 to 50,000 mg/kg of feed did not induce organotoxicity (including the heart, liver, kidneys, spleen, lungs, gastrointestinal tract, brain, testes, epididymides, ovaries, and uterus) at the tested doses.

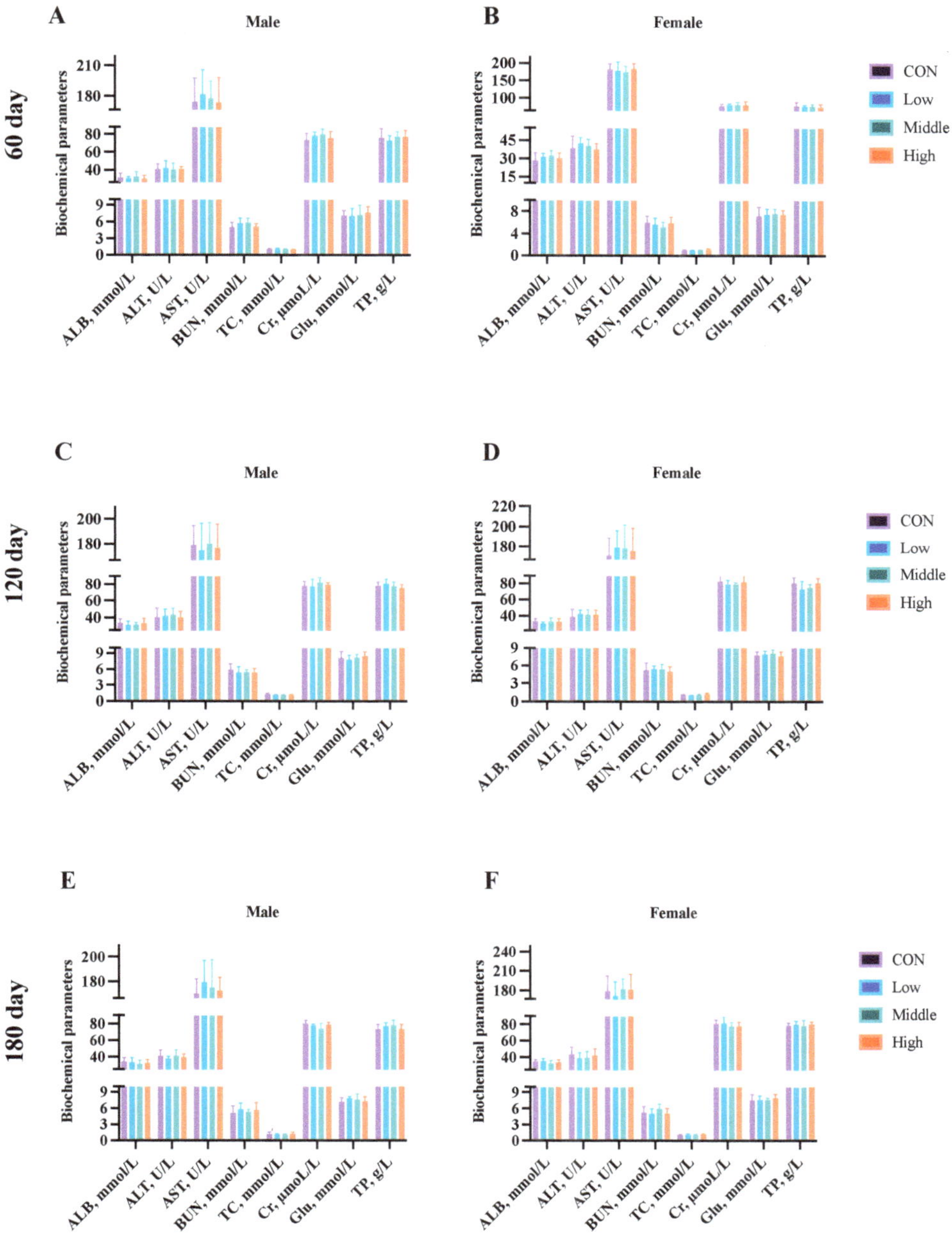

Figure 7. Effects of sublancin on the serum biochemical parameters of SD rats fed for 60 (**A**,**B**), 120 (**C**,**D**) and 180 (**E**,**F**) days during the chronic toxicity trial ($n = 5$/sex/group). Each point represents the mean ± SD. CON group, basal diet; Low group, basal diet + 2000 mg/kg; Middle group, basal diet + 10,000 mg/kg; High group, basal diet + 50,000 mg/kg. ALB, albumin; ALT, alanine aminotransferase; AST, aspartate aminotransferase; BUN, blood urea nitrogen; TC, total cholesterol; Cr, creatinine; Glu, glucose; TP, total protein.

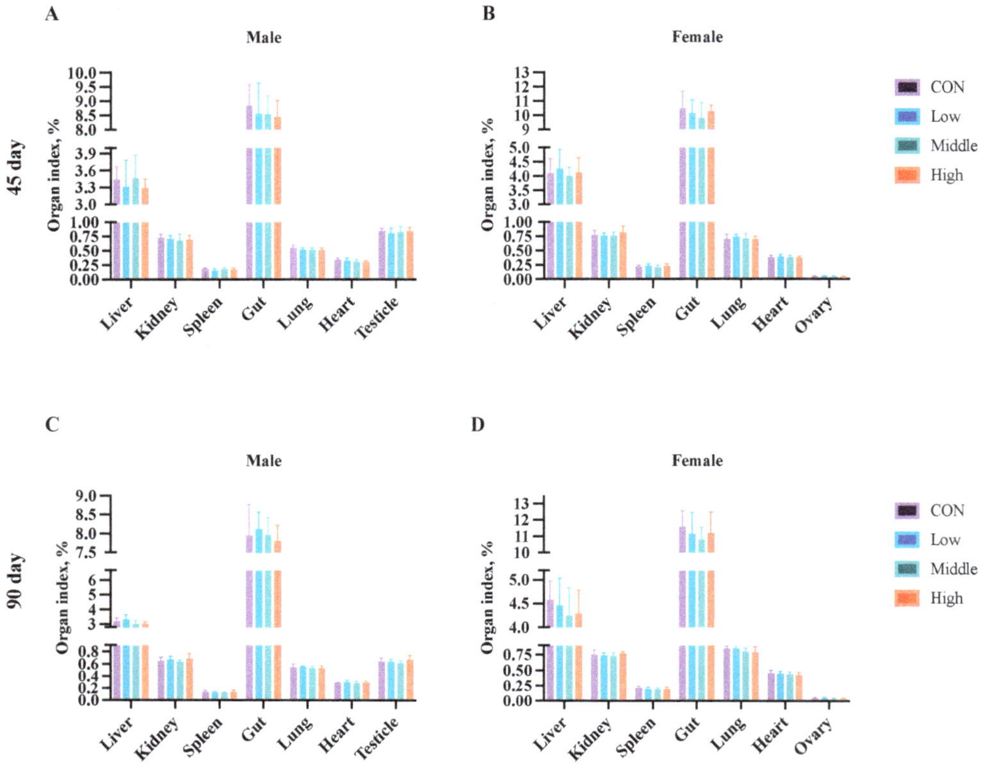

Figure 8. Effects of sublancin on the organ index of SD rats fed for 45 (**A**,**B**) and 90 (**C**,**D**) days during the subchronic toxicity trial (n = 5/sex/group). Each point represents the mean ± SD. CON group, basal diet; Low group, basal diet + 2000 mg/kg; Middle group, basal diet + 10,000 mg/kg; High group, basal diet + 50,000 mg/kg.

3.8. Histopathology Analysis

Throughout the 90-day subchronic and 180-day chronic toxicity studies, no treatment-related clinical signs or mortality were observed in any experimental groups. The gross necropsy examination revealed no abnormalities in either the sublancin-treated (2000–50,000 mg/kg feed) or control groups.

Histopathological analysis was conducted on the key organs (heart, liver, spleen, lungs, kidneys, duodenum, brain, and reproductive organs) from the high-dose (50,000 mg/kg) and control groups at the study's termination. A comprehensive evaluation of multiple parameters, including clinical observations, body weight, food/water consumption, hematology, clinical biochemistry, organ weights, and gross pathology, revealed no significant treatment-related effects ($p > 0.05$). Consequently, the histopathological comparison focused on the highest-dose group versus the controls.

As shown in Figures 10 and 11, microscopic examination demonstrated a normal tissue architecture in all examined organs, with no pathological alterations attributable to sublancin administration. These results conclusively demonstrate that chronic ad libitum exposure to sublancin at doses up to 50,000 mg/kg feed induced no adverse effects on major organ systems in SD rats.

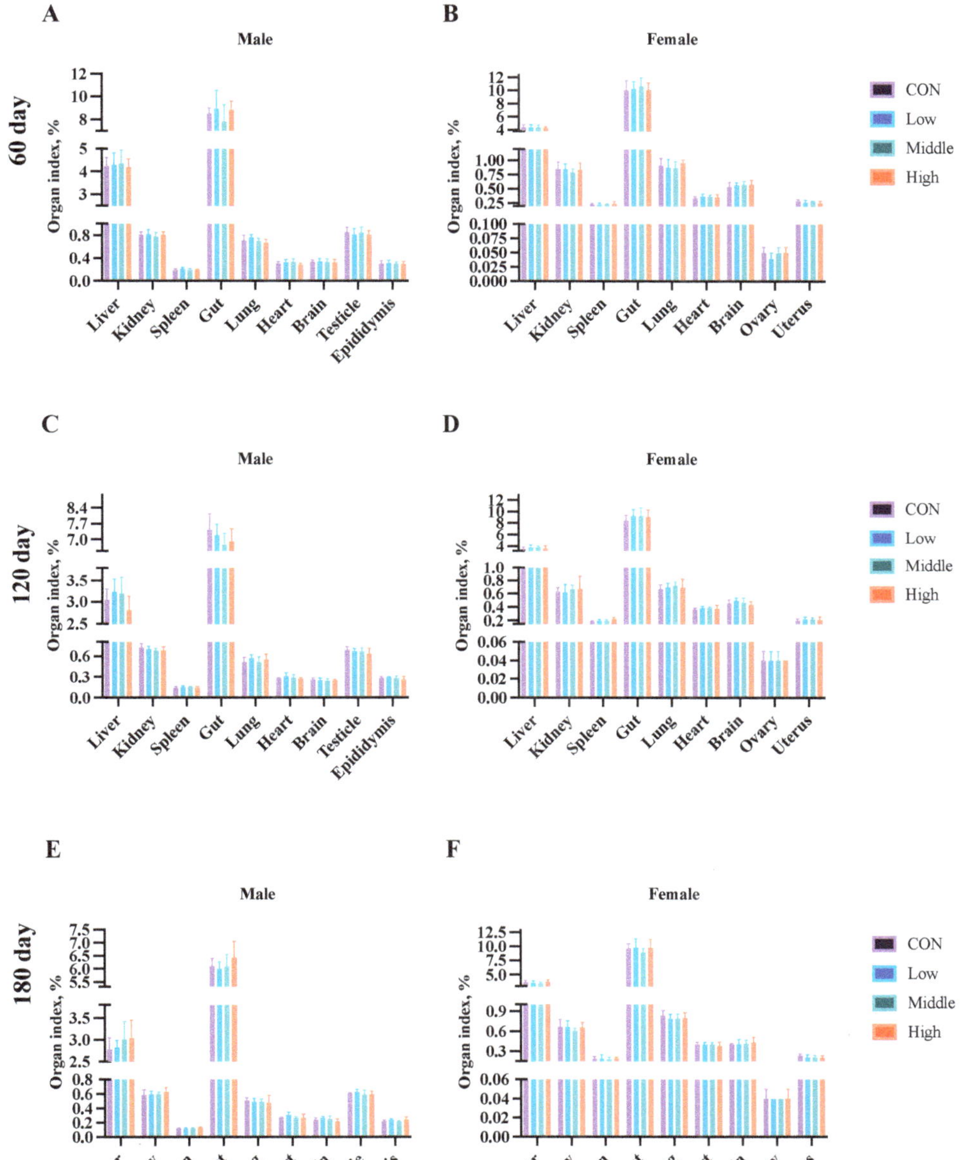

Figure 9. Effects of sublancin on the organ index of SD rats fed for 60 (**A**,**B**), 120 (**C**,**D**) and 180 (**E**,**F**) days during the chronic toxicity trial (n = 5/sex/group). Each point represents the mean ± SD. CON group, basal diet; Low group, basal diet + 2000 mg/kg; Middle group, basal diet + 10,000 mg/kg; High group, basal diet + 50,000 mg/kg.

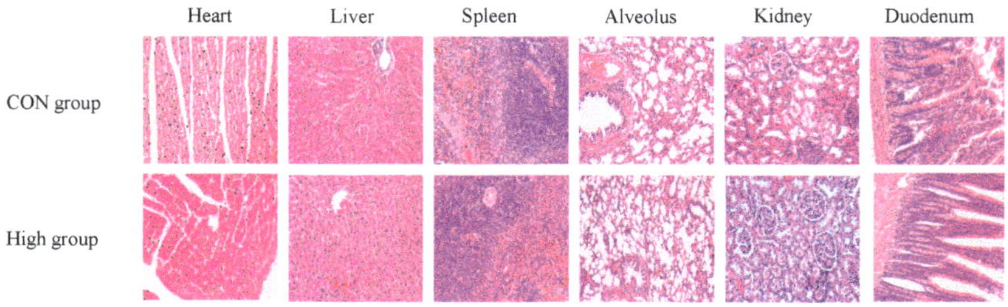

Figure 10. Histological sections of various tissues (heart, liver, spleen, alveolus, kidneys, and duodenum) from SD rats subjected to 90-day subchronic toxicity testing with sublancin (H&E staining, 40×). CON group, basal diet; High group, basal diet + 50,000 mg/kg.

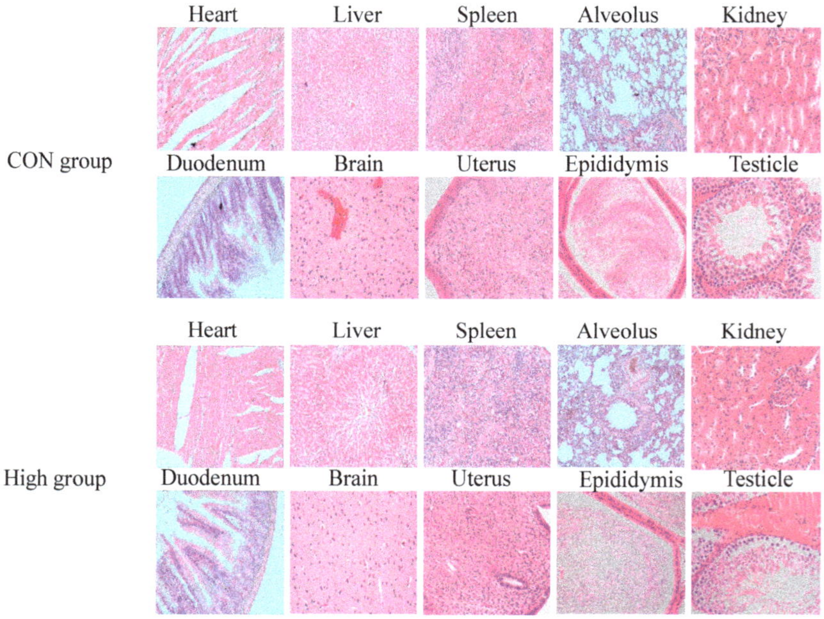

Figure 11. Histological sections of various tissues (heart, liver, spleen, alveolus, kidneys, duodenum, brain, and reproductive tissues) from SD rats subjected to 180-day chronic toxicity testing with sublancin (H&E staining, 40×). CON group, basal diet; High group, basal diet + 50,000 mg/kg.

In summary, administering sublancin at doses ranging from 2000 to 50,000 mg/kg of feed did not cause significant differences in appetite, water intake, or weight gain in SD rats ($p > 0.05$). Additionally, no significant differences were observed in hematological, biochemical, or pathological parameters in the sublancin-treated groups when compared with the control group ($p > 0.05$). These results indicate that sublancin does not exhibit subchronic and chronic toxicity within the dose range of 2000 to 50,000 mg/kg of feed. The NOEL was established at 50,000 mg/kg of feed based on both the 90-day subchronic and 180-day chronic toxicity studies.

4. Discussion

As antimicrobial resistance increasingly threatens human health, developing alternative compounds to replace conventional antibiotics has become a research imperative [21,22]. Antimicrobial peptides (AMPs) have broad-spectrum antimicrobial activity. They are highly effective against bacteria, fungi, parasites, and viruses, while showing a low tendency for developing resistance [23,24]. The unique pharmacokinetics and bactericidal mechanisms of AMPs position them as potential candidates to replace conventional antibiotics [25]. In animal husbandry, AMPs are recognized as ideal antibiotic alternatives [26], while in human medicine, they represent promising next-generation therapeutic agents for infectious diseases [27]. Sublancin is a cyclic (glycol) peptide made up of 37 amino acids with a molecular weight of 3877.642 Da [11]. The unique α-helical secondary structure has been confirmed through circular dichroism (CD) and nuclear magnetic resonance (NMR) studies [28,29]. Studies demonstrate that sublancin not only effectively inhibits the growth of diverse Gram-positive bacteria (e.g., *Staphylococcus aureus* and *Clostridium perfringens*) in vitro and in vivo [4–7], but also confers broad-spectrum protection against multidrug-resistant pathogen infections [8,30,31]. Given its unique antibacterial mechanism as an AMP and low resistance potential, sublancin may represent a promising candidate for innovative veterinary therapeutics [12,13].

In China, veterinary drug safety evaluations strictly adhere to the Regulations on Veterinary Drug Administration and Measures for Veterinary Drug Registration, in accordance with the Guidelines for Safety Pharmacology Studies of Veterinary Chemical Drugs issued by the Ministry of Agriculture and Rural Affairs (MARA). These regulations mandate that novel veterinary compounds (e.g., antibiotic alternatives) must undergo systematic acute, subchronic, and chronic toxicity testing, with particular emphasis on assessing the long-term effects on animal growth, hematology, organ function, and histopathology (MARA Announcement No. 1247). Our toxicology study evaluated sublancin using standardized 90-day and 180-day testing protocols at doses up to 50,000 mg/kg feed. This maximum dose represented 4166–5000× the established efficacious dose in rats (1.0–1.2 mg/kg) [6] and 83× the effective dose in broilers (equivalent to 60 mg/kg feed) [9] in vivo. The study design complied with GLP standards and regulatory requirements for veterinary drug registration.

Before initiating clinical trials to evaluate the efficacy of AMPs, acute and chronic toxicity tests in animal models are essential prerequisites. Chronic toxicity testing is crucial for establishing a comprehensive safety profile of investigational drugs, with particular emphasis on characterizing the long-term effects [14]. Subchronic toxicity studies, typically conducted over 90 days, provide detailed data on compound-induced changes in growth, development, physiological functions, and organ pathology [32]. These studies identify potential chronic toxicities and ensure that veterinary drugs do not significantly compromise animal health in practical applications [32]. Specifically, they determine the toxicological threshold and safe dosage by monitoring behavioral anomalies, appetite, water consumption, weight gain dynamics, hematological profiles, serum biochemistry, and organ coefficients, reflecting the potential impacts on fundamental physiological functions [33]. A 90-day trial identifies the maximum dose that does not significantly harm animals, providing critical information for subsequent chronic toxicity studies [34]. Subchronic toxicity studies constitute a pivotal element in drug safety evaluation, forming the foundation for nonclinical safety assessments and serving as a mandatory prerequisite prior to a clinical trial's initiation [32]. Chronic toxicity studies systematically characterize 180-day exposure effects in animals, focusing on chronic disease progression, parenchymal organ damage (liver, kidneys, and heart), and potential carcinogenicity. They determine the NOEL for long-term drug use, establishing a scientific basis for safe dosages [17,32]. These systematic toxicity tests ensures the safety of veterinary drugs used in practical applications and comply with national and international regulations [35]. Given China's policies prohibiting antibiotics in animal feed and limiting their use in veterinary medicine, this study's results are significant for developing and promoting novel veterinary drugs without residues [1]. These systematic toxicological evaluations provide essential data for establishing safety standards of investigational veterinary drugs, while supporting regulatory compliance with national and international guidelines for veterinary pharmaceutical development.

Assessing the safety of potential veterinary drugs is crucial for regulatory approval and usually involves a thorough review of various scientific documents, including monographs, peer-reviewed articles, and reports. This study investigates the safety of sublancin as a potential antibiotic alternative in rats, especially examining its long-term chronic toxicity to address a gap in the existing literature. Following the Guidelines for Chronic Toxicity and Carcinogenicity Testing of Veterinary Drugs from the Ministry of Agriculture and Rural Affairs of the People's Republic of China, we performed 90-day subchronic and 180-day chronic toxicity tests on SD rats with varying doses of sublancin to comprehensively assess its long-term feeding effects. The results indicated that sublancin, administered at doses of 2000–50,000 mg/kg feed, had no significant adverse effects on the growth and development, hematological parameters, serum biochemical parameters, or organ coefficients of rats, demonstrating a favorable safety profile. The trial involved continuous administration of the test substance for 60, 120, and 180 days. During this period, researchers observed adverse reactions and other indicators in the animals to assess the toxicity level and identify potential target organs. The evaluation criteria included changes in the food intake, water consumption, and body weight of the test animals, along with alterations in physiological and biochemical parameters and organ indices, and histopathological changes [17]. Neither the subchronic nor chronic toxicity tests revealed any abnormalities in the mental status, excreta, or diet of rats across all dose groups, and there were no recorded cases of poisoning or death. Collectively, these findings demonstrate sublancin's long-term safety profile and support its translational development as a veterinary therapeutic.

During both the 90-day and 180-day trials, sublancin did not negatively affect key physiological parameters like appetite, water consumption, or weight gain in rats, indicating that it does not alter the basic physiological functions of animals. Additionally, long-term feeding of sublancin to rats did not result in significant differences in appearance, behavior, or mental state compared with the control group, confirming that sublancin has no adverse effects on the daily behavior or quality of life of rats. Changes in body weight are among the most significant indicators of toxicity in animals [35]. Existing literature suggests that an initial weight loss of 10% or more in experimental animals may be a strong indicator of adverse side effects and could potentially endanger survival. In contrast, our study revealed no statistically significant variations in the body weight gain of rats. Although there are no direct reports on sublancin's effects on animal growth and production performance, numerous studies have demonstrated that adding various doses of AMPs to feed significantly promotes growth and improves production performance in broiler chickens [36,37], laying hens [38], weaned piglets [39], aquaculture [40,41], and ruminants [42]. Integrating all endpoints, administration of sublancin at doses up to 50,000 mg/kg demonstrated no significant alterations in fundamental physiological parameters, including appetite, water intake, and body weight dynamics, thereby establishing a potentially wide safety margin based upon efficacy data from the literature (the NOEL here was 50,000 mg/kg) for maintaining homeostatic growth and metabolic functions during chronic 180-day exposure.

Food intake and water consumption are important indicators of animal health. Systematic recording regularly enables prompt detection of behavioral abnormalities and variations in physiological parameters. If persistent abnormal changes in food intake or water consumption occur, particularly those independent of environmental factors or physiological conditions, the possibility of chronic toxicity must be systematically evaluated. Sublancin showed no significant effects on the water and food intake of rats. This finding provides evidence to support the long-term safety of sublancin in rodent models and support further translational development. In summary, AMPs promote livestock growth through direct growth-promoting effects and indirect effects by enhancing overall animal health and welfare, supporting their potential application in livestock growth.

Hematology and blood indices reflect the animal's ability to digest and absorb nutrients, its nutrient metabolism, and overall health. Monitoring these parameter changes enables a more precise toxicity evaluation of the test substance [43]. In this study, sublancin-treated rats showed no significant differences in hematological parameters (hemoglobin, red blood cell count, white blood cell count, and platelet count) compared with the control group, demonstrating no adverse effects on hematopoietic function. Furthermore, the liver function indicators (ALT, AST, ALP), kidney function (BUN, Cr), and blood lipid levels (TCH, TG) remained unaffected by sublancin, suggesting a low toxicity risk to the rats' major metabolic organs and systems. Published studies demonstrate that sublancin exerts significant therapeutic effects in animal models of specific pathological conditions. Wang et al. [12] demonstrated that oral sublancin administration (1 mg/kg) significantly upregulated *IL-1*, *IL-6*, and *TNF-α* gene expression in the macrophages of cyclophosphamide-induced immunosuppressed mice. In a murine methicillin-resistant *Staphylococcus aureus* (MRSA) model, sublancin exhibited anti-inflammatory activity, elevating chemokines (CXCL1 and MCP-1) and suppressing the production of pro-inflammatory cytokines (TNF-α) [8]. Wang et al. [7] further demonstrated that dietary sublancin supplementation reduced pro-inflammatory cytokines (IL-1β, IL-6, and TNF-α) in the ileum of broiler chickens with *Clostridium perfringens*-induced necrotic enteritis, confirming its anti-inflammatory efficacy. Liu et al. [9] also found that dietary sublancin significantly enhanced the immune response in specific-pathogen-free broiler chickens vaccinated against Newcastle disease, promoting lymphocyte proliferation; stimulating secretion of IFN-γ, IL-10, and IL-4; and elevating antibody titers. Notably, these effects were more pronounced at lower doses, suggesting its potential as a vaccine adjuvant. Overall, the high doses of sublancin in this study did not cause changes in serum biochemical parameters, indicating a low toxicity risk to the major metabolic organs and systems of animals under normal physiological conditions. Furthermore, the existing literature reports that sublancin has significant therapeutic effects on test animals under pathological conditions [9,12], further confirming its safety and providing scientific evidence for its therapeutic potential in practical applications.

Measuring organ coefficients serves as a sensitive indicator of a drug's impact on organs, and their fluctuations are recognized as markers for evaluating the effects of drug exposure on organ integrity [44]. This study found no significant differences in the organ coefficients of the heart, liver, spleen, lungs, and kidneys between sublancin-treated rats and the control group, indicating that sublancin does not cause significant damage or hypertrophy in these organs. Furthermore, the histopathological examination revealed no abnormal structural changes in the major organs of sublancin-treated rats, further confirming the safety of sublancin for these organs. Additionally, previous studies indicate that sublancin has protective and reparative effects on animal organs under pathological conditions. Wang et al. [5] found that adding sublancin to feed reduces intestinal damage caused by *Staphylococcus aureus* infection in young mice. Specifically, administering 4 mg/kg sublancin significantly improved intestinal repair and reduced mortality. Wang et al. [7] reported that dietary sublancin supplementation significantly increased jejunal villus height and duodenal crypt depth in broiler chickens with *Clostridium perfringens*-induced necrotic enteritis. Histopathological and functional analyses revealed no sublancin-induced abnormalities in renal structure or function, even at the maximum tested dose (50,000 mg/kg), despite kidneys being the principal detoxification organs [45]. Further histopathological examination revealed no signs of toxicity in other organs. Morphological observations of the major organs in the rats showed no necrosis, inflammation, or other pathological changes. This finding further demonstrates that high doses of sublancin have no subchronic toxicity and are considered highly safe. Collectively, the integrated subchronic (90-day) and chronic (180-day) toxicity data establish a NOEL of 50,000 mg/kg for sublancin, whichm together with the previously determined acute oral LD50 exceeding 5000 mg/kg, demonstrates systemic and organ-level safety across all evaluated endpoints, when considered against published efficacious dosing levels both in mice [6] and broilers [7,9]. These robust toxicological findings provide compelling experimental evidence to support further the preclinical development of sublancin as a veterinary therapeutic candidate.

It is noteworthy that published studies on sublancin administration in animal models have employed either intraperitoneal injection [4,5] or dietary/water supplementation [9,12]. As a peptide antimicrobial, sublancin shares the common delivery challenges of oral peptides [46], though its actual gastrointestinal stability needs empirical determination. While intraperitoneal injection introduces compounds directly into systemic circulation, risking acute toxicity from localized high concentrations (e.g., peritoneal irritation) [47], oral administration better mimics real-world exposure through gastrointestinal absorption [48]. Actually, our GLP-compliant study demonstrated an exceptional safety profile for sublancin: even at 50,000 mg/kg orally, no treatment-related toxicity was observed. Specifically, we found the absence of gastrointestinal symptoms, no clinically significant hematological changes, and normal histopathology scores. These findings, consistent with our following study of sublancin's gastric stability in SD rats, suggest that sublancin maintains structural integrity during digestion, with its degradation products exhibiting minimal toxic potential. Our completed pharmacokinetic studies in rats (data to be published) further support that oral administration of sublancin to rats results in minimal systemic absorption, primarily localizes in the gastrointestinal tract, and undergoes rapid excretion without evidence of tissue accumulation or plasma protein binding. The PK profile suggests that sublancin may retain partial stability in the gastrointestinal tract. Future investigations comparing PK/PD parameters across different administration routes may further optimize the dosing strategies.

5. Conclusions

This study systematically evaluated the safety of sublancin as a novel veterinary drug candidate through 90-day subchronic and 180-day chronic oral toxicity tests at doses of 2000, 10,000, and 50,000 mg/kg in feed. Comprehensive assessments revealed no adverse effects on physiological parameters (appetite, weight gain, water consumption), hematological indices, serum biochemistry, or organ histopathology in SD rats, establishing a NOEL of 50,000 mg/kg. These findings clearly support the long-term safety of oral administration of sublancin and its potential development as a veterinary therapeutic.

Supplementary Materials: The following supporting information can be downloaded at https://www.mdpi.com/article/10.3390/toxics13050413/s1. Table S1. Effects of sublancin on the daily food intake of rats fed for 45 and 90 days during the subchronic toxicity trial. Table S2-1. Effects of sublancin on the daily food intake of rats fed for 60 days during the chronic toxicity trial. Table S2-2. Effect of sublancin on the daily food intake of rats fed for 120 days during the chronic toxicity trial. Table S2-3. Effect of sublancin on the daily food intake of rats fed for 180 days during the chronic toxicity trial. Table S3. Effect of sublancin on the daily water intake of rats fed for 45 and 90 days during the subchronic toxicity trial. Table S4-1. Effect of sublancin on the daily water intake of rats fed for 60 days during the chronic toxicity trial. Table S4-2. Effect of sublancin on the daily water intake of rats fed for 120 days during the chronic toxicity trial. Table S4-3. Effect of sublancin on the daily water intake of rats fed for 180 days during the chronic toxicity trial. Table S5. Effect of sublancin on the average daily gain of rats fed for 45 and 90 days during the subchronic toxicity trial. Table S6-1.

Effect of sublancin on the average daily gain of rats fed for the 60 days during chronic toxicity trial. Table S6-2. Effect of sublancin on the average daily gain of rats fed for 120 days during the chronic toxicity trial. Table S6-3. Effect of sublancin on the average daily gain of rats fed for 180 days during the chronic toxicity trial. Table S7. Effect of sublancin on the hematology of rats fed for 45 and 90 days during the subchronic toxicity trial. Table S8. Effect of sublancin on the hematology of rats fed for 60, 120, and 180 days during the chronic toxicity trial. Table S9. Effect of sublancin on the biochemical parameters of rats fed for 45 and 90 days during the subchronic toxicity trial. Table S10. Effect of sublancin on the serum biochemical parameters of rats fed for 60, 120, and 180 days during the chronic toxicity trial. Table S11. Effect of sublancin on the organ index of rats fed for 45 and 90 days during the subchronic toxicity trial. Table S12. Effect of sublancin on the organ index of rats fed for 60, 120, and 180 days during the chronic toxicity trial.

Author Contributions: Investigation, writing—original draft, data curation, supervision, project administration, Y.G.; writing, methodology, software, data curation, Z.L.; methodology, investigation, P.X.; investigation, resources, G.G.; validation, investigation, resources T.H.; writing—review and editing, supervision, Y.L. All authors have read and agreed to the published version of the manuscript.

Funding: This study was supported by Wuhan Polytechnic University Industry-Academia-Research Project (Project No.: whpu-2023-kj-2709).

Institutional Review Board Statement: The animal study protocol was approved by the Animal Care and Use Committee of China Agricultural University approved all the animal experiments in this study (approval number AW42800202-3-5).

Informed Consent Statement: Not applicable.

Data Availability Statement: Data are available upon request.

Acknowledgments: This work was supported by Wuhan Textile University through the provision of research facilities and resources. The authors express their special appreciation to Chen Rong for his patient direction and careful review of the manuscript, which ensured its linguistic precision and clarity. The authors would like to acknowledge the support of China Agricultural University, Wuhan Institute of Technology, and Wuhan Polytechnic University for providing laboratory facilities and resources. During the preparation of this manuscript, the authors used Kimi, developed by Moonshot AI, for the purposes of polishing the grammar and fluency of the text.

Conflicts of Interest: Yong Guo, Gantong Guo and Tao He were employed by Sinagri YingTai Bio-peptide Co., Ltd. The remaining authors declare that the research was conducted in the absence of any commercial or financial relationships that could be construed as a potential conflict of interest. The funding sponsors had no role in the design of the study; in the collection, analyses, or interpretation of data; in the writing of the manuscript, and in the decision to publish the results.

Abbreviations

The following abbreviations are used in this manuscript.

SD	Sprague–Dawley
NOEL	No-observed-effect level
NOAEL	No-observed-adverse-effect level
LD	Linear dichroism
E	Glutamic acid
S	Serine
K	Lysine
I	Isoleucine
C	Cysteine
T	Threonine
W	Tryptophan

L	Leucine
G	Glycine
T	Threonine
Q	Glutamine
N	Asparagine
Y	Tyrosine
R	Arginine
N	Asparagine
HGB	Hemoglobin
RBC	Red blood cell count
WBC	White blood cell count
PLT	Platelet count
HCT	Hematocrit
EOS	Eosinophils
BAS	Basophils
MO	Monocytes
LY	Lymphocytes
TP	Total protein
ALB	Albumin
ALT	Alanine aminotransferase
AST	Aspartate aminotransferase
BUN	Blood urea nitrogen
TC	Total cholesterol
Cr	Creatinine
Glu	Glucose
TG	Triglycerides
H&E	Hematoxylin and eosin
AMPs	Antimicrobial peptides

References

1. Yang, L.; Zeng, X.; Qiao, S. Advances in research on solid-state fermented feed and its utilization: The pioneer of private customization for intestinal microorganisms. *Anim. Nutr.* **2021**, *7*, 905–916. [CrossRef]
2. Paik, S.H.; Chakicherla, A.; Hansen, J.N. Identification and Characterization of the Structural and Transporter Genes for, and the Chemical and Biological Properties of, Sublancin 168, a Novel Lantibiotic Produced by *Bacillus subtilis* 168. *J. Biol. Chem.* **1998**, *273*, 23134–23142. [CrossRef] [PubMed]
3. Biswas, S.; Garcia De Gonzalo, C.V.; Repka, L.M.; van der Donk, W.A. Structure-Activity Relationships of the S-Linked Glycocin Sublancin. *ACS Chem. Biol.* **2017**, *12*, 2965–2969. [CrossRef]
4. Wang, S.; Wang, Q.; Zeng, X.; Ye, Q.; Huang, S.; Yu, H.; Yang, T.; Qiao, S. Use of the Antimicrobial Peptide Sublancin with Combined Antibacterial and Immunomodulatory Activities To Protect against Methicillin-Resistant *Staphylococcus aureus* Infection in Mice. *J. Agric. Food Chem.* **2017**, *65*, 8595–8605. [CrossRef] [PubMed]
5. Wang, Q.; Zeng, X.; Wang, S.; Hou, C.; Yang, F.; Ma, X.; Thacker, P.; Qiao, S. The bacteriocin sublancin attenuates intestinal injury in young mice infected with *Staphylococcus aureus*. *Anat. Rec.* **2014**, *297*, 1454–1461. [CrossRef]
6. Wang, S.; Ye, Q.; Wang, K.; Zeng, X.; Huang, S.; Yu, H.; Ge, Q.; Qi, D.; Qiao, S. Enhancement of Macrophage Function by the Antimicrobial Peptide Sublancin Protects Mice from Methicillin-Resistant *Staphylococcus aureus*. *J. Immunol. Res.* **2019**, *2019*, 3979352. [CrossRef]
7. Wang, S.; Zeng, X.F.; Wang, Q.W.; Zhu, J.L.; Peng, Q.; Hou, C.L.; Thacker, P.; Qiao, S.Y. The antimicrobial peptide sublancin ameliorates necrotic enteritis induced by *Clostridium perfringens* in broilers. *J. Anim. Sci.* **2015**, *93*, 4750–4760. [CrossRef]
8. Li, J.; Chen, J.; Yang, G.; Tao, L. Sublancin protects against methicillin-resistant *Staphylococcus aureus* infection by the combined modulation of innate immune response and microbiota. *Peptides* **2021**, *141*, 170533. [CrossRef] [PubMed]
9. Liu, Y.; Zhang, J.; Wang, S.; Guo, Y.; He, T.; Zhou, R. A Novel Adjuvant "Sublancin" Enhances Immune Response in Specific Pathogen-Free Broiler Chickens Inoculated with Newcastle Disease Vaccine. *J. Immunol. Res.* **2019**, *2019*, 1016567. [CrossRef]
10. Costa, M.; Meirinhos, C.; Cunha, E.; Gomes, D.; Pereira, M.; Dias, R.; Tavares, L.; Oliveira, M. Nisin Mutant Prevention Concentration and the Role of Subinhibitory Concentrations on Resistance Development by Diabetic Foot Staphylococci. *Antibiotics* **2022**, *11*, 972. [CrossRef]

11. Garcia De Gonzalo, C.V.; Zhu, L.; Oman, T.J.; van der Donk, W.A. NMR structure of the S-linked glycopeptide sublancin 168. *ACS Chem. Biol.* **2014**, *9*, 796–801. [CrossRef] [PubMed]
12. Wang, S.; Huang, S.; Ye, Q.; Zeng, X.; Yu, H.; Qi, D.; Qiao, S. Prevention of Cyclophosphamide-Induced Immunosuppression in Mice with the Antimicrobial Peptide Sublancin. *J. Immunol. Res.* **2018**, *2018*, 4353580. [CrossRef]
13. Oman, T.J.; Boettcher, J.M.; Wang, H.; Okalibe, X.N.; van der Donk, W.A. Sublancin is not a lantibiotic but an S-linked glycopeptide. *Nat. Chem. Biol.* **2011**, *7*, 78–80. [CrossRef]
14. Nawabjan, S.A.; Singh, K.; Iswarya, G.S.M.; Au-Yeung, R.K.H.; Zhang, F.; Zhang, L.; El-Nezami, H.; Chow, B.K.C. Preclinical toxicity evaluation of the novel anti-hypertensive compound KSD179019. *Food Chem. Toxicol.* **2025**, *196*, 115195. [CrossRef]
15. Maache, S.; Tahraoui, A.; Soulo, N.; Benlabchir, A.A.; Nouioura, G.; Bouslamti, M.; Aldisi, D.; Saghrouchni, H.; Giesy, J.P.; Aboul-Soud, M.A.M.; et al. Sub-chronic and acute toxicity of aqueous extracts *Salvia blancoana* subsp. mesatlantica (Maire) Figuerola to rodents. *Toxicol. Rep.* **2024**, *13*, 101847. [CrossRef] [PubMed]
16. Li, Y.; Han, C.; Shi, L.; Chen, C.; Zhao, J.; Liu, T.; Zhuo, Q. Ninety-Day Feeding Test of Stacked DBN9936 × DBN9501 Maize on Sprague Dawley Rats. *J. Appl. Toxicol.* **2024**, *45*, 646–658. [CrossRef]
17. Su, B.C.; Li, C.C.; Liu, C.W.; Chen, J.Y. A Pilot Safety Assessment for Recombinant *Epinephelus lanceolatus* Piscidin Yeast Powder as a Drug Food Additive after Subacute and Subchronic Administration to SD Rats. *Mar. Drugs* **2020**, *18*, 586. [CrossRef]
18. Taychaworaditsakul, W.; Saenjum, C.; Lumjuan, N.; Chawansuntati, K.; Sawong, S.; Jaijoy, K.; Na Takuathung, M.; Sireeratawong, S. Safety of Oral Carica papaya L. Leaf 10% Ethanolic Extract for Acute and Chronic Toxicity Tests in Sprague Dawley Rats. *Toxics* **2024**, *12*, 198. [CrossRef] [PubMed]
19. Fleming, M.; Staveley, J.; Samel, A.; Aufderheide, J.; Blickley, T.M.; Bone, A.; Bruns, E.; Catron, T.; Edwards, D.; Gallagher, S.; et al. Critical review of chronic toxicity testing approaches with the saltwater mysid (*Americamysis bahia*) used in pesticide registration. *Environ. Toxicol. Chem.* **2025**, *44*, 916–925. [CrossRef]
20. Simancas-Herbada, R.; Fernández-Carballido, A.; Aparicio-Blanco, J.; Slowing, K.; Rubio-Retama, J.; López-Cabarcos, E.; Torres-Suárez, A.-I. Controlled Release of Highly Hydrophilic Drugs from Novel Poly(Magnesium Acrylate) Matrix Tablets. *Pharmaceutics* **2020**, *12*, 174. [CrossRef]
21. Mátis, G.; Tráj, P.; Hanyecz, V.; Mackei, M.; Márton, R.A.; Vörösházi, J.; Kemény, Á.; Neogrády, Z.; Sebők, C. Immunomodulatory properties of chicken cathelicidin-2 investigated on an ileal explant culture. *Vet. Res. Commun.* **2024**, *48*, 2527–2535. [CrossRef] [PubMed]
22. Velayatipour, F.; Tarrahimofrad, H.; Zamani, J.; Fotouhi, F.; Aminzadeh, S. In-vitro antimicrobial activity of AF-DP protein and in-silico approach of cell membrane disruption. *J. Biomol. Struct. Dyn.* **2024**, 1–18. [CrossRef] [PubMed]
23. Li, H.; Niu, J.; Wang, X.; Niu, M.; Liao, C. The Contribution of Antimicrobial Peptides to Immune Cell Function: A Review of Recent Advances. *Pharmaceutics* **2023**, *15*, 2278. [CrossRef] [PubMed]
24. Yang, L.; Luo, M.; Liu, Z.; Li, Y.; Lin, Z.; Geng, S.; Wang, Y. BamA-targeted antimicrobial peptide design for enhanced efficacy and reduced toxicity. *Amino Acids* **2023**, *55*, 1317–1331. [CrossRef]
25. Elbediwi, M.; Rolff, J. Metabolic pathways and antimicrobial peptide resistance in bacteria. *J. Antimicrob. Chemother.* **2024**, *79*, 1473–1483. [CrossRef]
26. García-Beltrán, J.M.; Arizcun, M.; Chaves-Pozo, E. Antimicrobial Peptides from Photosynthetic Marine Organisms with Potential Application in Aquaculture. *Mar. Drugs* **2023**, *21*, 290. [CrossRef]
27. Takada, M.; Ito, T.; Kurashima, M.; Matsunaga, N.; Demizu, Y.; Misawa, T. Structure-Activity Relationship Studies of Substitutions of Cationic Amino Acid Residues on Antimicrobial Peptides. *Antibiotics* **2022**, *12*, 19. [CrossRef]
28. Ji, S.; Li, W.; Baloch, A.R.; Wang, M.; Cao, B. Improved production of sublancin via introduction of three characteristic promoters into operon clusters responsible for this novel distinct glycopeptide biosynthesis. *Microb. Cell Factories* **2015**, *14*, 17. [CrossRef]
29. Hsieh, Y.S.; Wilkinson, B.L.; O'Connell, M.R.; Mackay, J.P.; Matthews, J.M.; Payne, R.J. Synthesis of the bacteriocin glycopeptide sublancin 168 and S-glycosylated variants. *Org. Lett.* **2012**, *14*, 1910–1913. [CrossRef]
30. Wu, C.; Biswas, S.; Garcia De Gonzalo, C.V.; van der Donk, W.A. Investigations into the Mechanism of Action of Sublancin. *ACS Infect. Dis.* **2019**, *5*, 454–459. [CrossRef]
31. Norris, G.E.; Patchett, M.L. The glycocins: In a class of their own. *Curr. Opin. Struct. Biol.* **2016**, *40*, 112–119. [CrossRef]
32. Yan, L.; Mao, J.; Shi, W.; Ren, L.; Li, J.; Geng, B.; Wang, H.; Zhang, J.; Tian, Y.; Zhang, B.; et al. Subchronic toxicity study of ferric oxide nanoparticles through intragastric administration: A 94-d, repeated dose study in Sprague Dawley rats. *Regul. Toxicol. Pharmacol.* **2023**, *140*, 105381. [CrossRef] [PubMed]
33. Villa, R.E.; Azimonti, G.; Bonos, E.; Christensen, H.; Durjava, M.; Dusemund, B.; Gehring, R.; Glandorf, B.; Kouba, M.; López-Alonso, M.; et al. Safety of a feed additive consisting of Duddingtonia flagransNCIMB 30336 (BioWorma®) for all grazing animals (International Animal Health Products Pty Ltd.). *EFSA J.* **2025**, *23*, e9366. [CrossRef] [PubMed]
34. Zhao, Q.; Yang, Z.S.; Cao, S.J.; Chang, Y.F.; Cao, Y.-Q.; Li, J.-B.; Yao, Z.-X.; Wen, Y.-P.; Huang, X.-B.; Wu, R.; et al. Acute oral toxicity test and assessment of combined toxicity of cadmium and aflatoxin B1 in kunming mice. *Food Chem. Toxicol.* **2019**, *131*, 110577. [CrossRef] [PubMed]

35. Gibb, S. Toxicity testing in the 21st century: A vision and a strategy. *Reprod. Toxicol.* **2008**, *25*, 136–138. [CrossRef]
36. Dai, Z.; Shang, L.; Wang, F.; Zeng, X.; Yu, H.; Liu, L.; Zhou, J.; Qiao, S. Effects of Antimicrobial Peptide Microcin C7 on Growth Performance, Immune and Intestinal Barrier Functions, and Cecal Microbiota of Broilers. *Front. Vet. Sci.* **2021**, *8*, 813629. [CrossRef]
37. Gao, S.; Zhang, Q.; Liu, C.; Shen, H.; Wang, J. Effects of maggot antimicrobial peptides on growth performance, immune function, and cecal flora of yellow-feathered broilers. *Front. Vet. Sci.* **2023**, *10*, 1156964. [CrossRef]
38. Roque-Borda, C.A.; Pereira, L.P.; Guastalli, E.A.L.; Soares, N.M.; Mac-Lean, P.A.B.; Salgado, D.D.; Meneguin, A.B.; Chorilli, M.; Vicente, E.F. HPMCP-Coated Microcapsules Containing the Ctx(Ile^{21})-Ha Antimicrobial Peptide Reduce the Mortality Rate Caused by Resistant Salmonella Enteritidis in Laying Hens. *Antibiotics* **2021**, *10*, 616. [CrossRef]
39. Liu, N.; Ma, X.; Jiang, X. Effects of Immobilized Antimicrobial Peptides on Growth Performance, Serum Biochemical Index, Inflammatory Factors, Intestinal Morphology, and Microbial Community in Weaning Pigs. *Front. Immunol.* **2022**, *13*, 872990. [CrossRef] [PubMed]
40. Wang, S.; Xie, S.; Zhou, A.; Zhang, C.; Wen, L.; Xu, G.; Zou, J. Effects of mixed antimicrobial peptide on the growth performance, antioxidant and immune responses and disease resistance of Pengze crucian carp (*Carassius auratus* var. Pengze). *Fish. Shellfish. Immunol.* **2021**, *114*, 112–118. [CrossRef]
41. An, Z.; Chen, F.; Hao, H.; Xiong, M.; Peng, H.; Sun, H.; Wang, K.J. Growth-promoting effect of antimicrobial peptide Scy-hepc on mariculture large yellow croaker *Larimichthys crocea* and the underlying mechanism. *Fish Shellfish. Immunol.* **2023**, *134*, 108649. [CrossRef] [PubMed]
42. Shi, J.; Lei, Y.; Wu, J.; Li, Z.; Zhang, X.; Jia, L.; Wang, Y.; Ma, Y.; Zhang, K.; Cheng, Q.; et al. Antimicrobial peptides act on the rumen microbiome and metabolome affecting the performance of castrated bulls. *J. Anim. Sci. Biotechnol.* **2023**, *14*, 31. [CrossRef]
43. Akmal, H.; Ahmad, S.; Abbasi, M.H.; Jabeen, F.; Shahzad, K. A study on assessing the toxic effects of ethyl paraben on rohu (*Labeo rohita*) using different biomarkers; hemato-biochemical assays, histology, oxidant and antioxidant activity and genotoxicity. *PLoS ONE* **2024**, *19*, e0302691. [CrossRef]
44. Zhang, G.; Li, Y.; Chen, T.; Gao, Y.; Sun, J.; Yang, W.; Song, L.; Su, P.; Ma, M.; Zhang, Z.; et al. Comparative study of the efficacy and pharmacokinetics of reduning injection and atomization inhalation. *Biomed. Pharmacother. Biomed. Pharmacother.* **2019**, *118*, 109226. [CrossRef]
45. Yadav, R.; Kumar, D.; Singh, J.; Jangra, A. Environmental toxicants and nephrotoxicity: Implications on mechanisms and therapeutic strategies. *Toxicology* **2024**, *504*, 153784. [CrossRef] [PubMed]
46. Yang, G.; Shang, L.; Liu, L.; Li, Z.; Zeng, X.; Ding, X.; Huang, J.; Qiao, S.; Yu, H. Engineering and Purification of Microcin C7 Variants Resistant to Trypsin and Analysis of Their Biological Activity. *Antibiotics* **2023**, *12*, 1346. [CrossRef] [PubMed]
47. Balog, S.; Jeong, S.; Asahina, K. Recruitment of large peritoneal macrophages to capsular fibrosis developed on the liver surface. *FASEB J.* **2024**, *38*, e23327. [CrossRef]
48. Wang, Y.; Ning, Z.H.; Tai, H.W.; Long, S.; Qin, W.C.; Su, L.M.; Zhao, Y.H. Relationship between lethal toxicity in oral administration and injection to mice: Effect of exposure routes. *Regul. Toxicol. Pharmacol.* **2015**, *71*, 205–212. [CrossRef]

International Journal of
Molecular Sciences

MDPI

Article

Therapeutic Potential of Quadrigemine I Against Lymphoma: Mechanistic Insights from Cell Lines and Xenograft Models Demonstrating DNA Damage, Oxidative Stress, and Pathway Regulation

Junmei Pu [1,2,3], Wenfeng Shi [1,2,3], Jiabao Cui [1,2,3], Hui Yang [1,2,3], Jianxin Cao [1,2,3], Yaping Liu [1,2,3], Shanshan Xiao [1,2,3,*] and Guiguang Cheng [1,2,3,*]

1 Faculty of Food Science and Engineering, Kunming University of Science and Technology, Kunming 650500, China; 20222125014@stu.kust.edu.cn (J.P.); swf821130135@163.com (W.S.); 15045227277@163.com (J.C.); 13595862451@163.com (H.Y.); 13313130@kust.edu.cn (J.C.); liuyaping@kust.edu.cn (Y.L.)
2 Yunnan Key Laboratory of Plateau Food Advanced Manufacturing, Kunming 650500, China
3 Yunnan International Joint Laboratory of Green Food Processing, Kunming 650500, China
* Correspondence: shine330@kust.edu.cn (S.X.); ggcheng@kmust.edu.cn (G.C.)

Abstract: Lymphoma is one of the malignant tumors that significantly threatens human health. Quadrigemine I, an indole alkaloid derived from the leaves of *Psychotria pilifera*, has been studied for its potential anti-tumor properties, but its mechanisms remain poorly understood. The CCK-8 assay was used to determine the cytotoxic effect of quadrigemine I on lymphoma cells. Flow cytometry was employed to analyze apoptosis and reactive oxygen species (ROS) levels in these cells. DNA damage was assessed by the comet assay, and the underlying mechanisms were investigated using RNA sequencing (RNA-seq) and real-time quantitative PCR (RT-qPCR). The anti-tumor activity of quadrigemine I was evaluated in tumor xenograft mice. Quadrigemine I suppressed lymphoma cell proliferation with an IC_{50} of 0.46 μM. It induced apoptosis, promoted ROS generation, and caused DNA damage in tumor cells. RNA-seq analysis revealed that the significantly differentially expressed genes were notably enriched in the ErbB, p53, and apoptosis signaling pathways. RT-qPCR demonstrated altered expression levels of key genes in the aforementioned pathways. In vivo, quadrigemine I significantly inhibited tumor growth in xenograft mice by increasing apoptosis in tumor tissues, with reduced Ki-67 and Bcl-2 expression and elevated cleaved caspase-3 levels. Quadrigemine I may serve as a novel anti-tumor agent for lymphoma therapy.

Keywords: lymphoma; quadrigemine I; xenograft mice; RNA-seq analysis

1. Introduction

Lymphoma is a malignant tumor originating in the lymphatic system, characterized by the abnormal proliferation and spread of lymphocytes, leading to swollen lymphatic tissue and impaired organ function [1]. Clinical symptoms include enlarged lymph nodes, systemic manifestations (e.g., fever, night sweats, malaise), anemia, and thrombocytopenia [2]. The prevalence of malignant lymphoma continues to rise, making it one of the major tumor types that pose a serious threat to human health [3]. Lymphoma exhibits significant individual variability, specific symptoms, and poor treatment prognosis, with a high likelihood of recurrence within 2–3 years [4]. Therefore, recurrence and drug resistance remain major challenges in current clinical treatment, seriously threatening patients' lives and health [5].

Common therapeutic drugs and strategies for lymphoma include chemotherapeutic agents (e.g., the CHOP regimen, ABVD regimen), targeted therapies (e.g., rituximab, belantamab mafodotin), immunotherapy (e.g., PD-1 inhibitors), and stem cell transplantation. Although these treatments are highly effective in many cases, they have certain drawbacks and limitations, such as high toxicity, susceptibility to drug resistance, high treatment costs, and long-term immune system impairment. Therefore, there is an urgent need to develop safe and effective therapeutic agents for the prevention and treatment of lymphoma [6].

Indole alkaloids are important medicinal compounds for drug exploration. To date, many indole alkaloids from plants, animals, and fungi have been developed as successful agents for clinical treatment of cancer, pain, diabetes, cardiovascular disease, and inflammatory diseases [7]. For lymphomas, vincristine, vinblastine, and camptothecin are important anti-lymphoma agents and are usually consumed with other chemotherapeutic agents [8]. Thus, indole alkaloids are important sources of new anti-lymphoma substances.

Quadrigemine I is an indole alkaloid isolated from *Psychotria pilifera*, a plant belonging to the *Rubiaceae* family. A previous study demonstrated that quadrigemine I has significant cytotoxic activity against various human cancer cells [9]. The IC_{50} of quadrigemine I was 0.56 μM for human acute promyelocytic leukemia cells (HL-60). While its effects in leukemia have been explored, its potential role in lymphoma therapy remains underinvestigated. Recent studies have suggested that indole alkaloids, such as quadrigemine I, can modulate various signaling pathways involved in cancer cell proliferation, apoptosis, and metastasis. However, the therapeutic effect of quadrigemine I on lymphomas has not yet been studied, and its possible mechanisms also need to be explored.

In this paper, the extraction, isolation, and identification of quadrigemine I were performed. The inhibitory effect of quadrigemine I on cell proliferation was evaluated in human large-cell immunoblastic lymphoma cells (SR) and Burkitt lymphoma cells (Raji) using the CCK-8 assay. In addition, the protective effect of quadrigemine I against lymphoma was assessed in an SR cell-induced lymphoma model in NOD/SCID mice. The effects of quadrigemine I on cell apoptosis, reactive oxygen species (ROS) generation, and DNA damage were systematically analyzed. Furthermore, RNA-seq analysis was conducted to identify key target genes and signaling pathways.

2. Results

2.1. Cytotoxic Effect of Indole Alkaloids on Human Lymphoma Cells and Normal Cells

We evaluated the effect of indole alkaloids on the viability of SR and Raji cells using the CCK-8 assay. Quadrigemine I exhibited concentration-dependent anti-proliferative activity in both SR and Raji cells. IC_{50} values of quadrigemine I on SR and Raji cells were 0.46 ± 0.12 μM and 1.99 ± 0.17 μM, respectively, while the positive control (CDDP) on SR and Raji cells was 7.29 ± 0.16 μM and 11.25 ± 0.31 μM, respectively (Figure 1B,C). We observed that the SR and Raji cell lines exhibited distinct responses to quadrigemine I. This difference may be attributed to their origins from different types of lymphomas—T cell and B cell, respectively. T cells and B cells possess distinct characteristics in terms of signal transduction, proliferation patterns, and apoptotic pathways, which may contribute to their differential sensitivity to quadrigemine I.

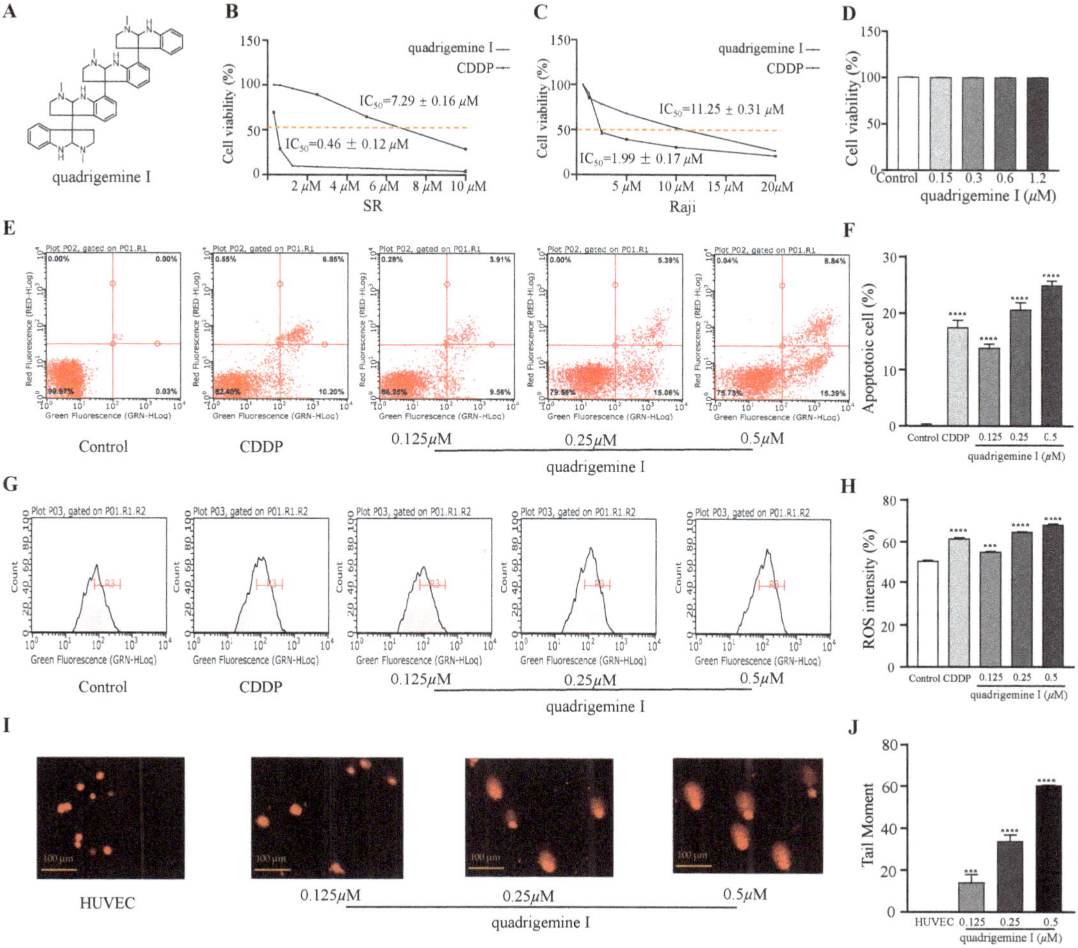

Figure 1. Effects of quadrigemine I on cell viability, apoptosis, ROS production, and DNA damage in SR cells. (**A**) The chemical structure of quadrigemine I. (**B–D**) SR cells (**B**), Raji cells (**C**), and HUVECs (**D**) were treated with different concentrations of quadrigemine I. Cell viability was evaluated by the CCK-8 assay. (**E,F**) Apoptosis on SR cells induced by quadrigemine I. (**G,H**) Intracellular ROS production in SR cells induced by quadrigemine I. (**I,J**) DNA comet images (scale bar: 100 μm) of SR cells induced by quadrigemine I. Data represent means ± SD from three independent experiments. *** $p < 0.001$ and **** $p < 0.0001$ compared with the control group.

Furthermore, the effect of quadrigemine I on normal cells was evaluated by the CCK-8 assay, which showed that quadrigemine I had no significant effect on the cell viability of HUVECs at 0.5 μM, 0.25 μM, and 0.125 μM. Collectively, these results demonstrate that quadrigemine I exerts significant anti-proliferative effects on SR cells at concentrations below the cytotoxic threshold (Figure 1D).

2.2. Effects of Quadrigemine I on SR Cell Apoptosis

To determine the percentage of mid- and late-stage apoptotic cells, flow cytometry was performed using Annexin V-FITC/PI staining. The results showed a significant increase in mid- and late-stage apoptotic cells after quadrigemine I treatment compared to the control group (Figure 1E). Treatment with quadrigemine I for 48 h dose dependently induced cell apoptosis, as evidenced by the increase in the proportion of Annexin V-positive cells from 3.90 ± 0.01% (control group) to 14.38 ± 0.07% (1.25 μM), 21.79 ± 0.26% (2.5 μM), and 30.89 ± 1.57% (5 μM) (Figure 1F). These findings suggest that quadrigemine I reduces cell viability through inducing cell apoptosis.

2.3. *Generation of Intracellular ROS Induced by Quadrigemine I*

Excessive generation of ROS is known to trigger cellular apoptosis and/or autophagy. Therefore, we analyzed the effect of quadrigemine I on intracellular ROS production in SR cells. After treatment with quadrigemine I at various concentrations (0.125, 0.25, and 0.5 μM), a significant increase in intracellular ROS levels was observed in SR cells (Figure 1G). Quantitative analysis showed that quadrigemine I at 0, 0.125, 0.25, and 0.5 μM significantly elevated ROS levels from approximately 50.42 ± 0.45% to 55.13 ± 0.18%, 64.85 ± 0.17%, and 68.31 ± 0.31%, respectively (Figure 1H). Compared with the control group, CDDP treatment elevated ROS levels from 50.42 ± 0.45% to 61.65 ± 0.39%. These results suggest that quadrigemine I (0.25 and 0.5 μM) may trigger apoptosis through increasing ROS levels in SR cells ($p < 0.05$).

2.4. *DNA Damage Induced by Quadrigemine I*

ROS can also induce DNA damage [10]. Therefore, we further investigated the role of quadrigemine I in DNA damage in SR cells. The comet assay demonstrated that the degree of DNA damage in SR cells treated with quadrigemine I was dose dependent compared to the control group (Figure 1I). Higher concentrations of quadrigemine I led to more pronounced DNA damage in SR cells, with tail DNA percentages of 14.03 ± 0.29% (1.25 μM), 33.78 ± 0.44% (0.25 μM), and 60.39 ± 0.22% (0.5 μM) (Figure 1J). The migrated cell nuclei with DNA strand breaks exhibited teardrop, pear-shaped, or large fan-like tails. In addition, treatment of HUVECs with the same concentrations of quadrigemine I did not cause DNA damage. These findings further suggest that quadrigemine I reduced cell viability through apoptotic mechanisms.

2.5. *RNA-Seq Analysis of Quadrigemine I-Regulated Gene Expression in SR Cells*

To further explore the mechanism of quadrigemine I-induced apoptosis, total RNA was extracted from SR cells treated with quadrigemine I for 48 h and used for RNA-seq analysis. We compared the gene expression profiles of the quadrigemine I-treated group with those of the control group. The results exhibited that SR cells treated with 0.5 μM quadrigemine I showed substantial changes in transcriptional profiles. Therefore, we selected the high-dose group for further analysis. RNA-seq analysis showed that a total of 1127 significantly differentially expressed genes (DEGs) were identified in quadrigemine I-treated SR cells, including 687 upregulated and 440 downregulated genes (Figure 2A–C). KEGG enrichment analysis of the significantly differentially expressed genes identified a total of 129 pathways. Notably, several pathways relevant to the focus of this study were identified, including the ErbB signaling pathway, p53 signaling pathway, TNF signaling pathway, T cell receptor (TCR) signaling pathway, mitogen-activated protein kinase (MAPK) signaling pathway, and apoptosis (Figure 2D).

To elucidate the specific mechanisms of action of quadrigemine I on the six candidate signaling pathways, we performed a differential analysis of the expression levels of key genes enriched in these pathways. In the ErbB signaling pathway (Figure 3A), which plays a key role in regulating cell proliferation and differentiation in various tissues and is crucial for maintaining tissue homeostasis, the ErbB family receptors are often aberrantly expressed or deregulated in tumors and play pivotal roles in cancer onset and metastatic progression. The results showed that the Erb-B2 receptor tyrosine kinase 3 (ERBB3) and transforming growth factor alpha (TGFA) genes were downregulated, suggesting that quadrigemine I treatment may block the activation of the ErbB signaling pathway, inhibit the growth and proliferation of lymphoma cells, and induce their apoptosis.

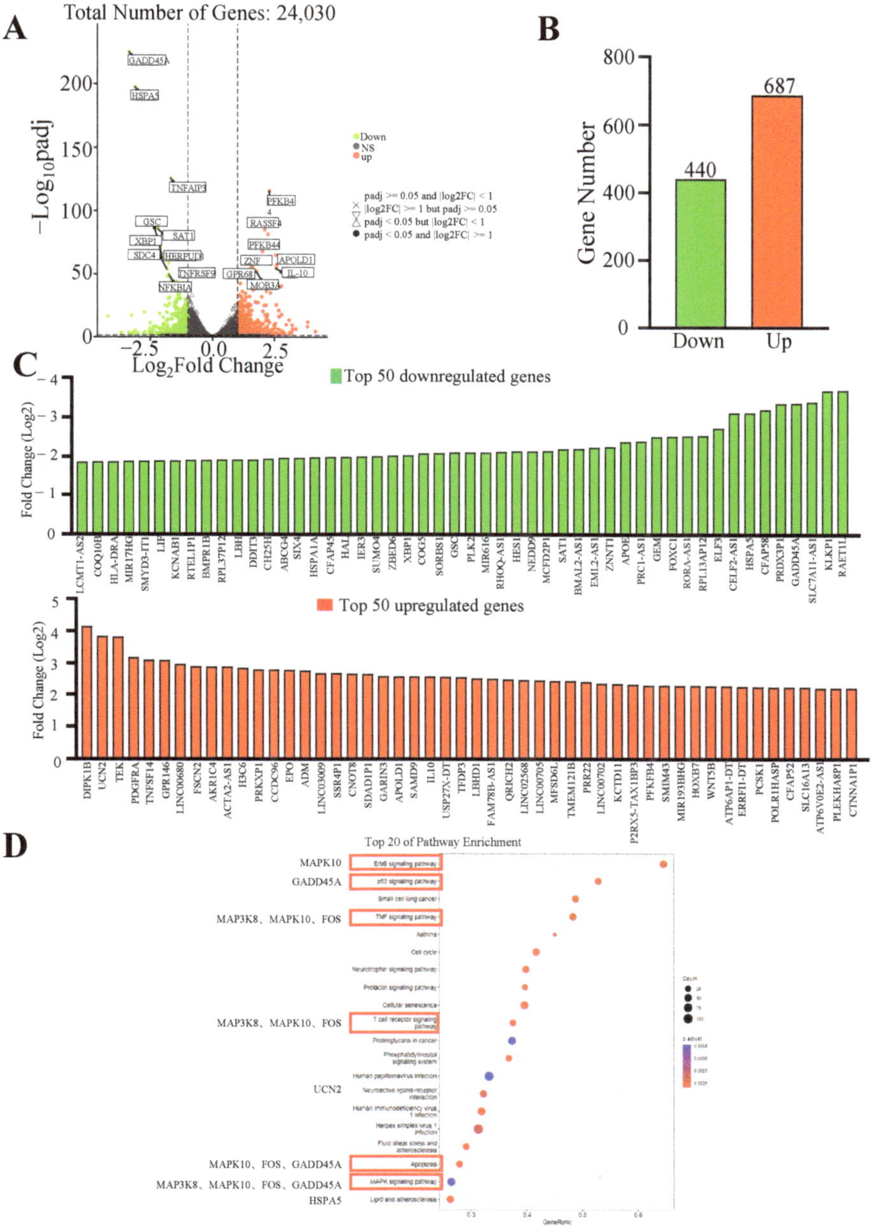

Figure 2. Transcriptional profiles of quadrigemine I-treated SR cells. (**A**) Volcano plots showed differentially expressed mRNAs. Green dots indicated downregulated mRNAs, red dots represented upregulated mRNAs, and gray dots showed mRNAs with nonsignificant differences. (**B**) The number of genes upregulated and downregulated by quadrigemine I. (**C**) Bar graphs showing the 18 DEGs that were most noticeably upregulated genes and downregulated genes in quadrigemine I-treated SR cells. (**D**) KEGG pathway enrichment (top 20).

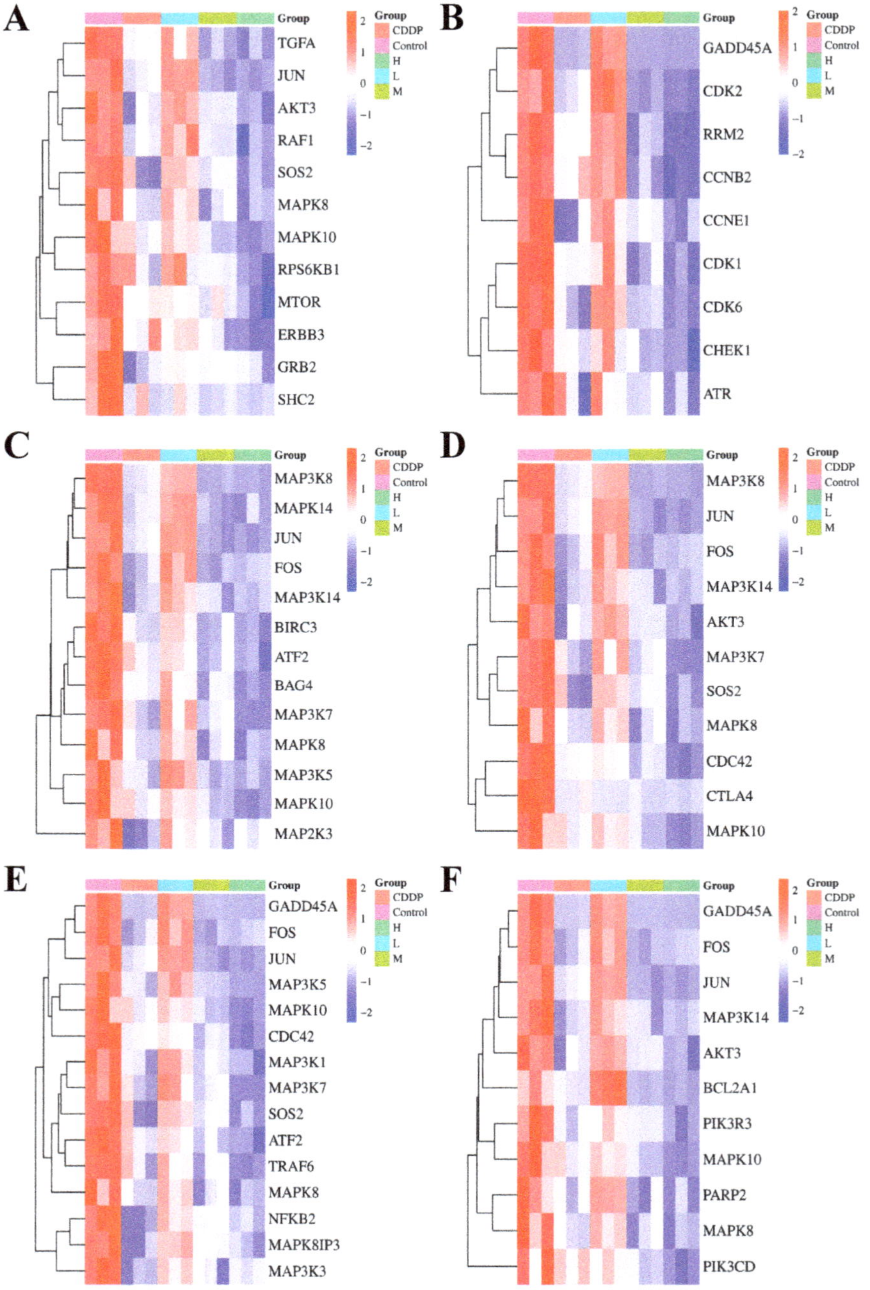

Figure 3. Quadrigemine I regulates the expression of genes in SR cells. (**A**) The heatmap of genes involved in the ErbB signaling pathway. (**B**) The heatmap of genes involved in the P53 signaling pathway. (**C**) The heatmap of genes involved in the TNF signaling pathway. (**D**) The heatmap of genes involved in the T cell receptor signaling pathway. (**E**) The heatmap of genes involved in the MAPK signaling pathway. (**F**) The heatmap of genes involved in the apoptosis signaling pathway. The heatmap showed all the differentially expressed mRNAs between quadrigemine I-treated SR cells (n = 3) and normal SR cells (n = 3) with adjusted $p < 0.05$ and $|\text{Log}_2\text{FC}| \geq 1.0$. Upregulated genes were presented as red dots, whereas downregulated ones were in blue.

The p53 signaling pathway (Figure 3B) maintains genetic stability by participating in DNA repair, chromosome recombination, and nucleotide excision in sister chromatids, and its downregulation may affect the expression of other DNA damage repair-related genes. In our study, the growth arrest and DNA damage inducible alpha (GADD45A) gene expression level was downregulated, leading to a decline in the overall DNA repair capacity of cells.

In the TNF signaling pathway (Figure 3C), which is a crucial intracellular signaling pathway involved in immune regulation, inflammatory responses, cell proliferation, differentiation, and apoptosis, the expression of FOS and JUN genes was significantly downregulated, which may result in decreased expression of genes associated with cell cycle progression, thereby slowing the proliferation of tumor cells.

In the TCR signaling pathway (Figure 3D), which plays a central and indispensable role in the adaptive immune response by enabling T cells to recognize specific antigens and initiate appropriate immune reactions, the downregulation of AKT serine/threonine kinase 3 (AKT3) and phosphatidylinositol-4,5-bisphosphate 3-kinase catalytic subunit delta (PIK3CD) gene expression impaired the transmission of TCR signals. This disrupted the balance between normal T cell proliferation and apoptosis.

In the MAPK signaling pathway (Figure 3E), 105 genes were upregulated and 141 genes were downregulated. This pathway is a crucial intracellular signaling transduction pathway that plays a vital role in various cellular physiological processes. The MAPK subfamilies, JNK and p38 MAPK, play crucial roles in cellular apoptosis. In this pathway, genes associated with these subfamilies, such as JUN and activating transcription factor 2 (ATF2), were significantly downregulated. This may affect the expression of anti-apoptotic genes, making SR cells more susceptible to apoptosis after quadrigemine I treatment.

In the apoptosis signaling pathway (Figure 3F), the downregulation of apoptosis-related genes, such as BCL2-related protein A1 (BCL2A1), may inhibit the survival and proliferation of tumor cells, thereby suppressing tumor growth.

RNA-seq analysis revealed that seven genes related to apoptosis were altered after treating with quadrigemine I. The expression of urocortin 2 (UCN2) was evidently upregulated, with a $\log_2$FoldChange of 3.83, and the expression of heat shock protein family A (Hsp70) member 5 (HSPA5), MAPK10, FOS, GADD45A, E74 Like ETS transcription factor 3 (ELF3), and mitogen-activated protein kinase kinase kinase 8 (MAP3K8) was evidently downregulated in quadrigemine I-treated SR cells. The transcriptome results showed that the treatment in the high-concentration group had the best effect and could effectively reverse the expression levels of these genes, which was consistent with the previous analysis results.

To validate these findings, the key genes involved in the aforementioned pathways—including MAP3K8, MAPK10, FOS, UCN2, ELF3, GADD45A, and HSPA5—were selected for RT-qPCR analysis. The RT-qPCR results showed that UCN2 expression was upregulated in a dose-dependent manner following quadrigemine I treatment. In contrast, the expression levels of GADD45A, MAP3K8, MAPK10, FOS, HSPA5, and ELF3 were downregulated in a dose-dependent manner. These results were consistent with the transcriptomic data (Figure 4).

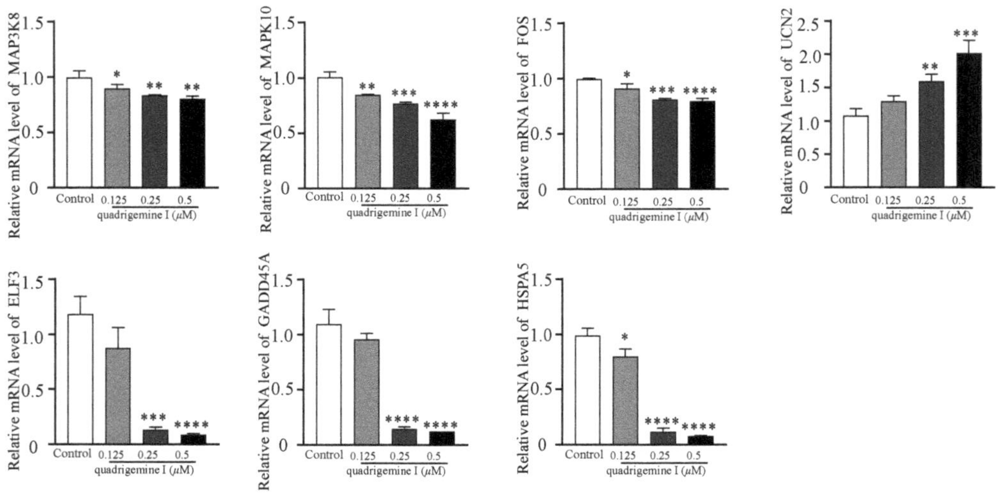

Figure 4. Quadrigemine I regulates the expression of genes in SR cells. Bar plots showing the expression level of the selected DEGs using RT-qPCR. n = 3; * $p < 0.05$, ** $p < 0.01$ *** $p < 0.001$, **** $p < 0.0001$.

2.6. *Safety Evaluation of Quadrigemine I in Mice*

Cell experiments demonstrated that quadrigemine I inhibited the growth of SR cells in vitro. To evaluate whether quadrigemine I is a potential agent for lymphoma treatment, we first assessed its safety in vivo. Figure S1 shows the body weight and organ-to-body weight ratio of mice following continuous intraperitoneal administration of 5 mg/kg quadrigemine I over a 14-day period. All mice survived, gained weight, and displayed normal activity and health. No noticeable abnormal behaviors, adverse toxic effects, or significant changes ($p > 0.05$) in organ weights (including heart, liver, spleen, lung, kidney, and brain) were observed. The hematological analysis results of mice after intraperitoneal injection of quadrigemine I for 14 days are shown in Table S1. The serum biochemical parameters are presented in Table S2. Both hematological and biochemical indices showed no significant differences between the groups. These results suggested that quadrigemine I had no effect on liver and kidney function in mice. H&E staining revealed no pathological changes in the tissue samples (Figure S2). Therefore, it can be concluded that quadrigemine I (5 mg/kg) is safe for intraperitoneal injection in mice and is suitable for subsequent experiments.

2.7. *Quadrigemine I Inhibits Tumor Growth in Tumor Xenograft Mice*

We conducted tumor xenograft assays in NOD/SCID mice to investigate the in vivo anti-lymphoma effects of quadrigemine I. Compared with the control group, the body weights of the mice did not show any significant differences throughout the experiment, except for the CDDP group (Figure 5A). Tumor growth was significantly suppressed in the quadrigemine I (5 mg/kg) group compared to the model group (Figure 5B–D). At the end of the experiment, we analyzed the organ-to-body weight ratio and found no significant differences in the heart, lung, kidney, or brain (Figure 5E–H). Notably, the organ indices of the liver and spleen in the model group were significantly increased, while treatment with quadrigemine I significantly reduced the organ indices of the liver and spleen to normal levels (Figure 5I,J).

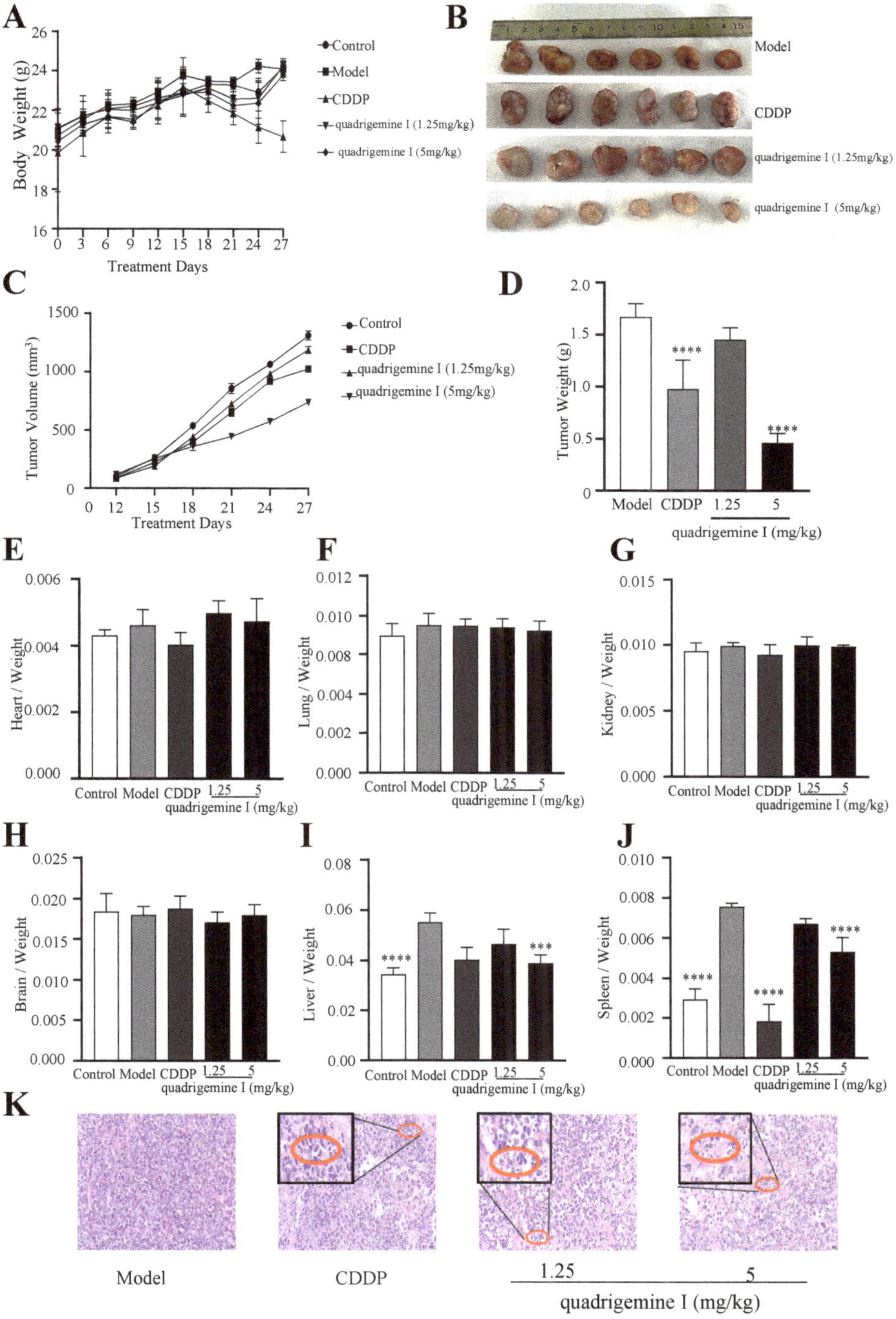

Figure 5. Effects of quadrigemine I on tumor growth in xenograft mice. NOD/SCID mice were subcutaneously injected with SR cells and had an intraperitoneal injection with quadrigemine I. (**A**) Body weights measured every 3 days. (**B**) Images of excised tumors. (**C**) Tumor volume. (**D**) Tumor weights. (**E–J**) The ratio of organs to body weight. (**K**) H&E staining of tumor tissues; the scale bar stands for 20 μm. The red circle marks apoptotic cells characterized by nuclear condensation and fragmentation. *** $p < 0.001$ and **** $p < 0.0001$ compared with the control group.

Furthermore, histopathological examination revealed that the model group exhibited high-density tumor cell clusters with an orderly arrangement and darker nuclear staining. In contrast, tumor tissues from the quadrigemine I dose groups and the CDDP group showed a sparse growth pattern with areas of increased apoptosis, characterized by nuclear condensation and fragmentation (Figure 5K, red circle). These results indicated that quadrigemine I reduced the tumor burden in mice.

Subsequently, to further investigate the mechanism, we performed TUNEL staining and analyzed the expression of apoptosis-related proteins in tumor tissues. We examined the expression of Ki-67, Bcl-2, and cleaved caspase-3 by immunofluorescence staining. As shown in Figure 6A, TUNEL staining of tumor tissues revealed a greater number of apoptotic cells in tumors treated with quadrigemine I compared to those from the model group. Quadrigemine I decreased Ki-67 and Bcl-2 expression while increasing cleaved caspase-3 expression in a dose-dependent manner (Figure 6B–D). These data suggested that quadrigemine I could inhibit tumor growth in tumor xenograft mice by regulating apoptosis pathways in vivo.

Additionally, we assessed the expression of inflammatory factors in tumor tissues (Figure 7). Compared to the control group, the levels of IL-1β showed a significant increase in the model group: from 50.48 ± 0.84 ng/L to 98.17 ± 0.54 ng/L in serum, from 48.00 ± 0.51 ng/L to 88.65 ± 0.83 ng/L in liver tissue, and from 53.07 ± 0.42 ng/L to 98.13 ± 0.53 ng/L in spleen tissue. However, treatment with quadrigemine I at 1.25 mg/kg and 5 mg/kg effectively inhibited the excessive secretion of inflammatory factors. Specifically, the inhibition rates for IL-1β were 10.79 ± 0.45% and 23.25 ± 0.29% in serum, 4.50 ± 1.52% and 19.92 ± 1.03% in liver tissue, 7.20 ± 0.18% and 18.26 ± 1.09% in spleen tissue, and 0.64 ± 0.60% and 18.43 ± 1.02% in tumor tissue.

Similarly, compared to the control group, the levels of IL-6 in the model group also significantly increased to 161.33 ± 1.49 ng/L, 135.84 ± 1.61 ng/L, 154.66 ± 0.70 ng/L, and 157.68 ± 0.73 ng/L, respectively. After treatment with quadrigemine I (1.25 and 5 mg/kg), the inhibition rates for IL-6 were 20.95 ± 0.30% and 30.00 ± 0.86% in serum, 2.13 ± 1.99% and 22.91 ± 0.71% in liver tissue, 5.10 ± 0.62% and 25.33 ± 0.98% in spleen tissue, and 2.74 ± 0.40% and 14.17 ± 0.09% in tumor tissue.

The levels of NO also showed a significant decrease after treatment with quadrigemine I, with inhibition rates of 9.07 ± 0.99% and 42.24 ± 1.02% in serum, 7.25 ± 1.21% and 36.53 ± 0.82% in liver tissue, 13.77 ± 1.21% and 24.84 ± 1.36% in spleen tissue, and 4.58 ± 1.35% and 19.32 ± 0.47% in tumor tissue.

The levels of TNF-α in the model group were initially 1013.51 ± 7.49 ng/L, 945.65 ± 5.99 ng/L, 890.07 ± 8.30 ng/L, and 992.28 ± 4.90 ng/L, respectively, but significantly decreased after treatment with quadrigemine I, with inhibition rates of 10.28 ± 0.64% and 49.83 ± 0.39% in serum, 2.49 ± 1.11% and 34.59 ± 1.11% in liver tissue, 2.22 ± 0.30% and 27.90 ± 0.09% in spleen tissue, and 3.79 ± 0.87% and 24.48 ± 1.05% in tumor tissue.

Overall, the levels of inflammatory factors in the model group were markedly elevated, while treatment with quadrigemine I resulted in a general reduction in these inflammatory factors across all tissues. The levels of IFN-γ in the model group significantly increased to 1126.89 ± 8.68 ng/L, 1088.81 ± 17.42 ng/L, 1075.39 ± 9.77 ng/L, and 773.86 ± 2.63 ng/L, respectively. After treatment with quadrigemine I, the inhibition rates for IFN-γ were 9.95 ± 1.01% and 49.42 ± 0.71% in serum, 6.35 ± 2.05% and 43.13 ± 0.60% in liver tissue, 21.21 ± 0.21% and 46.59 ± 0.66% in spleen tissue, and 3.97 ± 0.88% and 38.03 ± 0.22% in tumor tissue. These results indicated that quadrigemine I can alleviate tumor-induced inflammatory responses in vivo.

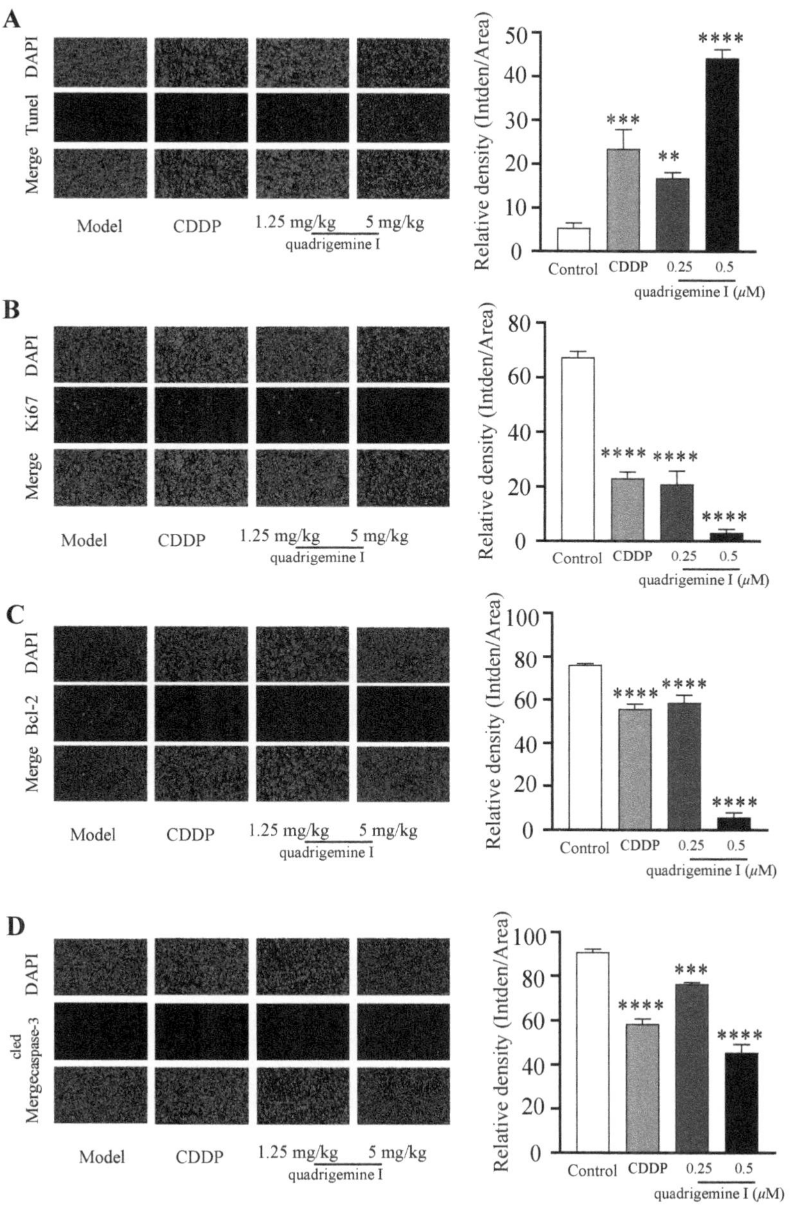

Figure 6. Effects of quadrigemine I on apoptosis in tumor xenograft mice. (**A**) The tumor apoptosis is examined by TUNEL staining, the blue fluorescence (DAPI) in the tumor represents tumor cells, red fluorescence (TUNEL, TMR red) represents apoptotic tumor cells, and the apoptosis index is calculated; the scale bar stands for 10 μm. (**B**) Representative images of immunofluorescence staining of Ki-67 on tumor tissues are shown, and the percentage of Ki-67-positive cells is calculated; the scale bar stands for 10 μm. (**C**) Representative images of immunofluorescence staining of Bcl-2 on tumor tissues are shown, and the percentage of Bcl-2-positive cells is calculated; the scale bar stands for 10 μm. (**D**) Representative images of immunofluorescence staining of caspase-3 on tumor tissues are shown, the blue fluorescence (DAPI) in tumor represents tumor cells, the red fluorescence (TUNEL, TMR red) represents apoptotic tumor cells, and the apoptosis index is calculated; the scale bar stands for 10 μm. The data are shown as the mean ± SD. ** $p < 0.01$, *** $p < 0.001$ and **** $p < 0.0001$ compared with the model group.

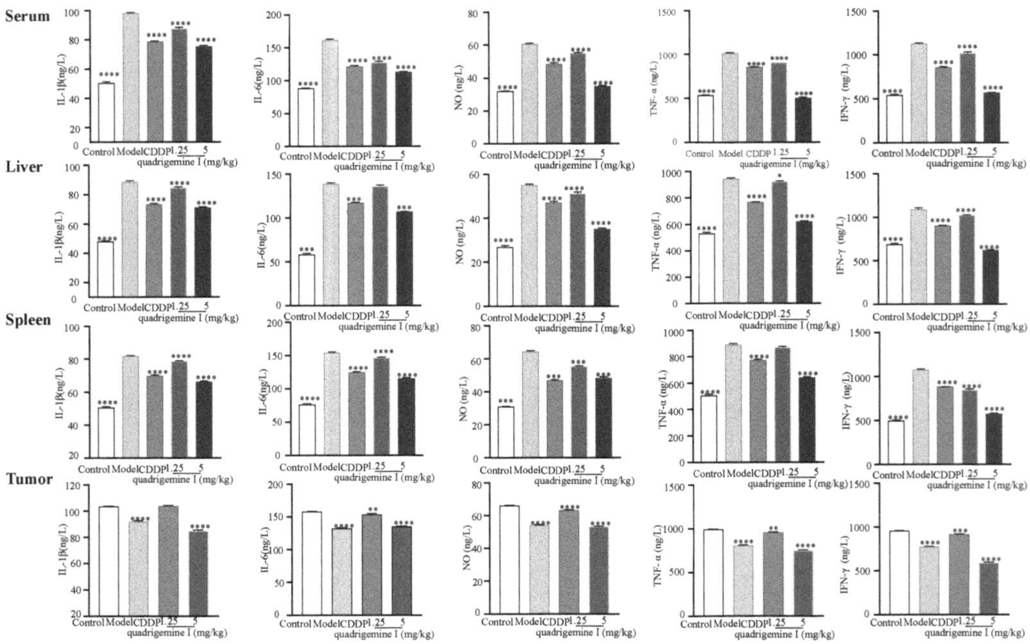

Figure 7. Effect of quadrigemine I on the production of IL-1β, IL-6, NO, TNF-α, and IFN-γ in serum, liver, spleen, and tumor tissue. * $p < 0.05$, ** $p < 0.01$, *** $p < 0.001$, and **** $p < 0.0001$ compared with the model control.

3. Discussion

Quadrigemine I is a natural indole alkaloid derived from *P. pilifera*. To our knowledge, this research represents, for the first time, the inhibitory effect of quadrigemine I on lymphoma cells. Our study investigated the effects of quadrigemine I on the proliferation, apoptosis, ROS generation, and DNA damage of SR cells. The results indicated that quadrigemine I inhibited proliferation, increased ROS generation, and induced cell apoptosis and DNA damage in SR cells. The CCK-8 assay showed that quadrigemine I inhibited the proliferation of SR cells with an IC_{50} of 0.46 ± 0.12 μM. Our results showed that quadrigemine I exhibited a low IC_{50} against SR cells and displayed no cytotoxicity toward HUVECs (Figure 1). One major limitation of this study is the use of only two lymphoma cell lines—SR cells representing T cell lymphoma and Raji cells representing B cell lymphoma. Although these lines provide initial insight into the effects of quadrigemine I on different lymphocyte-derived tumor types, they do not fully capture the heterogeneity and complexity of lymphomas. The differences observed may be influenced by intrinsic characteristics specific to these cell lines rather than representative of broader lymphoma subtypes. Future studies will aim to validate these findings across a larger panel of lymphoma cell lines, including those of various genetic backgrounds and clinical relevance, as well as primary patient-derived tumor cells.

Using flow cytometry, we found that quadrigemine I treatment dramatically promoted apoptosis in a dose-dependent manner. Notably, the results (Figure 1F) showed minimal effect of quadrigemine I on the proportion of early apoptotic cells, while a significant increase was observed in late apoptotic populations. This suggests that quadrigemine I may rapidly drive cells from a viable state to late apoptosis or secondary necrosis, bypassing the early apoptotic phase. One possible explanation is that the overproduction of ROS and the induction of DNA damage trigger a robust and acute pro-apoptotic response, leading to mitochondrial dysfunction and rapid activation of caspase cascades [11]. Alternatively, the selected time point for apoptosis detection may have primarily captured the late phase of apoptosis [12]. Future studies involving time course experiments will be necessary to further delineate the dynamics of quadrigemine I-induced apoptosis.

Additionally, the mechanism of quadrigemine I-induced cell apoptosis is not fully understood. Excessive cellular levels of ROS cause damage to proteins, nucleic acids, lipids, membranes, and organelles, leading to the activation of cell death processes [13]. DNA strand breaks (DSBs) are the most severe incidence of DNA damage, which can cause cell apoptosis [14]. An effective DNA repair process could restore DNA integrity, whereas unrepaired DNA damage might trigger cell death, or damaged cells might evade apoptosis, potentially leading to cancer [15]. In this study, we observed a significant increase in intracellular ROS levels following quadrigemine I treatment, which correlated with enhanced DNA damage, as indicated by the comet assay (Figure 1). Excessive ROS can cause oxidative modifications to DNA, including strand breaks and base damage, thereby compromising genomic integrity. This DNA damage activates the ATM/ATR-mediated DNA damage response (DDR), which in turn can trigger the intrinsic apoptotic pathway through mitochondrial dysfunction, cytochrome c release, and caspase activation. Our findings suggest that ROS-induced DNA damage is a critical upstream event leading to apoptosis in quadrigemine I-treated SR cells. This mechanistic cascade is consistent with previous reports [16].

Transcriptome analysis showed that quadrigemine I altered the expression of thousands of genes in SR cells (Figure 2). These DEGs were significantly enriched in multiple tumor-associated signaling pathways. These key genes, including MAP3K8, MAPK10, FOS, ELF3, GADD45A, HSPA5, and UCN2, have been shown in previous studies to be closely associated with apoptosis [14,15]. For example, HSPA5 (heat shock protein family A (Hsp70) member 5) can play diverse functional roles in cell viability, proliferation, apoptosis, attachment, and innate and adaptive immunity regulation, which can lead to various diseases, including cancers and coronavirus disease 2019 (COVID-19) [17]. HSPA5 expression was significantly upregulated in 14 types of cancer, including cholangiocarcinoma, colon adenocarcinoma, esophageal carcinoma, diffuse large B cell lymphoma, pancreatic adenocarcinoma, and others. These findings suggest that HSPA5 is a crucial cancer marker, highly expressed in most malignant tumors, and that targeted therapy targeting HSPA5 may be a feasible approach for cancer treatment [18]. Liu et al. found that after treating melanoma cells with CDDP, the expression of GADD45A was upregulated in a dose- and time-dependent manner. The inhibition of GADD45A made melanoma cells more sensitive to cisplatin, enhanced cisplatin-induced DNA damage, and induced apoptosis [19]. Studies have also shown that miR-335-5p can exert its suppressive effect on gastric cancer by targeting MAPK10. The underlying mechanism is that the downregulation of MAPK10 inhibits the proliferation, migration, and invasion of gastric cancer cells and induces apoptosis [20]. Studies have found that ELF3 is differentially expressed in non-small cell lung cancer. In lung cancer cells harboring K-ras mutations and EGFR L858R/T790M mutations, the knockdown of ELF3 significantly increases cell apoptosis [21]. Ucn2 is a specific ligand for the corticotropin-releasing factor (CRF) receptor CRFR2. Previous studies have shown that Ucn2 could suppress tumor growth predominantly by the inhibition of tumor vascularization, but also potentially through direct effects on tumor cell proliferation [22]. To the best of our knowledge, we report, for the first time, that quadrigemine I upregulated the expression of UCN2 and downregulated the expression of MAP3K8, MAPK10, FOS, ELF3, GADD45A, and HSPA5 on SR cells. These findings provide novel insights into the molecular mechanisms underlying the potential role of quadrigemine I in cellular stress response, survival, and apoptosis regulation (Figures 3 and 4). For these key and novel targets, we will further use RNA interference (RNAi) and overexpression techniques to validate the importance of these targets in SR cell apoptosis. Notably, transcriptomic enrichment analysis revealed that genes in the ERBB signaling pathway were significantly modulated in response to quadrigemine I treatment, particularly in SR cells. Although the ERBB pathway is not typically regarded as a major proliferative signal in normal lymphocytes, its altered activation in lymphoma cells may represent a disease-specific adaptation or a drug-induced response. Additionally, we will conduct in vivo studies to further investigate the role and mechanism of quadrigemine I in the treatment of lymphoma.

Using NOD/SCID xenograft mice as a model, it was demonstrated that quadrigemine I treatment significantly inhibited tumor growth without exhibiting toxicity (Figure 5). Furthermore, histopathological examination, immunofluorescence, and TUNEL staining analysis revealed that quadrigemine I treatment caused tumor cell apoptosis (Figure 6). However, the specific molecular targets of quadrigemine I and its detailed mechanism in the apoptosis signaling pathway require further investigation. Future research could explore the interaction of quadrigemine I with other apoptosis-regulating proteins (such as Bax, p53, etc.), as well as its long-term efficacy and safety in vivo and vitro [23]. In this study, we evaluated the anti-tumor efficacy of quadrigemine I in vivo using two dosage groups. The selection of these doses was based on preliminary in vivo safety evaluations and the limited availability of the compound at the time of extraction. While both doses demonstrated significant tumor-suppressive effects, we acknowledge that the current study does not fully address the dose–response relationship of quadrigemine I in vivo. Future studies will involve additional dose groups and expanded pharmacodynamic assessments to comprehensively determine the optimal dosing strategy for quadrigemine I in lymphoma.

Inflammation is closely related to the development of cancer, and inhibiting chronic inflammation or accelerating its resolution can improve the overall survival rate in cancer therapy [24]. The mechanisms by which inflammation promotes cancer include inflammatory cytokines activating intracellular signaling pathways to promote tumor cell proliferation and inhibit apoptosis; inflammatory mediators inducing the expression of angiogenic factors to promote tumor angiogenesis and facilitate tumor metastasis; and inflammatory processes in the tumor microenvironment causing immune cell dysfunction, the accumulation of immunosuppressive cells, and the inhibition of cytotoxic T cell activity, thereby assisting tumor cells in immune escape [25]. A decrease in key inflammatory biomarker (such as IL-6, TNF-α, IFN-γ, etc.) levels can reshape the tumor immune microenvironment. For example, Zhao et al. found that cancer-associated adipocytes (CAAs) secrete large amounts of IL-6, which activates STAT3 to induce macrophage polarization toward the M2 type, promoting breast cancer progression. This indicates that IL-6 plays an important role in inducing M2 polarization of tumor-associated macrophages (TAMs) [26]. Other studies have shown that IL-1β can promote MDSC accumulation, thereby driving tumor progression. Therefore, reducing inflammation may decrease the number of Myeloid-Derived Suppressor Cells (MDSCs) and slow down tumor progression [27]. TNF-α upregulation is involved in the angiogenesis process in the tumor microenvironment, affecting the tumor's nutrient supply and waste removal, creating favorable conditions for tumor growth, and promoting tumor cell migration and invasion [28]. Nitric oxide (NO) is involved in the process by which tumor cells acquire stem-cell-like abilities, helping them escape immune system recognition. NO can suppress the activity of cytotoxic T cells, natural killer cells, and other immune cells, reducing their ability to recognize and kill tumor cells. Therefore, downregulating NO expression may help inhibit tumor growth [29]. IFN-γ in the tumor microenvironment induces the production of immunosuppressive cells, causes immune cell functional exhaustion, and can also induce tumor cell metabolic reprogramming, enabling them to adapt to immune pressure, thereby promoting tumor cell immune escape [30].

These findings highlight the therapeutic potential of quadrigemine I in regulating both tumor growth and the associated inflammatory response. Quadrigemine I was observed to modulate the inflammatory microenvironment, a hallmark of tumor progression. Our data indicated that treatment with quadrigemine I significantly reduced the secretion of key pro-inflammatory cytokines, such as TNF-α, IL-6, and IL-1β, in various tissues, including serum, liver, spleen, and tumor tissues (Figure 7). This anti-inflammatory effect is consistent with the mechanisms proposed for other natural compounds that target inflammatory pathways in cancer [31].

Although our study demonstrates the therapeutic efficacy of quadrigemine I both in vitro and in vivo, we acknowledge that the lack of pharmacokinetic and biodistribution data remains a limitation. Given that quadrigemine I was administered systemically, it is essential to investigate its absorption, distribution, metabolism, and excretion (ADME) characteristics in future studies. These parameters will provide critical insights into its bioavailability, tissue targeting, and safety profile, all of which are crucial for optimizing dosing strategies and advancing quadrigemine I toward clinical application. Ongoing and future work in our laboratory will include detailed pharmacokinetic analyses and imaging-based biodistribution studies to address this gap and support the translational potential of quadrigemine I.

4. Materials and Methods

4.1. Chemicals and Reagents

^{1}D and ^{2}D NMR spectra were performed on a Bruker DRX–500 spectrometer (Bruker BioSpin GmBH, Rheinstetten, Germany). The fractions were analyzed using thin-layer chromatography (TLC, GF254, Qingdao Marine Chemical Ltd., Qingdao, China), and the spots were detected using Dragendorff's reagent and 10% H_2SO_4 in ethanol. 2′,7′-Dichlorofluorescin diacetate (DCFH-DA) was purchased from Sigma-Aldrich (Shanghai, China). The Cell Counting Kit-8 (CCK-8) was purchased from UElandy Biotech Co., Ltd. (Suzhou, China). The Annexin V fluorescein isothiocyanate (FITC)/propidium iodide (PI) apoptosis detection kit was purchased from 4A Biotech Co., Ltd. (Beijing, China). Phosphate-buffered saline (PBS) was obtained from Gibco (Grand Island, NY, USA).

4.2. Plant Material

The leaves of *P. pilifera* were collected in July 2022 from Mengna, Yunnan Province, China. The plant material with species number Cheng2022072001 was identified by Prof. Yaping Liu from the Kunming University of Science and Technology.

4.3. Extraction, Isolation, and Structure Characterization of Quadrigemine I

The air-dried leaves of *P. pilifera* (5.4 kg) were extracted with 90% ethanol aqueous solution three times. After filtration by nylon gauze, the extraction solutions were collected and evaporated by a rotary evaporator to yield an extract. Then, the extract was applied to a silica gel column, eluted by a $CHCl_3$-MeOH-$NH_3 \cdot H_2O$ solution system (10:1:0.3), and further purified by an MPLC column with RP–C18 chromatography (MeOH/H_2O, 50:50) to obtain quadrigemine I (63 mg). Finally, quadrigemine I (10 mg) was dissolved in a DMSO solution for NMR analysis. The ^{1}H and ^{13}C NMR spectra were recorded on a Bruker 500 MHz spectrometer. The chemical shifts (δ) were expressed in parts per million and coupling constants (J) in Hz.

^{1}H NMR (500 MHz, DMSO-D6) δ 1.82-3.31 (16H, m), 2.28, 2.33, 2.37, 2.49 (12 H, 4×N-CH3), 4.89 (1H, s), 4.84 (1H, s), 4.78 (1H, s), 4.58 (1H, s), 5.58 (1H, s), 5.69, (1H, s), 7.01 (1H, s), 7.08, (1H, s), 7.15, (4H, s), and 7.22 (1H, s). ^{13}C NMR (125 MHz, DMSO-D6) δ: 152.1, 151.5, 149.8, 148.6, 132.9, 132.1, 131.6, 127.8, 127.7, 125.7, 125.6, 124.8, 124.4, 123.5, 123.2, 117.2,

116.6, 116.4, 115.9, 107.0, 106.9, 86.4, 85.3, 83.1, 82.2, 62.9, 62.7, 59.8, 59.3, 51.9, 37.9, 37.6, 36.9, and 35.1. By comparison with the reported data, this compound was characterized as quadrigemine I (Figure 1A) [32].

4.4. Cell Culture

SR cells were purchased from the American Type Culture Collection (ATCC). Human umbilical vein endothelial cells (HUVECs) were acquired from Wuhan Servicebio technology Co., Ltd. (Wuhan, China), and Raji cells were obtained from Wuhan Servicebio technology Co., Ltd. (Wuhan, China). In this study, two lymphoma cell lines were used: SR cells, derived from T cell lymphoma, and Raji cells, derived from B cell lymphoma. These cell lines were selected to represent different lymphoma subtypes for investigating the effects of quadrigemine I on lymphocyte malignancy.

SR cells were cultured in an RPMI-1640 medium supplemented with 10% fetal bovine serum (FBS) and 1% penicillin/streptomycin (Gibco, Grand Island, NY, USA). HUVECs were cultured in a ScienCell ECM 1001 endothelial cell medium supplemented with 5% FBS, 1% endothelial cell growth supplement (ECGS), and 1% penicillin/streptomycin. Media and supplements were purchased from ScienCell research laboratories (Carlsbad, CA, USA). All cells were maintained at 37 °C with 5% CO_2.

4.5. Cell Viability Assay

The CCK-8 assay was used to measure cytotoxicity. Briefly, HUVECs (5×10^3 cells/well) were seeded in 96-well plates for 24 h and then treated with quadrigemine I for 20 h. SR cells were cultured in 96-well plates (5×10^4 cells/well) and treated with quadrigemine I or cisplatin (CDDP) for 44 h. Then, 10 μL of CCK-8 solution was added to each well and incubated at 37 °C for 4 h. Absorbance was measured at 450 nm using a microplate reader (SpectraMax M5, Molecular Devices, San Francisco, CA, USA).

4.6. Determination of Intracellular Reactive Oxygen Species (ROS)

SR cells (1×10^6 cells/well) were seeded into 6-well plates and treated with quadrigemine I (0, 0.125, 0.25, 0.5 μM) or CDDP (8 μM) for 48 h. After treatment, the cells were harvested, washed twice with cold PBS, and incubated with 10 μM DCFH-DA in the dark for 30 min. Following incubation, the cells were rinsed twice and analyzed by flow cytometry (Guava® easyCyte 6-2L, Millipore, Billerica, MA, USA).

4.7. Comet Assay

SR cells and HUVECs were treated with quadrigemine I in 6-well plates for 48 h and then harvested. Single-cell suspensions were prepared and collected for the assay. After gel preparation, the cells were lysed in a comet assay lysis buffer at 4 °C for 1 h. Single-cell electrophoresis was performed at 25 V and 300 mA for 20 min. Finally, the cells were stained with DAPI, and data analysis was conducted using CASP software (version 1.2.3 beta 1).

4.8. Cell Apoptosis Assays

SR cells (1×10^6 cells/well) were seeded into 6-well plates and cultured for 48 h. After the incubation period, the cells were harvested, rinsed three times with cold PBS, and then resuspended in 100 μL of binding buffer. Next, 5 μL of Annexin V-FITC was introduced, and the cells were incubated in the dark for 5 min. Thereafter, 10 μL of propidium iodide (PI, 20 μg/mL) and 200 μL of a binding buffer were added to the cell suspension. Apoptosis levels were determined by flow cytometry (Guava® easyCyte 6-2L, Millipore, Billerica, MA, USA).

4.9. RNA-Seq Analysis

SR cells (1 × 10^6 cells/well) in the logarithmic growth phase were treated with quadrigemine I (0.125, 0.25, 0.5 μM) or CDDP (8 μM) in 6-well plates for 48 h. The cells were then collected, washed once with PBS, and lysed using TRIzol reagent (Thermo Fisher Scientific, Waltham, MA, USA). The obtained cDNA libraries were sequenced by Beijing Yijian Testing Technology Co., Ltd. (Beijing, China).

4.10. Real-Time Quantitative PCR (RT-qPCR)

Total RNA was extracted using TRIzol reagent (Thermo Fisher Scientific, Waltham, MA USA). cDNA synthesis was conducted using the cDNA synthesis kit (TIANGEN, Beijing, China) in a 20 μL reaction volume. Quantitative PCR (qPCR) was carried out with SYBR Green Master Mix (TIANGEN, Beijing, China), and the reactions were performed and analyzed on a CFX Connect Real-Time System (Bio-Rad, Hercules, CA, USA). The relative expression levels of target genes were determined using the ΔΔCt method [33]. The qPCR primers (Table 1) were designed and synthesized by Sangon Biotech (Shanghai, China).

Table 1. The primers used for RT-qPCR analysis.

Gene	Species	Primer Sequences
MAP3K8	human	F:5′ AGCAACATTGTTTTCATGTCCAC3′ R:5′ AGCAACATTGTTTTCATGTCCAC3′
MAPK10	human	F:5′ AGCAACATTGTTTTCATGTCCAC3′ R:5′ AGCAACATTGTTTTCATGTCCAC3′
FOS	human	F:5′CTTCCTGGAGCAGTGTGGAG3′ R:5′ AGCAACATTGTTTTCATGTCCAC3′
UCN2	human	F:5′ AGAAGCAGCTGGTGGCG3′ R:5′ CACTGGGACAACCAGGACTC3′
ELF3	human	F:5′ AAGGCGTCTTCAAGTTCC3′ R:5′ TAGGTCATGTTGCTGTTCTT3′
GADD45A	human	F:5′ GACGAATCCACATTCATCTC3′ R:5′ GACGAATCCACATTCATCTC3′
HSPA5	human	F:5′ GTTACAATCAAGGTCTAT3′ R:5′ CATTCACATCTATCTCAA3′
GAPDH	human	F:5′ AGTGGCAAAGTGGAGATT3′ R:5′ GTGGAGTCATACTGGAACA3′

4.11. Safety Evaluation in Mice

Twelve ICR mice were randomly divided into two groups (5 mg/kg dose group and control group), with six mice in each group. Each mouse was intraperitoneally injected with 0.2 mL of the solution daily. During the experiment, the reactions of each animal were observed before and after administration. Body weight changes were measured every two days. At the end of the experimental period, organ indices, histopathological changes, and hematological parameters were evaluated in the mice.

4.12. In Vivo Anti-Tumor Evaluation

Female NOD/SCID mice aged 4 weeks were purchased from Sybef Biotechnology Co., Ltd. (Beijing, China). The mice were kept under specific pathogen-free (SPF) conditions at Yunnan University with a 12 h light/dark cycle. They were given sufficient food and water, and all experimental procedures followed the guidelines approved by the Yunnan University Laboratory Animal Care Committee. The Ethics Committee of Yunnan University approved the animal use and experimental protocols (permit number YNU20251207).

For xenograft tumor studies, SR cells (1 × 10^7) were subcutaneously injected into NOD/SCID mice at a volume of 100 μL. Tumor volume was calculated using the formula length × (width^2)/2. When the tumor volume reached 100 mm^3, the mice were randomly divided into groups (n = 6) and treated with intraperitoneal injections of normal saline (NS), CDDP (5 mg/kg), or quadrigemine I (1.25 and 5 mg/kg) for 14 days.

4.13. Hematoxylin and Eosin (H&E) Staining, Immunofluorescence, and Terminal Deoxynucleotidyl Transferase dUTP Nick End Labeling (TUNEL) Assays

Tumor tissues were removed from the xenograft mice and fixed in 4% paraformaldehyde for 24 h at room temperature, followed by dehydration and paraffin embedding. The tumor tissues were sectioned into 5 µM thick slices and stained with hematoxylin and eosin solution. The stained sections were scanned using a pathology slide scanner (Nikon, Tokyo,Japan), and the images were viewed using CaseViewer 2.4 software.

For immunofluorescence analysis, the sections were incubated overnight with primary antibodies, including cleaved cysteine-aspartic acid protease 3 (cleaved caspase-3, Affinity, San Francisco, CA, USA), B cell lymphoma 2 (Bcl-2, Servicebio, Wuhan, China), and Kiel 67 (Ki-67, Servicebio, Wuhan, China). After washing, the sections were incubated with fluorescent Cy3 goat anti-rabbit IgG secondary antibodies (Servicebio, Wuhan, China). Nuclei were counterstained with DAPI, and fluorescent signals were observed under a fluorescence microscope (Nikon, Tokyo, Japan). The images were analyzed using ImageJ software (version 1.8.0).

The TUNEL assay was performed according to the protocol provided with the TUNEL apoptosis detection kit (Servicebio, Wuhan, China). Briefly, paraffin-embedded tumor sections were deparaffinized and incubated with proteinase K to remove nuclear proteins. The sections were then incubated with TdT and dUTP enzyme reaction mixtures. Nuclei were stained with DAPI, and the sections were observed under a fluorescence microscope (Nikon, Tokyo, Japan). The percentage of TUNEL-positive cells was quantified using ImageJ software (version 1.8.0).

4.14. ELISA Analysis of Interleukin (IL)-1β, IL-6, Nitric Oxide (NO), Tumor Necrosis Factor (TNF)-α, and Interferon (IFN)-γ

The mice were euthanized, and the tumor, liver, and spleen tissues, along with serum samples, were harvested. Following this, the concentrations of IL-1β, IL-6, NO, TNF-α, and IFN-γ in the samples were quantified using ELISA kits, which were procured from Jiangsu Meimian Industrial Co., Ltd. (Yancheng, China).

4.15. Statistical Analysis

Data are presented as mean ± standard deviation (SD) from three independent experiments. Statistical analysis was performed using one-way ANOVA followed by Dunnett's multiple comparisons test. Significance levels were defined as * $p < 0.05$, ** $p < 0.01$, *** $p < 0.001$, and **** $p < 0.0001$.

5. Conclusions

In conclusion, our study suggests that quadrigemine I is effective in inhibiting lymphoma cell proliferation while remaining safe for normal human cells. At a dose of 5 mg/kg, quadrigemine I significantly suppressed tumor growth in xenograft NOD/SCID mice. Its mechanism of action may be associated with immune evasion, inflammation, and apoptosis pathways. These findings highlight quadrigemine I as a promising anti-proliferative agent for the treatment of lymphomas.

Supplementary Materials: The following supporting information can be downloaded at https://www.mdpi.com/article/10.3390/ijms26104848/s1.

Author Contributions: Conceptualization, W.S., J.C. (Jianxin Cao) and S.X.; methodology, J.P., J.C. (Jiabao Cui) and S.X.; software, J.P., W.S., J.C. (Jiabao Cui) and H.Y.; validation, H.Y.; formal analysis, J.P.; investigation, J.P., J.C. (Jiabao Cui) and H.Y.; resources, Y.L.; data curation, J.P., W.S., J.C. (Jiabao Cui) and H.Y.; writing—original draft preparation, J.P.; writing—review and editing, S.X. and G.C.; visualization, J.P.; supervision, S.X. and G.C.; project administration, J.C. (Jianxin Cao), S.X. and G.C.; funding acquisition, S.X. All authors have read and agreed to the published version of the manuscript.

Funding: This work was supported by the National Natural Science Foundation of China (32260108), the Yunnan Fundamental Research Projects (202501AT070352 and 202401CF070107), and the Scientific Research Fund Project of Yunnan Provincial Department of Education (2024J0081).

Institutional Review Board Statement: The animal study protocol was approved by the Laboratory Animal Welfare Ethics Committee of Yunnan University, which approved the animal use and experimental protocols (permit number YNU20251207; approved on 16 February 2025).

Informed Consent Statement: Not applicable.

Data Availability Statement: The original contributions presented in this study are included in the article/Supplementary Materials. Further inquiries can be directed to the corresponding author.

Acknowledgments: We sincerely thank Chengrui Wang (engineer, Wekemo Tech Group Co., Ltd. Shenzhen, China) for the instruction on transcriptome analysis.

Conflicts of Interest: The authors declare no conflicts of interest.

Abbreviations

The following abbreviations are used in this manuscript:

DCFH-DA	2′,7′-Dichlorofluorescin diacetate
2ATF2	Activating transcription factor
AKT3	AKT serine/threonine kinase 3
ATCC	American Type Culture Collection
BCL2A1	BCL2-related protein A1
Raji	Burkitt lymphoma cells
CCK-8	Cell Counting Kit-8
CDDP	Cisplatin
CRF	Corticotropin-releasing factor
DEGs	Differentially expressed genes
DSBs	DNA strand breaks
ELF3	E74-like ETS transcription factor 3
ERBB3	Erb-B2 receptor tyrosine kinase 3
FBS	Fetal bovine serum
FITC	Fluorescein isothiocyanate
GADD45A	Growth arrest and DNA damage inducible alpha
HSPA5	Heat shock protein family A (Hsp70) member 5
H&E	Hematoxylin and eosin
HL-60	Human acute promyelocytic leukemia cells
SR	Human large-cell immunoblastic lymphoma cells
HUVECs	Human umbilical vein endothelial cells
IFN	Interferon
IL	Interleukin
MAPK	Mitogen-activated protein kinase
MAP3K8	Mitogen-activated protein kinase 8
NO	Nitric oxide
PIK3CD	Phosphatidylinositol-4,5-bisphosphate 3-kinase catalytic subunit delta
PI	Propidium iodide
ROS	Reactive oxygen species

RT-qPCR	Real-time quantitative PCR
RNA-seq	RNA sequencing
SPF	Specific pathogen free
TCR	T cell receptor
TUNEL	Terminal deoxynucleotidyl transferase dUTP nick end labeling
TGFA	Transforming growth factor alpha
TNF	Tumor necrosis factor
UCN2	Urocortin 2

References

1. Hawkes, E.A.; Barraclough, A.; Sehn, L.H. Limited-Stage Diffuse Large B-Cell Lymphoma. *Blood* **2022**, *139*, 822–834. [CrossRef] [PubMed]
2. Sehn, L.H.; Salles, G. Diffuse Large B-Cell Lymphoma. *N. Engl. J. Med.* **2021**, *384*, 842–858. [CrossRef] [PubMed]
3. Siegel, R.L.; Miller, K.D.; Fuchs, H.E.; Jemal, A. Cancer Statistics, 2022. *CA Cancer J. Clin.* **2022**, *72*, 7–33. [CrossRef] [PubMed]
4. Islami, F.; Siegel, R.L.; Jemal, A. The Changing Landscape of Cancer in the USA—Opportunities for Advancing Prevention and Treatment. *Nat. Rev. Clin. Oncol.* **2020**, *17*, 631–649. [CrossRef]
5. Mugnaini, E.N.; Ghosh, N. Lymphoma. *Prim. Care* **2016**, *43*, 661–675. [CrossRef]
6. Pfreundschuh, M.; Kuhnt, E.; Trümper, L.; Österborg, A.; Trneny, M.; Shepherd, L.; Gill, D.S.; Walewski, J.; Pettengell, R.; Jaeger, U.; et al. CHOP-like Chemotherapy with or without Rituximab in Young Patients with Good-Prognosis Diffuse Large-B-Cell Lymphoma: 6-Year Results of an Open-Label Randomised Study of the MabThera International Trial (MInT) Group. *Lancet Oncol.* **2011**, *12*, 1013–1022. [CrossRef]
7. Sravanthi, T.V.; Manju, S.L. Indoles—A Promising Scaffold for Drug Development. *Eur. J. Pharm. Sci.* **2016**, *91*, 1–10. [CrossRef]
8. Chadha, N.; Silakari, O. Indoles as Therapeutics of Interest in Medicinal Chemistry: Bird's Eye View. *Eur. J. Med. Chem.* **2017**, *134*, 159–184. [CrossRef]
9. Luo, X.; Zhang, H.; Li, Y.; Li, X.; Cai, X. Indole Alkaloid Compound Quadrigemine Extracted from Psychotria Pilifera Hutch, and Its Salts and Analogs for Cancer Treatment. CN102617609A, 16 January 2012.
10. Srinivas, U.S.; Tan, B.W.Q.; Vellayappan, B.A.; Jeyasekharan, A.D. ROS and the DNA Damage Response in Cancer. *Redox Biol.* **2019**, *25*, 101084. [CrossRef]
11. Wei, Y.; Liu, F.; Zhu, X.; Liu, X.; Li, H.; Hou, L.; Ma, X.; Li, F.; Liu, H. Artesunate Disrupts Ribosome RNA Biogenesis and Inhibits Ovarian Cancer Growth by Targeting FANCA. *Phytomedicine* **2025**, *136*, 156333. [CrossRef]
12. Li, Z.; Mo, R.; Gong, J.; Han, L.; Wang, W.; Huang, D.; Xu, J.; Sun, Y.; Chen, S.; Han, G.; et al. Dihydrotanshinone I Inhibits Gallbladder Cancer Growth by Targeting the Keap1-Nrf2 Signaling Pathway and Nrf2 Phosphorylation. *Phytomedicine* **2024**, *129*, 155661. [CrossRef]
13. Jakubczyk, K.; Dec, K.; Kałduńska, J.; Kawczuga, D.; Kochman, J.; Janda, K. Reactive Oxygen Species—Sources, Functions, Oxidative Damage. *Pol. Merkur. Lek.* **2020**, *48*, 124–127.
14. Luan, S.; Yun, X.; Rao, W.; Xiao, C.; Xu, Z.; Lang, J.; Huang, Q. Emamectin Benzoate Induces ROS-Mediated DNA Damage and Apoptosis in Trichoplusia Tn5B1-4 Cells. *Chem.-Biol. Interact.* **2017**, *273*, 90–98. [CrossRef]
15. Zhang, Z.; Ren, Z.; Chen, S.; Guo, X.; Liu, F.; Guo, L.; Mei, N. ROS Generation and JNK Activation Contribute to 4-Methoxy-TEMPO-Induced Cytotoxicity, Autophagy, and DNA Damage in HepG2 Cells. *Arch. Toxicol.* **2018**, *92*, 717–728. [CrossRef]
16. Yuan, L.-Q.; Wang, C.; Lu, D.-F.; Zhao, X.-D.; Tan, L.-H.; Chen, X. Induction of Apoptosis and Ferroptosis by a Tumor Suppressing Magnetic Field through ROS-Mediated DNA Damage. *Aging* **2020**, *12*, 3662–3681. [CrossRef] [PubMed]
17. Li, T.; Fu, J.; Cheng, J.; Elfiky, A.A.; Wei, C.; Fu, J. New Progresses on Cell Surface Protein HSPA5/BiP/GRP78 in Cancers and COVID-19. *Front. Immunol.* **2023**, *14*, 1166680. [CrossRef]
18. Lee, A.S. GRP78 Induction in Cancer: Therapeutic and Prognostic Implications. *Cancer Res.* **2007**, *67*, 3496–3499. [CrossRef]
19. Liu, J.; Jiang, G.; Mao, P.; Zhang, J.; Zhang, L.; Liu, L.; Wang, J.; Owusu, L.; Ren, B.; Tang, Y.; et al. Down-Regulation of GADD45A Enhances Chemosensitivity in Melanoma. *Sci. Rep.* **2018**, *8*, 4111. [CrossRef] [PubMed]
20. Gao, Y.; Wang, Y.; Wang, X.; Zhao, C.; Wang, F.; Du, J.; Zhang, H.; Shi, H.; Feng, Y.; Li, D.; et al. miR-335-5p Suppresses Gastric Cancer Progression by Targeting MAPK10. *Cancer Cell Int.* **2021**, *21*, 71. [CrossRef]
21. Lee, J.-S.; Choi, Y.E.; Kim, S.; Han, J.-Y.; Goh, S.-H. ELF3 Is a Target That Promotes Therapeutic Efficiency in EGFR Tyrosine Kinase Inhibitor-Resistant Non-Small Cell Lung Cancer Cells via Inhibiting PKCί. *Int. J. Mol. Sci.* **2021**, *22*, 12287. [CrossRef]
22. Hao, Z.; Huang, Y.; Cleman, J.; Jovin, I.S.; Vale, W.W.; Bale, T.L.; Giordano, F.J. Urocortin2 Inhibits Tumor Growth via Effects on Vascularization and Cell Proliferation. *Proc. Natl. Acad. Sci. USA* **2008**, *105*, 3939–3944. [CrossRef] [PubMed]

23. Kandhavelu, J.; Subramanian, K.; Naidoo, V.; Sebastianelli, G.; Doan, P.; Konda Mani, S.; Yapislar, H.; Haciosmanoglu, E.; Arslan, L.; Ozer, S.; et al. A Novel EGFR Inhibitor, HNPMI, Regulates Apoptosis and Oncogenesis by Modulating BCL-2/BAX and P53 in Colon Cancer. *Br. J. Pharmacol.* **2024**, *181*, 107–124. [CrossRef] [PubMed]
24. Zhao, H.; Wu, L.; Yan, G.; Chen, Y.; Zhou, M.; Wu, Y.; Li, Y. Inflammation and Tumor Progression: Signaling Pathways and Targeted Intervention. *Signal Transduct. Target. Ther.* **2021**, *6*, 263. [CrossRef]
25. Crusz, S.M.; Balkwill, F.R. Inflammation and Cancer: Advances and New Agents. *Nat. Rev. Clin. Oncol.* **2015**, *12*, 584–596. [CrossRef]
26. Zhao, C.; Zeng, N.; Zhou, X.; Tan, Y.; Wang, Y.; Zhang, J.; Wu, Y.; Zhang, Q. CAA-Derived IL-6 Induced M2 Macrophage Polarization by Activating STAT3. *BMC Cancer* **2023**, *23*, 392. [CrossRef]
27. Bunt, S.K.; Yang, L.; Sinha, P.; Clements, V.K.; Leips, J.; Ostrand-Rosenberg, S. Reduced Inflammation in the Tumor Microenvironment Delays the Accumulation of Myeloid-Derived Suppressor Cells and Limits Tumor Progression. *Cancer Res.* **2007**, *67*, 10019–10026. [CrossRef]
28. Laha, D.; Grant, R.; Mishra, P.; Nilubol, N. The Role of Tumor Necrosis Factor in Manipulating the Immunological Response of Tumor Microenvironment. *Front. Immunol.* **2021**, *12*, 656908. [CrossRef] [PubMed]
29. PeÑarando, J.; Aranda, E.; RodrÍguez-Ariza, A. Immunomodulatory Roles of Nitric Oxide in Cancer: Tumor Microenvironment Says "NO" to Antitumor Immune Response. *Transl. Res.* **2019**, *210*, 99–108. [CrossRef]
30. Chen, Y.; Niu, N.; Xue, J. Double-Edged Roles of IFNγ in Tumor Elimination and Immune Escape. *J. Pancreatol.* **2023**, *6*, 8–17. [CrossRef]
31. Gouda, N.A.; Alshammari, S.O.; Abourehab, M.A.S.; Alshammari, Q.A.; Elkamhawy, A. Therapeutic Potential of Natural Products in Inflammation: Underlying Molecular Mechanisms, Clinical Outcomes, Technological Advances, and Future Perspectives. *Inflammopharmacology* **2023**, *31*, 2857–2883. [CrossRef]
32. Jannic, V.; Guéritte, F.; Laprévote, O.; Serani, L.; Martin, M.-T.; Sévenet, T.; Potier, P. Pyrrolidinoindoline Alkaloids from Psychotria o Leoides and Psychotria l Yciiflora. *J. Nat. Prod.* **1999**, *62*, 838–843. [CrossRef] [PubMed]
33. Yu, M.; Ren, L.; Liang, F.; Zhang, Y.; Jiang, L.; Ma, W.; Li, C.; Li, X.; Ye, X. Effect of Epiberberine from Coptis Chinensis Franch on Inhibition of Tumor Growth in MKN-45 Xenograft Mice. *Phytomedicine* **2020**, *76*, 153216. [CrossRef] [PubMed]

biomolecules

MDPI

Article

Baicalin Alleviates Piglet Immunosuppression Induced by *Glaesserella parasuis* via Promoting CD163/Tumor Necrosis Factor-like Weak Inducer of Apoptosis-Mediated Autophagy

Shulin Fu [1,2,*,†], Ronghui Luo [1,2,†], Jingyang Li [1,2], Yunjian Fu [1,2], Qiaoli Dong [1,2], Siyu Liu [1,2], Yamin Sun [1,2], Ling Guo [1,2], Jin Hu [1,2] and Yinsheng Qiu [1,2]

1 Wuhan Engineering and Technology Research Center of Animal Disease-Resistant Nutrition, School of Animal Science and Nutritional Engineering, Wuhan Polytechnic University, Wuhan 430023, China; 20230411035@whpu.edu.cn (R.L.); 20220411050@whpu.edu.cn (J.L.); 20230411033@whpu.edu.cn (Y.F.); 20230411045@whpu.edu.cn (Q.D.); 20220411030@whpu.edu.cn (S.L.); 20230412014@whpu.edu.cn (Y.S.); guoling1101@whpu.edu.cn (L.G.); hujin@whpu.edu.cn (J.H.); qiuyinsheng6405@whpu.edu.cn (Y.Q.)

2 Hubei Key Laboratory of Animal Nutrition and Feed Science, School of Animal Science and Nutritional Engineering, Wuhan Polytechnic University, Wuhan 430023, China

* Correspondence: shulinfu@whpu.edu.cn

† These authors contributed equally to the work.

Abstract: *Glaesserella parasuis* (*G. parasuis*) causes vascular inflammation in piglets, resulting in vascular damage. However, the mechanism causing vascular inflammation remains unclear. Baicalin possesses an anti-inflammatory function. Tumor necrosis factor-like weak inducer of apoptosis (TWEAK) has been implicated in immunosuppression. CD163, a scavenger receptor expressed on macrophages that acts as a decoy receptor for TWEAK, plays a crucial role in the regulation of autophagy and inflammation. This research investigated the efficacy of baicalin in reducing immunosuppression elicited by *G. parasuis* through the regulation of CD163/TWEAK-mediated autophagy. The data demonstrated that *G. parasuis* altered routine blood indicators and biochemical parameters, increased cytokine production, and induced blood vessel tissue damage. *G. parasuis* reduced the CD3+ T cell proportion, CD3+CD4+ T cell proportion, and CD3+CD8+ T cell proportion in piglet blood. The proteomic analysis revealed that CD163 was differentially expressed in the blood vessels of challenged piglets. Baicalin was found to regulate CD163/TWEAK axis expression, inhibit Notch/Wnt signaling pathway activation, promote autophagy, and reduce NLRP3/Caspase 1 signaling pathway activation. Baicalin also decreased cytokine production and alleviated pathological tissue damage in the blood vessels of *G. parasuis*-challenged piglets. Taken together, this study indicates that baicalin alleviates *G. parasuis*-induced immunosuppression and might promote CD163/TWEAK-mediated autophagy. This finding suggests that baicalin could serve as a potential therapeutic agent to control *G. parasuis infection* and related vascular inflammation.

Keywords: baicalin; *Glaesserella parasuis*; immunosuppression; CD163/TWEAK; autophagy

1. Introduction

Glaesserella parasuis (*G. parasuis*) is a Gram-negative bacterium that colonizes the pig upper respiratory tract [1]. *G. parasuis* is widely distributed in pig farms and is responsible for Glässer's disease [2]. This disease can cause significant morbidity and mortality, leading to serious economic losses in piglet raising. The typical characteristics of Glässer's disease are fibrinous polyserositis, pneumonia, arthritis, and meningitis [3]. According to the agar gel diffusion serotyping method, *G. parasuis* can be classified into 15 serotypes, but up to 20% of isolates cannot be typed [4]. Serovar 5 was considered the highly virulent strain [5]. The pathogenic mechanism of *G. parasuis* remains unclear at present, and thus, understanding the molecular mechanisms of *G. parasuis* infection is crucial for developing better prevention and control strategies.

G. parasuis infection in piglets can lead to immunosuppression of the host, but the immunosuppression mechanism remains unclear. CD163 is a scavenger receptor that is predominantly expressed on the surface of macrophages [6]. CD163 has important functions in hemoglobin–haptoglobin complex recognition and clearance [7]. Previous research reported that CD163 participates in various pathological processes, such as inflammation and immune responses [8]. Tumor necrosis factor-like weak inducer of apoptosis (TWEAK) is a member of the tumor necrosis factor superfamily [9]. CD163 can act as a decoy receptor for TWEAK, modulating TWEAK-induced signaling pathways [10]. The crosstalk of CD163 and TWEAK has been implicated in immunosuppression [11]. CD163-expressing macrophages could be affected by TWEAK, as they can internalize TWEAK-bound complexes and contribute to the immunosuppressive environment [12]. Studies have shown that the engagement of TWEAK with FN14 can regulate the efficacy of immune cells, such as macrophages and T cells, leading to the inhibition of immune responses and the progression of various diseases [13]. The TWEAK/Fn14 signaling pathway participates in multiple biological activities, such as angiogenesis and inflammatory responses [14,15]. However, whether CD163/TWEAK is involved in *G. parasuis*-induced immunosuppression has not been investigated.

Autophagy is a cellular process that involves the degradation and recycling of cellular components [16]. Recent studies have suggested that CD163 participates in modulating autophagy [17]. TWEAK regulates the expression of autophagy-related genes or proteins, leading to changes in autophagic flux [18]. However, whether CD163 interacts with TWEAK and regulates autophagy has not been explored in *G. parasuis*-infected piglets.

Baicalin was extracted from the root of *Scutellaria baicalensis* Georgi and has been reported to have important biological functions, such as antioxidant activity, antibacterial and antiviral effects, and the regulation of the immune system [19]. Baicalin has also shown great potential in the treatment of inflammatory diseases [20]. Research has reported that baicalin reduces pulmonary inflammation and injury by attenuating the TLR4/NF-κB pathway in Mycoplasma pneumoniae-infected mice [21]. Baicalin suppresses macrophage JNK-mediated adipose tissue inflammation [22]. Baicalin inhibits monosodium urate crystal-induced pyroptosis in a renal tubular epithelial cell line through Panx-1/P2 × 7 pathways [23]. In addition, baicalin reduces cell damage and affects the H6N6-induced autophagy level of RAW264.7 cells [24]. Baicalin induces the death of non-small cell lung cancer cells via MCOLN3-mediated lysosomal dysfunction and autophagy blockage [25]. Baicalin attenuates diabetic cardiomyopathy by inhibiting autophagy and cell death through SENP1/SIRT3 signaling pathway activation [25]. Thus, it is very interesting to investigate whether baicalin could regulate *G. parasuis*-induced autophagy.

In this study, we performed a detailed study of the efficacy of baicalin in the inhibition of *G. parasuis*-elicited immunosuppression by promoting CD163/TWEAK-mediated autophagy. It was reported that baicalin could regulate *G. parasuis*-induced autophagy, which might be regulated by CD163/TWEAK in piglets, and our results might provide new strategies to prevent *G. parasuis* infection in the clinic.

2. Materials and Methods

2.1. Ethics Statement

The animal study was approved by the Animal Care and Use Committee of Wuhan Polytechnic University, Hubei Province, China (WPU202308003), approval date 2 August 2023. All piglets used in this study were euthanized at the end of the experiment.

2.2. Bacterial

The serovar 5 *G. parasuis* SH0165 isolate used in this study is a highly virulent strain, which was isolated from the lung of a commercial pig with typical arthritis, fibrinous polyserositis, hemorrhagic pneumonia, and meningitis [26]. The SH0165 isolate was plated on TSA plates (Difco Laboratories, USA) or cultured in TSB culture (Difco Laboratories, USA). The culture was supplemented with 10 μg/mL of NAD (Biofroxx, Guangzhou, Guangzhou Saiguo biotech Co., Ltd.) and 10% fetal bovine serum (Sijiqing, Hangzhou, China) at 37 °C.

2.3. Drugs

Baicalin used in this study was obtained from Sichuan Xieli Pharmaceutical Co., Ltd. (Pengzhou, China). The drug levamisole was purchased from Beijing Solarbio Science & Technology Co., Ltd. (Beijing, China).

2.4. Animal Experiment Design 1

Twenty 30-day old naturally farrowed early-weaned (NFEW) Duroc × Landrace × Large White piglets weighing nine to ten kg were obtained from Wuhan Fenglongxin Breeding Professional Cooperative (Wuhan, China). Twenty piglets were randomly divided into control and infection groups, each comprising ten piglets. The infection group piglets were intraperitoneally infected with 1×10^8 CFU *G. parasuis* resuspended in 1 mL of TSB. The control group piglets were intraperitoneally challenged with an equivalent volume of TSB. All piglets from each group were monitored for three days following challenge by *G. parasuis*. After being challenged at 24 h, 48 h, and 72 h, blood samples were collected for routine hematological and biochemical parameter determination using commercial kits (Shanghai Kehua Bio-Engineering Co., Ltd., Shanghai, China) on a Hitec7100 automatic blood analyzer (Hitachi, Japan) [27].

2.5. Histopathological Analysis

After being challenged for three days by *G. parasuis*, the piglets were euthanized, and the aortae were obtained for histopathological analysis. The aortic samples were fixed using immersion in 10% neutral buffered formalin. The 4 μm tissue sections were stained using hematoxylin and eosin [28], then the stained sections were examined with a BX43 light microscope (Olympus, Tokyo, Japan).

2.6. Flow Cytometry

To explore the effect of *G. parasuis* on T cell differentiation, the blood was collected at 24 h, 48 h, and 72 h after *G. parasuis* challenge. The collected blood was treated with RBC lysis buffer (TaKaRa, Dalian, China) to remove erythrocytes prior to staining. Following filtration to a single-cell suspension using a 200 μm filter, the samples were stained with mouse anti-porcine CD3ε-FITC (Southern Biotech, Birmingham, AL, USA), mouse anti-porcine CD4-SPRD (Southern Biotech), and mouse anti-porcine CD8a-PE (Southern Biotech). The blood samples were repeated at least three times, and the data were analyzed using CytExpert SRT software.

2.7. Proteome Analysis

The piglet aorta tissue samples from the control and the infection group were obtained and immediately put into liquid nitrogen for proteome analysis [29]. Briefly, the blood vessel samples were added to lysis buffer. After collecting supernatant, the protein concentration of the supernatant was assessed using BCA protein assay. Sequence-grade modified trypsin (Promega, Madison, WI, USA) was used to obtain the peptide mixture. The obtained peptides were analyzed using LC-MS/MS with Orbitrap Fusion™ Lumos™ Tribrid™ coupled with the EASY-nLC 1200 system (Thermo Fisher Scientific, MA, USA). Differentially expressed proteins were filtered with the standard of fold change >1.5 and Q value < 0.05. The obtained genes were mapped to the KEGG database by employing FDR $\leq$ 0.05 as the threshold.

2.8. Animal Experiment Design 2

This animal experiment was reported in a previous study [27]. Briefly, sixty 30-day NFEW piglets weighing nine to ten kg were randomly divided into six groups, including the control, infection, levamisole, 25 mg/kg of baicalin, 50 mg/kg of baicalin, and 100 mg/kg of baicalin groups. Before *G. parasuis* challenge, the levamisole, 25 mg/kg of baicalin, 50 mg/kg of baicalin, and 100 mg/kg of baicalin groups were treated via intramuscular injection with 15 mg/kg of body weight (BW) of levamisole, 25 mg/kg of BW of baicalin, 50 mg/kg of BW of baicalin, and 100 mg/kg of BW of baicalin, respectively. Two hours after administration on the first day, all treatment groups except for the control group were intraperitoneally infected with 1×10^8 CFU *G. parasuis* resuspended in one mL of TSB. The control group received one mL of TSB only intraperitoneally. Hence, the five groups, except the control group, used intramuscular injection with the same drugs after being challenged for six hours. Each treatment was administered for two days and was adopted twice a day. The piglets were observed for seven days following *G. parasuis* infection.

2.9. RT-PCR(Reverse Transcription-Polymerase Chain Reaction)

The piglet aorta tissue samples were collected for gene expression level determination by using RT-PCR [30]. Briefly, the total RNA from the blood vessels was extracted by using TRIzol reagent (Invitrogen, USA). The concentration of the isolated RNA was measured on a Qubit 2.0 fluorometer (Thermo Fisher Scientific, USA). Following reverse transcription into complementary DNA (cDNA) using reverse transcriptase (TaKaRa, Dalian, China), the RT-PCR was carried out by using a SYBR Green PCR Kit (TaKaRa) based on the manufacturer's instructions. Each sample transcription was repeated at least three times. The internal gene adopted in this study was *GAPDH*. The relative expression levels of the gene were evaluated by employing the threshold cycle (CT) method. The gene relative expression fold changes were measured by employing the $2^{-\Delta\Delta CT}$ CT formula. The primers for RT-PCR used in this study are presented in Table 1.

Table 1. Primer sequences used for qRT-PCR analysis.

Gene		Nucleotide Sequence (5′-3′)	Tm (°C)	Length (bp)
IL-1β	Forward	TCTGCATGAGCTTTGTGCAAG	59.7	155
	Reverse	ACAGGGCAGACTCGAATTCAAC	60.9	
IL-6	Forward	CTTCTGGTGATGGCTACTG	52.7	134
	Reverse	TTGCCGAGGATGTACTTAA	50	
IL-8	Forward	ACAGCAGTAACAACAACAAG	50.2	117
	Reverse	GACCAGCACAGGAATGAG	53.2	
IL-10	Forward	CGTGGAGGAGGTGAAGAGTG	55.4	178
	Reverse	TTAGTAGAGTCGTCATCCTGGAAG	55.6	

Table 1. *Cont.*

Gene		Nucleotide Sequence (5′-3′)	Tm (°C)	Length (bp)
IL-18	Forward	AGTAACCATATCTGTGCAGTGT	54	155
	Reverse	TCTTATCACCATGTCCAGGAAC	53	
TNF-α	Forward	CGCTCTTCTGCCTACTGCACTTC	60.7	164
	Reverse	CTGTCCCTCGGCTTTGACATT	57.8	
Caspase1	Forward	TACAAGAATCCCAGGCGGTG	57.5	128
	Reverse	CCTTTGGGCTATGTCTGGGG	58.6	
NLRP3	Forward	CAGGCTTCTGGGACACCTTT	59.9	110
	Reverse	GTGCAGCCCTAGTCAGAGTC	58.8	
TWEAK	Forward	AGGCCAAGGCAGGCCAGCG	68.1	100
	Reverse	GTCGGAGCCAGAGGCGGAGG	66.8	
HES1	Forward	GTGAGTGCATGAACGAGGT	59.2	118
	Reverse	GTCATGGCGTTGATCTGGGT	60.4	
JAG1	Forward	TTTCAGGGCGACCTTGCATC	61	121
	Reverse	CCACACCACACCTTCGAGC	61	
RBP-J	Forward	CTTGAACTTACAGGACAGAAT	53.2	105
	Reverse	GACGACACAGAGCATACT	53.7	
GSK-3β	Forward	CAGAACCACCTCCTTTGCG	58.8	100
	Reverse	GGTCACCTTGCTGCCATCC	61.1	
GAPDH	Forward	GGCACAGTCAAGGCGGAGAAC	61.9	105
	Reverse	AGCACCAGCATCACCCCATTTG	61	
c-Myc	Forward	CTGCCAAGAGGGCTAAGTT	54.6	135
	Reverse	TCTGGCGTTCCAAGACATT	56	
WNT3A	Forward	TACTCCTCTGCAGCCTGAAGCA	63.4	322
	Reverse	ATGGCGTGGACAAAGGCCGAC	65.8	
Notch1	Forward	GATGGCATCAATTCCTTTAC	52.2	149
	Reverse	TGAGGGCAGGTACACTTGT	53	
β-catenin	Forward	AAGGCAATCCTGAAGAAGA	52	148
	Reverse	ATAGCAGCTCGTACCCTCT	51.4	
CD163	Forward	TGCTGTAGTCGCTGTTCT	55.9	117
	Reverse	ACTTTCACCTCCACTCTTC	54	

2.10. Western Blot

The levels of protein expression were assessed by using Western blot [31]. Briefly, aorta total proteins were isolated using RIPA lysis buffer. The concentration of proteins was determined by using a BCA protein assay kit (Beyotime Biotechnology, China). Following transfer onto PVDF membranes (Millipore, USA), the proteins were separated by utilizing 8 to 12% concentration of SDS-PAGE. The membranes were blocked with 15% nonfat milk and incubated with anti-CD163 (#16646-1, 1:1000, Protenitech), anti-JAG1 (#WL03261, 1:1000, Wanleibio), anti-Notch1 (#A19090, 1:1500, ABclonal), anti-HES1 (#A11718,1:1000, ABclonal), anti-WNT3A (#A0642, 1:1000, ABclonal), anti-GSK-3β (#A2081, 1:2000, ABclonal), anti-RBP-J (#WL04414, 1:1000, Wanleibio), anti-c-Myc (#HA721182, 1:1000, Huaan Biotechnology), anti-P62 (#A19700, 1:40,000, ABclonal), anti-LC3B (#A19665, 1:4000, ABclonal), anti-Beclin1 (#A21191, 1:10,000, ABclonal), anti-Caspase1 (#A0964, 1:1000, ABclonal), anti-NLRP3 (#A24294, 1:1000, ABclonal), and anti-β-catenin antibodies (#51067-2, 1:2000, Protenitech), respectively, overnight at 4 °C. After washing five times with TBST, the membranes were incubated with a second antibody, HRP Goat-Anti Rabbit IgG (1:10,000, Abbkine), for 60 min under 37 °C. The band signals of proteins were detected using an enhanced chemiluminescence detection kit (ABclonal, Wuhan, China). The protein expression level data were evaluated using a FluorChemFC2 AIC system (Alpha Innotech, USA).

2.11. Detection of CD163 Expression by Immunohistochemistry

The CD163 expression level was also measured using immunohistochemistry analysis. Briefly, aorta samples were prepared from paraffin-embedded blocks. The slices were cut and incubated with anti-CD163 (#16646-1, 1:500, protenitech) immunohistochemistry (IHC) at 4 °C overnight. CD163 was detected with HRP goat anti-rabbit secondary antibody and visualized using diaminobenzidine (DAB) (Servicebio, Wuhan, China). The negative control received only the second antibody. Following counterstaining with hematoxylin (Servicebio, Wuhan, China) for three min, microscopic examination was carried out with a BX43 light microscope, and Image 1.52a software was employed to determine the optical density sum of the brown area.

2.12. Statistical Analysis

The experimental data were displayed as mean ± SD. Differences were analyzed using analysis of variance (ANOVA). A value of $p < 0.05$ displays statistical significance.

3. Results

3.1. *G. parasuis* Altered the Routine Blood Indicators and Biochemical Parameters, Increased Cytokines Production, and Induced Blood Vessel Tissue Damage

Compared to the control group, for those infected by *G. parasuis* for 24 h, the white blood cells (WBCs), red blood cells (RBCs), platelets (PLTs), neutrophils (Neus), and lymphocytes (Lyms) were significantly reduced in the infection group, while the monocytes (MONs) were raised in the infection group ($p < 0.05$), as shown in Figure 1A. At 72 h, Lym and Neu continued to reduce in the infection group compared to the control group ($p < 0.05$).

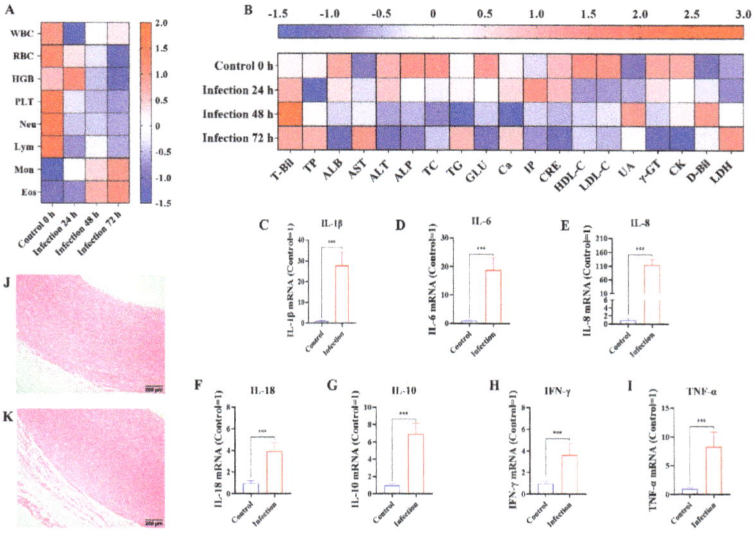

Figure 1. The effects of *G. parasuis* on the routine blood indicators, biochemical parameters, cytokine production, and blood vessel tissue damage of piglets. On 24 h, 48 h, 72 h infected by *G. parasuis*, the blood were collected and the routine blood indicators (**A**) and biochemical parameters (**B**) were determined. At 72 h after being challenged by *G. parasuis*, the mRNA levels of IL-1β (**C**), IL-6 (**D**), IL-8 (**E**), IL-18 (**F**), IL-10 (**G**), IFN-γ (**H**), and TNF-α (**I**) in the blood vessels were assessed using RT-PCR, and the blood vessels from the control group (**J**) and the infection group (**K**) were obtained for pathological analysis. Hemoglobin: HGB; eosinophils: Eos; total protein: TP; total cholesterol: TC; triglycerides: TG; calcium: Ca; uric acid: UA; *** $p < 0.001$.

At 24 h, albumin (ALB), glucose (GLU), high-density lipoprotein cholesterol (HDL-C), γ-glutamyl transpeptidase (γ-GT), and creatine kinase (CK) levels were reduced, whereas total bilirubin (T-Bil), aspartate aminotransferase (AST), inorganic phosphate (IP), and direct bilirubin (D-Bil) were raised in the infection group ($p < 0.05$), as shown in Figure 1B. At 72 h, T-Bil, AST, D-Bil, and lactate dehydrogenase (LDH) levels were increased, and the levels of ALB, alanine aminotransferase (ALT), alkaline phosphatase (ALP), GLU, creatinine (CRE), low-density lipoprotein cholesterol (LDL-C), γ-GT, and CK were decreased in the infection group compared to the control group ($p < 0.05$).

At 72 h after being infected by *G. parasuis*, the mRNA levels of IL-1β, IL-6, IL-8, IL-10, IL-18, IFN-γ, and TNF-α in the blood vessels were upregulated in the infection group compared to the control group ($p < 0.001$), as shown in Figure 1C–I. Pathological analysis of vessel tissue indicated that the vessels from the infection group underwent serious tissue damage with massive inflammatory cell infiltration and severe hemorrhage, while some minor tissue damage was present in the control group, as seen in Figure 1J,K.

3.2. G. parasuis Reduced the Proportion of CD3+, CD3+CD4+, and CD3+CD8+ T Cells in the Blood of Piglets

The results showed that compared to the control group at 24 h after infection with *G. parasuis*, the CD3+, CD3+CD4+, and CD3+CD8+ T cell proportions were reduced in the infection group ($p < 0.001$), as shown in Figure 2. The CD3+, CD3+ CD4+, and CD3+ CD8+ T cell proportions increased from 48 h to 72 h.

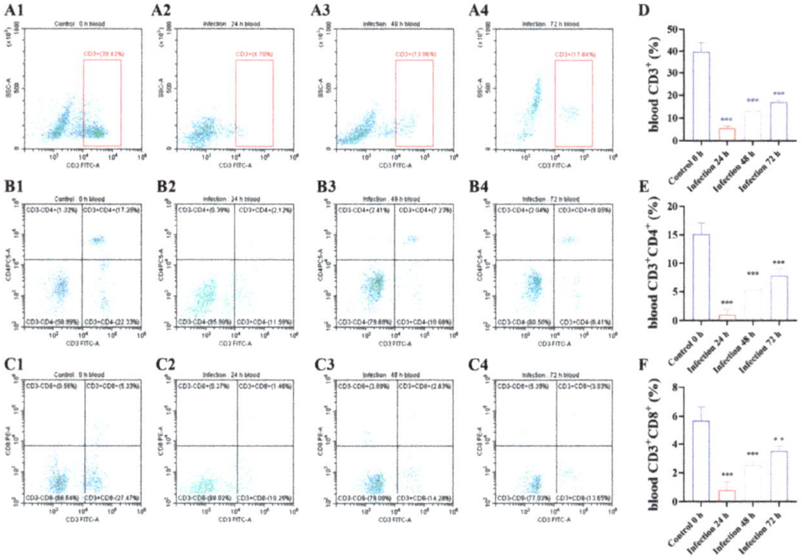

Figure 2. Following infected by *G. parasuis* at 24 to 72 h, piglet bloods were collected to determine CD3+, CD3+CD4+, and CD3+CD8+ T cell proportions. (**A1–A4**,**D**): the CD3+ T cells; (**B1–B4**,**E**): the CD3+ CD4+ T cells; (**C1–C4**,**F**): the CD3+ CD8+ T cells; ** $p < 0.01$; *** $p < 0.001$.

3.3. G. parasuis Infection Dysregulated Proteins in the Blood Vessels of Piglets

Proteomics was used to evaluate the host's immunosuppression mechanism triggered by *G. parasuis* in the blood vessels of piglets. The data indicated that 61,724 peptides were obtained, which corresponds to 5764 protein groups, as shown in Figure 3A. As shown in Figure 3B,C, 1358 differentially expressed proteins were identified from the blood vessels of the infection group, 760 differentially expressed proteins were upregulated, and 598 differentially expressed proteins were diminished according to the standard of | fold change | > 1.5 and Q value < 0.05. The CD163 expression level was significantly upregulated in the infection group, as shown in Figure 3B,F. Gene Ontology (GO) enrichment analysis demonstrated that the differentially expressed proteins were mainly enriched in the cellular process, metabolic process, and biological regulation from the biological processes category, as shown in Figure 3D. And the differentially expressed proteins were mainly enriched in the cellular anatomical entity and protein-containing complex in the cellular component. In the molecular function category, the differentially expressed proteins were dominantly enriched in binding, catalytic activity, and molecular function regulator, and KEGG analysis showed that the differentially expressed proteins were mainly enriched in spliceosome, necroptosis, cell adhesion molecules, Notch signaling pathway, and Wnt signaling pathway, as shown in Figure 3E.

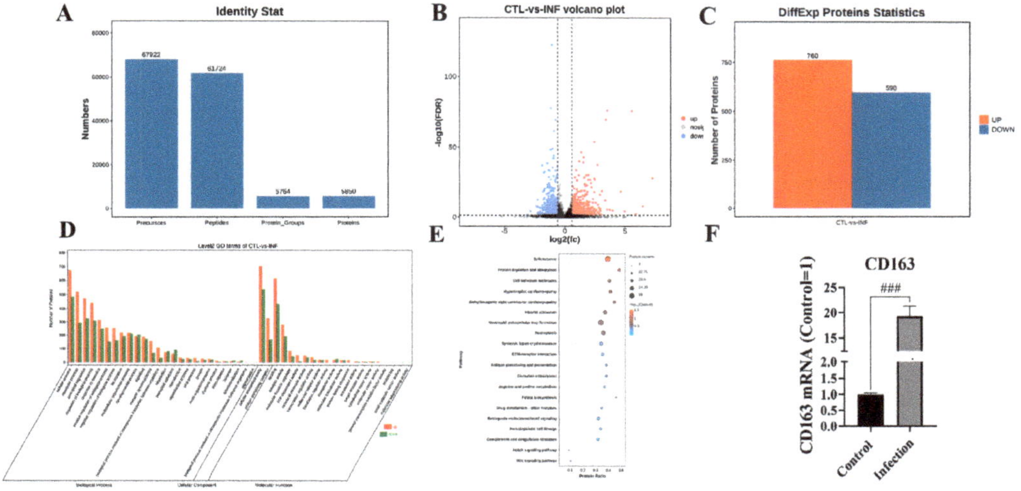

Figure 3. Proteomic analysis of piglet blood vessels after *G. parasuis* challenge. (**A**) The quantified proteins from the proteomic analysis; (**B**) volcano plot of the dysregulated proteins; (**C**) differentially expressed proteins (DEPs) in the infection group and the control group; (**D**) GO enrichment analysis of the DEPs; (**E**) KEGG analysis of the differentially expressed proteins; (**F**) detection of the CD163 expression level by RT-PCR; INF: the infection group; CTL: the control group; ### $p < 0.001$.

3.4. Baicalin Regulated CD163/TWEAK Axis Expression in Blood Vessels from Piglets Infected by *G. parasuis*

When the piglets were infected with *G. parasuis*, the CD163 mRNA expression level was significantly increased, and the mRNA level of TWEAK was significantly decreased in the infection group ($p < 0.001$), as shown in Figure 4A,F. Levamisole could attenuate the CD163 mRNA expression level and alter the TWEAK mRNA expression level compared to the infection group ($p < 0.001$). The 25, 50 and 100 mg/kg of baicalin diminished CD163 mRNA expression and increased the TWEAK mRNA expression level from the blood vessels compared to the infection group ($p < 0.001$).

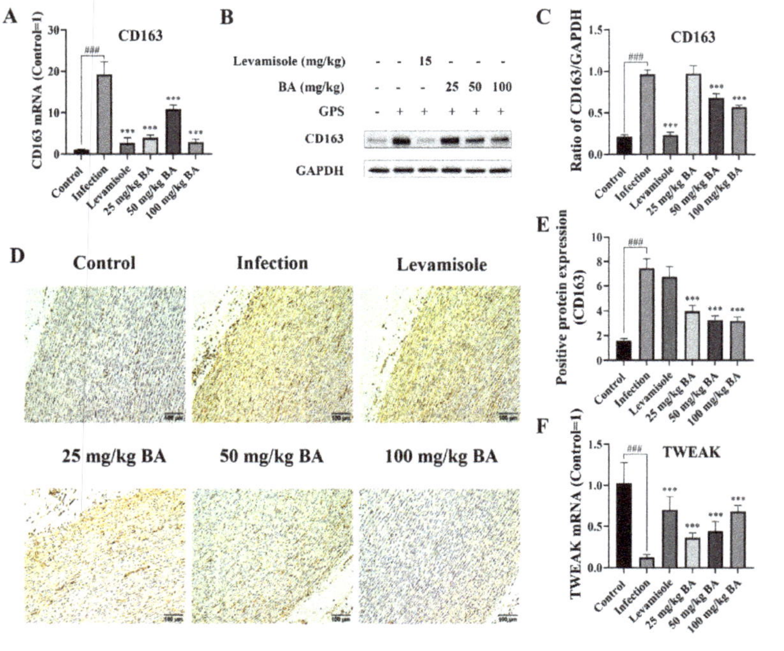

Figure 4. Determination of the function of baicalin on CD163/TWEAK expression in the blood vessels. When the piglets were challenged by *G. parasuis*, the CD163 expression level was assessed using RT-PCR (**A**), Western blot (**B**,**C**), and immunohistochemistry (**D**,**E**). The TWEAK expression level was determined using RT-PCR (**F**). BA: baicalin; ### $p < 0.001$ versus controls; *** significance at $p < 0.001$. Original images can be found in Supplementary Materials.

The CD163 protein expression level was determined using Western blot analysis. The data indicated that *G. parasuis* triggered upregulated CD163 protein expression compared to the control group ($p < 0.001$), as shown in Figure 4B,C. Levamisole and 50 to 100 mg/kg of baicalin inhibited the CD163 protein expression level compared to the infection group ($p < 0.001$).

The protein expression level of CD163 was also measured via immunohistochemistry. The results demonstrated that the CD163 protein expression level was raised in the infection group compared to the control group ($p < 0.001$), as shown in Figure 4D,E, while the protein expression level of CD163 was decreased by 25 to 100 mg/kg of baicalin compared to the infection group ($p < 0.001$).

3.5. Baicalin Inhibited Notch/Wnt Signaling Pathways Activation in the Blood Vessels from Piglets Challenged by G. parasuis

Previous research reported that CD163 could regulate Notch signaling pathway activation [10]. Notch 1 is a key receptor in the Notch 1 signaling pathway. After *G. parasuis* infection of piglets, the mRNA expression level of Notch 1 from the blood vessels was observed to decrease compared to the control group ($p < 0.001$), as shown in Figure 5A. The RBP-J could interact with Notch 1 and adjust the expression of downstream target genes. The blood vessels from the infection group showed a decrease in RBP-J mRNA expression compared to the control group ($p < 0.001$), as shown in Figure 5E. JAG1, a ligand in the Notch signaling pathway, was found to be downregulated at mRNA levels after *G. parasuis* infection ($p < 0.001$), as shown in Figure 5I, which could further enhance the Notch pathway activation. HES1, as the downstream target gene of the Notch signaling pathway, was found to show a decrease in mRNA expression from the blood vessels of the *G. parasuis*-challenged piglets ($p < 0.001$), as shown in Figure 5M, which indicated that the Notch pathway actively regulated its target genes. The levels of Notch 1, RBP-J, JAG1, and HES1 mRNA expression were significantly raised in the vessels from the levamisole group compared to the infection group ($p < 0.001$), as shown in Figure 5A,E,I,M. The use of 25 to 100 mg/kg of baicalin upregulated Notch 1, RBP-J, JAG1, and HES1 mRNA expression levels from the blood vessels compared to the infection group ($p < 0.01$). In the infection group, Notch 1, RBP-J, JAG1, and HES1 protein levels were significantly inhibited compared to the control group, while 25 to 50 mg/kg of baicalin increased the protein expression levels of Notch 1, RBP-J, JAG1, and HES1 compared to the infection group ($p < 0.05$), as shown in Figure 5B,F,J,N.

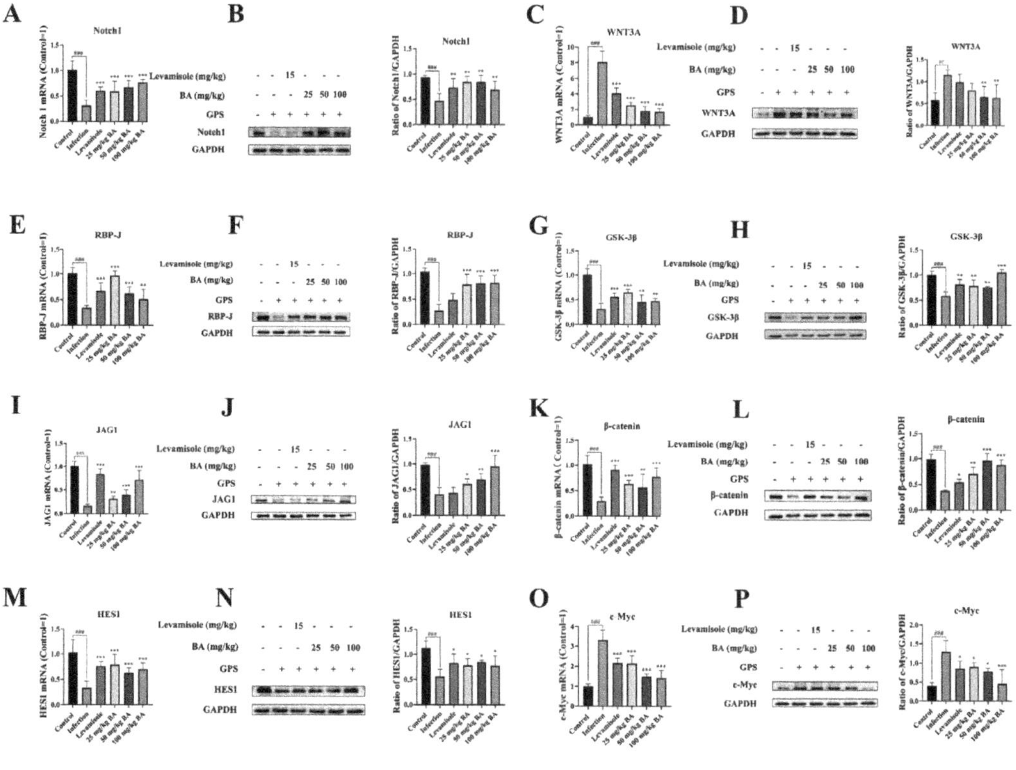

Figure 5. The effects of baicalin on Notch/Wnt signaling pathways' activation in piglets' blood vessels. Notch pathway-related molecule expression levels (Notch 1, RBP-J, JAG1, and HES1) and the expression levels of Wnt pathway-related molecules (WNT3A, GSK-3β, β-catenin, and c-Myc) were assessed using RT-PCR and Western blot. (**A**): Notch 1, (**E**): RBP-J, (**I**): JAG1, (**M**): HES1, (**C**): WNT3A, (**G**): GSK-3β, (**K**): β-catenin, (**O**): c-Myc at mRNA level; (**B**): Notch 1, (**F**): RBP-J, (**J**): JAG1, (**N**): HES1, (**D**): WNT3A, (**H**): GSK-3β, (**L**): β-catenin, (**P**): c-Myc at protein level; BA: baicalin; ## $p < 0.01$ versus controls; ### $p < 0.001$ versus controls; * significance at $p < 0.05$; ** significance at $p < 0.01$; *** significance at $p < 0.001$. Original images can be found in Supplementary Materials.

Compared to the control group, the WNT3A mRNA expression level was raised in the infection group ($p < 0.001$), as shown in Figure 5C. β-catenin is an effector of the Wnt signaling pathway. In the infection group, the mRNA expression level of β-catenin was significantly upregulated in the blood vessels compared to the control group ($p < 0.001$), as shown in Figure 5K. GSK-3β is a kinase that phosphorylates β-catenin, marking it for degradation. The expression of GSK-3β was found to be downregulated after infection

($p < 0.001$), as shown in Figure 5G. The mRNA expression level of c-Myc, as a well-known downstream target gene in the Wnt signaling pathway, was increased from the infection group ($p < 0.001$), as shown in Figure 5O. The mRNA levels of GSK-3β and β-catenin were significantly increased, and WNT3A and c-Myc mRNA levels were significantly downregulated in the levamisole group compared to the infection group ($p < 0.001$), as shown in Figure 5C,G,K,O. The 25 to 100 mg/kg of baicalin could enhance GSK-3β and β-catenin mRNA expressions and attenuate WNT3A and c-Myc mNRA expression levels compared to the infection group ($p < 0.01$).

The protein expressions of WNT3A, GSK-3β, β-catenin, and c-Myc were also assessed using Western blot. The data indicated that *G. parasuis* upregulated WNT3A and c-Myc protein levels and downregulated the GSK-3β and β-catenin protein levels in the infection group ($p < 0.01$), as shown in Figure 5D,H,L,P. Treatment with 25 mg/kg–50 mg/kg of baicalin raised the protein levels of GSK-3β and β-catenin and inhibited the protein levels of WNT3A and c-Myc compared to the infection group ($p < 0.05$).

3.6. Baicalin Promoted Autophagy in the Blood Vessels of *G. parasuis*-Infected Piglets

Following infection with *G. parasuis*, the protein levels of Beclin1 and LC3B in the blood vessels of piglets were decreased significantly compared to the control group ($p < 0.001$), as shown in Figure 6A–D. In contrast, the protein expression level of P62 was increased compared to the control group ($p < 0.001$), as shown in Figure 6E,F. When the piglets were treated with levamisole, the protein level of LC3B was upregulated compared to the infection group ($p < 0.001$), as shown in Figure 6C,D. Overall, 50 to 100 mg/kg baicalin could reduce p62 protein levels and promote Beclin1 and LC3B protein expression compared to the infection group ($p < 0.05$), as shown in Figure 6A–D.

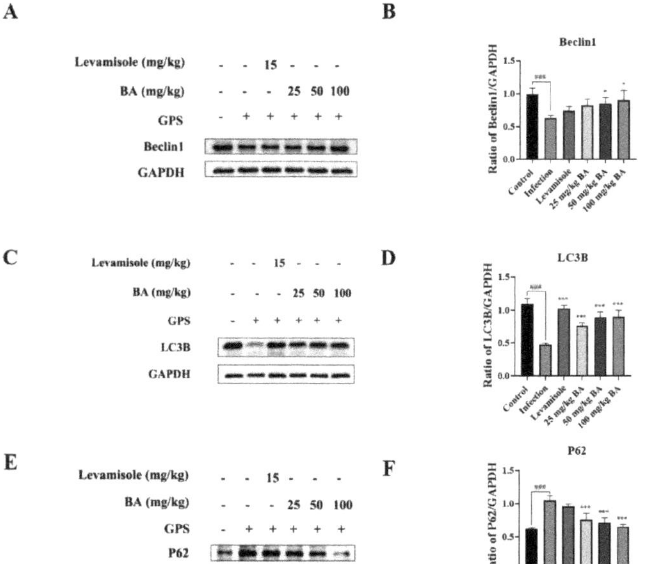

Figure 6. The function of baicalin on autophagy in the blood vessels of *G. parasuis*-infected piglets. Beclin1 (**A**,**B**), LC3B (**C**,**D**), and p62 (**E**,**F**) expression levels were assessed using Western blot. BA: baicalin; $^{###}$ $p < 0.001$ versus control; * significance present at $p < 0.05$; *** significance present at $p < 0.001$. Original images can be found in Supplementary Materials.

3.7. Baicalin Inhibited NLRP3/Caspase 1 Signaling Pathway Activation in the Blood Vessels of *G. parasuis*-Infected Piglets

As displayed in Figure 7A,D, the mRNA expression levels of NLRP3 and Caspase 1 in the infection group were raised compared to the control group ($p < 0.001$). Compared to the infection group, levamisole and 25 to 100 mg/kg of baicalin attenuated NLRP3 and Caspase 1 mRNA expression levels ($p < 0.001$). At the protein levels, NLRP3 and Caspase 1 were raised from the infection group compared to the control group ($p < 0.01$), as shown in Figure 7B,C,E,F. Compared to the infection group, treatment with levamisole and 25 mg/kg–50 mg/kg of baicalin reduced the NLRP3 and Caspase 1 expression levels ($p < 0.05$).

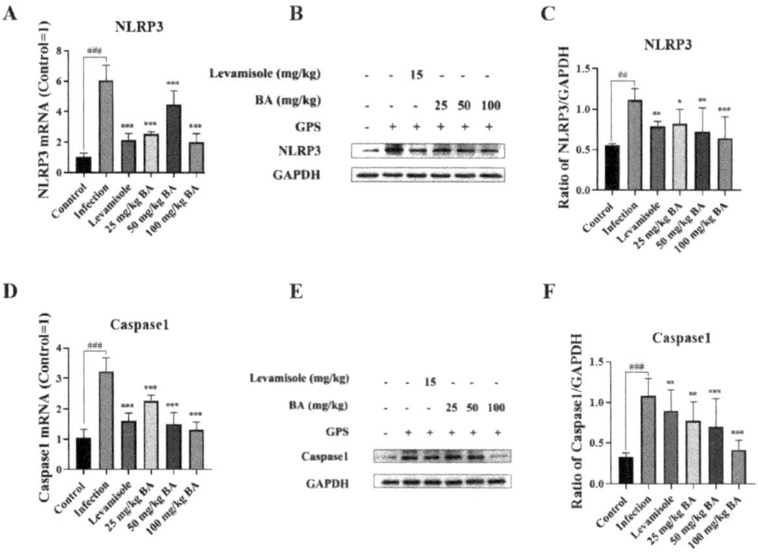

Figure 7. The function of baicalin on NLRP3/Caspase 1 signaling pathway activation in the blood vessels of *G. parasuis*-infected piglets. The NLRP3 and Caspase 1 expression levels were measured using RT-PCR method and Western blot. (**A**): NLRP3 mRNA level; (**B**,**C**): NLRP3 protein level; (**D**): Caspase 1 mRNA level; (**E**,**F**): Caspase 1 protein level; BA: baicalin; ## $p < 0.01$ versus control; ### $p < 0.001$ versus control; * significance present at $p < 0.05$; ** significance present at $p < 0.01$; *** significance present at $p < 0.001$. Original images can be found in Supplementary Materials.

3.8. Baicalin Decreased Cytokine Production and Diminished Pathological Tissue Damage in Blood Vessels of *G. parasuis*-Infected Piglets

The data showed that *G. parasuis* significantly induced the mRNA expression levels of IL-1β, IL-18, and TNF-α in the blood vessels ($p < 0.001$), as shown in Figure 8A–C. When the piglets were administered levamisole or baicalin, the mRNA expression levels of IL-1β, IL-18, and TNF-α in the blood vessels were weakened compared to the infection group ($p < 0.001$).

The alleviating effect of baicalin on blood vessel damage was also measured. The data indicated that in the infection group, inflammatory cell infiltration and hemorrhage were detected in the blood vessels (Figure 8E), while there was little injury displayed in the control group (Figure 8D). Only some minor damage was seen in the levamisole group or baicalin groups, as shown in Figure 8F–I.

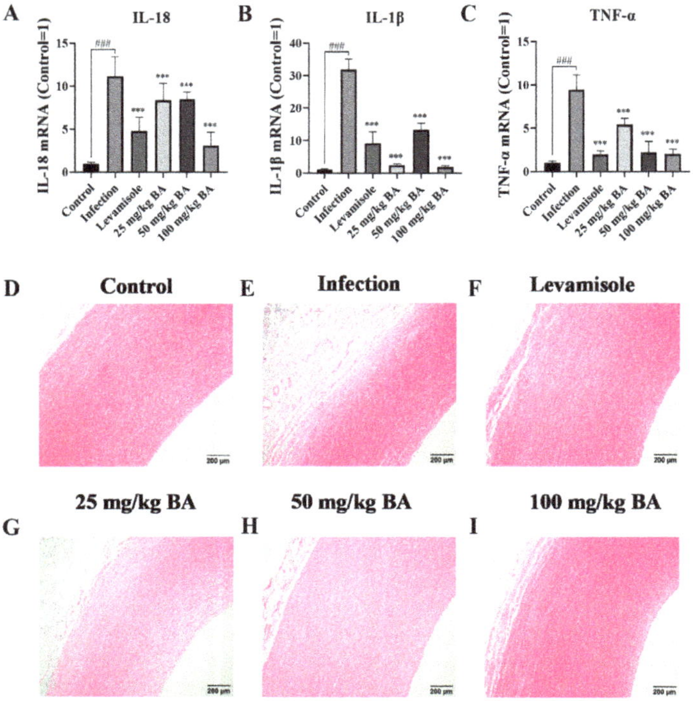

Figure 8. The effect of baicalin on cytokine production and pathological tissue injury in the blood vessels of *G. parasuis*-infected piglets. IL-18 (**A**), IL-1β (**B**), and TNF-α (**C**) expression levels were determined by RT-PCR. The pathological tissue damage assessment in blood vessels was carried out by histopathological analysis. (**D**): the control group; (**E**): the infection group; (**F**): the levamisole group; (**G**): 25 mg/kg of baicalin group; (**H**): 50 mg/kg of baicalin group; (**I**): 100 mg/kg of baicalin group; BA: baicalin; ### $p < 0.001$ versus control; *** significance present at $p < 0.001$.

4. Discussion

G. parasuis is a significant pathogen in the pig industry, causing Glässer's disease with significant clinical signs [32]. In this study, *G. parasuis* infection in piglets led to a significant decrease in the proportion of CD3+, CD3+CD4, and CD3+CD8+ T cells in the blood. This indicates that *G. parasuis* could elicit immunosuppression in the host, which has also been reported in previous work [27,29]. It has been reported that human immunodeficiency virus (HIV) primarily attacks CD4+ T cells, CD3+ T cells, and CD8+ T cells, resulting in a sharp decline in numbers and leading to immunosuppression [33]. The imbalance and abnormal function of CD4+ and CD8+ T cells have important functions in the regulation of the development of systemic lupus erythematosus [34]. In the tumor microenvironment, the distribution and function of CD3+, CD4+, and CD8+ T cells also change [35], and tumor cells might suppress T cell activity through immune checkpoint molecule production and immunosuppressive cytokine secretion [36]. It was therefore hypothesized that *G. parasuis* could directly affect the host's immune response's contribution to the progression of immunosuppression. The mechanism underlying immunosuppression remains unclear. However, it is crucial to understand for developing effective prevention and control strategies.

Our previous study showed that the PD-1/PD-L1 axis triggered immunosuppression via the PI3K/Akt/mTOR signaling pathway [29]. Whether there are any other molecules that can induce host immunosuppression needs to be explored. It was found that CD163 was significantly upregulated in the vessels of piglets infected by *G. parasuis* through vessel proteomics analysis. CD163 is a scavenger receptor predominantly expressed on macrophages [7]. CD163 acts as a decoy receptor for TWEAK [37], which is considered a multifunctional pro-inflammatory cytokine and is significantly expressed on macrophages [38]. The CD163/TWEAK axis is involved in various pathological processes of disease. The lipid droplet (LD)-laden macrophages (LLMs) were present with immunosuppressive phenotypes, accompanied by extensive expression of CD163, and reduced the antitumor activities of CD8+ T cells [39]. The CD163+ macrophages are thought to counteract tumor immunity by enhancing immunosuppressive mechanisms [40]. In this study, the CD163 expression level was increased, while the mRNA level of TWEAK was decreased in the infection group, suggesting that CD163 and TWEAK may be involved in *G. parasuis*-induced immunosuppression. Baicalin regulated the expression of the CD163/TWEAK axis, indicating that baicalin may alleviate *G. parasuis*-induced immunosuppression by modulating the CD163/TWEAK axis. Further studies are needed to elucidate the detailed mechanism of action of baicalin on the CD163/TWEAK axis, and other pig breeds or ages will also be used to study the function of baicalin on the regulation of CD163/TWEAK.

The CD163/TWEAK axis was reported to be one of the chief regulators of Notch signaling during the inflammation process [41]. The Notch signaling pathway was thought to be related to dextran sodium sulfate-induced colitis in mice [42]. Blocking the Notch signaling pathway reduced PRRSV infection both in vitro and in vivo [43], so the Notch signaling pathway can act as a potential novel therapeutic target for kidney and liver diseases [44,45]. The Wnt signaling has important roles in liver regeneration, tumorigenesis, and cardiovascular disease [46,47]; thus, regulation of Wnt signaling has become an attractive method to control liver disease and cardiovascular disease. However, the mechanism by which the CD163/TWEAK axis might regulate Notch and Wnt signaling pathways in response to *G. parasuis* infection remains largely unexplored. This study notably observed a significant regulation of key components of the Notch and Wnt signaling pathway. The use of levamisole and baicalin provided further insights into the modulation of these pathways. Levamisole, which is known to have immunomodulatory effects, significantly regulated the expression of key Notch and Wnt pathway components. These findings suggest that levamisole and baicalin might have therapeutic potential in modulating the Notch and Wnt signaling pathways in response to *G. parasuis* challenges.

Autophagy, a critical cellular process involving the degradation and recycling of cellular components, plays an important role in pathological conditions [48]. The crosstalk between autophagy and Wnt/Notch signaling controls cardiac differentiation [49]. Notch and Wnt signaling pathways can interact with autophagy and affect the survival and death of tumor cells [50]. Autophagy was involved in mouse kidney development and podocyte differentiation, which was regulated by Notch signaling [51]. Wnt signaling activation regulated autophagy due to a dietary magnesium deficiency in injury-induced osteoarthritis [52]. This study found that when piglets were infected with *G. parasuis*, the Notch and Wnt signaling pathways were altered, and baicalin could regulate Notch/Wnt signaling and autophagy, suggesting that Notch and Wnt signaling pathways may regulate autophagy in *G. parasuis* infection.

5. Conclusions

In conclusion, our study showed that baicalin regulated CD163/TWEAK axis expression, inhibited Notch/Wnt signaling pathways' activation, promoted autophagy, reduced NLRP3/Caspase 1 signaling pathway activation, decreased cytokine production, and alleviated pathological tissue damage in blood vessels. This study provides initial evidence that the CD163/TWEAK axis could induce host immunosuppression, and baicalin promotes autophagy in piglets challenged by *G. parasuis*. Understanding the role of autophagy in *G. parasuis* infection could offer a new view of the mechanism of disease and potential treatment targets to control *G. parasuis* infection in clinical settings. We will focus on exploring the relationship between autophagy and *G. parasuis*-induced immunosuppression in our future studies.

Supplementary Materials: The following supporting information can be downloaded at: https://www.mdpi.com/article/10.3390/biom15050722/s1, File S1: original western blot images.

Author Contributions: S.F. conceived and designed the experiments; S.F., R.L., J.L., Y.F., Q.D., S.L., and Y.S. performed the experiments; S.F., R.L., J.L., Y.F., L.G., J.H., and Y.Q. analyzed the data; S.F. authored the paper. All authors have read and agreed to the published version of the manuscript.

Funding: This work was supported by the Key Research and Development Plan of Hubei Province, China (2023BBB069), the National Natural Science Foundation of China (grant No. 32273067), and Hubei Provincial Natural Science Funding (2023AFB746).

Institutional Review Board Statement: The animal study was approved by the Animal Care and Use Committee of Wuhan Polytechnic University, Hubei Province, China (WPU202308003), approval date 2 August 2023.All piglets used in this study were euthanized at the end of the experiment.

Informed Consent Statement: Not applicable.

Data Availability Statement: The original contributions presented in this study are included in the article.

Conflicts of Interest: The authors declare no conflicts of interest.

References

1. Gong, X.; Cui, Q.; Zhang, W.; Shi, Y.; Zhang, P.; Zhang, C.; Hu, G.; Sahin, O.; Wang, L.; Shen, Z.; et al. Genomic insight into the diversity of *Glaesserella parasuis* isolates from 19 countries. *mSphere* **2024**, *9*, e0023124. [CrossRef]
2. González-Fernández, A.; Mencía-Ares, O.; García-Iglesias, M.J.; Petrocchi-Rilo, M.; Miguélez-Pérez, R.; Perelló-Jiménez, A.; Herencia-Lagunar, E.; Acebes-Fernández, V.; Gutiérrez-Martín, C.B.; Martínez-Martínez, S. TbpB-based oral mucosal vaccine provides heterologous protection against Glässer's disease caused by different serovars of Spanish field isolates of *Glaesserella parasuis*. *Porcine Health Manag.* **2024**, *10*, 57. [CrossRef] [PubMed]
3. Chen, Q.; Yu, B.; Su, F.; Ye, S.; Xu, L.; Yuan, X.; Wu, S.; Zhang, H.; Li, J. Ribosomal protein L32 contributes to the growth, antibiotic resistance and virulence of *Glaesserella parasuis*. *Front. Vet. Sci.* **2024**, *11*, 1361023. [CrossRef] [PubMed]
4. Wang, Z.; Zhao, Q.; Wei, H.; Wen, X.; Cao, S.; Huang, X.; Wu, R.; Yan, Q.; Huang, Y.; Wen, Y. Prevalence and seroepidemiology of *Haemophilus parasuis* in Sichuan province, China. *PeerJ* **2017**, *5*, e3379. [CrossRef] [PubMed]
5. Yan, P.; Jia, Y.C.; Zhang, X.L.; Zhou, Y.Y.; Guo, Y.; Yin, R.L.; Yuan, J.; Wang, L.X.; Guo, Z.B.; Wang, J.Y.; et al. Virulence assessment of four *Glaesserella parasuis* strains isolated in Liaoning province of China. *Res. Vet. Sci.* **2023**, *158*, 226–234. [CrossRef]
6. Siwan, E.; Wong, J.; Brooks, B.A.; Shinko, D.; Baker, C.J.; Deshpande, N.; McLennan, S.V.; Twigg, S.M.; Min, D. Deep Immune and RNA Profiling Revealed Distinct Circulating CD163+ Monocytes in Diabetes-Related Complications. *Int. J. Mol. Sci.* **2024**, *25*, 10094. [CrossRef]
7. Schleh, M.W.; Ameka, M.; Rodriguez, A.; Hasty, A.H. Deficiency of the hemoglobin-haptoglobin receptor, CD163, worsens insulin sensitivity in obese male mice. *Diabetes* **2024**, *73*, 1990–2002. [CrossRef]
8. Gobejishvili, L.; Vatsalya, V.; Avila, D.V.; Feygin, Y.B.; McClain, C.J.; Mokshagundam, S.; Barve, S. Association of Circulating Markers of Microbial Translocation and Hepatic Inflammation with Liver Injury in Patients with Type 2 Diabetes. *Biomedicines* **2024**, *12*, 1227. [CrossRef]
9. Zhu, Y.J.; Chen, H.L.; Huang, J.K.; Cai, X.J.; Zhan, B.L. TWEAK increases angiogenesis to promote diabetic skin wound healing by regulating Fn14/EGFR signaling. *J. Cosmet. Dermatol.* **2024**, *23*, 4230–4238. [CrossRef]
10. Akahori, H.; Karmali, V.; Polavarapu, R.; Lyle, A.N.; Weiss, D.; Shin, E.; Husain, A.; Naqvi, N.; Van Dam, R.; Habib, A.; et al. CD163 interacts with TWEAK to regulate tissue regeneration after ischaemic injury. *Nat. Commun.* **2015**, *6*, 7792. [CrossRef]
11. Kowal-Bielecka, O.; Bielecki, M.; Guiducci, S.; Trzcinska-Butkiewicz, B.; Michalska-Jakubus, M.; Matucci-Cerinic, M.; Brzosko, M.; Krasowska, D.; Chyczewski, L.; Kowal, K. High serum sCD163/sTWEAK ratio is associated with lower risk of digital ulcers but more severe skin disease in patients with systemic sclerosis. *Arthritis Res. Ther.* **2013**, *15*, R69. [CrossRef]

12. Gorvel, L.; Olive, D. Tumor associated macrophage in HPV(+) tumors: Between immunosuppression and inflammation. *Semin. Immunol.* **2023**, *65*, 101671. [PubMed]
13. Campbell, S.; Michaelson, J.; Burkly, L.; Putterman, C. The role of TWEAK/Fn14 in the pathogenesis of inflammation and systemic autoimmunity. *Front. Biosci.* **2004**, *9*, 2273–2284. [CrossRef]
14. Winkles, J.A. The TWEAK-Fn14 cytokine-receptor axis: Discovery, biology and therapeutic targeting. *Nat. Rev. Drug Discov.* **2008**, *7*, 411–425. [CrossRef] [PubMed]
15. Di Martino, L.; Dave, M.; Menghini, P.; Xin, W.; Arseneau, K.O.; Pizarro, T.T.; Cominelli, F. Protective Role for TWEAK/Fn14 in Regulating Acute Intestinal Inflammation and Colitis-Associated Tumorigenesis. *Cancer Res.* **2016**, *76*, 6533–6542. [CrossRef]
16. Shan, C.; Wang, Y.; Wang, Y. The Crosstalk between Autophagy and Nrf2 Signaling in Cancer: From Biology to Clinical Applications. *Int. J. Biol. Sci.* **2024**, *20*, 6181–6206. [CrossRef] [PubMed]
17. Zhang, D.; Lindstrom, A.; Kim, E.J.; Hwang, C.I.; Hall, M.L.; Lin, T.Y.; Li, Y. SEMA3C Supports Pancreatic Cancer Progression by Regulating the Autophagy Process and Tumor Immune Microenvironment. *Front. Oncol.* **2022**, *12*, 890154.
18. Carr, D.; Lau, R.; Hnatykiw, A.D.; Ward, G.C.D.; Daneshmand, M.; Cabrita, M.A.; Pratt, M.A.C. cIAP2 Is an Independent Signaling and Survival Factor during Mammary Lactational Involution and Tumorigenesis. *J. Mammary Gland. Biol. Neoplasia* **2018**, *23*, 109–123. [CrossRef]
19. Zhang, Q.; Guo, S.; Wang, H. The Protective Role of Baicalin in the Regulation of NLRP3 Inflammasome in Different Diseases. *Cell Biochem. Biophys.* **2024**. [CrossRef]
20. Wen, Y.; Wang, Y.; Zhao, C.; Zhao, B.; Wang, J. The Pharmacological Efficacy of Baicalin in Inflammatory Diseases. *Int. J. Mol. Sci.* **2023**, *24*, 9317.
21. Song, D.; Wei, W.; Zhang, J.; Zhang, L.; Huo, J.; Wang, W. The mechanism of baicalin in improving pulmonary inflammatory response and injury and regulating intestinal flora in Mycoplasma pneumoniae pneumonia mice. *Cell Signal* **2025**, *126*, 111530. [CrossRef] [PubMed]
22. Zhu, Z.; Yu, M.; Xu, M.; Ji, X.; Zong, X.; Zhang, Z.; Shang, W.; Zhang, L.; Fang, P. Baicalin suppresses macrophage JNK-mediated adipose tissue inflammation to mitigate insulin resistance in obesity. *J. Ethnopharmacol.* **2024**, *332*, 118355. [CrossRef] [PubMed]
23. Fu, W.; Liu, Z.; Wang, Y.; Li, X.; Yu, X.; Li, Y.; Yu, Z.; Qiu, Y.; Mei, Z.; Xu, L. Baicalin inhibits monosodium urate crystal-induced pyroptosis in renal tubular epithelial cell line through Panx-1/P2 $\times$ 7 pathways: Molecular docking, molecular dynamics, and in vitro experiments. *Chem. Biol. Drug Des.* **2024**, *103*, e14522. [CrossRef]
24. Yang, X.; Li, J.; Shan, C.; Song, X.; Yang, J.; Xu, H.; Ou, D. Baicalin reduced injury of and autophagy-related gene expression in RAW264.7 cells infected with H6N6 avian influenza virus. *Heliyon* **2024**, *10*, e32645. [CrossRef]
25. Dong, X.; Liu, X.; Lin, D.; Zhang, L.; Wu, Y.; Chang, Y.; Jin, M.; Huang, G. Baicalin induces cell death of non-small cell lung cancer cells via MCOLN3-mediated lysosomal dysfunction and autophagy blockage. *Phytomedicine* **2024**, *133*, 155872. [CrossRef]
26. Fu, S.; Liu, S.; Li, J.; Dong, Q.; Fu, Y.; Luo, R.; Sun, Y.; Tian, X.; Liu, W.; Zong, B.; et al. Baicalin and probenecid protect against *Glaesserella parasuis* challenge in a piglet model. *Vet. Res.* **2024**, *55*, 96. [CrossRef]
27. Fu, S.; Li, J.; You, J.; Liu, S.; Dong, Q.; Fu, Y.; Luo, R.; Sun, Y.; Tian, X.; Liu, W.; et al. Correction: Baicalin attenuates PD-1/PD-L1 axis-induced immunosuppression in piglets challenged with *Glaesserella parasuis* by inhibiting the PI3K/Akt/mTOR and RAS/MEK/ERK signalling pathways. *Vet. Res.* **2024**, *55*, 127. [CrossRef]
28. Huang, B.; Li, F.; You, D.; Deng, L.; Xu, T.; Lai, S.; Ai, Y.; Huang, J.; Zhou, Y.; Ge, L.; et al. Porcine reproductive and respiratory syndrome virus infects the reproductive system of male piglets and impairs development of the blood-testis barrier. *Virulence* **2024**, *15*, 2384564. [CrossRef]
29. Li, J.; Liu, S.; Dong, Q.; Fu, Y.; Sun, Y.; Luo, R.; Tian, X.; Guo, L.; Liu, W.; Qiu, Y.; et al. PD-1/PD-L1 axis induced host immunosuppression via PI3K/Akt/mTOR signalling pathway in piglets infected by *Glaesserella parasuis*. *BMC Vet. Res.* **2024**, *20*, 141. [CrossRef]
30. Fu, S.; Yin, R.; Zuo, S.; Liu, J.; Zhang, Y.; Guo, L.; Qiu, Y.; Ye, C.; Liu, Y.; Wu, Z.; et al. The effects of baicalin on piglets challenged with *Glaesserella parasuis*. *Vet. Res.* **2020**, *51*, 102. [CrossRef]
31. Hu, X.; Zheng, Y.; Fang, M.; Liang, Z.; Wen, C.; Lin, J.; Lin, Z.; Chen, S. Knockdown of the long noncoding RNA VSIG2-1:1 promotes the angiogenic ability of human pulmonary microvascular endothelial cells by activating the VEGF/PI3K/AKT pathway. *Respir. Res.* **2024**, *25*, 412. [CrossRef] [PubMed]
32. Lin, Y.; Zhu, N.; Liu, J.; Wen, S.; Xu, Y.; Xu, X.; Cai, X. The role of cytolethal distending toxin in *Glaesserella parasuis* JS0135 strain infection: Cytotoxicity, phagocytic resistance and pathogenicity. *Vet. Microbiol.* **2024**, *295*, 110168. [CrossRef]
33. Shenton, J.M.; Teranishi, M.; Abu-Asab, M.S.; Yager, J.A.; Uetrecht, J.P. Characterization of a potential animal model of an idiosyncratic drug reaction: Nevirapine-induced skin rash in the rat. *Chem. Res. Toxicol.* **2003**, *16*, 1078–1089. [CrossRef]
34. Yuan, S.; Zeng, Y.; Li, J.; Wang, C.; Li, W.; He, Z.; Ye, J.; Li, F.; Chen, Y.; Lin, X.; et al. Phenotypical changes and clinical significance of CD4(+)/CD8(+) T cells in SLE. *Lupus Sci. Med.* **2022**, *9*, e000660. [CrossRef]

35. Nagatani, Y.; Kiyota, N.; Imamura, Y.; Koyama, T.; Funakoshi, Y.; Komatsu, M.; Itoh, T.; Teshima, M.; Nibu, K.I.; Sakai, K.; et al. Different characteristics of the tumor immune microenvironment among subtypes of salivary gland cancer. *Asia Pac. J. Clin. Oncol.* **2024**, *20*, 779–788. [CrossRef]
36. Wu, Y.; Xiao, Y.; Ding, Y.; Ran, R.; Wei, K.; Tao, S.; Mao, H.; Wang, J.; Pang, S.; Shi, J.; et al. Colorectal cancer cell-derived exosomal miRNA-372-5p induces immune escape from colorectal cancer via PTEN/AKT/NF-κB/PD-L1 pathway. *Int. Immunopharmacol.* **2024**, *143*, 113261. [CrossRef]
37. Qian, J.K.; Ma, Y.; Huang, X.; Li, X.R.; Xu, Y.F.; Liu, Z.Y.; Gu, Y.; Shen, K.; Tian, L.J.; Wang, Y.T.; et al. The CD163/TWEAK/Fn14 axis: A potential therapeutic target for alleviating inflammatory bone loss. *J. Orthop. Translat* **2024**, *49*, 82–95. [CrossRef]
38. Liu, M.; Ren, Y.; Zhou, Z.; Yang, J.; Shi, X.; Cai, Y.; Arreola, A.X.; Luo, W.; Fung, K.M.; Xu, C.; et al. The crosstalk between macrophages and cancer cells potentiates pancreatic cancer cachexia. *Cancer Cell* **2024**, *42*, 885–903.e884. [CrossRef] [PubMed]
39. Wang, Y.; Chen, W.; Qiao, S.; Zou, H.; Yu, X.J.; Yang, Y.; Li, Z.; Wang, J.; Chen, M.S.; Xu, J.; et al. Lipid droplet accumulation mediates macrophage survival and Treg recruitment via the CCL20/CCR6 axis in human hepatocellular carcinoma. *Cell Mol. Immunol.* **2024**, *21*, 1120–1130. [CrossRef] [PubMed]
40. Davidsson, S.; Huotilainen, S.; Carlsson, J.; Sundqvist, P. Soluble Levels of CD163, PD-L1, and IL-10 in Renal Cell Carcinoma Patients. *Diagnostics* **2022**, *12*, 336. [CrossRef] [PubMed]
41. Ratajczak, W.; Atkinson, S.D.; Kelly, C. The TWEAK/Fn14/CD163 axis-implications for metabolic disease. *Rev. Endocr. Metab. Disord.* **2022**, *23*, 449–462. [CrossRef]
42. Xu, L.; Liu, W.; Huang, X.; Sun, T.; Mei, L.; Liu, M.; Ren, Z.; Wang, M.; Zheng, H.; Wang, Q.; et al. Sinomenine hydrochloride improves DSS-induced colitis in mice through inhibition of the Notch signaling pathway. *BMC Gastroenterol.* **2024**, *24*, 451. [CrossRef] [PubMed]
43. Ding, G.; Li, Y.; Li, D.; Dou, M.; Fu, C.; Chen, T.; Cui, X.; Zhang, Q.; Yang, P.; Hou, Y.; et al. SRCAP is involved in porcine reproductive and respiratory syndrome virus activated Notch signaling pathway. *J. Virol.* **2024**, *98*, e0121624. [CrossRef]
44. Adams, J.M.; Jafar-Nejad, H. The Roles of Notch Signaling in Liver Development and Disease. *Biomolecules* **2019**, *9*, 608. [CrossRef]
45. Mukherjee, M.; Fogarty, E.; Janga, M.; Surendran, K. Notch Signaling in Kidney Development, Maintenance, and Disease. *Biomolecules* **2019**, *9*, 692. [CrossRef] [PubMed]
46. Zou, G.; Park, J.I. Wnt signaling in liver regeneration, disease, and cancer. *Clin. Mol. Hepatol.* **2023**, *29*, 33–50. [CrossRef]
47. Akoumianakis, I.; Polkinghorne, M.; Antoniades, C. Non-canonical WNT signalling in cardiovascular disease: Mechanisms and therapeutic implications. *Nat. Rev. Cardiol.* **2022**, *19*, 783–797. [CrossRef]
48. Ren, S.Y.; Xu, X. Role of autophagy in metabolic syndrome-associated heart disease. *Biochim. Biophys. Acta* **2015**, *1852*, 225–231. [CrossRef]
49. Kolahdouzmohammadi, M.; Kolahdouz-Mohammadi, R.; Tabatabaei, S.A.; Franco, B.; Totonchi, M. Revisiting the Role of Autophagy in Cardiac Differentiation: A Comprehensive Review of Interplay with Other Signaling Pathways. *Genes* **2023**, *14*, 1328. [CrossRef]
50. Carballo, G.B.; Ribeiro, J.H.; Lopes, G.P.F.; Ferrer, V.P.; Dezonne, R.S.; Pereira, C.M.; Spohr, T. GANT-61 Induces Autophagy and Apoptosis in Glioblastoma Cells despite their heterogeneity. *Cell Mol. Neurobiol.* **2021**, *41*, 1227–1244. [CrossRef]
51. Zhang, C.; Li, W.; Wen, J.; Yang, Z. Autophagy is involved in mouse kidney development and podocyte differentiation regulated by Notch signalling. *J. Cell Mol. Med.* **2017**, *21*, 1315–1328. [CrossRef]
52. Bai, R.; Miao, M.Z.; Li, H.; Wang, Y.; Hou, R.; He, K.; Wu, X.; Jin, H.; Zeng, C.; Cui, Y.; et al. Increased Wnt/β-catenin signaling contributes to autophagy inhibition resulting from a dietary magnesium deficiency in injury-induced osteoarthritis. *Arthritis Res. Ther.* **2022**, *24*, 165. [CrossRef]

veterinary sciences

MDPI

Article

Multi-Omics Unveils Inflammatory Regulation of Fermented Sini Decoction Dregs in Broilers Infected with Avian Pathogenic *Escherichia coli*

Shuanghao Mo [1,2], Xin Fang [1,2], Wenxi Xiao [1,2], Bowen Huang [1,2], Chunsheng Li [1,2], Hui Yang [1,2], Yilin Wu [1], Yiming Wang [1,2,*] and Hongxia Ma [1,2,3,4]

1 College of Animal Science and Technology, College of Veterinary Medicine, Jilin Agricultural University, Changchun 130118, China; mosh0907@mails.jlau.edu.cn (S.M.); 18043281905@163.com (X.F.); huangbw1999@163.com (B.H.); jluyanghui163@163.com (H.Y.); ealin_1088@163.com (Y.W.); hongxia0731001@163.com (H.M.)
2 Jilin Provincial Key Laboratory of New Veterinary Drug R&D and Creation, Changchun 130118, China
3 College of Life Sciences, Jilin Agricultural University, Changchun 130118, China
4 The Engineering Research Center of Bioreactor and Drug Development, Ministry of Education, Jilin Agricultural University, Xincheng Street No. 2888, Changchun 130118, China
* Correspondence: wym27149@jlau.edu.cn; Tel.: +86-15943008568

Simple Summary: Avian colibacillosis, a common bacterial disease in poultry caused by avian pathogenic *Escherichia coli*, leads to significant economic losses in the agricultural industry and poses risks to human health through potential food safety issues. This study addressed the growing problem of antibiotic resistance, which complicates disease control, by seeking alternative strategies. Our goal was to develop a safe, effective treatment using fermented dregs from traditional Chinese medicine, processed with *Lactobacillus rhamnosus* strain, to combat the infection. Through analyzing the transcriptome data of the *GEO* database, we identified 11 key genes, including *TLR4*, which play a vital role in recognizing and fighting the bacteria by activating immune defenses. We also tested natural compounds, *Dioscin* and *Celastrol*, extracted from the fermented liquid, finding they reduced inflammation triggered by the infection. The fermented liquid itself proved more effective in live animals, likely due to the combined action of multiple beneficial components. This approach offers a sustainable method to manage avian colibacillosis, decrease reliance on antibiotics, and enhance poultry health. These advancements could support efforts to reduce the cost of poultry farming, improve animal welfare, and ensure society produces safer food.

Abstract: Avian colibacillosis causes significant economic losses and raises concerns for human health due to food safety risks, a problem exacerbated by the increase in antibiotic resistance. This study aimed to develop novel antibacterial strategies using fermented liquid of Sini decoction dregs to address these challenges. We analyzed the transcriptome of the chicken thymus sample *GSE69014* in the *GEO* database to identify immune-related genes, performed molecular docking to assess compound interactions, and experimental validation via Western blot and ELISA to evaluate anti-inflammatory effects. Results revealed 11 core genes, including *TLR4*, critical for immune responses against the infection, with *TLR4* activating key inflammatory pathways. Fermented liquid with probiotics enhanced bioactivity, and natural compounds *Dioscin* and *Celastrol* from the fermented liquid inhibited inflammation by targeting the *TLR4-MD2* complex. In animal models, fermented liquid outperformed individual compounds, likely due to synergistic effects, significantly reducing inflammatory markers. These findings demonstrate that fermented liquid of Sini decoction dregs offers a promising, sustainable approach to control avian colibacillosis,

 https://doi.org/10.3390/vetsci12050479

mitigate antibiotic resistance, and improve poultry health, providing a scientific foundation for its application in farming to reduce economic losses and enhance food safety.

Keywords: avian pathogenic *Escherichia coli* (APEC); fermented liquid of sini decoction dreg (FLSDD); anti-inflammatory

1. Introduction

Traditional Chinese medicine (TCM), developed over thousands of years, with a long history and remarkable pharmacological effects, has been extensively studied in the field of pharmacology and widely applied in various therapeutic approaches until now. TCM is generally categorized into single-herb and multi-herb formulas. Compared to single-herb formulations, multi-herb formulas exhibit higher efficacy and versatility in disease prevention and treatment [1,2].

The primary resources of TCM include three categories: herbal medicines, animal-derived medicines, and mineral medicines. Among these, herbal medicines account for 87% of the resources. However, in the production of TCM products, only an average of 5% of the active ingredients are extracted and utilized. Systematic studies on *Panax notoginseng* residues have shown that traditional TCM companies primarily extract single active ingredients, resulting in the discard of large amounts of organic nutrients, such as cellulose, hemicellulose, lignin, proteins, nucleic acids, and trace elements [3]. Moreover, some effective components are not fully extracted, leading to significant resource waste [4]. According to incomplete statistics, TCM manufacturers consume millions of tons of herbal materials annually, generating tens of millions of tons of medicinal wastewater and dregs [5].

To address this issue, researchers have implemented various effective strategies and methods to make full use of the medicinal dregs. These measures include using dregs as a substrate for edible mushrooms, converting dregs into organic fertilizers, and utilizing dregs as feed additives and fuel [6].

Avian colibacillosis, caused by specific serotypes of opportunistic pathogenic *Escherichia coli* [7], can occur throughout the year and affect chickens of all ages, to which chicks and broilers are particularly susceptible. This disease presents a range of typical symptoms, including acute septicemia, salpingitis, peritonitis, panophthalmitis, arthritis, omphalitis, coligranuloma, and early embryo mortality. The main routes of infection are the respiratory and digestive tracts [8]. Additionally, vertical transmission through eggs and feces-contaminated eggs is a key pathway for disease spread [9].

Probiotics are a group of active microorganisms that are beneficial to the host. They are introduced into the gastrointestinal and reproductive systems of humans and animals, where they produce active compounds proven to benefit health. These benefits include enhancing the hosts' microecological balance and promoting beneficial gut functions. Globally, research on probiotics has become a focus due to their ability not only to promote animal growth, regulate normal gut microbiome, maintain ecological balance and improve gastrointestinal function, but also enhance food digestibility and bioavailability, reduce serum cholesterol, control endotoxins, inhibit the growth of harmful bacteria in the gut, boost immunity and help prevent and treat various conditions including gastrointestinal syndromes and respiratory infections [10,11]. Among many probiotic strains, *Lactobacillus rhamnosus* is the best-known and most extensively studied one. It can adapt to the digestive environment, utilize prebiotics to stimulate growth, and play a key role in balancing and enhancing gut function, promoting the growth and activity of *Bifidobacterium* and *Lactobacillus acidophilus*, preventing diarrhea and reducing respiratory infections [12].

This study conducted an in-depth investigation and analysis of transcriptome data *GSE69014* from the Gene Expression Omnibus (*GEO*) database and utilized *Lactobacillus* derived from food sources to ferment the Sini decoction dregs (SDD), which were then fed to broilers. Additionally, Liquid Chromatography–Mass Spectrometry (LC–MS) was employed to identify the components of the fermented liquid of Sini decoction dregs (FLSDD), followed by bioinformatics analysis to determine its pharmacological activity. This approach aims to elucidate the key molecular mechanisms of avian pathogenic *Escherichia coli* infectious inflammation in broilers and explore potential feasible prevention and treatment strategies.

2. Materials and Methods

2.1. Transcriptome Sequencing Datasets

The transcriptome sequencing dataset *GSE69014* [13] was obtained from the *GEO* database. In the *GSE69014* dataset, male broilers were challenged with avian pathogenic *Escherichia coli* (APEC), and the thymuses were collected on days 1 and 5 post-infection. Based on autopsy scores of the liver, pericardium, and air sacs, the infected chickens were classified into mild and severe categories, representing resistant (R) and susceptible (S) phenotypes, respectively. Uninfected chickens were categorized as the healthy group (NC). All gene expression data measurements are available for download from the *GEO* database.

2.2. Differential Expression Genes (DEGs) Analysis

The dataset was read using R Studio (R 4.3.2) and preprocessed using the dplyr package. Subsequently, the ggplot2, tidyverse, and ggrepel packages were used to perform gene expression significance analysis and DEGs analysis on the *GSE69014* dataset, and a volcano plot was generated.

2.3. Weighted Gene Co-Expression Network Analysis (WGCNA)

Weighted Gene Co-expression Network Analysis (WGCNA) constructs a co-expression network from gene expression data and uses hierarchical clustering to group genes into functionally related modules, thereby revealing interactions between genes and their impact on biological traits. Using the WGCNA R package, highly co-expressed gene groups were identified in the *GSE69014* datasets, and genes were classified into different modules based on their expression patterns to elucidate potential biological pathways and pathogenic mechanisms of APEC infection in chickens. Furthermore, genes from modules highly correlated with external traits were evaluated and selected as candidate genes associated with WGCNA analysis results.

2.4. Protein–Protein Interaction (PPI) Network

For the key genes identified in the *GSE69014* dataset, the *STRING* database was used to predict and visualize the protein–protein interaction (PPI) network. The resulting network was then imported into Cytoscape v3.9.1, and the CytoHubba plugin was used to identify a set of hub genes as biomarkers for APEC infection.

2.5. CIBERSORT Immune Infiltration Analysis

In order to evaluate the composition and relative abundance of immune cells in each sample of *GSE69014*, we used the CIBERSORT (Cell-type Identification By Estimating Relative Subsets Of RNA Transcripts) algorithm to analyze the transcriptome data. CIBERSORT is a computational method based on support vector regression (SVR) that can use gene expression data to infer the relative proportions of 22 immune cell subsets in a sample. In order to explore the potential association between gene expression levels and immune cell infiltration, we used R to calculate the Spearman correlation coefficient between the expression levels of each gene and the proportion of immune cells inferred by CIBERSORT and used ggplot2 to draw stacked bar charts and correlation heat maps.

2.6. Gene Enrichment Analysis

Samples from the *GSE69014* dataset were divided into a healthy group and an APEC-infected group (susceptible(S) phenotypes). GSEA enrichment analysis was performed using R packages such as clusterProfiler and DESeq2, and key pathways were identified by examining the enrichment results. Key pathways related to APEC infection in broilers were identified by integrating bioinformatics databases (KEGG) and literature research.

2.7. Preparation of Sini Decoction Dregs Fermentation Liquid

Sini decoction (SD) is a compound traditional Chinese medicine formula composed of *Angelica sinensis, jujube, cinnamon, red peony root, Aristolochia debilis, licorice*, and *Asarum*. All medicinal herbs were provided by Beijing Ben Cao Fang Yuan (Bozhou) Pharmaceutical Technology Co., Ltd. (Bozhou, China), and were subjected to two rounds of water decoction to prepare the SDD before fermentation. The *Lactobacillus rhamnosus* strain used was isolated from kimchi and preserved at the China Center for Type Culture Collection (CCTCC), with strain number CCTCC M 2022083. First, 30 g of SDD, which had been subjected to two rounds of water decoction and dried, were mixed with 300 mL of distilled water and sterilized at 121 °C for 20 min, then cooled to room temperature to obtain the unfermented SDD dregs extract. Subsequently, under aseptic conditions, 1 mL of overnight-cultured (12 h) *L. rhamnosus* inoculum was added to the cooled dregs aqueous extract (AE) and incubated at 37 °C and 160 rpm for 12 h to produce the fermented liquid. After fermentation, the mixture was centrifuged at 10,000 rpm for 10 min to obtain the supernatant as the FLSDD, which was stored at −20 °C until use.

2.8. Metabolites Extraction

FLSDD samples were thawed at 4 °C and vortexed for 1 min. Then, an appropriate amount of the sample was accurately transferred to a 2 mL centrifuge tube, followed by the addition of 400 μL methanol, and vortexed for another minute. The mixture was then centrifuged at 12,000 rpm for 10 min at 4 °C, and the supernatant was transferred to a new 2 mL centrifuge tube and concentrated to dryness. The dried sample was reconstituted with 150 μL of 80% methanol solution containing 2-chloro-L-phenylalanine (4 ppm), and the reconstituted sample was filtered through a 0.22 μm membrane filter. The filtrate was collected into a vial for LC–MS analysis.

2.9. LC–MS/MS Analysis

Chromatographic conditions: A Thermo Scientific™ Vanquish™ Horizon UHPLC system (Thermo Fisher Scientific, Germering, Germany; address: Sommerstrasse 6, 82110 Germering, Germany; www.thermofisher.com, accessed on 13 May 2025) was used, equipped with an ACQUITY UPLC® HSS T3 column (2.1 × 100 mm, 1.8 μm) (Waters Corporation, Milford, MA, USA; address: 34 Maple Street, Milford, MA 01757, USA; www.waters.com, accessed on 13 May 2025). The column temperature was maintained at 40 °C, and the flow rate was set to 0.3 mL/min. The mobile phase consisted of acetonitrile with 0.1% formic acid and water. The gradient elution started with 8% acetonitrile, increased to 98% over 8 min, maintained at 98% until the 10th minute, then quickly dropped back to 8%, and continued until the 12th minute.

Mass spectrometry conditions: A Thermo Q Exactive Focus mass spectrometer was used, equipped with an electrospray ionization source. Data were collected in both positive and negative ion modes. The capillary temperature was set to 325 °C. In full scan mode, the m/z range was 100 to 1000, with a resolution of 70,000 and a collision energy of 30 eV.

2.10. Molecular Docking

The crystal structures of protein (*3VQ2*) were downloaded from the PDB database (https://www.rcsb.org/, accessed on 13 May 2025) and processed using AutoDockTools 1.5.6, which included removing ligands and water, adding hydrogen atoms, and saving the files in pdbqt format to be used as receptors for molecular docking. The selected ligand compounds (potential active components from FLSDD) were hydrogenated and converted into 3D coordinates using RDKit, and then converted to pdbqt format using Open Babel 2.4.1 to serve as ligands for molecular docking. Molecular docking was performed using AutoDock Vina, and the docking results were visualized using Pymol 2.5.7 and Discovery Studio 2024 Client.

2.11. Construction of Cell Inflammation Model

Chicken macrophages *HD11* were inoculated in 6-well plates at 1×10^6/well and cultured in a 37 °C incubator for 12 h. The control group received no treatment. AE, FLSDD, *Dioscin* (DIO) and *Celastrol* (CEL) were added to each experimental group. Except for the control group, the rest of the experimental groups were inoculated with APEC in the well plate at a multiplicity of infection MOI = 0.1 and cultured at 37 °C for 6 h. The specific treatment methods are shown in Table 1.

Table 1. Cell administration concentration and challenge dose.

ID	Group	Concentration	Challenge Dose
1	NC	—	—
2	AE	50 mg/mL	1×10^5/well
3	FLSDD	50 mg/mL	1×10^5/well
4	CEL	1 μg/mL	1×10^5/well
5	DIO	150 μg/mL	1×10^5/well
7	MG	—	1×10^5/well

All compounds were dissolved in DMSO, and the drug concentration in the FLSDD group was calculated based on the mass ratio of drug residue to water. In 100 mg/mL of FLSDD, the mass ratio of drug residue to water was 1:10. (m_{dregs}:m_{water} = 1:10).

2.12. Construction of Broiler Prevention Model

A total of 105 3-day-old Ross broilers were randomly divided into seven groups. The control group received no treatment. AE, FLSDD, DIO, and CEL were added to each experimental group. After 7 days of feeding, all treatment groups except the control group were challenged orally. The mental state of the birds was observed every 8 h, and after 24 h, blood samples were collected, and the birds were euthanized by cervical dislocation, followed by the collection of the duodenum. The detailed treatment methods are shown in Table 2.

Table 2. Chicken administration dose and challenge dose.

ID	Group	Treat	Dosage	Challenge Dose
1	NC	—	—	—
2	AE	Oral	100 mg/mL, 0.5 mL/day/bird	1×10^9 CFU
3	FLSDD	Oral	100 mg/mL, 0.5 mL/day/bird	1×10^9 CFU
4	CEL	Oral	0.1 mg/kg/day/bird	1×10^9 CFU
5	DIO	Oral	10 mg/kg/day/bird	1×10^9 CFU
7	MG	—	—	1×10^9 CFU

All compounds were dissolved in DMSO, and the drug concentration in the FLSDD group was calculated based on the mass ratio of drug residue to water. In 100 mg/mL of FLSDD, the mass ratio of drug residue to water was 1:10. (m_{dregs}:m_{water} = 1:10).

2.13. Total Protein Extraction and Western Blot

All experimental antibodies were purchased from Wuhan Proteintech Co., Ltd. (Proteintech Group, Wuhan, Hubei, China; address: 666 Gaoxin Avenue, East Lake High-tech Development Zone, Wuhan 430075, Hubei, China; www.ptgcn.com, accessed on 13 May 2025). Chicken macrophages *HD11* were washed with pre-chilled PBS, and 400 μL of RIPA lysis buffer was added. After lysis on ice for 10 min, the mixture was transferred to a 1.5 mL centrifuge tube and lysed on ice for an additional 30 min. It was then centrifuged at 4 °C and 12,000 r/min for 15 min, and the supernatant was collected. Chicken duodenal tissue was quick-frozen in liquid nitrogen, weighed, and thoroughly ground. For every 100 mg of solid tissue, 1 mL of pre-chilled lysis buffer (RIPA lysis buffer + PMSF) was added. The mixture was lysed on ice for 30 min, and the suspension was centrifuged at 12,000 r/min and 4 °C for 15 min, after which the supernatant was collected. The protein concentrations of all samples were determined using a BCA protein quantification kit and normalized. Based on the size of the target protein, SDS-PAGE gels were prepared. Protein samples and markers were sequentially loaded into the wells, electrophoresed at 80 V for 20 min, then switched to 120 V until the loading buffer migrated to the bottom. The proteins were transferred to a PVDF membrane at 200 mA for 60 min and blocked with 5% skim milk powder for 2 h. Primary antibodies were incubated at 4 °C for 12 h, followed by secondary antibodies for 2 h. Excess antibodies were washed off with TBST, and ECL luminescent solution was used for development in a protein imaging system, *β-actin* was used as an internal reference, and ImageJ (ImageJ bundled with 64-bit Java 8) was employed for grayscale analysis to quantify the expression levels of the target proteins.

2.14. Enzyme-Linked Immunosorbent Assay (ELISA)

The ELISA kit was purchased from Shanghai Enzyme-linked Biotechnology Co., Ltd. (Shanghai, China; address: 5500 Yuanjiang Road, Minhang District, Shanghai 201111, China; en.mlbio.cn, accessed on 13 May 2025). Blood samples were collected from the wing vein of broilers in each group to obtain serum, and the levels of *IL-1β*, *IL-4*, *IL-6*, *IL-8*, *IL-10*, *IL-12*, *TNF-α*, and *iNOS* were measured according to the manufacturer's instructions.

3. Results

3.1. Transcriptome Analysis Results

3.1.1. Identification of DEGs

After processing the transcriptome sequencing dataset GSE69014, genes with $|Log_2FC| \geq 0.6$ and $p \leq 0.05$ were selected as DEGs. As shown in Figure 1, A total of 1034 DEGs were identified in the GSE69014 dataset (943 upregulated, 91 downregulated). Compared to the resistant group (R), the susceptible phenotype (S) broilers showed more DEGs post-infection.

3.1.2. WGCNA and Identification of Key Modules

Data cleaning was performed on the 24 transcriptome samples from GSE69014: first, clustering analysis was conducted to exclude outlier samples (height $\geq$ 100,000) and genes with low expression and poor quality (fragments per kilobase of transcript per million mapped reads $\leq$ 100), Figure 2a. After identifying abnormal values or outliers, hierarchical clustering was conducted on the remaining genes in GSE69014. Based on the scale independence (Figure 2b) and mean connectivity (Figure 2c) of the network, a soft threshold of 20 was selected to ensure a scale-free topology. Dynamic tree cutting was used to identify gene modules, and dendrograms and module colors were visualized, merging highly correlated modules; see Figure 2d. By correlating with external traits, the relationship between module eigengenes and external sample traits was calculated, and a heatmap was plotted. As shown in Figure 2e, two gene modules related to the susceptible phenotype (S) were identified in GSE69014: *MEblack* and *MEbrown*. The 2129 genes in these modules were considered potential genes associated with APEC infection.

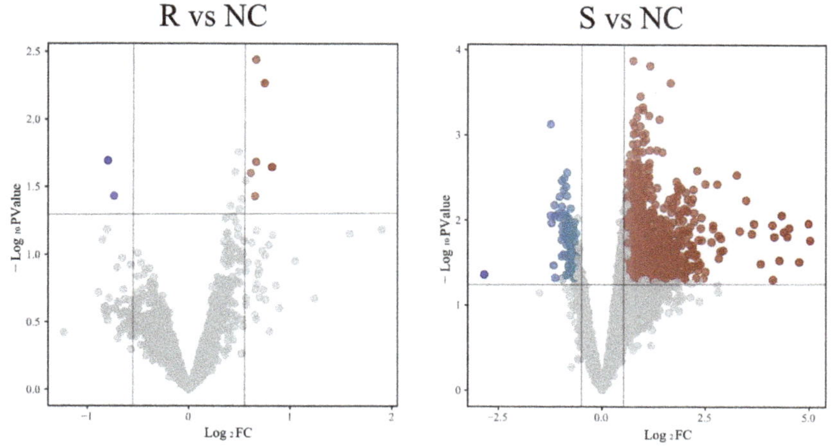

Figure 1. Volcano plot of DEGs group comparison. R represents the resistant phenotype, S represents the susceptible phenotype, and the healthy group is denoted by NC. Log_2 Fold Change (Log_2FC) represents the logarithm to base 2 of the ratio of gene expression levels under two conditions, indicating the magnitude and direction of differential gene expression. Red dots represent significantly up-regulated genes, and blue dots represent significantly down-regulated genes ($|Log_2FC| \geq 0.6$ and $p \leq 0.05$). Gray dots represent non-significantly changed genes ($|Log_2FC| < 0.6$ or $p > 0.05$).

3.1.3. Construction of PPI Network

A comprehensive analysis of the DEGs and WGCNA results from dataset GSE69014 identified 665 key genes associated with severe APEC infection in broilers. A PPI network was constructed using *STRING* and imported into Cytoscape, where *TLR4* was found at the core of the PPI network, suggesting its significant biological importance; see Figure 3.

3.1.4. GSEA Results

The GSEA enrichment analysis of dataset GSE69014 revealed that 23 pathways were significantly upregulated and 25 pathways were significantly downregulated (|Normalized Enrichment Score, NES| ≥ 1.5, $p \leq 0.05$), Figure 4a. In addition, single-gene enrichment analysis of *TLR4* revealed that among the six signaling pathways it is involved in, two were significantly activated. These are the Phagosome signaling pathway (NES = 1.86, $p < 0.0001$) and the Toll-like receptor signaling pathway (NES = 1.57, $p < 0.02$), Figure 4b.

3.1.5. CIBERSORT Immune Infiltration Results

The immune infiltration profiles of 24 thymic samples from the GSE69014 dataset were simulated using the CIBERSORT algorithm. The proportions of immune cells in each sample and their intergroup differences are presented in Figure 5a,c. Although the differences between groups for most immune cells were not statistically significant, correlation analysis between these cell proportions and core gene expression levels revealed significant associations for certain immune cell types. Notably, apart from thymic cells, *M1* and *M2 macrophage* subtypes exhibited significant positive correlations with most core genes, with *TLR4* showing a Pearson correlation coefficient of 0.52 ($p < 0.01$) with both subtypes, Figure 5b,d.

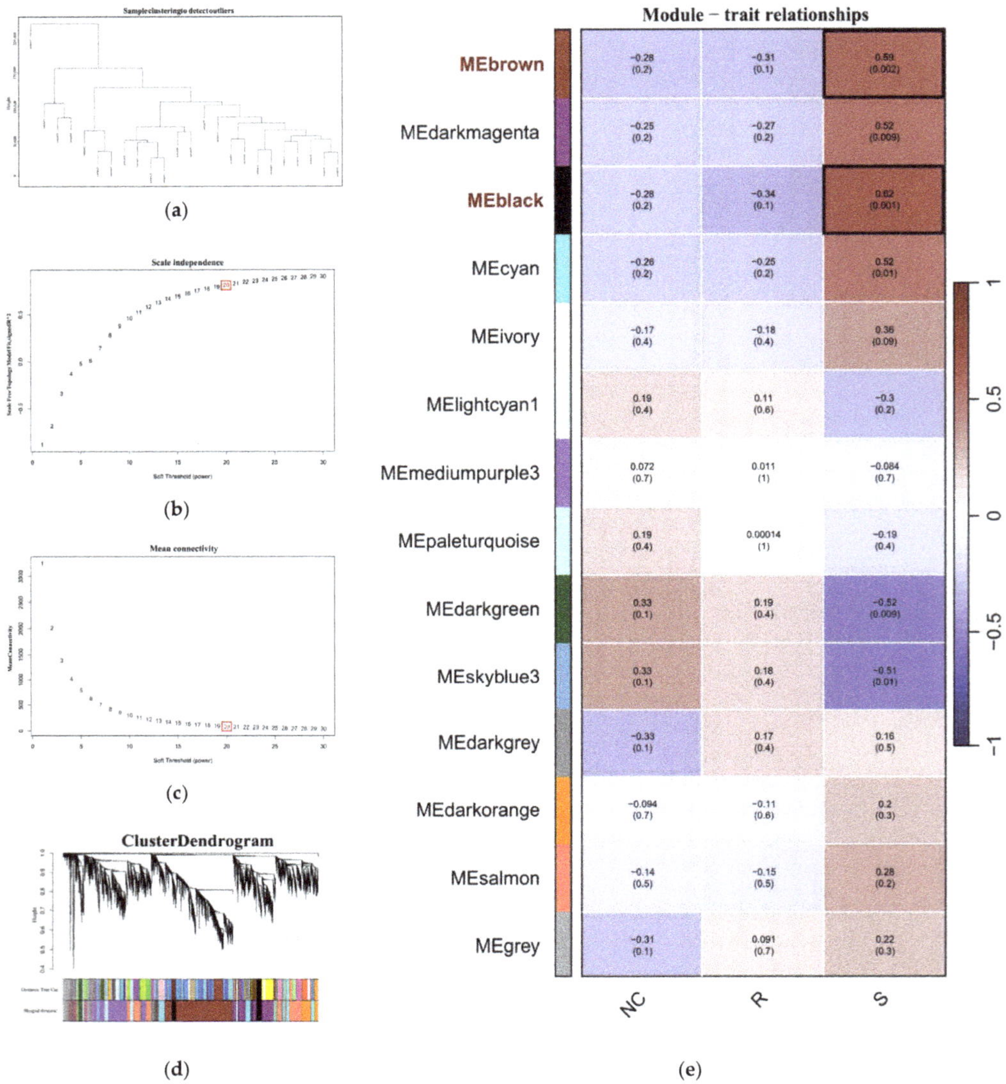

Figure 2. (**a**) The sample dendrogram, showing the similarity between samples through the dendrogram; (**b**,**c**) the soft-thresholding selection plot, which illustrates the network's topological properties under different soft-thresholding powers (soft threshold power, defined as the exponent β applied to the correlation matrix to enhance strong gene correlations and construct a scale-free network). (**b**) shows the scale independence, indicating how well the network fits a scale-free topology (R^2), while (**c**) displays the mean connectivity index of the scale-free topology model. By selecting an appropriate soft-thresholding power (β = 20 in this analysis), the network can maintain scale-free characteristics and preserve significant correlations between genes. (**d**) The cluster dendrogram, which displays the clustering of genes, with genes assigned to different modules based on their expression similarities. Different colors at the bottom represent gene modules identified through the dynamic tree cut method, where each color indicates a group of genes that tend to behave similarly. The dendrogram above shows the hierarchical clustering process, with the height on the y-axis reflecting the degree of dissimilarity between genes—taller branches indicate greater differences. The merging of color blocks at the bottom shows how modules are grouped or related to each other based on

their gene expression patterns. (**e**) The module-trait relationships heatmap, which displays the correlation between different gene modules and traits. Each module is represented with a correlation coefficient to the traits, indicated by both color and numerical values. Red represents positive correlation, green represents negative correlation, and the deeper the color, the stronger the correlation. R represents the resistant phenotype, S represents the susceptible phenotype, and the healthy group is denoted by NC.

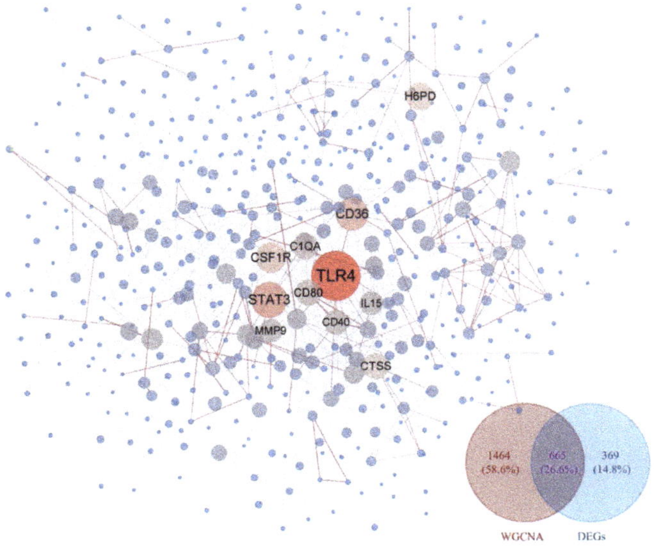

Figure 3. PPI network and Venn diagram. By comprehensively screening genes from the MEbrown and MEblack modules in WGCNA along with DEGs, a protein–protein interaction network was constructed using the 665 genes shared between them. In this network, larger dark red nodes indicate a higher degree of connectivity within the network, while the redder the edges between nodes, the more reliable the interaction between them (based on the combined score from the STRING database). Conversely, smaller nodes suggest poorer connectivity of that node within the network, and the bluer the edges between nodes, the less reliable their interactions are.

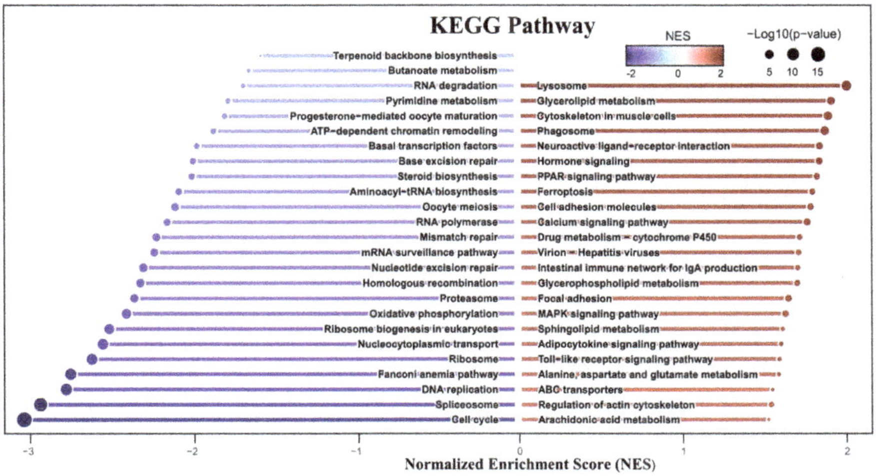

(**a**)

Figure 4. *Cont.*

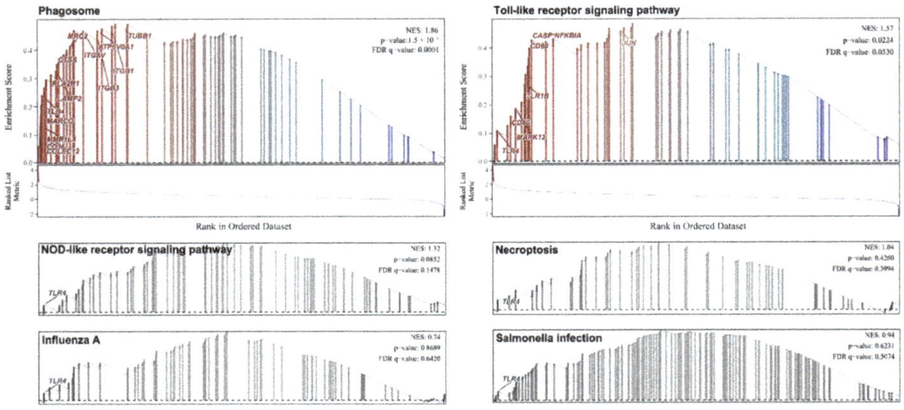

(**b**)

Figure 4. (**a**) Bidirectional column chart. Displays 48 signaling pathways from the GSEA enrichment analysis based on the KEGG database, with those significantly activated (red) and suppressed (blue) identified (| Normalized Enrichment Score, NES | $\geq 1.5, p \leq 0.05$); (**b**) Single-gene enrichment analysis result. GSEA results of the six signaling pathways involved in the *TLR4*, of which the gray ones are insignificant pathways ($p \geq 0.05$).

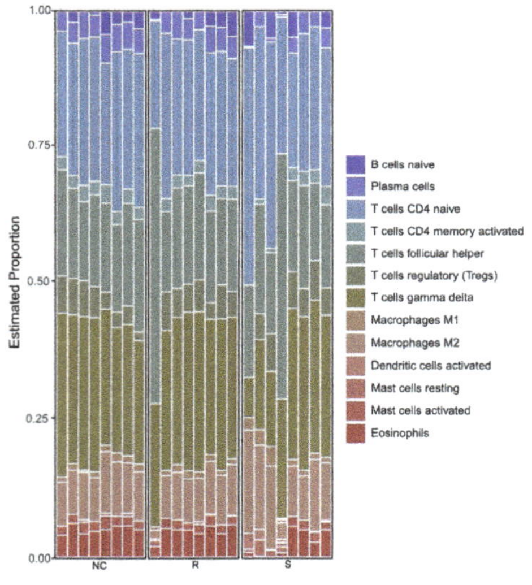

(**a**)

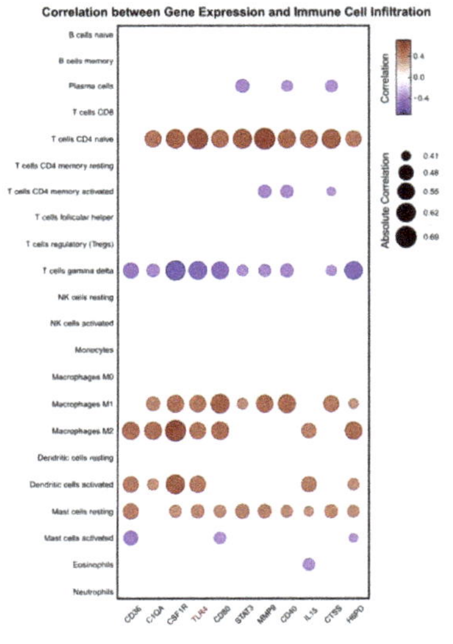

(**b**)

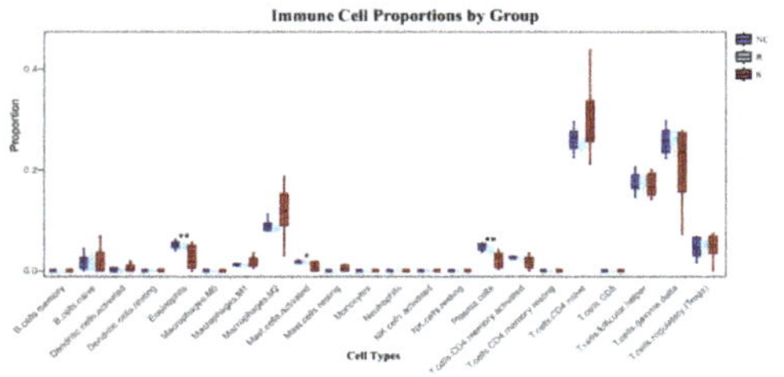

(**c**)

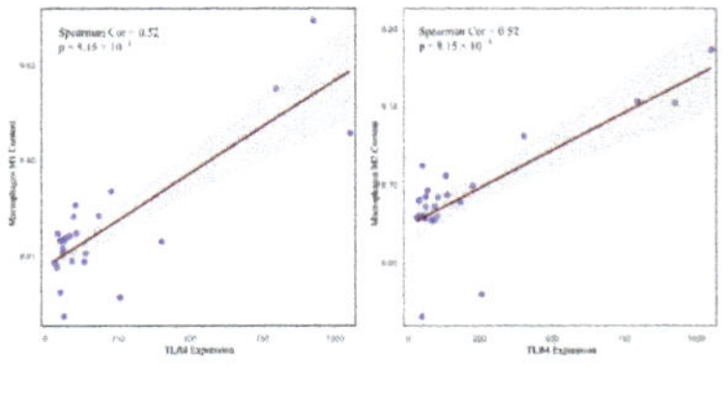

(**d**)

Figure 5. (**a**) The immune cell stacking chart displays the relative abundance of different types of immune cells in each thymus sample. R represents the resistant phenotype, S represents the susceptible phenotype, and the healthy group is denoted by NC. (**b**) The correlation heatmap analyzes the relationships between 11 core genes in the PPI network (Figure 3) and 22 types of immune cells using the Spearman correlation coefficient. The heatmap displays the absolute values of the correlation coefficients ($|Cor|$), highlighting only the significant correlations between genes and immune cells where $|Cor| \geq 0.4$ and $p \leq 0.05$; (**c**) shows the differences in the proportion of each group of immune cells. R represents the resistant phenotype, S represents the susceptible phenotype, and the healthy group is denoted by NC. * $p < 0.05$ and ** $p < 0.01$ indicate a statistically significant difference from the NC group. (**d**) Scatter plots showing the association of *TLR4* gene expression with M1 and M2 macrophage subsets (calculated using Spearman correlation).

3.1.6. Components of FLSDD

LC–MS identified a total of 7527 natural compounds and metabolites in the FLSDD. By querying *DrugBank*, *TCMSP*, and *HERB* databases, a total of eight compounds targeting *TLR4* were obtained; see Figure 6 and Table 3.

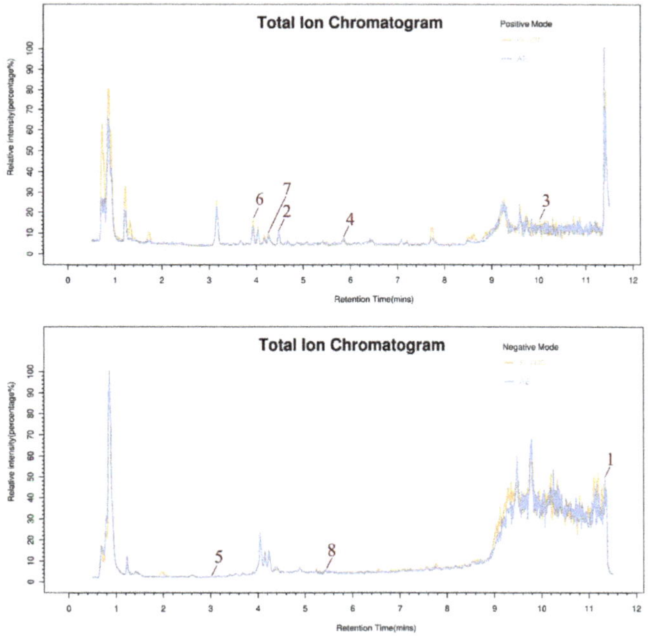

Figure 6. The total ion chromatograms of FLSDD and AE. By matching the results with a database, a total of eight natural compounds targeting TLR4 were identified in LC–MS positive and negative ion modes. As indicated by the numbers in the figure, detailed information of the compounds is shown in Table 3.

Table 3. List of natural compounds targeting *TLR4* in FLSDD and AE.

ID	Compounds Name	CAS	Structures	References
1	*Resveratrol*	501-36-0	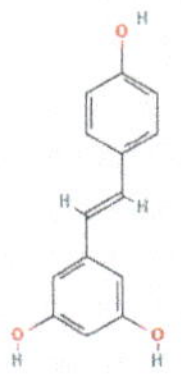	[14,15]

Table 3. *Cont.*

ID	Compounds Name	CAS	Structures	References
2	*Oleuropein*	32619-42-4	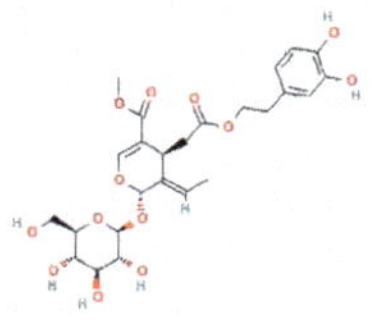	[16,17]
3	*Cinnamaldehyde*	104-55-2	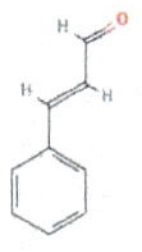	[18]
4	*Dioscin*	19057-60-4	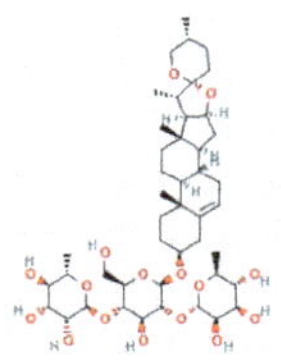	[19,20]
5	*Fisetin*	528-48-3	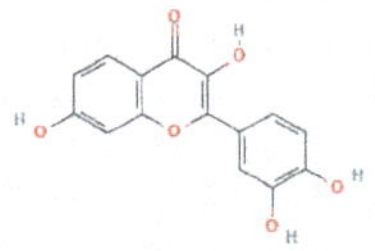	[21]
6	*Puerarin*	3681-99-0	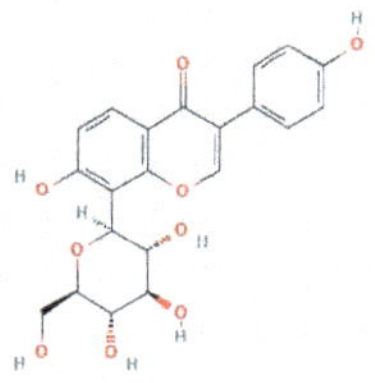	[22]
7	*Saikosaponin A*	20736-09-8	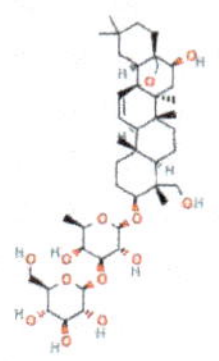	[23]
8	*Celastrol*	34157-83-0	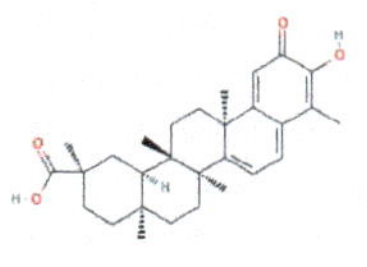	[24]

3.1.7. Molecular Docking Results

The molecular docking results of *TLR4* and eight components in the FLSDD are shown in Figure 7a. Among them, two compounds (*Dioscin* and *Celastrol*) exhibited high affinity for the *TLR4*-MD2 complex, with binding energies of −10.8 kcal/mol and −10.3 kcal/mol, respectively. Both compounds were able to occupy the hydrophobic pocket in *MD2* that binds to lipid A of lipopolysaccharide (LPS); see Figure 7b.

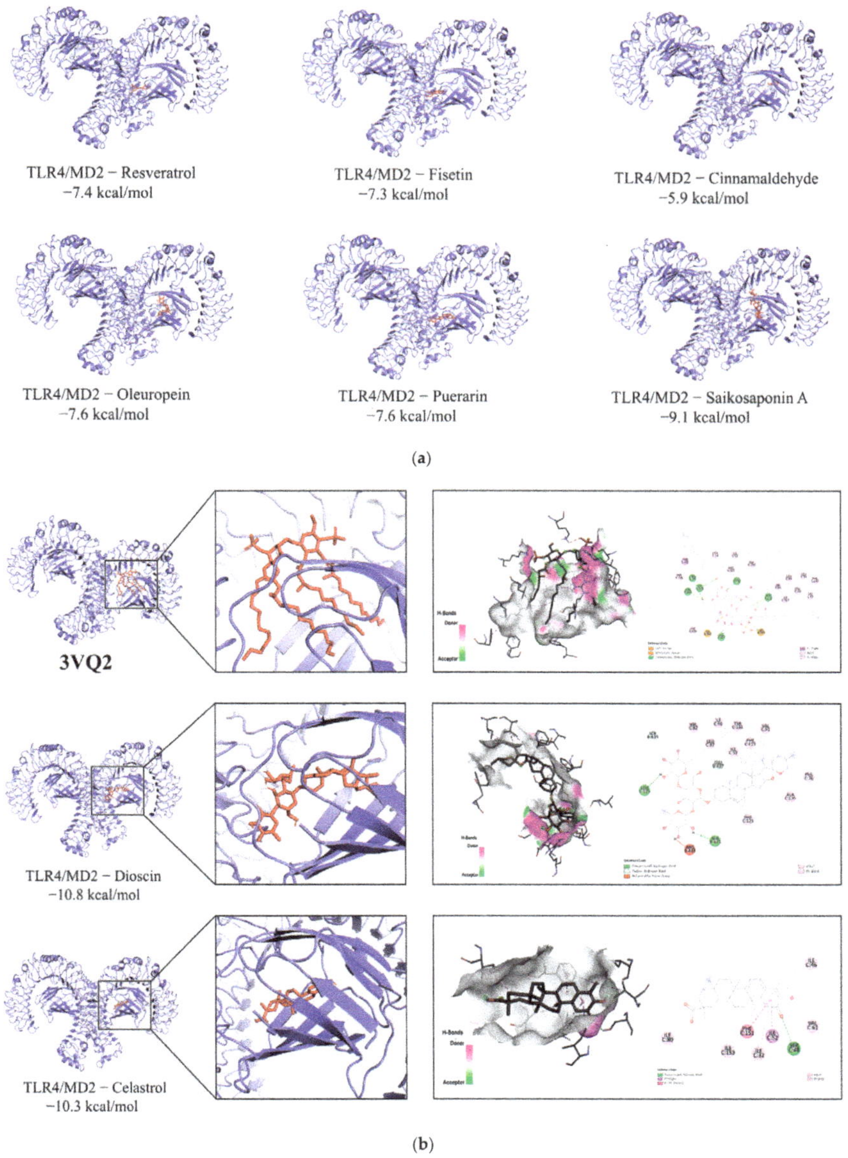

Figure 7. (**a**) The molecular docking results. The blue part is the protein receptor and the red part is the ligand. Show the binding modes of the *TLR4-MD2* complex dimer and six components. (**b**) Binding interactions of LPS and natural compounds with the *TLR4-MD2* complex dimer. The PDB entry *3VQ2* demonstrates the binding interaction between LPS and the *TLR4-MD2* complex dimer. Additionally, the binding modes of two natural compounds (*Dioscin* and *Celastrol*) with the same protein receptor from *3VQ2* are shown alongside it.

3.2. *Cell and Animal Experiments Results*

3.2.1. Western Blot

Western blot analysis was performed to assess key proteins in the *Toll-like receptor signaling pathway* and its downstream *NF-κB* and *MAPK* cascades in chicken macrophages *HD11*(Figure 8a) and duodenum(Figure 8b) across treatment groups. Results revealed that in the APEC infection group (IG), *TLR4* and *MyD88* levels were markedly elevated compared to the control group (NC) ($p \leq 0.01$). APEC infection also significantly increased the phosphorylation of *p38, JUN, IκBα,* and *p65,* with a consistent rise observed at both cellular and organismal levels. In cell models, *Dioscin* (DIO) and *Celastrol* (CEL) treatments showed slightly better inhibition of these proteins' elevated expression and phosphorylation compared to FLSDD and dregs aqueous extract (AE) groups. However, in animal models, their preventive effects were generally outperformed by the FLSDD treatment group.

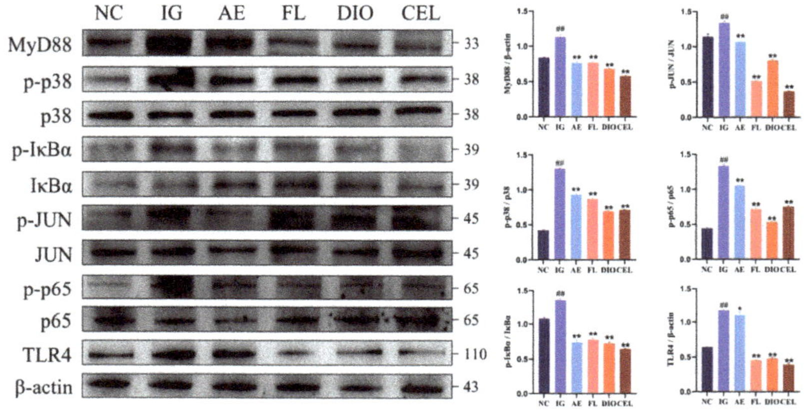

(**a**)

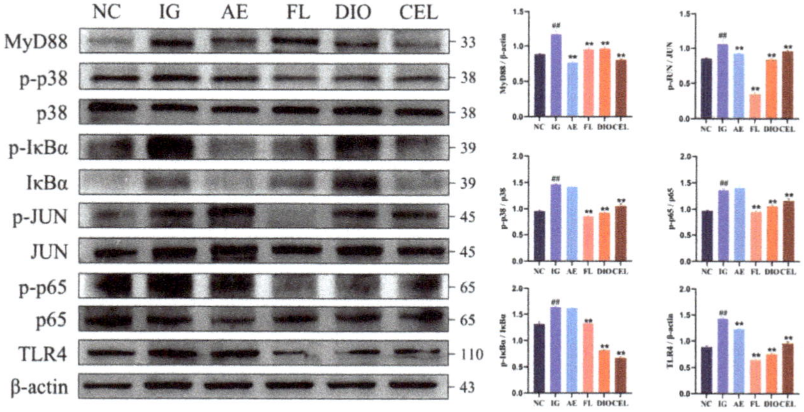

(**b**)

Figure 8. NC denotes the control group, IG denotes the APEC infection group, AE denotes the dregs aqueous extract treatment group, FL denotes the fermented liquid treatment group, and DIO and CEL denote the Dioscin and Celstrol treatment groups, respectively. (**a**) Protein expression levels in chicken macrophages *HD11*. (**b**) Protein expression levels in the chicken duodenum. ## $p < 0.01$ indicate a statistically significant difference from the NC group; * $p < 0.05$ and ** $p < 0.01$ indicate a statistically significant difference from the IG group. WB original figures see Supplementary Materials.

3.2.2. ELISA

The ELISA quantitative detection results of cytokines showed that, compared to the NC, all detected cytokine levels in the IG were significantly elevated, indicating that APEC infection induced a strong inflammatory response. This inflammatory response was effectively suppressed by AE. FLSDD, DIO, and CEL, as representative active components in FLSDD, showed some inhibitory effects on certain inflammatory cytokines, but their effects were not as strong as FLSDD; see Figure 9.

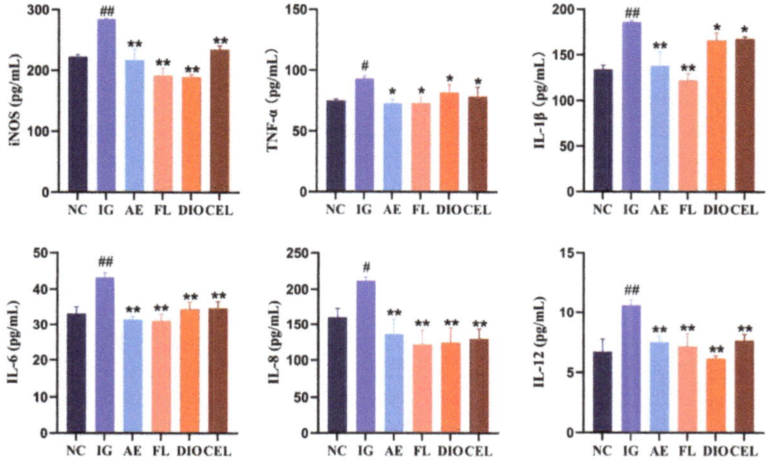

Figure 9. ELISA detection of serum inflammatory factor levels in each treatment group, # $p < 0.05$ and ## $p < 0.01$ indicate a statistically significant difference from the NC group; * $p < 0.05$ and ** $p < 0.01$ indicate a statistically significant difference from the IG group; NC denotes the control group, IG denotes the APEC infection group, AE denotes the dregs aqueous extract treatment group, FL denotes the fermented liquid treatment group, and DIO and CEL denote the Dioscin and Celstrol treatment groups, respectively.

4. Discussion

Avian colibacillosis is one of the most common bacterial diseases in livestock and poultry, causing substantial economic losses to the farming industry and posing potential threats to human health and food safety [25]. With the expansion of farming operations and environmental impacts, the incidence of avian colibacillosis has been increasing [26]. Current treatment mainly relies on antimicrobial drugs, but long-term misuse has led to an increase in resistant strains, making disease prevention and control more challenging [27]. Therefore, there is an urgent need to find new antibacterial agents and anti-infection strategies to address resistance issues and effectively control the disease. According to incomplete statistics, TCM manufacturers consume millions of tons of plant medicinal materials annually, generating tens of millions of tons of medicinal dregs [28]. TCM has various pharmacological activities, and fermentation of medicinal dregs using probiotics such as *Lactobacillus* can achieve comprehensive utilization of TCM waste while enhancing its bioactivity [29]. This process combines the advantages of TCM, enzymes, probiotics, and secondary metabolites, providing a safe and efficient method that facilitates animal absorption. It holds significant practical implications for preventing and controlling livestock and poultry diseases like avian colibacillosis, reducing resistance, and improving production performance [30].

By mining the transcriptomic data of 24 chicken thymus samples from GSE69014, we identified 11 core genes, including *TLR4* (*CD36, C1QA, CSF1R, TLR4, CD80, STAT3,*

MMP9, CD40, IL15, CTSS, H6PD), which collectively regulate immune and inflammatory responses during APEC infection. Among them, *CD36* and *TLR4*, as pattern recognition receptors, recognize bacterial LPS to initiate innate immune responses [31,32]; *C1QA* promotes bacterial clearance by activating the complement system [33]; *CSF1R* regulates macrophage differentiation and function, enhancing phagocytosis [34]; *CD80* and *CD40* play critical roles in antigen presentation and T-cell activation, facilitating adaptive immune responses [35]; *STAT3* and *IL15* maintain immune balance by regulating cytokine signaling and immune cell proliferation [36]; *MMP9* promotes immune cell migration by degrading the extracellular matrix [37]; *CTSS* participates in antigen processing and MHC-II molecule presentation [38]; and *H6PD* may indirectly influence immune responses by modulating oxidative stress [39]. Together, these genes orchestrate the immune defense mechanisms against APEC infection.

The *TLR4* gene encodes Toll-like receptor 4, a critical pattern recognition receptor that recognizes LPS from Gram-negative bacteria [40], playing a central role in phagosome signaling and Toll-like receptor signaling pathways. GSEA enrichment analysis revealed that in the susceptible (S) phenotype, the Toll-like receptor signaling pathway involving *TLR4*, along with *NF-κB* and *MAPK* cascades, exhibited significant activation. Located on the phagosome membrane, *TLR4* senses LPS and activates the *Toll-like receptor signaling pathway* via the adaptor protein *MyD88*, while also initiating signals within the phagosome to promote bacterial phagocytosis and degradation [41]. Downstream of the *Toll-like receptor signaling pathway*, *TLR4* activates *NF-κB* and *MAPK* cascades: *NF-κB* induces the expression of pro-inflammatory cytokines (e.g., *TNF-α*, *IL-6*) through *IκB* phosphorylation and nuclear translocation, thereby enhancing the inflammatory response; the *MAPK* pathway (e.g., *p38*) regulates the activity of the *AP-1* (*JUN*) transcription factor, further promoting the expression of inflammation- and immune-related genes [42]. These cascades collectively amplify immune signals, recruiting and activating immune cells such as macrophages and neutrophils to effectively clear bacterial infections. Additionally, immune infiltration analysis showed a positive correlation between *TLR4* expression levels and the proportions of M1 and M2 macrophage subtypes, suggesting that *TLR4* may play a significant role in macrophage polarization: by recognizing LPS from Gram-negative bacteria, *TLR4* activates signaling pathways that promote the polarization of pro-inflammatory M1 macrophages to enhance antibacterial phagocytosis and inflammation [43], while potentially inducing M2 macrophage generation through negative feedback or crosstalk signals, thereby facilitating tissue repair and inflammation resolution [44].

MD2 is a small glycoprotein closely associated with *TLR4*, featuring a hydrophobic pocket on its surface specifically designed to accommodate the fatty acid chains of lipid A in LPS [45]. In the *3VQ2* structure, the hydrophobic fatty acid chains of lipid A insert into *MD2*'s hydrophobic pocket through van der Waals forces and hydrophobic interactions, forming tight Alkyl and Pi–Alkyl interactions with hydrophobic amino acids such as PHE (C:119, C:121, C:151, etc.), LEU (C:54, C:94, etc.), ILE (C:52, C:124, etc.), and VAL (C:63, C:61, etc.). Meanwhile, the phosphate groups of lipid A interact with the positively charged LYS (A:263) via ionic interactions and salt bridges, enhancing binding stability, while its hydroxyl and oxygen atoms form hydrogen bonds with SER (C:120) and GLU (C:122), further stabilizing the complex. This binding mode induces a conformational change in *MD2*, promoting *TLR4* dimerization and thereby initiating downstream immune signaling [46]. Among the natural products identified in FLSDD, DIO possesses multiple cyclic structures and hydrophobic side chains, resembling the hydrophobic fatty acid chains of lipid A. Molecular docking results show that DIO competitively inserts into *MD2*'s hydrophobic pocket, forming Alkyl or Pi–Alkyl interactions with PHE (C:119, C:121), LEU (C:87), and others, while its hydroxyl and polar groups form hydrogen bonds with SER (B:413, B:439), GLU (B:437), and others, occupying key binding sites on *MD2*. This suggests that DIO may interfere with LPS binding to the *TLR4-MD2* complex through competitive binding or steric hindrance. Similarly, CEL, which also contains multiple cyclic frameworks and hydrophobic side chains, can insert into *MD2*'s hydrophobic pocket, forming Alkyl and Pi–Alkyl interactions with PHE (C:151), ILE (C:52, C:46), and others, while its hydroxyl or polar groups form hydrogen bonds with SER (C:48) and VAL (C:61), occupying critical binding sites on *MD2*, potentially disrupting the normal binding of LPS to the *TLR4-MD2* complex.

Western blot analysis revealed that in chicken macrophages *HD11* and duodenum, the IG exhibited significantly elevated expression levels of *TLR4* and *MyD88* compared to the NC ($p \leq 0.01$), alongside a marked increase in the phosphorylation levels of *p38*, *JUN*, *IκBα*, and *p65*. This trend was consistent at both cellular and organismal levels, indicating that APEC infection strongly induces a pro-inflammatory response by activating the *TLR4-MyD88* signaling pathway and its downstream *NF-κB* and *MAPK* cascades. Combined with ELISA results, all detected cytokines (e.g., *TNF-α*, *IL-1β*, *IL-6*, *IL-8*, *IL-12*) in the IG were significantly higher than in the NC, further confirming that APEC infection triggers a robust systemic inflammatory response. In cellular models, the DIO and CEL treatment groups showed slightly better inhibition of the expression and phosphorylation of proteins in the *Toll-like receptor signaling pathway* compared to the FL and AE, suggesting that DIO and CEL may interfere with LPS binding to the *TLR4-MD2* complex by competitively occupying *MD2*'s hydrophobic pocket, thereby suppressing pathway activation. However, in animal models, the FL demonstrated an overall superior preventive effect compared to the DIO and CEL groups, with ELISA results also showing that FLSDD more effectively suppressed *TNF-α*, *IL-1β*, *IL-6*, *IL-8*, and *IL-12* levels. This may indicate that other components in FLSDD, beyond DIO and CEL, exert synergistic effects in vivo, enhancing its anti-inflammatory activity; alternatively, FLSDD might indirectly suppress *TLR4* signaling pathway activation by modulating the gut microbiota and its metabolites, thus more effectively reducing systemic inflammation. The weaker performance of DIO and CEL in animal models compared to in vitro cell experiments may be attributed to their metabolic stability or bioavailability in vivo [47,48], potentially resulting in insufficient effective concentrations to sustain inflammation suppression. The anti-inflammatory efficacy of FLSDD may be linked to its enrichment with other potent anti-inflammatory components, though virtual drug screening based on molecular docking did not account for drug metabolism or absorption rates [49], and the specific components require further isolation and identification through additional methods.

While FLSDD demonstrates promising anti-inflammatory effects against APEC infection, its application as a natural product faces challenges related to component variability and standardization. The complex composition of FLSDD, influenced by factors such as raw material sources and fermentation conditions, may lead to batch-to-batch variations in the content and proportion of active compounds, potentially affecting therapeutic consistency [50]. Furthermore, the synergistic effects among multiple components, while contributing to FLSDD's efficacy, are difficult to systematically characterize and track, complicating quality control. The observed weaker in vivo performance of individual compounds like DIO and CEL compared to in vitro results suggests potential limitations in their metabolic stability and bioavailability, which may also apply to other FLSDD components.

To address these challenges, future research should focus on establishing robust quality control standards, such as quantitative analysis of key active compounds using LC–MS and chemical fingerprinting to ensure batch consistency [51]. Standardized production processes, including controlled raw material sourcing and fermentation conditions, are essential to minimize variability. Additionally, network pharmacology and metabolomics approaches can elucidate the mechanisms of synergistic effects [52], while pharmacokinetic studies will optimize the bioavailability of active components. These efforts will enhance the reliability and scalability of FLSDD as a novel anti-inflammatory agent, supporting its potential in combating avian colibacillosis and promoting sustainable poultry farming.

5. Conclusions

This study systematically elucidated the critical role of *TLR4* in APEC infection and its regulatory mechanisms through transcriptomic analysis, molecular docking, Western blot, and ELISA experiments, while also validating the potential of FLSDD and its active components, DIO and CEL, in suppressing inflammatory responses. FLSDD, through the synergistic action of multiple components, demonstrated superior anti-inflammatory effects in animal models compared to individual compounds, providing a scientific basis for developing novel antibacterial and anti-infection strategies using traditional Chinese medicine dregs. This not only offers new insights into addressing antimicrobial resistance in avian colibacillosis but also paves the way for the efficient utilization of Chinese medicinal resources and the prevention and control of livestock and poultry diseases. Future research can focus on further isolating and identifying other active components in FLSDD, combined with pharmacokinetic studies, to optimize its application strategies in preventing and treating avian colibacillosis, thereby better supporting the sustainable development of the farming industry, reducing economic losses, and ensuring food safety.

Supplementary Materials: The following supporting information can be downloaded at: https://www.mdpi.com/article/10.3390/vetsci12050479/s1.

Author Contributions: Conceptualization, S.M. and X.F.; methodology, W.X.; software, S.M.; validation, S.M., B.H. and Y.W. (Yilin Wu); formal analysis, Y.W. (Yiming Wang); investigation, C.L.; resources, Y.W. (Yiming Wang); data curation, W.X.; writing—original draft preparation, S.M.; writing—review and editing, X.F.; visualization, S.M. and X.F.; supervision, H.Y.; project administration, H.M.; funding acquisition, Y.W. (Yiming Wang). All authors have read and agreed to the published version of the manuscript.

Funding: This research was funded by the key technology of agricultural efficient and green production of Jilin Province science and technology development plan project, grant number 20230202085NC, supported by the Jilin Provincial Science and Technology Department.

Institutional Review Board Statement: The animal study protocol was approved by the Jilin Agricultural University Laboratory Animal Welfare and Ethics Committee (protocol code 20250313001 and date of approval 13 March 2025).

Informed Consent Statement: Not applicable.

Data Availability Statement: The transcriptome dataset GSE69014 can be downloaded from the GEO database (https://www.ncbi.nlm.nih.gov/geo/query/acc.cgi?acc=GSE69014, accessed on 13 May 2025).

Acknowledgments: We are grateful to Guangzhou Gene Denovo Biotechnology Co., Ltd. (Guangzhou, China) for assisting with sequencing, and other members who contributed to this study.

Conflicts of Interest: The authors declare no conflicts of interest.

Abbreviations

The following abbreviations are used in this manuscript:

BCA	Bicinchoninic Acid
ECL	Enhanced Chemiluminescence Luminescent Solution
PBS	Phosphate-Buffered Saline

PMSF	Phenylmethylsulfonyl Fluoride
PVDF	Polyvinylidene Difluoride
RIPA	Radioimmunoprecipitation Assay
SDS-PAGE	Sodium Dodecyl Sulfate-Polyacrylamide Gel Electrophoresis
TBST	Tris-Buffered Saline with Tween

References

1. Zhao, M.; Chen, Y.; Wang, C.; Xiao, W.; Chen, S.; Zhang, S.; Yang, L.; Li, Y. Systems pharmacology dissection of multi-scale mechanisms of action of Huo-Xiang-Zheng-Qi formula for the treatment of gastrointestinal diseases. *Front. Pharmacol.* **2019**, *9*, 1448. [CrossRef]
2. Yao, Y.; Zhang, X.; Wang, Z.; Zheng, C.; Li, P.; Huang, C.; Tao, W.; Xiao, W.; Wang, Y.; Huang, L. Deciphering the combination principles of Traditional Chinese Medicine from a systems pharmacology perspective based on Ma-huang Decoction. *J. Ethnopharmacol.* **2013**, *150*, 619–638. [CrossRef] [PubMed]
3. Zhang, S.; Ding, C.; Liu, X.; Zhao, Y.; Ding, Q.; Sun, S.; Zhang, J.; Yang, J.; Liu, W.; Li, W. Research progress on extraction, isolation, structural analysis and biological activity of polysaccharides from panax genus. *Molecules* **2023**, *28*, 3733. [CrossRef]
4. Yuan, D.; Cui, Y.; Kou, Y.; Guo, X.; Chen, B. Environmental risk and management of herbal-extraction residues induced by the composition and metal binding properties of DOM. *Front. Environ. Sci.* **2022**, *10*, 856308. [CrossRef]
5. Tao, W.; Jin, J.; Zheng, Y.; Li, S. Current advances of resource utilization of herbal extraction residues in China. *Waste Biomass Valorization* **2021**, *12*, 5853–5868. [CrossRef]
6. Cui, Y.; Li, J.; Deng, D.; Lu, H.; Tian, Z.; Liu, Z.; Ma, X. Solid-state fermentation by Aspergillus niger and Trichoderma koningii improves the quality of tea dregs for use as feed additives. *PLoS ONE* **2021**, *16*, e0260045. [CrossRef]
7. Apostolakos, I.; Laconi, A.; Mughini-Gras, L.; Yapicier, Ö.Ş.; Piccirillo, A. Occurrence of colibacillosis in broilers and its relationship with avian pathogenic *Escherichia coli* (APEC) population structure and molecular characteristics. *Front. Vet. Sci.* **2021**, *8*, 737720. [CrossRef] [PubMed]
8. Feng, A.; Akter, S.; Leigh, S.A.; Wang, H.; Pharr, G.T.; Evans, J.; Branton, S.L.; Landinez, M.P.; Pace, L.; Wan, X.-F. Genomic diversity, pathogenicity and antimicrobial resistance of *Escherichia coli* isolated from poultry in the southern United States. *BMC Microbiol.* **2023**, *23*, 15. [CrossRef]
9. Ibrahim, R.A.; Cryer, T.L.; Lafi, S.Q.; Basha, E.-A.; Good, L.; Tarazi, Y.H. Identification of *Escherichia coli* from broiler chickens in Jordan, their antimicrobial resistance, gene characterization and the associated risk factors. *BMC Vet. Res.* **2019**, *15*, 159. [CrossRef]
10. Geng, T.; He, F.; Su, S.; Sun, K.; Zhao, L.; Zhao, Y.; Bao, N.; Pan, L.; Sun, H. Probiotics Lactobacillus rhamnosus GG ATCC53103 and Lactobacillus plantarum JL01 induce cytokine alterations by the production of TCDA, DHA, and succinic and palmitic acids, and enhance immunity of weaned piglets. *Res. Vet. Sci.* **2021**, *137*, 56–67. [CrossRef]
11. Weill, F.S.; Cela, E.M.; Paz, M.L.; Ferrari, A.; Leoni, J.; Maglio, D.H.G. Lipoteichoic acid from Lactobacillus rhamnosus GG as an oral photoprotective agent against UV-induced carcinogenesis. *Br. J. Nutr.* **2013**, *109*, 457–466. [CrossRef] [PubMed]
12. Ahn, J.; Kyoung, H.; Park, K.I.; Kim, Y.; Lee, J.I.; Ha, S.; Nam, J.; Kim, K.; Kang, Y.; Park, Y. PSXIII-7 Effect of Dietary Supplementation of Lactobacillus Rhamnosus on Blood Profiles, Intestinal Morphology, Immune Responses, and Gene Expression of Tight Junction Proteins and Cytokines of Weaned Pigs. *J. Anim. Sci.* **2023**, *101* (Suppl. S3), 574–575. [CrossRef]
13. Sun, H.; Liu, P.; Nolan, L.K.; Lamont, S.J. Thymus transcriptome reveals novel pathways in response to avian pathogenic *Escherichia coli* infection. *Poult. Sci.* **2016**, *95*, 2803–2814. [CrossRef]
14. Guo, L.; Zhang, X.; Lv, N.; Wang, L.; Gan, J.; Jiang, X.; Wang, Y. Therapeutic role and potential mechanism of resveratrol in atherosclerosis: *TLR4*/NF-κB/HIF-1α. *Mediat. Inflamm.* **2023**, *2023*, 1097706. [CrossRef]
15. Okan, A.; Doğanyiğit, Z.; Yilmaz, S.; Uçar, S.; Arikan Söylemez, E.S.; Attar, R. Evaluation of the protective role of resveratrol against sepsis caused by LPS via *TLR4*/NF-κB/TNF-α signaling pathways: Experimental study. *Cell Biochem. Funct.* **2023**, *41*, 423–433. [CrossRef] [PubMed]
16. He, J.; Huang, L.; Sun, K.; Li, J.; Han, S.; Gao, X.; Wang, Q.-Q.; Yang, S.; Sun, W.; Gao, H. Oleuropein alleviates myocardial ischemia–reperfusion injury by suppressing oxidative stress and excessive autophagy via *TLR4*/MAPK signaling pathway. *Chin. Med.* **2024**, *19*, 59. [CrossRef]
17. Yakut, S.; Gelen, V.; Kara, H.; Özkanlar, S.; Yeşildağ, A. Silver Nanoparticles Loaded With Oleuropein Alleviates LPS-Induced Acute Lung Injury by Modulating the *TLR4*/P2X7 Receptor-Mediated Inflammation and Apoptosis in Rats. *Environ. Toxicol.* **2024**, *39*, 4960–4973. [CrossRef]
18. Chen, P.; Zhou, J.; Ruan, A.; Zeng, L.; Liu, J.; Wang, Q. Cinnamic Aldehyde, the main monomer component of Cinnamon, exhibits anti-inflammatory property in OA synovial fibroblasts via *TLR4/MyD88* pathway. *J. Cell. Mol. Med.* **2022**, *26*, 913–924. [CrossRef]
19. Han, J.; Shi, G.; Li, W.; Xie, Y.; Li, F.; Jiang, D. Preventive effect of *Dioscin* against monosodium urate-mediated gouty arthritis through inhibiting inflammasome NLRP3 and *TLR4*/NF-κB signaling pathway activation: An in vivo and in vitro study. *J. Nat. Med.* **2021**, *75*, 37–47. [CrossRef]

20. Jin, S.; Guan, T.; Wang, S.; Hu, M.; Liu, X.; Huang, S.; Liu, Y. *Dioscin* alleviates cisplatin-induced mucositis in rats by modulating gut microbiota, enhancing intestinal barrier function and attenuating *TLR4*/NF-κB signaling cascade. *Int. J. Mol. Sci.* **2022**, *23*, 4431. [CrossRef]
21. Huang, X.; Shen, H.; Liu, Y.; Qiu, S.; Guo, Y. Fisetin attenuates periodontitis through FGFR1/*TLR4*/NLRP3 inflammasome pathway. *Int. Immunopharmacol.* **2021**, *95*, 107505. [CrossRef] [PubMed]
22. Jian, J.; Wang, D.; Xiong, Y.; Wang, J.; Zheng, Q.; Jiang, Z.; Zhong, J.; Yang, S.; Wang, L. Puerarin alleviated oxidative stress and ferroptosis during renal fibrosis induced by ischemia/reperfusion injury via *TLR4*/Nox4 pathway in rats. *Acta Cir. Bras.* **2023**, *38*, e382523. [CrossRef] [PubMed]
23. Liu, X.-L.; Fan, L.; Yue, B.-H.; Lou, Z. Saikosaponin A mitigates the progression of Parkinson's disease via attenuating microglial neuroinflammation through *TLR4/MyD88*/NF-κB pathway. *Eur. Rev. Med. Pharmacol. Sci.* **2023**, *27*, 6956–6971.
24. Jin, G.-j.; Peng, X.; Chen, Z.-G.; Wang, Y.-l.; Liao, W.-J. Celastrol attenuates chronic constrictive injury-induced neuropathic pain and inhibits the *TLR4*/NF-κB signaling pathway in the spinal cord. *J. Nat. Med.* **2022**, *76*, 268–275. [CrossRef]
25. Gerasimova, A.; Novikova, O.; Savicheva, A. Avian colibacillosis—Current aspects. *Vet. Sci. Today* **2023**, *2023*, 284. [CrossRef]
26. Ahmed, A.A.; Salem, H.M.; Hamoud, M.M.; Amer, M.M. Avian Colibacillosis, Multidrug Resistance, Antibiotic Alternatives: An Updated Review. *Egypt. J. Vet. Sci.* **2025**, *56*, 1–21. [CrossRef]
27. Bhattarai, R.K.; Basnet, H.B.; Dhakal, I.P.; Devkota, B. Antimicrobial resistance of avian pathogenic *Escherichia coli* isolated from broiler, layer, and breeder chickens. *Vet. World* **2024**, *17*, 480. [CrossRef]
28. Lu, Q.; Li, C. Comprehensive utilization of Chinese medicine residues for industry and environment protection: Turning waste into treasure. *J. Clean. Prod.* **2021**, *279*, 123856. [CrossRef]
29. Yang, H.-Y.; Han, L.; Lin, Y.-Q.; Li, T.; Wei, Y.; Zhao, L.-H.; Tong, X.-L. Probiotic fermentation of herbal medicine: Progress, challenges, and opportunities. *Am. J. Chin. Med.* **2023**, *51*, 1105–1126. [CrossRef]
30. Hao, S.; Shi, W.; Chen, L.; Kong, T.; Wang, B.; Chen, S.; Guo, X. CATH-2-derived antimicrobial peptide inhibits multidrug-resistant *Escherichia coli* infection in chickens. *Front. Cell. Infect. Microbiol.* **2024**, *14*, 1390934. [CrossRef]
31. Gauthier, A.E.; Rotjan, R.D.; Kagan, J.C. Lipopolysaccharide detection by the innate immune system may be an uncommon defence strategy used in nature. *Open Biol.* **2022**, *12*, 220146. [CrossRef]
32. Baranova, I.N.; Bocharov, A.V.; Vishnyakova, T.G.; Chen, Z.; Ke, Y.; Birukova, A.A.; Yuen, P.S.; Tsuji, T.; Star, R.A.; Birukov, K.G. Class B Scavenger Receptor CD36 as a Potential Therapeutic Target in Inflammation Induced by Danger-Associated Molecular Patterns. *Cells* **2024**, *13*, 1992. [CrossRef]
33. Chen, G.-Y.; Xu, X.-D.; Feng, J.-X.; Zhang, J.; Li, X.-P. A single C1q A chain promotes leukocytes phagocytosis/chemotaxis and mediates bacterial clearance in Japanese flounder Paralichthys olivaceus. *Aquaculture* **2024**, *587*, 740840. [CrossRef]
34. Smirnova, T.; Spertini, C.; Spertini, O. CSF1R inhibition combined with GM-CSF reprograms macrophages and disrupts protumoral interplays with AML cells. *Cancers* **2021**, *13*, 5289. [CrossRef] [PubMed]
35. Wang, X.; Zhao, X.-Y. Transcription factors associated with IL-15 cytokine signaling during NK cell development. *Front. Immunol.* **2021**, *12*, 610789. [CrossRef] [PubMed]
36. Mazzoccoli, L.; Liu, B. Dendritic cells in shaping anti-tumor T cell response. *Cancers* **2024**, *16*, 2211. [CrossRef]
37. Zeng, Y.; Gao, M.; Lin, D.; Du, G.; Cai, Y. Prognostic and immunological roles of MMP-9 in Pan-cancer. *BioMed Res. Int.* **2022**, *2022*, 2592962. [CrossRef]
38. Zhao, K.; Sun, Y.; Zhong, S.; Luo, J.-L. The multifaceted roles of cathepsins in immune and inflammatory responses: Implications for cancer therapy, autoimmune diseases, and infectious diseases. *Biomark. Res.* **2024**, *12*, 165. [CrossRef]
39. De Angelis, M.; Amatore, D.; Checconi, P.; Zevini, A.; Fraternale, A.; Magnani, M.; Hiscott, J.; De Chiara, G.; Palamara, A.T.; Nencioni, L. Influenza virus down-modulates G6PD expression and activity to induce oxidative stress and promote its replication. *Front. Cell. Infect. Microbiol.* **2022**, *11*, 804976. [CrossRef]
40. Li, D.; Wu, M. Pattern recognition receptors in health and diseases. *Signal Transduct. Target. Ther.* **2021**, *6*, 291.
41. Wicherska-Pawłowska, K.; Wróbel, T.; Rybka, J. Toll-like receptors (TLRs), NOD-like receptors (NLRs), and RIG-I-like receptors (RLRs) in innate immunity. TLRs, NLRs, and RLRs ligands as immunotherapeutic agents for hematopoietic diseases. *Int. J. Mol. Sci.* **2021**, *22*, 13397. [CrossRef]
42. Li, X.; Xu, M.; Shen, J.; Li, Y.; Lin, S.; Zhu, M.; Pang, Q.; Tan, X.; Tang, J. Sorafenib inhibits LPS-induced inflammation by regulating Lyn-MAPK-NF-kB/AP-1 pathway and *TLR4* expression. *Cell Death Discov.* **2022**, *8*, 281. [CrossRef]
43. Sawoo, R.; Dey, R.; Ghosh, R.; Bishayi, B. *TLR4* and TNFR1 blockade dampen M1 macrophage activation and shifts them towards an M2 phenotype. *Immunol. Res.* **2021**, *69*, 334–351. [CrossRef]
44. Yan, L.; Wang, J.; Cai, X.; Liou, Y.C.; Shen, H.M.; Hao, J.; Huang, C.; Luo, G.; He, W. Macrophage plasticity: Signaling pathways, tissue repair, and regeneration. *MedComm* **2024**, *5*, e658. [CrossRef]
45. Oblak, A.; Jerala, R. The molecular mechanism of species-specific recognition of lipopolysaccharides by the MD-2/TLR4 receptor complex. *Mol. Immunol.* **2015**, *63*, 134–142. [CrossRef]

46. Hopp, M.T.; Holze, J.; Lauber, F.; Holtkamp, L.; Rathod, D.C.; Miteva, M.A.; Prestes, E.B.; Geyer, M.; Manoury, B.; Merle, N.S. Insights into the molecular basis and mechanism of heme-triggered *TLR4* signalling: The role of heme-binding motifs in *TLR4* and *MD2*. *Immunology* **2024**, *171*, 181–197. [CrossRef] [PubMed]
47. Bandopadhyay, S.; Anand, U.; Gadekar, V.S.; Jha, N.K.; Gupta, P.K.; Behl, T.; Kumar, M.; Radha; Shekhawat, M.S.; Dey, A. *Dioscin*: A review on pharmacological properties and therapeutic values. *BioFactors* **2022**, *48*, 22–55. [PubMed]
48. Wang, C.; Dai, S.; Zhao, X.; Zhang, Y.; Gong, L.; Fu, K.; Ma, C.; Peng, C.; Li, Y. *Celastrol* as an emerging anticancer agent: Current status, challenges and therapeutic strategies. *Biomed. Pharmacother.* **2023**, *163*, 114882. [CrossRef]
49. Opo, F.A.D.M.; Rahman, M.M.; Ahammad, F.; Ahmed, I.; Bhuiyan, M.A.; Asiri, A.M. Structure based pharmacophore modeling, virtual screening, molecular docking and ADMET approaches for identification of natural anti-cancer agents targeting XIAP protein. *Sci. Rep.* **2021**, *11*, 4049.
50. Bolmanis, E.; Dubencovs, K.; Suleiko, A.; Vanags, J. Model predictive control—A stand out among competitors for fed-batch fermentation improvement. *Fermentation* **2023**, *9*, 206. [CrossRef]
51. Yao, Z.; Yu, J.; Tang, Z.; Liu, H.; Ruan, K.; Song, Z.; Liu, Y.; Yan, K.; Liu, Y.; Tang, Y. Multi-evaluating strategy for Siji-kangbingdu mixture: Chemical profiling, fingerprint characterization, and quantitative analysis. *Molecules* **2019**, *24*, 3545. [CrossRef] [PubMed]
52. Mukherjee, P.K.; Banerjee, S.; Kar, A. Molecular combination networks in medicinal plants: Understanding synergy by network pharmacology in Indian traditional medicine. *Phytochem. Rev.* **2021**, *20*, 693–703. [CrossRef]

biomolecules

MDPI

Article

Baicalin Relieves *Glaesserella parasuis*-Triggered Immunosuppression Through Polarization via MIF/CD74 Signaling Pathway in Piglets

Shulin Fu [1,2,*,†], Qiaoli Dong [1,2,†], Yunjian Fu [1,2], Ronghui Luo [1,2], Jingyang Li [1,2], Yamin Sun [1,2], Siyu Liu [1,2], Yinsheng Qiu [1,2], Ling Guo [1,2] and Jin Hu [1,2]

1 Wuhan Engineering and Technology Research Center of Animal Disease-Resistant Nutrition, School of Animal Science and Nutritional Engineering, Wuhan Polytechnic University, Wuhan 430023, China; 20230411045@whpu.edu.cn (Q.D.); 20230411033@whpu.edu.cn (Y.F.); 20230411035@whpu.edu.cn (R.L.); 20220411050@whpu.edu.cn (J.L.); 20230412014@whpu.edu.cn (Y.S.); 20220411030@whpu.edu.cn (S.L.); qiuyinsheng6405@whpu.edu.cn (Y.Q.); guoling1101@whpu.edu.cn (L.G.); hujin@whpu.edu.cn (J.H.)

2 Hubei Key Laboratory of Animal Nutrition and Feed Science, School of Animal Science and Nutritional Engineering, Wuhan Polytechnic University, Wuhan 430023, China

* Correspondence: shulinfu@whpu.edu.cn

† These authors contributed equally to the work.

Abstract: *Glaesserella parasuis* (*G. parasuis*) infection is responsible for Glässer's disease in pigs. *G. parasuis* could trigger piglet immunosuppression, but the mechanism of inducing immunosuppression by *G. parasuis* remains unknown. Macrophage migration inhibitory factor (MIF)/CD74 axis has been shown to participate in inflammation response and immunosuppression, but the function of MIF/CD74 during immunosuppression elicited by *G. parasuis* has not been fully explored. This experiment explored the efficacy of baicalin on immunosuppression elicited by *G. parasuis* alleviation through regulating polarization via the MIF/CD74 signaling pathway. Our data indicated that baicalin reduced IL-1β, IL-6, IL-8, IL-18, TNF-α, and COX-2 expression, and regulated MIF/CD74 axis expression in the spleen. Immunohistochemistry analysis showed that baicalin enhanced CD74 protein levels in the spleen of piglets induced by *G. parasuis*. Baicalin regulated the PI3K/Akt/mTOR signaling pathway and RAF/MEK/ERK signaling activation, modified the expression of the autophagy-related proteins Beclin-1, P62, and LC3B, promoted M2 polarization to M1 polarization, and enhanced CD3, CD4, CD8, and TIM3 levels in the spleen of piglets elicited by *G. parasuis*. Our study reveals the important functions of the MIF/CD74 axis in *G. parasuis*-induced immunosuppression and may offer a new therapeutic method to control *G. parasuis* infection.

Keywords: baicalin; *Glaesserella parasuis*; inflammation; MIF/CD74; autophagy; polarization; immunosuppression

1. Introduction

Glaesserella parasuis (*G. parasuis*) is a significant pathogen responsible for Glässer's disease in pigs [1]. The symptoms caused by *G. parasuis* are fibrotic polyserositis, arthritis, and meningitis [2]. So far, 15 serovars have been verified by using gel immunodiffusion determination; however, more than 20% of isolates have not been serotyped using this method [3]. Serovars 4, 5, and 13 were considered as the highly virulent strains [4]. *G. parasuis* is a bacterium that colonizes the upper respiratory tract of pigs, which can spread rapidly within a herd [5]. Once inside the host, *G. parasuis* has the potential to invade and multiply within the tissues, leading to an acute systemic inflammatory response [6]. This infectious condition poses a considerable threat to swine production worldwide, contributing to substantial economic losses to the pig farms [7]. Prevention and control measures are crucial to mitigate the harmful effects of *G. parasuis*. However, the diversity of *G. parasuis* serovars lacks cross-serovar protection, contributing to limiting the effectiveness of vaccines [8]. Thus, it is urgent to find new drugs to prevent and control *G. parasuis* infection.

Infection of piglets with *G. parasuis* can result in immunosuppression in piglets [9]; however, the mechanism of immunosuppression elicited by *G. parasuis* in piglets is still unknown. Leukocyte differentiation antigen 74 (CD74), the MHC II-associated invariant chain, is a type II transmembrane glycoprotein [10]. CD74 promoted the formation of the tumor microenvironment through enhancing S100A8 and S100A9 production in the pancreatic cancer [11]. CD74 aided the accumulation and efficacy of regulatory T cells in tumors [12] and was involved in *Staphylococcus aureus*-induced dose-dependent immunosuppression [13]. In addition, CD74 was considered the receptor for the macrophage migration inhibitory factor (MIF) [14]. As a receptor for CD74, MIF can regulate the innate and adaptive immune response [15]. MIF has been shown to play a key role in immune system regulation and is involved in autoimmune disorders and tumorigenesis [16]. Furthermore, some studies indicate that MIF contributes to human melanoma myeloid-derived suppressor cell induction and T-cell immunosuppression [17]. Research has reported that MIF/CD74 damaged LECs' barrier via inhibiting tight junction proteins in an ERK1/2-dependent manner [18]. Interaction between CD74 and MIF enhances the secretion of proinflammatory mediators and promotes unactivated $CD4^+$ T cell infection, leading to viral spreading [19]. Blockade with immunosuppressive MIF/CD74 on macrophages and dendritic cells contributed to the downregulation of immunosuppressive factors from MOs and the upregulation of the capacity of DCs to activate cytotoxic T cells [20]. However, whether MIF/CD74 signaling was related to immunosuppression in piglets induced by *G. parasuis* has not been evaluated.

Bacterial antibiotic resistance is becoming more widespread and increasingly serious due to excessive utilization of antibiotics in healthcare and agriculture [21]. A Chinese veterinary drug was thought to be a good antibiotic alternative to relieve the negative impacts of infectious diseases [22]. Levamisole, a compound with established immunomodulatory properties, has been used as an immunostimulant and anti-cancer drug in humans and farm animals, making it an effective control in this study [23]. Baicalin, a flavonoid compound, is extracted from the root of *Scutellaria baicalensis* Georgi [24]. Baicalin possesses anti-inflammatory, antibacterial, antioxidant, and immunomodulatory properties [25]. It has been reported that baicalin suppressed macrophage JNK/SIRT1/p65-triggered adipose tissue inflammation [26] and diminished hepatocyte injury resulting from aflatoxin B_1 through TP53-related ferroptosis signaling [27]. In addition, baicalin reduced inflammatory responses to reduce lung cancer by acting on SOCS1/NF-κB/STAT3 axis [28], ameliorated *Mycoplasma gallisepticum*-elicited inflammatory damage by diminishing ceramide accumulation in DF-1 cells [29] and alleviated lung inflammatory responses and fibrosis elicited by silica through diminishing TLR4/NF-kB pathway in rats [30]. Baicalin also weakened porcine extraintestinal pathogenic *Escherichia coli* virulence through reducing the LuxS/AI-2 quorum-sensing system [31], and attenuated intestinal inflammatory damage triggered by LPS through diminishing NF-κB and NLRP3 pathways [32]. Baicalin inhibited PANX-1/P2Y6 signaling pathway activation in porcine aortic vascular endothelial cells infected by *G. parasuis* [33]. Although baicalin has important biological functions, the mechanism by which baicalin can alleviate *G. parasuis* elicited piglet immunosuppression via MIF/CD74 signaling has not been thoroughly studied.

In this study, the molecular mechanism of MIF/CD74-induced host immunosuppression via polarization in piglet spleen triggered by *G. parasuis* and whether baicalin could alleviate immunosuppression was investigated. This study deepens our knowledge about the molecular mechanism of inducing immunosuppression and provides new targets to prevent *G. parasuis* infection.

2. Materials and Methods

2.1. Ethics Statement

The animal study was approved by the Animal Care and Use Committee of Wuhan Polytechnic University, Hubei Province, China (WPU202308003), 2 August 2023. All experimental animals were euthanised at the end of the experiment.

2.2. Bacteria and Drugs

The *G. parasuis* SH0165 strain, serovar 5, used in this study, is highly virulent. It was isolated from the lung of a commercial pig which had the symptoms of arthritis, fibrinous polyserositis, hemorrhagic pneumonia, and meningitis [34]. The *G. parasuis* bacteria were grown on TSA or cultured in TSB at 37 °C. Simultaneously, 10 μg/mL of NAD (Guangzhou Saiguo Biotech Co., Ltd., Guangzhou, China) and 10% foetal bovine serum (Sijiqing, Hangzhou, China) were added to the culture.

Baicalin was obtained from Sichuan Xieli Pharmaceutical Co., Ltd. (Pengzhou, Sichuan, China). Levamisole was obtained from Beijing Solarbio Science & Technology Co., Ltd. (Beijing, China).

2.3. Design of the Experiment

The spleen tissue of piglets was obtained from our previous animal experiment [35]. Briefly, sixty 30-day-old naturally farrowed early-weaned piglets were randomly divided into the control group, the infection group, the levamisole group, the 25 mg/kg baicalin group, the 50 mg/kg baicalin group, and the 100 mg/kg baicalin group. All piglets were pretreated with an intramuscular injection of 15 mg/kg body weight (BW) levamisole, 25 mg/kg BW baicalin, 50 mg/kg BW baicalin, or 100 mg/kg BW baicalin, except the control group and the infection group. A total of 2 h later, after the first day of administration, all piglets were intraperitoneally infected with 1×10^8 CFU *G. parasuis* in 1 mL of TSB except the control group. The piglets from the control group were intraperitoneally injected with an equivalent volume of TSB. After being challenged by *G. parasuis* for 6 h, the groups were injected with the same drugs, except the control group. Every treatment was administered twice a day for 2 days. All groups were observed for 7 days after *G. parasuis* infection, and the spleens were collected for analysis.

2.4. RT-PCR

The spleens were collected from all groups for cytokine IL-1β, IL-6, IL-8, IL-10, IL-18, TNF-α, and COX-2 production measurement by RT-PCR [36]. Briefly, the RNAs from the spleens were extracted by using the TRIzol reagent (Invitrogen, Carlsbad, CA, USA). After analysis of the concentration and quality using a Qubit 2.0 fluorometer (Thermo Fisher Scientific, Waltham, MA, USA), the RNA was reverse transcribed to complementary DNA by using a reverse transcriptase (Abclonal, Wuhan, China) and PCR determination was implemented by employing a SYBR Green PCR Kit (Abclonal, Wuhan, China). The transcriptions of each sample were repeated at least three times. GAPDH was employed as the internal control. The relative expression levels of genes were quantified employing the threshold cycle (CT) method. The gene relative expression fold changes were performed employing the $2^{-\Delta\Delta CT}$ CT formula. The RT-PCR primers used in this study are listed in Table 1.

Table 1. Primer sequences for qRT-PCR analysis.

Gene		Nucleotide Sequence (5′-3′)	Tm (°C)	Length (bp)
IL-1β	Forward	TCTGCATGAGCTTTGTGCAAG	59.7	155
	Reverse	ACAGGGCAGACTCGAATTCAAC	60.9	
IL-6	Forward	CTTCTGGTGATGGCTACTG	52.7	134
	Reverse	TTGCCGAGGATGTACTTAA	50.0	
IL-8	Forward	ACAGCAGTAACAACAACAAG	50.2	117
	Reverse	GACCAGCACAGGAATGAG	53.2	
IL-10	Forward	CGTGGAGGAGGTGAAGAGTG	55.4	178
	Reverse	TTAGTAGAGTCGTCATCCTGGAAG	55.6	
IL-18	Forward	AGTAACCATATCTGTGCAGTGT	54.0	155
	Reverse	TCTTATCACCATGTCCAGGAAC	53.0	
TNF-α	Forward	CGCTCTTCTGCCTACTGCACTTC	60.7	164
	Reverse	CTGTCCCTCGGCTTTGACATT	57.8	
CD74	Forward	CTGGGAGTGTACCCGAAGC	58.6	104
	Reverse	CGCAGCCAGTTCTCAAAGAG	56.6	
MIF	Forward	GTGCGCCCTTTGCAGTCT	59.5	146
	Reverse	CCGCGTTCATGTCGTAGTAGTT	57.1	
COX-2	Forward	TGTGAAAGGGAGGAAAGA	49.7	133
	Reverse	CTGATGGGTGAAGTGCTG	53.5	
TIM-3	Forward	TCAAGCCTCATCACTTTGG	53.7	145
	Reverse	TGACGGAGCAGTAACACTC	50.8	
PI3K	Forward	TTGCTACAATCAATCGCCAGGAGAC	59.3	147
	Reverse	CTTCCCGTTGTTGCCATCGTTTG	59.7	
Akt	Forward	GGACGGGCACATCAAGATCACTG	60.8	126
	Reverse	TAGTCGTTGTCCTCCAGCACCTC	61.2	
mTOR	Forward	AGTACCTCCAGGACACCATGAACC	60.9	108
	Reverse	CAGACCTCACAGCCACAGAAAGC	61.0	
GAPDH	Forward	GGCACAGTCAAGGCGGAGAAC	61.9	105
	Reverse	AGCACCAGCATCACCCCATTTG	61.0	
LC3B	Forward	AGCCTTCTTCCTGTTAGTG	51.3	135
	Reverse	TTCATTCCGAAAGTCTCC	48.4	
P62	Forward	CCCCAATGTGATCTGTGATG	53.4	126
	Reverse	TTGCTGTGCTCCTTGTGAAT	54.5	
Beclin-1	Forward	TGAGGATACCCAAGCAAG	51.2	146
	Reverse	ATGTGGAGAAAGGCAAGA	50.4	
Inos	Forward	TGGAAGCGGTAACAAAGGA	53.7	482
	Reverse	CACGAGGTCAGGAGGGATT	56.8	
ARG1	Forward	GGACCTGTGCTTTGCTGAT	55.6	122
	Reverse	TTCCGTTCTTCTTGATTTCTG	49.7	
RAF	Forward	CAACACTGATGCTGCTGGTAA	55.4	113
	Reverse	CAGATGGCGACTTGGAATG	53.8	
MEK	Forward	CGTGAATGAGCCACCTCCC	60.7	149
	Reverse	CCACCTCGGACCGTTTGAT	60.2	
ERK	Forward	GCTCTTGAAGACGCAGCAC	58.6	108
	Reverse	CAGCAGGTTGGAAGGTTTG	58.5	
CD86	Forward	TGTGGGATGGTGTCCTTTG	55.0	106
	Reverse	GTTCACTCGCCTTCCTGTT	55.2	
CD3	Forward	GTTTGCTGATGGTGGTGTA	52.4	144
	Reverse	TGGGCTCATAGTCTGGATT	52.5	
CD4	Forward	AGCCTCAGTTACCGAGTT	52.7	138
	Reverse	ATCCTCTTGTCTTCCACTTC	51.1	
CD8	Forward	GTTACATCTCTGGTTACAAGG	49.9	139
	Reverse	AAGAAGACGGACATGAAGTT	50.7	
CD163	Forward	TGCTGTAGTCGCTGTTCT	55.9	117
	Reverse	ACTTTCACCTCCACTCTTC	54.0	

2.5. Determination of Baicalin on CD3 Expression by Immunofluorescence

Immunofluorescence analysis was employed to assess the function of baicalin and levamisole on the CD3 level in the spleen tissue [37]. Briefly, the spleen tissues were fixed with paraformaldehyde and then treated with 0.5% Triton X-100 (Sangon Biotech, Shanghai, China). Following blocking with 5% BSA at 37 °C for 30 min, the samples were incubated with a primary antibody against CD3 (SouthernBiotech, Birmingham, AL, USA) for 12 h at 4 °C. The nuclei were counterstained with DAPI (Servicebio, Wuhan, China). Afterwards, the tissues were labeled and washed, the fluorescent photographs were obtained using a Nikon Eclipse C1 microscope (Nikon, Tokyo, Japan). Fluorescence intensity and cell surface area were measured by utilizing the Case Viewer software (CaseViewer 3.3, 3DHISTECH, Budapest, Hungary).

2.6. Detection of CD74 Expression by Immunohistochemistry

The CD74 expression level was also determined by using immunohistochemistry analysis [38]. Briefly, spleen tissues were prepared from paraffin-embedded blocks. Then, slices were cut and incubated with an anti-CD74 antibody (Abclonal, Wuhan, China, 1:500, Cat: A24027) for immunohistochemistry (IHC) at 4 °C overnight. CD74 was reacted with goat anti-mouse antibody (1:200) and visualized using diaminobenzidine (DAB) (Servicebio, Wuhan, China). The negative control was only added with the second antibody. Following counterstaining with hematoxylin (Servicebio, Wuhan, China) for 3 min, the microscopic observation was carried out, and the Image 1.52a software was employed to determine the optical density sum of the brown area (Olympus BX43, Tokyo, Japan).

2.7. Western Blotting

The protein levels were estimated by the Western blotting method [39]. Briefly, the tissue proteins were extracted by employing the RIPA Lysis Buffer (Beyotime Biotechnology, Shanghai, China). The protein concentration was determined by employing an Enhanced BCA Protein Assay Kit (Beyotime Biotechnology, Shanghai, China), and then isolated by employing 10% SDS-PAGE and transferred onto a PVDF (polyvinylidene difluoride) membrane. Flowing being blocked by using BSA for 120 min and washed five times using TBST, the membrane was incubated with the primary antibody of MIF (Boster biological technology, Wuhan, China, 1:2000, Cat: BA2058), CD74 (ABclonal, 1:10,000, Cat: A24027), PI3K (ABclonal,1:500, Cat: A4992), p-PI3K (ABclonal, 1:1000, Cat: AP0427), Akt (proteintech, 1:2000, Cat: 10176-2), p-Akt (proteintech, 1:2000, Cat: 66444-1), mTOR (proteintech, 1:5000, Cat: 66888-1), p-mTOR (proteintech, 1:20,000, Cat: 67778-1), RAF (Abmart, Shanghai, China, 1:2000, Cat: T55225), MEK1/2 (ABclonal, 1:6000, Cat: A4868), p-MEK1/2 (ABclonal, 1:10,000, Cat: AP1349), ERK1/2 (ABclonal, 1:5000, Cat: A22447), p-ERK1/2 (ABclonal, 1:2000, Cat: AP0485), P62 (ABclonal, 1:40,000, Cat: A19700), LC3B (ABclonal, 1:4000, Cat: A19665), Beclin-1 (ABclonal, 1:10,000, Cat: A21191), or GAPDH (proteintech, 1:5000, Cat:10494-1-AP) for 12 h at 4 °C, respectively. The membrane was washed with TBST five times and incubated with corresponding HRP Goat Anti-Rabbit IgG (MIF, CD74, PI3K, p-PI3K, Akt, RAF, MEK1/2, p-MEK1/2, ERK1/2, p-ERK1/2, P62, LC3B, Beclin-1, GAPDH) (Abbkine, Wuhan, China, 1:10,000) or HRP Goat Anti-Mouse IgG (H+L) (p-Akt, mTOR, p-mTOR) (ABclonal, 1:5000) at 37 °C for 1 h, and were visualized with an ECL Enhanced Kit (ABclonal, Wuhan, China). The protein levels were quantified in a FluorChem™ FC2 AIC system (Alpha Innotech, Santa Clara, CA, USA) by employing ImageJ software (V1.8.0.112).

2.8. Statistical Analysis

The results are presented as the mean ± standard deviation. Statistical differences were carried out with one-way analysis of variance (ANOVA). ANOVA was used when three or more independent groups were compared. $p < 0.05$ was considered significantly different. ## $p < 0.01$ vs. controls; ### $p < 0.001$ vs. controls; * stands for $p < 0.05$; ** stands for $p < 0.01$; *** stands for $p < 0.001$.

3. Results

3.1. Baicalin Inhibited IL-1β, IL-6, IL-8, IL-18, TNF-α, and COX-2 Expression in Piglet Spleen Triggered by G. parasuis

Cytokines are multifunctional mediators of cellular communication that are involved in inflammatory responses [40]; thus, we analyzed the effect of baicalin on cytokine production in piglets infected with *G. parasuis*. The data indicated that IL-1β, IL-6, IL-8, IL-18, TNF-α, and COX-2 mRNA levels in the spleen in the infection group were significantly upregulated ($p < 0.001$) (Figure 1). Levamisole significantly inhibited IL-1β, IL-6, IL-8, IL-18, TNF-α, and COX-2 mRNA expressions in the spleen compared to the infection group ($p < 0.001$) (Figure 1). In addition, 25–100 mg/kg baicalin reduced the mRNA expression levels of IL-1β, IL-6, IL-8, IL-18, TNF-α, and COX-2 in the spleen compared to the infection group ($p < 0.05$) (Figure 1).

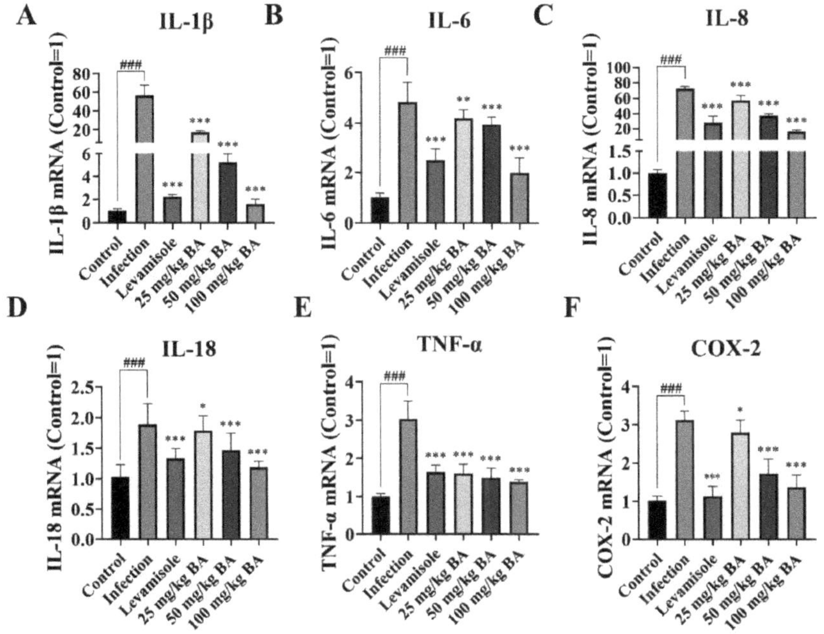

Figure 1. The effect of baicalin on cytokine production in piglet spleen triggered by *G. parasuis*. The IL-1β (**A**), IL-6 (**B**), IL-8 (**C**), IL-18 (**D**), TNF-α (**E**), and COX-2 (**F**) mRNA levels were determined by RT-PCR. BA: Baicalin; ### $p < 0.001$ vs. controls; * stands for $p < 0.05$; ** stands for $p < 0.01$; *** stands for $p < 0.001$.

3.2. Baicalin Improved MIF/CD74 Axis Levels in Piglet Spleen Elicited by G. parasuis

MIF-CD74 signaling blockade on macrophages reduced the immunosuppressive factors contributing to restoring the antitumor immune response in metastatic melanoma [20]. When the piglets were infected with *G. parasuis*, the mRNA expression levels of MIF and CD74 in the spleen were significantly reduced ($p < 0.001$) (Figure 2), while levamisole enhanced the MIF and CD74 mRNA levels in the spleen compared to the infection group ($p < 0.01$) (Figure 2). Doses of 25–100 mg/kg baicalin also increased the mRNA expression levels of MIF and CD74 in the spleen compared to the infection group ($p < 0.05$) (Figure 2). The MIF and CD74 protein expression levels in the spleen were determined using the western blotting method. The data indicated that the protein expression levels of MIF and CD74 in the spleen were downregulated in the infection group ($p < 0.001$), while levamisole and 25–100 mg/kg baicalin could increase CD74 protein level in the spleen compared to the infection group ($p < 0.05$) (Figure 2). Treatment with 50–100 mg/kg baicalin upregulated MIF protein level in the spleen compared to the infection group ($p < 0.001$) (Figure 2).

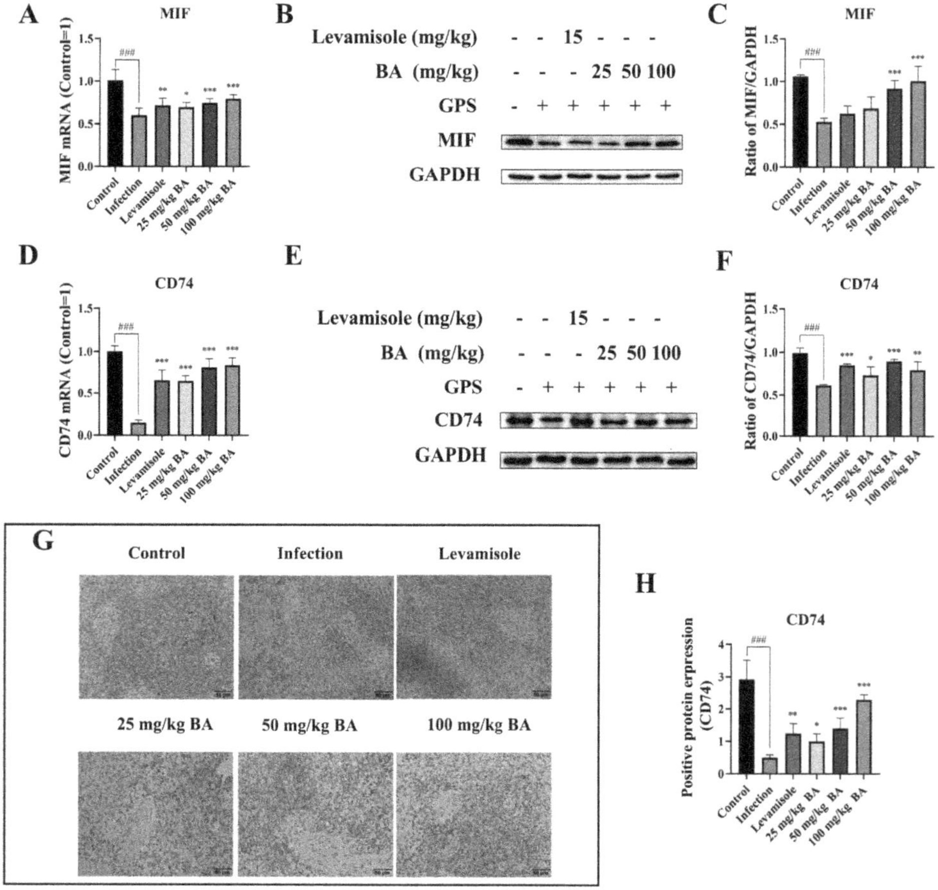

Figure 2. Detection of baicalin on MIF/CD74 axis expression in *G. parasuis* induced piglet spleen. (**A**): MIF mRNA expression level using RT-PCR; (**B**,**C**): MIF protein expression level using western blot; (**D**): CD74 mRNA level by RT-PCR; (**E**,**F**): CD74 protein expression level by western blot; (**G**,**H**): CD74 protein expression level by immunohistochemistry; BA: Baicalin; ### $p < 0.001$ vs. controls; * stands for $p < 0.05$; ** stands for $p < 0.01$; *** stands for $p < 0.001$. Original Western blot images are provided in the Supplementary Materials.

The CD74 protein expression level was also assessed by immunohistochemistry. The data demonstrated that the *G. parasuis* inhibited the positive protein expression level of CD74 in the spleen compared to the control group, while levamisole and 25–100 mg/kg baicalin promoted the positive protein level of CD74 in the spleen compared to the infection group ($p < 0.05$) (Figure 2).

3.3. Baicalin Promoted PI3K/Akt/mTOR Pathway Activation in G. parasuis Triggered Piglet Spleen

Pieces of evidence have shown that a large number of natural products were targeting PI3K/AKT/mTOR-induced autophagy, leading to diminishing tumor growth [41]; thus, we explored the efficacy of baicalin on PI3K/Akt/mTOR signaling activation. The data indicated that the PI3K, Akt, and mTOR mRNA levels were significantly decreased in the spleen when the piglets were challenged by *G. parasuis* ($p < 0.001$) (Figure 3). The PI3K, Akt, and mTOR mRNA levels treated with levamisole or baicalin were significantly increased compared to the infection group ($p < 0.05$) (Figure 3). In addition, p-PI3K, p-Akt, and p-mTOR protein levels were also declined in the infection group compared to the control group ($p < 0.01$) (Figure 3), while levamisole or 50–100 mg/kg baicalin alleviated the decrease in the protein levels of p-PI3Kand p-Akt ($p < 0.05$) (Figure 3). Dosages of 100 mg/kg baicalin increased p-mTOR protein levels, in contrast with the infection group ($p < 0.05$) (Figure 3).

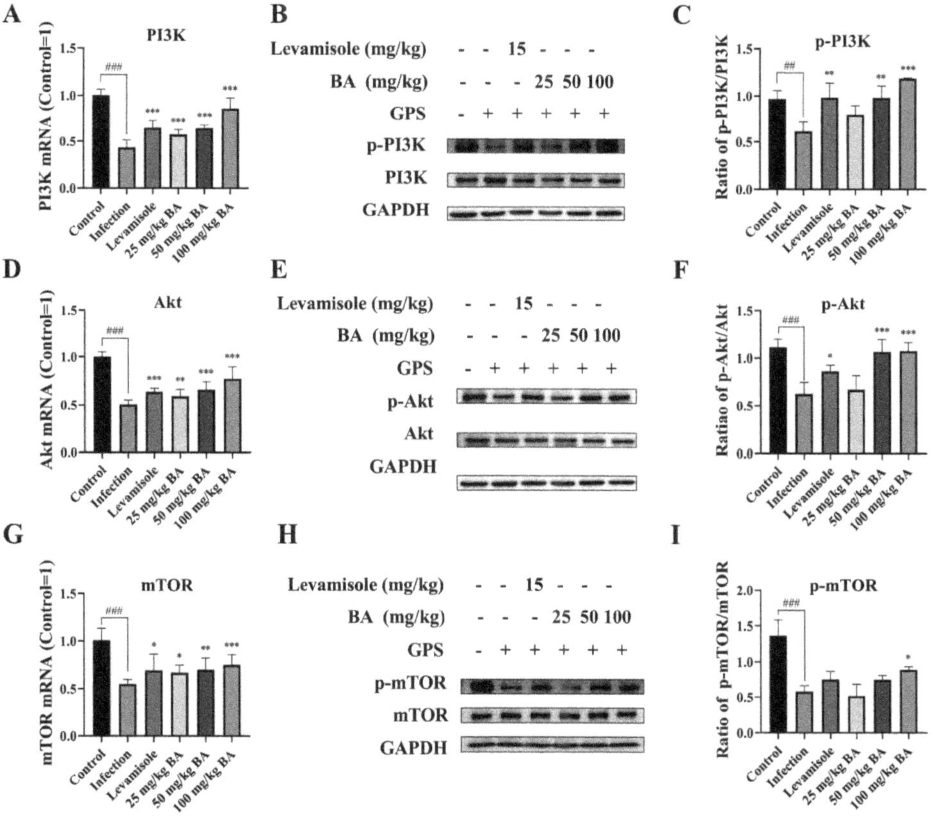

Figure 3. Measurement of the function of baicalin in PI3K/Akt/mTOR activation in *G. parasuis*-infected piglet spleen. (**A**): PI3K mRNA expression level using RT-PCR; (**B**,**C**): p-PI3K protein level using western blot; (**D**): Akt mRNA level using RT-PCR; (**E**,**F**): p-Akt protein expression level via western blot; (**G**): mTOR mRNA expression level via RT-PCR; (**H**,**I**): p-mTOR protein expression level via western blot; BA: Baicalin; ## $p < 0.01$ vs. controls; ### $p < 0.001$ vs. controls; * stands for $p < 0.05$; ** stands for $p < 0.01$; *** stands for $p < 0.001$. Original Western blot images are provided in the Supplementary Materials.

3.4. Baicalin Regulated the RAF/MEK/ERK Signaling Activation in G. parasuis Triggered Piglet Spleen

Previous research reported that MIF activated the CD74-dependent "sustained" activation of RAF/MEK/ERK involved in apoptosis and autophagy [42]. Thus, the RAF, MEK, and ERK levels were evaluated. The data demonstrated that *G. parasuis* challenge declined the RAF, MEK, and ERK mRNA levels in the spleen ($p < 0.001$) (Figure 4). Levamisole and 25–100 mg/kg baicalin promoted RAF, MEK, and ERK mRNA levels, in contrast with the infection group ($p < 0.01$) (Figure 4). The RAF, p-MEK, and p-ERK protein levels were suppressed, in contrast with the control group ($p < 0.01$) (Figure 4). Levamisole induced MEK and ERK phosphorylation levels in the spleen ($p < 0.001$) (Figure 4). Furthermore, 50–100 mg/kg baicalin also reversed the effect on RAF and p-ERK protein levels, in contrast with the infection group ($p < 0.05$) (Figure 4).

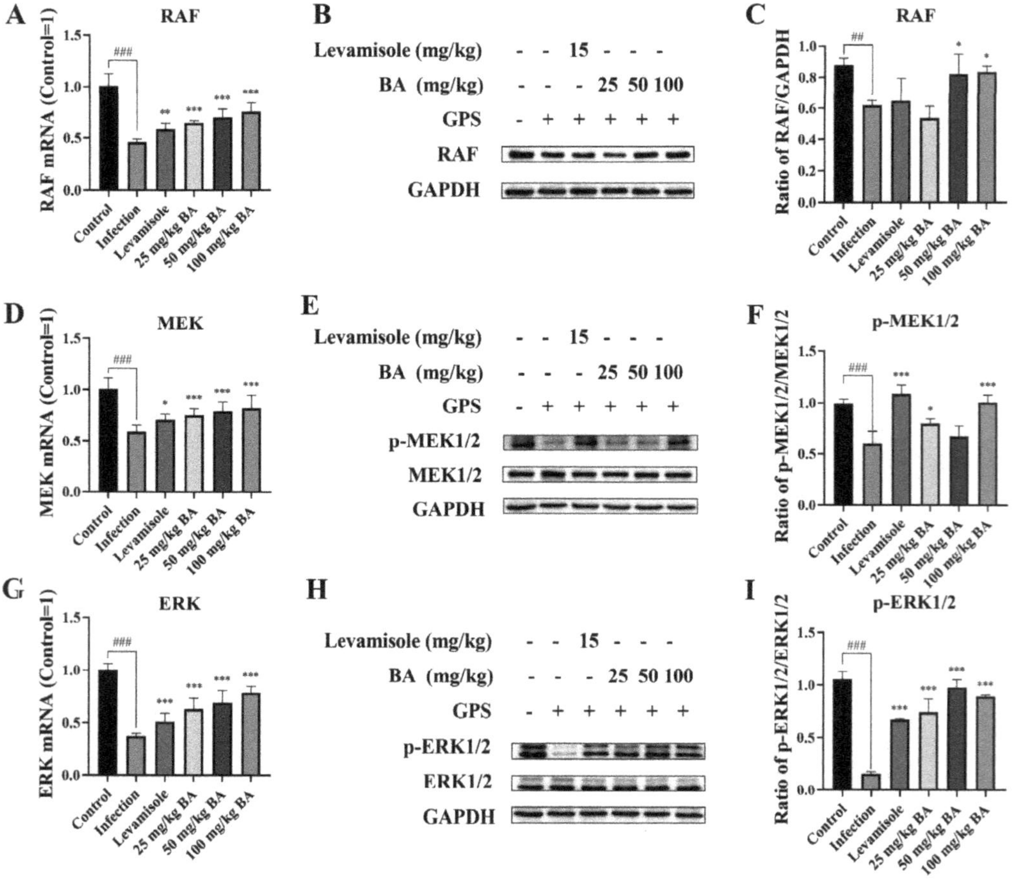

Figure 4. The efficacy of baicalin on RAF/MEK/ERK activation in *G. parasuis*-infected piglet spleen. (**A**): RAF mRNA expression level using RT-PCR; (**B**,**C**): RAF protein level using western blot; (**D**): MEK mRNA level using RT-PCR; (**E**,**F**): p-MEK1/2 protein level via western blot; (**G**): ERK mRNA level via RT-PCR; (**H**,**I**): p-ERK1/2 protein expression level via western blot; BA: Baicalin; ## $p < 0.01$ vs. controls; ### $p < 0.001$ vs. controls; * stands for $p < 0.05$; ** stands for $p < 0.01$; *** stands for $p < 0.001$. Original Western blot images are provided in the Supplementary Materials.

3.5. Baicalin Modified Autophagy-Related Proteins Beclin-1, P62, and LC3B Expression in G. parasuis-Triggered Piglet Spleens

PI3K/Akt/mTOR and RAF/MEK/ERK signaling activation could lead to autophagy [43]. The levels of the autophagy-related proteins Beclin-1, P62, and LC3B in the spleen were determined using the RT-PCR and western blotting methods. The data indicated that at the mRNA and protein level, *G. parasuis* induced the LC3B expression level and downregulated P62 and Beclin-1 expression levels in the spleen from the infection group, in contrast with the control group ($p < 0.001$) (Figure 5). Levamisole treatment reduced the LC3B mRNA and protein levels, upregulated Beclin-1 mRNA and protein levels, and increased P62 mRNA expression level in the spleen, in contrast with the infection group ($p < 0.01$) (Figure 5). Dosages of 25–100 mg/kg baicalin also suppressed the mRNA and protein levels of LC3B and triggered P62, Beclin-1 mRNA, and protein levels in the spleen ($p < 0.05$) (Figure 5).

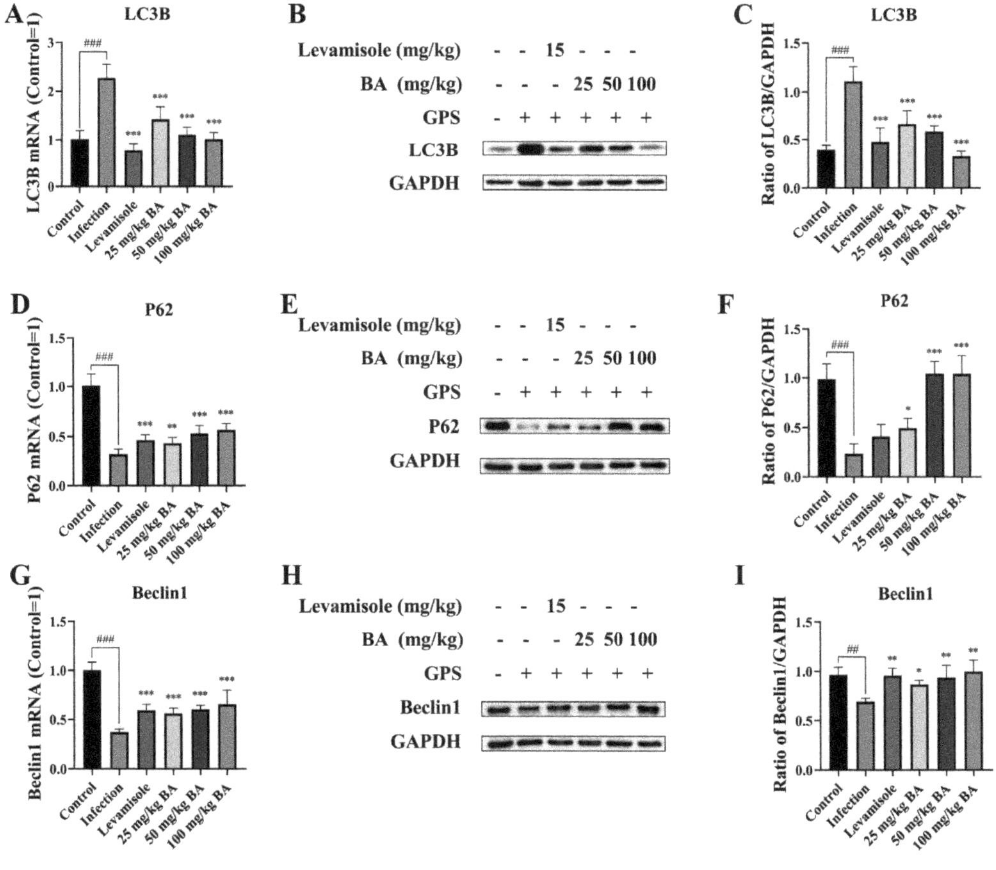

Figure 5. Assessment of the effect of baicalin on autophagy regulation in *G. parasuis*-infected piglet spleen. The levels of the autophagy-related proteins Beclin-1, P62, and LC3B were determined using RT-PCR and western blot. (**A**): LC3B mRNA expression level via RT-PCR; (**B**,**C**): LC3B protein expression level via western blot; (**D**): P62 mRNA expression level via RT-PCR; (**E**,**F**): P62 protein expression level via western blot; (**G**): Beclin-1 mRNA expression level via RT-PCR; (**H**,**I**): Beclin-1 protein level via western blot; BA: Baicalin; ## $p < 0.01$ vs. controls; ### $p < 0.001$ vs. controls; * stands for $p < 0.05$; ** stands for $p < 0.01$; *** stands for $p < 0.001$. Original Western blot images are provided in the Supplementary Materials.

3.6. Baicalin and Levamisole Promoted M2 Polarization to M1 Polarization in G. parasuis-Triggered Piglet Spleens

Targeting MIF-CD74 enhanced microglia M1 polarization in non-small cell lung cancer [44]. To investigate the immunomodulatory effect of the MIF/CD74 axis in the spleen of *G. parasuis*-infected piglets, the M1 and M2 phenotype markers were detected. The RT-PCR results demonstrated that the *G. parasuis* remarkably decreased iNOS and CD86 expression (M1 markers) and increased ARG1, IL-10, and CD163 expression (M2 markers) in the spleen ($p < 0.05$) (Figure 6). Levamisole and 25–100 mg/kg baicalin upregulated the mRNA expression of iNOS and CD86 and attenuated the levels of ARG1, IL-10, and CD163 in the spleen, in contrast with the infection group ($p < 0.05$) (Figure 6).

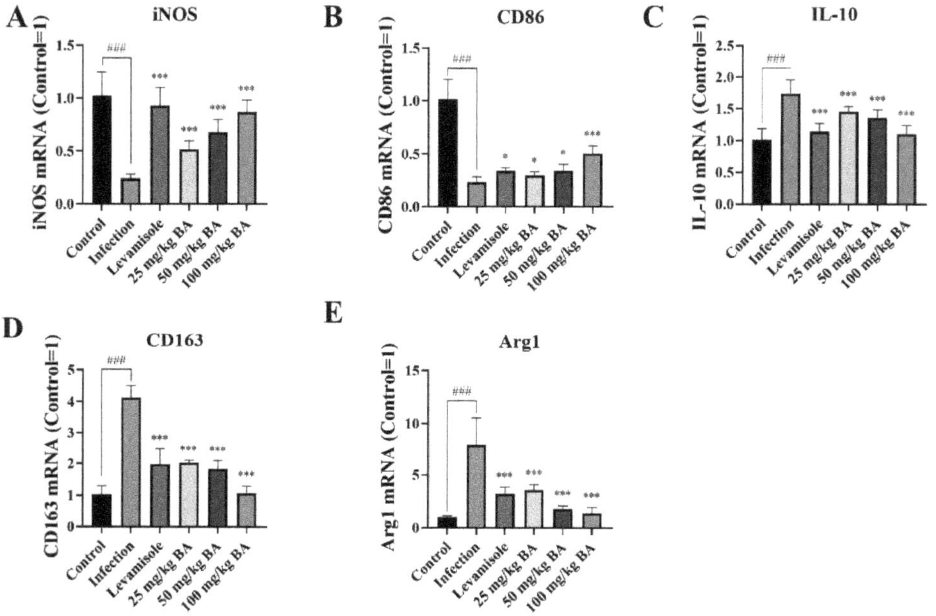

Figure 6. Determination of the effect of baicalin on polarization change in *G. parasuis*-infected piglet spleen. The iNOS (**A**), CD86 (**B**) (M1 markers), IL-10 (**C**), CD163 (**D**), and Arg1 (**E**) (M2 markers) mRNA levels were determined via RT-PCR. BA: Baicalin; ### $p < 0.001$ vs. controls; * stands for $p < 0.05$; *** stands for $p < 0.001$.

3.7. Baicalin Enhanced CD3, CD4, CD8, and TIM3 Levels in G. parasuis-Triggered Piglet Spleens

Previous research reported that tumor-associated M2 macrophages promoted tumor growth and suppressed immune response [45]. The decrease in the number of $CD3^+$, $CD4^+$, and $CD8^+$ T cells is one important reason causing host immunosuppression [46]; thus, we determined the CD3, CD4, CD8, and TIM3 levels in *G. parasuis*-infected spleen. The results showed that, after *G. parasuis* challenge, the CD3, CD4, CD8, and TIM3 mRNA levels were decreased, in contrast with the control group ($p < 0.001$) (Figure 7). When the piglets were pretreated with levamisole or 25–100 mg/kg baicalin, the CD3, CD4, CD8, and TIM3 mRNA levels were increased, in contrast with the infection group ($p < 0.05$) (Figure 7), suggesting that levamisole and baicalin could modify T cell differentiation.

We also detected CD3 expression using an immunofluorometric assay. The results demonstrated that the mean optical density of CD3 in the infection group was significantly decreased compared to that in the control group ($p < 0.001$) (Figure 7). Levamisole and

25–100 mg/kg baicalin enhanced the mean optical density of CD3 compared to the control group ($p < 0.001$) (Figure 7).

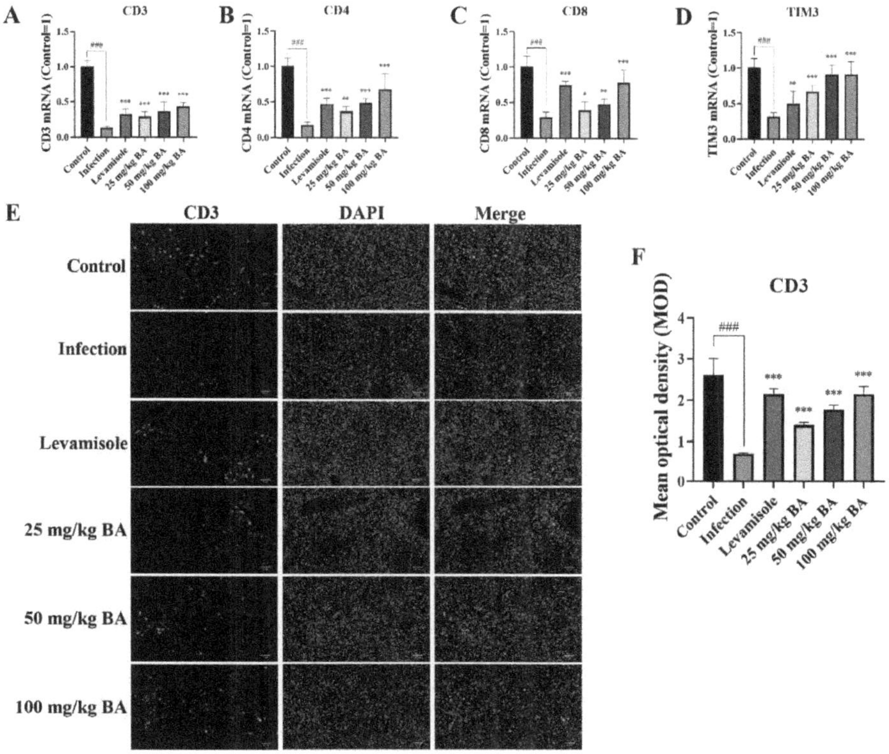

Figure 7. The effect of baicalin on CD3, CD4, CD8, and TIM3 levels in *G. parasuis*-infected piglet spleen. The CD3 (**A**), CD4 (**B**), CD8 (**C**), and TIM3 (**D**) levels were determined using RT-PCR. The CD3 protein expression level (**E**,**F**) was measured via immunofluorometric assay. BA: Baicalin; ### $p < 0.001$ vs. controls; * stands for $p < 0.05$; ** stands for $p < 0.01$; *** stands for $p < 0.001$.

4. Discussion

G. parasuis is a Gram-negative bacterium, and it is a common flora in the upper respiratory tract of pigs [47]. Under certain stress conditions, *G. parasuis* can cause systemic infections characterized by arthritis, meningitis, and polyserositis, known as Glässer's disease [48]. This disease is considered a major cause of pig deaths, resulting in serious losses to the global pig production [1]. *G. parasuis* infection can cause piglet immunosuppression; however, the immunosuppression mechanism remains unknown; thus, we used the piglet model to assess the molecular mechanism of immunosuppression caused by *G. parasuis*.

Our previous studies have shown that the activation of the PD-1/PD-L1 axis triggered by *G. parasuis* could lead to piglet immunosuppression [9]. In this study, we found that MIF/CD74 axis activation induced by *G. parasuis* could result in piglet immunosuppression. MIF/CD74 axis, as an immune checkpoint, has been confirmed to promote immunosuppression and evade immune surveillance [49]. MIF-CD74 signaling blockade on macrophages restored the antitumor immune response against metastatic melanoma [20]. Known as a macrophage migration inhibitor, MIF attracted and retained activated immune cells from the periphery to inflammatory tissue [50]. MIF is involved in immune regulation, inflammatory response, and cell proliferation [51]. The biological function of MIF is mainly mediated by its main receptor, CD74 [52]. As a receptor for MIF, CD74 is mainly expressed in macrophages [53]. MIF and the receptor CD74 can be used as a promising target to treat Ewing sarcoma, which is infiltrated by immunosuppressive myeloid populations [54]. CD74 promoted immunosuppressive tumor microenvironment formation in breast cancer in mice; therefore, CD74 might be considered a novel therapeutic target in breast cancer [55]. Therefore, in this study, we investigated the effect of baicalin on MIF/CD74 axis expression in the spleen triggered by *G. parasuis*. Our results indicated that baicalin regulated MIF/CD74 axis expression in the spleen triggered by *G. parasuis*, which suggested that MIF and CD74 might serve as new targets to treat *G. parasuis* infection.

Autophagy is a process by which cells degrade and recycle proteins and organelles to maintain intracellular homeostasis [30]. Autophagy is a key molecular pathway to preserve cellular and organismal homeostasis [56]. Autophagy offers an adaptive function to protect organisms against diverse pathologies, such as infections [57]. LC3B, Beclin-1, and p62 play key functions in the induction of autophagy during the disease state [58]. The autophagy-related protein LC3B accelerates PRMT1 mRNA decay, thereby enhancing autophagy [59]. P62 has been demonstrated to regulate autophagy [60]. This study showed that *G. parasuis* induced autophagy, and baicalin and levamisole treatment inhibited the autophagy phenomenon. Previous reports showed that MIF promoted cardiac stem cell proliferation and endothelial differentiation via PI3K/Akt/mTOR pathway activation [61]. The PI3K-Akt-mTOR pathway activation promoted necrotic cell death by suppressing autophagy [62]. Scoparone reduced hepatic inflammation and regulated autophagy in nonalcoholic steatohepatitis mice through PI3K/AKT/mTOR pathway regulation in macrophages [63]. Thus, we assessed the effect of baicalin on the PI3K/AKT/mTOR pathway and autophagy activation in the spleen of piglets challenged with *G. parasuis*. Our data demonstrated that baicalin regulated the PI3K/AKT/mTOR pathway and reduced autophagy in piglets challenged with *G. parasuis*, which enhances the new function of baicalin in autophagy regulation.

The RAF-MEK-ERK pathway is a characterized MAPK signaling pathway that participates in cell proliferation and survival [64]. Fibroblast GSK-3α promoted fibrosis via the RAF-MEK-ERK pathway in the injured heart [65]. The EIF3H-HAX1 increased RAF-MEK-ERK signaling activity to enhance colorectal cancer progression [66]. In addition, many studies reported that RAF/MEK/ERK signaling is associated with autophagy modification [67]. Geniposide alleviated lipopolysaccharide-induced murine kidney podocyte apoptosis via promoting RAF/MEK/ERK-mediated autophagy [68]. The use of 5-aminolevulinic acid photodynamic therapy diminished HPV viral load through autophagy by modifying RAF/MEK/ERK in HeLa cells [69]. Thus, we inferred that baicalin could modify autophagy through RAF/MEK/ERK pathway regulation, but the specific mechanism needs further research.

Macrophages are usually classified into two categories, which are named M1 macrophages and M2 macrophages [70]. Previous research reported that M1 macrophages and M2 macrophages are associated with inflammation responses [71]. M1 and M2 macrophage polarization imbalance might trigger pathological consequences and lead to diseases [72]. Previous research showed that CD5L promoted M2 macrophage polarization by an autophagy-mediated ID3 increase [73]. Puerarin induced macrophage M2 polarization to display antinon-alcoholic steatohepatitis activity through autophagy activation [74]. Baicalin induces M1 polarization to exert antiviral effects [75]. Thus, we explored the levels of M1 markers (iNOS, CD86) and M2 markers (ARG1, IL-10, CD163). Immunotherapy-activated T cells could recruit and skew late-stage activated M1-like macrophages, critical for cancer immunotherapy [76]. Wnt5a triggered the M2 polarization of tumor-associated macrophages by modifying CaKMII-ERK1/2-STAT3-mediated IL-10 production, leading to the induction of tumor growth and the metastasis of colorectal cancer [77]. The results indicated that M1 was downregulated and M2 was upregulated in the *G. parasuis*-infected spleen, suggesting that infection with *G. parasuis* in piglets induces macrophage polarization towards the M2 phenotype, while baicalin could reverse the M2 phenotype to the M1 phenotype.

Previous research reported that nicotinamide phosphoribosyltransferase-driven tumor-associated macrophage M2 polarization results in an immunosuppressive microenvironment in colorectal cancer [78]. Macrophages activated by the histamine receptor H1 could polarize toward the M2-like immunosuppressive phenotype with the dysfunction of T cells [79]. T cells have an important effect in modifying the immune system balance [80]. Sepsis-associated immunosuppression was correlated to a reduction in the number of T cells [81]. CD4+ T cells have important functions in immunosuppression in inflammatory diseases [82]. Metformin improved cancer immunotherapy by remedying tumor-infiltrating CD8 T lymphocytes from the immunosuppression triggered by hypoxia [83]. T-cell exhaustion is a major reason for immunosuppression in the tumor microenvironment [84]. The T cell immunoglobulin domain and mucin domain 3 (TIM3) served as targets for CD8+ T cell exhaustion [7]. Therefore, the expression levels of CD3, CD4, CD8, and TIM3 in *G. parasuis*-infected piglet spleens were measured, and our results confirmed that infection with *G. parasuis* in piglets could lead to host immunosuppression, and baicalin could alleviate the immunosuppression in piglets caused by *G. parasuis* [35].

Taken together, our study showed that *G. parasuis* infection could promote the MIF/CD74 axis and regulate the PI3K/Akt/mTOR and RAF/MEK/ERK signaling pathways induced autophagy, thereby transforming M1 towards M2 polarization in the spleen, contributing to host immunosuppression, whereas baicalin could reverse this signaling activation in order to reduce host immunosuppression. Our study is the first to report that baicalin could alleviate host immunosuppression by regulating the MIF/CD74 axis-induced polarization of macrophages, which might provide some novel targets to relieve the immunosuppression and inflammation caused by *G. parasuis*. Our future study will explore the therapeutic potential of baicalin in combination with other treatments.

Supplementary Materials: The following supporting information can be downloaded at: https://www.mdpi.com/article/10.3390/biom15050640/s1. Original Western blot images.

Author Contributions: S.F. conceived and designed the experiments; S.F., Q.D., Y.F., R.L., J.L., Y.S. and S.L. performed the experiments; S.F., Q.D., Y.Q., L.G. and J.H. analyzed the data; S.F. wrote this paper. All authors have read and agreed to the published version of the manuscript.

Funding: This work was supported by the National Natural Science Foundation of China (grant no. 32273067) and Hubei Provincial Natural Science Funding (2023AFB746).

Institutional Review Board Statement: The animal study was approved by the Animal Care and Use Committee of Wuhan Polytechnic University, Hubei Province, China (WPU202308003), approval date 2 August 2023.

Informed Consent Statement: Not applicable.

Data Availability Statement: All relevant data are within the manuscript.

Conflicts of Interest: The authors declare no conflicts of interest.

References

1. Lin, Y.; Zhu, N.; Liu, J.; Wen, S.; Xu, Y.; Xu, X.; Cai, X. The role of cytolethal distending toxin in *Glaesserella parasuis* JS0135 strain infection: Cytotoxicity, phagocytic resistance and pathogenicity. *Vet. Microbiol.* **2024**, *295*, 110168. [CrossRef] [PubMed]
2. Zhou, Y.; Jiang, D.; Yao, X.; Luo, Y.; Yang, Z.; Ren, M.; Zhang, G.; Yu, Y.; Lu, A.; Wang, Y. Pan-genome wide association study of *Glaesserella parasuis* highlights genes associated with virulence and biofilm formation. *Front. Microbiol.* **2023**, *14*, 1160433. [CrossRef]

3. Wu, C.F.; Hsu, C.Y.; Chou, C.C.; Wang, C.M.; Huang, S.W.; Kuo, H.C. Serotypes, virulence factors and multilocus sequence typing of *Glaesserella parasuis* from diseased pigs in Taiwan. *PeerJ* **2023**, *11*, e15823. [CrossRef]
4. Macedo, N.; Gottschalk, M.; Strutzberg-Minder, K.; Van, C.N.; Zhang, L.; Zou, G.; Zhou, R.; Marostica, T.; Clavijo, M.J.; Tucker, A.; et al. Molecular characterization of *Glaesserella parasuis* strains isolated from North America, Europe and Asia by serotyping PCR and LS-PCR. *Vet. Res.* **2021**, *52*, 68. [CrossRef]
5. Mao, W.; Wang, Z.; Wen, S.; Lin, Y.; Gu, J.; Sun, J.; Wang, H.; Cao, Q.; Xu, Y.; Xu, X.; et al. LRRC8A promotes *Glaesserella parasuis* cytolethal distending toxin-induced p53-dependent apoptosis in NPTr cells. *Virulence* **2023**, *14*, 2287339. [CrossRef]
6. Liu, F.; Gao, Y.; Jiao, J.; Zhang, Y.; Li, J.; Ding, L.; Zhang, L.; Chen, Z.; Song, X.; Yang, G.; et al. Upregulation of TLR4-Dependent ATP Production Is Critical for *Glaesserella parasuis* LPS-Mediated Inflammation. *Cells* **2023**, *12*, 751. [CrossRef]
7. Chen, C.; Zhao, F.; Peng, J.; Zhao, D.; Xu, L.; Li, H.; Ma, S.; Peng, X.; Sheng, X.; Sun, Y.; et al. Soluble Tim-3 serves as a tumor prognostic marker and therapeutic target for $CD8^+$ T cell exhaustion and anti-PD-1 resistance. *Cell Rep. Med.* **2024**, *5*, 101686. [CrossRef]
8. Silva, G.F.R.; Moreno, L.Z.; Matajira, C.E.C.; Silva, A.P.S.; Araújo, K.M.; Gomes, V.T.M.; Barbosa, M.R.F.; Sato, M.I.Z.; Moreno, A.M. Serotyping and Antimicrobial Susceptibility Profiling of *Glaesserella parasuis* Isolated from Diseased Swine in Brazil. *Pathogens.* **2022**, *11*, 1443. [CrossRef]
9. Li, J.; Liu, S.; Dong, Q.; Fu, Y.; Sun, Y.; Luo, R.; Tian, X.; Guo, L.; Liu, W.; Qiu, Y.; et al. PD-1/PD-L1 axis induced host immunosuppression via PI3K/Akt/mTOR signalling pathway in piglets infected by *Glaesserella parasuis*. *BMC Vet. Res.* **2024**, *20*, 141. [CrossRef]
10. Tanese, K.; Ogata, D. The role of macrophage migration inhibitory factor family and CD74 in the pathogenesis of melanoma. *Exp. Dermatol.* **2024**, *33*, e15122. [CrossRef]
11. Hong, W.C.; Lee, D.E.; Kang, H.W.; Kim, M.J.; Kim, M.; Kim, J.H.; Fang, S.; Kim, H.J.; Park, J.S. CD74 Promotes a Pro-Inflammatory Tumor Microenvironment by Inducing S100A8 and S100A9 Secretion in Pancreatic Cancer. *Int. J. Mol. Sci.* **2023**, *24*, 12993. [CrossRef] [PubMed]
12. Bonnin, E.; Rodrigo Riestra, M.; Marziali, F.; Mena Osuna, R.; Denizeau, J.; Maurin, M.; Saez, J.J.; Jouve, M.; Bonté, P.E.; Richer, W.; et al. CD74 supports accumulation and function of regulatory T cells in tumors. *Nat. Commun.* **2024**, *15*, 3749. [CrossRef] [PubMed]
13. Rasmussen, G.; Cajander, S.; Bäckman, A.; Källman, J.; Söderquist, B.; Strålin, K. Expression of HLA-DRA and CD74 mRNA in whole blood during the course of complicated and uncomplicated Staphylococcus aureus bacteremia. *Microbiol. Immunol.* **2017**, *61*, 442–451. [CrossRef] [PubMed]
14. Li, Q.L.; Tang, J.; Zhao, L.; Ruze, A.; Shan, X.F.; Gao, X.M. The role of CD74 in cardiovascular disease. *Front. Cardiovasc. Med.* **2022**, *9*, 1049143. [CrossRef]
15. Merk, M.; Mitchell, R.A.; Endres, S.; Bucala, R. D-dopachrome tautomerase (D-DT or MIF-2): Doubling the MIF cytokine family. *Cytokine* **2012**, *59*, 10–17. [CrossRef]
16. Bach, J.P.; Rinn, B.; Meyer, B.; Dodel, R.; Bacher, M. Role of MIF in inflammation and tumorigenesis. *Oncology* **2008**, *75*, 127–133. [CrossRef]
17. Yaddanapudi, K.; Rendon, B.E.; Lamont, G.; Kim, E.J.; Al Rayyan, N.; Richie, J.; Albeituni, S.; Waigel, S.; Wise, A.; Mitchell, R.A. MIF Is Necessary for Late-Stage Melanoma Patient MDSC Immune Suppression and Differentiation. *Cancer Immunol. Res.* **2016**, *4*, 101–112. [CrossRef]
18. Shu, W.; Wang, Y.; Deji, Z.; Li, C.; Chen, C.; Ding, W.; Du, P.; Wang, X. Infliximab modifies CD74-mediated lymphatic abnormalities and adipose tissue alterations in creeping fat of Crohn's disease. *Inflamm. Res.* **2024**, *73*, 1157–1172. [CrossRef]
19. Trifone, C.; Salido, J.; Ruiz, M.J.; Leng, L.; Quiroga, M.F.; Salomón, H.; Bucala, R.; Ghiglione, Y.; Turk, G. Interaction Between Macrophage Migration Inhibitory Factor and CD74 in Human Immunodeficiency Virus Type I Infected Primary Monocyte-Derived Macrophages Triggers the Production of Proinflammatory Mediators and Enhances Infection of Unactivated $CD4^+$ T Cells. *Front. Immunol.* **2018**, *9*, 1494. [CrossRef]
20. Figueiredo, C.R.; Azevedo, R.A.; Mousdell, S.; Resende-Lara, P.T.; Ireland, L.; Santos, A.; Girola, N.; Cunha, R.; Schmid, M.C.; Polonelli, L.; et al. Blockade of MIF-CD74 Signalling on Macrophages and Dendritic Cells Restores the Antitumour Immune Response Against Metastatic Melanoma. *Front. Immunol.* **2018**, *9*, 1132. [CrossRef]
21. Tripathy, A.; Sen, P.; Su, B.; Briscoe, W.H. Natural and bioinspired nanostructured bactericidal surfaces. *Adv. Colloid Interface Sci.* **2017**, *248*, 85–104. [CrossRef] [PubMed]
22. Abdelnour, S.A.; Ghazanfar, S.; Abdel-Hamid, M.; Abdel-Latif, H.M.R.; Zhang, Z.; Naiel, M.A.E. Therapeutic uses and applications of bovine lactoferrin in aquatic animal medicine: An overview. *Vet. Res. Commun.* **2023**, *47*, 1015–1029. [CrossRef] [PubMed]
23. Gholami, M.H.; Rassouli, A.; Mirzaei, S.; Hashemi, F. The potential immunomodulatory effect of levamisole in humans and farm animals. *J. Adv. Vet. Anim. Res.* **2023**, *10*, 620–629. [CrossRef] [PubMed]

24. Jin, X.; Liu, M.Y.; Zhang, D.F.; Zhong, X.; Du, K.; Qian, P.; Yao, W.F.; Gao, H.; Wei, M.J. Baicalin mitigates cognitive impairment and protects neurons from microglia-mediated neuroinflammation via suppressing NLRP3 inflammasomes and TLR4/NF-κB signaling pathway. *CNS Neurosci. Ther.* **2019**, *25*, 575–590. [CrossRef]
25. Wen, Y.; Wang, Y.; Zhao, C.; Zhao, B.; Wang, J. The Pharmacological Efficacy of Baicalin in Inflammatory Diseases. *Int. J. Mol. Sci.* **2023**, *24*, 9317. [CrossRef]
26. Zhu, Z.; Yu, M.; Xu, M.; Ji, X.; Zong, X.; Zhang, Z.; Shang, W.; Zhang, L.; Fang, P. Baicalin suppresses macrophage JNK-mediated adipose tissue inflammation to mitigate insulin resistance in obesity. *J. Ethnopharmacol.* **2024**, *332*, 118355. [CrossRef]
27. Zhang, H.J.; Luo, J.Z.; Lan, C.L.; Teng, X.; Ge, B.; Liu, J.Q.; Xie, H.X.; Yang, K.J.; Qin, C.J.; Zhou, X.; et al. Baicalin protects against hepatocyte injury caused by aflatoxin B_1 via the TP53-related ferroptosis Pathway. *Ecotoxicol. Environ. Saf.* **2024**, *281*, 116661. [CrossRef]
28. Guo, L.; Yue, M.; Ma, C.; Wang, Y.; Hou, J.; Li, H. Baicalin reduces inflammation to inhibit lung cancer via targeting SOCS1/NF-κB/STAT3 axis. *Heliyon* **2024**, *10*, e29361. [CrossRef]
29. Chen, X.; Ishfaq, M.; Wang, J. Baicalin ameliorates Mycoplasma gallisepticum-induced inflammatory injury via inhibiting STIM1-regulated ceramide accumulation in DF-1 cells. *Poult. Sci.* **2023**, *102*, 102687. [CrossRef]
30. Zhang, Y.; Liu, F.; Jia, Q.; Zheng, L.; Tang, Q.; Sai, L.; Zhang, W.; Du, Z.; Peng, C.; Bo, C.; et al. Baicalin alleviates silica-induced lung inflammation and fibrosis by inhibiting TLR4/NF-?B pathway in rats. *Physiol. Res.* **2023**, *72*, 221–233. [CrossRef]
31. Zong, B.; Xiao, Y.; Wang, P.; Liu, W.; Ren, M.; Li, C.; Fu, S.; Zhang, Y.; Qiu, Y. Baicalin Weakens the Virulence of Porcine Extraintestinal Pathogenic Escherichia coli by Inhibiting the LuxS/AI-2 Quorum-Sensing System. *Biomolecules* **2024**, *14*, 452. [CrossRef] [PubMed]
32. Lu, Q.; Wang, N.; Wen, D.; Guo, P.; Liu, Y.; Fu, S.; Ye, C.; Wu, Z.; Qiu, Y. Baicalin attenuates lipopolysaccharide-induced intestinal inflammatory injury via suppressing PARP1-mediated NF-κB and NLRP3 signalling pathway. *Toxicon* **2024**, *239*, 107612. [CrossRef] [PubMed]
33. Fu, S.; Tian, X.; Peng, C.; Zhang, D.; Zhou, L.; Yuan, Y.; He, J.; Guo, L.; Qiu, Y.; Ye, C.; et al. Baicalin inhibited PANX-1/P2Y6 signaling pathway activation in porcine aortic vascular endothelial cells infected by *Glaesserella parasuis*. *Heliyon* **2024**, *10*, e23632. [CrossRef] [PubMed]
34. Sun, P.; Yang, Y.; Yang, L.; Qian, Y.; Liang, M.; Chen, H.; Zhang, J.; Qiu, Y.; Guo, L.; Fu, S. Quercetin Protects Blood-Brain Barrier Integrity via the PI3K/Akt/Erk Signaling Pathway in a Mouse Model of Meningitis Induced by *Glaesserella parasuis*. *Biomolecules* **2024**, *14*, 696. [CrossRef]
35. Fu, S.; Li, J.; You, J.; Liu, S.; Dong, Q.; Fu, Y.; Luo, R.; Sun, Y.; Tian, X.; Liu, W.; et al. Baicalin attenuates PD-1/PD-L1 axis-induced immunosuppression in piglets challenged with *Glaesserella parasuis* by inhibiting the PI3K/Akt/mTOR and RAS/MEK/ERK signalling pathways. *Vet. Res.* **2024**, *55*, 95. [CrossRef]
36. Liu, Y.; Kim, J.H.; Lim, H.K.; Huang, L.; Choi, W.; Kopalli, S.R.; Lee, S.; Lee, B.H.; Lee, J.H.; Ju, Y.; et al. Millingtonia hortensis L. f. ethanol extract exerts in vivo and in vitro anti-inflammatory activities through inhibition of Syk kinase in NF-κB pathway. *J. Ethnopharmacol.* **2024**, *332*, 118386. [CrossRef]
37. Yang, K.; Shan, X.; Songru, Y.; Fu, M.; Zhao, P.; Guo, W.; Xu, M.; Chen, H.; Lu, R.; Zhang, C. Network pharmacology integrated with experimental validation to elucidate the mechanisms of action of the Guizhi-Gancao Decoction in the treatment of phenylephrine-induced cardiac hypertrophy. *Pharm. Biol.* **2024**, *62*, 456–471. [CrossRef]
38. Chen, Y.; Zeng, Z.; Chen, J. Role of BACH1 in multiple myeloma. *Hematology* **2024**, *29*, 2352687. [CrossRef]
39. Tian, X.; Li, J.; Liu, S.; Dong, Q.; Fu, Y.; Luo, R.; Sun, Y.; Guo, L.; Lu, Q.; Ye, C.; et al. Anemoside B4 attenuates necrotic enteritis of laying hens induced by Clostridium perfringens via inhibiting NF-κB and PI3K/Akt/mTOR signalling pathways. *Heliyon* **2024**, *10*, e33161. [CrossRef]
40. Campbell, I.L. Cytokine-mediated inflammation, tumorigenesis, and disease-associated JAK/STAT/SOCS signaling circuits in the CNS. *Brain Res. Brain Res. Rev.* **2005**, *48*, 166–177. [CrossRef]
41. Xu, Z.; Han, X.; Ou, D.; Liu, T.; Li, Z.; Jiang, G.; Liu, J.; Zhang, J. Targeting PI3K/AKT/mTOR-mediated autophagy for tumor therapy. *Appl. Microbiol. Biotechnol.* **2020**, *104*, 575–587. [CrossRef] [PubMed]
42. Lue, H.; Kapurniotu, A.; Fingerle-Rowson, G.; Roger, T.; Leng, L.; Thiele, M.; Calandra, T.; Bucala, R.; Bernhagen, J. Rapid and transient activation of the ERK MAPK signalling pathway by macrophage migration inhibitory factor (MIF) and dependence on JAB1/CSN5 and Src kinase activity. *Cell. Signal.* **2006**, *18*, 688–703. [CrossRef] [PubMed]
43. Yun, Z.S.; Zhihua, S.; Xuelian, T.; Min, X.; Rongjing, H.; Mei, L. Rosmarinic acid activates the Ras/Raf/MEK/ERK signaling pathway to regulate $CD8^+$ T cells and autophagy to clear Chlamydia trachomatis in reproductive tract-infected mice. *Mol. Immunol.* **2024**, *171*, 105–114. [CrossRef] [PubMed]
44. Liu, L.; Wang, J.; Wang, Y.; Chen, L.; Peng, L.; Bin, Y.; Ding, P.; Zhang, R.; Tong, F.; Dong, X. Blocking the MIF-CD74 axis augments radiotherapy efficacy for brain metastasis in NSCLC via synergistically promoting microglia M1 polarization. *J. Exp. Clin. Cancer Res. CR* **2024**, *43*, 128. [CrossRef]

45. Petrovic, M.; Majchrzak, O.B.; Marecar, R.; Laingoniaina, A.C.; Walker, P.R.; Borchard, G.; Jordan, O.; Tankov, S. Combining antimiR-25 and cGAMP Nanocomplexes Enhances Immune Responses via M2 Macrophage Reprogramming. *Int. J. Mol. Sci.* **2024**, *25*, 12787. [CrossRef]
46. Huang, S.; Zhao, Y.; Liao, P.; Wang, J.; Li, Z.; Tan, J.; Zha, X.; Chen, S.; Li, Y.; Zhong, L. Different expression patterns of VISTA concurrent with PD-1, Tim-3, and TIGIT on T cell subsets in peripheral blood and bone marrow from patients with multiple myeloma. *Front. Oncol.* **2022**, *12*, 1014904. [CrossRef]
47. Gong, X.; Cui, Q.; Zhang, W.; Shi, Y.; Zhang, P.; Zhang, C.; Hu, G.; Sahin, O.; Wang, L.; Shen, Z.; et al. Genomic insight into the diversity of *Glaesserella parasuis* isolates from 19 countries. *mSphere* **2024**, *9*, e0023124. [CrossRef]
48. Yang, Z.; Zhang, Y.; Zhao, Q.; Du, S.; Huang, X.; Wu, R.; Yan, Q.; Han, X.; Wen, Y.; Cao, S.J. HbpA from *Glaesserella parasuis* induces an inflammatory response in 3D4/21 cells by activating the MAPK and NF-κB signalling pathways and protects mice against *G. parasuis* when used as an immunogen. *Vet. Res.* **2024**, *55*, 93. [CrossRef]
49. Ghoochani, A.; Schwarz, M.A.; Yakubov, E.; Engelhorn, T.; Doerfler, A.; Buchfelder, M.; Bucala, R.; Savaskan, N.E.; Eyüpoglu, I.Y. MIF-CD74 signaling impedes microglial M1 polarization and facilitates brain tumorigenesis. *Oncogene* **2016**, *35*, 6246–6261. [CrossRef]
50. Meza-Romero, R.; Benedek, G.; Leng, L.; Bucala, R.; Vandenbark, A.A. Predicted structure of MIF/CD74 and RTL1000/CD74 complexes. *Metab. Brain Dis.* **2016**, *31*, 249–255. [CrossRef]
51. Calandra, T.; Bucala, R. Macrophage Migration Inhibitory Factor (MIF): A Glucocorticoid Counter-Regulator within the Immune System. *Crit. Rev. Immunol.* **2017**, *37*, 359–370. [CrossRef] [PubMed]
52. Simons, D.; Grieb, G.; Hristov, M.; Pallua, N.; Weber, C.; Bernhagen, J.; Steffens, G. Hypoxia-induced endothelial secretion of macrophage migration inhibitory factor and role in endothelial progenitor cell recruitment. *J. Cell. Mol. Med.* **2011**, *15*, 668–678. [CrossRef] [PubMed]
53. Ezaki, A.; Yano, H.; Pan, C.; Fujiwara, Y.; Anami, T.; Ibe, Y.; Motoshima, T.; Yatsuda, J.; Esumi, S.; Miura, Y.; et al. Potential protumor function of CD74 in clear cell renal cell carcinoma. *Hum. Cell* **2024**, *37*, 1535–1543. [CrossRef] [PubMed]
54. He, F.; Xu, J.; Zeng, F.; Wang, B.; Yang, Y.; Xu, J.; Sun, X.; Ren, T.; Tang, X. Integrative analysis of Ewing's sarcoma reveals that the MIF-CD74 axis is a target for immunotherapy. *Cell Commun. Signal. CCS* **2025**, *23*, 23. [CrossRef]
55. Pellegrino, B.; David, K.; Rabani, S.; Lampert, B.; Tran, T.; Doherty, E.; Piecychna, M.; Meza-Romero, R.; Leng, L.; Hershkovitz, D.; et al. CD74 promotes the formation of an immunosuppressive tumor microenvironment in triple-negative breast cancer in mice by inducing the expansion of tolerogenic dendritic cells and regulatory B cells. *PLoS Biol.* **2024**, *22*, e3002905. [CrossRef]
56. Klionsky, D.J.; Petroni, G.; Amaravadi, R.K.; Baehrecke, E.H.; Ballabio, A.; Boya, P.; Bravo-San Pedro, J.M.; Cadwell, K.; Cecconi, F.; Choi, A.M.K.; et al. Autophagy in major human diseases. *EMBO J.* **2021**, *40*, e108863. [CrossRef]
57. Levine, B.; Kroemer, G. Autophagy in the pathogenesis of disease. *Cell* **2008**, *132*, 27–42. [CrossRef]
58. Vishnupriya, S.; Priya Dharshini, L.C.; Sakthivel, K.M.; Rasmi, R.R. Autophagy markers as mediators of lung injury-implication for therapeutic intervention. *Life Sci.* **2020**, *260*, 118308. [CrossRef]
59. Hwang, H.J.; Ha, H.; Lee, B.S.; Kim, B.H.; Song, H.K.; Kim, Y.K. LC3B is an RNA-binding protein to trigger rapid mRNA degradation during autophagy. *Nat. Commun.* **2022**, *13*, 1436. [CrossRef]
60. Jeong, S.J.; Zhang, X.; Rodriguez-Velez, A.; Evans, T.D.; Razani, B. p62/SQSTM1 and Selective Autophagy in Cardiometabolic Diseases. *Antioxid. Redox Signal.* **2019**, *31*, 458–471. [CrossRef]
61. Cui, J.; Zhang, F.; Wang, Y.; Liu, J.; Ming, X.; Hou, J.; Lv, B.; Fang, S.; Yu, B. Macrophage migration inhibitory factor promotes cardiac stem cell proliferation and endothelial differentiation through the activation of the PI3K/Akt/mTOR and AMPK pathways. *Int. J. Mol. Med.* **2016**, *37*, 1299–1309. [CrossRef] [PubMed]
62. Wu, Y.T.; Tan, H.L.; Huang, Q.; Ong, C.N.; Shen, H.M. Activation of the PI3K-Akt-mTOR signaling pathway promotes necrotic cell death via suppression of autophagy. *Autophagy* **2009**, *5*, 824–834. [CrossRef] [PubMed]
63. Liu, B.; Deng, X.; Jiang, Q.; Li, G.; Zhang, J.; Zhang, N.; Xin, S.; Xu, K. Scoparone improves hepatic inflammation and autophagy in mice with nonalcoholic steatohepatitis by regulating the ROS/P38/Nrf2 axis and PI3K/AKT/mTOR pathway in macrophages. *Biomed. Pharmacother.* **2020**, *125*, 109895. [CrossRef]
64. Ullah, R.; Yin, Q.; Snell, A.H.; Wan, L. RAF-MEK-ERK pathway in cancer evolution and treatment. *Semin. Cancer Biol.* **2022**, *85*, 123–154. [CrossRef]
65. Umbarkar, P.; Tousif, S.; Singh, A.P.; Anderson, J.C.; Zhang, Q.; Tallquist, M.D.; Woodgett, J.; Lal, H. Fibroblast GSK-3α Promotes Fibrosis via RAF-MEK-ERK Pathway in the Injured Heart. *Circ. Res.* **2022**, *131*, 620–636. [CrossRef]
66. Jin, H.; Huang, X.; Pan, Q.; Ma, N.; Xie, X.; Wei, Y.; Yu, F.; Wen, W.; Zhang, B.; Zhang, P.; et al. The EIF3H-HAX1 axis increases RAF-MEK-ERK signaling activity to promote colorectal cancer progression. *Nat. Commun.* **2024**, *15*, 2551. [CrossRef]
67. Zhang, L.; Chen, L.; Qi, M.; Yu, F.; Ni, X.; Hong, H.; Xu, H.; Xu, S. Glyphosate induces autophagy in hepatic L8824 cell line through NO-mediated activation of RAS/RAF/MEK/ERK signaling pathway and energy metabolism disorders. *Fish Shellfish Immunol.* **2023**, *137*, 108772. [CrossRef]

68. Li, X.; Ma, A.; Liu, K. Geniposide alleviates lipopolysaccharide-caused apoptosis of murine kidney podocytes by activating Ras/Raf/MEK/ERK-mediated cell autophagy. *Artif. Cells Nanomed. Biotechnol.* **2019**, *47*, 1524–1532. [CrossRef]
69. Xie, J.; Wang, S.; Li, Z.; Ao, C.; Wang, J.; Wang, L.; Peng, X.; Zeng, K. 5-aminolevulinic acid photodynamic therapy reduces HPV viral load via autophagy and apoptosis by modulating Ras/Raf/MEK/ERK and PI3K/AKT pathways in HeLa cells. *J. Photochem. Photobiol. B Biol.* **2019**, *194*, 46–55. [CrossRef]
70. Bardi, G.T.; Smith, M.A.; Hood, J.L. Melanoma exosomes promote mixed M1 and M2 macrophage polarization. *Cytokine* **2018**, *105*, 63–72. [CrossRef]
71. Yunna, C.; Mengru, H.; Lei, W.; Weidong, C. Macrophage M1/M2 polarization. *Eur. J. Pharmacol.* **2020**, *877*, 173090. [CrossRef] [PubMed]
72. Cutolo, M.; Campitiello, R.; Gotelli, E.; Soldano, S. The Role of M1/M2 Macrophage Polarization in Rheumatoid Arthritis Synovitis. *Front. Immunol.* **2022**, *13*, 867260. [CrossRef] [PubMed]
73. Sanjurjo, L.; Aran, G.; Téllez, É.; Amézaga, N.; Armengol, C.; López, D.; Prats, C.; Sarrias, M.R. CD5L Promotes M2 Macrophage Polarization through Autophagy-Mediated Upregulation of ID3. *Front. Immunol.* **2018**, *9*, 480. [CrossRef]
74. Fang, X.; Lan, X.; Zhu, M.; He, M.; Sun, M.; Cao, Y.; Zhu, D.; Guo, D.; Luo, H. Puerarin Induces Macrophage M2 Polarization to Exert Antinonalcoholic Steatohepatitis Pharmacological Activity via the Activation of Autophagy. *J. Agric. Food Chem.* **2024**, *72*, 7187–7202. [CrossRef]
75. Geng, P.; Zhu, H.; Zhou, W.; Su, C.; Chen, M.; Huang, C.; Xia, C.; Huang, H.; Cao, Y.; Shi, X. Baicalin Inhibits Influenza A Virus Infection via Promotion of M1 Macrophage Polarization. *Front. Pharmacol.* **2020**, *11*, 01298. [CrossRef]
76. van Elsas, M.J.; Middelburg, J.; Labrie, C.; Roelands, J.; Schaap, G.; Sluijter, M.; Tonea, R.; Ovcinnikovs, V.; Lloyd, K.; Schuurman, J.; et al. Immunotherapy-activated T cells recruit and skew late-stage activated M1-like macrophages that are critical for therapeutic efficacy. *Cancer Cell* **2024**, *42*, 1032–1050.e10. [CrossRef]
77. Liu, Q.; Yang, C.; Wang, S.; Shi, D.; Wei, C.; Song, J.; Lin, X.; Dou, R.; Bai, J.; Xiang, Z.; et al. Wnt5a-induced M2 polarization of tumor-associated macrophages via IL-10 promotes colorectal cancer progression. *Cell Commun. Signal. CCS* **2020**, *18*, 51. [CrossRef]
78. Hong, S.M.; Lee, A.Y.; Kim, B.J.; Lee, J.E.; Seon, S.Y.; Ha, Y.J.; Ng, J.T.; Yoon, G.; Lim, S.B.; Morgan, M.J.; et al. NAMPT-Driven M2 Polarization of Tumor-Associated Macrophages Leads to an Immunosuppressive Microenvironment in Colorectal Cancer. *Adv. Sci.* **2024**, *11*, e2303177. [CrossRef]
79. Li, H.; Xiao, Y.; Li, Q.; Yao, J.; Yuan, X.; Zhang, Y.; Yin, X.; Saito, Y.; Fan, H.; Li, P.; et al. The allergy mediator histamine confers resistance to immunotherapy in cancer patients via activation of the macrophage histamine receptor H1. *Cancer Cell* **2022**, *40*, 36–52.e9. [CrossRef]
80. Yeung, M.Y.; Grimmig, T.; Sayegh, M.H. Costimulation Blockade in Transplantation. *Adv. Exp. Med. Biol.* **2019**, *1189*, 267–312.
81. Yuan, Y.; Hua, L.; Zhou, J.; Liu, D.; Ouyang, F.; Chen, X.; Long, S.; Huang, Y.; Liu, X.; Zheng, J.; et al. The effect of artesunate to reverse CLP-induced sepsis immunosuppression mice with secondary infection is tightly related to reducing the apoptosis of T cells via decreasing the inhibiting receptors and activating MAPK/ERK pathway. *Int. Immunopharmacol.* **2023**, *124 Pt A*, 110917. [CrossRef]
82. Wang, X.; Zhou, H.; Liu, Q.; Cheng, P.; Zhao, T.; Yang, T.; Zhao, Y.; Sha, W.; Zhao, Y.; Qu, H. Targeting regulatory T cells for cardiovascular diseases. *Front. Immunol.* **2023**, *14*, 1126761. [CrossRef] [PubMed]
83. Finisguerra, V.; Dvorakova, T.; Formenti, M.; Van Meerbeeck, P.; Mignion, L.; Gallez, B.; Van den Eynde, B.J. Metformin improves cancer immunotherapy by directly rescuing tumor-infiltrating CD8 T lymphocytes from hypoxia-induced immunosuppression. *J. Immunother. Cancer* **2023**, *11*, e005719. [CrossRef] [PubMed]
84. Naoi, Y.; Morinaga, T.; Nagasaki, J.; Ariyasu, R.; Ueda, Y.; Yamashita, K.; Zhou, W.; Kawashima, S.; Kawase, K.; Honobe-Tabuchi, A.; et al. CD106 in Tumor-Specific Exhausted $CD8^+$ T Cells Mediates Immunosuppression by Inhibiting TCR Signaling. *Cancer Res.* **2024**, *84*, 2109–2122. [CrossRef] [PubMed]

veterinary sciences

MDPI

Article

Euphorbia humifusa Willd. ex Schltdl. Mitigates Liver Injury via KEAP1-NFE2L2-Mediated Ferroptosis Regulation: Network Pharmacology and Experimental Validation

Hongxu Du [1,2,*], Kunzhao Yang [1], Jingyi Yang [1], Junjie Wan [1], Yu Pan [1], Weijie Song [1], Shuang Xu [1], Cheng Chen [1] and Jiahui Li [1]

[1] Department of Traditional Chinese Veterinary Medicine, College of Veterinary Medicine, Southwest University, Chongqing 402460, China
[2] Institute of Traditional Chinese Veterinary Medicine, Southwest University, Chongqing 402460, China
* Correspondence: hongxudu@swu.edu.cn

Simple Summary: Liver injury, a prevalent disorder in animals with a multifactorial etiology, is critically influenced by ferroptosis—an iron-dependent cell death mechanism. Although *Euphorbia humifusa* Willd. ex Schltdl. (EHW) exhibits pharmacological properties, its role in modulating ferroptosis-associated liver injury remains underexplored. This study evaluated EHW's therapeutic effects using a CCl_4-induced murine hepatotoxicity model. EHW administration significantly attenuated liver dysfunction, mitigated ferroptosis, and restored the regulatory pathways linked to iron metabolism and lipid peroxidation. These findings position EHW as a promising therapeutic agent for veterinary liver diseases, primarily through ferroptosis pathway regulation.

Abstract: Liver injury poses major health risks in livestock, necessitating effective therapeutic interventions. This study elucidates the hepatoprotective mechanisms of *Euphorbia humifusa* Willd. ex Schltdl. (EHW) by integrating network pharmacology, molecular docking, and experimental validation. Using a CCl_4-induced liver injury model mimicking veterinary clinical scenarios, EHW markedly alleviated hepatic damage, demonstrated by reduced liver index, serum ALT and AST levels, histopathological lesions, iron accumulation, inflammatory cytokines, and ferroptosis-associated gene expression. Network pharmacology identified EHW's core bioactive components (quercetin, kaempferol, and β-sitosterol) and critical targets (*IL-6*, *STAT3*, *HIF-1α*, *PTGS2*, *NFE2L2*, and *KEAP1*) which were linked to ferroptosis and oxidative stress. Molecular docking revealed robust binding affinities between these compounds and ferroptosis-related proteins. *In vivo* validation confirmed that EHW inhibited *KEAP1*, activated *NFE2L2*-mediated antioxidant defenses (upregulating *SOD1* and *NQO1*), restored iron homeostasis (lowering *TFR1*, elevating *FTH1*), and attenuated phospholipid peroxidation by suppressing *ACSL4* and *ALOX12*. These results indicate that EHW mitigates ferroptosis-driven liver injury via KEAP1-NFE2L2 signaling to restore iron homeostasis and reduce oxidative stress, offering a mechanistic foundation for its clinical application in veterinary hepatoprotection.

Keywords: *Euphorbia humifusa* Willd. ex Schltdl.; hepatic injury; ferroptosis; network pharmacology

1. Introduction

The liver is one of the most vital organs in livestock, playing a crucial role in key physiological functions such as synthesis, metabolism, and immunity [1]. Once liver damage occurs, it severely disrupts these essential physiological processes [2]. In veterinary practice, liver injury is commonly caused by factors such as poor feed quality [3], pathogenic microbial infections [4], the ingestion of toxic substances [5], or improper management practices [6]. When liver damage occurs, it can lead to significant harm to the animal's health. If timely interventions are not implemented, liver injury may rapidly progress, resulting in liver failure and potentially leading to death [7]. Therefore, there is an urgent need to identify effective therapeutic agents to address the common issue of liver damage in veterinary clinical settings.

As a key organ for iron storage and metabolism in the body, maintaining normal iron homeostasis is crucial for the liver to preserve its proper function [8]. Ferroptosis is a recently emerging form of iron-dependent programmed cell death, characterized by the peroxidation of membrane lipids [9]. Various pathways, including redox balance, lipid metabolism, and iron metabolism, can influence the progression of ferroptosis in cells [10]. Research has demonstrated that ferroptosis exerts a dual function in liver diseases. Specifically, in the context of hepatocellular carcinoma, the induction of ferroptosis has been established as a significant strategy to retard tumor progression [11]. However, in the case of non-alcoholic fatty liver disease, alcoholic liver injury, and liver fibrosis, inhibiting ferroptosis has demonstrated beneficial effects in alleviating the condition [12–14]. Furthermore, in liver injury studies, Yang et al. [15] found that Maresin 1 could inhibit ferroptosis by activating the *NFE2L2/HMOX1/GPX4* pathway, thereby reducing liver damage. Similarly, Wei et al. [16] found that Gan Dan Kang, a traditional Chinese medicine (TCM) compound, could alleviate CCL_4-induced hepatocellular ferroptosis in mice through the activation of the *KEAP1/NFE2L2* pathway. In veterinary clinical practice, inhibiting ferroptosis in liver tissue has also been widely recognized as an important approach to mitigating hepatic injury in livestock. For instance, Zhao et al. [17] found that Phlorotannins could alleviate heat-stress-induced liver damage in broiler chickens by reducing hepatic ferroptosis. Similarly, Huang et al. [18] demonstrated that the total flavonoids of *Rhizoma Drynaria* could reverse AFB_1-induced liver injury in broilers by inhibiting ferroptosis.

Euphorbia humifusa Willd. ex Schltdl. (EHW) is an annual herbaceous plant widely distributed in China. As a TCM, it is also known as Dijincao and has been documented in the *Compendium of Materia Medica* (*Ben-Cao-Gang-Mu* in Chinese) since ancient times [19]. According to TCM theory, EHW is considered to have a pungent and neutral nature, with properties that include clearing heat, detoxifying and cooling the blood, and stopping bleeding, as well as promoting diuresis and alleviating jaundice. Modern medical studies have shown that EHW possesses anti-inflammatory [20], antiviral [21], hypoglycemic [22], and anti-tumor effects [23], and its therapeutic efficacy in treating intestinal and hepatic inflammation is widely acknowledged [24]. However, the potential of EHW to alleviate liver damage through the ferroptosis pathway has not yet been reported. Therefore, this study aims to construct a liver injury model in mice using CCL_4, a classic hepatotoxic substance, and to employ network pharmacology to predict potential targets, followed by experimental validation. The goal is to explore whether EHW can mitigate liver damage through the regulation of ferroptosis and to investigate the underlying mechanisms. The findings from this study may contribute to a deeper understanding of the mechanisms of EHW's action and provide a theoretical basis for its clinical application.

2. Materials and Methods

2.1. Reagents and Materials

EHW was purchased from Kangyiyin Biotechnology Co., Ltd. (Bozhou, China). CCL_4 was obtained from Guangdong Puhui Chemical Technology Co., Ltd. (Shaoguan, China).

The reagent kits for aspartate aminotransferase (AST) and alanine aminotransferase (ALT) were purchased from Nanjing Jiancheng Bioengineering Institute (Nanjing, China). Hematoxylin and eosin staining kits were from BKMAM Biology Technology Co., Ltd. (Changde, China), and the Prussian blue staining kit and 2 × Universal Blue SYBR Green qPCR Master Mix were from Wuhan Servicebio Technology Co., Ltd. (Wuhan, China). RNAiso Plus was purchased from TaKaRa Bio Lnc. (Tokyo, Japan), and the All-in-One 5 × RT MasterMix reverse transcription kit was obtained from Applied Biological Materials Inc. (Vancouver, BC, Canada).

2.2. Preparation of the Drug

Distilled water was added to a total of 50 g of EHW at a ratio of 15:1. After soaking for 50 min, the mixture was brought to a boil over a high heat and then simmered for 45 min. The liquid was filtered through gauze, and the filtrate was collected. The remaining drug residue was then re-extracted with an equal amount of distilled water, brought to a boil over high heat, and simmered for 45 min. This process was repeated three times. Then, the combined liquid was concentrated using a rotary evaporator to a final concentration of 0.4 g/mL. Finally, a portion of this concentrated liquid was diluted with distilled water to achieve a concentration of 0.2 g/mL. Both concentrations were aliquoted and stored in a refrigerator at 4 °C for later use.

2.3. Animal Grouping and Treatment

A total of 24 SPF-grade male KM mice (purchased from Southwest Medical University) were randomly assigned into four groups: the control group (CON), model group (CCL_4), high-dose EHW group (EHW-H), and low-dose EHW group (EHW-L), with 6 mice in each group. After 3 d of acclimatization, all mice were treated with daily oral gavage for 7 consecutive days at a dose of 0.1 mL/10 g of EHW decoction. The high-dose group received EHW at 4 g/kg, and the low-dose group received 2 g/kg. The CON and CCL_4 groups were given an equivalent volume of sterile water. One hour after the final dose, the mice were injected intraperitoneally with CCL_4 solution (0.1% CCL_4 in olive oil) at a dose of 0.1 mL/10 g. Analogously, the CON group received an equivalent volume of olive oil. Eighteen hours later, the mice were gavaged once more and then anesthetized for blood collection from the ocular sinus. Finally, the mice were euthanized via cervical dislocation, and their liver tissues were harvested. Part of the liver was fixed in 4% paraformaldehyde, and another portion was rapidly frozen in liquid nitrogen and stored at −80 °C for long-term preservation. The experimental procedure and treatment flow was created with the BioGDP website (https://biogdp.com/ (accessed on 17 October 2024)) [25], as shown in Figure 1. All animal experiments were conducted in accordance with the guidelines approved by the Ethics Committee of Southwest University for Animal Welfare (Ethics No. LAC2024-1-0346).

2.4. Measurement of Liver Organ Index

The liver was carefully removed from the mice, and the surface moisture was gently blotted with bibulous paper. The organs were then precisely weighed using an electronic balance, and the organ index was calculated using the following formula:

$$\text{Liver Index} = \text{Organ Weight (g)}/\text{Body Weight (g)}$$

2.5. Serum AST and ALT Activity Assays in Mice

Serum samples from mice were used to measure the activities of AST and ALT using the respective test kits, following the instructions provided by the manufacturers. These assays were performed to evaluate the liver function status of the mice.

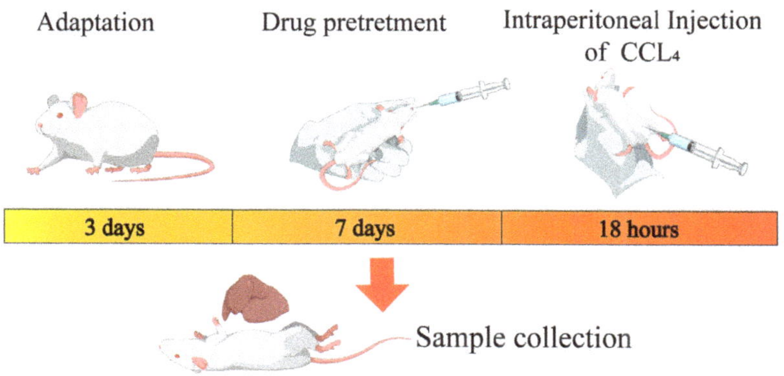

Figure 1. Animal experimental flowchart.

2.6. Hematoxylin and Eosin (HE) Staining

Fresh liver tissue samples were fixed in 4% paraformaldehyde and processed through gradient ethanol dehydration. The samples were then embedded in paraffin, sectioned, and stained with hematoxylin and eosin (HE). After mounting, the pathological changes in the liver tissue of the mice from each group were observed under a microscope (CX31RTSF, OLYMPUS).

2.7. Prussian Blue Staining

The process of tissue embedding and sectioning was performed as described in Section 2.6. For Prussian blue staining, the protocol provided by the manufacturer was strictly followed. After mounting, the liver tissue sections were examined for iron deposition in the livers of the mice across different groups.

2.8. Collection of EHW Chemical Constituents

The primary active components of EHW were retrieved from the TCMSP database (https://www.tcmsp-e.com/tcmsp.php (accessed on 3 November 2024)) by searching for "Diercao". The active ingredients were selected based on the criteria of oral bioavailability (OB) $\geq$ 30% and drug-likeness (DL) $\geq$ 0.18.

2.9. Screening of Active Ingredient Targets

SMILES identifiers for all compounds of EHW were obtained from the PubChem database (https://pubchem.ncbi.nlm.nih.gov/ (accessed on 3 November 2024)). Potential targets were identified by using the Swiss Target Prediction database (http://www.swisstargetprediction.ch/ (accessed on 3 November 2024)), with targets with a probability greater than 0 considered. Additionally, the Super-PRED database (https://prediction.charite.de/ (accessed on 3 November 2024)) was used to further identify potential targets of EHW active components. The active ingredient targets were then combined with those from the TCMSP database and duplicates were removed, resulting in a comprehensive list of drug targets.

2.10. Screening of Liver Injury and Ferroptosis Targets, and Identification of Potential EHW Targets for Liver Injury and Ferroptosis Treatment

Liver-injury-related targets were collected using the GeneCards database (https://www.genecards.org/ (accessed on 18 October 2024)) with the keyword "liver injury," selecting those with a relevance score > 8.0. Additional liver-injury-related targets were retrieved from the DisGeNET database (https://www.disgenet.org/ (accessed on 2 March 2024)). The two target sets were combined and duplicates removed, forming the disease target set. Ferroptosis-related targets were compiled from the FerrDb database (http://www.zhounan.org/ferrdb (accessed on 27 September 2024)), and duplicates were similarly removed to generate the ferroptosis target set. The three sets of drug active ingredient targets, disease targets, and ferroptosis targets were then intersected using Venn diagrams created by https://www.bioinformatics.com.cn (accessed on 25 December 2024), an online platform to identify potential therapeutic targets for EHW in liver injury through ferroptosis modulation.

2.11. Construction of the "Drug–Disease–Active Ingredient–Intersecting Target" Network

The intersecting targets obtained in Section 2.10 were used to construct network and type files, which were then imported into Cytoscape 3.9.1 software to create a "drug–disease–active ingredient–intersecting target" network diagram.

2.12. Protein–Protein Interaction (PPI) Network Construction for Intersecting Targets

The intersecting targets identified in Section 2.10 were imported into the String platform (https://string-db.org/ (accessed on 3 November 2024)) to retrieve protein–protein interaction (PPI) data while removing isolated targets. The resulting PPI data were then imported into the Cytoscape 3.9.1 software for topological analysis and visualization.

2.13. GO Enrichment and KEGG Pathway Enrichment Analysis

The intersecting targets obtained in Section 2.10 were imported into the DAVID database (https://david.ncifcrf.gov/ (accessed on 3 November 2024)) for Gene Ontology (GO) functional and Kyoto Encyclopedia of Genes and Genomes (KEGG) pathway enrichment analysis. This provided insights into the potential biological processes (BPs), cellular components (CCs), molecular functions (MFs), and signaling pathways involved in the treatment of liver injury through ferroptosis by EHW. Enriched GO functions and KEGG pathways were visualized using bubble charts generated by https://www.bioinformatics.com.cn (accessed on 3 November 2024), an online platform for data analysis and visualization.

2.14. Molecular Docking

The SDF format files for the active ingredients of EHW were obtained from the PubChem database. Core protein PDB numbers were determined using Uniprot (https://www.uniprot.org/ (accessed on 5 December 2024)), and the PDB format files for these proteins were downloaded from the PDB database (https://www.rcsb.org/ (accessed on 5 December 2024)). Molecular docking and visualization were performed using the MOE 2022.02 software.

2.15. RNA Extraction and Quantitative PCR (qPCR) Analysis

Total RNA was extracted strictly according to the manufacturer's instructions for the Trizol reagent kit. RNA concentration and quality were assessed using a NanoDrop 200 ultra-micro spectrophotometer. cDNA was synthesized following the guidelines provided by the reverse transcription kit. GAPDH was used as the internal reference gene, and RT-qPCR reactions were performed using a fluorescence quantitative PCR machine (FQD-96X, BIOER). The relative expression of target genes was analyzed using the $2^{-\Delta\Delta Ct}$ method [26]. Primer sequences for the relevant genes were synthesized by Shanghai Sangon Biotech Co., Ltd. (Shanghai, China), with the sequence details provided in Table 1.

2.16. Statistical Analysis

Data were analyzed using IBM SPSS Statistics 26, and the results were expressed as the mean ± standard error of the mean (SEM), and statistical significance was determined using one-way ANOVA followed by Duncan's multiple range test. GraphPad Prism 9 was utilized for data visualization. In all statistical analyses, a p-value of less than 0.05 was considered to indicate a statistically significant difference.

Table 1. Sequences of primers for RT-qPCR.

Gene	Primer (5′→3′)	Length (bp)
IL-1β	F: CACTACAGGCTCCGAGATGAACAAC R: TGTCGTTGCTTGGTTCTCCTTGTAC	145
IL-6	F: CTTCTTGGGACTGATGCTGGTGAC R: TCTGTTGGGAGTGGTATCCTCTGTG	91
TNF-α	F: CGCTCTTCTGTCTACTGAACTTCGG R: GTGGTTTGTGAGTGTGAGGGTCTG	113
PTGS2	F: TCTGGTGCCTGGTCTGATGATG R: CTATGAGTATGAGTCTGCTGGTTTGG	134
FSP-1	F: AGAACCGGATGGTGTTGCTAC R: CACCTCGTTAAACTTGCCAGG	104
TFR1	F: AGCCAGATCAGCATTCTCTAACTTG R: CTCCACATGACTGTTATCTCCATCTAC	100
FTH1	F: CAAGTGCGCCAGAACTACCA R: GCCACATCATCTCGGTCAAAA	122
FPN	F: ACCAAGGCAAGAGATCAAACC R: AGACACTGCAAAGTGCCACAT	138
GPX4	F: CTCCGAGTTCCTGGGCTTGTG R: CCGTCGATGTCCTTGGCTGAG	87
ACSL4	F: CTCACCATTATATTGCTGCCTGT R: TCTCTTTTTGCCATAGCGTTTTTCT	116
LPCAT3	F: CCCCACATCACAGACGACTATC R: TCTCACGGTCCCATTTTCATC	192
ALOX12	F: ACCTCAGACAATAGCAGCGGA R: TCAACGTCCATTCAAAGTCCAG	92
P62	F: GAACTCGCTATAAGTGCAGTGT R: AGAGAAGCTATCAGAGAGGTGG	131
NFE2L2	F: GTTGCCACCGCCAGGACTAC R: AAACTTGTACCGCCTCGTCTGG	88
KEAP1	F: CTGGAGGATCATACCAAGCAGG R: GGATACCCTCAATGGACACCAC	220
NQO1	F: TGGCCGAACACAAGAAGCTG R: GCTACGAGCACTCTCTCAAACC	112
HMOX1	F: CTGGAGATGACACCTGAGGTCAA R: CTGACGAAGTGACGCCATCTG	150
SOD1	F: GGAACCATCCACTTCGAGCAG R: ACAGCCTTGTGTATTGTCCCC	126
GAPDH	F: GTCGTGGAGTCTACTGGTGTCTTC R: AGTTGTCATATTTCTCGTGGTTCA	145

3. Results

3.1. Liver Organ Index and Liver Function Analysis

The results of the liver organ index for each group of mice are shown in Figure 2A. Compared to the CON group, the liver organ index in the CCL_4 group was significantly elevated ($p < 0.05$). However, in comparison to the CCL_4 group, the liver organ index in both the EHW-H and EHW-L groups significantly decreased ($p < 0.05$). The liver function indicators for the EHW treatment groups are presented in Figure 2B,C. Compared to the CON group, serum AST and ALT activities were markedly increased in the CCL_4 group ($p < 0.01$). Treatment with EHW reversed this trend, as the serum ALT activity in the EHW-H group significantly decreased compared to the CCL_4 group ($p < 0.05$), and there was a downward trend in the serum ALT activity in the EHW-L group. Additionally, serum AST activity in both the EHW-H and EHW-L groups was significantly reduced ($p < 0.01$), with the EHW-H group showing a more pronounced effect than the EHW-L group.

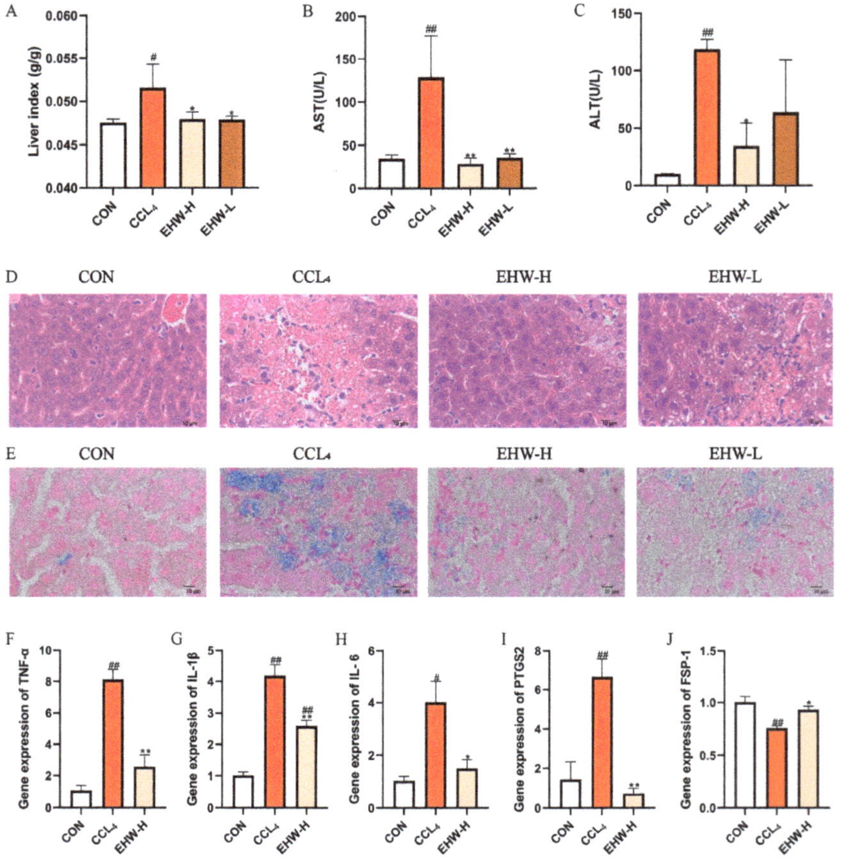

Figure 2. (**A**) Changes in the organ index of mouse livers. (**B**) Changes in serum AST levels. (**C**) Changes in serum ALT levels. (**D**) HE staining results of mouse liver tissue. (**E**) Prussian blue staining results of mouse liver tissue. (**F**) Relative mRNA expression of *TNF-α* in mouse liver. (**G**) Relative mRNA expression of *IL-1β* in mouse liver. (**H**) Relative mRNA expression of *IL-6* in mouse liver. (**I**) Relative mRNA expression of *PTGS2* in mouse liver. (**J**) Relative mRNA expression of *FSP-1* in mouse liver. Note: * indicates a statistically significant difference compared to the CCL_4 group, # indicates a statistically significant difference compared to the CON group, * $p < 0.05$, ** $p < 0.01$; # $p < 0.05$, ## $p < 0.01$.

3.2. Pathological Histological Examination of HE Staining

The HE staining results of the liver tissues from the mice in each experimental group are shown in Figure 2D. In the CON group, the liver tissue exhibited a normal structure, with intact hepatocytes and no significant pathological damage or inflammatory cell infiltration. In the CCL_4 group, hepatic cords were disorganized, and prominent fatty degeneration of the liver was observed, along with extensive hepatocyte necrosis and inflammatory cell infiltration. In contrast, the liver architecture in both the EHW-H and EHW-L groups was partially restored, fatty degeneration was alleviated, hepatocyte necrosis was notably improved, and the number of inflammatory cells infiltrating the liver was significantly reduced. The EHW-H group showed a better therapeutic effect compared to the EHW-L group.

3.3. Prussian Blue Staining Examination

The Prussian blue staining results for the liver tissues from each experimental group are shown in Figure 2E. No significant blue iron ferricyanide deposits were observed in the liver tissues of the CON group. In the CCL_4 group, noticeable blue deposits were present in the areas of liver necrosis, indicating iron overload. In the EHW-H and EHW-L groups, the liver necrosis areas were significantly reduced, and the blue deposition at the necrotic sites was markedly lower than in the CCL_4 group. Furthermore, the EHW-H group demonstrated a more pronounced reduction in iron deposition compared to the EHW-L group.

3.4. Expression of Liver Inflammatory Cytokines and Ferroptosis-Related Genes

The relative expression levels of liver inflammatory cytokines and ferroptosis-related genes in mice are shown in Figure 2F–J. Compared to the CON group, the relative mRNA expression levels of *TNF-α*, *IL-1β*, and *PTGS2* were significantly elevated in the CCL_4 group ($p < 0.01$), and *IL-6* mRNA expression was also significantly increased ($p < 0.05$). In contrast, the mRNA levels of *TNF-α*, *IL-1β*, and *PTGS2* were significantly reduced in the EHW-H group compared to the CCL_4 group ($p < 0.01$), and *IL-6* expression also decreased significantly ($p < 0.05$). Additionally, the mRNA expression of *FSP-1* was significantly increased in the EHW-H group ($p < 0.05$), while it was significantly decreased in the CCL_4 group ($p < 0.01$).

3.5. Major Active Components of EHW

Based on the criteria of OB $\geq$ 30% and DL $\geq$ 0.18 in the TCMSP database, six major active components of EHW were identified. Detailed information on these compounds is presented in Table 2.

Table 2. Information on the major active components of EHW.

Molid	Active Ingredient	OB (%)	DL
MOL001002	ellagic acid	43.06	0.43
MOL000359	sitosterol	36.91	0.75
MOL000422	kaempferol	41.88	0.24
MOL006326	ensaculin	45.76	0.86
MOL006331	4′,5-dihydroxyflavone	48.55	0.19
MOL000098	quercetin	46.43	0.28

3.6. Screening of Drug Active Component Targets

Using the TCMSP, Swiss Target Prediction, and SuperPred databases, the targets of the screened active components of EHW were collected and analyzed. After removing duplicates, a total of 578 potential targets for the active components of EHW were obtained.

3.7. Potential Targets of EHW for Improving Ferroptosis-Related Liver Injury

By searching the GeneCards database with the keyword "liver injury" and applying the filtering criterion of having a relevance score > 8, the target information was merged with data from the DisGeNET database. After removing duplicate entries, 2360 disease-related targets for liver injury were identified. In the FerrDb database, 564 ferroptosis-related disease targets were obtained. By intersecting the 578 targets of EHW with the 564 ferroptosis-related targets and the 2360 liver injury-related targets, 52 potential targets for EHW in improving ferroptosis-related liver injury were identified. The Venn diagram illustrating these results is shown in Figure 3A.

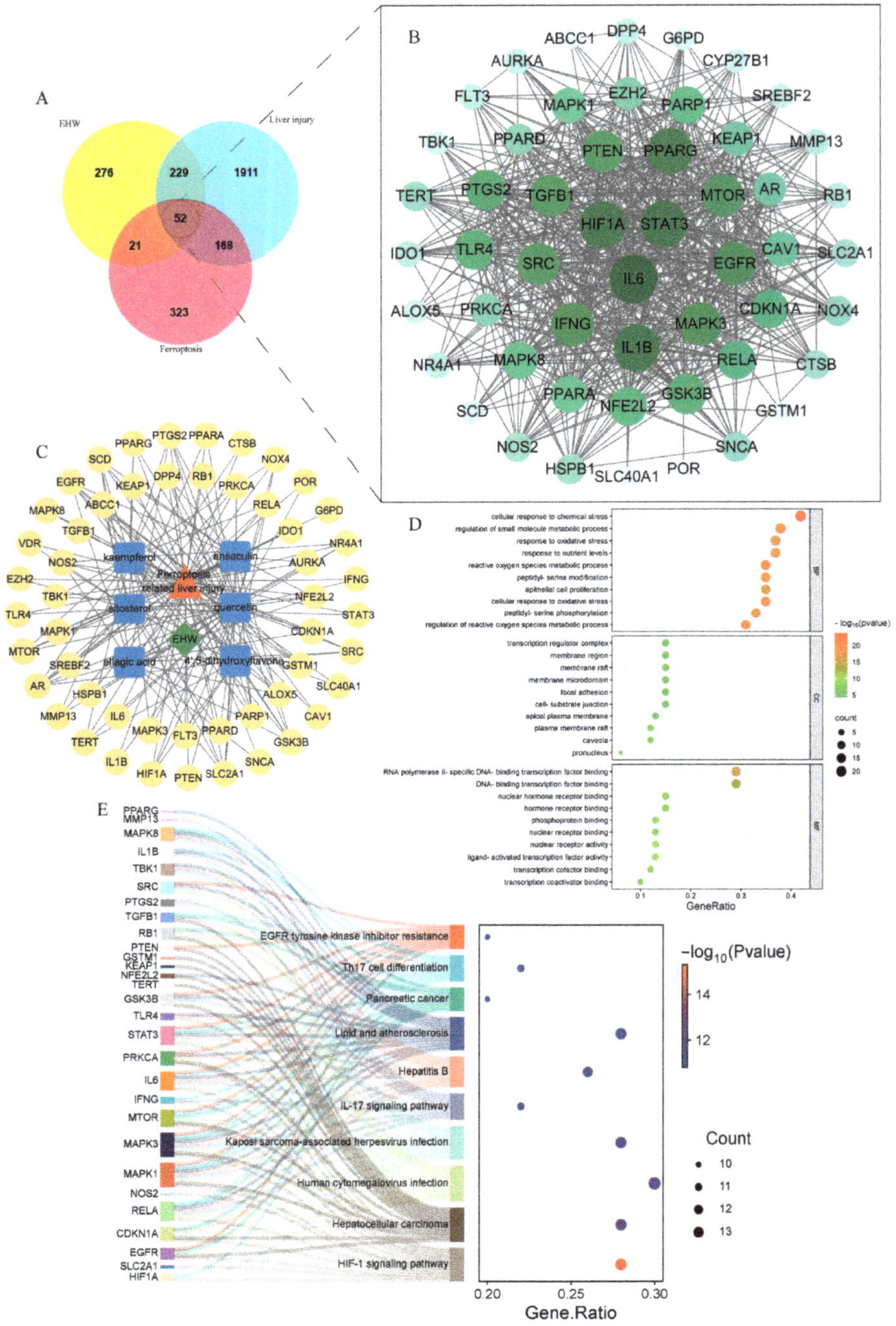

Figure 3. (**A**) Venn diagram of EHW component targets, liver injury targets, and ferroptosis targets. (**B**) PPI network diagram of the intersecting targets. (**C**) "Drug–active component–intersecting target" diagram. (**D**) GO enrichment analysis of EHW's effects on ferroptosis-related liver injury. (**E**) KEGG enrichment analysis of EHW's effects on ferroptosis-related liver injury.

3.8. Construction of PPI Network for the Intersecting Targets

The potential targets of EHW for improving ferroptosis-related liver injury were input into the STRING database to construct a PPI network, and the data were visualized using Cytoscape 3.9.1 software (Figure 3B). By calculating the degree values, the top core targets were found to be *IL-6, STAT3, HIF1A, PTGS2, NFE2L2,* and *KEAP1*, among others.

3.9. "Drug–Disease–Active Component–Intersecting Target" Network Construction

As shown in Figure 3C, the "drug–disease–active component–intersecting target" relationship network was constructed. The network contained 60 nodes and 184 edges. Based on the degree values, the core active components of EHW for improving ferroptosis-related liver injury were identified. The top-ranked active components included quercetin, kaempferol, and sitosterol.

3.10. GO Enrichment Analysis and KEGG Pathway Enrichment Analysis

The intersecting targets were subjected to GO functional enrichment analysis and KEGG pathway enrichment analysis ($p < 0.05$). A total of 2214 BPs, 75 CCs, 132 MFs, and 156 signaling pathways were identified. As shown in Figure 3D,E, the potential targets of EHW for alleviating ferroptosis-related liver injury mainly involved BPs such as response to oxidative stress, the regulation of reactive oxygen species, metabolic processes, and ROS metabolic processes. These targets were predominantly located in cellular components like the transcription regulator complex, membrane regions, and membrane rafts. Molecular functions such as RNA polymerase II-specific DNA-binding transcription factor binding, transcription coactivator binding, and DNA-binding transcription factor binding were the most prominent. The KEGG pathway enrichment results indicated that the enriched pathways were primarily associated with Hepatitis B, the *HIF-1* signaling pathway, and the *IL-17* signaling pathway.

3.11. Molecular Docking Results

Figure 4 presents the molecular docking results between the main active components of EHW and the potential key target proteins associated with ferroptosis-related liver injury. The key target proteins identified through topological analysis, including *NFE2L2, KEAP1, LPCAT3, ALOX12,* and *ACSL4*, were pre-processed and subjected to molecular docking alongside the core active components. The results showed that the top three active components, based on their degree value, had binding energies with ferroptosis-related proteins *NFE2L2, KEAP1, LPCAT3, ALOX12,* and *ACSL4* which were lower than −4.5 kcal/mol. Based on the binding energies, the stronger binding component–target combinations were selected for visualization (Figure 5). The results demonstrated that quercetin interacts with the target *ACSL4* at the Tyr468 and ASP573 sites, sitosterol interacts with *ALOX12* at the Asn287 site, sitosterol interacts with *KEAP1* at the Arg326 site, quercetin interacts with *LPCAT3* at the Cys383 site, and quercetin interacts with *NFE2L2* at the Val606 and Val463 sites.

3.12. Effect of EHW on the KEAP1-NFE2L2 Signaling Pathway

As shown in Figure 6, compared to the CON group, the mRNA expression level of *KEAP1* in the CCL_4 group was significantly elevated ($p < 0.05$), while the mRNA expression of *SOD-1* was significantly reduced ($p < 0.01$). Additionally, the mRNA expression levels of *P62, NFE2L2,* and *NQO1* were significantly decreased ($p < 0.05$), and the mRNA expression level of *HMOX1* showed a downward trend. However, in comparison to the CCL_4 group, the EHW-H group exhibited a significant reduction in *KEAP1* mRNA expression ($p < 0.05$). Moreover, the mRNA expression of *P62* and *SOD-1* was significantly increased ($p < 0.01$), and the mRNA levels of *NFE2L2*, *NQO1*, and *HMOX1* were all significantly elevated ($p < 0.05$).

Figure 4. Heatmap of the binding affinity between the key targets related to ferroptosis and liver injury and the active components of EHW.

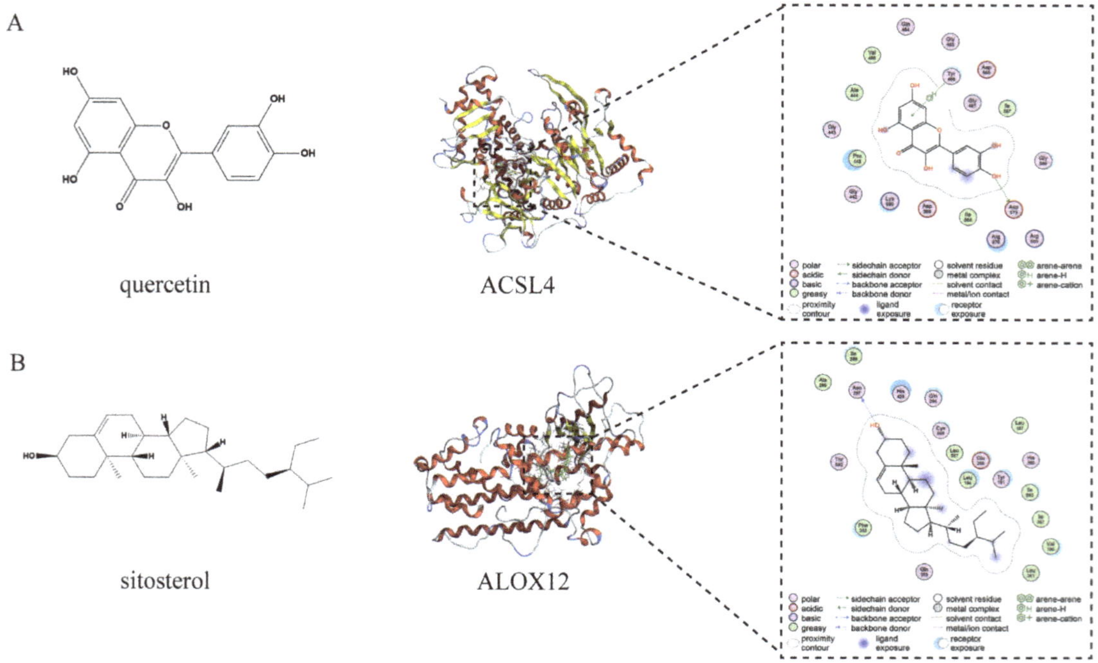

Figure 5. *Cont.*

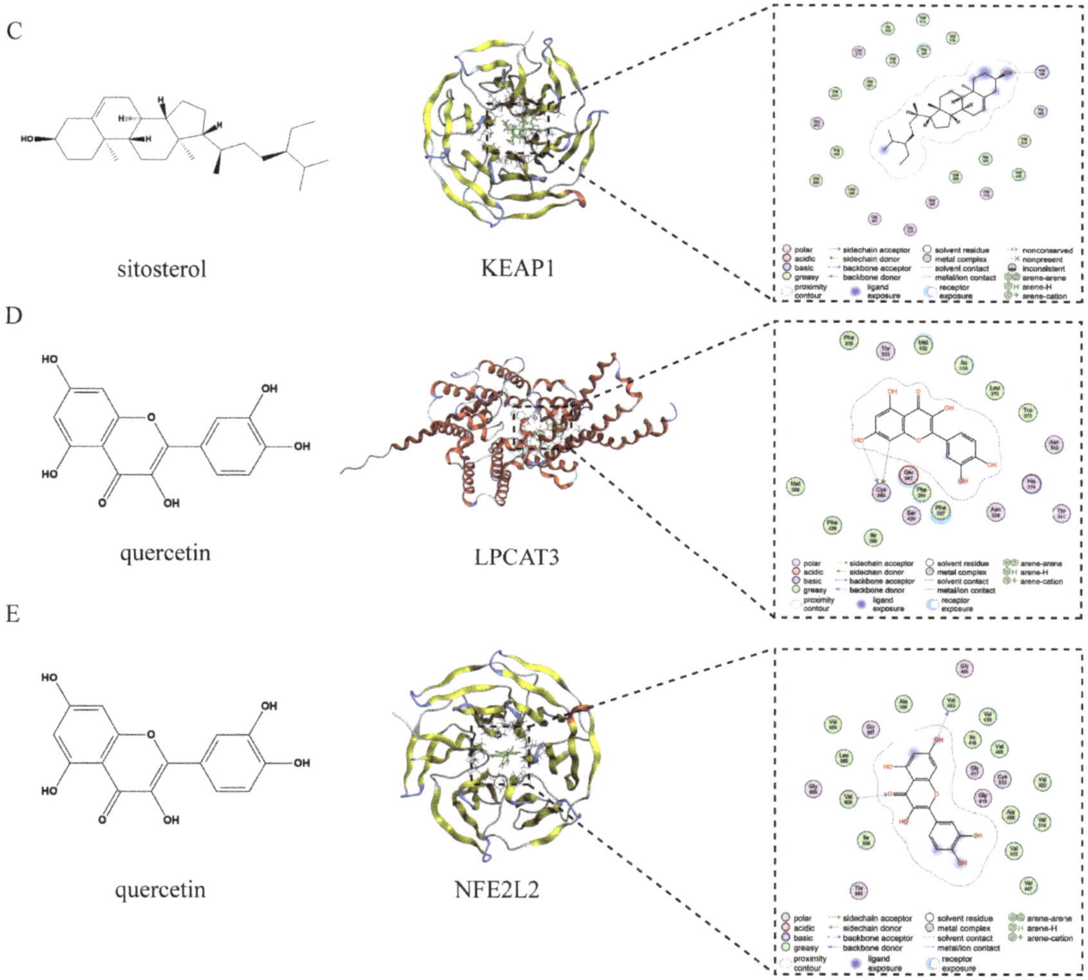

Figure 5. (**A**) Molecular docking visualization of quercetin and *ACSL4*. (**B**) Molecular docking visualization of sitosterol and *ALOX12*. (**C**) Molecular docking visualization of sitosterol and *KEAP1*. (**D**) Molecular docking visualization of quercetin and *LPCAT3*. (**E**) Molecular docking visualization of quercetin and *NFE2L2*.

3.13. Effect of EHW on the Expression of Iron Homeostasis-Related Genes

As shown in Figure 7, compared to the CON group, the mRNA expression of *TFR1* was significantly elevated in the CCL_4 group ($p < 0.01$), while the mRNA expression of *FTH1* was significantly reduced ($p < 0.01$). Additionally, the mRNA expression level of *FPN* was significantly decreased ($p < 0.05$). In contrast, when compared to the CCL_4 group, the EHW-H group showed a significant reduction in *TFR1* mRNA expression ($p < 0.05$), while *FTH1*'s mRNA expression was significantly increased ($p < 0.01$), and the mRNA expression of *FPN* exhibited an upward trend.

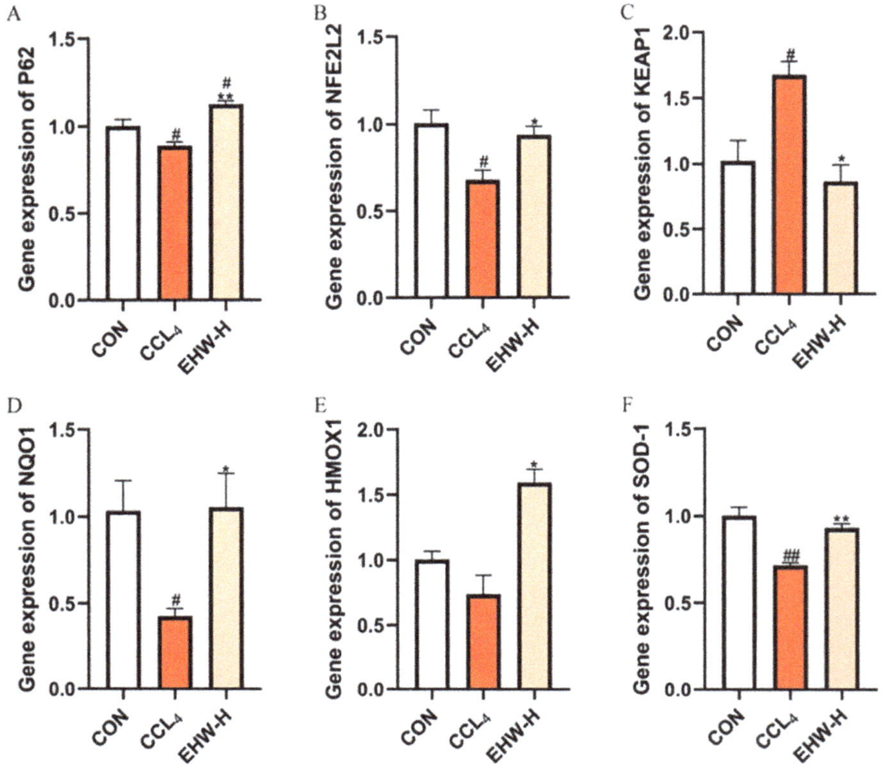

Figure 6. The effect of EHW on the KEAP1-NFE2L2 signaling pathway. (**A**) Relative *P62* mRNA expression. (**B**) Relative *NFE2L2* mRNA expression. (**C**) Relative *KEAP1* mRNA expression. (**D**) Relative *NQO1* mRNA expression. (**E**) Relative *HMOX1* mRNA expression. (**F**) Relative *SOD-1* mRNA expression. Note: * indicates a statistically significant difference compared to the CCL_4 group, # indicates a statistically significant difference compared to the CON group, * $p < 0.05$, ** $p < 0.01$; # $p < 0.05$, ## $p < 0.01$.

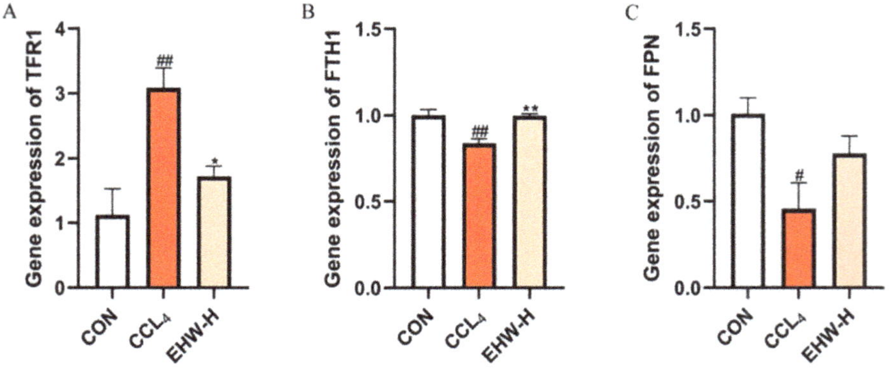

Figure 7. The effect of EHW on iron homeostasis-related genes.(**A**) Relative *TFR1* mRNA expression. (**B**) Relative *FTH1* mRNA expression. (**C**) Relative *FPN* mRNA expression. Note: * indicates a statistically significant difference compared to the CCL_4 group, # indicates a statistically significant difference compared to the CON group, * $p < 0.05$, ** $p < 0.01$; # $p < 0.05$, ## $p < 0.01$.

3.14. Effect of EHW on the Expression of Genes in the PLOOH Metabolism Pathway

As shown in Figure 8, compared to the CON group, the mRNA expression levels of *ACSL4* and *ALOX12* were significantly increased in the CCL_4 group ($p < 0.01$), and the mRNA expression of *LPCAT3* was also significantly elevated ($p < 0.05$). On the other hand, the mRNA expression of *GPX4* was significantly decreased ($p < 0.01$). In contrast, when compared to the CCL_4 group, the EHW-H group exhibited a significant reduction in the mRNA expression of *ACSL4* and *ALOX12* ($p < 0.01$), as well as a decrease in the mRNA expression of *LPCAT3* ($p < 0.05$). Moreover, the mRNA expression of *GPX4* was significantly increased in the EHW-H group ($p < 0.01$).

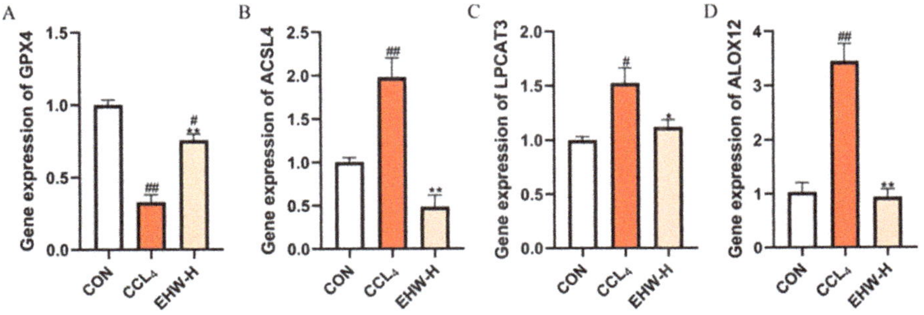

Figure 8. The effect of EHW on gene expression in the PLOOH metabolism pathway. (**A**) Relative *GPX4* mRNA expression. (**B**) Relative *ACSL4* mRNA expression. (**C**) Relative *LPCAT3* mRNA expression. (**D**) Relative *ALOX12* mRNA expression. Note: * indicates a statistically significant difference compared to the CCL_4 group, # indicates a statistically significant difference compared to the CON group, * $p < 0.05$, ** $p < 0.01$; # $p < 0.05$, ## $p < 0.01$.

4. Discussion

Liver injury is a common condition in veterinary clinical practice and poses a significant threat to livestock production [27]. Inhibiting excessive hepatocyte death is an effective approach to protect the liver and treat liver diseases [28]. Increasing evidence suggests that ferroptosis plays a crucial role in the occurrence and progression of liver-injury-related diseases [29,30]. Fortunately, with the rapid development of traditional Chinese (veterinary) medicine research, its potential use in liver injury treatment demonstrating safety, reliability, and multi-target effects has garnered increasing attention [31]. In this study, we established a mouse model of liver injury using CCL_4, a classic hepatotoxic substance. The results showed that, compared to the CON group, the CCL_4 group had a significantly increased liver index, and serum AST and ALT activities were markedly elevated. Histological examination via HE staining revealed disorganized hepatic cords, hepatocyte fatty degeneration, and considerable hepatocyte necrosis with inflammatory cell infiltration in the CCL_4 group. However, EHW successfully reversed these trends. Compared to the CCL_4 group, both the EHW-H and EHW-L groups exhibited a significant reduction in the liver index, and serum AST activity was markedly decreased. Serum ALT activity in the EHW-L group showed a decreasing trend. Consistently, in the EHW-H group, serum ALT activity significantly decreased. HE staining also revealed that liver architecture was restored in both the EHW-H and EHW-L groups, hepatocyte fatty degeneration was alleviated, hepatocyte necrosis was notably improved, and the area of inflammatory cell infiltration was substantially reduced, with the EHW-H group showing a better therapeutic effect than the EHW-L group. The excessive production and release of pro-inflammatory cytokines is one of the major characteristics of liver injury, and *TNF-α* generation is considered an important early marker of CCL_4-induced liver injury [32]. In this study, compared to the

CON group, the relative mRNA expression levels of *TNF-α* and *IL-1β* in the CCL_4 group were significantly increased, and *IL-6* mRNA expression was also significantly elevated. In contrast, compared to the CCL_4 group, the EHW-H group showed a significant decrease in the relative mRNA expression levels of *TNF-α* and *IL-1β*, and *IL-6* mRNA expression was significantly reduced. These results suggest that EHW has a protective effect against CCL_4-induced liver injury, and this protective effect appears to be dose-dependent.

Ferroptosis is an iron-dependent form of programmed cell death, and abnormal iron accumulation is a key feature in the onset of ferroptosis [33]. *PTGS2* is considered an important marker of ferroptosis [34], and *FSP-1* is a recently identified inhibitor that plays a crucial role in suppressing ferroptosis [35]. To further investigate whether the hepatoprotective effect of EHW is related to the alleviation of ferroptosis in hepatocytes, we first performed Prussian blue staining. The results revealed that in the CCL_4 group, there was a significant increase in blue precipitates in the areas of liver necrosis, indicating the presence of iron overload. In contrast, in the EHW treatment groups, the areas of liver necrosis were significantly reduced, and the extent of blue precipitate staining in the liver tissue was markedly less than in the CCL_4 group. Furthermore, we measured the relative mRNA expression levels of *PTGS2* and *FSP-1* in the liver. Compared to the CON group, the relative mRNA expression of *PTGS2* was significantly elevated in the CCL_4 group, while the expression of *FSP-1* was significantly reduced. In contrast, in the EHW-H group, the relative mRNA expression of *PTGS2* was significantly decreased, while the expression of *FSP-1* was notably increased compared to the CCL_4 group. These results suggest that CCL_4-induced liver injury is closely associated with the excessive occurrence of ferroptosis, and that EHW may alleviate ferroptosis, thereby exerting its hepatoprotective effects.

Network pharmacology, an interdisciplinary research approach that integrates knowledge from fields such as systems biology and pharmacology, enables the analysis of the systemic effects of multi-component drugs on the human body [36]. It has significant advantages in exploring the therapeutic targets of TCM and investigating disease mechanisms [37,38]. In this study, by constructing an intersecting target PPI network, we identified the core targets with higher degree values, including *PTGS2*, *NFE2L2*, *KEAP1*, and others. These targets exhibited frequent interactions with the major active compounds of EHW, such as quercetin, kaempferol, and sitosterol, suggesting that EHW may exert its hepatoprotective effects by targeting these core molecules. GO enrichment analysis, which linked the genes identified in this study to functional categories in the Gene Ontology database, revealed statistically significant differences in specific biological processes, cellular components, and molecular functions [39]. KEGG analysis aids in understanding the regulatory processes of various metabolic pathways and cellular processes in biological systems, thus shedding light on key biological mechanisms and interactions [40]. In this study, GO enrichment analysis indicated that the potential targets of EHW in alleviating ferroptosis-induced liver injury were mainly involved in BPs such as the response to oxidative stress, regulation of reactive oxygen species (ROS) metabolic processes, and ROS metabolic processes. These targets were predominantly localized in cellular components, including the transcription regulator complex, membrane regions, and membrane rafts. In terms of molecular functions, they were primarily associated with RNA polymerase II-specific DNA-binding transcription factor binding, transcription coactivator binding, and DNA-binding transcription factor binding. KEGG pathway enrichment analysis revealed that the enriched pathways were primarily related to the Hepatitis B, *HIF-1* signaling, and *IL-17* signaling pathways. Molecular docking is a computational biology method typically used to study the interactions between drugs and their targets. In this study, molecular docking of the three top-ranking active compounds with key ferroptosis targets (*KEAP1*, *NFE2L2*, *LPCAT3*, *ALOX12*, and *ACSL4*) demonstrated that the binding energies of the target small molecule complexes were all significantly lower than −4.5 kcal/mol, suggesting a strong affinity between the active components and the targets. The visualization results further indicated that the active components and core targets were closely connected in a specific spatial structure, implying that the major active compounds of EHW have a strong biological affinity with the ferroptosis-related core targets. These findings suggest that EHW may exert its hepatoprotective effects by acting on the KEAP1-NFE2L2 signaling pathway and reducing the accumulation of PLOOH, thereby mitigating hepatocyte ferroptosis.

Ferroptosis is a form of cell death closely associated with oxidative stress [41]. The KEAP1-NFE2L2 pathway is a crucial regulator of cellular oxidative stress [42]. Upon activation, *NFE2L2* exerts potent antioxidant effects and plays a key role in the regulation of ferroptosis [43]. *KEAP1* negatively regulates *NFE2L2* [44], while *P62* can interact with *KEAP1* at the *NFE2L2* binding site, competing with *KEAP1* and thereby activating *NFE2L2* [45]. Once activated, *NFE2L2* induces an increase in *P62*, creating a positive feedback loop [46]. More importantly, when *NFE2L2* is activated, the downstream antioxidant genes such as *HMOX1*, *NQO2*, and *SOD-1* are activated, directly exerting antioxidant cell protection and alleviating ferroptosis [47]. In this study, compared to the CON group, the CCL_4 group showed a significant increase in *KEAP1* mRNA expression, a marked decrease in *SOD-1* mRNA expression, and significant reductions in the mRNA expression levels of *P62*, *NFE2L2*, and *NQO1*, with *HMOX1* mRNA expression also showing a decreasing trend. In contrast, compared to the CCL_4 group, the EHW-H group showed a significant decrease in *KEAP1* mRNA expression, a dramatic increase in *P62* and *SOD-1* mRNA expression, and significant increases in the mRNA expression levels of *NFE2L2*, *NQO1*, and *HMOX1*. These results suggest that EHW can activate *NFE2L2*, upregulate the expression of antioxidant genes, reduce cellular oxidative stress and ferroptosis, and thereby exert a protective effect against liver injury.

Studies have shown that the dysregulation of iron metabolism is a critical trigger for ferroptosis [48]. The excessive accumulation of iron in the labile iron pool (LIP) is a key factor in the initiation of ferroptosis [49]. The iron content in the LIP is tightly regulated by various molecules. Under normal physiological conditions, iron ions typically interact with transferrin, after which they are internalized into the cell via *TFR1* [50]. Once inside the cell, iron is either stored in the iron storage protein complex or, alternatively, exported by *FPN* [51,52]. Notably, *FTH1* serves as a critical component of the iron storage protein complex [53]. When *TFR1* is overexpressed [54], *FTH1* is insufficient [55] or *FPN* is suppressed [56], and iron accumulates abnormally in the LIP, leading to the Fenton reaction, which triggers lipid peroxidation and ultimately induces ferroptosis [57]. Interestingly, recent studies have discovered that the activation of the KEAP1-NFE2L2 pathway not only regulates the body's antioxidant response but also plays a critical role in iron metabolism. Numerous studies have confirmed that key regulators of iron metabolism, such as *FTH1* and *FPN*, are controlled by *NFE2L2* [58]. In this study, compared to the CON group, the CCL_4 group showed a significant increase in *TFR1* mRNA expression, a marked decrease in *FTH1* mRNA expression, and a significant reduction in *FPN* mRNA expression. In contrast, the EHW-H group exhibited a significant decrease in *TFR1* mRNA expression, a dramatic increase in *FTH1* mRNA expression, and an upward trend in *FPN* mRNA expression compared to the CCL_4 group. These results suggest that EHW may regulate iron metabolism by activating the KEAP1-NFE2L2 pathway, restoring cellular iron homeostasis, and inhibiting excessive ferroptosis, thereby alleviating liver injury.

The accumulation of phospholipid hydroperoxides (PLOOHs) is the most direct cause of ferroptosis [59]. Timely clearance of PLOOHs or the inhibition of the synthesis of their precursors, phospholipid-containing polyunsaturated fatty acid chains (PUFA-PL), can exert a positive effect on ferroptosis [60]. *GPX4* is the most well-known reductase of PLOOHs, as it reduces PLOOHs to their corresponding alcohols [34]. Notably, the expression of this gene is also regulated by *NFE2L2* [61]. Another critical pathway to prevent ferroptosis is the inhibition of PUFA-PL synthesis. *ACSL4* is a fatty acid synthetase that specifically catalyzes the activation of long-chain polyunsaturated fatty acids (PUFAs) [62]. Under the catalysis of *ACSL4*, these long-chain PUFAs are conjugated with coenzyme A (CoA) to form PUFA-CoA intermediates [63]. Subsequently, *LPCAT3* esterifies these intermediates to generate PUFA-PL [64], which is ultimately converted into PLOOHs through the action of lipoxygenases such as *ALOX12* [65]. In this study, compared to the CON group, the CCL_4 group showed a dramatic increase in the mRNA expression of *ACSL4* and *ALOX12*, a significant increase in *LPCAT3* mRNA expression, and a marked decrease in *GPX4* mRNA expression. In contrast, the EHW-H group exhibited a significant reduction in the mRNA expression of *ACSL4* and *ALOX12*, a notable decrease in *LPCAT3* mRNA expression, and a dramatic increase in *GPX4* mRNA expression compared to the CCL_4 group. These results suggest that EHW may reduce the generation of PLOOHs and accelerate their reduction, thereby decreasing the accumulation of PLOOHs, inhibiting ferroptosis, and alleviating liver injury.

5. Conclusions

In summary, EHW may exert its hepatoprotective effects by regulating the KEAP1-NFE2L2 signaling pathway to alleviate oxidative stress, restore cellular iron homeostasis, reduce the accumulation of PLOOHs, and ultimately mitigate ferroptosis in the liver. On the other hand, despite promising results, the direct applicability of EHW in veterinary practice requires further pharmacokinetic, toxicity, and large-scale clinical trials to confirm its safety and effectiveness across different livestock breeds.

Author Contributions: Conceptualization, H.D.; methodology, K.Y.; software, J.Y. and Y.P.; validation, H.D., J.W. and W.S.; formal analysis, S.X., C.C. and J.L.; resources, S.X., C.C. and H.D.; data curation, K.Y.; writing—original draft preparation, K.Y. and J.Y.; writing—review and editing, H.D.; visualization, K.Y.; supervision, H.D.; project administration, H.D.; funding acquisition, H.D. All authors have read and agreed to the published version of the manuscript.

Funding: This work was supported by the National Natural Science Foundation of China (32402934), the Natural Science Foundation of Chongqing (CSTB2022NSCQ-MSX0470), and the College Students' Innovative Entrepreneurial Training Plan Program of Chongqing (S202410635371).

Institutional Review Board Statement: All animal experiments were conducted in accordance with the guidelines approved by the Ethics Committee of Southwest University for Animal Welfare (Ethics No. LAC2024-1-0346; date: 31 October 2024).

Informed Consent Statement: Not applicable.

Data Availability Statement: The data presented in this study are available on request from the corresponding author.

Acknowledgments: We are grateful to all other staff in the Department of Traditional Chinese Veterinary Medicine of Southwest University for their assistance in the experiments.

Conflicts of Interest: The authors declare no conflicts of interest.

Abbreviations

The following abbreviations are used in this manuscript:

EHW	*Euphorbia humifusa* Willd. ex Schltdl.
TCM	Traditional Chinese Medicine
AST	Aspartate aminotransferase
ALT	Alanine aminotransferase

HE	Hematoxylin–eosin staining
OB	Oral bioavailability
DL	Drug-likeness
PPI	Protein–protein interaction
GO	Gene Ontology
KEGG	Kyoto Encyclopedia of Genes and Genomes
BP	Biological process
CC	Cellular component
MF	Molecular function
SEM	Standard error of the mean
LIP	Labile iron pool
PLOOHs	Phospholipid hydroperoxides
PUFA-PL	Phospholipid-containing polyunsaturated fatty acid chains
PUFAs	Polyunsaturated fatty acids
CoA	Coenzyme A

References

1. Peleman, C.; Hellemans, S.; Veeckmans, G.; Arras, W.; Zheng, H.; Koeken, I.; Van San, E.; Hassannia, B.; Walravens, M.; Kayirangwa, E.; et al. Ferroptosis is a targetable detrimental factor in metabolic dysfunction-associated steatotic liver disease. *Cell Death Differ.* **2024**, *31*, 1113–1126. [CrossRef] [PubMed]
2. Kubes, P.; Jenne, C. Immune Responses in the Liver. *Annu. Rev. Immunol.* **2018**, *36*, 247–277. [CrossRef] [PubMed]
3. Liu, S.; Kang, W.; Mao, X.; Ge, L.; Du, H.; Li, J.; Hou, L.; Liu, D.; Yin, Y.; Liu, Y.; et al. Melatonin mitigates aflatoxin B1-induced liver injury via modulation of gut microbiota/intestinal FXR/liver TLR4 signaling axis in mice. *J. Pineal. Res.* **2022**, *73*, e12812. [CrossRef]
4. Twedt, D.C.; Cullen, J.; Mccord, K.; Janeczko, S.; Dudak, J.; Simpson, K. Evaluation of fluorescence in situ hybridization for the detection of bacteria in feline inflammatory liver disease. *J. Feline. Med. Surg.* **2014**, *16*, 109–117. [CrossRef]
5. Weingarten, M.A.; Sande, A.A. Acute liver failure in dogs and cats. *J. Vet. Emerg. Crit. Care* **2015**, *25*, 455–473. [CrossRef]
6. Dedeaux, A.M.; Flesner, B.K.; Reinhart, J.M.; Langohr, I.M.; Husnik, R.; Geraci, S.N.; Taboada, J.; Rademacher, N.; Thombs, L.A.; Bryan, J.N.; et al. Biochemical, functional, and histopathologic characterization of lomustine-induced liver injury in dogs. *Am. J. Vet. Res.* **2020**, *81*, 810–820. [CrossRef]
7. Wu, Y.; He, X.; Chen, H.; Lin, Y.; Zheng, C.; Zheng, B. Extraction and characterization of hepatoprotective polysaccharides from Anoectochilus roxburghii against CCL_4-induced liver injury via regulating lipid metabolism and the gut microbiota. *Int. J. Biol. Macromol.* **2024**, *277 Pt 3*, 134305. [CrossRef]
8. Xu, W.; Hu, Z.; Zhang, J.; Tang, Y.; Xing, H.; Xu, P.; Ma, Y.; Niu, Q. Cross-talk between autophagy and ferroptosis contributes to the liver injury induced by fluoride via the mtROS-dependent pathway. *Ecotoxicol. Environ. Saf.* **2023**, *250*, 114490. [CrossRef] [PubMed]
9. Dixon, S.J.; Olzmann, J.A. The cell biology of ferroptosis. *Nat. Rev. Mol. Cell Biol.* **2024**, *25*, 424–442. [CrossRef]
10. Liu, J.; Kang, R.; Tang, D. Signaling pathways and defense mechanisms of ferroptosis. *FEBS J.* **2022**, *289*, 7038–7050. [CrossRef]
11. Yang, R.; Gao, W.; Wang, Z.; Jian, H.; Peng, L.; Yu, X.; Xue, P.; Peng, W.; Li, K.; Zeng, P. Polyphyllin I induced ferroptosis to suppress the progression of hepatocellular carcinoma through activation of the mitochondrial dysfunction via Nrf2/HO-1/GPX4 axis. *Phytomedicine* **2024**, *122*, 155135. [CrossRef]
12. Yin, X.; Mi, Y.; Wang, X.; Li, Y.; Zhu, X.; Bukhari, I.; Wang, Q.; Zheng, P.; Xue, X.; Tang, Y. Exploration and Validation of Ferroptosis-Associated Genes in ADAR1 Deletion-Induced NAFLD through RNA-seq Analysis. *Int. Immunopharmacol.* **2024**, *134*, 112177. [CrossRef] [PubMed]
13. Zhou, L.; Xiao, M.; Li, Y.; Chitrakar, B.; Sheng, Q.; Zhao, W. Ursolic Acid Ameliorates Alcoholic Liver Injury through Attenuating Oxidative Stress-Mediated Ferroptosis and Modulating Gut Microbiota. *J. Agric. Food Chem.* **2024**, *72*, 21181–21192. [CrossRef] [PubMed]
14. Wu, A.; Feng, B.; Yu, J.; Yan, L.; Che, L.; Zhuo, Y.; Luo, Y.; Yu, B.; Wu, D.; Chen, D. Fibroblast growth factor 21 attenuates iron overload-induced liver injury and fibrosis by inhibiting ferroptosis. *Redox Biol.* **2021**, *46*, 102131. [CrossRef]
15. Yang, W.; Wang, Y.; Zhang, C.; Huang, Y.; Yu, J.; Shi, L.; Zhang, P.; Yin, Y.; Li, R.; Tao, K. Maresin1 Protect Against Ferroptosis-Induced Liver Injury Through ROS Inhibition and Nrf2/HO-1/GPX4 Activation. *Front. Pharmacol.* **2022**, *13*, 865689. [CrossRef] [PubMed]

16. Wei, Y.Y.; Wang, H.R.; Fan, Y.M.; Gu, J.H.; Zhang, X.Y.; Gong, X.H.; Hao, Z.H. Acute liver injury induced by carbon tetrachloride reversal by Gandankang aqueous extracts through nuclear factor erythroid 2-related factor 2 signaling pathway. *Ecotoxicol. Environ. Saf.* **2023**, *251*, 114527. [CrossRef]
17. Zhao, Z.X.; Yuan, Y.M.; Zhao, Z.H.; Yao, Q.H.; Ye, X.Q.; Wang, Y.Y.; Liu, H.M.; Jha, R.; Balasubramanian, B.; Liu, W.C. Phlorotannin Alleviates Liver Injury by Regulating Redox Balance, Apoptosis, and Ferroptosis of Broilers under Heat Stress. *Antioxidants* **2024**, *13*, 1048. [CrossRef]
18. Huang, S.; Lin, L.; Wang, S.; Ding, W.; Zhang, C.; Shaukat, A.; Xu, B.; Yue, K.; Zhang, C.; Liu, F. Total Flavonoids of Rhizoma Drynariae Mitigates Aflatoxin B1-Induced Liver Toxicity in Chickens via Microbiota-Gut-Liver Axis Interaction Mechanisms. *Antioxidants* **2023**, *12*, 819. [CrossRef]
19. Wu, X.; Ma, G.L.; Chen, H.W.; Zhao, Z.Y.; Zhu, Z.P.; Xiong, J.; Yang, G.X.; Hu, J.F. Antibacterial and antibiofilm efficacy of the preferred fractions and compounds from *Euphorbia humifusa* (herba euphorbiae humifusae) against Staphylococcus aureus. *J. Ethnopharmacol.* **2023**, *306*, 116177. [CrossRef]
20. Luyen, B.T.; Tai, B.H.; Thao, N.P.; Thao, N.P.; Eun, K.J.; Cha, J.Y.; Xin, M.J.; Lee, Y.M.; Kim, Y. Anti-inflammatory components of *Euphorbia humifusa* Willd. *Bioorg. Med. Chem. Lett.* **2014**, *24*, 1895–1900. [CrossRef]
21. Chang, S.Y.; Park, J.H.; Kim, Y.H.; Kang, J.S.; Min, J.Y. A natural component from *Euphorbia humifusa* Willd displays novel, broad-spectrum anti-influenza activity by blocking nuclear export of viral ribonucleoprotein. *Biochem. Biophys. Res. Commun.* **2016**, *471*, 282–289. [CrossRef] [PubMed]
22. Rakotondrabe, T.F.; Fan, M.; Guo, M. Exploring potential antidiabetic and anti-inflammatory flavonoids from *Euphorbia humifusa* with an integrated strategy. *Front. Pharmacol.* **2022**, *13*, 980945. [CrossRef] [PubMed]
23. Shin, S.Y.; Kim, C.G.; Jung, Y.J.; Jung, Y.; Jung, H.; Im, J.; Lim, Y.; Lee, Y.H. *Euphorbia humifusa* Willd exerts inhibition of breast cancer cell invasion and metastasis through inhibition of TNFα-induced MMP-9 expression. *BMC Complement. Altern. Med.* **2016**, *16*, 413. [CrossRef]
24. The State Pharmacopoeia Commission of the People's Republic of China. *Pharmacopoeia of the People's Republic of China*; Chemical Industry Press: Beijing, China, 2005; p. 84.
25. Jiang, S.; Li, H.; Zhang, L.; Mu, W.; Zhang, Y.; Chen, T.; Wu, J.; Tang, H.; Zheng, S.; Liu, Y.; et al. Generic Diagramming Platform (GDP): A comprehensive database of high-quality biomedical graphics. *Nucleic Acids Res.* **2025**, *53*, D1670–D1676. [CrossRef] [PubMed]
26. Livak, K.J.; Schmittgen, T.D. Analysis of relative gene expression data using real-time quantitative PCR and the 2(-Delta Delta C(T)) Method. *Methods.* **2001**, *25*, 402–408. [CrossRef]
27. Wang, Y.; Wu, J.; Wang, L.; Yang, P.; Liu, Z.; Rajput, S.A.; Hassan, M.; Qi, D. Epigallocatechin Gallate and Glutathione Attenuate Aflatoxin B-Induced Acute Liver Injury in Ducklings via Mitochondria-Mediated Apoptosis and the Nrf2 Signalling Pathway. *Toxins* **2022**, *14*, 876. [CrossRef]
28. Zhang, J.; Guo, J.; Yang, N.; Huang, Y.; Hu, T.; Rao, C. Endoplasmic reticulum stress-mediated cell death in liver injury. *Cell Death Dis.* **2022**, *13*, 1051. [CrossRef]
29. Ge, Y.; Yang, S.; Zhang, T.; Gong, S.; Wan, X.; Zhu, Y.; Fang, Y.; Hu, C.; Yang, F.; Yin, L.; et al. Ferroptosis participated in inhaled polystyrene nanoplastics-induced liver injury and fibrosis. *Sci. Total Environ.* **2024**, *916*, 170342. [CrossRef]
30. Wu, J.; Wang, Y.; Jiang, R.; Xue, R.; Yin, X.; Wu, M.; Meng, Q. Ferroptosis in liver disease: New insights into disease mechanisms. *Cell Death Discov.* **2021**, *7*, 276. [CrossRef]
31. Sun, Y.K.; Zhang, Y.F.; Xie, L.; Rong, F.; Zhu, X.Y.; Xie, J.; Zhou, H.; Xu, T. Progress in the treatment of drug-induced liver injury with natural products. *Pharmacol. Res.* **2022**, *183*, 106361. [CrossRef]
32. Kim, H.Y.; Park, J.; Lee, K.H.; Lee, D.U.; Kwak, J.H.; Kim, Y.S.; Lee, S.M. Ferulic acid protects against carbon tetrachloride-induced liver injury in mice. *Toxicology.* **2011**, *282*, 104–111. [CrossRef] [PubMed]
33. Rittase, W.B.; Slaven, J.E.; Suzuki, Y.J.; Muir, J.M.; Lee, S.H.; Rusnak, M.; Brehm, G.V.; Bradfield, D.T.; Symes, A.J.; Day, R.M. Iron Deposition and Ferroptosis in the Spleen in a Murine Model of Acute Radiation Syndrome. *Int. J. Mol. Sci.* **2022**, *23*, 11029. [CrossRef]
34. Yang, W.S.; Sriramaratnam, R.; Welsch, M.E.; Shimada, K.; Skouta, R.; Viswanathan, V.S.; Cheah, J.H.; Clemons, P.A.; Shamji, A.F.; Clish, C.B.; et al. Regulation of ferroptotic cancer cell death by GPX4. *Cell* **2014**, *156*, 317–331. [CrossRef] [PubMed]
35. Doll, S.; Freitas, F.P.; Shah, R.; Aldrovandi, M.; Da Silva, M.C.; Ingold, I.; Goya Grocin, A.; Xavier da Silva, T.N.; Panzilius, E.; Scheel, C.H.; et al. FSP1 is a glutathione-independent ferroptosis suppressor. *Nature* **2019**, *575*, 693–698. [CrossRef] [PubMed]
36. Shang, L.; Wang, Y.; Li, J.; Zhou, F.; Xiao, K.; Liu, Y.; Zhang, M.; Wang, S.; Yang, S. Mechanism of Sijunzi Decoction in the treatment of colorectal cancer based on network pharmacology and experimental validation. *J. Ethnopharmacol.* **2023**, *302 (Pt A)*, 115876. [CrossRef]
37. Zhao, L.; Zhang, H.; Li, N.; Chen, J.; Xu, H.; Wang, Y.; Liang, Q. Network pharmacology, a promising approach to reveal the pharmacology mechanism of Chinese medicine formula. *J. Ethnopharmacol.* **2023**, *309*, 116306. [CrossRef] [PubMed]

38. Wang, Z.Y.; Li, M.Z.; Li, W.J.; Ouyang, J.F.; Gou, X.J.; Huang, Y. Mechanism of action of Daqinjiao decoction in treating cerebral small vessel disease explored using network pharmacology and molecular docking technology. *Phytomedicine.* **2023**, *108*, 154538. [CrossRef]
39. Cai, B.; Qi, M.; Zhang, X.; Zhang, D. Integrating Network Pharmacology with in vitro Experiments to Validate the Efficacy of Celastrol Against Hepatocellular Carcinoma Through Ferroptosis. *Drug Des. Devel. Ther.* **2024**, *18*, 3121–3141. [CrossRef]
40. Zhang, S.; Cai, X.; Khan, G.J.; Cheng, J.; He, J.; Zhai, K.; Mao, Y. Exploring the molecular mechanism of Artemisia rupestris L. for the treatment of hepatocellular carcinoma via PI3K/AKT pathway. *J. Ethnopharmacol.* **2024**, *322*, 117572. [CrossRef]
41. Yu, Y.; Yan, Y.; Niu, F.; Wang, Y.; Chen, X.; Su, G.; Liu, Y.; Zhao, X.; Qian, L.; Liu, P.; et al. Ferroptosis: A cell death connecting oxidative stress, inflammation and cardiovascular diseases. *Cell Death Discov.* **2021**, *7*, 193. [CrossRef]
42. Adinolfi, S.; Patinen, T.; Deen, A.J.; Pitkänen, S.; Härkönen, J.; Kansanen, E.; Küblbeck, J.; Levonen, A.L. The KEAP1-NRF2 pathway: Targets for therapy and role in cancer. *Redox Biol.* **2023**, *63*, 102726. [CrossRef] [PubMed]
43. Kerins, M.J.; Ooi, A. The Roles of NRF2 in Modulating Cellular Iron Homeostasis. *Antioxid. Redox Signal.* **2018**, *29*, 1756–1773. [CrossRef]
44. Baird, L.; Yamamoto, M. The Molecular Mechanisms Regulating the KEAP1-NRF2 Pathway. *Mol. Cell Biol.* **2020**, *40*, e00099-20. [CrossRef] [PubMed]
45. Komatsu, M.; Kurokawa, H.; Waguri, S.; Taguchi, K.; Kobayashi, A.; Ichimura, Y.; Sou, Y.S.; Ueno, I.; Sakamoto, A.; Tong, K.I.; et al. The selective autophagy substrate p62 activates the stress responsive transcription factor Nrf2 through inactivation of Keap1. *Nat. Cell Biol.* **2010**, *12*, 213–223. [CrossRef] [PubMed]
46. Lee, D.H.; Park, J.S.; Lee, Y.S.; Han, J.; Lee, D.K.; Kwon, S.W.; Han, D.H.; Lee, Y.H.; Bae, S.H. SQSTM1/p62 activates NFE2L2/NRF2 via ULK1-mediated autophagic KEAP1 degradation and protects mouse liver from lipotoxicity. *Autophagy* **2020**, *16*, 1949–1973. [CrossRef]
47. Song, W.; Zhang, L.; Cui, X.; Wang, R.; Ma, J.; Xu, Y.; Jin, Y.; Wang, D.; Lu, Z. Nobiletin alleviates cisplatin-induced ototoxicity via activating autophagy and inhibiting NRF2/GPX4-mediated ferroptosis. *Sci. Rep.* **2024**, *14*, 7889. [CrossRef]
48. Tang, D.; Chen, X.; Kang, R.; Kroemer, G. Ferroptosis: Molecular mechanisms and health implications. *Cell Res.* **2021**, *31*, 107–125. [CrossRef]
49. Lei, L.; Yuan, J.; Dai, Z.; Xiang, S.; Tu, Q.; Cui, X.; Zhai, S.; Chen, X.; He, Z.; Fang, B.; et al. Targeting the Labile Iron Pool with Engineered DFO Nanosheets to Inhibit Ferroptosis for Parkinson's Disease Therapy. *Adv. Mater.* **2024**, *36*, e2409329. [CrossRef]
50. Gao, M.; Monian, P.; Quadri, N.; Ramasamy, R.; Jiang, X. Glutaminolysis and Transferrin Regulate Ferroptosis. *Mol. Cell* **2015**, *59*, 298–308. [CrossRef]
51. Cui, J.; Chen, Y.; Yang, Q.; Zhao, P.; Yang, M.; Wang, X.; Mang, G.; Yan, X.; Wang, D.; Tong, Z.; et al. Protosappanin A Protects DOX-Induced Myocardial Injury and Cardiac Dysfunction by Targeting ACSL4/FTH1 Axis-Dependent Ferroptosis. *Adv. Sci.* **2024**, *11*, e2310227. [CrossRef]
52. Li, Y.; Zeng, X.; Lu, D.; Yin, M.; Shan, M.; Gao, Y. Erastin induces ferroptosis via ferroportin-mediated iron accumulation in endometriosis. *Hum. Reprod.* **2021**, *36*, 951–964. [CrossRef] [PubMed]
53. Fang, Y.; Chen, X.; Tan, Q.; Zhou, H.; Xu, J.; Gu, Q. Inhibiting Ferroptosis through Disrupting the NCOA4-FTH1 Interaction: A New Mechanism of Action. *ACS Cent. Sci.* **2021**, *7*, 980–989. [CrossRef] [PubMed]
54. Wang, D.; Liang, W.; Huo, D.; Wang, H.; Wang, Y.; Cong, C.; Zhang, C.; Yan, S.; Gao, M.; Su, X.; et al. SPY1 inhibits neuronal ferroptosis in amyotrophic lateral sclerosis by reducing lipid peroxidation through regulation of GCH1 and TFR1. *Cell Death Differ.* **2023**, *30*, 369–382. [CrossRef]
55. Kong, N.; Chen, X.; Feng, J.; Duan, T.; Liu, S.; Sun, X.; Chen, P.; Pan, T.; Yan, L.; Jin, T.; et al. Baicalin induces ferroptosis in bladder cancer cells by downregulating FTH1. *Acta Pharm. Sin. B* **2021**, *11*, 4045–4054. [CrossRef]
56. Bao, W.D.; Zhou, X.T.; Zhou, L.T.; Wang, F.; Yin, X.; Lu, Y.; Zhu, L.Q.; Liu, D. Targeting miR-124/Ferroportin signaling ameliorated neuronal cell death through inhibiting apoptosis and ferroptosis in aged intracerebral hemorrhage murine model. *Aging Cell.* **2020**, *19*, e13235. [CrossRef]
57. Gan, B. Mitochondrial regulation of ferroptosis. *J. Cell Biol.* **2021**, *220*, e202105043. [CrossRef] [PubMed]
58. Dodson, M.; Castro-portuguez, R.; Zhang, D.D. NRF2 plays a critical role in mitigating lipid peroxidation and ferroptosis. *Redox Biol.* **2019**, *23*, 101107. [CrossRef]
59. Jiang, X.; Stockwell, B.R.; Conrad, M. Ferroptosis: Mechanisms, biology and role in disease. *Nat. Rev. Mol. Cell Biol.* **2021**, *22*, 266–282. [CrossRef]
60. Zheng, J.; Conrad, M. The Metabolic Underpinnings of Ferroptosis. *Cell Metab.* **2020**, *32*, 920–937. [CrossRef]
61. Yang, S.; Wang, L.; Zeng, Y.; Wang, Y.; Pei, T.; Xie, Z.; Xiong, Q.; Wei, H.; Li, W.; Li, J.; et al. Salidroside alleviates cognitive impairment by inhibiting ferroptosis via activation of the Nrf2/GPX4 axis in SAMP8 mice. *Phytomedicine* **2023**, *114*, 154762. [CrossRef]
62. Doll, S.; Proneth, B.; Tyurina, Y.Y.; Panzilius, E.; Kobayashi, S.; Ingold, I.; Irmler, M.; Beckers, J.; Aichler, M.; Walch, A.; et al. ACSL4 dictates ferroptosis sensitivity by shaping cellular lipid composition. *Nat. Chem. Biol.* **2017**, *13*, 91–98. [CrossRef] [PubMed]

63. Soupene, E.; Kuypers, F.A. Mammalian long-chain acyl-CoA synthetases. *Exp. Biol. Med.* **2008**, *233*, 507–521. [CrossRef] [PubMed]
64. Dixon, S.J.; Winter, G.E.; Musavi, L.S.; Lee, E.D.; Snijder, B.; Rebsamen, M.; Superti-Furga, G.; Stockwell, B.R. Human Haploid Cell Genetics Reveals Roles for Lipid Metabolism Genes in Nonapoptotic Cell Death. *ACS Chem. Biol.* **2015**, *10*, 1604–1609. [CrossRef] [PubMed]
65. Kuhn, H.; Banthiya, S.; Van Leyen, K. Mammalian lipoxygenases and their biological relevance. *Biochim. Biophys. Acta.* **2015**, *1851*, 308–330. [CrossRef]

biomolecules

MDPI

Article

Baicalin, Amoxicillin, and Probenecid Provide Protection in Mice Against *Glaesserella parasuis* Challenge

Jingyang Li [1,2], Ronghui Luo [1,2], Yunjian Fu [1,2], Siyu Liu [1,2], Qiaoli Dong [1,2], Yamin Sun [1,2], Xinyue Tian [1,2], Yi Zhu [1,2], Peiyi Wang [1,2], Ling Guo [1,2], Qirong Lu [1,2], Chun Ye [1,2], Shulin Fu [1,2,*] and Yinsheng Qiu [1,2,*]

1 Hubei Key Laboratory of Animal Nutrition and Feed Science, Wuhan Polytechnic University, Wuhan 430023, China; lijingyang026@163.com (J.L.); 18971562947@163.com (R.L.); mopanfyj@163.com (Y.F.); lusy617@163.com (S.L.); 15838785140@163.com (Q.D.); 18137892679@163.com (Y.S.); tianxinyue111@126.com (X.T.); 18272347532@163.com (Y.Z.); peiyi1220@163.com (P.W.); guoling1101@whpu.edu.cn (L.G.); qirongluvet@whpu.edu.cn (Q.L.); yechun@whpu.edu.cn (C.Y.)

2 Hubei Collaborative Innovation Center for Animal Nutrition and Feed Safety, Wuhan 430023, China

* Correspondence: shulinfu@whpu.edu.cn (S.F.); qiuyinsheng6405@whpu.edu.cn (Y.Q.)

Abstract: *Glaesserella parasuis* (*G. parasuis*) causes Glässer's disease and systemic inflammatory responses in the host. The currently available therapies have limited efficacy and fail to achieve a balance between anti-inflammatory and antibacterial effects. In this study, we investigated the effects of baicalin, amoxicillin, and probenecid on blood biochemical parameters, routine blood indicators, survival rate, bacterial burden, and pathological tissue damage in *G. parasuis*-challenged mice. Treatment with baicalin, amoxicillin, and probenecid significantly modified the blood biochemical parameters and routine blood test indicators, increased the survival rate, attenuated the bacterial burden, and alleviated pathological tissue damage in *G. parasuis*-challenged mice. Treatment with baicalin, amoxicillin, and probenecid also increased the number of $CD3^+$, $CD3^+CD4^+$, and $CD3^+CD8^+$ T cells as measured by flow cytometry, and restored the intensity of the CD3, CD4, and CD8 protein expression in the blood vessels of *G. parasuis*-challenged mice by immunohistochemistry. These compounds reduced interleukin 1β (IL-1β), IL-18, tumor necrosis factor alpha (TNF-α), and high mobility group box 1 protein (HMGB1) expression in the spleen of *G. parasuis*-challenged mice. Furthermore, baicalin, amoxicillin, and probenecid inhibited activation of the family pyrin domain containing 3 (NLRP3) inflammasome and apoptosis in the spleen of *G. parasuis*-challenged mice. This study showed the important roles of baicalin, amoxicillin, and probenecid in the modulation of the inflammatory response of Glässer's disease. The findings might provide new strategies for combination therapy using antibiotics and anti-inflammatory drugs to control *G. parasuis* infection.

Keywords: baicalin; amoxicillin; probenecid; *Glaesserella parasuis*; anti-inflammatory; antibacterial

1. Introduction

Glaesserella parasuis (*G. parasuis*), one of the most common bacterial upper respiratory pathogens, is a serious threat to porcine industry development [1]. *G. parasuis* is responsible for Glässer's disease of the pig and could lead to huge economic losses [2]. The typical clinical signs of Glässer's disease are fibrinous polyserositis, polyarthritis, and meningitis [3]. To date, 15 serotypes of *G. parasuis* have been verified by the gel immunodiffusion test, but up to 20% of the isolates cannot be classified [4]. Serovar 5 is thought to be the highly virulent and epidemic strain in China [5]. So far, the pathogenic mechanism of *G. parasuis* remains unclear. Previous research has shown that the *htrA* gene is involved in adhesion and related to survival and pathogenicity of *G. parasuis* [6]. PilA of *G. parasuis* mediates adhesion to porcine kidney-15 (PK-15) cells and swine tracheal epithelial cells (STECs) through fibronectin and laminin [7]. Deletion of QseBC weakens the virulence of *G. parasuis*, adhesion to pulmonary alveolar macrophage (PAM) cells, and host inflammation [8]. Due to the unclear pathogenesis of *G. parasuis*, it is important to study its infection mechanisms to identify effective treatments.

In clinical practice, antibiotics are generally used to control bacterial infections in pigs [9]. Amoxicillin belongs to the aminopenicillin family of penicillin; it can inhibit the synthesis of bacterial cell walls and has efficient broad-spectrum antibacterial effects [10]. *G. parasuis* is sensitive to amoxicillin; thus, it has been widely used to control *G. parasuis* infection on pig farms [11]. However, the clinical management of *G. parasuis* infections is increasingly constrained by emerging antimicrobial resistance [12], which underscores the critical need for implementing combination therapeutic strategies to optimize antimicrobial stewardship while maintaining clinical efficacy. In the present study, amoxicillin was used as the positive control.

Probenecid is lipid soluble and used as a putative neurotherapeutic [13]. Probenecid has been shown to have an anti-inflammatory function via inhibiting OAT1/3 mediated transport of kynurenic acid in the brain [14]. It can modulate myocardial contractility and displays anti-inflammatory properties that might inhibit inflammasomes, modify purinergic-pannexin-1 signaling and TRPV2 channels [15]. In the previous study, mice were used to establish the *G. parasuis* infection model [16]. However, the anti-inflammatory efficacy of probenecid in *G. parasuis*-challenged mice has not been investigated.

Due to the excessive use of antibiotics, the phenomenon of antibiotic resistance has become increasingly severe [17]. Traditional Chinese medicine could alleviate the problem of antibiotic resistance. Baicalin, extracted from *Scutellaria baicalensis* Georgi, has important anti-inflammatory [18] and anti-microbial effects [19]. Baicalin suppresses interleukin-1β (IL-1β)-induced apoptosis, inflammation, oxidative stress, and extracellular matrix degradation in human nucleus pulposus cells [20]. Baicalin reduces renal inflammation in mesangial proliferative glomerulonephritis through the nuclear factor erythroid 2-related factor 2 (Nrf2)/antioxidant response element (ARE) and phosphoinositide 3-kinase (PI3K)/AKT pathways [21]. Baicalin relieves lung inflammation triggered by lipopolysaccharide (LPS) in mice via the nuclear factor kappa B (NF-κB) and mitogen-activated protein kinase (MAPK) pathways [22]. However, there has been no investigation of antibiotic and traditional Chinese medicine combination therapy to control *G. parasuis* infection. Therefore, in this study, we investigated the ability of amoxicillin, probenecid, the combination of amoxicillin and probenecid, and baicalin to ameliorate *G. parasuis* infection in mice.

2. Results

2.1. The Effects of Baicalin, Probenecid, Amoxicillin, and the Combination of Probenecid and Amoxicillin on the Body Weight (BW), Survival Rate, and Bacterial Burden of G. parasuis-Challenged Mice

After *G. parasuis* challenge, the weight of the mice from the infection, Pro10, Pro20, Pro40, Amo20, the Pro20 + Amo20, and the Bai100 groups decreased (Figure 1A; see Section 4.4 for details on the groups). The mice from the control group showed an increase in BW, and the mice from the infection group grew the slowest (Figure 1A). Mice treated with baicalin exhibited the fastest BW recovery (Figure 1A).

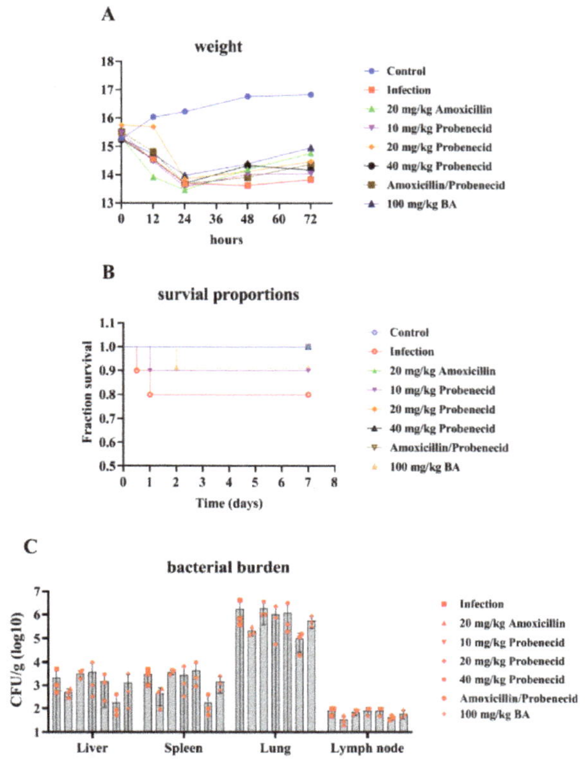

Figure 1. Detection of the effects of baicalin on the body weight (**A**), survival rate (**B**), and bacterial burden (**C**) of *G. parasuis*-challenged mice. When the mice were pretreated with drugs and challenged by *G. parasuis*, the body weight, survival rate, and the bacterial burdens in the liver, spleen, lung, and lymph node were determined.

Following challenged with 2×10^9 colony-forming units (CFUs) of *G. parasuis*, the infection group showed a mortality rate of 20% (Figure 1B). The Bai100 group provided 90% protection against the *G. parasuis* challenge compared to the infection group ($p < 0.05$) (Figure 1B). The Pro10, Pro20, Pro40, Amo20, and Pro20 + Amo20 groups provided 90%, 100%, 100%, 100%, and 100% protection, respectively, when the mice were challenged by *G. parasuis* (Figure 1B). The Pro10, Pro20, Pro40, Amo20, and Pro20 + Amo20 groups showed a higher survival rate than the infection group ($p < 0.05$) (Figure 1B). Furthermore, the Pro20, Pro40, Amo20, and Pro20 + Amo20 groups showed a higher survival rate than the Pro10 and Bai100 groups ($p < 0.05$) (Figure 1B).

We examined the bacterial burden in the liver, spleen, lung, and lymph node. The moribund mice from the infection group had detectable bacteria in the liver, spleen, lung, and lymph node. The surviving mice had a variable bacterial burden in the tissue specimens. The surviving mice from the Amo20 and Pro20 + Amo20 groups had less bacterial burden in the liver, spleen, lung, and lymph node (Figure 1C). The mice from the Pro10 group had a higher bacterial burden in the lung (Figure 1C).

2.2. Baicalin, Probenecid, Amoxicillin, and the Combination of Probenecid and Amoxicillin Altered the Blood Biochemical Parameters and Routine Blood Test Indicators in G. parasuis-Challenged Mice

We explored the blood biochemical parameters and routine blood test indicators 12, 24, and 72 h after challenging mice with *G. parasuis*. Total bilirubin (TB), alanine aminotransferase (ALT), cholesterol (CHOL), and triglycerides (TG) were significantly upregulated, and albumin (ALB), alkaline phosphatase (ALP), glucose (GLU), calcium (Ca), and lactate dehydrogenase (LDH) were decreased at 12, 24, and 72 h after *G. parasuis* challenge in the infection group compared with the control group ($p < 0.05$) (Tables 1–3). We found that 20 mg/kg BW amoxicillin reduced the ALT and CHOL levels and increased the Ca level 12 h after the *G. parasuis* challenge compared with the infection group ($p < 0.01$) (Table 1). Treatment with 20 or 40 mg/kg BW probenecid or 20 mg/kg BW probenecid and amoxicillin increased the ALB and LDH level and reduced the phosphorus (P) level compared with the infection group ($p < 0.05$) (Table 1). When the mice were pretreated with 100 mg/kg BW baicalin for 12 h, the CHOL and P levels were decreased compared with the infection group ($p < 0.05$) (Table 1). At 72 h after *G. parasuis* challenge, the aspartate aminotransferase (AST), CHOL, TG, and high-density lipoprotein (HDL) levels had returned to normal levels (Table 3).

Table 1. The blood biochemical parameters for 12 h.

Item	Control	GPS	Amo20	Pro10	Pro20	Pro40	Pro20 + Amo20	Bai100	SEM	*p* Value						
	(A)	(B)	(C)	(D)	(E)	(F)	(G)	(H)		BvsA	CvsB	DvsB	EvsB	FvsB	GvsB	HvsB
TB (mg/dL)	0.13	1.20	0.60	1.00	0.90	0.70	0.40	1.08	0.10	0.005	0.089	0.556	0.378	0.146	0.028	0.734
TP (g/dL)	30.99	30.16	28.40	28.18	33.48	30.97	30.84	29.27	0.48	0.626	0.311	0.255	0.065	0.633	0.689	0.603
ALB (g/dL)	21.73	16.01	17.76	17.36	19.67	18.85	19.41	17.54	0.46	0.001	0.248	0.369	0.024	0.070	0.034	0.309
AST (U/L)	124.00	142.00	113.00	221.00	268.00	344.00	258.00	208.00	17.29	0.621	0.415	0.039	0.002	<0.001	0.005	0.377
ALT (U/L)	26.00	62.00	38.00	40.00	60.00	52.00	50.00	51.00	2.92	<0.001	0.009	0.015	0.872	0.251	0.170	0.221
ALP (U/L)	185.00	97.00	95.00	99.00	76.00	96.00	107.00	106.00	6.80	<0.001	0.881	0.811	0.078	0.928	0.360	0.424
CHOL (mg/dL)	1.19	2.00	1.68	1.74	1.94	1.85	1.83	1.51	0.06	<0.001	0.005	0.015	0.566	0.149	0.098	<0.001
TG (mg/dL)	1.33	3.10	2.02	1.54	3.47	2.23	1.98	2.69	0.20	0.015	0.118	0.030	0.570	0.203	0.106	0.544
GLU (mg/dL)	3.53	0.70	1.03	1.57	0.72	0.83	1.57	1.05	0.19	<0.001	0.302	0.014	0.958	0.676	0.014	0.284
Ca (mg/dL)	2.11	1.55	1.99	1.77	1.54	1.72	1.79	1.56	0.05	<0.001	0.002	0.084	0.956	0.174	0.065	0.956
P (mg/dL)	2.42	1.74	1.73	1.53	1.65	1.69	1.48	1.62	0.07	<0.001	0.053	0.004	0.019	0.032	0.002	0.014
HDL (mg/dL)	0.81	0.76	0.77	0.83	0.89	0.87	0.87	0.71	0.02	0.001	0.985	0.237	0.606	0.788	0.153	0.505
LDL (mg/dL)	0.11	0.07	0.06	0.08	0.10	0.08	0.09	0.08	0.01	0.079	0.777	0.572	0.168	0.777	0.399	0.777
LDH (U/L)	1191.70	620.40	613.10	800.20	1094.70	1159.30	875.60	647.20	53.60	<0.001	0.951	0.142	0.001	<0.001	0.043	0.821

Table 2. The blood biochemical parameters for 24 h.

Item	Control	GPS	Amo20	Pro10	Pro20	Pro40	Pro20 + Amo20	Bai100	SEM	*p* Value						
	(A)	(B)	(C)	(D)	(E)	(F)	(G)	(H)		BvsA	CvsB	DvsB	EvsB	FvsB	GvsB	HvsB
TB (mg/dL)	0.17	0.56	0.33	0.62	0.14	0.42	0.26	0.31	0.04	<0.001	0.001	0.246	<0.001	0.022	<0.001	<0.001
TP (g/dL)	32.18	32.15	31.59	32.63	31.97	29.72	29.43	30.09	0.44	0.989	0.750	0.786	0.918	0.179	0.135	0.252
ALB (g/dL)	22.15	16.44	17.51	18.62	18.33	17.67	20.07	19.99	0.41	<0.001	0.311	0.048	0.082	0.245	0.003	0.003
AST (U/L)	119.00	135.00	120.00	159.00	158.00	218.00	196.00	154.00	7.72	0.361	0.402	0.167	0.184	<0.001	0.002	0.265
ALT (U/L)	21.00	34.00	25.00	32.00	38.00	40.00	32.00	35.00	1.51	0.003	0.029	0.615	0.360	0.143	0.615	0.801
ALP (U/L)	183.00	68.00	77.00	102.00	100.00	86.00	103.00	83.00	7.24	<0.001	0.329	0.002	0.002	0.053	0.001	0.099
CHOL (mg/dL)	1.21	1.92	1.57	1.71	1.88	1.62	1.55	1.75	0.05	<0.001	0.018	0.141	0.754	0.042	0.014	0.220
TG (mg/dL)	1.32	1.59	0.90	0.81	1.20	0.98	0.96	1.15	0.07	0.275	0.011	0.005	0.125	0.022	0.019	0.084
GLU (mg/dL)	3.47	1.10	0.93	1.50	0.90	1.10	1.71	1.97	0.22	0.003	0.809	0.564	0.772	1.000	0.382	0.220
Ca (mg/dL)	2.09	1.47	1.31	1.97	1.92	1.68	1.44	2.05	0.07	0.002	0.347	0.007	0.014	0.206	0.855	0.003
P (mg/dL)	2.24	1.01	1.22	1.50	1.37	1.27	1.67	1.22	0.09	<0.001	0.460	0.092	0.210	0.350	0.028	0.460
HDL (mg/dL)	0.70	0.74	0.70	0.76	0.82	0.78	0.81	0.84	0.02	0.968	0.609	0.480	0.179	0.330	0.204	0.111
LDL (mg/dL)	0.12	0.10	0.09	0.10	0.10	0.08	0.08	0.09	0.01	0.274	0.852	0.710	0.710	0.461	0.359	0.852
LDH (U/L)	1221.30	621.20	636.13	820.00	867.20	817.30	855.20	657.50	40.07	<0.001	0.813	0.006	0.001	0.006	0.001	0.568

Table 3. The blood biochemical parameters for 72 h.

Item	Control	GPS	Amo20	Pro10	Pro20	Pro40	Pro20 + Amo20	Bai100	SEM	*p* Value						
	(A)	(B)	(C)	(D)	(E)	(F)	(G)	(H)		BvsA	CvsB	DvsB	EvsB	FvsB	GvsB	HvsB
TB (mg/dL)	0.15	0.57	0.18	0.25	0.21	0.41	0.29	0.22	0.03	<0.001	<0.001	<0.001	<0.001	0.045	0.002	<0.001
TP (g/dL)	31.07	28.84	31.39	30.91	30.36	30.00	29.14	30.38	0.36	0.150	0.104	0.181	0.319	0.441	0.839	0.312
ALB (g/dL)	20.09	15.02	17.01	16.68	17.39	18.19	18.37	19.76	0.39	<0.001	0.075	0.133	0.038	0.008	0.006	<0.001
AST (U/L)	123.00	84.00	119.00	106.00	117.00	128.00	107.00	110.00	3.62	0.004	0.007	0.073	0.011	0.002	0.066	0.036
ALT (U/L)	21.00	20.00	13.00	20.00	16.00	23.00	12.00	14.00	0.95	0.622	0.030	1.000	0.227	0.274	0.008	0.061
ALP (U/L)	197.00	59.00	65.00	75.00	66.00	53.00	76.00	87.00	10.65	<0.001	0.813	0.544	0.784	0.823	0.520	0.304
CHOL (mg/dL)	1.25	1.08	1.29	1.44	1.48	1.50	1.40	1.53	0.03	0.051	0.021	<0.001	<0.001	<0.001	0.001	<0.001
TG (mg/dL)	1.29	0.53	0.85	0.76	0.81	0.85	0.78	0.84	0.04	<0.001	<0.001	0.004	0.001	<0.001	0.002	<0.001
GLU (mg/dL)	3.44	1.23	1.30	1.87	1.43	1.60	1.33	1.23	0.16	<0.001	0.776	0.011	0.377	0.115	0.656	1.000
Ca (mg/dL)	2.14	2.29	1.96	1.50	2.03	1.90	1.89	1.93	0.06	0.379	0.058	<0.001	0.130	0.028	0.023	0.041
P (mg/dL)	2.31	1.11	1.18	1.38	1.09	1.04	1.26	2.08	0.11	<0.001	0.778	0.307	0.959	0.808	0.566	0.002
HDL (mg/dL)	0.75	0.45	0.52	0.58	0.63	0.57	0.56	0.60	0.02	<0.001	0.256	0.063	0.014	0.077	0.111	0.029
LDL (mg/dL)	0.10	0.15	0.23	0.22	0.23	0.22	0.20	0.21	0.01	0.073	0.014	0.023	0.011	0.018	0.073	0.058
LDH (U/L)	1435.10	755.10	1130.20	805.30	1069.60	1022.30	723.00	895.00	56.03	<0.001	0.025	0.745	0.055	0.097	0.835	0.368

Regarding the routine blood test results, the white blood cell (WBC), red blood cell (RBC), platelet (PLT), and lymphocyte (LYM) levels were significantly lower in the infection group compared with the control group at 12, 24, and 72 h after *G. parasuis* challenge

($p < 0.05$) (Supplemental Tables S1–S3). Compared with the control group, neutrophils (NE) were significantly increased in the infection group at 24 h after the *G. parasuis* challenge ($p < 0.001$) (Supplemental Table S2) and then decreased at 72 h ($p < 0.001$) (Supplemental Table S3). Treatment with 20 mg/kg BW amoxicillin increased the PLT and monocyte (MONO) levels at 12 h after the *G. parasuis* challenge compared with the infection group ($p < 0.01$) (Supplemental Table S2). The RBC, MONO, and eosinophil (EOS) levels at 12 h after the *G. parasuis* challenge were significantly higher in the Pro40, Pro20+Pro20, and Bai100 groups compared with the infection group ($p < 0.05$) (Supplemental Table S2). At 72 h after the *G. parasuis* challenge, most of the routine blood indicators had returned to normal levels (Supplemental Table S3).

2.3. Baicalin, Probenecid, Amoxicillin, and the Combination of Probenecid and Amoxicillin Inhibited Inflammatory Cytokines and High Mobility Group Box 1 Protein (HMGB1) Production in G. parasuis-Challenged Mice

We evaluated the expression of the inflammatory cytokines IL-1β, IL-18, and tumor necrosis factor α (TNF-α) after mice were challenged by *G. parasuis* for 72 h by using quantitative real-time reverse transcription PCR (RT-qPCR) and Western blotting. *G. parasuis* infection increased IL-1β, IL-18, and TNF-α messenger RNA (mRNA) levels compared with the control group ($p < 0.001$) (Figure 2A–C). The IL-1β, IL-18, and TNF-α mRNA levels were decreased in the Pro10, Pro20, Pro40, Amo20, Pro20 + Amo20, and Bai100 groups compared with the infection group ($p < 0.001$) (Figure 2A–C). The IL-1β, IL-18, and TNF-α protein levels were increased in the infection group compared with the control group ($p < 0.001$) (Figure 2E–H). The IL-1β, IL-18, and TNF-α protein levels were reduced in the Pro10, Pro20, Pro40, Amo20, Pro20 + Amo20, and Bai100 groups compared with the infection group ($p < 0.05$) (Figure 2E–H) (except TNF-α in the Pro10 group).

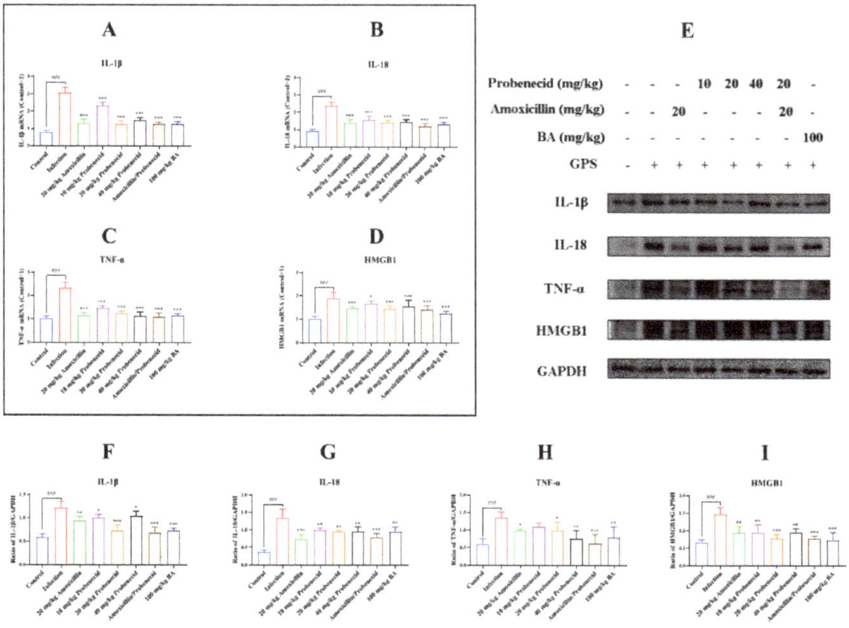

Figure 2. Measurement of the inflammatory cytokines and HMGB1 production in *G. parasuis*-challenged mice. After the mice were challenged, the spleens were collected and IL-1β (**A**,**E**,**F**), IL-18 (**B**,**E**,**G**), TNF-α (**C**,**E**,**H**), and HMGB1 (**D**,**E**,**I**) expressions of mRNA and protein levels were determined by RT-PCR and Western blotting. ### $p < 0.001$ versus control group; * $p < 0.05$, ** $p < 0.01$, and *** $p < 0.001$ versus the infection group.

The HMGB1 mRNA and protein levels in the spleen were increased in the infection group compared with the control group ($p < 0.001$) (Figure 2D). The HMGB1 mRNA and protein levels were decreased in the Pro10, Pro20, Pro40, Amo20, Pro20 + Amo20, and Bai100 groups compared with the infection group ($p < 0.05$) (Figure 2D,E).

2.4. Baicalin, Probenecid, Amoxicillin, and the Combination of Probenecid and Amoxicillin Modified the Proportion of CD3$^+$, CD4$^+$, and CD8$^+$ T Cells in the Blood and Spleen of G. parasuis-Challenged Mice

Immune cells, including T cells, play important roles in regulating innate and adaptive immunity. We used flow cytometry to determine the proportions of T cells in the blood and spleen of *G. parasuis*-challenged mice for 72 h (Figures 3–5, Supplemental Figures S1–S3). In the blood, the proportion of CD3$^+$, CD3$^+$CD4$^+$, and CD3$^+$CD8$^+$ T cells were significantly deceased in the infection group compared with the control group ($p < 0.001$) (Figures 3–5). The Pro10, Pro20, Amo20, and Pro20 + Amo20 groups, but not the Pro40 group, showed higher proportions of CD3$^+$ and CD3$^+$CD4$^+$ T cells compared with the infection group ($p < 0.05$) (Figures 3 and 4), while the proportions of CD3$^+$CD8$^+$ T cells were increased in the Pro20 and Amo20 groups ($p < 0.05$) (Figure 5). When the mice were pretreated with 100 mg/kg BW baicalin, the proportions of CD3$^+$ and CD3$^+$CD4$^+$ T cells were upregulated compared with the infection group ($p < 0.05$) (Figures 3 and 4).

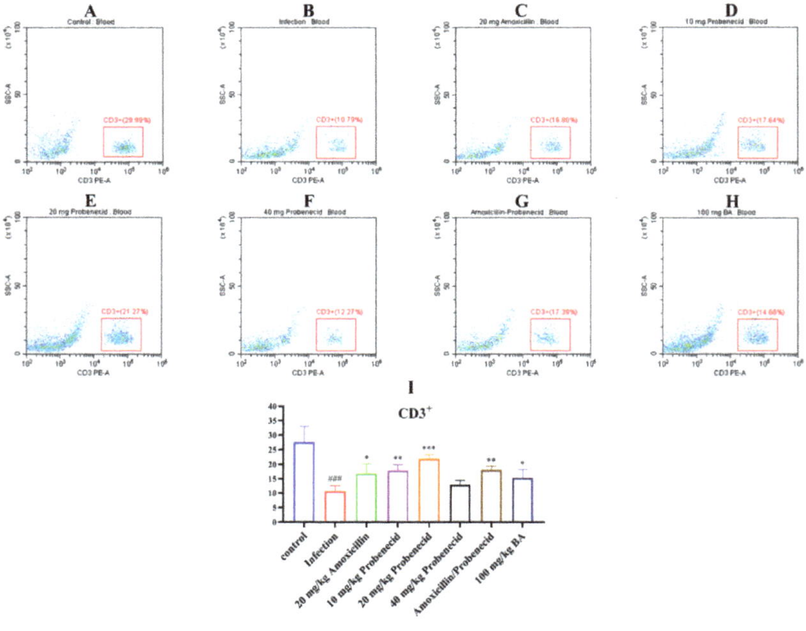

Figure 3. Flow immune cells of CD3$^+$ T cells from the blood were analyzed using cytometry. (**A**) The control group; (**B**) the infection group; (**C**) the Amo20 group; (**D**) the Pro10 group; (**E**) the Pro20 group; (**F**) the Pro40 group; (**G**) the Pro20 + Amo20 group; (**H**) the Bai100 group; (**I**) the results of CD3$^+$ T cells in each group. $^{###}$ $p < 0.001$ versus the control group; * $p < 0.05$, ** $p < 0.01$, and *** $p < 0.001$ versus the infection group.

In the spleen, the proportions of CD3$^+$, CD3$^+$CD4$^+$, and CD3$^+$CD8$^+$ T cells were increased in the infection group compared with the control group ($p < 0.001$) (Supplemental Figures S1–S3). The Pro10, Pro20, Pro40, Amo20, and Pro20 + Amo20 groups showed increased proportions of CD3$^+$, CD3$^+$CD4$^+$, and CD3$^+$CD8$^+$ T cells compared with the infection group ($p < 0.01$) (Supplemental Figures S1–S3). Finally, 100 mg/kg BW baicalin upregulated the proportions of $CD3^+$, $CD3^+CD4^+$, and $CD3^+CD8^+$ T cells compared with the infection group ($p < 0.001$) (Supplemental Figures S1–S3).

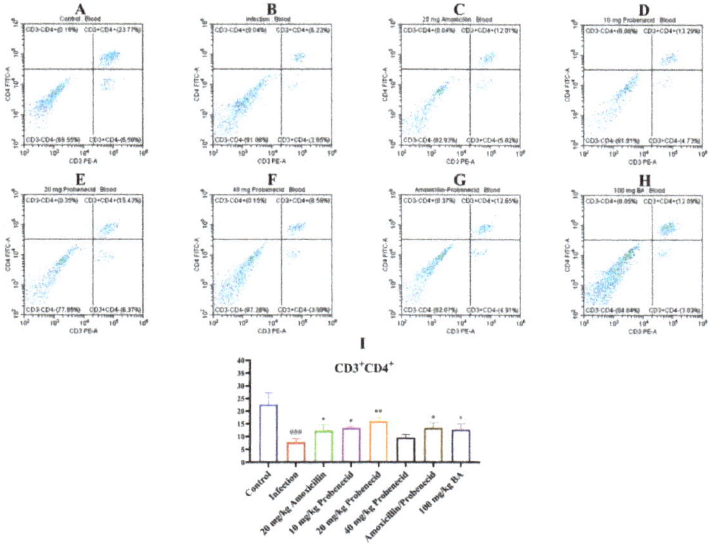

Figure 4. The proportion of $CD3^+CD4^+$ T cells from the blood were analyzed using cytometry. (**A**) The control group; (**B**) the infection group; (**C**) the Amo20 group; (**D**) the Pro10 group; (**E**) the Pro20 group; (**F**) the Pro40 group; (**G**) the Pro20 + Amo20 group; (**H**) the Bai100 group; (**I**) the results of $CD3^+CD4^+$ T cells in each group. ### $p < 0.001$ versus the control group; * $p < 0.05$ and ** $p < 0.01$ versus the infection group.

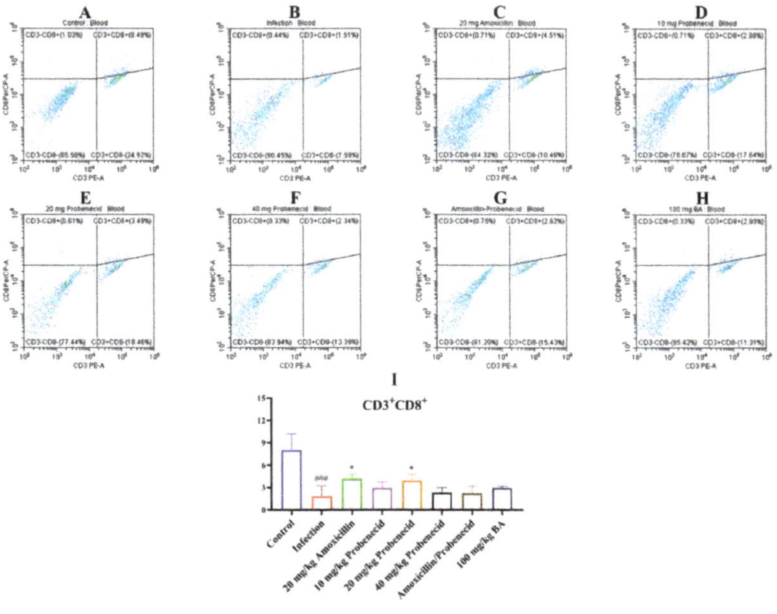

Figure 5. The proportion of $CD3^+CD8^+$ T cells from the blood were analyzed using cytometry. (**A**) The control group; (**B**) the infection group; (**C**) the Amo20 group; (**D**) the Pro10 group; (**E**) the Pro20 group; (**F**) the Pro40 group; (**G**) the Pro20 + Amo20 group; (**H**) the Bai100 group; (**I**) the results of $CD3^+CD8^+$ T cells in each group. ### $p < 0.001$ versus the control group; * $p < 0.05$ versus the infection group.

We also detected the CD3 (Figure 6), CD4 (Supplemental Figure S4), and CD8 (Supplemental Figure S5) expression in the blood vessels by immunohistochemistry. The results showed that the expression of CD3, CD4, and CD8 was intense in the blood vessels of the control group, while in the infection group, the CD3, CD4, and CD8 protein immunostaining was decreased ($p < 0.001$) (Figure 6, Supplemental Figures S4 and S5). Treatment with 20 mg/kg probenecid, 20 mg/kg amoxicillin, the combination of 20 mg/kg probenecid and 20 mg/kg amoxicillin, and 100 mg/kg baicalin could restore the intensity of the CD3, CD4, and CD8 protein expression in the blood vessels of mice compared to the infection group ($p < 0.05$) (Figure 6, Supplemental Figures S4 and S5).

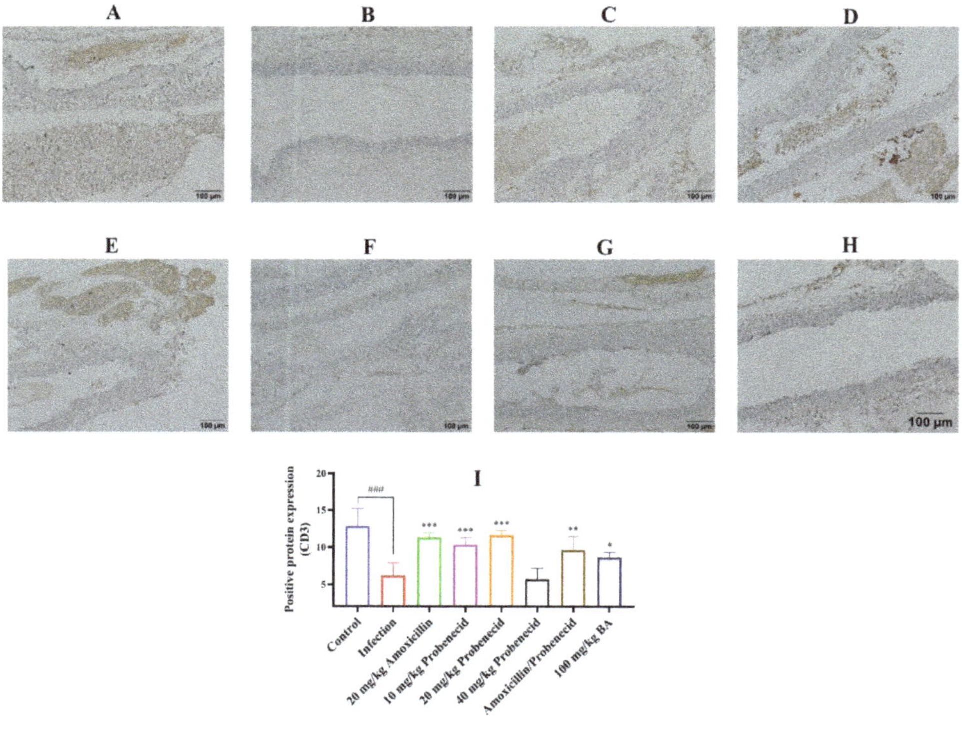

Figure 6. Detection of the CD3 expression in the blood vessels by immunohistochemistry presented in yellow color. Data are expressed as mean ± SD (I). (**A**) the control group; (**B**) the infection group; (**C**) the Amo20 group; (**D**) the Pro10 group; (**E**) the Pro20 group; (**F**) the Pro40 group; (**G**) the Pro20 + Amo20 group; (**H**) the Bai100 group; (**I**) the CD3 expression in the blood vessels. ### $p < 0.001$ versus the control group; * $p < 0.05$, ** $p < 0.01$, and *** $p < 0.001$ versus the infection group.

2.5. Baicalin, Probenecid, Amoxicillin, and the Combination of Probenecid and Amoxicillin Attenuated NLR Family Pyrin Domain Containing 3 (NLRP3) Inflammasome Activation in Spleen of G. parasuis-Challenged Mice

We assessed NLRP3 inflammasome activation in the spleen by measuring NLRP3, ASC, and caspase-1 expression. The NLRP3 and caspase-1 mRNA levels were increased in the infection group compared with the control group ($p < 0.001$) (Figure 7A,C). However, the mRNA levels of NLRP3 and caspase-1 were decreased in the Pro10, Pro20, Pro40, Amo20, Pro20 + Amo20, and Bai100 groups compared with the infection group ($p < 0.01$) (Figure 7A,C).

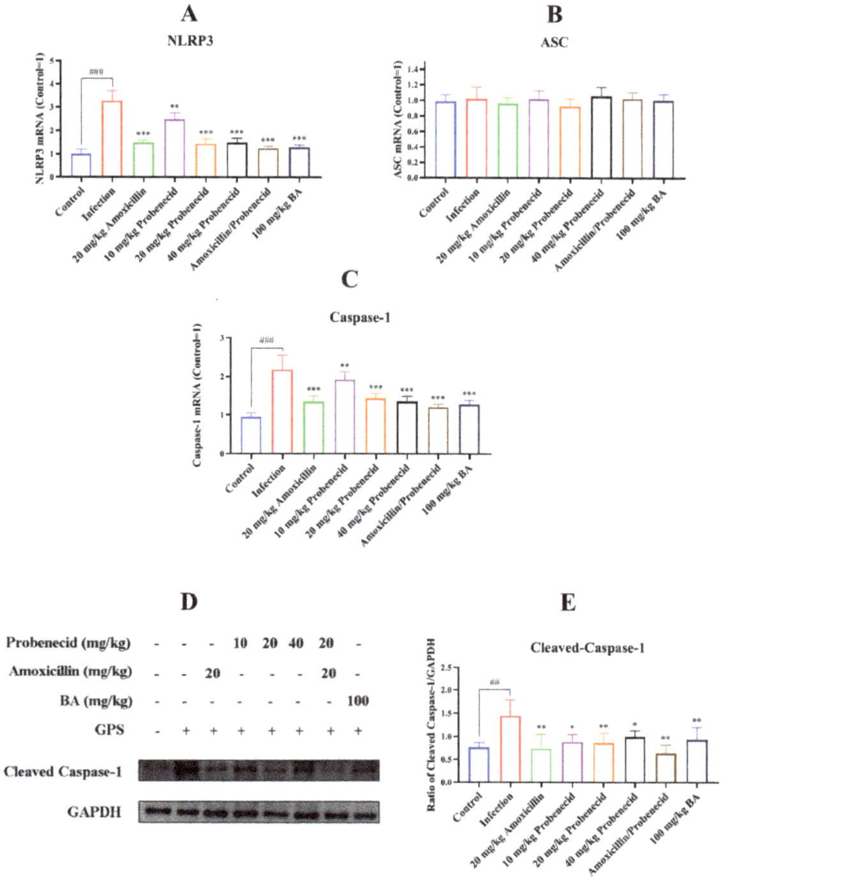

Figure 7. Determination of the effect of baicalin on NLRP3 activation in the spleen of *G. parasuis*-challenged mice. NLRP3 (**A**), ASC (**B**), and caspase-1 (**C**) expressions at mRNA levels were determined by RT-PCR and cleaved caspase-1 protein (**D**,**E**) expression level was evaluated by Western blotting. ## $p < 0.01$ and ### $p < 0.001$ versus the control group; * $p < 0.05$, ** $p < 0.01$, and *** $p < 0.001$ versus the infection group.

We also used Western blotting to determine the cleaved caspase-1 protein level in the spleen after *G. parasuis* infection. The level was increased in the infection group compared with the control group ($p < 0.01$) (Figure 7D,E), while amoxicillin and probenecid treatment reduced cleaved caspase-1 expression compared with the infection group ($p < 0.05$) (Figure 7D,E). Treatment with 100 mg/kg BW baicalin also attenuated cleaved caspase-1 protein expression compared with the infection group ($p < 0.01$) (Figure 7D,E).

2.6. Baicalin, Probenecid, Amoxicillin, and the Combination of Probenecid and Amoxicillin Inhibited Apoptosis in the Spleen of G. parasuis-Challenged Mice

To explore the role of caspase-3 in *G. parasuis*-induced apoptosis in the spleen, we used Western blotting to determine the activated caspase-3 protein level. It was increased in the infection group compared with the control group ($p < 0.001$) (Figure 8A,B). Its expression was decreased in Pro10, Pro20, Pro40, Amo20, Pro20 + Amo20, and Bai100 groups compared with the infection group ($p < 0.01$) (Figure 8A,B).

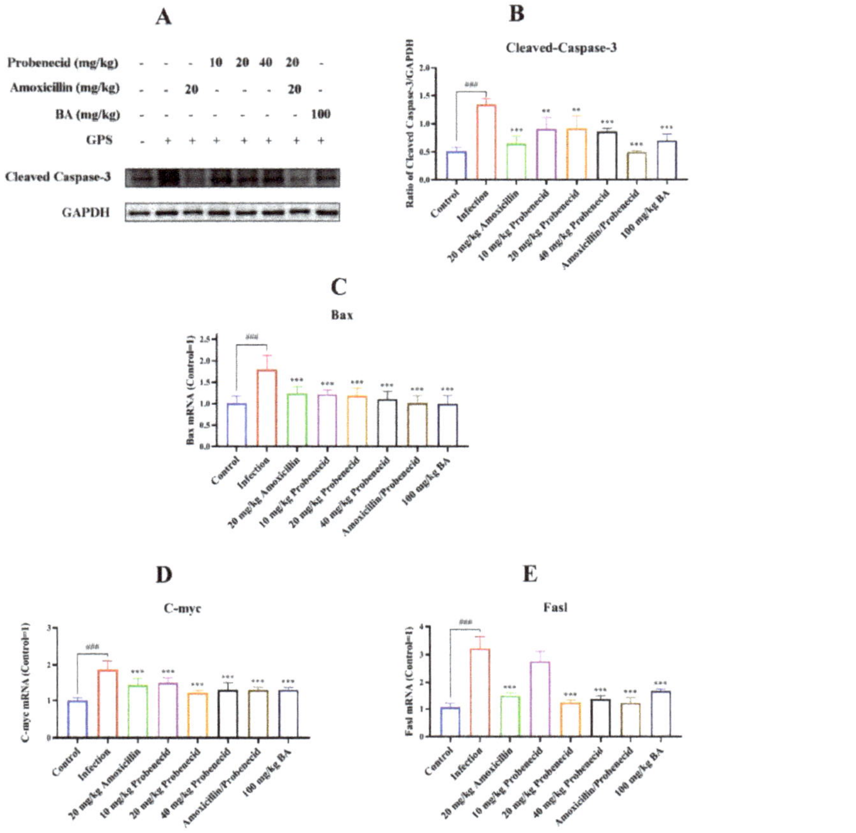

Figure 8. Evaluation of the efficacy of baicalin on cell apoptosis in the spleen of *G. parasuis*-challenged mice. When the mice were infected by *G. parasuis*, the cell apoptosis in the spleen was determined. The activated caspase-3 (**A**,**B**) expression was determined by Western blotting. The Bax (**C**), C-myc (**D**), and Fasl (**E**) expression levels were measured by RT-PCR. ### $p < 0.001$ versus the control group; ** $p < 0.01$ and *** $p < 0.001$ versus the infection group.

We also assessed the mitochondria-dependent apoptosis pathway induced by *G. parasuis* by examining the mRNA levels of apoptotic genes (Bax, C-myc, and Fasl). The infection group showed elevated Bax, C-myc, and Fasl mRNA levels compared with the control group ($p < 0.001$) (Figure 8C–E). All levels were decreased in the Pro20, Pro40, Amo20, Pro20 + Amo20, and Bai100 groups compared to the infection group ($p < 0.001$) (Figure 8C–E).

2.7. Baicalin, Probenecid, Amoxicillin, and the Combination of Probenecid and Amoxicillin Relieved Pathological Lung and Spleen Tissue Damage in G. parasuis-Challenged Mice

We assessed pathological changes in the lung and spleen. In the control group, there was no obvious pathological damage, while in the infection group, there was serious pathological damage, including hemorrhage and inflammatory infiltration in the lung, and necrosis and white pulp lymphocytes reduction in the spleen (Figure 9). In the Pro10, Pro20, Pro40, Amo20, Pro20 + Amo20, and Bai100 groups, there was only minor pathological damage (Figure 9), suggesting that baicalin, amoxicillin, and probenecid relieve pathological tissue damage caused by *G. parasuis* challenge.

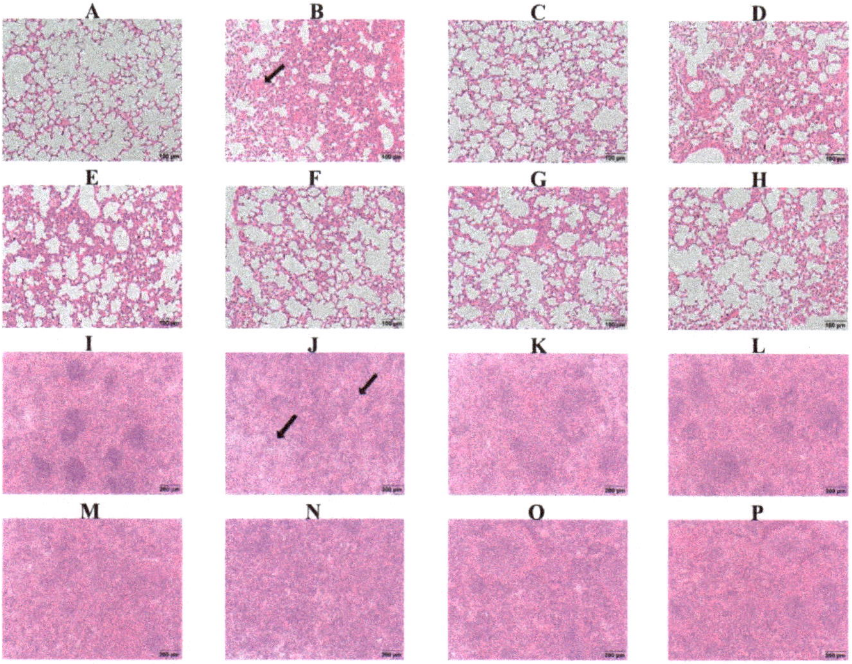

Figure 9. Histopathological analysis of mice lung (**A–H**) and spleen (**I–P**) in the control group (**A**,**I**), the infection group (**B**,**J**), the Amo20 group (**C**,**K**), the Pro10 group (**D**,**L**), the Pro20 group (**E**,**M**), the Pro40 group (**F**,**N**), the Pro20 + Amo20 group (**G**,**O**), and the Bai100 group (**H**,**P**). Lung and spleen tissues were fixed by immersion and embedded in paraffin. Tissue sections were cut and stained by hematoxylin and eosin, and examined by light microscopy. The arrow represents the pathological change.

3. Discussion

In this study, we compared the efficacy of baicalin, probenecid, and amoxicillin to protect against *G. parasuis* infection in mice. We found that these compounds could provide protection against *G. parasuis* infection. Our study is the first report regarding the ability of baicalin and probenecid to protect against *G. parasuis*. These findings might provide a new approach for the control of *G. parasuis* infection through combination therapy, but the use of a mouse model may not fully replicate the conditions of *G. parasuis* infection in pigs; thus, we will use a piglet model to explore the protection of baicalin, amoxicillin, and probenecid in piglets against a *G. parasuis* challenge in the future study.

A bacterial infection can elicit the host's innate and adaptive immune responses [23]. T cells recognize pathogens, including bacteria, that infect host cells [24]. CD3 is a T cell surface marker [25], and CD4+ T helper cells are key regulators of host health and disease [26]. $CD8^+$T cells, known as cytotoxic T cells, are the key component of the host's adaptive immune system. They are responsible for killing cells infected by bacteria or viruses and form memory T cells to provide long-term protection for the host [27]. In chronic infections (e.g., HIV) and the tumor microenvironment, the auxiliary function of $CD4^+$ T cells is dysregulated, and they cannot provide effective assistance to $CD8^+$ T cells [28].

Although *G. parasuis* is known as the causative pathogen of Glässer's disease, it is still unclear how the adaptive immune system responds to this bacterium. *G. parasuis* could elicit immunosuppression, but the exact mechanism remains unclear. We found that the

WBC and LYM levels decreased sharply from 12 to 72 h after the *G. parasuis* challenge. We also assessed the $CD3^+$, $CD3^+CD4^+$, and $CD3^+CD8^+$ T cell levels in the blood and spleen. The proportion of each population was significantly decreased in the infection group compared with the control group. Previous research reported that a decrease in $CD4^+$ T cells reduces macrophage phagocytosis and the ability to kill pathogens [29]. This might be one of the mechanisms by which *G. parasuis* induces immunosuppression. We found that amoxicillin, probenecid, and baicalin could reverse the reduction in $CD3^+$, $CD3^+CD4^+$, and $CD3^+CD8^+$ T cell levels. These findings shed light on how the adaptive immune system responds to *G. parasuis* infection, and how amoxicillin, probenecid, and baicalin exert an immunomodulatory effect on T cell differentiation.

HMGB1 is a prototypical damage-associated molecular pattern (DAMP) and promotes inflammation [30]. HMGB1 plays a pivotal role in inflammatory diseases by binding to its corresponding receptors [31]. In mice, HMGB1 enables bacterial lipids to trigger receptor-interacting protein kinase 3 (RIPK3)-mediated necroptosis and apoptosis, leading to tissue damage [32]. HMGB1 is involved in LPS-induced hyperinflammation in mice [33]. The deletion of HMGB1 in airway epithelial cells exaggerates the inflammatory response in mice with muco-obstructive airway disease [34]. We found that *G. parasuis* promoted HMGB1 production in the spleen, which might result in tissue damage, but the mechanism needs to be investigated more deeply. Probenecid provides protection against oxygen–glucose deprivation injury in primary astrocytes by inhibiting HMGB1 production [35]. Probenecid also protects against transient focal cerebral ischemic injury by inhibiting HMGB1 release in mice [36]. Baicalin protected piglets from a *G. parasuis* challenge by reducing HMGB1 production [37]. In the present study, baicalin, probenecid, and amoxicillin attenuated HMGB1 release, which relieved tissue damage, but the exact mechanism requires further elucidation.

The NLRP3 inflammasome is an intracellular multiprotein complex that can be activated by a series of pathogenic microbes [38]. NLRP3 inflammasome activation in bone marrow-derived macrophages leads to liver fibrosis [39]. Moreover, the NLRP3 inflammasome is involved in Alzheimer's disease-related neuroinflammation [40]. Researchers have reported that the NLRP3 inflammasome can mediate apoptosis [41,42]. We found that *G. parasuis* challenge activated the NLRP3 inflammasome signaling pathway and apoptosis in the spleen.

Many traditional Chinese medicines are anti-inflammatory. Shizhifang inhibits hyperuricemia-induced renal tubular epithelial cell pyroptosis through targeting NLRP3 to induce tubular inflammation [43]. QijiShujiang granules alleviate the dopaminergic neuronal injury of Parkinson's disease by inhibiting the NLRP3/caspase-1 pathway [44]. Yunvjian decoction attenuates LPS-induced periodontitis in alveolar bone by suppressing the NLRP3 pathway [45]. We found that amoxicillin, probenecid, and baicalin attenuated NLRP3 pathway activation and apoptosis in the spleen triggered by *G. parasuis*.

NLRP3 inflammasome activation results in the release of cytokines such as IL-1β, IL-18, and TNF-α [38]. TNF-α and IL-1β are the key drivers of retinal endothelial dysfunction in non-infectious uveitis, leading to macular oedema and vascular leakage [46]. IL-1β disrupts blood–brain barrier development by inhibiting endothelial Wnt/β-catenin signaling during neuroinflammation [47]. IL-1β induces inflammatory responses in human aortic valve interstitial cells (AVICs) and is involved in the pathogenesis of aortic stenosis [48]. IL-18 induces inflammatory and cytotoxic immune cell activities, contributing to autoimmunity [49]. Exaggerated cytokine responses could cause severe tissue damage [50]. We found that amoxicillin, probenecid, and baicalin could reduce the tissue damage induced by *G. parasuis* and attenuated IL-1β, IL-18, and TNF-α expression.

4. Materials and Methods

4.1. Ethics Statement

This animal study was approved by the Animal Care and Use Committee of Wuhan Polytechnic University, Hubei Province, China (WPU202307002). All experimental animals were euthanized via intraperitoneal injection with 1% pentobarbital sodium (50 mg/kg) at the end of the experiment.

4.2. Bacterial and Culture Conditions

G. parasuis SH0165 serovar 5 was originally isolated from the lung of a commercially produced pig that had the typical Glässer's disease characteristics of arthritis, fibrinous polyserositis, hemorrhagic pneumonia, and meningitis [37]. The SH0165 strain was cultured in tryptic soy broth (TSB) (Difco Laboratories, Detroit, MI, USA) or tryptic soy agar (TSA) (Difco Laboratories) supplemented with 10 μg/mL nicotinamide adenine dinucleotide (NAD) (Ameresco, Framingham, MA, USA) and 10% fetal bovine serum (Sijiqing, Jiangsu Sijiqing Biotechnology Co., Ltd., Nanjing, China) at 37 °C.

4.3. Drugs and the Mice

Baicalin was purchased from Sichuan Taikang Pharmaceutical Co., Ltd. (Yuechi, China). When used, baicalin was dissolved and diluted in TSB. Probenecid and amoxicillin were obtained from MedChemexpress (Monmouth Junction, NJ, USA).

One hundred and sixty 4–5-week old female BALB/c mice were purchased from Hubei Provincial Center for Disease Control and Prevention (China).

4.4. Experimental Design

The mice were randomly divided into eight groups (20 mice/group): the control group, the infection group, the 10 mg/kg BW probenecid group (Pro10), the 20 mg/kg BW probenecid group (Pro20), the 40 mg/kg BW probenecid group (Pro40), the amoxicillin group (Amo20), the probenecid and amoxicillin group (Pro20 + Amo20), and the baicalin group (Bai100). Before the *G. parasuis* challenge, the mice from the Pro10, Pro20, and Pro40 groups were injected intramuscularly with 10, 20, and 40 mg/kg BW probenecid, respectively. The mice in the Amo20 group were injected intramuscularly with 20 mg/kg BW amoxicillin. The mice in the Pro20 + Amo20 group were injected intramuscularly with 20 mg/kg BW probenecid and 20 mg/kg BW amoxicillin. The mice in the Bai100 group were injected intramuscularly with 100 mg/kg BW baicalin. After 2 h, the infection, Pro10, Pro20, Pro40, Amo20, Pro20 + Amo20, and Bai100 groups were challenged with an intraperitoneal injection of 2×10^9 CFU *G. parasuis* in 1 mL of TSB. The mice from the control group were injected intraperitoneally with an equivalent volume of TSB. The drugs were administered twice a day for 2 days. All mice were monitored for 7 days after the *G. parasuis* challenge, during which time their BW, morbidity, and mortality were recorded.

4.5. Blood Biochemical Parameters and Routine Blood Test Indicators

At 12, 24, and 72 h after the *G. parasuis* challenge, blood was collected to assess blood biochemical parameters and routine blood test indicators. The blood biochemical parameters were explored as described in a previous study [51]. Briefly, 500 μL of blood from the caudal vein was collected in the tubes containing ethylenediaminetetraacetic acid (EDTA). The blood samples were centrifuged at 3000 rpm for 30 min under 4 °C to obtain plasma samples, which were analyzed with commercially available kits (Shanghai KehuaBioengineering Co., Ltd., Shanghai, China). The biochemical parameters detected included TB, total protein (TP), ALB, AST, ALT, ALP, CHOL, TG, GLU, Ca, P, HDL, low-density lipoprotein (LDL), and LDH. Anticoagulated blood samples were used to determine the

WBC, RBC, hemoglobin (HGB), PLT, NE, LYM, MONO, and EOS levels with an automatic blood analyzer (Hitachi HITEC 7100, Tokyo, Japan).

4.6. Flow Cytometry

Anticoagulated whole blood samples and spleens from mice 72 h after the *G. parasuis* challenge were used to prepare single-cell suspensions. Splenocytes were isolated from spleens as described previously [52]. Briefly, the spleen samples were filtered through strainers (70 μm) to obtain splenocytes, which were rinsed and resuspended in cold phosphate-buffered saline (PBS), before being lysed with lysis buffer. Afterwards, the blood cells and splenocytes were labelled with the following monoclonal antibodies: PE anti-mouse CD3ε, FITC anti-mouse CD4, PerCP/Cyanine5.5 anti-mouse CD8α (BioLegend, San Diego, CA, USA). The samples were analyzed with flow cytometry (Beckman, Shanghai, China).

4.7. Bacterial Burden

After challenging the mice with *G. parasuis* for 24 h, the bacterial burden in the liver, spleen, lung, and lymph node were aseptically collected, tissues were homogenized in sterile PBS (1 mL) using a mechanical homogenizer, and serial 10-fold dilutions of the homogenates were prepared in PBS and cultured in triplicate in TSA at 37 °C for 24 h. Bacterial colonies were counted, and colony-forming units (CFUs) per gram of tissue were calculated.

4.8. ***RT**-qPCR*

RT-qPCR was used to determine the expression of inflammatory cytokines (IL-1β, IL-18, and TNF-α) and HMGB1 [53]. Briefly, total RNA was extracted from spleen samples with the TRIzol reagent (Invitrogen, Waltham, MA, USA) and reverse transcribed to complementary DNA (cDNA) using reverse transcriptase (TaKaRa, Beijing, China) according to the manufacturer's protocol. RT-qPCR was performed by using the SYBR Green PCR Kit (TaKaRa), following the manufacturer's protocol, and a 7500 Real-Time PCR System (Applied Biosystems, Waltham, MA, USA). The specific amplification of each gene was verified by using melting curve analysis. The relative fold change was calculated by using the $2^{-\Delta\Delta Ct}$ method. The primers used in this study for RT-PCR are presented in Table 4.

Table 4. Primer sequences for qRT-PCR analysis.

Gene		Nucleotide Sequence (5′–3′)	Tm (°C)	Length (bp)
NLRP3	Forward	TTTATTTGTACCCAAGGCTGCTATC	54.7	148
	Reverse	CAACGGACACTCGTCATCTTCA	57.1	
ASC	Forward	CAGTTCGTGCAGAGACCACCA	60.2	140
	Reverse	CTGCTCCAGGTCCATCACCA	60.0	
Caspase-1	Forward	TCCGAGGGTTGGAGCTCAAG	59.9	137
	Reverse	CTGGCCAGGCAGCAAATTC	57.8	
HMGB1	Forward	ATGACAAGCAGCCCTATGAGAA	55.9	82
	Reverse	CCTTTAGCTCTGTAGGCAGCAA	57.1	
Fasl	Forward	GCCATCACAACCACTCCC	56.8	107
	Reverse	CCAGAGCCACCAGAACCA	57.9	
Bax	Forward	AGCTGCAGAGGATGATTGCTG	57.9	178
	Reverse	CTGATCAGCTCGGGCATTTA	54.8	
C-myc	Forward	TCGCCCAAATCCTGTACCTC	57.2	193
	Reverse	TCTCCACAGACACCACATCAAT	55.8	

Table 4. *Cont.*

Gene		Nucleotide Sequence (5′–3′)	Tm (°C)	Length (bp)
IL-1β	Forward	TGCCACCTTTTGACAGTGATG	55.6	138
	Reverse	TGATGTGCTGCTGCGAGATT	57.3	
IL-18	Forward	GGACACTTTCTTGCTTGCCA	56.0	166
	Reverse	CAGCCTCGGGTATTCTGTTATG	55.4	
TNF-α	Forward	GCCCCCAGTCTGTATCCTTCTA	58.4	209
	Reverse	TTCGGAAAGCCCATTTGAGT	54.3	
GAPDH	Forward	AAGCCCATCACCATCTTCCA	56.4	88
	Reverse	CACCAGTAGACTCCACGACA	56.4	

4.9. Western Blotting

Western blotting was performed as described previously [37] with some minor modifications. Briefly, total protein was extracted from spleens with a total protein extraction kit (Beyotime Biotechnology, Shanghai, China). The protein concentration was determined by a BCA protein assay kit (Sigma, Bellevue, WA, USA). Proteins were resolved with 12% sodium dodecyl sulphate–polyacrylamide gel electrophoresis (SDS-PAGE) and transferred onto polyvinylidene difluoride membranes. The membranes were incubated in 5% non-fat milk for 90 min to block nonspecific protein binding and then washed five times with Tris-buffered saline containing Tween 20 (TBST). The membranes were incubated with primary antibody of IL-1β (Cat. No. A1112, Polyclonal, 1:1000, ABclonal, Woburn, MA, USA), IL-18 (Cat. No. A23076, Monoclonal, 1:5000, ABclonal), TNF-α (Cat. No. 60291-1, Monoclonal, 1:1000, Proteintech, Tokyo, Japan), HMGB1 (Cat. No. PAB12414, Polyclonal, 1:1000, Abnova, Taipei City, Taiwan), cleaved caspase-1 (Cat. No. ABP0172, Polyclonal, 1:1000, Abbkine, Atlanta, GA, USA), activated caspase-3 (Cat. No. 9661S, Monoclonal, 1:1000, Cell Signaling, Danvers, MA, USA), or GAPDH (Cat. No. 10494-1-AP, Polyclonal, 1:5000, Proteintech, Rosemont, IL, USA) for 12 h at 4 °C. After washing the membranes five times with TBST, they were incubated with horseradish peroxidase-linked goat anti-rabbit antibody (Proteintech, Rosemont, IL, USA) for 30 min at 37 °C. The colored bands were analyzed by utilizing ImageJ 1.52a software to measure the gray values with a FluorChemFC2 AIC system (Alpha Innotech, San Leandro, CA, USA). The protein expression levels were measured by comparing the gray values of the colored bands with the gray values of the internal controls.

4.10. Immunohistochemistry

Immunohistochemistry was performed as described previously [54] with some minor modifications. Briefly, blood vessel samples from mice were collected 72 h after the *G. parasuis* challenge, and following rehydration in a graded alcohol series, the slides were heated in an appropriate buffer solution and incubated with blocking solution to reduce nonspecific background staining. After incubation with the primary antibody for 1 h, the slides were washed with buffer solution to remove unbound primary antibody. Then, the slides were incubated with secondary antibody for 1 h. After washing in buffer solution, substrate or a fluorophore was added to the slides to visualize the antigen–antibody complex. Following counterstaining with hematoxylin, the slides were dehydrated through a graded alcohol series and cleared with xylene.

4.11. Histopathological Analysis

Lung and spleen tissues were fixed by immersion in 10% neutral buffered formalin and then embedded in paraffin. After embedding, 4 μm tissue sections were cut and stained with hematoxylin and eosin using a standard protocol. The stained sections were examined with a light microscope.

4.12. Statistical Analysis

The experimental data are presented as the mean ± standard deviation. Statistical differences were determined with analysis of variance (ANOVA). The log-rank test was used for survival analysis. $p < 0.05$ was considered to be a statistically significant difference.

5. Conclusions

Taken together, our results demonstrated that amoxicillin, probenecid, and baicalin improved the blood biochemical parameters, routine blood test indicators, and survival rate, and reduced the bacterial burden and pathological tissue damage caused by a *G. parasuis* challenge. These compounds modified CD3+, CD3+CD4+, and CD3+CD8+ T cell differentiation in the blood and splenocytes; attenuated IL-1β, IL-18, TNF-α, and HMGB1 release; and inhibited NLRP3 inflammasome activation and apoptosis in the spleen. Our research provides new insights regarding the combined use of antibiotics and anti-inflammatory agents to control *G. parasuis* infection.

Supplementary Materials: The following supporting information can be downloaded at https://www.mdpi.com/article/10.3390/biom15040507/s1.

Author Contributions: S.F. and Y.Q. conceived and designed the experiments; J.L., R.L., Y.F., S.L., Q.D., Y.S., X.T., Y.Z. and P.W. performed the experiments; J.L., S.F., L.G., Q.L. and C.Y. analyzed the data; S.F. wrote the paper. All authors have read and agreed to the published version of the manuscript.

Funding: This work was supported by the National Natural Science Foundation of China (grant no. 32072917; no. 32273067), Key Research and Development Plan of Hubei Province, China (2023BBB069).

Institutional Review Board Statement: Not applicable.

Informed Consent Statement: Not applicable.

Data Availability Statement: All relevant data are within the manuscript.

Conflicts of Interest: The authors declare no conflicts of interest.

References

1. Hua, K.; Li, T.; He, Y.; Guan, A.; Chen, L.; Gao, Y.; Xu, Q.; Wang, H.; Luo, R.; Zhao, L.; et al. Resistin secreted by porcine alveolar macrophages leads to endothelial cell dysfunction during Haemophilus parasuis infection. *Virulence* **2023**, *14*, 2171636. [PubMed]
2. Mugabi, R.; Silva, A.; Hu, X.; Gottschalk, M.; Aragon, V.; Macedo, N.R.; Sahin, O.; Harms, P.; Main, R.; Tucker, A.W.; et al. Molecular characterization of Glaesserella parasuis strains circulating in North American swine production systems. *BMC Vet. Res.* **2023**, *19*, 135.
3. Duan, Y.; Hao, Y.; Feng, H.; Shu, J.; He, Y. Research progress on Haemophilus parasuis vaccines. *Front. Vet. Sci.* **2025**, *12*, 1492144.
4. Kielstein, P.; Rapp-Gabrielson, V.J. Designation of 15 serovars of Haemophilus parasuis on the basis of immunodiffusion using heat-stable antigen extracts. *J. Clin. Microbiol.* **1992**, *30*, 862–865.
5. Yan, P.; Jia, Y.C.; Zhang, X.L.; Zhou, Y.Y.; Guo, Y.; Yin, R.L.; Yuan, J.; Wang, L.X.; Guo, Z.B.; Wang, J.Y.; et al. Virulence assessment of four Glaesserella parasuis strains isolated in Liaoning province of China. *Res. Vet. Sci.* **2023**, *158*, 226–234.
6. Zhang, X.; Lin, Y.; Xu, X.; Wen, S.; Wang, Z.; Gu, J.; He, Q.; Cai, X. HtrA is involved in stress response and adhesion in Glaesserella parasuis serovar 5 strain Nagasaki. *Vet. Microbiol.* **2023**, *282*, 109748.
7. An, J.; Cai, J.; Zhang, B.; Li, Y. Pili Subunit PilA Contributes to the Cytoadhesion of Glaesserella Parasuis to Host Cells and Provides Immunoprotection. *Appl. Environ. Microbiol.* **2023**, *89*, e0200222.
8. Yan, X.; Dai, K.; Gu, C.; Yu, Z.; He, M.; Xiao, W.; Zhao, M.; He, L. Deletion of two-component system QseBC weakened virulence of Glaesserella parasuis in a murine acute infection model and adhesion to host cells. *PeerJ* **2022**, *10*, e13648.
9. Rochegüe, T.; Haenni, M.; Mondot, S.; Astruc, C.; Cazeau, G.; Ferry, T.; Madec, J.Y.; Lupo, A. Impact of Antibiotic Therapies on Resistance Genes Dynamic and Composition of the Animal Gut Microbiota. *Animals* **2021**, *11*, 3280. [CrossRef]

10. Elsbroek, L.; Amiteye, D.; Schreiber, S.; Herrmann, F. Molecular Imaging of Isolated Escherichia coli DH5α Peptidoglycan Sacculi Identifies the Mechanism of Action of Cell Wall-Inhibiting Antibiotics. *ACS Chem. Biol.* **2023**, *18*, 848–860.
11. El Garch, F.; de Jong, A.; Simjee, S.; Moyaert, H.; Klein, U.; Ludwig, C.; Marion, H.; Haag-Diergarten, S.; Richard-Mazet, A.; Thomas, V.; et al. Monitoring of antimicrobial susceptibility of respiratory tract pathogens isolated from diseased cattle and pigs across Europe, 2009–2012: VetPath results. *Vet. Microbiol.* **2016**, *194*, 11–22. [PubMed]
12. An, J.; Zhang, C.; Li, Y. Outer Membrane Vesicles Associated with the Delivery of Amoxicillin Resistance in Glaesserella parasuis. *Microb. Drug Resist.* **2023**, *29*, 263–270. [PubMed]
13. Clark, R.S.B.; Empey, P.E.; Kochanek, P.M.; Bell, M.J. N-Acetylcysteine and Probenecid Adjuvant Therapy for Traumatic Brain Injury. *Neurotherapeutics* **2023**, *20*, 1529–1537. [PubMed]
14. Wirthgen, E.; Hoeflich, A.; Rebl, A.; Günther, J. Kynurenic Acid: The Janus-Faced Role of an Immunomodulatory Tryptophan Metabolite and Its Link to Pathological Conditions. *Front. Immunol.* **2017**, *8*, 1957.
15. Onódi, Z.; Koch, S.; Rubinstein, J.; Ferdinandy, P.; Varga, Z.V. Drug repurposing for cardiovascular diseases: New targets and indications for probenecid. *Br. J. Pharmacol.* **2023**, *180*, 685–700.
16. Fu, S.; Zhang, M.; Ou, J.; Liu, H.; Tan, C.; Liu, J.; Chen, H.; Bei, W. Construction and immune effect of Haemophilus parasuis DNA vaccine encoding glyceraldehyde-3-phosphate dehydrogenase (GAPDH) in mice. *Vaccine* **2012**, *30*, 6839–6844.
17. Singh, H.P.; Wilkinson, S.; Kamran, S. Decreasing Antibiotic Use in a Community Neonatal Intensive Care Unit: A Quality Improvement Initiative. *Am. J. Perinatol.* **2024**, *41* (Suppl. 1), e2767–e2775.
18. Gao, L.; Zhao, J.X.; Qin, X.M.; Zhao, J. The ethanol extract of Scutellaria baicalensis Georgi attenuates complete Freund's adjuvant (CFA)-induced inflammatory pain by suppression of P2X3 receptor. *J. Ethnopharmacol.* **2023**, *317*, 116762.
19. Du, Z.; Han, J.; Luo, J.; Bi, G.; Liu, T.; Kong, J.; Chen, Y. Combination effects of baicalin with linezolid against Staphylococcus aureus biofilm-related infections: In vivo animal model. *N. Microbiol.* **2023**, *46*, 258–263.
20. Bai, X.; Yao, M.; Zhu, X.; Lian, Y.; Zhang, M. Baicalin suppresses interleukin-1β-induced apoptosis, inflammatory response, oxidative stress, and extracellular matrix degradation in human nucleus pulposus cells. *Immunopharmacol. Immunotoxicol.* **2023**, *45*, 433–442.
21. Ning, X.; Luo, D.; Chen, Y.; Shao, Y.; Xu, J. Baicalin Reduces Renal Inflammation in Mesangial Proliferative Glomerulonephritis through Activation of Nrf2/ARE and PI3K/AKT Pathways. *Discov. Med.* **2023**, *35*, 372–382. [CrossRef] [PubMed]
22. Shen, B.; Zhang, H.; Zhu, Z.; Ling, Z.; Zeng, F.; Wang, Y.; Wang, J. Baicalin Relieves LPS-Induced Lung Inflammation via the NF-κB and MAPK Pathways. *Molecules* **2023**, *28*, 1873. [CrossRef] [PubMed]
23. Lin, Z.; Meng, F.; Ma, Y.; Zhang, C.; Zhang, Z.; Yang, Z.; Li, Y.; Hou, L.; Xu, Y.; Liang, X.; et al. In situ immunomodulation of tumors with biosynthetic bacteria promote anti-tumor immunity. *Bioact. Mater.* **2024**, *32*, 12–27. [CrossRef] [PubMed]
24. Xu, Y.; Wang, Z.; Gao, W.; Wang, H.; Wang, Z.; Tian, C.; Chen, G. Detection of the CD8(+) T cell immune response in mice infected with OVA-Listeriamonocytogenes. *STAR Protoc.* **2023**, *4*, 102582. [CrossRef]
25. Schardey, J.; Lu, C.; Neumann, J.; Wirth, U.; Li, Q.; Jiang, T.; Zimmermann, P.; Andrassy, J.; Bazhin, A.V.; Werner, J.; et al. Differential Immune Infiltration Profiles in Colitis-Associated Colorectal Cancer versus Sporadic Colorectal Cancer. *Cancers* **2023**, *15*, 4743. [CrossRef]
26. Reynolds, G.; Cliff, E.R.S.; Mohyuddin, G.R.; Popat, R.; Midha, S.; Liet Hing, M.N.; Harrison, S.J.; Kesselheim, A.S.; Teh, B.W. Infections following bispecific antibodies in myeloma: A systematic review and meta-analysis. *Blood Adv.* **2023**, *7*, 5898–5903. [CrossRef]
27. Chen, Y.; Zander, R.A.; Wu, X.; Schauder, D.M.; Kasmani, M.Y.; Shen, J.; Zheng, S.; Burns, R.; Taparowsky, E.J.; Cui, W. BATF regulates progenitor to cytolytic effector CD8(+) T cell transition during chronic viral infection. *Nat. Immunol.* **2021**, *22*, 996–1007. [CrossRef]
28. Smyk-Pearson, S.; Tester, I.A.; Klarquist, J.; Palmer, B.E.; Pawlotsky, J.M.; Golden-Mason, L.; Rosen, H.R. Spontaneous recovery in acute human hepatitis C virus infection: Functional T-cell thresholds and relative importance of CD4 help. *J. Virol.* **2008**, *82*, 1827–1837. [CrossRef]
29. Merlo, A.; Saverino, D.; Tenca, C.; Grossi, C.E.; Bruno, S.; Ciccone, E. CD85/LIR-1/ILT2 and CD152 (cytotoxic T lymphocyte antigen 4) inhibitory molecules down-regulate the cytolytic activity of human CD4+ T-cell clones specific for Mycobacterium tuberculosis. *Infect. Immun.* **2001**, *69*, 6022–6029. [CrossRef]
30. Trivedi, A.; Bose, D.; Saha, P.; Roy, S.; More, M.; Skupsky, J.; Klimas, N.G.; Chatterjee, S. Prolonged Antibiotic Use in a Preclinical Model of Gulf War Chronic Multisymptom-Illness Causes Renal Fibrosis-like Pathology via Increased micro-RNA 21-Induced PTEN Inhibition That Is Correlated with Low Host Lachnospiraceae Abundance. *Cells* **2023**, *13*, 56. [CrossRef]
31. Tan, L.; Chan, W.; Zhang, J.; Wang, J.; Wang, Z.; Liu, J.; Li, J.; Liu, X.; Wang, M.; Hao, L.; et al. Regulation of RIP1-Mediated necroptosis via necrostatin-1 in periodontitis. *J. Periodontal Res.* **2023**, *58*, 919–931. [PubMed]
32. Meng, R.; Gu, L.; Lu, Y.; Zhao, K.; Wu, J.; Wang, H.; Han, J.; Tang, Y.; Lu, B. High mobility group box 1 enables bacterial lipids to trigger receptor-interacting protein kinase 3 (RIPK3)-mediated necroptosis and apoptosis in mice. *J. Biol. Chem.* **2019**, *294*, 8872–8884. [CrossRef] [PubMed]
33. Zhang, H.; Xu, G.; Wu, X.; Xu, Y.; Xu, L.; Zou, Y.; Yang, X.; Pan, L.; Lei, B.; Mu, J.; et al. Fei-Yan-Qing-Hua decoction decreases hyperinflammation by inhibiting HMGB1/RAGE signaling and promotes bacterial phagocytosis in the treatment of sepsis. *J. Ethnopharmacol.* **2024**, *321*, 117553. [PubMed]

34. Mao, Y.; Patial, S.; Saini, Y. Airway epithelial cell-specific deletion of HMGB1 exaggerates inflammatory responses in mice with muco-obstructive airway disease. *Front. Immunol.* **2022**, *13*, 944772.
35. Jian, Z.; Ding, S.; Deng, H.; Wang, J.; Yi, W.; Wang, L.; Zhu, S.; Gu, L.; Xiong, X. Probenecid protects against oxygen-glucose deprivation injury in primary astrocytes by regulating inflammasome activity. *Brain Res.* **2016**, *1643*, 123–129.
36. Xiong, X.X.; Gu, L.J.; Shen, J.; Kang, X.H.; Zheng, Y.Y.; Yue, S.B.; Zhu, S.M. Probenecid protects against transient focal cerebral ischemic injury by inhibiting HMGB1 release and attenuating AQP4 expression in mice. *Neurochem. Res.* **2014**, *39*, 216–224. [CrossRef]
37. Fu, S.; Yin, R.; Zuo, S.; Liu, J.; Zhang, Y.; Guo, L.; Qiu, Y.; Ye, C.; Liu, Y.; Wu, Z.; et al. The effects of baicalin on piglets challenged with Glaesserella parasuis. *Vet. Res.* **2020**, *51*, 102. [CrossRef]
38. Que, X.; Zheng, S.; Song, Q.; Pei, H.; Zhang, P. Fantastic voyage: The journey of NLRP3 inflammasome activation. *Genes Dis.* **2024**, *11*, 819–829.
39. Ping, D.; Qi, J.; Li, M.; Sun, X.; Peng, Y.; Liu, C. Fuzheng Huayu recipe alleviates liver fibrosis via inhibiting NLRP3 inflammasome activation in macrophages. *J. Ethnopharmacol.* **2024**, *318*, 117001. [CrossRef]
40. Zhao, Y.; Liu, X.; Liu, X.; Zhang, J.; Zhang, Y.; Wen, Y.; Yang, G. Salvianolic acid B exerts protective effects against Aβ-induced neuroinflammation through the inhibition of NLRP3 inflammasome activation and switching of M1/M2 polarization. *Tissue Cell* **2023**, *85*, 102260.
41. Zhou, S.; Liu, Y.; Xue, B.; Yuan, P. Low-dose Esketamine suppresses NLRP3-mediated apoptotic and pyroptotic cell death in microglial cells to ameliorate LPS-induced depression via ablating GSK-3β. *Behav. Brain Res.* **2024**, *459*, 114782. [PubMed]
42. He, L.; Zhou, W.; Wang, L.; Tang, N.; Wang, Y.; Zhong, H.; Tang, Y.; Xi, D.; He, F. Murine Cytomegalovirus Infection Induced miR-1929-3p Down-Regulation Promotes the Proliferation and Apoptosis of Vascular Smooth Muscle Cells in Mice by Targeting Endothelin A Receptor and Downstream NLRP3 Activation Pathway. *Mol. Biotechnol.* **2023**, *65*, 1954–1967. [PubMed]
43. Zhou, J.; Wang, C.; Zhang, X.; Wu, Z.; Wu, Y.; Li, D.; Gao, J. Shizhifang ameliorates pyroptosis of renal tubular epithelial cells in hyperuricemia through inhibiting NLRP3 inflammasome. *J. Ethnopharmacol.* **2023**, *317*, 116777.
44. Huan, P.; Sun, X.; He, Z.; Yang, S.; Wang, X.; Xie, H.; Wang, L.; He, J. Qiji Shujiang granules alleviates dopaminergic neuronal injury of parkinson's disease by inhibiting NLRP3/Caspase-1 pathway mediated pyroptosis. *Phytomed. Int. J. Phytother. Phytopharm.* **2023**, *120*, 155019.
45. Ye, Q.; Lin, B.; Xu, P.; Zhang, F.; Wang, N.; Shou, D. Yunvjian decoction attenuates lipopolysaccharide-induced periodontitis by suppressing NFκB/NLRP3/IL-1β pathway. *J. Ethnopharmacol.* **2024**, *319*, 117279.
46. Barros Ferreira, L.; Ashander, L.M.; Ma, Y.; Appukuttan, B.; Williams, K.A.; Best, G.; Smith, J.R. Effects of tumor necrosis factor-α and interleukin-1β on human retinal endothelial cells. *Cytokine* **2024**, *173*, 156407.
47. Fetsko, A.R.; Sebo, D.J.; Budzynski, L.B.; Scharbarth, A.; Taylor, M.R. IL-1β disrupts blood-brain barrier development by inhibiting endothelial Wnt/β-catenin signaling. *bioRxiv* **2023**. *preprint*.
48. Nadlonek, N.; Lee, J.H.; Reece, T.B.; Weyant, M.J.; Cleveland, J.C., Jr.; Meng, X.; Fullerton, D.A. Interleukin-1 Beta induces an inflammatory phenotype in human aortic valve interstitial cells through nuclear factor kappa Beta. *Ann. Thorac. Surg.* **2013**, *96*, 155–162.
49. Ihim, S.A.; Abubakar, S.D.; Zian, Z.; Sasaki, T.; Saffarioun, M.; Maleknia, S.; Azizi, G. Interleukin-18 cytokine in immunity, inflammation, and autoimmunity: Biological role in induction, regulation, and treatment. *Front. Immunol.* **2022**, *13*, 919973.
50. Saxena, K.; Roverato, N.D.; Reithmann, M.; Mah, M.M.; Schregle, R.; Schmidtke, G.; Silbern, I.; Urlaub, H.; Aichem, A. FAT10 is phosphorylated by IKKβ to inhibit the antiviral type-I interferon response. *Life Sci. Alliance* **2023**, *7*, e202101282.
51. Zhao, X.; Ali, S.; Hassan, M.F.; Bashir, M.A.; Ni, X.; Lv, C.; Yang, H.; Danzeng, B.; Quan, G. Effects of graded levels of dietary protein supplementation on milk yield, body weight gain, blood biochemical parameters, and gut microbiota in lactating ewes. *Front. Vet. Sci.* **2023**, *10*, 1223450. [CrossRef] [PubMed]
52. Lu, X.; Wang, X.; Cheng, H.; Wang, X.; Liu, C.; Tan, X. Anti-triple-negative breast cancer metastasis efficacy and molecular mechanism of the STING agonist for innate immune pathway. *Ann. Med.* **2023**, *55*, 2210845. [CrossRef] [PubMed]
53. Fu, D.; Wu, J.; Wu, X.; Shao, Y.; Song, X.; Tu, J.; Qi, K. The two-component system histidine kinase EnvZ contributes to Avian pathogenic Escherichia coli pathogenicity by regulating biofilm formation and stress responses. *Poult. Sci.* **2023**, *102*, 102388. [CrossRef] [PubMed]
54. Li, H.; Yang, H.; Lu, S.; Wang, X.; Shi, X.; Mao, P. Autophagy-dependent ferroptosis is involved in the development of endometriosis. *Gynecol. Endocrinol. Off. J. Int. Soc. Gynecol. Endocrinol.* **2023**, *39*, 2242962. [CrossRef]

metabolites

MDPI

Review

Recent Advances in Pretreatment Methods and Detection Techniques for Veterinary Drug Residues in Animal-Derived Foods

Qing Dai [1,2,3], Shusheng Tang [1,2,*] and Chongshan Dai [1,2,*]

[1] National Key Laboratory of Veterinary Public Health and Safety, College of Veterinary Medicine, China Agricultural University, Beijing 100193, China
[2] Technology Innovation Center for Food Safety Surveillance and Detection (Hainan), Sanya Institute of China Agricultural University, Sanya 572025, China
[3] China Institute of Veterinary Drug Control, Beijing 100081, China
* Correspondence: tssfj@cau.edu.cn (S.T.); daichongshan@cau.edu.cn (C.D.)

Abstract: Veterinary drugs are extensively employed in livestock, poultry, and aquaculture, playing a crucial role in preventing and treating animal diseases, facilitating growth, and enhancing feed conversion rates. Nevertheless, veterinary drug residues in animal-derived foods pose a direct or potential threat to human life and health. Precise detection of these residues in animal-derived foods to ensure their safety has become an important mission. In this review, we sum up the current progress of applied pretreatment methods and detection techniques for veterinary drug residues in animal-derived foods. At present, sample pretreatment methods mainly consist of the following: liquid–liquid extraction; solid-phase extraction; immunoaffinity chromatography; Quick, Easy, Cheap, Effective, Rugged, and Safe (QuEChERS) method; and molecular imprinting technology. Detection techniques mainly involve chromatographic techniques, immunoassay techniques, fluorescence polarization immunoassay, and surface-enhanced Raman scattering. We also discussed the advantages and limitations of these technologies. Moreover, we point out the development direction and tendency of detection techniques in the future, providing references for the detection of veterinary drug residues in animal-derived foods.

Keywords: animal derived foods; veterinary drug residues; pretreatment methods; detection techniques

1. Introduction

Veterinary drugs utilized for disease treatment and prevention might linger in animal-derived foods, like milk, eggs, honey, and meat, which could endanger the public's health [1–3]. However, some illegal enterprises, for the sake of swift profit seeking, fail to abide by the medication instructions and use veterinary drugs or even prohibited drugs beyond their scope, dosage, and treatment duration. These actions will exert complex and far-reaching impacts on target animal safety, contact safety, environmental and ecological safety, animal food safety, and public health safety [4,5]. Veterinary drug residues refer to the residual substances associated with drugs that accumulate in an animal's body post-treatment with medication, encompassing drug prototypes and their metabolites [6,7]. Veterinary drug residues do not remain at the food processing stage; instead, they accumulate along the food chain. They not only contaminate the environment but also inflict substantial harm to the human body when amassed in a certain quantity over a long period [8,9]. Consequently, strict supervision of veterinary drug residues is essential to control the safety and quality of animal-derived foods and guarantee consumer safety [10,11].

Recent advancements in the detection of veterinary drug residues have predominantly focused on edible animals and their by-products, including organs, eggs, bee products, and aquatic species such as fish and shrimp [12–15]. The complexity of animal-derived food matrices, characterized by numerous interfering substances, necessitates the development of both efficient pretreatment methodologies and highly sensitive and specific detection technologies [16–18]. In the global context, various nations have adopted divergent approaches to veterinary drug residue detection, influenced by their respective food safety regulations and technological infrastructures. In China, liquid chromatography–mass spectrometry (LC-MS) is recognized as the "golden standard" due to its superior sensitivity and precision. Additionally, China is actively advancing and implementing rapid detection technologies such as Gold Immunochromatographic Assay (GICA) and Enzyme-Linked Immunosorbent Assay (ELISA) [19]. The trajectory of veterinary drug residue detection technology in China is increasingly oriented towards high-throughput and high-sensitivity methodologies. The United States extensively utilizes LC-MS for veterinary drug residue detection, leveraging its remarkable sensitivity and accuracy [20,21]. Concurrently, the U.S. is actively researching and applying biosensor and biochar technologies, which are gaining significant attention for their rapid and high-throughput capabilities [22]. European countries also place substantial emphasis on the application of LC-MS in veterinary drug residue detection [23]. There has been extensive research and implementation of pretreatment methods such as Quick, Easy, Cheap, Effective, Rugged, and Safe (QuEChERS) and solid-phase extraction (SPE) [24]. Microbiological testing methods are widely employed in Europe, particularly for the rapid screening of antibiotic residues [25]. Given the stringent food safety regulations in Europe, the region boasts advanced and diverse veterinary drug residue detection technologies.

For this review, the most important articles published from 1 January 2000 to 1 January 2025 relating to pretreatment methods and detection techniques for veterinary drug residues in animal-derived foods were selected from the Scopus, PubMed, and Web of Science databases. The keyword combinations used for the literature search are as follows: 'veterinary drug residues and pretreatment methods', 'veterinary drug residues and animal-derived foods', or 'drug residues and animal-derived foods'. For each pretreatment method or analysis method, some representative publications were further selected.

This paper provides a comprehensive review of recent pretreatment methods and detection techniques for animal-derived foods, intending to contribute to the future development of veterinary drug residue detection technologies. By examining the current landscape and identifying key trends, we aim to offer valuable insights and recommendations for enhancing the efficacy and applicability of these critical methodologies.

2. Pretreatment Methods

Animal-derived foods possess complex matrices, high concentrations of interfering substances, and low residue levels, which render their separation and purification arduous [26]. They fall into the category of detection technologies for trace components in complex matrix samples. In a comprehensive detection experiment, pretreatment is the most time-consuming and crucial task, directly affecting the precision and accuracy of the experiment [27]. Pretreatment methods for veterinary drug residues mainly consist of extraction, separation, purification, concentration, and enrichment. Extraction and purification are fundamental steps since they can remove matrix interference and safeguard instruments [28,29]. Currently, establishing a rapid and efficient sample pretreatment method has become the development trend and challenge for veterinary drug residue analysis techniques.

2.1. Liquid–Liquid Extraction (LLE) Technology

LLE is a separation technique based on the principle of differential distribution coefficients of solutes in two immiscible solvent systems [30]. When a sample extract containing the target veterinary drug is brought into contact with a selected organic solvent, the drug molecules distribute themselves between the organic and aqueous phases according to their solubility, which is determined by their chemical properties [30]. This distribution, governed by the partition coefficient, enables the initial separation and migration of the veterinary drug toward the organic phase, where it is more soluble [30]. LLE is widely recognized for its operational simplicity and effectiveness in extracting target analytes from complex matrices. However, it consumes a significant volume of organic solvents, which not only increases operational costs but also raises safety and environmental concerns due to the flammable, explosive, and toxic nature of many solvents [31,32]. Additionally, LLE is rarely used alone for the pretreatment of animal-derived foods with complex matrices, and often requires combining it with other pretreatment methods to achieve optimal results [33].

The advantages of LLE include its excellent selectivity, which leverages the chemical properties of veterinary drugs to choose compatible organic solvents, such as chloroform for lipophilic drugs. This enables precise separation from complex matrices like animal tissues and milk, effectively reducing interference from hydrophilic impurities [30]. The method is also operationally simple, requiring minimal equipment—common separatory funnels are sufficient—and involves straightforward steps such as manual shaking and phase separation, making it accessible for routine use [34,35]. Furthermore, LLE is cost-effective, utilizing low-cost organic solvents like acetonitrile and ethyl acetate, combined with simple equipment, resulting in a low initial investment for detection [35].

Despite its advantages, LLE has notable limitations. The high consumption of organic solvents not only increases costs but also poses safety risks and environmental concerns due to their hazardous nature [36]. Emulsification is another common issue, particularly in samples containing proteins or surfactants, where the organic and aqueous phases may form stable emulsions that resist separation, necessitating additional demulsification steps and complicating the process [35]. Additionally, the extraction efficiency of LLE can be limited, especially for drugs with similar partition coefficients. A single extraction may be insufficient, requiring multiple extraction steps that are time-consuming and labor-intensive, thereby reducing overall efficiency [37].

Several studies have demonstrated the application of LLE in veterinary drug residue analysis. For instance, Sun et al. employed Na_2EDTA-Mcllvaine buffer and acetonitrile as extraction solvents, along with anhydrous $MgSO_4$ and NaCl as extraction salts, in a salt-precipitation-assisted LLE method combined with high-performance liquid chromatography–tandem mass spectrometry (HPLC-MS/MS). This approach enabled the detection of 22 veterinary drug residues in aquatic products, with detection limits ranging from 0.5 to 1.0 μg/kg and accuracy between 71.4% and 120% [38]. Similarly, Wang et al. utilized methanol and acetonitrile as extraction solvents, performing static extraction three times. The preliminarily purified extract was further refined using magnetic materials and analyzed by ultra-performance liquid chromatography–tandem mass spectrometry (UPLC-MS/MS). Under optimized conditions, this method identified 33 types of antibiotics and 37 types of pesticides [39]. In summary, while LLE offers significant advantages in terms of selectivity, simplicity, and cost-effectiveness, its limitations, including high solvent consumption, emulsification risks, and limited extraction efficiency, necessitate careful consideration and often require complementary techniques to achieve reliable and efficient results in veterinary drug residue analysis [40].

2.2. SPE

SPE is a widely used sample preparation technique based on the principles of adsorption and desorption. It is also a technique designed for rapid and selective sample preparation and purification prior to chromatographic analysis [41]. When a sample extract passes through an extraction column packed with a solid-phase adsorbent, veterinary drugs are selectively adsorbed onto the active sites of the adsorbent surface due to their specific chemical structures and properties. The adsorption strength varies depending on the polarity and functional groups of the veterinary drugs, allowing for selective retention of target analytes [42]. During this process, impurities with weaker adsorption affinities are washed away with the eluent, achieving preliminary purification. Subsequently, an appropriate elution solvent is used to desorb the target veterinary drugs from the adsorbent, and the eluate is collected for subsequent analysis [39,40]. Since its development, SPE has been extensively applied in various fields due to its simplicity, high efficiency, and rapidity. Compared to traditional LLE, SPE is faster, more solvent efficient, and exhibits better reproducibility, making it the preferred method for sample pretreatment [43,44].

SPE offers several notable advantages. First, it provides effective enrichment of trace amounts of veterinary drugs. For example, in water samples, SPE can concentrate extremely low drug concentrations, significantly enhancing detection sensitivity and enabling compliance with stringent regulatory limits [45]. Second, it exhibits high selectivity. By choosing appropriate adsorbents, target analytes can be precisely adsorbed, effectively excluding a large number of interfering impurities and achieving excellent purification [46]. Third, SPE consumes significantly less solvent compared to LLE, reducing operational costs, environmental pollution, and safety risks [39,40]. Several studies have demonstrated the effectiveness of SPE in veterinary drug residue analysis. For instance, Zhu et al. employed an ethylene diamine tetraacetic acid (pH 7)-acetonitrile (30:70) solvent system combined with a one-step SPE cleanup method. Using UPLC-MS/MS, they simultaneously determined multiple veterinary drugs with excellent sensitivity, achieving limits of quantification ranging from 0.2 μg/kg to 3.0 μg/kg [47]. Similarly, Mehl et al. developed a high-throughput planar SPE cleaning method to screen 81 veterinary drugs from six different groups (glucocorticoids, dulcimers, antiparasitic drugs, antibacterial drugs, non-steroidal anti-inflammatory drugs, and antibiotics) in four matrices (honey, pig muscle, milk, and eggs). This method consumed 13 times less solvent, was more environmentally friendly, and was five times faster than conventional methods [48].

However, SPE also has some limitations. The high cost of specialized, high-performance adsorbents can increase the initial investment for detection. Additionally, the operational procedure is relatively complex, involving multiple steps such as column packing, conditioning, sample loading, washing, and elution [49]. Each step requires strict control of conditions to avoid affecting recovery rates, demanding a high level of technical proficiency from operators [50] Furthermore, SPE is susceptible to matrix effects, particularly in complex sample matrices such as animal tissue extracts. Matrix components may occupy adsorption sites or compete with veterinary drugs for adsorption, reducing extraction efficiency and interfering with the accuracy of detection results. As a result, additional optimization steps are often required to mitigate these effects [43,44].

In summary, SPE is a powerful and versatile sample preparation technique that offers significant advantages in terms of enrichment, selectivity, and solvent efficiency [51]. However, its limitations, including high adsorbent costs, complex operational procedures, and susceptibility to matrix effects, necessitate careful optimization and skilled execution to ensure reliable and accurate results in veterinary drug residue analysis.

2.3. Immunoaffinity Chromatography (IAC)

IAC is a highly sophisticated analytical technique that exploits the specific immunological interactions between antigens and antibodies for the selective extraction and purification of target analytes [52–54]. This method involves immobilizing monoclonal or polyclonal antibodies specific to veterinary drugs onto a solid-phase matrix to create an immunoaffinity column [55]. When a sample extract containing veterinary drug residues passes through the column, the target drug, acting as an antigen, binds specifically to the immobilized antibodies through precise molecular recognition mechanisms, forming stable antigen–antibody complexes. Concurrently, other components in the sample matrix that lack affinity for the antibodies are eluted with the mobile phase, achieving initial separation and purification. Subsequently, the environmental conditions are altered using an elution buffer to disrupt the antigen–antibody interactions, allowing for the collection of the eluted veterinary drug for precise downstream analysis [52–54,56].

IAC offers several notable advantages. First, it exhibits exceptional specificity due to the precise molecular recognition capability of antibodies for their target veterinary drugs [57]. This enables efficient extraction of drugs from complex biological matrices such as animal tissue homogenates, blood, and urine, significantly reducing matrix interference and enhancing detection accuracy. Second, IAC demonstrates high sensitivity, capable of capturing even trace amounts of veterinary drug residues, making it particularly suitable for stringent regulatory requirements in food safety monitoring. Third, this technique is operationally efficient; once the immunoaffinity column is prepared, the loading and elution procedures are straightforward and time-efficient, facilitating high-throughput analysis [52–54]. Several studies have demonstrated the effectiveness of IAC in veterinary drug residue analysis. For example, Peng et al. developed an IAC method for the selective purification of 3-methylquinoxaline-2-carboxylic acid from pig muscle and liver, coupled with high-performance liquid chromatography–ultraviolet detection (HPLC-UV). The method achieved detection limits of 1.0–3.0 μg/kg and quantification limits of 4.0–10.0 μg/kg, with an average recovery rate of 80.1–87.7% and a relative standard deviation of less than 8.5% [58]. Similarly, Wang et al. developed an IAC clean-up method combined with HPLC-MS/MS for the simultaneous determination of chloramphenicol, zearalanone, α-zearalanol, β-zearalanol, zearalenone, α-zearalenol, and β-zearalenol in pig muscle. The method demonstrated excellent linearity ($r \geq 0.9990$), precision (relative standard deviation < 2.9%), average recovery (74.5–105.0%), and detection limits (0.04–0.10 μg/kg), making it rapid, reliable, sensitive, and highly applicable to real samples [59].

However, IAC also presents certain limitations. The development of specific antibodies, whether it is monoclonal or polyclonal, requires significant technical expertise and financial investment [60]. Additionally, antibody stability is sensitive to environmental factors necessitating strict storage conditions to maintain functionality. Another limitation is the constrained lifespan of immunoaffinity columns. Repeated use leads to decreased antibody activity and reduced binding capacity, requiring frequent column replacements and increasing operational costs [61]. Furthermore, cross-reactivity with structural analogs in sample matrices may interfere with antibody recognition, potentially leading to false positives [57]. Therefore, supplementary analytical methods are often required to validate results and ensure data reliability.

In short, IAC is a powerful tool for veterinary drug residue analysis, offering exceptional specificity, sensitivity, and operational efficiency. However, its application requires careful consideration of its limitations, including the complexity of antibody development, column lifespan constraints, and potential cross-reactivity [62]. By addressing these challenges, IAC can be effectively integrated into analytical workflows to ensure accurate and reliable detection of veterinary drug residues in complex matrices.

2.4. QuEChERS

The QuEChERS method has emerged as a highly regarded pretreatment technique for the detection of veterinary drug residues, primarily based on dispersive solid-phase extraction (d-SPE) technology [61]. The process begins with the addition of an organic solvent, such as acetonitrile, to the sample to dissolve and extract target veterinary drugs from complex matrices like animal muscles, organs, and feed [61]. Concurrently, desiccants like anhydrous magnesium sulfate are introduced to facilitate rapid phase separation between the aqueous and organic layers, minimizing moisture interference in subsequent analyses. Following this, a dispersive solid-phase sorbent is added, which selectively adsorbs impurities such as organic acids, pigments, fats, and sugars from the sample extract, while the target veterinary drugs remain in the organic phase [63,64]. This step ensures a high degree of purification, making the method particularly effective for complex sample matrices.

The QuEChERS method offers several notable advantages. First, it is simple and fast, requiring only basic laboratory equipment such as centrifuge tubes and vortex mixers. This allows for the completion of multiple steps in a short timeframe, significantly enhancing pretreatment efficiency. Second, it is cost-effective, as the organic solvents, desiccants, and sorbents used are relatively inexpensive, making the method suitable for large-scale sample analysis. Third, it provides effective purification by efficiently removing various interfering impurities, leading to more accurate and reliable detection results. These characteristics make QuEChERS particularly well-suited for analyzing complex matrices in veterinary drug residue detection. Several studies have demonstrated the effectiveness of the QuEChERS method in veterinary drug residue analysis. For instance, Yang et al. employed an improved QuEChERS method to determine the residues of 192 veterinary drugs. This method involved efficient extraction with 90% acetonitrile, dehydration with sodium sulfate and acetic acid, and subsequent cleanup using a dispersive solid-phase extraction sorbent, primary secondary amine (PSA). The method was suitable for samples with varying fat contents, providing detection limits ranging from 0.5 to 10 μg/kg, high recovery rates (60–120%), and a low relative standard deviation (<20%) [65]. Similarly, Wang et al. established quantitative methods for 146 types of veterinary drug residues in beef and chicken samples. After enzymatic hydrolysis, the samples were treated with an optimized QuEChERS method. The calibration curves exhibited good linearity, with correlation coefficients ranging from 0.9921 to 0.9994. The recovery rates were within 52.1–138.2%, and the relative standard deviation was 0.4–17.7%. The limits of detection and quantification were 0.15–3.03 μg/kg and 0.5–10 μg/kg, respectively [66]. Mu et al. established a rapid method for the simultaneous determination of 14 antiviral drugs and related metabolites, including amantadine, rimantadine, oseltamivir, oseltamivir carboxylate, memantine, abidol, moxifloxacin, acyclovir, ganciclovir, famciclovir, penciclovir, ribavirin, and immunomodulators, in chicken meat using QuEChERS and liquid chromatography–mass spectrometry techniques. The improved QuEChERS method was used, and the determination time of the target compound was less than 11 min, with a recovery rate of 56.2% to 113.4% [67].

However, the QuEChERS method also has certain limitations. The selectivity of the sorbents is not absolute, meaning that for certain structurally unique impurities or veterinary drugs, the desired adsorption or retention may not be achieved, potentially compromising detection accuracy [68]. Additionally, optimization can be challenging, as different types of veterinary drugs and sample matrices require adjustments in the types and quantities of desiccants and sorbents. This process can be time-consuming and labor-intensive, often relying heavily on empirical knowledge. Furthermore, residual matrix components may still interfere with subsequent analyses even after purification, necessitating additional measures such as the use of internal standards to mitigate these effects and ensure the precision of the detection results [58,59].

In short, the QuEChERS method provides a robust, efficient, and cost-effective solution for the pretreatment of samples in veterinary drug residue analysis. Its simplicity, speed, and effective purification make it highly suitable for complex matrices [69]. However, its limitations, including sorbent selectivity issues, optimization challenges, and residual matrix effects, necessitate careful optimization and supplementary techniques to ensure accurate and reliable detection results [70]. By addressing these challenges, the QuEChERS method can be effectively integrated into analytical workflows for veterinary drug residue detection.

2.5. Molecular Imprinting Technology (MIT)

MIT has emerged as a cutting-edge strategy for detecting veterinary drug residues, leveraging its unique molecular recognition capabilities and pretreatment efficiency [71]. At its core, MIT relies on the formation of reversible complexes between template molecules (target veterinary drugs) and functional monomers in solution, driven by covalent or non-covalent interactions. Polymerization is subsequently initiated by cross-linkers, producing a rigid three-dimensional polymeric network—termed Molecularly Imprinted Polymers (MIPs)—that mirrors the template's structural and chemical features. Following template removal via elution, MIPs retain complementary cavities that exhibit high specificity for the target molecules. During analysis, these cavities selectively adsorb veterinary drug residues from complex matrices through spatial and chemical affinity, enabling efficient separation from interfering substances [72,73]. This molecular recognition mechanism positions MIPs as powerful tools for sample pretreatment in residue analysis.

Recent studies highlight MIT's applicability in veterinary drug monitoring. For instance, Song et al. synthesized a tetracycline-imprinted MIP via precipitation polymerization using methacrylic acid as a functional monomer. Coupled with dispersive SPE and high-performance liquid chromatography–mass spectrometry HPLC-MS, the method enabled the simultaneous detection of seven macrolide antibiotics in pork. At spiked concentrations of 5–100 mg/kg, recoveries ranged from 68.6% to 95.5%, with intra- and inter-day precision (relative standard deviations, RSDs) below 8%. The method achieved detection limits (LODs) of 0.2–0.5 mg/kg and quantification limits (LOQs) of 0.5–2.0 mg/kg, demonstrating both sensitivity and reproducibility [74]. In another study, Tang et al. developed a clenbuterol-imprinted MIP-coated stirrer for β-receptor agonist analysis in animal-derived foods. Integrating stir bar sorptive extraction with HPLC-MS/MS, the method achieved recoveries of 75.8–97.9% in pork and liver samples, with RSDs of 2.6–5.3%. The technique exhibited exceptional sensitivity, with LODs and LOQs as low as 0.05–0.15 μg/kg and 0.10–0.30 μg/kg, respectively [75].

MIT offers three principal advantages in veterinary drug detection. First, MIPs exhibit customized specificity: their tailored cavities enable precise recognition of target analytes even in complex biological matrices (e.g., tissues, blood, and milk), minimizing matrix interference and enhancing sensitivity. Second, MIPs demonstrate robust stability, retaining structural integrity across diverse temperatures, pH ranges, and organic solvents, which permits reuse and reduces operational costs. Third, the technology is highly adaptable: by modulating template molecules, monomers, and cross-linkers, MIPs can be engineered for diverse veterinary drugs, broadening their applicability [76,77]. However, MIT faces notable challenges. The synthesis process remains labor-intensive, requiring meticulous optimization of polymerization conditions and post-treatment steps, which demands specialized expertise and prolongs development timelines. Additionally, while MIPs exhibit high specificity, cross-reactivity with structurally analogous compounds may occur, potentially yielding false positives [78]. Thus, confirmatory techniques (e.g., mass spectrometry) are often indispensable for validation [79].

In short, MIT represents a transformative approach for veterinary drug residue analysis, combining selectivity, stability, and versatility. While its current limitations necessitate complementary methods for verification, ongoing advancements in polymer design and automation hold promise for streamlining MIP synthesis and enhancing reliability in complex matrices [80].

A comprehensive systematic comparison of these pretreatment methods, encompassing LOD, accuracy, advantages, limitations, feasibility, and cost-effectiveness for veterinary drug residue analysis in animal-derived food matrices, is presented in Table 1. For example, LLE demonstrates particular efficacy for high-fat matrices [81], albeit with prolonged processing times (typically 2–4 h) and substantial solvent-related expenses. SPE enables faster sample preparation (30–60 min) through diverse adsorbent selections, yet incurs significant material costs. IAC, while providing exceptional post-purification efficiency, faces challenges in antibody development costs and cross-reactivity risks and cross-reactivity risks for complex matrices. The QuEChERS method provides rapid and cost-efficient analysis for complex matrices, whereas MIT achieves remarkable specificity (85–95% recognition accuracy) despite complex polymer synthesis requirements [82].

Table 1. A comprehensive systematic comparison among various pretreatment methods for veterinary drug residues in animal-derived foods.

Method	LOD (μg/kg)	Accuracy (Recovery Rate)	Applicability	Advantages	Disadvantages	Feasibility	Cost-Effectiveness
LLE	0.5–1.0	71.4–120%	Suitable for lipophilic drugs in aquatic products, milk, and tissues.	Excellent selectivity for lipophilic drugs. Simple operation with minimal equipment. Low cost.	High solvent consumption. Prone to emulsification. Limited extraction efficiency for polar analytes.	Simple but requires solvent management.	Low
SPE	0.2–3.0	60–120%	Broad applicability for diverse matrices (honey, muscle, milk, and eggs).	Effective enrichment of trace residues. High selectivity with tailored adsorbents. Reduced solvent use.	High adsorbent costs. Complex optimization for matrix effects. Susceptible to column clogging.	Requires skilled optimization.	Moderate to high
IAC	0.04–0.10	74.5–105%	Ideal for specific targets (e.g., chloramphenicol and β-agonists) in muscle/liver.	Exceptional specificity via antibody–antigen binding. High sensitivity for trace residues. High-throughput potential.	Antibody development is costly/time-consuming. Limited column lifespan. Cross-reactivity risks.	Antibody-dependent and storage-sensitive.	High
QuEChERS	0.15–3.03	52.1–138.2%	Effective for high-fat matrices (beef and chicken) and multi-residue analysis.	Rapid and simple. Cost-effective with minimal solvents. Effective impurity removal.	Sorbent selectivity limitations. Residual matrix interference. Optimization challenges for diverse analytes.	Easy to implement with standard lab tools.	Low to moderate

Table 1. *Cont.*

Method	LOD (μg/kg)	Accuracy (Recovery Rate)	Applicability	Advantages	Disadvantages	Feasibility	Cost-Effectiveness
MIT	0.05–0.5	68.6–95.5%	Customizable for antibiotics (e.g., tetracyclines and β-agonists) in complex matrices.	Tailored specificity via imprinting. Reusable and stable. Adaptable to diverse targets.	Labor-intensive synthesis. Cross-reactivity with structural analogs. Requires confirmatory methods.	Specialized expertise needed for polymer design.	Moderate

3. Chromatographic Detection Techniques

Animal-derived food contains relatively low levels of veterinary drug residues and large quantities of interfering substances, which makes it a challenge to detect trace components in complex matrix samples [83,84]. At present, the detection of veterinary drug residues in animal-based foods mainly relies on chromatography–mass spectrometry technology. Commonly used techniques encompass gas chromatography–mass spectrometry (GS-MS), LC-MS, and capillary electrophoresis–mass spectrometry (CE-MS). In brief, the traditional approaches for detecting veterinary drug residues are typically based on laboratory tests using large-scale instrument analysis [85]. Although these methods offer abundant qualitative and quantitative techniques and relatively high detection accuracy, they have drawbacks such as necessitating complex pre-processing methods, requiring professional training for testers, and having relatively expensive equipment, which are not favorable for large-scale and rapid daily risk monitoring. As science and technology progress, convenient and highly sensitive rapid detection methods for veterinary drug residues have gradually become the chief means of daily risk monitoring [86].

3.1. GC-MS

GC-MS is a robust analytical technique that integrates the separation capabilities of GC with the detection features of MS [87]. This process begins with the gasification of the sample, where the analyte is introduced into the GC column. The stationary phase in the GC column selectively retains components based on their physicochemical properties, primarily their volatility and interaction with the stationary phase. Following separation in the GC column, the analytes are introduced into the MS, where they undergo ionization, and their molecular ions are detected based on their mass-to-charge ratio (m/z) [88]. This dual functionality of GC-MS enables both separation and identification, making it highly suitable for detecting and quantifying trace volatile organic compounds (VOCs) and other small molecules such as antibiotics in complex matrices, such as food and pharmaceutical samples [89,90]. For instance, Liu et al. developed a novel pre-column derivatization GC-MS/MS method to determine residual decanoate in chicken tissues (muscle, liver, and kidney) [91]. The sample was extracted and purified through liquid–liquid extraction combined with solid-phase extraction and derivatized using acetic anhydride and pyridine. The recovery rate of decanoic acid ester ranged from 77.38% to 89.65%, with intra-day and inter-day relative standard deviations of 1.63–5.74% and 2.27–8.06%, respectively. This method was applied to analyze the tissues of 60 chickens purchased from nearby supermarkets, revealing that only one sample contained 15.6 μg/kg of residual decanoic acid [91]. In another study, Wang et al. developed a rapid and sensitive solvent extraction method for the determination of spectinomycin and lincomycin residues in chicken samples. After derivatization of the target compounds, GC-MS/MS detection was utilized. This method achieved a recovery rate of 80.0% to 95.7%, a relative standard deviation of 1.0% to 3.4%, a detection limit of 2.3 to 4.3 μg/kg, and a quantification limit of 5.6 to 9.5 μg/kg [92]. Guo et al. successfully developed a rapid and sensitive solvent extraction coupled with the GC-MS/MS method for the determination of spectinomycin and lincomycin residues in chicken samples. Following derivatization of the target compounds, the method demonstrated recovery rates of 80.0–95.7%, RSDs of 1.0–3.4%, with LODs ranging from 2.3 to 4.3 μg/kg and LOQs between 5.6 and 9.5 μg/kg [93]. Yu et al. established a GC-MS method to detect and identify amitraz and its metabolite, 2,4-dimethylaniline, in liver and kidney tissues of food-producing animals. At spiked concentrations of 50–300 μg/kg, the method achieved recovery rates of 72.4–101.3%, with RSDs below 11.5% [94].

GC-MS offers several advantages that make it an indispensable tool in analytical chemistry. For example, GC-MS excels at separating and identifying complex mixtures, making it ideal for analyzing pharmaceutical residues in biological matrices [8,95]. GC-MS provides high specificity due to the precise molecular recognition during the ionization process and it is applicable to a wide range of analytes, including organic pollutants, heavy metals, and pharmaceuticals with higher sensitivity [89].

In short, GC-MS is a powerful analytical tool for detecting and quantifying trace volatile organic compounds in complex matrices. Its ability to provide both separation and identification makes it indispensable in food safety monitoring and pharmaceutical residue analysis. However, its application requires careful optimization of the sample preparation and instrumental settings to ensure accurate and reliable results. Despite its limitations, advancements in analytical chemistry continue to enhance its utility and reduce its dependency on expensive equipment, making it more accessible for routine use.

3.2. *Liquid Chromatography Quadrupole-Time-of-Flight Mass Spectrometry (LC-QTOF-MS)*

The core principle of LC-QTOF-MS lies in its seamless integration of the separation capabilities of liquid chromatography (LC) with high-resolution, accurate mass measurement of quadrupole time-of-flight mass spectrometry (QTOF-MS) [96]. After sample injection into the liquid chromatography system, different veterinary drug components are separated based on their distribution coefficients between the stationary and mobile phases, as well as their polarity and molecular weight characteristics, as they are eluted sequentially from the chromatographic column driven by the mobile phase. Subsequently, the eluted components enter the QTOF-MS, where they are ionized under the influence of the ion source. The resulting ion beams undergo preliminary screening in the quadrupole to eliminate some interfering ions before entering the time-of-flight analyzer. In the flight tube, ions travel at different speeds depending on their mass-to-charge ratio (m/z), with the flight time inversely proportional to the m/z ratio. The detector accurately records the ion flight times and converts them into mass spectral data, enabling precise determination of ion mass with ultra-high resolution. This generates highly accurate mass spectra, which, through database comparison or analysis of characteristic fragment ions, allow for the precise identification and quantification of trace veterinary drug residues [97,98]. A simple working mode diagram is shown in Figure 1.

LC-QTOF-MS is recognized as a high-resolution mass spectrometer capable of achieving fine resolution for compounds with similar mass-to-charge ratios, making it particularly advantageous for differentiating and characterizing similar ions [99,100]. For instance, Li et al. developed a comprehensive screening method using HPLC-QTOF-MS for the simultaneous determination of 141 veterinary drugs and their metabolites in pork. Most of the tested compounds exhibited a detection limit of 0.5 μg/kg, with recovery rates exceeding 70% [101]. Similarly, Vardali et al. established a UPLC-QTOF-MS method for the concurrent determination of 20 veterinary drug residues and metabolites (including tetracyclines, quinolones, sulfonamides, and diaminopyrimidines) in the edible muscle and skin tissues of European sea bass. The recovery rates for all target compounds in both muscle and skin tissues ranged from 93.8% to 107.5%, with detection and quantification limits spanning 2.22 to 15.00 μg/kg and 6.67 to 45.46 μg/kg, respectively [102].

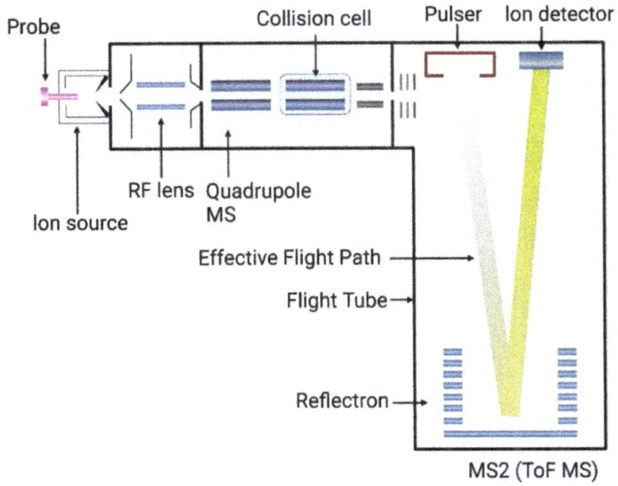

Figure 1. Schematic illustration of the principle of LC-QTOF-MS.

This technology boasts several notable advantages [95]. Firstly, it excels in the deep separation and precise recognition of veterinary drugs within complex biological matrices. This capability effectively mitigates interference from structurally similar substances, thereby providing reliable and accurate results. Secondly, the technology exhibits exceptional sensitivity. It can detect extremely low concentrations of veterinary drug residues, ensuring compliance with stringent food safety standards and offering a high level of assurance in the detection process. Thirdly, it possesses robust qualitative capabilities. The acquisition of high-resolution mass spectral data yields rich structural information, which aids in the identification of unknown substances [103]. This broadens the scope of detection and enhances the overall analytical power of the technology [104].

However, it is important to acknowledge certain limitations, particularly in terms of cost and data processing. For instance, the acquisition and maintenance of the required instrumentation entail significant expenses. Additionally, substantial investment in laboratory infrastructure and the employment of highly skilled personnel are necessary to operate and maintain the technology effectively [105]. Moreover, the vast amount of high-resolution data generated necessitates the use of specialized software for interpretation. This process can be both time-consuming and labor-intensive. Furthermore, improper handling of the data may introduce errors, which could compromise the accuracy of the detection results [106]. Therefore, meticulous attention to detail and rigorous quality control measures are essential to ensure the reliability and precision of the technology.

In summary, LC-QTOF-MS represents a powerful analytical tool in the field of veterinary drug residue analysis, offering unparalleled precision and sensitivity. Nevertheless, its application demands careful consideration of the associated costs and technical challenges.

3.3. Liquid Chromatography-Tandem Mass Spectrometry (LC-MS/MS)

LC-MS/MS is widely used in the field of veterinary drug residue detection [107]. Its working principle is as follows [46]: First, the sample is injected into the liquid chromatography system. Utilizing the differences in distribution coefficients between the mobile phase and the stationary phase, based on properties such as polarity and molecular weight, the various components are eluted from the chromatography column in a specific order, achieving initial separation. Subsequently, the eluted components enter the mass spectrometer, where the molecules are ionized to form ion beams under the action of the ion source. The ions first undergo primary mass spectrometry analysis, where they are separated and screened based on differences in the mass-to-charge ratio (m/z). After screening out the target ions, these ions further collide with an inert gas in the collision chamber, breaking apart to produce characteristic fragment ions, which are then analyzed by secondary mass spectrometry. By comprehensively analyzing the mass-to-charge ratio and abundance information of the parent ions from the primary mass spectrometry and the fragment ions from the secondary mass spectrometry, and comparing them with known databases, precise qualitative and quantitative determination of veterinary drugs is achieved [108].

In recent years, as LC-MS/MS has been developing, it has turned into a commonly utilized method for analyzing trace organic compounds. This technology can identify trace organic compounds at a concentration of 1 μg/kg [109,110]. This technology offers the following significant advantages: It boasts extremely high sensitivity, is capable of detecting trace or even ultra-trace levels of veterinary drug residues, meets increasingly stringent food safety standards, and provides strong support to ensure the quality and safety of animal-derived food products. It exhibits excellent selectivity; through the dual selection and analysis of parent ions and fragment ions, it effectively eliminates many interfering components in complex sample matrices, accurately identifies target veterinary drugs, and reduces the likelihood of false-positive results. Its qualitative and quantitative accuracy is strong, with rich mass spectrometry information combined with professional software algorithms ensuring reliable detection results, making it suitable for the simultaneous detection of multiple veterinary drug residues in complex biological samples [111]. For instance, Lehotay et al. carried out a study regarding the regulatory analysis of veterinary drug residues in liquid and powdered eggs by means of UPLC-MS/MS and evaluated the validation of 169 veterinary drugs in those eggs. When it comes to the extraction and injection sample preparation method, for 139–141 (82–83%) out of 169 different drug analytes added to powder and liquid eggs at three supervisory levels, it achieved acceptable recovery rates ranging from 70 to 120% and a relative standard deviation of less than 25% [112]. Additionally, Na et al. adopted improved QuEChERS and UPLC-MS/MS to confirm the methods for 197 pesticides, 56 veterinary drug components, and 5 toxin components. In the improved analysis method, all components demonstrated good linearity ($r2 \geq 0.98$). For most compounds, the limit of quantification is 0.05 mg/kg. And the recovery rate lies between 70.09 and 119.76% with a relative standard deviation of less than 18.91% [113]. Wittenberg et al. established an LC-MS/MS method for the determination of 52 veterinary drug residues in milk powder. The LOQs ranged from 0.02 to 82 ng/g. In skimmed milk powder spiked at three concentration levels, 40 out of the 52 target compounds exhibited satisfactory recovery rates (70–120%) and precision (relative standard deviations, RSDs < 20%). Comparable results were also achieved in whole milk powder, whole milk protein, whey protein concentrate, and whey protein isolate [114].

On the contrary, LC-MS/MS technology also has certain limitations. Firstly, the instruments are expensive, with high acquisition costs, and subsequent maintenance and upkeep require skilled technicians and stringent laboratory environments, increasing detection costs and operational difficulty. Secondly, the operation is complex, as sample preparation, instrument parameter optimization, and data processing all require specialized knowledge and extensive experience, making it challenging for beginners to master [115,116].

Despite these limitations, LC-MS/MS remains a vital tool for ensuring food safety and regulatory compliance.

3.4. Liquid Chromatography Coupled to Ion Trap Mass Spectrometry (LC-IT-MS)

LC-IT-MS is grounded in the synergistic operation of liquid chromatography and ion trap mass spectrometry [117]. In this technique, the sample is first introduced into the liquid chromatography system, where different veterinary drug components are sequentially separated based on their distinct distribution coefficients between the mobile and stationary phases. This separation is facilitated by differences in polarity, molecular weight, and other physicochemical properties within the chromatographic column. Subsequently, the separated components are eluted into the ion trap mass spectrometer [118].

Within the ion source, the molecules of veterinary drugs are ionized, and these ions are then introduced into the ion trap. The ion trap employs a radiofrequency electric field to capture, store, and manipulate the ions, allowing for the precise selection of ions with specific mass-to-charge ratios (m/z). By adjusting the parameters of the electric field, the ions are induced to collide with a buffer gas, leading to fragmentation and the generation of multiple-stage fragment ions [117]. Through in-depth analysis of these multi-stage mass spectra, coupled with the use of specialized software and reference databases, accurate identification and quantitative determination of veterinary drugs are achieved [117].

LC-IT-MS addresses the limitations of LC-MS/MS in terms of compound retrieval breadth and is capable of simultaneously confirming and screening multiple trace compounds [119–121]. For instance, Kang et al. developed a multi-residue method using ion trap LC-IT-MS to screen for 110 veterinary drug residues. Among these, 108 (98%) were detected at concentrations below 10 ppm. The method demonstrated an average accuracy ranging from 63% to 122% in measuring drug residues in muscle, liver, and kidney tissues, with repeatability and reproducibility ranging from 5% to 22% and 7% to 23%, respectively [122]. Similarly, Casado et al. employed Ultra-Performance Liquid Chromatography-Ion Trap Mass Spectrometry (UPLC-IT-MS) for the simultaneous determination of 23 veterinary drug residues, including beta-blockers, beta-agonists, and non-steroidal anti-inflammatory drugs, in meat samples. The detection limits of this method ranged from 0.01 to 18.75 μg/kg, while the quantification limits spanned from 0.02 to 62.50 μg/kg [123]. Furthermore, analysis of actual beef samples revealed traces of propranolol, ketoprofen, and diclofenac in some instances.

The technique exhibits several notable advantages: (1) It can provide exceptional qualitative capability. The rich fragment ion information provided by multi-stage mass spectrometry effectively distinguishes structurally similar veterinary drugs, significantly enhancing qualitative accuracy and supporting complex sample analysis. (2) LC-IT-MS is highly sensitive to trace levels of veterinary drug residues, meeting stringent food safety standards and ensuring the quality and safety of animal-derived products. (3) Compared to high-end hyphenated instruments, the ion trap mass spectrometer is compact and imposes less stringent requirements on laboratory infrastructure and funding, facilitating widespread adoption [124]. Despite its advantages, LC-IT-MS has certain limitations. (1) The complex motion of ions within the ion trap and uncertainties during collision processes can lead to variations in quantitative results, potentially affecting detection accuracy. (2) The method's scanning speed is comparatively slow, making it less efficient for rapid data acquisition in multi-component analyses or high-throughput screening scenarios, thereby extending the overall detection cycle and somewhat limiting its application in large-scale, high-throughput detection contexts [125].

In conclusion, LC-IT-MS represents a powerful analytical tool with significant advantages in qualitative analysis and sensitivity, though its limitations in precision and scanning speed warrant consideration in specific applications [126].

3.5. CE-MS

The fundamental principle of CE-MS involves the following sequential processes: Initially, a sample solution containing veterinary drug residues is introduced into the capillary electrophoresis system. Under the influence of a high-voltage electric field, the various components of veterinary drugs are separated within the capillary based on differences in their charge, size, and shape. Components with higher electrophoretic mobility reach the end of the capillary first. Subsequently, the separated components enter the mass spectrometer immediately, where they are ionized by the ion source, converting them into ions with specific mass-to-charge ratios (m/z). These ions then move through the mass analyzer according to their m/z values under the influence of magnetic or electric fields, ultimately being detected by the detector to generate a mass spectrum. By comparing the obtained spectra with known standard libraries and analyzing characteristic ion peaks, precise qualitative and quantitative detection of veterinary drugs is achieved [127,128].

The development of CE-MS has addressed the sensitivity limitations traditionally associated with CE. CE is characterized by high separation efficiency, rapid analysis speed, low reagent consumption, and straightforward operation. In the context of detecting residues in animal-based foods, CE-MS is particularly suitable for analyzing both weakly and strongly ionized metabolites, as well as their three-dimensional structures [129–131]. For instance, Blasco et al. successfully utilized CE-MS/MS to simultaneously detect eight residual veterinary drugs (ciprofloxacin, enrofloxacin, sulfacetamide, sulfabenzamide, sulfachloropyridazine, sulfaquinoxaline, sulfathiazole, and sulfamethoxazole) in milk. Acetonitrile was employed as the extraction solvent, and the Oasis HLB solid-phase extraction (SPE) column was used for purification. All eight veterinary drugs exhibited recovery rates exceeding 78%. The detection limits of this method ranged from 1 to 9 μg/kg, indicating the feasibility of CE-MS/MS for the initial analysis of veterinary drugs in milk [132].

This technique offers several notable advantages in separation efficiency, sensitivity, and sample requirement [133]. (1) The capillary electrophoresis allows for the fine separation of veterinary drugs based on their molecular properties in a short period, effectively addressing the coexistence of multiple components in complex sample matrices. This reduces peak broadening and enhances resolution. (2) CE-MS can detect trace amounts of veterinary drug residues, meeting increasingly stringent food safety standards and providing robust protection for the quality and safety of animal-derived foods. (3) This technique requires only a small amount of sample, significantly reducing detection costs. This is particularly advantageous for precious samples or situations with limited sample availability. However, CE-MS has several limitations [134]; for instance, in the required instrumentation, the interfacing technology between capillary electrophoresis and mass spectrometry is challenging, requiring precise calibration to ensure their synergistic operation. This places high demands on the technical expertise of the operator. Additionally, in methodological instability, the electrophoretic process is susceptible to variations in temperature and buffer composition, leading to poor reproducibility of migration times and peak areas. Frequent calibration and optimization are necessary, which increases the complexity and uncertainty of the detection process.

4. Rapid Detection Techniques

In recent years, a novel detection concept named rapid detection has been put forward. It represents a technical means that can generate detection outcomes within a brief time frame, and this also encompasses the preparation of test samples. When contrasted with traditional detection techniques like common physical–chemical testing and large-scale instrument testing, the rapid detection technologies such as Immunoassay analysis techniques, fluorescence polarization immunoassay, and surface-enhanced Raman scattering, showed superiority in terms of short detection time, uncomplicated instrument equipment, and convenience for on-site online detection [135–137].

4.1. *Immunoassay Analysis Techniques*

Immunological methods of veterinary drug residue analysis are based on the highly specific reaction of the substance with antibodies. Currently, various immune-based screening assays for antibiotic residue detection have been developed, such as GICA, ELISA, chemiluminescence immunoassay (CLIA), radioimmunoassay (RIA), and fluorescence immunoassay (FIA), and others [138]. The immunoassays are characterized by their high specificity and sensitivity. In general, the LODs achieved are lower and the analysis time is significantly shorter than that required for some of the microbiological tests. Based on the manipulation of assay formats and the use of broad-specificity antibodies, the multi-analysis of several targets in one single test is available [139].

4.1.1. GICA

The GICA method is a qualitative and semi-quantitative analysis approach that employs colloidal gold as a tracer. It is intercepted at the sites where antigens and antibodies specifically react and then shows color. This method is characterized by high-speed operation and great accuracy [140]. Ziyang et al. created a rapid analysis method which combines ultrasound-assisted SPE and colloidal gold test paper and utilized it to detect the residues of veterinary drugs like chloramphenicol, malachite green, cryptomalachite green, and semicarbazide in common aquatic products. Under the optimized conditions, the detection limit of this method is between 0.01 and 0.5 μg/kg, the absolute recovery rate is within 84.2–112.9%, and the average detection time is 15 min [141]. Na et al. produced a specific mouse monoclonal antibody against nitriles and developed a colloidal gold immunochromatographic test strip based on mouse monoclonal antibodies for screening nitriles in milk. The minimum detection limit was 1.146 ng/mL, and the average detection time was 10 min [142]. GICA possesses the following significant advantages: it is extremely simple to operate, requires no specialized instruments, and the results can be interpreted visually based on color development. This makes it suitable for on-site rapid testing in settings such as farms, slaughterhouses, and farmers' markets, significantly improving detection efficiency [143]. The assay is fast, typically completed within 5 to 15 min, allowing for timely feedback on veterinary drug residues and facilitating quick decision-making. Additionally, it is cost-effective, with minimal consumables and relatively simple test strip preparation, reducing overall detection costs and enabling large-scale application [144]. However, the sensitivity of GICA is relatively limited, making it less capable of detecting trace amounts of veterinary drug residues, which may not meet the increasingly stringent requirements for low residue limits in food safety standards. Additionally, the qualitative accuracy is suboptimal, as it is susceptible to interference from sample matrices, such as fats and proteins in animal tissues, which can lead to false-positive or false-negative results [145]. Therefore, confirmatory testing using other methods is often necessary to ensure result reliability. Furthermore, quantitative analysis is challenging, as the color intensity of the test strips only provides a rough estimate of the veterinary drug concentration range, making it less suitable for scenarios requiring precise quantification [142,146].

4.1.2. ELISA

ELISA is a widely used analytical technique in veterinary residue [147]. Its principle is grounded in the specific immune reaction between antigens and antibodies. In this method, a known antigen or antibody is immobilized onto the surface of a solid-phase carrier, such as a microplate. When the tested sample is added, if it contains the target veterinary drug (antigen), it will specifically bind to the immobilized antibody, forming an antigen–antibody complex. Following this initial binding, unbound substances are washed away to ensure specificity. Subsequently, an enzyme-labeled secondary antibody is introduced. This secondary antibody is designed to specifically recognize and bind to the antigen that has already been captured by the primary antibody. After this step, another washing process is performed to remove any unbound enzyme-labeled antibodies. The final stage involves the addition of an enzyme substrate. The enzyme catalyzes a chemical reaction with the substrate, leading to the generation of a colored product [148]. The intensity of this color is directly proportional to the amount of antigen present in the sample. The absorbance of this colored product is then measured using a microplate reader. By comparing the absorbance values to a standard curve, the concentration of the veterinary drug in the sample can be accurately determined [149]. A schematic illustration of the principle is shown in Figure 2.

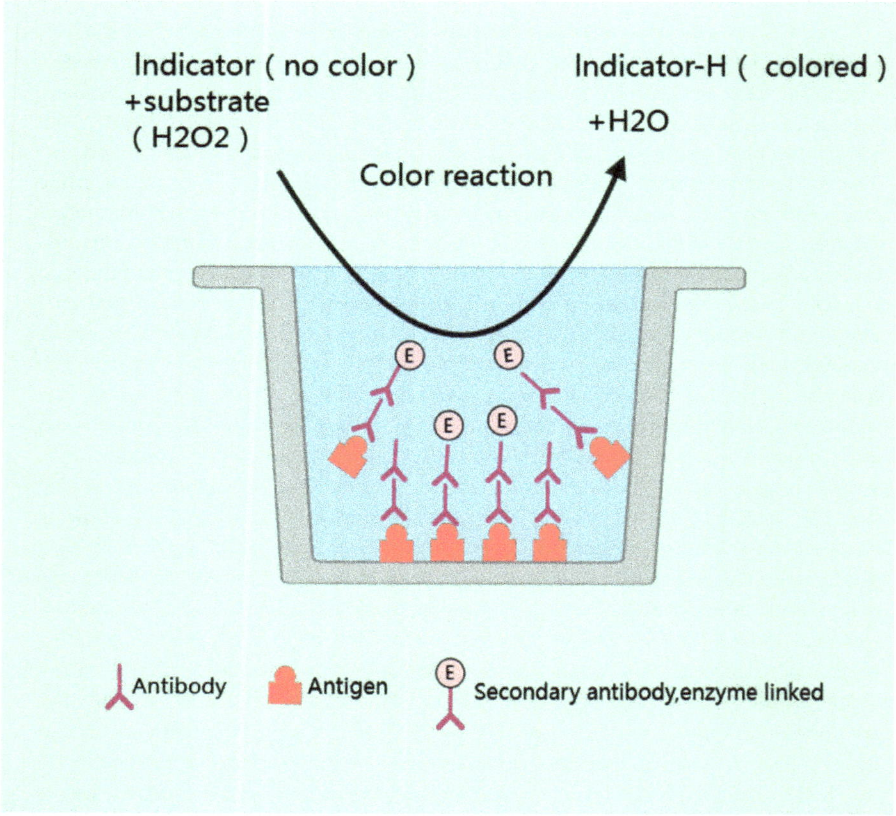

Figure 2. Schematic illustration of the principle of ELISA.

The ELISA method can produce a color change in enzyme reaction substrates via the specific binding of antigens and antibodies [150]. Fu et al. developed an indirect competitive enzyme-linked immunosorbent assay based on specific monoclonal antibodies for quickly screening estradiol in animal-derived foods. The detection limit and quantification limit are 68.1–103.1 ng/L and 100.8–12.7 ng/L, respectively. The recovery rate of estradiol varies from 70.1% to 103.1%, with a coefficient of variation of less than 13.9% [151]. Yin et al. prepared a monoclonal antibody with high specificity and sensitivity and developed an indirect competitive enzyme-linked immunosorbent assay to detect praziquantel. The limit of detection calculated from the analysis of broccoli, cabbage, wheat, corn, rice, chicken, fish, and crayfish samples was 1.56–2.72 mug/kg, and the average recovery rate was 81.25–103.19% [152]. The ELISA method usually has several advantages. In the sensitivity, it can detect trace amounts of veterinary drug residues, following the stringent food safety standards and provide strong support for ensuring the quality and safety of animal-derived foods. Additionally, its specificity is strong, as the precise binding of antigens and antibodies ensures highly selective recognition of the target veterinary drug, effectively excluding interference from other impurities in the sample and ensuring accurate results. Moreover, its operation is relatively simple, requiring no complex instruments; conventional laboratory equipment such as microplates and pipettes can be used. Multiple samples can be processed simultaneously, making it suitable for batch testing and significantly improving detection efficiency [153,154]. However, the ELISA method also has some limitations. There is a risk of cross-reactivity, where antibodies may misidentify structurally similar veterinary drugs or their metabolites, leading to false-positive results and affecting detection accuracy. Enzyme activity is susceptible to environmental factors such as temperature and pH, requiring strict control of operating conditions. Improper control can reduce enzyme activity, thereby affecting the reliability of the results [155].

4.2. Fluorescence Polarization Immunoassay (FPIA)

FPIA exhibits distinct advantages in veterinary drug residue detection, leveraging the principle of fluorescence polarization changes induced by specific antigen–antibody binding reactions [156]. In this method, a fluorophore-labeled veterinary drug antigen is introduced into a reaction system containing specific antibodies along with the sample. In the solution state, if the sample is devoid of veterinary drug residues, the fluorophore-labeled antigen can fully bind with the antibodies, forming large molecular complexes. This results in a reduction in rotational speed, thereby increasing the degree of fluorescence polarization. Conversely, if the sample contains the target veterinary drug, due to the competitive binding mechanism, it will preferentially bind with the antibodies, reducing the chance for the fluorophore-labeled antigen to bind with the antibodies [157]. Consequently, fewer complexes are formed, and the number of fluorophore-labeled antigens in small molecular states increases, leading to faster rotational speeds and a decrease in fluorescence polarization. By utilizing a fluorescence polarization analyzer to measure the fluorescence polarization value and comparing it with a standard curve, the amount of veterinary drug residue can be determined [158,159]. A schematic illustration of the principle of FPIA is shown in Figure 3.

FPIA usually provides rapid detection, typically completing the measurement of a single sample within minutes, making it suitable for on-site rapid screening. The technique exhibits considerable sensitivity, capable of detecting trace amounts of veterinary drugs, which is crucial for ensuring food safety and controlling the misuse of veterinary drugs, aligning with current stringent regulatory limits [160]. Additionally, FPIA is straightforward to operate, requiring no complex sample pretreatment steps. The process involves simple mixing and measurement, and it does not demand advanced technical skills from operators, making it user-friendly [161]. For example, Duan et al. created a rapid FPIA for the measurement of erythromycin in milk. To attain high sensitivity, five erythromycin tracers having diverse fluorescent structures were synthesized and paired with three monoclonal antibodies for detecting erythromycin in milk. The detection limit reached 14.08 μg/L and the recovery rate ranged from 96.08 to 107.77%. The overall testing time from adding the sample to obtaining the result reading was less than 5 min [162]. Duan et al. also devised a uniform high-throughput screening dual-wavelength FPIA for determining sulfamethoxazole and trimethoprim in milk, with recovery rates of 81.7–97.2% and 78.6–103.6%, respectively, and a coefficient of variation under 18.9%. Including sample pretreatment, it could be accomplished within 15 min [163].

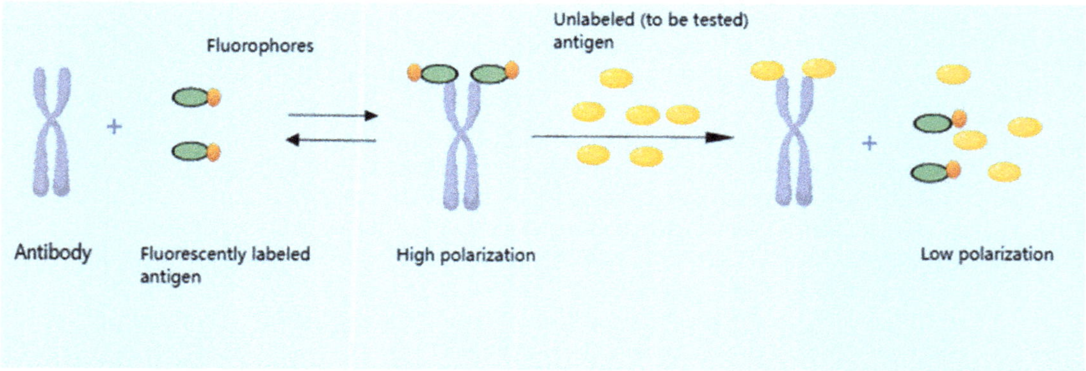

Figure 3. Schematic illustration of the principle of FPIA.

FPIA has notable limitations. The instrumentation, such as fluorometers with polarization modules, involves significant upfront costs, which can deter small-scale laboratories. Its detection scope is restricted by the availability of high-affinity antibodies and fluorescent tracers, particularly for veterinary drugs with complex structures or low immunogenicity, limiting multi-residue analysis [164]. Additionally, matrix components (e.g., fats or proteins in animal tissues) can interfere with fluorescence polarization signals, necessitating pre-treatment steps like solid-phase extraction or dilution to ensure accuracy [165]. While FPIA offers rapid, homogeneous detection, these challenges highlight the need for complementary methods in complex residue analysis.

4.3. Surface-Enhanced Raman Scattering (SERS)

SERS is founded on the Raman scattering principle (shown in Figure 4). When pyridine molecules are adsorbed on rough silver electrodes, the Raman scattering signal can be enhanced by 4–6 orders of magnitude. Through the combination of SERS technology and various materials, the detection of small molecules in veterinary drug residues becomes possible [166–168]. Dhakal et al. worked out a straightforward surface-enhanced Raman spectroscopy approach for on-the-spot screening of tetracycline residues in milk and water. In water–tetracycline solutions, tetracycline vibration modes were spotted at 1285, 1317, and 1632 cm^{-1}, while in milk–tetracycline solutions, they were seen at 1322 and 1621 cm^{-1}. A tetracycline residue concentration as minuscule as 0.01 ppm was detected in both solutions. The peak intensities at 1285 and 1322 cm^{-1} were utilized to calculate the tetracycline concentrations in water and milk, with correlation coefficients of 0.92 for water and 0.88 for milk [169]. Fan et al. devised a lateral flow immunosensor depending on dual SERS for the concurrent detection of tetracycline and penicillin residues in milk. Two nanoprobes are capable of recognizing tetracycline-BSA and ampicillin-BSA, respectively, which makes it easier to detect these two kinds of antibiotics simultaneously on a single test line. The detection limits for tetracycline and penicillin are as low as 0.015 ng/mL and 0.010 ng/mL correspondingly, with recovery rates varying from 88.8% to 111.3% and relative standard deviation less than 16% [170].

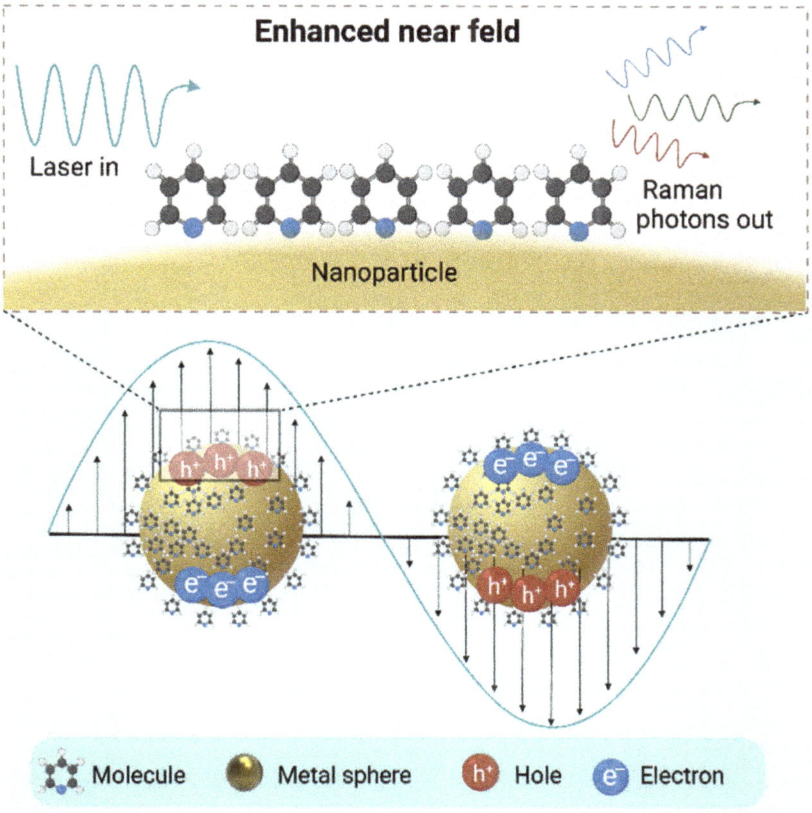

Figure 4. Schematic illustration of the principle of SERS.

SERS technology offers significant advantages such as it exhibits extremely high sensitivity, enabling the detection of trace and even ultra-trace levels of veterinary drug residues. This capability is crucial for meeting stringent food safety standards and ensuring the quality and safety of animal-derived products, as it effectively monitors the use of veterinary drugs at very low concentration levels [171]. The detection process is rapid, with spectral acquisition and analysis typically completed in a short time, making it suitable for on-site rapid screening scenarios such as farmers' markets and customs quarantine [172]. This allows for the timely identification of problematic products, thereby enhancing regulatory efficiency. Additionally, SERS requires minimal sample pretreatment, as it can directly analyze raw or simply processed samples, reducing errors and time associated with extensive preprocessing and simplifying the detection workflow [173]. Finally, SERS testing requires Raman spectrometers and suitable SERS active substrates, and has low-cost manufacturability and commercial accessibility.

However, SERS is mostly confined to academic laboratories and is not recognized as a gold-standard analytical technique, especially in the field of veterinary drug residue detection. This is due to several certain limitations, including its reproducibility is relatively poor. The Raman enhancement effect is highly susceptible to factors such as the size, shape, and aggregation state of metal nanoparticles, as well as the adsorption mode between the sample and nanoparticles. These variables can lead to significant fluctuations in results across different experimental batches or detection positions, compromising data reliability [174]. Quantitative analysis is challenging; although SERS can qualitatively identify veterinary drugs, the complex enhancement mechanisms, and matrix interferences make it difficult to establish precise quantitative models. The linear relationship between spectral signal intensity and veterinary drug concentration is hard to control accurately, posing challenges for the precise determination of residue levels [175,176].

5. Comparative Analysis of Merits and Limitations in Contemporary Veterinary Drug Residue Analytical Techniques for Animal-Derived Food Products

We systematically compared the advantages and disadvantages of the current main methods for detecting veterinary drug residues (as illustrated in Table 2). Of note, GC-MS requires meticulous preparation and optimization, presents certain cost limitations, and is particularly effective for volatile compounds [177]. LC-QTOF-MS offers high-resolution analysis but is constrained by elevated operational costs and complex data processing requirements, making it suitable for complex matrices. LC-MS/MS, while highly sensitive and applicable to numerous food types, is both expensive and technically demanding to operate [177]. LC-IT-MS, despite its slower scan speed, provides a more economical option for preliminary screening purposes. CE-MS achieves high separation efficiency but faces technical implementation challenges, though it is particularly effective for ionized metabolites [178]. Taken together, these established chromatographic platforms, such as GC-MS (LOD 0.1–5 μg/kg for volatile compounds) and LC-MS/MS, demonstrate unparalleled sensitivity and accuracy (85–115% recovery rates), albeit constrained by prolonged analysis cycles, substantial operational costs, and stringent laboratory requirements. In contrast, rapid immunoassays and SERS-based approaches enable on-site screening within 5 min at <10% of instrumental costs, particularly valuable for high-throughput surveillance (200 samples/hour) and resource-limited settings, despite relatively compromised sensitivity and occasional cross-reactivity issues. This technological dichotomy underscores its importance in developing integrated detection paradigms. A hierarchical strategy combining rapid preliminary screening (5–10% presumptive positives) with subsequent instrumental confirmation achieves 300–500% efficiency enhancement while maintaining regulatory-grade reliability. Such synergistic systems address critical industry challenges posed by quick growth in global veterinary drug consumption (for example, global antibiotic sales are expected to increase from 93,309 tons in 2017 to 104,079 tons by 2030) [179], particularly enabling real-time monitoring across decentralized supply chains and rapid response to contamination emergencies. This dual-track approach not only optimizes resource allocation but also establishes an adaptive framework for emerging veterinary pharmacopeia, positioning itself as an indispensable component in modern food safety ecosystems. Future research should prioritize the standardization of validation protocols and the development of multiplexed detection platforms to fully realize the transformative potential of rapid screening technologies in global food surveillance networks.

Table 2. A comprehensive systematic comparison among various detection techniques for veterinary drug residues in animal-derived foods.

Method	LOD (μg/kg)	Accuracy (Recovery Rate)	Applicability	Advantages	Disadvantages	Feasibility	Costing
GC-MS	2.3–4.3	77.38–95.7%	Volatile/semi-volatile compounds.	High specificity for volatile analytes. Robust qualitative capabilities. Wide applicability for small molecules.	Requires derivatization for non-volatilecompounds. Limited to thermally stable analytes.	Requires derivatization expertise.	Moderate to high
LC-QTOF-MS	0.5	More than 70%	High-resolution multi-residue screening.	Ultra-high resolution for accurate mass identification. Broad-spectrum detection. Rich structural data.	High equipment/maintenance costs. Demands advanced data analysis skills.	Requires high-end infrastructure.	Very high
LC-MS/MS	0.02–82	70–120%	Gold standard for trace-level quantification.	High sensitivity and selectivity. Reliable for multi-residue analysis. Robust quantitative accuracy.	Expensive instrumentation. Complex sample preparation. Matrix effects require mitigation.	Skilled operation and maintenance needed.	High
LC-IT-MS	0.01–18.75	63–122%	Multi-stage fragmentation for structural elucidation.	Mul-ti-stage mass for structural insights. Compact and cost-effective.	Slower scanning speeds. Moderate resolution limits complex mixture analysis.	Suitable for targeted analysis.	Moderate
CE-MS	1–9	More than 78%	Ionizable metabolites.	High separation efficiency. Minimal sample/reagent consumption. Fast analysis.	Poor reproducibility due to buffer/temperature sensitivity.	Technically demanding for calibration.	Moderate
GICA	0.01–0.5	84.2–112.9%	Rapid on-site screening.	Equipment-free, rapid results. Low cost and user-friendly.	Qualitative/semi-quantitative only. Limited sensitivity for trace residues. Matrix interference risks.	Ideal for field testing.	Low
ELISA	1.56–2.72	70.1–103.1%	High-throughput screening.	High throughput and specificity. Cost-effective for batch analysis. Minimal instrumentation.	Cross-reactivity with analogs. Enzyme activity affected by environmental factors.	Requires antibody development.	Low to moderate
FPIA	0.01	78.6–107.77%	Homogeneous assays.	Rapid and homogeneous. Minimal sample pretreatment. Moderate sensitivity.	Limited by antibody/tracer availability. Matrix interference in complex samples.	Suitable for simple matrices.	Moderate
SERS	0.01–0.015	88.8–111.3%	Ultra-trace detection.	Ultra-high sensitivity. Rapid and minimal pretreatment. Multiplexing potential.	Poor reproducibility due to nanoparticle variability.	Requires nanoparticle optimization.	High

6. Conclusions and Further Prospects

Recent advancements in pre-treatment methods and detection technologies for veterinary drug residues in animal-derived foods have significantly bolstered food safety and regulatory adherence. Innovations in sample preparation, such as SPE, QuEChERS, and MIPs, have streamlined the extraction and purification of complex matrices. These methods have reduced both time and solvent usage while enhancing selectivity [180].

Simultaneously, cutting-edge detection technologies, including LC-IT-MS, LC-MS/MS, biosensors, and nanomaterial-based assays, now facilitate ultrasensitive, multi-residue analysis with high specificity. These technological strides address critical challenges such as low residue concentrations, matrix interference, and the necessity for rapid, high-throughput screening. The integration of automation, artificial intelligence (AI) for data analysis, and portable devices has further revolutionized the field, enabling real-time monitoring and on-site testing. Such progress not only ensures compliance with stringent global regulations (e.g., EU Maximum Residue Limits [MRLs], FDA guidelines) but also mitigates public health risks associated with antibiotic resistance, allergic reactions, and chronic toxicity from veterinary drug residues [181,182].

In resource-limited regions such as China and other developing nations, the development of veterinary drug residue monitoring systems requires strategic methodology selection that prioritizes cost-efficiency, technical accessibility, and large-scale implementation ability. For sample pretreatment, the QuEChERS approach demonstrates superior practicality through its solvent-minimized workflow (<10 mL acetonitrile/sample), equipment independence (requiring only basic centrifugation), and broad matrix compatibility spanning meat, dairy, and aquatic products [183]. Field-deployable detection platforms, particularly GICA and ELISA, achieve reliable on-site screening with rapid turnaround (≤30 min/test) and operator-friendly protocols, maintaining 89–94% concordance with reference methods in multi-laboratory validations [184]. However, advanced techniques like LC-MS/MS provide unparalleled sensitivity and multi-residue capabilities (detection limits 0.1–5 μg/kg, >200 simultaneous analytes), their reliance on expensive instrumentation (about USD500,000+ instrumentation costs) and specialized expertise limits their feasibility in underfunded regions. Therefore, a hybrid strategy combining QuEChERS-based pretreatment with GICA/ELISA screening, supplemented by targeted LC-MS/MS confirmation in centralized facilities, offers a balanced approach to ensure food safety without compromising accessibility [185,186]. This framework aligns with the socioeconomic realities of third-world countries, where cost-efficient, field-deployable solutions are critical for large-scale monitoring and regulatory compliance.

The next-generation paradigm for veterinary drug residue analysis in animal-derived foods necessitates synergistic advancements in pretreatment and detection technologies. In sample preparation, intelligent precision extraction strategies are emerging, such as biomimetic molecularly imprinted nanoreactors that achieve a higher recovery rate through size-exclusion mechanisms, coupled with AI-driven automated platforms capable of processing multiple samples in a short time. Additionally, ultrasensitive detection breakthroughs including CRISPR-Cas12a-mediated cascades and plasmonic metasurface-enhanced Raman spectroscopy utilizing gold nanorod arrays with deep learning deconvolution, can further enhance detection accuracy and speed [187–189]. Additionally, developing universal protocols for the simultaneous detection of diverse veterinary drugs (antibiotics, hormones, and anthelmintics) remains crucial, especially with the emergence of novel drug formulations. Harmonizing detection thresholds and methodologies across regions will reduce trade barriers and ensure equitable food safety standards. Enhanced surveillance of veterinary drug residues in food chains is vital to combat antimicrobial resistance, requiring robust collaboration between governments, industries, and researchers. Public education on residue risks and the implementation of transparent labeling systems (e.g., blockchain traceability) can empower consumers to make informed choices and foster trust in food systems. By addressing these areas, the scientific community can continue to advance the safety and quality of animal-derived foods, ensuring they meet the highest standards of public health and regulatory compliance.

Author Contributions: Conceptualization, Q.D. and C.D.; methodology, Q.D. and C.D.; software, Q.D. and C.D.; formal analysis, S.T.; investigation, Q.D. and C.D.; writing—original draft preparation, Q.D. and C.D.; writing—review and editing, Q.D., S.T. and C.D.; supervision, S.T. and C.D.; project administration, S.T. and C.D.; funding acquisition, S.T. and C.D. All authors have read and agreed to the published version of the manuscript.

Funding: This study was supported by Beijing Natural Science Foundation (6252017). It was also supported by National Natural Science Foundation of China (32473083).

Institutional Review Board Statement: Not applicable.

Informed Consent Statement: Not applicable.

Data Availability Statement: All data were included in this manuscript.

Conflicts of Interest: The authors declare no conflicts of interest.

References

1. Wang, B.; Xie, K.; Lee, K. Veterinary drug residues in animal-derived foods: Sample preparation and analytical methods. *Foods* **2021**, *10*, 555. [CrossRef] [PubMed]
2. Khalifa, H.O.; Shikoray, L.; Mohamed, M.I.; Habib, I.; Matsumoto, T. Veterinary drug residues in the food chain as an emerging public health threat: Sources, analytical methods, health impacts, and preventive measures. *Foods* **2024**, *13*, 1629. [CrossRef] [PubMed]
3. Pratiwi, R.; Ramadhanti, S.P.; Amatulloh, A.; Megantara, S.; Subra, L. Recent advances in the determination of veterinary drug residues in food. *Foods* **2023**, *12*, 3422. [CrossRef]
4. Mesfin, Y.M.; Mitiku, B.A.; Tamrat Admasu, H. Veterinary drug residues in food products of animal origin and their public health consequences: A review. *Vet. Med. Sci.* **2024**, *10*, e70049. [CrossRef]
5. Zhang, H.; Chen, Q.; Niu, B. Risk assessment of veterinary drug residues in meat products. *Curr. Drug Metab.* **2020**, *21*, 779–789. [CrossRef]
6. Bacanlı, M.; Başaran, N. Importance of antibiotic residues in animal food. *Food Chem. Toxicol.* **2019**, *125*, 462–466. [CrossRef]
7. Parmar, J.K.; Chaubey, K.K.; Gupta, V.; Bharath, M.N. Assessment of various veterinary drug residues in animal originated food products. *Vet. World* **2021**, *14*, 1650–1664. [CrossRef] [PubMed]
8. Colopi, A.; Guida, E.; Cacciotti, S.; Fuda, S.; Lampitto, M.; Onorato, A.; Zucchi, A.; Balistreri, C.R.; Grimaldi, P.; Barchi, M. Dietary exposure to pesticide and veterinary drug residues and their effects on human fertility and embryo development: A global overview. *Int. J. Mol. Sci.* **2024**, *25*, 9116. [CrossRef]
9. Chicoine, A.; Erdely, H.; Fattori, V.; Finnah, A.; Fletcher, S.; Lipp, M.; Sanders, P.; Scheid, S. Assessment of veterinary drug residues in food: Considerations when dealing with sub-optimal data. *Regul. Toxicol. Pharmacol.* **2020**, *118*, 104806. [CrossRef]
10. Rana, M.S.; Lee, S.Y.; Kang, H.J.; Hur, S.J. Reducing veterinary drug residues in animal products: A review. *Food Sci. Anim. Resour.* **2019**, *39*, 687–703. [CrossRef]
11. Atta, A.H.; Atta, S.A.; Nasr, S.M.; Mouneir, S.M. Current perspective on veterinary drug and chemical residues in food of animal origin. *Environ. Sci. Pollut. Res. Int.* **2022**, *29*, 15282–15302. [CrossRef]
12. Lozano, A.; Hernando, M.D.; Ucles, S.; Hakme, E.; Fernandez-Alba, A.R. Identification and measurement of veterinary drug residues in beehive products. *Food Chem.* **2019**, *274*, 61–70. [CrossRef]
13. Xia, S.; Niu, B.; Chen, J.; Deng, X.; Chen, Q. Risk analysis of veterinary drug residues in aquatic products in the Yangtze river delta of China. *J. Food Prot.* **2021**, *84*, 1228–1238. [CrossRef] [PubMed]
14. Guo, X.; Tian, H.; Yang, F.; Fan, S.; Zhang, J.; Ma, J.; Ai, L.; Zhang, Y. Rapid determination of 103 common veterinary drug residues in milk and dairy products by ultra performance liquid chromatography tandem mass spectrometry. *Front. Nutr.* **2022**, *9*, 879518. [CrossRef]
15. Li, F.; Lv, H.; Zhu, F.; Zhang, Q.; Xu, Q.; Ji, W. High throughput detection of veterinary drug residues in chicken and eggs. *Food Chem.* **2025**, *463*, 141267. [CrossRef] [PubMed]

16. Rahman, M.M.; Lee, D.J.; Jo, A.; Yun, S.H.; Eun, J.B.; Im, M.H.; Shim, J.H.; Abd El-Aty, A.M. Onsite/on-field analysis of pesticide and veterinary drug residues by a state-of-art technology: A review. *J. Sep. Sci.* **2021**, *44*, 2310–2327. [CrossRef]
17. Liu, L.; Gao, Y.; Liu, J.; Li, Y.; Yin, Z.; Zhang, Y.; Pi, F.; Sun, X. Sensitive techniques for poct sensing on the residues of pesticides and veterinary drugs in food. *Bull. Environ. Contam. Toxicol.* **2021**, *107*, 206–214. [CrossRef]
18. Li, M.; Zhang, X. Nanostructure-based surface-enhanced raman spectroscopy techniques for pesticide and veterinary drug residues screening. *Bull. Environ. Contam. Toxicol.* **2021**, *107*, 194–205. [CrossRef]
19. Zhang, B.; Lang, Y.; Guo, B.; Cao, Z.; Cheng, J.; Cai, D.; Shentu, X.; Yu, X. Indirect competitive enzyme-linked immunosorbent assay based on broad-spectrum antibody for simultaneous determination of thirteen fluoroquinolone antibiotics in *Rana catesbeianus*. *Foods* **2023**, *12*, 2530. [CrossRef]
20. Park, D.; Choi, Y.S.; Kim, J.Y.; Choi, J.D.; Moon, G.I. Determination of flunixin and 5-hydroxy flunixin residues in livestock and fishery products using liquid chromatography-tandem mass spectrometry (lc-ms/ms). *Food Sci. Anim. Resour.* **2024**, *44*, 873–884. [CrossRef]
21. Gens, K.D.; Singer, R.S.; Dilworth, T.J.; Heil, E.L.; Beaudoin, A.L. Antimicrobials in animal agriculture in the united states: A multidisciplinary overview of regulation and utilization to foster collaboration: On behalf of the society of infectious diseases pharmacists. *Open Forum Infect. Dis.* **2022**, *9*, ofac542. [CrossRef]
22. Li, R.; Wen, Y.; Wang, F.; He, P. Recent advances in immunoassays and biosensors for mycotoxins detection in feedstuffs and foods. *J. Anim. Sci. Biotechnol.* **2021**, *12*, 108. [CrossRef]
23. Park, H.; Kim, E.; Lee, T.H.; Park, S.; Choi, J.D.; Moon, G. Multiclass method for the determination of anthelmintic and antiprotozoal drugs in livestock products by ultra-high-performance liquid chromatography-tandem mass spectrometry. *Food Sci. Anim. Resour.* **2023**, *43*, 914–937. [CrossRef]
24. Beguiristain, I.; Alongi, A.; Rubies, A.; Granados, M. Analysis of corticosteroids in samples of animal origin using QuEChERS and ultrahigh-performance liquid chromatography coupled to high-resolution mass spectrometry. *Anal. Bioanal. Chem.* **2019**, *411*, 449–457. [CrossRef]
25. Okerman, L.; Noppe, H.; Cornet, V.; De Zutter, L. Microbiological detection of 10 quinolone antibiotic residues and its application to artificially contaminated poultry samples. *Food Addit. Contam.* **2007**, *24*, 252–257. [CrossRef] [PubMed]
26. Jia, Q.; Liao, G.Q.; Chen, L.; Qian, Y.Z.; Yan, X.; Qiu, J. Pesticide residues in animal-derived food: Current state and perspectives. *Food Chem.* **2024**, *438*, 137974. [CrossRef] [PubMed]
27. Lin, S.; Zhao, Z.; Lv, Y.K.; Shen, S.; Liang, S.X. Recent advances in porous organic frameworks for sample pretreatment of pesticide and veterinary drug residues: A review. *Analyst* **2021**, *146*, 7394–7417. [CrossRef] [PubMed]
28. Jia, M.; E, Z.; Zhai, F.; Bing, X. Rapid multi-residue detection methods for pesticides and veterinary drugs. *Molecules* **2020**, *25*, 3590. [CrossRef] [PubMed]
29. Tao, X.; Peng, Y.; Liu, J. Nanomaterial-based fluorescent biosensors for veterinary drug detection in foods. *J. Food Drug Anal.* **2020**, *28*, 575–594. [CrossRef]
30. Zhang, J.; Hu, B. Liquid-liquid extraction (lle). In *Separation and Purification Technologies in Biorefineries*; John Wiley & Sons: Hoboken, NJ, USA, 2013; pp. 61–78.
31. Tshepho, R.; Dube, S.; Nindi, M.M. Ionic liquid-based dispersive liquid-liquid microextraction of anthelmintic drug residues in small-stock meat followed by LC-ESI-MS/MS detection. *Food Sci. Nutr.* **2023**, *11*, 6288–6302. [CrossRef]
32. Araujo, P.; Iqbal, S.; Arno, A.; Espe, M.; Holen, E. Validation of a liquid-liquid extraction method to study the temporal production of d-series resolvins by head kidney cells from atlantic salmon (*Salmon salar*) exposed to docosahexaenoic acid. *Molecules* **2023**, *28*, 4728. [CrossRef] [PubMed]
33. Zheng, X.; Xie, Y.; Chen, Z.; Cao, M.; Lei, X.; Le, T. A comprehensive review on the pretreatment and detection methods of nitrofurans and their metabolites in animal-derived food and environmental samples. *Food Chem. X* **2024**, *24*, 101928. [CrossRef] [PubMed]
34. Doria, S.; Yost, J.; Gagnon, Z. Free-flow biomolecular concentration and separation of proteins and nucleic acids using tef-chophoresis. *Talanta* **2023**, *255*, 124198. [CrossRef]
35. Kannouma, R.E.; Hammad, M.A.; Kamal, A.H.; Mansour, F.R. Miniaturization of liquid-liquid extraction; the barriers and the enablers. *Microchem. J.* **2022**, *182*, 107863. [CrossRef]
36. Faraji, H. Advancements in overcoming challenges in dispersive liquid-liquid microextraction: An overview of advanced strategies. *TrAC Trends Anal. Chem.* **2024**, *170*, 117429. [CrossRef]
37. Cantwell, F.F.; Losier, M. Chapter 11 Liquid—Liquid extraction. In *Comprehensive Analytical Chemistry*; Elsevier: Amsterdam, The Netherlands, 2002; Volume 37, pp. 297–340.
38. Sun, Q.; Zhang, S.; Chen, Q.; Cao, L. Salting-out assisted liquid-liquid extraction for the simple and rapid determination of veterinary antibiotic residues in aquatic products by HPLC-MS/MS. *Food Chem.* **2024**, *460*, 140775. [CrossRef] [PubMed]

39. Wang, J.; Xu, J.; Ji, X.; Wu, H.; Yang, H.; Zhang, H.; Zhang, X.; Li, Z.; Ni, X.; Qian, M. Determination of veterinary drug/pesticide residues in livestock and poultry excrement using selective accelerated solvent extraction and magnetic material purification combined with ultra-high-performance liquid chromatography-tandem mass spectrometry. *J. Chromatogr. A* **2020**, *1617*, 460808. [CrossRef]
40. Pang, M.; Xu, J.; Tang, Y.; Guo, Y.; Ding, H.; Wang, R.; Zhang, T.; Zhang, G.; Guo, X.; Dai, G.; et al. Combining gc-ms/mS with a lle-spe sample pretreatment step to simultaneously analyse enrofloxacin and ofloxacin residues in chicken tissues and pork. *Food Chem.* **2024**, *456*, 139972. [CrossRef]
41. Zhang, Y.; Xue, X.; Su, S.; Guo, Z.; Wang, J.; Ding, L.; Liu, Y.; Zhu, J. A multi-class, multi-residue method for detection of veterinary drugs in multiple meat using a pass-through cleanup spe technique and uplc-ms/mS analysis. *Food Anal. Methods* **2018**, *11*, 2865–2884. [CrossRef]
42. Badawy, M.E.I.; El-Nouby, M.A.M.; Kimani, P.K.; Lim, L.W.; Rabea, E.I. A review of the modern principles and applications of solid-phase extraction techniques in chromatographic analysis. *Anal. Sci. Int. J. Jpn. Soc. Anal. Chem.* **2022**, *38*, 1457–1487. [CrossRef]
43. Cheng, R.; Mao, X.; Yu, J.; Liu, F.; Guo, L.; Luo, D.; Wan, Y. A dispersive solid-phase extraction method for the determination of Aristolochic acids in Houttuynia cordata based on MIL-101(Fe): An analytes-oriented adsorbent selection design. *Food Chem.* **2023**, *407*, 135074. [CrossRef] [PubMed]
44. Zhang, Q.; Okoli, C.P. Development of a solid-phase extraction method based on biocompatible starch polyurethane polymers for gc-ms analysis of polybrominated diphenyl ethers in ambient water samples. *Molecules* **2022**, *27*, 3253. [CrossRef]
45. Islas, G.; Ibarra, I.S.; Hernandez, P.; Miranda, J.M.; Cepeda, A. Dispersive solid phase extraction for the analysis of veterinary drugs applied to food samples: A review. *Int. J. Anal. Chem.* **2017**, *2017*, 8215271. [CrossRef] [PubMed]
46. Brandi, J.; Siragusa, G.; Robotti, E.; Marengo, E.; Cecconi, D. Analysis of veterinary drugs and pesticides in food using liquid chromatography-mass spectrometry. *TrAC Trends Anal. Chem.* **2024**, *179*, 117888. [CrossRef]
47. Zhu, Y.; Jiang, X.; Shen, D.; Mao, J.; Cao, Y.; Zhang, K.; Peng, J.; Dong, F.; Wang, N.; He, K. A one-step solid-phase extraction with uhplc-ms/ms for fast and accurate determination of multi-class veterinary drugs in animal muscles. *Food Chem.* **2023**, *428*, 136712. [CrossRef]
48. Mehl, A.; Hudel, L.; Bucker, M.; Morlock, G.E. Validated screening method for 81 multiclass veterinary drug residues in food via online-coupling high-throughput planar solid-phase extraction to high-performance liquid chromatography-orbitrap tandem mass spectrometry. *J. Agric. Food Chem.* **2022**, *70*, 10886–10898. [CrossRef]
49. Rodriguez-Mozaz, S.; Lopez de Alda, M.J.; Barceló, D. Advantages and limitations of on-line solid phase extraction coupled to liquid chromatography–mass spectrometry technologies versus biosensors for monitoring of emerging contaminants in water. *J. Chromatogr. A* **2007**, *1152*, 97–115. [CrossRef]
50. Talib, N.; Mohd-Setapar, S.; Khamis, A. The benefits and limitations of methods development in solid phase extraction: Mini review. *J. Teknol. (Sci. Eng.)* **2014**, *69*, 69–72. [CrossRef]
51. Zheng, M.; Tang, S.; Bao, Y.; Daniels, K.D.; How, Z.T.; El-Din, M.G.; Wang, J.; Tang, L. Fully-automated spe coupled to uhplc-ms/ms method for multiresidue analysis of 26 trace antibiotics in environmental waters: SPE optimization and method validation. *Environ. Sci. Pollut. Res. Int.* **2022**, *29*, 16973–16987. [CrossRef]
52. Sharmeen, S.; Suh, K.; Kyei, I.; Jones, J.; Olupathage, H.; Campbell, A.; Hage, D.S. Immunoaffinity chromatography for protein purification and analysis. *Curr. Protoc.* **2023**, *3*, e867. [CrossRef]
53. Ren, J.; Xiong, H.; Huang, C.; Ji, F.; Jia, L. An engineered peptide tag-specific nanobody for immunoaffinity chromatography application enabling efficient product recovery at mild conditions. *J. Chromatogr. A* **2022**, *1676*, 463274. [CrossRef] [PubMed]
54. Poddar, S.; Sharmeen, S.; Hage, D.S. Affinity monolith chromatography: A review of general principles and recent developments. *Electrophoresis* **2021**, *42*, 2577–2598. [CrossRef]
55. Fitzgerald, J.; Leonard, P.; Darcy, E.; Sharma, S.; O'Kennedy, R. Immunoaffinity chromatography: Concepts and applications. *Methods Mol. Biol.* **2017**, *1485*, 27–51. [CrossRef] [PubMed]
56. Subramanian, A. Immunoaffinity chromatography. *Mol. Biotechnol.* **2002**, *20*, 41–47. [CrossRef]
57. Moser, A.C.; Hage, D.S. Immunoaffinity chromatography: An introduction to applications and recent developments. *Bioanalysis* **2010**, *2*, 769–790. [CrossRef] [PubMed]
58. Peng, D.; Zhang, X.; Wang, Y.; Pan, Y.; Liu, Z.; Chen, D.; Sheng, F.; Yuan, Z. An immunoaffinity column for the selective purification of 3-methyl-quinoxaline-2-carboxylic acid from swine tissues and its determination by high-performance liquid chromatography with ultraviolet detection and a colloidal gold-based immunochromatographic assay. *Food Chem.* **2017**, *237*, 290–296. [CrossRef]
59. Wang, Q.; Zhao, H.; Xi, C.; Wang, G.; Chen, D.; Ding, S. Determination of chloramphenicol and zeranols in pig muscle by immunoaffinity column clean-up and LC-MS/MS analysis. *Food Addit. Contam. Part A Chem. Anal. Control Expo. Risk Assess.* **2014**, *31*, 1177–1186. [CrossRef]
60. Nelson, M.A.; Reiter, W.S.; Hage, D.S. Chromatographic competitive binding immunoassays: A comparison of the sequential and simultaneous injection methods. *Biomed. Chromatogr.* **2003**, *17*, 188–200. [CrossRef]

61. Thompson, N.E.; Burgess, R.R. Immunoaffinity chromatography: Advantages and limitations. In *Analytical Separation Science*; Wiley: Hoboken, NJ, USA, 2015; pp. 483–502. [CrossRef]
62. Kavianpour, A.; Ashjari, M.; Hosseini, S.N.; Khatami, M. Quantitative assessment of LPS-HBsAg interaction by introducing a novel application of immunoaffinity chromatography. *Prep. Biochem. Biotechnol.* **2023**, *53*, 672–682. [CrossRef]
63. Gonzalez-Curbelo, M.A.; Varela-Martinez, D.A.; Riano-Herrera, D.A. Pesticide-residue analysis in soils by the QuEChERS method: A Review. *Molecules* **2022**, *27*, 4323. [CrossRef]
64. Kim, K.; Choi, Y.; Mok, S.; Moon, H.B.; Jeon, J. Optimization of the QuEChERS method for multi-residue analysis of pharmaceuticals and pesticides in aquaculture products. *Food Chem.* **2023**, *399*, 133958. [CrossRef] [PubMed]
65. Yang, G.; Zhang, J.; Tang, Y.; Kong, C.; Li, S.; Wang, S.; Ding, S.; Gu, L.; Shen, X.; Martin, A.A.; et al. Development and validation of rapid screening of 192 veterinary drug residues in aquatic products using HPLC-HRMS coupled with QuEChERS. *Food Chem. X* **2024**, *22*, 101504. [CrossRef]
66. Wang, H.; Tian, H.; Ai, L.F.; Liang, S.X. Screening and quantification of 146 veterinary drug residues in beef and chicken using QuEChERS combined with high performance liquid chromatography-quadrupole orbitrap mass spectrometry. *Food Chem.* **2023**, *408*, 135207. [CrossRef]
67. Mu, P.; Xu, N.; Chai, T.; Jia, Q.; Yin, Z.; Yang, S.; Qian, Y.; Qiu, J. Simultaneous determination of 14 antiviral drugs and relevant metabolites in chicken muscle by uplc-ms/ms after QuEChERS preparation. *J. Chromatogr. B Analyt. Technol. Biomed. Life Sci.* **2016**, *1023–1024*, 17–23. [CrossRef]
68. Kaufmann, A.; Butcher, P.; Maden, K.; Walker, S.; Widmer, M. Assessment and validation of the p-QuEChERS sample preparation methodology for the analysis of >200 veterinary drugs in various animal-based food matrices. *Food Addit. Contam. Part A Chem. Anal. Control Expo. Risk Assess.* **2023**, *40*, 356–372. [CrossRef] [PubMed]
69. Hu, Q.; Sun, J.; Yu, H.; Feng, R.; Zhang, J.; Zhou, H.; Ji, S. Simultaneous Screening of 172 Veterinary drugs by modified QuEChERS-LC-MS/MS in TCM galli gigerii endothelium corneum. *J. Chromatogr. Sci.* **2024**, *62*, 439–443. [CrossRef] [PubMed]
70. Sun, Q.; Liu, J.; Gou, Y.; Chen, T.; Shen, X.; Wang, T.; Li, Y.; He, H.; Deng, H.; Hua, Y. Determination of veterinary drugs in foods of animal origin by QuEChERS coupled with ultra performance liquid chromatography-tandem mass spectrometry (UPLC-MS/MS). *J. Chromatogr. A* **2025**, *1744*, 465726. [CrossRef]
71. Chen, L.; Wang, X.; Lu, W.; Wu, X.; Li, J. Molecular imprinting: Perspectives and applications. *Chem. Soc. Rev.* **2016**, *45*, 2137–2211. [CrossRef]
72. Ye, K.; Xu, S.; Zhou, Q.; Wang, S.; Xu, Z.; Liu, Z. Advances in molecular imprinting technology for the extraction and detection of quercetin in plants. *Polymers* **2023**, *15*, 2107. [CrossRef]
73. Xu, S.; Xu, Z.; Liu, Z. Paper-based molecular-imprinting technology and its application. *Biosensors* **2022**, *12*, 595. [CrossRef]
74. Song, X.; Zhou, T.; Li, J.; Su, Y.; Xie, J.; He, L. Determination of macrolide antibiotics residues in pork using molecularly imprinted dispersive solid-phase extraction coupled with LC-MS/MS. *J. Sep. Sci.* **2018**, *41*, 1138–1148. [CrossRef] [PubMed]
75. Tang, J.; Wang, J.; Shi, S.; Hu, S.; Yuan, L. Determination of beta-agonist residues in animal-derived food by a liquid chromatography-tandem mass spectrometric method combined with molecularly imprinted stir bar sorptive extraction. *J. Anal. Methods Chem.* **2018**, *2018*, 9053561. [CrossRef]
76. Elfadil, D.; Lamaoui, A.; Della Pelle, F.; Amine, A.; Compagnone, D. Molecularly Imprinted polymers combined with electrochemical sensors for food contaminants analysis. *Molecules* **2021**, *26*, 4607. [CrossRef] [PubMed]
77. Farooq, S.; Xu, L.; Ullah, S.; Li, J.; Nie, J.; Ping, J.; Ying, Y. Advancements and greenification potential of magnetic molecularly imprinted polymers for chromatographic analysis of veterinary drug residues in milk. *Compr. Rev. Food Sci. Food Saf.* **2024**, *23*, e13399. [CrossRef]
78. Tang, Y.; Meng, H.; Wang, W.; Song, Y.; Wang, S.; Li, Z.; Wang, X.; Hu, X. Off-line magnetic Fe_3O_4@SiO_2@MIPs-based solid phase dispersion extraction coupling with HPLC for the simultaneous determination of olaquindox and its metabolite in fish muscle and milk samples. *Food Chem. X* **2023**, *17*, 100611. [CrossRef]
79. Norman, R.L.; Singh, R.; Muskett, F.W.; Parrott, E.L.; Rufini, A.; Langridge, J.I.; Runau, F.; Dennison, A.; Shaw, J.A.; Piletska, E.; et al. Mass spectrometric detection of KRAS protein mutations using molecular imprinting. *Nanoscale* **2021**, *13*, 20401–20411. [CrossRef]
80. Li, D.; Luo, K.; Zhang, L.; Gao, J.; Liang, J.; Li, J.; Pan, H. Research and application of highly selective molecular imprinting technology in chiral separation analysis. *Crit. Rev. Anal. Chem.* **2023**, *53*, 1066–1079. [CrossRef] [PubMed]
81. Khatibi, S.A.; Hamidi, S.; Siahi-Shadbad, M.R. Application of liquid-liquid extraction for the determination of antibiotics in the foodstuff: Recent trends and developments. *Crit. Rev. Anal. Chem.* **2022**, *52*, 327–342. [CrossRef]
82. Wang, J.; Yang, H.Y.; Wang, X.D.; Lv, Y.F.; Wei, N. Application of QuEChERS for analysis of contaminants in dairy products: A review. *J. Food Prot.* **2025**, *88*, 100453. [CrossRef]
83. Xiao, Z.; Wang, J.; Cao, Y.; Yao, T.; Wang, S.; Liu, J.; Suo, D.; Tian, J.; Jia, Z.; Li, Y.; et al. Quick and high-throughput quantification of 22 beta-agonists residues in animal-derived foods using enzymatic probe sonication. *Food Chem.* **2023**, *408*, 135262. [CrossRef]

84. Zhang, Q.; Du, H.; Zhang, Y. Recent progress on the detection of animal-derived food stimulants using mass spectrometry-based techniques. *Front. Nutr.* **2023**, *10*, 1226530. [CrossRef] [PubMed]
85. Hajrulai-Musliu, Z.; Uzunov, R.; Jovanov, S.; Musliu, D.; Dimitrieska-Stojkovikj, E.; Stojanovska-Dimzoska, B.; Angeleska, A.; Stojkovski, V.; Sasanya, J.J. Multi-class/residue method for determination of veterinary drug residues, mycotoxins and pesticide in urine using LC-MS/MS technique. *BMC Vet. Res.* **2023**, *19*, 156. [CrossRef]
86. Li, G.; Li, H.; Zhai, J.; Guo, J.; Li, Q.; Wang, C.F.; Chen, S. Microfluidic fluorescent platform for rapid and visual detection of veterinary drugs. *RSC Adv.* **2022**, *12*, 8485–8491. [CrossRef]
87. Mondello, L.; Cordero, C.; Janssen, H.-G.; Synovec, R.E.; Zoccali, M.; Tranchida, P.Q. Comprehensive two-dimensional gas chromatography–mass spectrometry. *Nat. Rev. Methods Primers* **2025**, *5*, 7. [CrossRef]
88. Brettell, T.A.; Lum, B.J. Analysis of drugs of abuse by gas chromatography-mass spectrometry (gc-ms). *Methods Mol. Biol.* **2018**, *1810*, 29–42. [CrossRef]
89. Gruber, B.; David, F.; Sandra, P. Capillary gas chromatography-mass spectrometry: Current trends and perspectives. *TrAC Trends Anal. Chem.* **2020**, *124*, 115475. [CrossRef]
90. Welke, J.E.; Hernandes, K.C.; Lago, L.O.; Silveira, R.D.; Marques, A.T.B.; Zini, C.A. Flavoromic analysis of wines using gas chromatography, mass spectrometry and sensory techniques. *J. Chromatogr. A* **2024**, *1734*, 465264. [CrossRef]
91. Liu, S.; Tang, Y.; Lu, Y.; Guo, Y.; Xie, K.; Guan, F.; Gao, P.; Zhu, Y.; Dong, Y.; Zhang, T.; et al. Qualitative and quantitative determination of decoquinate in chicken tissues by gas chromatography tandem mass spectrometry. *Molecules* **2023**, *28*, 3875. [CrossRef]
92. Wang, B.; Wang, Y.; Xie, X.; Diao, Z.; Xie, K.; Zhang, G.; Zhang, T.; Dai, G. Quantitative analysis of spectinomycin and lincomycin in poultry eggs by accelerated solvent extraction coupled with gas chromatography tandem mass spectrometry. *Foods* **2020**, *9*, 651. [CrossRef]
93. Guo, Y.; Xie, X.; Diao, Z.; Wang, Y.; Wang, B.; Xie, K.; Wang, X.; Zhang, P. Detection and determination of spectinomycin and lincomycin in poultry muscles and pork by ASE-SPE-GC–MS/MS. *J. Food Compos. Anal.* **2021**, *101*, 103979. [CrossRef]
94. Yu, H.; Tao, Y.; Le, T.; Chen, D.; Ishsan, A.; Liu, Y.; Wang, Y.; Yuan, Z. Simultaneous determination of amitraz and its metabolite residue in food animal tissues by gas chromatography-electron capture detector and gas chromatography-mass spectrometry with accelerated solvent extraction. *J. Chromatogr. B Analyt. Technol. Biomed. Life Sci.* **2010**, *878*, 1746–1752. [CrossRef] [PubMed]
95. Zhu, Z.-J.; Schultz, A.W.; Wang, J.; Johnson, C.H.; Yannone, S.M.; Patti, G.J.; Siuzdak, G. Liquid chromatography quadrupole time-of-flight mass spectrometry characterization of metabolites guided by the metlin database. *Nat. Protoc.* **2013**, *8*, 451–460. [CrossRef] [PubMed]
96. Glish, G.L.; Burinsky, D.J. Hybrid mass spectrometers for tandem mass spectrometry. *J. Am. Soc. Mass Spectrom.* **2008**, *19*, 161–172. [CrossRef]
97. Chernushevich, I.V.; Loboda, A.V.; Thomson, B.A. An introduction to quadrupole-time-of-flight mass spectrometry. *J. Mass Spectrom. JMS* **2001**, *36*, 849–865. [CrossRef]
98. Allen, D.R.; McWhinney, B.C. Quadrupole Time-of-Flight Mass Spectrometry: A paradigm shift in toxicology screening applications. *Clin. Biochem. Rev.* **2019**, *40*, 135–146. [CrossRef] [PubMed]
99. Cui, X.; Du, M.; Wei, K.; Dai, C.; Yang, R.Y.H.; Zhou, B.; Luo, Z.; Yang, X.; Yu, Y.; Lin, W.; et al. Study of Xuanhuang Pill in protecting against alcohol liver disease using ultra-performance liquid chromatography/time-of-flight mass spectrometry and network pharmacology. *Front. Endocrinol.* **2023**, *14*, 1175985. [CrossRef]
100. Zhang, H.; Chang, Q.; Li, J.; Qiu, G.; Wu, F.; Zhu, R.; Wang, X.; Su, M. A liquid chromatography-time-of-flight/mass spectrometry method for analysis of pesticides and transfer behavior in Radix Codonopsis and Angelica sinensis decoctions. *Anal. Methods* **2023**, *15*, 2121–2131. [CrossRef]
101. Li, X.; Chi, Q.; Xia, S.; Pan, Y.; Chen, Y.; Wang, K. Untargeted multi-residue method for the simultaneous determination of 141 veterinary drugs and their metabolites in pork by high-performance liquid chromatography time-of-flight mass spectrometry. *J. Chromatogr. A* **2020**, *1634*, 461671. [CrossRef] [PubMed]
102. Vardali, S.C.; Samanidou, V.F.; Kotzamanis, Y.P. Development and validation of an ultra performance liquid chromatography-quadrupole time of flight-mass spectrometry (in MS(E) mode) method for the quantitative determination of 20 antimicrobial residues in edible muscle tissue of European sea bass. *J. Chromatogr. A* **2018**, *1575*, 40–48. [CrossRef]
103. Choi, B.S.; Lee, D.U.; Kim, W.S.; Park, C.W.; Choe, W.J.; Moon, M.J. Simultaneous screening of 322 residual pesticides in fruits and vegetables using gc-ms/ms and deterministic health risk assessments. *Foods* **2023**, *12*, 3001. [CrossRef]
104. Petrarca, M.H.; Braga, P.A.C.; Reyes, F.G.R.; Bragotto, A.P.A. Exploring miniaturized sample preparation approaches combined with LC-QToF-MS for the analysis of sulfonamide antibiotic residues in meat- and/or egg-based baby foods. *Food Chem.* **2022**, *366*, 130587. [CrossRef] [PubMed]
105. Stachniuk, A.; Kozub, A.; Czeczko, R.; Montowska, M.; Fornal, E. LC-QTOF-MS evaluation of rabbit-specific peptide markers for meat quantitation. *J. Food Drug Anal.* **2022**, *30*, 223–233. [CrossRef] [PubMed]

106. Yaseen, S.B.; Akram, A.; Musharraf, S.G.; Wajidi, M.; Tabassum, N.; Nazir, N.; Shah, S.M.Z. The lc-qtof-ms/ms analysis of acid degradation products of rifaximin, an antibiotic. *MethodsX* **2022**, *9*, 101735. [CrossRef]
107. Zhang, M.; Li, E.; Su, Y.; Zhang, Y.; Xie, J.; He, L. Quick Multi-Class Determination of Residues of Antimicrobial Veterinary Drugs in Animal Muscle by LC-MS/MS. *Molecules* **2018**, *23*, 1736. [CrossRef] [PubMed]
108. Hayashi, M.; Kanda, M.; Yoshikawa, S.; Nakajima, T.; Hayashi, H.; Matsushima, Y.; Ohba, Y.; Koike, H.; Nagano, C.; Otsuka, K.; et al. Single-laboratory Validation Study and Surveillance Using an Improved Multiresidue Analytical Method for Veterinary Drugs in Livestock Products by lc-ms/ms. *Shokuhin Eiseigaku Zasshi* **2023**, *64*, 53–60. [CrossRef]
109. Zhang, S.; Yang, Z.; Yu, H.; Chen, Y.; Yuan, F.; Zhang, X.; Fang, S. Furazolidone and nitrofurazone metabolic studies in crucian carp by ultra-performance liquid chromatography Tandem Mass Spectrometry. *J. Chromatogr. Sci.* **2022**, *60*, 963–969. [CrossRef]
110. Cheng, Y.H.; Lee, C.H.; Wang, S.Y.; Chou, C.Y.; Yang, Y.J.; Kao, C.C.; Wu, H.Y.; Dong, Y.; Hung, W.Y.; Su, C.Y.; et al. Multiplexed antibody glycosylation profiling using dual enzyme digestion and liquid chromatography-triple quadrupole mass spectrometry method. *Mol. Cell Proteom.* **2024**, *23*, 100710. [CrossRef]
111. Jung, Y.S.; Kim, D.B.; Nam, T.G.; Seo, D.; Yoo, M. Identification and quantification of multi-class veterinary drugs and their metabolites in beef using LC-MS/MS. *Food Chem.* **2022**, *382*, 132313. [CrossRef]
112. Lehotay, S.J. Comparison of analyte identification criteria and other aspects in triple quadrupole tandem mass spectrometry: Case study using UHPLC-MS/MS for regulatory analysis of veterinary drug residues in liquid and powdered eggs. *Anal. Bioanal. Chem.* **2022**, *414*, 287–302. [CrossRef]
113. Na, T.W.; Seo, H.J.; Jang, S.N.; Kim, H.; Yun, H.; Kim, H.; Ahn, J.; Cho, H.; Hong, S.H.; Kim, H.J.; et al. Multi-residue analytical method for detecting pesticides, veterinary drugs, and mycotoxins in feed using liquid- and gas chromatography coupled with mass spectrometry. *J. Chromatogr. A* **2022**, *1676*, 463257. [CrossRef]
114. Wittenberg, J.B.; Simon, K.A.; Wong, J.W. Targeted multiresidue analysis of veterinary drugs in milk-based powders using liquid chromatography-tandem mass spectrometry (lc-ms/ms). *J. Agric. Food Chem.* **2017**, *65*, 7288–7293. [CrossRef] [PubMed]
115. Desmarchelier, A.; Bessaire, T.; Savoy, M.C.; Tarres, A.; Mujahid, C.; Beck, A.; Mottier, P.; Delatour, T. Screening of 154 veterinary drug residues in foods of animal origin using lc-ms/ms: First action 2020.04. *J. AOAC Int.* **2021**, *104*, 650–681. [CrossRef]
116. Jang, S.; Seo, H.; Kim, H.; Kim, H.; Ahn, J.; Cho, H.; Hong, S.; Lee, S.; Na, T. Development of a quantitative method for detection of multiclass veterinary drugs in feed using modified quppe extraction and lc-ms/ms. *Molecules* **2022**, *27*, 4483. [CrossRef] [PubMed]
117. Khadim, A.; Yaseen Jeelani, S.U.; Khan, M.N.; Kumari, S.; Raza, A.; Ali, A.; Zareena, B.; Zaki Shah, S.M.; Musharraf, S.G. Targeted analysis of veterinary drugs in food samples by developing a high-resolution tandem mass spectral library. *J. Agric. Food Chem.* **2023**, *71*, 12839–12848. [CrossRef] [PubMed]
118. Al-Shakliah, N.S.; Kadi, A.A.; Aljohar, H.I.; AlRabiah, H.; Attwa, M.W. Profiling of in vivo, in vitro and reactive zorifertinib metabolites using liquid chromatography ion trap mass spectrometry. *RSC Adv.* **2022**, *12*, 20991–21003. [CrossRef]
119. Berendsen, B.J.A.; Meijer, T.; Mol, H.G.J.; van Ginkel, L.; Nielen, M.W.F. A global inter-laboratory study to assess acquisition modes for multi-compound confirmatory analysis of veterinary drugs using liquid chromatography coupled to triple quadrupole, time of flight and orbitrap mass spectrometry. *Anal. Chim. Acta* **2017**, *962*, 60–72. [CrossRef]
120. Guo, Y.; Mao, R.; Zhang, Y.; Li, R.; Oduro, P.K.; Si, D.; Han, L.; Huang, Y.; Pan, G. An integrated strategy for the systematic chemical characterization of Salvianolate lyophilized injection using four scan modes based on the ultra-high performance liquid chromatography-triple quadrupole-linear ion trap mass spectrometry. *J. Pharm. Biomed. Anal.* **2022**, *215*, 114769. [CrossRef]
121. Chen, G.Y.; Zhang, Q. Quantification of plasma oxylipins using solid-phase extraction and reversed-phase liquid chromatography-triple quadrupole mass spectrometry. *Methods Mol. Biol.* **2021**, *2306*, 171–186. [CrossRef]
122. Kang, J.; Park, S.J.; Park, H.C.; Hossain, M.A.; Kim, M.A.; Son, S.W.; Lim, C.M.; Kim, T.W.; Cho, B.H. Multiresidue screening of veterinary drugs in meat, milk, egg, and fish using liquid chromatography coupled with ion trap time-of-flight mass spectrometry. *Appl. Biochem. Biotechnol.* **2017**, *182*, 635–652. [CrossRef]
123. Casado, N.; Morante-Zarcero, S.; Perez-Quintanilla, D.; Sierra, I. Application of a hybrid ordered mesoporous silica as sorbent for solid-phase multi-residue extraction of veterinary drugs in meat by ultra-high-performance liquid chromatography coupled to ion-trap tandem mass spectrometry. *J. Chromatogr. A* **2016**, *1459*, 24–37. [CrossRef]
124. Sabei, F.Y.; Khardali, I.; Al-Kasim, M.A.; Shaheen, E.S.; Oraiby, M.; Alamir, A.; David, B.; Alshahrani, S.; Jali, A.M.; Attafi, M.; et al. Disposition kinetics of cathinone and its metabolites after oral administration in rats. *Curr. Drug Metab.* **2024**, *25*, 220–226. [CrossRef] [PubMed]
125. Sun, Q.; Wang, B.; Xu, L.; Xiong, L.; Guo, D.; Cui, W.; Xu, X.; Lin, Y. Quality consistency evaluation on three species of North Patrininae herba by high-performance liquid chromatography coupled with electrospray ion trap time-of-flight mass spectrometry and network pharmacology approaches. *J. Sep. Sci.* **2022**, *45*, 3852–3865. [CrossRef]
126. Chen, Z.; Zhu, X.; Geng, Y.; Dai, J.; Tang, S.; Adams, E.; Chen, D.; Yuan, Y. Analysis of impurity profiling of arbekacin sulfate by ion-pair liquid chromatography coupled with pulsed electrochemical detection and online ion suppressor-ion trap-time off light mass spectrometry. *J. Pharm. Biomed. Anal.* **2022**, *221*, 115061. [CrossRef]

127. Saade, J.; Biacchi, M.; Giorgetti, J.; Lechner, A.; Beck, A.; Leize-Wagner, E.; Francois, Y.N. Analysis of monoclonal antibody glycopeptides by capillary electrophoresis-mass spectrometry coupling (ce-ms). *Methods Mol. Biol.* **2021**, *2271*, 97–106. [CrossRef] [PubMed]
128. Ferre, S.; Boccard, J.; Rudaz, S.; Gonzalez-Ruiz, V. Evaluation of prototype ce-ms interfaces. *Methods Mol. Biol.* **2022**, *2531*, 1–13. [CrossRef]
129. Latosinska, A.; Siwy, J.; Mischak, H.; Frantzi, M. Peptidomics and proteomics based on ce-ms as a robust tool in clinical application: The past, the present, and the future. *Electrophoresis* **2019**, *40*, 2294–2308. [CrossRef]
130. Fang, P.; Pan, J.Z.; Fang, Q. A robust and extendable sheath flow interface with minimal dead volume for coupling ce with esi-ms. *Talanta* **2018**, *180*, 376–382. [CrossRef]
131. Sanchez-Lopez, E.; Marina, M.L. Applications of capillary electrophoresis-mass spectrometry to chiral analysis. *Methods Mol. Biol.* **2022**, *2531*, 211–225. [CrossRef]
132. Blasco, C.; Pico, Y.; Andreu, V. Analytical method for simultaneous determination of pesticide and veterinary drug residues in milk by CE-MS. *Electrophoresis* **2009**, *30*, 1698–1707. [CrossRef]
133. Seyfinejad, B.; Jouyban, A. Capillary electrophoresis-mass spectrometry in pharmaceutical and biomedical analyses. *J. Pharm. Biomed. Anal.* **2022**, *221*, 115059. [CrossRef]
134. Shamsi, S.A.; Akter, F. Capillary electrophoresis mass spectrometry: Developments and applications for enantioselective analysis from 2011–2020. *Molecules* **2022**, *27*, 4126. [CrossRef] [PubMed]
135. Li, H.; Lu, Y.; Zhang, L.; Qin, L.; Wen, H.; Fan, X.; Peng, D. Highly sensitive magnetic-nanoparticle-based immunochromatography assay for rapid detection of amantadine in chicken and eggs. *Biosensors* **2023**, *14*, 23. [CrossRef] [PubMed]
136. Zhang, L.; Yin, X.; Yang, H.; Wen, H.; Han, S.; Pan, X.; Li, H.; Peng, D. A sensitive and specific monoclonal antibody based enzyme-linked immunosorbent assay for the rapid detection of pretilachlor in grains and the environment. *Foods* **2023**, *13*, 12. [CrossRef]
137. Lee, M.J.; Aliya, S.; Lee, E.S.; Jeon, T.J.; Oh, M.H.; Huh, Y.S. Simple and rapid detection of ractopamine in pork with comparison of LSPR and LFIA sensors. *J. Food Drug Anal.* **2022**, *30*, 590–602. [CrossRef]
138. Ahmed, S.; Ning, J.; Peng, D.; Chen, T.; Ahmad, I.; Ali, A.; Lei, Z.; Abu bakr Shabbir, M.; Cheng, G.; Yuan, Z. Current advances in immunoassays for the detection of antibiotics residues: A review. *Food Agric. Immunol.* **2020**, *31*, 268–290. [CrossRef]
139. Calidonio, J.M.; Hamad-Schifferli, K. Biophysical and biochemical insights in the design of immunoassays. *Biochim. Biophys. Acta Gen. Subj.* **2023**, *1867*, 130266. [CrossRef]
140. Sun, B.; Chen, Z.; Feng, B.; Chen, S.; Feng, S.; Wang, Q.; Niu, X.; Zhang, Z.; Zheng, P.; Lin, M.; et al. Development of a colloidal gold-based immunochromatographic assay for rapid detection of nasal mucosal secretory IgA against SARS-CoV-2. *Front. Microbiol.* **2024**, *15*, 1386891. [CrossRef]
141. Ziyang, H.; Shaoen, Z.; Qingzhou, C.; Hong, L.; Jianxin, S.; Kaiqiang, W.; Huiying, W.; Limin, C. Novel solid-phase extraction promotes simultaneous colloidal gold immunochromatographic assay of malachite green, leuco-malachite green, chloramphenicol, and semi-carbazone metabolites in aquatic products. *Food Chem.* **2025**, *465*, 142118. [CrossRef] [PubMed]
142. Na, G.; Hu, X.; Yang, J.; Sun, Y.; Kwee, S.; Tang, L.; Xing, G.; Xing, Y.; Zhang, G. A rapid colloidal gold-based immunochromatographic strip assay for monitoring nitroxynil in milk. *J. Sci. Food Agric.* **2020**, *100*, 1860–1866. [CrossRef]
143. Chen, L.; Sun, Y.; Hu, X.; Xing, Y.; Kwee, S.; Na, G.; Zhang, G. Colloidal gold-based immunochromatographic strip assay for the rapid detection of diminazene in milk. *Food Addit. Contam. Part A Chem. Anal. Control Expo. Risk Assess.* **2020**, *37*, 1667–1677. [CrossRef]
144. Lu, Y.; Wang, S.; Cai, X.; Cao, M.; Lu, Q.; Hu, D.; Chen, Q.; Xiong, X. A rapid colloidal gold immunochromatographic assay based on polyclonal antibodies against HtpsC protein for the detection of Streptococcus suis. *Front. Microbiol.* **2023**, *14*, 1294368. [CrossRef] [PubMed]
145. Na, G.; Hu, X.; Yang, J.; Sun, Y.; Kwee, S.; Tang, L.; Xing, G.; Xing, Y.; Zhang, G. Colloidal gold-based immunochromatographic strip assay for the rapid detection of bacitracin zinc in milk. *Food Chem.* **2020**, *327*, 126879. [CrossRef] [PubMed]
146. Liu, F.; Cao, Y.; Yan, M.; Sun, M.; Zhang, Q.; Wang, J.; Fu, G.; Liu, R.; Huang, Y.; Su, J. Development of a colloidal gold immunochromatographic assay for duck enteritis virus detection using monoclonal antibodies. *Pathogens* **2021**, *10*, 365. [CrossRef] [PubMed]
147. Shen, J.; Qian, C.; Jiang, H.; Yang, H. Development of an enzyme-linked immunosorbent assay for the determination of maduramicin in broiler chicken tissues. *J. Agric. Food Chem.* **2001**, *49*, 2697–2701. [CrossRef]
148. Chen, R.; Shang, H.; Niu, X.; Huang, J.; Miao, Y.; Sha, Z.; Qin, L.; Huang, H.; Peng, D.; Zhu, R. Establishment and evaluation of an indirect ELISA for detection of antibodies to goat *Klebsiella pneumonia*. *BMC Vet. Res.* **2021**, *17*, 107. [CrossRef]
149. Qin, L.; Xiao, J.; Yang, H.; Liang, J.; Li, L.; Wu, S.; Peng, D. Rapid immunoassays for the detection of quinoxalines and their metabolites residues in animal-derived foods: A review. *Food Chem.* **2024**, *443*, 138539. [CrossRef]

150. Wernike, K.; Aebischer, A.; Michelitsch, A.; Hoffmann, D.; Freuling, C.; Balkema-Buschmann, A.; Graaf, A.; Muller, T.; Osterrieder, N.; Rissmann, M.; et al. Multi-species ELISA for the detection of antibodies against SARS-CoV-2 in animals. *Transbound. Emerg. Dis.* **2021**, *68*, 1779–1785. [CrossRef]
151. Fu, B.; Gao, H.; Fang, C.; Cheng, G.; Wang, H.; Wang, Y.; Hao, H.; Wang, X.; Huang, L.; Peng, D. Development of a monoclonal antibody-based indirect competitive enzyme-linked immunosorbent assay for screening of diethylstilbestrol in animal-derived foods. *Heliyon* **2024**, *10*, e39769. [CrossRef]
152. Yin, X.; Li, H.; Wu, S.; Lu, Y.; Yang, Y.; Qin, L.; Li, L.; Xiao, J.; Liang, J.; Si, Y.; et al. A sensitive and specific enzyme-linked immunosorbent assay for the detection of pymetrozine in vegetable, cereal, and meat. *Food Chem.* **2023**, *418*, 135949. [CrossRef]
153. Wang, W.; Zhang, Y.; Yang, H. Development of a nucleocapsid protein-based blocking elisa for the detection of porcine deltacoronavirus antibodies. *Viruses* **2022**, *14*, 1815. [CrossRef]
154. He, D.; Sun, M.; Jiang, X.; Zhang, S.; Wei, F.; Wu, B.; Diao, Y.; Tang, Y. Development of an indirect competitive ELISA method based on ORF2 detecting the antibodies of novel goose astrovirus. *J. Virol. Methods* **2023**, *311*, 114643. [CrossRef] [PubMed]
155. Xu, J.; Wang, J.; Li, Z.; He, X.; Zhao, S.; Ma, Q.; Li, X.; Liu, J.; Liu, A.; Li, Y.; et al. A universal ELISA assay for detecting six strains of ovine Babesia species in China. *Vet. Parasitol.* **2021**, *300*, 109616. [CrossRef]
156. Guo, L.; Liu, M.; Li, Q.; Dong, B.; Li, H.; Mari, G.M.; Liu, R.; Yu, W.; Yu, X.; Wang, Z.; et al. Synthesis and characterization of tracers and development of a fluorescence polarization immunoassay for amantadine with high sensitivity in chicken. *J. Food Sci.* **2021**, *86*, 4754–4767. [CrossRef] [PubMed]
157. He, S.; Liang, D.; Xiong, J.; Wang, Z.; Zheng, P.; Zhang, H.; Ren, Z.; Jiang, H. Development of a sensitive and rapid fluorescence polarization immunoassay for high throughput screening eight glucocorticoids in beef. *J. Pharm. Biomed. Anal.* **2022**, *214*, 114719. [CrossRef]
158. Mukhametova, L.I.; Eremin, S.A.; Arutyunyan, D.A.; Goryainova, O.S.; Ivanova, T.I.; Tillib, S.V. Fluorescence Polarization Immunoassay of Human Lactoferrin in Milk Using Small Single-Domain Antibodies. *Biochemistry* **2022**, *87*, 1679–1688. [CrossRef] [PubMed]
159. Hendrickson, O.D.; Mukhametova, L.I.; Zvereva, E.A.; Zherdev, A.V.; Eremin, S.A. A sensitive fluorescence polarization immunoassay for the rapid detection of okadaic acid in environmental waters. *Biosensors* **2023**, *13*, 477. [CrossRef]
160. Smith, D.S.; Eremin, S.A. Fluorescence polarization immunoassays and related methods for simple, high-throughput screening of small molecules. *Anal. Bioanal. Chem.* **2008**, *391*, 1499–1507. [CrossRef]
161. Mukhametova, L.I.; Eremin, S.A. Fluorescence polarization assays for organic compounds in food safety. *Front. Biosci. (Elite Ed.)* **2024**, *16*, 4. [CrossRef]
162. Duan, C.; Zhang, H.; Zhang, Y.; Li, Q.; Li, P.; Mari, G.M.; Eremin, S.A.; Shen, J.; Wang, Z. A robust homogeneous fluorescence polarization immunoassay for rapid determination of erythromycin in milk. *Foods* **2023**, *12*, 1581. [CrossRef]
163. Duan, C.; Zhang, Y.; Li, P.; Li, Q.; Yu, W.; Wen, K.; Eremin, S.A.; Shen, J.; Yu, X.; Wang, Z. Dual-wavelength fluorescence polarization immunoassay for simultaneous detection of sulfonamides and antibacterial synergists in milk. *Biosensors* **2022**, *12*, 1053. [CrossRef]
164. Zhang, Q.; Zou, M.; Wang, W.; Li, J.; Liang, X. Design, synthesis, and characterization of tracers and development of a fluorescence polarization immunoassay for rapid screening of 4,4′-dinitrocarbanilide in chicken muscle. *Foods* **2021**, *10*, 1822. [CrossRef] [PubMed]
165. Zhou, L.; Yang, J.; Tao, Z.; Eremin, S.A.; Hua, X.; Wang, M. Development of fluorescence polarization immunoassay for imidacloprid in environmental and agricultural samples. *Front. Chem.* **2020**, *8*, 615594. [CrossRef]
166. Chowdhury, J.; Ranc, V.; Sarkar, S. Editorial: Surface enhanced raman scattering: Theory and applications. *Front. Chem.* **2023**, *11*, 1176495. [CrossRef]
167. Liu, H.L.; Zhan, K.; Wang, K.; Xia, X.H. Nanopore-based surface-enhanced Raman scattering technologies. *Sci. Bull.* **2022**, *67*, 1539–1541. [CrossRef]
168. Tanwar, S.; Kim, J.H.; Bulte, J.W.M.; Barman, I. Surface-enhanced Raman scattering: An emerging tool for sensing cellular function. *Wiley Interdiscip. Rev. Nanomed. Nanobiotechnol.* **2022**, *14*, e1802. [CrossRef]
169. Dhakal, S.; Chao, K.; Huang, Q.; Kim, M.; Schmidt, W.; Qin, J.; Broadhurst, C.L. A simple surface-enhanced raman spectroscopic method for on-site screening of tetracycline residue in whole milk. *Sensors* **2018**, *18*, 424. [CrossRef]
170. Fan, R.; Tang, S.; Luo, S.; Liu, H.; Zhang, W.; Yang, C.; He, L.; Chen, Y. Duplex surface enhanced raman scattering-based lateral flow immunosensor for the low-level detection of antibiotic residues in milk. *Molecules* **2020**, *25*, 5249. [CrossRef]
171. Zhao, Y.; Xu, Y.; Jing, X.; Ma, W. SERS-active plasmonic metal NP-CsPbX(3) films for multiple veterinary drug residues detection. *Food Chem.* **2023**, *412*, 135420. [CrossRef]
172. Xu, M.L.; Gao, Y.; Han, X.X.; Zhao, B. Innovative Application of sers in food quality and safety: A brief review of recent trends. *Foods* **2022**, *11*, 5249. [CrossRef]
173. Wang, L.; Ma, P.; Chen, H.; Chang, M.; Lu, P.; Chen, N.; Yuan, Y.; Chen, N.; Zhang, X. Rapid determination of mixed pesticide residues on apple surfaces by surface-enhanced raman spectroscopy. *Foods* **2022**, *11*, 1089. [CrossRef]

174. Lee, K.M.; Yarbrough, D.; Kozman, M.M.; Herrman, T.J.; Park, J.; Wang, R.; Kurouski, D. A rapid and convenient screening method for detection of restricted monensin, decoquinate, and lasalocid in animal feed by applying SERS and chemometrics. *Food Chem. Toxicol.* **2020**, *144*, 111633. [CrossRef] [PubMed]
175. Zhang, N.; Lv, H.; Wang, J.; Yang, Z.; Ding, Y.; Zhao, B.; Tian, Y. An aptamer-based colorimetric/SERS dual-mode sensing strategy for the detection of sulfadimethoxine residues in animal-derived foods. *Anal. Methods* **2023**, *15*, 1047–1053. [CrossRef]
176. Xu, X.; Zhao, S.; Xie, Y.; Zhang, L.; Shao, Y.; Lin, J.; Wu, A. Advances in SERS detection method combined with microfluidic technology for bio-analytical applications. *Spectrochim. Acta A Mol. Biomol. Spectrosc.* **2025**, *332*, 125797. [CrossRef] [PubMed]
177. Vanaja, P.; Moorthy, N.; Rajpoot, V.S.; Rao, H.; Goswami, R.K.; Subash, P.; Khute, S.; Rao, K.S. Metabolite profiling, antimalarial potentials of Schleichera oleosa using LC-MS and GC-MS: In vitro, molecular docking and molecular dynamics. *Front. Mol. Biosci.* **2025**, *12*, 1543939. [CrossRef]
178. Johnson, K.R.; Gao, Y.; Gregus, M.; Ivanov, A.R. On-capillary cell lysis enables top-down proteomic analysis of single mammalian cells by ce-ms/ms. *Anal. Chem.* **2022**, *94*, 14358–14367. [CrossRef]
179. Tiseo, K.; Huber, L.; Gilbert, M.; Robinson, T.P.; Van Boeckel, T.P. Global Trends in Antimicrobial Use in Food Animals from 2017 to 2030. *Antibiotics* **2020**, *9*, 918. [CrossRef]
180. Rapacz, D.; Smolinska-Kempisty, K.; Wolska, J. Preparation and characterization of SPE column with smart green molecularly imprinted polymers materials for selective determination of S-metolachlor herbicide. *Sci. Rep.* **2025**, *15*, 3153. [CrossRef]
181. Zad, N.; Tell, L.A.; Ampadi Ramachandran, R.; Xu, X.; Riviere, J.E.; Baynes, R.; Lin, Z.; Maunsell, F.; Davis, J.; Jaberi-Douraki, M. Development of machine learning algorithms to estimate maximum residue limits for veterinary medicines. *Food Chem. Toxicol.* **2023**, *179*, 113920. [CrossRef]
182. Wu, P.Y.; Chou, W.C.; Wu, X.; Kamineni, V.N.; Kuchimanchi, Y.; Tell, L.A.; Maunsell, F.P.; Lin, Z. Development of machine learning-based quantitative structure-activity relationship models for predicting plasma half-lives of drugs in six common food animal species. *Toxicol. Sci.* **2025**, *203*, 52–66. [CrossRef]
183. Kaufmann, A.; Butcher, P.; Maden, K.; Walker, S.; Widmer, M. Improving the QuEChERS liquid/liquid extraction of analytes with widely varying physicochemical properties: Example of 201 veterinary drugs in milk. *J. AOAC Int.* **2022**, *105*, 1030–1042. [CrossRef]
184. Fodor, A.; Cseppento, D.C.N.; Badea, G.E.; Petrehele, A.I.; Groze, A.; Tit, D.M.; Bungau, S.G. Colloidal gold immunochromatography and elisa traceability of tetracycline residues from raw milk to its dairy products. *In Vivo* **2023**, *37*, 1619–1627. [CrossRef] [PubMed]
185. Chang, Q.; Ge, L.; Li, J.; Qiu, G.; Wu, F.; Zhang, H.; Xu, F.; Zhu, R.; Qi, P.; Bai, R.; et al. Automated QuEChERS for the determination of 482 pesticide residues in Radix codonopsis by GC-Q-TOF/MS and LC-Q-TOF/MS. *Anal. Methods* **2021**, *13*, 5660–5669. [CrossRef] [PubMed]
186. Zhou, Z.; Zhang, S.; Chen, J.; Luo, W.; Kang, F.; Ren, Y.; Zhou, W. Development and application of a new QuEChERS method coupled with uplc-qtof-ms/ms for analysis of tiafenacil and its photolysis products in water. *J. Agric. Food Chem.* **2024**, *72*, 27007–27018. [CrossRef] [PubMed]
187. Zhang, Y.; Tang, H.; Chen, W.; Zhang, J. Nanomaterials used in fluorescence polarization based biosensors. *Int. J. Mol. Sci.* **2022**, *23*, 8625. [CrossRef]
188. Park, J.; Kim, Y.W.; Jeon, H.J. Machine learning-driven innovations in microfluidics. *Biosensors* **2024**, *14*, 613. [CrossRef]
189. Zhang, D.; He, S.; Wang, Z.; Yan, P.; Li, H.; Xu, X.; Zeng, Q.; Wang, N.; Chen, X. Recent advances in CRISPR/Cas systems-associated surface enhanced Raman spectroscopy biosensor towards nucleic acid detection. *Med. Nov. Technol. Devices* **2024**, *24*, 100336. [CrossRef]

International Journal of
Molecular Sciences

MDPI

Review

The Intestinal Macrophage–Intestinal Stem Cell Axis in Inflammatory Bowel Diseases: From Pathogenesis to Therapy

Tao Quan, Ran Li and Ting Gao *

College of Veterinary Medicine, China Agricultural University, Beijing 100083, China; s20243051072@cau.edu.cn (T.Q.); 18246269672@163.com (R.L.)
* Correspondence: gt1994@cau.edu.cn

Abstract: The gut plays a crucial role in digestion and immunity, so its balance is essential to overall health. This balance relies on dynamic interactions between intestinal epithelial cells, immune cells, and crypt stem cells. Inflammatory bowel disease (IBD), which consists of ulcerative colitis and Crohn's disease, is a chronic relapsing inflammatory disease of the gastrointestinal tract closely related to immune dysfunction. Stem cells, known for their ability to self-renew and differentiate, play an important role in repairing damaged intestinal epithelium and maintaining homeostasis in vivo. Macrophages are key gatekeepers of intestinal immune homeostasis and have a significant impact on IBD. Current research has focused on the link between epithelial cells and stem cells, but interactions with macrophages, which have been recognized as attractive targets for the development of new therapeutic approaches to disease, have been less explored. Recently, the developing field of immunometabolism has reinforced that metabolic reprogramming is a key determinant of macrophage function and subsequent disease progression. The aim of this review is to explore the role of the macrophage–stem cell axis in the maintenance of intestinal homeostasis and to summarize potential approaches to treating IBD by manipulating the cellular metabolism of macrophages, as well as the main opportunities and challenges faced. In summary, our overview provides a framework for understanding the critical role of macrophage immunometabolism in maintaining gut health and potential therapeutic targets.

Keywords: intestinal stem cells; inflammatory bowel disease; macrophage; immunometabolism

1. Introduction

The intestinal tract is a large and complex organ that is responsible for the digestion and absorption of nutrients, elimination of wastes, and immunomodulation. Coordination between multiple cells, including intestinal epithelial cells, immune cells, and stem cells, plays a key role in regulating homeostasis in the gut. Inflammatory bowel disease (IBD), including ulcerative colitis (UC) and Crohn's disease (CD), is a chronic and recurrent multifactorial disease that originates within the gastrointestinal tract. Currently, IBD has become a major healthcare burden globally and is also recognized as a major pathological risk factor for the induction of colorectal cancer. Therefore, better parsing the pathophysiologic basis and molecular or biochemical processes of IBD may provide new perspectives for our early diagnosis and discovery of appropriate therapeutic options [1].

Macrophages are first-line cells of innate immunity, located in the gut lamina propria and mesenchymal tissues, and are responsible for early inflammation onset and late inflammation relief. In the last decade or so, with the burgeoning field of immunometabolism, and the mechanisms of interactions between immunity and metabolism, researchers have emphasized that cellular metabolism has emerged as a key factor in the activation of macrophages and the regulation of their function [2,3]. Notably, dysregulation of macrophage immunometabolism is not only recognized as a key determinant of intestinal diseases such as IBD but also plays an important role in a variety of diseases including cancer, cardiovascular disease, atherosclerosis, and lipid metabolism imbalance. Therefore, targeting macrophage immune metabolism with small molecules and metabolic intermediates may provide new perspectives for the treatment of IBD [4–6]; however, the metabolic profile of macrophages in the context of IBD has not been fully revealed.

LGR5+ stem cells are located in the basal layer of the intestine and possess a strong capacity for self-renewal and differentiation. Intestinal stem cells maintain homeostasis in the gut by continuously generating intestinal epithelial cells and a variety of other cell types to replenish cell death caused by normal or abnormal factors in the gut. It has been shown that a complex network of signaling pathways, including BMP, Wnt/β-linker protein, Notch, and Hedgehog, facilitates communication between crypt stem cells and epithelial cells that work together to maintain intestinal tissue homeostasis [7]. However, these do not fully elucidate the regulatory mechanisms that maintain stem cell homeostasis.

In this review, we generalize and summarize the role of the macrophage–stem cell axis in regulating intestinal homeostasis and the interactions between these two to elucidate their roles in maintaining the intestinal mucosal barrier, modulating immune responses and promoting tissue repair. In addition, we provide a framework to understand the metabolic profile of macrophages in IBD. We also highlight the salient features of macrophage immune metabolic pathways in IBD and summarize the potentially effective therapeutic options proposed for targeting macrophage immune metabolism in IBD, as well as the opportunities and challenges for future development. Our goal is to provide new insights and strategies for the prevention and treatment of gut-related diseases.

2. Intestinal Homeostasis and Inflammatory Bowel Disease

The intestinal tract is anatomically divided into the small intestine, which includes the duodenum, jejunum, and ileum, and the large intestine, which includes the cecum, colon, and rectum, and these segments of the intestinal tract collectively perform the roles of nutrient digestion and absorption, waste elimination, and immune regulation. In addition, the intestinal tract is also the first barrier for the host to play a defense role. The small intestine relies on the intestinal epithelium to digest and absorb nutrients, while the intestinal epithelium can also be regarded as a mucosal barrier against the invasion of toxic and harmful substances from the outside world. Below the epithelium is a thin layer of connective tissue known as the lamina propria, which plays a major role in intestinal immunity. Various types of cells in the intestine, immune cells (including macrophages, dendritic cells, T cells and B cells, etc.) and intestinal epithelial cells (IEC) are independent of each other and closely linked, and together they maintain intestinal homeostasis [8,9]. In addition, intestinal health is also closely related to the overall health, nutritional status, intracirculatory homeostasis, and psychological status of the body [10,11]. Therefore, an imbalance in intestinal homeostasis cannot induce a variety of intestinal diseases, such as IBD (CD and UC), diarrhea, intestinal cancer, etc., and can easily lead to problems in other tissues and organs of the organism [12].

IBD, including CD and UC, is one of the most common intestinal diseases, and has been referred to as the "green cancer" due to its unknown etiology and difficulty in treating the disease [13]. IBD, which is typically characterized by diarrhea, abdominal pain, and intestinal bleeding, is a pathological process caused by damage to the intestinal mucosal barrier, an imbalance of intestinal microbiota, and an immune disorder [14,15].

IBD is a widespread disease with a global prevalence, especially in Western countries, and its incidence is increasing [16]. Although the predisposing factors and pathologic process of IBD are not clear, genetic factors, dietary habits, and environmental factors are believed to play important roles in the pathogenesis of IBD [17,18]. Among them, intestinal immune dysfunction is regarded as one of the important factors that induce IBD. Specifically, intestinal immune cells, including macrophages, dendritic cells, T cells, and B cells, can maintain intracellular homeostasis by secreting a variety of cytokines [19], and alterations in these cytokines can induce intestinal inflammation, which is closely related to the pathogenesis of IBD [20,21]. In addition, intestinal flora can provide key signaling molecules during the development and functioning of immune functions. Therefore, intestinal flora and their metabolites are not only involved in the regulation of intestinal immune processes but can also influence the susceptibility of the host to multiple immune-mediated diseases, including IBD [22,23], despite the gut microbiota–immunity–IBD axis and its complexity. The relationship between the triad of gut microbes, host immunity, and IBD has been discussed and summarized in some detail in several refereed journals.

New research suggests that improving the inflammatory response may be a new and effective target for treating patients with IBD [24–26]. Data consistent with this study show that a biologic agent, an inhibitor of tumor necrosis factor (TNF)-α, effectively accelerated the rate of mucosal healing in patients with IBD, signaling great progress in the treatment of IBD [19]. However, there are some patients who do not respond significantly to such agents, which means that new therapies need to be sought to repair mucosal damage in IBD patients. Earlier studies have suggested that transplantation or infusion of mesenchymal stem cells (MSCs) and hematopoietic stem cells (HSCs) into IBD patients can have beneficial effects, and a series of trials have been conducted to validate the effectiveness of this therapy [27,28]. Not only that, but recent clinical data also suggest that allogeneic bone marrow MSCs are still capable of treating patients with celiac disease in the cavities [29]. Thus, whether stem cells can be a good alternative treatment for patients with IBD remains controversial, but great potential exists.

3. How Immune Cells Modulate Intestinal Stem Cells to Ameliorate Intestinal Inflammation

It is well known that intestinal stem cells (ISC) have a strong proliferative and differentiation capacity, which allows them to continuously replenish dead and detached intestinal epithelial cells and to generate organoids by in vitro culture [30]. The resulting organoids are structurally very similar to intestinal epithelial cells, with intestinal villi and crypts and contain a variety of cell types from the intestinal epithelium [31]. Transplantation of this intestinal organoid to mucosa suffering from inflammation is expected to accelerate healing of the injured mucosa [32]. In addition, there are also data suggesting that organoids cultured from intestinal stem cells extracted from fetal intestine or adult small intestine can also be successfully transplanted to damaged sites in the colon, but they are slightly less able to adapt to the environment [33]. Recent studies have confirmed that human intestinal organoids accelerate healing of damaged mucosa in immunodeficient mice [34]. Therefore, we believe that restoration of intestinal stem cell activity may provide force an effective therapeutic approach to re-establish the mucosal barrier in IBD and, thus, improve the clinical symptoms of IBD.

Multiple mucosal surfaces of the organism, such as the respiratory and gastrointestinal tracts, are in constant contact with each other and with a variety of commensal microorganisms and pathogens from inside and outside the body. In the intestine, a dense monolayer of epithelial cells bound by tight junction proteins prevents these toxic and harmful substances from entering the host. As a result of the continuous attack by harmful substances, this epithelial cell needs to renew itself continuously, approximately every 5–7 days [35].

However, when the renewal process of intestinal epithelial cells is impeded, it can lead to the development of a variety of intestinal diseases, such as IBD and/or colon/rectal cancer.

The differentiation of intestinal stem cells into different daughter intestinal epithelial cells is in a dynamic equilibrium process. During this process, various signaling pathways coordinate intercellular communication and work together to maintain intestinal homeostasis. Dysregulation of these pathways also predisposes to intestinal diseases. Currently, a large number of studies have highlighted several key signaling pathways by which intestinal stem cells promote repair and proliferation of intestinal epithelial cells, including the BMP signaling pathway, Wnt/β-catenin signaling, Notch signaling pathway, Hedgehog signaling pathway, Hippo-YAP signaling pathway, and EGF signaling pathway, but these studies have not been able to elaborate the specific mechanisms of regulating intestinal stem cells [36].

Previous data have demonstrated the importance of immune function in maintaining homeostasis in the gut [37]. New studies have found that immune cells and intestinal epithelial cells communicate with each other despite their relative independence, and that normal communication between these two is a key mediator in maintaining intestinal function [11,38,39]. Not only that, but immune cells are also involved in the regeneration process of IEC [11,40,41]. Increasing evidence has found that immune cells play an important role in the self-renewal of ISC. The in vitro intestinal stem cell-immune cell co-culture system provides effective evidence to argue the validity of this view.

Therefore, this section summarizes the important roles of multiple immune cells, including T cells, B cells, and macrophages, in regulating the ecological niche of ISC and their pathways of action.

3.1. T Cells

T cells present in the intestinal lamina propria are not only involved in intestinal immune function but have also emerged as effective targets for the treatment of immune dysregulation-mediated intestinal injury. Recent data suggest that both $CD4^+$ and $CD8^+$ T cells have the potential to mediate ISC luminal injury in graft-versus-host disease models. Following bone marrow transplantation (BMT)-mediated gastrointestinal tract injury, diseased T cells identified the ISC ecotone as the site of intestinal origin in three-dimensional imaging examinations [42]. In addition, recruitment of pathogenic T cells to the basal region of the crypt resulted in the loss of ISC expressing both major histocompatibility complex class (MHC) I and II [42]. In a similar report, another study explored changes in the ISC ecological niche using a T-cell-mediated in vivo model of graft-versus-host disease (GVHD) and in vitro intestinal organoid-T-cell co-cultures [43]. This study demonstrated that both ISC and Paneth cell numbers per crypt were significantly reduced in mice treated with bone marrow or allogeneic T cells. In contrast, these mice had significantly increased densities of both lamina propria- and intraepithelial-associated T cells in the crypt region [43].

Recently, researchers developed methods for co-culturing gut-like organs with T cells in human and mouse models to reveal the molecular mechanisms underlying the loss of ISC activity in IBD. It was found that mouse heterologous reactive T cells significantly reduced the number of intestinal-like organs, whereas human-derived $CD8^+$ cytotoxic T cells significantly reduced the efficiency of intestinal-like organs. Meanwhile, in the absence of Pan cells, IFN-γ secreted by $CD4^+$ T cells induced apoptosis in ISC by activating the JAK/STAT signaling pathway [41]. Overall, these studies have not only expanded our understanding of immune cell–intestinal stem cell communication, but based on this, have also suggested that T cells could be an effective target for ameliorating IBD by acting on

ISC. In future studies, considering that T cells may act on ISC by secreting IFN-γ, inhibition of the JAK/STAT signaling pathway could be considered as a new ameliorative pathway for effective treatment of IBD.

Meanwhile, recent studies have utilized fetal intestinal organoids to explore the interactions between epithelial and immune cells during fetal development [44]. Recent single-cell RNA sequencing data identified a specific group of CD4$^+$ effector memory T cells that secrete large amounts of TNF-α [45]. On this basis, the investigators used a co-culture model of organoids from fetal intestinal cultures with CD4$^+$ Tem cells to explore the effect of TNF-α on intestinal epithelial cells. The results showed that a small number of CD4$^+$Tem cells from fetal intestines mediated TNF-α to promote the proliferation of intestinal stem cells; conversely, a large number of CD4$^+$Tem cells from fetal intestines disrupted the function of intestinal stem cells [45]. The latter result is consistent with observations in infants with necrotizing small bowel colitis [45].

Interaction between T helper (TH) cells and intestinal stem cells can help restore intestinal stem cell activity and rebuild the intestinal epithelial barrier system [40]. The current study utilized single-cell RNA sequencing to identify a subpopulation of Lgr5+ ISC that can highly express MHC II molecules. In vitro culturing of intestinal organoids revealed that MHC II-expressing intestinal stem cells have the ability to present antigens in the functioning of intestinal epithelial cells. Moreover, MHC II mediated the contact between TH cells and intestinal stem cells, which in turn induced the activation of TH cells in an antigen-presenting manner [38]. Additional studies have also observed the effects on intestinal stem cells after co-culture of intestinal organoids with different TH cell subsets or after specific labeling with cytokines to better reveal the interactions between intestinal stem cells and TH cells. It was found that the addition of TH2 cells or IL-14 with IL-13 cytokines depleted the intestinal stem cell pool. Consistent with this, TH1 and TH17 cells or their secreted cytokines IFN-γ and IL-17 significantly reduced the number of intestinal stem cells but upregulated the number of transit-expanding cells. In contrast, regulatory T (Treg) cells or IL-10 upregulated the number of intestinal stem cells [40]. Another study found that co-culturing human T cells with ISC promoted the maturation of human pluripotent stem cell-derived intestinal organoids (hIOs) in vitro, suggesting that IL-2 promotes the growth of human T cells by activating the STAT3 signaling pathway [46].

Together, the above findings suggest that interactions between intestinal stem cells and T cells are involved in regulating intestinal stem cell fate. Notably, intestinal stem cells can sense T cells via MHC II molecules and, thus, activate T cells. However, more studies are needed in the future to further reveal the relationship between intestinal immunity and the intestinal epithelial barrier.

3.2. *Innate Lymphoid Cells*

Innate lymphoid cells (ILCs) are regarded as important regulators of intestinal mucosal immunity and play an important role in the maintenance of intestinal homeostasis and the integrity of the intestinal mucosal barrier [47]. ILCs can be categorized into three major groups, namely, ILC1, ILC2, and ILC3, based on the cytokines they produce, their phenotype, and the developmental pathway [48]. It has been suggested that the cytokines secreted by ILCs are closely related to intestinal stem cell activity and function. Among them, IL-22 has been shown to be involved in regulating the maintenance of intestinal stem cell activity and differentiation and to protect against DNA damage [49,50]. Several studies have demonstrated that IL-22 promotes intestinal epithelial repair in DSS-induced colitis [51–53]. However, the exact mechanism of how IL-22 is involved in regulating the fate of intestinal stem cells remains to be further explored, although we already know that IL-22 receptors exist in intestinal stem cells [54,55]. In addition, IL-22 is closely associated with the repair of intestinal epithelial damage in GVHD and BMT, in which allogeneic T cells can inhibit ISC function [55,56]. Mechanistically, IL-22 can activate the STAT3 signaling pathway in Lgr5+ intestinal stem cells independently of PAN cells to maintain intestinal stem cell activity and promote their differentiation [56]. IL-22 not only enhances the survival of intestinal stem cells but also their proliferative viability, the latter of which is critical for the regeneration of damaged epithelial cells [55]. Together, the above studies suggest that the IL-22–STAT3 axis may be one of the effective targets for the treatment of IBD. However, a recent study seems to contradict this conclusion [57]. This study showed that in the colonic epithelium, IL-22 induces endoplasmic reticulum stress response with transcriptional programs. Specifically, high expression of the IL-22-responsive transcriptional module and endoplasmic reticulum (ER) stress response was found in CD patients. In contrast, blocking IL-22 alleviated colitis by inhibiting epithelial ER stress [57]. The above studies suggest that IL-22 acts as a double-edged sword in the pathogenesis of chronic intestinal inflammation.

Environmental genotoxic silver can also lead to mutations in intestinal stem cells and will furthermore induce cancer [58,59]. As an evolutionarily conserved response pathway in ISC, DNA damage response (DDR) can be involved in maintaining genome integrity at the cellular level [60]. Recent studies have suggested that IL-22, produced by ILC3 and gδ T cells in the intestinal mucosa, could be regarded as a potent regulator of the DDR pathway in colonic stem cells (CSCs) [61]. Specifically, after DNA damage, IL-22 can overexpress the IL-22 receptor, which in turn effectively activates the DDR in CSCs. Alternatively, when IL-22 signaling is inhibited, intestinal stem cells exposed to carcinogens show significant mutations, which can exacerbate colon cancer development and progression [61]. In addition, a previous study explored the protective effect of Lactobacillus royale on intestinal mucosal integrity by establishing a co-culture system of intestinal organoids and lamina propria lymphocytes, and the results demonstrated that the Lactobacillus royale metabolite indole-3-aldehyde furthered enteric stem cell-mediated regeneration of intestinal epithelium by stimulating the secretion of IL-22 by lamina propria lymphocytes [50].

Group 2 innate lymphoid-like cells (ILC2) are also involved in the regulation of intestinal stem cell fate. Among them, IL-13 secreted by ILC2 has been shown to promote the differentiation of cup cells in intestinal organoid models. Also, IL-33 secreted by intestinal epithelial cells promotes IL-13 secretion by ILC2 [62]. In addition, it has been shown that in the intestine, IL-4 and IL-13 can be involved in the differentiation of tufted cells that regulate type II immune responses [63–65]. It is interesting to know about the relationship between the tufted cell-mediated intestinal immune system and epithelial cells. For example, tufted cells may act as immune sentinels and respond accordingly to parasites present in the gut. In addition, succinic acid from helminths induces IL-25 secretion by tufted cells [65,66], whereas IL-22 recruits ILC2 in the intestinal mucosa and secretes large amounts of IL-4 with IL-13, inflammatory factors that, in turn, clear parasites by promoting mucus production by cuprocytes [65]. Although these studies suggest that IL-4 and IL-13 secreted by ILC2 can induce more cell differentiation, including cup and tuft cells, the interactions between ILC2 and intestinal stem cells and their specific molecular mechanisms still need to be further explored.

Currently, a range of anti-inflammatory drugs (corticosteroids and aminosalicylates), immunosuppressive agents (azathioprine, methotrexate, and mercaptopurine), and biologic drugs (anti-TNFα drugs such as uterumab and vedolizumab; and the interleukin-12/23 antagonist uterumab) have been developed for the treatment of intestinal inflammation, targeting these above immune cells [17,67]. However, there are some limitations of these drugs that restrict their general use, such as limited efficacy, severe side effects, and unresponsiveness to immunosuppressive agents [17,67]. Over the past decades, targeting macrophages has been recognized as one of the measures that can effectively treat intestinal inflammation [4].

4. The Role of Macrophages in IBD

IBD is a major case of intestinal homeostatic imbalance and further pathological changes. Activation of the immune system with increased inflammatory response is an important hallmark of IBD, and macrophages are thought to play an important role in this process [68].

Furthermore, intestinal macrophages are involved in the regulation of intestinal stem cell renewal, homeostasis maintenance and differentiation. Studies have shown that macrophage depletion significantly reduces the number of intestinal stem cells and impairs the differentiation process of intestinal stem cells. Unlike macrophages in other tissues, intestinal macrophages are not only abundant but also play the first line of defense in the intestinal tract, and, therefore, intestinal macrophages have distinct anti-inflammatory characteristics and exhibit high phagocytic and bactericidal activity and high tolerance to foreign substances [69,70].

4.1. Origin, Phenotype and Function of Intestinal Macrophages

Macrophages are derived from circulating monocytes in the blood or from a pool of resident macrophages in the gut and are one of the important immune cells against foreign pathogens. Results of in vitro studies targeting monocyte-derived macrophages have shown that macrophages can be categorized into classically activated M1 macrophages and selectively activated M2 macrophages [71–73]. Although the use of M1 versus M2 macrophages can effectively differentiate macrophage response types, it does not fully summarize all phenotypes of macrophages in vivo [74–76]. Nevertheless, this categorization still contributes to our understanding of the metabolic programming of different macrophage functions. Therefore, in the description of this review, we refer to those that stimulate inflammation/tissue injury/hinder wound healing and trigger Th1/Th2 immune responses as M1 macrophages, and those that antagonize inflammation/promote tissue repair and promote Th2 immune responses as M2 macrophages (Figure 1).

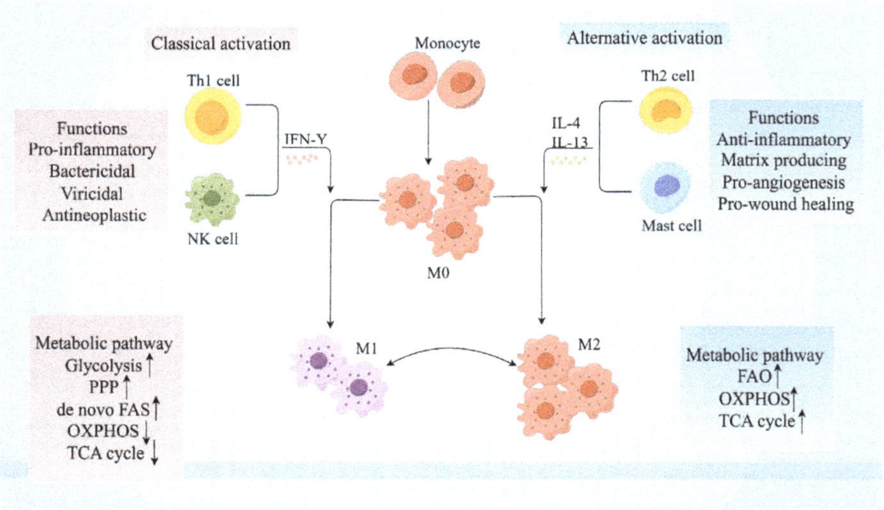

Figure 1. Characteristics of polarized macrophages M0 macrophages derived from monocyte polarizes into M1 macrophages with IFN-γ secreted from Th1 cell and NK cell stimuli. M1 macrophages

perform pro-inflammatory, bactericidal, viricidal, and antineoplastic activity. In M1 macrophages, glycolysis, PPP, and de novo FAS are upregulated, and OXPHOS and the TCA cycle are downregulated. M0 macrophages polarizes= into M2 macrophages with IL-4, IL-10, and IL-13 secreted from Th2 cell and mast cell. M2 macrophages preform anti-inflammatory, matrix producing, pro-angiogenesis and pro-wound healing function. In M2 macrophages, FAO, OXPHOS, and the TCA cycle are upregulated. In special cases, M1 can be polarized toward M2. But whether M2 can be polarized to M1 is still debated. M0 macrophages: unactuated macrophages; M1 macrophage: classically activated macrophages; M2 macrophages: alternatively activated macrophages; IFN: interferon; IL: interleukin; PPP: pentose phosphate pathway; OXPHOS: oxidative phosphorylation; FAS: fatty acid synthesis; FAO: fatty acid oxidation; TCA: tricarboxylic acid cycle; the dots under the arrows are inflammatory factors; A rising arrow indicates a strengthening of the pathway and a falling arrow indicates a weakening of the pathway.

4.2. *Differentiation of Macrophages During IBD*

In the maintenance of organismal homeostasis, monocytes derived from mouse circulating Ly6Chi and human CD14hi are continuously recruited into the intestine and further differentiate into mature CX3CR1hiF4/80+ macrophages to complement resident macrophages. These blood-borne macrophages exhibit an M2-like macrophage phenotype due to the high expression of CD206 and CD163, which can exert anti-inflammatory effects by secreting the anti-inflammatory factors IL-10 and TGF-β and decreasing the response to TLR [77]. In addition, this cell class is involved in the maintenance of intestinal homeostasis by effectively clearing apoptotic epithelial cells and translocated bacteria [78]. In addition, M2 macrophages maintain the integrity of the intestinal epithelial barrier by promoting IL-10 secretion from $CD4^{+}CD25^{+}$ Treg [79], which in turn promotes intestinal stem cell renewal [80,81].

Cellular metabolism can be involved in macrophage activation and function by generating energy and building blocks required for cell maintenance and proliferation [82,83]. The major metabolic pathways involved in macrophage immune metabolism include glycolysis, tricarboxylic acid cycle (TCA), pentose-phosphate pathway (PPP), fatty acid oxidation (FAO), fatty acid synthesis, and amino acid metabolism [82–84]. Notably, macrophage subpopulations present different metabolic and energy requirements, as well as the flexibility to respond to different environments and cytokines [85] (Figure 2).

Specifically, the metabolism of M1 macrophages activated by lipopolysaccharide is characterized by an impaired TCA and mitochondrial oxidative phosphorylation (OXPHOS), in contrast to enhanced glycolytic pathways and increased PPP flux [82,86–88]. These metabolic pathways rapidly produce ATP and accumulate large amounts of immunomodulatory molecules, including lactate, citrate, and succinate [82]. In contrast, IL-4-activated M2 macrophages maintain an intact TCA cycle and full OXPHOS [89–91]. Whereas amino acid metabolic pathways are also involved in macrophage polarization, e.g., glutamine is required for IL-1β secretion by M1 macrophages [88,92], M2 macrophage polarization requires the glutamine-UDP-N-acetylglucosamine pathway and α-ketoglutarate produced by glutamine catabolism for maintenance [92,93]. In addition, arginine is required for the substrate of iNOS and arginase in M1 versus M2 macrophages, respectively [90,94]. Importantly, macrophage polarization in vivo is also stimulated by a variety of nutrients, but since the supply of nutrients is very limited, macrophages also need to compete with other cells for nutrients [95]. Therefore, when nutrient metabolism is altered, macrophages must promptly adjust their immune response to meet the needs of the organism. In the next section, we will generalize and summarize how perturbations in intracellular nutrient metabolism affect macrophage polarization and function during the pathological process of IBD.

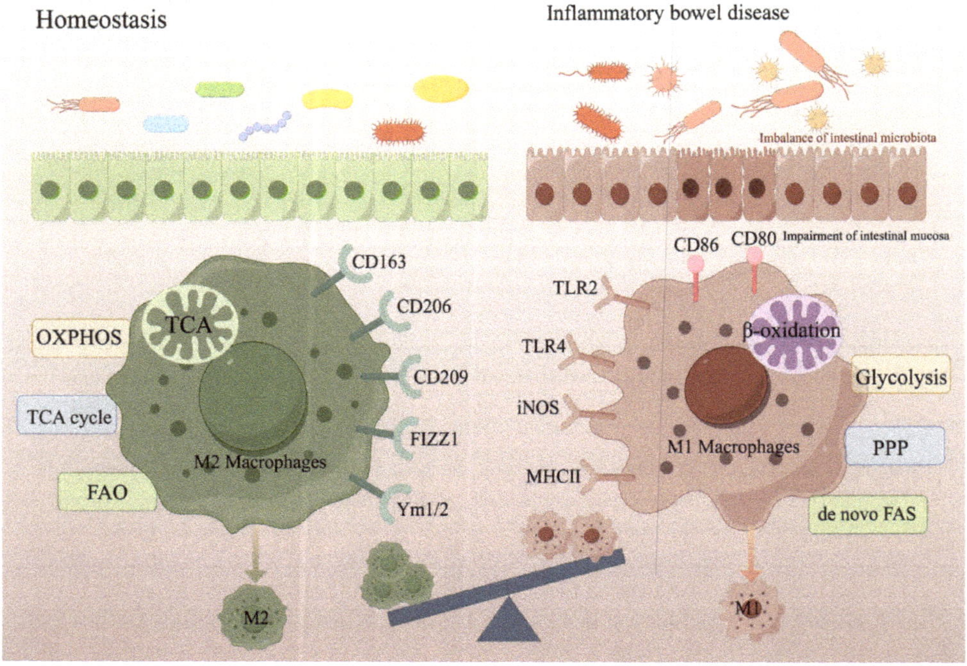

Figure 2. Macrophage status in the healthy gut versus under IBD. In the context of intestinal homeostasis, the intestinal mucosal barrier is intact, and the intestinal flora is balanced. In the lamina propria, M0 macrophages are predominantly polarized to M2 macrophages. Its surface markers include CD163, CD206, CD209, FIZZ1, and Ym1/2, and the metabolic mode of M2 macrophages is dominated by OXPHOS, the TCA cycle, and FAO. On the background of inflammatory bowel disease, the intestinal mucosal barrier is damaged and the intestinal flora is imbalanced. In the lamina propria, it is dominated by M1 macrophages, whose surface markers include CD86, CD80, TLR2, TLR4, iNOS, and MHC II, and whose metabolism is dominated by glycolysis, PPP and de novo FAS; CD163, CD206, CD209, FIZZ1, Ym1/2: Marker proteins of M2 macrophages; CD80, CD86, TLR2, TLR4, iNOS, MHCII: Marker proteins of M1 macrophages.

5. How Macrophages Regulate Intestinal Stem Cells to Maintain Intestinal Homeostasis

The successful establishment of an intestinal organoid and macrophage co-culture system confirmed the involvement of macrophages in the functional regulation of ISC. Macrophages rely on a variety of cytokines secreted by themselves to regulate intestinal stem cell function, including IL-6, IL-8, IFN-γ, and TGFβ-1, which are mediated by macrophages to promote the maturation of intestinal epithelial cells and the establishment of the mucosal barrier [96]. In the healthy intestine, macrophages are distributed at the crypts. Hyaluronic acid in the extracellular matrix can promote proliferation of intestinal stem cells by binding to TLR4 receptors on macrophage membranes first, which in turn promotes macrophage secretion of prostaglandin E2 (PGE2), which subsequently promotes proliferation of intestinal stem cells by activating EGFR signaling in intestinal stem cells, ultimately lengthening the length of the early intestine. When macrophages were depleted with chlorophosphate, the proliferative vigor of intestinal stem cells was significantly reduced, but no increase in the number of SOX9+ reserve stem cells was observed [97]. In contrast, the results of another study showed a significant decrease in the number and proliferative capacity of Lgr5+ intestinal stem cells but a significant increase in the proliferative viability of SOX9++ Bmil+ reserve stem cells after depletion of macrophages by long-term administration of CSF1R antibody in mice. This may be due to the difference in age between neonatal and adult mice.

Macrophages are also involved in the functional regulation of intestinal stem cells in the MSCs population. MSCs can ameliorate IBD by replacing damaged cells by homing to damaged tissues and differentiating into functional cells [98]. A study in mice showed that oral administration of *Lactobacillus rhamnosus* was protective of the gut and that *Lactobacillus rhamnosus* GG (LGG) was dependent on TLR2 and COX2 for its protective effect. It was associated with a shift in COX-2+ MSCs from the lamina propria of the villi to the lamina propria near crypt epithelial stem cells. This shift attenuated radiation-induced intestinal damage and accelerated the proliferation of intestinal stem cells [99].

Using a mouse model of IBD, macrophages were found to accelerate intestinal mucosal barrier repair by activating the Wnt signaling pathway [100]. Recent data further suggest that colony-stimulating factor (CSF1)-dependent macrophages in the intestinal wall can be used to maintain the ecological niche of intestinal stem cells in the small intestine [97], a result that suggests that CSF1 is expected to be a potent target for improving the intestinal epithelial barrier after inflammatory or chemotherapeutic injury. In addition, the researchers established a human intestinal-like-macrophage co-culture model to explore the interactions between the organism and intestinal pathogens [96]. The results of this study suggest that macrophages may play a role in promoting intestinal epithelial maturation and rebuilding the intestinal mucosal barrier by mediating their secretion of various cytokines such as IL-6, IL-8, IFNg, and TGFβ1 [96]. These studies have greatly expanded our understanding of the immune cell–intestinal stem cell interactions, but the reality is far more complex than we can comprehend.

In this review, we draw a framework for a better understanding of the metabolic signature of macrophages in IBD and summarize and generalize how this metabolic reprogramming process is triggered by microenvironmental factors, including signals originating from the host cell as well as sensing by microorganisms and their metabolites. We also outline the salient features of macrophage immune metabolism in IBD and summarize the therapeutic tools developed to target macrophage immune metabolism in IBD. Finally, we have also attempted to present the current problems and future developable directions in the field of studying macrophage immune metabolism and IBD.

In IBD, the process of M2-like macrophage polarization is inhibited, whereas the process of M1-like macrophage polarization is promoted [70,101]. A variety of inflammatory factors secreted by M1 macrophages, such as IL-12, IL-23, and IL-1β, mediate the inflammatory response triggered by Th1 and Th17, which further contribute to intestinal epithelial injury [102,103]. In particular, although M1 macrophages play a key role in the pathological colonization of colitis, they can still exert anti-inflammatory effects by promoting M2 macrophage polarization thereby accelerating the healing of mucosal injury [77,80]. Notably, it has been found that bone marrow-derived M2 macrophages significantly ameliorate colitis in mice by transplantation or intraperitoneal injection [104–106]. Since the balance between M2-M2 macrophages plays a critical role in intestinal health, unraveling the regulatory mechanisms of macrophage polarization will provide effective protocols for developing targeted macrophages to ameliorate intestinal inflammation (Figure 3).

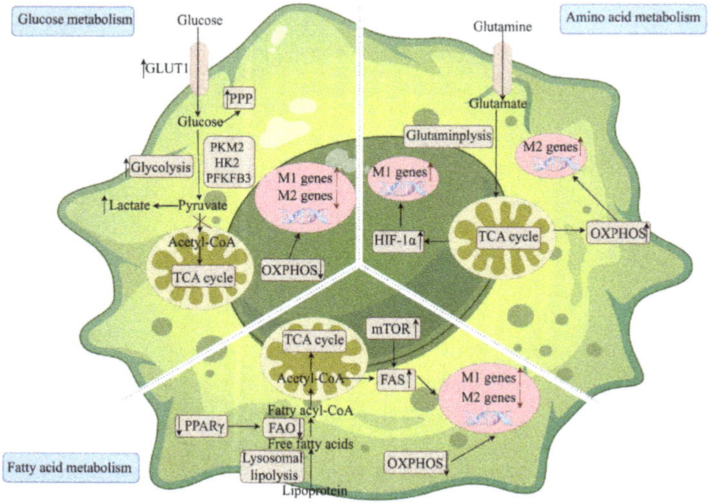

Figure 3. Metabolic reprogramming of macrophages in IBD. During IBD, three types of metabolism involved in macrophages, including glucose metabolism, amino acid metabolism, and fatty acid metabolism, are altered. In glucose metabolism, the entry of pyruvate into mitochondria is reduced and pyruvate is mainly converted to lactate, leading to a truncated TCA cycle, further promoting M1 macrophage-associated gene expression and inhibits M2 macrophage-associated gene expression. In fatty acid metabolism, the anti-inflammatory PPAR-γ and FAO are downregulated during IBD, which causes impaired OXPHOS and M2 macrophage polarization; supplementation of unsaturated fatty acids such as conjugated linoleic acid prevents IBD by targeting at PPAR-γ. In amino acid metabolism, glutamine is another key metabolic fuel for macrophages, and glutamine can enhance M2 polarization by fueling TCA cycle activity and OXPHOS or M1 polarization by activating succinate-HIF-1α pathway. FAO, fatty acid oxidization; FAS, fatty acid synthesis; GLUT1, glucose transporter 1; HIF-1α, hypoxia-inducible factor-1 alpha; HK2, hexokinase II; mTOR, mammalian target of rapamycin complex; OXPHOS, oxidative phosphorylation; PFKFB3, 6-phosphofructo-2-kinase/fructose-2, 6-bisphosphatase; PKM2, pyruvate kinase 2; PPAR-γ, peroxisome proliferator–activated receptor gamma; TCA, tricarboxylic acid; the dots under the arrows are inflammatory factors; A rising arrow indicates a strengthening of the pathway and a falling arrow indicates a weakening of the pathway.

5.1. Macrophage-Derived Factors Could Regulate Stem Cell Fate

Macrophages play a critical role in mammalian development and exhibit important homeostatic activity in virtually all organ systems of the body through the production of growth factors and other mediators that provide nutritional support to the tissues in which they reside [107]. Resident tissue macrophages are important drivers of inflammatory and tissue regenerative responses to infection, autoimmune disease, mechanical or toxic injury, and a variety of other causes of organ damage [108]. Following tissue injury, the basement membrane is further disrupted by macrophage-derived matrix metalloproteinases, which help to promote the movement of inflammatory cells toward the site of tissue injury. They also produce chemokines that direct the initial recruitment of inflammatory cells [109]. Macrophages can also act as scavenger cells, phagocytizing cellular debris, invasive biocytes, neutrophils, and other apoptotic cells present after tissue injury [110,111]. Thus, if macrophage recruitment or activation is disrupted after tissue injury, the early inflammatory response tends to be reduced [112]. However, impaired macrophage recruitment and activation may also reduce wound debridement and prolong exposure to pro-inflammatory stimuli, leading to incomplete tissue regeneration [113]. Thus, the repair response of macrophages after tissue injury must be well coordinated.

In addition to supporting the initial inflammatory response following injury, local tissue macrophages respond to stimuli found in the local environment and rapidly transition to a wound repair phenotype. Wound repair macrophages are characterized by enhanced production of multiple growth factors, including platelet-derived (PDGF), insulin-like (IGF-1), and vascular endothelial (VEGF-α) growth factors, which promote cellular proliferation and vascular development, collectively contributing to the relief of local hypoxia that occurs after injury [114–118]. They also produce soluble mediators, such as TGF-β1, which stimulate local and recruited tissue fibroblasts to differentiate into myofibroblasts, promoting wound contraction and closure, as well as synthesis of extracellular matrix components [119]. Proliferation and expansion of many neighboring parenchymal and stromal cells are also enhanced by macrophages, which may activate stem cells and local progenitor cell populations involved in the tissue repair response if the injury is severe or chronic [120].

During the final stages of tissue repair, macrophages exhibit an anti-inflammatory or pro-catabolic phenotype [121]. These macrophages preferentially secrete anti-inflammatory mediators such as IL-10 and TGF-β1 and express cell surface receptors such as PD-L2 to suppress the immune system and ameliorate inflammation, which, if not effectively controlled, leads to collateral cell death and prolongs or significantly impairs the repair process [122–125]. In some tissues, they may also replace or supplement the resident tissue macrophage population. Thus, the three major phases of tissue repair must be carefully regulated, with macrophages playing unique and critical roles at each stage [108]. Indeed, interruption of macrophage recruitment or activation at any point in the process may result in inefficient or incomplete repair, with uncontrolled production of macrophages by inflammatory mediators and growth factors and defective wound healing or macrophages both leading to an inefficient repair response. In some cases, dysregulation of the wound repair response may lead to the development of pathologic fibrosis or scar formation, which can further compromise normal tissue function and ultimately lead to organ failure and death [126]. Although tissue repair is orchestrated by multiple cell types since macrophages are involved in all stages of the wound repair response [71], they have emerged as potentially important therapeutic targets [127,128].

5.2. Macrophage Glucose Metabolism and IBD

During the pathologic process of IBD, one of the key pathological changes in IBD occurs due to an imbalance between oxygen supply and demand in the intestinal microenvironment, which further leads to severe hypoxia at the site of inflammation [129]. In this hypoxic environment, in order to meet the high energy demands of the gut, macrophages undergo a shift in the metabolic pathway from oxidative phosphorylation to aerobic glycolysis. This transition is mediated by hypoxia-inducible factor 1α (HIF-1α) and key glycolytic proteins including glucose transporter 1 (GLUT1), hexokinase II, PKM2, and 6-phosphofructo-2-kinase/fructose-2,6-bisphosphatase (PFKFB3) [89,130]. The HIF-1α-mediated glycolytic metabolic pathway plays a key role in the pro-inflammatory effects exerted by macrophages [131]. In addition, activation of HIF-1α in myeloid cells exacerbates the manifestation of colitis [132], whereas specific knockdown of the gene in myeloid cells significantly ameliorates colitis [133]. In addition, the glycolytic reprogramming process can be regulated through the formation of PKM2-HIF-1α complexes in the nucleus [134]. Recent findings showed that PKM2 de-adhesion of SIRT5 significantly reduced the levels of IL-1β produced by macrophages, which in turn ameliorated dextran sulfate sodium salt (DSS)-induced colitis in mice [130], suggesting that PKM2 could be an effective target for the treatment of IBD. Collectively, these findings explain that metabolic reorganization to aerobic glycolysis is a key feature that determines macrophage function and subsequent IBD progression.

In addition to enhanced glycolysis, M1 macrophages exhibit interrupted tricarboxylic acid cycle pathways, which can further lead to a high accumulation of succinate and citrate [86,88]. Succinate is not only an intermediate product of the TCA but also a receptor ligand for succinate receptor 1 (SUCNR1) [135]. It is important to note that IBD patients have significantly higher fecal and serum concentrations of succinate [136,137], and their intestinal expression of SUCNR1 is also significantly increased [138]. Importantly, large amounts of succinic acid promote the stabilization of intracellular HIF-1α, which in turn enhances glycolytic metabolism in M1 macrophages and ultimately induces the secretion of the pro-inflammatory factor IL-1β [88,139]. Studies on colitis have demonstrated that resting peritoneal macrophages from SUCNR1-deficient mice secrete less pro-inflammatory factors IL-1β, IL-6 and TNF-α and, thus, exert a protective effect against colitis. The second breakpoint in the TCA cycle occurs at isocitrate dehydrogenase, which leads to the accumulation of intracellular citrate, which subsequently moves from the mitochondria into the cytoplasm, where it is then cleaved by ATP-citrate lyase [140] This export and catabolism of mitochondrial citrate is essential for the secretion of prostaglandins, NO, and ROS in pro-inflammatory macrophages [140,141]. In addition, metabolomics analyses have mentioned that the levels of lactate, pyruvate, and citrate are all significantly increased in the serum of mice with spontaneous colitis [142], suggesting that the glycolytic process is enhanced with the blockage of the TCA cycling process, which is tightly linked to the pathogenesis of IBD. Therefore, succinate and citrate are considered as key metabolites that differentiate between IBD patients and healthy individuals.

5.3. Macrophage Fatty Acid Metabolism and IBD

Studies have also identified changes in lipid metabolism that are also present in IBD patients, which in turn could lead to the use of inflammatory lipids as a potential biomarker for IBD [143,144], where peroxisome proliferator-activated receptor gamma (PPAR-γ) could serve as a key conjugate of lipids with innate immunity [145,146]. Studies have shown that the activation of PPAR-γ in macrophages significantly ameliorates experimental colitis in mice [147], whereas specific knockdown of the PPAR-γ gene in macrophages exacerbates colitis pathology [148]. This result could partially explain the PPAR-γ-dependent lipid metabolism in macrophages due to the fact that activation of PPAR-γ could ultimately contribute to the metabolic reorganization of the anti-inflammatory phenotype of macrophages through upregulation of Arg1 expression, promotion of β-oxidized fatty acids, and mitochondrial biogenesis processes [149]. However, inactivation of the mTOR-Semaphorin 6D-PPAR-γ signaling pathway can lead to impaired fatty acid uptake and FAO processes, which in turn induces an imbalance in macrophage polarization and worsens the pathological changes in colitis [149]. Consistent with these findings, the utilization of dietary or probiotic-derived conjugated linoleic acid can further ameliorate IBD in both humans and mice through the activation of PPAR-γ [150]. Thus, together, these findings suggest that lipid metabolism is involved in macrophage polarization process and PPAR-γ can be considered as one of the effective targets for the treatment of IBD.

5.4. Macrophage Amino Acid Metabolism and IBD

In addition to protein building blocks, some amino acids are also involved in immunoregulation during macrophage metabolism. Not only that, disorders of amino acid metabolism have been observed in some IBD patients, and the degree of disorders is positively correlated with the severity of the disease [151,152]. Glutamine is not only the most abundant free amino acid in human plasma but also an essential amino acid required in the body during injury or disease processes [92], whereas patients with IBD often show altered metabolism of glutamine [153,154], including decreased glutamine concentration, decreased uptake, and decreased activity of glutaminase [154,155]. Glutamine metabolism plays a dual role during macrophage activation. It not only promotes M2 macrophage polarization through the glutamine-UDP-acetylglucosamine pathway and utilizes α-ketoglutarate produced by glutamine catabolism but also mediates glutamine-dependent aging-free and/or γ-aminobutyric acid shunting processes to induce M1 macrophage polarization [88,92,93]. Different polarization processes lead to different effects on macrophage immune function; therefore, glutamine can be pro- or anti-inflammatory in regulating the pathological process of IBD, depending on the different pathological context. The use of alanyl-glutamine therapy was found to inhibit the secretion of Th17-related cytokines by reducing the number of macrophages infiltrating the peritoneal cavity, ultimately ameliorating the inflammatory response [156–158]. In contrast, findings have shown that glutamine supplementation up-regulates pro-inflammatory cytokines including IL-1β, IL-17, and TNF-α [159]. The above different experimental results may be due to the different doses of glutamine used, hence the data suggesting that low doses of glutamine (0.5% basal diet) significantly promote the expression of pro-inflammatory factors, whereas high doses of glutamine (2% basal diet) inhibit the expression of pro-inflammatory factors [159]. Recent studies have found that the alpha-ketoglutarate/succinate ratio is a key factor in the regulation of macrophage immune metabolism by glutamine. A high value of this ratio promotes M2 macrophage polarization and vice versa for M1 macrophages [93]. Therefore, it is difficult to determine the optimal glutamine dose to improve IBD. In this process, factors such as the pathological course of IBD, the supplemental dose of glutamine, and the metabolic balance between glutamine and other amino acids should be included in the study.

6. Therapeutic Approaches Targeting Macrophage Immunometabolism in IBD

Macrophage metabolism is a dynamic process that changes in response to alterations in the tissue microenvironment, and these changes also reflect the different pathologic processes of a given disease. As mentioned earlier, macrophage metabolic processes are significantly altered in IBD, including enhanced glycolysis, impaired FAO processes with altered amino acid metabolic functions.

Tofacitinib is an oral pan-JAK inhibitor that has been approved for the treatment of UC [160,161], and it promotes M2 macrophage polarization in colitis mice by inducing macrophage OXPHOS processes [162,163]. However, some patients have experienced adverse side effects with tofacitinib, such as severe infections (e.g., herpes zoster) and thromboembolism [164,165]. To circumvent the occurrence of these side effects, researchers have developed selective JAK1 inhibitors such as filgotinib [166,167] and ubadatinib [168,169] as alternatives to tofacitinib, and the results of a phase II clinical study have shown that both of these drugs have demonstrated good results in relieving celiac disease and moderate to severe UC [166–170]. However, the mechanism of how these two drugs activate intestinal macrophages and participate in the regulation of their metabolism remains to be further investigated.

Recent studies have also identified other glycolytic regulators other than JAK that can be used to activate macrophages and treat IBD, such as mTORC1, HIF-1α, and AMPK. Mmong these, d-mannose, a naturally occurring C-2 glucose exopolymer, alleviates ulcerative colitis in mice by inhibiting glucose catabolism and succinate-mediated activation of HIF-1α in macrophages [171]. Similarly, it was also shown that spermidine, a natural polyamine, could enhance oxidative phosphorylation and FAO processes through activation of the AMPK/HIF-1α pathway, causing macrophages to exhibit anti-inflammatory properties, ultimately effectively ameliorating colitis in mice [172]. In addition to the endogenous molecules mentioned above, some chemicals present in natural plants can also target the immune metabolism of macrophages to exert ameliorative effects on IBD. Zhuang et al. have clarified that the small molecule adenosine can restore the balance between the M1 and M2 types of macrophages by inhibiting the HIF-1α-dependent glycolysis pathway and, thus, exerts an ameliorative effect on DSS-induced colitis in mice [173]. In addition, Dioscin, a steroidal saponin extracted from the roots of Chrysanthemum pastorale, simultaneously inhibits M1 macrophage polarization by modulating the mTORC1/HIF-1α-dependent glycolytic pathway and promotes M2 macrophage polarization by regulating the mTORC2/PPAR-γ-dependent FAO process, which ultimately ameliorates DSS-induced colitis in mice [174]. Didymin, a dietary citrus flavonoid, ameliorates ulcerative colitis by rebalancing M1/M2 macrophages through 4-hydroxyacyl coenzyme a dehydrogenase β-mediated FAO rather than glycolysis [175].

In addition, reviewing studies of approved drugs with metabolic targets could deepen our understanding of the metabolic adaptations of these drugs and provide new opportunities for better treatment of IBD. It is well known that metformin, a well-established hypoglycemic agent, ameliorates colitis in mice by inhibiting the endogenous fatty acid synthesis/AKT pathway and its downstream MAPK pathway in macrophages [176]. Overall, the above studies suggest that the manipulation of macrophage metabolic pathways and related transcriptional regulators may provide a viable strategy for finding effective targets and potent drugs for the treatment of IBD. It is important to note that many of the metabolic targets found in macrophages may have already been identified in many other cell types. Therefore, combining macrophage-targeted drug delivery technologies, such as nanotechnology, will greatly accelerate the effectiveness of new drug discovery (Table 1).

Table 1. Therapeutic approaches targeting macrophage immunometabolism in experimental IBD.

Drug	Target Point	Effect	Machine	References
Tofacitinib	A pan-JAK inhibitor	Increases M2-like macrophages by promoting OXPHOS	Suppressing JAK1 and JAK3	[163]
GLPG0555	A JAK1 selective inhibitor	Increases M2-like macrophages by promoting OXPHOS	Suppressing JAK1	[163]
D-mannose	Monosaccharide	Inhibits pro-inflammatory macrophage IL-1β production by impairing glucose catabolism and succinate-mediated HIF-1α activation	By phosphomannose isomerase	[171]
Spermidine	Polyamine	Promoting M2 polarization by enhancing OXPHOS and FAO fluxes	Blocking AMPK/HIF-1α pathway	[172]
Tiliroside	Flavonoid	Restores M1/M2 balance by blocking glycolysis	Suppressing HIF-1α	[173]
Dioscin	Steroidal saponin	Inhibits M1 by inhibiting aerobic glycolysis; promotes M2 polarization by facilitating FAO	Suppressing mTORC1/HIF-1α pathway and promoting mTORC2/PPAR-γ pathway	[174]
Didymin	Flavonoid	Restoration of M1/M2 balance via promoting FAO	Promoting 4 hydroacyl-CoA dehydrogenase β	[175]
Metformin	A hypoglycemic drug	Inhibition of pro-inflammatory macrophage activation through inhibition of fatty acid synthesis	Suppressing fatty acid synthase/Akt/MAPK pathway	[176]

Abbreviations: OXPHOS, oxidative phosphorylation; JAK, Janus Kinase; HIF-1α, hypoxia-inducible factor-1 alpha; M1: Classical activated macrophages; M2: Alternatively activated macrophages; AMPK, adenosine monophosphate-activated protein kinase; FAO, fatty acid oxidization; mTORC1, mammalian target of rapamycin complex 1; mTORC2, mammalian target of rapamycin complex 2; Akt, protein kinase B; MAPK, mitogen-activated protein kinase; PPAR-γ, peroxisome proliferator-activated receptor gamma.

7. Specific Goals and Potential Challenges of Treatment

Research on immune metabolism over the past decade has made significant progress and has clearly demonstrated that the foundation of life is constituted by the interplay of immune metabolism, which is also at the core of numerous diseases such as IBD. In the above description, we highlight the importance of the metabolic fate of nutrients versus microenvironmental stimuli during macrophage polarization. We also summarize potentially effective drugs that target macrophage immune metabolism and thereby treat IBD. As mentioned earlier, recent advances in research on immune metabolism have deepened our understanding of macrophage polarization processes and suggested new ideas for effective therapeutic tools for IBD.

Despite the continuous new advances around these topics, there are still important unanswered questions targeting macrophage immune metabolism in IBD. The first problem is that most of the molecular mechanisms currently used to unravel macrophage immune metabolism have been obtained through in vitro studies, whereas in vivo, the intestinal microenvironment in which macrophages reside is much more complex. For example, various changing cellular components (nutrient concentrations, metabolites, cytokines, transcriptional regulators, etc.) and extracellular factors (oxygen tension, pH, stimuli from other cells or tissues) affect macrophage polarization and their metabolism, but in vitro studies do not mimic this complex tissue microenvironment. The resolution of these issues contributes to a broader analysis of macrophage immunometabolism over tissue/organ or even systemic water. The advanced technologies that are currently available, such as mass spectrometry imaging, multi-omics coanalysis, and single-cell sequencing technologies, will help to piece together a complete picture of macrophage immunometabolism in IBD.

Another issue that exists is that dynamic cellular metabolic processes can respond rapidly to environmental changes under disease. Since the intestinal microenvironment changes differently during the different pathological processes of IBD (e.g., inflammatory or remission phases), the study of macrophage immune metabolism during the multiple phases of IBD would accelerate our understanding of the dynamic metabolic profiles of macrophages and analyze whether these profiles could be effective targets for disease amelioration. Finally, different sources of cells bring different metabolic profiles, and most of the studies carried out on macrophage immune metabolism have been performed with bone marrow or monocyte-derived macrophages. Therefore, in order to better characterize the different metabolic profiles of macrophage immunity in normal and IBD, studies using multiple sources of macrophages, such as monocyte-derived or macrophages from the lamina propria or myofibroblasts, are needed and will be the focus of future research.

8. Barriers to Clinical Translation

Although the various types of drugs developed to target macrophages for the treatment of intestinal inflammation have made considerable progress, drug development has been severely constrained by a number of factors such as biological complexity, drug-biological interactions, scale-up effects in the translational process, and the complexity of quality control, which have been the main barriers to low clinical translation rates.

8.1. Biological Barriers

Drugs undergo a complex multi-step cascade process in the body to exert their efficacy, including entry into the blood circulation, accumulation to the site of inflammation, penetration into the interior of inflamed tissues, endocytosis, intracellular transport, and drug release. Inefficiency in any step reduces the overall therapeutic effect. However, there are a range of biological barriers present in the body that can prevent the drug from moving efficiently through each of these processes, limiting therapeutic efficacy.

8.1.1. Nano-Biological Interactions in the Bloodstream

The presence of a large number of proteins in the blood may bind to the surface of the drug to form a "protein crown", thereby altering its physicochemical characteristics and stability and preventing the specific binding of the target molecule to the receptor. In addition, blood protein fractions vary from patient to patient, leading to heterogeneity of the "protein crowns" and unpredictability of their fate in vivo.

8.1.2. Clearance by the Mononuclear Phagocytic System (MPS)

Most drugs are taken up and cleared by macrophages or endothelial cells in the liver or spleen, preventing further delivery to inflammatory tissues. At the same time, non-specific accumulation in healthy tissues can lead to long-term toxicity.

8.2. Clinical Safety Issues

Many patients develop severe immune-related adverse reactions after taking the drug, leading to the termination of clinical trials. There is an urgent need to supplement the means of evaluating the safety of drugs with new methods.

8.3. Physicochemical Characteristics and Scale-Up Effects

Physicochemical parameters are crucial for the biological effects of drugs, including immune escape, drug penetration, and cellular uptake. The complexity of drug microstructures and compositions, and the multi-step or complex technologies involved in the construction process, result in poor reproducibility of preparation and extreme difficulty in scale-up, which in turn affects in vivo biological effects. Therefore, the production of drugs requires strict control of the physical and chemical properties between batches, with higher requirements for chemistry, manufacturing and control.

8.4. Animal Models

The lack of animal models capable of accurately simulating precise human inflammatory bowel disease is also one of the shortcomings, leading to poor correlation between preclinical studies and clinical trial results (pharmacokinetics, biodistribution and safety, etc.) [177].

9. Concluding Remarks and Future Perspectives

Intestinal homeostasis is a finely orchestrated process in which intestinal epithelial cells, macrophages, and stem cells play vital roles. Intestinal epithelial cells are responsible for maintaining immune tolerance by upholding the integrity of the intestinal barrier and regulating immune responses. Macrophages, on the other hand, are pivotal in the clearance of pathogens and the control of inflammatory reactions. They engage in dynamic interactions with intestinal epithelial cells and other immune cells to maintain a delicate balance within the intestinal immune system while also overseeing the regulation of intestinal stem cells, thus, ensuring the overall stability of intestinal tissue. Stem cells exhibit the remarkable capacity for self-renewal and differentiation along multiple lineages. This continuous generation of intestinal epithelial cells and various other cell types within the intestinal environment is vital for the ongoing repair and regeneration of intestinal tissue. The interactions between intestinal epithelial cells and macrophages serve as a means to regulate the intestinal immune response and modulate the state of inflammation, thereby impacting the integrity of the intestinal barrier function. This intricate web of mutual regulation between macrophages and stem cells, as well as between stem cells and the intestinal epithelium, represents a pivotal step in the repair and regeneration of intestinal tissue (Figure 4).

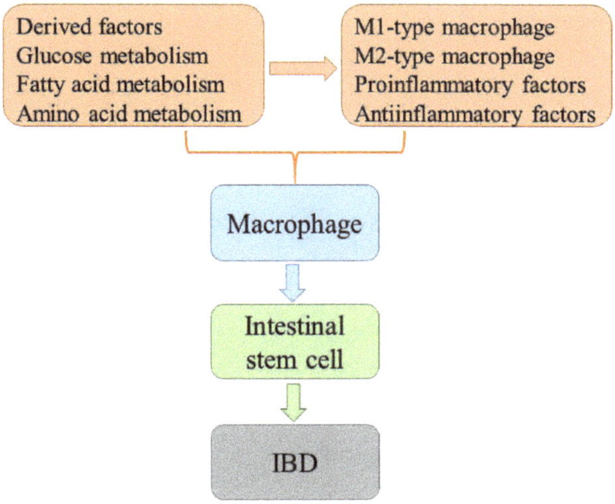

Figure 4. Illustrates the interaction patterns among IBD, macrophages, and crypt stem cells. Beyond establishing a vital barrier, these cells also exert control over macrophage function, modulating the immune response by influencing macrophage polarization and participating in defense against pathogens as well as the secretion of cytokines. Macrophages contribute positively to the promotion of intestinal stem cell proliferation through their regulatory role. Stem cells, endowed with self-renewal and differentiation capabilities, play a pivotal role in repairing damaged intestinal epithelium and maintaining homeostasis.

Author Contributions: Conceptualization, methodology, formal analysis, writing—original draft preparation, visualization, T.Q.; resources, supervision, project administration, funding acquisition, R.L.; conceptualization, formal analysis, validation, writing—review and editing, project administration, funding acquisition, supervision, T.G. All authors have read and agreed to the published version of the manuscript.

Funding: This work was supported by the Chinese National Natural Science Foundation (32402845, 32172801, 32372954) and Beijing Natural Science Foundation (5244041, 6222019).

Data Availability Statement: Data will be made available on request.

Conflicts of Interest: The authors declare that they have no known competing financial interests or personal relationships that could have appeared to influence the work reported in this paper.

Abbreviations

IBD: Inflammatory bowel disease; UC: Ulcerative colitis; CD: Crohn's disease; IEC: Intestinal epithelial cells; MSCs: Mesenchymal stem cells; HSCs: Hematopoietic stem cells; ISC: Intestinal stem cells; BMT: Bone marrow transplantation; MHC: Histocompatibility complex class; GVHD: Graft-versus-host disease; TNF: Tumor necrosis factor; TH: T helper; Treg: Regulatory T; hIOs: Human pluripotent stem cell-derived intestinal organoids; ILCs: Innate lymphoid cells; ER: Endoplasmic reticulum; DDR: DNA damage response; CSCs: Colonic stem cells; ILC2: Group 2 innate lymphoid-like cells; TCA: Tricarboxylic acid cycle; PPP: Pentose-phosphate pathway; FAO: Fatty acid oxidation; OXPHOS: Oxidative phosphorylation; PGE2: Prostaglandin E2; LGG: *Lactobacillus rhamnosus* GG; CSF1: Colony-stimulating factor; GLUT1: Glucose transporter 1; PFKFB3: 6-phosphofructo-2-kinase/fructose-2,6-bisphosphatase; DSS: Dextran sulfate sodium salt; SUCNR1: Succinate receptor 1; PPAR-γ: Peroxisome proliferator-activated receptor gamma.

References

1. Chelakkot, C.; Ghim, J.; Ryu, S.H. Mechanisms regulating intestinal barrier integrity and its pathological implications. *Exp. Mol. Med.* **2018**, *50*, 1–9. [PubMed]
2. Nguyen, H.D.; Aljamaei, H.M.; Stadnyk, A.W. The production and function of endogenous interleukin-10 in intestinal epithelial cells and gut homeostasis. *Cell. Mol. Gastroenterol. Hepatol.* **2021**, *12*, 1343–1352. [CrossRef] [PubMed]
3. Ma, J.; Piao, X.; Mahfuz, S.; Long, S.; Wang, J. The interaction among gut microbes, the intestinal barrier and short chain fatty acids. *Anim. Nutr.* **2021**, *9*, 159–174. [PubMed]
4. Na, Y.R.; Stakenborg, M.; Seok, S.H.; Matteoli, G. Macrophages in intestinal inflammation and resolution: A potential therapeutic target in IBD. *Nat. Rev. Gastroenterol. Hepatol.* **2019**, *16*, 531–543. [CrossRef]
5. Delfini, M.; Stakenborg, N.; Viola, M.F.; Boeckxstaens, G. Macrophages in the gut: Masters in multitasking. *Immunity* **2022**, *55*, 1530–1548.
6. Wang, E.J.; Wu, M.Y.; Ren, Z.Y.; Zheng, Y.; Ye, R.D.; Tan, C.S.H.; Wang, Y.; Lu, J.H. Targeting macrophage autophagy for inflammation resolution and tissue repair in inflammatory bowel disease. *Burn. Trauma* **2023**, *11*, tkad004.
7. Zheng, L.; Duan, S.L. Molecular regulation mechanism of intestinal stem cells in mucosal injury and repair in ulcerative colitis. *World J. Gastroenterol.* **2023**, *29*, 2380–2396.
8. Eri, R.; Chieppa, M. Messages from the Inside. The dynamic environment that favors intestinal homeostasis. *Front. Immunol.* **2013**, *4*, 323. [CrossRef]
9. Saenz, S.A.; Taylor, B.C.; Artis, D. Welcome to the neighborhood: Epithelial cell-derived cytokines license innate and adaptive immune responses at mucosal sites. *Immunol. Rev.* **2008**, *226*, 172–910.
10. Carding, S.R.; Davis, N.; Hoyles, L. Review article: The human intestinal virome in health and disease. *Aliment. Pharmacol. Ther.* **2017**, *46*, 800–815. [CrossRef]
11. Peterson, L.W.; Artis, D. Intestinal epithelial cells: Regulators of barrier function and immune homeostasis. *Nat. Rev. Immunol.* **2014**, *14*, 141–153. [CrossRef] [PubMed]
12. Garrett, W.S.; Gordon, J.I.; Glimcher, L.H. Homeostasis and inflammation in the intestine. *Cell* **2010**, *140*, 859–870. [CrossRef]
13. Hansen, J.J. Immune responses to intestinal microbes in inflammatory bowel diseases. *Curr. Allergy Asthma Rep.* **2015**, *15*, 61. [CrossRef]
14. Podolsky, D.K. Inflammatory bowel disease. *N. Engl. J. Med.* **2002**, *347*, 417–429. [CrossRef] [PubMed]
15. Kostic, A.D.; Xavier, R.J.; Gevers, D. The Microbiome in inflammatory bowel disease: Current status and the future ahead. *Gastroenterology* **2014**, *146*, 1489–1499. [CrossRef]
16. Ng, S.C. Emerging leadership lecture: Inflammatory bowel disease in Asia: Emergence of a "Western" disease. *J. Gastroenterol. Hepatol.* **2015**, *30*, 440–445. [CrossRef]
17. Ungaro, R.; Mehandru, S.; Allen, P.B.; Peyrin-Biroulet, L.; Colombel, J.F. Ulcerative colitis. *Lancet* **2017**, *389*, 1756–1770. [CrossRef]
18. Torres, J.; Mehandru, S.; Colombel, J.F.; Peyrin-Biroulet, L. Crohn's disease. *Lancet* **2017**, *389*, 1741–1755. [CrossRef]
19. Neurath, M.F. Cytokines in inflammatory bowel disease. *Nat. Rev. Immunol.* **2014**, *14*, 329–342. [CrossRef] [PubMed]
20. Neurath, M.F. New targets for mucosal healing and therapy in inflammatory bowel diseases. *Mucosal Immunol.* **2014**, *7*, 6–19. [CrossRef]
21. Strober, W.; Fuss, I.J.; Blumberg, R.S. The immunology of mucosal models of inflammation. *Annu. Rev. Immunol.* **2002**, *20*, 495–549. [CrossRef]
22. Rooks, M.G.; Garrett, W.S. Gut microbiota, metabolites and host immunity. *Nat. Rev. Immunol.* **2016**, *16*, 341–352. [CrossRef]
23. Lavelle, A.; Sokol, H. Gut microbiota-derived metabolites as key actors in inflammatory bowel disease. *Nat. Rev. Gastroenterol. Hepatol.* **2020**, *17*, 223–237. [CrossRef] [PubMed]
24. Neurath, M.F.; Travis, S.P. Mucosal healing in inflammatory bowel diseases: A systematic review. *Gut* **2012**, *61*, 1619–1635. [CrossRef]
25. Colombel, J.F.; Rutgeerts, P.; Reinisch, W.; Esser, D.; Wang, Y.; Lang, Y.; Marano, C.W.; Strauss, R.; Oddens, B.J.; Feagan, B.G.; et al. Early mucosal healing with infliximab is associated with improved long-term clinical outcomes in ulcerative colitis. *Gastroenterology* **2011**, *141*, 1194–1201. [CrossRef]
26. Pineton de Chambrun, G.; Peyrin-Biroulet, L.; Lemann, M.; Colombel, J.F. Clinical implications of mucosal healing for the management of IBD. *Nat. Rev. Gastroenterol. Hepatol.* **2010**, *7*, 15–29. [CrossRef]
27. Cho, Y.B.; Park, K.J.; Yoon, S.N.; Song, K.H.; Kim, D.S.; Jung, S.H.; Kim, M.; Jeong, H.Y.; Yu, C.S. Long-term results of adipose-derived stem cell therapy for the treatment of Crohn's fistula. *Stem Cells Transl. Med.* **2015**, *4*, 532–537. [CrossRef]
28. Ciccocioppo, R.; Gallia, A.; Sgarella, A.; Kruzliak, P.; Gobbi, P.G.; Corazza, G.R. Long-term follow-up of Crohn disease fistulas after local injections of bone marrow-derived mesenchymal stem cells. *Mayo Clin. Proc.* **2015**, *90*, 747–755. [CrossRef]
29. Barnhoorn, M.C.; Wasser, M.; Roelofs, H.; Maljaars, P.W.J.; Molendijk, I.; Bonsing, B.A.; Oosten, L.E.M.; Dijkstra, G.; van der Woude, C.J.; Roelen, D.L.; et al. Long-term evaluation of allogeneic bone marrow-derived mesenchymal stromal cell therapy for Crohn's disease perianal fistulas. *J. Crohns Colitis* **2020**, *14*, 64–70. [CrossRef] [PubMed]

30. Sato, T.; Vries, R.G.; Snippert, H.J.; van de Wetering, M.; Barker, N.; Stange, D.E.; van Es, J.H.; Abo, A.; Kujala, P.; Peters, P.J.; et al. Single Lgr5 stem cells build crypt-villus structures in vitro without a mesenchymal niche. *Nature* **2009**, *459*, 262–265. [CrossRef] [PubMed]
31. Sato, T.; van Es, J.H.; Snippert, H.J.; Stange, D.E.; Vries, R.G.; van den Born, M.; Clevers, H. Paneth cells constitute the niche for Lgr5 stem cells in intestinal crypts. *Nature* **2011**, *469*, 415–418. [CrossRef]
32. Yui, S.; Nakamura, T.; Sato, T.; Nemoto, Y.; Mizutani, T.; Zheng, X.; Ichinose, S.; Nagaishi, T.; Okamoto, R.; Tsuchiya, K.; et al. Functional engraftment of colon epithelium expanded in vitro from a single adult $Lgr5^+$ stem cell. *Nat. Med.* **2012**, *18*, 618–623. [CrossRef]
33. Fukuda, M.; Mizutani, T.; Mochizuki, W.; Matsumoto, T.; Nozaki, K.; Sakamaki, Y.; Ichinose, S.; Okada, Y.; Tanaka, T.; Watanabe, M.; et al. Small intestinal stem cell identity is maintained with functional Paneth cells in heterotopically grafted epithelium onto the colon. *Genes Dev.* **2014**, *28*, 1752–1757. [CrossRef]
34. Sugimoto, S.; Ohta, Y.; Fujii, M.; Matano, M.; Shimokawa, M.; Nanki, K.; Date, S.; Nishikori, S.; Nakazato, Y.; Nakamura, T.; et al. Reconstruction of the human colon epithelium In Vivo. *Cell Stem Cell* **2018**, *22*, 171–176.e5. [CrossRef]
35. Zhu, P.; Lu, T.; Wu, J.; Fan, D.; Liu, B.; Zhu, X.; Guo, H.; Du, Y.; Liu, F.; Tian, Y.; et al. Gut microbiota drives macrophage-dependent self-renewal of intestinal stem cells via niche enteric serotonergic neurons. *Cell Res.* **2022**, *32*, 555–569. [CrossRef]
36. Hu, D.; Yan, H.; He, X.C.; Li, L. Recent advances in understanding intestinal stem cell regulation. *F1000Research* **2019**, *8*, F1000 Faculty Rev-72. [CrossRef]
37. Date, S.; Sato, T. Mini-gut organoids: Reconstitution of the stem cell niche. *Annu. Rev. Cell Dev. Biol.* **2015**, *31*, 269–289. [CrossRef]
38. Onyiah, J.C.; Colgan, S.P. Cytokine responses and epithelial function in the intestinal mucosa. *Cell Mol. Life Sci.* **2016**, *73*, 4203–4212. [CrossRef]
39. Andrews, C.; McLean, M.H.; Durum, S.K. Cytokine tuning of intestinal epithelial function. *Front. Immunol.* **2018**, *9*, 1270. [CrossRef]
40. Biton, M.; Haber, A.L.; Rogel, N.; Burgin, G.; Beyaz, S.; Schnell, A.; Ashenberg, O.; Su, C.W.; Smillie, C.; Shekhar, K.; et al. T Helper cell cytokines modulate intestinal stem cell renewal and differentiation. *Cell* **2018**, *175*, 1307–1320.e22. [CrossRef] [PubMed]
41. Saha, S.; Aranda, E.; Hayakawa, Y.; Bhanja, P.; Atay, S.; Brodin, N.P.; Li, J.; Asfaha, S.; Liu, L.; Tailor, Y.; et al. Macrophage-derived extracellular vesicle-packaged WNTs rescue intestinal stem cells and enhance survival after radiation injury. *Nat. Commun.* **2016**, *7*, 13096. [PubMed]
42. Fu, Y.Y.; Egorova, A.; Sobieski, C.; Kuttiyara, J.; Calafiore, M.; Takashima, S.; Clevers, H.; Hanash, A.M. T cell recruitment to the intestinal stem cell compartment drives immune-mediated intestinal damage after allogeneic transplantation. *Immunity* **2019**, *51*, 90–103.e3.
43. Takashima, S.; Martin, M.L.; Jansen, S.A.; Fu, Y.; Bos, J.; Chandra, D.; O'Connor, M.H.; Mertelsmann, A.M.; Vinci, P.; Kuttiyara, J.; et al. T cell-derived interferon-gamma programs stem cell death in immune-mediated intestinal damage. *Sci. Immunol.* **2019**, *4*, aay8556.
44. Fordham, R.P.; Yui, S.; Hannan, N.R.; Soendergaard, C.; Madgwick, A.; Schweiger, P.J.; Nielsen, O.H.; Vallier, L.; Pedersen, R.A.; Nakamura, T.; et al. Transplantation of expanded fetal intestinal progenitors contributes to colon regeneration after injury. *Cell Stem Cell* **2013**, *13*, 734–744. [CrossRef]
45. Schreurs, R.; Baumdick, M.E.; Sagebiel, A.F.; Kaufmann, M.; Mokry, M.; Klarenbeek, P.L.; Schaltenberg, N.; Steinert, F.L.; van Rijn, J.M.; Drewniak, A.; et al. Human fetal TNF-alpha-cytokine-producing $CD4^+$ effector memory T cells promote intestinal development and mediate inflammation early in life. *Immunity* **2019**, *50*, 462–476.e8. [CrossRef] [PubMed]
46. Jung, K.B.; Lee, H.; Son, Y.S.; Lee, M.O.; Kim, Y.D.; Oh, S.J.; Kwon, O.; Cho, S.; Cho, H.S.; Kim, D.S.; et al. Interleukin-2 induces the in vitro maturation of human pluripotent stem cell-derived intestinal organoids. *Nat. Commun.* **2018**, *9*, 3039. [CrossRef]
47. Diefenbach, A.; Gnafakis, S.; Shomrat, O. Innate lymphoid cell-epithelial cell modules sustain intestinal homeostasis. *Immunity* **2020**, *52*, 452–463.
48. Panda, S.K.; Colonna, M. Innate lymphoid cells in mucosal immunity. *Front. Immunol.* **2019**, *10*, 861.
49. Zha, J.M.; Li, H.S.; Lin, Q.; Kuo, W.T.; Jiang, Z.H.; Tsai, P.Y.; Ding, N.; Wu, J.; Xu, S.F.; Wang, Y.T.; et al. Interleukin 22 expands transit-amplifying cells while depleting $Lgr5^+$ stem cells via inhibition of Wnt and Notch signaling. *Cell. Mol. Gastroenterol. Hepatol.* **2019**, *7*, 255–274.
50. Hou, Q.; Ye, L.; Liu, H.; Huang, L.; Yang, Q.; Turner, J.R.; Yu, Q. Lactobacillus accelerates ISCs regeneration to protect the integrity of intestinal mucosa through activation of STAT3 signaling pathway induced by LPLs secretion of IL-22. *Cell Death Differ.* **2018**, *25*, 1657–1670.
51. Huber, S.; Gagliani, N.; Zenewicz, L.A.; Huber, F.J.; Bosurgi, L.; Hu, B.; Hedl, M.; Zhang, W.; O'Connor, W.; Murphy, A.J.; et al. IL-22BP is regulated by the inflammasome and modulates tumorigenesis in the intestine. *Nature* **2012**, *491*, 259–263. [CrossRef]
52. Satoh-Takayama, N.; Vosshenrich, C.A.; Lesjean-Pottier, S.; Sawa, S.; Lochner, M.; Rattis, F.; Mention, J.J.; Thiam, K.; Cerf-Bensussan, N.; Mandelboim, O.; et al. Microbial flora drives interleukin 22 production in intestinal $NKp46^+$ cells that provide innate mucosal immune defense. *Immunity* **2008**, *29*, 958–970. [CrossRef]

53. Zenewicz, L.A.; Yancopoulos, G.D.; Valenzuela, D.M.; Murphy, A.J.; Stevens, S.; Flavell, R.A. Innate and adaptive interleukin-22 protects mice from inflammatory bowel disease. *Immunity* **2008**, *29*, 947–957. [CrossRef]
54. Barker, N.; van Es, J.H.; Kuipers, J.; Kujala, P.; van den Born, M.; Cozijnsen, M.; Haegebarth, A.; Korving, J.; Begthel, H.; Peters, P.J.; et al. Identification of stem cells in small intestine and colon by marker gene Lgr5. *Nature* **2007**, *449*, 1003–1007. [CrossRef]
55. Hanash, A.M.; Dudakov, J.A.; Hua, G.; O'Connor, M.H.; Young, L.F.; Singer, N.V.; West, M.L.; Jenq, R.R.; Holland, A.M.; Kappel, L.W.; et al. Interleukin-22 protects intestinal stem cells from immune-mediated tissue damage and regulates sensitivity to graft versus host disease. *Immunity* **2012**, *37*, 339–350. [CrossRef]
56. Lindemans, C.A.; Calafiore, M.; Mertelsmann, A.M.; O'Connor, M.H.; Dudakov, J.A.; Jenq, R.R.; Velardi, E.; Young, L.F.; Smith, O.M.; Lawrence, G.; et al. Interleukin-22 promotes intestinal-stem-cell-mediated epithelial regeneration. *Nature* **2015**, *528*, 560–564. [CrossRef]
57. Powell, N.; Pantazi, E.; Pavlidis, P.; Tsakmaki, A.; Li, K.; Yang, F.; Parker, A.; Pin, C.; Cozzetto, D.; Minns, D.; et al. Interleukin-22 orchestrates a pathological endoplasmic reticulum stress response transcriptional programme in colonic epithelial cells. *Gut* **2020**, *69*, 578–590. [CrossRef]
58. Blokzijl, F.; de Ligt, J.; Jager, M.; Sasselli, V.; Roerink, S.; Sasaki, N.; Huch, M.; Boymans, S.; Kuijk, E.; Prins, P.; et al. Tissuespecific mutation accumulation in human adult stem cells during life. *Nature* **2016**, *538*, 260–264. [CrossRef]
59. Wu, S.; Powers, S.; Zhu, W.; Hannun, Y.A. Substantial contribution of extrinsic risk factors to cancer development. *Nature* **2016**, *529*, 43–47. [CrossRef]
60. Blanpain, C.; Mohrin, M.; Sotiropoulou, P.A.; Passegue, E. DNA-damage response in tissue-specific and cancer stem cells. *Cell Stem Cell* **2011**, *8*, 16–29. [CrossRef]
61. Gronke, K.; Hernandez, P.P.; Zimmermann, J.; Klose, C.S.N.; Kofoed-Branzk, M.; Guendel, F.; Witkowski, M.; Tizian, C.; Amann, L.; Schumacher, F.; et al. Interleukin-22 protects intestinal stem cells against genotoxic stress. *Nature* **2019**, *566*, 249–253. [CrossRef]
62. Waddell, A.; Vallance, J.E.; Hummel, A.; Alenghat, T.; Rosen, M.J. IL-33 induces murine intestinal goblet cell differentiation indirectly via innate lymphoid cell IL-13 secretion. *J. Immunol.* **2019**, *202*, 598–607. [CrossRef]
63. Gerbe, F.; Sidot, E.; Smyth, D.J.; Ohmoto, M.; Matsumoto, I.; Dardalhon, V.; Cesses, P.; Garnier, L.; Pouzolles, M.; Brulin, B.; et al. Intestinal epithelial tuft cells initiate type 2 mucosal immunity to helminth parasites. *Nature* **2016**, *529*, 226–230. [CrossRef]
64. Howitt, M.R.; Lavoie, S.; Michaud, M.; Blum, A.M.; Tran, S.V.; Weinstock, J.V.; Gallini, C.A.; Redding, K.; Margolskee, R.F.; Osborne, L.C.; et al. Tuft cells, taste-chemosensory cells, orchestrate parasite type 2 immunity in the gut. *Science* **2016**, *351*, 1329–1333. [CrossRef]
65. von Moltke, J.; Ji, M.; Liang, H.E.; Locksley, R.M. Tuft-cell-derived IL-25 regulates an intestinal ILC2-epithelial response circuit. *Nature* **2016**, *529*, 221–225. [CrossRef]
66. Schneider, C.; O'Leary, C.E.; von Moltke, J.; Liang, H.E.; Ang, Q.Y.; Turnbaugh, P.J.; Radhakrishnan, S.; Pellizzon, M.; Ma, A.; Locksley, R.M. A metabolite-triggered tuft cell-ILC2 circuit drives small intestinal remodeling. *Cell* **2018**, *174*, 271–284.e14. [CrossRef]
67. Greuter, T.; Rieder, F.; Kucharzik, T.; Peyrin-Biroulet, L.; Schoepfer, A.M.; Rubin, D.T.; Vavricka, S.R. Emerging treatment options for extraintestinal manifestations in IBD. *Gut* **2021**, *70*, 796–802. [CrossRef]
68. Grainger, J.R.; Konkel, J.E.; Zangerle-Murray, T.; Shaw, T.N. Macrophages in gastrointestinal homeostasis and infammation. *Pflügers Arch.-Eur. J. Physiol.* **2017**, *469*, 527–539. [CrossRef]
69. Bain, C.C.; Bravo-Blas, A.; Scott, C.L.; Perdiguero, E.G.; Geissmann, F.; Henri, S.; Malissen, B.; Osborne, L.C.; Artis, D.; Mowat, A.M. Constant replenishment from circulating monocytes maintains the macrophage pool in the intestine of adult mice. *Nat. Immunol.* **2014**, *15*, 929–937. [CrossRef]
70. Bain, C.C.; Scott, C.L.; Uronen-Hansson, H.; Gudjonsson, S.; Jansson, O.; Grip, O.; Guilliams, M.; Malissen, B.; Agace, W.W.; Mowat, A.M. Resident and pro-inflammatory macrophages in the colon represent alternative context-dependent fates of the same Ly6Chi monocyte precursors. *Mucosal Immunol.* **2013**, *6*, 498–510. [CrossRef]
71. Mosser, D.M.; Edwards, J.P. Exploring the full spectrum of macrophage activation. *Nat. Rev. Immunol.* **2008**, *8*, 958–969. [CrossRef]
72. Murray, P.J.; Allen, J.E.; Biswas, S.K.; Fisher, E.A.; Gilroy, D.W.; Goerdt, S.; Gordon, S.; Hamilton, J.A.; Ivashkiv, L.B.; Lawrence, T.; et al. Macrophage activation and polarization: Nomenclature and experimental guidelines. *Immunity* **2014**, *41*, 14–20. [CrossRef]
73. Shapouri-Moghaddam, A.; Mohammadian, S.; Vazini, H.; Taghadosi, M.; Esmaeili, S.A.; Mardani, F.; Seifi, B.; Mohammadi, A.; Afshari, J.T.; Sahebkar, A. Macrophage plasticity, polarization, and function in health and disease. *J. Cell Physiol.* **2018**, *233*, 6425–6440. [CrossRef]
74. Han, X.; Ding, S.; Jiang, H.; Liu, G. Roles of Macrophages in the Development and Treatment of Gut Inflammation. *Front. Cell Dev. Biol.* **2021**, *9*, 625423. [CrossRef]
75. Martinez, F.O.; Gordon, S. The M1 and M2 paradigm of macrophage activation: Time for reassessment. *F1000Prime Rep.* **2014**, *6*, 13. [CrossRef]

76. Hu, X.; Yuan, X.; Zhang, G.; Song, H.; Ji, P.; Guo, Y.; Liu, Z.; Tian, Y.; Shen, R.; Wang, D. The intestinal epithelial-macrophage-crypt stem cell axis plays a crucial role in regulating and maintaining intestinal homeostasis. *Life Sci.* **2024**, *344*, 122452. [CrossRef]
77. Isidro, R.A.; Appleyard, C.B. Colonic macrophage polarization in homeostasis, inflammation, and cancer. *Am. J. Physiol. Gastrointest. Liver Physiol.* **2016**, *311*, G59–G73. [CrossRef]
78. Medina-Contreras, O.; Geem, D.; Laur, O.; Williams, I.R.; Lira, S.A.; Nusrat, A.; Parkos, C.A.; Denning, T.L. CX3CR1 regulates intestinal macrophage homeostasis, bacterial translocation, and colitogenic Th17 responses in mice. *J. Clin. Investig.* **2011**, *121*, 4787–4795. [CrossRef]
79. Murai, M.; Turovskaya, O.; Kim, G.; Madan, R.; Karp, C.L.; Cheroutre, H.; Kronenberg, M. Interleukin 10 acts on regulatory T cells to maintain expression of the transcription factor Foxp3 and suppressive function in mice with colitis. *Nat. Immunol.* **2009**, *10*, 1178–1184. [CrossRef] [PubMed]
80. Bain, C.C.; Mowat, A.M. Macrophages in intestinal homeostasis and inflammation. *Immunol. Rev.* **2014**, *260*, 102–117. [CrossRef] [PubMed]
81. Bain, C.C.; Schridde, A. Origin, Differentiation, and Function of Intestinal Macrophages. *Front. Immunol.* **2018**, *9*, 2733.
82. Ryan, D.G.; O'Neill, L.A.J. Krebs Cycle Reborn in Macrophage Immunometabolism. *Annu. Rev. Immunol.* **2020**, *38*, 289–313. [CrossRef] [PubMed]
83. Voss, K.; Hong, H.S.; Bader, J.E.; Sugiura, A.; Lyssiotis, C.A.; Rathmell, J.C. A guide to interrogating immunometabolism. *Nat. Rev. Immunol.* **2021**, *21*, 637–652. [PubMed]
84. O'Neill, L.A.; Kishton, R.J.; Rathmell, J. A guide to immunometabolism for immunologists. *Nat. Rev. Immunol.* **2016**, *16*, 553–565.
85. Gaber, T.; Strehl, C.; Buttgereit, F. Metabolic regulation of inflammation. *Nat. Rev. Rheumatol.* **2017**, *13*, 267–279.
86. Jha, A.K.; Huang, S.C.; Sergushichev, A.; Lampropoulou, V.; Ivanova, Y.; Loginicheva, E.; Chmielewski, K.; Stewart, K.M.; Ashall, J.; Everts, B.; et al. Network integration of parallel metabolic and transcriptional data reveals metabolic modules that regulate macrophage polarization. *Immunity* **2015**, *42*, 419–430.
87. Lauterbach, M.A.; Hanke, J.E.; Serefidou, M.; Mangan, M.S.J.; Kolbe, C.C.; Hess, T.; Rothe, M.; Kaiser, R.; Hoss, F.; Gehlen, J.; et al. Toll-like receptor signaling rewires macrophage metabolism and promotes histone acetylation via ATP-citrate lyase. *Immunity* **2019**, *51*, 997–1011.e7.
88. Tannahill, G.M.; Curtis, A.M.; Adamik, J.; Palsson-McDermott, E.M.; McGettrick, A.F.; Goel, G.; Frezza, C.; Bernard, N.J.; Kelly, B.; Foley, N.H.; et al. Succinate is an inflammatory signal that induces IL-1β through HIF-1α. *Nature* **2013**, *496*, 238–242. [CrossRef]
89. Castegna, A.; Gissi, R.; Menga, A.; Montopoli, M.; Favia, M.; Viola, A.; Canton, M. Pharmacological targets of metabolism in disease: Opportunities from macrophages. *Pharmacol. Ther.* **2020**, *210*, 107521.
90. Van den Bossche, J.; O'Neill, L.A.; Menon, D. Macrophage Immunometabolism: Where Are We (Going)? *Trends Immunol.* **2017**, *38*, 395–406.
91. Yan, J.; Horng, T. Lipid metabolism in regulation of macrophage functions. *Trends Cell Biol.* **2020**, *30*, 979–989. [CrossRef] [PubMed]
92. Ren, W.; Xia, Y.; Chen, S.; Wu, G.; Bazer, F.W.; Zhou, B.; Tan, B.; Zhu, G.; Deng, J.; Yin, Y. Glutamine Metabolism in Macrophages: A Novel Target for Obesity/Type 2 Diabetes. *Adv. Nutr.* **2019**, *10*, 321–330. [CrossRef] [PubMed]
93. Liu, P.S.; Wang, H.; Li, X.; Chao, T.; Teav, T.; Christen, S.; Di Conza, G.; Cheng, W.C.; Chou, C.H.; Vavakova, M.; et al. α-ketoglutarate orchestrates macrophage activation through metabolic and epigenetic reprogramming. *Nat. Immunol.* **2017**, *18*, 985–994. [CrossRef]
94. Rath, M.; Müller, I.; Kropf, P.; Closs, E.I.; Munder, M. Metabolism via Arginase or Nitric Oxide Synthase: Two Competing Arginine Pathways in Macrophages. *Front. Immunol.* **2014**, *5*, 532. [CrossRef] [PubMed]
95. Kedia-Mehta, N.; Finlay, D.K. Competition for nutrients and its role in controlling immune responses. *Nat. Commun.* **2019**, *10*, 2123. [CrossRef]
96. Noel, G.; Baetz, N.W.; Staab, J.F.; Donowitz, M.; Kovbasnjuk, O.; Pasetti, M.F.; Zachos, N.C. A primary human macrophage-enteroid co-culture model to investigate mucosal gut physiology and host-pathogen interactions. *Sci. Rep.* **2017**, *7*, 45270.
97. Sehgal, A.; Donaldson, D.S.; Pridans, C.; Sauter, K.A.; Hume, D.A.; Mabbott, N.A. The role of CSF1R-dependent macrophages in control of the intestinal stem-cell niche. *Nat. Commun.* **2018**, *9*, 1272. [CrossRef]
98. Griffin, M.D.; Elliman, S.J.; Cahill, E.; English, K.; Ceredig, R.; Ritter, T. Concise review: Adult mesenchymal stromal cell therapy for inflammatory diseases: How well are we joining the dots? *Stem Cells* **2013**, *31*, 2033–2041. [CrossRef]
99. Riehl, T.E.; Alvarado, D.; Ee, X.; Zuckerman, A.; Foster, L.; Kapoor, V.; Thotala, D.; Ciorba, M.A.; Stenson, W.F. *Lactobacillus rhamnosus* GG protects the intestinal epithelium from radiation injury through release of lipoteichoic acid, macrophage activation and the migration of mesenchymal stem cells. *Gut* **2019**, *68*, 1003–1013. [CrossRef]
100. Cosin-Roger, J.; Ortiz-Masia, D.; Calatayud, S.; Hernandez, C.; Esplugues, J.V.; Barrachina, M.D. The activation of Wnt signaling by a STAT6-dependent macrophage phenotype promotes mucosal repair in murine IBD. *Mucosal Immunol.* **2016**, *9*, 986–998. [CrossRef]

101. Yip, J.L.K.; Balasuriya, G.K.; Spencer, S.J.; Hill-Yardin, E.L. The role of intestinal macrophages in gastrointestinal homeostasis: Heterogeneity and implications in disease. *Cell. Mol. Gastroenterol. Hepatol.* **2021**, *12*, 1701–1718. [CrossRef] [PubMed]
102. Friedrich, M.; Pohin, M.; Powrie, F. Cytokine Networks in the Pathophysiology of Inflammatory Bowel Disease. *Immunity* **2019**, *50*, 992–1006. [CrossRef]
103. Kamada, N.; Hisamatsu, T.; Okamoto, S.; Sato, T.; Matsuoka, K.; Arai, K.; Nakai, T.; Hasegawa, A.; Inoue, N.; Watanabe, N.; et al. Abnormally differentiated subsets of intestinal macrophage play a key role in Th1-dominant chronic colitis through excess production of IL-12 and IL-23 in response to bacteria. *J. Immunol.* **2005**, *175*, 6900–6908. [CrossRef] [PubMed]
104. Ackermann, M.; Mucci, A.; McCabe, A.; Frei, S.; Wright, K.; Snapper, S.B.; Lachmann, N.; Williams, D.A.; Brendel, C. Restored macrophage function ameliorates disease pathophysiology in a mouse model for IL10 receptor-deficient very early onset inflammatory bowel disease. *J. Crohn's Colitis* **2021**, *15*, 1588–1595. [CrossRef] [PubMed]
105. Hunter, M.M.; Wang, A.; Parhar, K.S.; Johnston, M.J.; Van Rooijen, N.; Beck, P.L.; McKay, D.M. In vitro-derived alternatively activated macrophages reduce colonic inflammation in mice. *Gastroenterology* **2010**, *138*, 1395–1405. [CrossRef]
106. Leung, G.; Wang, A.; Fernando, M.; Phan, V.C.; McKay, D.M. Bone marrow-derived alternatively activated macrophages reduce colitis without promoting fibrosis: Participation of IL-10. *Am. J. Physiol. Gastrointest. Liver Physiol.* **2013**, *304*, G781–G792. [CrossRef]
107. Wynn, T.A.; Chawla, A.; Pollard, J.W. Macrophage biology in development, homeostasis and disease. *Nature* **2013**, *496*, 445–455. [CrossRef]
108. Wynn, T.A.; Vannella, K.M. Macrophages in tissue repair, regeneration, and fibrosis. *Immunity* **2016**, *44*, 450–462. [CrossRef]
109. Wynn, T.A.; Barron, L. Macrophages: Master regulators of inflammation and fibrosis. *Semin. Liver Dis.* **2010**, *30*, 245–257. [CrossRef]
110. Peiser, L.; Mukhopadhyay, S.; Gordon, S. Scavenger receptors in innate immunity. *Curr. Opin. Immunol.* **2002**, *14*, 123–128. [CrossRef]
111. Gordon, S. Phagocytosis: An immunobiologic process. *Immunity* **2016**, *44*, 463–475. [CrossRef] [PubMed]
112. Duffield, J.S.; Forbes, S.J.; Constandinou, C.M.; Clay, S.; Partolina, M.; Vuthoori, S.; Wu, S.; Lang, R.; Iredale, J.P. Selective depletion of macrophages reveals distinct, opposing roles during liver injury and repair. *J. Clin. Investig.* **2005**, *115*, 56–65. [CrossRef]
113. Zhang, M.Z.; Yao, B.; Yang, S.; Jiang, L.; Wang, S.; Fan, X.; Yin, H.; Wong, K.; Miyazawa, T.; Chen, J.; et al. CSF-1 signaling mediates recovery from acute kidney injury. *J. Clin. Investig.* **2012**, *122*, 4519–4532. [CrossRef] [PubMed]
114. Rappolee, D.A.; Mark, D.; Banda, M.J.; Werb, Z. Wound macrophages express TGF-α and other growth factors in vivo: Analysis by mRNA phenotyping. *Science* **1988**, *241*, 708–712. [CrossRef] [PubMed]
115. Berse, B.; Brown, L.F.; Van de Water, L.; Dvorak, H.F.; Senger, D.R. Vascular permeability factor (vascular endothelial growth factor) gene is expressed differentially in normal tissues, macrophages, and tumors. *Mol. Biol. Cell* **1992**, *3*, 211–220. [CrossRef]
116. Shimokado, K.; Raines, E.W.; Madtes, D.K.; Barrett, T.B.; Benditt, E.P.; Ross, R. A significant part of macrophage-derived growth factor consists of at least two forms of PDGF. *Cell* **1985**, *43*, 277–286. [CrossRef]
117. Chujo, S.; Shirasaki, F.; Kondo-Miyazaki, M.; Ikawa, Y.; Takehara, K. Role of connective tissue growth factor and its interaction with basic fibroblast growth factor and macrophage chemoattractant protein-1 in skin fibrosis. *J. Cell. Physiol.* **2009**, *220*, 189–195. [CrossRef]
118. Willenborg, S.; Lucas, T.; van Loo, G.; Knipper, J.A.; Krieg, T.; Haase, I.; Brachvogel, B.; Hammerschmidt, M.; Nagy, A.; Ferrara, N.; et al. CCR2 recruits an inflammatory macrophage subpopulation critical for angiogenesis in tissue repair. *Blood* **2012**, *120*, 613–625. [CrossRef]
119. Akhurst, R.J.; Hata, A. Targeting the TGFβ signalling pathway in disease. *Nat. Rev. Drug Discov.* **2012**, *11*, 790–811. [CrossRef]
120. Stappenbeck, T.S.; Miyoshi, H. The role of stromal stem cells in tissue regeneration and wound repair. *Science* **2009**, *324*, 1666–1669. [CrossRef]
121. Ramachandran, P.; Iredale, J.P.; Fallowfield, J.A. Resolution of liver fibrosis: Basic mechanisms and clinical relevance. *Semin. Liver Dis.* **2015**, *35*, 119–131. [PubMed]
122. Zigmond, E.; Bernshtein, B.; Friedlander, G.; Walker, C.R.; Yona, S.; Kim, K.W.; Brenner, O.; Krauthgamer, R.; Varol, C.; Müller, W.; et al. Macrophage-restricted interleukin-10 receptor deficiency, but not IL-10 deficiency, causes severe spontaneous colitis. *Immunity* **2014**, *40*, 720–733.
123. Shouval, D.S.; Biswas, A.; Goettel, J.A.; McCann, K.; Conaway, E.; Redhu, N.S.; Mascanfroni, I.D.; Al Adham, Z.; Lavoie, S.; Ibourk, M.; et al. Interleukin-10 receptor signaling in innate immune cells regulates mucosal immune tolerance and anti-inflammatory macrophage function. *Immunity* **2014**, *40*, 706–719.
124. Khalil, N.; Bereznay, O.; Sporn, M.; Greenberg, A.H. Macrophage production of transforming growth factor beta and fibroblast collagen synthesis in chronic pulmonary inflammation. *J. Exp. Med.* **1989**, *170*, 727–737.
125. Said, E.A.; Dupuy, F.P.; Trautmann, L.; Zhang, Y.; Shi, Y.; El-Far, M.; Hill, B.J.; Noto, A.; Ancuta, P.; Peretz, Y.; et al. Programmed death-1-induced interleukin-10 production by monocytes impairs CD4+ T cell activation during HIV infection. *Nat. Med.* **2010**, *16*, 452–459. [PubMed]

126. Wynn, T.A.; Ramalingam, T.R. Mechanisms of fibrosis: Therapeutic translation for fibrotic disease. *Nat. Med.* **2012**, *18*, 1028–1040.
127. Getts, D.R.; Terry, R.L.; Getts, M.T.; Deffrasnes, C.; Muller, M.; van Vreden, C.; Ashhurst, T.M.; Chami, B.; McCarthy, D.; Wu, H.; et al. Therapeutic inflammatory monocyte modulation using immune-modifying microparticles. *Sci. Transl. Med.* **2014**, *6*, 219ra7. [CrossRef]
128. Happle, C.; Lachmann, N.; Skuljec, J.; Wetzke, M.; Ackermann, M.; Brennig, S.; Mucci, A.; Jirmo, A.C.; Groos, S.; Mirenska, A.; et al. Pulmonary transplantation of macrophage progenitors as effective and long-lasting therapy for hereditary pulmonary alveolar proteinosis. *Sci. Transl. Med.* **2014**, *6*, 250ra113. [CrossRef]
129. Van Welden, S.; Selfridge, A.C.; Hindryckx, P. Intestinal hypoxia and hypoxia-induced signalling as therapeutic targets for IBD. *Nat. Rev. Gastroenterol. Hepatol.* **2017**, *14*, 596–611. [CrossRef]
130. Wang, F.; Wang, K.; Xu, W.; Zhao, S.; Ye, D.; Wang, Y.; Xu, Y.; Zhou, L.; Chu, Y.; Zhang, C.; et al. SIRT5 desuccinylates and activates pyruvate kinase M2 to block macrophage IL-1β production and to prevent DSS-induced colitis in mice. *Cell Rep.* **2017**, *19*, 2331–2344.
131. Wang, T.; Liu, H.; Lian, G.; Zhang, S.Y.; Wang, X.; Jiang, C. HIF1α-Induced Glycolysis Metabolism Is Essential to the Activation of Inflammatory Macrophages. *Mediat. Inflamm.* **2017**, *2017*, 9029327. [CrossRef] [PubMed]
132. Kim, Y.E.; Lee, M.; Gu, H.; Kim, J.; Jeong, S.; Yeo, S.; Lee, Y.J.; Im, S.H.; Sung, Y.C.; Kim, H.J.; et al. HIF-1α activation in myeloid cells accelerates dextran sodium sulfate-induced colitis progression in mice. *Dis. Models Mech.* **2018**, *11*, dmm033241. [CrossRef]
133. Bäcker, V.; Cheung, F.Y.; Siveke, J.T.; Fandrey, J.; Winning, S. Knockdown of myeloid cell hypoxia-inducible factor-1α ameliorates the acute pathology in DSS-induced colitis. *PLoS ONE* **2017**, *12*, e0190074.
134. Palsson-McDermott, E.M.; Curtis, A.M.; Goel, G.; Lauterbach, M.A.; Sheedy, F.J.; Gleeson, L.E.; van den Bosch, M.W.; Quinn, S.R.; Domingo-Fernandez, R.; Johnston, D.G.; et al. Pyruvate kinase M2 regulates Hif-1α activity and IL-1β induction and is a critical determinant of the warburg effect in LPS-activated macrophages. *Cell Metab.* **2015**, *21*, 65–80. [CrossRef]
135. Krzak, G.; Willis, C.M.; Smith, J.A.; Pluchino, S.; Peruzzotti-Jametti, L. Succinate receptor 1: An emerging regulator of myeloid cell function in inflammation. *Trends Immunol.* **2021**, *42*, 45–58. [CrossRef]
136. Fremder, M.; Kim, S.W.; Khamaysi, A.; Shimshilashvili, L.; Eini-Rider, H.; Park, I.S.; Hadad, U.; Cheon, J.H.; Ohana, E. A transepithelial pathway delivers succinate to macrophages, thus perpetuating their pro-inflammatory metabolic state. *Cell Rep.* **2021**, *36*, 109521. [PubMed]
137. Macias-Ceja, D.C.; Ortiz-Masiá, D.; Salvador, P.; Gisbert-Ferrándiz, L.; Hernández, C.; Hausmann, M.; Rogler, G.; Esplugues, J.V.; Hinojosa, J.; Alós, R.; et al. Succinate receptor mediates intestinal inflammation and fibrosis. *Mucosal Immunol.* **2019**, *12*, 178–187.
138. Bauset, C.; Lis-Lopez, L.; Coll, S.; Gisbert-Ferrándiz, L.; Macias-Ceja, D.C.; Seco-Cervera, M.; Navarro, F.; Esplugues, J.V.; Calatayud, S.; Ortiz-Masia, D.; et al. SUCNR1 mediates the priming step of the inflammasome in intestinal epithelial cells: Relevance in ulcerative colitis. *Biomedicines* **2022**, *10*, 532. [CrossRef]
139. Fernández-Veledo, S.; Ceperuelo-Mallafré, V.; Vendrell, J. Rethinking succinate: An unexpected hormone-like metabolite in energy homeostasis. *Trends Endocrinol. Metab.* **2021**, *32*, 680–692.
140. Williams, N.C.; O'Neill, L.A.J. A Role for the Krebs cycle intermediate citrate in metabolic reprogramming in innate immunity and inflammation. *Front. Immunol.* **2018**, *9*, 141.
141. Infantino, V.; Convertini, P.; Cucci, L.; Panaro, M.A.; Di Noia, M.A.; Calvello, R.; Palmieri, F.; Iacobazzi, V. The mitochondrial citrate carrier: A new player in inflammation. *Biochem. J.* **2011**, *438*, 433–436. [CrossRef] [PubMed]
142. Schicho, R.; Shaykhutdinov, R.; Ngo, J.; Nazyrova, A.; Schneider, C.; Panaccione, R.; Kaplan, G.G.; Vogel, H.J.; Storr, M. Quantitative metabolomic profiling of serum, plasma, and urine by ^{1}H NMR spectroscopy discriminates between patients with inflammatory bowel disease and healthy individuals. *J. Proteome Res.* **2012**, *11*, 3344–3357. [CrossRef]
143. Guan, S.; Jia, B.; Chao, K.; Zhu, X.; Tang, J.; Li, M.; Wu, L.; Xing, L.; Liu, K.; Zhang, L.; et al. UPLC-QTOF-MS-based plasma lipidomic profiling reveals biomarkers for inflammatory bowel disease diagnosis. *J. Proteome Res.* **2020**, *19*, 600–609. [CrossRef] [PubMed]
144. Scoville, E.A.; Allaman, M.M.; Brown, C.T.; Motley, A.K.; Horst, S.N.; Williams, C.S.; Koyama, T.; Zhao, Z.; Adams, D.W.; Beaulieu, D.B.; et al. Alterations in lipid, amino acid, and energy metabolism distinguish crohn's disease from ulcerative colitis and control subjects by serum metabolomic profiling. *Metabolomics* **2018**, *14*, 17. [CrossRef] [PubMed]
145. Chawla, A.; Barak, Y.; Nagy, L.; Liao, D.; Tontonoz, P.; Evans, R.M. PPAR-gamma dependent and independent effects on macrophage-gene expression in lipid metabolism and inflammation. *Nat. Med.* **2001**, *7*, 48–52. [CrossRef]
146. Wahli, W.; Michalik, L. PPARs at the crossroads of lipid signaling and inflammation. *Trends Endocrinol. Metab.* **2012**, *23*, 351–363. [CrossRef]
147. Shah, Y.M.; Morimura, K.; Gonzalez, F.J. Expression of peroxisome proliferator-activated receptor-gamma in macrophage suppresses experimentally induced colitis. *Am. J. Physiol. Gastrointest. Liver Physiol.* **2007**, *292*, G657–G666. [CrossRef]
148. Hontecillas, R.; Horne, W.T.; Climent, M.; Guri, A.J.; Evans, C.; Zhang, Y.; Sobral, B.W.; Bassaganya-Riera, J. Immunoregulatory mechanisms of macrophage PPAR-γ in mice with experimental inflammatory bowel disease. *Mucosal Immunol.* **2011**, *4*, 304–313. [CrossRef]

149. Kang, S.; Nakanishi, Y.; Kioi, Y.; Okuzaki, D.; Kimura, T.; Takamatsu, H.; Koyama, S.; Nojima, S.; Nishide, M.; Hayama, Y.; et al. Semaphorin 6D reverse signaling controls macrophage lipid metabolism and anti-inflammatory polarization. *Nat. Immunol.* **2018**, *19*, 561–570. [CrossRef]
150. Bassaganya-Riera, J.; Viladomiu, M.; Pedragosa, M.; De Simone, C.; Carbo, A.; Shaykhutdinov, R.; Jobin, C.; Arthur, J.C.; Corl, B.A.; Vogel, H.; et al. Probiotic bacteria produce conjugated linoleic acid locally in the gut that targets macrophage PPAR γ to suppress colitis. *PLoS ONE* **2012**, *7*, e31238. [CrossRef]
151. Bosch, S.; Struys, E.A.; van Gaal, N.; Bakkali, A.; Jansen, E.W.; Diederen, K.; Benninga, M.A.; Mulder, C.J.; de Boer, N.K.H.; de Meij, T.G.J. Fecal amino acid analysis can discriminate de novo treatment-naïve pediatric inflammatory bowel disease from controls. *J. Pediatr. Gastroenterol. Nutr.* **2018**, *66*, 773–778. [CrossRef] [PubMed]
152. Kolho, K.L.; Pessia, A.; Jaakkola, T.; de Vos, W.M.; Velagapudi, V. Faecal and serum metabolomics in paediatric inflammatory bowel disease. *J. Crohn's Colitis* **2017**, *11*, 321–334.
153. Ooi, M.; Nishiumi, S.; Yoshie, T.; Shiomi, Y.; Kohashi, M.; Fukunaga, K.; Nakamura, S.; Matsumoto, T.; Hatano, N.; Shinohara, M.; et al. GC/MS-based profiling of amino acids and TCA cycle-related molecules in ulcerative colitis. *Inflamm. Res.* **2011**, *60*, 831–840.
154. Sido, B.; Seel, C.; Hochlehnert, A.; Breitkreutz, R.; Dröge, W. Low intestinal glutamine level and low glutaminase activity in Crohn's disease: A rational for glutamine supplementation? *Dig. Dis. Sci.* **2006**, *51*, 2170–2179.
155. Hu, X.; Deng, J.; Yu, T.; Chen, S.; Ge, Y.; Zhou, Z.; Guo, Y.; Ying, H.; Zhai, Q.; Chen, Y.; et al. ATF4 deficiency promotes intestinal inflammation in mice by reducing uptake of glutamine and expression of antimicrobial peptides. *Gastroenterology* **2019**, *156*, 1098–1111.
156. Chu, C.C.; Hou, Y.C.; Pai, M.H.; Chao, C.J.; Yeh, S.L. Pretreatment with alanyl-glutamine suppresses T-helper-cell-associated cytokine expression and reduces inflammatory responses in mice with acute DSS-induced colitis. *J. Nutr. Biochem.* **2012**, *23*, 1092–1099. [CrossRef]
157. Hou, Y.C.; Liu, J.J.; Pai, M.H.; Tsou, S.S.; Yeh, S.L. Alanyl-glutamine administration suppresses Th17 and reduces inflammatory reaction in dextran sulfate sodium-induced acute colitis. *Int. Immunopharmacol.* **2013**, *17*, 1–8.
158. Xu, Q.; Hu, M.; Li, M.; Hou, J.; Zhang, X.; Gao, Y.; Chachar, B.; Li, X. Dietary bioactive peptide alanyl-glutamine attenuates dextran sodium sulfate-induced colitis by modulating gut microbiota. *Oxidative Med. Cell. Longev.* **2021**, *2021*, 5543003.
159. Ren, W.; Liu, S.; Chen, S.; Zhang, F.; Li, N.; Yin, J.; Peng, Y.; Wu, L.; Liu, G.; Yin, Y.; et al. Dietary L-glutamine supplementation increases Pasteurella multocida burden and the expression of its major virulence factors in mice. *Amino Acids* **2013**, *45*, 947–955. [CrossRef]
160. Digby-Bell, J.L.; Atreya, R.; Monteleone, G.; Powell, N. Interrogating host immunity to predict treatment response in inflammatory bowel disease. *Nat. Rev. Gastroenterol. Hepatol.* **2020**, *17*, 9–20.
161. Sandborn, W.J.; Su, C.; Sands, B.E.; D'Haens, G.R.; Vermeire, S.; Schreiber, S.; Danese, S.; Feagan, B.G.; Reinisch, W.; Niezychowski, W.; et al. Tofacitinib as induction and maintenance therapy for ulcerative colitis. *N. Engl. J. Med.* **2017**, *376*, 1723–1736. [CrossRef] [PubMed]
162. Cordes, F.; Lenker, E.; Spille, L.J.; Weinhage, T.; Bettenworth, D.; Kessel, C.; Schmidt, H.H.; Foell, D.; Varga, G. Tofacitinib reprograms human monocytes of IBD patients and healthy controls toward a more regulatory phenotype. *Inflamm. Bowel Dis.* **2020**, *26*, 391–406. [CrossRef] [PubMed]
163. De Vries, L.C.S.; Duarte, J.M.; De Krijger, M.; Welting, O.; Van Hamersveld, P.H.P.; Van Leeuwen-Hilbers, F.W.M.; Moerland, P.D.; Jongejan, A.; D'Haens, G.R.; De Jonge, W.J.; et al. A JAK1 selective kinase inhibitor and tofacitinib affect macrophage activation and function. *Inflamm. Bowel Dis.* **2019**, *25*, 647–660. [CrossRef]
164. Deepak, P.; Alayo, Q.A.; Khatiwada, A.; Lin, B.; Fenster, M.; Dimopoulos, C.; Bader, G.; Weisshof, R.; Jacobs, M.; Gutierrez, A.; et al. Safety of tofacitinib in a real-world cohort of patients with ulcerative colitis. *Clin. Gastroenterol. Hepatol.* **2021**, *19*, 1592–1601.e3. [CrossRef] [PubMed]
165. Sandborn, W.J.; Panés, J.; D'Haens, G.R.; Sands, B.E.; Su, C.; Moscariello, M.; Jones, T.; Pedersen, R.; Friedman, G.S.; Lawendy, N.; et al. Safety of tofacitinib for treatment of ulcerative colitis, based on 4.4 years of data from global clinical trials. *Clin. Gastroenterol. Hepatol.* **2019**, *17*, 1541–1550. [CrossRef]
166. Feagan, B.G.; Danese, S.; Loftus, E.V., Jr.; Vermeire, S.; Schreiber, S.; Ritter, T.; Fogel, R.; Mehta, R.; Nijhawan, S.; Kempiński, R.; et al. Filgotinib as induction and maintenance therapy for ulcerative colitis (SELECTION): A phase 2b/3 double-blind, randomised, placebo-controlled trial. *Lancet* **2021**, *397*, 2372–2384. [CrossRef]
167. Vermeire, S.; Schreiber, S.; Petryka, R.; Kuehbacher, T.; Hebuterne, X.; Roblin, X.; Klopocka, M.; Goldis, A.; Wisniewska-Jarosinska, M.; Baranovsky, A.; et al. Clinical remission in patients with moderate-to-severe Crohn's disease treated with filgotinib (the FITZROY study): Results from a phase 2, double-blind, randomised, placebo-controlled trial. *Lancet* **2017**, *389*, 266–275. [CrossRef]
168. Sandborn, W.J.; Feagan, B.G.; Loftus, E.V., Jr.; Peyrin-Biroulet, L.; Van Assche, G.; D'Haens, G.; Schreiber, S.; Colombel, J.F.; Lewis, J.D.; Ghosh, S.; et al. Efficacy and safety of upadacitinib in a randomized trial of patients with crohn's disease. *Gastroenterology* **2020**, *158*, 2123–2138.e8. [CrossRef]

169. Sandborn, W.J.; Ghosh, S.; Panes, J.; Schreiber, S.; D'Haens, G.; Tanida, S.; Siffledeen, J.; Enejosa, J.; Zhou, W.; Othman, A.A.; et al. Efficacy of upadacitinib in a randomized trial of patients with active ulcerative colitis. *Gastroenterology* **2020**, *158*, 2139–2149.e14. [CrossRef]
170. Holmer, A.K.; Singh, S. Expanding targeted immune modulators in ulcerative colitis. *Lancet* **2021**, *397*, 2313–2315. [CrossRef]
171. Torretta, S.; Scagliola, A.; Ricci, L.; Mainini, F.; Di Marco, S.; Cuccovillo, I.; Kajaste-Rudnitski, A.; Sumpton, D.; Ryan, K.M.; Cardaci, S. D-mannose suppresses macrophage IL-1β production. *Nat. Commun.* **2020**, *11*, 6343. [CrossRef] [PubMed]
172. Liu, R.; Li, X.; Ma, H.; Yang, Q.; Shang, Q.; Song, L.; Zheng, Z.; Zhang, S.; Pan, Y.; Huang, P.; et al. Spermidine endows macrophages anti-inflammatory properties by inducing mitochondrial superoxide-dependent AMPK activation, Hif-1α upregulation and autophagy. *Free Radic. Biol. Med.* **2020**, *161*, 339–350. [CrossRef]
173. Zhuang, H.; Lv, Q.; Zhong, C.; Cui, Y.; He, L.; Zhang, C.; Yu, J. Tiliroside ameliorates ulcerative colitis by restoring the M1/M2 macrophage balance via the HIF-1α/glycolysis pathway. *Front. Immunol.* **2021**, *12*, 649463. [CrossRef] [PubMed]
174. Wu, M.M.; Wang, Q.M.; Huang, B.Y.; Mai, C.T.; Wang, C.L.; Wang, T.T.; Zhang, X.J. Dioscin ameliorates murine ulcerative colitis by regulating macrophage polarization. *Pharmacol. Res.* **2021**, *172*, 105796. [CrossRef]
175. Lv, Q.; Xing, Y.; Liu, Y.; Chen, Q.; Xu, J.; Hu, L.; Zhang, Y. Didymin switches M1-like toward M2-like macrophage to ameliorate ulcerative colitis via fatty acid oxidation. *Pharmacol. Res.* **2021**, *169*, 105613. [CrossRef] [PubMed]
176. Xiong, W.; Sun, K.Y.; Zhu, Y.; Zhang, X.; Zhou, Y.H.; Zou, X. Metformin alleviates inflammation through suppressing FASN-dependent palmitoylation of Akt. *Cell Death Dis.* **2021**, *12*, 934. [CrossRef]
177. Katsandegwaza, B.; Horsnell, W.; Smith, K. Inflammatory bowel disease: A Review of pre-clinical murine models of human disease. *Int. J. Mol. Sci.* **2022**, *23*, 9344. [CrossRef]

International Journal of
Molecular Sciences

MDPI

Review

Role of Mitochondrial Dynamics in Skin Homeostasis: An Update

Tao Quan, Ran Li and Ting Gao *

College of Veterinary Medicine, China Agricultural University, Beijing 100083, China;
s20243051072@cau.edu.cn (T.Q.); 18246269672@163.com (R.L.)
* Correspondence: gt1994@cau.edu.cn

Abstract: Skin aging is the most prominent phenotype of host aging and is the consequence of a combination of genes and environment. Improving skin aging is essential for maintaining the healthy physiological function of the skin and the mental health of the human body. Mitochondria are vital organelles that play important roles in cellular mechanisms, including energy production and free radical balance. However, mitochondrial metabolism, mitochondrial dynamics, biogenesis, and degradation processes vary greatly in various cells in the skin. It is well known that mitochondrial dysfunction can promote the aging and its associated diseases of the skin, resulting in the damage of skin physiology and the occurrence of skin pathology. In this review, we summarize the important role of mitochondria in various skin cells, review the cellular responses to vital steps in mitochondrial quality regulation, mitochondrial dynamics, mitochondrial biogenesis, and mitochondrial phagocytosis, and describe their importance and specific pathways in skin aging.

Keywords: ultraviolet radiation; skin aging; mitochondria; oxidative stress

1. Introduction

Cellular aging is a complex physiological process characterized by an eternal inhibition of proliferation [1]. Numerous data show that cellular senescence is accompanied by a gradual decrease in the ability to proliferate and differentiate, eventually becoming physiologically nonfunctional [2]. It is well known that skin aging is the most obvious feature of organismal aging. In addition to physiological aging caused by endogenous factors such as aging, exogenous factors including ultraviolet radiation further aggravate skin aging [3]. Clinically, the phenotypes of skin aging include skin thinning, wrinkle appearance, hyperpigmentation, and capillary dilation with enlarged pores [4]. Skin aging not only changes the host's appearance and brings serious psychological burden, which in turn hinders normal work and social interactions, but also leads to the occurrence of related diseases and even skin cancer [5,6]. Current research has confirmed the importance of mitochondrial functional homeostasis in skin health and the key role that mitochondrial dysfunction plays in skin inflammation, skin diseases, and skin cancers. Mitochondrial dysfunction in the skin and other tissues is mechanistically based on excessive ROS production and its induction of oxidative stress, although some specific skin injuries are mediated by other mitochondrial pathways [7]. Although scientists have made many breakthroughs in revealing mitochondria-associated genes and proteins that play important roles in health- and disease-related environments, it is still difficult to unify views of how mitochondrial dysfunction drives the changes associated with skin aging [8].

Here, we not only discuss the critical role mitochondria play in healthy skin physiology versus premature skin aging and the associated diseases that result, but we also discuss

 https://doi.org/10.3390/ijms26051803

the newly proposed idea that mitochondrial dysfunction promotes the pathogenesis of age-related skin diseases. Specifically, therapeutic strategies to ameliorate mitochondrial dysfunction and thereby slow skin aging by targeting mitochondrial dynamics, mitochondrial biogenesis, and mitochondrial autophagy are argued to be beneficial to humans in terms of healthy aging, prevention of age-associated diseases, and longevity.

2. Skin Structure

The skin is the outer sheath of the body, which is in contact with the external environment. It is a complex tissue that consists of several layers of tissue, each of which is made up of different cells. It has important functions such as sensory, thermal insulation, and protection against water loss, and is a physical barrier against pathogens [7].

The skin is made up of more than 20 different types of cells and is divided into three layers: epidermis, dermis, and subcutaneous. The epidermis forms the outermost layer, offering a defensive barrier to the skin [9]. The epidermis is mainly differentiated from keratinocytes. Originating from the ectoderm during embryogenesis, keratinocytes form the main cell type in the epidermis and form the protective layer of the skin. Upon differentiation as they move from the basal layer to the surface of the skin, keratinocytes undergo morphological changes and produce keratin, which contributes to skin strength and waterproofing [10,11].

The dermis is located under the epidermis and is connected to the epidermis by the dermal-epidermal connection and contains a variety of structures, including blood vessels, sebaceous glands, sweat glands, etc. Dermal-epidermal junction (DEJ), which consists of an abundance of extracellular matrix (ECM) and various functional structures. In the compartments of the skin, the dermis has the richest cellular richness. Among them, fibroblasts are the main cell type in the dermis and exert a central role in the synthesis and secretion of the ECM, which is composed of structurally related proteins including collagen and elastin, glycoproteins, and glycosaminoglycans. Overall, these components maintain the elasticity, fullness, and resilience of the skin [10,12–14]. Other cells are also present in the dermis, including mesenchymal stem cells (MSCs), fat cells, and immune cells, the latter of which are less numerous due to the presence of blood vessels [15]. In addition, the dermal chamber is home to additional functionally complex ingredients, including hair follicles, sweat glands, sebaceous glands, and sensory nerve endings [14].

Finally, the hypodermis is the innermost layer of skin [14]. It is mainly constituted of adipocytes, preadipocytes, and stem cells derived from adipose but is also composed of fibroblasts, macrophages, T cells, and erythrocytes, forming a network of interstitial vascular cells in nerves, muscles, and hair follicles [16]. The central role of the subcutaneous tissue is to link the skin to substructures, including muscles and bones, offering a pillar and insulation to the host and being regarded as a permanent energy storage chamber [17].

3. Skin Aging

Acts as a barrier between the body and the external environment, the vital location of the skin determines the important function of keeping balance in the body and maintaining the survival of the host [18]. This leads to the skin, the organ most susceptible to environmental elements, including pollutants, temperature, and sun exposure. Physical and chemical responses induced by a variety of internal and external factors can lead to a gradual deficiency of skin structure and function, ultimately resulting in skin aging [19]. Intrinsic aging is mainly determined by genes and is the natural result of alterations in the physiological system over time; however, external aging is caused by a combination of external environments [20], including ultraviolet (UV) exposure, smoke, pollutants, nutrients, and exercise (Figure 1). Clinical phenotypes are associated with skin aging, including dry and rough, wrinkle formation, decreased elasticity, no regular pigment distribution, and a pronounced vascular system [21].

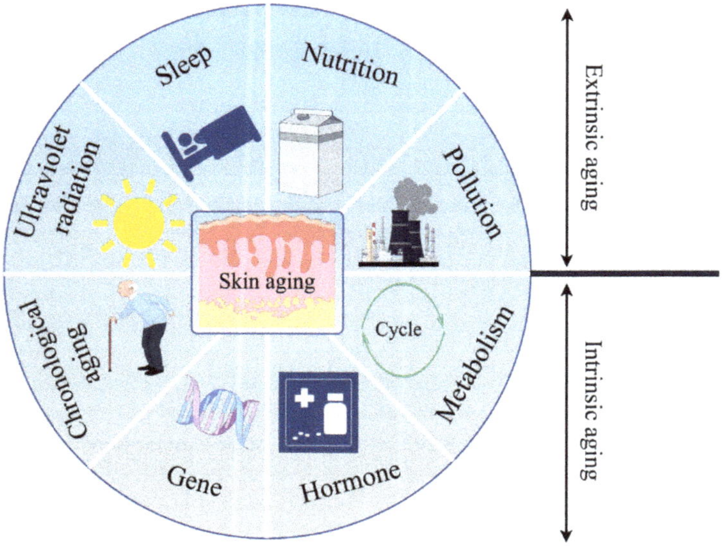

Figure 1. Skin aging. Skin aging is divided into endogenous and exogenous aging. Intrinsic skin aging is mainly determined by chronological aging, genes, hormones, and metabolism and is the natural result of changes in the physiological system over time, while external aging is caused by a combination of external environments, including ultraviolet radiation, sleep, nutrition, and pollution.

Ultraviolet exposure is the most common external factor that induces skin aging [22,23]. Ultraviolet radiation (UVR) exposure is an important environmental element that induces apoptosis and skin barrier damage [24]. Of the three subcategories of UVR, UVA (320–400 nm) and UVB (280–320 nm) reach earth in large numbers, while UVC (200–280 nm) does not [25]. As a result, UVA and UVB bear responsibility for most UVR-related health damages [25]. UVA is a more extensive ingredient that penetrates the dermis, while UVB, which typically does not exceed 10% of UVR, has high energy and is mainly assimilated by epidermal cells [25,26].

Senescence is often defined as the slow decline of an organism's functions over time, which can be induced by the synchronous deterioration of interrelated cellular capacities. Although often mistaken for being found only in modern humans and animals in function, in fact, most metazoans indicate various markers of aging during their lifetime [27]. There is a great deal of interest in understanding aging and aging-related conditions and their associated molecular mechanisms in the hope of delaying these alterations. Researchers have long struggled to explore the process of aging. Especially if aging becomes apparent long after the reduced fertility, and therefore the power of nature has long since been weakened, how does the genetic process of aging work?

4. Mitochondria and Skin

Mitochondria are found in almost all eukaryotic cells, and they are multifunctional organelles. Mitochondria exert a vital role in signaling, including bioenergetics, reactive oxygen species (ROS) production, catabolism and anabolism, iron-sulfur clusters and heme biosynthesis, calcium and iron balance, and apoptosis, and are central to the regulation of cellular metabolism and homeostasis [28–32]. These organelles are essential for life, dynamics, and cellular stress responses [29,33]. Mitochondrial dysfunction is closely associated with various phenotypes of aging, such as impaired oxidative phosphorylation (OXPHOS) processes, increased oxidative damage, decreased mitochondrial quality control, decreased metabolic enzyme activity, and changes in mitochondrial morphology, kinetics, and biogenesis [27,34]. Mitochondrial dysfunction can affect host multi-organ systems, resulting in a pathological state known as "mitochondrial disease". In addition, mitochondrial dysfunction is also associated with many age-related pathological conditions, including neurodegenerative diseases, metabolic diseases, and cardiovascular diseases [35–40]. Cellular homeostasis and mitochondrial health require tight regulation between the generation of new cells and the clearance of damaged mitochondria. Recent studies on skin homeostasis and aging have identified the indispensable functions of mitochondria [41] (Figure 2).

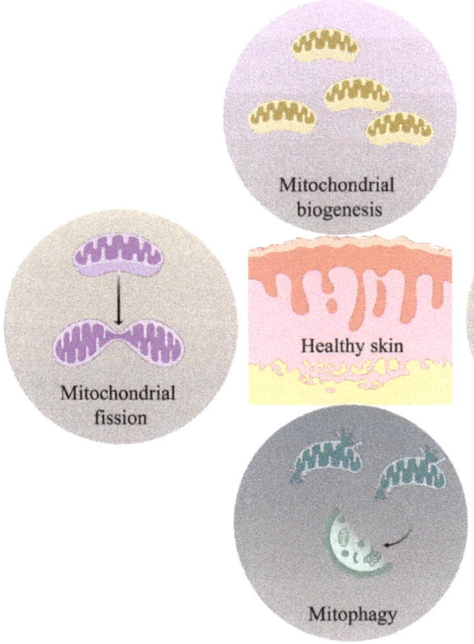

Figure 2. Skin health is tied to mitochondrial homeostasis. The state of healthy skin is associated with dynamic mitochondrial quality control, including mitochondrial biogenesis, mitochondrial fusion, mitochondrial division, and mitochondrial autophagy.

A growing body of research has shown that mitochondria have become important players in maintaining skin homeostasis [7]. Mitochondria are at key locations in the skin. While the skin does not have a huge energy requirement compared to other organs such as skeletal muscle, it is vital for effects such as hair growth, immune homeostasis, hormone regulation, pigmentation, and wound healing [42]. UV-caused oxidative stress and its capacity for signaling during extrinsic-induced skin aging have been well documented [43]. However, the importance of mitochondrial damage in UVR-caused skin injury or associated skin manifestations remains to be further clarified. Our review aimed to illustrate the mitochondrial mechanism. Here, we summarize how mitochondria are involved in skin aging and how the opinions can usher in novel insights of mitochondria-targeted approaches that potentially slow down the rate of aging.

4.1. Effects of Mitochondria on the Epidermis

It is important to note that the ability required for the sustained regeneration of the skin's epidermis derives from the adenosine triphosphate (ATP) produced by the mitochondria [44]. Moreover, as keratinocytes gradually differentiate from the basal layer and move up to the basil, mitochondrial structure and function also undergo dynamic changes. In addition, mitochondria-derived ROS can also be regarded as a signal to regulate epidermal differentiation [45]. In terms of metabolism, mitochondria are intermediates that approve the biosynthetic pathways associated with keratin synthesis and lipid metabolism, so mitochondrial homeostasis within keratinocytes is essential for the maintenance of the skin barrier and even host homeostasis [46].

When keratinocytes and slow-proliferating fibroblasts are compared, the accumulation of functional damages ends in keratinocytes for a less pronounced time, and in addition to mitochondrial damage, other mechanisms can affect epidermal integrity [47–49]. Alterations in mitochondrial structure are seen during keratinocyte differentiation and persist throughout the skin aging process. The mitochondria number in keratinocytes is significantly higher in the granular layer than in keratinocytes in the spinous layer, and the mitochondria within these different layers form a compact mitochondrial network system [50]. In addition to persistently strong inter-individual variation in mitochondrial proliferation keratinocytes, no age-related differences in mitochondrial numbers were observed to compare cell isolation from young and elderly subjects in vitro, while mitochondrial junction decreased in older keratinocytes with younger ones [50,51].

4.2. Effects of Mitochondria on the Dermis

Fibroblasts within the dermis exert an important role in maintaining the skin morphology and function. They are beneficial for the synthesis and keeping of ECM and are positively involved in distinct skin-associated processes, including collagen synthesis and repair. As energy-hungry cells, fibroblasts rely heavily on mitochondria to perform key features, such as ECM synthesis, the synthesis of growth factors, and cell signaling. In addition, mitochondrial function is essential for maintaining fibroblast viability, facilitating proliferation, and enhancing collagen formation, which is vital for keeping skin integrity [52–54].

4.3. Effects of Mitochondria on the Skin Accessories

The sebaceous gland is an exocrine gland that is beneficial for the release of lipid-abundant sebum and is a key constituent of epidermal defenses, capacity, temperature maintenance, skin strength, and skin microbiota balance [55]. The sebaceous glands are usually linked to hair follicles, a structure consisting of a single pipe leading to the hair follicle, a cluster of secretory cells present at the base of the duct. Glandular cells are mainly made up of a special type of adipocyte that can synthesize and accumulate lipids and participate in other components that constitute sebum [56,57]. Taking into account the active metabolic activity of sebaceous lipids, these cells are rich in highly active mitochondria in their cytoplasm [58].

Mitochondrial dynamics has a vital influence on melanocytes [59]. Studies have shown that genetic and chemical-physical suppression of the mitochondrial fission protein Drp1 results in upregulated melanin generation and mitochondrial lengthening, while reduction of the mitochondrial fusion protein dynein-like GTPase 1 inhibits melanin production, resulting in mitochondrial fragmentation. Correspondingly, mitochondrial fragmentation caused by carbonyl cyanide 3-chlorophenylhydrazone (CCCP) treatment also decreased melanin production. In addition, mitochondrial fission activates the ROS-ERK loop, which leads to the proteasome breakdown of microphthalmia-related transcription elements, further promoting the inhibition of melanin production, favoring the speculation that mitochondrial dynamics exert a vital role in melanin production [60].

Perhaps no other structure is so closely related to the energy of the young and the reduction of the elderly. These associations of the complicated and antagonistic capacities of mitochondria have slowly changed the way we think about the sub-organelles. Mitochondria are seen as more than just energy factories; they are also thought of as a connector for intracellular signal transduction, controllers of innate immunity, and regulators of stem cell activity. These properties, in turn, offer clues to how mitochondria modulate skin aging and skin age-associated disorders [34].

5. Structure and Function of Mitochondria

Mitochondria are dynamic organelles within eukaryotic cells and are encased by two layers of membranes. They contain unique DNA that plays a decisive role in maintaining the energy and metabolic requirements of cells. The main task of mitochondria is to produce ATP through cellular metabolism, thus serving as an important source of energy. In addition to energy production, mitochondria exert a key role in regulating cell proliferation, differentiation, calcium, iron balance, and programmed death of cells [61,62]. In addition, mitochondria are associated with keeping electrolyte balance, activating inflammatory responses, and coordinating redox responses [43].

Mitochondria are surrounded by two membranes, inner and outer. This morphological structure constitutes two independent water compartments: the intermembrane space and the mitochondrial matrix, which exert a key role in keeping the cellular function of this cell. The outer membrane exhibits high permeability and consists of various large protein-based pores and positive transporters that are collectively responsible for the transport of proteins, nucleotides, ions, and metabolites between the cytoplasm and the intermembrane space. By contrast, the permeability of the inner membrane is more restricted, and resembles the plasma membrane of a cell. The inner membrane constitutes a large number of folds (cristae) and extends into the matrix of the organelle, in which proteins associated with electron transport and ATP generation. The intermembranous space is located between the two layers of membranes, and its constitution is similar in structure to the cytoplasm. This matrix is encapsulated by an inner membrane and consists of enzymes, mitochondrial DNA (mtDNA), and ribosomes, which are vital for ATP generation. Mitochondrial DNA is encapsulated in a nucleoid, which is translated into vital mitochondrial proteins, which constitute the oxidative phosphorylation complex [61].

The mitochondrial dynamics homeostasis is essential for keeping mitochondrial and cellular balance, and it is regulated by mitochondrial dynamics, mitochondrial biogenesis, and mitophagy. These processes determine the morphology, number, distribution, and location of mitochondria, allowing for optimal cellular energy generation, calcium balance, and other important mitochondrial activity.

Mitochondrial dynamics, mitochondrial biogenesis, and mitophagy are coordinatedly governing and interacting. The mitochondrial fission cleaves the injured mitochondria and cuts them into smaller pieces that can be enclosed by vesicle membranes that provide autophagic content for the mitophagy [63]. The mitochondrial fusion process indirectly governs mitophagy, and when fusion is suppressed, mitochondrial fission further facilitates mitophagy [64]. Meanwhile, mitophagy also influences the fission pathway of mitochondria via binding receptors to related molecules related to mitochondrial dynamics [65,66]. Mitochondrial biogenesis is a closely controlled process that is implicated in a variety of signaling loops and signaling transcription elements. Mitochondrial biogenesis permits cells to satisfy up-regulated energy requirements under different stimuli, eliminating fragmented mitochondria, and is important for cells to adapt to challenges [67]. Once the mitochondria are injured, the mitochondrial network is unable to keep an abundant energy provision [68].

5.1. Mitophagy

Mitophagy is a highly selective type of autophagy that permits lysosomal degradation of damaged mitochondria. Mitophagy is associated with fission and fusion, and together they protect mitochondrial homeostasis [69]. Most studies have elucidated the PTEN-caused kinase 1 (PINK1) and Parkin-mediated mitophagy loops [70]. PINK1 localizes to IMM under healthy conditions and is sustainably broken down through proteases. However, when the body is stressed, ΔΨm is depolarized, which in turn leads to the inactivity of proteases and the PINK1 stability. PINK1 is transported by translocase to OMM and accumulates in OMM, where it is autophosphorylated and hires Parkin, subsequently ubiquitinating the substrates, such as Mfn1/2 and voltage-dependent anion channel-1 (VDAC1). The ubiquitinated proteins are then marked and combined with the adaptor protein p62 and eventually undergo light chain 3 (LC3)-mediated clearance of mitochondrial phagocytosis, resulting in degradation of injured mitochondria [71]. Research suggests that Nrf2 also has a vital influence in regulating mitochondrial phagocytosis and maintaining mitochondrial balance. Nrf2 could facilitate mitochondrial phagocytosis and clear damaged mitochondria through the upregulation of p62 and PINK [72]. At the same time, the diminished ability to clear injured mitochondria leads to a large amount of ROS, the unleashing of damage-associated molecular patterns (DAMPs), and the occurrence of inflammation [73,74].

5.2. Mitochondrial Dynamics

In order to adapt to alterations in the cellular environment and keep the mitochondria balanced, this process in which mitochondria constantly undergo fission and fusion is known as "mitochondrial dynamics". Mitochondria maintain the healthy structure and function through a homeostasis of fission and fusion, which is a prerequisite for it to keep physiological capacity. The homeostasis is mediated by various dynamically associated GTPases, which maintain the quantity and quality of mitochondria in parental and daughter cells [75].

Mitochondrial fusion, as a complementary measure, favors the exchange of content between adjacent mitochondria, while fission allows the isolation of damaged mitochondria for the follow-up clear by mitophagy. Mitochondrial proteins (MFN) 1 and 2 are held responsible for the fusion of the outer mitochondrial membrane (OMM); however, the inner mitochondrial membrane (IMM) fusion is dependent on the (optic atrophy 1 (OPA1) protein [76]. In contrast, dyne-associated protein 1 (Drp-1) could be recruited from the cytoplasm to the mitochondrial surface as a central regulator that coordinates fission events [77,78]. The homeostasis of mitochondrial fusion and fission, by regulating mitochondrial morphology, facilitating material exchange, keeping the stability and integrity of mtDNA, and isolating damaged mitochondria, is very important for mitochondrial function [79].

5.3. Mitochondrial Biogenesis

Mitochondrial biogenesis is essential for the synthesis of new mitochondria within cells, which is able to be activated by various distinct signals under stress situations and is influenced by toxin increases and mtDNA mutations [80]. Mitochondrial biogenesis is essential for maintaining the quantity and quality of mitochondria involved in the nuclear and mitochondrial genomes. Some transcription elements are associated with the process of mitochondrial biogenesis regulated by physiological stimuli, such as physical activity, nutrient intake, and temperature, etc. Peroxisome proliferation-activating receptor (PPAR)-γ coactivator-1α (1α)

regulates the process of co-transcriptional regulators with various transcription elements, including nuclear respiration factors (NRF-1 and NRF-2), mitochondrial transcription factors (Tfam), uncoupling proteins (UCP2), PPARs, thyroid hormones, glucocorticoids, estrogen, and estrogen-associated receptors (ERRs) α and γ [81–83]. NRF-1, NRF-2, and Tfam mainly regulate mitochondrial enzyme transcription and mtDNA synthesis [84]. In addition to the transcription elements, two other vital enzymes that are thought to be metabolic receptors that control mitochondrial biogenesis are AMP-activated protein kinase (AMPK) and the mammalian counterpart silencing information regulator 1 (SIRT1). In the case of energy deprivation, AMPK and SIRT1 regulate PGC-1α by phosphorylation and deacetylation, respectively [85]. Some clinical research have indicated that the morphology and mitochondrial numbers in tissues, including the heart, skeletal muscle, and liver tissue, are altered under pathogenic conditions [86].

6. Mitochondrial Dysfunction and Aging

Mitochondrial damage and cell aging are two signs of aging and are linked to each other [87]. Mitochondrial dysfunction is characterized as a reduction in the respiration ability of each mitochondrion and a decrease in mitochondrial membrane potential, a process always accompanied by an excess of ROS level, which is thought to be both a consequence and a result of cell aging. This damage has a vital influence on various feedback pathways that cause and maintain the aging manifestations [88]. During cell aging, the increases of damaged mitochondria are caused by elements including excess ROS and mitophagy disorder, which selectively remove dysfunctional mitochondria. As the accumulation of mitochondrial damage gradually increases, mitochondrial dysfunction is further exacerbated, ultimately resulting in downregulated cell bioenergy and increased ROS [89]. This interaction between mitochondrial damage and aging not only leads to cell aging but also has vital influences, affecting cell and organ capacity to play a central role in the broader manifestations of aging and age-associated disorders in organs.

As we age, mitochondria change in morphology, quantity, and OXPHOS activity [90]. Senescence is closely linked to alterations in mitochondrial structure, including hyperfusion or upregulated fragmentation and injury of the mitochondrial network [7,91]. In addition, with age, the oxidative phosphorylation efficiency of mitochondria decreases, resulting in a decrease in ATP production and a decrease in $\Delta\Psi M$, which impairs cellular energy metabolism. These changes are associated with upregulation of ROS levels during aging, the cumulation of which leads to oxidative stress to mitochondria and cellular ingredients, leading to cell aging [92]. In addition, mitochondrial turnover decreases during aging, which illustrates impaired mitochondrial quality control processes, including downregulation of mitochondrial biogenesis and mitophagy [7]. The mitochondria's response to stressful events depends on the stimulus and cell type. Notably, the overdose of somatic mtDNA mutations due to internal or external damage impairs mitochondrial capabilities and leads to upregulated ROS formation [93,94].

6.1. The Regulatory Roles of Mitochondrial Biogenesis in Aging

As we age, mitochondrial biogenesis gradually downregulates. Mitochondrial biogenesis relies on different signaling cascades to produce normal and intact mitochondria. Generally, PGC-1α is regarded as the intermediate connector of mitochondrial biogenesis. Research has confirmed the importance of PGC-1α in the skin aging process. UVB-irradiated human keratinocyte cells (HaCaT) underwent senescence compared to cells not exposed to UV light, which is characterized by significant downregulation of AMPK phosphorylation levels, SIRT-1, and PGC-1α contents [95]. Guo et al. found that cryptotanshinone (CTS) can promote mitochondrial production in skin cells and ameliorate skin aging via regulating the

AMPK-PGC-1α signaling pathway [96]. Similarly, the standardized kaempferia parviflora extract inhibits the intrinsic aging process in human dermal fibroblasts and hairless mice by up-regulating the level of PGC-1α, therefore facilitating mitochondrial biogenesis and ultimately inhibiting cellular senescence and mitochondrial dysfunction [97]. Increased PGC-1α promotes mitochondriogenesis, ameliorates mitochondrial capacities, and delays intrinsic and extrinsic element-caused skin aging.

6.2. The Regulatory Roles of Mitophagy in Aging

In senescent cells, mitophagy is activated [12]. Decreased mitophagy and impaired autophagy pathways have been implicated in aging and various age-associated disorders, such as neurodegenerative diseases, cardiac and autoimmune disorders, glucose-metabolic disorders, and tumors [98]. Mitochondrial depolarization is a mark of mitophagy [99]. Activation of DRP1 leads to abnormalities in the morphology and function of mitochondria [100]. Parkin and PINK1 are two protein signs of mitochondrial phagocytosis, and they have a vital influence on mitochondrial capacities [101]. PINK1 continuously accumulates in impaired mitochondria and regulates Parkin, promoting mitophagy [102]. Research showed that when exposed to ultraviolet light, cells initiate mitophagy to clear injured mitochondria, decrease ROS levels, keep photoaging cell balance, suppress mitophagy, disturb photoaging cellular homeostasis, and activate the apoptotic response.

6.3. The Regulatory Roles of Mitochondrial Dynamics in Aging

The decrease in mitochondrial biogenesis with age may be the result of an imbalance between mitochondrial fission and fusion pathways, as well as mitochondrial phagocytosis suppression, which favors the elimination of impaired mitochondria [103]. ROS is a family of free radicals that include superoxide anions, hydroxyl groups, peroxyl radicals, and other non-free radicals capable of producing free radicals [104]. While the production of ROS within the cell is an unavoidable process in itself, the cell simultaneously possesses many defense systems to eliminate it to maintain homeostasis. The overproduction of reactive oxygen species is associated with the oxidative damage caused by lipids, DNA, and proteins [105,106]. Previous studies have shown that oxidative stress is linked to a variety of pathophysiological conditions, including aging and age-related diseases [107].

Mitochondria have been the focus of aging research since the introduction of the free radical theory of aging and its later improved version, the mitochondrial aging theories [108–110]. Several papers have been introduced in which aged mice genetically deficient in mitochondrial NAD(P)+ transhydrogenase, an enzyme involved in mitochondrial NADPH supply, show impaired motor behavior. According to these papers, it was shown that NNT deficiency interacts with the effects of aging to exacerbate motor dysfunction and cause impaired mitochondrial respiration in oxidative muscles [111–114]. Among the studies on mitochondrial aging, an increase in ROS can cause cellular deterioration with age, with the theory of mitochondrial aging emphasizing mitochondria as the primary producer of ROS. While the accumulation of oxidative damage occurs with age, evidence on the causal relationship of ROS or mitochondrial ROS as a driver behind aging remains lacking [115].

7. Mitochondrial Dysfunction and Skin Aging

Skin aging is an intricate process that results in a decrease in the balance and regenerative ability of the organ. Mitochondria are vital organelles that have an important influence on cell mechanisms, including energy synthesis and free radical sustainment. However, mitochondrial dynamics, biogenesis, and degradation processes vary widely in different cells of the skin. It is well known that mitochondrial dysfunction can lead to skin aging and inflammation, further inducing damage to skin physiological functions and the occurrence of related skin diseases. Next, we will summarize the importance of mitochondria in various skin cell species and how damage to mitochondrial structure, functions, and metabolism in the cell contents contributes to the skin aging process (Figure 3).

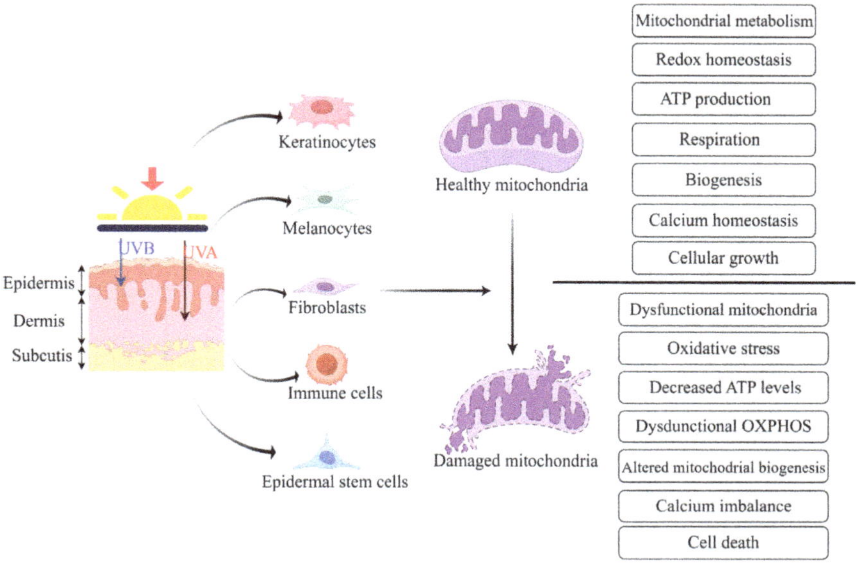

Figure 3. Mitochondrial dysfunction is closely related to skin aging. Ultraviolet exposure (UVB and UVA) induces senescence in different types of cells within the skin, including melanocytes, fibroblasts, immune cells, and epidermal stem cells. Mitochondria in these cells undergo functional and morphological changes, including dysfunctional mitochondria, oxidative stress, decreased ATP levels, dysfunctional OXPHOS, altered mitochondrial biogenesis, calcium imbalance, and cell death.

7.1. Mitochondrial Dysfunction and Epidermal Aging

Ultraviolet radiation directly influences the structure and function of the epidermis. Excessive sun exposure can damage keratinocytes, resulting in a decline in the function of skin tissue, which in turn can lead to an imbalance in skin homeostasis [18]. Massive UV exposure can result in necrosis of some epidermal keratinocytes. The damage to cell membrane integrity leads to the emission of cellular contents into the surrounding environment. Adjacent intact keratinocytes absorb the generated contents, such as double-stranded RNA, which promotes toll-like receptors 3 in the endosomes. This process triggers a series of processes of increased inflammatory response and lipid-associated processes [116,117].

UVB exposure of spontaneously transformed aneuploidy immortal keratinocyte cell line cells results in the emission of mitochondria and nuclear-injured DNA, culminating in the promotion of cyclic guanosine monophosphate-adenosine monophosphate synthase/protein stimulator of interferon genes DNA sensor and causing an endogenous immune effect [118]. In addition, long-lasting UV irradiation of spontaneously transformed aneuploidy immortal keratinocytes resulted in a reduction in ΔΨM, a downregulation in mitochondrial mass, and mitochondrial remodeling associated with mitochondria [119,120]. An age-dependent study of mitochondrial networks in youth and older adults showed that keratinocytes in older skin build an obviously more dispersive network with smaller, more compact clusters of mitochondria compared to keratinocytes in younger skin [50].

From the perspective of functional metabolism, aging is linked to alterations in mitochondrial metabolism within capital keratinocytes, including a decrease in

OXPHOS [121,122]. The amount of ROS in keratinocytes accumulates with age, while the mitochondrial membrane potential decreases, leading to a metabolic shift from OXPHOS to anaerobic glycolysis in mitochondria. Glycolysis is a compensatory mechanism designed to regulate membrane potential depolarization and gradual decline in cardiolipin levels [121]. The coordination between OXPHOS and glycolysis also controls the homeostasis between stem cell proliferation and differentiation, which in turn determines the fate of epidermal stem cells [123]. An imbalance of low nicotinamide adenine dinucleotide/nicotinamide (NAD+/NAM) ratios also creates keratinocyte differentiation disorders and is linked to the acquisition of premature keratinocyte aging manifestations [122]. They may lead to new therapeutic advances for skin conditions characterized by dysregulated epidermal homeostasis and premature skin aging, such as photoaging.

7.2. Mitochondrial Dysfunction and Dermal Aging

During aging, fibroblasts accumulate mitochondrial injury with the loss of mtDNA, which results in structural and functional changes in the ECM and induces an inflammatory response, which in turn promotes the skin wrinkles formation [52,124]. Fibroblasts from aging skin show a number of features linked to aging manifestations and mitochondrial injury. The features include damaged mitochondrial respiration, which reflects upregulated baseline respiration and increased proton leakage, as well as decreased ATP-related respiration and upregulated mitochondrial mass and ROS level [125]. Notably, chronologically aging fibroblasts show a decrease in both the level and activity of mitochondrial complex 2 [126,127]. The latest study by Wedel et al. (2023) reveals a change in mitochondrial fission and fusion pathways in skin fibroblasts during skin aging. The research indicates that a decreased incidence of mitochondrial fission processes or an upregulated perfusion propensity may lead to premature prolongation of mitochondria in aging skin fibroblasts with reduced growth differentiation factor 15 (GDF15) expression [128]. These data were compared to research, in which they elucidated that fibroblasts from centenarian skin maintain mitochondrial bioenergetic capacities via reshaping the mitochondrial network to perfusion. These results highlight the mitochondrial dynamics of senescent skin fibroblasts, suggesting the implications for further investigation to uncover the intricate mechanisms of mitochondrial dynamics during skin aging [129].

7.3. Mitochondrial Dysfunction and Adipose Tissue Aging

During the process of skin aging, there are noticeable changes in the quality and distribution of adipose tissue, which is characterized by an upregulation in visceral adipose and a reduction in subcutaneous adipose. Visceral fat tissue is closely linked to impaired metabolic influences, while subcutaneous fat tissue is thought to be metabolically useful. Age-induced peripheral adipose loss may be associated with flaws in lipogenesis and an exacerbation of the inflammatory response to subcutaneous adipose. Mitochondrial damage in adipose tissue during host aging can lead to cell aging, chronic inflammatory response, decreased stem cell activity, and obesity and insulin resistance [130,131]. Aging adipose progenitor cells exhibit injury differentiation ability, which is connected to overdose expression of p53 and p16INK4a, decreased insulin-dependent glucose transport, and dilated adipocyte hypertrophy [132,133].

8. Mechanisms of Mitochondrial Dysfunction Leading to Skin Aging

8.1. Mitophagy and Skin Aging

Mitophagy is central to mitochondrial quality and quantity control and balance. Once mitophagy is impaired, mitochondrial damage will lead to damaged energy balance and cell injury, which will eventually induce aging and various age-associated disorders, including cancer, cardiovascular disease, metabolic disease, etc. Similarly, mitophagy has been indicated to have a vital influence on the skin aging process. Studies showed that aging keratin-forming cells with high expression of miR-30a showed BNIP3L-dependent mitophagy defects in the final differentiation stage, reflecting that senescent skin development is associated with impaired mitophagy [51]. However, there have been some different voices regarding the negative or positive effects of mitophagy on skin aging. Research showed that mitochondrial dysfunction can be reversed by enhancing mitophagy by restoring imbalance in intracellular homeostasis, further protecting human skin fibroblasts from UVA-caused photoaging [134]. Research also emphasized that the Polygoni Multiflori Radix Preparat (PMRP) enhanced mitophagy by promoting the PINK/parkin signaling loop. However, studies also suggested that mitophagy does not inevitably preserve cells but may induce negative influences and cause apoptotic reactions. In research investigating the protective mechanism of a metabolite derived from intestine microflora, urolithin A in skin aging, centering on the mitophagy process, their data indicated that urolithin A lessened the accumulation of intracellular reactive oxygen species, which promoted the phosphorylation and afterwards nuclear translocation of NRF2, subsequently driving the activation of downstream antioxidative enzymes and restoring mitochondrial function by induction of mitophagy, which was regulated by the SIRT3-FOXO3-PINK1-PARKIN network [135].

8.2. Mitochondrial Dynamics and Skin Aging

The process of mitochondrial fusion and fission is involved in the regulation of cellular homeostasis in states of high energy demand, and the interruption of fusion results in the damage of respiratory capacity, which results in mitochondrial damage [136]. The effect of mitochondrial kinetic imbalance on aging is unquestionable and exerts an important role in the skin aging process. Recently, much research has concerned the significance of Drp1 in skin aging, and studies have also shown that inhibiting the expression of Drp1 in cells of skin may slow skin aging. Latest data have found that essential oils from *C. laurie foras* and *C. monspelis* can upregulate the level of the anti-aging gene SIRT1 in normally cultured HaCaT cells. The results are thought to be induced by a decrease of Drp1 level in HaCaT cells [137].

The agonist of the adiponectin receptor, AdipoRon (AR), upregulates phosphorylation of Drp1 Serine 637 via promoting AMP-activated protein kinase (AMPK), preventing mitochondrial translocation, and improving skin inflammatory response-induced skin aging. UV exposure promotes the translocation of Drp1 to mitochondria, causing mitochondrial division and promoting the aging of healthy human keratinocytes (NHK), while suppression of Drp1 delays mitochondrial division. However, Drp1 knockdown did not thoroughly inhibit the likelihood of mitochondrial fission, indicating that Drp1-induced mitochondrial translocation may not be the only pathway by which mitochondrial fission happens in the cells. As of now, there are no clear data on the alterations of mitochondrial fusion protein (Mfn) during skin aging, while Mfn is obviously increased in both skin cancer and melanoma, suggesting that excessive mitochondrial fission caused by Mfn accelerates skin damage [138]. The study revealed that a shift from fusion to a fission network happened in aging skin and that the homeostasis between fission and fusion was damaged [50].

8.3. Mitochondrial Biogenesis and Skin Aging

Mitochondrial biogenesis is an important part of mitochondrial quality and quantity monitoring, and AMPK/SIRT1/PGC-1α is a key signaling pathway controlling mitochondrial biogenesis. In a study of myocardial infarction, Guo indicated that the SIRT1/PGC-1α loop plays a vital role in mitochondrial biogenesis [136]. Meanwhile, Sirt1 has been suggested to exhibit anti-aging influences in many cellular models, which are linked to a weakened intracellular response to oxidative stress. Sirt1 modulates the transcriptional viability of PGC-1α by deacetylation. The levels of p-AMPK, Sirt1, and PGC-1α were significantly reduced in HaCaT cells and human foreskin fibroblasts (HFFs) irradiated with ultraviolet light. Excess ultraviolet radiation suppresses the expression of nuclear DNA by damaging it, thereby inhibiting mitochondrial biogenesis. Therefore, these mitochondrial biogenesis-associated proteins have important clinical application value in the diagnosis of skin aging [80].

9. Therapeutic Targeting of Skin Mitochondria

In this context, therapeutic drugs that directly aim to mitochondrial function have a wide range of applications in dermatology. They can broadly block harmful products or facilitate cell-specific signaling loops. Topical therapy with a proper vehicle could facilitate permeability of the epidermal layer and dermal-linked tissue, and mitochondria-targeted vehicles or peptides could assist the treatment to achieve its intended location of action to maximize effect.

Given the harmful influences of massively ROS DNA injury and cellular aging [7,139], ROS clearers—naturally occurring or synthetic—have long been regarded as promising treatments to defend mitochondria from injury [140]. For example, CoQ10, a natural ROS clearer that is also regarded as an electron shuttle between Complexes I/II and III of the ETC, is connected to skin aging [141]. New research in mice has indicated that important CoQ10 applications inhibited UV-caused skin aging [142] and upregulated collagen levels and the fibroblast cell quantity [141]. Ubiquitous throughout the body, CoQ10 is an endogenous lipophilic molecule with a central role in cellular energy metabolism—more specifically, the mitochondrial electron transport chain. Additionally, CoQ10 functions as an effective antioxidant molecule involved in the neutralization of reactive species and other free radicals [143]. CoQ10 levels in the skin decrease with age and exposure to UV radiation [141]. Topical supplementation of CoQ10 has been shown to ameliorate signs of skin aging both by stabilizing mitochondrial activity and by exerting antioxidant properties, which typically translates to the regeneration of skin processes involved in cutaneous skin aging [127]. Numerous studies have demonstrated the anti-aging effects of CoQ10 on cultured human dermal fibroblasts [144]. It is a critical ingredient in many anti-aging and regenerative skin creams, and as shown by Knott et al. [145], topical application of two different CoQ10-containing formulas significantly replenished the levels of this antioxidant in the dermal and epidermal layers of older skin (>60 years). Photo-stressed skin benefitted from CoQ10 application on account of a reduction in in situ free radical levels. Schniertshauer et al. [146] further showed that CoQ10 could restore ATP production, prevent mitophagy, and alleviate oxidative stress in aged skin cells. In addition, CoQ10 supplied to the media of the human skin-on-a-chip system resulted in upregulated epidermal and dermal thickness and upregulated keratinocyte marker and collagen I expression [147], a reversal of manifestations of skin aging.

While various natural products have been strengthened for their antioxidant capacity to protect or restore skin aging, as shown in various cosmeceuticals, synthetic substances have also been surveyed for their ROS-clearing viability. For example, MHY553, a PPARα agonist, dose-dependently cleared ROS and suppressed inflammatory factor expression via PPARα expression in the skin [148]. Exhilaratingly, there are over 250 clinical experiments of antioxidants on skin or skin-associated treatments (clinicaltrials.gov, January 2022). As mitochondria are the capital generators of cellular ROS, mitochondria-derived antioxidants, including MitoQ (up to 1000× more beneficial mitochondrial uptake than CoQ10) and tiron (spontaneously localized to the mitochondria without the demand for engineered targeting), have indicated important therapies [149].

10. Conclusions and Future Prospects

Skin aging is the noticeable performance of host aging, which is the consequence of the connection of genetic and environmental elements. The effectiveness of skin aging is urgent and vital to keep healthy physiological capacities of the skin and humans' mental health. The review summarizes the molecular pathways of the vital steps in mitochondrial quality and quantity regulation: mitochondrial dynamics, mitochondrial biogenesis, and mitophagy, and clarifies their influences and underlying mechanisms in skin aging. Recently, the improvement of skin aging induced by environmental elements mainly depended on physical shielding. Antioxidants and enzymes connected to DNA restoration could be supplied to sunblock as ancillary ingredients to partially improve the skin from UV exposure. The supplementation of skin repair or nutritional treatments, including various antioxidants, pharmacological reagents, and active compounds that could obviously ameliorate mitochondrial capacity and improve the metabolic function of the skin, seems to be an effective way. However, basic and clinical studies to prove whether these supplements can be applied in skin anti-aging are still very limited and deserve further investigation.

Author Contributions: T.Q.: Conceptualization, Methodology, Formal analysis, Writing—original draft, Visualization; R.L.: Resources, Supervision, Project administration, Funding acquisition; T.G.: Conceptualization, Formal analysis, Validation, Writing—review and editing, Project administration, Funding acquisition, Supervision. All authors have read and agreed to the published version of the manuscript.

Funding: This work was supported by the Chinese National Natural Science Foundation (32402845, 32172801, 32372954) and Beijing Natural Science Foundation (5244041, 6222019).

Institutional Review Board Statement: Not applicable.

Informed Consent Statement: Not applicable.

Data Availability Statement: Data will be made available on request.

Conflicts of Interest: The authors declare that they have no known competing financial interests or personal relationships that could have appeared to influence the work reported in this paper.

Abbreviations

DEJ: dermal-epidermal junction; ECM: extracellular matrix; MSCs: mesenchymal stem cells; UV: ultraviolet; UVR: ultraviolet radiation; ROS: reactive oxygen species; OXPHOS: oxidative phosphorylation; CCCP: carbonyl cyanide 3-chlorophenylhydrazone; mtDNA: mitochondrial DNA; PINK1: PTEN-caused kinase 1; VDAC1: voltage-dependent anion channel-1; LC3: light chain 3; DAMPs: damage-associated molecular patterns; MFN: mitochondrial proteins; OMM: outer mitochondrial membrane; IMM: inner mitochondrial membrane; OPA1: optic atrophy 1; Drp-1: dyne-associated protein 1; PPAR-1α: peroxisome proliferation-activating receptor-γ coactivator-1α; NRF: nuclear respiration factors; UCP2: uncoupling proteins; ERRs: estrogen-associated receptors; AMPK: AMP-activated protein kinase; SIRT1: silencing information regulator 1; $\Delta\Psi$M: mitochondrial membrane potential; HaCaT: human keratinocytes cells; CTS: cryptotanshinone; NAD+/NAM: nicotinamide adenine dinucleotide/nicotinamide; GDF15: growth differentiation factor 15; PMRP: polygoni multiflori radix preparat; AMPK: AMP-activated protein kinase; NHK: normal human keratinocytes; Mfn: mitochondrial fusion protein; HFFs: human foreskin fibroblasts.

References

1. Ogrodnik, M. Cellular aging beyond cellular senescence: Markers of senescence prior to cell cycle arrest in vitro and in vivo. *Aging Cell* **2021**, *20*, e13338. [CrossRef] [PubMed]
2. Summer, R.; Shaghaghi, H.; Schriner, D.; Roque, W.; Sales, D.; Cuevas-Mora, K.; Desai, V.; Bhushan, A.; Ramirez, M.I.; Romero, F. Activation of the mTORC1/PGC-1 axis promotes mitochondrial biogenesis and induces cellular senescence in the lung epithelium. *Am. J. Physiol. Cell. Mol. Physiol.* **2019**, *316*, L1049–L1060. [CrossRef] [PubMed]
3. Franco, A.C.; Aveleira, C.; Cavadas, C. Skin senescence: Mechanisms and impact on whole-body aging. *Trends Mol. Med.* **2022**, *28*, 97–109. [CrossRef]
4. Hu, S.C.-S.; Lin, C.-L.; Yu, H.-S. Dermoscopic assessment of xerosis severity, pigmentation pattern and vascular morphology in subjects with physiological aging and photoaging. *Eur. J. Dermatol.* **2019**, *29*, 274–280. [CrossRef]
5. D'Uva, G.; Baci, D.; Albini, A.; Noonan, D.M. Cancer chemoprevention revisited: Cytochrome P450 family 1B1 as a target in the tumor and the microenvironment. *Cancer Treat. Rev.* **2018**, *63*, 1–18. [CrossRef]
6. Banach, K.; Kowalska, J.; Rzepka, Z.; Beberok, A.; Rok, J.; Wrze'sniok, D. The role of UVA radiation in ketoprofen-mediated BRAF-mutant amelanotic melanoma cells death—A study at the cellular and molecular level. *Toxicol. In Vitro* **2021**, *72*, 105108. [CrossRef]
7. Sreedhar, A.; Aguilera-Aguirre, L.; Singh, K.K. Mitochondria in skin health, aging, and disease. *Cell Death Dis.* **2020**, *11*, 444. [CrossRef]
8. Popov, L.D. Mitochondrial biogenesis: An update. *J. Cell. Mol. Med.* **2020**, *24*, 4892–4899. [CrossRef]
9. Martic, I.; Papaccio, F.; Bellei, B.; Cavinato, M. Mitochondrial dynamics and metabolism across skin cells: Implications for skin homeostasis and aging. *Front. Physiol.* **2023**, *14*, 1284410. [CrossRef]
10. Zhang, Z.; Michniak-Kohn, B.B. Tissue engineered human skin equivalents. *Pharmaceutics* **2012**, *4*, 26–41. [CrossRef]
11. Hill, D.S.; Robinson, N.D.; Caley, M.P.; Chen, M.; O'Toole, E.A.; Armstrong, J.L.; Przyborski, S.; Lovat, P.E. A novel fully humanized 3D skin equivalent to model early melanoma invasion. *Mol. Cancer Ther.* **2015**, *14*, 2665–2673. [CrossRef] [PubMed]
12. Wang, H.; Guo, B.; Hui, Q.; Lin, F.; Tao, K. CO_2 lattice laser reverses skin aging caused by UVB. *Aging* **2020**, *12*, 7056–7065. [CrossRef] [PubMed]
13. Kim, J.H.; Jeong, H.D.; Song, M.J.; Lee, D.H.; Chung, J.H.; Lee, S.T. SOD3 suppresses the expression of MMP-1 and increases the integrity of extracellular matrix in fibroblasts. *Antioxidants* **2022**, *11*, 928. [CrossRef] [PubMed]
14. Hofmann, E.; Schwarz, A.; Fink, J.; Kamolz, L.P.; Kotzbeck, P. Modelling the complexity of human skin in vitro. *Biomedicines* **2023**, *11*, 794. [CrossRef]
15. Abdallah, F.; Mijouin, L.; Pichon, C. Skin immune landscape: Inside and outside the organism. *Mediat. Inflamm.* **2017**, *2017*, 5095293. [CrossRef]
16. Trevor, L.V.; Riches-Suman, K.; Mahajan, A.L.; Thornton, M.J. Adipose tissue: A source of stem cells with potential for regenerative therapies for wound healing. *J. Clin. Med.* **2020**, *9*, 2161. [CrossRef]
17. Mancuso, P.; Bouchard, B. The impact of aging on adipose function and adipokine synthesis. *Front. Endocrinol.* **2019**, *10*, 137. [CrossRef]
18. Slominski, A.; Wortsman, J. Neuroendocrinology of the skin. *Endocr. Rev.* **2000**, *21*, 457–487. [CrossRef]
19. Naidoo, K.; Hanna, R.; Birch-Machin, M.A. What is the role of mitochondrial dysfunction in skin photoaging? *Exp. Dermatol.* **2018**, *27*, 124–128. [CrossRef]
20. Farage, M.A.; Miller, K.W.; Elsner, P.; Maibach, H.I. Intrinsic and extrinsic factors in skin ageing: A review. *Int. J. Cosmet. Sci.* **2008**, *30*, 87–95. [CrossRef]
21. Yaar, M.; Eller, M.S.; Gilchrest, B.A. Fifty years of skin aging. *J. Investig. Dermatol. Symp. Proc.* **2002**, *7*, 51–58. [CrossRef] [PubMed]
22. Kim, D.J.; Iwasaki, A.; Chien, A.L.; Kang, S. UVB-mediated DNA damage induces matrix metalloproteinases to promote photoaging in an AhR- and SP1-dependent manner. *JCI Insight* **2022**, *7*, e156344. [CrossRef] [PubMed]
23. Kim, H.; Jang, J.; Song, M.J.; Park, C.H.; Lee, D.H.; Lee, S.H.; Chung, J.H. Inhibition of matrix metalloproteinase expression by selective clearing of senescent dermal fibroblasts attenuates ultraviolet-induced photoaging. *Biomed. Pharmacother.* **2022**, *150*, 113034. [CrossRef] [PubMed]
24. Moriyama, M.; Moriyama, H.; Uda, J.; Kubo, H.; Nakajima, Y.; Goto, A.; Morita, T.; Hayakawa, T. BNIP3 upregulation via stimulation of ERK and JNK activity is required for the protection of keratinocytes from UVB-induced apoptosis. *Cell Death Dis.* **2017**, *8*, e2576. [CrossRef]
25. Brand, R.M.; Wipf, P.; Durham, A.; Epperly, M.W.; Greenberger, J.S.; Falo, L.D., Jr. Targeting mitochondrial oxidative stress to mitigate UV-induced skin damage. *Front. Pharmacol.* **2018**, *9*, 920. [CrossRef]
26. Valejo Coelho, M.M.; Matos, T.R.; Apetato, M. The dark side of the light: Mechanisms of photocarcinogenesis. *Clin. Dermatol.* **2016**, *34*, 563–570. [CrossRef]
27. Kauppila, T.E.S.; Kauppila, J.H.K.; Larsson, N.G. Mammalian mitochondria and aging: An update. *Cell Metab.* **2017**, *25*, 57–71. [CrossRef]

28. Nicholls, D.G. Mitochondria and calcium signaling. *Cell Calcium.* **2005**, *38*, 311–731. [CrossRef]
29. Liu, Z.; Butow, R.A. Mitochondrial retrograde signaling. *Annu. Rev. Genet.* **2006**, *40*, 159–185. [CrossRef]
30. Nilsson, R.; Schultz, I.J.; Pierce, E.L.; Soltis, K.A.; Naranuntarat, A.; Ward, D.M.; Baughman, J.M.; Paradkar, P.N.; Kingsley, P.D.; Culotta, V.C.; et al. Discovery of genes essential for heme biosynthesis through large-scale gene expression analysis. *Cell Metab.* **2009**, *10*, 119–130. [CrossRef]
31. Wallace, D.C.; Fan, W.; Procaccio, V. Mitochondrial energetics and therapeutics. *Annu. Rev. Pathol.* **2010**, *5*, 297–348. [CrossRef] [PubMed]
32. Srivastava, S. Emerging therapeutic roles for NAD^+ metabolism in mitochondrial and age-related disorders. *Clin. Transl. Med.* **2016**, *5*, 25. [CrossRef] [PubMed]
33. Butow, R.A.; Avadhani, N.G. Mitochondrial signaling: The retrograde response. *Mol. Cell* **2004**, *14*, 1–15. [CrossRef]
34. Sun, N.; Youle, R.J.; Finkel, T. The mitochondrial basis of aging. *Mol. Cell* **2016**, *61*, 654–666. [CrossRef]
35. Petersen, K.F.; Befroy, D.; Dufour, S.; Dziura, J.; Ariyan, C.; Rothman, D.L.; DiPietro, L.; Cline, G.W.; Shulman, G.I. Mitochondrial dysfunction in the elderly: Possible role in insulin resistance. *Science* **2003**, *300*, 1140–1142. [CrossRef]
36. Lowell, B.B.; Shulman, G.I. Mitochondrial dysfunction and type 2 diabetes. *Science* **2005**, *307*, 384–387. [CrossRef]
37. Wallace, D.C. Mitochondria and cancer. *Nat. Rev. Cancer* **2012**, *12*, 685–698. [CrossRef]
38. Lane, R.K.; Hilsabeck, T.; Rea, S.L. The role of mitochondrial dysfunction in age-related diseases. *Biochim. Biophys. Acta* **2015**, *1847*, 1387–1400. [CrossRef]
39. Tocchi, A.; Quarles, E.K.; Basisty, N.; Gitari, L.; Rabinovitch, P.S. Mitochondrial dysfunction in cardiac aging. *Biochim. Biophys. Acta* **2015**, *1847*, 1424–1433. [CrossRef]
40. Li, Q.O.Y.; Soro-Arnaiz, I.; Aragonés, J. Age-dependent obesity and mitochondrial dysfunction. *Adipocyte* **2017**, *6*, 161–166. [CrossRef]
41. Ahn, C.S.M.; Etallo, C.M. Mitochondria as biosynthetic factories for cancer proliferation. *Cancer Metab.* **2015**, *3*, 1. [CrossRef] [PubMed]
42. Stout, R.; Birch-Machin, M. Mitochondria's role in skin ageing. *Biology* **2019**, *8*, 29. [CrossRef] [PubMed]
43. Lin, S.J.; Kaeberlein, M.; Andalis, A.A.; Sturtz, L.A.; Defossez, P.A.; Culotta, V.C.; Fink, G.R.; Guarente, L. Calorie restriction extends Saccharomyces cerevisiae lifespan by increasing respiration. *Nature* **2002**, *418*, 344–348. [CrossRef]
44. Kirkinezos, I.G.; Moraes, C.T. Reactive oxygen species and mitochondrial diseases. *Semin. Cell Dev. Biol.* **2001**, *12*, 449–457. [CrossRef]
45. Hamanaka, R.B.; Chandel, N.S. Mitochondrial metabolism as a regulator of keratinocyte differentiation. *Cell Logist.* **2013**, *3*, e25456. [CrossRef]
46. Trompette, A.; Pernot, J.; Perdijk, O.; Alqahtani, R.A.A.; Domingo, J.S.; Camacho-Muñoz, D.; Wong, N.C.; Kendall, A.C.; Wiederkehr, A.; Nicod, L.P.; et al. Gut-derived short-chain fatty acids modulate skin barrier integrity by promoting keratinocyte metabolism and differentiation. *Mucosal. Immunol.* **2022**, *15*, 908–926. [CrossRef]
47. Yaar, M.; Gilchrest, B.A. Ageing and photoageing of keratinocytes and melanocytes. *Clin. Exp. Dermatol.* **2001**, *26*, 583–591. [CrossRef]
48. Farage, M.A.; Miller, K.W.; Elsner, P.; Maibach, H.I. Characteristics of the aging skin. *Adv. Wound Care* **2013**, *2*, 5–10. [CrossRef]
49. Khalid, K.A.; Nawi, A.F.M.; Zulkifli, N.; Barkat, M.A.; Hadi, H. Aging and wound healing of the skin: A review of clinical and pathophysiological hallmarks. *Life* **2022**, *12*, 2142. [CrossRef]
50. Mellem, D.; Sattler, M.; Pagel-Wolff, S.; Jaspers, S.; Wenck, H.; Rübhausen, M.A.; Fischer, F. Fragmentation of the mitochondrial network in skin in vivo. *PLoS ONE* **2017**, *12*, e0174469. [CrossRef]
51. Chevalier, F.P.; Rorteau, J.; Ferraro, S.; Martin, L.S.; Gonzalez-Torres, A.; Berthier, A.; El Kholti, N.; Lamartine, J. MiR-30a-5p alters epidermal terminal differentiation during aging by regulating BNIP3L/NIX-dependent mitophagy. *Cells* **2022**, *11*, 836. [CrossRef] [PubMed]
52. Krutmann, J.; Schroeder, P. Role of mitochondria in photoaging of human skin: The defective powerhouse model. *J. Investig. Dermatol. Symp. Proc.* **2009**, *14*, 44–49. [CrossRef] [PubMed]
53. Katsuyama, Y.; Yamawaki, Y.; Sato, Y.; Muraoka, S.; Yoshida, M.; Okano, Y.; Masaki, H. Decreased mitochondrial function in UVA-irradiated dermal fibroblasts causes the insufficient formation of type I collagen and fibrillin-1 fibers. *J. Dermatol. Sci.* **2022**, *108*, 22–29. [CrossRef] [PubMed]
54. Yanes, B.; Rainero, E. The interplay between cell-extracellular matrix interaction and mitochondria dynamics in cancer. *Cancers* **2022**, *14*, 1433. [CrossRef]
55. Ahmed, N.S.; Foote, J.B.; Singh, K.K. Impaired mitochondria promote aging-associated sebaceous gland dysfunction and pathology. *Am. J. Pathol.* **2022**, *192*, 1546–1558. [CrossRef]
56. Lovászi, M.; Szegedi, A.; Zouboulis, C.C.; Törőcsik, D. Sebaceous-immunobiology is orchestrated by sebum lipids. *Derm.-Endocrinol.* **2017**, *9*, e1375636. [CrossRef]

57. Wikramanayake, T.C.; Nicu, C.; Gherardini, J.; Mello, A.C.G.C.V.; Chéret, J.; Paus, R. Mitochondrially localized MPZL3 functions as a negative regulator of sebaceous gland size and sebocyte proliferation. *J. Investig. Dermatol.* **2022**, *142*, 2524–2527.e7. [CrossRef]
58. Schneider, M.R.; Paus, R. Sebocytes, multifaceted epithelial cells: Lipid production and holocrine secretion. *Int. J. Biochem. Cell Biol.* **2010**, *42*, 181–185. [CrossRef]
59. Yu, R.; Lendahl, U.; Nistér, M.; Zhao, J. Regulation of mammalian mitochondrial dynamics: Opportunities and challenges. *Front. Endocrinol.* **2020**, *11*, 374. [CrossRef]
60. Kim, E.S.; Park, S.J.; Goh, M.J.; Na, Y.J.; Jo, D.S.; Jo, Y.K.; Shin, J.H.; Choi, E.S.; Lee, H.K.; Kim, J.Y.; et al. Mitochondrial dynamics regulate melanogenesis through proteasomal degradation of MITF via ROS-ERK activation. *Pigment Cell Melanoma Res.* **2014**, *27*, 1051–1062. [CrossRef]
61. Vasileiou, P.V.S.; Evangelou, K.; Vlasis, K.; Fildisis, G.; Panayiotidis, M.I.; Chronopoulos, E.; Passias, P.G.; Kouloukoussa, M.; Gorgoulis, V.G.; Havaki, S. Mitochondrial homeostasis and cellular senescence. *Cells* **2019**, *8*, 686. [CrossRef] [PubMed]
62. Lee, S.E.; Kang, Y.C.; Kim, Y.; Kim, S.; Yu, S.H.; Park, J.H.; Kim, I.H.; Kim, H.Y.; Han, K.; Lee, H.K.; et al. Preferred migration of mitochondria toward cells and tissues with mitochondrial damage. *Int. J. Mol. Sci.* **2022**, *23*, 15734. [CrossRef] [PubMed]
63. Li, W.; Li, H.; Zheng, L.; Xia, J.; Yang, X.; Men, S.; Yuan, Y.; Fan, Y. Ginsenoside CK improves skeletal muscle insulin resistance by activating DRP1/PINK1-mediated mitophagy. *Food Funct.* **2023**, *14*, 1024–1036. [CrossRef]
64. Ma, F.; Li, H.; Huo, H.; Han, Q.; Liao, J.; Zhang, H.; Li, Y.; Pan, J.; Hu, L.; Guo, J.; et al. N-acetyl-L-cysteine alleviates FUNDC1-mediated mitophagy by regulating mitochondrial dynamics in type 1 diabetic nephropathy canine. *Life Sci.* **2023**, *313*, 121278. [CrossRef]
65. Al-Bari, M.A.A.; Xu, P. Molecular regulation of autophagy machinery by mTOR-dependent and -independent pathways. *Ann. N. Y. Acad. Sci.* **2020**, *1467*, 3–20. [CrossRef]
66. Shirihai, O.S.; Song, M.; Dorn, G.W., 2nd. How mitochondrial dynamism orchestrates mitophagy. *Circ. Res.* **2015**, *116*, 1835–1849. [CrossRef]
67. Shi, R.; Guberman, M.; Kirshenbaum, L.A. Mitochondrial quality control: The role of mitophagy in aging. *Trends Cardiovasc. Med.* **2018**, *28*, 246–260. [CrossRef]
68. Zhu, N.; Xu, M.H.; Li, Y. Bioactive oligopeptides from ginseng (*Panax ginseng* Meyer) suppress oxidative stress-induced senescence in fibroblasts via NAD+/SIRT1/PGC-1α signaling pathway. *Nutrients* **2022**, *14*, 5289. [CrossRef]
69. Pickles, S.; Vigié, P.; Youle, R.J. Mitophagy and quality control mechanisms in mitochondrial maintenance. *Curr. Biol.* **2018**, *28*, R170–R185. [CrossRef]
70. Matsuda, N.; Sato, S.; Shiba, K.; Okatsu, K.; Saisho, K.; Gautier, C.A.; Sou, Y.S.; Saiki, S.; Kawajiri, S.; Sato, F.; et al. PINK1 stabilized by mitochondrial depolarization recruits parkin to damaged mitochondria and activates latent parkin for mitophagy. *J. Cell Biol.* **2010**, *189*, 211–221. [CrossRef]
71. Geisler, S.; Holmström, K.M.; Skujat, D.; Fiesel, F.C.; Rothfuss, O.C.; Kahle, P.J.; Springer, W. PINK1/parkin-mediated mitophagy is dependent on VDAC1 and p62/SQSTM1. *Nat. Cell Biol.* **2010**, *12*, 119–131. [CrossRef] [PubMed]
72. Ryoo, I.G.; Kwak, M.K. Regulatory crosstalk between the oxidative stress-related transcription factor Nfe2l2/Nrf2 and mitochondria. *Toxicol. Appl. Pharmacol.* **2018**, *359*, 24–33. [CrossRef] [PubMed]
73. Niu, Q.; Chen, J.; Xia, T.; Li, P.; Zhou, G.; Xu, C.; Zhao, Q.; Dong, L.; Zhang, S.; Wang, A. Excessive ER stress and the resulting autophagic flux dysfunction contribute to fluoride-induced neurotoxicity. *Environ. Pollut.* **2018**, *233*, 889–899. [CrossRef]
74. Meyer, J.N.; Leuthner, T.C.; Luz, A.L. Mitochondrial fusion, fission, and mitochondrial toxicity. *Toxicology* **2017**, *391*, 42–53. [CrossRef]
75. Morton, H.; Kshirsagar, S.; Orlov, E.; Bunquin, L.E.; Sawant, N.; Boleng, L.; George, M.; Basu, T.; Ramasubramanian, B.; Pradeepkiran, J.A.; et al. Defective mitophagy and synaptic degeneration in Alzheimer's disease: Focus on aging, mitochondria and synapse. *Free Radic. Biol. Med.* **2021**, *172*, 652–667. [CrossRef]
76. Jugé, R.; Breugnot, J.; Da Silva, C.; Bordes, S.; Closs, B.; Aouacheria, A. Quantification and characterization of UVB-induced mitochondrial fragmentation in normal primary human keratinocytes. *Sci. Rep.* **2016**, *6*, 35065. [CrossRef]
77. Pagliuso, A.; Cossart, P.; Stavru, F. The ever-growing complexity of the mitochondrial fission machinery. *Cell. Mol. Life Sci.* **2018**, *75*, 355–374. [CrossRef]
78. Gupta, D.; Abdullah, T.S. Regulation of mitochondrial dynamics in skin: Role in pathophysiology. *Int. J. Dermatol.* **2022**, *61*, 541–547. [CrossRef]
79. Chan, D.C. Mitochondrial dynamics and its involvement in disease. *Annu. Rev. Pathol. Mech. Dis.* **2020**, *15*, 235–259. [CrossRef]
80. Zhang, C.; Gao, X.; Li, M.; Yu, X.; Huang, F.; Wang, Y.; Yan, Y.; Zhang, H.; Shi, Y.; He, X. The role of mitochondrial quality surveillance in skin aging: Focus on mitochondrial dynamics, biogenesis and mitophagy. *Ageing Res. Rev.* **2023**, *87*, 101917. [CrossRef]
81. Ventura-Clapier, R.; Garnier, A.; Veksler, V. Transcriptional control of mitochondrial biogenesis: The central role of PGC-1alpha. *Cardiovasc. Res.* **2008**, *79*, 208–217. [CrossRef] [PubMed]
82. Jornayvaz, F.R.; Shulman, G.I. Regulation of mitochondrial biogenesis. *Essays Biochem.* **2010**, *47*, 69–84. [PubMed]

83. Hock, M.B.; Kralli, A. Transcriptional control of mitochondrial biogenesis and function. *Annu. Rev. Physiol.* **2009**, *71*, 177–203. [CrossRef] [PubMed]
84. Virbasius, J.V.; Scarpulla, R.C. Activation of the human mitochondrial transcription factor A gene by nuclear respiratory factors: A potential regulatory link between nuclear and mitochondrial gene expression in organelle biogenesis. *Proc. Natl. Acad. Sci. USA* **1994**, *91*, 1309–1313. [CrossRef]
85. Reznick, R.M.; Shulman, G.I. The role of AMP-activated protein kinase in mitochondrial biogenesis. *J. Physiol.* **2006**, *574*, 33–39. [CrossRef]
86. Ren, J.; Pulakat, L.; Whaley-Connell, A.; Sowers, J.R. Mitochondrial biogenesis in the metabolic syndrome and cardiovascular disease. *J. Mol. Med.* **2010**, *88*, 993–1001. [CrossRef]
87. Miwa, S.; Kashyap, S.; Chini, E.; von Zglinicki, T. Mitochondrial dysfunction in cell senescence and aging. *J. Clin. Investig.* **2022**, *132*, e158447. [CrossRef]
88. Chapman, J.; Fielder, E.; Passos, J.F. Mitochondrial dysfunction and cell senescence: Deciphering a complex relationship. *FEBS Lett.* **2019**, *593*, 1566–1579. [CrossRef]
89. Ghosh-Choudhary, S.K.; Liu, J.; Finkel, T. The role of mitochondria in cellular senescence. *FASEB J.* **2021**, *35*, e21991. [CrossRef]
90. Shigenaga, M.K.; Hagen, T.M.; Ames, B.N. Oxidative damage and mitochondrial decay in aging. *Proc. Natl. Acad. Sci. USA* **1994**, *91*, 10771–10778. [CrossRef]
91. Liesa, M.; Shirihai, O.S. Mitochondrial dynamics in the regulation of nutrient utilization and energy expenditure. *Cell Metab.* **2013**, *17*, 491–506. [CrossRef] [PubMed]
92. Gao, X.; Yu, X.; Zhang, C.; Wang, Y.; Sun, Y.; Sun, H.; Zhang, H.; Shi, Y.; He, X. Telomeres and mitochondrial metabolism: Implications for cellular senescence and age-related diseases. *Stem Cell Rev. Rep.* **2022**, *18*, 2315–2327. [CrossRef]
93. Zorov, D.B.; Vorobjev, I.A.; Popkov, V.A.; Babenko, V.A.; Zorova, L.D.; Pevzner, I.B.; Silachev, D.N.; Zorov, S.D.; Andrianova, N.V.; Plotnikov, E.Y. Lessons from the discovery of mitochondrial fragmentation (fission): A review and update. *Cells* **2019**, *8*, 175. [CrossRef]
94. Cavinato, M.; Madreiter-Sokolowski, C.T.; Büttner, S.; Schosserer, M.; Zwerschke, W.; Wedel, S.; Grillari, J.; Graier, W.F.; Jansen-Dürr, P. Targeting cellular senescence based on interorganelle communication, multilevel proteostasis, and metabolic control. *FEBS J.* **2021**, *288*, 3834–3854. [CrossRef]
95. Jing, R.; Guo, K.; Zhong, Y.; Wang, L.; Zhao, J.; Gao, B.; Ye, Z.; Chen, Y.; Li, X.; Xu, N.; et al. Protective effects of fucoidan purified from Undaria pinnatifida against UV-irradiated skin photoaging. *Ann. Transl. Med.* **2021**, *9*, 1185. [CrossRef]
96. Guo, K.; Liu, R.; Jing, R.; Wang, L.; Li, X.; Zhang, K.; Fu, M.; Ye, J.; Hu, Z.; Zhao, W.; et al. Cryptotanshinone protects skin cells from ultraviolet radiation-induced photoaging via its antioxidant effect and by reducing mitochondrial dysfunction and inhibiting apoptosis. *Front. Pharmacol.* **2022**, *13*, 1036013. [CrossRef]
97. Park, J.E.; Woo, S.W.; Kim, M.B.; Kim, C.; Hwang, J.K. Standardized kaempferia parviflora extract inhibits intrinsic aging process in human dermal fibroblasts and hairless mice by inhibiting cellular senescence and mitochondrial dysfunction. *Evid. Based Complement. Alternat. Med.* **2017**, *2017*, 6861085. [CrossRef]
98. Palikaras, K.; Daskalaki, I.; Markaki, M.; Tavernarakis, N. Mitophagy and age-related pathologies: Development of new therapeutics by targeting mitochondrial turnover. *Pharmacol. Ther.* **2017**, *178*, 157–174. [CrossRef]
99. Mengus, C.; Neutzner, M.; Bento, A.C.P.F.; Bippes, C.C.; Kohler, C.; Decembrini, S.; Häusel, J.; Hemion, C.; Sironi, L.; Frank, S.; et al. VCP/p97 cofactor UBXN1/SAKS1 regulates mitophagy by modulating MFN2 removal from mitochondria. *Autophagy* **2022**, *18*, 171–190. [CrossRef]
100. Ko, H.J.; Tsai, C.Y.; Chiou, S.J.; Lai, Y.L.; Wang, C.H.; Cheng, J.T.; Chuang, T.H.; Huang, C.F.; Kwan, A.L.; Loh, J.K.; et al. The phosphorylation status of Drp1-Ser637 by PKA in mitochondrial fission modulates mitophagy via PINK1/Parkin to exert multipolar spindles assembly during mitosis. *Biomolecules* **2021**, *11*, 424. [CrossRef]
101. Barazzuol, L.; Giamogante, F.; Brini, M.; Calì, T. PINK1/Parkin mediated mitophagy, Ca^{2+} signalling, and ER-mitochondria contacts in Parkinson's disease. *Int. J. Mol. Sci.* **2020**, *21*, 1772. [CrossRef] [PubMed]
102. Sarraf, S.A.; Sideris, D.P.; Giagtzoglou, N.; Ni, L.; Kankel, M.W.; Sen, A.; Bochicchio, L.E.; Huang, C.H.; Nussenzweig, S.C.; Worley, S.H.; et al. PINK1/Parkin influences cell cycle by sequestering TBK1 at damaged mitochondria, inhibiting mitosis. *Cell Rep.* **2019**, *29*, 225–235.e5. [CrossRef] [PubMed]
103. Chistiakov, D.A.; Sobenin, I.A.; Revin, V.V.; Orekhov, A.N.; Bobryshev, Y.V. Mitochondrial aging and age-related dysfunction of mitochondria. *BioMed Res. Int.* **2014**, *2014*, 238463. [CrossRef]
104. Halliwell, B. Reactive species and antioxidants. Redox biology is a fundamental theme of aerobic life. *Plant Physiol.* **2006**, *141*, 312–322. [CrossRef]
105. Halliwell, B.; Gutteridge, J.M.C. *Free Radicals in Biology and Medicine*, 4th ed.; Oxford University Press: Oxford, UK, 2007.
106. Finkel, T.; Holbrook, N.J. Oxidants, oxidative stress and the biology of ageing. *Nature* **2000**, *408*, 239–247. [CrossRef]
107. Thanan, R.; Oikawa, S.; Hiraku, Y.; Ohnishi, S.; Ma, N.; Pinlaor, S.; Yongvanit, P.; Kawanishi, S.; Murata, M. Oxidative stress and its significant roles in neurodegenerative diseases and cancer. *Int. J. Mol. Sci.* **2015**, *16*, 193–217. [CrossRef]

108. Harman, D. Aging: A theory based on free radical and radiation chemistry. *J. Gerontol.* **1956**, *11*, 298–300. [CrossRef]
109. Harman, D. The biologic clock: The mitochondria? *J. Am. Geriatr. Soc.* **1972**, *20*, 145–147. [CrossRef]
110. Alexeyev, M.F. Is there more to aging than mitochondrial DNA and reactive oxygen species? *FEBS J.* **2009**, *276*, 5768–5787. [CrossRef]
111. Francisco, A.; Engel, D.F.; Figueira, T.R.; Rogério, F.; de Bem, A.F.; Castilho, R.F. Mitochondrial NAD(P)$^+$ transhydrogenase is unevenly distributed in different brain regions, and its loss causes depressive-like behavior and motor dysfunction in mice. *Neuroscience* **2020**, *440*, 210–229. [CrossRef]
112. Navarro, C.D.C.; Francisco, A.; Costa, E.F.D.; Dalla Costa, A.P.; Sartori, M.R.; Bizerra, P.F.V.; Salgado, A.R.; Figueira, T.R.; Vercesi, A.E.; Castilho, R.F. Aging-dependent mitochondrial bioenergetic impairment in the skeletal muscle of NNT-deficient mice. *Exp. Gerontol.* **2024**, *193*, 112465. [CrossRef] [PubMed]
113. Kadoguchi, T.; Takada, S.; Yokota, T.; Furihata, T.; Matsumoto, J.; Tsuda, M.; Mizushima, W.; Fukushima, A.; Okita, K.; Kinugawa, S. Deletion of NAD(P)H oxidase 2 prevents angiotensin II-induced skeletal muscle atrophy. *BioMed Res. Int.* **2018**, *2018*, 3194917. [CrossRef] [PubMed]
114. Figueira, T.R.; Francisco, A.; Ronchi, J.A.; Dos Santos, G.R.R.M.; Santos, W.D.; Treberg, J.R.; Castilho, R.F. NADPH supply and the contribution of NAD(P)$^+$ transhydrogenase (NNT) to H_2O_2 balance in skeletal muscle mitochondria. *Arch. Biochem. Biophys.* **2021**, *707*, 108934. [CrossRef]
115. Bokov, A.; Chaudhuri, A.; Richardson, A. The role of oxidative damage and stress in aging. *Mech. Ageing Dev.* **2004**, *125*, 811–826. [CrossRef]
116. Bernard, J.J.; Cowing-Zitron, C.; Nakatsuji, T.; Muehleisen, B.; Muto, J.; Borkowski, A.W.; Martinez, L.; Greidinger, E.L.; Yu, B.D.; Gallo, R.L. Ultraviolet radiation damages self noncoding RNA and is detected by TLR3. *Nat. Med.* **2012**, *18*, 1286–1290. [CrossRef]
117. Borkowski, A.W.; Gallo, R.L. UVB radiation illuminates the role of TLR3 in the epidermis. *J. Investig. Dermatol.* **2014**, *134*, 2315–2320. [CrossRef]
118. Li, C.; Liu, W.; Wang, F.; Hayashi, T.; Mizuno, K.; Hattori, S.; Fujisaki, H.; Ikejima, T. DNA damage-triggered activation of cGAS-STING pathway induces apoptosis in human keratinocyte HaCaT cells. *Mol. Immunol.* **2021**, *131*, 180–190. [CrossRef]
119. Kelly, J.; Murphy, J.E. Mitochondrial tolerance to single and repeat exposure to simulated sunlight in human epidermal and dermal skin cells. *J. Photochem. Photobiol. B* **2016**, *165*, 298–304. [CrossRef]
120. Bellei, B.; Picardo, M. Premature cell senescence in human skin: Dual face in chronic acquired pigmentary disorders. *Ageing Res. Rev.* **2020**, *57*, 100981. [CrossRef]
121. Dumas, M.; Maftah, A.; Bonte, F.; Ratinaud, M.H.; Meybeck, A.; Julien, R. Flow cytometric analysis of human epidermal cell ageing using two fluorescent mitochondrial probes. *Comptes Rendus L'academie Sci. Ser. III* **1995**, *318*, 191–197.
122. Tan, C.L.; Chin, T.; Tan, C.Y.R.; Rovito, H.A.; Quek, L.S.; Oblong, J.E.; Bellanger, S. Nicotinamide metabolism modulates the proliferation/differentiation balance and senescence of human primary keratinocytes. *J. Investig. Dermatol.* **2019**, *139*, 1638–1647.e3. [CrossRef] [PubMed]
123. Folmes, C.D.; Dzeja, P.P.; Nelson, T.J.; Terzic, A. Metabolic plasticity in stem cell homeostasis and differentiation. *Cell Stem Cell* **2012**, *11*, 596–606. [CrossRef] [PubMed]
124. Ho, C.Y.; Dreesen, O. Faces of cellular senescence in skin aging. *Mech. Ageing Dev.* **2021**, *198*, 111525. [CrossRef]
125. Rorteau, J.; Chevalier, F.P.; Bonnet, S.; Barthélemy, T.; Lopez-Gaydon, A.; Martin, L.S.; Bechetoille, N.; Lamartine, J. Maintenance of chronological aging features in culture of normal human dermal fibroblasts from old donors. *Cells* **2022**, *11*, 858. [CrossRef] [PubMed]
126. Bowman, A.; Birch-Machin, M.A. Age-dependent decrease of mitochondrial complex II activity in human skin fibroblasts. *J. Investig. Dermatol.* **2016**, *136*, 912–919. [CrossRef] [PubMed]
127. Schniertshauer, D.; Gebhard, D.; Bergemann, J. Age-dependent loss of mitochondrial function in epithelial tissue can be reversed by coenzyme Q10. *J. Aging Res.* **2018**, *2018*, 6354680. [CrossRef]
128. Wedel, S.; Martic, I.; Guerrero Navarro, L.; Ploner, C.; Pierer, G.; Jansen-Dürr, P.; Cavinato, M. Depletion of growth differentiation factor 15 (GDF15) leads to mitochondrial dysfunction and premature senescence in human dermal fibroblasts. *Aging Cell* **2023**, *22*, e13752. [CrossRef]
129. Sgarbi, G.; Matarrese, P.; Pinti, M.; Lanzarini, C.; Ascione, B.; Gibellini, L.; Dika, E.; Patrizi, A.; Tommasino, C.; Capri, M.; et al. Mitochondria hyperfusion and elevated autophagic activity are key mechanisms for cellular bioenergetic preservation in centenarians. *Aging* **2014**, *6*, 296–310. [CrossRef]
130. Woo, C.Y.; Jang, J.E.; Lee, S.E.; Koh, E.H.; Lee, K.U. Mitochondrial dysfunction in adipocytes as a primary cause of adipose tissue inflammation. *Diabetes Metab. J.* **2019**, *43*, 247–256. [CrossRef]
131. Ou, M.Y.; Zhang, H.; Tan, P.C.; Zhou, S.B.; Li, Q.F. Adipose tissue aging: Mechanisms and therapeutic implications. *Cell Death Dis.* **2022**, *13*, 300. [CrossRef]
132. Krstic, J.; Reinisch, I.; Schupp, M.; Schulz, T.J.; Prokesch, A. p53 functions in adipose tissue metabolism and homeostasis. *Int. J. Mol. Sci.* **2018**, *19*, 2622. [CrossRef] [PubMed]

133. de Lange, P.; Lombardi, A.; Silvestri, E.; Cioffi, F.; Giacco, A.; Iervolino, S.; Petito, G.; Senese, R.; Lanni, A.; Moreno, M. Physiological approaches targeting cellular and mitochondrial pathways underlying adipose organ senescence. *Int. J. Mol. Sci.* **2023**, *24*, 11676. [CrossRef] [PubMed]
134. Liu, W.; Yan, F.; Xu, Z.; Chen, Q.; Ren, J.; Wang, Q.; Chen, L.; Ying, J.; Liu, Z.; Zhao, J.; et al. Urolithin A protects human dermal fibroblasts from UVA-induced photoaging through NRF2 activation and mitophagy. *J. Photochem. Photobiol. B.* **2022**, *232*, 112462. [CrossRef]
135. Chen, L.; Qin, Y.; Liu, B.; Gao, M.; Li, A.; Li, X.; Gong, G. PGC-1α-mediated mitochondrial quality control: Molecular mechanisms and implications for heart failure. *Front. Cell Dev. Biol.* **2022**, *10*, 871357. [CrossRef]
136. Guo, Z.; Fan, D.; Liu, F.Y.; Ma, S.Q.; An, P.; Yang, D.; Wang, M.Y.; Yang, Z.; Tang, Q.Z. NEU1 Regulates Mitochondrial Energy Metabolism and Oxidative Stress Post-myocardial Infarction in Mice via the SIRT1/PGC-1 Alpha Axis. *Front. Cardiovasc. Med.* **2022**, *9*, 821317. [CrossRef]
137. Ledrhem, M.; Nakamura, M.; Obitsu, M.; Hirae, K.; Kameyama, J.; Bouamama, H.; Gadhi, C.; Katakura, Y. Essential oils derived from cistus species activate mitochondria by inducing SIRT1 expression in human keratinocytes, leading to senescence inhibition. *Molecules* **2022**, *27*, 2053. [CrossRef]
138. Soares, C.D.; de Lima Morais, T.M.; Mariano, F.V.; Altemani, A.; Corrêa, M.B.; Reis, R.R.D.D.; Amorim, L.S.; Ferreira, S.M.S.; de Almeida, O.P.; Carlos, R.; et al. Expression of mitochondrial dynamics markers during melanoma progression: Comparative study of head and neck cutaneous and mucosal melanomas. *J. Oral Pathol. Med.* **2019**, *48*, 373–381. [CrossRef]
139. Chen, G.; Kroemer, G.; Kepp, O. Mitophagy: An emerging role in aging and age-associated diseases. *Front. Cell Dev. Biol.* **2020**, *8*, 200. [CrossRef]
140. Russell, O.M.; Gorman, G.S.; Lightowlers, N.; Turnbull, D.M. Mitochondrial diseases: Hope for the future. *Cell* **2020**, *181*, 168–188. [CrossRef]
141. Ayunin, Q.; Miatmoko, A.; Soeratri, W.; Erawati, T.; Susanto, J.; Legowo, D. Improving the anti-ageing activity of coenzyme Q10 through protransfersome-loaded emulgel. *Sci. Rep.* **2022**, *12*, 906. [CrossRef]
142. Wu, H.; Zhong, Z.; Lin, S.; Qiu, C.; Xie, P.; Lv, S.; Cui, L.; Wu, T. Coenzyme Q_{10} sunscreen prevents progression of ultraviolet-induced skin damage in mice. *BioMed Res. Int.* **2020**, *2020*, 9039843. [CrossRef] [PubMed]
143. Sguizzato, M.; Mariani, P.; Spinozzi, F.; Benedusi, M.; Cervellati, F.; Cortesi, R.; Drechsler, M.; Prieux, R.; Valacchi, G.; Esposito, E. Ethosomes for Coenzyme Q10 cutaneous administration: From design to 3D skin tissue evaluation. *Antioxidants* **2020**, *9*, 485. [CrossRef] [PubMed]
144. Hoppe, U.; Bergemann, J.; Diembeck, W.; Ennen, J.; Gohla, S.; Harris, I.; Jacob, J.; Kielholz, J.; Mei, W.; Pollet, D.; et al. Coenzyme Q10, a cutaneous antioxidant and energizer. *Biofactors* **1999**, *9*, 371–378. [CrossRef]
145. Knott, A.; Achterberg, V.; Smuda, C.; Mielke, H.; Sperling, G.; Dunckelmann, K.; Vogelsang, A.; Krüger, A.; Schwengler, H.; Behtash, M.; et al. Topical treatment with coenzyme Q10-containing formulas improves skin's Q10 level and provides antioxidative effects. *Biofactors* **2015**, *41*, 383–390. [CrossRef]
146. Schniertshauer, D.; Müller, S.; Mayr, T.; Sonntag, T.; Gebhard, D.; Bergemann, J. Accelerated Regeneration of ATP Level after Irradiation in Human Skin Fibroblasts by Coenzyme Q10. *Photochem. Photobiol.* **2016**, *92*, 488–494. [CrossRef]
147. Kim, J.; Kim, K.; Sung, G.Y. Coenzyme Q10 efficacy test for human skin equivalents using a pumpless skin-on-a-chip system. *Int. J. Mol. Sci.* **2020**, *21*, 8475. [CrossRef]
148. Jung, H.J.; Kim, S.M.; Kim, D.H.; Bang, E.; Kang, D.; Lee, S.; Chun, P.; Moon, H.R.; Chung, H.Y. 2,4-Dihydroxyphenyl-benzo[d] thiazole (MHY553), a synthetic PPARα agonist, decreases age-associated inflammatory responses through PPARα activation and RS scavenging in the skin. *Exp. Gerontol.* **2021**, *143*, 111153. [CrossRef]
149. Oyewole, A.O.; Birch-Machin, M.A. Mitochondria-targeted antioxidants. *FASEB J.* **2015**, *29*, 4766–4771. [CrossRef]

antioxidants

MDPI

Article

Isorhamnetin Ameliorates Non-Esterified Fatty Acid-Induced Apoptosis, Lipid Accumulation, and Oxidative Stress in Bovine Endometrial Epithelial Cells via Inhibiting the MAPK Signaling Pathway

Haimiao Lv [1], Lijuan Liu [1], Wenna Zou [1], Ying Yang [1], Yuan Li [1], Shengji Yang [1], Aixin Liang [1,2,*] and Liguo Yang [1,2,*]

1 Key Laboratory of Agricultural Animal Genetics, Breeding and Reproduction of Ministry of Education, College of Animal Science and Technology, Huazhong Agricultural University, Wuhan 430070, China; lvhaimiao0310@webmail.hzau.edu.cn (H.L.); 2022302110088@webmail.hzau.edu.cn (L.L.); qiuyi@webmail.hzau.edu.cn (W.Z.); hzyy666@webmail.hzau.edu.cn (Y.Y.); li_yuan@webmail.hzau.edu.cn (Y.L.); 149248@webmail.hzau.edu.cn (S.Y.)

2 National Center for International Research on Animal Genetics, Breeding and Reproduction, Huazhong Agricultural University, Wuhan 430070, China

* Correspondence: lax.pipi@mail.hzau.edu.cn (A.L.); ylg@mail.hzau.edu.cn (L.Y.)

Abstract: High concentrations of non-esterified fatty acids (NEFA) in the blood contribute to various metabolic disorders and are linked to endometritis in dairy cows. Isorhamnetin (ISO), a flavonoid found in many plants, is known for its antioxidant, anti-inflammatory, and anti-obesity properties. This study systematically assessed NEFA-induced damage in bovine endometrial epithelial cells (bEECs) and investigated whether ISO alleviates NEFA-induced cell damage and its underlying molecular mechanisms. Our observations revealed that excessive NEFA inhibited proliferation and induced apoptosis in bEECs, accompanied by an increase in the expression of BAX and cleaved caspase-3. We further observed that NEFA could induce lipid accumulation, reactive oxygen species (ROS) generation, and the release of pro-inflammatory factors IL-1β, IL-6, and TNF-α in bEECs. RNA sequencing and Western blot analysis revealed that NEFA induced damage in bEECs by activating MAPK signaling pathway. Notably, ISO treatment ameliorated these effects induced by NEFA, as evidenced by decreased protein levels of BAX, cleaved caspase-3, and PPAR-γ, along with reductions in triglyceride content, ROS generation, and levels of IL-1β, IL-6, and TNF-α. Mechanistically, our experimental results demonstrated that ISO inhibited NEFA-induced activation of MAPK signaling. Overall, ISO shows promise for therapeutic development to address NEFA-related endometritis in dairy cows.

Keywords: non-esterified fatty acids; isorhamnetin; endometrial epithelial cells; MAPK signaling; bovine

1. Introduction

Ketosis is a common metabolic disease that occurs in high-producing dairy cows during early lactation and is associated with the incidence of endometritis [1–4]. Dairy cows experience a negative energy balance (NEB) due to the increased nutrient demands of lactation and a decrease in dry matter intake (DMI) during the postpartum transition period [5,6]. To compensate for the energy shortage in early lactation, the body elevates fat mobilization, resulting in elevated concentrations of non-esterified fatty acids (NEFA) in the blood. Previous research has indicated that serum NEFA levels in subclinical ketosis

https://doi.org/10.3390/antiox14020156

cows can reach approximately 0.6 mM [2,7,8]. By comparison, NEFA concentrations can reach approximately 1.2 mM in clinical ketosis or fatty liver [7,9–11]. When the antioxidant defenses in these animals are insufficient to counteract the generation of ROS and reactive nitrogen species (RNS) triggered by elevated NEFA levels, these free radicals can adversely affect physiological processes, leading to cellular damage and dysfunction [12]. Recent studies have shown evidence that high concentrations of NEFA (1.2 mM), similar to the NEFA levels observed in dairy cows experiencing fatty liver during the transition period, induce apoptotic cell death in hepatocytes through triggered mitochondrial dysfunction in bovine hepatocytes [10]. Additionally, the high levels of NEFA can lead to the accumulation of lipids and mitochondrial dysfunction in bovine mammary epithelial cells [13–15]. Metabolic diseases, including ketosis, can adversely affect the reproductive performance of dairy cows. Recent studies have shown that exposure to high levels of NEFA in vitro leads to fatty acid accumulation in preantral follicles and induces inflammatory responses in the ovarian cortex [16]. Elevated NEFA levels in follicular fluid impair follicular growth in cattle [17]. Furthermore, NEFA induces granulosa cell apoptosis in cattle through the PI3K/AKT/FoxO1 pathway activated by ROS [18]. However, the effect and precise mechanism of high levels of NEFA on the bovine endometrial epithelial cells remain unclear.

Isorhamnetin (ISO), also recognized as 3-O-methylquercetin, is a flavonoid that occurs naturally in various vegetables and fruits, particularly in *Hippophae rhamnoides* L. and the leaves of *Ginkgo biloba* L. [19,20]. ISO has been reported to exhibit a range of beneficial effects, including antioxidant, anti-inflammatory, anti-tumor, and anti-adipogenic activities [21–24]. Peroxisome proliferator-activated receptor gamma (PPAR-γ) is a ligand-dependent transcription factor that regulates adipogenesis and energy homeostasis [25]. The transcriptional activation of PPAR-γ in the liver induces the adipogenic program, facilitating the storage fatty acids in lipid droplets, as observed in adipocytes [26]. A recent study reported that dietary ISO acts as PPAR-γ antagonist, providing protective effects against obesity and hepatic steatosis in diet-induced obese mice [27]. Interestingly, recent studies have indicated that circulating NEFA in cattle induces dose-dependent activation of PPARs in various tissue cell types, particularly PPAR-δ and PPAR-γ [28]. Furthermore, Lee et al. demonstrated that isorhamnetin inhibits adipogenesis by downregulating PPAR-γ and C/EBP-α in 3T3-L1 adipocytes [22]. In addition, isorhamnetin mitigates cognitive impairments and neuroinflammation induced by high-fat and high-fructose diets by modulating the NF-κB and MAPK signaling pathways [29]. Nevertheless, the impact of isorhamnetin on NEFA-induced damage and lipid accumulation remains unexplored.

In this study, we aimed to explore the full potential effects and molecular mechanisms of NEFA on apoptosis, lipid accumulation, oxidative stress, and inflammation in bovine endothelial cells (bEECs). Importantly, we sought to examine the protective role of isorhamnetin on NEFA-induced damage in bEECs. Our findings may enhance the applicability of isorhamnetin in improving the health and production performance of transitional dairy cows, thereby significantly reducing veterinary costs and increasing the economic benefits for the dairy industry.

2. Materials and Methods

2.1. Preparation of Non-Esterified Fatty Acids and Isorhamnetin

A stock solution of NEFA (52.7 mM) was prepared by dissolving a mixture of 2.8 mM palmitoleic acid, 7.6 mM stearic acid, 2.6 mM linoleic acid, 16.8 mM palmitic acid, and 22.9 mM oleic acid (MedChemExpress, Shanghai, China) in 0.1 M KOH at 60 °C, as previously described [10,13,14]. Isorhamnetin was provided by ApexBio (Houston, TX, USA, purity $\geq$ 99.64%), and its chemical structural formula is shown in Figure S1. A stock solution of isorhamnetin was dissolved in dimethyl sulfoxide at a concentration of 10 mM and stored at −20 °C. Prior to use, different concentrations of isorhamnetin were prepared by dissolving it in the medium to achieve the appropriate dosage.

2.2. Cell Culture

The bovine endometrial epithelial cell line (bEECs) was purchased from Yaji Biological (Shanghai, China). Following thawing, bEECs were cultured in complete DMEM/F12 (Hyclone, Logan, CA, USA) culture medium. For identification purposes, the cells were immunoblotted with antibodies against cytokeratin 18 (Figure S2), which serves as an epithelial cell marker. After 2 to 3 passages, the cells were seeded into culture plates containing DMEM/F12 with 10% FBS, and the medium was replaced every 24 h. To evaluate the effects of varying NEFA concentrations on bEECs, the cells were cultured in DMEM/F12 medium supplemented with 0.11% fatty acid-free bovine serum albumin (BSA, Solebao, Beijing, China) and 0.1% insulin–transferrin–selenium (ITS, Sigma-Aldrich, Saint Louis, MO, USA). NEFA was applied for 24 h to cells at 0, 0.3, 0.6, or 1.2 mM. To assess the pharmacological effects of isorhamnetin on NEFA-treated cells, the cells were pretreated with isorhamnetin at concentrations of 1, 2.5, 5, or 10 μM for 12 h, followed by exposure to 1.2 mM NEFA for 24 h in serum-free DMEM/F12 medium (containing 0.11% BSA and 0.1% ITS).

2.3. Immunofluorescence

Cytokeratin 18 immunofluorescence staining was used to assess the purity of bEECs. Fixed cells underwent treatment with 4% paraformaldehyde for a duration of 30 min. Afterwards, they were subjected to permeabilization using 0.1% Triton X-100, which was maintained for a period of 1 h. Following washing with PBS, anti-cytokeratin 18 (Abcam, Cambridge, UK) was incubated overnight at 4 °C on the fixed cells. Following primary antibody incubation, the cells were treated with FITC-conjugated goat anti-rabbit IgG (Sigma-Aldrich, Saint Louis, MO, USA) for 45 min, followed by nuclear staining with 4′,6-diamidino-2-phenylindole (DAPI, Service bio, Wuhan, China) for 10 min. Fluorescence images were captured using a laser confocal microscope (Leica, Mannheim, Germany).

2.4. CCK-8 and EdU Assay

A CCK-8 kit (Dojindo Laboratories, Kumamoto, Japan) was used to measure cell viability according to the manufacturer's instructions. For the EdU incorporation assay, cell proliferation was assessed using a 5-ethynyl-2′-deoxyuridine (EdU) assay kit (Med Chem Express, Shanghai, China) following the manufacturer's instructions. Details are provided in Supplementary Materials and Methods.

2.5. Cell Cycle and Apoptosis Assay

After incubation with the indicated treatment, cell cycle analysis was performed using the Cell Cycle Kit (KeyGEN BioTECH, Nanjing, China). After fixation overnight at 4 °C in 70% ethanol, the cells were stained following the kit instructions, and the results were detected via flow cytometry. Apoptosis was assessed using the Annexin V-FITC Apoptosis Detection Kit (KeyGEN BioTECH, Nanjing, China). Briefly, drug-treated cells were harvested and resuspended in 500 μL of binding buffer containing the appropriate reagents. Subsequently, 5 μL of propidium iodide (PI) and 5 μL of Annexin V-FITC were sequentially added and mixed thoroughly. The cell suspension was incubated for 15 min at room temperature in the dark. Samples were then analyzed by flow cytometry, with data processed using FlowJo software (FlowJo, Version 10.4).

2.6. Triglyceride Measurement

Following treatment, the cells were digested with trypsin and subsequently collected by centrifugation. The concentrations of intracellular triglycerides were determined using a Triglyceride Assay Kit (Applygen Technologies Inc., Beijing, China) following the manufacturer's instructions. The results were normalized to the concentration of protein. A BCA Protein Assay Kit (HY CEZMBIO, Wuhan, China) was used to measure protein content.

2.7. Detection of ROS

ROS levels in bEECs were measured using an ROS detection kit (Beyotime, Shanghai, China). Cells were incubated with 2 mL of 10 μM DCFH-DA (diluted in serum-free medium) at 37 °C for 20 min in the dark. After incubation, the cells were washed three times with serum-free medium to remove any excess probes, collected, and analyzed for ROS production via flow cytometry.

2.8. Oil Red O and BODIPY Staining

Lipid droplet formation was assessed using Oil Red O (Solarbio, Beijing, China) and BODIPY493/503 (MedChemExpress, Shanghai, China) staining techniques. Details are provided in Supplementary Materials and Methods.

2.9. Enzyme-Linked Immunosorbent Assay (ELISA)

Cytokine production (IL-1β, IL-6, and TNF-α) in the collected supernatant was measured using ELISA kits (Enzyme-Linked Biosystems, Shanghai, China). The optical density was then determined at 450 nm with a multimode microplate reader (Tecan, Männedorf, Switzerland).

2.10. Western Blotting Assay

RIPA lysis buffer was employed for protein extraction from bEECs. Protein concentration was assessed using the BCA kit. Following the addition of loading buffer (Servicebio, Wuhan, China) and boiling for 10 min, equivalent amounts of protein (10–20 μg) were separated by electrophoresis on a 10–15% SDS-PAGE gel and transferred to a PVDF membrane (Millipore, Billerica, MA, USA). The PVDF membrane was blocked with 5% skim milk for 2 h and then incubated overnight at 4 °C with primary antibodies against PPAR-γ (1:1000, ZEN-BIOSCIENCE, Chengdu, China), cleaved caspase-3 (1:1000, Abmart, Shanghai, China), BAX (1:1000, Affinity Biosciences, Liyang City, China), CyclinD (1:1000, Affinity Biosciences, Liyang City, China), p38 (1:1000, Affinity Biosciences, Liyang City, China), phospho-p38 (Thr180/Tyr182, 1:1000, Affinity Biosciences, Liyang City, China), phospho-JNK (Thr183/Tyr185, 1:1000, Abmart, Shanghai, China), JNK (1:1000, Abmart, Shanghai, China), phospho-ERK (Thr202/Tyr204, 1:1000, Abmart, Shanghai, China), ERK (1:1000, Abclone, Wuhan, China) and GAPDH (1:1000, Servicebio, Wuhan, China). After washing with TBST, the membrane was incubated with goat anti-rabbit/mouse IgG (H + L)-HRP (1:10,000, Abclone, Wuhan, China) at room temperature for 2 h. Following three washes with TBST, the protein bands were detected by an ECL developer (Biosharp, Anhui, China) and chemiluminescence imager (Tanon, Shanghai, China).

2.11. RNA-Seq and Transcriptome Analysis

Transcriptome sequences were obtained from three samples each of the control and 1.2 mM NEFA treatment groups. Differential gene expression analysis was performed using DESeq2 (v1.38.3), identifying genes with |log2FoldChange| > 1 and a p-value < 0.05 as differentially expressed genes (DEGs). A significant enrichment in GO and KEGG pathways was deemed significant by p-values ≤ 0.05 based on hypergeometric distribution tests. Details are provided in Supplementary Materials and Methods.

2.12. RNA Isolation and Quantitative Real-Time PCR

Total RNA isolated from cells was performed following the instructions of the total RNA extraction kit (Vazyme, Nanjing, China). Details are provided in Supplementary Materials and Methods.

2.13. Statistical Analysis

Data were analyzed and plotted by GraphPad (Version 6.0) and presented as mean ± standard error of the mean (SEM). The experiments were performed in three replicates ($n = 3$). For multiple comparisons, one-way ANOVA was conducted followed by Tukey's post-hoc test.

3. Results

3.1. NEFA Inhibits Cell Viability and Induces Apoptosis in bEECs

To determine whether NEFA induces cytotoxicity in bEECs, its effect on cell viability was examined. The results indicated that high concentrations of NEFA significantly reduced cell viability compared to the control group. Of note, a dosage of 1.2 mM of NEFA led to a decrease in cell viability to 77.3% (Figure 1A). Flow cytometry analysis revealed a dose-dependent increase in apoptosis rates following NEFA treatment (Figure 1B,C). High concentration NEFA treated cells showed higher levels of cleaved caspase-3 and the pro-apoptotic protein BAX compared to the control group (Figure 1D–F). These findings indicate that excess NEFA inhibits the cell viability and induces the apoptosis of bEECs.

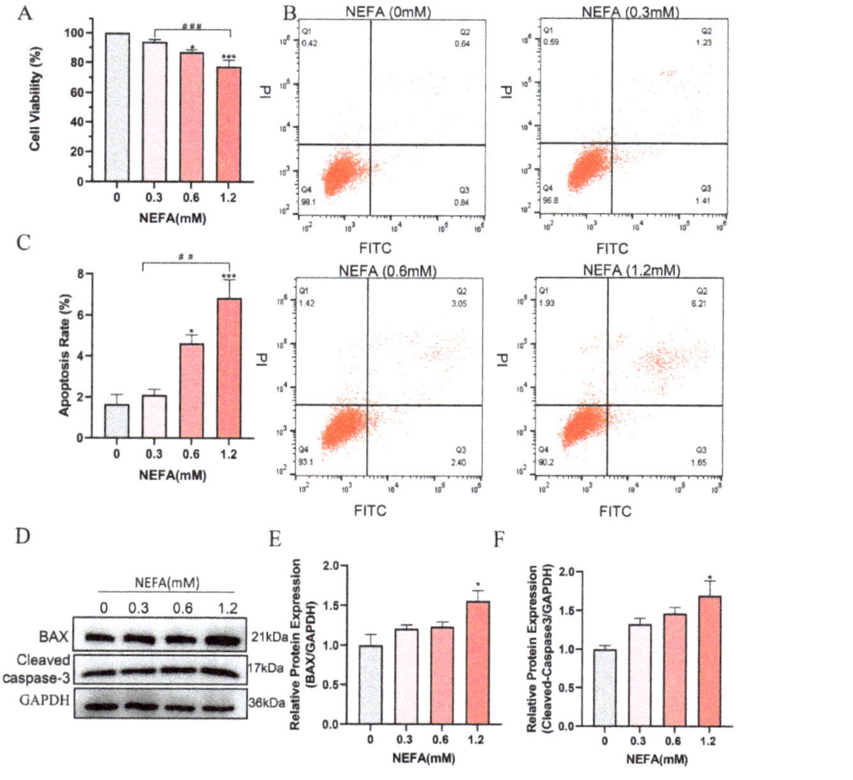

Figure 1. NEFA decreases cell proliferation and induces apoptosis in bEECs. BEECs were exposed to NEFA at the different concentrations (0, 0.3, 0.6, 1.2 mM) for 24 h. (**A**) Cell viability of bEECs was

assessed using the CCK-8 assay. (**B**) Flow cytometry results illustrating the apoptotic rates of bEECs under various treatment conditions. (**C**) Quantification was presented by the ratio of apoptotic cells/total cells. (**D**) Western blot analysis of BAX and cleaved caspase3 in the different treatments. (**E**) Quantification of normalized protein expression intensity of BAX. (**F**) Quantification of normalized protein expression intensity of cleaved caspase-3. The experiments were performed in three replicates ($n = 3$). * $p < 0.05$, *** $p < 0.001$ indicate statistically significant differences between the treatment group and the control group;, ## $p < 0.01$, ### $p < 0.001$ indicate statistically significant differences between different treatment groups. No marking indicates no significant difference.

3.2. NEFA Causes Lipid Accumulation in bEECs

NEFA has been demonstrated to induce excessive lipid accumulation in various non-adipose cells of dairy cows, resulting in cellular dysfunction and even cell death [13,30]. To investigate whether NEFA can cause lipid accumulation in bEECs, the cells were stained with BODIPY (green) after treatment with various concentrations of NEFA for 24 h. As shown in Figure 2A,B, lipid accumulation within bEECs increased gradually with the increasing NEFA concentrations. Consistent with the BODIPY staining data, intercellular TG content in the various concentrations of NEFA-treated groups was markedly enhanced compared with the control group (Figure 2C). The maximum intercellular TG content was observed in 1.2 mM NEFA treatment group, which was 10-fold higher than that in the control group. In addition, Western blot results showed that NEFA significantly upregulated PPAR-γ protein expression, a key regulator of adipogenesis (Figure 2D,E). These findings indicate that NEFA induces lipid accumulation in bEECs.

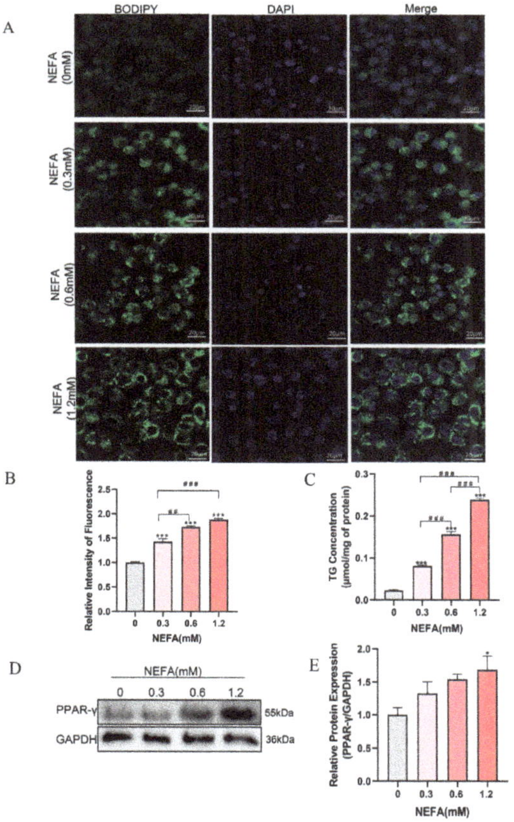

Figure 2. NEFA induces lipid accumulation in bEECs. BEECs were exposed to NEFA at different concentrations (0, 0.3, 0.6, 1.2 mM) for 24 h. (**A**) Representative BODIPY staining images of bEECs;

scale bar: 20 μm. (**B**) Comparison of relative BODIPY fluorescence intensity in bEECs. (**C**) Intercellular TG content in bEECs. (**D**) Western blot analysis of PPAR-γ expression across various treatment groups. (**E**) Quantification of normalized protein expression intensity of PPAR-γ. The experiments were performed in three replicates ($n = 3$). * $p < 0.05$, *** $p < 0.001$ indicate statistically significant differences between the treatment group and the control group; ## $p < 0.01$, ### $p < 0.001$ indicate statistically significant differences between different treatment groups. No marking indicates no significant difference.

3.3. NEFA Induces ROS Accumulation and the Release of Inflammatory Factors in bEECs

The intracellular ROS levels in bEECs exhibited a gradually increase with increasing NEFA concentrations, showing significant elevation in the treatment groups of 0.3, 0.6, and 1.2 mM NEFA when compared to the control group (Figure 3A,B). Subsequently, an ELISA assay was performed to investigate the concentrations of inflammatory cytokines, including IL-1β, IL-6, and TNF-α. The results indicated that as NEFA concentration increased, there was a marked rise in the release of pro-inflammatory cytokines IL-1β, IL-6, and TNF-α, which was significantly enhanced only in the groups treated with 0.6 mM and 1.2 mM NEFA when compared to the control group (Figure 3C–E). These findings suggest that excessive NEFA induces oxidative stress and inflammation in bEECs.

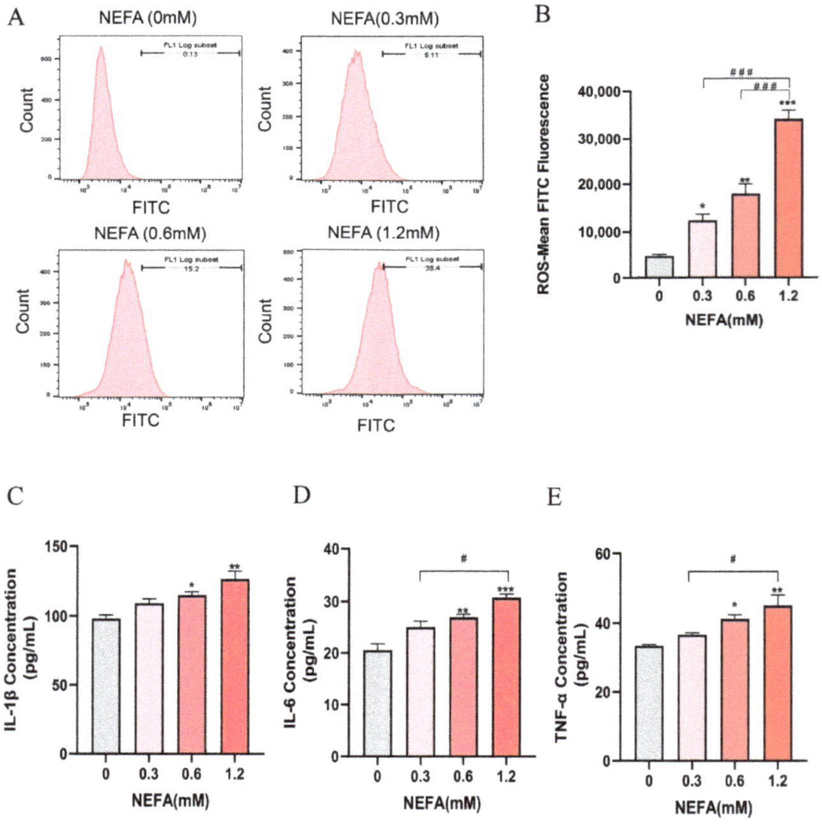

Figure 3. NEFA stimulates ROS accumulation and the release of inflammatory factors in bEECs. BEECs were exposed to NEFA at the different concentrations (0, 0.3, 0.6, 1.2 mM) for 24 h. (**A**) The level of ROS of bEECs were detected by flow cytometry. (**B**) Quantification of the level of ROS of bEECs. (**C**) IL-1β levels in the supernatant of bEECs treated with different concentrations of NEFA.

(**D**) IL-6 levels in the supernatant of bEECs treated with different concentrations of NEFA. (**E**) TNF-α levels in the supernatant of bEECs treated with different concentrations of NEFA. The experiments were performed in three replicates (n = 3). * $p < 0.05$, ** $p < 0.01$, *** $p < 0.001$ indicate statistically significant differences between the treatment group and the control group; # $p < 0.05$, ### $p < 0.001$ indicate statistically significant differences between different treatment groups. No marking indicates no significant difference.

3.4. NEFA Alters a Variety of Biological Processes and Signaling Pathways in bEECs

To further investigate the mechanisms behind NEFA-induced apoptosis, lipid accumulation, oxidative stress, and inflammation in bEECs, we next performed RNA sequencing (RNA-seq) analysis on cells after treatment with 1.2 mM NEFA. Heat map representation of these genes shows differential expression between the NEFA group and control group (Figure 4A). Compared to the control group, a total of 2394 DEGs were identified after NEFA treatment, including 1193 upregulated and 1201 downregulated genes (Figure 4B). The top 20 DEGs with the maximal fold changes (up- and downregulated) induced by NEFA are presented in Figure 4C. Several genes from DEGs were selected for RT-qPCR validation analysis, such as *CD36*, *CXCL8*, *IL-36*, *GRO1*, and *PDGFB* (Figure S3). A total of 1758 significant GO-BP terms, 154 significant GO-CC terms, and 251 significant GO-MF terms were identified. The significantly enriched GO terms mainly included the apoptotic process, cell death, MAPK cascade, response to oxidative stress, lipid droplet formation, and inflammatory response (Figure 4D). Chord diagrams revealed functional interconnections between multiple key biological processes and critical genes (Figure 4E). The scatter plot of KEGG pathway enrichment analysis revealed that the top 25 enrichment pathways primarily comprised the PI3K-Akt signaling pathway, MAPK signaling pathway, and ECM–receptor interaction (Figure 4F).

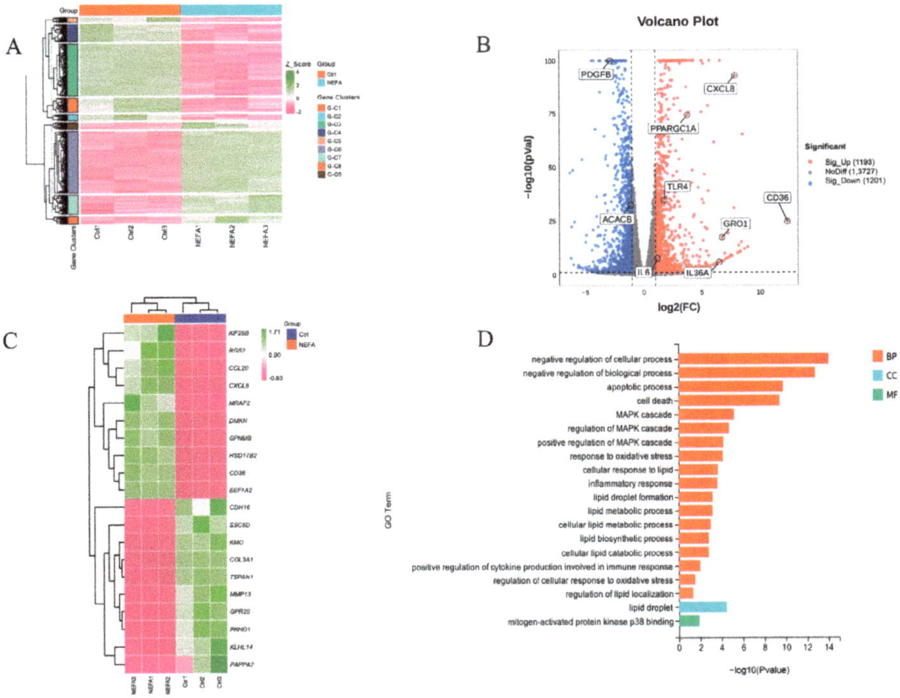

Figure 4. *Cont.*

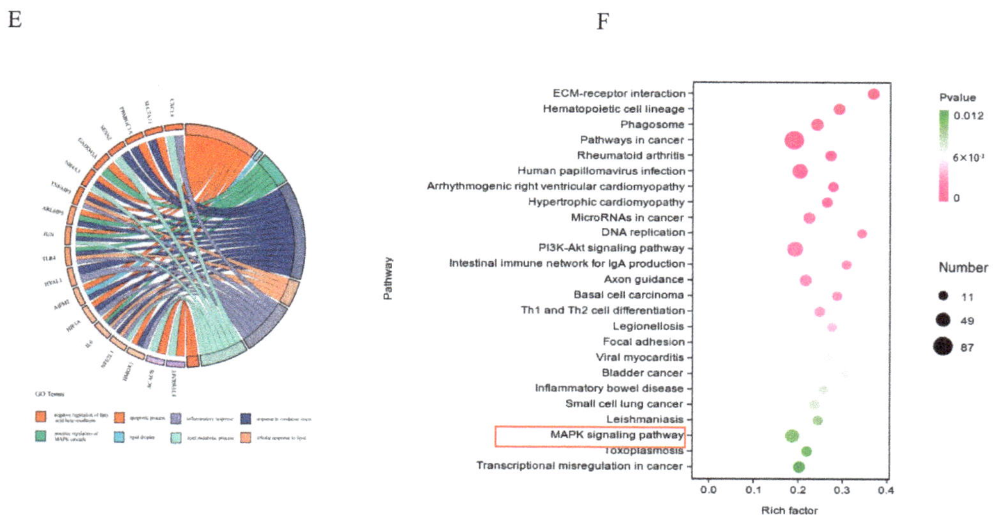

Figure 4. The results of RNA-seq data. BEECs were treated with 1.2 mM NEFA for 24 h. (**A**) The heatmap shows the expression of DEGs between the control group and NEFA group. (**B**) Volcano-plot representation of DEGs in control group and NEFA group. (**C**) The top 20 DEGs with the maximal fold changes (up- and downregulated) induced by NEFA. (**D**) Bar chart of the GO analysis of significant DEGs. (**E**) Chord diagram of the biological process analysis of significant DEGs. (**F**) Top 25 KEGG pathways with the highest enrichment; the red box represents the pathways we validated in subsequent experiments.

3.5. NEFA Activates MAPK Signaling Pathway in bEECs

Given that both GO and KEGG pathway analyses enriched the MAPK signaling pathway, which is closely linked to inflammation, cellular growth, and oxidative stress [31], we speculated that the MAPK signaling pathway may be involved in the adverse effects of NEFA on bEECs. Therefore, we further determined the expression changes in MAPK pathways including ERK, p38, and JNK by Western blot. As shown in Figure 5A–D, the phosphorylation levels of ERK and p38 were markedly activated in 0.6 mM and 1.2 mM NEFA-stimulated bEECs when compared to control, whereas the JNK phosphorylation was significantly activated in 1.2 mM NEFA-treated bEECs (Figure 5E,F). These findings demonstrate that excess NEFA can activate the MAPK signaling pathway in bEECs.

3.6. Isorhamnetin Promotes the Proliferation of bEECs

Isorhamnetin has been confirmed to possess multiple effects, such as anti-inflammatory, antioxidant, and anti-apoptotic properties [19,32]. In this study, we evaluated the effect of ISO on the proliferation and cell cycle of bEECs after treatment with various concentrations of ISO for 24 h. The results of CCK8 assay indicated that 2.5 μM, 5 μM and 10 μM isorhamnetin markedly raised the cell viability of bEECs when compared to the control group (Figure 6A). The quantification of EdU staining results showed a marked increase in the percentage of EdU-incorporating cells following treatment with 1, 2.5, 5, and 10 μM isorhamnetin (Figure 6B,C). In addition, flow cytometry analysis revealed that 1, 2.5, 5, and 10 μM isorhamnetin significantly enhanced the ratio of bEECs in the S-phase of the cell cycle (Figure 6D,E). Concurrently, western blot results revealed that the protein expression of Cyclin D significantly increased at 5μM and 10 μM isorhamnetin compared to the control group (Figure 6F,G). These findings indicate that isorhamnetin has a positive impact on the proliferation of bEECs.

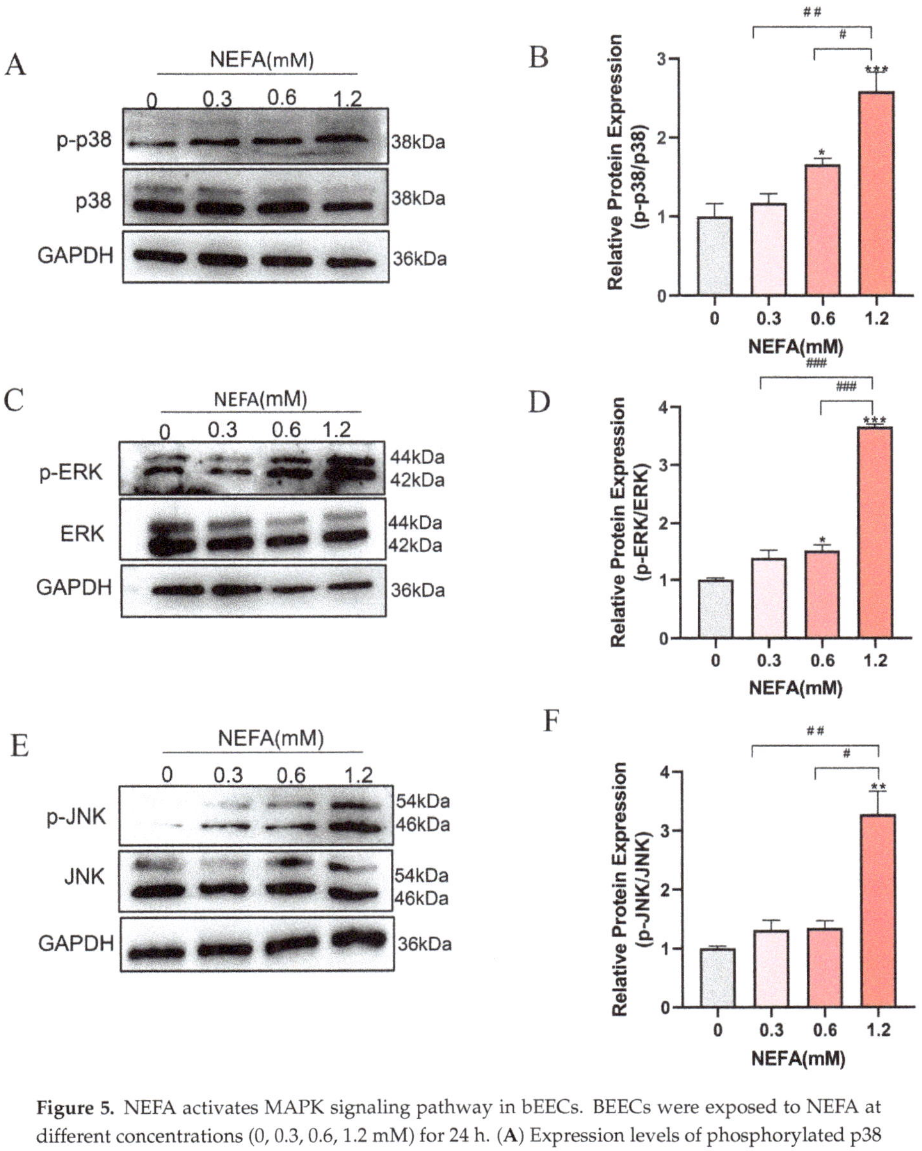

Figure 5. NEFA activates MAPK signaling pathway in bEECs. BEECs were exposed to NEFA at different concentrations (0, 0.3, 0.6, 1.2 mM) for 24 h. (**A**) Expression levels of phosphorylated p38 and p38 in bEECs treated with varying concentrations of NEFA. (**B**) Quantification of the ratio of p-p38/p38. (**C**) Expression levels of phosphorylated ERK and ERK in bEECs treated with varying concentrations of NEFA. (**D**) Quantification of the ratio of p-ERK/ERK. (**E**) Expression levels of phosphorylated JNK and JNK in bEECs treated with varying concentrations of NEFA. (**F**) Quantification of the ratio of p-JNK/ JNK. The experiments were performed in three replicates ($n = 3$). * $p < 0.05$, ** $p < 0.01$, *** $p < 0.001$ indicate statistically significant differences between the treatment group and the control group; # $p < 0.05$, ## $p < 0.01$, ### $p < 0.001$ indicate statistically significant differences between different treatment groups. No marking indicates no significant difference.

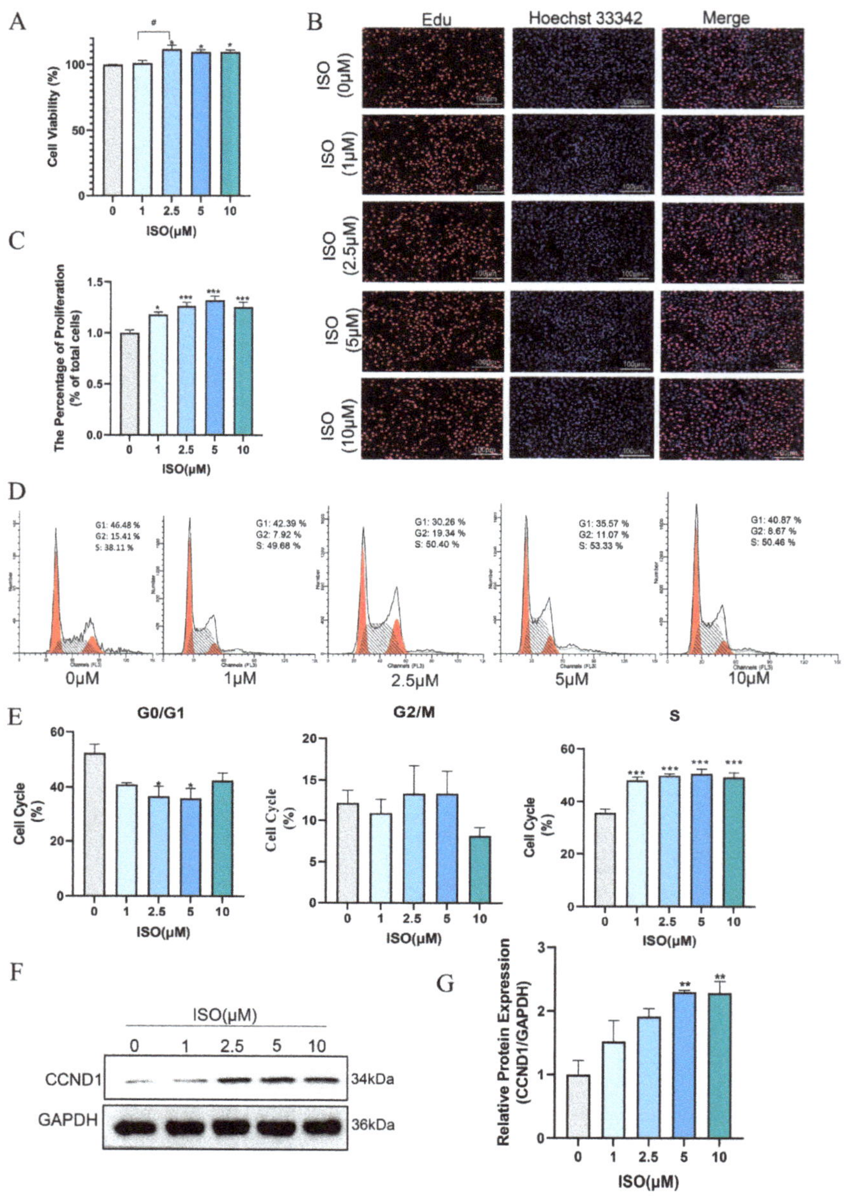

Figure 6. Isorhamnetin increases bEEC proliferation. BEECs were treated with varying concentrations of ISO (0, 1, 2.5, 5, and 10 μM) for 24 h. (**A**) Cell viability of bEECs was assessed using the CCK-8 assay. (**B**) Image of bEECs stained with EdU. EdU stained red and nuclei stained blue. Scale bar: 100 μm. (**C**) The bar chart displays the percentage of EdU-positive cells. (**D**) Flow cytometry results illustrating cell cycle distribution in the control and ISO-treated groups (1, 2.5, 5, and 10 μM). (**E**) The bar chart illustrates the cell cycle phase distributions (G0/G1, G2/M, and S) of bEECs across the various treatment groups. (**F**) Representative images of Western blot analysis of CCND1 in the different treatments. (**G**) Quantification of CCND1 protein levels in the different treatments. The experiments were performed in three replicates ($n = 3$). * $p < 0.05$, ** $p < 0.01$, *** $p < 0.001$ indicate statistically significant differences between the treatment group and the control group; # $p < 0.05$, indicate statistically significant differences between different treatment groups. No marking indicates no significant difference.

3.7. Isorhamnetin Attenuates NEFA-Induced Apoptosis in bEECs

To explore the protective effect of ISO on NEFA-induced apoptosis in bEECs, the cells were pretreated with varying concentrations of ISO for 12 h, then exposed to 1.2 mM NEFA for 24 h. The CCK-8 assay indicated that the cell viability of bEECs in the combined group of 2.5 μM ISO and NEFA increased by 12.9% compared to the NEFA alone treatment group (Figure 7A). Consequently, a concentration of 2.5 μM ISO was selected for subsequent experiments. Furthermore, isorhamnetin (2.5 μM) remarkably reduced the cell apoptosis rate compared to the NEFA alone treatment group (Figure 7B,C), which was accompanied by a significant decrease in the apoptotic biomarkers BAX and cleaved caspase-3 (Figure 7D–F). Taken together, these results indicate that ISO ameliorates the impaired cell viability and apoptosis induced by NEFA in bEECs.

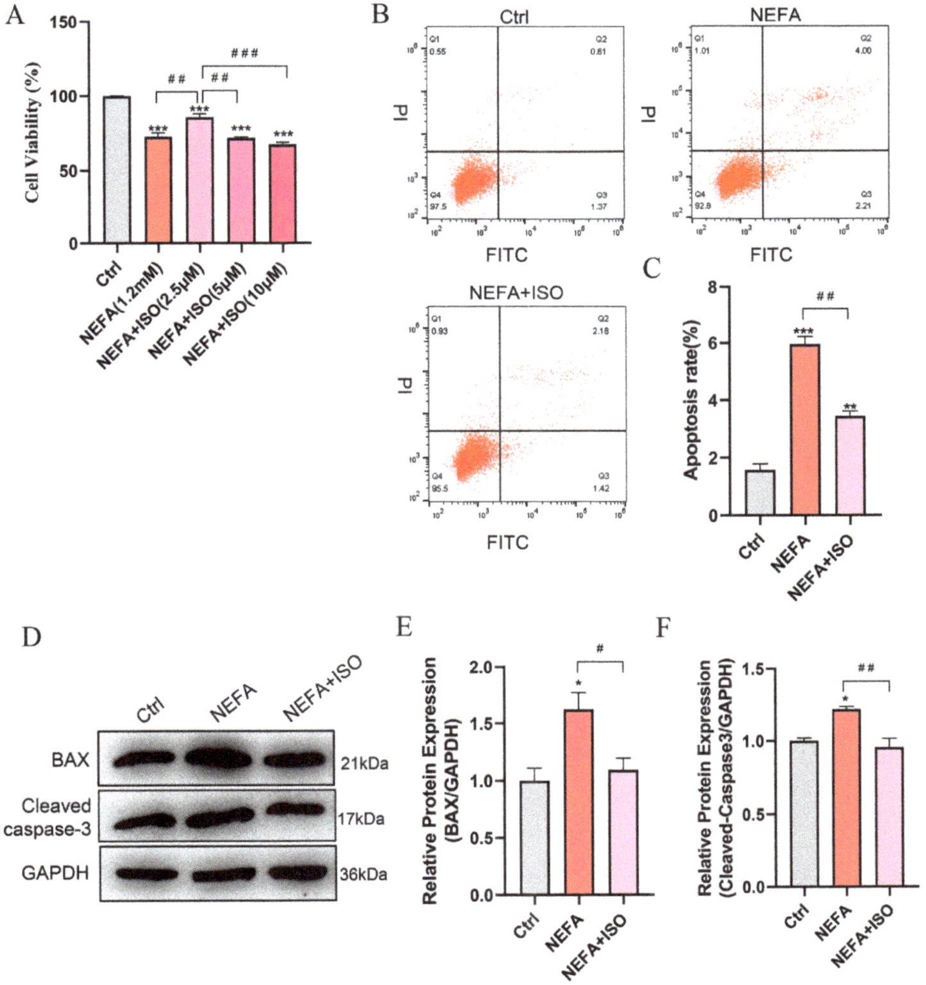

Figure 7. Isorhamnetin attenuates NEFA-induced apoptosis in bEECs. BEECs were pretreated with or without ISO at different concentrations (2.5, 5, 10 μM) or only 2.5 μM for 12 h and subsequently treated with NEFA (1.2 mM) for 24 h. (**A**) Cell viability of bEECs was assessed using the CCK-8 assay. (**B**) Flow cytometry results illustrating the apoptotic rates of bEECs under various treatment conditions. (**C**) Quantification was presented by the ratio of apoptotic cells/total cells. (**D**) Western

blot analysis of BAX and cleaved caspase-3 in the different treatments. (**E**) Quantification of BAX protein levels in the different treatments. (**F**) Quantification of cleaved caspase3 protein levels in the different treatments. The experiments were performed in three replicates (n = 3). * $p < 0.05$, ** $p < 0.01$, *** $p < 0.001$ indicate statistically significant differences between the treatment group and the control group; # $p < 0.05$, ## $p < 0.01$, ### $p < 0.001$ indicate statistically significant differences between different treatment groups. No marking indicates no significant difference.

3.8. Isorhamnetin Mitigates NEFA-Induced Lipid Accumulation in bEECs

BODIPY and Oil Red O staining revealed that lipid droplet (LD) content was remarkably reduced in the isorhamnetin pretreatment plus NEFA group compared to the NEFA alone treatment group (Figure 8A–C). Similarly, NEFA alone treatment increased the TG content to 8.47 times that of control group, while pretreatment with isorhamnetin (2.5 μM) recovered it to 6.33 times that of control group (Figure 8D). In addition, ISO significantly reduced the protein expression of PPAR-γ compared to the NEFA alone treatment group (Figure 8E,F). These findings suggest that ISO alleviates lipid accumulation caused by NEFA in bEECs.

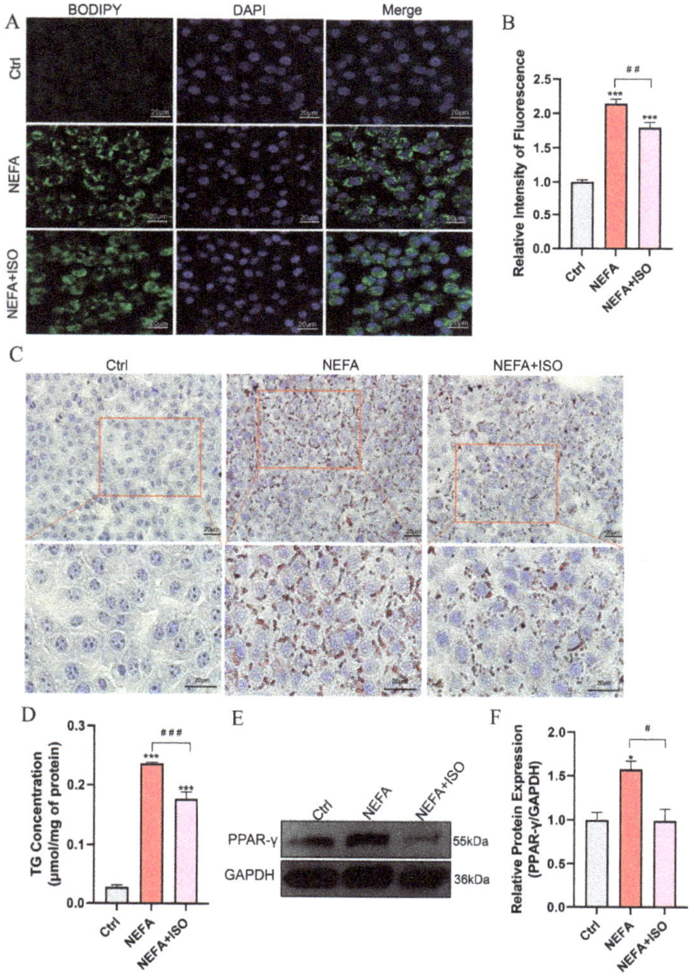

Figure 8. Isorhamnetin reduces lipid accumulation in NEFA-induced bEECs. BEECs were pretreated with or without ISO at a concentration of 2.5 μM for 12 h, followed by exposure to NEFA (1.2 mM)

for an additional 24 h. (**A**) Representative BODIPY staining images of bEECs; scale bar: 20 μm. (**B**) Comparison of relative BODIPY fluorescence intensity in bEECs. (**C**) Representative Oil Red O staining images of bEECs; scale bar: 20 μm. (**D**) Intercellular TG content in bEECs. (**E**) Western blot analysis of PPAR-γ expression across various treatment groups. (**F**) Quantification of normalized protein expression intensity of PPAR-γ. The experiments were performed in three replicates ($n = 3$). * $p < 0.05$, *** $p < 0.001$ indicate statistically significant differences between the treatment group and the control group; # $p < 0.05$, ## $p < 0.01$, ### $p < 0.001$ indicate statistically significant differences between different treatment groups. No marking indicates no significant difference.

3.9. Isorhamnetin Decreases NEFA-Induced ROS Accumulation and the Release of Inflammatory Factors in bEECs

ISO is reported to exhibit potent anti-inflammatory and antioxidant properties [19], but whether it can reduce ROS and the release of inflammatory factors in bEECs remains unknown. As illustrated in Figure 9A,B, ISO effectively attenuated the levels of ROS induced by NEFA treatment. Furthermore, ELISA results showed that ISO could remarkedly decreased the levels of IL-1β, IL-6, and TNF-α in cell culture supernatant and reverse the inflammatory response induced by NEFA (Figure 9C–E). Our findings reveal that ISO effectively reduces ROS generation and suppresses the inflammatory response in NEFA-treated bEECs.

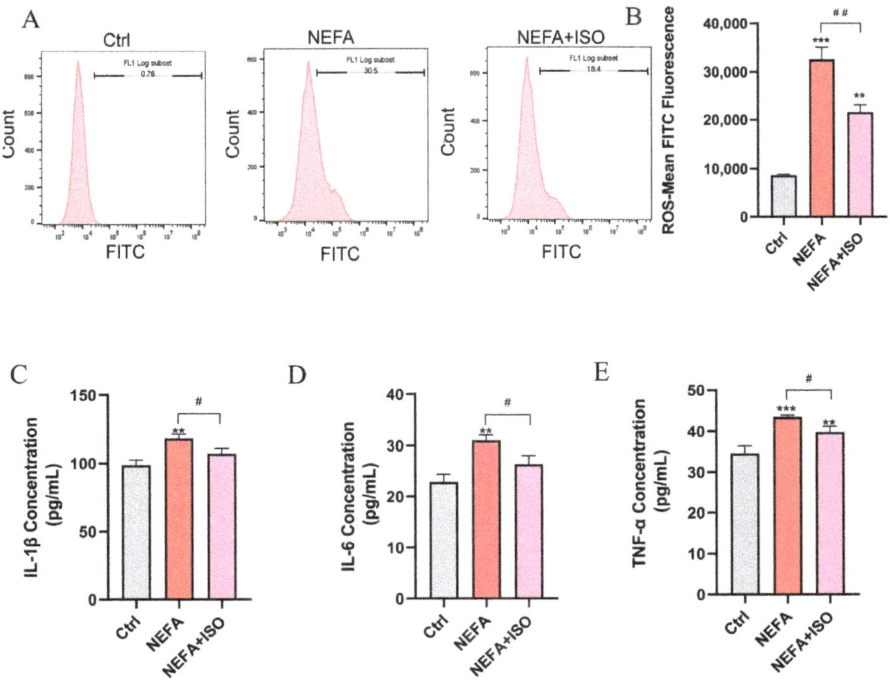

Figure 9. Isorhamnetin decreases NEFA-induced ROS accumulation and the release of inflammatory factors in bEECs. BEECs were pretreated with or without ISO at a concentration of 2.5 μM for 12 h, followed by exposure to NEFA (1.2 mM) for an additional 24 h. (**A**) The level of ROS in bEECs was detected by flow cytometry. (**B**) Quantification of the level of ROS in bEECs. (**C**) IL-1β levels in the supernatant of bEECs. (**D**) IL-6 levels in the supernatant of bEECs. (**E**) TNF-α levels in the supernatant of bEECs. The experiments were performed in three replicates ($n = 3$). ** $p < 0.01$, *** $p < 0.001$ indicate statistically significant differences between the treatment group and the control group; # $p < 0.05$, ## $p < 0.01$, indicate statistically significant differences between different treatment groups. No marking indicates no significant difference.

3.10. Isorhamnetin Reduces NEFA-Induced Activation of MAPK Signaling Pathway

To further investigate the therapeutic impact of ISO on NEFA-induced cellular injury, the MAPK signaling pathway was examined through Western blot. The results demonstrated that pretreatment with 2.5 μM ISO remarkedly reduced the phosphorylation levels of ERK, JNK, and p38 relative to the NEFA alone treatment group (Figure 10A–F). These findings suggest that ISO may attenuate the NEFA-induced bEECs damage through the MAPK signaling pathway.

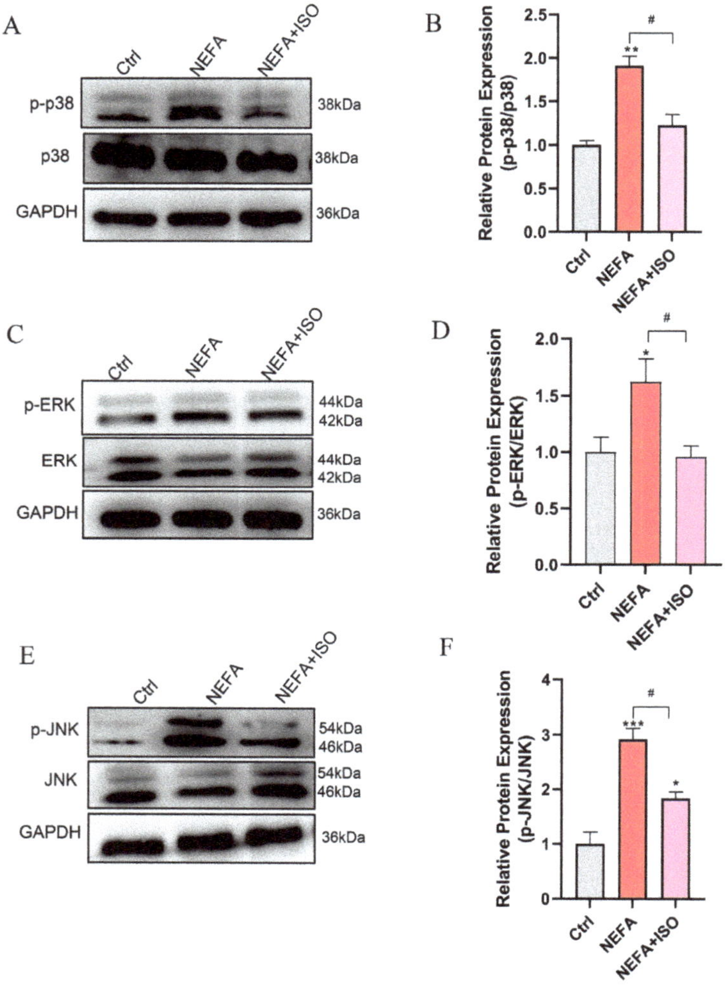

Figure 10. Isorhamnetin reduces NEFA-induced activation of MAPK pathway. BEECs were pretreated with or without ISO at a concentration of 2.5 μM for 12 h, followed by exposure to NEFA (1.2 mM) for an additional 24 h. (**A**) Western blot analysis of p38 and p-p38 in the different treatments. (**B**) Quantification of the ratio of p-p38/p38. (**C**) Western blot analysis of ERK and p-PERK in the different treatments. (**D**) Quantification of the ratio of p-ERK/ERK. (**E**) Western blot analysis of JNK and p-JNK in the different treatments. (**F**) Quantification of the ratio of p-JNK/JNK. The experiments were performed in three replicates ($n = 3$). * $p < 0.05$, ** $p < 0.01$, *** $p < 0.001$ indicate statistically significant differences between the treatment group and the control group; # $p < 0.05$, indicates statistically significant differences between different treatment groups. No marking indicates no significant difference.

4. Discussion

High blood concentrations of NEFA are considered as a significant risk factor for ketosis in dairy cows during the transition period. In this study, we observed that elevated levels of NEFA lead to damage in bEECs, as evidenced by decreased cell viability, activation of apoptosis, lipid accumulation, ROS generation, and the release of inflammatory factors. Alleviating these NEFA-induced negative effects is crucial for maintaining endometrial health, establishing pregnancy, and improving embryo development in cattle. Isorhamnetin, a flavonoid present in various medicinal plants, has been demonstrated in numerous studies to possess considerable anti-inflammatory and anti-obesity properties. Our findings confirm that ISO may mitigate NEFA-induced damage in bEECs by modulating the MAPK signaling pathways. These findings provide novel evidence that ISO could serve as a therapeutic agent to protect the endometrium of dairy cows during a transitional period.

Cows with higher serum NEFA concentrations tended to suffer from subclinical endometritis [33]; however, the precise regulatory effects of NEFA on endometrium cells remain to be elucidated. In this study, we evaluated the effects of different concentrations of NEFA (0.3, 0.6, and 1.2 mM) on the proliferation, apoptosis, lipid accumulation, ROS generation, and inflammation in bEECs. These varying concentrations of NEFA may correspond to different clinical conditions. Additionally, we primarily focused on shorter exposure times (12 h and 24 h), as our objective was to investigate the acute effects of NEFA exposure, which are relevant to the pathophysiological conditions of clinical ketosis. However, we also recognize that exploring longer exposure durations, especially at lower NEFA concentrations, could provide valuable insights into the chronic cytotoxic effects and represents an important direction for future research. This study found that NEFA reduced bEEC viability in a dose-dependent manner, accompanied by a dose-dependent increase in cell apoptosis. These findings are consistent with previous studies conducted by Chankeaw's team [34]. The apoptotic processes induced by NEFA have been reported through various mechanisms, including the induction of endoplasmic reticulum (ER) stress [35] and the activation of caspase-3 [36]. Here, we demonstrated that NEFA exerts pro-apoptotic effects by upregulating the expression of BAX and cleaved caspase-3, suggesting that NEFA causes mitochondrial dysfunction in bEECs. At the cellular level, previous studies have reported that the negative effects on cell viability and apoptosis may be related to the lipotoxicity induced by NEFA [13,34]. Our research further demonstrated that NEFA treatment increased the BODIPY fluorescence intensity and intracellular TG content, indicating NEFA induces lipid accumulation in bEECs. Concurrently, this effect is strongly supported by our RNA-seq data, which demonstrate that DEGs are mainly enriched in lipid-related biological processes, including cellular response to lipid, lipid droplet formation, lipid biosynthetic process, and lipid metabolic process. Notably, NEFA-stimulated lipid accumulation may arise from the expression upregulation of PPAR-γ, a key regulator responsible for triglyceride formation and storage in lipid droplets [37]. Recent studies have shown that an overload of lipids can induce oxidative stress [38]. Several studies have demonstrated that elevated NEFA levels are linked to increased mitochondrial ROS production and oxidative stress across various cell types, including bovine hepatocytes and granulosa cells [18,39]. Similarly, our experiments demonstrated that ROS levels increased in a NEFA dose-dependent manner in bEECs. In addition, ROS production is often associated with the occurrence of cellular inflammation [40]. Consistently, we observed that excessive NEFA stimulated the secretion of inflammatory factors IL-1β, TNF-α, and IL-6 in bEECs. This observation was further confirmed by GO enrichment analysis.

Recently, isorhamnetin, as a flavonoid active compound, has been widely studied. ISO has been reported to protect porcine oocytes from apoptosis, oxidative stress, and endoplasmic reticulum stress induced by zearalenone through the PI3K/AKT signaling pathway [41]. ISO pretreatment can reduce the loss of mitochondrial membrane potential, improve cellular morphological damage, and inhibit H_2O_2-induced apoptosis in HaCaT cells [42]. According to the published data, isorhamnetin raised the ratio of S-phase cells in response to GC proliferation by increasing Cyclin family protein expression [32]. Consistent with these findings, we observed that isorhamnetin treatment increased the ratio of cells in S-phase and Cyclin D expression significantly. Meanwhile, we also proved that ISO treatment promoted the proliferation of bEECs, as indicated by EdU-positive cells. These findings suggest that ISO is beneficial for bEEC growth. Besides the pro-proliferation effect, ISO is a novel antagonist of PPAR-γ that inhibits adipocyte differentiation induced by the PPAR-γ agonist rosiglitazone, thereby obstructing obesity development and alleviating hepatic steatosis caused by high-fat diet and leptin deficiency [27]. Lee et al. demonstrated that low concentrations of ISO (0.1, 0.5, 1, and 10 μM) can inhibit adipogenesis and promote mitochondrial biogenesis in 3T3-L1 cells [43]. These results suggest that isorhamnetin may have a protective effect against NEFA-induced damage and lipid accumulation in bEECs.

Interestingly, we found that 2.5 μM ISO is more effective than 5 μM or 10 μM in protecting bEECs from the effects of NEFA. Based on these beneficial effects, we further investigated the potential impact of 2.5 μM ISO in alleviating damage to bEECs induced by NEFA. Compared to the NEFA alone treatment group, the addition of ISO significantly enhanced cell viability and reduced the apoptosis rate by downregulating the expression of BAX and cleaved caspase-3 in bEECs. Furthermore, our findings demonstrated that ISO significantly alleviated oxidative stress and the inflammatory response induced by NEFA in bEECs, as evidence by a decrease in the production of ROS, IL-1β, IL-6, and TNF-α. Our results are consistent with former studies showing that isorhamnetin can alleviate oxidative stress and inflammation in paracetamol-induced liver damage [44] and heat stroke-affected lung injury [45]. As the antagonist of PPAR-γ, ISO expectedly downregulated the expression of PPAR-γ, thus reducing lipid accumulation in NEFA-treated bEECs. Notably, the protective effect of ISO observed in bEECs in this study may be a subsequent effect of reduced lipid accumulation. For instance, isorhamnetin-mediated reduction in lipid accumulation may attributed to decreased ROS production and inflammatory response, thereby alleviating cell apoptosis induced by NEFA. To our knowledge, this study is the first to report the inhibitory effect of isorhamnetin on lipotoxicity in bEECs.

The MAPK pathways relay, amplify, and integrate signals from various stimuli, triggering physiological responses such as cellular proliferation, differentiation, development, inflammatory reactions, and apoptosis in mammalian cells [46]. It has been proved that NEFA induced hepatocyte apoptosis mediating via the ROS-ERK/JNK signaling pathway and oxidative stress [39]. Likewise, NEFA increased reactive oxygen species (ROS) levels in bovine mammary epithelial cells by activating the MAPK signaling pathway [47]. In the current study, KEGG and GO analyses revealed that the MAPK signaling pathways were significantly enriched in NEFA treatment cells, indicating MAPK may play a significant role in NEFA-induced damage to bEECs. Subsequently, Western blot results confirmed a dose-dependent increase in the phosphorylation levels of JNK, ERK, and p38 following NEFA treatment. Of note, isorhamnetin attenuates the NEFA-induced activation of the MAPK signaling pathway in bEECs, as evidenced by the reduction in the phosphorylation expression of JNK, ERK, and p38. These findings align with previous observations that ISO inhibits LPS-induced phosphorylation of ERK and JNK in vivo by modulating the MAPK signaling pathways [48].

5. Conclusions

According to these findings, medium and high concentrations of NEFA induce lipid accumulation in bEECs and trigger apoptosis, oxidative stress, and inflammatory responses via activation of the MAPK signaling pathways, including p38, ERK, and JNK. Notably, ISO, as a PPAR-γ antagonist, may alleviate NEFA-induced lipid accumulation in bEECs by downregulating the expression of PPAR-γ. In addition, we confirmed that ISO mitigates apoptosis, oxidative stress, and inflammation in bEECs by attenuating NEFA-activated MAPK signaling (Figure 11). Collectively, these findings illuminate that isorhamnetin may be a promising candidate for pharmaceutical development or as part of a dietbased strategy to support bovine uterine health during the transition period.

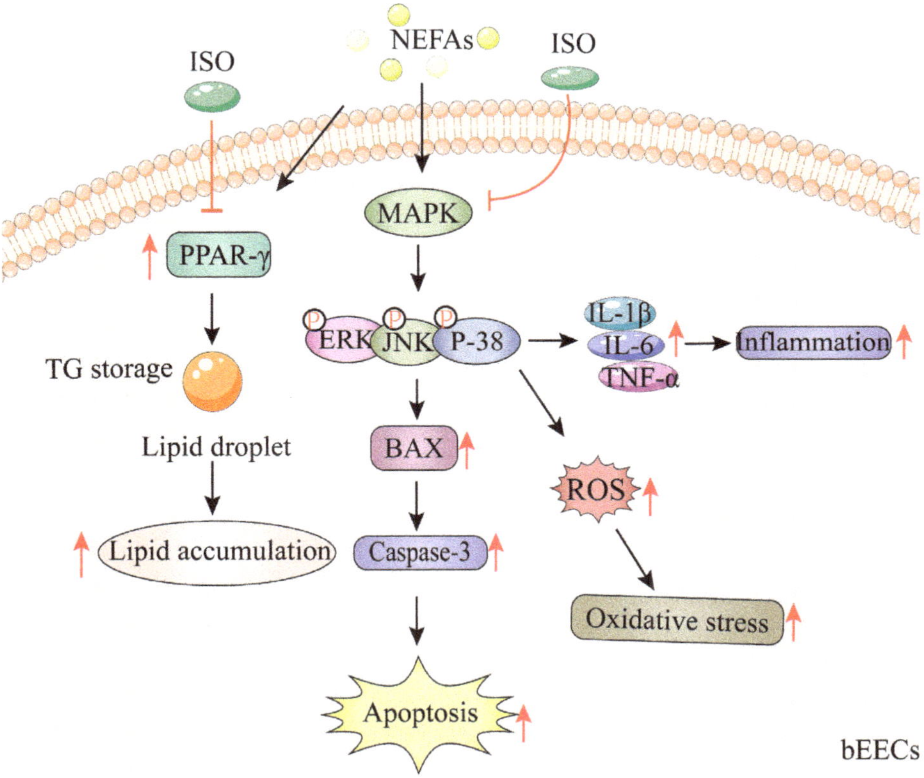

Figure 11. Isorhamnetin ameliorates NEFA-induced damage in bovine endometrial epithelial cells via inhibiting MAPK signaling pathway. High concentrations of NEFA induce lipid accumulation in bEECs and trigger apoptosis, oxidative stress, and inflammatory responses via activation of the MAPK pathway. Notably, ISO, as a PPAR-γ antagonist, alleviate NEFA-induced lipid accumulation in bEECs by downregulating PPAR-γ. In addition, ISO mitigates apoptosis, oxidative stress, and inflammation in bEECs by attenuating NEFA-activated MAPK signaling. The red "T" arrow represents "inhibition"; the black arrow represents "promotion".

Supplementary Materials: The following supporting information can be downloaded at: https://www.mdpi.com/article/10.3390/antiox14020156/s1, Figure S1: Chemical structure of isorhamnetin. Figure S2: Cytokeratin 18 (CK 18) immunofluorescence staining. Nuclei (blue) and CK18 (green); scale bar represents 10 μM. Figure S3: mRNA expression of CD36, CXCL8, IL-36, GRO1, and PDGFB.

Author Contributions: Conceptualization, H.L., A.L. and L.Y.; Data curation, H.L.; Formal analysis, H.L., L.L., W.Z., Y.Y., Y.L., A.L. and L.Y.; Investigation, L.L.; Methodology, H.L.; Resources, L.Y.; Software, H.L., L.L., W.Z., Y.Y., Y.L. and S.Y.; Visualization, S.Y.; Writing—original draft, H.L.; Writing—review and editing, A.L. and L.Y. All authors have read and agreed to the published version of the manuscript.

Funding: This research was funded by the Biological Breeding-Major Projects in National Science and Technology (2023ZD04049), the National Natural Science Foundation of China (NO. 32172731) and the National Key R&D Program of China (NO. 2023YFD1300604).

Institutional Review Board Statement: Not applicable.

Informed Consent Statement: Not applicable.

Data Availability Statement: The RNA-seq data reported in this article are available in the NCBI database (Bioproject PRJNA1163307).

Conflicts of Interest: The authors declare no conflicts of interest.

References

1. Braga Paiano, R.; Becker Birgel, D.; Harry Birgel Junior, E. Uterine Involution and Reproductive Performance in Dairy Cows with Metabolic Diseases. *Animals* **2019**, *9*, 93. [CrossRef] [PubMed]
2. Shin, E.K.; Jeong, J.K.; Choi, I.S.; Kang, H.G.; Hur, T.Y.; Jung, Y.H.; Kim, I.H. Relationships among ketosis, serum metabolites, body condition, and reproductive outcomes in dairy cows. *Theriogenology* **2015**, *84*, 252–260. [CrossRef] [PubMed]
3. Wathes, D.C.; Cheng, Z.; Chowdhury, W.; Fenwick, M.A.; Fitzpatrick, R.; Morris, D.G.; Patton, J.; Murphy, J.J. Negative energy balance alters global gene expression and immune responses in the uterus of postpartum dairy cows. *Physiol. Genom.* **2009**, *39*, 1–13. [CrossRef] [PubMed]
4. Varhidi, Z.; Csiko, G.; Bajcsy, A.C.; Jurkovich, V. Uterine Disease in Dairy Cows: A Comprehensive Review Highlighting New Research Areas. *Vet. Sci.* **2024**, *11*, 66. [CrossRef] [PubMed]
5. Perez-Baez, J.; Risco, C.A.; Chebel, R.C.; Gomes, G.C.; Greco, L.F.; Tao, S.; Thompson, I.M.; do Amaral, B.C.; Zenobi, M.G.; Martinez, N.; et al. Association of dry matter intake and energy balance prepartum and postpartum with health disorders postpartum: Part II. Ketosis and clinical mastitis. *J. Dairy Sci.* **2019**, *102*, 9151–9164. [CrossRef] [PubMed]
6. Shahzad, K.; Lopreiato, V.; Liang, Y.; Trevisi, E.; Osorio, J.S.; Xu, C.; Loor, J.J. Hepatic metabolomics and transcriptomics to study susceptibility to ketosis in response to prepartal nutritional management. *J. Anim. Sci. Biotechnol.* **2019**, *10*, 96. [CrossRef] [PubMed]
7. Brisken, C.; O'Malley, B. Hormone action in the mammary gland. *Cold Spring Harb. Perspect. Biol.* **2010**, *2*, a003178. [CrossRef]
8. Singh, K.; Erdman, R.A.; Swanson, K.M.; Molenaar, A.J.; Maqbool, N.J.; Wheeler, T.T.; Arias, J.A.; Quinn-Walsh, E.C.; Stelwagen, K. Epigenetic regulation of milk production in dairy cows. *J. Mammary Gland. Biol. Neoplasia* **2010**, *15*, 101–112. [CrossRef] [PubMed]
9. Duffield, T.F.; Sandals, D.; Leslie, K.E.; Lissemore, K.; McBride, B.W.; Lumsden, J.H.; Dick, P.; Bagg, R. Efficacy of monensin for the prevention of subclinical ketosis in lactating dairy cows. *J. Dairy Sci.* **1998**, *81*, 2866–2873. [CrossRef]
10. Gao, W.; Du, X.; Lei, L.; Wang, H.; Zhang, M.; Wang, Z.; Li, X.; Liu, G.; Li, X. NEFA-induced ROS impaired insulin signalling through the JNK and p38MAPK pathways in non-alcoholic steatohepatitis. *J. Cell. Mol. Med.* **2018**, *22*, 3408–3422. [CrossRef]
11. Hayirli, A. The role of exogenous insulin in the complex of hepatic lipidosis and ketosis associated with insulin resistance phenomenon in postpartum dairy cattle. *Vet. Res. Commun.* **2006**, *30*, 749–774. [CrossRef]
12. Khan, M.Z.; Ma, Y.; Xiao, J.; Chen, T.; Ma, J.; Liu, S.; Wang, Y.; Khan, A.; Alugongo, G.M.; Cao, Z. Role of Selenium and Vitamins E and B9 in the Alleviation of Bovine Mastitis during the Periparturient Period. *Antioxidants* **2022**, *11*, 657. [CrossRef]
13. Li, J.; Zhao, C.; Liu, M.; Chen, L.; Zhu, Y.; Gao, W.; Du, X.; Song, Y.; Li, X.; Liu, G.; et al. Nuciferine Ameliorates Nonesterified Fatty Acid-Induced Bovine Mammary Epithelial Cell Lipid Accumulation, Apoptosis, and Impaired Migration via Activating LKB1/AMPK Signaling Pathway. *J. Agric. Food Chem.* **2023**, *71*, 443–456. [CrossRef] [PubMed]
14. Loor, J.J.; Bionaz, M.; Drackley, J.K. Systems physiology in dairy cattle: Nutritional genomics and beyond. *Annu. Rev. Anim. Biosci.* **2013**, *1*, 365–392. [CrossRef] [PubMed]
15. Unger, R.H.; Clark, G.O.; Scherer, P.E.; Orci, L. Lipid homeostasis, lipotoxicity and the metabolic syndrome. *Biochim. Biophys. Acta* **2010**, *1801*, 209–214. [CrossRef] [PubMed]
16. Pedroza, G.H.; Lanzon, L.F.; Rabaglino, M.B.; Walker, W.L.; Vahmani, P.; Denicol, A.C. Exposure to non-esterified fatty acids in vitro results in changes in the ovarian and follicular environment in cattle. *Anim. Reprod. Sci.* **2022**, *238*, 106937. [CrossRef] [PubMed]
17. Ferst, J.G.; Missio, D.; Bertolin, K.; Gasperin, B.G.; Leivas, F.G.; Bordignon, V.; Goncalves, P.B.; Ferreira, R. Intrafollicular injection of nonesterified fatty acids impaired dominant follicle growth in cattle. *Anim. Reprod. Sci.* **2020**, *219*, 106536. [CrossRef]
18. Lei, Z.; Ali, I.; Yang, M.; Yang, C.; Li, Y.; Li, L. Non-Esterified Fatty Acid-Induced Apoptosis in Bovine Granulosa Cells via ROS-Activated PI3K/AKT/FoxO1 Pathway. *Antioxidants* **2023**, *12*, 434. [CrossRef]
19. Gong, G.; Guan, Y.Y.; Zhang, Z.L.; Rahman, K.; Wang, S.J.; Zhou, S.; Luan, X.; Zhang, H. Isorhamnetin: A review of pharmacological effects. *Biomed. Pharmacother.* **2020**, *128*, 110301. [CrossRef]

20. Navarro-Hortal, M.D.; Varela-Lopez, A.; Romero-Marquez, J.M.; Rivas-Garcia, L.; Speranza, L.; Battino, M.; Quiles, J.L. Role of flavonoids against adriamycin toxicity. *Food Chem. Toxicol.* **2020**, *146*, 111820. [CrossRef]
21. Boesch-Saadatmandi, C.; Loboda, A.; Wagner, A.E.; Stachurska, A.; Jozkowicz, A.; Dulak, J.; Doring, F.; Wolffram, S.; Rimbach, G. Effect of quercetin and its metabolites isorhamnetin and quercetin-3-glucuronide on inflammatory gene expression: Role of miR-155. *J. Nutr. Biochem.* **2011**, *22*, 293–299. [CrossRef] [PubMed]
22. Lee, J.; Jung, E.; Lee, J.; Kim, S.; Huh, S.; Kim, Y.; Kim, Y.; Byun, S.Y.; Kim, Y.S.; Park, D. Isorhamnetin represses adipogenesis in 3T3-L1 cells. *Obesity* **2009**, *17*, 226–232. [CrossRef]
23. Manu, K.A.; Shanmugam, M.K.; Ramachandran, L.; Li, F.; Siveen, K.S.; Chinnathambi, A.; Zayed, M.E.; Alharbi, S.A.; Arfuso, F.; Kumar, A.P.; et al. Isorhamnetin augments the anti-tumor effect of capecitabine through the negative regulation of NF-κB signaling cascade in gastric cancer. *Cancer Lett.* **2015**, *363*, 28–36. [CrossRef] [PubMed]
24. Seo, K.; Yang, J.H.; Kim, S.C.; Ku, S.K.; Ki, S.H.; Shin, S.M. The antioxidant effects of isorhamnetin contribute to inhibit COX-2 expression in response to inflammation: A potential role of HO-1. *Inflammation* **2014**, *37*, 712–722. [CrossRef]
25. Stump, M.; Guo, D.F.; Lu, K.T.; Mukohda, M.; Cassell, M.D.; Norris, A.W.; Rahmouni, K.; Sigmund, C.D. Nervous System Expression of PPARγ and Mutant PPARγ Has Profound Effects on Metabolic Regulation and Brain Development. *Endocrinology* **2016**, *157*, 4266–4275. [CrossRef] [PubMed]
26. Lee, Y.K.; Park, J.E.; Lee, M.; Hardwick, J.P. Hepatic lipid homeostasis by peroxisome proliferator-activated receptor gamma 2. *Liver Res.* **2018**, *2*, 209–215. [CrossRef] [PubMed]
27. Zhang, Y.; Gu, M.; Cai, W.; Yu, L.; Feng, L.; Zhang, L.; Zang, Q.; Wang, Y.; Wang, D.; Chen, H.; et al. Dietary component isorhamnetin is a PPARγ antagonist and ameliorates metabolic disorders induced by diet or leptin deficiency. *Sci. Rep.* **2016**, *6*, 19288. [CrossRef]
28. Busato, S.; Bionaz, M. The interplay between non-esterified fatty acids and bovine peroxisome proliferator-activated receptors: Results of an in vitro hybrid approach. *J. Anim. Sci. Biotechnol.* **2020**, *11*, 91. [CrossRef] [PubMed]
29. Mulati, A.; Zhang, X.; Zhao, T.; Ren, B.; Wang, L.; Liu, X.; Lan, Y.; Liu, X. Isorhamnetin attenuates high-fat and high-fructose diet induced cognitive impairments and neuroinflammation by mediating MAPK and NFκB signaling pathways. *Food Funct.* **2021**, *12*, 9261–9272. [CrossRef] [PubMed]
30. Shi, Z.; Li, X.B.; Peng, Z.C.; Fu, S.P.; Zhao, C.X.; Du, X.L.; Fang, Z.Y.; Wang, Z.; Liu, G.W.; Li, X.W. Berberine Protects against NEFA-Induced Impairment of Mitochondrial Respiratory Chain Function and Insulin Signaling in Bovine Hepatocytes. *Int. J. Mol. Sci.* **2018**, *19*, 1691. [CrossRef] [PubMed]
31. Sun, J.; Nan, G. The Mitogen-Activated Protein Kinase (MAPK) Signaling Pathway as a Discovery Target in Stroke. *J. Mol. Neurosci. MN* **2016**, *59*, 90–98. [CrossRef] [PubMed]
32. Li, X.; Chen, H.; Zhang, Z.; Xu, D.; Duan, J.; Li, X.; Yang, L.; Hua, R.; Cheng, J.; Li, Q. Isorhamnetin Promotes Estrogen Biosynthesis and Proliferation in Porcine Granulosa Cells via the PI3K/Akt Signaling Pathway. *J. Agric. Food Chem.* **2021**, *69*, 6535–6542. [CrossRef] [PubMed]
33. Yáñez, U.; Herradón, P.G.; Becerra, J.J.; Peña, A.I.; Quintela, L.A. Relationship between Postpartum Metabolic Status and Subclinical Endometritis in Dairy Cattle. *Animals* **2022**, *12*, 242. [CrossRef]
34. Chankeaw, W.; Guo, Y.Z.; Båge, R.; Svensson, A.; Andersson, G.; Humblot, P. Elevated non-esterified fatty acids impair survival and promote lipid accumulation and pro-inflammatory cytokine production in bovine endometrial epithelial cells. *Reprod. Fertil. Dev.* **2018**, *30*, 1770–1784. [CrossRef]
35. Zhang, Y.; Xue, R.; Zhang, Z.; Yang, X.; Shi, H. Palmitic and linoleic acids induce ER stress and apoptosis in hepatoma cells. *Lipids Health Dis.* **2012**, *11*, 1. [CrossRef]
36. Ohtsu, A.; Tanaka, H.; Seno, K.; Iwata, H.; Kuwayama, T.; Shirasuna, K. Palmitic acid stimulates interleukin-8 via the TLR4/NF-κB/ROS pathway and induces mitochondrial dysfunction in bovine oviduct epithelial cells. *Am. J. Reprod. Immunol.* **2017**, *77*, e12642. [CrossRef] [PubMed]
37. Montaigne, D.; Butruille, L.; Staels, B. PPAR control of metabolism and cardiovascular functions. *Nat. Rev. Cardiol.* **2021**, *18*, 809–823. [CrossRef] [PubMed]
38. Chae, S.Y.; Kim, Y.; Park, C.W. Oxidative Stress Induced by Lipotoxicity and Renal Hypoxia in Diabetic Kidney Disease and Possible Therapeutic Interventions: Targeting the Lipid Metabolism and Hypoxia. *Antioxidants* **2023**, *12*, 2083. [CrossRef] [PubMed]
39. Li, Y.; Ding, H.; Liu, L.; Song, Y.; Du, X.; Feng, S.; Wang, X.; Li, X.; Wang, Z.; Li, X.; et al. Non-esterified Fatty Acid Induce Dairy Cow Hepatocytes Apoptosis via the Mitochondria-Mediated ROS-JNK/ERK Signaling Pathway. *Front. Cell Dev. Biol.* **2020**, *8*, 245.
40. Mittal, M.; Siddiqui, M.R.; Tran, K.; Reddy, S.P.; Malik, A.B. Reactive oxygen species in inflammation and tissue injury. *Antioxid. Redox Signal.* **2014**, *20*, 1126–1167. [CrossRef]
41. Li, X.; Duan, J.; Wang, S.; Cheng, J.; Chen, H.; Zhang, Z.; Yang, L.; Hua, R.; Li, Q. Isorhamnetin protects porcine oocytes from zearalenone-induced reproductive toxicity through the PI3K/Akt signaling pathway. *J. Anim. Sci. Biotechnol.* **2023**, *14*, 22. [CrossRef]

42. Hu, W.; Zhang, J.; Wang, H.; Guan, M.; Dai, L.; Li, J.; Kang, X. Protective effects of isorhamnetin against H_2O_2-induced oxidative damage in HaCaT cells and comprehensive analysis of key genes. *Sci. Rep.* **2023**, *13*, 2498. [CrossRef] [PubMed]
43. Lee, M.S.; Kim, Y. Effects of Isorhamnetin on Adipocyte Mitochondrial Biogenesis and AMPK Activation. *Molecules* **2018**, *23*, 1853. [CrossRef]
44. Rousta, A.M.; Mirahmadi, S.M.; Shahmohammadi, A.; Mehrabi, Z.; Fallah, S.; Baluchnejadmojarad, T.; Roghani, M. Therapeutic Potential of Isorhamnetin following Acetaminophen-Induced Hepatotoxicity through Targeting NLRP3/NF-κB/Nrf2. *Drug Res.* **2022**, *72*, 245–254. [CrossRef] [PubMed]
45. Dong, X.; Li, J.J.; Ma, N.; Liu, A.Z.; Liu, J.W. Anti-Inflammatory and Anti-Oxidative Effects of Isorhamnetin for Protection Against Lung Injury in a Rat Model of Heatstroke in a Dry-Heat Environment. *Med. Sci. Monit.* **2022**, *28*, e935426. [CrossRef]
46. Zhang, W.; Liu, H.T. MAPK signal pathways in the regulation of cell proliferation in mammalian cells. *Cell Res.* **2002**, *12*, 9–18. [CrossRef]
47. Yan, Y.; Huang, J.; Huan, C.; Li, L.; Li, C. Non-Esterified Fatty Acid Induces ER Stress-Mediated Apoptosis via ROS/MAPK Signaling Pathway in Bovine Mammary Epithelial Cells. *Metabolites* **2022**, *12*, 803. [CrossRef] [PubMed]
48. Chi, G.; Zhong, W.; Liu, Y.; Lu, G.; Lu, H.; Wang, D.; Sun, F. Isorhamnetin protects mice from lipopolysaccharide-induced acute lung injury via the inhibition of inflammatory responses. *Inflamm. Res.* **2016**, *65*, 33–41. [CrossRef]

toxics

MDPI

Communication

Tea Polyphenols Relieve the Fluoride-Induced Oxidative Stress in the Intestinal Porcine Epithelial Cell Model

Chunyan Xie [1,†], Shuyi Niu [2,†] and Wen Tian [2,*]

1 Tianjin Key Laboratory of Animal Molecular Breeding and Biotechnology, Tianjin Livestock and Poultry Health Breeding Technology Engineering Center, Institute of Animal Science and Veterinary, Tianjin Academy of Agricultural Sciences, Tianjin 300381, China; xie.chunyan@foxmail.com
2 Key Laboratory of Agro-ecological Processes in Subtropical Region, Institute of Subtropical Agriculture, Chinese Academy of Sciences, Changsha 410125, China; niu461130365@foxmail.com
* Correspondence: tianw2022@isa.ac.cn
† These authors contributed equally to this work.

Abstract: Prolonged excessive intake of fluoride (F) can result in fluorosis, leading to a range of tissue oxidative damages. Therefore, mitigating the oxidative stress induced by fluorosis has become a significant research concern. Consequently, how to relieve oxidative stress caused by fluorosis is an urgent matter. In the present study, intestinal porcine epithelial (IPEC-J2) cells were chosen to explore the underlying mechanism of tea polyphenols (TPs) on F-induced oxidative stress. The results show that the cytotoxicity of IPEC-J2 cells induced by F presented a dose-dependent manner according to cell viability. Additionally, F treatment inhibited the activity of T-SOD, CAT, and GSH-Px as well as their transcription levels, increased the reactive oxygen (ROS) formation and cell damage rates, and then promoted cell apoptosis through the results of TUNEL and mitochondrial membrane potential detection when compared with the IPEC-J2 cells from the control group. As the main antioxidant ingredient in tea, TPs alleviated F-induced cell oxidation and apoptosis via blocking F-induced ROS generation and LDH's release, as well as promoting the transcription of tight junction (TJ) proteins and the activities of antioxidant enzymes in IPEC-J2 cells. These results provide a new treatment strategy for F-induced intestinal oxidative impairment.

Keywords: fluorosis; TPs; oxidative stress; ROS; IPEC-J2 cells

1. Introduction

Humans and animals can be exposed to F through drinking water, air, and food supplements. However, long-term excessive F intake would cause fluorosis, which might disrupt the redox equilibrium, produce more free radicals, and then cause significant oxidative damage in bones, teeth, the gastrointestinal tract, etc. [1,2]. Many studies showed that excessive F impairs antioxidant defense systems [3,4]. F accumulation could increase ROS formation [5,6] and reduce antioxidant enzyme activity [3,6]. It has been reported that NaF treatment significantly increases the hepatic MDA level in mice and induces the generation of free radicals, thus leading to intracellular mitochondrial damage [7,8]. F overexposure in the diet could reduce the serum antioxidant capacity, destroy the intestinal barrier, and increase the intestinal permeability of mice [9]. However, there are still no oxidation-free and effective treatment methods for fluorosis, which has become a global public health problem in several nations, including China [10].

It is well known that antioxidants could be an effective strategy for maintaining intestinal health [11]. As one of the most popular drinks worldwide, tea is rich in polyphenols,

 https://doi.org/10.3390/toxics13020083

which are the major active compounds present in teas [12]. It has been proven many times that TPs have anti-oxidant activity and stress resistance properties thanks to their ability to limit ROS formation by inhibiting the activity of oxidative enzymes, improve the activity of endogenic antioxidants, and sweep ROS, indicating they play a key role in regulating free radicals [13,14]. TPs could activate different antioxidant enzymes and attenuate the oxidative damage of DNA. Apart from this, TPs are also regarded as a potent stress regulator in mediating the stress induced by metal ions via ROS scavenging [11,15]. It has been reported that TPs in green tea could prevent some metal-induced oxidative stress and damage [16]. However, few studies are available linking TPs to the occurrence of fluorosis, and we believe that this issue should be looked into, given the proven oxidation resistance-promoting effect of TPs.

Generally, F is transported to various tissues of the body after being absorbed by the gastrointestinal tract. As the main organ exposed to nutrients and toxic food contaminants, the defenses of the intestine are not enough to respond adequately against the oxidative stress and contribute to developing intestinal pathologies under conditions which exacerbate ROS production [17,18]. Moreover, aside from residents, pigs breeding in most fluoride-rich areas still have to face the F threat from local underground water. Therefore, based on the fact that IPEC-J2 cells are widely used to study the impacts of nutrients or toxins on intestinal epithelial cell permeability and intestinal function [19,20], we aimed to explore the underlying mechanism by which TPs eliminate intestinal oxidative stress induced by F exposure in IPEC-J2 cells in this study.

2. Materials and Methods

This study was conducted in accordance with the Basic & Clinical Pharmacology & Toxicology policy for experimental and clinical studies (2023) [21].

2.1. Cells and Reagents

IPEC-J2 cells (BeNa Culture Collection, Beijing, China) were cultured in DMEM + 10% FBS + 100 U/mL penicillin and 0.1 mg/mL streptomycin (Gibico, New York, NY, USA) at 37 °C with 5% CO_2 throughout the experiment. A cell culture was performed using 12 well Costar Snapwell inserts (Corning Inc., Corning, NY, USA) from 2.5×10^5 to 4.0×10^5 per well in a 0.5 mL volume. The IPEC-J2 cells were allowed to grow for about 36 h to reach an average cell density of 2.5×10^5 per well. Then, the subsequent analysis was conducted.

TPs (purity $\geq$ 98%) were purchased from Jiangshu Ruixiang Biotechnology Co., Ltd. (Changzhou, China), and NaF (purity $\geq$ 98%) was purchased from Maclin Biochemical Technology Co., Ltd. (Shanghai, China).

2.2. Establishment of NaF-Treated Cell Model and Determination of Working Concentration of TPs

IPEC-J2 cells were cultured in 96 well Costar Snapwell inserts (Corning Inc., Corning, NY, USA), and the remaining procedures followed the same protocol described in Section 2.1. Based on previous studies, we selected the approximate concentrations of the NaF (0 mM, 1 mM, 2 mM, 3 mM, and 4 mM) and TPs (0, 100, 200, 300, and 400 $mg \cdot L^{-1}$) to establish the F-induced IPEC-J2 cell model protected by TPs [22–24]. After the IPEC-J2 cells were incubated with a series of NaF for 24 h, we determined the working concentration of NaF according to the cell viability (about 50%), which was found by testing with Cell Counting Kit-8 (CCK8, Dojindo Molecular Technologies, Inc., Kumamoto, Japan). The TP working solution was conformed via the cell viability of F-induced cells.

2.3. Cell Treatment

According to studies about the well-known role of TPs in maintaining intestinal health as antioxidants [11–15,18,20], we just divided the IPEC-J2 cells into three groups without the TP group alone—a control group (Control, cultured with medium without FBS for 24 h), NaF group (NaF, treated with 4 mM NaF for 24 h), and NaF +TPs group (NaF +TPs, treated with 4 mM NaF and 200 mg·L^{-1} TPs for 24 h)—after the IPEC-J2 cells were cultured in 12 well Costar Snapwell inserts until the cell confluency reached 80%. Next, we collected cell samples and stored them at −80 °C for the following analysis.

2.4. Determination of Antioxidant-Related Parameters

The activities of the total superoxide dismutase (T-SOD), catalase (CAT), and glutathione peroxidase (GSH-Px) were measured with SOD activity assay kit, CAT activity assay kit and GPX activity assay kit from Beijing Boxbio Science & Technology Co.,Ltd. (Beijing, China). After incubation as described in Section 2.3, the cells were harvested into the extract solution of the corresponding kit after being washed with ice-cold and sterile PBS. Then, the supernatant was collected after being centrifuged at 8000× *g* at 4 °C for 10 min for subsequent analysis according to the instructions of the kits. The absorbance was measured with a microplate reader (Infinite M200 PRO, Tecan Trading AG, Shanghai, China).

2.5. Intracellular ROS, Cell Apoptosis, and Mitochondrial Membrane Potential (MMP) Detection

The intracellular ROS, cell apoptosis, and MMP of the IPEC-J2 cells were detected using an ROS assay kit, a one-step TUNEL apoptosis assay kit, and an MMP assay kit with JC-1 (Beyotime Biotechnology, Shanghai, China) after being collected as described in Section 2.4. Then, they were detected in accordance with their introductions. The fluorescence intensity is analyzed by Image J software (version 1.53a, National Institutes of Health, Bethesda, MD, USA).

2.6. Lactate Dehydrogenase (LDH) Detection

The relative cell damage rate was evaluated via the LDH in the cytoplasm releasing into the medium when the cell membrane was damaged. The LDH level in the medium was detected in accordance with the instructions of the Cytotoxicity LDH Assay Kit-WST® (Dojindo Molecular Technologies, Inc., Kumamoto, Japan) to evaluate the cell membrane permeability via determining the relative cell damage rate.

2.7. Quantitative Real-Time PCR (RT-qPCR)

The total RNA of the IPEC-J2 cells was extracted using Trizol reagent (Tiangen Biotech Co., Ltd., Beijing, China), and reverse transcribed with a prime script RT reagent kit (Takara Biotechnology Co., Ltd., Dalian, China). The primer sequences for GPx-1, GPx-2, GPx-3, GPx-4, CAT, SOD, Occludin, Claudin-1, ZO-1, and GAPDH are listed in Table 1. The relative mRNA levels of the genes were calculated with the $2^{-\Delta\Delta Ct}$ method in the case of GAPDH as an internal control.

Table 1. The primer sequences for RT-qPCR.

Gene	ID	Primer Sequence (5′-3′)	Length (bp)
GAPDH	NM_007393.5	F: TGTCCACCTTCCAGCAGATGT	101
		R: AGCTCAGTAACAGTCCGCCTAGA	
GPX1	NM_214201.1	F: TGGGGAGATCCTGAATTG	184
		R: GATAAACTTGGGGTCGGT	
GPX2	NM_001115136.1	F: TGCAACCAATTTGGACATCAG	122
		R: TTCACGTCACACTTCTGGATAAGG	
GPX3	NM_001115155.1	F: AAACAGGAACCGGGAGACAA	156
		R: AGGACAGGCGTTCTTCAGGAA	

Table 1. *Cont.*

Gene	ID	Primer Sequence (5′-3′)	Length (bp)
GPX4	NM_214407.1	F: GATTCTGGCCTTCCCTTGC	173
		R: TCCCCTTGGGCTGGACTTT	
CAT	XM_021081498.1	F: CGAAGGCGAAGGTGTTTG	370
		R: AGTGTGCGATCCATATCC	
SOD	XM_021100523.1	F: ATGCTGACGCTGCTCTGTGCTTA	143
		R: TCCTGCCAGATCTCCGTCACTTT	
Occludin	NM_001360536.1	F: CCCCTCTTTCCTTAGGCGAC	168
		R: TTCAAAAGGCCTCACGGACA	
Claudin-1	NM_008770.3	F: TGGTGGACATCCTCATCCTT	190
		R: GCCAGCAGAATAAGGAGCAC	
ZO-1	NM_001417374.1	F: GAGCAGGCTTTGGAGGAGAC	162
		R: TGGGACAAAAGTCCGGGAAG	

2.8. Statistical Analysis

Data were analyzed by one-way ANOVA in SPSS statistics 21 (SPSS Institute, Inc., New York, NY, USA) and presented as the mean ± SEM. LSD was used for multiple comparisons, where $p < 0.05$ was considered to have a statistical difference and *, **, and *** represent $p < 0.05$, $p < 0.01$, and $p < 0.001$, respectively. GraphPad prism 8.3 (GraphPad Software, San Diego, CA, USA) was utilized to create bar charts and cell survival curves.

3. Results

3.1. TPs Increased the Viability of IPEC-J2 Cells Induced by F

The IPEC-J2 cells were exposed to NaF at different concentrations (0 mM, 1 mM, 2 mM, 3 mM, and 4 mM). As shown in Figure 1A, 4 mM NaF was chosen as the inducing solution when the viability of the cells dropped to about 50% ($p < 0.001$). Then, we found that 200 mg·L^{-1} of TPs had the best effect on increasing the viability of F-induced cells ($p < 0.001$) (Figure 1B). Therefore, we used 4 mM NaF and 200 mg·L^{-1} of TPs as the working concentrations in the next series of cell experiments.

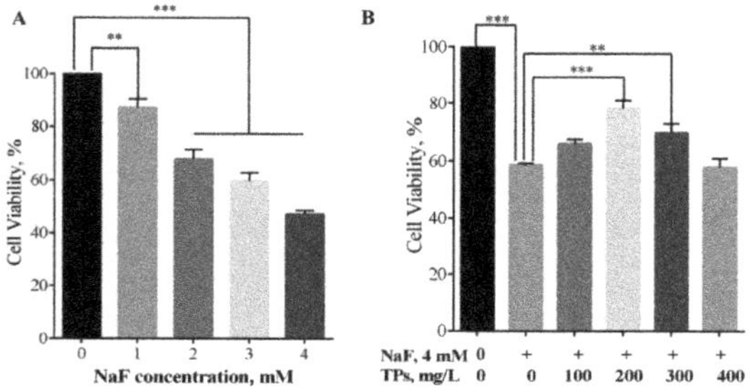

Figure 1. The viability of IPEC-J2 cells treated with different concentrations of NaF or TPs. (**A**) Effect of different concentrations of NaF on the viability of IPEC-J2 cells ($n = 3$). (**B**) Protect effect of different concentrations of TPs on the viability of IPEC-J2 cells ($n = 3$). Data were presented as mean ± SEM. ** $p < 0.01$. *** $p < 0.001$.

3.2. Effect of TPs on REDOX-Related Indexes in F-Induced IPEC-J2 Cells

As shown in Figure 2A, NaF induction inhibited ($p < 0.01$) the activity of CAT, T-SOD, and GSH-Px in cells when compared with the control group, but TP supplementation improved ($p < 0.001$) the activity of CAT and GSH-Px in the F-induced IPEC-J2 cells. The RT-qPCR results also showed a similar trend in the mRNA levels of CAT, SOD, GPx-2, GPx-3, and GPx-4. However, GPx-1 expression was significantly increased ($p < 0.001$) by the NaF treatment and not altered by TP supplementation in the NaF-induced IPEC-J2 cells (Figure 2B), indicating that TPs did not relieve F-induced oxidative stress in the IPEC-J2 cells via regulating the level of GPx-1.

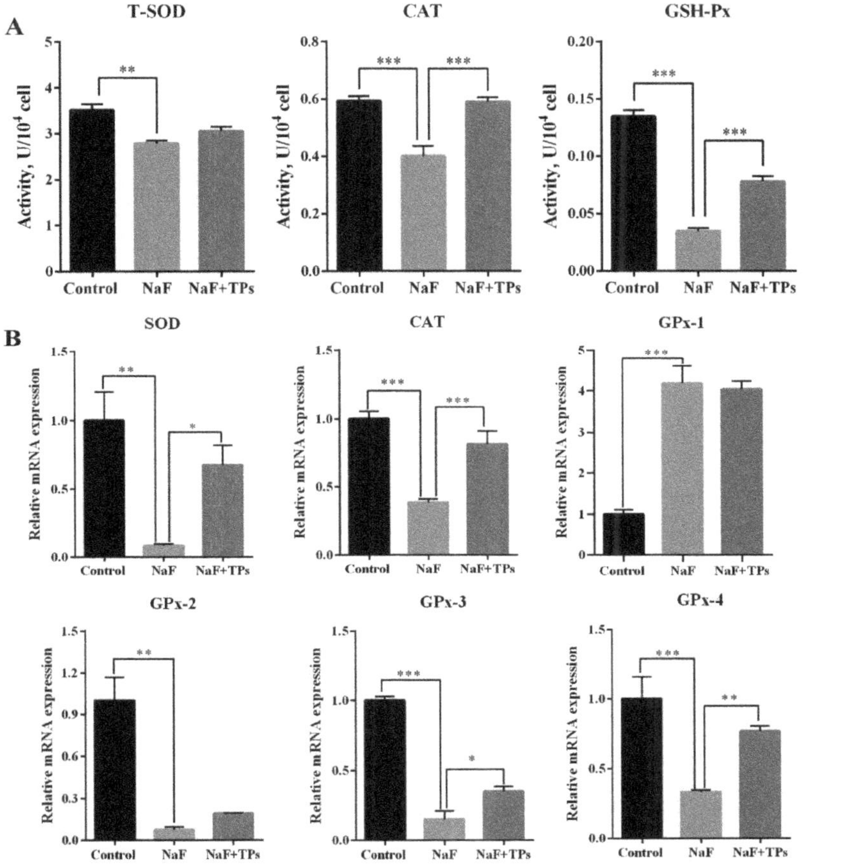

Figure 2. Determination of REDOX-related indexes in IPEC-J2 cells. (**A**) Activity of antioxidant related enzymes ($n = 3$). (**B**) Relative mRNA expression of genes associated with antioxidant enzymes in IPEC-J2 cells ($n = 3$). * $p < 0.05$. ** $p < 0.01$. *** $p < 0.001$.

The ROS in the IPEC-J2 cells were detected by immunofluorescence after treatment with TPs and NaF for 24 h, which was highly relevant to the level of antioxidant enzymes. The ROS level significantly increased after NaF induction ($p < 0.001$) when compared with the control group but almost reduced to a normal level under TP supplementation of the NaF-induced cells (Figure 3).

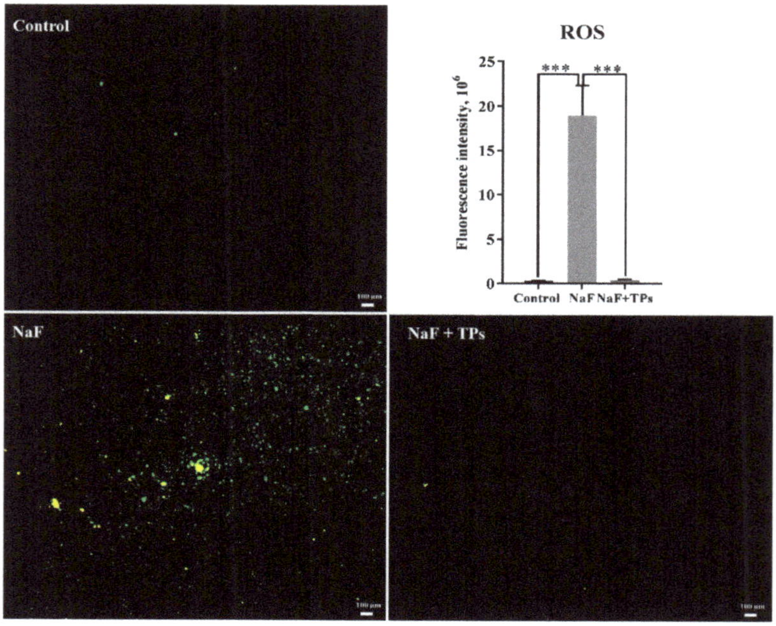

Figure 3. ROS levels in IPEC-J2 cells ($n = 3$). Data were presented as mean ± SEM. *** $p < 0.001$.

3.3. *Effects of TPs on the Cell Injury-Related Indexes in F-Induced IPEC-J2 Cells*

Both TUNEL and MMP detection are key indicators in evaluating cell apoptosis. According to the introduction of the TUNEL kit, apoptosis was determined by the relative fluorescence intensity (red/green). As shown in Figure 4A (1,2), NaF induction alone significantly reduced the red fluorescence intensity ($p < 0.01$) and increased the green fluorescence intensity ($p < 0.001$), which resulted in an obvious decrease in the relative fluorescence intensity (red/green) ($p < 0.001$). As expected, TP supplementation notably relieved the downward trend in the relative fluorescence intensity (red/green) ($p < 0.01$) when compared with that in the NaF group. Aside from MMP detection, TUNEL also showed similar results in apoptosis. As shown in Figure 4B, the green fluorescence in the TPs + NaF group was significantly reduced when compared with that in the NaF group ($p < 0.001$) and almost returned to the level in the control group, from which it was inferred that TPs have a significant effect in alleviating the proportion of apoptosis induced by F exposure.

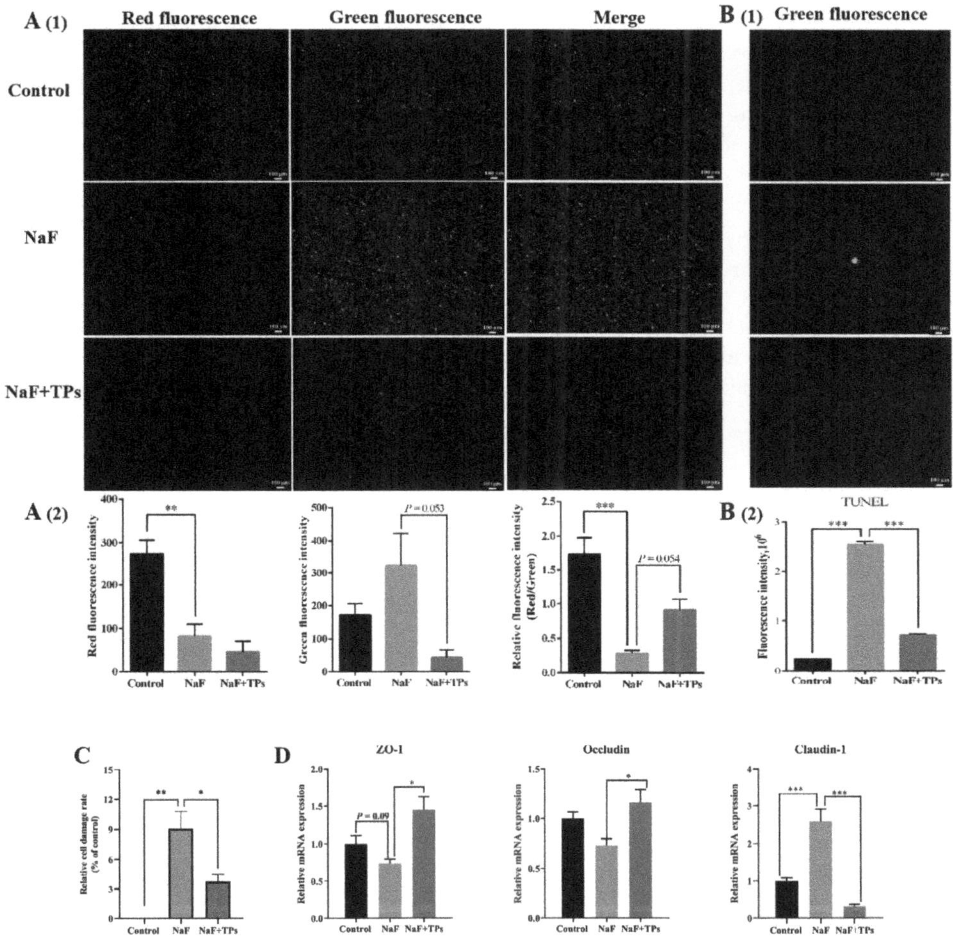

Figure 4. The cell injury-related indexes in IPEC-J2 cells. (**A**) Mitochondrial membrane potential detection with JC-1 in IPEC-J2 cells ($n = 4$): (**1**) Red fluorescence and green fluorescence reflect the high and low MMP of mitochondria, respectively; (**2**) The relative ratio of red fluorescence to green fluorescence in IPEC-J2 cells, and the decrease in the red/green fluorescence ratio indicates that mitochondria are damaged or the cells are in the early apoptosis state. (**B**) TUNEL detection ($n = 3$): (**1**) Fluorescent imagines show apoptotic IPEC-J2 cells, (**2**) Fluorescence intensity analysis in IPEC-J2 cells (**C**) The relative IPEC-J2 cell damage rate according to the LDH level in the medium ($n = 3$). (**D**) The relative mRNA levels of TJ proteins in IPEC-J2 cells ($n = 3$). Data presented as mean ± SEM. * $p < 0.05$. ** $p < 0.01$. *** $p < 0.001$.

In addition, we also found that TP supplementation decreased the IPEC-J2 cell damage rate induced by NaF via inhibiting LDH's release from the cytoplasm to the medium ($p < 0.05$) (Figure 4C). As for the level of TJ proteins in the IPEC-J2 cells, TP supplementation improved ($p < 0.05$) the relative mRNA levels of ZO-1 and Occludin when compared with those in the NaF group but significantly inhibited the Claudin-1 level ($p < 0.001$) (Figure 4D). Combined with the above results in Figure 4A–D, it was supposed that TPs contributed to dealing with the cell injury induced by excessive F exposure.

4. Discussion

In the absence of effective medication, tissue damage from fluorosis has become a key public health issue. It is well known that long-term high F intake increases oxidative stress levels, inflammatory-related factor secretion, apoptosis, and organelle damage [25,26]. In this study, we found significant dose–response changes in IPEC-J2 cell viability due to the content of F added, which demonstrated that the toxic effects of F increased with the exposure concentration. Many studies have shown that tea or its functional components could inhibit cell proliferation of cancer cells via regulating some cell survival pathways [27–29], protecting neurocytes or hepatocytes against oxidative stress-mediated cell death [30,31]. The present study found that the viability of F-exposed cells was increased by 200 mg/L of TP supplementation, which further demonstrated that TPs could promote cell proliferation against apoptosis induced by heavy metals.

The major antioxidant enzymes in the intestine, such as SOD, CAT, and GSH-Px, could protect cell from oxidative stress induced by superoxides [32]. Actually, F could inhibit the activity of antioxidant-related enzymes and decrease some antioxidant contents, all of which lead to excessive oxygen free radical generation and greater oxidative stress [33]. Likewise, F induction not only inhibited the activities of T-SOD, CAT, and GSH-Px but also their transcription levels, as presented in this study. ROS mainly occur due to enzymatic reactions or the byproducts of cellular respiration. Accordingly, antioxidant defenses decline when the redox state is altered by ROS accumulation [18]. To maintain the ROS at an appropriate level, tissues usually use antioxidant components to reduce free radical cytotoxicity and provide major antioxidant defenses against the ROS threat, especially SOD and CAT [34]. The antioxidant activity of TPs partially depends on their activities as ROS scavengers [35]. In this study, it was further verified that TPs are efficient at resisting intestinal oxidative stress induced by F exposure, according to the levels of ROS and the antioxidant-related enzymes of IPEC-J2 cells.

Abnormal oxidation can interfere with reversible oxidation-reduction reactions, which physiologically function in cell proliferation and apoptosis [36]. Mitochondria participate in oxidative-mediated apoptosis [37]. Loss of some antioxidant-related substances from mitochondria would increase ROS generation and lead to colonic epithelial cell apoptosis [38]. It has been reported that ROS generation would lead to mitochondria-mediated apoptosis [39]. Consequently, the available evidence in this study shows that TPs relieved IPEC-J2 cell apoptosis caused by F induction, referring to the results for the TUNEL and MMP, which may have directly reduced LDH's release from the cytoplasm to the medium, thus attenuating the cell damage rate. Furthermore, it has been proven that F disrupted the intestinal epithelial tight junction integrity [40,41] and TPs help improve intestinal barrier function [42], which confirms that TP supplements may contribute to restoring an intestinal barrier disrupted by F exposure through increasing TJ protein expression in IPEC-J2 cells. However, more research is still needed to determine exactly how TPs lead to susceptibility in IPEC-J2 cells exposed to excessive F.

5. Conclusions

Collectively, our results highlight the important role of TPs in protecting intestinal epithelial cells from F-induced oxidative stress via improving the activity of antioxidant enzymes and limiting ROS generation to modify oxidative balance and cell apoptosis. These findings would provide more evidence supporting that TPs could be used as a valuable candidate for preventing F-induced gastrointestinal injury in F-rich areas.

Author Contributions: C.X. designed and supervised the experiment, analyzed the data, and wrote the manuscript; S.N. conducted the experiments and analyzed the data; W.T. supervised the experiment and edited the manuscript. All authors have read and agreed to the published version of the manuscript.

Funding: This work was supported by a grant from the Tianjin Agriculture and Rural Committee (2024ZYCX012).

Institutional Review Board Statement: Not applicable.

Informed Consent Statement: Not applicable.

Data Availability Statement: The data presented in this study are available upon request from the corresponding author.

Acknowledgments: We thank Xin Wu for the valuable scientific suggestions and discussion.

Conflicts of Interest: The authors declare that they have no known competing financial interests or personal relationships that could have appeared to influence the work reported in this paper.

References

1. Ferreira, M.K.M.; Aragão, W.A.B.; Bittencourt, L.O.; Puty, B.; Dionizio, A.; de Souza, M.P.C.; Buzalaf, M.A.R.; de Oliveira, E.H.; Crespo-Lopez, M.E.; Lima, R.R. Fluoride exposure during pregnancy and lactation triggers oxidative stress and molecular changes in hippocampus of offspring rats. *Ecotoxicol. Environ. Saf.* **2021**, *208*, 111437. [CrossRef]
2. Srivastava, S.; Flora, S.J.S. Fluoride in Drinking Water and Skeletal Fluorosis: A Review of the Global Impact. *Curr. Environ. Health Rep.* **2020**, *7*, 140–146. [CrossRef]
3. Das, S.S.; Maiti, R.; Ghosh, D. Fluoride-induced immunotoxicity in adult male albino rat: A correlative approach to oxidative stress. *J. Immunotoxicol.* **2006**, *3*, 49–55. [CrossRef]
4. Reddy, Y.P.; Tiwari, S.K.; Shaik, A.P.; Alsaeed, A.; Sultana, A.; Reddy, P.K. Effect of sodium fluoride on neuroimmunological parameters, oxidative stress and antioxidative defenses. *Toxicol. Mech. Methods* **2014**, *24*, 31–36. [CrossRef]
5. Goschorska, M.; Gutowska, I.; Baranowska-Bosiacka, I.; Piotrowska, K.; Metryka, E.; Safranow, K.; Chlubek, D. Influence of Acetylcholinesterase Inhibitors Used in Alzheimer's Disease Treatment on the Activity of Antioxidant Enzymes and the Concentration of Glutathione in THP-1 Macrophages Under Fluoride-Induced Oxidative Stress. *Int. J. Environ. Res. Public Health* **2019**, *16*, 10. [CrossRef]
6. Wang, J.J.; Wei, Z.-K.; Han, Z.; Liu, Z.-Y.; Zhang, Y.; Zhu, X.-Y.; Li, X.-W.; Wang, K.; Yang, Z.-T. Sodium fluoride exposure triggered the formation of neutrophil extracellular traps. *Environ. Pollut.* **2020**, *257*, 113583. [CrossRef]
7. Song, C.; Zhao, J.; Fu, B.; Li, D.; Mao, T.; Peng, W.; Wu, H.; Zhang, Y. Melatonin-mediated upregulation of Sirt3 attenuates sodium fluoride-induced hepatotoxicity by activating the MT1-PI3K/AKT-PGC-1alpha signaling pathway. *Free Radic. Biol. Med.* **2017**, *112*, 616–630. [CrossRef]
8. Li, H.; Hao, Z.; Wang, L.; Yang, J.; Zhao, Y.; Cheng, X.; Yuan, H.; Wang, J. Dietary Calcium Alleviates Fluorine-Induced Liver Injury in Rats by Mitochondrial Apoptosis Pathway. *Biol. Trace Elem. Res.* **2022**, *200*, 271–280. [CrossRef]
9. Li, M.; Wang, J.; Wu, P.; Manthari, R.K.; Zhao, Y.; Li, W.; Wang, J. Self-recovery study of the adverse effects of fluoride on small intestine: Involvement of pyroptosis induced inflammation. *Sci. Total Environ.* **2020**, *742*, 140533. [CrossRef]
10. Dai, Y.-H.; Wei, J.-R.; Chen, X.-Q. Interactions between tea polyphenols and nutrients in food. *Compr. Rev. Food Sci. Food Saf.* **2023**, *18* (Suppl. S1), 21. [CrossRef]
11. Prasanth, M.I.; Sivamaruthi, B.S.; Chaiyasut, C.; Tencomnao, T. A Review of the Role of Green Tea (*Camellia sinensis*) in Antiphotoaging, Stress Resistance, Neuroprotection, and Autophagy. *Nutrients* **2019**, *11*, 474. [CrossRef]
12. Khan, N.; Mukhtar, H. Tea Polyphenols in Promotion of Human Health. *Nutrients* **2019**, *11*, 39. [CrossRef]
13. Qi, G.; Mi, Y.; Wang, Y.; Li, R.; Huang, S.; Li, X.; Liu, X. Neuroprotective action of tea polyphenols on oxidative stress-induced apoptosis through the activation of the TrkB/CREB/BDNF pathway and Keap1/Nrf2 signaling pathway in SH-SY5Y cells and mice brain. *Food Funct.* **2017**, *8*, 4421–4432. [CrossRef]
14. Simos, Y.V.; Verginadis, I.I.; Toliopoulos, I.K.; Velalopoulou, A.P.; Karagounis, I.V.; Karkabounas, S.C.; Evangelou, A.M. Effects of catechin and epicatechin on superoxide dismutase and glutathione peroxidase activity, in vivo. *Redox Rep.* **2012**, *17*, 181–186. [CrossRef]
15. Winiarska-Mieczan, A. Protective effect of tea against lead and cadmium-induced oxidative stress-a review. *Biometals* **2018**, *31*, 909–926. [CrossRef]
16. Garcia-Rodriguez, M.D.C.; Hernández-Cortés, L.M.; Mendoza-Núñez, V.M.; Arenas-Huertero, F. Effects of green tea polyphenols against metal-induced genotoxic damage: Underlying mechanistic pathways. *J. Toxicol. Environ. Health B Crit. Rev.* **2023**, *26*, 371–386. [CrossRef]

17. Londero, A.S.; Arana, M.R.; Perdomo, V.G.; Tocchetti, G.N.; Zecchinati, F.; Ghanem, C.I.; Ruiz, M.L.; Rigalli, J.P.; Mottino, A.D.; García, F.; et al. Intestinal multidrug resistance-associated protein 2 is down-regulated in fructose-fed rats. *J. Nutr. Biochem.* **2017**, *40*, 178–186. [CrossRef]
18. Diaz de Barboza, G.; Guizzardi, S.; Moine, L.; Tolosa de Talamoni, N. Oxidative stress, antioxidants and intestinal calcium absorption. *World J. Gastroenterol.* **2017**, *23*, 2841–2853. [CrossRef]
19. Hermes, R.G.; Manzanilla, E.G.; Martín-Orúe, S.M.; Pérez, J.F.; Klasing, K.C. Influence of dietary ingredients on in vitro inflammatory response of intestinal porcine epithelial cells challenged by an enterotoxigenic Escherichia coli (K88). *Comp. Immunol. Microbiol. Infect. Dis.* **2011**, *34*, 479–488. [CrossRef]
20. Kang, R.; Li, R.; Dai, P.; Li, Z.; Li, Y.; Li, C. Deoxynivalenol induced apoptosis and inflammation of IPEC-J2 cells by promoting ROS production. *Environ. Pollut.* **2019**, *251*, 689–698. [CrossRef]
21. Tveden-Nyborg, P.; Bergmann, T.K.; Jessen, N.; Simonsen, U.; Lykkesfeldt, J. BCPT 2023 policy for experimental and clinical studies. *Basic Clin. Pharmacol. Toxicol.* **2023**, *133*, 391–396. [CrossRef]
22. Yuksek, V.; Dede, S.; Çetin, S.; Usta, A.; Taşpınar, M. Vitamin D may assist the UPR against sodium fluoride-induced damage by reducing RIPK1, ATG5, BECN1, oxidative stress and increasing caspase-3 in the osteoblast MC3T3-E1 cell line. *J. Trace Elem. Med. Biol.* **2023**, *80*, 127293. [CrossRef]
23. Korkmaz, R.; Yuksek, V.; Dede, S. The Effects of Sodium Fluoride (NaF) Treatment on the PI3K/Akt Signal Pathway in NRK-52E Cells. *Biol. Trace Elem. Res.* **2022**, *200*, 3294–3302. [CrossRef]
24. Yang, F.; Oz, H.S.; Barve, S.; De Villiers, W.J.; McClain, C.J.; Varilek, G.W. The green tea polyphenol (-)-epigallocatechin-3-gallate blocks nuclear factor-kappa B activation by inhibiting I kappa B kinase activity in the intestinal epithelial cell line IEC-6. *Mol. Pharmacol.* **2001**, *60*, 528–533. [CrossRef]
25. Wang, F.; Li, Y.; Tang, D.; Yang, B.; Tian, T.; Tian, M.; Meng, N.; Xie, W.; Zhang, C.; He, Z.; et al. Exploration of the SIRT1-mediated BDNF-TrkB signaling pathway in the mechanism of brain damage and learning and memory effects of fluorosis. *Front. Public Health* **2023**, *11*, 1247294. [CrossRef]
26. Souza-Monteiro, D.; Ferreira, M.K.M.; Bittencourt, L.O.; Aragão, W.A.B.; Oliveira, I.G.d.; Maia, C.S.F.; Freire, M.A.M.; Zohoori, F.V.; Buzalaf, M.A.R.; Lima, R.R. Intrauterine and Postnatal Exposure to High Levels of Fluoride Is Associated with Motor Impairments, Oxidative Stress, and Morphological Damage in the Cerebellum of Offspring Rats. *Int. J. Mol. Sci.* **2022**, *23*, 8556. [CrossRef]
27. Fan, X.; Xiao, X.; Mao, X.; Chen, D.; Yu, B.; Wang, J.; Yan, H. Tea bioactive components prevent carcinogenesis via anti-pathogen, anti-inflammation, and cell survival pathways. *IUBMB Life* **2021**, *73*, 328–340. [CrossRef]
28. Liu, L.Q.; Nie, S.P.; Shen, M.Y.; Hu, J.L.; Yu, Q.; Gong, D.; Xie, M.Y. Tea Polysaccharides Inhibit Colitis-Associated Colorectal Cancer via Interleukin-6/STAT3 Pathway. *J. Agric. Food Chem.* **2018**, *66*, 4384–4393. [CrossRef]
29. Yang, J.; Chen, B.; Gu, Y. Pharmacological evaluation of tea polysaccharides with antioxidant activity in gastric cancer mice. *Carbohydr. Polym.* **2012**, *90*, 943–947. [CrossRef]
30. Almajano, M.P.; Vila, I.; Gines, S. Neuroprotective effects of white tea against oxidative stress-induced toxicity in striatal cells. *Neurotox. Res.* **2011**, *20*, 372–378. [CrossRef]
31. Braud, L.; Peyre, L.; De Sousa, G.; Armand, M.; Rahmani, R.; Maixent, J.-M. Effect of Brewing Duration on the Antioxidant and Hepatoprotective Abilities of Tea Phenolic and Alkaloid Compounds in a t-BHP Oxidative Stress-Induced Rat Hepatocyte Model. *Molecules* **2015**, *20*, 14985–15002. [CrossRef] [PubMed]
32. Ivanov, A.V.; Bartosch, B.; Isaguliants, M.G. Oxidative Stress in Infection and Consequent Disease. *Oxidative Med. Cell. Longev.* **2017**, *2017*, 3496043. [CrossRef] [PubMed]
33. Babu, S.; Manoharan, S.; Ottappilakkil, H.; Perumal, E. Role of oxidative stress-mediated cell death and signaling pathways in experimental fluorosis. *Chem. Biol. Interact.* **2022**, *365*, 110106. [CrossRef]
34. Bhattacharyya, A.; Chattopadhyay, R.; Mitra, S.; Crowe, S.E. Oxidative stress: An essential factor in the pathogenesis of gastrointestinal mucosal diseases. *Physiol. Rev.* **2014**, *94*, 329–354. [CrossRef]
35. Perez-Torres, I.; Castrejón-Téllez, V.; Soto, M.E.; Rubio-Ruiz, M.E.; Manzano-Pech, L.; Guarner-Lans, V. Oxidative Stress, Plant Natural Antioxidants, and Obesity. *Int. J. Mol. Sci.* **2021**, *22*, 1786. [CrossRef]
36. Jones, D.P. Radical-free biology of oxidative stress. *Am. J. Physiol. Cell Physiol.* **2008**, *295*, C849–C868. [CrossRef]
37. Redza-Dutordoir, M.; Averill-Bates, D.A. Activation of apoptosis signalling pathways by reactive oxygen species. *Biochim. Biophys. Acta* **2016**, *1863*, 2977–2992. [CrossRef]
38. Lash, L.H. Mitochondrial GSH transport and intestinal cell injury: A commentary on "Contribution of mitochondrial GSH transport to matrix GSH status and colonic epithelial cell apoptosis". *Free Radic. Biol. Med.* **2008**, *44*, 765–767. [CrossRef]
39. Rivoira, M.A.; Marchionatti, A.M.; Centeno, V.A.; de Barboza, G.E.D.; López, M.E.P.; de Talamoni, N.G.T. Sodium deoxycholate inhibits chick duodenal calcium absorption through oxidative stress and apoptosis. *Comp. Biochem. Physiol. A Mol. Integr. Physiol.* **2012**, *162*, 397–405. [CrossRef]

40. Li, L.; Xin, J.; Wang, H.; Wang, Y.; Peng, W.; Sun, N.; Huang, H.; Zhou, Y.; Liu, X.; Lin, Y.; et al. Fluoride disrupts intestinal epithelial tight junction integrity through intracellular calcium-mediated RhoA/ROCK signaling and myosin light chain kinase. *Ecotoxicol. Environ. Saf.* **2023**, *257*, 114940. [CrossRef]
41. Jin, Y.; Gao, X.Y.; Zhao, J.; Tian, W.S.; Zhang, Y.L.; Tian, E.J.; Zhou, B.H.; Wang, H.W. Estrogen deficiency aggravates fluoride-induced small intestinal mucosa damage and junctional complexes proteins expression disorder in rats. *Ecotoxicol. Environ. Saf.* **2022**, *246*, 114181. [CrossRef] [PubMed]
42. Chen, C.; Wang, H.; Hong, T.; Huang, X.; Xia, S.; Zhang, Y.; Chen, X.; Zhong, Y.; Nie, S. Effects of tea polysaccharides in combination with polyphenols on dextran sodium sulfate-induced colitis in mice. *Food Chem. X* **2022**, *13*, 100190. [CrossRef] [PubMed]

International Journal of
Molecular Sciences

MDPI

Article

Extraction and Identification of Polysaccharide from *Lentinus edodes* and Its Effect on Immunosuppression and Intestinal Barrier Injury Induced by Cyclophosphamide

Xiaodi Jin [1,†], Zhiyong Wu [1,2,†], Hao Chen [1], Weiqi Liu [1], Fuhua Gu [1] and Jichang Li [1,*]

1 College of Veterinary Medicine, Northeast Agricultural University, Harbin 150030, China; jxd19950925@163.com (X.J.); wuzhiyong@neau.edu.cn (Z.W.); chenhao980915@163.com (H.C.); 17621867263@163.com (W.L.); fhua666@126.com (F.G.)
2 Heilongjiang Academy of Chinese Medicine Sciences, Harbin 150030, China
* Correspondence: lijichang@neau.edu.cn
† These authors contributed equally to this work.

Abstract: *Lentinus edodes* serves as a significant source of both medicine and food, with its key component, lentinan (LNT), recognized as an effective immunomodulator. However, the mechanisms by which it regulates immune and intestinal functions under conditions of immunosuppression remain unclear. This study aims to investigate the components of lentinan and examine its potential effects on countering cyclophosphamide (CP)-induced immunosuppression, intestinal barrier damage, and dysregulation of gut microbiota. In this study, the effects of LNT were evaluated by serological indicators, histopathological changes in ileum, tight-junction-related protein expression, cytokine expression levels, and gut microbiota 16S rRNA gene sequencing. We found that LNT was effective in mitigating the abnormalities in body weight, immune organ index, and serum levels of IL-6, IL-2, IFN-γ, and IgG in mice induced by CP ($p < 0.05$). Furthermore, LNT demonstrated the ability to alleviate intestinal barrier damage induced by CP by increasing the mRNA levels of TNF-α, IL-1β, IFN-γ, Occludin, and ZO-1 ($p < 0.05$). Additionally, 16S rRNA gene sequencing revealed that LNT also normalized the disrupted abundance of *Firmicutes*, *Proteobacteria*, and *Bacteroidets* caused by CP. This restoration brought the gut microbiota back to normal levels and increased the abundance of certain tumor-inhibiting bacteria, such as *Alistipes*. Overall, lentinan demonstrated the ability to reverse the immunosuppressive effects induced by cyclophosphamide and modulate gut microbiota to restore a healthy microbial balance.

Keywords: lentinan; immunosuppression; intestinal barrier; gut microbiota

1. Introduction

In recent times, attention has increasingly turned to the advancement of active compounds within traditional Chinese medicine, especially concerning research on anti-inflammatory properties, antioxidant effects, and immune system functionalities. The immune system is essential for the body, helping to fend off pathogen attacks and maintain internal balance. Once the immune system is compromised, the body becomes more sensitive to pathogens [1]. Cyclophosphamide (CP), a chemotherapy agent, can influence apoptosis mediated by the Fas/Fas L death receptor signal transduction pathway [2] and can also affect cytokine secretion through the NF-κB pathway [3,4]. Furthermore, recent studies have demonstrated that CP can induce an imbalance in gut microbiota intestinal flora [5]. Therefore, it is often used to establish immunosuppressive models [6,7]. At present, the immunomodulators used in clinical practice have certain side effects, such as Levamisole [8]. Therefore, the development of safer immunomodulators is of great significance for the prevention and treatment of immunosuppressive diseases. Contemporary studies in medicine have established that traditional Chinese medicine can exert an influence on the immune system by affecting both innate and acquired immunity [9–11].

Over the past decades, immunostimulation has been one of the most exploited biological functions of polysaccharides [12]. *Lentinus edodes*, which originated in China, is the second-largest mushroom produced in the world and is also a precious medical mushroom with a long reputation in China. Lentinan (LNT), the principal active component extracted from *Lentinus edodes*, exists as sugar chains interconnected by glycosidic bonds. It is a high-molecular-weight polymer composed of various monosaccharides. Previous research has been demonstrated to have a strong effect on immunomodulatory [13–15], antitumor [16], and antiviral effects [17]. Research indicates that LNT may serve as a natural protective agent against damage in bovine mammary epithelial cells, which are mediated by the modulation of the Nrf2 pathway, owing to its anti-inflammation [18]. Moreover, LNT could improve antibiotic-induced gut microbiota imbalance in mice [19]. Gut microbiota has been extensively studied, as it not only reflects the normal physiological functions of the body but also influences overall health. An imbalance in gut microbiota can lead to chronic immune diseases. Research has demonstrated that such imbalances may also impact neurological conditions, including Alzheimer's disease, via the gut–brain axis [20]. Consequently, flora plays a crucial part in maintaining healthy immune regulation and intestinal barrier function [21]. Further investigation is necessary to determine if lentinan can reduce the side effects associated with cyclophosphamide, enhance immune function, and modulate gut microbiota.

Therefore, CP was utilized in this study to create an immunosuppressive mouse model for examining the modulation of immune function and the therapeutic impact of LNT on intestinal function and gut microbiota in mice. The aim of this study is to extract and identify the key components of lentinan, investigate its regulatory effects on immunosuppression induced by CP, analyze its impact on intestinal flora, and explore the drug mechanism of lentinan. The findings aim to offer insights into the potential combined use of immunomodulators and microbial agents in future applications.

2. Results

2.1. Characterization of LNT

LNT appeared as a hygroscopic powder with a brownish-yellow hue, with a total carbohydrate content of 73.43% (*w*/*w*) determined using the phenol–sulfuric acid method. Additionally, the composition included 17.17% protein content (*w*/*w*) as measured by a protein assay kit and 3.03% galacturonic acid content. The result of monosaccharide components in Table 1 shows that it was mainly composed of glucose (76.10%), galactose (11.67%), and D-mannose (4.73%). It also contained small amounts of fucose (2.96%), ribose (1.82%), and D-glucuronic acid (6.23%) (Figure 1C). The following supporting information can be found in Supplementary Materials File S1.

Table 1. Identification of monosaccharides by HPLC.

Number	Retention Time	Compound	Area	Concentration μg/mL	Percentage %
1	15.768	D-Mannose	326,336	11.219	4.728
2	20.149	Ribose	125,820	3.898	1.823
3	20.918	L-Rhamnose	4197	0.199	0.061
4	25.249	D-Glucuronic acid	122,023	4.705	1.768
5	33.552	Glucose	5,252,654	202.539	76.104
6	38.492	Galactose	805,757	24.538	11.674
7	41.975	L-Arabinose	60,595	1.539	0.878
8	47.026	Fucose	204,572	6.873	2.964

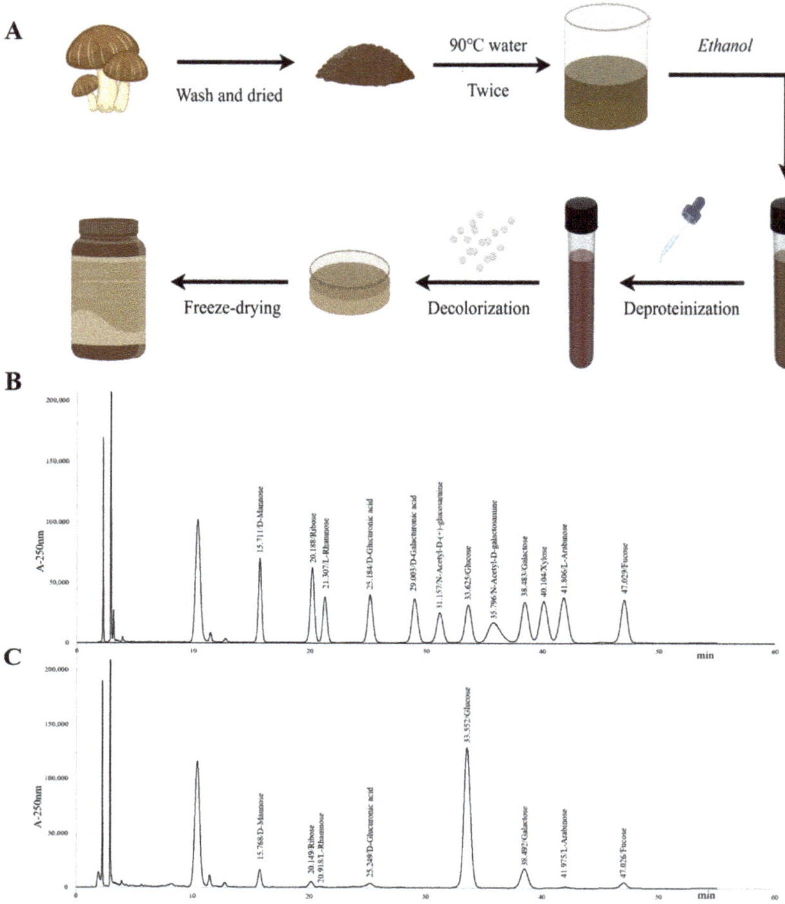

Figure 1. Extraction and HPLC identification of lentinan: (**A**) Extraction process of lentinan. The figure was designed by Figdraw 2.0. (**B**) A chromatogram of standard mixed solution. (**C**) A chromatogram of sample solution.

The results from SEM illustrate the surface morphology of LNT at four distinct magnifications: 500 μm, 100 μm, 50.0 μm, and 20.0 μm (Figure 2B). SEM analysis indicated that LNT exhibited an irregular large layer structure, characterized by a minimal presence of fibrotic structures interwoven with a lamellar arrangement. Atomic force microscopy (AFM) is an effective method for observing the surface morphology of polysaccharides [22]. Atomic force microscopy results in Figure 2C revealed that, in the 2D image, LNT displayed a sheet-like and fibrous structure, with heights ranging from −1.1 nm to 2.0 nm, and the value of Image Ra was 0.273 nm. The 3D image shown in Figure 2D reproduces the same structure as that in SEM.

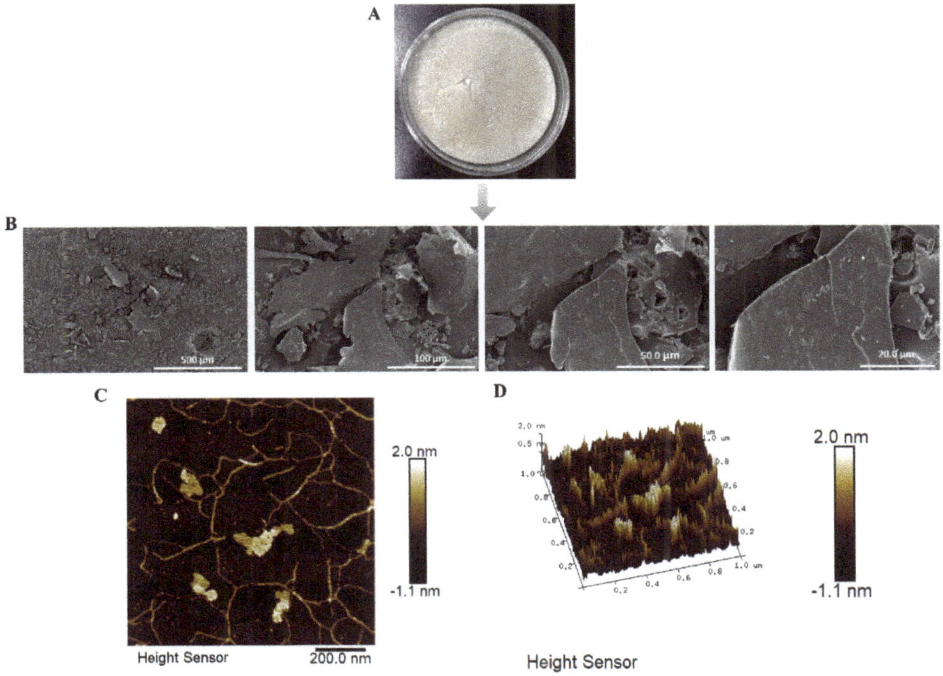

Figure 2. Identification of surface characteristics of lentinan: (**A**) Freeze-dried lentinan. (**B**) The SEM images of LNT were obtained in the order of 500–20 μm. (**C**,**D**) Atomic force microscopy (AFM) 2D, phase and 3D images, respectively.

2.2. LNT Regulates Body Weight and Index of Immune Organ

To assess the impact of LNT on the immunosuppressive effects caused by CP, an experiment utilizing mice was conducted, illustrated in Figure 3. CP has been shown to have an effect on body weight in mice after intraperitoneal injection [23,24]. Figure 3A, following five consecutive days of CP administration, there was a notable decline in the body weight of the mice subjected to CP. By the fifth day, the body weight in the CP group was significantly lower than that in the C group ($p < 0.05$). After 9 days, an observed gradual increase in body weight was noted in the CL group. On day 17, the body weight of C-group mice was significantly higher than that of the CP group ($p < 0.05$). No notable difference in body weight was found between the mice treated with LNT and those in the control group.

Immune organs can reflect the immune function of the body [25]. After the administration of CP, both the thymus and spleen displayed significant changes. In Figure 3B, compared with the C group, the thymus index was notably decreased, while the spleen index was increased ($p < 0.05$). Conversely, the thymus index in the CL group showed a significant increase, and the spleen index exhibited a significant decrease compared with the CP group after treatment with LNT (Figure 3B) ($p < 0.05$). Furthermore, no significant difference was observed between the C and L groups ($p > 0.05$). These findings suggest that LNT prevented weight loss and restored injury to the thymus and spleen in mice induced by CP.

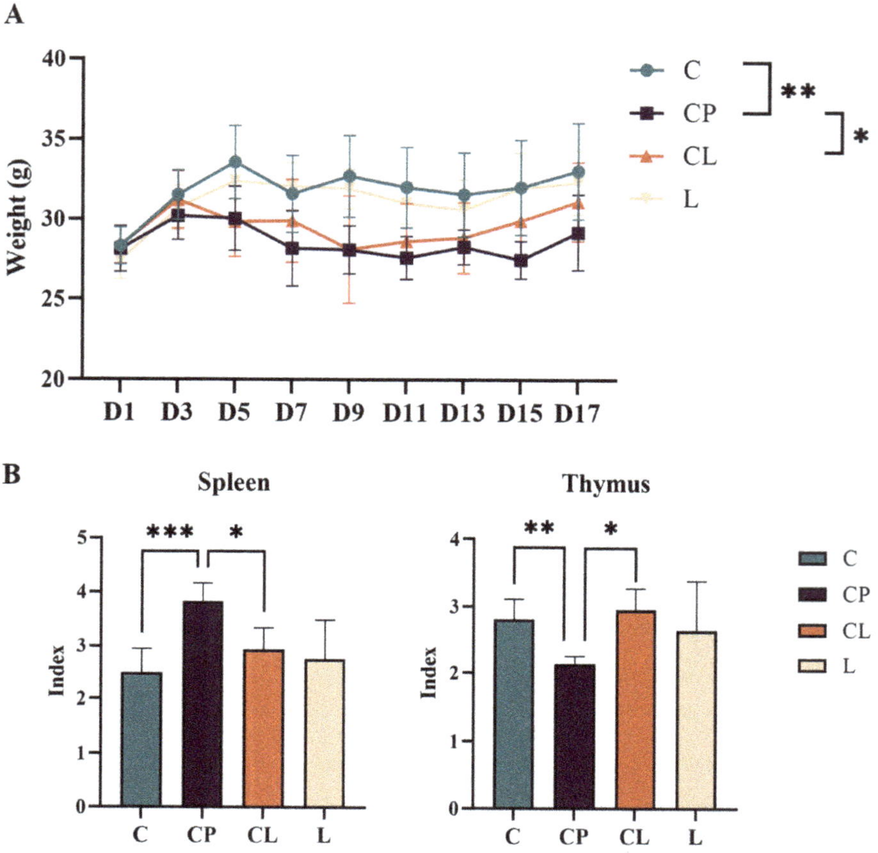

Figure 3. Physical indexes of experimental animals: Effects of CP and LNT on body weight and organ index after administration. (**A**) Body weight curves of four groups. (**B**) The organ indices of mice. Data are presented as the mean ± SD and analyzed by one-way ANOVA with Tukey's multiple comparisons. * $p < 0.05$, ** $p < 0.01$, *** $p < 0.001$.

2.3. Effects of LNT on Serum Cytokine Contents in LNT-Treated Mice

The levels of IL-2, IL-6, TNF-α, IFN-γ, and IgG in serum were assessed using ELISA. Additionally, sIgA levels in the ileum were also measured using the same method (Figure 4). However, in the CP group, a contrasting effect was observed: serum levels of IL-2, IL-6, and IFN-γ were found to be reduced compared with those in the C group. ($p < 0.05$). Besides, the level of IgG in serum and sIgA in the ileum had a downward trend, but the difference was not significant. The CL group significantly increased serum IL-6, IL-2, and IFN-γ levels as compared with the CP group ($p < 0.05$). The level of IgG in serum and sIgA in the ileum had an upward trend, though not statistically significant. The level of TNF-α in serum was increased remarkably in the CP group compared with the C group, and the CL group could alleviate the phenomenon could alleviate the above-mentioned phenomenon ($p > 0.05$).

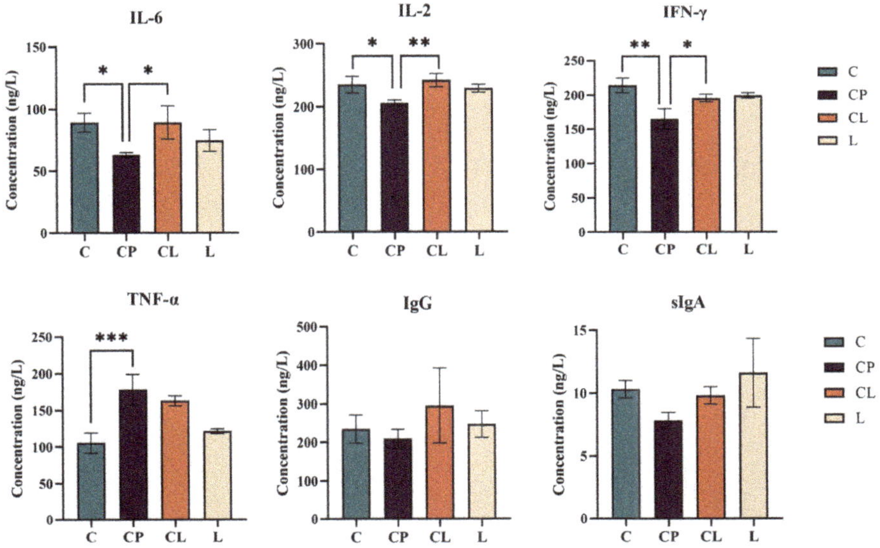

Figure 4. Cytokines in serum and *sIgA* in the ileum: The concentration of IL-6, IL-2, IFN-γ, TNF-α, and IgG in serum and sIgA in intestinal tract. Data are presented as the mean ± SD and analyzed by one-way ANOVA with Tukey's multiple comparisons. * $p < 0.05$, ** $p < 0.01$, *** $p < 0.001$.

2.4. The Results of Histopathology

The intestinal mucosa was examined using a light microscope on the tissue of the terminal ileum (Figure 5A). As is shown in Figure 5A, the intestinal villi in the C group appeared normal and were well organized, with the structure of the intestinal mucosa and epithelial cells intact in the small intestine. In comparison with the C group, after the administration of CP, the CP group exhibited significant damage to the intestinal wall, destruction of the structure of intestinal villi, and a looser arrangement. Following the intervention of LNT, the morphology of intestinal villi in the CL group gradually reverted to a palisade-like structure. The clarity and structure of the intestinal villi improved, with a closer arrangement of the villi. There was no significant difference in ileum morphology between the L group and the C group. Our results suggest that LNT could alleviate the intestinal damage that was caused by CP and protect the integrity of the intestinal structure.

As is shown in Figure 5B, the ileum was further assessed using a scanning electron microscope (SEM). At a scale of 300 μm, the integrity of intestinal villi in the CP group was compromised, exhibiting signs of damage to their surfaces. Compared with the CP group, the CL group has better cell completeness and narrower intercellular spaces. The intestinal microvilli were sparser and smoother in the CL group than in the CP group, meaning that treatment with LNT reduced this disruption. In comparison with the C group, the intestinal villi surface structure of the L group exhibited no significant changes, and the integrity of the intestinal villi remained intact. This demonstrates that LNT alone did not induce mechanical damage to the intestinal tract of normal mice. At a scale of 20.0 μm, it was observed that the CP group had reduced colonization of *Segmented Filamentous Bacteria* (SFB) due to immunosuppression, leaving more pores on the intestinal mucosal surface. The CL group significantly restored the colonization of this bacterium (Figure 5B).

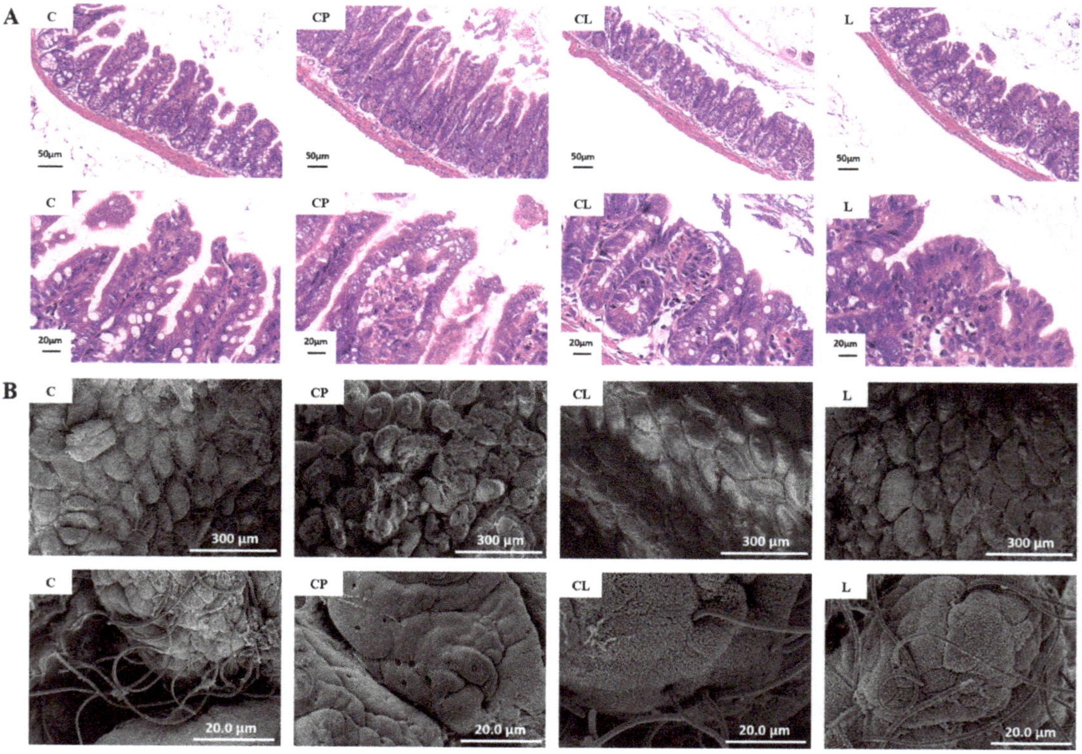

Figure 5. Pathological examination of ileum: (**A**) Representative histological images of ileum tissues by H&E staining. Scale bar: 50 μm/20 μm. The intestinal villi of the C group were normal and neatly arranged. The CP group exhibited significant damage to the intestinal wall, destruction of the structure of intestinal villi, and a looser arrangement. The clarity and structure of the intestinal villi in the CL group improved, with a closer arrangement of the villi. (**B**) SEM images of ileum tissues. Scale bar: 300 μm/20 μm.

2.5. Effect of LNT on Gene Expression in CP-Treated Mice

In Figure 6, RT-qPCR analysis displayed that CP-induced mice had a lower expression of related genes, such as TNF-α, IL-1β, IFN-γ, and Claudin 1, than that in the normal group ($p < 0.05$). The CL group increased the related gene expression in comparison with the CP group ($p < 0.05$). Besides, the expression of GPR41, Occludin, and ZO-1 increased in the CP group and reversed in the CL group compared with the CP group ($p > 0.05$). Surprisingly, no matter which group, they changed the expression of GPR43 and MUC2 to a higher level. In the L group, intestinal mRNA expression levels of GPR41, GPR43, and MUC2 were also increased after administration of LNT alone. In general, LNT could promote the related genes of cytokines (TNF-α, IL-1β, and IFN-γ) to change intestinal function. Meanwhile, both CP and LNT could change the expression of GPR43 and MUC2.

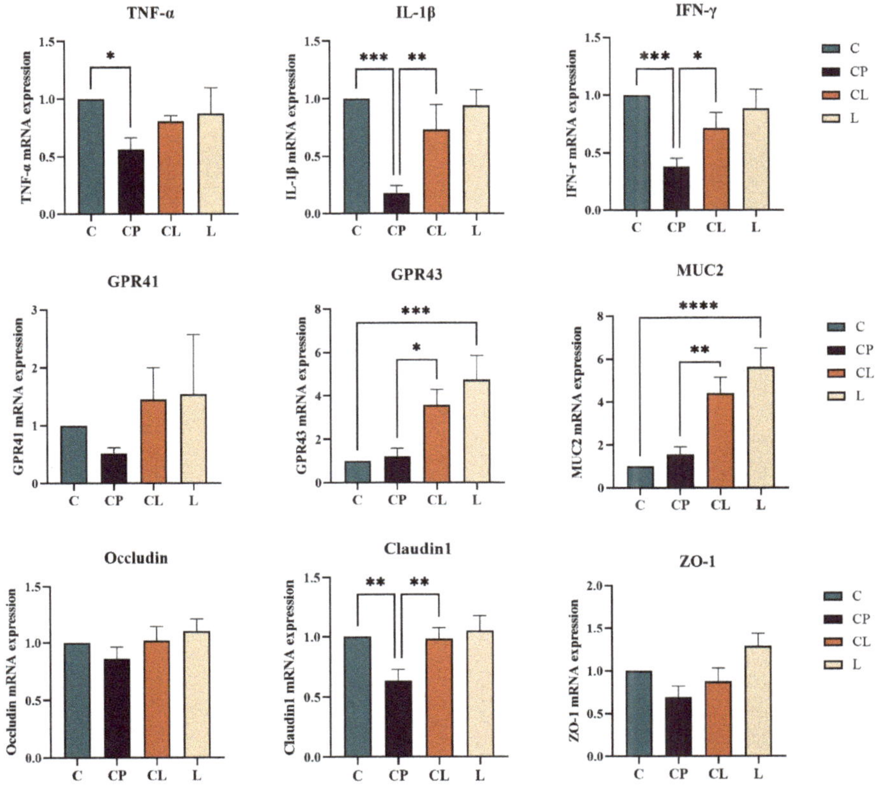

Figure 6. Rt-qPCR results of ileum: The mRNA expression of inflammatory cytokines in ileum, including TNF-α, IL-1β, and IFN-γ. The mRNA expression of GPR43, GPR41, and MUC2. The mRNA expression of intestinal tight junction proteins Occludin, Claudin1, and ZO-1. Data are presented as the mean ± SD and analyzed by one-way ANOVA with Tukey's multiple comparisons. * $p < 0.05$, ** $p < 0.01$, *** $p < 0.001$, **** $p < 0.0001$.

2.6. LNT Modulates the Structure of the Gut Microbiota in CP-Treated Mice

Bioinformatics analyses of the microbiome were conducted using QIIME2 version 2019.4 [26], with minor adjustments made in accordance with the official tutorials (https://docs.qiime2.org/2019.4/tutorials/, accessed on 15 August 2022). In this experiment, CP treatment significantly changed the gut microbiota of mice. As shown in Figure 7A, we enumerate the changes in species composition at the top 10 phylum levels. The results indicate that the abundance of *Bacteroidetes* has decreased, whereas the abundances of *Firmicutes* and *Proteobacteria* have increased following treatment with CP. Conversely, under LNT treatment, *Bacteroidetes* increased, while both *Firmicutes* and *Proteobacteria* decreased compared with the CP group (Figure 7A). In Figure 7B, we present the species composition of flora at the top 20 genus levels. Following CP treatment, *Lactobacillus*, *Bifidobacterium*, and *Roseburia* exhibited a downward trend, which was restored by LNT treatment. Other bacterial groups demonstrated an upward trend after CP administration, while the composition of intestinal flora was partially restored following LNT treatment (Figure 7B). In comparison with group C, the relative abundance of both phylum and genus in group LNT exhibited an upward trend, indicating that LNT may enhance the abundance of gut microbiota.

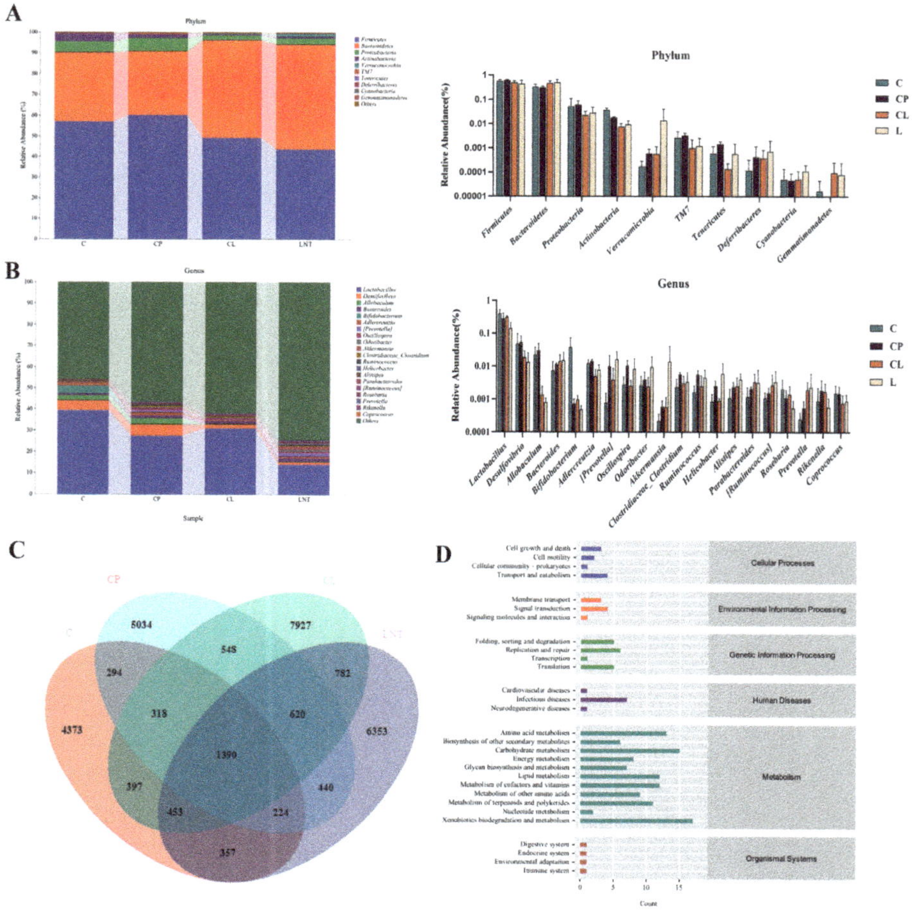

Figure 7. Gut microbiota structure changes based on 16S rRNA gene sequencing: (**A**,**B**) Species diversity composition of intestinal flora at phylum and genus levels. (**C**) Venn diagram of species differences between the four groups. (**D**) Enrichment analysis of KEGG metabolic pathways related to changes in gut microbiota.

In the CP group, the Chao1, Shannon, and Simpson indices were lower than those in the C group. The CL group adjusted the microflora structure, restoring the typical α-diversity indices to levels comparable to those of the C group (Figure 8A). The alterations in gut microbiota structure were further validated by the distinct clustering of the four groups observed in the principal coordinates analysis (PCoA) scatter diagram and the nonmetric multidimensional scaling (NMDS) analysis (Figure 8B,C). Additionally, orthogonal partial least squares discrimination analysis (OPLS-DA) at both the phylum and genus levels revealed distinct clusters within the gut microbiota (Figure 8D). Following the administration of CP, a shift in the microbial community structure was observed, which was partially reversed by LNT treatment. Notably, the LNT group also induced changes in the colony structure of the C group.

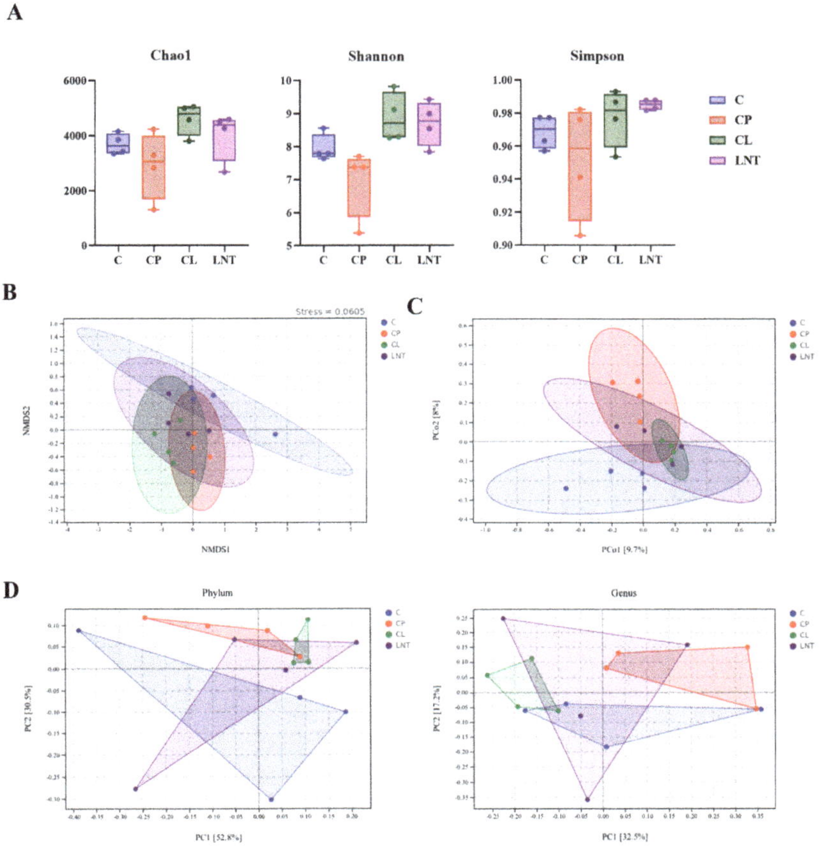

Figure 8. Diversity index results based on 16S rRNA gene sequencing: (**A**) The Alpha diversity index analysis of gut microbiota included Chao1, Shannon, and Simpson. (**B**) Nonmetric multidimensional scaling (NMDS) analysis. (**C**) Principal coordinate analysis (PCoA) of the microbiota based on Bray–Curtis distance. (**D**) Orthogonal partial least squares discrimination analysis (OPLS-DA) at both the phylum and genus levels.

3. Discussions

Cyclophosphamide, a frequently used chemotherapeutic agent in clinical settings, is commonly employed in the treatment of tumors and autoimmune conditions. Despite its therapeutic benefits, the drug is associated with notable side effects, most notably immunosuppression and intestinal dysfunction. Previous research has demonstrated that lentinan polysaccharide possesses antioxidant, anti-radiation, and immune-enhancing properties [18,27]. This study investigated the immunomodulatory effects of LNT on CP-induced immunosuppression, with a particular focus on the impact of LNT on CP-induced changes in intestinal flora and function.

LNT was obtained from *Lentinus edodes* by water extraction and alcohol precipitation. The carbohydrate content of LNT was measured at 73.43%, with glucose, galactose, and D-mannose identified as the top three monosaccharides in descending order. Additionally, the protein content was found to be 17.17%, and the galacturonic acid content was recorded at 3.03%. Consequently, the next step is to investigate the bioactive role of carbohydrates as the primary components of LNT. The spleen and thymus play crucial roles as immune organs, and the indices of these immune organs reflect the functioning of innate immunity to some degree [28]. In our study, we found that CP can reduce the thymus index in mice, which is consistent with previous studies [29]. Furthermore, the administration of cyclophosphamide results in a higher spleen index, while the LNT treatment reversed the indices induced by CP. Our research results are consistent with those in previous research reports [30].

Cytokines are crucial in both mediating and regulating the immune response. IL-6, a multifunctional cytokine, significantly influences the regulation of immune activity [31]. IFN-γ stimulates the production of T cells, activates macrophages, and regulates the balance between Th1 and Th2 cells [32]. On the other hand, IL-2 is a multifunctional cytokine that supports T cell proliferation and triggers the development of Treg cells [33]. TNF-α is a potent pro-inflammatory cytokine that plays a crucial role in the immune response to inflammation. Our study revealed that treatment with LNT resulted in increased levels of IL-6, IFN-γ, and IL-2 in the serum supernatants of mice when compared with CP. Additionally, there was a noticeable upward trend in serum IgG levels following LNT treatment. Secretory IgA (sIgA) is produced by intestinal plasma cells and plays a crucial role in maintaining the integrity of the intestinal barrier [34]. It serves as an indicator of the functionality of the intestinal immune barrier to a certain extent [35]. Selective colonization of E. coli in the gut of germ-free animals has been shown to produce sIgA [36]. In instances of immunosuppression in our research, both the composition and the quantity of gut microbiota are altered, leading to a reduction in sIgA levels. Following treatment with LNT, these alterations were reversed, and similar functionality was observed in the unmolded group. These findings suggested that LNT has the potential to effectively alleviate the immunosuppression caused by CP.

The integrity of the intestinal mucosa plays an important role in the maintenance of intestinal mucosal barrier function, which prevents gut microbes and their toxins from migrating to extra-intestinal tissues and organs [37]. The analysis of our HE results revealed that mechanical injury occurred in the intestine following CP modeling, with LNT demonstrating a significant improvement in reducing this injury. Additionally, SEM results indicated a notable decrease in *SFB* colonization in the ileum of mice, accompanied by the presence of significant cavities on the surface of the villi, leading to mechanical damage in the ileum. However, administration of LNT successfully restored *SFB* colonization and mitigated the mechanical damage. Notably, SFB has been identified as a potent inducer of T helper 17 (Th17) cells, which play a crucial role in host defense against microbial pathogens [38].

Rt-qPCR results revealed that CP significantly decreased the expression levels of TNF-α, IL-1β, IFN-γ, and Claudin1 in the ileum when compared with the C group. Following LNT intervention, there was a notable increase in the expression of TNF-a, IL-1β, IFN-r, and Claudin1. In comparison with group C, CP was found to lower the expression of GPR41, Occludin, and ZO-1 in the ileum. LNT could restore these abnormal indexes, though the difference was not statistically significant. TNF-α, IL-1β, and IFN-γ are critical cytokines that play significant roles in the immune response. During pathogen invasion, TNF-α and IL-1β facilitate the recruitment and activation of inflammatory cells, while IFN-γ enhances cellular immune responses by promoting the activation of macrophages and their antigen presentation capabilities. In the context of CP-induced immunosuppression, the cytotoxic effects of CP lead to a reduction in the secretion of these cytokines, which significantly increases the body's susceptibility to pathogens. The application of LNT makes up for the above cytokine defects in the application of CP. Occludin and Claudin-1, as tight junction proteins on the intestinal surface, directly affect the permeability of the intestinal tract and play an important role in maintaining the normal barrier function of the intestinal tract [39]. We propose that Occludin, Claudin-1, and sIgA play significant roles in enhancing the recognition of and immunity to pathogen invasion, with Occludin and Claudin-1 contributing to the mechanical barrier, while sIgA is associated with the mucosal immune barrier. In our study, MUC2 always showed an upward trend. MUC2, a significant gel-forming mucin, is secreted by intestinal epithelial goblet cells and plays a crucial role in transmitting immunomodulatory signals to reduce the immunogenicity of intestinal microorganisms. Interestingly, certain strains, like *Shigella* and *Sobacterium nucleatum*, have been found to increase the expression of MUC2 [40–42]. This means there is a two-way regulation between the MUC2 and the gut bacteria. In our 16S sequence detection results, the abundance of *Shigella* bacteria in the CP, CL, and LNT groups was significantly higher than that in the C group, resulting in a notable increase in MUC2 mRNA expression. Changes in the quantity and proportion of flora stimulated alterations in the short-chain fatty acid (SCFAs) content in the intestine, leading to increased mRNA expressions of the SCFAs receptors GPR41 and GPR43. Simultaneously, the small intestine's epithelial cells respond to short-chain fatty acids (SCFAs) through the activation of the extracellular signal-regulated kinase 1/2 and p38 mitogen-activated protein kinase signaling pathways by GPR41 and GPR43. This activation subsequently leads to the production of chemokines and cytokines, which play a crucial role in immune responses [43].

This research employed 16S rRNA gene sequencing to assess the diversity and composition of microbiota. To evaluate the complexity of species diversity in gut microbiota, α-diversity was measured using various indices, such as Chao1 (https://www.genescloud.cn/home, accessed on 20 October 2022), Shannon (https://www.genescloud.cn/home, accessed on 20 October 2022), and Simpson (https://www.genescloud.cn/home, accessed on 20 October 2022). The findings indicated that CP led to a reduction in α-diversity, implying a decline in both the abundance and diversity of the gut microbial community, aligning with the observations made by Chen et al. [44]. KEGG level 2 analysis of the changes in the gut microbiota showed that the alterations were associated with various anabolic pathways.

As is widely recognized, numerous microbial communities within the intestines function as a biological barrier for the gut [45]. Intestinal $CD4^+$ T lymphocytes can be differentiated into functional subtypes with regulatory or effector functions through the *Bacteroidetes*, thus playing a role in protecting the gut [46]. The *Proteobacteria* phylum contains various pathogenic microbes like *Helicobacter* and *Salmonella*. An increase in *Proteobacteria* levels can indicate a disruption in the gut microbiota and serve as a potential diagnostic factor for diseases [47]. We found that, at the phylum level, following LNT treatment, there was a decrease in the abundance of *Firmicutes* and *Proteobacteria*, while *Bacteroidetes* increased. This result indicates that LNT may enhance the dysbiosis of microbiota in mice with suppressed immunity and elevate the concentration of probiotics. At the genus level, the composition of *Lactobacillus*, *Bifidobacterium*, and *Roseburia* in the intestinal tract of the CP group decreased following the administration of CP. In contrast, the bacterial composition in the CL group showed recovery after the administration of LNT. It is well established that *Lactobacillus* and *Bifidobacterium*, which are among the most commonly utilized probiotic agents [48], play significant roles in mitigating intestinal inflammation and enhancing intestinal immunity and barrier function [49]. Recent studies have indicated that *Roseburia* may protect against colorectal tumorigenesis through the production of butyrate [50]. Additionally, the treatment of LNT further enhances the abundance of bacteria associated with cancer treatment; for instance, the increased presence of *Alistipes* is likely to have a beneficial effect on cancer therapy [51].

In conjunction with these findings, we demonstrate that LNT can not only mitigate CP-induced immunosuppression and intestinal dysfunction but also regulate the imbalance of intestinal microbiota induced by CP while enhancing the abundance of bacteria associated with adjuvant cancer therapy and treatment.

4. Materials and Methods

4.1. Materials and Reagents

The fresh *Lentinus edodes* were collected from Harbin (Harbin, Heilongjiang, China). CP was purchased from Suri Pharmaceutical Chemical Co., LTD (Suzhou, Jiangsu, China). ELISA kits of interferon-γ (*IFN-γ*), interleukin-2 (*IL-2*), interleukin-6 (*IL-6*), tumor necrosis factor-α (*TNF-α*), immunoglobulin G and sIgA (*IgG* and *sIgA*) were purchased from Andy Huatai Technology Co., Ltd. (Beijing, China). TRIzol® Reagent (9108Q), PrimeScript RT reagent Kit (RR047A), and TB Green® Premix Ex Taq™ (RR820A) were purchased from Takara Baori Doctor Material Technology Co., Ltd. (Beijing, China).

4.2. Polysaccharide Preparation and Characterization

4.2.1. Extraction of LNT

Based on the previous literature on the extraction methods of fungal plants [52], we adjusted the partial extraction methods. The specific methods are described below. The extraction process of lentinan is shown in Figure 1A. The *Lentinus edodes* were crushed into powder after being dried. The residue was extracted twice with 90 °C water for 2 h each time. After filtration and concentration by rotary evaporator, the filtrate was precipitated with 75% ethanol (final concentration) overnight. The precipitate was collected after centrifugation at 4000× *g* for 15 min and redissolved using sterilized distilled water. Deproteinization was performed with Sevage reagent and decolorized with macroporous resin. The crude product was obtained after vacuum freeze-drying, named LNT.

4.2.2. Determination of Total Carbohydrate, Protein, and Galacturonic Acid Contents

The total carbohydrate contents were analyzed by the Phenolsulfuric acid method using dextrose as a standard [53]. Specific methods are as follows. Accurately weigh the dried D-glucose in a test tube and dilute it with deionized water to achieve concentrations of 0.01, 0.02, 0.04, 0.06, and 0.08 mg/mL. For each test tube solution, take 1 mL and add 1.0 mL of 5% phenol solution and 5.0 mL of concentrated sulfuric acid. After vortexing, allow the reaction to proceed at room temperature for half an hour. The absorbance at 490 nm was measured using the solutions without D-glucose as the control. Each sample is repeated three times.

M-hydroxybiphenyl colorimetric method was utilized to quantify the Galacturonic acid (GalA) content in LNT [54]. M-hydroxybiphenyl colorimetric method was utilized to quantify GalA in LNT. Weigh 0.15 g m-hydroxybiphenyl, dissolve it in 5 mg/mL sodium hydroxide solution, and adjust the concentration to 1.5 mg/mL. Then, add 20 μL of LNT solutions into a 1 mL centrifuge tube, add 60 μL of sodium tetraborate/sulfuric acid solution, mix it with a vortex mixer, heat it in a boiling water bath for 5 min, and put it in an ice water bath. Then, 100 μL of the 1.5 mg/mL m-hydroxybiphenyl solution was added to the mixture and shaken for an additional 5 min. A total of 200 μL of the mixture was added to a 96-well plate, and the absorbance was measured at 520 nm. Prepare a standard GalA solution of 0–60 μmol/mL in the same way and measure the absorbance at 520 nm according to the above method. Each sample is repeated three times.

Protein content was detected using the BCA protein assay kit (BL521A, Biosharp) according to the manufacturer's instructions. Briefly, liquid A and liquid B in the reagent are mixed at a ratio of 50:1. The standard solution is then diluted to a concentration ranging from 0.025 mg/mL to 2 mg/mL. In a 96-well plate, 20 μL of each of the standard diluents and sample solutions was added to each well, followed by the addition of 200 μL of the working solution. After incubating at 37 °C for 30 min, the absorbance was measured at 562 nm. Each sample is repeated three times.

4.2.3. Determination of Monosaccharides Composition

LNT (5 mg) was hydrolyzed with 2 M trifluoroacetic acid (TFA, 3 mL) at 120 °C for 4 h. Next, the hydrolysate was derivatized with 3-methyl-1phenyl-2-pyrazolin-5-one (PMP) and then analyzed by a previously reported HPLC method, with some modification [55,56]. The samples or monosaccharide standards were mixed with an equal volume of 0.6 M NaOH aqueous solution (250 μL) and 0.4 M PMP methanol solution. The derivatization proceeded at 70 °C for 1 h after vortex blending. Then, the reaction mixture was neutralized with 500 μL of an HCl solution (0.3 M), and the excess PMP was removed by chloroform three times. Then, the filtrate of the aqueous layer was evaluated by a Daojin LC-20AD HPLC instrument with an Xtimate C18 column (4.6 × 200 mm 5 μm). A potassium dihydrogen phosphate solution (50 mM) and acetonitrile in a ratio of 83:17 (*v*/*v*) were used as eluent, and the flow rate was maintained at 1.0 mL/min.

4.2.4. Morphological Observations

The LNT was affixed to the sample stage using double-sided conductive tape, and images were captured at various magnifications with a Phenom Pro scanning electron microscope (SEM, HITACHI, Tokyo, Japan). Atomic force microscopy (AFM) shooting was improved based on previous research methods [57]. The LNT was fully dissolved in pure water at a concentration of 30 μg/mL, and 30 μL of this solution was deposited onto a freshly cleaved mica surface to obtain atomic force microscopy (AFM) images. Following air drying, the morphology was examined using a Dimension Edge AFM (Bruker, Penang, Malaysia).

4.3. *Animal Experiments*

4.3.1. Animals and Experimental Design

Twenty-four male Kunming mice (inbred, 6–8 weeks old), weighing 18~22 g, were provided by the Liaoning Changsheng Biotechnology Co., Ltd. (Benxi, Liaoning, China, license number SCXK (Liao)2020-0001). After fasting for 12h, the mice were anesthetized with isoflurane in an anesthesia box and then euthanized with eye blood collection. Throughout the whole experiment, the mice were reared at room temperature (25 ± 2 °C), humidity (55–60%), and a 12 h light–dark cycle with arbitrary water and food. The present study was conducted under the approval of the Institutional Animal Care and Use Committee of Northeast Agricultural University (Heilongjiang province, China) (SYXK (Hei) 2012–2067) in accordance with the Laboratory Animal Guideline for ethical review of animal welfare.

Following a week of adaptive breeding, the mice were split into four groups at random, with each group containing 6 mice. From the 1st day to the 5th day, the intraperitoneal injection of mice was as below: Model group (CP) and LNT treatment group (CL): intraperitoneal injection of 80 mg/kg CP; NC group (C) and LNT administration alone group (L): intraperitoneal injection of 0.9% NS. From the 6th day to the 15th day, there were two LNT groups (CL and L): intragastric administration of 200 mg/kg LNT [19], and the other groups were as follows: intragastric administration of 0.9%NS (Figure 9). The mice were euthanized on the 17th day, and serum, spleen, thymus, and ileum were collected for subsequent experiments.

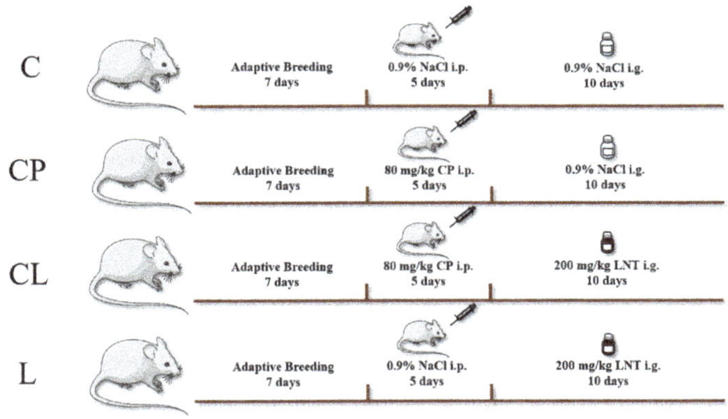

Figure 9. Animal and experimental design. Animals (n = 24) were divided into four experimental groups. From the 1st day to the 5th day, the intraperitoneal injection of mice was as below: Model group (CP) and LNT treatment group (CL): intraperitoneal injection of 80 mg/kg CP; NC group (C) and LNT administration alone group (L): intraperitoneal injection of 0.9% NS. From the 6th day to the 15th day, there were two LNT groups (CL and L): intragastric administration of 200 mg/kg LNT, and the other groups were as follows: intragastric administration of 0.9%NS.

4.3.2. Immune Organs Indices

After administration, the mice were anesthetized with isoflurane and euthanized by eye blood collection. The thymus gland and spleen were removed and weighed. Immune organ index was determined using the following formula: Thymus or spleen index = weight of thymus or spleen/body weight × 1000.

4.3.3. Measurements of Cytokines in Serum and sIgA in the Ileum

Serum was obtained by centrifugation after the blood was left for 2 h at 4 °C. The ileum samples were crushed and centrifuged at 4 °C 3000 rpm for 5 min to obtain the supernatant. The concentrations of cytokines in the serum and sIgA in the ileum were determined by the ELISA kit according to the instructions of the manufacturer.

4.3.4. Real Time-qPCR (RT-qRCR) Assay

The total RNA was extracted from the ileum using the Trizol reagent. The RNA was performed at 37 °C for 15 min and 85 °C for 5 sec to prepare cDNA, which was then stored at −80 °C for use. Real-time PCR was carried out using a LightCycle 96 real-time PCR system and a real-time PCR master mix (SYBR Green) reagent kit (Takara, China). Custom-made primers are present in Table 2. The conditions for real-time PCR were 95 °C for 5 min and then 45 cycles at 95 °C for 10 s, 60 °C for 30 s, and 95 °C for 15 s, 60 °C for 1 min, 95 °C for 15 s. Relative mRNA expression was calculated by the $2^{-\triangle\triangle}$CT method.

Table 2. Primers used for Rt-PCR analysis of genes.

Genes	Forward (5′-3′)	Reverse (5′-3′)
IL-1β	CACTACAGGCTCCGAGATGAACAAC	TGTCGTTGCTTGGTTCTCCTTGTAC
IFN-γ	CTTGAAAGACAATCAGGCCATC	CTTGGCAATACTCATGAATGCA
TNF-α	ATGTCTCAGCCTCTTCTCATTC	GCTTGTCACTCGAATTTTGAGA
GPR41	CCACACTGCTCATCTTCTTCGTCTG	ACGGACTCTCACGGCTGACATAG
GPR43	CTGTATGGAGTGATCGCTGCTCTG	CTGCTCTTGGGTGAAGTTCTCGTAG
MUC2	CGAGCACATCACCTACCACATCATC	TCCAGAATCCAGCCAGCCAGTC
Claudin-1	AGATACAGTGCAAAGTCTTCGA	CAGGATGCCAATTACCATCAAG
Occludin	TGCTTCATCGCTTCCTTAGTAA	GGGTTCACTCCCATTATGTACA
ZO-1	CTGGTGAAGTCTCGGAAAAATG	CATCTCTTGCTGCCAAACTATC
GAPDH	CATCTTCCAGGAGCGAGACC	TCCACCACCCTGTTGCTGTA

4.3.5. Histopathology

The ileum was collected after washing and fixed in a 10% formalin solution. They were then sectioned and stained with H&E.

4.3.6. Scanning Electron Microscopy (SEM)

Scanning electron microscopy (SEM) was performed as previously described. Briefly, samples were fixed with 2.5% glutaraldehyde and rinsed in 0.1 M sodium phosphate buffer. After dehydration and replacement with ethanol and tert-butyl alcohol, the mixture was dried at −20 °C for 30 min, and then dried in ES-2030 (HITACHI) for 4 h. Samples were metal coated and examined with SU8010 field emission scanning electron microscope (HITACHI, Tokyo, Japan).

4.3.7. 16S rRNA Gene Sequencing of the Gut Microbiota in Fecal Samples

The total DNA was extracted from the fecal samples following a previously reported method, with certain modifications [58]. Briefly, the OMEGA Soil DNA Kit (M5636-02) (Omega Bio-Tek, Norcross, GA, USA) was used following the manufacturer's recommendations. The bacterial 16S rRNA gene V3-V4 region was amplified by PCR using the forward primer 338F (5′-ACTCCTACGGGAGGCAGCA-3′) and the reverse primer 806R (5′-GGACTACHVGGGTWTCTAAT-3′). PCR amplicons were purified with Vazyme VAHTSTM DNA Clean Beads (Vazyme, Nanjing, China) and quantified using the Quant-iT

PicoGreen dsDNA Assay Kit (Invitrogen, Carlsbad, CA, USA). Sequence data analyses were mainly performed using QIIME2 version 2019.4 and R packages (v3.2.0).

4.4. Statistical Analysis

All results were confirmed in at least three independent experiments. Results were expressed as mean ± SD. All data were analyzed by one-way ANOVA with Tukey's post-test. Significance was defined as $p < 0.05$.

5. Conclusions

In summary, the LNT we extracted is a brown-yellow, hygroscopic powder with a total carbohydrate content of 73.43% (*w*/*w*), a protein content of 17.17% (*w*/*w*), and a galacturonic acid content of 3.03%. In vivo experiments confirmed that LNT can regulate cyclophosphamide (CP)-induced immunosuppression, restore cytokine levels in serum, and reinforce the intestinal mucosal barrier. Additionally, LNT has been shown to normalize the gut microbiota structure affected by CP, particularly by enhancing the growth of bacteria associated with cancer treatment. These findings establish a foundation for the potential development of LNT as functional foods and drugs with immunomodulatory properties. However, the precise mechanism by which LNT exerts its immunomodulatory effects through gut microbiota remains unclear and warrants further investigation.

Supplementary Materials: The following supporting information can be downloaded at https://www.mdpi.com/article/10.3390/ijms252212432/s1.

Author Contributions: J.L., conceptualization and supervision; X.J., methodology, investigation, and writing; H.C., resources and data curation; W.L., methodology and data curation; F.G., methodology and validation; Z.W., investigation and writing. All data were generated in-house, and no paper mill was used. All authors agree to be accountable for all aspects of this work, ensuring integrity and accuracy. All authors have read and agreed to the published version of the manuscript.

Funding: This work is supported by the Natural Science Foundation of Heilongjiang Province (LH2022C038) and Special Funding for Postdoctoral of Heilongjiang Province (LBH-TZ2119).

Institutional Review Board Statement: Animal experiments were conducted in compliance with the regulations of the Laboratory Animal Ethics Committee of Northeastern Agricultural University (approved No.: NEAUEC20220319, approved date: 10 March 2022).

Informed Consent Statement: No applicable.

Data Availability Statement: All relevant data are within the manuscript and its Supplementary Materials.

Acknowledgments: We extend our sincere gratitude to Li Jichang for supplying the experimental conditions. We also thank FigDraw for providing independent painting resources. Additionally, we appreciate the financial support from the Natural Science Foundation of Heilongjiang Province and the Special Funding for Postdoctoral Research of Heilongjiang Province.

Conflicts of Interest: The authors declare no conflicts of interest.

Abbreviations

LNT: lentinan; CP, cyclophosphamide; GalA, galacturonic acid; IL-6, interleukin 6; IL-2, interleukin 2; IFN-γ, interferon-γ; TNF-α, tumor necrosis factor α; IgG, immunoglobulin G; sIgA, secretory immunoglobulin A; IL-1β, interleukin 1β; GPR41, G protein-coupled receptor 41; GPR43, G protein-coupled receptor 43; MUC2, Mucin2; ZO-1, zonula occludens-1; SEM, scanning electron microscopy; SFB, *Segmented Filamentous Bacteria*; NMDS, nonmetric multidimensional scaling; PCoA, principal coordinates analysis; OPLS-DA, orthogonal partial least squares discrimination analysis.

References

1. Wang, H.; Xu, L.; Yu, M.; Wang, Y.; Jiang, T.; Yang, S.; Lv, Z. Glycosaminoglycan from *Apostichopus japonicus* induces immunomodulatory activity in cyclophosphamide-treated mice and in macrophages. *Int. J. Biol. Macromol.* **2019**, *130*, 229–237. [CrossRef] [PubMed]
2. Mahiou, J.; Walter, U.; Lepault, F.; Godeau, F.; Bach, J.F.; Chatenoud, L. In vivo blockade of the Fas-Fas ligand pathway inhibits cyclophosphamide-induced diabetes in NOD mice. *J. Autoimmun.* **2001**, *16*, 431–440. [CrossRef] [PubMed]
3. Fang, J.; Wang, Y.; Lv, X.; Shen, X.; Ni, X.; Ding, K. Structure of a β-glucan from *Grifola frondosa* and its antitumor effect by activating Dectin-1/Syk/NF-κB signaling. *Glycoconj. J.* **2012**, *29*, 365–377. [CrossRef] [PubMed]
4. An, H.; Yu, Y.; Zhang, M.; Xu, H.; Qi, R.; Yan, X.; Liu, S.; Wang, W.; Guo, Z.; Guo, J.; et al. Involvement of ERK, p38 and NF-kappaB signal transduction in regulation of TLR2, TLR4 and TLR9 gene expression induced by lipopolysaccharide in mouse dendritic cells. *Immunology* **2002**, *106*, 38–45. [CrossRef]
5. Xu, X.; Zhang, X. Effects of cyclophosphamide on immune system and gut microbiota in mice. *Microbiol. Res.* **2015**, *171*, 97–106. [CrossRef]
6. Zhou, R.; He, D.; Xie, J.; Zhou, Q.; Zeng, H.; Li, H.; Huang, L. The Synergistic Effects of Polysaccharides and Ginsenosides from American Ginseng (*Panax quinquefolius* L.) Ameliorating Cyclophosphamide-Induced Intestinal Immune Disorders and Gut Barrier Dysfunctions Based on Microbiome-Metabolomics Analysis. *Front. Immunol.* **2021**, *12*, 665901. [CrossRef]
7. Zhang, J.; Zhou, H.C.; He, S.B.; Zhang, X.F.; Ling, Y.H.; Li, X.Y.; Zhang, H.; Hou, D.D. The immunoenhancement effects of sea buckthorn pulp oil in cyclophosphamide-induced immunosuppressed mice. *Food Funct.* **2021**, *12*, 7954–7963. [CrossRef]
8. Kim, J.W.; Choi, J.S.; Seol, D.J.; Choung, J.J.; Ku, S.K. Immunomodulatory Effects of Kuseonwangdogo-Based Mixed Herbal Formula Extracts on a Cyclophosphamide-Induced Immunosuppression Mouse Model. *Evid.-Based Complement. Altern. Med.* **2018**, *2018*, 6017412. [CrossRef]
9. Liu, H.; Ma, J.; Gong, F.; Wei, F.; Zhang, X.; Wu, H. Technology, Structural characterisation and immunomodulatory effects of polysaccharides isolated from *Dendrobium aphyllum*. *Int. J. Food Sci. Technol.* **2018**, *53*, 1185–1194. [CrossRef]
10. Zou, M.; Chen, Y.; Sun-Waterhouse, D.; Zhang, Y.; Li, F. Immunomodulatory acidic polysaccharides from *Zizyphus jujuba* cv. Huizao: Insights into their chemical characteristics and modes of action. *Food Chem.* **2018**, *258*, 35–42. [CrossRef]
11. Du, J.; Chen, X.; Wang, C.; Sun, H. Pathway analysis of global gene expression change in dendritic cells induced by the polysaccharide from the roots of *Actinidia eriantha*. *J. Ethnopharmacol.* **2018**, *214*, 141–152. [CrossRef] [PubMed]
12. Tabarsa, M.; Jafari, A.; You, S.; Cao, R. Immunostimulatory effects of a polysaccharide from *Pimpinella anisum* seeds on RAW264.7 and NK-92 cells. *Int. J. Biol. Macromol.* **2022**, *213*, 546–554. [CrossRef] [PubMed]
13. Shirasaka, T.; Shimamoto, Y.; Fukushima, M. Inhibition by oxonic acid of gastrointestinal toxicity of 5-fluorouracil without loss of its antitumor activity in rats. *Cancer Res.* **1993**, *53*, 4004–4009. [PubMed]
14. Spears, C.P.; Shahinian, A.H.; Moran, R.G.; Heidelberger, C.; Corbett, T.H. In vivo kinetics of thymidylate synthetase inhibition of 5-fluorouracil-sensitive and -resistant murine colon adenocarcinomas. *Cancer Res.* **1982**, *42*, 450–456.
15. Mansour, E.G.; Gray, R.; Shatila, A.H.; Tormey, D.C.; Cooper, M.R.; Osborne, C.K.; Falkson, G. Survival advantage of adjuvant chemotherapy in high-risk node-negative breast cancer: Ten-year analysis—An intergroup study. *J. Clin. Oncol.* **1998**, *16*, 3486–3492. [CrossRef]
16. Shirasaka, T.; Nakano, K.; Takechi, T.; Satake, H.; Uchida, J.; Fujioka, A.; Saito, H.; Okabe, H.; Oyama, K.; Takeda, S.; et al. Antitumor activity of 1 M tegafur-0.4 M 5-chloro-2,4-dihydroxypyridine-1 M potassium oxonate (S-1) against human colon carcinoma orthotopically implanted into nude rats. *Cancer Res.* **1996**, *56*, 2602–2606.
17. Ren, G.; Xu, L.; Lu, T.; Yin, J. Structural characterization and antiviral activity of lentinan from *Lentinus edodes* mycelia against infectious hematopoietic necrosis virus. *Int. J. Biol. Macromol.* **2018**, *115*, 1202–1210. [CrossRef]
18. Meng, M.; Huo, R.; Wang, Y.; Ma, N.; Shi, X.; Shen, X.; Chang, G. Lentinan inhibits oxidative stress and alleviates LPS-induced inflammation and apoptosis of BMECs by activating the Nrf2 signaling pathway. *Int. J. Biol. Macromol.* **2022**, *222 Pt B*, 2375–2391. [CrossRef]
19. Ji, X.; Su, L.; Zhang, P.; Yue, Q.; Zhao, C.; Sun, X.; Li, K.; Liu, X.; Zhang, S.; Zhao, L. Lentinan improves intestinal inflammation and gut dysbiosis in antibiotics-induced mice. *Sci. Rep.* **2022**, *12*, 19609. [CrossRef]
20. Chen, C.; Ahn, E.H.; Kang, S.S.; Liu, X.; Alam, A.; Ye, K. Gut dysbiosis contributes to amyloid pathology, associated with C/EBPβ/AEP signaling activation in Alzheimer's disease mouse model. *Sci. Adv.* **2020**, *6*, eaba0466. [CrossRef]
21. Round, J.L.; Mazmanian, S.K. The gut microbiota shapes intestinal immune responses during health and disease. *Nat. Rev. Immunol.* **2009**, *9*, 313–323. [CrossRef]
22. Gong, Y.; Ma, Y.; Cheung, P.C.; You, L.; Liao, L.; Pedisić, S.; Kulikouskaya, V. Structural characteristics and anti-inflammatory activity of UV/H(2)O(2)-treated algal sulfated polysaccharide from *Gracilaria lemaneiformis*. *Food Chem. Toxicol.* **2021**, *152*, 112157. [CrossRef] [PubMed]
23. Ding, Y.; Yan, Y.; Chen, D.; Ran, L.; Mi, J.; Lu, L.; Jing, B.; Li, X.; Zeng, X.; Cao, Y. Modulating effects of polysaccharides from the fruits of *Lycium barbarum* on the immune response and gut microbiota in cyclophosphamide-treated mice. *Food Funct.* **2019**, *10*, 3671–3683. [CrossRef] [PubMed]
24. Huang, J.; Huang, J.; Li, Y.; Wang, Y.; Wang, F.; Qiu, X.; Liu, X.; Li, H. Sodium Alginate Modulates Immunity, Intestinal Mucosal Barrier Function, and Gut Microbiota in Cyclophosphamide-Induced Immunosuppressed BALB/c Mice. *J. Agric. Food Chem.* **2021**, *69*, 7064–7073. [CrossRef] [PubMed]

25. Sun, X.; Liu, N.; Wu, Z.; Feng, Y.; Meng, X. Anti-tumor activity of a polysaccharide from blueberry. *Molecules* **2015**, *20*, 3841–3853. [CrossRef] [PubMed]
26. Bolyen, E.; Rideout, J.R.; Dillon, M.R.; Bokulich, N.A.; Abnet, C.C.; Al-Ghalith, G.A.; Alexander, H.; Alm, E.J.; Arumugam, M.; Asnicar, F.; et al. Reproducible, interactive, scalable and extensible microbiome data science using QIIME 2. *Nat. Biotechnol.* **2019**, *37*, 852–857. [CrossRef]
27. Liu, Y.H.; Ma, S.D.; Fu, Q.J.; Zhao, L.Y.; Li, Y.; Wang, H.Q.; Li, M.C. Effect of lentinan on membrane-bound protein expression in splenic lymphocytes under chronic low-dose radiation. *Int. Immunopharmacol.* **2014**, *22*, 505–514. [CrossRef]
28. Ding, G.B.; Nie, R.H.; Lv, L.H.; Wei, G.Q.; Zhao, L.Q. Preparation and biological evaluation of a novel selenium-containing exopolysaccharide from *Rhizobium* sp. N613. *Carbohydr. Polym.* **2014**, *109*, 28–34. [CrossRef]
29. Becker, J.C.; Schrama, D. The dark side of cyclophosphamide: Cyclophosphamide-mediated ablation of regulatory T cells. *J. Investig. Dermatol.* **2013**, *133*, 1462–1465. [CrossRef]
30. Zhang, Q.; Cong, R.; Hu, M.; Zhu, Y.; Yang, X. Immunoenhancement of Edible Fungal Polysaccharides (Lentinan, Tremellan, and Pachymaran) on Cyclophosphamide-Induced Immunosuppression in Mouse Model. *Evid.-Based Complement. Altern. Med.* **2017**, *2017*, 9459156. [CrossRef]
31. Lambert, K.; Diggins, K.E.; Jones, B.E.; Hundhausen, C.; Maerz, M.D.; Hocking, A.M.; Sanda, S.; Greenbaum, C.J.; Linsley, P.S.; Cerosaletti, K.; et al. IL-6-Driven pSTAT1 Response Is Linked to T Cell Features Implicated in Early Immune Dysregulation. *Front. Immunol.* **2022**, *13*, 935394. [CrossRef] [PubMed]
32. Cai, Z.; Li, W.; Wang, H.; Yan, W.; Zhou, Y.; Wang, G.; Cui, J.; Wang, F. Anti-tumor and immunomodulating activities of a polysaccharide from the root of *Sanguisorba officinalis* L. *Int. J. Biol. Macromol.* **2012**, *51*, 484–488. [CrossRef]
33. Liu, N.; Dong, Z.; Zhu, X.; Xu, H.; Zhao, Z. Characterization and protective effect of *Polygonatum sibiricum* polysaccharide against cyclophosphamide-induced immunosuppression in Balb/c mice. *Int. J. Biol. Macromol.* **2018**, *107 Pt A*, 796–802. [CrossRef]
34. Xiang, Q.; Wu, X.; Pan, Y.; Wang, L.; Cui, C.; Guo, Y.; Zhu, L.; Peng, J.; Wei, H. Early-Life Intervention Using Fecal Microbiota Combined with Probiotics Promotes Gut Microbiota Maturation, Regulates Immune System Development, and Alleviates Weaning Stress in Piglets. *Int. J. Mol. Sci.* **2020**, *21*, 503. [CrossRef] [PubMed]
35. Diehl, G.E.; Longman, R.S.; Zhang, J.X.; Breart, B.; Galan, C.; Cuesta, A.; Schwab, S.R.; Littman, D.R. Microbiota restricts trafficking of bacteria to mesenteric lymph nodes by CX(3)CR1(hi) cells. *Nature* **2013**, *494*, 116–120. [CrossRef] [PubMed]
36. Hapfelmeier, S.; Lawson, M.A.; Slack, E.; Kirundi, J.K.; Stoel, M.; Heikenwalder, M.; Cahenzli, J.; Velykoredko, Y.; Balmer, M.L.; Endt, K.; et al. Reversible microbial colonization of germ-free mice reveals the dynamics of IgA immune responses. *Science* **2010**, *328*, 1705–1709. [CrossRef] [PubMed]
37. Nakamoto, N.; Amiya, T.; Aoki, R.; Taniki, N.; Koda, Y.; Miyamoto, K.; Teratani, T.; Suzuki, T.; Chiba, S.; Chu, P.S.; et al. Commensal *Lactobacillus* Controls Immune Tolerance during Acute Liver Injury in Mice. *Cell Rep.* **2017**, *21*, 1215–1226. [CrossRef] [PubMed]
38. Ivanov, I.I.; Atarashi, K.; Manel, N.; Brodie, E.L.; Shima, T.; Karaoz, U.; Wei, D.; Goldfarb, K.C.; Santee, C.A.; Lynch, S.V.; et al. Induction of intestinal Th17 cells by segmented filamentous bacteria. *Cell* **2009**, *139*, 485–498. [CrossRef]
39. Yang, H.; Rao, J.N.; Wang, J.Y. Posttranscriptional Regulation of Intestinal Epithelial Tight Junction Barrier by RNA-binding Proteins and microRNAs. *Tissue Barriers* **2014**, *2*, e28320. [CrossRef]
40. Prakash, R.; Bharathi Raja, S.; Devaraj, H.; Devaraj, S.N. Up-regulation of MUC2 and IL-1β expression in human colonic epithelial cells by *Shigella* and its interaction with mucins. *PLoS ONE* **2011**, *6*, e27046. [CrossRef]
41. Dharmani, P.; Strauss, J.; Ambrose, C.; Allen-Vercoe, E.; Chadee, K. *Fusobacterium nucleatum* infection of colonic cells stimulates MUC2 mucin and tumor necrosis factor alpha. *Infect. Immun.* **2011**, *79*, 2597–2607. [CrossRef] [PubMed]
42. Gao, Y.; Li, S.; Wang, J.; Luo, C.; Zhao, S.; Zheng, N. Modulation of Intestinal Epithelial Permeability in Differentiated Caco-2 Cells Exposed to Aflatoxin M1 and Ochratoxin A Individually or Collectively. *Toxins* **2017**, *10*, 13. [CrossRef] [PubMed]
43. Kim, M.H.; Kang, S.G.; Park, J.H.; Yanagisawa, M.; Kim, C.H. Short-chain fatty acids activate GPR41 and GPR43 on intestinal epithelial cells to promote inflammatory responses in mice. *Gastroenterology* **2013**, *145*, 396–406.e10. [CrossRef] [PubMed]
44. Chen, D.; Chen, G.; Ding, Y.; Wan, P.; Peng, Y.; Chen, C.; Ye, H.; Zeng, X.; Ran, L. Polysaccharides from the flowers of tea (*Camellia sinensis* L.) modulate gut health and ameliorate cyclophosphamide-induced immunosuppression. *J. Funct. Foods* **2019**, *61*, 103470. [CrossRef]
45. Ferguson, A.S.; Huang, W.E.; Lawson, K.A.; Doherty, R.; Gibert, O.; Dickson, K.W.; Whiteley, A.S.; Kulakov, L.A.; Thompson, I.P.; Kalin, R.M.; et al. Microbial analysis of soil and groundwater from a gasworks site and comparison with a sequenced biological reactive barrier remediation process. *J. Appl. Microbiol.* **2007**, *102*, 1227–1238. [CrossRef]
46. Bousbaine, D.; Fisch, L.I.; London, M.; Bhagchandani, P.; Rezende de Castro, T.B.; Mimee, M.; Olesen, S.; Reis, B.S.; VanInsberghe, D.; Bortolatto, J.; et al. A conserved Bacteroidetes antigen induces anti-inflammatory intestinal T lymphocytes. *Science* **2022**, *377*, 660–666. [CrossRef]
47. Shin, N.R.; Whon, T.W.; Bae, J.W. Proteobacteria: Microbial signature of dysbiosis in gut microbiota. *Trends Biotechnol.* **2015**, *33*, 496–503. [CrossRef]
48. Azad, M.A.K.; Sarker, M.; Li, T.; Yin, J. Probiotic Species in the Modulation of Gut Microbiota: An Overview. *BioMed Res. Int.* **2018**, *2018*, 9478630. [CrossRef]

49. Ni, Y.; Zhang, Y.; Zheng, L.; Rong, N.; Yang, Y.; Gong, P.; Yang, Y.; Siwu, X.; Zhang, C.; Zhu, L.; et al. *Bifidobacterium* and *Lactobacillus* improve inflammatory bowel disease in zebrafish of different ages by regulating the intestinal mucosal barrier and microbiota. *Life Sci.* **2023**, *324*, 121699. [CrossRef]
50. Kang, X.; Liu, C.; Ding, Y.; Ni, Y.; Ji, F.; Lau, H.C.H.; Jiang, L.; Sung, J.J.; Wong, S.H.; Yu, J. *Roseburia intestinalis* generated butyrate boosts anti-PD-1 efficacy in colorectal cancer by activating cytotoxic CD8(+) T cells. *Gut* **2023**, *72*, 2112–2122. [CrossRef]
51. Iida, N.; Dzutsev, A.; Stewart, C.A.; Smith, L.; Bouladoux, N.; Weingarten, R.A.; Molina, D.A.; Salcedo, R.; Back, T.; Cramer, S.; et al. Commensal bacteria control cancer response to therapy by modulating the tumor microenvironment. *Science* **2013**, *342*, 967–970. [CrossRef] [PubMed]
52. Luo, Z.; Gao, Q.; Li, Y.; Bai, Y.; Zhang, J.; Xu, W.; Xu, J. *Flammulina velutipes* Mycorrhizae Attenuate High Fat Diet-Induced Lipid Disorder, Oxidative Stress and Inflammation in the Liver and Perirenal Adipose Tissue of Mice. *Nutrients* **2022**, *14*, 3830. [CrossRef]
53. Dubois, M.; Gilles, K.; Hamilton, J.K.; Rebers, P.A.; Smith, F. A colorimetric method for the determination of sugars. *Nature* **1951**, *168*, 167. [CrossRef] [PubMed]
54. Gharibzahedi, S.M.T.; Smith, B.; Guo, Y. Ultrasound-microwave assisted extraction of pectin from fig (*Ficus carica* L.) skin: Optimization, characterization and bioactivity. *Carbohydr. Polym.* **2019**, *222*, 114992. [CrossRef] [PubMed]
55. Jayamanohar, J.; Devi, P.B.; Kavitake, D.; Priyadarisini, V.B.; Shetty, P.H. Prebiotic potential of water extractable polysaccharide from red kidney bean (*Phaseolus vulgaris* L.). *LWT* **2019**, *101*, 703–710. [CrossRef]
56. Yang, T.H.; Chiu, C.Y.; Lu, T.J.; Liu, S.H.; Chiang, M.T. The Anti-Obesity Effect of Polysaccharide-Rich Red Algae (*Gelidium amansii*) Hot-Water Extracts in High-Fat Diet-Induced Obese Hamsters. *Mar. Drugs* **2019**, *17*, 532. [CrossRef]
57. Zhang, X.; Hong, L.; Zhu, B.J.; Yuan, Y.; Li, S.P.; Zhao, J. Atomic force microscopy based conformation and immunological activity of Lentinan injections. *Int. J. Biol. Macromol.* **2023**, *253 Pt 3*, 126901. [CrossRef]
58. Lv, W.; Liu, C.; Ye, C.; Sun, J.; Tan, X.; Zhang, C.; Qu, Q.; Shi, D.; Guo, S. Structural modulation of gut microbiota during alleviation of antibiotic-associated diarrhea with herbal formula. *Int. J. Biol. Macromol.* **2017**, *105 Pt 3*, 1622–1629. [CrossRef]

antioxidants

MDPI

Article

Rosmarinic Acid Attenuates *Salmonella enteritidis*-Induced Inflammation via Regulating TLR9/NF-κB Signaling Pathway and Intestinal Microbiota

Dandan Yi [1], Menghui Wang [1], Xia Liu [1], Lanqian Qin [1], Yu Liu [1], Linyi Zhao [1], Ying Peng [1], Zhengmin Liang [1,2,3,*] and Jiakang He [1,2,3,*]

[1] College of Animal Science and Technology, Guangxi University, Nanning 530004, China; 2118402010@st.gxu.edu.cn (D.Y.); 2218302032@st.gxu.edu.cn (M.W.); 2118302028@st.gxu.edu.cn (X.L.); 2018302030@st.gxu.edu.cn (L.Q.); 2018402003@st.gxu.edu.cn (Y.L.); 2218302046@st.gxu.edu.cn (L.Z.); 2118302033@st.gxu.edu.cn (Y.P.)

[2] Guangxi Key Laboratory of Animal Breeding, Disease Control and Prevention, Nanning 530004, China

[3] Guangxi Zhuang Autonomous Region Engineering Research Center of Veterinary Biologics, Nanning 530004, China

[*] Correspondence: liangzm@gxu.edu.cn (Z.L.); jkhe@gxu.edu.cn (J.H.)

Abstract: *Salmonella enteritidis* (*SE*) infection disrupts the homeostasis of the intestinal microbiota, causing an intestinal inflammatory response and posing a great threat to human and animal health. The unreasonable use of antibiotics has led to an increase in the prevalence of drug-resistant *SE*, increasing the difficulty of controlling *SE*. Therefore, new drug strategies and research are urgently needed to control *SE*. Rosmarinic acid (RA) is a natural phenolic acid with various pharmacological activities, including antioxidant, anti-inflammatory and antibacterial properties. However, the protective effects and mechanism of RA on intestinal inflammation and the gut microbial disorders caused by *SE* have not been fully elucidated. In this study, RAW264.7 cells, MCECs and BALB/c mice were challenged with *SE* to assess the protective effects and mechanisms of RA. The results showed that RA enhanced the phagocytic ability of RAW264.7 cells, reduced the invasion and adhesion ability of *SE* in MCECs, and inhibited *SE*-induced inflammation in cells. Moreover, RA inhibited the activation of the NF-κB signaling pathway by upregulating TLR9 expression. Importantly, we found that RA provided protection against *SE* and increased the diversity and abundance of the intestinal microbiota in mice. Compared with infection control, RA significantly increased the abundance of *Firmicutes* and *Acidibacteria* and decreased the abundance of *Proteobacteria*, *Epsilonbacteraeota* and *Bacteroidota*. However, RA failed to alleviate *SE*-induced inflammation and lost its regulatory effects on the TLR9/NF-κB signaling pathway after destroying the gut microbiota with broad-spectrum antibiotics. These results indicated that RA attenuated *SE*-induced inflammation by regulating the TLR9/NF-κB signaling pathway and maintaining the homeostasis of the gut microbiota. Our study provides a new strategy for preventing *SE*-induced intestinal inflammation.

Keywords: Rosmarinic acid; *Salmonella enteritidis*; TLR9; NF-κB; Gut microbiota

1. Introduction

Salmonella invasion and intracellular replication within host cells result in gastroenteritis, bacteremia, enteric fever and focal infections [1]. *Salmonella* is prone to contaminating animal foods. It is frequently detected in seafood [2], beef [3], chicken and pork [4], causes significant economic losses in animal husbandry and seriously endangers public health [5,6]. *Salmonella* invades intestinal epithelial cells, inducing an excessive inflammatory response in the gut and disrupting intestinal mucosa barrier function [7]. The virulence factor T3SS-2 of *Salmonella* enhances its colonization in the intestinal lumen by achieving chemotactic-dependent utilization of intestinal inflammation [7]. Maintaining the intestinal barrier integrity and intestinal microbiota homeostasis offers colonization resistance against *Salmonella*, limiting bacterial expansion and transmission [8]. Moreover, enhancing the antioxidant capacity of epithelial cells and inhibiting excessive inflammatory responses in the intestine contribute to the control of *Salmonella* infection [9].

Macrophages are recruited to the site of infection to exert their phagocytic function after infection with *Salmonella*. Moreover, this response has been studied intensively in epithelial cells, the target of *Salmonella* during gastrointestinal infections [10]. The ability of immune cells, particularly macrophages, to eliminate bacteria is often associated with pattern recognition receptors (PRRs), especially TLRs. TLRs activate signaling pathways that provide specific immunological responses tailored to microbes expressing PAMPs [11]. TLRs include cell surface TLRs and endosomal TLRs [12]. TLR9 is an intracellular TLR that recognizes unmethylated CpG motifs in the double-stranded DNA of bacteria. TLR9 activation amplifies the T-cell response and promotes bacterial clearance; thus, TLR9 activation may be an effective strategy for resisting pathogen entry [13]. Previous studies have proven that TLR9 is important for the host defense against *Salmonella typhimurium* invasion in mice [14]. Persistent infection with *Salmonella* leads to an inflammatory response in the body, and the NF-κB signaling pathway plays an important role in *Salmonella*-induced inflammation. TLR9 negatively regulates the NF-κB-NLRP3-IL-1β pathway and plays a vital role in defense against *Salmonella typhimurium* and the protection of the intestinal integrity. Moreover, CpG treatment induced increased TLR9 expression and enhanced the host resistance to *Salmonella* [15]. These studies indicated that activating TLR9 and inhibiting the NF-κB pathway are beneficial for the host to clear *Salmonella* and alleviate intestinal inflammatory damage.

The gut microbiota promotes the integrity of the intestinal mucosa, provides essential nutrients such as vitamins and enzymes, protects hosts from pathogen infections and regulates innate and adaptive immunity [16]. The metabolites produced by the symbiotic gut microbiota affect the host's health through recognition by the immune system [17]. The disturbance of the gut microbiota leads to cardiac metabolism disorders, type 2 diabetes, obesity, and various chronic inflammatory diseases [18–20]. Moreover, the gut microbiota is crucial for resisting the colonization of exogenous microorganisms [21]. Mice harboring low-complexity gut microbiota are unable to resist *Salmonella* infection [22]. Maintaining normal gut microbiota can reduce the colonization of *Salmonella* and maintain intestinal barrier function [23].

Salmonella enteritidis (*SE*), a common serotype of *Salmonella*, is one of the main pathogens causing acute gastroenteritis. Currently, the main prevention and treatment drugs for *SE* infection are antibiotics, and the unreasonable use of antibiotics has led to the emergence of *SE* resistance and damage to intestinal barrier function. Residues of veterinary drugs in animal food impair public health and safety, and because of their powerful biological activity and nontoxicity and the advantage of having multiple targets, the effective monomeric components of Chinese herbal medicine are an outstanding choice for controlling *Salmonella* [24]. Rosmarinic acid (RA) is a bioactive phenolic compound commonly found in plants of the *Lamiaceae* and *Boraginaceae* families, and possesses anti-inflammatory, anticancer and antibacterial activities [25–27]. RA has been widely used in the beverage, food and skincare industries [28,29], provides protection against colitis and inhibits the growth of *Escherichia coli*, *Proteus mirabilis* and *Bacillus cereus* [28,30,31]. Additionally, RA has direct antibacterial and bactericidal effects on *Salmonella* in vitro [32]. However, the role of RA on *SE* infection in vivo and the regulatory mechanism of RA against *SE* are unclear.

In this study, we investigated the protective effects of RA on *Salmonella*-challenged RAW264.7 cells, MCECs and BALB/c mice, and the regulatory effects of RA on the TLR9/NF-κB signaling pathway and the gut microbiota. Overall, this study revealed that RA has a good protective effect on *SE*-induced inflammation and intestinal damage in mice by regulating the TLR9/NF-κB signaling pathway and gut microbiota. These findings indicate that RA is a mild, effective and potentially effective agent with few side effects that can be used to prevent *SE* infection.

2. Materials and Methods

2.1. Reagents

RA was purchased from McLean Biochemical Technology Co., Ltd. (Shanghai, China); LB nutrient agar and nutrient broth were purchased from Beijing Land Bridge Co., Ltd. (Beijing, China); DMEM and FBS were purchased from Thermo Fisher Scientific (Waltham, MA, USA); RNAlater was purchased from Genstar (Guangzhou, China); FITC was purchased from Sigma-Aldrich (Shanghai, China); rabbit monoclonal antibodies against p-p65, p65, p-IκB-α, IκB-α and β-actin were purchased from Abmart (Shanghai, China); ELISA kits were purchased from Mlbio (Shanghai, China); SOD, MDA, GPT and GOT biochemical indicator detection kits were purchased from Nanjing Jiancheng Biology Co., Ltd. (Nanjing, China); and E6446 dihydrochloride was purchased from Selleck (Chongqing, China).

2.2. Animal

BALB/c mice (female, 6–8 weeks, 18–22 g) were purchased from Beijing Si Bei Fu Biotechnology Co., Ltd. (Beijing, China). All mice received food and water ad libitum. The laboratory temperature was maintained at 22–24 °C, with a relative humidity of 40–50%. This study was conducted according to the recommendations of the Academy of Animal Research Guidelines and approved by the Animal Ethics Committee of Guangxi University (protocol number: GXU-2022-336). All animal experiments involving mice protocols and procedures were performed according to the protocols for the care of laboratory animals, Ministry of Science and Technology People's Republic of China. The mice were euthanized using anesthesia.

2.3. Bacteria and Cell Culture

The *Salmonella enterica* (*SE*) serovar *Enteritidis* strain was isolated from chickens and maintained by the Pharmacology and Pathology Laboratory of the School of Animal Science and Technology, Guangxi University. *SE* was grown in LB broth at 37 °C and shaken at 180 rpm for 6 h, centrifuged at 3000 rpm for 10 min, and resuspended with PBS for challenge.

RAW264.7 cells and MCECs were cultured in high-glucose Dulbecco's modified Eagle's medium (DMEM) supplemented with 10% fetal bovine serum and penicillin streptomycin, and incubated with 5% CO_2 at 37 °C.

2.4. Minimum Inhibitory Concentration (MIC)

RA was dissolved and diluted with LB broth. Then, 100 μL of RA at different concentrations (0.0157–8 mg/mL) was added to the 96-well plate. Next, 100 μL of *SE* was added, and the final concentration of RA was 0.00785–4 mg/mL. The experiment was conducted three times in parallel.

2.5. CCK-8 Assay

A CCK-8 kit was used to evaluate the impact of RA on the viability of RAW 264.7 cells and MCECs. Briefly, cells (10^5/well) in antibiotic-free medium were grown on 96-well plates at 37 °C for 12 h. Except for the control wells, the cells were treated with different concentrations of RA (0–480 μg/mL) and incubated at 37°C in 5% CO_2 for 6 h, 12 h, 24 h or 48 h. Then, 10 μL of CCK-8 reagent was added. After incubation at 37 °C for 2 h, the optical density (OD) was determined at 450 nm via a microplate reader. The experiment was conducted six times in parallel. Survival rate data in the blank control group was compared with other groups at the same time point for statistical analysis. The percentage survival was determined using the following formula:

$$\text{Viability (\%)} = (OD_{\text{treated cells}} - OD_{\text{control}})/(OD_{\text{control cells}} - OD_{\text{control}}) \times 100\%$$

2.6. CFU Detection

RAW264.7 cells and MCECs (10^5/well) were cultured in antibiotic-free medium on 24-well plates for 12 h, followed by pretreatment with different concentrations of RA (0–60 μg/mL) for 6 h. Then, *SE* was challenged with an MOI of 100, the supernatant was removed 2 h after the challenge, and the cells were washed with PBS three times to remove unattached bacteria. For the adhesion assay, the cells were incubated with 0.1% Triton X-100, and the suspensions were serially diluted with PBS and then plated on LB plates. The plates were cultured at 37°C for 16–18 h. Finally, the number of *SE* colonies that adhered to the cells was counted. For the invasion assay, the cells were washed with PBS three times after incubation with 0.1% Triton X-100 and then incubated in DMEM supplemented with penicillin and streptomycin (100 μg/mL) for 60 min. Furthermore, the cells were lysed with 1% Triton X-100, and the number of intracellular bacteria was determined via an adhesion assay. Finally, the number of *SE* colonies that invaded the cells was counted.

2.7. Phagocytosis Test

The cells were pretreated with RA (0–60 μg/mL) for 6 h and washed with PBS after pretreatment. Then, the cells were challenged with FITC-labeled *SE* at an MOI of 100 for 2 h, washed with PBS three times and fixed with 4% formaldehyde for 15 min. Finally, the cells were treated with DAPI for 10 min and observed using a fluorescence microscope.

2.8. Inflammatory Factor Detection

Cells were pretreated with RA (0–60 μg/mL) for 6 h and then challenged with *SE* (MOI: 100) for 2 h. The supernatant was collected to measure cytokine levels. The cells were lysed with TRIzol or RIPA buffer and then stored at −80 °C for RT–qPCR or Western blot detection.

2.9. Molecular Docking

The crystal structure of the target protein TLR9 was retrieved from the PDB database (PDB-No. 3WPF). The molecular structure of RA was obtained from the PubChem compound database. We prepared and converted the protein and ligand files into PDBQT format. Water molecules were removed from the protein structure, and hydrogen atoms were added accordingly. The grid box dimensions were determined based on the protein domain. PyMol 3.0 software was used to visualize the docking model.

2.10. RA Pretreatment and SE Challenge in Mice

To evaluate the effects of RA on the survival rate of *SE*-challenged mice, the mice were randomly divided into two groups: the *SE* infection control group and the RA + *SE* group. The mice in the RA + *SE* group were orally administered RA (20 mg/ kg·bw) daily for 7 days, and the mice in the *SE* infection control group were orally administered the same dose of PBS. Then, the mice were intraperitoneally challenged with *SE* (2.5×10^8 CFUs/mL, 0.2 mL) and observed for 15 days. Body weight and mortality were recorded daily.

To evaluate the effects of RA on *SE*-induced enteritis, the mice were randomly divided into four groups: the blank control group, the RA control group, the *SE* infection control group and the RA + *SE* group. Mice in the RA control group and RA + *SE* group were orally administered with RA (20 mg/ kg·bw) daily for 7 days, and mice in *SE* group and the blank control group were orally administered an equivalent volume of PBS. On the 7th day, 12 h after RA pretreatment, the mice were intraperitoneally injected with *SE* (2.5×10^8 CFUs/mL, 0.2 mL). The mice were euthanized 24 h after the challenge. Serum, intestinal tissues and feces were collected for further analysis.

To analyze the effects of the gut microbiota on RA-induced protection, the mice were randomly divided into four groups: the blank control group, the *SE* infection control group, the antibiotic control group and the RA + *SE* + antibiotic group. Mice in the antibiotic control group and the RA + *SE*+ antibiotic group were orally administered with broad-spectrum antibiotics. Six hours after antibiotic treatment, mice in the RA + *SE* + antibiotics group were orally administered with RA (20 mg/ kg·bw) daily for 7 days, and the other mice were orally administered an equivalent volume of PBS. On the 7th day, 12 h after RA pretreatment, the mice were intraperitoneally injected with *SE* (2.5 × 10^8 CFUs/mL, 0.2 mL). The mice were euthanized 24 h after the challenge. Serum, intestinal tissues and feces were collected for further analysis. Broad-spectrum antibiotics included neomycin (100 mg/L), streptomycin (50 mg/L), penicillin (100 mg/L), vancomycin (50 mg/L) and metronidazole (100 mg/L).

2.11. *Measurement of Biochemical Indicators*

The levels of glutamic oxaloacetic transaminase (GOT), glutamic pyruvic transaminase (GPT), malondialdehyde (MDA) and SOD in mouse serum were detected with a biochemical reagent kit according to the manufacturer's instructions.

2.12. *Measurement of Cytokine Levels*

The levels of TNF-α, IL-1β and IL-6 were detected via ELISA kits according to the manufacturer's instructions.

2.13. *Histological Evaluation*

Duodenal and colon tissues were collected, fixed in 4% paraformaldehyde and embedded in paraffin. Tissues were cut into 3 μm series slices, dewaxed in xylene, rehydrated in an ethanol gradient and then stained with hematoxylin and eosin. Histological pathology was scored according to a previously established scoring system on the basis of the degree of epithelial damage, inflammatory infiltrate in the mucosa, submucosa and muscularis/serosa, crypt damage, severity of inflammation, intestinal villus injury and damage to intestinal glands [33].

2.14. *RT-qPCR*

Total RNA was extracted from cells and intestinal tissues using a TRIzol kit according to the manufacturer's instructions. Reverse transcription was performed with an ABM reagent kit according to the manufacturer's instructions. The primers used for real-time PCR are listed in the Appendix A.1 (Table A1). All samples were analyzed via the Roche 96 Light Cycle 96 system (Roche, Sweden) and programmed to undergo one cycle (95 °C for 2 min) and 40 cycles (95 °C for 15 s, 60 °C for 30 s, and 72 °C for 20 s). The mRNA expression levels of the targeted genes were normalized to that of β-actin. All experiments were performed in triplicate.

2.15. *Western Blot*

Cells and intestinal tissues were lysed in RIPA lysis buffer. Protein concentrations were determined using the Bradford or bicinchoninic acid (BCA) kit. Aliquots containing 20 μg of protein were resolved by 10% sodium dodecyl sulfate-polyacrylamide gel electrophoresis (SDS-PAGE) and transferred to PVDF membranes. The membranes were blocked with 5% BSA for 2 h at room temperature before being incubated overnight with primary antibodies diluted 1:1500 in 5% BSA. The membranes were then washed five times in 1× TBST, incubated with a secondary antibody conjugated to horseradish peroxidase at a dilution of 1:3000 with BSA at 37 °C for 1 h, and processed with an enhanced chemiluminescence (ECL) kit.

2.16. *Analysis of 16S rDNA Sequencing*

Intestinal contents were collected and stored at −80 °C. Fecal genomic DNA extraction and the evaluation of DNA purity and concentration were performed by Lianchuan Biotechnology Co., Ltd. (Hangzhou, China). PCR amplification of 16S rDNA, library construction, ion S5 sequencing and bioinformatics analysis were performed on the V3-V4 domain of 16S rDNA in the gut microbiota.

2.17. Data Analysis

All data were analyzed uisng GraphPad Prism (version 8. 0) to test for normal distributions. For normally distributed data, one-way analysis of variance (ANOVA) followed by Tukey's multiple comparison test was used to analyze significance. For non-normally distributed data, the Kruskal-Wallis test was used to analyze the body weight of mice, and the Mantle-Cox test was used to analyze the survival rates of mice, and the results were expressed as the mean ± standard deviation (SD); a *p*-value less than 0.05 was considered statistically significant.

3. Results

3.1. RA Alleviated Salmonella Enteritidis (SE) -Induced Inflammatory Responses In Vitro

The MIC of RA against *SE* was 1 mg/mL (Appendix A.2 (Table A2)). Moreover, pretreatment with 60 μg/mL of RA for 6 h had no significant effect on the viability of RAW264.7 cells and MCECs. Thus, 60 μg/mL RA was used for subsequent experiments (Figure 1A,B). To investigate the effects of RA on the phagocytic ability of RAW264.7 cells and the ability of *SE* to adhere to and invade MCECs, cells were challenged with FITC-labeled *SE*. Compared with the untreated control, RA significantly enhanced the RAW264.7 cell phagocytosis of *SE* in a dose-dependent manner (Figure 1C,D). Moreover, RA significantly reduced the invasion and adhesion of *SE* in MCECs (Figure 1E,F). *SE* infection causes the excessive secretion of proinflammatory cytokines. The RA pretreatment significantly decreased the levels of TNF-α, IL-1β and IL-6 (Figure 1G,H), indicating that RA inhibited the *SE*-induced inflammatory response. Overall, these results indicated that RA promoted macrophage phagocytosis, inhibited the adhesion and invasion of *SE* and alleviated the *SE*-induced inflammatory response in vitro.

3.2. RA Regulated TLR9/NF-κB Signaling Pathway

TLRs and the NF-κB signaling pathway play significant roles in bacterial invasion. In this study, we evaluated the effects of RA on the expression of TLR9 and the NF-κB signaling pathway. To assess the binding energy and interaction mode between RA and TLR9, we used Python 3.13.0, Yasara 24.10.5 and DS4.0 software to determine the protein–ligand docking and binding energy for each interaction. The results showed that RA could bind to the TLR9 protein, forming hydrogen bonds with the amino acid residues HIS394, ARG470 and ASP469 (Figure 2A). In addition, the free binding energy of RA for the TLR9 protein was −8.61 kcal/mol (Appendix A.3 (Table A3)), indicating that RA could spontaneously and stably bind to TLR9. The *SE* challenge inhibited the mRNA expression of TLR9 and increased the mRNA expression of TNF-α and IL-6, and RA significantly reversed these changes (Figure 2B–G). Furthermore, we found that RA pretreatment significantly increased the protein expression level of TLR9 and inhibited the protein expression levels of p-p65 and p-IκB-α compared to the *SE* infection control group (Figure 2H–O). Collectively, these results revealed that RA regulated TLR9 and inhibited the phosphorylation of p65 and IκB-α to limit the activation of the NF-κB signaling pathway against *SE* infection.

3.3. RA Inhibited NF-κB Signaling Pathway by Upregulating TLR9

To confirm the relationship between TLR9 and the NF-κB signaling pathway in RA-mediated anti-inflammatory effects, the cells were pretreated with the TLR9 inhibitor E664 dihydrochloride. In the presence of E664 dihydrochloride, *SE* infection significantly increased the levels of TNF-α, IL-1β and IL-6 and reduced the mRNA expression level of TLR9. After pretreatment with E664 dihydrochloride, RA pretreatment failed to reverse these effects (Figure 3A–H). The protein expression levels of p-p65 and p-IκB-α in the inhibitor-treated group were significantly greater than those in the blank group, indicating that the suppression of TLR9 led to the activation of the NF-κB signaling pathway and caused the loss of the regulatory effect of RA on the NF-κB signaling pathway (Figure 3I–P). Overall, these results suggested that RA alleviated the *SE*-induced inflammatory response by promoting TLR9 expression to inhibit the NF-κB signaling pathway.

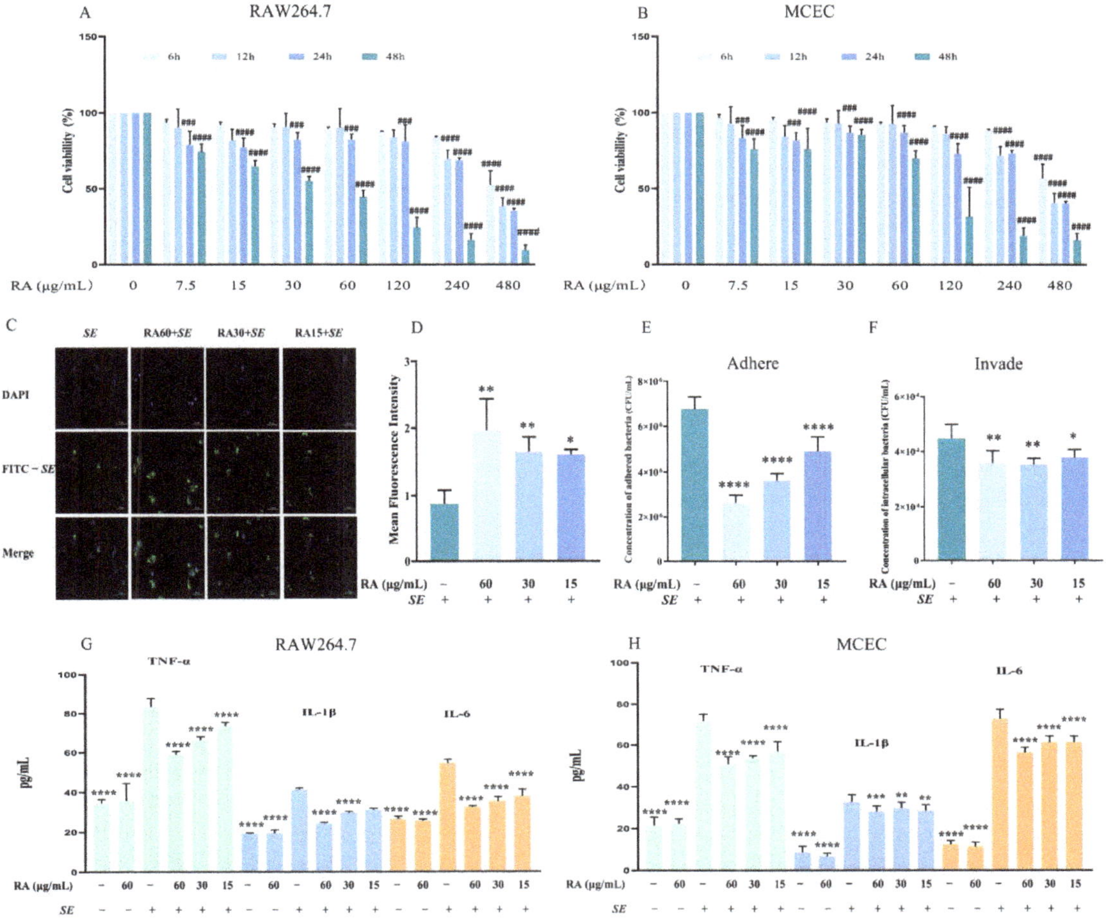

Figure 1. RA alleviated the cellular inflammation caused by *SE*. (**A**,**B**) Effects of RA on the viability of RAW264.7 cells and MCECs. The cells were treated with different concentrations of RA for 6 h, 12 h, 24 h, or 48 h, and a CCK-8 assay was used to detect cell viability. The cell survival rate data at the same time were obtained for statistical analysis ($n = 6$). (**C**,**D**) The fluorescence images and intensity of the phagocytic ability of RAW264.7 cells. The cells were pretreated with different concentrations of RA and challenged with FITC-labeled *SE* (MOI = 100) for 2 h ($n = 3$). (**E**) The bacteria adhering to MCECs. The cells were pretreated with different concentrations of RA, challenged with *SE* (MOI = 100) for 2 h and lysed with PBS containing 0.1% Triton X-100. The bacteria adhering to the cells were detected ($n = 6$). (**F**) The intracellular bacteria in MCECs. The cells were treated with RA and challenged with *SE* (MOI = 100) for 2 h, and then treated with DMEM containing 1% penicillin and streptomycin for 1 h. The cells were subsequently lysed with PBS containing 1% Triton X-100 and the intracellular bacteria were detected ($n = 6$). (**G**) Levels of TNF-α, IL-1β and IL-6 in RAW264.7 cells ($n = 6$). (**H**) Levels of TNF-α, IL-1β and IL-6 in MCECs ($n = 6$). Cells were pretreated with different concentrations of RA for 6 h, and challenged with *SE* (MOI = 100) for 2 h. Cytokine levels in the cell supernatant were detected via ELISA. Statistical analysis was performed by one-way ANOVA followed by Tukey's multiple comparison test, and expressed as the mean ± SD, * $p < 0.05$, ** $p < 0.01$, *** $p < 0.001$, **** $p < 0.0001$ versus the *SE* infection control group, ### $p < 0.001$, #### $p < 0.0001$ versus the blank control group.

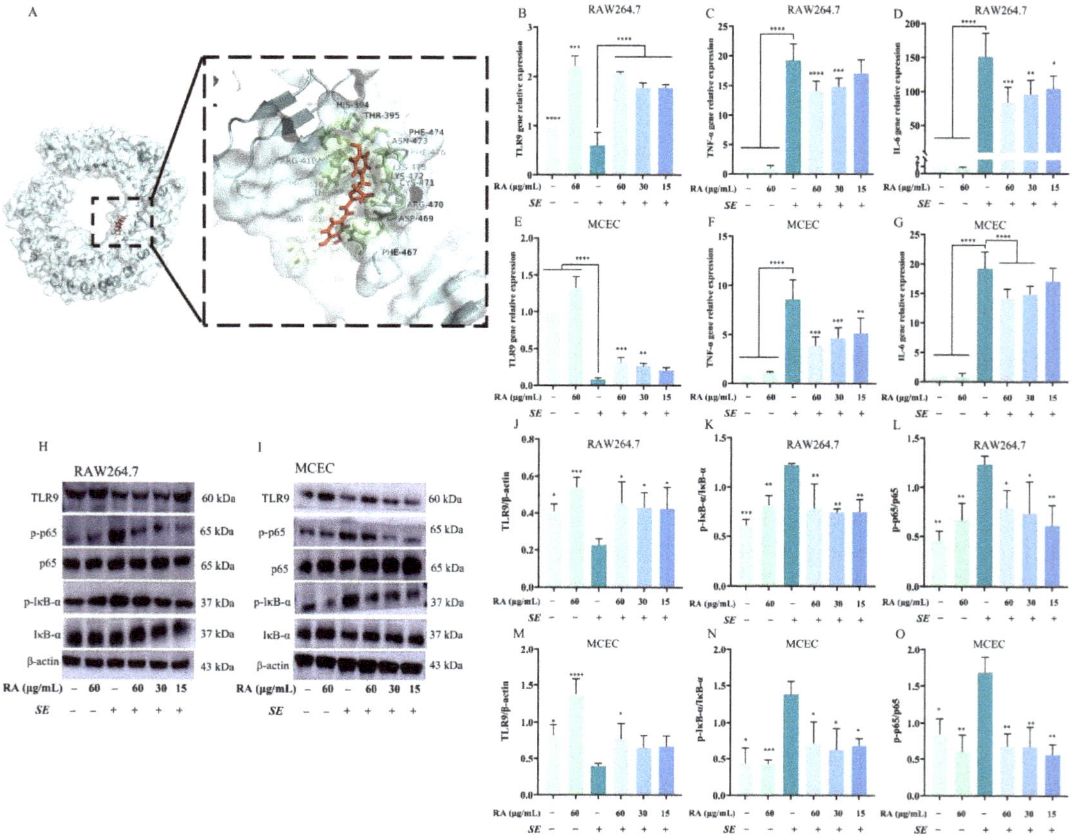

Figure 2. RA alleviated inflammation caused by *SE* via regulating the TLR9/NF-κB signaling pathway. Cells were pretreated with different concentrations of RA for 6 h and challenged with *SE* (MOI = 100) for 2 h; cells were collected for RT-qPCR or Western blot detection. (**A**) Molecular docking analysis of the binding mode and affinity of RA for TLR9. (**B–D**) The mRNA expression levels of TLR9, TNF-α and IL-6 in RAW264.7 cells (n = 6). (**E–G**) The mRNA expression levels of TLR9, TNF-α and IL-6 in MCECs (n = 6). (**H–O**) The protein expression levels of TLR9, p65, p-p65, IκB-α and p-IκB-α in RAW264.7 cells and MCECs (n = 3). Statistical analysis was performed by one-way ANOVA followed by Tukey's multiple comparison test, and expressed as the mean ± SD, * $p < 0.05$, ** $p < 0.01$, *** $p < 0.001$, **** $p < 0.0001$ versus the *SE* infection control group.

3.4. RA Provided Protection against SE in Mice

To investigate the protective effect and mechanism of RA on *SE* infection in vivo, the mice were orally treated with RA for 7 days. We found that RA reduced the mortality rate caused by *SE* (Figure 4A,B). Moreover, compared with the *SE* infection control group, RA significantly alleviated body weight loss, decreased the DAI, increased the length of the colon (Figure 4C–G) and decreased the organ indices of the liver and other organs (Appendix B.1, Figure A1). Consistently, we found that *SE* infection led to the enlargement of the intestinal cavity, damage to and detachment of the duodenal villi, reduction in colonic glands and thinning of the muscular layer (Figure 4H). Moreover, the RA pretreatment significantly improved the pathological damage to the duodenum and colon (Figure 4I,J). These results indicate that RA provided protection against *SE* in mice.

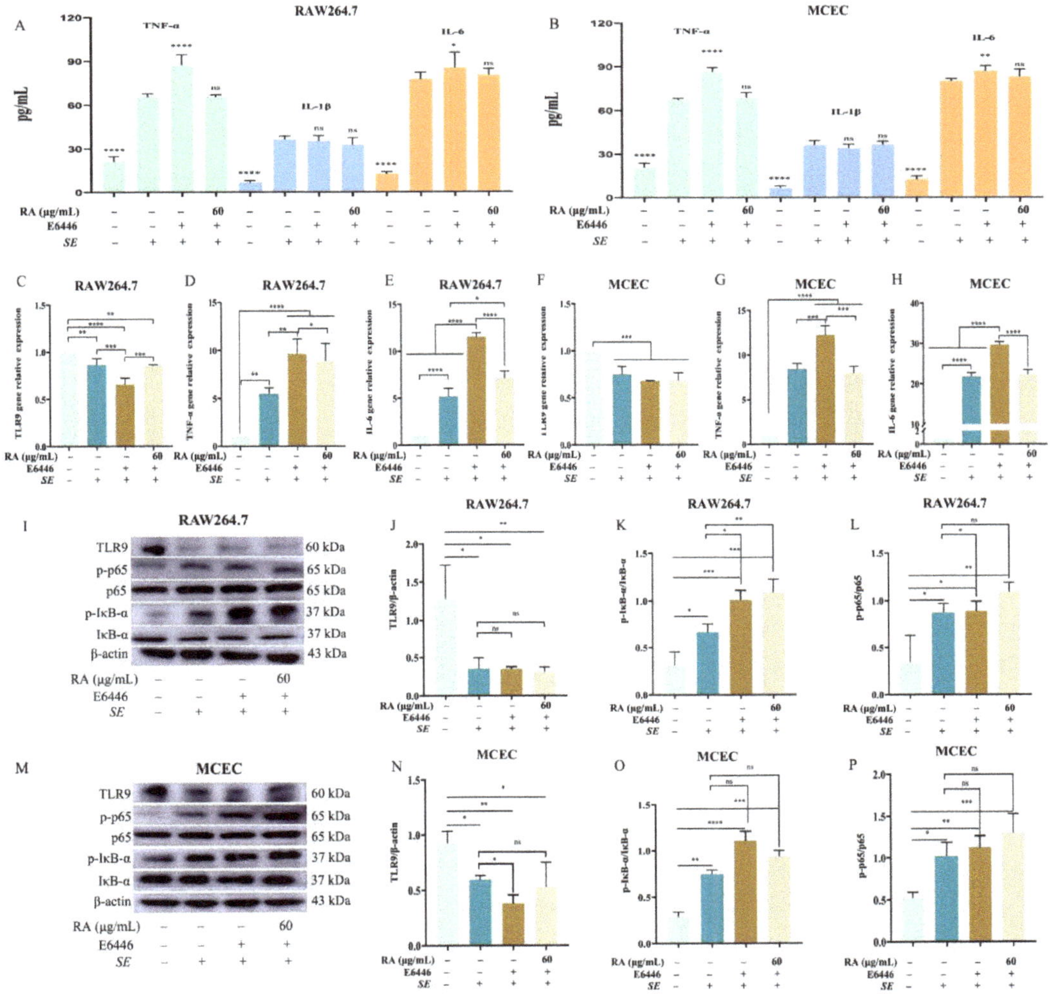

Figure 3. RA inhibited the activation of NF-κB signaling pathway by upregulating TLR9. The cells were pretreated with 10 μM E6446 (TLR9 inhibitor dihydrochloride) for 2 h, and co-incubated with RA for 6 h and then challenged with *SE* (MOI = 100) for 2 h. The cell supernatants were collected to detect cytokines, and the cells were collected for RT–qPCR or Western blot detection. (**A**,**B**) Levels of proinflammatory cytokines in RAW264.7 cells and MCECs ($n = 6$). (**C**–**E**) The mRNA expression levels of TLR9, TNF-α and IL-6 in RAW264.7 cells ($n = 6$). (**F**–**H**) The mRNA expression levels of TLR9, TNF-α and IL-6 in MCECs ($n = 6$). (**I**–**L**) The protein expression levels of TLR9, p65, p-p65, IκB-α and p-IκB-α in RAW264.7 cells ($n = 3$). (**M**–**P**) The protein expression levels of TLR9, p65, p-p65, IκB-α and p-IκB-α in MCECs ($n = 3$). For (**A**,**B**), statistical analysis was performed by one-way ANOVA followed by Tukey's multiple comparison test, and expressed as the mean ± SD, ^{ns} $p > 0.05$, * $p < 0.05$, ** $p < 0.01$, *** $p < 0.001$, **** $p < 0.0001$ versus the *SE* infection control group. For (**C**–**P**), statistical analysis was performed by one-way ANOVA followed by Tukey's multiple comparison test, and expressed as the mean ± SD, ^{ns} $p > 0.05$, * $p < 0.05$, ** $p < 0.01$, *** $p < 0.001$, **** $p < 0.0001$.

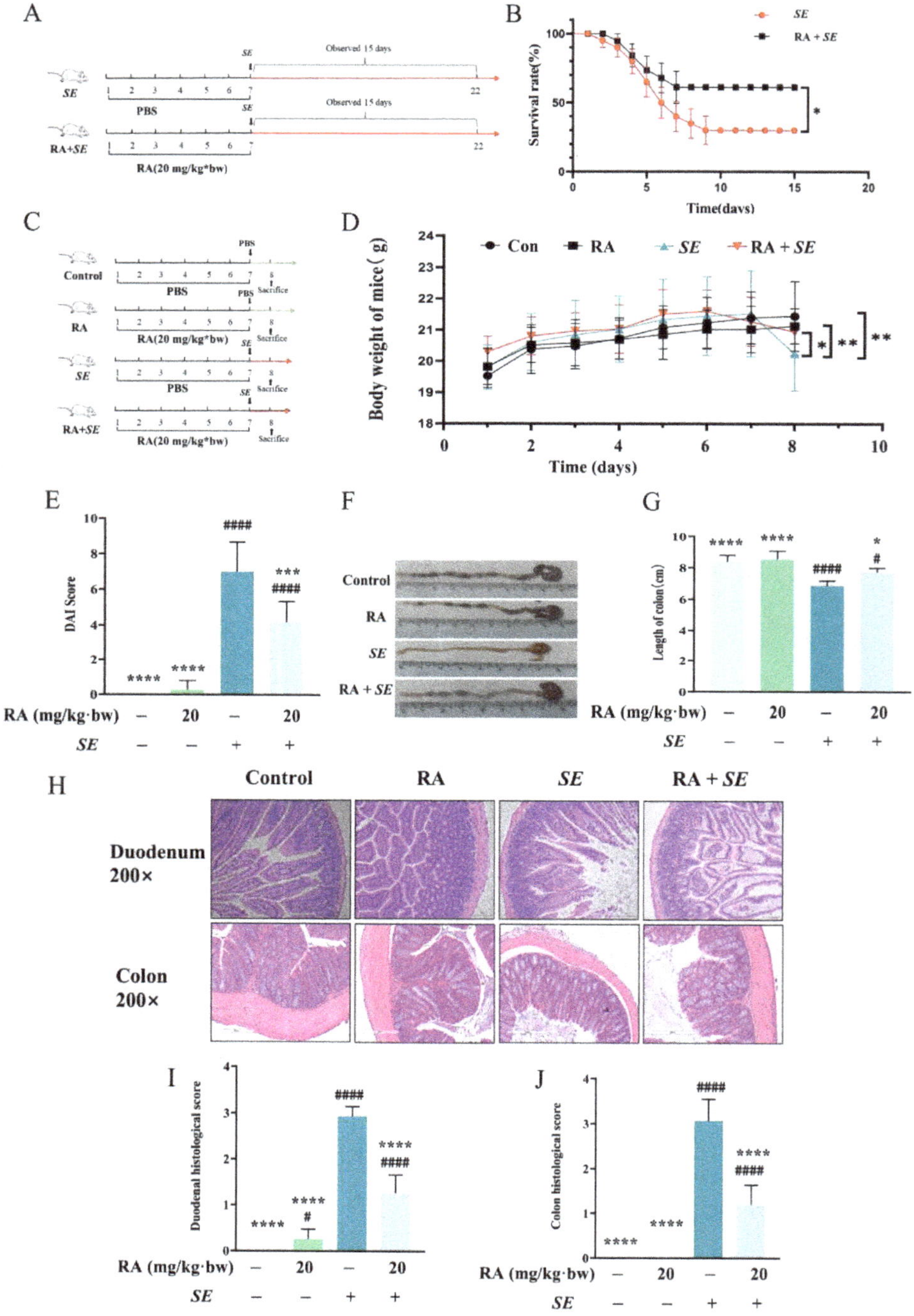

Figure 4. Protective effects of RA on *SE*-challenged mice. (**A**) Experimental design for evaluating the effect of RA on the survival rate of mice challenged with *SE*. Mice were orally pretreated with RA (20 mg/kg·bw) for 7 days and then intraperitoneally injected with 0.2 mL of *SE* at a concentration of 2.5×10^8 CFUs/mL. The mortality of mice was recorded. (**B**) Survival rate of the mice ($n = 20$).

(**C**) Experimental design for testing the protective effect of RA against *SE*. The treatment and challenge of the mice were the same as those in (**A**), and samples were collected 24 h after challenge. (**D**) Changes in the body weights of the mice (n = 10). (**E**) Disease activity index (n = 10). (**F**,**G**) Measurement of colon length (n = 6). (**H**) The histopathological changes in duodenum and colon (n = 3). (**I**) Duodenum histopathological score (n = 3). (**J**) Colon histopathological score (n = 3). For (**B–D**), statistical analysis was performed by Mantle-Cox test and Kruskal-Wallis test, respectively, and *SE* infection control group was compared with other groups. In (**G–J**), statistical analysis was performed by one-way ANOVA followed by Tukey's multiple comparison test, and expressed as the mean ± (SD), * $p < 0.05$, ** $p < 0.01$, *** $p < 0.001$, **** $p < 0.0001$ versus the *SE* infection control group, # $p < 0.05$, #### $p < 0.0001$ versus the blank control group.

3.5. RA Alleviated SE-Induced Intestinal Inflammation and Oxidative Damage

Excessive inflammation and oxidative stress cause damage to organs and tissues. In this study, we found that *SE* infection caused a significant increase in the levels of GOT, GPT and MDA in serum, and led to a significant decrease in the SOD level in serum compared with the blank control group, suggesting *SE* infection led to oxidative damage, but RA significantly decreased the production of GOT, GPT and MDA, and significantly increased the SOD production compared with that in the *SE* infection control group, indicating that RA pretreatment may alleviate oxidative damage caused by *SE* infection in mice (Figure 5A–D). Moreover, *SE* infection significantly increased the production of TNF-α, IL-1β and IL-6 and inhibited TLR9 expression compared with the blank control group; however, the RA pretreatment alleviated the overproduction of inflammatory cytokines and the suppression of TLR9 compared with the *SE* infection control group (Figure 5E–H). Moreover, in comparison with the blank control group, *SE* infection significantly increased the protein expression levels of p-p65 and p-IκB-α, while RA downregulated *SE*-caused p-p65 and p-IκB-α expression (Figure 5I), which was consistent with the findings of the cell experiments. Collectively, these results indicated that RA improved *SE*-induced inflammatory and oxidative damage in mice, which may be associated with upregulating TLR9 and inhibiting the NF-κB signaling pathway.

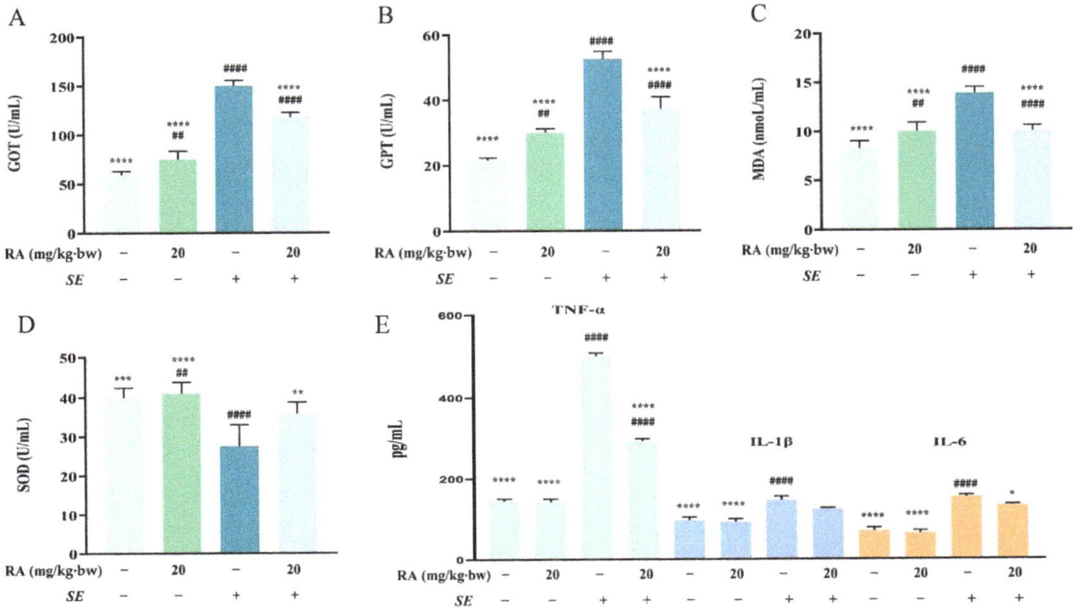

Figure 5. *Cont.*

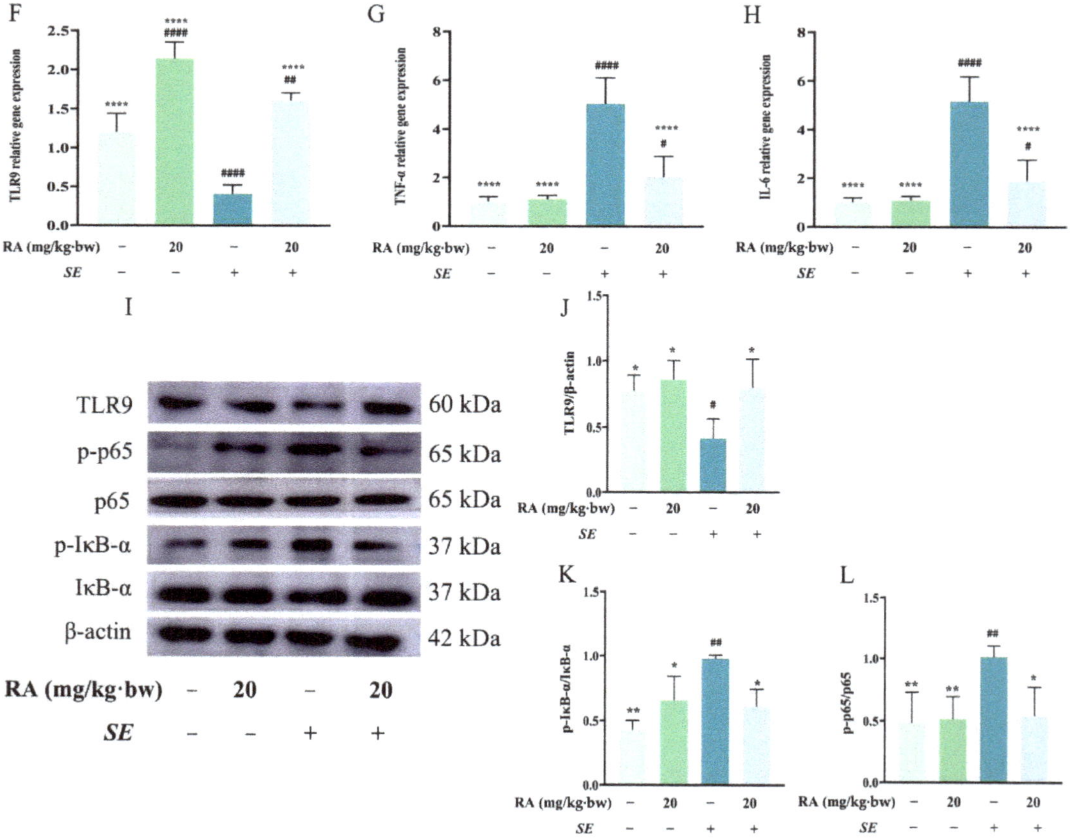

Figure 5. RA alleviated *SE*-induced intestinal inflammation by regulating TLR9 and inhibiting the activation of the NF-κB signaling pathway. The mice were orally pretreated with RA (20 mg/kg·bw) for 7 days, 0.2 mL of *SE* at a concentration of 2.5×10^8 CFUs/mL was intraperitoneally injected, and samples were collected 24 h after challenge. (**A–D**) Serum levels of GOT, GPT, MDA and SOD ($n = 6$). (**E**) Levels of TNF-α, IL-1β and IL-6 in the serum ($n = 6$) (**F–H**). The mRNA expression levels of TLR9, TNF-α and IL-6 in colon tissues ($n = 6$). (**I–L**) The protein expression levels of TLR9, p65, p-p65, IκB-α and p-IκB-α in colon tissues ($n = 3$). Statistical analysis was performed by one-way ANOVA followed by Tukey's multiple comparison test, and expressed as the mean ± SD, * $p < 0.05$, ** $p < 0.01$, *** $p < 0.001$, **** $p < 0.0001$ versus the *SE* infection control group, # $p < 0.05$, ## $p < 0.01$, #### $p < 0.0001$ versus the blank control group.

3.6. RA Improved SE-Caused Intestinal Microbiota Disorder

The gut microbiota plays a crucial role in maintaining intestinal homeostasis. We analyzed the composition and structure of the gut microbiota to investigate the effect of RA on the gut microbiota of mice. A total of 16,027 significant features were obtained for the 16S rRNA genes of the four bacterial groups (Figure A2A). The rarefaction curve of the OTUs of the mouse gut bacterial community tended to flatten, and the sequencing depth basically covered most of the microbial species in the sample (Figure A3B). A Venn diagram of the overlapping and shared OTU data revealed that 766 OTUs in the RA pretreatment group were shared with those in the control group, and 733 OTUs in the infection control group were shared with those in the control group (Figure A3C). A Circos diagram revealed that after the RA pretreatment, *Lachnospiraceae_NK4A136* dominated the intestinal microbiota in the mice at the genus level (Figure 6A). According to the α-diversity analysis, the species abundance of the gut microbiota in the *SE* infection control group was significantly lower than that in the control group, suggesting that *SE* could decrease the species diversity of the intestinal microbiota in mice. Compared with that in the *SE* infection control group, the α-diversity of the gut microbiota in the RA pretreatment group was significantly greater, indicating that RA had an effect on the abundance of bacterial species in the gut microbiota in the context of *SE*-induced intestinal inflammation (Figure 6B–D). A β-diversity analysis was performed via a principal coordinate analysis and non-metric multidimensional scaling analysis to explore the differences in the microbiota composition between the groups (Figure 6E and Appendix B.2, Figure A2). Clearly, a marked separation between the *SE* infection control group and the blank control group was observed, indicating that the microbiota composition was altered after modeling with *SE*. There was no overlap between the *SE* infection control group and the RA pretreatment group, indicating that RA altered the gut microbiota structure in mice infected with *SE*. These results suggested that the gut microbiota of the mice was uniquely altered after the RA supplementation. A taxonomic analysis at the phylum level revealed that *SE* infection caused an increase in the abundance of *Proteobacteria*, *Epsilonbacteraeota* and *Bacteroidota* as well as a decrease in the abundance of *Fiemicutes* and *Acidibacteria*; the RA pretreatment reversed this trend (Figure 6F–K). A taxonomic analysis at the genus level revealed that *SE* infection significantly increased the abundance of *Muribaculaceae*, *Bilophila* and *Tyzzerella* but decreased the abundance of *Lachnospiraceae_NK4A136*, *Acetatifactor* and *Intestinimonas*, and RA reversed the changes in the gut microbiota caused by *SE* (Figure 6L–R). The use of the linear discriminant analysis effect size (LEfSe) to display the microbial characteristics is the most likely to explain the differences between categories and to reveal the taxonomic groups with the greatest differences in abundance between different groups (Figure A3E). These results indicated that the RA pretreatment increased the abundance and diversity of the gut microbiota, upregulated the abundance of *Lachnospiraceae_NK4A136*, *Acetatifactor* and *Intestinimonas*, and reduced the abundance of *Muribaculaceae*, *Bilophila* and *Tyzzerella* at the genus level to improve the intestinal microbiota disorders caused by *SE*.

3.7. RA Failed to Provide Protection in Pseudo Germ-Free Mice

To verify the relationship between the RA-induced protection and the gut microbiota. We use of broad-spectrum antibiotics cleared the gut microbiota of the mice (Appendix B.3, Figure A3). In this study, we found that the elimination of the intestinal microbiota aggravated *SE* infection, and the mice without normal gut microbiota showed significant weight loss, increased DAI scores, intestinal enlargement and an increased organ coefficient of the liver and other organs (Figure 7A–F); the RA pretreatment did not reverse these changes. Moreover, the mice treated with broad-spectrum antibiotics presented a high level of TNF-α in serum after challenge with *SE*; RA did not alleviate the increase in TNF-α (Figure 7G). The pathological results of the duodenum showed that RA did not improve the *SE*-induced intestinal enlargement or intestinal villus damage in the duodenum (Figure 7H,I). These results suggested that the protective effect of RA against *SE* infection depends on the gut microbiota.

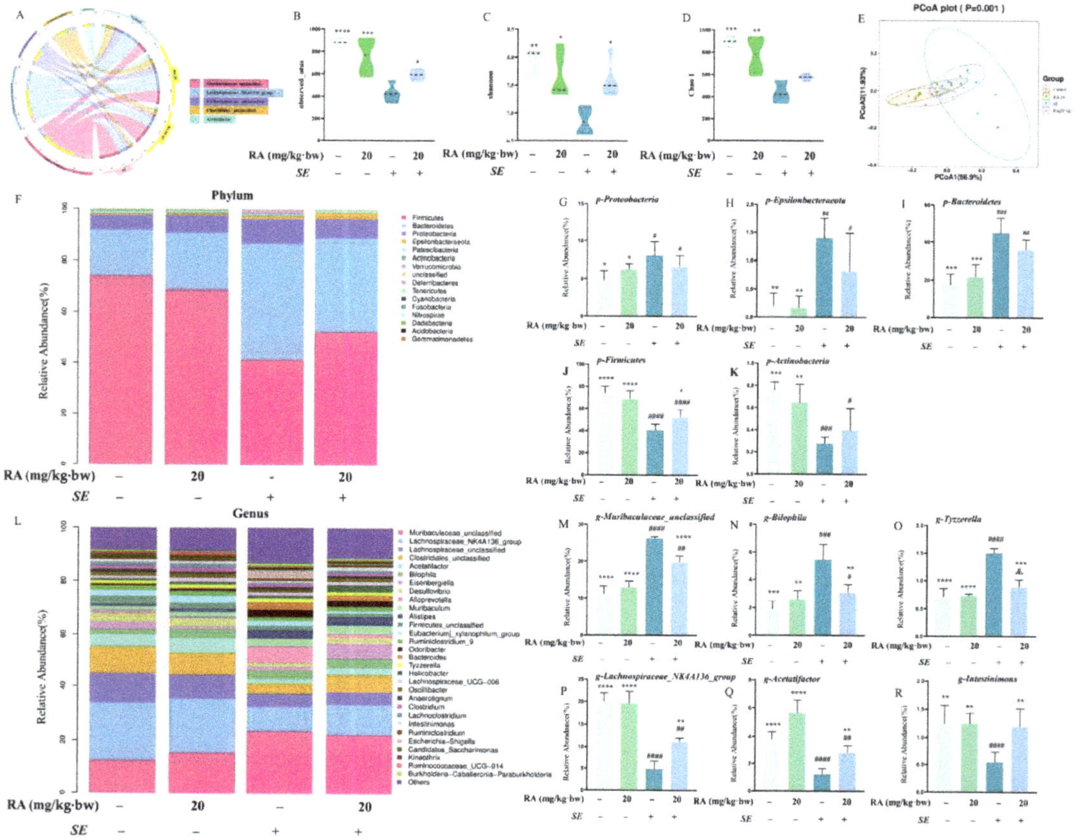

Figure 6. RA regulated the gut microbiota in *SE*-challenged mice (n = 6). The mice were orally pretreated with RA (20 mg/kg·bw) for 7 days, then intraperitoneally injected with 0.2 mL of *SE* at a concentration of 2.5×10^8 CFUs/mL. (**A**) Circos diagram. (**B**) Alpha diversity analysis: observed OTUs. (**C**) Alpha diversity analysis: Shannon index. (**D**) Alpha diversity analysis: Chao 1. (**E**) β-diversity analysis: PCoA. (**F**) Relative abundance of the gut microbiota at the genus level. (**G–K**) Relative abundance of *p__Proteobacteria*, *p__Epsilonbacteraeota*, *p__Bacteroidota*, *p__Firmicutes* and *p__Actinobacteria*. (**L**) Relative abundance of the gut microbiota at the genus level. (**M–R**) Relative abundance of g__*Muribaculaceae*_unclassified, g__*Bilophila*, g__*Tyzzerella*, g__*Lachnospiraceae_NK4A136*_group, g__*Acetatifactor* and g__*Intestinimonas*. Statistical analysis was performed by one-way ANOVA followed by Tukey's multiple comparison test, and expressed as the mean ± SD, * $p < 0.05$, ** $p < 0.01$, *** $p < 0.001$, **** $p < 0.0001$ versus the *SE* infection control group, # $p < 0.05$, ## $p < 0.01$, ### $p < 0.001$, #### $p < 0.0001$ versus the blank control group.

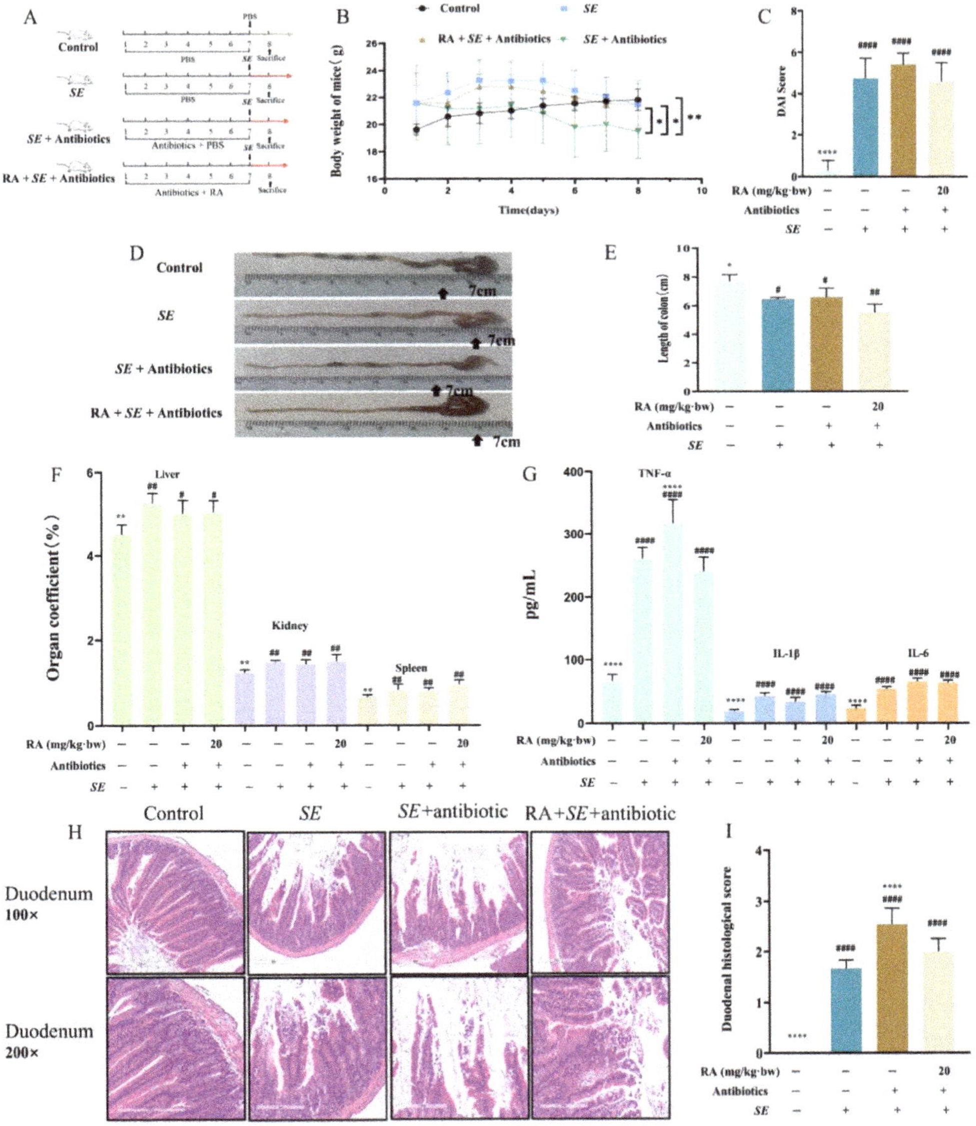

Figure 7. RA lost its protective effect against *SE* after clearance of the microbiota. (**A**) Experimental design in mice. The mice were orally pretreated with RA (20 mg/kg·bw) or antibiotics for 7 days, 0.2 mL of *SE* at a concentration of 2.5×10^8 CFUs/mL was intraperitoneally injected, and samples were collected 24 h after challenge (n = 10). (**B**) Changes in body weight (n = 10). (**C**) Disease activity index (n = 10). (**D**,**E**) Measurement of colon length (n = 10). (**F**) Organ indices of the liver, kidney and spleen (n = 10). (**G**) Levels of TNF-α, IL-1β and IL-6 in the serum (n = 6). (**H**) Histopathological changes in the duodenum (n = 3). (**I**) Duodenal histopathological score (n = 3). For (**B**), statistical analysis was performed using Kruskal-Wallis test, and the *SE* infection control group was compared with other groups. For (**C–I**), statistical analysis was performed by one-way ANOVA followed by Tukey's multiple comparison test, and expressed as the mean ± SD, * $p < 0.05$, ** $p < 0.01$, **** $p < 0.0001$ versus the *SE* infection control group, # $p < 0.05$, ## $p < 0.01$, #### $p < 0.0001$ versus the blank control group.

3.8. The Regulation of TLR9/NF-κB Signaling Pathway by RA Was Dependent on Gut Microbiota

To determine the relationship between the TLR9/NF-κB signaling pathway and gut microbiota regulated by RA, this study detected the expression of the TLR9/NF-κB signaling pathway in pseudo sterile mice. The *SE* challenge significantly decreased the mRNA expression of TLR9, but significantly increased the mRNA expression of TNF-α and IL-6 (Figure 8A–C). Moreover, the *SE* challenge significantly decreased the protein expression of TLR9 and significantly increased the protein expression of p-p65 and p-IκB-α, but RA did not reverse these effects (Figure 8D–G). These results indicated that the loss of the regulatory effect of RA on TLR9 led to the increased activation of the NF-κB signaling pathway and exacerbated intestinal inflammation in the pseudo sterile mice, which suggested that the gut microbiota is important for the protective effect of RA on *SE* infection.

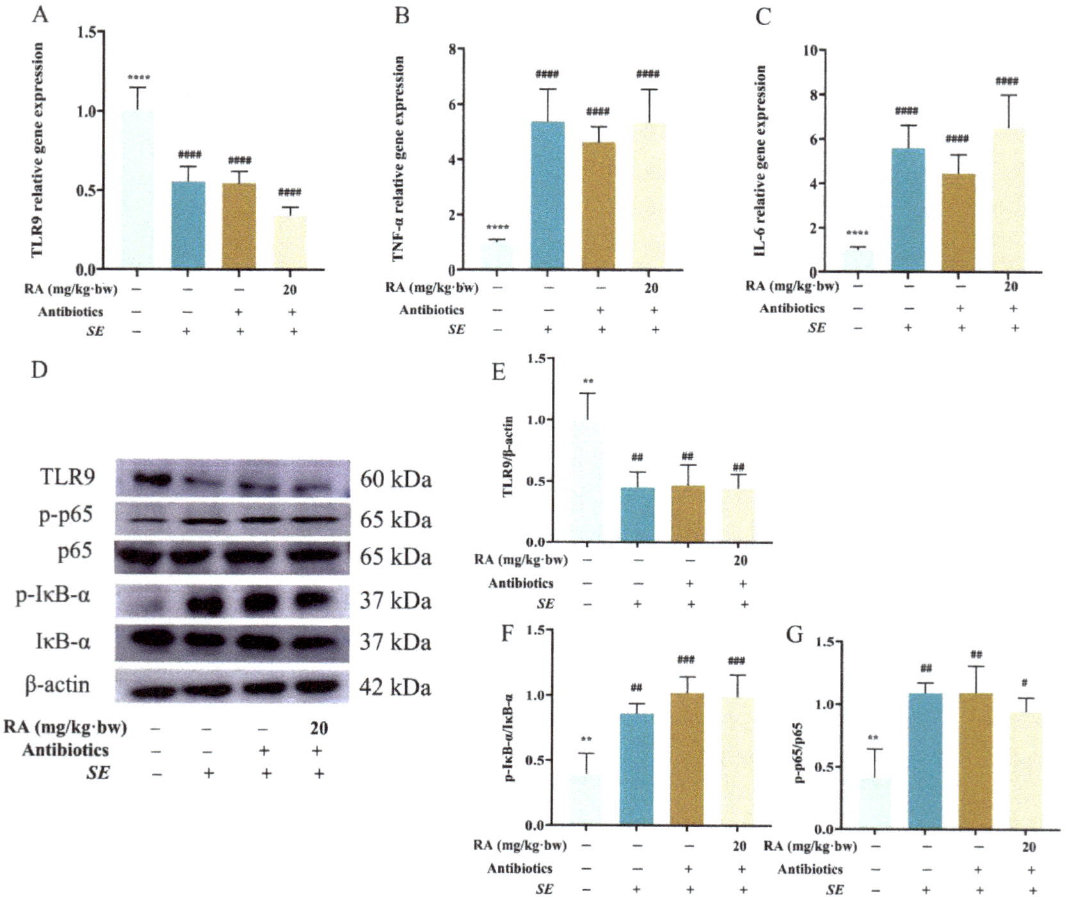

Figure 8. RA failed to inhibit NF-κB signaling pathway by regulating TLR9 after clearance of the microbiota. The mice were orally pretreated with RA (20 mg/ kg·bw) and antibiotics for 7 days and then intraperitoneally injected with 0.2 mL of *SE* at a concentration of 2.5×10^8 CFUs/mL. Samples were collected 24 h after challenge. (**A–C**) The mRNA expression levels of TLR9, TNF-α and IL-6 in colon tissues (n = 6). (**D–G**) The protein expression levels of TLR9, p65, p-p65, IκB-α and p-IκB-α in colon tissues (n = 3). Statistical analysis was performed by one-way ANOVA followed by Tukey's multiple comparison test, and expressed as the mean ± SD, ** $p < 0.01$, **** $p < 0.0001$ versus the *SE* infection control group, $^{\#}$ $p < 0.05$, $^{\#\#}$ $p < 0.01$, $^{\#\#\#}$ $p < 0.001$, $^{\#\#\#\#}$ $p < 0.0001$ versus the blank control group.

4. Discussion

Salmonella enteritidis (*SE*) is a zoonotic pathogen that poses a great threat to human and animal health, causing significant economic losses to the livestock industry [28]. In modern society, the excessive intake of refined grains and insufficient intake of dietary fiber damage the gut microbiota, leading to a decrease in resistance to intestinal pathogen infections [34,35]. Antibiotics are often used to control *SE* infection, but their excessive use leads to the production of drug-resistant *SE*. Additionally, the excessive use of antibiotics disturbs the gut microbiota and damages the intestinal barrier [36]. Moreover, the inflammation induced by *Salmonella typhimurium* accelerated the transfer of plasmids and bacteriophages, promoting the spread of antibiotic resistance [37]. Therefore, there is an urgent need to find new drug strategies that can reduce bacterial resistance and prevent *SE* infection.

The control of inflammation and balance of the gut microbiota balance is important for resisting *SE* infection. The effective active ingredients of natural plants exhibit anti-inflammatory and antibacterial effects as well as low toxicity, and maintain the balance of the gut microbiota. Importantly, they are not prone to drug resistance. Like Ganoderma lucidum polysaccharide [38] and Astragalus polysaccharide [39], they are candidate drugs for limiting inflammation and regulating the gut microbiota caused by *SE*. RA, a compound derived from plants in the *Lamiaceae* family, has anti-inflammatory and antioxidant effects. However, its protective effects and mechanisms against *SE* challenge in vitro and in vivo are not yet clear. In this study, we found that the MIC of RA against *SE* was 1 mg/mL, which was consistent with the findings of a previous study [32]. Intestinal epithelial cells are the main target cells for *SE* infection, and *SE* invades intestinal epithelial cells and causes systemic infection and inflammatory reactions [40]. During infection, macrophages strongly phagocytose *SE* to reduce its adhesion and invasion to intestinal epithelial cells. Furthermore, we found that RA enhanced the ability of RAW264.7 cells to phagocytose *SE*, reduced the ability of *SE* to adhere and invade MCECs and alleviated the *SE*-induced overproduction of TNF-α, IL-1β and IL-6. Our study revealed that RA offered protection against *SE* in vivo. We found that RA reduced the mortality rate of the *SE*-challenged mice, significantly alleviated the *SE*-induced decreases in body weight and DAI, increased the organ index and shortened the colon. Additionally, the RA pretreatment significantly improved the pathological changes in the duodenum and colon in the *SE*-infected mice. Previous studies have shown that *SE* infection significantly decreased the SOD and GSH activities in Caco-$_2$ cells and mice, which was consistent with our findings [9,41]. These alterations may be attributed to the active defense against the oxidative damage to intestinal epithelial cells after *Salmonella* infection [42]. Additionally, our data revealed that *SE* infection increased the TNF-α, IL-1β and IL-6 and MDA levels. MDA, a sensitive index assessing the degree of oxidative injury, was linked to chronic inflammation in the gut [43]. High levels of TNF-α, IL-1β and IL-6 and MDA in the *SE*-stimulated mice indicated that *SE* infection caused an imbalance in the redox status, resulting in oxidative stress and inflammatory responses. However, the mice pretreated with RA displayed a remarkable increase in SOD activity in serum, and a notable reduction in the levels of TNF-α, IL-1β and IL-6 and MDA in the serum, indicating that RA could enhance the antioxidant and anti-inflammatory ability to provide protection against *SE*.

TLRs are pattern recognition receptors that recognize a broad variety of structurally conserved molecules derived from microbes [44]. TLRs play a crucial role in preventing host invasion in pathogenic infections, and TLR2, TLR4 and TLR9 are involved in the defense against *Salmonella* in vivo [45]. TLR9 is vital for resistance to *Salmonella* infection, and its absence leads to severe *Salmonella* hepatitis, a high bacterial load, sustained inflammation and tissue damage [14,15,46]. In our study, we discovered that RA and TLR9 had multiple binding sites, and that RA bound well to the TLR9 protein, forming hydrogen bonds with the amino acid residues HIS394, ARG470 and ASP469. Consistently, we found that the RA pretreatment significantly upregulated the mRNA and protein expression of TLR9. After blocking TLR9 expression, RA failed to protect against *SE* infection, suggesting that the protection induced by RA was dependent on TLR9, which was consistent with the findings of a previous study, showing that TLR9 negatively regulates the NF-κB signaling pathway to alleviate IEC inflammation caused by *Salmonella typhimurium* [15]. The NF-κB signaling pathway is activated by various stimuli to regulate the inflammatory response [47]. The phosphorylation and nuclear translocation of p65 and IκB-α can activate the NF-κB signaling pathway, which plays a crucial role in regulating TNF-α, IL-1β and IL-6 [48]. Our study revealed that RA significantly inhibited the NF-κB signaling pathway to decrease the expression of TNF-α, IL-1β and IL-6, exerting a protective effect on *Salmonella*-induced intestinal and cellular inflammation. However, TLR9 suppression led to the loss of the ability of RA to inhibit the secretion of proinflammatory cytokines and the phosphorylation of p65 and IκB-α in vitro. Overall, our study suggested that RA inhibited the activation of the NF-κB signaling pathway by increasing TLR9 expression to alleviate *SE*-induced intestinal inflammation.

The gut is colonized by a large and diverse microbiota, and the microbiota composition is vital to the intestinal barrier [49]. High levels of fiber, unsaturated fatty acids and polyphenols in food are beneficial for physical health [50]. As a polyphenolic acid compound with medicinal and edible homology, RA is an effective active ingredient in plants rich in dietary fiber [51], and it has a protective effect on the gut microbiota [52]. The invasion of exogenous microorganisms leads to a decrease in the microbial diversity of the intestinal flora. Our study revealed that *SE* infection resulted in a decrease in the abundance and diversity of the gut microbiota in mice, whereas RA effectively increased the abundance and diversity of the gut microbiota. In the present study, the RA pretreatment increased the abundance of *Firmicutes* and *Acidibacteria* and decreased the abundance of *Proteobacteria*, *Epsilonbacteraeota* and *Bacteroidota* at the phylum level. *Lachnospiraceae_NK4A136* is a representative bacterium that produces butyric acid and contributes to maintaining intestinal barrier function [53]. After the RA pretreatment, *Lachnospiraceae_NK4A136* dominated the mouse gut microbiota at the genus level, suggesting that RA alleviated *Salmonella*-induced intestinal damage and maintained intestinal health through *Lachnospiraceae_NK4A136*. The long-term use of broad-spectrum antibiotics can greatly interfere with or eliminate gut microbiota, and thus weaken the resistance of the gut microbiota to colonization by exogenous pathogens [54]. After the clearance of the gut microbiota, the susceptibility of the intestine to *SE* increased [55]. Therefore, the gut microbiota plays an extremely important role in resisting *SE* infection. When the intestinal microbiota was cleared with broad-spectrum antibiotics, RA failed to alleviate inflammation and duodenum damage. Furthermore, RA also failed to regulate the expression of p-p65, p-IκB-α, proinflammatory cytokines and TLR9 in the *SE*-challenged mice, revealing that the gut microbiota played an important role in the protective effect of RA. The gut microbiota is a prerequisite for resisting exogenous pathogen infection; however, further research is needed on how RA upregulates TLR9 to inhibit the activation of the NF-κB signaling pathway through the gut microbiota to resist *SE* infection, and whether it directly affects the metabolites of the gut microbiota.

5. Conclusions

In summary, our study demonstrated that RA alleviated *SE*-induced inflammation by upregulating TLR9 and inhibiting the NF-κB signaling pathway, which is closely related to maintaining gut microbiota homeostasis in mice. Supplementing with RA may be a preventive strategy to alleviate the damage caused by *SE* in humans and animals. Our study provided theoretical support for the application and development of RA in the prevention and treatment of *SE* infection.

Author Contributions: Original draft, review editing, methodology, visualization and data curation, D.Y.; methodology and software, L.Q.; formal analysis and investigation, X.L.; conceptualization and validation, Y.L.; data curation and validation, L.Z.; formal analysis and visualization, M.W.; methodology and supervision, Y.P.; formal analysis, methodology and review editing, Z.L.; methodology, resources and review editing, J.H. All data were generated in-house, and no paper mill was used. All authors have read and agreed to the published version of the manuscript.

Funding: This research was funded by the National Natural Science Foundation grant number 31960717 and the Innovation Project of Guangxi Graduate Education YCBZ number 2021007.

Institutional Review Board Statement: This study was conducted according to the recommendations of the Academy of Animal Research Guidelines and approved by the Animal Ethics Committee of Guangxi University (protocol number: GXU-2022-336).

Informed Consent Statement: Not applicable.

Data Availability Statement: The datasets used and analyzed during the current study are available from the corresponding author upon reasonable request.

Conflicts of Interest: The authors declare that they have no conflicts of interest.

Abbreviations

TLR9: toll-like receptor-9; NF-κB, nuclear factor kappa-light-chain-enhancer of activated B cells; *SE* (used in the chart), *Salmonella enterica;* GOT, glutamic oxaloacetic transaminase; GPT, glutamic-pyruvic transaminase; IL-1β, interleukin-1β; IL-6, interleukin-6; LB, Luria-Bertani culture medium; MDA, malondialdehyde; p-p65, phosphorylated p65; p-IκB-α, phosphorylated IκB-α; RA, rosmarinic acid; SOD, superoxide dismutase; TNF-α, tumour necrosis factor α; PAMPs, pathogen-associated molecular patterns; E6446 (used in the chart), E6446 dihydrochloride; MIC, minimum inhibitory concentration; BSA, bovine serum albumin; 20 mg/kg·bw, 20 mg/ kg·body weight.

Appendix A

Appendix A.1

Table A1. Primers for RT-qPCR.

Target Gene	Forward (5′–3′)	Reverse (5′–3′)	Genbank Number
TNF-α	CGCTGAGGTCAATCTGC	GGCTGGGTAGAGAATGGA	XM_042684006.1
IL-6	ACAGAAGGAGTGGCTAAGGA	AGGCATAACGCACTAGGTTT	XM_032905335.1
TLR9	CAGGAGCGGTGAAGGTG	GAAGGGAGGTCAGATTGGC	NM_031178.2
β-actin	CCTCACTGTCCACCTTCC	GGGTGTAAAACGCAGCTC	XM_058569360.1

Appendix A.2

The minimum inhibitory concentration of RA against *SE*. The minimum inhibitory concentration of RA against *SE* is 1 mg/mL.

Table A2. The minimum inhibitory concentration of RA against *SE*.

Drug Concentration (mg/mL)	4.00	2.00	1.00	0.500	0.250	0.125	0.0625	0.0313	0.0157	0.00785	Positive Control	Negative Control
1	−	−	−	+	+	+	+	+	+	+	+	−
2	−	−	−	+	+	+	+	+	+	+	+	−
3	−	−	−	+	+	+	+	+	+	+	+	−

+ represents positive, indicating the presence of bacterial precipitation, − represents negative, and no bacterial precipitation appears.

Appendix A.3

The free binding energy of RA and the TLR9 protein is −8.6100 kcal/mL.

Table A3. The free binding energy of RA and TLR9 protein.

Run	Bind. Energy [kcal/mol]	Dissoc. Constant [pM]	Contacting Receptor Residues
001	000008.6100	00000000488412.0000	HIS A 394 THR A 395 ARG A 418 PHE A 419 PHE A 467 ASP A 469 ARG A 470 CYS A 471 LYS A 472 ASN A 473 LYS A 475 THR A 477 ASP A 479 SER A 481 ARG A 482 SER A 503 SER A 505 HIS A 506 ASP A 528
002	000008.6100	00000000488412.0000	HIS A 394 THR A 395 ARG A 418 PHE A 419 PHE A 467 ASP A 469 ARG A 470 CYS A 471 LYS A 472 ASN A 473 LYS A 475 THR A 477 ASP A 479 SER A 481 ARG A 482 SER A 503 SER A 505 HIS A 506 ASP A 528
003	000008.4100	00000000684523.3125	ARG A 418 PHE A 419 PHE A 467 ASP A 469 CYS A 471 LYS A 472 ASN A 473 LYS A 475 THR A 477 ASP A 479 SER A 481 SER A 503 SER A 505 ASP A 528 SER A 530
004	000008.4100	00000000684523.3125	ARG A 418 PHE A 419 PHE A 467 ASP A 469 CYS A 471 LYS A 472 ASN A 473 LYS A 475 THR A 477 ASP A 479 SER A 481 SER A 503 SER A 505 ASP A 528 SER A 530
005	000008.3810	00000000718862.0000	HIS A 394 THR A 395 HIS A 397 ARG A 418 PHE A 419 PHE A 467 ASP A 469 ARG A 470 CYS A 471 LYS A 472 ASN A 473 LYS A 475 THR A 477 ASP A 479 SER A 503
006	000008.3810	00000000718862.0000	HIS A 394 THR A 395 HIS A 397 ARG A 418 PHE A 419 PHE A 467 ASP A 469 ARG A 470 CYS A 471 LYS A 472 ASN A 473 LYS A 475 THR A 477 ASP A 479 SER A 503
007	000008.3720	00000000729865.1875	PRO A 30 ALA A 31 PHE A 32 LEU A 33 CYS A 35 GLU A 36 LEU A 37 LYS A 38 PRO A 39 GLN A 779 THR A 780 VAL A 782 PRO A 783 GLY A 784 LEU A 785 ALA A 786
008	000008.3720	00000000729865.1875	PRO A 30 ALA A 31 PHE A 32 LEU A 33 CYS A 35 GLU A 36 LEU A 37 LYS A 38 PRO A 39 GLN A 779 THR A 780 VAL A 782 PRO A 783 GLY A 784 LEU A 785 ALA A 786

Appendix B

Appendix B.1

The organ coefficient is an important indicator for evaluating the degree of bacterial infection. *SE* infection leads to a significant decrease in the weight and food and water intake of mice, as well as pathological phenomena such as organ edema and congestion. Therefore, the increase in the organ coefficient was caused by *SE* infection. After the pretreatment with RA, the increase in the organ coefficient in the liver, kidney and spleen caused by *SE* was significantly reduced.

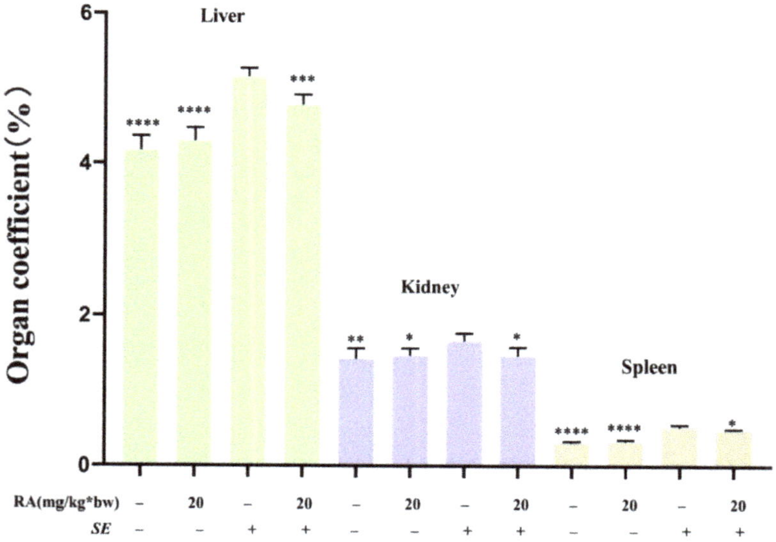

Figure A1. The protective effect of RA on *SE*-infected mice. Organ indices of the liver, kidney and spleen ($n = 6$). Statistical analysis was performed by one-way ANOVA followed by Tukey's multiple comparison test, and expressed as the mean ± SD, * $p < 0.05$, ** $p < 0.01$, *** $p < 0.001$, **** $p < 0.0001$ versus the *SE* infection control group.

Appendix B.2

RA regulated the gut microbiota in the *SE*-infected mice.

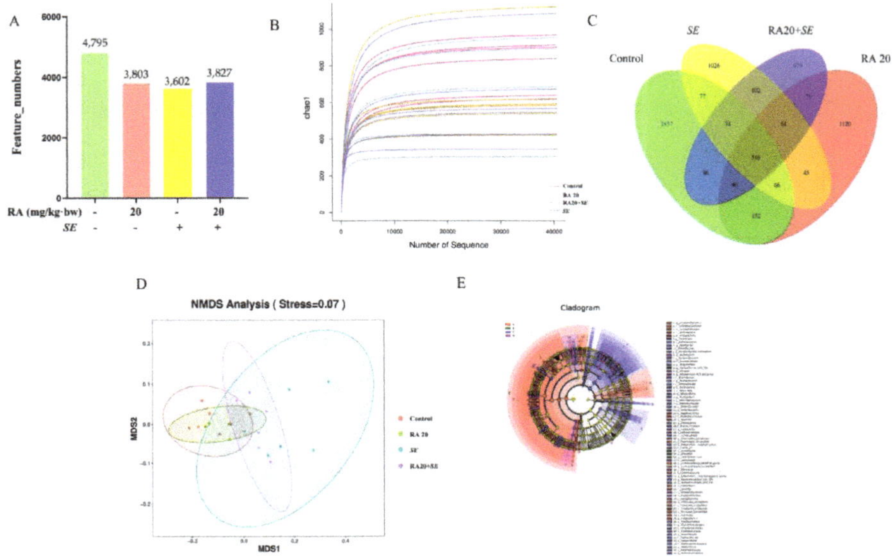

Figure A2. RA regulated gut microbiota in *SE*-infected mice. (**A**): Feature numbers; (**B**): rare curve of the OTU level; (**C**): Venn diagram; (**D**): non-metric multi-dimensional scaling analysis; (**E**): linear discriminant analysis effect size.

Appendix B.3

The broad-spectrum antibiotic pretreatment significantly reduced the number of intestinal bacteria in the mice.

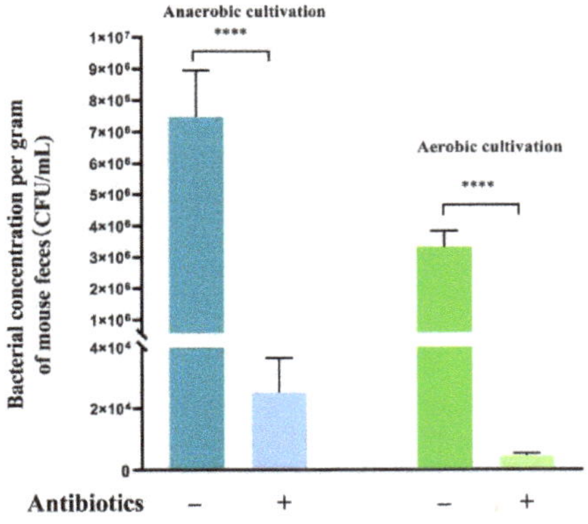

Figure A3. Broad-spectrum antibiotic pretreatment significantly reduced the number of intestinal bacteria in mice. The broad-spectrum antibiotics used included neomycin (100 mg/L), streptomycin (50 mg/L), penicillin (100 mg/L), vancomycin (50 mg/L), and metronidazole (100 mg/L). The mice were orally administered with 0.5 mL of broad-spectrum antibiotics by gavage for 7 d ($n = 6$). After 7 d, 10 g of mouse feces was collected under sterile conditions. The feces were diluted with 1 g/mL sterile water and cultured anaerobically and aerobically on LB medium via the dilution coating plate method. The bacterial colonies on the plate were counted after 12 h at 37 °C. Student's *t*-test was used for statistical analysis. **** for $p < 0.0001$ versus the blank control group.

References

1. LaRock, D.L.; Chaudhary, A.; Miller, S.I. Salmonellae interactions with host processes. *Nat. Rev. Microbiol.* **2015**, *13*, 191–205. [CrossRef] [PubMed]
2. Karp, B.E.; Leeper, M.M.; Chen, J.C.; Tagg, K.A.; Francois Watkins, L.K.; Friedman, C.R. Multidrug-Resistant Salmonella Serotype Anatum in Travelers and Seafood from Asia, United States. *Emerg. Infect. Dis.* **2020**, *26*, 1030–1033. [CrossRef] [PubMed]
3. Oueslati, W.; Rjeibi, M.R.; Mhadhbi, M.; Jbeli, M.; Zrelli, S.; Ettriqui, A. Prevalence, virulence and antibiotic susceptibility of Salmonella spp. strains, isolated from beef in Greater Tunis (Tunisia). *Meat Sci.* **2016**, *119*, 154–159. [CrossRef] [PubMed]
4. Lin, D.; Yan, M.; Lin, S.; Chen, S. Increasing prevalence of hydrogen sulfide negative Salmonella in retail meats. *Food Microbiol.* **2014**, *43*, 1–4. [CrossRef] [PubMed]
5. Elbediwi, M.; Pan, H.; Biswas, S.; Li, Y.; Yue, M. Emerging colistin resistance in Salmonella enterica serovar Newport isolates from human infections. *Emerg. Microbes Infect.* **2020**, *9*, 535–538. [CrossRef]
6. Shi, S.; Wu, S.; Shen, Y.; Zhang, S.; Xiao, Y.; He, X.; Gong, J.; Farnell, Y.; Tang, Y.; Huang, Y.; et al. Iron oxide nanozyme suppresses intracellular Salmonella Enteritidis growth and alleviates infection in vivo. *Theranostics* **2018**, *8*, 6149–6162. [CrossRef]
7. Zhang, K.; Griffiths, G.; Repnik, U.; Hornef, M. Seeing is understanding: Salmonella's way to penetrate the intestinal epithelium. *Int. J. Med. Microbiol.* **2018**, *308*, 97–106. [CrossRef]
8. Jacobson, A.; Lam, L.; Rajendram, M.; Tamburini, F.; Honeycutt, J.; Pham, T.; Van Treuren, W.; Pruss, K.; Stabler, S.R.; Lugo, K.; et al. A Gut Commensal-Produced Metabolite Mediates Colonization Resistance to Salmonella Infection. *Cell Host Microbe* **2018**, *24*, 296–307.e297. [CrossRef]
9. Guo, F.; Liu, H.; Li, X.; Hu, Z.; Huang, J.; Bi, R.; Abbas, W.; Guo, Y.; Wang, Z. Sophy β-Glucan from the Black Yeast Aureobasidium pullulans Attenuates Salmonella-Induced Intestinal Epithelial Barrier Injury in Caco-2 Cell Monolayers via Exerting Anti-Oxidant and Anti-Inflammatory Properties. *Antioxidants* **2023**, *13*, 48. [CrossRef]
10. Masud, S.; Prajsnar, T.K.; Torraca, V.; Lamers, G.E.M.; Benning, M.; Van Der Vaart, M.; Meijer, A.H. Macrophages target Salmonella by Lc3-associated phagocytosis in a systemic infection model. *Autophagy* **2019**, *15*, 796–812. [CrossRef]
11. Kawai, T.; Akira, S. Toll-like receptors and their crosstalk with other innate receptors in infection and immunity. *Immunity* **2011**, *34*, 637–650. [CrossRef] [PubMed]
12. Khanmohammadi, S.; Rezaei, N. Role of Toll-like receptors in the pathogenesis of COVID-19. *J. Med. Virol.* **2021**, *93*, 2735–2739. [CrossRef] [PubMed]
13. Vaknin, I.; Blinder, L.; Wang, L.; Gazit, R.; Shapira, E.; Genina, O.; Pines, M.; Pikarsky, E.; Baniyash, M. A common pathway mediated through Toll-like receptors leads to T- and natural killer-cell immunosuppression. *Blood* **2008**, *111*, 1437–1447. [CrossRef] [PubMed]
14. Zhan, R.; Han, Q.; Zhang, C.; Tian, Z.; Zhang, J. Toll-Like receptor 2 (TLR2) and TLR9 play opposing roles in host innate immunity against Salmonella enterica serovar Typhimurium infection. *Infect. Immun.* **2015**, *83*, 1641–1649. [CrossRef] [PubMed]
15. Li, Y.; Liu, M.; Zuo, Z.; Liu, J.; Yu, X.; Guan, Y.; Zhan, R.; Han, Q.; Zhang, J.; Zhou, R.; et al. TLR9 Regulates the NF-κB-NLRP3-IL-1β Pathway Negatively in Salmonella-Induced NKG2D-Mediated Intestinal Inflammation. *J. Immunol.* **2017**, *199*, 761–773. [CrossRef] [PubMed]
16. Zhang, Y.; Chen, R.; Zhang, D.; Qi, S.; Liu, Y. Metabolite interactions between host and microbiota during health and disease: Which feeds the other? *Biomed. Pharmacother.* **2023**, *160*, 114295. [CrossRef]
17. Brown, E.M.; Clardy, J.; Xavier, R.J. Gut microbiome lipid metabolism and its impact on host physiology. *Cell Host Microbe* **2023**, *31*, 173–186. [CrossRef]
18. Malesza, I.J.; Malesza, M.; Walkowiak, J.; Mussin, N.; Walkowiak, D.; Aringazina, R.; Bartkowiak-Wieczorek, J.; Mądry, E. High-Fat, Western-Style Diet, Systemic Inflammation, and Gut Microbiota: A Narrative Review. *Cells* **2021**, *10*, 3164. [CrossRef]
19. Chen, C.; Liao, J.; Xia, Y.; Liu, X.; Jones, R.; Haran, J.; McCormick, B.; Sampson, T.R.; Alam, A.; Ye, K. Gut microbiota regulate Alzheimer's disease pathologies and cognitive disorders via PUFA-associated neuroinflammation. *Gut* **2022**, *71*, 2233–2252. [CrossRef]
20. Yang, G.; Wei, J.; Liu, P.; Zhang, Q.; Tian, Y.; Hou, G.; Meng, L.; Xin, Y.; Jiang, X. Role of the gut microbiota in type 2 diabetes and related diseases. *Metab. Clin. Exp.* **2021**, *117*, 154712. [CrossRef]
21. Ducarmon, Q.R.; Zwittink, R.D.; Hornung, B.V.H.; van Schaik, W.; Young, V.B.; Kuijper, E.J. Gut Microbiota and Colonization Resistance against Bacterial Enteric Infection. *Microbiol. Mol. Biol. Rev.* **2019**, *83*, e00007-19. [CrossRef] [PubMed]
22. Lubin, J.B.; Green, J.; Maddux, S.; Denu, L.; Duranova, T.; Lanza, M.; Wynosky-Dolfi, M.; Flores, J.N.; Grimes, L.P.; Brodsky, I.E.; et al. Arresting microbiome development limits immune system maturation and resistance to infection in mice. *Cell Host Microbe* **2023**, *31*, 554–570.e557. [CrossRef] [PubMed]
23. Liu, J.; Liu, H.; Liu, H.; Teng, Y.; Qin, N.; Ren, X.; Xia, X. Live and pasteurized Akkermansia muciniphila decrease susceptibility to Salmonella Typhimurium infection in mice. *J. Adv. Res.* **2023**, *52*, 89–102. [CrossRef] [PubMed]
24. Tsou, L.K.; Lara-Tejero, M.; RoseFigura, J.; Zhang, Z.J.; Wang, Y.C.; Yount, J.S.; Lefebre, M.; Dossa, P.D.; Kato, J.; Guan, F.; et al. Antibacterial Flavonoids from Medicinal Plants Covalently Inactivate Type III Protein Secretion Substrates. *J. Am. Chem. Soc.* **2016**, *138*, 2209–2218. [CrossRef] [PubMed]
25. Jin, B.R.; Chung, K.S.; Hwang, S.; Hwang, S.N.; Rhee, K.J.; Lee, M.; An, H.J. Rosmarinic acid represses colitis-associated colon cancer: A pivotal involvement of the TLR4-mediated NF-κB-STAT3 axis. *Neoplasia* **2021**, *23*, 561–573. [CrossRef]

26. Chung, C.H.; Jung, W.; Keum, H.; Kim, T.W.; Jon, S. Nanoparticles Derived from the Natural Antioxidant Rosmarinic Acid Ameliorate Acute Inflammatory Bowel Disease. *ACS Nano* **2020**, *14*, 6887–6896. [CrossRef]
27. Elian, C.; Andaloussi, S.A.; Moilleron, R.; Decousser, J.W.; Boyer, C.; Versace, D.L. Biobased polymer resources and essential oils: A green combination for antibacterial applications. *J. Mater. Chem. B* **2022**, *10*, 9081–9124. [CrossRef]
28. Righi, N.; Boumerfeg, S.; Fernandes, P.A.R.; Deghima, A.; Baali, F.; Coelho, E.; Cardoso, S.M.; Coimbra, M.A.; Baghiani, A. Thymus algeriensis Bioss & Reut: Relationship of phenolic compounds composition with in vitro/in vivo antioxidant and antibacterial activity. *Food Res. Int.* **2020**, *136*, 109500. [CrossRef]
29. Huerta-Madroñal, M.; Caro-León, J.; Espinosa-Cano, E.; Aguilar, M.R.; Vázquez-Lasa, B. Chitosan–Rosmarinic acid conjugates with antioxidant, anti-inflammatory and photoprotective properties. *Carbohydr. Polym.* **2021**, *273*, 118619. [CrossRef]
30. Dahchour, A. Anxiolytic and antidepressive potentials of rosmarinic acid: A review with a focus on antioxidant and anti-inflammatory effects. *Pharmacol. Res.* **2022**, *184*, 106421. [CrossRef]
31. Noor, S.; Mohammad, T.; Rub, M.A.; Raza, A.; Azum, N.; Yadav, D.K.; Hassan, M.I.; Asiri, A.M. Biomedical features and therapeutic potential of rosmarinic acid. *Arch. Pharmacal Res.* **2022**, *45*, 205–228. [CrossRef] [PubMed]
32. Zhang, J.; Cui, X.; Zhang, M.; Bai, B.; Yang, Y.; Fan, S. The antibacterial mechanism of perilla rosmarinic acid. *Biotechnol. Appl. Biochem.* **2022**, *69*, 1757–1764. [CrossRef] [PubMed]
33. Singla, B.; Holmdahl, R.; Csanyi, G. Editorial: Oxidants and Redox Signaling in Inflammation. *Front. Immunol.* **2019**, *10*, 545. [CrossRef] [PubMed]
34. Schnupf, P.; Gaboriau-Routhiau, V.; Cerf-Bensussan, N. Modulation of the gut microbiota to improve innate resistance. *Curr. Opin. Immunol.* **2018**, *54*, 137–144. [CrossRef] [PubMed]
35. Desai, M.S.; Seekatz, A.M.; Koropatkin, N.M.; Kamada, N.; Hickey, C.A.; Wolter, M.; Pudlo, N.A.; Kitamoto, S.; Terrapon, N.; Muller, A.; et al. A Dietary Fiber-Deprived Gut Microbiota Degrades the Colonic Mucus Barrier and Enhances Pathogen Susceptibility. *Cell* **2016**, *167*, 1339–1353.e1321. [CrossRef]
36. Li, J.; Cha, R.; Zhao, X.; Guo, H.; Luo, H.; Wang, M.; Zhou, F.; Jiang, X. Gold Nanoparticles Cure Bacterial Infection with Benefit to Intestinal Microflora. *ACS Nano* **2019**, *13*, 5002–5014. [CrossRef]
37. Wotzka, S.Y.; Nguyen, B.D.; Hardt, W.D. Salmonella Typhimurium Diarrhea Reveals Basic Principles of Enteropathogen Infection and Disease-Promoted DNA Exchange. *Cell Host Microbe* **2017**, *21*, 443–454. [CrossRef]
38. Li, M.; Yu, L.; Zhai, Q.; Chu, C.; Wang, S.; Zhao, J.; Zhang, H.; Tian, F.; Chen, W. Combined Ganoderma lucidum polysaccharide and ciprofloxacin therapy alleviates Salmonella enterica infection, protects the intestinal barrier, and regulates gut microbiota. *Food Funct.* **2023**, *14*, 6896–6913. [CrossRef]
39. Dong, N.; Li, X.; Xue, C.; Wang, C.; Xu, X.; Bi, C.; Shan, A.; Li, D. Astragalus polysaccharides attenuated inflammation and balanced the gut microflora in mice challenged with Salmonella typhimurium. *Int. Immunopharmacol.* **2019**, *74*, 105681. [CrossRef]
40. Yuan, H.; Zhou, L.; Chen, Y.; You, J.; Hu, H.; Li, Y.; Huang, R.; Wu, S. Salmonella effector SopF regulates PANoptosis of intestinal epithelial cells to aggravate systemic infection. *Gut Microbes* **2023**, *15*, 2180315. [CrossRef]
41. Shi, Z.; Nan, Y.; Zhou, X.; Zhang, W.; Zhang, Z.; Zhang, C.; Duan, H.; Ge, J.; Zhao, L. Molecular Mechanisms of Intestinal Protection by *Levilactobacillus brevis 23017* against *Salmonella typhimurium* C7731-Induced Damage: Role of Nrf2. *Microorganisms* **2024**, *12*, 1135. [CrossRef] [PubMed]
42. Ookawara, T.; Imazeki, N.; Matsubara, O.; Kizaki, T.; Oh-Ishi, S.; Nakao, C.; Sato, Y.; Ohno, H. Tissue distribution of immunoreactive mouse extracellular superoxide dismutase. *Am. J. Physiol.* **1998**, *275*, C840–C847. [CrossRef] [PubMed]
43. Sottero, B.; Rossin, D.; Poli, G.; Biasi, F. Lipid Oxidation Products in the Pathogenesis of Inflammation-related Gut Diseases. *Curr. Med. Chem.* **2018**, *25*, 1311–1326. [CrossRef] [PubMed]
44. Wesch, D.; Peters, C.; Oberg, H.H.; Pietschmann, K.; Kabelitz, D. Modulation of $\gamma\delta$ T cell responses by TLR ligands. *Cell. Mol. Life Sci.* **2011**, *68*, 2357–2370. [CrossRef] [PubMed]
45. Arpaia, N.; Godec, J.; Lau, L.; Sivick, K.E.; McLaughlin, L.M.; Jones, M.B.; Dracheva, T.; Peterson, S.N.; Monack, D.M.; Barton, G.M. TLR signaling is required for Salmonella typhimurium virulence. *Cell* **2011**, *144*, 675–688. [CrossRef]
46. Celhar, T.; Yasuga, H.; Lee, H.Y.; Zharkova, O.; Tripathi, S.; Thornhill, S.I.; Lu, H.K.; Au, B.; Lim, L.H.K.; Thamboo, T.P.; et al. Toll-Like Receptor 9 Deficiency Breaks Tolerance to RNA-Associated Antigens and Up-Regulates Toll-Like Receptor 7 Protein in Sle1 Mice. *Arthritis Rheumatol.* **2018**, *70*, 1597–1609. [CrossRef]
47. Yu, H.; Lin, L.; Zhang, Z.; Zhang, H.; Hu, H. Targeting NF-κB pathway for the therapy of diseases: Mechanism and clinical study. *Signal Transduct. Target. Ther.* **2020**, *5*, 209. [CrossRef]
48. Feng, Z.; Li, X.; Lin, J.; Zheng, W.; Hu, Z.; Xuan, J.; Ni, W.; Pan, X. Oleuropein inhibits the IL-1β-induced expression of inflammatory mediators by suppressing the activation of NF-κB and MAPKs in human osteoarthritis chondrocytes. *Food Funct.* **2017**, *8*, 3737–3744. [CrossRef]
49. Fung, T.C.; Olson, C.A.; Hsiao, E.Y. Interactions between the microbiota, immune and nervous systems in health and disease. *Nat. Neurosci.* **2017**, *20*, 145–155. [CrossRef]
50. Perler, B.K.; Friedman, E.S.; Wu, G.D. The Role of the Gut Microbiota in the Relationship Between Diet and Human Health. *Annu. Rev. Physiol.* **2023**, *85*, 449–468. [CrossRef]
51. Oliveira-Alves, S.C.; Vendramini-Costa, D.B.; Betim Cazarin, C.B.; Maróstica Júnior, M.R.; Borges Ferreira, J.P.; Silva, A.B.; Prado, M.A.; Bronze, M.R. Characterization of phenolic compounds in chia (*Salvia hispanica* L.) seeds, fiber flour and oil. *Food Chem.* **2017**, *232*, 295–305. [CrossRef] [PubMed]

52. Zhou, Z.; He, W.; Tian, H.; Zhan, P.; Liu, J. Thyme (*Thymus vulgaris* L.) polyphenols ameliorate DSS-induced ulcerative colitis of mice by mitigating intestinal barrier damage, regulating gut microbiota, and suppressing TLR4/NF-κB-NLRP3 inflammasome pathways. *Food Funct.* **2023**, *14*, 1113–1132. [CrossRef] [PubMed]
53. Ma, L.; Ni, Y.; Wang, Z.; Tu, W.; Ni, L.; Zhuge, F.; Zheng, A.; Hu, L.; Zhao, Y.; Zheng, L.; et al. Spermidine improves gut barrier integrity and gut microbiota function in diet-induced obese mice. *Gut Microbes* **2020**, *12*, 1832857. [CrossRef] [PubMed]
54. Li, D.T.; Feng, Y.; Tian, M.L.; Ji, J.F.; Hu, X.S.; Chen, F. Gut microbiota-derived inosine from dietary barley leaf supplementation attenuates colitis through PPARγ Signaling activation. *Microbiome* **2021**, *9*, 83. [CrossRef]
55. Fang, D.; Xu, T.Q.; Sun, J.Y.; Shi, J.R.; Li, F.L.; Yin, Y.Q.; Wang, Z.Q.; Liu, Y. Nicotinamide Mononucleotide Ameliorates Sleep Deprivation-Induced Gut Microbiota Dysbiosis and Restores Colonization Resistance against Intestinal Infections. *Adv. Sci.* **2023**, *10*, e2207170. [CrossRef]

antioxidants

MDPI

Article

Ozone Therapy in the Integrated Treatment of Female Dogs with Mammary Cancer: Oxidative Profile and Quality of Life

Laís Pereira Silva [1], Ricardo Wagner Portela [2,*], Marília Carneiro Machado [1,3], Gisele André Baptista Canuto [4], João Moreira Costa-Neto [1,5], Vitor de Moraes Pina de Carvalho [1], Hanna Carvalho de Sá [4], Karine Araújo Damasceno [1,6], Vinicius Ricardo Cuña de Souza [7], Clarisse Simões Coelho [7,8,9] and Alessandra Estrela-Lima [1,5,*]

1 Research Center on Mammary Oncology (NPqOM), Federal University of Bahia, Salvador 40170-110, Brazil; laisps@ufba.br (L.P.S.); marilia.carneiro@ufrb.edu.br (M.C.M.); jmcn@ufba.br (J.M.C.-N.); vitor.moraes@ufba.br (V.d.M.P.d.C.); karine.damasceno@fiocruz.br (K.A.D.)
2 Laboratory of Immunology and Molecular Biology (LABIMUNO), Institute of Health Sciences, Federal University of Bahia, Salvador 40110-100, Brazil
3 Center for Agricultural, Environmental and Biological Sciences, Federal University of Reconcavo of Bahia, Cruz das Almas 44380-000, Brazil
4 Department of Analytical Chemistry, Institute of Chemistry, Federal University of Bahia, Salvador 40170-110, Brazil; gisele.canuto@ufba.br (G.A.B.C.); carvalho.hanna@ufba.br (H.C.d.S.)
5 Department of Anatomy, Pathology and Veterinary Clinics, Federal University of Bahia, Salvador 40170-110, Brazil
6 Experimental Pathology Laboratory (LAPEX), Gonçalo Moniz Institute, Oswaldo Cruz Foundation (FIOCRUZ), Salvador 40296-710, Brazil
7 Faculty of Veterinary Medicine, Lusofona University, 1749-024 Lisbon, Portugal; vinicius.souza@ulusofona.pt (V.R.C.d.S.); clarisse.coelho@ulusofona.pt (C.S.C.)
8 Veterinary and Animal Research Centre (CECAV), Faculty of Veterinary Medicine, Lusofona University, 1749-024 Lisbon, Portugal
9 Mediterranean Institute for Agriculture, Environment and Development, Universidade de Évora, 7004-516 Évora, Portugal
* Correspondence: rwportela@ufba.br (R.W.P.); aestrela@ufba.br (A.E.-L.)

Abstract: Considering the high frequency of malignant breast tumors, there is a growing search for new therapeutic strategies that control neoplastic growth and dissemination, combined with fewer adverse reactions. Therefore, this study evaluated the effects of ozone therapy in female dogs with mammary cancer undergoing chemotherapy treatment. Twenty-five canines diagnosed with malignant mammary neoplasia were divided into two groups: one treated with carboplatin alone ($n = 11$) and the other with carboplatin associated with ozone therapy ($n = 14$). Clinical and laboratory evaluations, mastectomy, analysis of the oxidative profile based on total antioxidant capacity (TAC) and serum concentrations of malondialdehyde (MDA), survival rate, and quality of life were performed. Animals in the ozone therapy group had higher concentrations of red blood cells and platelets, significantly improving the survival rate and quality of life. Furthermore, adverse reactions were less intense and frequent in this group, which was associated with an increase in TAC and a reduction in MDA. These results indicate that the combination of carboplatin and ozone therapy represents a promising complementary treatment for female dogs with mammary cancer, as it was associated with fewer adverse reactions and a better oxidative profile.

Keywords: breast cancer; chemotherapy-induced side effects; integrative medicine; ozone; reactive oxygen species

1. Introduction

Breast cancer is one of the most prevalent cancers in women worldwide [1]. Despite substantial advancements in conventional treatments, there is a growing demand for innovative therapeutic strategies that can control neoplastic growth and dissemination along with fewer adverse reactions, especially in patients at advanced stages of the disease [2,3]. In this context, adjuvant therapy using the gas mixture of oxygen–ozone (O_2–O_3) emerges as a promising alternative for the treatment of various types of cancer due to its immunomodulatory, anti-inflammatory, and oxidative stress-inducing properties [4,5].

In breast cancer, oxidative stress mechanisms play a fundamental role in activating cellular signaling pathways, proliferation, and apoptosis, influencing tumor progression and aggressiveness [6,7]. In vitro and preclinical studies demonstrate the ozone's ability to damage tumor cells directly, enhancing the effects of radiotherapy and chemotherapy and reducing adverse reactions to oncological treatments [8–11]. Additionally, indirect effects have been observed in animal models, such as immunomodulation and tissue oxygenation improvement, as additional benefits during cancer treatment [9,12–14].

In women undergoing mastectomy, ozone–oxygen therapy has been shown to accelerate the healing process, enabling radiotherapy application [4]. Preliminary results also suggest that ozone therapy may provide effective therapeutic support in relieving pain and fatigue in patients with various types of cancer during palliative therapies [13,15]. Recent studies investigated the potential of ozone therapy to inhibit the growth of human breast cancer cells, with promising results in inducing apoptosis and inhibiting proliferation [16]. However, the efficacy of ozone therapy as an adjuvant option in oncological treatment requires targeted clinical trials for specific tumors and well-defined circumstances [17].

In veterinary medicine, the use of ozone–oxygen mixture is widespread in various species [18,19], but its application as adjuvant antineoplastic therapy in female dogs, which are considered models for comparative studies on human breast cancer, remains limited. Considering this knowledge gap, the importance and the necessity of new research in this field are evident. In this way, the present study aimed to evaluate the effects of medicinal ozone as an integrative therapy for female dogs with mammary cancer undergoing conventional chemotherapy with carboplatin.

2. Materials and Methods

2.1. Ethical Aspects

The Ethics Committee for the Use of Experimental Animals of the School of Veterinary Medicine and Zootechny of the Federal University of Bahia approved this research project (protocol n° 18/2021). All the procedures were conducted according to the Brazilian College of Animal Experimentation (COBEA). All animals were domiciliated, and dog guardians were informed about the details of the research project by means of a signed Free and Informed Consent Form.

2.2. Clinical Evaluation and Mastectomy

The anamnesis was accomplished by a detailed evaluation of physiological parameters, clinical history, and reproductive records. Preoperative clinical evaluation included a complete blood count; serum urea, creatinine, alkaline phosphatase, alanine aminotransferase, ionized calcium, and glucose concentrations; thoracic radiological examination (anterolateral right, anterolateral left, and ventral-dorsal); and a total abdominal ultrasound to assess for metastasis. Clinical stage classification was performed based on tumor size (T), involvement of regional lymph nodes (N), and presence or absence of distant metastases (M) based on the TNM system [20].

The macroscopic evaluation of inguinal and axillary lymph nodes was performed by palpation. Neoplastic involvement was confirmed by histopathological examination of lymph nodes following mastectomy. All the animals underwent a total unilateral mastectomy to remove the inguinal lymph node. All dogs used compressive bandages and post-surgical clothing until the skin stitches were removed. During this period, physiological and behavioral parameters and possible surgical wound complications were evaluated.

2.3. Histological Classification and Grade

Fragments of the affected mammary gland, including skin and subcutaneous tissues, were fixed in phosphate-buffered 10% neutral formalin and processed by the routine paraffin embedding technique. Histological sections (4 μm) were stained by the hematoxylin–eosin (HE) method. In all cases, two veterinarian pathologists analyzed duplicate slides. Tumor samples were categorized according to the histopathological diagnosis following the World Health Organization (WHO) criteria and complemented by the Consensus regarding the diagnosis, prognosis, and treatment of canine and feline mammary tumors, 2019 [20]. Tumors initially classified as solid carcinomas were reclassified following the proposal by Nakagaki et al. [21].

The histological grades of the tumors were described by the Nottingham system as modified by Elston and Ellis [22], which evaluates the percentage of tubule formation, nuclear pleomorphism, and the mitotic index. The female dogs with a histopathological diagnosis of malignancy, histopathological Grades II, III, or regional lymph node metastasis were referred for chemotherapy treatment. Based on histopathological diagnoses, the animals were allocated to groups for treatment in a paired manner, seeking to obtain equivalence of histological types between the groups.

2.4. Chemotherapy Treatment

Twenty days post-surgery, carboplatin chemotherapy commenced in six sessions, with intervals of 21–28 days between each session. The dogs received a 5 min infusion of carboplatin via the cephalic vein, with doses ranging from 240 to 300 mg/m^2 per session, depending on the animal's size [23]. The primary acute dose-limiting toxicity of carboplatin is myelosuppression, particularly neutropenia, and thrombocytopenia, occurring approximately 14 days post-treatment. However, some dogs, especially those under 10 kg, may reach their nadir at 21 days following carboplatin administration [23].

Fourteen days after each carboplatin session (roughly 48 h before the next session), biological samples were collected for urinalysis, complete blood counts, and serum biochemistry analysis. These laboratory tests confirmed the animals' suitability for the carboplatin-based chemotherapy procedure. Ondansetron (0.2 mg/kg), ranitidine (2 mg/kg), and promethazine (0.1 mg/kg) were administered subcutaneously. All animals in this study followed this protocol, and these medications were prescribed orally for three consecutive days post-chemotherapy, except for promethazine, which was given as a single dose.

2.5. Chemotherapy Treatment and Ozone Therapy

Ozone Life's specific medical ozone generation equipment was used for ozone therapy. The O&L Portable model was coupled to a medical oxygen cylinder, achieving an ozone concentration with more than 40 values for different oxygen flows (1, ¾ = 0.75, ½ =0.5, ¼ = 0.25, and 215B1/8 = 0.125) L/min, ensuring precise levels of ozone in the gas mixture. This device is certified by the Brazilian National Agency of Sanitary Surveillance (ANVISA 81509100001).

During the entire cycle of chemotherapy with carboplatin, ozone therapy was performed every seven days in the minor autohemotherapy modality (Ahm—20 μg/ O_2-O_3), in the proportion of 1 mL of whole blood (without anticoagulant): 1 mL of the mixture (O_3–O_2) with intramuscular (IM) application to the VG 14 acupoint, except on the day of chemotherapy treatment. The VG 14 point, called Governor Vessel 14 (Da-Shui), is in a depression in the dorsal midline, between the seventh cervical vertebra and the first thoracic vertebra, in the cranial direction to the highest point of the inter-scapular. This acupoint (VG 14) is commonly used in hemopuncture, with the objective of modulating the immune system, having indications for its use in cases of immunological disorders, thrombocytopenia, anemia, infectious diseases, and chronic diseases. Indeed, the VG14 point was chosen due to its easy access to the application of ozonated autohemotherapy and its history of stimulating the immune system [24,25].

2.6. Chemotherapy Adverse Reactions

The animals were monitored weekly via phone calls during chemotherapy and assessed biweekly through clinical exams focusing on potential adverse drug reactions. To evaluate these reactions, we followed the Veterinary Cooperative Oncology Group—Common Terminology Criteria for Adverse Events (VCOG-CTCAE) guidelines [26]. This consensus outlines a grading scale (from 1 to 5) for the severity of chemotherapy-induced adverse reactions (such as vomiting and diarrhea) in dogs and cats, with the following distinctions: Grade 1 (mild): asymptomatic or mild symptoms, only diagnostic observations, no need for therapeutic intervention; Grade 2 (moderate): moderate limitation of daily activities, requires minimal or non-invasive outpatient intervention; Grade 3 (severe or clinically significant, but not immediately fatal): significant restriction in daily activities, hospitalization indicated; Grade 4 (life-threatening): requires urgent therapeutic intervention; and Grade 5: death related to adverse drug reactions [26]. Dog owners provided information and completed weekly reports indicating potential adverse drug reactions.

2.7. Assessment of Hematological and Biochemical Parameters

Blood samples, approximately 5 mL, were obtained by puncture of the jugular vein and previous fasting for 10 h and immediately placed in tubes containing EDTA (ethylenediaminetetraacetic acid) for complete blood counts and in tubes containing coagulant gel for determination of serum biochemical profile. The blood counts were performed using an automated blood cell analyzer (ADVIA 60, Bayer HealthCare, Tarrytown, NY, USA), determining the total counts of red blood cells, leukocytes, hemoglobin, packed cell volume, and platelets. The differential leukocyte counts were performed by optical microscopy evaluation of the blood smears, which were prepared at the time of sample collection, and stained using the Rosenfeld technique.

The gel tubes were centrifuged at 3000 rpm at 4 °C for 10 min to obtain the serum and to perform serum biochemistry tests (liver and kidney function analysis) on a Smart 200 automatic biochemical analyzer. The liver function profile included alkaline phosphatase (FA) and alanine aminotransferase (ALT) serum levels. For renal function evaluation, serum urea and creatinine concentrations were determined. The manufacturer's recommendations were followed. Hematological and biochemical evaluations were carried out pre-surgery. Pre-chemotherapy assessments consisted of blood count, ALT, and creatinine. Patients whose results were not within the reference values for the species were not subjected to chemotherapy.

2.8. Sample Collection for Oxidative Profile Evaluation

To analyze the oxidative status, blood was collected from female dogs undergoing therapeutic protocols at three different times: 48 h before the first chemotherapy session (moment A), 48 h before the fourth chemotherapy session (moment B), and 45 days after completion of the treatment with carboplatin (moment C). From each animal, five milliliters (mL) of blood was collected in tubes containing clot activator and centrifuged at 3000 rpm at 4 °C for 10 min. Then, serum aliquots were placed in microtubes and stored at −80 °C until analysis.

2.9. Total Antioxidant Capacity (TAC)

The Trolox Equivalent Antioxidant Capacity assay was used to determine total antioxidant capacity (TAC), following the method described by Erel [27]. Briefly, the 2,2′-Azinobis-(3-ethylbenzothiazoline-6-sulfonic acid) (ABTS) radical was generated by the reaction between the diammonium salt ABTS (7.0 mmol L^{-1}) and potassium persulfate (2.45 mmol L^{-1}) and was kept in the dark at room temperature for 16 h. The ABTS solution was subsequently diluted with Phosphate-Buffered Saline (PBS) solution until it reached an absorbance of 0.70 nm ± 0.05 nm at 734 nm. The TAC determination was performed on a spectrophotometer (VersaMax™ Microplate Reader, Thermo Fisher Scientific, Waltham, USA) at 734 nm, after 6 min of the reaction of 198 μL ABTS diluted in PBS with

2 µL of standard solutions, samples, or blank. Ethanolic solutions of 6-hydroxy-2,5,7,8-tetramethylchroman-2-carboxylic acid (Trolox), in the range 0.25 to 2.00 mmol L^{-1}, were used for calibration in triplicate. PBS was used as a blank sample to correct absorbance. The TAC was calculated by determining the decrease in the absorbance and was expressed as relative to the Trolox standard (in mmol L^{-1}).

2.10. Determination of Malondialdehyde (MDA) Concentration

MDA serum concentrations were determined following the method described by Tang et al. [28]. MDA was produced from the reaction between 1:100 (v/v) malondialdehyde bis (dimethyl acetal) and 1% (v/v) sulfuric acid, kept at room temperature for 2 h. The solution was subsequently diluted ten times with 1% (v/v) sulfuric acid solution. The calibration curve was prepared from a stock solution of MDA (165 mmol L^{-1}) in triplicate and within the concentration range of 0.013–6.63 nmol. The samples were prepared by mixing 20 µL of serum samples with 500 µL of sulfuric acid (42 mmol L^{-1}) and adding 125 µL of phosphotungstic acid. The mixture was homogenized by vortexing and maintained at room temperature for 5 min, followed by centrifugation (15,000× g for 5 min). Amounts of 102 µL of butylated hydroxytoluene (BHT) (2.5 mmol L^{-1}) on ice and 98 µL of deionized water were added to the pellet. Two hundred microliters of samples and standard solutions were mixed with 600 µL of thiobarbituric acid (TBA) (prepared in glacial acetic acid). The mixture was then incubated at 95 °C for 1 h, followed by an ice bath for 10 min. The absorbance was measured at 532 nm on a spectrophotometer (VersaMax™ Microplate Reader, Thermo Fisher Scientific, Waltham, USA). The total concentration of MDA was expressed in nmol mL^{-1}.

2.11. Assessment of Quality of Life

During the chemotherapy treatment and for 45 days following its conclusion, the quality of life of the female dogs was evaluated using a specific questionnaire completed by their owners. This questionnaire, developed according to the guidelines of Yazbek and Fantoni [29], comprised 12 questions with four possible response options, each ranging from zero to three points, culminating in a maximum score of 36 points. A score of zero indicated the poorest quality of life, while a score of 36 represented the best. The questions focused on behavioral aspects, interaction with the owner, and assessments of pain, appetite, sleep disturbances, vomiting, diarrhea, urinary incontinence, or repletion.

2.12. Follow-Up and Survival

Clinical and laboratory tests were conducted every 15 days, while thoracic radiological examinations and total abdominal ultrasounds were performed monthly. One month after completing the conventional chemotherapy cycle with carboplatin (six sessions), the animals underwent clinical evaluations through laboratory tests, which showed that their neutrophil, platelet, and red cell counts were within normal ranges. This follow-up continued until the end of this study or the animal's death. Specific survival (SS) was defined as the period (in days) from mastectomy to death due to disease progression. In contrast, overall survival time was defined as the period (in days) from the surgical removal of the primary tumor to the date of death (from any cause) or this study's end date. The medical team evaluated the necessity for humane euthanasia and performed it with the owners' consent [30]. Animals that died during the follow-up period were necropsied to determine the cause of death and to identify any potential metastases or chemotherapy-induced lesions.

2.13. Statistical Analysis

The Student's t-test was applied for independent samples when age and weight means were compared between groups. Analysis of variance for repeated measures (one-way ANOVA) followed by the Tukey test were used to compare different time moments. The Chi-square test and the linear-by-linear association were used for categorical variables to compare proportions between groups. Kaplan–Meier analysis was used to construct the survival curves, and the log-rank test was used to compare the curves from different groups. The Spearman correlation analysis was applied to analyze possible correlations between the evaluated parameters. Binary logistic regression analysis served to compare multiple parameters. Cox regression was used for multivariate survival assessment. The significance level adopted was $p < 0.05$ at a 95% confidence interval, with a two-tailed analysis. Analyses were performed using Prism 8.0 (GraphPad, San Diego, CA, USA) and SPPS 26.0 for Windows (SPSS Inc., Chicago, IL, USA) software.

3. Results

3.1. Clinical and Pathological Features

From 50 eligible animals with mammary tumors, 25 female dogs diagnosed with malignant mammary neoplasms were included in the current investigation. The age of female dogs in this study ranged from three to 14 years, with a higher frequency of female dogs between 6 and 10 years (14/25—56%) and an average age of 10.2 ± 2.5 years. Weight ranged from 3 to 22 kg with an average of 9.9 ± 4.9, with a predominance of mixed breed animals (SRD) regardless of the group (9/25—36%), followed by Dachshund dogs (4/25—16%) and Poodles, respectively (4/25—16%).

Female dogs with T2 (between 3–5 cm/32% (8/25) and T3 (greater than 5 cm/ 32% (8/25) tumors represented together 64% (16/25) of cases, while animals with metastasis on regional lymph nodes represented 20% (5/25) and with distant metastasis only 8% (2/25). Despite the different types and histological grades of mammary cancer studied, a balance can be seen between the groups evaluated (Table 1). The female dogs that died presented confirmed metastatic foci after necropsy, and the histopathological evaluations of the organs were reclassified to stage V due to a worse prognosis, regardless of the regional metastasis.

Table 1. Clinical and pathological characteristics of the female dogs with mammary neoplasms included in this study.

	Carboplatin (CG)	Carboplatin + O_3 (O_3G)	Total	
	Mean (±SD)	**Mean (±SD)**	**Mean (±SD)**	***p*-Value**
Weight(Kg)	10.8 (±1.9)	9.2 (±3.7)	9.9 (±4.9)	
Minimum	3.9	3	3	0.428
Maximum	22	15.3	22	
Variable	**n (%)**	**n (%)**	**n (%)**	
Size				
Mini	2 (18.2)	2 (14.3)	4 (16.0)	
Small	4 (36.4)	9 (64.3)	13 (52.0)	0.468
Average	5 (45.5)	3 (21.4)	8 (32.0)	
Breed				
Mixed breed	5 (45.5)	4 (28.6)	9 (36.0)	
Beagle	1 (9.1)	1 (7.1)	2 (8.0)	
Poodle	2 (18.2)	2 (14.3)	4 (16.0)	
Dachshund	1 (9.1)	3 (21.4)	4 (16.0)	0.231
Pinscher	1 (9.1)	1 (7.1)	2 (8.0)	
Yorkshire Terrier	1 (9.1)	1 (7.1)	2 (8.0)	
Maltese	0	1 (7.1)	1 (4.0)	
Shih Tzu	0	1 (7.1)	1 (4.0)	
Reproductive Status				
Neutered	2 (18.2)	6 (42.9)	8 (32.0)	0.198
Not neutered	9 (81.8)	8 (57.1)	17 (68.0)	
Affected Mammary Chain				
Unilateral	6 (54.5)	6 (42.9)	12 (48.0)	0.695
Bilateral	5 (45.5)	8 (57.1)	13 (52.0)	

Table 1. *Cont.*

	Carboplatin (CG)	Carboplatin + O_3 (O_3G)	Total	
	Mean (±SD)	**Mean (±SD)**	**Mean (±SD)**	***p*-Value**
Involvement				
Single	5 (45.5)	3 (21.4)	8 (32.0)	0.21
Multiple	6 (54.5)	11 (78.6)	17 (68.0)	
TNM Stage				
I	0	2 (14.3)	2 (8.0)	
II	5 (45.5)	3 (21.4)	8 (32.0)	0.802
III	3 (27.3)	5 (35.7)	8 (32.0)	
IV	3 (27.3)	2 (14.3)	5 (20.0)	
V	0	2 (14.3)	2 (8.0)	
Lymph Node				
Metastasis	3 (27.3)	3 (21.4)	6 (24.0)	0.739
No metastasis	8 (72.7)	11 (78.6)	19 (76.0)	
Histological Grade				
Grade I	0	1 (10.0)	1 (5.9)	0.388
Grade II	7 (100)	9 (90.0)	16 (94.1)	

The animals were divided into two treatment groups: one treated only with carboplatin and the second one treated with carboplatin and ozone. Statistical tests: Student's *t*-test, Chi-square, and linear-by-linear association. Significance set at $p < 0.05$.

For the histopathological grading of the tumors, areas of invasiveness were used, with 16 tumors classified as Grade II and one tumor as Grade I, the latter being a carcinoma in a mixed tumor with behavior considered as atypical and aggressive, since the dog, at the time of diagnosis, already had distant metastasis (Table 2).

Table 2. Histopathological diagnosis and grading of mammary tumors in dogs were included in this study.

	Carboplatin (CG)	Carboplatin + O_3 (O_3G)	Total	
	n (%)	**n (%)**	**n (%)**	
Histological Types				0.146
Malignant adenomyoepithelioma	3 (27.3)	1 (7.1)	4 (16.0)	
Carcinoma in a mixed tumor (Grade II)	4 (36.4)	5 (35.7)	9 (36.0)	
Papillary carcinoma (Grade II)	1 (9.1)	2 (14.2)	3 (12.0)	
Solid papillary carcinoma (Grade II)	1 (9.1)	1 (7.1)	2 (8.0)	
Carcinosarcoma	1 (9.1)	2 (14.3)	3 (12.0)	
Malignant myoepithelioma	0	1 (7.1)	1 (4.0)	
Carcinoma in a mixed tumor (Grade I)	0	1 (7.1)	1 (4.0)	
Carcinoma with solid pattern (Grade II)	1 (9.1)	1 (7.1)	2 (8.0)	
Histological Grade				0.388
Grade I	0	1 (10.0)	1 (5.9)	
Grade II	7 (100)	9 (90.0)	16 (94.1)	

The animals were divided into two treatment groups: one treated only with carboplatin and the second one treated with carboplatin and ozone. Statistical tests: Chi-square and linear-by-linear association. Significance at $p < 0.05$.

3.2. *Chemotherapy Adverse Reactions*

Among the adverse effects/reactions of chemotherapy, regardless of the group, 28% (7/25) of the dogs presented vomiting and 20% (5/25) diarrhea. In the group treated only with carboplatin, Grade 1 adverse reactions were predominant, that is, asymptomatic animals or mild signs, with only diagnostic observations and no indication of therapeutic intervention. However, some animals in the first five CT sessions also presented adverse reactions classified as moderate (Grade 2), with moderate limitation of daily activities and the need for minimal or non-invasive outpatient intervention. In the O_3G, only reactions classified as mild (Grade 1) were identified (Figure 1). The Tukey test was applied to compare the medians between groups for each CT session. No significant difference was observed between the degrees of adverse reactions between the groups. However, when the Chi-square test was applied, comparing the effects frequencies between the two groups and considering all sessions, a difference was found between the groups (Table 3).

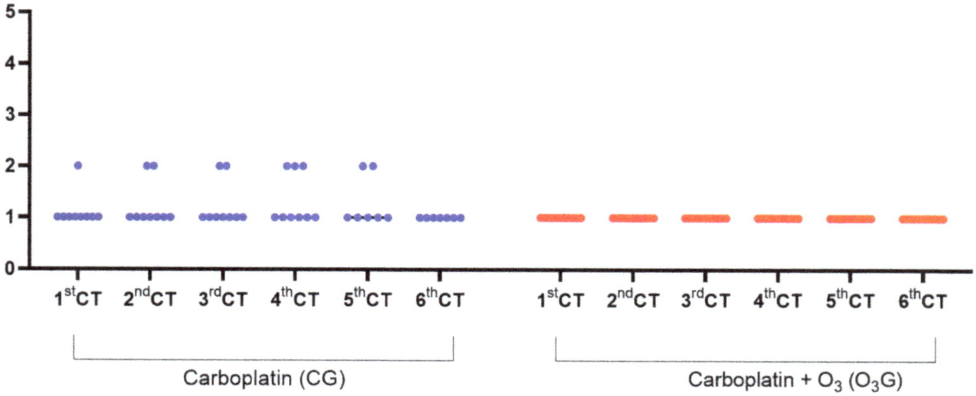

Figure 1. Comparison of degrees of adverse effects in each chemotherapy session according to the type of chemotherapy protocol established in female dogs with malignant mammary neoplasms. Mild, (2) moderate, (3) severe or clinically significant but not immediately fatal, (4) life-threatening, and (5) death related to an adverse drug reaction. The animals were treated with carboplatin alone or with the association of carboplatin and ozone therapy during six chemotherapy sessions. The Tukey test was applied to compare the medians between groups for each CT session and showed no significant difference ($p < 0.05$) between groups in all chemotherapy sessions.

Table 3. Comparison of adverse reaction degrees between experimental groups.

	Carboplatin (CG)	Carboplatin + O_3 (O_3G)	*p*-Value
Adverse reaction	n (%)	n (%)	
Mild	47 (73.4)	84 (100)	<0.001
Moderate	17 (26.6)	0	

The Chi-square statistical test was used to compare the results between the groups, with significance at $p < 0.05$.

3.3. Laboratory Parameters

Based on hematological parameters, the main effects of chemotherapy were thrombocytopenia and leukopenia due to neutropenia, which were significantly more intense in the group treated with carboplatin alone. The erythrogram of the animals in the O_3 group revealed a significant increase in red blood cells ($p = 0.017$) after the second CT session, remaining within normal parameters for the species throughout the cycle, and in platelets after the second ($p = 0.046$), third ($p = 0.009$), and sixth ($p = 0.004$) CT sessions when compared to the CG (Figure 2A,B).

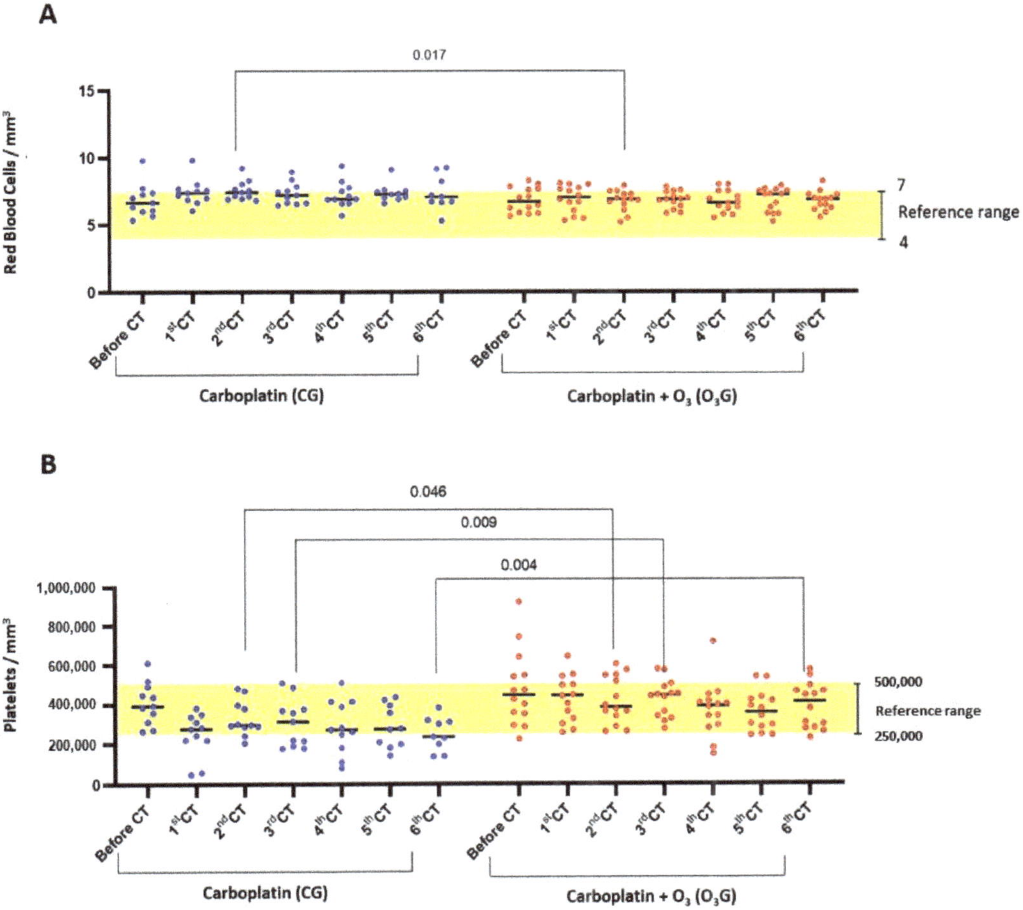

Figure 2. Comparison of red blood cell and platelet counts throughout the chemotherapy treatment of the animals from the control and ozone groups. (**A**) Total red blood cells; (**B**) platelets. The one-way ANOVA statistical test was used, with significance at $p < 0.05$.

Regarding the leukogram, leukopenia due to neutropenia was observed at different times, with a significant difference and lower values found in the carboplatin group (CT). Total leukocytes were reduced after the first ($p = 0.001$), third ($p = 0.038$), fourth ($p = 0.030$), fifth ($p = 0.018$), and sixth CT sessions ($p = 0.006$). Except after the fourth chemotherapy session, there was a significant difference in segmented neutrophil counts throughout the treatment, namely after the first ($p = 0.001$), second ($p = 0.018$), third ($p = 0.049$), fifth ($p = 0.006$), and sixth ($p = 0.016$) CT sessions (Figure 3A,B). Given these results in the group of animals treated only with carboplatin, extending the time between chemotherapy sessions to 28 days was necessary, which was initially established at 21 days. In this study, the biochemical parameters evaluated (creatinine and ALT) in both treated groups were similar.

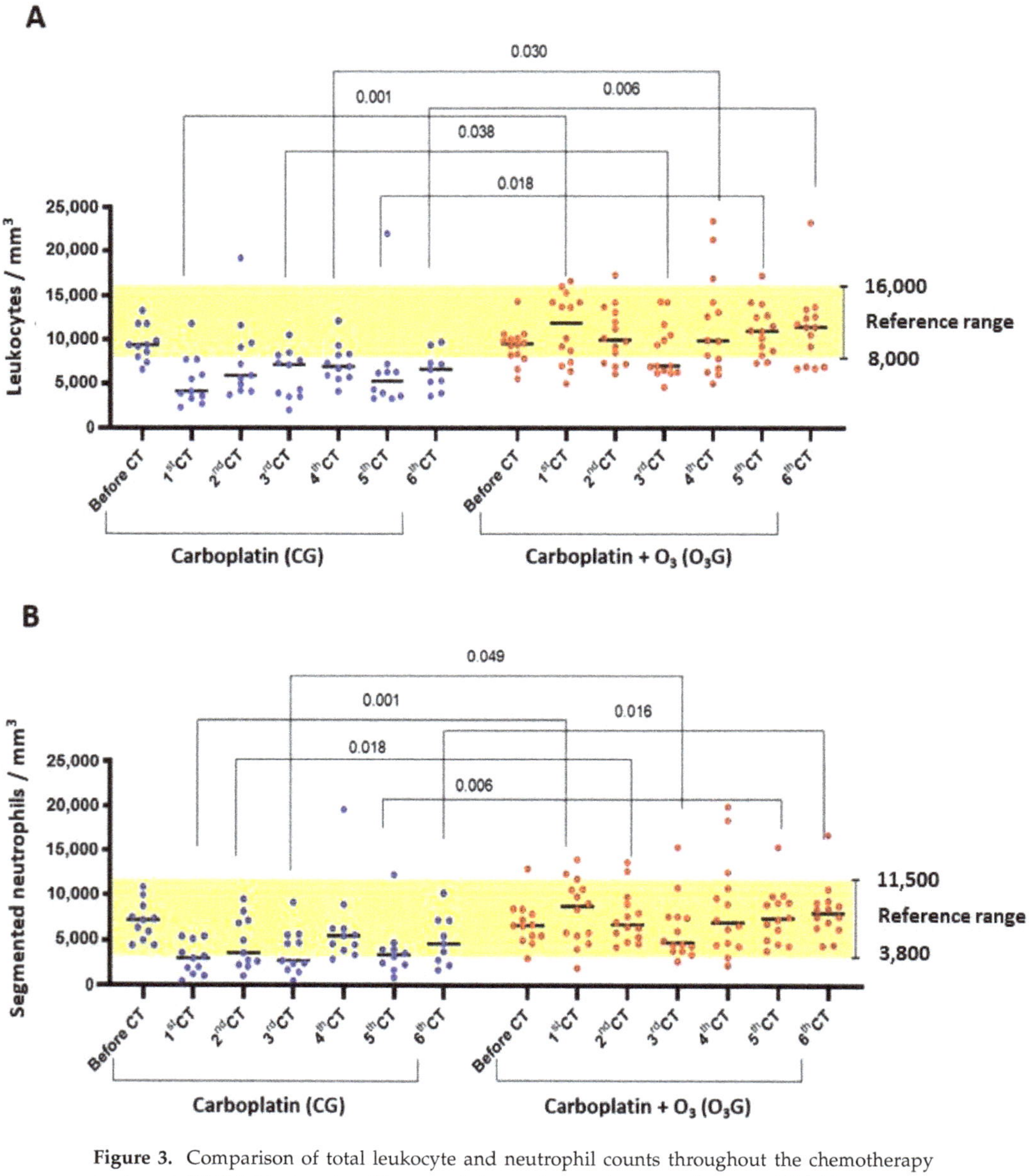

Figure 3. Comparison of total leukocyte and neutrophil counts throughout the chemotherapy treatment. (**A**) Total leukocytes; (**B**) neutrophil granulocytes (segmented neutrophils). The one-way ANOVA statistical test was used, with significance at $p < 0.05$.

3.4. Quality of Life

After the chemotherapy was carried out on 25 animals, 16 (64%) had their quality of life improved after the start of treatment, five (20%) showed no difference, and two (8%) had the quality of life worsened. In the group undergoing chemotherapy treatment associated with ozone, ten owners (10/14) thought there was an improvement in quality of life (72%), and four owners (28%) were unable to answer the question.

At the beginning of treatment, there was no significant difference between the treatment groups, but after the first chemotherapy session, there was a difference between the groups. In the second session, a difference in quality of life was observed between the groups, and this situation was also seen in the fourth, fifth, and sixth chemotherapy sessions (Table 4). Therefore, comparing the groups separately in terms of the initial and final moments, significant differences in quality of life were observed within the same group in different moments and between groups in the same moment.

Table 4. Comparison of quality of life between treated groups and in each chemotherapy (CT) session.

	Start	1st CT	2nd CT	3rd CT	4th CT	5th CT	6th CT
Quality of life							
CG	29.5 (±2.4) a	31.9 (±2.1) b	32.4 (±1.7) Ab	33.2 (±1.4) b	32.8 (±1.9) Ab	32.3 (±1.2) Ab	31.9 (±1.4) Ab
O3G	29.5 (±2.4) a	33.0 (±1.7) b	34.3 (±1.3) Bb	33.7 (±1.7) b	34.7 (±1.2) Bb	34.6 (±1.6) Bb	35.6 (±1.1) Bb

The results are expressed as means ± standard deviations, and different capital letters demonstrate statistical differences between groups (comparison of results in different lines). In addition, different lowercase letters demonstrate statistical differences between moments (comparison of results in different columns). The Chi-square statistical test was used, with significance at $p < 0.05$.

An increase in the quality of life was observed between the two treatment groups, with greater evidence and significance in the O_3G when compared to the CG in the second ($p = 0.002$), fourth ($p = 0.005$), fifth ($p = 0.005$), and sixth ($p < 0.001$) CT sessions (Figure 4).

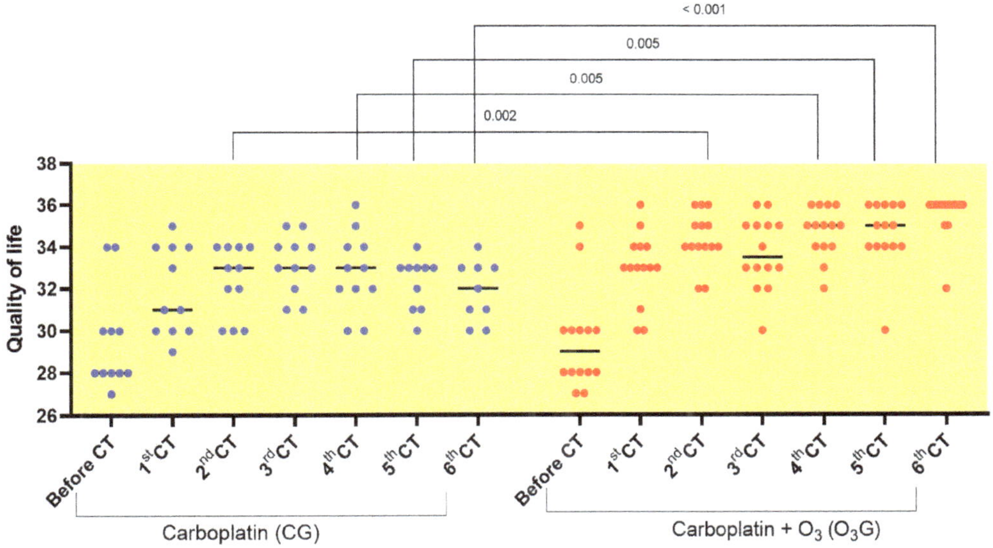

Figure 4. Comparison of quality of life scores between groups in each chemotherapy session. The one-way ANOVA statistical test was used, with significance set at $p < 0.05$.

3.5. *Oxidative Profile—Total Antioxidant Capacity (TAC)*

The mean serum concentrations of total antioxidant capacity (TAC) (mmol/L) between the CG and O_3G were significantly different ($p < 0.05$). There was no significant difference between the groups in moment A (start of treatment—before the first chemotherapy session). However, at moment B (middle of treatment—moments before the fourth chemotherapy session) and moment C (45 days after the end of therapy), a significant difference was observed ($p = 0.017$ and $p = 0.022$, respectively), with greater total antioxidant capacity in the O_3G ($p < 0.05$) (Table 5). Additionally, the graph showed an increase in total antioxidant capacity in the group treated with ozone (O_3G) when compared to the group treated with carboplatin alone (CG), as the treatment progressed (Figure 5A).

Table 5. Total antioxidant capacity (TAC) (mmol/L) serum concentration means and standard deviations in the CG and O_3G at different moments (A, B, and C).

TAC (mmol/L)			
	Carboplatin (CG)	Carboplatin + O_3 (O_3G)	*p*-Value
Moments			
A	78.2 (±31.7)	69.6 (±33.0)	0.517
B	55.8 (±24.7)	79.2 (±20.7)	0.017 *
C	63.5 (±32.7)	98.6 (±37.4)	0.022 *

The symbol "*" expresses statistically significant differences. The one-way ANOVA statistical test was used, followed by the Student's *t*-test, with significance set at $p < 0.05$. Moment A: 48 h before the first chemotherapy session; moment B: 48 h before the fourth chemotherapy session; and moment C: 45 days after completion of the treatment with carboplatin.

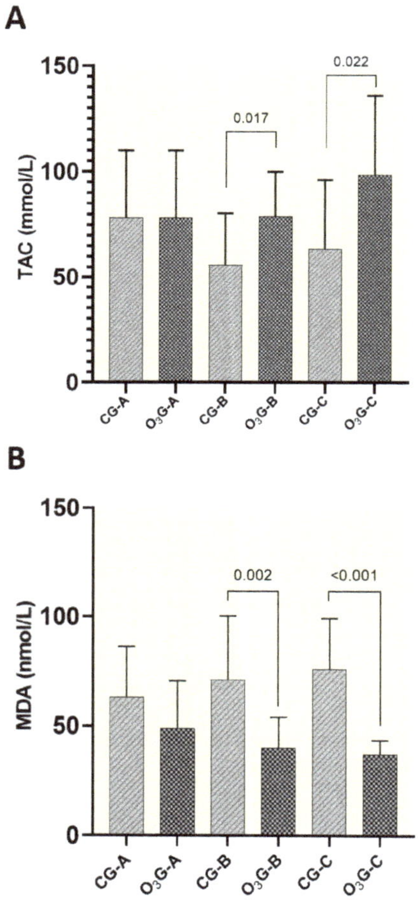

Figure 5. Total antioxidant capacity (TAC) (**A**) and mean serum malondialdehyde (MDA) (**B**) concentrations (nmol/L) of the animals treated with carboplatin alone (CG) or associated with carboplatin and O_3 (O_3G) at different times (A, B, and C). Moment A: 48 h before the first chemotherapy session; moment B: 48 h before the fourth chemotherapy session; and moment C: 45 days after completion of the treatment with carboplatin. The one-way ANOVA statistical test was used, with significance at $p < 0.05$.

3.6. *Oxidative Profile—Malondialdehyde (MDA)*

The mean serum concentrations of malondialdehyde (MDA) (nmol/L) between the CG and O3G showed significant differences ($p < 0.05$), except in moment A (start of treatment—before the first chemotherapy session). At moment B (middle of treatment—moments before the fourth chemotherapy session) and moment C (45 days after the end of treatment), there was a significant difference ($p = 0.002$ and $p < 0.001$, respectively) (Table 6). The graph shows a reduction in the concentration of malondialdehyde in the group treated with ozone (O3G) when compared to the group treated only with carboplatin (CG) (Figure 5B).

Table 6. Estimated means and standard deviations of serum concentrations of MDA (nmol/mL) in the CG and O3G in the different moments (A, B, and C).

MDA (mmol/L)			
	Carboplatin (CG)	**Carboplatin + O_3 (O3G)**	***p*-Value**
Moments			
A	62.8 (±23.4)	48.9 (21.6)	0.136
B	71.1 (±29.1)	40.2 (±13.9)	0.002 *
C	75.9 (±23.4)	37.3 (±6.4)	<0.001 *

Moment A: 48 h before the first chemotherapy session; moment B: 48 h before the fourth chemotherapy session; and moment C: 45 days after completion of the treatment with carboplatin. The symbol * expresses statistically significant differences. The one-way ANOVA statistical test was used, followed by the Student's *t*-test, with significance at $p < 0.05$.

3.7. *Comparison of Survival Curves*

The maximum follow-up period was approximately 28 months. The minimum survival time was 150 days, attributed to a dog in the carboplatin group with a histological diagnosis of carcinosarcoma that died due to disease progression after the fifth session of chemotherapy. The maximum survival time was 841 days, attributed to a dog from O3G with a histological diagnosis of Grade II papillary carcinoma and without lymph node metastasis. This dog was alive and under follow-up at the end of this study. Only the carboplatin group reached the median (563 days), as more than 50% of the population died. Based on the survival of the evaluated groups, significantly higher values can be seen in the O3G when compared to the CG ($p = 0.0224$, HR 6.294, and 95% CI 1.252–31.6) (Figure 6).

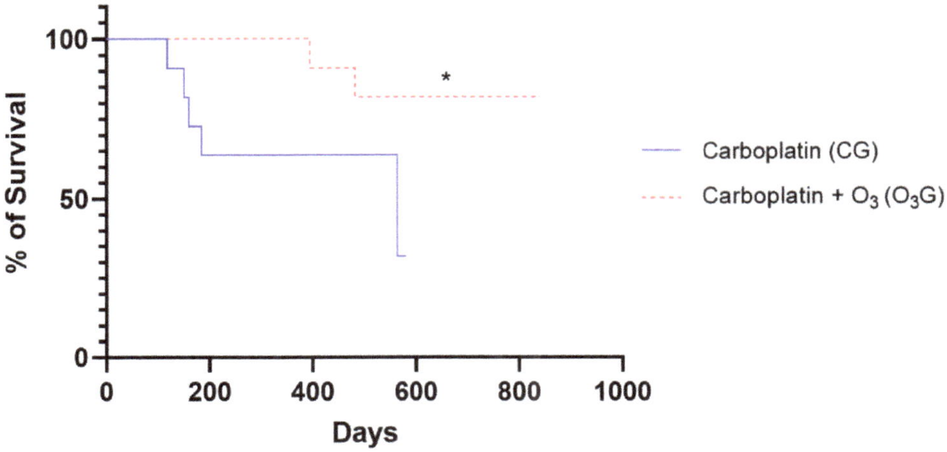

Figure 6. Survival curves of female dogs from the carboplatin group (CG) and carboplatin + O_3 group (O3G) after the chemotherapy sessions. The log-rank (Mantel–Cox) statistical test was used to compare the curves, with significance set at $p < 0.05$, and a significant difference was found ($p = 0.0224$), represented by the symbol "*".

3.8. Univariate and Multivariate Analyses

The analysis of the influence of pathological and clinical parameters associated with survival revealed a significant negative correlation between quality of life and survival in the univariate analysis ($p = 0.022$), indicating that a better quality of life may be associated with a higher chance of survival. However, the multivariate analysis did not maintain this correlation ($p = 0.170$), suggesting that other factors may influence the relationship between quality of life and survival. Ozone therapy ($p = 0.044$) and total antioxidant capacity (TAC) ($p = 0.039$) showed a significant positive correlation with survival only in the multivariate analysis. Meanwhile, malondialdehyde (MDA) showed a significant correlation with survival in both univariate and multivariate analyses (Table 7). This situation suggests that this parameter related to oxidative stress may play an important role in determining survival in female dogs with malignant mammary tumors.

Table 7. Univariate and multivariate analyses of clinical, pathological, and hematological parameters in female dogs with malignant mammary tumors considering the association with survival.

Parameters	Univariate [a]		Multivariate [b]	
	Spearman's Rho (CI 95%)	*p*-Value	Odds Ratio (CI 95%)	*p*-Value
Ozone therapy	0.268 (−0.081–0.582)	0.196	0.177 (0.033–0.955)	0.044 *
Alanine aminotransferase	−0.173 (−0.529–0.188)	0.506	0.926 (0.834–1.029)	0.152
Creatinine	0.224 (−0.276–0.591)	0.387	2.445 (0.410–14.574)	0.326
Red cells	0.087 (−0.479–0.428)	0.992	0.599 (0.190–1.888)	0.382
Platelets	−0.410 (−0.744–0.155)	0.102	1.006 (0.989–1.011)	0.132
Total leukocytes	−0.284 (−0.652–0.156)	0.270	1.009 (0.995–1.014)	0.179
Segmented neutrophils	−0.409 (−0.716–0.257)	0.103	1.002 (0.968–1.023)	0.081
Quality of life	−0.551 (−0.825–−0.195)	0.022 *	0.732 (0.469–1.143)	0.170
Adverse effects	0.310 (−0.169–0.669)	0.226	3.344 (0.342–32.736)	0.300
Clinical staging	0.278 (−0.038–0.628)	0.280	1.200 (0.637–2.260)	0.572
Tumor grade	0.116 (0.063–0.281)	0.658	2.303 (0.087–12.145)	0.43
TAC	−0.472 (−0.743–−0.158)	0.055	0.971 (0.944–0.998)	0.039 *
MDA	0.567 (0.262–0.796)	0.018*	1.038 (1.004–1.073)	0.026 *

Significant statistical associations are expressed by the symbol * when using the Spearman correlation test ([a]) or the Cox regression test ([b]), with significance at $p < 0.05$.

Furthermore, the evaluation of clinical, pathological, and hematologic parameters of ozone therapy demonstrated significant influences on total leukocytes ($p = 0.017$), segmented neutrophils ($p = 0.046$), quality of life ($p < 0.001$), survival ($p = 0.007$), and MDA levels ($p < 0.001$) in both univariate and multivariate analyses. In contrast, total antioxidant capacity (TAC) ($p = 0.039$) exhibited a significant positive correlation with the treatment only in the multivariate analysis (Table 8).

Table 8. Univariate and multivariate analyses of clinical, pathological, and hematologic parameters associated with carboplatin + ozone treatment in female dogs with malignant mammary tumors.

Parameters	Univariate [a]		Multivariate [b]	
	Spearman's Rho (CI 95%)	*p*-Value	Odds Ratio (CI 95%)	*p*-Value
Alanine aminotransferase	−0.383 (−0.761–0.103)	0.143	0.955 (0.907–1.005)	0.078
Creatinine	−0.111 (−0.629–0.449)	0.681	0.246 (0.008–7.491)	0.421
Red cells	−0.342 (−0.754–0.193)	0.195	0.413 (0.141–1.209)	0.106
Platelets	0.410 (−0.104–0.790)	0.114	1.004 (0.947–1.627)	0.115
Total leukocytes	0.588 (0.192–0.855)	0.017 *	1.208 (1.004–1.470)	0.020 *
Segmented neutrophils	0.506 (0.032–0.820)	0.046 *	1.015 (0.992–1.133)	0.037 *
Quality of life	0.923 (0.823–0.988)	<0.001 *	4.594 (1.445–14.607)	0.010 *
Adverse effects	−0.293 (−0.620–−0.174)	0.271	0.290 (0.021–4.026)	0.357
Clinical staging	0.043 (−0.486–0.529)	0.875	1.260 (0.555–2.858)	0.580
Tumor grade	−0.209 (−0.436–−0.139)	0.420	1.778 (0.257–15.273)	0.388
Survival (days)	0.525 (0.191–0.784)	0.007 *	1.005 (1.001–1.010)	0.027 *
Condition (live)	−0.345 (−0.678–0.050)	0.092	0.200 (0.030–1.351)	0.099
TAC	0.451 (−0.010–0.216)	0.080	1.022 (1.001–1.038)	0.035 *
MDA	−0.861 (0.047–−0.877)	<0.001 *	0.789 (0.627–0.993)	0.043 *

Significant statistical associations are expressed by the symbol * when using the Spearman correlation test ([a]) or the Cox regression test ([b]), with significance at $p < 0.05$.

4. Discussion

Many desirable effects of chemotherapy arise from increased free radicals and reactive oxygen species (ROS) in the tumor microenvironment. However, they can also mediate chemotherapy-induced toxicity in healthy cells, resulting in important adverse reactions and reduced response to treatment [31]. In this context, this work evaluated the effects of using medicinal ozone as an integrative therapy on the oxidative profile, survival, and quality of life of female dogs with mammary cancer undergoing conventional chemotherapy with carboplatin.

Carboplatin is one of the drugs that are used for the adjuvant treatment of mammary neoplasm in female dogs, resulting in a significant increase in survival rates and quality of life [30,32,33]. It is less nephrotoxic than cisplatin, and vomiting and diarrhea are less frequent and intense during a carboplatin treatment [34]. Machado and collaborators [33], when evaluating the effect of metronomic cyclophosphamide on the tolerability, efficacy, and pharmacokinetics of carboplatin in bitches with mammary carcinoma, reported Grade 1 and 2 adverse reactions in the group of female dogs treated only with carboplatin. Similar results were observed in our study in relation to the carboplatin group; however, in the group of bitches treated with carboplatin associated with ozone therapy, only adverse reactions considered mild (Grade 1) and self-limiting without the need for drug intervention were observed. In addition, a significant reduction in the incidence of nausea and vomiting was detected. It is known that using ozone in adequate doses promotes a reduction in chemoresistance and enhances the action of chemotherapy [10]. Furthermore, ozone can induce adequate doses of antioxidants, reducing or preventing oxidative damage caused by chemotherapy, and providing a clinical benefit for patients undergoing oncological treatment [15,31].

Leukopenia, anemia, and thrombocytopenia of varying intensities are expected adverse reactions in animals undergoing chemotherapy with carboplatin [30,33]. However, interestingly, in the group treated with carboplatin and ozone, a significant increase in the average values of red blood cells, platelets, and leukocytes was observed, including during the nadir period of the drug [23]. In humans, repeated and controlled therapeutic ozone sessions, as major autohemotherapy, induce mild and transient controlled oxidative stress, which results in an increase in the number of red blood cells with improved biochemical and functional characteristics [35,36]. More specifically, in red blood cells stimulated by ozone, there is a transient increase in the production of intracellular ATP with an increase in the rate of glycolysis, leading to the stimulation of the Krebs cycle through the production of 3-phosphoglycerate by 2,3 diphosphoglycerate (2, 3 DPG) and glucose 6 phosphate dehydrogenase (G6PDH), with changes in the hemoglobin dissociation curve, increased oxygen supply to tissues, in addition to higher concentrations of antioxidant enzymes [36,37].

Also, ozone can influence the rheological properties of blood; that is, it can improve the flexibility and elasticity of erythrocyte membranes and reduce blood viscosity [37,38]. There is also induction of nitric oxide production by vascular endothelial cells and, thus, promotion of vasodilation at the microcirculation level. In this way, activated erythrocytes enter the circulation gradually and continuously, replacing old erythrocytes adapted to chronic oxidative stress in a process known as oxidative preconditioning, as they regulate the antioxidant system [9,39]. Although Bocci et al. [9] suggest that 15 sessions of autohemotherapy are necessary to stimulate the production of new erythrocytes with biochemical improvements induced by ozone, in the present study, a significant increase in erythrocyte values was observed in the second session and, consequently, in tissue oxygenation in animals in the O_3G.

The assessment of tumor hypoxia was not performed in this study; however, it is noteworthy that increased concentrations of red blood cells in O_3G can result in better blood flow, greater transfer of oxygen to hypoxemic tissues, and a consequent better response to treatment. It is known that tumor hypoxia is an adverse factor for chemotherapy and radiotherapy, as it induces a physiological selection of tumor cells with reduced apoptotic potential, resulting in resistance to treatment and greater potential for tumor aggressiveness [40,41].

Another interesting result refers to the significant increase in the number of platelets observed after the first chemotherapy session in animals from the carboplatin and ozone group. The effects caused by ozone therapy on platelet induction are still controversial, and data in the literature regarding this process are scarce [35,42]. Nevertheless, Valacchi and Bocci [43] reported an increase in the number of platelets, reduced platelet aggregation, significantly greater amounts of platelet-derived growth factor (PDGF), and transforming growth factor-beta 1 (TGF-beta1) and interleukin-8 (IL-8), which are dose-dependent and released after ozonation, at a concentration of 20 $\mu g\ O_3/mL$ to 80 $\mu g\ O_3/mL$, by heparinized platelets in human blood samples. The ozone concentration in the present study is within the therapeutic window proposed by the aforementioned authors and justifies the increase in the number of platelets and lower platelet aggregation in the group that received weekly minor ozonated autohemotherapy.

Regarding the leukocyte counts, leukopenia was found in the CG animals, as expected, while a significant increase in total leukocytes and segmented neutrophils was identified in the animals that underwent weekly minor ozonated autohemotherapy. Ozone, through the transient action of H_2O_2, has a long-lasting and delayed action on polyunsaturated fatty acids (PUFAs) and other antioxidants, acting as a modulator of the immune system by inducing the production of cytokines in neutrophils, lymphocytes, and monocytes, which can cause leukocytosis and activation of leukocytes [9,44,45]. However, it is worth highlighting that, in the present study, the increase in total leukocyte and segmented neutrophil counts did not characterize leukocytosis, given that the values observed were within the reference range for the species, even during the nadir of the drug, when evident leukopenia was expected.

The therapeutic toxicity biochemical markers (ALT and creatinine) evaluated throughout the treatment remained within physiological parameters for the species for most of the treatment, regardless of the group. These results were expected and suggest that carboplatin chemotherapy did not affect the morphological integrity of hepatocytes and renal function. In fact, according to the literature, despite the renal excretion of the chemotherapy drug, carboplatin is a drug considered safe and can be used in nephropathic patients with discretion [46,47].

In the liver function assessment, the maintenance of ALT values is also justified, as carboplatin does not cause liver toxicity [48]. Aslaner and collaborators [49] evaluated the protective effect of ozone therapy on hepatotoxicity induced by methotrexate, a chemotherapy drug with the potential to cause hepatic steatosis, cholestasis, fibrosis, and cirrhosis, and concluded that the association with ozone resulted in a decrease in ALT levels. Absence of changes in serum biochemistry (creatinine and ALT) and hematological changes restricted to the nadir period of the drug in female dogs with breast cancer treated with carboplatin were also reported by Machado et al. [30], Machado et al. [33], and Lavalle et al. [32].

Oxidative stress was evaluated through assays on total antioxidant capacity (TAC) and lipid peroxidation product, measuring the malondialdehyde (MDA) serum levels. The analysis of TAC and MDA revealed linear, inverse, and significantly different profiles between the CG and O_3G at moments B (48 h before the fourth chemotherapy session) and C (45 days after completion of the treatment with carboplatin). In the group treated only with carboplatin, TAC values were decreasing, and those of MDA increased during chemotherapy treatment; the opposite was observed in the O_3G as treatment progressed, that is, TAC with an increasing linear profile and decreasing MDA. The increase in MDA in cancer patients, despite being controversial, has been associated with excessive production of reactive oxygen species (ROS) by neoplastic cells, antioxidant deficiency, induction of oxidative stress followed by molecular damage, and lipid peroxidation in different types of cancer, including human and canine breast cancer patients [10,11,50–52]. However, the increase in ROS and consequent genetic instability and increase in MDA can also be induced by surgery and mainly by chemotherapy. Some chemotherapy drugs, such as platinum derivatives, including carboplatin, induce apoptosis of neoplastic cells through an increase in ROS, which is one of their main mechanisms of action [13], justifying the increasing and significant MDA values observed in the group treated with carboplatin alone.

Regarding the high TAC and reduced MDA levels in the O_3G, the results demonstrated a reduction in oxidative stress status with ozone therapy. It is known that ozone acts as a modulator or pro-drug and enhances subsequent adaptive responses by inducing secondary messengers. Hydrogen peroxide, one of these secondary messengers, promotes the nuclear factor 2-related factor erythroid 2 (Nrf2) pathway and the synthesis of proteins that favor cell survival [53–55]. Low doses of O_3 generate mild oxidative stress, which activates the Keap1-dependent pathway mediated by Nrf2, which, in turn, stimulates gene expression of antioxidant response elements (AREs) [55]. Under basal conditions, Nrf2 is close to the Keap 1 protein; when there are stressful stimuli, Nrf2 moves from Keap1 and enters the cell nucleus, inducing the transcription of genes triggered by antioxidant response elements [56]. Ozonation prevents the degradation of Nrf2 and promotes its translocation to the nucleus, activating antioxidant response elements [55]. Experimental results demonstrated that ex vivo or in vivo ozone can activate Nrf2 [53,55]. This mechanism may explain the genomic target of ozone, which induces a response based on protein synthesis, such as the antioxidant enzymes HO-1, SOD, and CAT, providing more effective cellular protection against the harmful effects of free radicals, including lipid peroxidation/elevation of MDA resulting from chemotherapy treatment [53]. The results agree with other studies described in the literature and may suggest that a controlled application of O_3 during chemotherapy can protect against prolonged and important oxidative stress [31].

An increase in quality of life was observed in the O_3G, with an average quality of life score close to 36, which is considered the best possible index. The results obtained from the analysis of the questionnaires were confirmed by the satisfactory clinical response and reports from the tutors regarding the better general condition of the animals after ozone therapy. In a previous study carried out by our group, female dogs submitted only to chemotherapy with carboplatin presented quality of life scores ranging from 20 to 23, while those treated with chemotherapy and low-dose naltrexone (LDN) ranged from 31 to 36 [30]. Therefore, the results of this study support the idea that quality of life is directly related to the beneficial effects of ozone. Similarly, human clinical studies have demonstrated improved quality of life, reduced levels of reactive oxygen species (ROS), and increased antioxidant potential in patients who received chemotherapy combined with ozone therapy [15,31].

Of the seven dogs that died because of the neoplastic disease, five belonged to the group treated with carboplatin alone, and two belonged to the group treated with carboplatin associated with ozone therapy. The higher survival rate observed in the O_3G is a reflection, among other factors, of the better quality of life. Recently, the effect of ozone treatment on health-related quality of life and toxicity induced by radiotherapy and chemotherapy in symptomatic cancer survivors was also evaluated [57]. The results demonstrated that ozone treatment has the potential to positively impact not only the treatment, as observed in our study, but also the care of cancer survivors undergoing radiotherapy and chemotherapy by significantly improving quality of life, reducing toxicity, and offering a promising therapeutic option for managing chronic side effects.

We acknowledge the inherent limitations of our study, particularly regarding the small sample size and the heterogeneity of the tumors in terms of diagnosis and histopathological grading. These limitations precluded a more robust comparison with previously reported results observed in women, especially concerning a specific type of tumor. However, considering that the analyzed tumors were spontaneous and there was a balance between the evaluated groups, this same heterogeneity may have contributed to demonstrating the potential of ozone therapy as a complementary therapy in clinical practice. This approach is pertinent, given that the diversity of malignant mammary tumor types is a reality in medical practice.

5. Conclusions

The promising results of this study, although preliminary, encourage further exploration of ozone therapy as a complementary therapeutic alternative in the treatment of breast cancer. Ozone treatment has demonstrated beneficial or potentially beneficial effects by reducing adverse reactions to chemotherapy, promoting positive modulation of the oxidative profile, and improving quality of life, leading to increased survival rates in dogs undergoing treatment for mammary cancer. However, clinical trials involving more animals are necessary to validate the results observed herein and confirm the efficacy and safety of ozone therapy within the context of veterinary, comparative, and translational oncology.

Author Contributions: Conceptualization, A.E.-L.; methodology, L.P.S., M.C.M., V.d.M.P.d.C. and A.E.-L.; formal analysis, L.P.S., R.W.P. and A.E.-L.; investigation, L.P.S., G.A.B.C., H.C.d.S., J.M.C.-N. and K.A.D.; writing—original draft preparation, L.P.S., R.W.P., A.E.-L., G.A.B.C. and M.C.M.; writing—review and editing, L.P.S., V.R.C.d.S., C.S.C., M.C.M., V.d.M.P.d.C., R.W.P. and A.E.-L.; visualization, L.P.S., R.W.P. and A.E.-L.; supervision, R.W.P. and A.E.-L. All authors have read and agreed to the published version of the manuscript.

Funding: We thank the Coordenação de Aperfeiçoamento de Pessoal de Nível Superior (CAPES), Brazil, for the scholarship to LPS (Proc. 453702/2019-3). AE-L (Proc. 310058/2022-8) and RDP (Proc. 310058/2022-8) are supported by research and technological development fellowships of the National Council for Scientific and Technological Development (CNPq). The funders had no role in this study's design, in the collection, analysis, and interpretation of data, or in the writing of the manuscript.

Institutional Review Board Statement: This study was conducted according to the guidelines of the Declaration of Helsinki and approved by the Ethics Committee for the Use of Experimental Animals of the School of Veterinary Medicine and Zootechny of the Federal University of Bahia (protocol no. 18/2021). All the procedures were conducted according to the Brazilian College of Animal Experimentation (COBEA).

Informed Consent Statement: All animals were domiciliated, and dog guardians were informed about the details of the research project by means of a signed Free and Informed Consent Form.

Data Availability Statement: The raw data supporting this article's conclusions are available upon request.

Acknowledgments: We appreciate the logistic support of the Federal University of Bahia and the Postgraduate Program in Animal Science in the Tropics.

Conflicts of Interest: The authors declare that the research was conducted without any commercial or financial relationships that could be potential conflicts of interest.

References

1. Sung, H.; Ferlay, J.; Siegel, R.L.; Laversanne, M.; Soerjomataram, I.; Jemal, A.; Bray, F. Global cancer statistics 2020: GLOBOCAN estimates of incidence and mortality worldwide for 36 cancers in 185 countries. *CA Cancer J. Clin.* **2021**, *71*, 209–249. [CrossRef] [PubMed]
2. Harbeck, N.; Penault-Llorca, F.; Cortes, J.; Gnant, M.; Houssami, N.; Poortmans, P.; Ruddy, K.; Tsang, J. Breast cancer (Primer). *Nat. Rev. Dis. Primers* **2019**, *5*, 66. [CrossRef] [PubMed]
3. DeSantis, C.E.; Ma, J.; Gaudet, M.M.; Newman, L.A.; Miller, K.D.; Goding Sauer, A.; Jemal, A.; Siegel, R.L. Breast cancer statistics, 2019. *CA A Cancer J. Clin.* **2019**, *69*, 438–451. [CrossRef] [PubMed]
4. Clavo, B.; Santana-Rodríguez, N.; Llontop, P. Ozone Therapy as Adjuvant for Cancer Treatment: Is Further Research Warranted? *Evid. Based Complement Altern. Med.* **2018**, *2018*, 7931849. [CrossRef] [PubMed]
5. Agarwal, D.B. Role of ozone therapy and cancer: Myth or reality? *J. Med. Sci. Clin. Res.* **2019**, *7*, 720–727. [CrossRef]
6. Klaunig, J.E.; Kamendulis, L.; Hocevar, B.A. Oxidative Stress and Oxidative Damage in Carcinogenesis. *Toxicol. Pathol.* **2010**, *38*, 96–109. [CrossRef] [PubMed]
7. Waris, G.; Ahsan, H. Reactive oxygen species: Role in developing cancer and various chronic conditions. *J. Carcinog.* **2006**, *5*, 14. [CrossRef] [PubMed]
8. Bocci, V. Mechanisms of action and chemical-biological interactions between ozone and body compartments: A critical appraisal of the different administration routes. *Current Drug Therapy.* **2009**, *4*, 159–173. [CrossRef]
9. Bocci, V.; Zanardi, I.; Travagli, V. Ozone acting on human blood yields a hormetic dose-response relationship. *J. Transl. Med.* **2011**, *9*, 66. [CrossRef] [PubMed]
10. Simonetti, V.; Quagliariello, V.; Franzini, M.; Iaffaioli, R.V.; Maurea, N.; Valdenassi, L. Ozone exerts cytoprotective and anti-inflammatory effects in cardiomyocytes and skin fibroblastos after incubation with doxorubicin. *Evid. -Based Complement. Altern. Med.* **2019**, *2019*, 2169103. [CrossRef] [PubMed]
11. Luongo, M.; Marinelli, O.; Zeppa, L. Cannabidiol and Oxygen-Ozone Combination Induce Cytotoxicity in Human Pancreatic Ductal Adenocarcinoma Cell Lines. *Cancers* **2020**, *12*, 2774. [CrossRef] [PubMed]
12. Rossmann, A.; Mandic, R.; Heinis, J.; Höffken, H.; Küssner, O.; Kinscherf, R.; Weihe, E.; Bette, M. Intraperitoneal oxidative stress in rabbits with papillomavirus-associated head and neck cancer induces tumoricidal immune response that is adoptively transferable. *Clin. CancerResearch* **2014**, *20*, 4289–4301. [CrossRef] [PubMed]
13. Clavo, B.; Martínez-Sánchez, G.; Rodríguez-Esparragón, F.; Rodríguez-Abreu, D.; Galván, S.; Aguiar-Bujanda, D.; Díaz-Garrido, J.A.; Cañas, S.; Torres-Mata, L.B.; Fabelo, H.; et al. Modulation by Ozone Therapy of Oxidative Stress in Chemotherapy-Induced Peripheral Neuropathy: The Background for a Randomized Clinical Trial. *Int. J. Mol. Sci.* **2021**, *22*, 2802. [CrossRef] [PubMed]
14. Clavo, B.; Santana-Rodriguez, N.; Llontop, P.; Gutierrez, D.; Ceballos, D.; Méndez, C.; Rovira, G.; Suarez, G.; Rey-Baltar, D.; Garcia-Cabrera, L.; et al. Ozone therapy in the management of persistent radiation-induced rectal bleeding in prostate cancer patients. *Evid.-Based Complement. Altern. Med.* **2015**, *2015*, 480369. [CrossRef] [PubMed]
15. Tirelli, U.; Cirrito, C.; Pavanello, M.; Del Pup, L.; Lleshi, A.; Berretta, M. Oxygen-ozone therapy as support and palliative therapy in 50 cancer patients with fatigue: A short report. *Eur. Rev. Med. Pharmacol. Sci.* **2018**, *22*, 8030–8033. [CrossRef] [PubMed]
16. Yıldırım, M.; Erkişi, S.; Yılmaz, H.; Ünsal, N.; İnaç, E.; Tanrıver, Y.; Koçak, P. The apoptotic effect of ozone therapy on mitochondrial activity of highly metastatic breast cancer cell line MDA-MB-231 using in vitro approaches. *J. Interv. Med.* **2022**, *5*, 64–71. [CrossRef] [PubMed]
17. Li, Y.; Pu, R. Ozone Therapy for Breast Cancer: An Integrative Literature Review. *Integr. Cancer Ther.* **2024**, *23*, 15347354241226667. [CrossRef] [PubMed]
18. Sciorsci, R.L.; Lillo, E.; Occhiogrosso, L.; Rizzo, A. Ozone therapy in veterinary medicine: A review. *Res. Vet. Sci.* **2020**, *130*, 240–246. [CrossRef] [PubMed]
19. Orlandin, J.R.; Machado, L.C.; Ambrósio, C.E.; Travagli, V. Ozone and its derivatives in veterinary medicine: A careful appraisal. *Vet. Anim. Sci.* **2021**, *13*, 100191. [CrossRef] [PubMed]

20. Cassali, G.; Jark, P.; Gamba, C.; Damasceno, K.; Estrela-Lima, A.; Nardi, A.; Ferreira, E.; Horta, R.S.; Firmo, B.F.; Sueiro, F.A.R.; et al. Consensus Regarding the Diagnosis, Prognosis and Treatment of Canine and Feline Mammary Tumors—2019. *Braz. J. Vet. Pathol.* **2020**, *13*, 555–574.
21. Nakagaki, K.Y.; Nunes, M.M.; Garcia, A.P.V.; Nunes, F.C.; Schmitt, F.; Cassali, G.D. Solid Carcinoma of the Canine Mammary Gland: A Histological Type or Tumour Cell Arrangement? *J. Comp. Pathol.* **2022**, *190*, 1–12. [CrossRef] [PubMed]
22. Elston, E.W.; Ellis, I.O. Method for Grading Breast Cancer. *J. Clin. Pathol.* **1993**, *46*, 189–190. [CrossRef] [PubMed]
23. Coffee, C.; Roush, J.K.; Higginbotham, M.L. Carboplatin-induced myelosuppression as related to body weight in dogs. *Veter. Comp. Oncol.* **2020**, *18*, 804–810. [CrossRef] [PubMed]
24. Liu, Z.; Jiao, Y.; Yu, T.; Wang, H.; Zhang, Y.; Liu, D.; Xu, Y.; Guan, Q.; Lu, M. A Review on the Immunomodulatory Mechanism of Acupuncture in the Treatment of Inflammatory Bowel Disease. *Evid. -Based Complement. Altern. Med.* **2022**, *2022*, 8528938. [CrossRef] [PubMed]
25. Wang, S.; Yu, G.; Wang, Z.; Jacobson, O.; Lin, L.S.; Yang, W.; Deng, H.; He, Z.; Liu, Y.; Chen, Z.Y.; et al. Enhanced Antitumor Efficacy by a Cascade of Reactive Oxygen Species Generation and Drug Release. *Angewandte Chem. Int. Ed.* **2019**, *58*, 14758–14763. [CrossRef] [PubMed]
26. Leblanc, A.K.; Atherton, M.; Bentley, R.T.; Boudreau, C.E.; Burton, J.H.; Curran, K.M.; Dow, S.; Giuffrida, M.A.; Kellihan, H.B.; Mason, N.J.; et al. Veterinary cooperative oncology group—Common terminology criteria for adverse events (VCOG-CTCAE v2) following investigational therapy in dogs and cats. *Vet. Comp. Oncol.* **2021**, *19*, 311–352. [CrossRef] [PubMed]
27. Erel, O. A novel automated direct measurement method for total antioxidant capacity using a new generation, more stable ABTS radical cation. *Clin. Biochem.* **2004**, *37*, 277–285. [CrossRef]
28. Tang, Q.; Su, Y.W.; Xian, C.J. Determining Oxidative Damage by Lipid Peroxidation Assay in Rat Serum. *Bio. Protoc.* **2019**, *9*, e3263. [CrossRef] [PubMed]
29. Yazbek, K.V.; Fantoni, D.T. Validity of a health-related quality-of-life scale for dogs with signs of pain secondary to cancer. *JAVMA* **2005**, *226*, 1354–1358. [CrossRef] [PubMed]
30. Machado, M.C.; Costa-Neto, J.M.; Portela, R.D.; D'assis, M.J.M.H.; Martins-Filho, A.O.; Barrouin-Melo, S.M.; Borges, N.F.; Silva, F.L.; Estrela-Lima, A. The Effect of Naltrexone as a Carboplatin Chemotherapy-Associated Drug on the Immune Response, Quality of Life and Survival of Dogs with Mammary Carcinoma. *PLoS ONE* **2018**, *13*, e0204830. [CrossRef] [PubMed]
31. Clavo, B.; Rodríguez-Esparragón, F.; Rodríguez-Abreu, D.; Martínez-Sánchez, G.; Llontop, P.; Aguiar-Bujanda, D.; Fernández-Pérez, L.; Santana-Rodríguez, N. Modulation of Oxidative Stress by Ozone Therapy in the Prevention and Treatment of Chemotherapy-Induced Toxicity: Review and Prospects. *Antioxidants* **2019**, *8*, 588. [CrossRef]
32. Lavalle, G.E.; De Campos, C.B.; Bertagnolli, A.C.; Cassali, G.D. Canine malignant mammary gland neoplasms with advanced clinical staging treated with carboplatin and cyclooxygenase inhibitors. *In Vivo* **2012**, *26*, 375–379. [PubMed]
33. Machado, M.C.; Yamamoto, P.A.; PippA, L.F.; De Moraes, N.V.; Neves, F.M.F.; Portela, R.D.; Barrouin-Melo, S.M.; Hielm-Björkman, A.; Godoy, A.L.P.C.; Estrela-Lima, A. Pharmacokinetics of Carboplatin in Combination with Low-Dose Cyclophosphamide in Female Dogs with Mammary Carcinoma. *Animals* **2022**, *12*, 3109. [CrossRef]
34. Villarino, N.; Cox, S.; Yarbrough, J.; Martiãn-Jimãnez, T. Determination of carboplatin in canine plasma by high-performance liquid chromatography. *Biomed. Chromatogr.* **2009**, *24*, 908–913. [CrossRef] [PubMed]
35. Erken, G.; Erken, H.A.; Bilen, C.; Gencer, N. The effects of ex vivo ozone treatment on human erythrocyte carbonic anhydrase enzyme. *Arch. Physiol. Biochem.* **2017**, *142*, 171–174. [CrossRef] [PubMed]
36. Tricarico, G.; Travagli, V. The Relationship between Ozone and Human Blood in the Course of a Well-Controlled, Mild, and Transitory Oxidative Eustress. *Antioxidants* **2021**, *10*, 1946. [CrossRef]
37. Akbudak, I.H.; Kucukatay, V.; Kilic-Erkek, O.; Ozdemir, Y.; Bor-Kucukatay, M. Investigation of the effects of major ozone autohemotherapy application on erythrocyte deformability and aggregation. *Clin. Hemorheol. Microcirc.* **2019**, *71*, 365–372. [CrossRef] [PubMed]
38. Baieth, H.-S.; Elashmawi, I.S. Influence of ozone on the rheological and electrical properties of stored human blood. *J. Biomed. Res.* **2012**, *26*, 185–192. [CrossRef] [PubMed]
39. Ojeda-Chi, M.M.; Rodriguez-Vivas, R.I.; Esteve-Gasent, M.D.; Pérez de León, A.A.; Modarelli, J.J.; Villegas-Perez, S.L. Ehrlichia canis in dogs of Mexico: Prevalence, incidence, co-infection and factors associated. *Comp. Immunol. Microbiol. Infect. Dis.* **2019**, *67*, 1011351. [CrossRef]
40. Manalo, D.J.; Rowan, A.; Lavoie, T.; Natarajan, L.; Kelly, B.D.; Ye, S.Q.; Garcia, J.G.; Semenza, G.L. Transcriptional regulation of vascular endothelial cell responses to hypoxia by HIF-1. *Blood* **2005**, *105*, 659–669. [CrossRef] [PubMed]
41. Mole, D.R.; Blancher, C.; Copley, R.R.; Pollard, P.J.; Gleadle, J.M.; Ragoussis, J.; Ratcliffe, P.J. Genome-wide association of hypoxia-inducible factor (HIF)-1alpha and HIF-2alpha DNA binding with expression profiling of hypoxia-inducible transcripts. *J. Biol. Chem.* **2009**, *284*, 16767–16775. [CrossRef]
42. Bocci, V.; Luzzi, E.; Corradeschi, F.; Silvestri, S. Studies on the biological effects of ozone: 6. Production of transforming growth factor 1 by human blood after ozone treatment. *J. Biol. Regul. Homeost. Agents* **1994**, *8*, 108–112. [PubMed]
43. Valacchi, G.; Bocci, V. Studies on the biological effects of ozone: Release of factors from ozonated human platelets. *Mediat. Inflamm.* **1999**, *8*, 205–209. [CrossRef] [PubMed]

44. Chirumbolo, S.; Valdenassi, L.; Tirelli, U.; Ricevuti, G.; Pandolfi, S.; Vaiano, F.; Galoforo, A.; Loprete, F.; Simonetti, V.; Chierchia, M.; et al. The Oxygen-Ozone Adjunct Medical Treatment According to the Protocols from the Italian Scientific Society of Oxygen-Ozone Therapy: How Ozone Applications in the Blood Can Influence Clinical Therapy Success via the Modulation of Cell Biology and Immunity. *Biology* **2023**, *12*, 1512. [CrossRef] [PubMed]
45. Milillo, C.; Falcone, L.; Di Carlo, P.; Aruffo, E.; Del Boccio, P.; Cufaro, M.C.; Patruno, A.; Pesce, M.; Ballerini, P. Ozone effect on the inflammatory and proteomic profile of human macrophages and airway epithelial cells. *Respir. Physiol. Neurobiol.* **2023**, *307*, 103979. [CrossRef] [PubMed]
46. Gupta, S.; Portales-Castillo, I.; Daher, A.; Kitchlu, A. Conventional Chemotherapy Nephrotoxicity. *Adv. Chronic. Kidney Dis.* **2021**, *28*, 402–414. [CrossRef] [PubMed]
47. Duan, Z.Y.; Liu, J.Q.; Yin, P.; Li, J.J.; Cai, G.Y.; Chen, X.M. Impact of aging on the risk of platinum-related renal toxicity: A systematic review and meta-analysis. *Cancer Treat Rev.* **2018**, *69*, 243–253. [CrossRef] [PubMed]
48. Stockhaus, C.; Kohn, B.; Rudolph, R.; Brunnberg, L.; Giger, U. Correlation of haemostatic abnormalities with tumor stage and characteristics in dogs with mammary carcinoma. *J. Small Anim. Pract.* **1999**, *40*, 326–331. [CrossRef] [PubMed]
49. Aslaner, A.; Çakır, T.; Çelik, B.; Doğan, U.; Akyüz, C.; Baştürk, A.; Polat, C.; Gündüz, U.; Mayir, B.; Şehirli, A.Ö. The protective effect of intraperitoneal medical ozone preconditioning and treatment on hepatotoxicity induced by methotrexate. *Int. J. Clin. Exp. Med.* **2015**, *8*, 13303.
50. do Val Carneiro, J.L.; Nixdorf, S.L.; Mantovani, M.S.; da Silva do Amaral, H.A.C.; Aoki, M.N.; Amarante, M.K.; Fabris, B.A.; Pelegrinelli, F.M.H.; Ehara, W.M.A. Plasma malondialdehyde levels and CXCR4 expression in peripheral blood cells of breast cancer patients. *J. Cancer Res. Clin. Oncol.* **2009**, *135*, 997–1004. [CrossRef]
51. Kumaraguruparan, R.; Balachandran, C.; Manohar, B.M.; Nagini, S. Altered oxidant-antioxidant profile in canine mammary tumours. *Vet. Res. Commun.* **2005**, *29*, 287–296. [CrossRef]
52. Macotpet, A.; Suksawat, F.; Sukon, P.; Pimpakdee, K.; Pattarapanwichien, E.; Tangrassameeprasert, R.; Boonsiri, P. Oxidative stress in cancer-bearing dogs assessed by measuring serum malondialdehyde. *BMC Vet. Res.* **2013**, *9*, 101–106. [CrossRef] [PubMed]
53. Re, L.; Martínez-Sánchez, G.; Bordicchia, M.; Malcangi, G.; Pocognoli, A.; Morales-Segura, M.A.; Rothchild, J.; Rojas, A. Is ozone pre-conditioning effect linked to Nrf2/EpRE activation pathway in vivo? A preliminary result. *Eur. J. Pharmacol.* **2014**, *742*, 158–162. [CrossRef] [PubMed]
54. Siniscalco, D.; Trotta, M.C.; Brigida, A.L.; Maisto, R.; Luongo, M.; Ferraraccio, F.; D'Amico, M.; Di Filippo, C. Intraperitoneal administration of oxygen/ozone to rats reduces the pancreatic damage induced by atreptozotocin. *Biology* **2018**, *7*, 10. [CrossRef] [PubMed]
55. Galiè, M.; Costanzo, M.; Nodari, A.; Boschi, F.; Calderan, L.; Mannucci, S.; Covi, V.; Tabaracci, G.; Malatesta, M. Mild ozonisation activates antioxidant cell response by the Keap1/Nrf2 dependent pathway. *Free Radic. Biol. Med.* **2018**, *124*, 114–121. [CrossRef] [PubMed]
56. Pandey, P.; Singh, A.K.; Singh, M.; Tewari, M.; Shukla, H.S.; Gambhir, I.S. The see-saw of Keap1-Nrf2 pathway in cancer. *Crit. Rev. Oncol. Hematol.* **2017**, *116*, 89–98. [CrossRef] [PubMed]
57. Clavo, B.; Cánovas-Molina, A.; Ramallo-Fariña, Y.; Federico, M.; Rodríguez-Abreu, D.; Galván, S.; Ribeiro, I.; Marques da Silva, S.C.; Navarro, M.; González-Beltrán, D.; et al. Effects of Ozone Treatment on Health-Related Quality of Life and Toxicity Induced by Radiotherapy and Chemotherapy in Symptomatic Cancer Survivors. *Int. J. Environ. Res. Public Health* **2023**, *20*, 1479. [CrossRef] [PubMed]

MDPI AG
Grosspeteranlage 5
4052 Basel
Switzerland
Tel.: +41 61 683 77 34

MDPI Books Editorial Office
E-mail: books@mdpi.com
www.mdpi.com/books